Lecture Notes in Computer Scier

T0238660

Commenced Publication in 1973
Founding and Former Series Editors:
Gerhard Goos, Juris Hartmanis, and Jan van Leeuwen

Editorial Board

David Hutchison, UK
Josef Kittler, UK
Alfred Kobsa, USA
John C. Mitchell, USA
Oscar Nierstrasz, Switzerland
Bernhard Steffen, Germany
Demetri Terzopoulos, USA
Gerhard Weikum, Germany

Takeo Kanade, USA
Jon M. Kleinberg, USA
Friedemann Mattern, Switzerland
Moni Naor, Israel
C. Pandu Rangan, India
Madhu Sudan, USA
Doug Tygar, USA

Advanced Research in Computing and Software Science

Subline of Lectures Notes in Computer Science

Subline Series Editors

Giorgio Ausiello, *University of Rome 'La Sapienza', Italy*
Vladimiro Sassone, *University of Southampton, UK*

Subline Advisory Board

Susanne Albers, *University of Freiburg, Germany*
Benjamin C. Pierce, *University of Pennsylvania, USA*
Bernhard Steffen, *University of Dortmund, Germany*
Madhu Sudan, *Microsoft Research, Cambridge, MA, USA*
Deng Xiaotie, *City University of Hong Kong*
Jeannette M. Wing, *Carnegie Mellon University, Pittsburgh, PA, USA*

Artur Czumaj   Kurt Mehlhorn
Andrew Pitts   Roger Wattenhofer (Eds.)

Automata, Languages, and Programming

39th International Colloquium, ICALP 2012
Warwick, UK, July 9-13, 2012
Proceedings, Part II

 Springer

Volume Editors

Artur Czumaj
University of Warwick
Department of Computer Science and
Centre for Discrete Mathematics and its Applications
Warwick, UK
E-mail: a.czumaj@warwick.ac.uk

Kurt Mehlhorn
Max-Planck-Institut für Informatik
Saarbrücken, Germany
E-mail: mehlhorn@mpi-inf.mpg.de

Andrew Pitts
University of Cambridge
Computer Laboratory
Cambridge, UK
E-mail: andrew.pitts@cl.cam.ac.uk

Roger Wattenhofer
ETH Zurich, Switzerland
E-mail: wattenhofer@ethz.ch

ISSN 0302-9743   e-ISSN 1611-3349
ISBN 978-3-642-31584-8   e-ISBN 978-3-642-31585-5
DOI 10.1007/978-3-642-31585-5
Springer Heidelberg Dordrecht London New York

Library of Congress Control Number: 2012940794

CR Subject Classification (1998): F.2, F.1, C.2, H.3-4, G.2, I.2

LNCS Sublibrary: SL 1 – Theoretical Computer Science and General Issues

© Springer-Verlag Berlin Heidelberg 2012
This work is subject to copyright. All rights are reserved, whether the whole or part of the material is
concerned, specifically the rights of translation, reprinting, re-use of illustrations, recitation, broadcasting,
reproduction on microfilms or in any other way, and storage in data banks. Duplication of this publication
or parts thereof is permitted only under the provisions of the German Copyright Law of September 9, 1965,
in its current version, and permission for use must always be obtained from Springer. Violations are liable
to prosecution under the German Copyright Law.
The use of general descriptive names, registered names, trademarks, etc. in this publication does not imply,
even in the absence of a specific statement, that such names are exempt from the relevant protective laws
and regulations and therefore free for general use.

Typesetting: Camera-ready by author, data conversion by Scientific Publishing Services, Chennai, India

Printed on acid-free paper

Springer is part of Springer Science+Business Media (www.springer.com)

Preface

This volume contains the papers presented at the 39th International Colloquium on Automata, Languages and Programming (ICALP 2012), held during July 9–13, 2012 at the University of Warwick, UK. ICALP is the main conference and annual meeting of the European Association for Theoretical Computer Science (EATCS) and first took place in 1972. This year the ICALP program consisted of three tracks:

- Track A: Algorithms, Complexity and Games
- Track B: Logic, Semantics, Automata and Theory of Programming
- Track C: Foundations of Networked Computation

In response to the call for papers, the three Program Committees received a total of 432 submissions: 248 for Track A, 105 for Track B, and 79 for Track C. Each submission was reviewed by three or more Program Committee members, aided by sub-reviewers. The committees decided to accept 123 papers for inclusion in the scientific program: 71 papers for Track A, 30 for Track B, and 22 for Track C. The selection was made by the Program Committees based on originality, quality, and relevance to theoretical computer science. The quality of the submissions was very high indeed, and many deserving papers could not be selected.

The EATCS sponsored awards for both a best paper and a best student paper (to qualify for which, all authors must be students) for each of the three tracks, selected by the Program Committees.

The best paper awards were given to the following papers:

Track A: Leslie Ann Goldberg and Mark Jerrum for their paper "The Complexity of Computing the Sign of the Tutte Polynomial (and Consequent #P-hardness of Approximation)"

Track B: Volker Diekert, Manfred Kufleitner, Klaus Reinhardt, and Tobias Walter for their paper "Regular Languages are Church-Rosser Congruential"

Track C: Piotr Krysta and Berthold Vöcking for their paper "Online Mechanism Design (Randomized Rounding on the Fly)"

The best student paper awards were given to the following papers:

Track A: jointly, Shelby Kimmel for her paper "Quantum Adversary (Upper) Bound" and Anastasios Zouzias for his paper "A Matrix Hyperbolic Cosine Algorithm and Applications"

Track B: Yaron Velner for his paper "The Complexity of Mean-Payoff Automaton Expression"

Track C: Leonid Barenboim for his paper "On the Locality of Some NP-Complete Problems"

In addition to the contributed papers, the conference included six invited lectures, by Gilles Dowek (INRIA Paris), Kohei Honda (Queen Mary London), Stefano Leonardi (Sapienza University of Rome), Daniel Spielman (Yale), Berthold Vöcking (RWTH Aachen University), and David Harel (The Weizmann Institute of Science). David Harel's talk was in honor of Alan Turing, since the conference was one of the Alan Turing Centenary Celebration events, celebrating the life, work, and legacy of Alan Turing.

The following workshops were held as satellite events of ICALP 2012 on July 8, 2012:

- Workshop on Applications of Parameterized Algorithms and Complexity (APAC)
- 4th International Workshop on Classical Logic and Computation (CL&C)
- Third Workshop on Realistic models for Algorithms in Wireless Networks (WRAWN)

We wish to thank all the authors who submitted extended abstracts for consideration, the members of the three Program Committees for their scholarly efforts, and all sub-reviewers who assisted the Program Committees in the evaluation process. We thank the sponsors Microsoft Research, Springer-Verlag, EATCS, and the Centre for Discrete Mathematics and its Applications (DIMAP) for their support, and the University of Warwick for hosting ICALP 2012. We are also grateful to all members of the Organizing Committee and to their support staff. The conference-management system EasyChair was used to handle the submissions, to conduct the electronic Program Committee meeting, and to assist with the assembly of the proceedings.

May 2012

Artur Czumaj
Kurt Mehlhorn
Andrew Pitts
Roger Wattenhofer

Organization

Program Committee

Susanne Albers	Humboldt-Universität zu Berlin, Germany
Albert Atserias	Universitat Politecnica de Catalunya, Spain
Andrei Brodnik	University of Ljubljana, Slovenia
Harry Buhrman	University of Amsterdam, The Netherlands
Bernard Chazelle	Princeton University, USA
James Cheney	University of Edinburgh, UK
Siu-Wing Cheng	HKUST, Hong Kong, SAR China
Bob Coecke	Oxford University, UK
Amin Coja-Oghlan	University of Warwick, UK
Pierre-Louis Curien	CNRS, France
Ugo Dal Lago	Università di Bologna, Italy
Benjamin Doerr	Max Planck Institute for Informatics, Germany
Stefan Dziembowski	University of Warsaw, Poland
Javier Esparza	Technische Universität München, Germany
Michal Feldman	Hebrew University, Israel
Antonio Fernandez Anta	Institute IMDEA Networks, Spain
Paola Flocchini	University of Ottawa, Canada
Fedor Fomin	University of Bergen, Norway
Pierre Fraigniaud	CNRS and University of Paris 7, France
Philippa Gardner	Imperial College London, UK
Philipp Gibbons	Intel Labs Pittsburgh, USA
Erich Graedel	RWTH Aachen University, Germany
Magnus Halldorsson	Reykjavik University, Iceland
Moritz Hardt	IBM Almaden Research, USA
Maurice Herlihy	Brown University, USA
Daniel Hirschkoff	ENS Lyon, France
Nicole Immorlica	Northwestern University, USA
Bart Jacobs	Radboud University Nijmegen, The Netherlands
Valerie King	University of Victoria, Canada/Microsoft Research SVC, USA
Bartek Klin	University of Warsaw, Poland
Elias Koutsoupias	University of Athens, Greece
Dariusz Kowalski	University of Liverpool, UK
Piotr Krysta	University of Liverpool, UK

Amit Kumar   IIT Delhi, India
Stefano Leonardi   Sapienza University of Rome, Italy
Pinyan Lu   Microsoft Research Asia, Shanghai, China
Nancy Lynch   MIT CSAIL, USA
Kazuhisa Makino   University of Tokyo, Japan
Dahlia Malkhi   Microsoft Research, Silicon Valley, USA
Laurent Massoulie   Thomson Research, Paris, France
Kurt Mehlhorn   Max Planck Institut fur Informatik
                                Saarbrücken, Germany
Julian Mestre   University of Sydney, Australia
Aleksandar Nanevski   IMDEA-Software, Madrid, Spain
Flemming Nielson   Technical University of Denmark, Denmark
Rasmus Pagh   IT University of Copenhagen, Denmark
Alessandro Panconesi   Sapienza University of Rome, Italy
Rafael Pass   Cornell University, USA
Boaz Patt-Shamir   Tel Aviv University, Israel
Andrew Pitts   University of Cambridge, UK
Harald Raecke   Technische Universität München, Germany
Arend Rensink   University of Twente, The Netherlands
Peter Sanders   University of Karlsruhe, Germany
Davide Sangiorgi   University of Bologna, Italy
Piotr Sankowski   University of Warsaw, Poland
Saket Saurabh   The Institute of Mathematical Sciences,
                                Chennai, India
Carsten Schuermann   IT University of Copenhagen, Denmark
Nicole Schweikardt   Goethe-Universität Frankfurt am Main,
                                Germany
Angelika Steger   ETH Zurich, Switzerland
Andrzej Tarlecki   Warsaw University, Poland
Patrick Thiran   EPFL, Switzerland
Suresh Venkatasubramanian   University of Utah, USA
Bjorn Victor   Uppsala University, Sweden
Igor Walukiewicz   CNRS, LaBRI, France
Roger Wattenhofer   ETH Zurich, Switzerland
Thomas Wilke   University of Kiel, Germany
James Worrell   Oxford University, UK
Masafumi Yamashita   Kyushu University, Japan

Additional Reviewers

Aaronson, Scott
Abdalla, Michel
Abdullah, Amirali
Abu Zaid, Faried
Aceto, Luca
Adamczyk, Marek
Adler, Isolde
Agmon, Noa
Ahn, Jae Hyun
Alaei, Saeed
Albers, Susanne
Alessi, Fabio
Alglave, Jade
Almagor, Shaull
Altenkirch, Thorsten
Althaus, Ernst
Ambainis, Andris
Ambos-Spies, Klaus
Anand, S.
Andoni, Alexandr
Anh Dung, Phan
Anshelevich, Elliot
Antoniadis, Antonios
Arjona, Jordi
Arrighi, Pablo
Arvind, Vikraman
Asarin, Eugene
Asharov, Gilad
Aspnes, James
Ateniese, Giuseppe
Atig, Mohamed Faouzi
Azar, Yossi
Aziz, Haris
Bachrach, Yoram
Bae, Sang Won
Bansal, Nikhil
Barceló, Pablo
Bateni, Mohammadhossein
Batu, Tuğkan
Bauer, Andreas
Beame, Paul
Becchetti, Luca
Becker, Florent

Bei, Xiaohui
Benedikt, Michael
Bengtson, Jesper
Berenbrink, Petra
Bernáth, Attila
Berwanger, Dietmar
Bezhanishvili, Nick
Bhaskara, Aditya
Bhattacharyya, Arnab
Bhawalkar, Kshipra
Bidkhori, Hoda
Bienkowski, Marcin
Bienvenu, Laurent
Bierman, Gavin
Bille, Philip
Bilo', Vittorio
Bingham, Brad
Birget, Jean-Camille
Birkedal, Lars
Björklund, Henrik
Björklund, Johanna
Bloem, Roderick
Blondin Massé, Alexandre
Blumensath, Achim
Blömer, Johannes
Bonchi, Filippo
Bonifaci, Vincenzo
Bonsma, Paul
Boreale, Michele
Borgstrom, Johannes
Bosnacki, Dragan
Bosque, Jose Luis
Bouajjani, Ahmed
Boyle, Elette
Brach, Paweł
Brautbar, Michael
Bremer, Joachim
Briet, Jop
Bringmann, Karl
Broadbent, Anne
Brodal, Gerth Stølting
Bucciarelli, Antonio
Buchbinder, Niv

Bukh, Boris
Bulatov, Andrei
Cai, Leizhen
Caires, Luis
Calzavara, Stefano
Canetti, Ran
Capecchi, Sara
Carayol, Arnaud
Carbone, Marco
Caskurlu, Bugra
Celis, Laura Elisa
Censor-Hillel, Keren
Chakrabarti, Amit
Chakraborty, Sourav
Chakraborty, Tanmoy
Chan, Sze-Hang
Chatterjee, Krishnendu
Chatzigiannakis, Ioannis
Chekuri, Chandra
Chen, Ning
Chen, Xi
Childs, Andrew
Christodoulou, George
Chung, Kai-Min
Churchill, Martin
Chuzhoy, Julia
Ciancia, Vincenzo
Cicerone, Serafino
Cittadini, Luca
Clairambault, Pierre
Clemente, Lorenzo
Cohen, David
Colcombet, Thomas
Corbo, Jacomo
Cormode, Graham
Cornejo, Alejandro
Courcelle, Bruno
Cygan, Marek
Czerwinski, Wojciech
Czumaj, Artur
D'Souza, Deepak
Dadush, Daniel
David, Alexandre
De Leoni, Massimiliano
De Liguoro, Ugo

de Wolf, Ronald
Degorre, Aldric
Dell, Holger
Demangeon, Romain
Deng, Xiaotie
Deng, Yuxin
Deshpande, Amit
Devanur, Nikhil
Diekert, Volker
Dinitz, Michael
Dinsdale-Young, Thomas
Dittmann, Christoph
Downey, Rod
Doyen, Laurent
Drange, Pål
Drewes, Frank
Dreyer, Derek
Drucker, Andrew
Duflot, Marie
Dughmi, Shaddin
Durnoga, Konrad
Dyagilev, Kirill
Dyer, Martin
Efthymiou, Charilaos
Eisentraut, Christian
Elbassioni, Khaled
Ellen, Faith
Englert, Matthias
Epstein, Leah
Ergun, Funda
Fahrenberg, Uli
Feige, Uriel
Fekete, Sándor
Ferns, Norman
Filipiuk, Piotr
Finocchi, Irene
Fischer, Johannes
Fogarty, Seth
Forej, Vojtech
Fortnow, Lance
Fotakis, Dimitris
Frandsen, Gudmund Skovbjerg
Freydenberger, Dominik
Frieze, Alan
Fu, Hu

Fujito, Toshihiro
Fukunaga, Takuro
Gabbay, Murdoch
Gacs, Peter
Gal, Anna
Gamarnik, David
Gambin, Anna
Gamzu, Iftah
Ganguly, Sumit
Ganty, Pierre
Garcia Saavedra, Andres
García, Álvaro
Garg, Naveen
Gargano, Luisa
Gaspers, Serge
Gauwin, Olivier
Gavoille, Cyril
Gawrychowski, Pawel
Gay, Simon
Ge, Rong
Ghaffari, Mohsen
Ghica, Dan
Giachino, Elena
Gimbert, Hugo
Godskesen, Jens Chr.
Goerdt, Andreas
Goergiou, Chryssis
Goldberg, Leslie Ann
Goldhirsh, Yonatan
Golovach, Petr
Gorla, Daniele
Gottesman, Daniel
Goubault-Larrecq, Jean
Gould, Victoria
Grabmayer, Clemens
Grandoni, Fabrizio
Gravin, Nikolay
Grenet, Bruno
Griffin, Christopher
Grigorieff, Serge
Grohe, Martin
Gugelmann, Luca
Guha, Sudipto
Guillon, Pierre
Gupta, Anupam

Guruswami, Venkatesan
Gutin, Gregory
Gutkovas, Ramunas
Göller, Stefan
Haghpanah, Nima
Hague, Matthew
Hajiaghayi, Mohammadtaghi
Halevi, Shai
Halldorsson, Magnus M.
Hallgren, Sean
Hamano, Masahiro
Han, Xin
Hansen, Helle Hvid
Harwath, Frederik
Haviv, Ishay
Hay, David
Herbreteau, Frédéric
Hernich, André
Hertel, Philipp
Herzen, Julien
Hildebrandt, Thomas
Hodkinson, Ian
Hoefer, Martin
Hoenicke, Jochen
Hofheinz, Dennis
Hohenberger, Susan
Holzer, Markus
Horn, Florian
Hsu, Tsan-Sheng
Huang, Chien-Chung
Huang, Zengfeng
Hunter, Paul
Husfeldt, Thore
Høyer, Peter
Hüffner, Falk
Indyk, Piotr
Iosif, Radu
Ishii, Toshimasa
Ito, Takehiro
Jain, Abhishek
Jain, Rahul
Jancar, Petr
Jansen, Bart
Jansen, Klaus
Jeż, Łukasz

Jones, Neil
Jowhari, Hossein
Jurdzinski, Marcin
Jurdzinski, Tomasz
Kaiser, Lukasz
Kakimura, Naonori
Kaminski, Marcin
Kamiyama, Naoyuki
Kanté, Mamadou
Kapron, Bruce
Kari, Lila
Kartzow, Alexander
Kashefi, Elham
Katoen, Joost-Pieter
Katoh, Naoki
Katz, Jonathan
Kaufman, Tali
Kawachi, Akinori
Kawamura, Akitoshi
Kawase, Yasushi
Kazana, Tomasz
Kesner, Delia
Khabbazian, Majid
Kiefer, Stefan
Kijima, Shuji
Kirsten, Daniel
Kiwi, Marcos
Kiyomi, Masashi
Klasing, Ralf
Kleinberg, Robert
Kliemann, Lasse
Kobayashi, Yusuke
Koenemann, Jochen
Koenig, Barbara
Kolliopoulos, Stavros
Konrad, Christian
Kopczynski, Eryk
Kopparty, Swastik
Kortsarz, Guy
Korula, Nitish
Kosowski, Adrian
Koutavas, Vasileios
Kozen, Dexter
Kratsch, Dieter
Krauthgamer, Robert

Krcal, Jan
Krcal, Pavel
Kretinsky, Jan
Krishnaswamy, Ravishankar
Krugel, Johannes
Kucherov, Gregory
Kufleitner, Manfred
Kullmann, Oliver
Kupferman, Orna
Kuske, Dietrich
Kutzkov, Konstantin
Könemann, Jochen
Łącki, Jakub
Laird, Jim
Lammersen, Christiane
Land, Kati
Lasota, Slawomir
Laurent, Monique
Lee, Troy
Lelarge, Marc
Lengrand, Stephane
Leroux, Jerome
Levavi, Ariel
Ley-Wild, Ruy
Li, Shi
Lin, Anthony Widjaja
Lin, Huijia Rachel
Loff, Bruno
Lotker, Zvi
Lu, Chi-Jen
Lui, Edward
Luttenberger, Michael
Löffler, Maarten
M.S., Ramanujan
Macedonio, Damiano
Madet, Antoine
Madry, Aleksander
Maffei, Matteo
Magniez, Frederic
Mahajan, Meena
Maheshwari, Anil
Mahini, Hamid
Mahmoody, Mohammad
Mairesse, Jean
Makarychev, Konstantin

Malacaria, Pasquale

Malec, David

Malekian, Azarakhsh

Malkis, Alexander

Mandemaker, Jorik

Maneth, Sebastian

Maneva, Elitza

Manthey, Bodo

Marchetti-Spaccamela, Alberto

Mardare, Radu

Martens, Wim

Martin, Greg

Martini, Simone

Marx, Dániel

Massar, Serge

Matisziv, Tim

Mayr, Richard

Mazza, Damiano

McBride, Conor

Mcgregor, Andrew

Medina, Moti

Megow, Nicole

Mei, Alessandro

Mellies, Paul-Andre

Merro, Massimo

Methrayil Varghese, Praveen Thomas

Meyer, Roland

Michail, Othon

Milani, Alessia

Milius, Stefan

Mitra, Pradipta

Miyazaki, Shuichi

Mohar, Bojan

Momigliano, Alberto

Monaco, Gianpiero

Monmege, Benjamin

Montanari, Angelo

Moseley, Benjamin

Mostefaoui, Achour

Mosteiro, Miguel A.

Mozes, Shay

Mucha, Marcin

Mulligan, Dominic

Munagala, Kamesh

Møgelberg, Rasmus Ejlers

Nagarajan, Viswanath

Nandy, Subhas

Nanz, Sebastian

Navarra, Alfredo

Neis, Georg

Nesme, Vincent

Nestmann, Uwe

Neumann, Adrian

Nguyen, Huy

Nicolaou, Nicolas

Niehren, Joachim

Nielson, Hanne Riis

Nies, Andre

Nishide, Takashi

Nussbaum, Yahav

Nutov, Zeev

O Dunlaing, Colm

Obdrzalek, Jan

Obremski, Maciej

Onak, Krzysztof

Ono, Hirotaka

Ooshita, Fukuhito

Oren, Sigal

Orlandi, Alessio

Orlandi, Claudio

Oshman, Rotem

Otachi, Yota

Otto, Friedrich

Otto, Martin

Ouaknine, Joel

Oveis Gharan, Shayan

Padovani, Luca

Paes Leme, Renato

Pakusa, Wied

Panagiotou, Konstantinos

Parker, Matthew

Parrow, Joachim

Parys, Pawel

Patitz, Matthew

Paturi, Ramamohan

Paulusma, Daniel

Pavlovic, Dusko

Perdrix, Simon

Person, Yury

Peter, Ueli

Petit, Barbara
Philip, Geevarghese
Philippou, Anna
Phillips, Jeff
Pientka, Brigitte
Pietrzak, Krzysztof
Piliouras, Georgios
Pilipczuk, Marcin
Pilipczuk, Michal
Pin, Jean-Eric
Pinkau, Chris
Plandowski, Wojciech
Poll, Erik
Poplawski, Laura
Pountourakis, Emmanouil
Pretnar, Matija
Pruhs, Kirk
Puppis, Gabriele
Pyrga, Evangelia
Rabinovich, Roman
Radeva, Tsvetomira
Radoszewski, Jakub
Rafiey, Arash
Rafnsson, Willard
Raghavendra, Prasad
Rahmati, Zahed
Raman, Parasaran
Rao B.V., Raghavendra
Raskin, Jean-Francois
Rawitz, Dror
Reichardt, Ben
Reichman, Daniel
Renaud, Fabien
Reynier, Pierre-Alain
Riba, Colin
Ribichini, Andrea
Richard, Gaétan
Riveros, Cristian
Roland, Jérémie
Rosamond, Frances
Rosulek, Mike
Rybicki, Bartosz
Röglin, Heiko
Sachdeva, Sushant
Sadakane, Kunihiko

Saha, Chandan
Sahai, Amit
Salavatipour, Mohammad
Salvati, Sylvain
Samborski-Forlese, Julian
Sangnier, Arnaud
Sankur, Ocan
Santha, Miklos
Santhanam, Rahul
Santi, Paolo
Santos, Agustin
Sastry, Srikanth
Satti, Srinivasa Rao
Sauerwald, Thomas
Saxena, Nitin
Scafuro, Alessandra
Scarpa, Giannicola
Schaffner, Christian
Schieferdecker, Dennis
Schnoor, Henning
Schudy, Warren
Schwartz, Roy
Schweitzer, Pascal
Schwoon, Stefan
Scott, Philip
Segev, Danny
Seki, Shinnosuke
Senizergues, Geraud
Serre, Olivier
Seshadhri, C.
Seth, Karn
Sevilla, Andres
Sewell, Peter
Seyalioglu, Hakan
Shah, Chintan
Shapira, Asaf
Shen, Alexander
Shioura, Akiyoshi
Sikdar, Somnath
Silva, Alexandra
Silvestri, Riccardo
Singh, Mohit
Sivan, Balasubramanian
Skrzypczak, Michał
Smid, Michiel

Sobocinski, Pawel
Sommer, Christian
Soto, Jose
Spalek, Robert
Spieksma, Frits
Spirakis, Paul
Spöhel, Reto
Srinivasan, Aravind
Srinivasan, Srikanth
Srivastava, Gautam
Stacho, Juraj
Stark, Ian
Staton, Sam
Steurer, David
Stoddard, Greg
Stoelinga, Marielle I.A.
Straubing, Howard
Sun, He
Sun, Xiaoming
Sun, Xiaorui
Sviridenko, Maxim
Swamy, Chaitanya
Szegedy, Mario
Sørensen, Troels Bjerre
Tabareau, Nicolas
Takamatsu, Mizuyo
Takazawa, Kenjiro
Takimoto, Eiji
Tamaki, Suguru
Tamir, Tami
Tanaka, Keisuke
Tanigawa, Shin-Ichi
Taraz, Anusch
Tasson, Christine
Tautschnig, Michael
Telelis, Orestis
Terepeta, Michał
Terui, Kazushige
Tesson, Pascal
Thomas, Henning
Thraves, Christopher
Tokuyama, Takeshi
Toninho, Bernardo
Torenvliet, Leen
Toruńczyk, Szymon

Torán, Jacobo
Touili, Tayssir
Tredan, Gilles
Trevisan, Luca
Tulsiani, Madhur
Turrini, Andrea
Törmä, Ilkka
Uchizawa, Kei
Ueno, Kenya
Ueno, Shuichi
Ullman, Jon
Uno, Takeaki
Urzyczyn, Paweł
Vaidya, Nitin
Van Breugel, Franck
van Stee, Rob
Van Zuylen, Anke
Vassilvitskii, Sergei
Venema, Yde
Venkitasubramaniam,
     Muthuramakrishnan
Venturini, Rossano
Vereshchagin, Nikolay
Versari, Cristian
Verschae, Jose
Vidick, Thomas
Vijayaraghavan, Aravindan
Vilenchik, Dan
Vilenchik, Danny
Villard, Jules
Viola, Emanuele
Visconti, Ivan
Vishkin, Uzi
Vishnoi, Nisheeth
Vitanyi, Paul
Vondrak, Jan
Wachter, Björn
Wahlström, Magnus
Wang, Yajun
Ward, Justin
Wasowski, Andrzej
Weber, Tjark
Wee, Hoeteck
Wei, Zhewei
Weihmann, Jeremias

Weimann, Oren
Welch, Jennifer
Whittle, Geoff
Wiedijk, Freek
Wiese, Andreas
Williams, Ryan
Wimmer, Karl
Winzen, Carola
Wong, Prudence W.H.
Woodruff, David
Worrell, James
Wu, Yi
Wulff-Nilsen, Christian
Xia, Mingji
Xiao, David
Xie, Ning
Yamamoto, Masaki
Yamauchi, Yukiko
Yannakakis, Mihalis
Ye, Deshi
Yekhanin, Sergey
Yi, Ke

Yin, Yitong
Yokoo, Makoto
Yoshida, Yuichi
Young, Neal
Zadimoghaddam, Morteza
Zając, Michał
Zavattaro, Gianluigi
Zavou, Elli
Zdeborova, Lenka
Zelke, Mariano
Zhang, Chihao
Zhang, Fuyuan
Zhang, Jialin
Zhang, Jiawei
Zhang, Qin
Zhang, Shengyu
Zheng, Colin
Zheng, Jia
Zhong, Ning
Zhou, Yuan
Zohar, Aviv

Table of Contents – Part II

Invited Talks

Track B – Logic Semantics, Automata and Theory of Programming

Track C – Foundations of Networked Computation

Table of Contents – Part I

Track A – Algorithms, Complexity and Games

On Multiple Keyword Sponsored Search Auctions with Budgets

Riccardo Colini-Baldeschi[1,*], Monika Henzinger[2,**],
Stefano Leonardi[1,*], and Martin Starnberger[2,**]

[1] Sapienza University of Rome, Italy
[2] University of Vienna, Austria

Abstract. We study *multiple keyword* sponsored search auctions with budgets. Each keyword has *multiple ad slots* with a click-through rate. The bidders have additive valuations, which are linear in the click-through rates, and budgets, which are restricting their overall payments. Additionally, the number of slots per keyword assigned to a bidder is bounded.

We show the following results: (1) We give the first mechanism for multiple keywords, where click-through rates differ among slots. Our mechanism is incentive compatible in expectation, individually rational in expectation, and Pareto optimal. (2) We study the combinatorial setting, where each bidder is only interested in a subset of the keywords. We give an incentive compatible, individually rational, Pareto optimal, and deterministic mechanism for identical click-through rates. (3) We give an impossibility result for incentive compatible, individually rational, Pareto optimal, and deterministic mechanisms for bidders with diminishing marginal valuations.

1 Introduction

In *sponsored search* (or *adwords*) auctions advertisers bid on *keywords*. Such auctions are used by firms such as Google, Yahoo, and Microsoft [11]. The search result page for each keyword contains multiple slots for ads and each bidder is assigned to a limited number of slots on a search result page. The slots have a click-through rate (CTR), which is usually decreasing by the position of the slot on the search result page. The true valuation of the bidders is private knowledge and is assumed to depend linearly on the CTR. Moreover, valuations are assumed to be additive, i.e., the total valuation of a bidder is equal to the sum of his valuations for all the slots that are assigned to him.

A further key ingredient of an adwords auction is that bidders specify a budget on the payment charged for the ads, effectively linking the different keywords. The deterministic Vickrey auction [18] was designed to maximize social welfare

---

* This work was partially supported from ERC Starting Grant 259515: Practical Approximation Algorithms.
** This work has been funded by the Vienna Science and Technology Fund (WWTF) through project ICT10-002 and by the University of Vienna through IK I049-N.

A. Czumaj et al. (Eds.): ICALP 2012, Part II, LNCS 7392, pp. 1–12, 2012.
© Springer-Verlag Berlin Heidelberg 2012

in this and more general settings without budget restrictions. However, the introduction of budgets dramatically changes the nature of the problem. The Vickrey auction may charge more than the budget and is no longer feasible. Moreover, bidders might get assigned to slots even though their budget is arbitrary small and other bidders are interested in those slots. Thus, as was observed before [7,8], maximizing social welfare is not the right optimality criterion to use. In a seminal paper by Dobzinski et al. [7,8], they considered the multi-unit case with additive valuations, which in the sponsored search setting corresponds to each keyword having only one slot and all slots having identical CTR. They gave an incentive compatible (IC) auction based on Ausubel's ascending clinching auction [3] that produces a Pareto optimal (PO) and individually rational (IR) allocation if budgets are *public*. They also showed that this assumption is strictly needed, i.e., that no deterministic mechanism for *private* budgets exists if we insist on incentive compatibility, individual rationality, and on obtaining an allocation that is Pareto optimal. This impossibility result for *deterministic* mechanisms was strengthened for our setting to *public* budgets in Dütting et al. [9]. The question was open what optimality result can be achieved for *randomized* mechanisms. Due to the impossibility results for deterministic mechanisms it is unlikely that "strong" optimality criteria, such as bidder optimality, are achievable. Thus, the first question to study is whether Pareto optimality, which is a basic notion of optimality, can be achieved with *randomized* mechanisms. Note that if an allocation is Pareto optimal then it is impossible to make a bidder better off without making another bidder or the auctioneer worse off, and is therefore the least one should aim for.

Our Results. We give a positive answer to the above question and also present two further related results. Specifically, the paper contains the following three results: (1) *Multiple keywords with multiple slots:* We show that the multi-unit auction of Dobzinski et al. [7,8] can be adapted to study adwords auctions with multiple keywords having multiple slots, and budget limits for each bidder. We specifically model the case of several slots with different CTR, available for each keyword, and a bound on the number of slots per keyword (usually one) that can be allocated to a bidder. We first provide an IC, IR, and PO deterministic auction that provides a fractional allocation for the case of one keyword with divisible slots. Note that the impossibility result in [9] does not hold for divisible slots. In contrast, the impossibility result in [7,8] for multi-unit auctions applies also to this setting, and achieving IC, IR, and PO deterministic auctions is only possible if budgets are *public*. Thus, we restrict ourselves to the public budget case. Our auction is a one-shot auction, i.e., each bidder interacts only once with the auction. We then show how to probabilistically round this fractional allocation for the divisible case to an integer allocation for the indivisible case with multiple keywords (i.e., the adwords setting) and get an auction that is IC in expectation, IR in expectation, and PO.

(2) *Multiple keywords with combinatorial constraints and multiple slots:* So far we assumed that every bidder is interested in every keyword. In the second part of the paper we study the case that bidders are interested in only a subset of the

keywords, i.e., bidders have a non-zero identical valuation only on a *subset* of the keywords. The valuations are additive and each bidder is assigned at most one slot for a given keyword. We restrict the model by allowing only identical slots for each keyword, i.e., we require that all slots have the same CTR. This setting extends the combinatorial one-slot per keyword model considered by Fiat et al. [12] to multiple slots. We present a variation of the clinching auction that is deterministic, IC, IR, and PO.

(3) Finally, we also study non-additive valuations, namely valuations with diminishing marginal valuations. Diminishing marginal valuation (also called submodular) functions are widely used to model auction settings with marginal utilities being positive functions that are non-increasing in the number of items already allocated to the bidders. We show that even in the multi-unit (one slot per keyword) case there is no deterministic, IC, IR, and PO auction for private diminishing marginal valuations and public budgets. This shows how budgets complicate mechanism design: For the non-budgeted version of this setting Ausubel [3] gave his deterministic mechanism.

Related Work. Ascending clinching auctions are used in the FCC spectrum auctions, see [16,4,3]. For a motivation of adwords auctions see [17] on Google's auction for TV ads.

We first compare our results with those of a recent, unpublished work by Goel et al. [14] that was developed independently at the same time. They studied IC auctions with feasible allocations that must obey public polymatroid constraints and agents with identical or separable valuations (see their Lemma 3.10) and public budgets. The problem of auctions with polymatroid constraints was first studied by Bikhchandani et al. [6] for unbudgeted bidders and concave utilities. The auction in [14] is an adaption of the ascending auction in [6] to the case of budgeted bidders. The polymatroid constraints generalize on one hand the the multi-unit case in [7,8] and the multiple slots with different CTRs model presented in this paper. On the other hand, the PO ascending auction in [14] only returns allocations for divisible items whereas in Sect. 4 of this paper we demonstrate that these allocations can be rounded to *allocations for indivisible items* if we allow the auction to yield incentive compatibility in expectation. In Sect. 5, we present an IC, IR, and PO *deterministic* auction with feasible *allocations of indivisible slots* that obey matching constraints for the case of multiple identical slots.

There are three extensions of Dobzinski et al. [7,8]: (1) Fiat et al. [12] studied an extension to a combinatorial setting, where items are distinct and different bidders may be interested in different items. The auction presented in [12] is IC, IR, and PO for additive valuations and single-valued bidders (i.e., every bidder does not distinguish between the keywords in his public interest set). This is a special case of our combinatorial setting in Sect. 5 with multiple keywords but only one slot per keyword. (2) Bhattacharya et al. [5] dealt with private budgets, and gave an auction for one infinitely divisible item, where bidders cannot improve their utility by underreporting their budget. This leads to a randomized IC in expectation auction for one infinitely divisible item with both private valuations and budgets. (3) Several papers [1,2,10,13] studied *envy-free* outcomes that are bidder optimal, respectively PO, in an one-keyword adwords auction. In this setting they give (under certain conditions on the input) an IC auction with both private valuations and budgets.

Our impossibility result in Sect. 6 is related to two impossibility results: Lavi and May [15] show that there is no IC, IR, and PO deterministic mechanism for indivisible items and bidders with *monotone* valuations. Our result for indivisible items is stronger as it applies to bidders with non-negative and diminishing marginal valuations. In [14] the same impossibility result for *divisible* items and bidders with monotone and concave utility functions was given. Note that neither their result nor ours implies the other.

2   Problem Statement and Definitions

We have n bidders and m slots. We call the set of bidders $I := \{1, \ldots, n\}$ and the set of slots $J := \{1, \ldots, m\}$. Each bidder $i \in I$ has a private *valuation* v_i, a public *budget* b_i, and a public *slot constraint* κ_i, which is a positive integer. Each slot $j \in J$ has a public *quality* $\alpha_j \in \mathbb{Q}_{\geq 0}$. The slots are ordered such that $\alpha_j \geq \alpha_{j'}$ if $j > j'$, where ties are broken in some arbitrary but fixed order. We assume in Sect. 3 and 4 that the number of slots m fulfills $m = \sum_{i \in I} \kappa_i$ as the general case can be easily reduced to this setting.

Divisible case: In the divisible case we assume that there is only one keyword with infinitely divisible slots. Thus the goal is to assign each bidder i a fraction $x_{i,j} \geq 0$ of each slot j and charge him a payment p_i. A matrix $X = (x_{i,j})_{(i,j) \in I \times J}$ and a payment vector p are called an *allocation* (X, p). We call $c_i = \sum_{j \in J} \alpha_j x_{i,j}$ the *weighted capacity* allocated to bidder i. An allocation is *feasible* if it fulfills the following conditions: (1) the sum of the fractions assigned to a bidder does not exceed his *slot constraint* ($\sum_{j \in J} x_{i,j} \leq \kappa_i \ \forall i \in I$); (2) each of the slots is fully assigned to the bidders ($\sum_{i \in I} x_{i,j} = 1 \ \forall j \in J$); and (3) the payment of a bidder does not exceed his budget limit ($b_i \geq p_i \ \forall i \in I$).

Indivisible case: We additionally have a set R of *keywords*, where $|R|$ is public. The goal is to assign each slot $j \in J$ of keyword $r \in R$ to one bidder $i \in I$ while obeying various constraints. An assignment $X = (x_{i,j,r})_{(i,j,r) \in I \times J \times R}$ where $x_{i,j,r} = 1$ if slot j is assigned to bidder i in keyword r, and $x_{i,j,r} = 0$ otherwise, and a payment vector p form an *allocation* (X, p). We call $c_i = \sum_{j \in J} \frac{\alpha_j}{|R|} (\sum_{r \in R} x_{i,j,r})$ the *weighted capacity* allocated to bidder i. An allocation is *feasible* if it fulfills the following conditions: (1) the number of slots of a keyword that are assigned to a bidder does not exceed his *slot constraint* ($\sum_{j \in J} x_{i,j,r} \leq \kappa_i \ \forall i \in I, \forall r \in R$); (2) each slot is assigned to exactly one bidder ($\sum_{i \in I} x_{i,j,r} = 1 \ \forall j \in J, \forall r \in R$); and (3) the payment of a bidder does not exceed his budget limit ($b_i \geq p_i \ \forall i \in I$).

Combinatorial indivisible case: In the combinatorial case not all keywords are identical. Every bidder $i \in I$ has a publicly known set of interest $S_i \subseteq R$, and valuation v_i for all keywords in S_i and a valuation of zero for all other keywords. We model this case by imposing $x_{i,j,r} = 0 \ \forall r \notin S_i$.

Note that in all cases the budgets are bounds on *total* payments across keywords and *not* bounds on prices of individual keywords.

Properties of the auctions: The *utility* u_i of bidder i for a feasible allocation (X, p) is $c_i v_i - p_i$, the *utility* of the auctioneer (or mechanism) is $\sum_{i \in I} p_i$. We study auctions that select feasible allocations obeying the following conditions: (*Bidder rationality*) $u_i \geq 0$ for all bidders $i \in I$, (*Auctioneer rationality*) the utility of the auctioneer fulfills $\sum_{i \in I} p_i \geq 0$, and (*No-positive-transfer*) $p_i \geq 0$ for all bidders $i \in I$. An auction that on all inputs outputs an allocation that is both bidder rational and auctioneer rational is called *individually rational (IR)*. A feasible allocation (X, p) is *Pareto optimal* (PO) if there is no other feasible allocation (X', p') such that (1) the utility of no bidder in (X, p) is less than his utility in (X', p'), (2) the utility of the auctioneer in (X, p) is no less than his utility in (X', p'), and (3) at least one bidder or the auctioneer is better off in (X', p') compared with (X, p). An auction is *incentive compatible* (IC) if it is a dominant strategy for all bidders to reveal their true valuation. An auction is said to be *Pareto optimal* (PO) if the allocation it produces is PO. A randomized auction is IC in expectation, IR in expectation, respectively PO in expectation if the above conditions hold in expectation. We show that our randomized mechanism for indivisible slots is PO in expectation and that each realized allocation is PO.

3 Deterministic Clinching Auction for the Divisible Case

3.1 Characterization of Pareto Optimality

In this section we present a novel characterization of PO allocations that allows to address the divisible case of multiple slots with different CTRs. Given a feasible allocation (X, p), a *swap* between two bidders i and i' is a fractional exchange of slots, i.e., if there are slots j and j' and a constant $\tau > 0$ with $x_{i,j} \geq \tau$ and $x_{i',j'} \geq \tau$ then a swap between i and i' gives a new feasible (X', p) with $x'_{i,j} = x_{i,j} - \tau$, $x'_{i',j'} = x_{i',j'} - \tau$, $x'_{i,j'} = x_{i,j'} + \tau$, and $x'_{i',j} = x_{i',j} + \tau$. If $\alpha_j < \alpha_{j'}$ then the swap increases i's weighted capacity. We assume throughout this section that $\alpha_j \neq \alpha_{j'}$ for $j \neq j'$, the general case requires a small modification presented in the full version of our paper. To characterize PO allocations we first define for each bidder i the set N_i of bidders such that for every bidder a in N_i there exists a swap between i and a that increases i's weighted capacity. Given a feasible allocation (X, p) we use $h(i) := \max\{j \in J | x_{i,j} > 0\}$ for the slot with the highest quality that is assigned to bidder i and $l(i) := \min\{j \in J | x_{i,j} > 0\}$ for the slot with the lowest quality that is assigned to bidder i. Now, $N_i = \{a \in I | h(a) > l(i)\}$ is the set of all the bidders a such that i could increase his weighted capacity (and a could decrease his weighted capacity) if i traded with a, for example, if i received part of a's share of slot $h(a)$. To model sequences of swaps we define furthermore $N_i^k = N_i$ for $k = 1$ and $N_i^k = \bigcup_{a \in N_i^{k-1}} N_a$ for $k > 1$. Since we have only n bidders, $\bigcup_{k=1}^n N_i^k = \bigcup_{k=1}^{n'} N_i^k$ for all $n' \geq n$. We define $\tilde{N}_i := \bigcup_{k=1}^n N_i^k \setminus \{i\}$ as the set of *desired (recursive) trading partners* of i. See Fig. 3.1 for an example with four bidders. The bidders a in $\tilde{N}_i$ are all the bidders such that through a sequence of trades that "starts" with i and "ends" with a, bidder i could increase his weighted capacity, bidder a could decrease his weighted capacity, and the capacity of the remaining bidders involved in the swap would been unchanged. Now let $\tilde{v}_i = \min_{a \in \tilde{N}_i}(v_a)$ if $\tilde{N}_i \neq \emptyset$ and $\tilde{v}_i = \infty$ else.

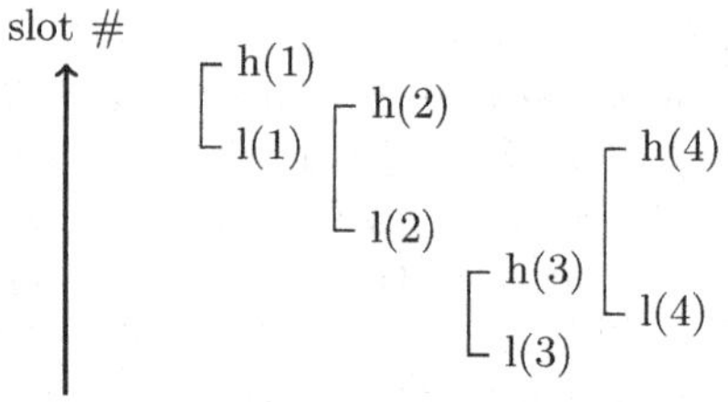

slot #

$$N_1 = \{1,2\} \qquad N_2 = \{1,2,4\}$$
$$N_3 = \{1,2,3,4\} \qquad N_4 = \{1,2,3,4\}$$

$$N_1^2 = \{1,2,4\} \qquad N_2^2 = \{1,2,3,4\}$$

$$N_1^3 = \{1,2,3,4\}$$

$$\tilde{N}_1 = \{2,3,4\} \qquad \tilde{N}_2 = \{1,3,4\}$$
$$\tilde{N}_3 = \{1,2,4\} \qquad \tilde{N}_4 = \{1,2,3\}$$

Fig. 1. Example of desired trading partners

Given a feasible allocation (X,p) we use $B := \{i \in I \mid b_i > p_i\}$ to denote the set of bidders who have a positive remaining budget. As we show below if for a given assignment we know $\tilde{v}_i$ for every bidder $i \in B$ then we can immediately decide whether the assignment is PO or not.

We say that a feasible allocation (X,p) contains a *trading swap sequence* (for short *trading swap*) if there exists a feasible allocation (X',p') and two bidders $u, w \in I$ such that

1. bidder w is a desired trading partner of u, i.e., $w \in \tilde{N}_u$,
2. for all $i \in I \setminus \{u,w\}$ it holds that the weighted capacity of i and the payment of i are unchanged by the swap, i.e., $\sum_{j \in J} \alpha_j x_{i,j} = \sum_{j \in J} \alpha_j x'_{i,j}$ and $p_i = p'_i$,
3. the weighted capacity of u increases by $\delta > 0$ and the weighted capacity of w decreases by δ, i.e., $\delta := \sum_{j \in J} \alpha_j (x'_{u,j} - x_{u,j}) = \sum_{j \in J} \alpha_j (x_{w,j} - x'_{w,j}) > 0$,
4. $v_u > v_w$, u pays after the swap exactly that amount more that w's weighted valuation decreases (i.e., $p'_u - p_u = v_w \delta$), and w pays exactly that amount less (i.e., $p_w - p'_w = v_w \delta$), and
5. u has a high enough budget to pay what is required by (X',p'), i.e., $b_u \geq p'_u$.

We say that the allocation (X',p') *results from the trading swap*. The existence of a trading swap is related to the $\tilde{v}_i$ of each bidder i with remaining budget.

Theorem 1. *A feasible allocation (X,p) contains no trading swaps if and only if $\tilde{v}_i \geq v_i$ for each bidder $i \in B$.*

The following theorem shows that the absence of trading swaps characterizes Pareto optimality. We will use exactly this fact to prove that the mechanism of the next section outputs a PO allocation.

Theorem 2. *A feasible allocation (X,p) is Pareto optimal if and only if it contains no trading swaps.*

Hence, the feasible allocation (X, p) is PO if and only if $\tilde{v}_i \geq v_i \; \forall i \in B$. This novel characterization of Pareto optimality is interesting, as the payment does not affect the values $\tilde{v}_i$, the payment only influences which bidders belong to B.

3.2 Multiple Keyword Auction for the Divisible Case

We describe next our deterministic clinching auction for divisible slots and show that it is IC, IR, and PO. The auction repeatedly increases a price "per capacity" and gives different weights to different slots depending on their CTRs. To perform the check whether all remaining unsold weighted capacity can still be sold we solve suitable linear programs. We will show that if the allocation of the auction did not fulfill the characterization of Pareto optimality given in Sect. 3.1, i.e., if it contained a trading swap, then one of the linear programs solved by the auction would not have computed an optimal solution. Since this is not possible, it will follow that the allocation is PO. A formal description of the auction is given in the procedures AUCTION and SELL. The input values of AUCTION are the bids, budget limits, and slot constraints that the bidders communicate to the auctioneer on the beginning of the auction, and information about the qualities of the slots. The auction is a so called "one-shot auction", the bidders are asked once for the valuations at the beginning of the auction and then they *cannot* input *any* further data.

Algorithm 1. Clinching auction for divisible slots

1: **procedure** AUCTION$(I, J, \alpha, \kappa, v, b)$
2:     $A \leftarrow I; \; \pi \leftarrow 0; \; \pi^+ \leftarrow 1$
3:     $c_i \leftarrow 0, \; p_i \leftarrow 0, \; d_i \leftarrow \infty \; \forall i \in I$
4:     **while** $\sum_{i \in I} c_i < \sum_{j \in J} \alpha_j$ **do**   ▷ unsold weighted capacity exists
5:         $E \leftarrow \{i \in A | \pi^+ > v_i\}$ ▷ bidders become exiting bidders
6:         **for** $i \in E$ **do**
7:             $(X, s) \leftarrow$ SELL$(I, J, \alpha, \kappa, c, d, i)$ ▷ sell to exiting bidder
8:             $(c_i, p_i, d_i) \leftarrow (c_i + s, p_i + s\pi, 0)$
9:         **end for**
10:         $A \leftarrow A \setminus E$ ▷ exiting bidders leave auction
11:         $d_i^+ \leftarrow \frac{b_i - p_i}{\pi^+} \; \forall i \in A$
12:         **while** $\exists i \in A$ with $d_i \neq d_i^+$ **do**   ▷ bidders with price π exist
13:             $i' \leftarrow \min(\{i \in A | d_i \neq d_i^+\})$ ▷ select bidder with price π
14:             $(X, s) \leftarrow$ SELL$(I, J, \alpha, \kappa, c, d, i')$ ▷ sell to bidder
15:             $(c_{i'}, p_{i'}) \leftarrow (c_{i'} + s, p_{i'} + s\pi)$
16:             $d_{i'}^+ \leftarrow \frac{b_{i'} - p_{i'}}{\pi^+}; \; d_{i'} \leftarrow d_{i'}^+$ ▷ increase bidder's price to π^+
17:         **end while**
18:         $\pi \leftarrow \pi^+; \; \pi^+ \leftarrow \pi^+ + 1$ ▷ increase price
19:     **end while**
20:     **return** (X, p)
21: **end procedure**

Algorithm 2. Determination of the weighted capacity that bidder i' clinches

---

1: **procedure** SELL$(I, J, \alpha, \kappa, c, d, i')$
2:     compute an optimal solution of the following linear program
       that is a vertex of the polytope defined by its constraints:
       minimize $\qquad\qquad\qquad \gamma_{i'}$
       s.t.: (a) $\qquad\quad \sum_{i \in I} x_{i,j} = 1 \ \forall j \in J$ $\qquad\qquad\qquad$ ▷ assign all slots
           (b) $\qquad\quad \sum_{j \in J} x_{i,j} = \kappa_i \ \forall i \in I$ $\qquad\qquad\qquad$ ▷ slot constraint
           (c) $\quad \sum_{j \in J} x_{i,j}\, \alpha_j - \gamma_i = c_i \ \forall i \in I$ $\qquad$ ▷ assign value to γ_i
           (d) $\qquad\qquad\quad \gamma_i \leq d_i \ \forall i \in I$ $\qquad\qquad\qquad$ ▷ demand constraint
           (e) $\qquad\qquad\quad x_{i,j} \geq 0 \ \forall i \in I, \forall j \in J$
           (f) $\qquad\qquad\qquad \gamma_i \geq 0 \ \forall i \in I$
3:     **return** $(X, \gamma_{i'})$
4: **end procedure**

The demand of the bidders for weighted capacity is computed by the mechanism based on their remaining budget and the current price. We assume throughout this section that $v_i \in \mathbb{N}_+$ and $b_i \in \mathbb{Q}_+$ for all $i \in I$.[1] The state of the auction is defined by the current price π, the next price π^+, the weighted capacity c_i that bidder $i \in I$ has clinched so far, and the payment p_i that has been charged so far to bidder i. We define the set of *active* bidders $A \subseteq I$ which are all those $i \in I$ with $\pi \leq v_i$, and the subset E of A of *exiting* bidders which are all those $i \in A$ with $\pi^+ > v_i$. The auction does not increase the price that a bidder $i \in I$ has to pay from π to π^+ for *all* bidders at the same time. Instead, it calls SELL each time before it increases the price for a single bidder. If the price that bidder $i \in A$ has to pay for weighted capacity is π then his demand is $d_i = \frac{b_i - p_i}{\pi}$. If the price he has to pay was already increased to π^+ then his demand is $d_i = \frac{b_i - p_i}{\pi^+} < \frac{b_i - p_i}{\pi}$. In this case, the demand corresponds to d_i^+, that is always equal to $\frac{b_i - p_i}{\pi^+}$. Different from the auction in [7,8,5] a bidder with $d_i = d_i^+$ is also charged the increased price π^+ if he receives additional weighted capacity. Since our price is incremented by one in each round and is not continuously increasing as in prior work, this is necessary for proving the Pareto optimality of the allocation.

The crucial point of the auction is that it sells only weighted capacity s to bidder i at a certain price π or π^+ if it cannot sell s to the other bidders. It computes s by solving a linear program in SELL. We use a linear program as there are two types of constraints to consider: The slot constraint in line (b) of the LP, which constraints "unweighted" capacity, and the demand constraint in line (d) of the LP, which is implied by the budget limit and constraints weighted capacity. In the homogeneous item setting in [7,8,5] there are no slot constraints and the demand constraints are unweighted, i.e., $\alpha_j = 1 \ \forall j \in J$. Thus, no linear program is needed to decide what amount to sell to whom.

---

[1] All the arguments go through if we simply assume that $v_i \in \mathbb{Q}_+$ for all $i \in I$ and there exists a publicly known value $z \in \mathbb{R}_+$ such that for all bidders i and i' either $v_i = v_{i'}$ or $|v_i - v_{i'}| \geq z$.

For each iteration of the outer while-loop the auction first calls SELL for each exiting bidder i and sells him s for price π. This is the last time when he can gain weighted capacity. Afterward, he is no longer an active bidder. Next, it calls SELL for one of the remaining active bidders who has $d_i \neq d_i^+$. It sells him the respective s and increases his price to π^+. It continues the previous step until the price of each active bidder is increased to π^+. Then it sets π to π^+ and π^+ to $\pi^+ + 1$.

It is crucial for the progress and the correctness of the mechanism that there is a feasible solution for the linear program in SELL every time that SELL is called. This is proved in the full version. It follows that the final assignment X is a feasible solution of the linear program in SELL. Thus it fulfills conditions (1) and (2) for a feasible allocation. Condition (3) is also fulfilled as by the definition of the demand of a bidder, the auction guarantees that $b_i \geq p_i$ for all i. Thus, the allocation (X, p) computed by the auction is a feasible allocation. As no bidder is assigned weighted capacity if his price is above his valuation and the mechanism never pays the bidders, the auction is IR. As it is an increasing price auction, it is also IC.

Proposition 1. *The auction is individually rational and incentive compatible and the allocation (X, p) it outputs has only rational entries.*

We show finally that the allocation (X, p) our auction computes does not contain any trading swap, and thus, by Theorem 2 it is PO. The proof shows that every trading swap in (X, p) would lead to a superior solution to one of the linear programs solved by the mechanism. Since the mechanism found an optimal solution this leads to a contradiction.

Theorem 3. *The allocation (X, p) returned by our auction is Pareto optimal.*

4   Randomized Clinching Auction for the Indivisible Case

We will now use the allocation computed by the deterministic auction for *divisible* slots to give a randomized auction for multiple keywords with *indivisible* slots that ensures that bidder i receives at most κ_i slots for each keyword. The randomized auction has to assign to every slot $j \in J$ exactly one bidder $i \in I$ for each keyword $r \in R$. We call a distribution over allocations for the indivisible case *Pareto superior* to another such distribution if the expected utility of a bidder or the auctioneer is higher, while all other expected utilities are at least as large. If a distribution has no Pareto superior distribution, we call it *Pareto optimal*. The basic idea is as follows: Given the PO solution for the *divisible* case, we construct a *distribution over allocations* of the *indivisible* case such that the expected utility of every bidder and of the auctioneer is the same as the utility of the bidder and the auctioneer in the *divisible* case. To be precise, we do not explicitly construct this distribution but instead we give an algorithm that can sample from this distribution. The mechanism for the *indivisible* case would, thus, first call the mechanism for the *divisible* case (with the same input) and then convert the resulting allocation (X^d, p^d) into a representation of a PO distribution over allocations for the *indivisible* case. It then samples from this representation to receive the allocation that it outputs. During all these steps the (expected) utility of the bidders and the auctioneer remains unchanged. As the mechanism for the *divisible* case is IR and IC this implies immediately that the mechanism for the *indivisible* case is IR in expectation and IC in expectation. To show that the final allocation is PO in expectation and also PO ex post we use the following lemma.

Lemma 1. *For every probability distribution over feasible allocations in the indivisible case there exists a feasible allocation (X^d, p^d) in the divisible case, where the utility of the bidders and the auctioneer equals their expected utility using this probability distribution.*

Lemma 1 implies that any probability distribution over feasible allocations in the *indivisible* case that is Pareto superior to the distribution generated by our auction would lead to a feasible allocation for the *divisible* case that is Pareto superior to (X^d, p^d). This is not possible as (X^d, p^d) is PO. Additionally, each realized allocation is ex-post Pareto optimal: if in the *indivisible* case there existed a Pareto superior allocation to one of the allocations that gets chosen with a positive probability in our auction, then a Pareto superior allocation would exist in the *divisible* case. By the same argument as above this would lead to a contradiction.

We still need to explain how to use the PO allocation (X^d, p^d) for the *divisible* case to give a probability distribution for the *indivisible* case with expected utility for every bidder equal to the utility in the divisible case and how to sample efficiently from this distribution. Given an input for the indivisible case we use it *as is* as an input for the algorithm for the divisible case, ignoring the number of keywords. Based on the allocation (X^d, p^d) for the divisible problem we construct a matrix M' of size $|J| \times \lambda$, where λ is the least common denominator of all the $x^d_{i,j}$ values and where each column of M' corresponds to a feasible assignment for the indivisible one-keyword case. Note that the same assignment can occur in multiple columns of M'. The matrix M' is our representation of the distribution over allocations in the indivisible case. To sample from the distribution we pick for each $r \in R$ a column uniformly at random from the columns of M'. The r-th choice gives the assignment of bidders to the slots of keyword r. The payments are set equal to p^d. In the full version, we give the construction of M' such that after the above sampling step the expected weighted capacity allocation to bidder $i \in I$ equals $\sum_{j \in J} \alpha_j x^d_{i,j}$, i.e., its weighted capacity in the divisible case. Additionally, all of the slots are fully assigned to the bidders, and hence, the stated properties are fulfilled by the randomized auction.

5   The Combinatorial Case with Multiple Slots

We consider single-valued combinatorial auctions with multiple identical slots in multiple keywords. Every bidder $i \in I$ has valuation v_i on all keywords of his interest set S_i. All other keywords are valued zero. The interest sets S_i and the budgets b_i are public knowledge. We further restrict to the case where at most one slot per keyword is allocated to a single bidder (i.e., $\kappa_i = 1$). We require that at least m bidders are interested in each keyword, where m is the number of slots for a keyword.

In our auction, we extend the B-matching based approach and the concept of trading alternating paths in the bidder/keyword bipartite graph by Fiat et al. [12] for their single-slot per keyword setting to our multi-slot per keyword setting.

We characterize a feasible allocation (H, p) by a tuple $H = (H_1, H_2, \ldots, H_n)$, where $H_i \subseteq S_i$ represents the set of keywords that are allocated to bidder i, and by a vector of payments $p = (p_1, p_2, \ldots, p_n)$ with $p_i \leq b_i$ for all $i \in I$. The utility of bidder i is defined by $u_i := v_i|H_i| - p_i$, and the utility of the auctioneer is $\sum_{i=1}^{n} p_i$. We base the allocation of the items in the clinching auction on B-matchings computed on a bipartite graph G with the union of keywords and bidders $(I \cup R)$ as vertex set and the preferences $\{(i, t) \in I \times R | t \in S_i\}$ as edge set. The vertices have degree constraints, which represent the demand constraints for the bidders and the number of unsold slots for the keywords. The B-matchings are the subgraphs of G, which fulfill the constraints, and have a maximal number of edges. The idea of the auction is to sell slots at the highest possible price such that all slots are sold and there exists no competition between bidders. We define the auction and give the proof of the following theorem in the full version.

Theorem 4. *The allocation (H^*, p^*) produced by the combinatorial clinching auction is incentive compatible, individually rational, and Pareto optimal.*

6   Impossibility for Diminishing Marginal Valuations

We assume in this section that we have multiple homogeneous indivisible items and bidders with private diminishing marginal valuations and public budgets. We show that there is no IC, IR, and PO deterministic mechanism for this case.

Bidder i's marginal valuation for obtaining a further item when k items are already assigned to him is $v_i(k + 1)$. His valuation for obtaining $k + 1$ items is therefore $\sum_{j=1}^{k+1} v_i(j)$. The marginal valuations have to fulfill $v_i(k) \geq v_i(k + 1)$ for $k \geq 1$. The initial clinching auction in [3] was indeed proposed for the case of diminishing marginal valuations but without budget limits.

We use that the case of additive valuations, which was studied by Dobzinski et al. [7,8], is a special case of ours, and that they showed that their auction is the only IC, IR, and PO deterministic auction for that case. We study bidders with diminishing marginal valuations that report additive valuations in order to raise the price paid by the other bidders and consequently decrease their demand. A possible decrease of the price charged to the non-truth telling bidders follows.

Theorem 5. *There is no incentive compatible, individually rational, Pareto optimal, and deterministic mechanism for multiple homogeneous indivisible items and agents with private diminishing marginal valuations and public budget limits.*

References

1. Aggarwal, G., Muthukrishnan, S., Pál, D., Pál, M.: General auction mechanism for search advertising. In: WWW 2009: Proceedings of the 18th International Conference on World Wide Web, pp. 241–250. ACM (2009)
2. Ashlagi, I., Braverman, M., Hassidim, A., Lavi, R., Tennenholtz, M.: Position auctions with budgets: Existence and uniqueness. The B.E. Journal of Theoretical Economics 10(1) (2010)
3. Ausubel, L.M.: An efficient ascending-bid auction for multiple objects. American Economic Review 94(5), 1452–1475 (2004)
4. Ausubel, L.M., Milgrom, P.R.: Ascending auctions with package bidding. Frontiers of Theoretical Economics 1(1), 1019 (2002)
5. Bhattacharya, S., Conitzer, V., Munagala, K., Xia, L.: Incentive compatible budget elicitation in multi-unit auctions. CoRR abs/0904.3501 (2009)
6. Bikhchandani, S., de Vries, S., Schummer, J., Vohra, R.V.: Ascending auctions for integral (poly)matroids with concave nondecreasing separable values. In: Teng, S.H. (ed.) SODA, pp. 864–873. SIAM (2008)
7. Dobzinski, S., Lavi, R., Nisan, N.: Multi-unit auctions with budget limits. In: FOCS, pp. 260–269. IEEE Computer Society (2008)
8. Dobzinski, S., Lavi, R., Nisan, N.: Multi-unit auctions with budget limits (2011), http://ie.technion.ac.il/~ronlavi/papers/budget-constraints.pdf
9. Dütting, P., Henzinger, M., Starnberger, M.: Auctions with heterogeneous items and budget limits (2012)
10. Dütting, P., Henzinger, M., Weber, I.: An expressive mechanism for auctions on the web. In: Srinivasan, S., Ramamritham, K., Kumar, A., Ravindra, M.P., Bertino, E., Kumar, R. (eds.) WWW, pp. 127–136. ACM (2011)
11. Edelman, B., Ostrovsky, M., Schwarz, M.: Internet advertising and the generalized second price auction: Selling billions of dollars worth of keywords. American Economic Review 97(1), 242–259 (2005)
12. Fiat, A., Leonardi, S., Saia, J., Sankowski, P.: Single valued combinatorial auctions with budgets. In: Shoham, Y., Chen, Y., Roughgarden, T. (eds.) ACM Conference on Electronic Commerce, pp. 223–232. ACM (2011)
13. Fujishige, S., Tamura, A.: A two-sided discrete-concave market with possibly bounded side payments: An approach by discrete convex analysis. Mathematics of Operations Research 32(1), 136–155 (2007)
14. Goel, G., Mirrokni, V.S., Leme, R.P.: Polyhedral clinching auctions and the adwords polytope. CoRR abs/1201.0404 (2012), to appear in 44th ACM Symposium on Theory of Computing (STOC 2012), New York (May 2012)
15. Lavi, R., May, M.: A Note on the Incompatibility of Strategy-Proofness and Pareto-Optimality in Quasi-Linear Settings with Public Budgets - Working Paper. In: Chen, N., Elkind, E., Koutsoupias, E. (eds.) WINE 2011. LNCS, vol. 7090, p. 417. Springer, Heidelberg (2011)
16. Milgrom, P.: Putting auction theory to work: The simulteneous ascending auction. Journal of Political Economy 108(2), 245–272 (2000)
17. Nisan, N., Bayer, J., Chandra, D., Franji, T., Gardner, R., Matias, Y., Rhodes, N., Seltzer, M., Tom, D., Varian, H., Zigmond, D.: Google's Auction for TV Ads. In: Albers, S., Marchetti-Spaccamela, A., Matias, Y., Nikoletseas, S., Thomas, W. (eds.) ICALP 2009, Part II. LNCS, vol. 5556, pp. 309–327. Springer, Heidelberg (2009)
18. Vickrey, W.: Counterspeculation, auctions, and competitive sealed tenders. The Journal of Finance 16(1), 8–37 (1961)

A Theory Independent Curry-De Bruijn-Howard Correspondence

Gilles Dowek

INRIA, 23 avenue d'Italie, CS 81321, 75214 Paris Cedex 13, France
gilles.dowek@inria.fr

Brouwer, Heyting, and Kolmogorov have proposed to define constructive proofs as algorithms, for instance, a proof of $A \Rightarrow B$ as an algorithm taking proofs of A as input and returning proofs of B as output. Curry, De Bruijn, and Howard have developed this idea further. First, they have proposed to express these algorithms in the lambda-calculus, writing for instance $\lambda f^{A \Rightarrow A \Rightarrow B} \lambda x^A (f\ x\ x)$ for the proof of the proposition $(A \Rightarrow A \Rightarrow B) \Rightarrow A \Rightarrow B$ taking a proof f of $A \Rightarrow A \Rightarrow B$ and a proof x of A as input and returning the proof of B obtained by applying f to x twice. Then, they have remarked that, as proofs of $A \Rightarrow B$ map proofs of A to proofs of B, their type $proof(A \Rightarrow B)$ is $proof(A) \rightarrow proof(B)$. Thus the function $proof$ mapping propositions to the type of their proofs is a morphism transforming the operation $\Rightarrow$ into the operation $\rightarrow$. In the same way, this morphism transforms cut-reduction in proofs into beta-reduction in lambda-terms.

This expression of proofs as lambda-terms has been extensively used in proof processing systems: Automath, Nuprl, Coq, Elf, Agda, etc. Lambda-calculus is a more compact representation of proofs, than natural deduction or sequent calculus proof-trees. This representation is convenient, for instance to store proofs on a disk and to communicate them through a network.

This has lead to the development of several typed lambda-calculi: Automath, the system F, the system Fω, the lambda-Pi-calculus, Martin-Löf intuitionistic type theory, the Calculus of Constructions, the Calculus of Inductive Constructions, etc. And we may wonder why so many different calculi are needed.

In some cases, the differences in the lambda-calculi reflect differences in the logic where proofs are expressed: some calculi, for instance, express constructive proofs, others classical ones. In other cases, they reflect differences in the inductive rules used to define proofs: some calculi are based on natural deduction, others on sequent calculus. But most of the times, the differences reflect differences in the theory where the proofs are expressed: arithmetic, the theory of classes—a.k.a. second-order logic—, simple type theory—a.k.a. higher-order logic—, predicative type theory, etc.

Instead of developing a customized typed lambda-calculus for each specific theory, we may attempt to design a general parametric calculus that permits to express the proofs of any theory. This way, the problem of expressing proofs in the lambda-calculus would be completely separated from that of choosing a theory.

A. Czumaj et al. (Eds.): ICALP 2012, Part II, LNCS 7392, pp. 13–15, 2012.
© Springer-Verlag Berlin Heidelberg 2012

A way to do this is to start from the lambda-Pi-calculus, that is designed to express proofs in minimal predicate logic and to define a theory in an axiomatic way, declaring a variable, or a constant, for each axiom. This is the approach of the *Logical framework* [8]. Yet, a limit of this approach is that the beta-reduction is too weak in presence of axioms, and we need to add axiom-specific proof-reduction rules, such as the rules of Gödel system T for the induction axiom, to emulate cut-reduction in specific theories.

We have proposed in [5] a different approach, where a theory is expressed, not with axioms, but with rewrite rules, as in Deduction modulo [6,7]. This has lead to the *lambda-Pi-calculus modulo*, and its implementation, the system *Dedukti* [2].

Although it is just a proof-checker, Dedukti is a universal proof-checker [3]. By choosing appropriate rewrite rules, the lambda-Pi-calculus modulo can be parametrized to express proofs of any theory that can be expressed in Deduction modulo, such as arithmetic, the theory of classes, simple type theory, some versions of set theory, etc. By choosing appropriate rewrite rules, the lambda-Pi-calculus can also emulate the system F, the system Fω, the Calculus of Constructions [5], the Calculus of Inductive Constructions [4], etc. This has lead to the development of systems to translate proofs from the system Coq to Dedukti [4] and from the system HOL to Dedukti [1].

This universal proof-checker opens new research directions that still remain to be investigated. First, what happens if we prove the proposition $A \Rightarrow B$ in a theory $\mathcal{T}_1$ and the proposition A in a theory $\mathcal{T}_2$? Is there a theory in which we can deduce B? Of course, if the theories $\mathcal{T}_1$ and $\mathcal{T}_2$ are incompatible—such as set theory with the axiom of choice and set theory with the negation of the axiom of choice—, it makes no sense to deduce B anywhere. But, there are also cases where one of the rewrite systems expressing $\mathcal{T}_1$ and $\mathcal{T}_2$ in the lambda-Pi-calculus modulo is a subset of the other, or where the union of these two systems defines a consistent theory, or where propositions and proofs of one theory may be translated into the other, and in all these cases, it makes sense to deduce B from the proofs of $A \Rightarrow B$ and A, even if these proofs have been developed in different theories and different systems.

More generally, although most proof processing systems are based on strong theories—simple type theory, the Calculus of Inductive Constructions, etc.—we know that many proofs developed in these systems use only a small part of this strength. Making explicit the axioms or rewrite rules defining these theories permits to identify which axiom, or which rule, is used in which proof, in a similar way as we, more or less, know which part of informal mathematics depends on the axiom of choice and which part does not.

Such an analysis may be a first step towards the development of libraries of proofs, where proofs would not be classified in function of the system in which they have been developed, but in function of the axioms and rules they use, i.e. to a true interoperability between proof systems.

References

1. Assaf, A.: Translating HOL in the lambda-Pi-calculus modulo, Master thesis (in preparation, 2012)
2. Boespflug, M.: Conception d'un noyau de vérification de preuves pour le lambda-Pi-calcul modulo, Doctoral thesis, École polytechnique (2011)
3. Boespflug, M., Carbonneaux, Q., Hermant, O.: The lambda-Pi calculus modulo as a universal proof language. In: Second International Workshop on Proof Exchange for Theorem Proving (2012)
4. Boespflug, M., Burel, G.: CoqInE: Translating the Calculus of inductive constructions into the lambda-Pi-calculus modulo. In: Second International Workshop on Proof Exchange for Theorem Proving (2012)
5. Cousineau, D., Dowek, G.: Embedding Pure Type Systems in the Lambda-Pi-Calculus Modulo. In: Della Rocca, S.R. (ed.) TLCA 2007. LNCS, vol. 4583, pp. 102–117. Springer, Heidelberg (2007)
6. Dowek, G., Hardin, T., Kirchner, C.: Theorem proving modulo. Journal of Automated Reasoning 31, 33–72 (2003)
7. Dowek, G., Werner, B.: Proof normalization modulo. The Journal of Symbolic Logic 68(4), 1289–1316 (2003)
8. Harper, R., Honsell, F., Plotkin, G.: A framework for defining logics. The Journal of the ACM 40(1) (1993)

Standing on the Shoulders of a Giant

One Persons Experience of Turings Impact

(Summary of the Alan M. Turing Lecture*)

David Harel

The Weizmann Institute of Science, Rehovot, 76100, Israel

A quote attributed to Isaac Newton says "If I have seen a little further it's because I stand on the shoulders of giants". This was indeed stated by Newton, but the general metaphor of a dwarf standing on a giant goes back many, many years earlier. I would recommend the wonderful 1965 book by Robert K. Merton, referred to fondly as OTSOG (on the shoulders of giants) [1] .

Alan M. Turing (1912-1954) can almost be said to be the Mozart of computer science. He died at a young age, under well-known tragic circumstances, leaving an incredibly brilliant, diverse, and versatile legacy. And we are left with the tantalizing question of what would he have achieved had he lived another 30 or 40 years. Turing conceived of the simplest yet most powerful models of computation, but also contributed the design of some of the most complex computers of his time. He proved that there are many things that computers *cannot* do (already in 1936!), but also taught us that there some amazing things that they *can* do. He carried out pioneering work on the idea of mathematical and computational modeling of biology, but also dealt with the question whether computers can be of human-like intelligence. I feel that Alan Turing will become recognized as one of the most important and influential scientists of all time, possibly alongside other giants like Galileo, Newton, and Einstein.

Many computers scientists feel that large parts of their research are rooted in Turing's work and I am one of them. This talk will discuss briefly three of Turing's main lines of work, and the crucial way in which they impacted one person's research — my own. I will not be discussing a central part of my work, that related to software and systems engineering and executable visual languages, but the talk will cover several of the other topics I have worked in over the years.

Computability

As is well known, Turing was a pioneering force behind the notion of computability and non-solvability [2], and thus, alongside the likes of Church, Gödel, Post

* Versions of this talk were presented in a number of conferences and symposia during 2012, to celebrate the Alan M. Turing Centennial year. Parts of the more recent research reported upon here were supported by an Advanced Research Grant to DH from the European Research Council (ERC) under the European Community's FP7 Programme.

A. Czumaj et al. (Eds.): ICALP 2012, Part II, LNCS 7392, pp. 16–22, 2012.
© Springer-Verlag Berlin Heidelberg 2012

and Kleene, was a central figure in establishing the limits of computing. In particular, Turing showed the halting problem to be undecidable, and essentially invented the notion of universal computing, via the universal Turing machine. At this point, one should definitely mention Rice's theorem [3], which can be viewed as a grand extension of Turing's undecidability result for the halting problem, and which establishes the dramatic fact that no non-trivial problem about computation can be decidable! This includes correctness, equivalence, efficiency, and many others... Thus, as a profession, computing constitutes the ultimate example of the barefoot shoemaker aphorism. See [4] for a detailed exposition of these and other results on the limitations of computing.

My own humble extensions of Turing's work in this area include three topics, which will be discussed briefly below: computability on finite structures, computability on infinite recursive structures, and high non-solvability/undecidability.

In a joint 1979 STOC paper with Ashok Chandra, we addressed the problem of defining the computable functions over finite structures [5]. The issue is the following. Say you want to compute a function on graphs, or on relational databases. These are really sets of tuples (a graph is a set of pairs of vertices), and there is no order on their elements. A function that takes advantage, so to speak, of some ordering that could be the result of a particular representation of the graph (e.g., on a Turing machine tape), should be outlawed. So what is the appropriate notion of computability over such structures? How should one extend Church-Turing computability from words or numbers to general (unordered, or partially unordered) structures? The answer offered in our 1979 paper is that a computable function over a structure has to be partial recursive, in the classical sense of Turing and co., and in addition has to be *consistent* (or *generic*, as the notion was later called), which means that the function has to preserve isomorphisms. Genericity captures the idea that the function should not use any information that is not present in the structure itself, such as an ordering on the tuples.

This definition, however, would have been worthless unless accompanied with a complete language for the computable functions — a sort of analogue for structures of Turing machines or the lambda calculus, or, for that matter, of any programming language over numbers or words. This, of course, raises the question of why not simply take as the complete language the set of generic Turing machines, i.e., exactly all those that preserve isomorphisms? The answer is that by Rice's theorem even the syntax of such a language in non-effective/non-computable: you cannot tell whether a string of symbols is a legal program in the language, because it is undecidable to tell whether a function preserves isomorphisms. What Chandra and I did in our paper [5] was to define a simple query language QL over relations (which is really a variant of the first order relational algebra enhanced by a *while-do* looping construct), and prove it complete. The subtle part of the proof was to show how, given an input relational structure S, we are able to use QL to program a relation that represents the set of automorphisms of S.

The paper itself (which appeared as a journal version in 1980) was written in the setting of relational databases, but it can be viewed as establishing computability over general finite relational structures. Indeed, its main result has been extended over the years to much more complex structures. We believe that this is a good example of the standing-on-the-shoulders-of-the-giant phenomenon, consisting of a natural and modest, yet basic, extension to general structures of Turing's original notion of computability .

A side remark worth including here concerns the second paper written with Chandra, which appeared in the 1980 FOCS conference [6]. There we continued the work on computable queries/functions on (unordered) structures, and defined the structural and the complexity-theoretic basics over such structures. These have been shown to be underlie many issues in descriptive complexity and finite model theory. In that paper, we posed the question of "does QPTIME have an effective enumeration?" (see p. 118 of the 1982 journal version). It can be viewed as asking whether there can be any effective language/logic capturing the polynomial time functions over general unordered structures, such as graphs. This problem (which was popularized in later writings of Gurevich and others) has been open now for over 30 years.

Many years later, with my PhD student Tirza Hirst [7], we extended the notion of computable functions on structures to deal with infinite recursive structures; e.g., relations whose set of tuples is computable. For example, a recursive graph is one whose vertex and edge sets are effective; the simplest case is a recursive binary relation over the natural numbers. We were able to obtain several results on recursive structures, including a completeness result for an appropriate variant of the QL language of [5] in the style of the proof given there [7,8]. This has to be done on a suitably restricted class of recursive structures, because the general class is not closed even under projection.

The third topic that interested me for many years, and which can also be viewed as directly extending the work of Turing (and in this case, that of Stephen Kleene too), is *high undecidability* or high unsolvability. Of specific interest are problems that can be shown to be complete for the lowest level of the analytic hierarchy, the so-called Σ_1^1/Π_1^1 level. Such problems are thus infinitely many levels of undecidability worse than the halting problem.

Here are some of the results I was able to obtain. First, the halting problem of Turing is extended to the halting problem for programs with countably infinite nondeterminism, and also the halting problem for parallel and concurrent programs under the assumption of fairness — that is, fair halting. Actually, these two problems turn out to be closely related, and both were shown to be highly undecidable [9].

Second, with various co-authors and over a period of several years, I was able to show that many satisfiability/validity problems for logics of programs are also highly undecidable, including variants of non-regular propositional dynamic logic, temporal logical in two dimensions, and more [10,11]. Many of these satisfiability results are proved by a reduction from another problem I was able to show highly undecidable — a recurring version of the tiling problem originating with Wang, in which there is an additional requirement that a specific tile has to occur infinitely often in a tiling of the plane [10].

Finally, I was able to show that asking whether a recursive graph has a Hamiltonian path is also highly undecidable, i.e., complete for the Σ_1^1/Π_1^1 level [12]. This problem was known to be undecidable but had not given rise to an upper bound residing in the arithmetical hierarchy. The proof in [12], establishing that the problem is actually outside the arithmetical hierarchy, is a rather picturesque reduction from the halting problem with countable non-determinism (which is essentially the unfoundedness of recursive trees).

In a second paper with Hirst [13], we were able to link high undecidability of properties over recursive structures to the approximability of finitary versions of those problems on the NP level.

Biological Modeling

As is well known, Turing pioneered the mathematical basis for modeling biological growth, or in more technical terms, *morphogenesis* [14]. He was thus one of the first to consider computational and mathematical means for modeling biological processes, specifically pattern-formation and growth.

My modest follow-up to that work involves modeling the development of the pancreas, carried out with Yaki Setty and Yaron Cohen. We used Statecharts and other computer science and software engineering techniques to build a dynamic executable model of a cell, which is "scheduled" to become a pancreatic cell [15,16]. When thousands of such cell statecharts are simulated together, the result is an interactive model that mimics the growth process (organogenesis) of the pancreas, ultimately obtaining its cauliflower- or broccoli-shaped final form. See also [17].

In more recent ongoing work, carried out with Naama Bloch, and also building on Turing's morphogenesis work, we are in the process of modeling the growth of a cancerous tumor. This is done in a manner similar to that used in the pancreas model, and again we model the dynamics of a cancer cell (and the surrounding blood vessels), trying to capture the crux of biological growth.

In the context of our use of Statecharts to model biological growth, the following quote from Turing's 1952 paper, 50 years earlier, is particularly illuminating: "...one proceeds as with a physical theory and defines an entity called 'the state of the system'. One then describes how that state is to be determined from the state at a moment very shortly before" [14].

The Turing Test

As is also very well known, Turing spent a lot of energy in his later years thinking about whether computers can think... The most famous outcome of this process involves Turing's imitation game, which has come to be called the *Turing test*, for determining whether a computer or a piece of software is intelligent [18].

The test involves a human sitting is one room and a computer in another, and a human interrogator in a third room who tries to distinguish one from the other. the computer passes the test if the interrogator is not able to tell the difference. An enormous amount of material has been written about this test and its significance, so there is no reason to recall any of it here.

My modest follow-up, once again as the dwarf standing on the shoulders of the giant, is a proposal for a Turing-like test for modeling nature. Continuing the subject matter of the previous section, one of the things I've been talking about for many years is a grand challenge in the area of systems biology for comprehensive and realistic modeling. The challenge is to construct a full, correct, true-to-all-known-facts, four-dimensional model of a multi-cellular organism. In short, the challenge is to build an interactive executable model of an animal [19]. As a second-level part of this proposal, I suggested using the *C. elegans* nematode as the model organism for tackling the challenge.

However, the question arises as to when you know that such a model is complete, so that you can satisfy yourself with its validity. This is to be contrasted with conventional modeling and analysis work in bioinformatics or systems biology, where you start out with a question or a set of questions, in which case the model is complete and valid when it can be shown to provide answers in full accordance with the answers that one gets in the laboratory. One can then go on to the next project and the next set of questions. In contrast, modeling an entire biological system — an organism such as a worm, a fly, or an elephant, or even just a complete organ such as a pancreas, a liver, a heart or a brain — the question arises as to when the project ends and the model can be claimed to be valid.

My 2005 suggestion for addressing this question is a Turing-like test, but with a Popperian twist [20]. In line with Turing's original test, one places the computer with its model in one room, and in the other one sets up an advanced laboratory researching the actual organism being modeled. If we are to take the *C. elegans* worm as the modeled organism, we should have an advanced *C. elegans* laboratory in the second room. Then, a team of interrogators, well-versed in the subject matter, is given time to probe the laboratory and the model, trying to tell the difference. Of course, the buffering between the interrogators and the two rooms has to be more elaborate than that in Turing's original test for intelligence, since, for example, many questions can be answered in a split second by a computer but could take months (or could be actually impossible) to answer in the laboratory. But this is a technicality I will not get into here.

The interesting twist present in this new version of the Turing test is that a model passing it can indeed be claimed to be a true model of an elephant, a *C. elegans*, or a liver, yet such a conclusion should be taken to be temporary, and should be stamped on the computer in non-permanent ink. The reason is that, as Carl Popper has taught us, this kind of definitive statement will no doubt be refuted in the future when more is discovered about the organism or the organ being modeled. Hence, such a model is really a theory of the biological system being modeled, and the theory will have to be revised when new facts are discovered, which is actually exactly what we want! See the discussion in [20].

*   *   *

In conclusion, all these are but very personal and idiosyncratic examples of the work of one particular computer scientist over a period of almost 35 years. They can all be viewed as modest extensions and generalizations of Turing's pioneering work in three different areas.

The emerging picture is definitely that of a dwarf standing on the shoulders of a true giant, and as a result perhaps being able to see just a tiny bit further.

References

1. Merton, R.K.: On The Shoulders of Giants: A Shandean Postscript. Free Press (1965)
2. Turing, A.M.: On Computable Numbers with an Application to the Entscheidungsproblem. Proc. London Math. Soc. 42, 230–265 (1936); Corrections appeared in: ibid 43, 544–546 (1937)
3. Rice, H.G.: Classes of recursively enumerable sets and their decision problems. Trans. Amer. Math. Soc. 74, 358–366 (1953)
4. Harel, D.: Computers Ltd.: What They Really Can't Do. Oxford University Press (2000); Revised paperback edition, 2003. Special Turing Centennial printing (2012)
5. Chandra, A.K., Harel, D.: Computable Queries for Relational Data Bases. J. Comput. System Sciences 21, 156–178 (1980); Also, Proc. ACM 11th Symp. on Theory of Computing, Atlanta, Georgia, pp. 309–318 (April 1979)
6. Chandra, A.K., Harel, D.: Structure and Complexity of Relational Queries. J. Comput. System Sci. 25, 99–128 (1982); Also, Proc. 21st IEEE Symp. on Foundations of Computer Science, Syracuse, New York (October 1980)
7. Hirst, T., Harel, D.: Completeness Results for Recursive Databases. J. Comput. System Sci. 52(3), 522–536 (1996); Also, Proc. 12th ACM Symp. on Principles of Database Systems, pp. 244–252. ACM Press, New York (1993)
8. Harel, D.: Towards a Theory of Recursive Structures. In: Enjalbert, P., Mayr, E.W., Wagner, K.W. (eds.) STACS 1994. LNCS, vol. 775, pp. 633–645. Springer, Heidelberg (1994)
9. Harel, D.: Effective Transformations on Infinite Trees, with Applications to High Undecidability, Dominoes and Fairness. J. Assoc. Comput. Mach. 33, 224–248 (1986)
10. Harel, D.: Recurring Dominoes: Making the Highly Undecidable Highly Understandable. Ann. Disc. Math. 24, 51–72 (1985); Also, Karpinski, M. (ed.) FCT 1983. LNCS, vol. 158, pp. 177–194. Springer, Heidelberg (1983)
11. Harel, D.: How Hard is it to Reason About Propositional Programs? In: Broy, M. (ed.) Program Design Calculi. NATO ASI Series, vol. F-118, pp. 165–184. Springer, Berlin (1993)
12. Harel, D.: Hamiltonian Paths in Infinite Graphs. Israel J. Math. 76(3), 317–336 (1991); Also, Proc. 23rd ACM Symp. on Theory of Computing, New Orleans, pp. 220–229 (1991)
13. Hirst, T., Harel, D.: Taking it to the Limit: On Infinite Variants of NP-Complete Problems. J. Comput. System Sci. 53(2), 180–193 (1996); Also, Proc. 8th IEEE Structure in Complexity Theory, pp. 292–304. IEEE Press, New York (1993)

14. Turing, A.M.: Computing Machinery and Intelligence. Mind 59, 433–460 (1950)
15. Setty, Y., Cohen, I.R., Dor, Y., Harel, D.: Four-Dimensional Realistic Modeling of Pancreatic Organogenesis. Proc. Natl. Acad. Sci. 105(51), 20374–20379 (2008)
16. Setty, Y., Cohen, I.R., Harel, D.: Executable Modeling of Morphogenesis: A Turing-Inspired Approach. Fundamenta Informaticae (in press)
17. Fisher, J., Harel, D., Henzinger, T.A.: Biology as Reactivity. Comm. Assoc. Comput. Mach. 54(10), 72–82 (2011)
18. Turing, A.M.: The chemical basis of morphogenesis. Philosophical Transactions of the Royal Society of London B 327, 37–72 (1952)
19. Harel, D.: A Grand Challenge for Computing: Full Reactive Modeling of a Multi-Cellular Animal. Bulletin of the EATCS, European Association for Theoretical Computer Science (81), 226–235 (2003); Early version prepared for the UK Workshop on Grand Challenges in Computing Research (November 2002), Reprinted in: Paun, Rozenberg, Salomaa (eds.) Current Trends in Theoretical Computer Science: The Challenge of the New Century, Algorithms and Complexity, vol. I, pp. 559–568. World Scientific (2004)
20. Harel, D.: A Turing-Like Test for Biological Modeling. Nature Biotechnology 23, 495–496 (2005)

Session Types and Distributed Computing

Kohei Honda

School of Electronic Engineering and Computer Science
Queen Mary, University of London
Mile End Road, London E1 4NS, UK

Abstract. Session types offer type-based structuring principles for distributed communicating processes. They were born from process encodings of various data structures in an asynchronous version of the π-calculus. Its theory is supported and enriched by several foundations, including Linear Logic, automata theory, typed π-calculi and typed λ-calculi. Unlike types for sequential computing, where types abstract and specify data and functions, session types abstract and specify structures of interactions in sessions – they describe *protocols* with which processes interact with each other in possibly interleaved sessions. In this type abstraction, the division of interactions into sessions is important, classifying each set of related interactions into a distinct session, an idea which has been known and practiced in networking research since early days. This division enables tractable type abstraction, making it easy to specify protocols and to validate programs against them, statically and at runtime.

In this talk we introduce central ideas of session types through illustrative examples, identify different properties of concurrent and distributed systems which session types and associated theories can specify and guarantee, present several key technical results underpinning theories and applications of session types, and discuss open research topics. If time remains, we shall report our recent experience of the use of session types for building and verifying large-scale distributed systems.

A. Czumaj et al. (Eds.): ICALP 2012, Part II, LNCS 7392, p. 23, 2012.
© Springer-Verlag Berlin Heidelberg 2012

Algorithms, Graph Theory, and the Solution of Laplacian Linear Equations*

Daniel A. Spielman

Departments of Computer Science and Mathematics
Program in Applied Mathematics
Yale University

In this talk, we survey major developments in the design of algorithms for solving Laplacian linear equations, by which we mean systems of linear equations in the Laplacian matrices of graphs and their submatrices. We begin with a few examples of where such equations arise, including the analysis of networks of resistors, the analysis of networks of springs, and the solution of maximum flow problems by interior point methods.

We solve these systems of linear equations by using Vaidy's idea [1] of preconditioning these systems of linear equations by the Laplacian matrices of subgraphs of the original graph. A preconditioner is essentially a matrix that is supposed to approximate the original matrix. If it is a good approximation and if it is easy to solve systems of equations in the preconditioner, then one can use the preconditioner to quickly solve systems of equations in the original matrix via algorithms such as the preconditioned conjugate gradient or the preconditioned chebyshev method. Vaidya's algorithm allows one to obtain ϵ-accurate solutions to Laplacian systems with m non-zero entries in time essentially $O(m^{1.75} \log \epsilon^{-1})$. For more background on Vaidya's work, we refer the reader to one of [2–5].

Vaidya initially suggested approximating a graph by a maximum spanning tree. However, it was a decade later before Boman and Hendrickson realized [6, 7] low-stretch spanning trees [8] provided much better results. The use of these trees reduced the running time of Laplacian solvers to $\widetilde{O}(m^{1.5+o(1)} \log \epsilon^{-1})$.

In fact, the best bound on the time required to obtain an ϵ-accurate solution using a maximum spanning tree is $O(mn \log \epsilon^{-1})$. Vaidya obtained the better bound mentioned above by adding a few edges of the original graph back to the maximum spanning tree. One can obtain much better preconditioners by adding edges back to low-stretch spanning trees.

Spielman and Teng [9, 10] found a way to add edges back to low-stretch spanning trees that resulted in Laplacian solvers that run in time $O(m \log^c n \log \epsilon^{-1})$, where c is a constant close to 100. This construction relied on many ingredients, including spectral sparsifiers, local graph partitioning algorithms, fast graph partitioning algorithms, and low-stretch spanning trees. Improvements in these ingredients (see [11–19]) have steadily decreased the exponent c. The most recent constructions of Koutis, Miller and Peng [20, 21] greatly simplify the whole process, and reduces the constant c to 1.

---

* This work is supported in part by the National Science Foundation under Grant Nos. 0915487 and 1111257.

A. Czumaj et al. (Eds.): ICALP 2012, Part II, LNCS 7392, pp. 24–26, 2012.
© Springer-Verlag Berlin Heidelberg 2012

References

1. Vaidya, P.M.: Solving linear equations with symmetric diagonally dominant matrices by constructing good preconditioners. Unpublished manuscript UIUC 1990. A talk based on the manuscript was presented at the IMA Workshop on Graph Theory and Sparse Matrix Computation, October 1991, Minneapolis (1990)
2. Joshi, A.: Topics in Optimization and Sparse Linear Systems. PhD thesis, UIUC (1997)
3. Chen, D., Toledo, S.: Vaidya's preconditioners: implementation and experimental study. Electronic Transactions on Numerical Analysis 16, 30–49 (2003)
4. Bern, M., Gilbert, J., Hendrickson, B., Nguyen, N., Toledo, S.: Support-graph preconditioners. SIAM J. Matrix Anal. & Appl. 27(4), 930–951 (2006)
5. Gremban, K.: Combinatorial Preconditioners for Sparse, Symmetric, Diagonally Dominant Linear Systems. PhD thesis, Carnegie Mellon University, CMU-CS-96-123 (1996)
6. Boman, E.G., Hendrickson, B.: Support theory for preconditioning. SIAM Journal on Matrix Analysis and Applications 25(3), 694–717 (2003)
7. Boman, E., Hendrickson, B.: On spanning tree preconditioners. Manuscript, Sandia National Lab. (2001)
8. Alon, N., Karp, R.M., Peleg, D., West, D.: A graph-theoretic game and its application to the k-server problem. SIAM Journal on Computing 24(1), 78–100 (1995)
9. Spielman, D.A., Teng, S.H.: Nearly-linear time algorithms for graph partitioning, graph sparsification, and solving linear systems. In: Proceedings of the Thirty-Sixth Annual ACM Symposium on Theory of Computing, pp. 81–90 (2004), Full version available at http://arxiv.org/abs/cs.DS/0310051
10. Spielman, D.A., Teng, S.H.: Nearly-linear time algorithms for preconditioning and solving symmetric, diagonally dominant linear systems. CoRR abs/cs/0607105 (2008), http://www.arxiv.org/abs/cs.NA/0607105 (submitted to SIMAX)
11. Elkin, M., Emek, Y., Spielman, D.A., Teng, S.H.: Lower-stretch spanning trees. SIAM Journal on Computing 32(2), 608–628 (2008)
12. Abraham, I., Bartal, Y., Neiman, O.: Nearly tight low stretch spanning trees. In: Proceedings of the 49th Annual IEEE Symposium on Foundations of Computer Science, pp. 781–790 (October 2008)
13. Abraham, I., Neiman, O.: Using petal-decompositions to build a low stretch spanning tree. In: Proceedings of The Fourty-Fourth Annual ACM Symposium on The Theory of Computing (STOC 2012) (to appear, 2012)
14. Andersen, R., Chung, F., Lang, K.: Local graph partitioning using pagerank vectors. In: Proceedings of the 47th Annual Symposium on Foundations of Computer Science, pp. 475–486 (2006)
15. Andersen, R., Peres, Y.: Finding sparse cuts locally using evolving sets. In: STOC 2009: Proceedings of the 41st Annual ACM Symposium on Theory of Computing, pp. 235–244. ACM, New York (2009)
16. Orecchia, L., Sachdeva, S., Vishnoi, N.K.: Approximating the exponential, the lanczos method and an $\tilde{O}(m)$-time spectral algorithm for balanced separator. In: Proceedings of The Fourty-Fourth Annual ACM Symposium on The Theory of Computing (STOC 2012) (to appear, 2012)
17. Spielman, D.A., Srivastava, N.: Graph sparsification by effective resistances. In: Proceedings of the 40th Annual ACM Symposium on Theory of Computing, pp. 563–568 (2008)

18. Batson, J.D., Spielman, D.A., Srivastava, N.: Twice-Ramanujan sparsifiers. In: Proceedings of the 41st Annual ACM Symposium on Theory of Computing, pp. 255–262 (2009)
19. Levin, A., Koutis, I., Peng, R.: Improved spectral sparsification and numerical algorithms for sdd matrices. In: Proceedings of the 29th Symposium on Theoretical Aspects of Computer Science (STACS) (to appear, 2012)
20. Koutis, I., Miller, G., Peng, R.: Approaching optimality for solving sdd linear systems. In: 2010 51st Annual IEEE Symposium on Foundations of Computer Science (FOCS), pp. 235–244 (2010)
21. Koutis, I., Miller, G., Peng, R.: A nearly-mlogn time solver for sdd linear systems. In: 2011 52nd Annual IEEE Symposium on Foundations of Computer Science, FOCS (2011)

Randomized Mechanisms for Multi-unit Auctions
(Extended Abstract)

Berthold Vöcking

Department of Computer Science
RWTH Aachen University
voecking@cs.rwth-aachen.de

Combinatorial auctions are a common abstraction of many complex resource allocation problems: A large number of items (goods, resources) should be assigned to a number of agents (bidders) with different valuations on bundles of items. They are the central representative problem for the field of algorithmic mechanism design. In this field, algorithmic problems are studied in a game theoretic setting in which the input of the algorithm is not publicly known but distributed among a set of selfish agents which would possibly lie about their private information if this would give an advantage to them. A mechanism is called *incentive compatible* or *truthful* if it allocates the goods and sets payments for the bidders in such a way that it is a dominant strategy for each bidder to report his/her valuations for different bundles of items in a truthful manner.

Multi-unit auctions are arguably the most basic variant of combinatorial auctions: m identical items shall be allocated to n bidders such that social welfare is maximized. Each bidder i has a valuation function $v_i : \{0, \ldots, m\} \to \mathbb{R}_{\geq 0}$. The valuation functions v_i are non-decreasing and $v_i(0) = 0$. The set of feasible allocations is

$$A = \left\{ s = (s_1, \ldots, s_n) \in \{0, \ldots, m\}^n \ \middle| \ \sum_{i=1}^{n} s_i \leq m \right\}.$$

The valuation of bidder i for allocation s is denoted by $v_i(s)$. There are *no externalities*, i.e., v_i depends only on the number s_i of items assigned to bidder i and not on the allocation of the remaining items to other bidders. The classical objective is to find an allocation $s \in A$ maximizing the *social welfare* $v(s) = \sum_{i=1}^{n} v_i(s_i)$.

In the single-minded case in which the valuation functions correspond to step functions with only one step, the multi-unit auction problem corresponds to the knapsack problem. Under this major simplification, multi-unit auctions admit a truthful fully polynomial-time approximation scheme (FPTAS) [2], which is a variation of the well-known (non-truthful) approximation scheme by Ibarra and Kim [7].

Multi-unit auctions with general valuations are a significantly richer and much harder problem. It is assumed that the valuation functions are not given explicitly but in form of an oracle that can be queried by the mechanism. The oracle answers *value queries*: Given $i \in [n]$, $k \in [m]$, what is the value $v_i(k)$? The challenge is to find an approximately optimal solution without querying the valuation functions completely. An efficient algorithm is supposed to run in time polynomial in $n \log m$, which is the established notion of *input length* in this context.

The problem is already challenging when there are only 2 bidders. First investigations are concerned with mechanisms using standard approaches corresponding to a

A. Czumaj et al. (Eds.): ICALP 2012, Part II, LNCS 7392, pp. 27–29, 2012.
© Springer-Verlag Berlin Heidelberg 2012

well-known characterization of Roberts [8]. In particular, it is known that affine maximizers with VCG-based payments require an exponential number of queries in the input length in order to achieve an approximation factor better than 2 for multi-unit auctions [4]. Only very recently, the exponential lower bound on the number of queries has even been extended towards a class of mechanisms beyond Roberts' characterization [5]. The conclusion that one can draw from these analyses is that getting an approximation ratio better than 2 for multi-unit auctions with deterministic mechanisms is either impossible or requires to develop completely new ideas.

Our talk surveys recent studies [3,6,10] showing that one can get much better approximation ratios when using randomization. Two different notions of truthfulness are distinguished:

- *Universal truthfulness:* A universally truthful mechanism is a probability distribution over deterministically truthful mechanisms.
- *Truthfulness in expectation:* A mechanism is truthful in expectation if a bidder always maximizes his/her expected utility by bidding truthfully.

Universal truthfulness is a very restrictive and strong concept: If the random bits used by the mechanism are known, a universally truthful algorithm corresponds to a deterministically truthful algorithm. Truthfulness in expectation is a somewhat weaker concept. In particular, it assumes that bidders are risk neutral.

Dobzinski and Dughmi [3] present a randomized FPTAS for multi-unit auctions with general valuations that is truthful in expectation. A more general result is shown by Dughmi and Roughgarden [6] who show that every packing problem that admits an FPTAS also admits a truthful-in-expectation randomized FPTAS. They employ techniques from the smoothed analysis of Pareto-optimal solutions presented in [1,9].

More generally, Dobzinski and Dughmi [3] study the difference in the computational power between mechanisms that are truthful in expectation and mechanisms that are universally truthful. They separate these two notions of truthfulness by showing that there is a variant of multi-unit auctions (with certain additional constraints) for which there exists an FPTAS being truthful in expectation but there does not exist a polynomial-time universally truthful algorithm with an approximation factor better than 2. They state that they ideally would like to prove this negative result for original multi-unit auctions rather than only for a technical variant of these auctions.

In [10], we present a universally truthful, randomized polynomial-time approximation scheme (PTAS). This disproves the implicit conjecture of Dobzinski and Dughmi [3] that a 2-approximation is best possible for multi-unit auctions. Our mechanism employs VCG payments in a non-standard way: The deterministic mechanisms underlying our universally truthful approximation scheme do not belong to the class of affine maximizers. Instead each of the deterministic mechanisms is composed of a collection of affine maximizers, one for each bidder. In particular, we introduce a subjective variant of VCG in which payments for different bidders are defined on the basis of possibly different affine maximizers.

References

1. Beier, R., Vöcking, B.: Typical properties of winners and losers in discrete optimization. SIAM J. Comput. 35(4), 855–881 (2006)

2. Briest, P., Krysta, P., Vöcking, B.: Approximation techniques for utilitarian mechanism design. In: Proc. of the 37th ACM Symp. on Theory of Computing (STOC), pp. 39–48 (2005)
3. Dobzinski, S., Dughmi, S.: On the power of randomization in algorithmic mechanism design. In: Proc. of the 50th IEEE Symp. on Foundations of Computer Science (FOCS), pp. 505–514 (2009)
4. Dobzinski, S., Nisan, N.: Mechanisms for multi-unit auctions. In: Proc. of the 8th ACM Conference on Electronic Commerce (EC) (2007)
5. Dobzinski, S., Nisan, N.: Multi-unit auctions: beyond Roberts. In: Proc. of the 12th ACM Conference on Electronic Commerce (EC), pp. 233–242 (2011)
6. Dughmi, S., Roughgarden, T.: Black-box randomized reductions in algorithmic mechanism design. In: Proc. of the 51st IEEE Symp. on Foundations of Computer Science (FOCS), pp. 775–784 (2010)
7. Ibarra, O.H., Kim, C.E.: Fast approximation algorithms for the knapsack and sum of subset problems. Journal of the ACM 22(4), 463–468 (1975)
8. Roberts, K.: The characterization of implementable choice rules. In: Laffont, J.-J. (ed.) Aggregation and Revelation of Preferences. Papers Presented at the First European Summer Workshop of the Economic Society, pp. 321–349. North-Holland (1979)
9. Röglin, H., Teng, S.-H.: Smoothed analysis of multiobjective optimization. In: Proc. of the 50th IEEE Symp. on Foundations of Computer Science (FOCS), pp. 681–690 (2009)
10. Vöcking, B.: A universally-truthful approximation scheme for multi-unit auctions. In: Proc. of the 23rd ACM Symp. on Discrete Algorithms (SODA), pp. 846–855 (2012)

Algebraic Synchronization Trees and Processes*

Luca Aceto[1], Arnaud Carayol[2], Zoltán Ésik[3], and Anna Ingólfsdóttir[1]

[1] ICE-TCS, School of Computer Science, Reykjavik University, Iceland
[2] LIGM, Université Paris-Est, CNRS, France
[3] Institute of Informatics, University of Szeged, Hungary

Abstract. We study algebraic synchronization trees, i.e., initial solutions of algebraic recursion schemes over the continuous categorical algebra of synchronization trees. In particular, we investigate the relative expressive power of algebraic recursion schemes over two signatures, which are based on those for Basic CCS and Basic Process Algebra, as a means for defining synchronization trees up to isomorphism as well as modulo bisimilarity and language equivalence. The expressiveness of algebraic recursion schemes is also compared to that of the low levels in the Caucal hierarchy.

1 Introduction

The study of recursive program schemes is one of the classic topics in programming language semantics. (See, e.g., [4,16,20,27] for some of the early references.) In this paper, we study recursion schemes from a process-algebraic perspective and investigate the expressive power of *algebraic* recursion schemes over the signatures of Basic CCS [24] and of Basic Process Algebra (BPA) [3] as a way of defining possibly infinite synchronization trees [23], which are essentially edge-labelled trees with a distinguished exit label ex. As depicted here, this exit label can only occur on edges whose target is a leaf. Both these signatures allow one to describe every finite synchronization tree and include a binary choice operator +. The difference between them is that the signature for Basic CCS, which is denoted by Γ in this paper,

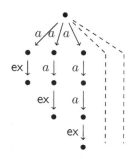

* A full version of this paper may be found at http://www.ru.is/faculty/luca/PAPERS/algsynch.pdf. Luca Aceto and Anna Ingólfsdóttir have been partially supported by the project 'Meta-theory of Algebraic Process Theories' (nr. 100014021) of the Icelandic Research Fund. Arnaud Carayol has been supported by the project AMIS (ANR 2010 JCJC 0203 01 AMIS). Zoltán Ésik has been partially supported by the project TÁMOP-4.2.1/B-09/1/KONV-2010-0005 'Creating the Center of Excellence at the University of Szeged', supported by the European Union and co-financed by the European Regional Fund, and by the National Foundation of Hungary for Scientific Research, grant no. K 75249. Zoltán Ésik's work on this paper was also partly supported by grant T10003 from Reykjavik University's Development Fund and a chair from the LabEx Bézout.

A. Czumaj et al. (Eds.): ICALP 2012, Part II, LNCS 7392, pp. 30–41, 2012.
© Springer-Verlag Berlin Heidelberg 2012

contains a unary action prefixing operation $a._-$ for each action a, whereas the signature for BPA, which we denote by Δ, has one constant a for each action that may label the edge of a synchronization tree and offers a full-blown sequential composition, or sequential product, operator. Intuitively, the sequential product $t \cdot t'$ of two synchronization trees is obtained by appending a copy of t' to the leaves of t that describe successful termination of a computation. In order to distinguish successful and unsuccessful termination, both the signatures Γ and Δ contain constants 0 and 1, which denote unsuccessful and successful termination, respectively. An example of a regular recursion scheme over the signature Δ is

$$X = (X \cdot a) + a, \tag{1}$$

and an example of an algebraic recursion scheme over the signature Γ is

$$F_1 = F_2(a.1), \quad F_2(v) = v + F_2(a.v). \tag{2}$$

The synchronization tree defined by these two schemes is depicted on page 30. In the setting of process algebras such as CCS [24] and ACP [3], synchronization trees are a classic model of process behaviour. They arise as unfoldings of labelled transition systems (LTSs) that describe the operational semantics of process terms and have been used to give denotational semantics to process description languages—see, for instance, [1]. Regular synchronization trees over the signature Γ are unfoldings of processes that can be described in the regular fragment of CCS, which is obtained by adding to the signature Γ a facility for the recursive definition of processes. On the other hand, regular synchronization trees over the signature Δ are unfoldings of processes that can be described in Basic Process Algebra (BPA) [3] augmented with constants for the deadlocked and the empty process as well as recursive definitions. For example, the tree that is defined by. (1) is Δ-regular.

As is well known, the collection of regular synchronization trees over the signature Δ strictly includes that of regular synchronization trees over the signature Γ even up to language equivalence. Therefore, the notion of regularity depends on the signature. But what is the expressiveness of algebraic recursion schemes over the signatures Γ and Δ? The aim of this paper is to begin the analysis of the expressive power of those recursion schemes as a means for defining synchronization trees, and their bisimulation or language equivalence classes.

In order to characterize the expressive power of algebraic recursion schemes defining synchronization trees, we interpret such schemes in continuous categorical Γ- and Δ-algebras of synchronization trees. Continuous categorical Σ-algebras are a categorical generalization of the classic notion of continuous Σ-algebra that underlies the work on algebraic semantics [19,20], and have been used in [8,9,17] to give semantics to recursion schemes over synchronization trees and words. (We refer the interested reader to [21] for a recent discussion of category-theoretic approaches to the solution of recursion schemes.) In this setting, the Γ-regular (respectively, Γ-algebraic) synchronization trees are those that are initial solutions of regular (respectively, algebraic) recursion schemes over the signature Γ. Δ-regular and Δ-algebraic synchronization trees are defined in similar fashion.

Our first contribution in the paper is therefore to provide a categorical semantics for first-order recursion schemes that define processes, whose behaviour is represented by synchronization trees. The use of continuous categorical Σ-algebras allows us to deal with arbitrary first-order recursion schemes; there is no need to restrict oneself to, say, 'guarded' recursion schemes, as one is forced to do when using a metric semantics (see, for instance, [10] for a tutorial introduction to metric semantics), and this categorical approach to giving semantics to first-order recursion schemes can be applied even when the order-theoretic framework either fails because of the lack of a 'natural' order or leads to undesirable identities.

As a second contribution, we provide a comparison of the expressive power of regular and algebraic recursion schemes over the signatures Γ and Δ, as a formalism for defining processes described by their associated synchronization trees up to isomorphism, bisimilarity [24,26] and language equivalence. Moreover, we compare the expressiveness of those recursion schemes to that of the low levels in the Caucal hierarchy. (As a benefit of the comparison with the Caucal hierarchy, we obtain structural properties and decidability of Monadic Second-Order Logic [29].) In the setting of language equivalence, the notion of Γ-regularity corresponds to the regular languages, the one of Δ-regularity corresponds to the context-free languages and Δ-algebraicity corresponds to the macro languages [18], which coincide with the languages generated by Aho's indexed grammars [2]. We present a pictorial summary of our expressiveness results on Figure 1. Moreover, we prove that the synchronization tree that is the unfolding of the bag (also known as multiset) over a binary alphabet depicted on Figure 2 is not Γ-algebraic, even up to language equivalence, and that it is not Δ-algebraic up to bisimilarity. These results are a strengthening of a classic theorem from the literature on process algebra proved by Bergstra and Klop in [5].

In order to obtain a deeper understanding of Γ-algebraic recursion schemes, as a final main contribution of the paper, we characterize their expressive power by following the lead of Courcelle [15,16]. In those references, Courcelle proved that a term tree is algebraic if, and only if, its branch language is a deterministic context-free language. In our setting, we associate with each synchronization tree with bounded branching a family of branch languages and we show that a synchronization tree with bounded branching is Γ-algebraic if, and only if, the family of branch languages associated with it contains a deterministic context-free language (Theorem 2). In conjunction with standard tools from formal language theory, this result can be used to show that certain synchronization trees are not Γ-algebraic.

The paper is organized as follows. In Section 2, we recall the notion of continuous categorical Σ-algebra. Synchronization trees are defined in Section 3, together with the signatures Γ and Δ that contain the operations on those trees that we use in this paper. We introduce regular and algebraic recursion schemes,

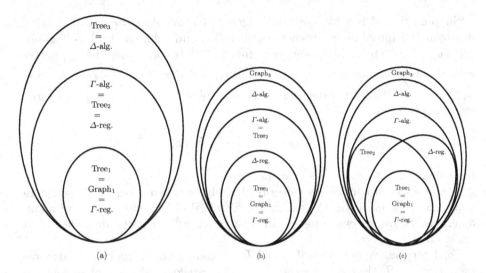

Fig. 1. The expressiveness hierarchies up to language equivalence (a), up to bisimilarity (b) and up to isomorphism (c)

as well as their initial solutions, in Section 4. Section 5 studies the expressive power of regular and algebraic recursion schemes over the signatures Γ and Δ. In Section 6, following Courcelle, we characterize the expressive power of Γ-algebraic recursion schemes by studying the branch languages of synchronization trees whose vertices have bounded outdegree.

2 Continuous Categorical Algebras

In this section, we recall the notion of continuous categorical Σ-algebra. Continuous categorical Σ-algebras were used in [8,9,17] to give semantics to recursion schemes over synchronization trees and words.

Let $\Sigma = \bigcup_{n \geq 0} \Sigma_n$ be a ranked set (or 'signature'). A *categorical Σ-algebra* is a small category A equipped with a functor $\sigma^A : A^n \to A$ for each $\sigma \in \Sigma_n$, $n \geq 0$. A *morphism* between categorical Σ-algebras A and B is a functor $h : A \to B$ such that, for each $\sigma \in \Sigma_n$, the diagram

$$
\begin{array}{ccc}
A^n & \xrightarrow{\ \sigma^A\ } & A \\
{\scriptstyle h^n}\big\downarrow & & \big\downarrow{\scriptstyle h} \\
B^n & \xrightarrow[\ \sigma^B\ ]{} & B
\end{array}
$$

commutes *up to a natural isomorphism* π_σ. Here, the functor $h^n : A^n \to B^n$ maps each object and morphism $(x_1, \dots, x_n)$ in A^n to $(h(x_1), \dots, h(x_n))$ in B^n. A morphism h is *strict* if, for all $\sigma \in \Sigma$, the natural isomorphism π_σ is the identity.

Suppose that A is a categorical Σ-algebra. We call A *continuous* if A has a distinguished initial object (denoted $\perp^A$ or 0^A) and colimits of all ω-diagrams $(f_k : a_k \to a_{k+1})_{k \geq 0}$. Moreover, each functor σ^A is continuous, i.e., preserves colimits of ω-diagrams. Thus, if $\sigma \in \Sigma_2$, say, and if $x_0 \overset{f_0}{\to} x_1 \overset{f_1}{\to} x_2 \overset{f_2}{\to} \ldots$ and $y_0 \overset{g_0}{\to} y_1 \overset{g_1}{\to} y_2 \overset{g_2}{\to} \ldots$ are ω-diagrams in A with colimits $(x_k \overset{\phi_k}{\to} x)_k$ and $(y_k \overset{\psi_k}{\to} y)_k$, respectively, then

$$\sigma^A(x_0, y_0) \overset{\sigma^A(f_0, g_0)}{\to} \sigma^A(x_1, y_1) \overset{\sigma^A(f_1, g_1)}{\to} \sigma^A(x_2, y_2) \overset{\sigma^A(f_2, g_2)}{\to} \ldots$$

has colimit $(\sigma^A(x_k, y_k) \overset{\sigma^A(\phi_k, \psi_k)}{\to} \sigma^A(x, y))_k$.

A morphism of continuous categorical Σ-algebras is a categorical Σ-algebra morphism that preserves the distinguished initial object and colimits of all ω-diagrams. Below we will often write just σ for σ^A, in particular when A is understood.

For later use, we note that if A and B are continuous categorical Σ-algebras then so is $A \times B$. Moreover, for each $k \geq 0$, the category $[A^k \to A]$ of all continuous functors $A^k \to A$ is also a continuous categorical Σ-algebra, where, for each $\sigma \in \Sigma_n$, $\sigma^{[A^k \to A]}(f_1, \ldots, f_n) = \sigma^A \circ \langle f_1, \ldots, f_n \rangle$, with $\langle f_1, \ldots, f_n \rangle$ standing for the target tupling of the continuous functors $f_1, \ldots, f_n : A^k \to A$. On natural transformations, $\sigma^{[A^k \to A]}$ is defined in a similar fashion. In $[A^k \to A]$, colimits of ω-diagrams are formed pointwise.

3   Synchronization Trees

A *synchronization tree* $t = (V, v_0, E, l)$ over an alphabet $\mathcal{A}$ of 'action symbols' consists of a finite or countably infinite set V of 'vertices' and an element $v_0 \in V$ (the 'root'), a set $E \subseteq V \times V$ of "edges" and a 'labelling function' $l : E \to \mathcal{A} \cup \{\text{ex}\}$. These data obey the following restrictions.

- (V, v_0, E) is a rooted tree: for each $u \in V$, there is a unique path $v_0 \rightsquigarrow u$.
- If $e = (u, v) \in E$ and $l(e) = \text{ex}$, then v is a leaf, and u is called an *exit vertex*.

A *morphism* $\phi : t \to t'$ of synchronization trees is a function $V \to V'$ that preserves the root, the edges and the labels, so that if (u, v) is an edge of t, then $(\phi(u), \phi(v))$ is an edge of t', and $l'(\phi(u), \phi(v)) = l(u, v)$. Morphisms are therefore functional *simulations* [22,26]. It is clear that the trees and tree morphisms form a category. The tree that has a single vertex and no edges is initial. It is known that the category of trees has colimits of all ω-diagrams, see [7]. (It also has binary coproducts.) In order to make the category of trees small, we may require that the vertices of a tree form a subset of some fixed infinite set.

The category $\mathsf{ST}(\mathcal{A})$ of synchronization trees over $\mathcal{A}$ is equipped with two binary operations: $+$ (sum) and $\cdot$ (sequential product or sequential composition), and either with a unary operation or a constant associated with each letter $a \in \mathcal{A}$.

The *sum* $t + t'$ of two trees is obtained by taking the disjoint union of the vertices of t and t' and identifying the roots. The edges and labelling are inherited. The *sequential product* $t \cdot t'$ of two trees is obtained by replacing each edge of t labelled ex by a copy of t'. With each letter $a \in \mathcal{A}$, we can either associate a constant, or a unary *prefixing operation*. As a constant, a denotes the tree with vertices v_0, v_1, v_2 and two edges: the edge (v_0, v_1), labelled a, and the edge (v_1, v_2), labelled ex. As an operation, $a(t)$ is the tree $a \cdot t$, for any tree t. Let 0 denote the tree with no edges and 1 the tree with a single edge labelled ex. On morphisms, all operations are defined in the expected way. For example, if $h : t \to t'$ and $h' : s \to s'$, then $h + h'$ is the morphism that agrees with h on the nonroot vertices of t and that agrees with h' on the nonroot vertices of s. The root of $t + s$ is mapped to the root of $t' + s'$.

In the sequel we will consider two signatures for synchronization trees, Γ and Δ. The signature Γ contains $+, 0, 1$ and each letter $a \in \mathcal{A}$ as a *unary* symbol. In contrast, Δ contains $+, \cdot, 0, 1$ and each letter $a \in \mathcal{A}$ as a *nullary* symbol. It is known that, for both signatures, $\mathsf{ST}(\mathcal{A})$ is a continuous categorical algebra. See [7] for details.

Two synchronization trees $t = (V, v_0, E, l)$ and $t' = (V', v_0', E', l')$ are *bisimilar* or *bisimulation equivalent* [24,26] if there is some symmetric relation $R \subseteq (V \times V') \cup (V' \times V)$ that relates their roots, and such that if $(v_1, v_2) \in R$ and there is some edge (v_1, v_1'), then there is an equally-labelled edge (v_2, v_2') with $(v_1', v_2') \in R$. The *path language* of a synchronization tree is composed of the words in $\mathcal{A}^*$ that label a path from the root to the source of an exit edge. Two trees are *language equivalent* if they have the same path language.

4   Algebraic Objects and Functors

When n is a non-negative integer, we denote the set $\{1, \ldots, n\}$ by $[n]$.

Definition 1. *Let Σ be a signature. A Σ-recursion scheme, or recursion scheme over Σ, is a sequence E of equations*

$$F_1(v_1, \ldots, v_{k_1}) = t_1, \ldots, F_n(v_1, \ldots, v_{k_n}) = t_n,$$

where each t_i is a term over the signature $\Sigma_\Phi = \Sigma \cup \Phi$ in the variables $v_1, \ldots, v_{k_i}$, and Φ contains the symbols F_i (sometimes called 'functor variables') of rank k_i, $i \in [n]$. A Σ-recursion scheme is regular *if $k_i = 0$, for each $i \in [n]$.*

Suppose that A is a continuous categorical Σ-algebra, and consider a Σ-recursion scheme of the form given above. Define

$$A^{r(\Phi)} = [A^{k_1} \to A] \times \cdots \times [A^{k_n} \to A].$$

Then $A^{r(\Phi)}$ is a continuous categorical Σ-algebra, as noted in Section 2.

When each F_i, $i \in [n]$, is interpreted as a continuous functor $f_i : A^{k_i} \to A$, each term over the extended signature $\Sigma_\Phi = \Sigma \cup \Phi$ in the variables $v_1, \ldots, v_m$ induces a continuous functor $A^m \to A$ that we denote by $t^A(f_1, \ldots, f_n)$. In fact, t^A is a continuous functor $t^A : A^{r(\Phi)} \to [A^m \to A]$. More precisely, we define t^A as follows. Let f_i, g_i denote continuous functors $A^{k_i} \to A$, $i \in [n]$, and let α_i be a natural transformation $f_i \to g_i$ for each $i \in [n]$. When t is the variable v_i, say, then $t^A(f_1, \ldots, f_n)$ is the ith projection functor $A^m \to A$, and $t^A(\alpha_1, \ldots, \alpha_n)$ is the identity natural transformation corresponding to this projection functor. Suppose now that t is of the form $\sigma(t_1, \ldots, t_k)$, where $\sigma \in \Sigma_k$ and $t_1, \ldots, t_k$ are terms. Then $t^A(f_1, \ldots, f_n) = \sigma^A \circ \langle h_1, \ldots, h_k \rangle$ and $t^A(\alpha_1, \ldots, \alpha_n) = \sigma^A \circ \langle \beta_1, \ldots, \beta_k \rangle$, where $h_j = t_j^A(f_1, \ldots, f_n)$ and $\beta_j = t_j^A(\alpha_1, \ldots, \alpha_n)$ for all $j \in [k]$. (Here, we use the same notation for a functor and the corresponding identity natural transformation.) Finally, when t is of the form $F_i(t_1, \ldots, t_{k_i})$, then $t^A(f_1, \ldots, f_n) = f_i \circ \langle h_1, \ldots, h_{k_i} \rangle$, and the corresponding natural transformation is $\alpha_i \circ \langle \beta_1, \ldots, \beta_{k_i} \rangle$, where the h_j and β_j, $j \in [k_i]$, are defined similarly as above.

Note that if each $\alpha_i : f_i \to f_i$ is an identity natural transformation (so that $f_i = g_i$, for all $i \in [n]$), then $t^A(\alpha_1, \ldots, \alpha_n)$ is the identity natural transformation $t^A(f_1, \ldots, f_n) \to t^A(f_1, \ldots, f_n)$.

In any continuous categorical Σ-algebra A, by target-tupling the functors t_i^A, we obtain a continuous functor

$$E^A : A^{r(\Phi)} \to A^{r(\Phi)}.$$

Indeed, we have that $t_i^A : A^{r(\Phi)} \to [A^{k_i} \to A]$, for $i \in [n]$, so that

$$E^A = \langle t_1^A, \ldots, t_n^A \rangle : A^{r(\Phi)} \to A^{r(\Phi)}.$$

Thus, E^A has an initial fixed point in $A^{r(\Phi)}$, unique up to natural isomorphism, that we denote by

$$|E^A| = (|E|_1^A, \ldots, |E|_n^A),$$

so that, in particular, $|E|_i^A = t_i^A(|E|_1^A, \ldots, |E|_n^A)$, at least up to isomorphism, for each $i \in [n]$.

Definition 2. *Suppose that A is a continuous categorical Σ-algebra. We say that $f : A^m \to A$ is Σ-algebraic, if there is a recursion scheme E such that f is isomorphic to $|E|_1^A$, the first component of the above-mentioned initial solution of E. When $m = 0$, we identify a Σ-algebraic functor with a Σ-algebraic object. Last, a Σ-regular object is an object isomorphic to the first component of the initial solution of a Σ-regular recursion scheme.*

In particular, we get the notions of Γ-algebraic and Γ-regular trees, and Δ-algebraic and Δ-regular trees.

Example 1. The Δ-regular recursion scheme (1) and the Γ-algebraic one (2) have the infinitely branching tree $\sum_{i \geq 1} a^i$ depicted on page 30 as their initial solutions. That tree is therefore both Δ-regular and Γ-algebraic. So Δ-regular and Γ-algebraic recursion schemes can be used to define infinitely branching trees that have an infinite number of subtrees, even up to language equivalence.

5 Comparing the Expressiveness of Classes of Recursion Schemes

In this section, we interpret recursion schemes over the continuous categorical algebra $\mathsf{ST}(\mathcal{A})$, viewed either as a Γ-algebra or as a Δ-algebra, and study the expressive power of classes of recursion schemes over the signatures Γ and Δ. It is clear that every Γ-regular tree is Δ-regular and that the inclusion is proper, since every Γ-regular tree has, up to isomorphism, only a finite number of subtrees, see [7,23], while there exist Δ-regular and Γ-algebraic trees that do not have this property (see Example 1). The strict inclusion also holds with respect to strong bisimulation equivalence or language equivalence. It is well-known that the languages of synchronization trees defined by Γ-regular schemes are the regular languages. On the other hand, modulo language equivalence, Δ-regular schemes are nothing but context-free grammars and have the same expressive power as Γ-algebraic schemes (see Theorem 1(1) and (4) below).

The Δ-regular trees that can be defined using regular Δ-recursion schemes that do not contain occurrences of the constants 0 and 1 correspond to unfoldings of the labelled transition systems denoted by terms in Basic Process Algebra (BPA) with recursion, see, for instance, [3,5]. Indeed, the signature of BPA contains one constant symbol a for each action as well as the binary + and · operation symbols, denoting nondeterministic choice and sequential composition, respectively. (Below, we write BPA for 'BPA with recursion'.) Alternatively, following [25], one may view BPA as the class of labelled transition systems associated with context-free grammars in Greibach normal form in which only leftmost derivations are permitted. The class of Basic Parallel Processes (BPP) is a parallel counterpart of BPA introduced by Christensen [14]. We refer the interested readers to [25] for the details of the formal definitions, which are not needed to appreciate the results to follow, and further pointers to the literature.

In the results to follow, we will compare the expressiveness of recursion schemes to that of the low levels in the Caucal hierarchy [12]. For the sake of completeness, following [11], we recall that Tree_0 and Graph_0 denote the collections of finite, edge-labelled trees and graphs, respectively. Moreover, for each $n \geq 0$, Tree_{n+1} stands for the collection of unfoldings of graphs in Graph_n, and the graphs in Graph_{n+1} are those that can be obtained from the trees in Tree_{n+1} by applying a monadic interpretation (or transduction). It is well known that Graph_1 is the class of all prefix-recognizable graphs [13].

The following theorem collects our main results on the expressiveness of recursion schemes over the signatures Δ and Γ. A pictorial summary of all our expressiveness results may be found on Figure 1. All the inclusions on that figure are strict, with the possible exception of the inclusion of the collection of the Δ-algebraic trees in Graph_3 up to bisimilarity and up to isomorphism. To the best of our knowledge, it is open whether those inclusions are strict. The fact that the path language of every synchronization tree in Tree_3 (respectively, Tree_2) is an indexed language (respectively, context-free language) is known from [11, Theorem 4].

Theorem 1.

1. *Every Δ-regular tree is Γ-algebraic.*
2. *There is a Γ-algebraic synchronization tree that is not bisimilar to any Δ-regular tree. Moreover, there is a Γ-algebraic synchronization tree that is neither definable in BPA modulo bisimilarity nor in BPP modulo language equivalence.*

3. *Each synchronization tree in* Tree$_2$ *is Γ-algebraic, but there is a Δ-regular (and hence Γ-algebraic) synchronization tree that is not in* Tree$_2$.

4. *Every Γ-algebraic synchronization tree is bisimilar to a tree in* Tree$_2$. *Therefore, modulo bisimilarity, the Γ-algebraic synchronization trees coincide with those in* Tree$_2$. *Moreover, each Γ-algebraic synchronization tree is language equivalent to a Δ-regular one.*

5. *Each Δ-algebraic synchronization tree is in* Graph$_3$ *and hence has a decidable monadic second-order theory. Moreover, there is a Δ-algebraic synchronization tree that does not belong to* Tree$_3$.

6. *The synchronization tree t_{bag} associated with the bag over a binary alphabet depicted on Figure 2 has an undecidable monadic second-order theory (even without the root being the source of an exit edge). Hence, it is not in the Caucal hierarchy and is therefore not Δ-algebraic, even up to bisimilarity. Moreover, t_{bag} is not Γ-algebraic up to language equivalence.*

7. *There exists a Γ-algebraic synchronization tree whose minimization with respect to bisimilarity does not have a decidable monadic second-order theory and hence is not in the Caucal hierarchy.*

Statement 6 in the above theorem is a strengthening of a classic result from the literature on process algebra proved by Bergstra and Klop in [5]. Indeed, in Theorem 4.1 in [5], Bergstra and Klop showed that the bag over a domain of values that contains at least two elements is not expressible in BPA, and the collection of synchronization trees that are expressible in BPA is strictly included in the Δ-algebraic synchronization trees.

Thomas showed in [28, Theorem 10] that the monadic second-order theory of the infinite two-dimensional grid is undecidable. However, we cannot use that result to prove that the synchronization tree t_{bag} has an undecidable monadic second-order theory. Indeed, the unfolding of the infinite two-dimensional grid is the full binary tree, which has a decidable monadic second-order theory.

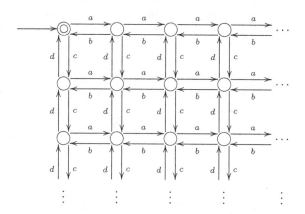

Fig. 2. An LTS whose unfolding is not a Δ-algebraic synchronization tree

Finally, we remark that Theorem 1(7) yields that the collection of synchronization trees in the Caucal hierarchy is *not* closed under quotients with respect to bisimilarity. Indeed, there is a Γ-algebraic tree whose quotient with respect to bisimilarity is not in the Caucal hierarchy. Nevertheless the result is sort of folklore.

6   Branch Languages of Bounded Synchronization Trees

Call a synchronization tree *bounded* if there is a constant k such that the outdegree of each vertex is at most k. Our aim in this section is to offer a language-theoretic characterization of the expressive power of Γ-algebraic recursion schemes defining synchronization trees. We shall do so by following Courcelle—see, e.g., [16]—and studying the branch languages of bounded synchronization trees. More precisely, we assign a family of branch languages to each bounded synchronization tree over an alphabet $\mathcal{A}$ and show that a bounded tree is Γ-algebraic if, and only if, the corresponding language family contains a deterministic context-free language (DCFL). Throughout this section, we will call Γ-algebraic trees just algebraic trees, and similarly for regular trees.

Definition 3. *Suppose that $t = (V, v_0, E, l)$ is a bounded synchronization tree over the alphabet $\mathcal{A}$. Denote by k the maximum of the outdegrees of the vertices of t. Let $\mathcal{B}$ denote the alphabet $\mathcal{A} \times [k]$. A determinization of t is a tree $t' = (V, v_0, E, l')$ over the alphabet $\mathcal{B}$ which differs from t only in the labelling as follows. Suppose that $v \in V$ with outgoing edges $(v, v_1), \ldots, (v, v_\ell)$ labelled $a_1, \ldots, a_\ell \in \mathcal{A} \cup \{\mathrm{ex}\}$ in t. Then there is some permutation π of the set $[\ell]$ such that the label of each (v, v_i) in t' is $(a_i, \pi(i))$.*

Consider a determinization t' of t. Let $v \in V$ and let $v_0, v_1, \ldots, v_m = v$ denote the vertices on the unique path from the root to v. The branch word *corresponding to v in t' is the alternating word*

$$k_0(a_1, i_1)k_1 \ldots k_{m-1}(a_m, i_m)k_m$$

where $k_0, \ldots, k_m$ denote the outdegrees of the vertices $v_0, \ldots, v_m$, and for each $j \in [m]$, (a_j, i_j) is the label of the edge (v_{j-1}, v_j) in t'. The branch language *$L(t')$ corresponding to a determinization t' of t consists of all the branch words of t'. Finally, the family of branch languages corresponding to t is:*

$$\mathcal{L}(t) = \{L(t') : t' \text{ is a determinization of } t\}.$$

By way of example, consider the LTS depicted in Figure 2. The synchronization tree t_{bag} that is obtained by unfolding this LTS from its start state is bounded. In fact, the outdegree of each non-leaf node is three. The branch words corresponding to the nodes of any determinization of the tree t_{bag} have the form

$$3(a_1, i_1)3 \ldots 3(a_m, i_m)k_m,$$

where k_m is either 3 or 0, $i_1, \ldots, i_m \in [3]$ and $a_1 \ldots a_m$ is a word with the property that, in any of its prefixes, the number of occurrences of the letter a

is greater than, or equal to, the number of occurrences of the letter b, and the number of occurrences of the letter c is greater than, or equal to, the number of occurrences of the letter d. Moreover, for each $j \in [m]$, $a_j = $ ex if and only if $j = m$ and $k_m = 0$. (Note that, when $a_m = $ ex, the number of a's in $a_1 \ldots a_{m-1}$ equals the number of b's, and similarly for c and d.)

Theorem 2.

1. *A bounded synchronization tree t is algebraic (respectively, regular) if, and only if, $\mathcal{L}(t)$ contains a DCFL (respectively, regular language).*
2. *The bounded synchronization trees in Tree_2 are the bounded Γ-algebraic synchronization trees.*

Using statement 2 in the above theorem, we can show that Figure 1(b) also applies for bounded synchronization trees.

The language-theoretic characterization of the class of bounded algebraic synchronization trees offered in Theorem 2 can be used to prove that certain trees are *not* algebraic. For example, consider the following Δ-algebraic scheme:

$$F_0 = F(1), \quad F(v) = a \cdot F(b \cdot v) + v \cdot c \cdot v \cdot 0.$$

Given any determinization of the synchronization tree t defined by this scheme, the non-context-free language $\{a^n b^n c b^n : n \geq 0\}$ is a homomorphic image of the intersection of its branch language with a regular language. Thus t is not Γ-algebraic.

References

1. Abramsky, S.: A Domain Equation for Bisimulation. Inf. Comput. 92(2), 161–218 (1991)
2. Aho, A.V.: Indexed Grammars — an Extension of Context-Free Grammars. J. ACM 15, 647–671 (1968)
3. Baeten, J.C.M., Basten, T., Reniers, M.A.: Process Algebra: Equational Theories of Communicating Processes. Cambridge University Press (2009)
4. de Bakker, J.W.: Recursive Procedures. Mathematical Centre Tracts, vol. 24. Mathematisch Centrum, Amsterdam (1971)
5. Bergstra, J.A., Klop, J.W.: The Algebra of Recursively Defined Processes and the Algebra of Regular Processes. In: Paredaens, J. (ed.) ICALP 1984. LNCS, vol. 172, pp. 82–94. Springer, Heidelberg (1984)
6. Bloom, S.L., Ésik, Z., Taubner, D.: Iteration Theories of Synchronization Trees. Inf. Comput. 102(1), 1–55 (1993)
7. Bloom, S.L., Ésik, Z.: Iteration Theories. Springer (1993)
8. Bloom, S.L., Ésik, Z.: The Equational Theory of Regular Words. Inf. Comput. 197, 55–89 (2005)
9. Bloom, S.L., Ésik, Z.: A Mezei-Wright Theorem for Categorical Algebras. Theor. Comput. Sci. 411, 341–359 (2010)
10. van Breugel, F.: An Introduction to Metric Semantics: Operational and Denotational Models for Programming and Specification Languages. Theor. Comput. Sci. 258, 1–98 (2001)

11. Carayol, A., Wöhrle, S.: The Caucal Hierarchy of Infinite Graphs in Terms of Logic and Higher-Order Pushdown Automata. In: Pandya, P.K., Radhakrishnan, J. (eds.) FSTTCS 2003. LNCS, vol. 2914, pp. 112–123. Springer, Heidelberg (2003)

12. Caucal, D.: On Infinite Terms Having a Decidable Monadic Theory. In: Diks, K., Rytter, W. (eds.) MFCS 2002. LNCS, vol. 2420, pp. 165–176. Springer, Heidelberg (2002)

13. Caucal, D.: On Infinite Transition Graphs Having a Decidable Monadic Theory. Theor. Comput. Sci. 290, 79–115 (2003)

14. Christensen, S.: Decidability and Decomposition in Process Algebras. PhD thesis ECS-LFCS-93-278, Department of Computer Science, Univ. of Edinburgh (1983)

15. Courcelle, B.: A Representation of Trees by Languages I and II. Theor. Comput. Sci. 6, 255–279 (1978); Theor. Comput. Sci. 7, 25–55

16. Courcelle, B.: Fundamental Properties of Infinite Trees. Theor. Comput. Sci. 25, 69–95 (1983)

17. Ésik, Z.: Continuous Additive Algebras and Injective Simulations of Synchronization Trees. J. Log. Comput. 12, 271–300 (2002)

18. Fischer, M.J.: Grammars with Macro-like Productions. In: 9th Annual Symp. Switching and Automata Theory, pp. 131–142. IEEE Press (1968)

19. Goguen, J.A., Thatcher, J.W., Wagner, E.G., Wright, J.B.: Initial algebra semantics and continuous algebras. J. ACM 24, 68–95 (1977)

20. Guessarian, I.: Algebraic Semantics. LNCS, vol. 99. Springer, Heidelberg (1981)

21. Milius, S., Moss, L.: The Category-Theoretic Solution of Recursive Program Schemes. Theor. Comput. Sci. 366, 3–59 (2006)

22. Milner, R.: An Algebraic Definition of Simulation Between Programs. In: Proceedings 2nd Joint Conference on Artificial Intelligence, pp. 481–489. BCS (1971)

23. Milner, R.: A Calculus of Communication Systems. LNCS, vol. 92. Springer, Heidelberg (1980)

24. Milner, R.: Communication and Concurrency. Prentice Hall (1989)

25. Moller, F.: Infinite Results. In: Sassone, V., Montanari, U. (eds.) CONCUR 1996. LNCS, vol. 1119, pp. 195–216. Springer, Heidelberg (1996)

26. Park, D.M.R.: Concurrency and Automata on Infinite Sequences. In: Deussen, P. (ed.) GI-TCS 1981. LNCS, vol. 104, pp. 167–183. Springer, Heidelberg (1981)

27. Scott, D.S.: The Lattice of Flow Diagrams. In: Symposium on Semantics of Algorithmic Languages 1971. Lecture Notes in Mathematics, vol. 188, pp. 311–366. Springer (1971)

28. Thomas, W.: A Short Introduction to Infinite Automata. In: Kuich, W., Rozenberg, G., Salomaa, A. (eds.) DLT 2001. LNCS, vol. 2295, pp. 130–144. Springer, Heidelberg (2002)

29. Thomas, W.: Constructing Infinite Graphs with a Decidable MSO-Theory. In: Rovan, B., Vojtáš, P. (eds.) MFCS 2003. LNCS, vol. 2747, pp. 113–124. Springer, Heidelberg (2003)

Streaming Tree Transducers*

Rajeev Alur and Loris D'Antoni

University of Pennsylvania

Abstract. Theory of tree transducers provides a foundation for understanding expressiveness and complexity of analysis problems for specification languages for transforming hierarchically structured data such
as XML documents. We introduce *streaming tree transducers* as an ana-
lyzable, executable, and expressive model for transforming unranked ordered trees (and hedges) in a single pass. Given a linear encoding of the
input tree, the transducer makes a single left-to-right pass through the input, and computes the output using a finite-state control, a visibly pushdown stack, and a finite number of variables that store output chunks
that can be combined using the operations of string-concatenation and
tree-insertion. We prove that the expressiveness of the model coincides
with transductions definable using monadic second-order logic (MSO).
We establish complexity upper bounds of EXPTIME for *type-checking* and
NEXPTIME for checking *functional equivalence* for our model. We consider variations of the basic model when inputs/outputs are restricted to
strings and ranked trees, and in particular, present the model of *bottom-
up ranked-tree transducers*, which is the first known MSO-equivalent
transducer model that processes trees in a bottom-up manner.

1   Introduction

Finite-state machines and logics for specifying tree transformations offer a suitable theoretical foundation for studying expressiveness and complexity of analysis problems for languages for processing and transforming XML documents.
Representative formalisms for specifying tree transductions include finite-state
top-down and bottom-up tree transducers, Macro tree transducers (MTT), attribute grammars, MSO (monadic second-order logic) definable graph transductions, and specialized programming languages such as XSLT and XDUCE [1–8].
    In this paper, we propose the model of *streaming tree transducers* (STT)
which has the following three properties: (1) *Single-pass processing:* an STT is a
deterministic machine that computes the output using a single left-to-right pass
through the linear encoding of the input tree; (2) *Expressiveness:* STTs specify
exactly the class of MSO-definable transductions; and (3) *Analyzability:* decision
problems such as type checking and checking functional equivalence of two STTs,
are decidable. The last two features indicate that our model has the commonly
accepted trade-off between analyzability and expressiveness in formal language

---

* A more detailed version of this paper is available at http://www.cis.upenn.edu/~
  alur/stt12.pdf.

A. Czumaj et al. (Eds.): ICALP 2012, Part II, LNCS 7392, pp. 42–53, 2012.
© Springer-Verlag Berlin Heidelberg 2012

theory. The motivation for designing streaming algorithms that can process a document in a single pass has led to streaming models for checking membership in a regular tree language and for querying [5, 9–11], but the problem of computing all MSO-definable transformations in a single pass has not been studied previously.

The transducer model integrates features of *visibly pushdown automata*, equivalently *nested word automata* [12], and *streaming string transducers* [13, 14]. In our model, the input tree is encoded as a *nested word*, which is a string over alphabet symbols, tagged with open/close brackets (or equivalently, call/return types) to indicate the hierarchical structure [9, 12]. The streaming tree transducer reads the input nested word left-to-right in a single pass. It uses finitely many states, together with a stack, but the type of operation applied to the stack at each step is determined by the hierarchical structure of the tags in the input. The output is computed using a finite set of variables that range over output nested words, possibly with *holes* that are used as place-holders for inserting subtrees. At each step, the transducer reads the next symbol of the input. If the symbol is an internal symbol, then the transducer updates its state and the output variables. If the symbol is a call symbol, then the transducer pushes a stack symbol, along with updated values of variables, updates the state, and reinitializes the variables. While processing a return symbol, the stack is popped, and the new state and new values for the variables are determined using the current state, current variables, popped symbol, and popped values from the stack. In each type of transition, the variables are updated using expressions that allow adding new symbols, *string concatenation*, and *tree insertion* (simulated by replacing the hole with another expression). A key restriction is that variables are updated in a manner that ensures that each value can contribute at most once to the eventual output, without duplication. This *single-use-restriction* is enforced via a binary *conflict* relation over variables: no output term combines conflicting variables, and variable occurrences in right-hand sides during each update are consistent with the conflict relation. The transformation computed by the model can be implemented as a single-pass linear-time algorithm.

We show that the model can be simplified in natural ways if we want to restrict either the input or the output, to either strings or ranked trees. For example, to compute transformations that output strings it suffices to consider variable updates that allow only concatenation, and to compute transformations that output ranked trees it suffices to consider variable updates that allow only tree insertion. The restriction to the case of ranked trees as inputs gives the model of *bottom-up ranked-tree transducers*. As far as we know, this is the only transducer model that processes trees in a bottom-up manner, and can compute all MSO-definable transformations.

The main technical result in the paper is that the class of transductions definable using streaming tree transducers is exactly the class of MSO-definable transductions. The starting point for our result is the known equivalence of MSO-definable transductions and Macro Tree Transducers with regular lookahead and single-use restriction, over ranked trees [3]. Our proof proceeds by establishing two key properties of STTs: the model is closed under *regular look ahead* and under *functional composition*. These proofs are challenging due to the requirement that a transducer can use only a fixed number of variables that can be updated by assignments that obey the single-use-restriction rules.

While decidability for a variety of analysis questions for our transducer model follows from corresponding results for MSO-definable transformations [15], we are interested in understanding the complexity bounds for a variety of analysis questions. Given a regular language L_1 of input trees and a regular language L_2 of output trees, the *type checking* problem is to determine if the output of the transducer on an input in L_1 is guaranteed to be in L_2. We establish an EXP-TIME upper bound on type checking. For checking functional equivalence of two streaming tree transducers, we show that if the two transducers are inequivalent, then we can construct a pushdown automaton A over the alphabet $\{0,1\}$ such that A accepts a word with equal number of 0's and 1's exactly when there is an input on which the two transducers compute different outputs. Using known techniques for computing the Parikh images of context-free languages [15–17], this leads to a NEXPTIME upper bound for checking functional inequivalence of two STTs. Assuming a bounded number of variables, the upper bound on the parametric complexity becomes NP. Improving the NEXPTIME bound remains a challenging open problem.

2   Transducer Model

Nested Words: Data with both linear and hierarchical structure can be encoded using nested words [12]. Given a set Σ of symbols, the *tagged alphabet* $\hat{\Sigma}$ consists of the symbols a, $\langle a$, and $a \rangle$, for each $a \in \Sigma$. A *nested word* over Σ is a finite sequence over $\hat{\Sigma}$. For a nested word $a_1 \cdots a_k$, a position j, for $1 \leq j \leq k$, is said to be a *call* position if the symbol a_j is of the form $\langle a$, a *return* position if the symbol a_j is of the form $a \rangle$, and an *internal* position otherwise. The tags induce a natural matching relation between call and return positions, and in this paper, we are interested only in *well-matched* nested words in which all calls/returns have matching returns/calls. A string over Σ is a nested word with only internal positions. Nested words naturally encode ordered trees. The empty tree is encoded by the empty string ε. The tree with a-labeled root with subtrees $t_1, \ldots t_k$ as children, in that order, is encoded by the nested word $\langle a \, \langle\!\langle t_1 \rangle\!\rangle \cdots \langle\!\langle t_k \rangle\!\rangle \, a \rangle$, where $\langle\!\langle t_i \rangle\!\rangle$ is the encoding of the subtree t_i. This transformation can be viewed as an inorder traversal of the tree. The encoding extends to *hedges* also.

Nested Words with Holes: A key operation that our transducer model relies on is *insertion* of one nested word within another. In order to define this, we consider nested words with holes, where a hole is represented by the special symbol ?. We require that a nested word can contain at most one hole, and we use a binary type to keep track of whether a nested word contains a hole or not. A type-0 nested word does not contain any holes, while a type-1 nested word contains one hole. We can view a type-1 nested word as a unary function from nested words to nested words. The set $W_0(\Sigma)$ of type-0 nested words over the alphabet Σ is defined by the grammar $W_0 := \varepsilon \mid a \mid \langle a\,W_0\,b \rangle \mid W_0\,W_0$, for $a, b \in \Sigma$. The set $W_1(\Sigma)$ of type-1 nested words over the alphabet Σ is defined by the grammar $W_1 := ? \mid \langle a\,W_1\,b \rangle \mid W_1\,W_0 \mid W_0\,W_1$, for $a, b \in \Sigma$. A *nested-word language* over Σ is a subset L of $W_0(\Sigma)$, and a *nested-word transduction* from an input alphabet Σ to an output alphabet Γ is a *partial* function f from $W_0(\Sigma)$ to $W_0(\Gamma)$.

Nested Word Expressions: In our transducer model, the machine maintains a set of variables that range over output nested words with holes. Each variable has an associated binary type: a type-k variable has type-k nested words as values, for $k = 0, 1$. The variables are updated using typed expressions, where variables can appear on the right-hand side, and we also allow substitution of the hole symbol by another expression. Formally, a set X of typed variables is a set that is partitioned into two sets X_0 and X_1 corresponding to the type-0 and type-1 variables. Given an alphabet Σ and a set X of typed variables, a *valuation* α is a function that maps X_0 to $W_0(\Sigma)$ and X_1 to $W_1(\Sigma)$. Given an alphabet Σ and a set X of typed variables, we define the sets $E_k(X, \Sigma)$, for $k = 0, 1$, of type-k expressions by the grammars: $E_0 := \varepsilon \mid a \mid x_0 \mid \langle a\,E_0\,b \rangle \mid E_0\,E_0 \mid E_1[E_0]$ and $E_1 :=? \mid x_1 \mid \langle a\,E_1\,b \rangle \mid E_0\,E_1 \mid E_1\,E_0 \mid E_1[E_1]$, where $a, b \in \Sigma$, $x_0 \in X_0$ and $x_1 \in X_1$. The clause $e[e']$ corresponds to substitution of the hole in a type-1 expression e by another expression e'. A valuation α for the variables X naturally extends to a type-consistent function that maps the expressions $E_k(X, \Sigma)$ to values in $W_k(\Sigma)$, for $k = 0, 1$.

Single Use Restriction: The transducer updates variables X using type-consistent assignments. To achieve the desired expressiveness, we need to restrict the reuse of variables in right-hand sides. In particular, we want to disallow the assignment $x := xx$ (which would double the length of x), but allow the assignment $(x, y) := (x, x)$, provided the variables x and y are guaranteed not to be combined later. For this purpose, we assume that the set X of variables is equipped with a binary relation η: if $\eta(x, y)$, then x and y cannot be combined. This "conflict" relation is required to be reflexive and symmetric (but need not be transitive). Two conflicting variables cannot occur in the same expression used in the right-hand side of an update or as an output. During an update, two conflicting variables can occur in multiple right-hand sides for updating conflicting variables. Thus, the assignment $(x, y) := (\langle a\,xa \rangle[y], a?)$ is allowed, provided $\eta(x, y)$ does not hold; the assignment $(x, y) := (ax[y], y)$ is not allowed; and the assignment $(x, y) := (ax, x[b])$ is allowed, provided $\eta(x, y)$ holds. Formally, given a set X of typed variables with a reflexive symmetric binary conflict relation η, and an alphabet Σ, an expression e in $E(X, \Sigma)$ is said to be *consistent* with η, if (1) each variable x occurs at most once in e, and (2) if $\eta(x, y)$ holds, then e does not contain both x and y. Given sets X and Y of typed variables, a conflict relation η, and an alphabet Σ, a *single-use-restricted assignment* is a function ρ that maps each type-k variable x in X to a right-hand side expression in $E_k(Y, \Sigma)$, for $k = 0, 1$, such that (1) each expression $\rho(x)$ is consistent with η, and (2) if $\eta(x, y)$ holds, and $\rho(x')$ contains x, and $\rho(y')$ contains y, then $\eta(x', y')$ must hold. The set of such single-use-restricted assignments is denoted $\mathcal{A}(X, Y, \eta, \Sigma)$.

At a return, the transducer assigns the values to its variables X using the values popped from the stack as well as the values returned. For each variable x, we will use x_p to refer to the popped value of x. Thus, each variable x is updated using an expression over the variables $X \cup X_p$. The conflict relation η extends naturally to variables in X_p: $\eta(x_p, y_p)$ holds exactly when $\eta(x, y)$ holds. Then, the update at a return is specified by assignments in $\mathcal{A}(X, X \cup X_p, \eta, \Sigma)$.

When the conflict relation η is the purely reflexive relation $\{(x, x) \mid x \in X\}$, the single-use-restriction means that a variable x can appear at most once in at most one right-hand side. We refer to this special case as "copyless".

STT Syntax: A *deterministic streaming tree transducer* (STT) S from input alphabet Σ to output alphabet Γ consists of a finite set of states Q; a finite set of stack symbols P; an initial state $q_0 \in Q$; a finite set of typed variables X with a reflexive symmetric binary conflict relation η; a partial output function $F : Q \mapsto E_0(X, \Gamma)$ such that each expression $F(q)$ is consistent with η; an internal state-transition function $\delta_i : Q \times \Sigma \mapsto Q$; a call state-transition function $\delta_c : Q \times \Sigma \mapsto Q \times P$; a return state-transition function $\delta_r : Q \times P \times \Sigma \mapsto Q$; an internal variable-update function $\rho_i : Q \times \Sigma \mapsto \mathcal{A}(X, X, \eta, \Gamma)$; a call variable-update function $\rho_c : Q \times \Sigma \mapsto \mathcal{A}(X, X, \eta, \Gamma)$; and a return variable-update function $\rho_r : Q \times P \times \Sigma \mapsto \mathcal{A}(X, X \cup X_p, \eta, \Gamma)$. An STT S with variables X is called *copyless* if the conflict relation η equals $\{(x, x) \mid x \in X\}$.

STT Semantics: To define the semantics of a streaming tree transducer, we consider configurations of the form (q, Λ, α), where $q \in Q$ is a state, α is a type-consistent valuation from variables X to typed nested words over Γ, and Λ is a sequence of pairs (p, β) such that $p \in P$ is a stack symbol and β is a type-consistent valuation from variables in X to typed nested words over Γ. The initial configuration is $(q_0, \varepsilon, \alpha_0)$ where α_0 maps each type-0 variable to ε and each type-1 variable to ?. The transition function δ over configurations is defined by:

1. **Internal transitions:** $\delta((q, \Lambda, \alpha), a) = (\delta_i(q, a), \Lambda, \alpha \cdot \rho_i(q, a))$.
2. **Call transitions:** $\delta((q, \Lambda, \alpha), \langle a) = (q', (p, \alpha \cdot \rho_c(q, a))\Lambda, \alpha_0)$, where $\delta_c(q, a) = (q', p)$.
3. **Return transitions:** $\delta((q, (p, \beta)\Lambda, \alpha), a\rangle) = (\delta_r(q, p, a), \Lambda, \alpha \cdot \beta_p \cdot \rho_r(q, p, a))$, where β_p is the valuation for variables X_p defined by $\beta_p(x_p) = \beta(x)$ for $x \in X$.

For an input word $w \in W_0(\Sigma)$, if $\delta^*((q_0, \varepsilon, \alpha_0), w) = (q, \varepsilon, \alpha)$ then if $F(q)$ is undefined then so is $[\![S]\!](w)$, otherwise $[\![S]\!](w) = \alpha(F(q))$. We say that a nested word transduction f from input alphabet Σ to output alphabet Γ is *STT-definable* if there exists an STT S such that $[\![S]\!] = f$.

Example: Tree Swap: Streaming tree transducers can easily implement standard tree-edit operations such as insertion, deletion, and relabeling. We show how swapping can be implemented as a copyless STT that mirrors the corresponding natural single-pass algorithm. Figure 1 shows the transduction that transforms the input tree by swapping the first (in inorder traversal) b-rooted subtree t_1 with the next (in inorder traversal) b-rooted subtree t_2, not contained in t_1, For clarity of presentation, let us assume that the input word encodes a tree: it does not contain any internal symbols and if a call position is labeled $\langle a$ then its matching return is labeled $a\rangle$.

The initial state is q_0 which means that the transducer has not yet encountered a b-label. In state q_0, the STT records the tree traversed so far using a type-0 variable x: upon an a-labeled call, x is stored on the stack, and is reset to ε; and upon an a-labeled return, x is updated to $x_p\langle a\,x\,a\rangle$. In state q_0, upon a b-labeled

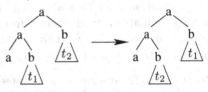

Fig. 1. Tree Swap

call, the STT pushes q_0 along with current x on the stack, resets x to ε, and updates its state to q'. In state q', the STT constructs the first b-labeled subtree t_1 in variable x: as long as it does not pop stack symbol q_0, at a call it pushes q' and x, and at a return, updates x to $x_p\langle a\,x\,a\rangle$ or $x_p\langle b\,x\,b\rangle$, depending on whether the current return symbol is a or b. When it pops q_0, it updates x to $\langle b\,x\,b\rangle$ (at this point, x contains the tree t_1, and its value will be propagated), sets another type-1 variable x' to x_p?, and changes its state to q_1. In state q_1, the STT is searching for the next b-labeled call, and processes a-labeled calls and returns exactly as in state q_0, but now using the type-1 variable x'. At a b-labeled call, it pushes q_1 along with x' on the stack, resets x to ε, and updates the state to q'. Now in state q', the STT constructs the second b-labeled subtree t_2 in variable x as before. When it pops q_1, the subtree t_2 corresponds to $\langle b\,x\,b\rangle$. The transducer updates x to $x'_p[\langle b\,x\,b\rangle]x_p$ capturing the desired swapping of the two subtrees t_1 and t_2 (the variable x' is no longer needed and is reset to ε to ensure copyless restriction), and switches to state q_2. In state q_2, the remainder of the tree is traversed adding it to x. The output function is defined only for the state q_2 and maps q_2 to x.

3 Properties and Variants

In this section, we note some properties and variants of streaming tree transducers aimed at understanding their expressiveness. First, STTs compute *linearly-bounded* outputs, that is, the length of the output word is within at most a constant factor of the length of the input word. The single-use-restriction ensures that at every step of the execution of the transducer on an input word, the sum of the sizes of all the variables that contribute to the output term at the end of the execution, can increase only by an additive constant.

Regular Nested-Word Languages: A streaming tree transducer with empty sets of string variables can be viewed as an *acceptor* of nested words: the input is accepted if the output function is defined in the terminal state, and rejected otherwise. In this case, the definition coincides with (deterministic) nested word automata (NWA). A language $L \subseteq W_0(\Sigma)$ of nested words is *regular* if it is accepted by such an automaton. This class includes all regular word languages, regular tree languages, and is a subset of deterministic context-free languages [12]. Given a nested-word transduction f from input alphabet Σ to output alphabet Γ, the *domain* of f is the set $Dom(f) \subseteq W_0(\Sigma)$ of input nested words w for which $f(w)$ is defined, and the *image* of f is the set $Img(f) \subseteq W_0(\Gamma)$ of output nested words w' such that $w' = f(w)$ for some w. It is easy to establish that: for an STT-definable transduction f from Σ to Γ, $Dom(f)$ is a regular language of nested words over Σ, and there exists an STT-definable transduction f from Σ to Γ, such that $Img(f)$ is not a regular language of nested words over Γ.

Bottom-Up Transducers: A nested-word automaton is called *bottom-up* if it resets its state along the call transition: if $\delta_c(q, a) = (q', p)$ then $q' = q_0$. The well-matched nested word sandwiched between a call and its matching return is processed by a bottom-up NWA independent of the outer context. It is known that bottom-up NWAs are as expressive as NWAs over well-matched words [12]. We show that a similar result holds for transducers also: there is no loss of expressiveness if the STT is disallowed to propagate information at a call to the linear successor. Note than every STT reinitializes all its variables at a call. An STT S is said to be a *bottom-up STT* if for every state $q \in Q$ and symbol $a \in \Sigma$, if $\delta_c(q, a) = (q', p)$ then $q' = q_0$. We can prove that: every STT-definable transduction is definable by a bottom-up STT.

Regular Look Ahead: Now we consider an extension of the STT model in which the transducer can make its decisions based on whether the remaining (well-matched) suffix of the input word belongs to a regular language of nested words. Such a test is called *regular look ahead*. Given a nested-word $w = a_1 a_2 \ldots a_k$, for each position $1 \le i \le k$, let $\text{WMS}(w, i)$ be the (well-matched) nested word $a_i, \ldots a_j$, where j is the maximal index l such that $a_i, \ldots a_l$ is well-matched. Then, a look-ahead test at step i can test a regular property of the word $\text{WMS}(w, i)$. Let L be a regular language of nested words, and let A be a (deterministic) bottom-up NWA for $reverse(L)$. Then, while processing a nested word, testing whether the word $\text{WMS}(w, i)$ belongs to L corresponds to testing whether the state of A after processing $reverse(\text{WMS}(w, i))$ is an accepting state of A. Since regular languages of nested words are closed under intersection, the state of a single bottom-up NWA A reading the input word in reverse can be used to test membership of the well-matched suffix at each step in different languages. This motivates the following formalization. Let $w = a_1 \ldots a_k$ be a nested word over Σ, and let A be a bottom-up NWA with states R processing nested words over Σ. Given a state $r \in R$, we define the r-look-ahead labeling of w to be the nested word $w_r = r_1 r_2 \ldots r_k$ over the alphabet R such that for each position $1 \le j \le k$, the call/return/internal type of r_j is the same as the type of a_j, and the corresponding symbol is the state of the NWA A after reading $reverse(a_j \ldots a_k)$ starting in state r. Then the *A-look-ahead labeling of w*, is the nested word $w_A = w_{r_0}$. An *STT-with-regular-look-ahead* consists of a bottom-up NWA A over Σ with states R, and an STT S from R to Γ. Such a transducer defines a streaming tree transduction from Σ to Γ: for an input word $w \in W(\Sigma)$, the output $[\![S, A]\!](w)$ is defined to be $[\![S]\!](w_A)$. The critical closure property for STTs is captured by the following result: the transductions definable by STTs with regular look-ahead are STT-definable. For the proof, given an NWA A with states R, and a bottom-up STT S over R, we construct an equivalent STT S'. Intuitively, at every step, for every possible look-ahead type $r \in R$ of the suffix, S' needs to maintain a copy of the state and variables of S. Updating these copies consistently while obeying single-use-restriction is the challenging part of the construction. To simplify the proof, we define an intermediate equivalent model of STTs, *multi-parameter STTs*, in which values of variables can contain multiple parameters.

Copyless STTs with RLA: Recall that an STT is said to be copyless if η only contains the reflexive relation. In an STT, an assignment of the form $(x, y) := (z, z)$ is allowed if x and y are guaranteed not to be combined, and thus, if only one of x and y contributes to the final output. In presence of regular-look-ahead test, the STT can check which variables contribute to the final output, and avoid redundant updates, and can thus be copyless: a nested-word transduction f is STT-definable iff it is definable by a copyless STT with regular-look-ahead.

Closure under Composition: Many of our results rely on the crucial property that STTs are closed under sequential composition: given two STT-definable transductions, f_1 from Σ_1 to Σ_2 and f_2 from Σ_2 to Σ_3, the composite transduction $f_2 \cdot f_1$ from Σ_1 to Σ_3 is STT-definable. In the constructive proof, given a copyless STT S_1 with regular-look-ahead, and a bottom-up STT S_2, we construct a multi-parameter STT S_3 with regular-look-ahead. The core idea of the construction is that S_3 maintains all possible "summaries" of executions of S_2 on each nested word stored in a variable of S_1.

Mapping Strings: A nested word captures both linear and hierarchical structure. There are two natural classes of nested words: strings are nested words with only linear structure, and ranked trees are nested words with only hierarchical structure. A transducer without a stack cannot use the hierarchical structure of the input nested word, and processes the input as a string of symbols. Since the transducer does not interpret the tags in a special manner, we will assume that the input is a string over Σ. This restricted transducer can still map strings to nested words (or trees) over Γ with interesting hierarchical structure, and hence, is called a *string-to-tree* transducer. This leads to the following definition: a *streaming string-to-tree transducer* (SSTT) S from input alphabet Σ to output alphabet Γ consists of a finite set of states Q; an initial state $q_0 \in Q$; a finite set of typed variables X; a partial output function $F : Q \mapsto E_0(X, \Gamma)$ such that for each state q, a variable x appears at most once in $F(q)$; a state-transition function $\delta_i : Q \times \Sigma \mapsto Q$; and a variable-update function $\rho_i : Q \times \Sigma \mapsto \mathcal{A}(X, X, \Gamma)$. Configurations of such a transducer are of the form (q, α), where $q \in Q$ is a state, and α is a type-consistent valuation for the variables X. The semantics $[\![S]\!]$ of such a transducer is a partial function from Σ^* to $W_0(\Gamma)$. In this case, the copyless restriction is enough to capture MSO completeness due to the following closure property: the transductions definable by SSTTs with regular look-ahead are SSTT-definable.

Mapping Ranked Trees: We consider how the definition can be simplified for the case of inputs restricted to ranked trees. The set $B(\Sigma)$ of *binary trees* over the alphabet Σ is then the subset of nested words defined by the grammar $T := \mathbf{0} \mid \langle a\, T\, T\, a \rangle$, for $a \in \Sigma$. The definition of an STT can be simplified in the following way if we know that the input is a binary tree. First, we do not need to worry about processing of internal symbols. Second, we restrict to bottom-up STTs due to their similarity to bottom-up tree transducers, where the transducer returns, along with the state, values for variables ranging over output nested words, as a result of processing a subtree. Finally, at a call, we know that there are exactly two subtrees, and hence, the propagation of information across matching calls and returns using a stack can be combined into a unified combinator: the transition function computes the result corresponding to a tree $a\langle t_l\, t_r \rangle$ based on the symbol a, and the results of processing subtrees t_l and t_r.

A *bottom-up ranked-tree transducer* (BRTT) S from binary trees over Σ to nested words over Γ consists of a finite set of states Q; an initial state $q_0 \in Q$; a finite set of typed variables X equipped with a conflict relation η; a partial output function $F : Q \mapsto E_0(X, \Gamma)$ such that for each state q, the expression $F(q)$ is consistent with η; a state-combinator function $\delta : Q \times Q \times \Sigma \mapsto Q$; and a variable-combinator function $\rho : Q \times Q \times \Sigma \mapsto \mathcal{A}(X_l \cup X_r, X, \eta, \Gamma)$, where X_l denotes the set of variables $\{x_l \mid x \in X\}$, X_r denotes the set of variables $\{x_r \mid x \in X\}$, and conflict relation η extends to these sets naturally. The state-combinator extends to trees in $B(\Sigma)$: $\delta^*(\mathbf{0}) = q_0$ and $\delta^*(a\langle t_l\, t_r \rangle) = \delta(\delta^*(t_l), \delta^*(t_r), a)$. The variable-combinator is used to map trees to valuations for X: $\alpha^*(\mathbf{0}) = \alpha_0$, where α_0 maps each type-0 variable to ε and each type-1 variable to ?, and $\alpha^*(a\langle t_l\, t_r \rangle) = \rho(\delta^*(t_l), \delta^*(t_r), a)[X_l \mapsto \alpha^*(t_l)][X_r \mapsto \alpha^*(t_r)]$. Given a tree $t \in B(\Sigma)$, let $\delta^*(t)$ be q and let $\alpha^*(t)$ be α. Then, if $F(q)$ is undefined then $[\![S]\!](t)$ is undefined, else $[\![S]\!](t)$ equals $\alpha(F(q))$ obtained by evaluating the expression $F(q)$ according to valuation α. We then prove that: a partial function from $B(\Sigma)$ to $W_0(\Gamma)$ is STT-definable iff it is BRTT-definable.

4   Expressiveness

The goal of this section is to prove that the class of nested-word transductions definable by STTs coincides with the class of transductions definable using Monadic Second Order logic (MSO). Our proof relies on the known equivalence between MSO and Macro Tree Transducers over *ranked trees* [3].

MSO for Nested Word Transductions: Formulas in monadic second-order logic (MSO) can be used to define functions from (labeled) graphs to graphs [2]. We adapt this general definition for our purpose of defining transductions over nested words. By adapting the simulation of string transducers by MSO [13, 18], we show that the computation of an STT can be encoded by MSO, and thus, every transduction computable by an STT is MSO definable.

Nested Words as Binary Trees: Nested words can be encoded as binary trees. This encoding is analogous to encoding of *unranked* trees as binary trees. Such an encoding increases the depth of the tree by imposing unnecessary hierarchical structure, and thus, is not suitable for processing of inputs. However, it is useful to simplify proofs of subsequent results about expressiveness. The desired transduction nw_bt from $W_0(\Sigma)$ to $B(\Sigma)$ is defined by $nw\_bt(\varepsilon) = \mathbf{0}$; $nw\_bt(aw) = a\langle\, nw\_bt(w)\, \mathbf{0}\,\rangle$; and $nw\_bt(\langle a\, w_1\, b\rangle\, w_2) = a\langle\, nw\_bt(w_1)\, b\langle\, nw\_bt(w_2)\, \mathbf{0}\,\rangle\,\rangle$. Observe that nw_bt is a one-to-one function. We can define the inverse partial function bt_nw from binary trees to nested words as follows: given $t \in B(\Sigma)$, if t equals $nw\_bt(w)$, for some $w \in W_0(\Sigma)$ (and if so, the choice of w is unique), then $bt\_nw(t) = w$, and otherwise $bt\_nw(t)$ is undefined. The next proposition shows that both these mappings can be implemented as STTs: the transductions $nw\_bt : W_0(\Sigma) \mapsto B(\Sigma)$ and $bt\_nw : B(\Sigma) \mapsto W_0(\Sigma)$ are STT-definable.

For a nested-word transduction f from $W_0(\Sigma)$ to $W_0(\Gamma)$, we can define another transduction $\tilde{f}$ that maps binary trees over Σ to binary trees over Γ: given a binary tree $t \in B(\Sigma)$, if t equals $nw\_bt(w)$, then $\tilde{f}(t) = nw\_bt(f(w))$, and otherwise $\tilde{f}(t)$ is undefined. The following proposition can be proved easily from the definitions of the encodings: if f is an MSO-definable transduction from $W_0(\Sigma)$ to $W_0(\Gamma)$, then the transduction $\tilde{f} : B(\Sigma) \mapsto B(\Gamma)$ is an MSO-definable binary-tree transduction and $f = bt\_nw \cdot \tilde{f} \cdot nw\_bt$. Since STT-definable transductions are closed under composition, to establish that every MSO-definable transduction is STT-definable, it suffices to consider MSO-definable transductions from binary trees to binary trees.

Macro Tree Transducers: A Macro Tree Transducer (MTT) [3, 4] is a tree transducer in which the translation of a tree may not only depend on its subtrees but also on its context. While the subtrees are represented by input variables, the context information is handled by parameters. We only need to consider deterministic MTTs with regular look ahead (MTTR) that map binary trees to binary trees. In general, MTTs are more expressive than MSO. The restrictions needed to limit the expressiveness rely on the *single-use-restriction* for using the parameters (SURP) and *finite copying* restriction in processing of the input (FCI). The following theorem is proved in [3]: a ranked-tree transduction f is MSO-definable iff there exists an MTTR M with SURP/FCI such that $f = [\![M]\!]$.

MSO Equivalence: We first show that bottom-up ranked-tree transducers are as expressive as MTTs with regular-look-ahead, with single-use restriction for the parameters, and finite-copying restriction for the input: if a ranked-tree transduction $f : B(\Sigma) \mapsto B(\Gamma)$ is definable by an MTTR with SURP/FCI, then it is BRTT-definable. Now, we can put together all the results to obtain the main technical result of the paper:

Theorem 1 (MSO Equivalence). *A nested-word transduction* $f : W_0(\Sigma) \mapsto W_0(\Gamma)$ *is STT-definable iff it is MSO-definable.*

5   Decision Problems

In this section, we show that a number of analysis problems for our model are decidable.

Output Analysis: Given an input nested word w over Σ, and an STT S from Σ to Γ, consider the problem of computing the output $[\![S]\!](w)$. To implement the operations of the STT efficiently, we can store the nested words corresponding to variables in linked lists with reference variables pointing to positions that correspond to holes. To process each symbol in w, the copyless update of variables can be executed by changing only a constant number of pointers. This leads to: given an input nested word w and an STT S with k variables, the output word $[\![S]\!](w)$ can be computed in time $O(k|w|)$.

The second problem we consider corresponds to *type checking*: given regular languages L_{pre} and L_{post} of nested words over Σ, and an STT S from Σ to Γ, the type checking problem is to determine if $[\![S]\!](L_{pre}) \subseteq L_{post}$ (that is, if for every $w \in L_{pre}$, $[\![S]\!](w) \in L_{post}$). This form of type checking is useful in checking consistency of XML schemas. For STTs, type checking can be solved in ExpTime: given an STT S from Σ to Γ, an NWA A accepting nested words over Σ, and an NWA B accepting nested words over Γ, checking $[\![S]\!](L(A)) \subseteq L(B)$ is solvable in time $O(|A|^3 \cdot |S|^3 \cdot n^{kn^2})$ where n is the number of states of B, and k is the number of variables in S.

As noted earlier, the image of an STT is not necessarily regular. However, the pre-image of a given regular language is regular, and can be computed. Given an STT S from input alphabet Σ to output alphabet Γ, and a language $L \subseteq W_0(\Gamma)$ of output words, the set $PreImg(L, S)$ consists of input nested words w such that $[\![S]\!](w) \in L$. We show that: given an STT S from Σ to Γ, and an NWA B over Γ, there is an algorithm to compute an NWA A over Σ such that $L(A) = PreImg(L(B), S)$. It follows that given an STT S and a regular language L of output nested words, there is an ExpTime algorithm to test whether $Img(S) \cap L$ is non-empty.

Functional Equivalence: Finally, we consider the problem of checking *functional equivalence* of two STTs: given two streaming tree transducers S and S', we want to check if they define the same transduction. Given two *streaming string transducers* S and S', [13, 14] shows how to construct an NFA A over the alphabet $\{0, 1\}$ such that the two transducers are *inequivalent* exactly when A accepts some word w such that w has equal number of 0's and 1's. The idea can be adopted for the case of STTs, but A now will be a nondeterministic pushdown automaton. The size of A is polynomial in the number of states of the input STTs, but exponential in the number of variables of the STTs. Results in [16, 17] can be adopted to check whether this pushdown automaton accepts a word with the same number of 0's and 1's.

Theorem 2 (Checking Equivalence). *Given two STTs S and S', the problem of checking whether $[\![S]\!] \neq [\![S']\!]$ is solvable in* NExpTime.

If the number of variables is bounded, then the size of A is polynomial, and this gives an upper bound of NP. For the transducers that map strings to nested words, that is, for streaming string-to-tree transducers (SSTT), the above construction yields a PSPACE bound.

Acknowledgements. We thank Joost Engelfriet for his valuable feedback.

References

1. Comon, H., Dauchet, M., Gilleron, R., Lugiez, D., Tison, S., Tommasi, M.: Tree automata techniques and applications. Draft (2002),
   http://www.grappa.univ-lille3.fr/tata/
2. Courcelle, B.: Monadic second-order definable graph transductions: A survey. Theor. Comput. Sci. 126(1), 53–75 (1994)
3. Engelfriet, J., Maneth, S.: Macro tree transducers, attribute grammars, and MSO definable tree translations. Information and Computation 154, 34–91 (1999)
4. Engelfriet, J., Vogler, H.: Macro tree transducers. J. Comput. System Sci. 31, 71–146 (1985)
5. Milo, T., Suciu, D., Vianu, V.: Typechecking for XML transformers. In: Proceedings of the 19th ACM Symposium on PODS, pp. 11–22 (2000)
6. Hosoya, H., Pierce, B.C.: XDuce: A statically typed XML processing language. ACM Trans. Internet Techn. 3(2), 117–148 (2003)
7. Martens, W., Neven, F.: On the complexity of typechecking top-down XML transformations. Theor. Comput. Sci. 336(1), 153–180 (2005)
8. Hosoya, H.: Foundations of XML Processing: The Tree-Automata Approach. Cambridge University Press (2011)
9. Segoufin, L., Vianu, V.: Validating streaming XML documents. In: Proceedings of the 21st ACM Symposium on PODS, pp. 53–64 (2002)
10. Neven, F., Schwentick, T.: Query automata over finite trees. Theor. Comput. Sci. 275(1-2), 633–674 (2002)
11. Madhusudan, P., Viswanathan, M.: Query Automata for Nested Words. In: Královič, R., Niwiński, D. (eds.) MFCS 2009. LNCS, vol. 5734, pp. 561–573. Springer, Heidelberg (2009)
12. Alur, R., Madhusudan, P.: Adding nesting structure to words. Journal of the ACM 56(3) (2009)
13. Alur, R., Cerný, P.: Expressiveness of streaming string transducers. In: IARCS Annual Conference on Foundations of Software Technology and Theoretical Computer Science. LIPIcs, vol. 8, pp. 1–12 (2010)
14. Alur, R., Cerný, P.: Streaming transducers for algorithmic verification of single-pass list-processing programs. In: Proceedings of 38th ACM Symposium on POPL, pp. 599–610 (2011)
15. Engelfriet, J., Maneth, S.: The equivalence problem for deterministic MSO tree transducers is decidable. Inf. Process. Lett. 100(5), 206–212 (2006)
16. Seidl, H., Schwentick, T., Muscholl, A., Habermehl, P.: Counting in Trees for Free. In: Díaz, J., Karhumäki, J., Lepistö, A., Sannella, D. (eds.) ICALP 2004. LNCS, vol. 3142, pp. 1136–1149. Springer, Heidelberg (2004)
17. Esparza, J.: Petri nets, commutative context-free grammars, and basic parallel processes. Fundam. Inform. 31(1), 13–25 (1997)
18. Engelfriet, J., Hoogeboom, H.: MSO definable string transductions and two-way finite-state transducers. ACM Trans. Comput. Log. 2(2), 216–254 (2001)

Causal Graph Dynamics*

Pablo Arrighi[1,2] and Gilles Dowek[3]

[1] Université de Grenoble, LIG, 220 rue de la chimie, 38400 SMH, France
[2] École Normale Supérieure de Lyon, LIP, 46 allée d'Italie, 69008 Lyon, France
parrighi@imag.fr
[3] INRIA, 23 avenue d'Italie, CS 81321, 75214 Paris Cedex 13, France.
gilles.dowek@inria.fr

Abstract. We extend the theory of Cellular Automata to arbitrary, time-varying graphs.

1   Introduction

A question. There are countless situations in which some agents (e.g. physical systems [19], computer processes [28], biochemical agents [27], economical agents [22], users of social networks, etc.) interact with their neighbours, leading to a global dynamics, the state of each agent evolving through the interactions. In most of these situations, the topology, i.e. who is next to whom also varies in time (e.g. agents become physically connected, get to exchange contact details, move around, etc.). The general concept of a dynamics caused by neighbour-to-neighbour interactions and with a time-varying neighbourhood, is therefore quite natural.

At the mathematical level, however, this general concept turns out to be rather difficult to formalize. There are at least three difficulties. The first is that the neighbourhood relation plays a double role in this story, as it is both a constraint upon the global dynamics, and a subject of the global dynamics, which modifies it. The second is that, as agents get created and deleted, the notion of who has become whom is not so obvious, but this notion is needed in order to state the causality property that only neighbours may communicate in one step of time. The third is to express that the global dynamics should 'act everywhere the same', a property akin to translation-invariance... but arbitrary graphs do not admit such translations.

Two approaches. Cellular Automata research lies at the cross-point between Physics, Mathematics, and Computer Science. Cellular Automata consist of a grid of identical square cells, each of which may take one of a finite number of possible states. The entire array evolves in discrete time steps. The dynamics is required to be translation-invariant (it commutes with translations of the grid) and causal (information cannot be transmitted faster than a fixed number of cells per time step). Whilst Cellular Automata are usually defined as exactly the functions having those physics-like symmetries, it turns out that they can also be

---

* The full version of this paper is available as arXiv:1202.1098.

A. Czumaj et al. (Eds.): ICALP 2012, Part II, LNCS 7392, pp. 54–66, 2012.
© Springer-Verlag Berlin Heidelberg 2012

characterized in purely mathematical terms as the set of translation-invariant continuous functions [16] for a certain metric. Moreover in a more Computer Science oriented-view, Cellular Automata can be seen as resulting from the synchronous application of the same local rule throughout the grid. These three complementary ways (physical causality, mathematical continuity and constructive locality) of addressing Cellular Automata are what anchors this field in rigorous foundations. Still, restricting to a fixed grid has been perceived to be a limitation. As a consequence Cellular Automata definitions have been extended from grids to Cayley or hyperbolic graphs, where most of the theory carries though [29,8,17]. But these graphs are quite regular in many respects, for instance they are self-similar under translations. More recently Cellular Automata definitions have been extended to graphs [28,9], in order to describe certain distributed algorithms. In these extensions, the topology remained fixed: they would be unable to account for evolving mobile networks, nor discrete formulations of general relativity [36]. *This paper aims at extending Cellular Automata theory to arbitrary, time-varying graphs. The theorems we prove are mainly inspired by those at the foundation of Cellular Automata theory. Two of them show the equivalence of causality with local rule constructions (Theorem 1) and continuity (Theorem 3), and two others are closure properties by composition (Theorem 2) and inversion (Theorem 4).*

The second, related line of work is that of Graph Rewriting. The idea of rewriting graphs by applying some replacement rules has arisen as a natural generalization of term rewriting, and is now widespread in Computer Science [30,12]. Whilst Cellular Automata theory focuses on changing states, Graph Rewriting focuses on changing the topology. But there are other fundamental differences. Whilst Cellular Automata theory focuses on the global dynamics resulting from the synchronous application of the same local rule throughout the graph, Graph Rewriting theory usually focuses on asynchronous applications of a rule which need not be local, this leading to an undefined global dynamics (hence the emphasis on properties such as confluence, non-interference, etc.). Amalgamated Graph Transformations [6,26] and Parallel Graph Transformations [13,37,38] are noticeable exceptions in this respect, as they work out rigorous ways to apply local rewriting rules synchronously throughout the graph. Still the properties of the resulting global dynamics are not examined. *This paper aims at extending the focus of Graph Rewriting to changing states, as well as to deduce aspects of Amalgamated/Parallel Graph Transformations from the axiomatic properties of the global dynamics.*

Third way. The idea of a rigorous model of computation in which both the states and the topology evolve is certainly not new, and can be attributed to Kolmogorov and Upsenskii [20], see also [33]. These models, as well as the more recent [31,7], are again asynchronous in a sense. There is no spatial parallelism, although it may be simulated [42]. Lately, several groups have taken a more practical approach to this problem, and have started to develop simulation environments [15,43,24] based on different programming paradigms, all of them implementing the idea of rewriting both the graph and its states via repeated applications of some replacement rules. These systems offer the possibility to apply the replacement rules simultaneously in different non-conflicting places. Such an evaluation strategy does provide some weak form of synchronism. Sometimes, when a set of rule is non-conflicting, this evaluation strategy happens to coincide with full synchronism. This pragmatic approach to extending Cellular Automata to time-varying graphs is advocated in [41,23,19], and has led to some advanced algorithmic constructions [40,39]. *This paper aims at proposing simple formalizations of the notions of translation-invariance and causality when referring to functions from labelled graphs to labelled graphs. It aims at deducing, from these two notions, what is the most general form of fully synchronous application of a local rule throughout a graph, as well as to achieve a full characterization of what are the suitable local rules.*

2   Graphs Dynamics

Graphs. We fix an uncountable infinite set V of names. The *vertices* of the graphs we consider in this paper are uniquely identified by a name u in V. Vertices may also have a *state* $\sigma(u)$ in Σ. Each vertex has several *ports*, numbered between 1 and a natural number π. A vertex and its port are written $u\!:\!i$. An *edge* is a pair $(u\!:\!i, v\!:\!j)$. Edges also have a *state* $\delta(u\!:\!i, v\!:\!j)$ in Δ. In fact, the presence of an edge is held in the domain of the function δ.

Definition 1 (Graph). *A graph G with states Σ, Δ and degree π is given by*

- *An at most countable subset $V(G)$ of V whose elements are called* vertices.
- *A set $1..\pi$ whose elements are called* ports.
- *A partial function σ from $V(G)$ to Σ giving the state of the vertices.*
- *A partial function δ from $(V(G)\!:\!1..\pi) \times (V(G)\!:\!1..\pi)$ to Δ giving the states of the edges, such that each $u:i$ in $(V(G):1..\pi)$ appears at most once in $dom(\delta)$.*

The set of all graphs with states Σ, Δ and degree π is written $\mathcal{G}_{\Sigma,\Delta,\pi}$. To ease notations, we sometimes write $v \in G$ for $v \in V(G)$.

The definition is tailored so that we are able to cut the graph around some vertices, whilst keeping the information about the connectivity with the surrounding vertices. The choice of V uncountable but $V(G)$ countable allows us to always pick fresh names. The edges are oriented, but for our dynamics to be compositional we need to define neighbours regardless of edge orientation:

Definition 2 (Neighbours). *We write $u \frown v$ if there exists ports $i, j \in 1..\pi$ such that either $\delta(u\!:\!i, v\!:\!j)$ or $\delta(v\!:\!j, u\!:\!i)$ is defined. We write $u \frown^k v$ if there exists $w_1, \ldots, w_{k-1}$ such that $u \frown w_1 \frown \ldots \frown w_{k-1} \frown v$. We write $u \frown^{\leq r} v$ if there exists $k \leq r$ such that $u \frown^k v$. The set of neighbours of radius r (a.k.a. of diameter $2r + 1$) of a set A with respect to a graph G is the set of the vertices v in $V(G)$ such that $u \frown^{\leq r} v$, for u in A.*

Moving on, a *pointed graph* is just a graph where one, or sometimes several, vertices are privileged.

Definition 3 (Pointed graph). *A* pointer set *of G is a subset of $V(G)$. A* pointed graph *is given by a graph G and pointer set A of G. The* set of pointed graphs *with states Σ, Δ and degree π is written $\mathcal{P}_{\Sigma,\Delta,\pi}$.*

Dynamics. Consider an arbitrary function from graphs to graphs. The difficulty we now address is that of expressing the condition that this function 'acts everywhere the same' — a property similar to that of translation-invariance in the realm of Cellular Automata. Of course arbitrary graphs do not admit translations; the first idea is that translation-invariance becomes an invariance under *isomorphisms*. In other words, the names of the vertices are somewhat immaterial, they can be renamed, unlike states and ports.

Definition 4 (Isomorphism). *An* isomorphism R *is a function from $\mathcal{G}_{\Sigma,\Delta,\pi}$ to $\mathcal{G}_{\Sigma,\Delta,\pi}$ which is specified by a bijection $R(.)$ from V to V. The image of a graph G under the isomorphism R is a graph RG whose set of vertices is $R(V(G))$, and whose partial functions σ_{RG} and δ_{RG} are the compositions $\sigma_G \circ R^{-1}$ and $\delta_G \circ R^{-1}$ respectively. When G and H are isomorphic we write $G \approx H$. Similarly, the image of a pointed graphs $P = (G, A)$ is the pointed graph $RP = (RG, R(A))$. When P and Q are isomorphic we write $P \approx Q$.*

It would seem that the graph *dynamics* we are interested in must commute with isomorphisms, as in [14]. Unfortunately, demanding straightforward commutation with isomorphisms would enter in conflict with the possibility to introduce new vertices.

Proposition 1 (Commuting forbids new vertices). *Let F be a function from $\mathcal{G}_{\Sigma,\Delta,\pi}$ to $\mathcal{G}_{\Sigma,\Delta,\pi}$, which commutes with isomorphisms, i.e. such that for any isomorphism R, $F \circ R = R \circ F$. Then for any graph G, $V(F(G)) \subseteq V(G)$.*

The question of circumventing this limitation has been pointed out to be a difficulty [35]. We propose to focus on the following, weaker property instead:

Definition 5 (Dynamics). *A* dynamics F *is a function from $\mathcal{G}_{\Sigma,\Delta,\pi}$ to $\mathcal{G}_{\Sigma,\Delta,\pi}$, such that the following two conditions are met:*

(i) Conjugacy. *For any isomorphism R there exists an isomorphism R', called a* conjugate *of R through F, such that $F \circ R = R' \circ F$, i.e. for all G, $F(RG) = R'F(G)$.*
(ii) Freshness. *For any family of graphs $(G^{(i)})$, $\left[\bigcap V(G^{(i)}) = \emptyset \Rightarrow \bigcap V(F(G^{(i)})) = \emptyset \right]$.*

The definition extends to functions from $\mathcal{P}_{\Sigma,\Delta,\pi}$ to $\mathcal{G}_{\Sigma,\Delta,\pi}$ in the obvious way.

Note that both conditions in the above definition would have been entailed by straightforward commutation of F with every R. Moreover, condition *(ii)* still has a natural interpretation: to generate a fresh common vertex (inside $\bigcap V(F(G^{(i)}))$) two parties (inside the $(G^{(i)})$) must use a common resource (inside $\bigcap V(G^{(i)})$). Even two persons bumping into each other in the street share a common medium: the street. Note that, due to this condition, dynamics send the empty graph to the empty graph.

3 Causal Dynamics

Our goal in this Section is to define the notion of causality in an axiomatic way. Intuitively a dynamics is *causal* if the state and connectivity of each image vertex is determined by a small graph describing the state and connectivity of the neighbours of some antecedent vertex. Moreover, not only each image vertex is generated by an antecedent vertex, but also each antecedent vertex generates only a bounded number of image vertices. In order to make this into a formal definition we clearly need a notion of *disk* around a set of vertices, as well as a notion of *antecedent*.

Definition 6 (Induced subgraph). *The* induced subgraph *of a graph G around a set U is a graph G_U:*

- *whose vertices $V(G_U)$ are given by the neighbours of radius one of the set $U' = (V(G) \cap U)$.*
- *whose partial function σ_{G_U} is the restriction of σ to U'.*
- *whose partial function δ_{G_U} is the restriction of δ to*

$$((V(G_U):1..\pi) \times (U':1..\pi)) \cup ((U':1..\pi) \times (V(G_U):1..\pi)).$$

Definition 7 (Disk). *The* envelope of radius r *(a.k.a. of diameter $2r + 1$) of a set A with respect to a graph G is the pointed graph whose graph is the induced subgraph of G around the neighbours of radius r of A, and whose pointer set is A itself. It is denoted G_A^r.*

The set of disks of radius r (a.k.a. of diameter $2r + 1$) of the set of graph $\mathcal{G}_{\Sigma,\Delta,\pi}$ is the set of envelopes $\{G_v^r \mid G \in \mathcal{G}_{\Sigma,\Delta,\pi}\}$, i.e. those centered on a single vertex. It is denoted $\mathcal{D}_{\Sigma,\Delta,\pi}^r$. The set of all disks of the set of graph $\mathcal{G}_{\Sigma,\Delta,\pi}$ is $\bigcup_r \mathcal{D}_{\Sigma,\Delta,\pi}^r$. It is denoted $\mathcal{D}_{\Sigma,\Delta,\pi}$.

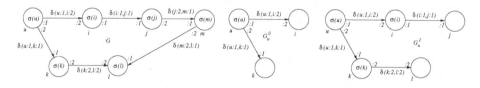

Dynamics make vertex names somewhat immaterial, but still do not prevent from using these names in order to identify the antecedent vertices of an image vertex.

Definition 8 (Antecedent codynamics). *Let F be a dynamics from $\mathcal{G}_{\Sigma,\Delta,\pi}$ to $\mathcal{G}_{\Sigma,\Delta,\pi}$. We define the antecedent codynamics $a()$ from V to subsets of V such that $v \in a(v')$ if and only if:*

$$\forall G, [v' \in F(G) \Rightarrow v \in G].$$

The following is our main definition.

Definition 9 (Causality). *A dynamics F from $\mathcal{G}_{\Sigma,\Delta,\pi}$ to $\mathcal{G}_{\Sigma,\Delta,\pi}$ is causal if and only if there exists a radius r and a bound b, such that the following two conditions are met:*

(i) Uniform continuity.

$$\forall v', v \in a(v'), \forall G, H, \; \left[G_v^r = H_v^r \Rightarrow F(G)_{v'} = F(H)_{v'} \right]$$

(ii) Boundedness.

$$\forall G, \forall v \in G, \; |\{v' \in F(G) \,|\, v \in a(v')\}| \leq b.$$

with $a(.)$ the antecedent codynamics of F.

This definition captures, in a formal way, the physical idea that information propagate at a bounded velocity.

4   Localizable Dynamics

The definition of causality does not provide us with a concrete way to construct such causal dynamics. We now introduce the more concrete notion of *localizable dynamics*, i.e. a dynamics which is induced by a local rule. Hence let us thus shift our focus towards bottom-up, local mechanisms for rewriting graphs in a parallel manner. This construction is reminiscent of [6,26]. We define the notion of *consistency* between two graphs as the fact that they do not disagree on their intersections. The *union* of two consistent graphs is itself a graph. See the long version of the paper for more details. Roughly speaking the result of a localizable dynamics is the union of the result of the parallel application of a *local rule*, which we now define.

Definition 10 (Consistent function). *A function f from $\mathcal{D}_{\Sigma,\Delta,\pi}^r$ to $\mathcal{G}_{\Sigma,\Delta,\pi}$ is consistent if and only if for any graph G, for any pair of disks G_u^r, G_v^r, $f(G_u^r)$ is consistent with $f(G_v^r)$.*

Definition 11 (Bounded function). *A subset of graphs $\mathcal{G}_{\Sigma,\Delta,\pi}$ is bounded if there exists b such that for any graph G in the set, $|V(G)|$ is less or equal to b. A partial function from $\mathcal{G}_{\Sigma,\Delta,\pi}$, or $\mathcal{P}_{\Sigma,\Delta,\pi}$, to $\mathcal{G}_{\Sigma,\Delta,\pi}$ is bounded if its co-domain is bounded.*

Definition 12 (Local rule). *A function from $\mathcal{D}_{\Sigma,\Delta,\pi}^r$ to $\mathcal{G}_{\Sigma,\Delta,\pi}$ is a local rule or radius r if it is a consistent bounded dynamics.*

Here is the natural way to parallelize the application of the above-described local rules into a global dynamics.

Definition 13 (localizable dynamics). *A dynamics F from $\mathcal{G}_{\Sigma,\Delta,\pi}$ to $\mathcal{G}_{\Sigma,\Delta,\pi}$ is localizable if and only if there exists r a radius and f a local rule from $\mathcal{D}_{\Sigma,\Delta,\pi}^r$ to $\mathcal{G}_{\Sigma,\Delta,\pi}$ such that for every graph G in $\mathcal{G}_{\Sigma,r}$,*

$$F(G) = \bigcup_{v \in G} f(G_v^r).$$

There are many, more specific-purpose models which a posteriori can be viewed as instances of the notion of causal graph dynamics developed in this paper [19,28,25,27]. We now show via two examples that our model subsumes, but is not restricted to, Cellular Automata.

Cellular automata. For some localizable dynamics, the graph is finite and does not change. Such dynamics are called *bounded cellular automata*. Some others slightly expand their border. Such dynamics are called *finite unbounded cellular automata*. One-dimensional finite unbounded cellular automata are usually defined as follows.

Definition 14 (Finite unbounded cellular automata). *An alphabet Σ is a finite set of symbols, with a distinguished quiescent symbol q. A configuration c is a function $\mathbb{Z} \to \Sigma$, that is equal to q almost everywhere. A finite unbounded cellular automaton is a function H from configurations to configurations such that $H(c)_{i+1} = h(c_i, c_{i+1})$, with h a function from Σ^2 to Σ such that $h(q, q) = q$.*

Note that the use of quiescent states is an artifact to express that configurations are finite but unbounded, i.e. that they may grow arbitrarily large.

A configuration c of such a Cellular Automaton can be represented as a finite graph as follows: there is an interval $I = [n, p]$ such that $c_i = q$ whenever $i \notin I$. We take $V(G) = I$, $\pi = 2$, $\delta(x\!:\!2, (x+1)\!:\!1)$ defined for $x \in [n, p)$ and undefined otherwise. Note that the geometry is not expressed by the names of the cells, with 1 next to 2, etc., but by the edges of the graph. The local rule f is defined on disks of radius one as follows (only the significant cases are shown; and the dashed nodes may be present or not):

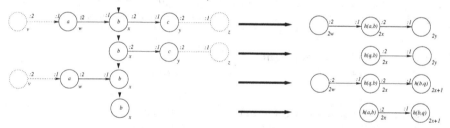

Take $\Sigma = \{0, 1\}$, $q = 0$ and $h(a, b) = (a + b) \mod 2$. Then $c = 10011$ is mapped to $c' = 110101$. Consider a coding of c, modulo isomorphism, e.g.

There are five vertices and hence five disks:

Taking the union, we obtain:

$$\underset{8}{\overset{1}{\bigcirc}} \xrightarrow[\;]{:2\;\;:1} \underset{18}{\overset{1}{\bigcirc}} \xrightarrow[\;]{:2\;\;:1} \underset{22}{\overset{0}{\bigcirc}} \xrightarrow[\;]{:2\;\;:1} \underset{26}{\overset{1}{\bigcirc}} \xrightarrow[\;]{:2\;\;:1} \underset{32}{\overset{0}{\bigcirc}} \xrightarrow[\;]{:2\;\;:1} \underset{33}{\overset{1}{\bigcirc}}$$

which is indeed a representation of c'.

The inflating grid. In another extreme case, the graph gets radically modified as each vertex v gives rise to four vertices $4v$, $4v + 1$, $4v + 2$, and $4v + 3$. The general case of the local rule, defined on disks of radius zero, is the following:

but we have to include fifteen other cases for vertices that do not have neighbours in all directions. This inflating grid may also be viewed as a way to generate smaller and smaller structures in a fixed size system.

If we now include states in the guise of colours on vertices, start with a grey vertex and rewrite

- a black vertex to a cluster of four black vertices,
- and a grey vertex to a cluster of grey, grey, grey, and black vertices,

we get the picture on the left. If, on the other hand, we rewrite

- a black vertex to a cluster of four black vertices,
- a white vertex to a cluster of white, white, white, and a black vertex,
- and a grey vertex to a cluster of four white vertices,

we get the picture on the right:

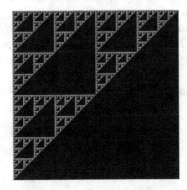
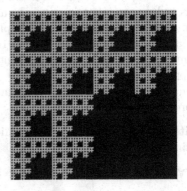

We now establish our main theorem: that a dynamics is causal if and only if it is localizable. On the one hand this shows that there is always a concrete way to construct a causal dynamics. On the other hand, this shows that the notion of a localizable dynamic is grounded on physical principles such as the bounded velocity of the propagation of information. Thus the Physical and the Computer Science-oriented notions coincide. A similar theorem is trivial for classical cellular automata, but much more challenging for instance for reversible cellular automata [18,10,3], quantum cellular automata [4], and graph dynamics. In contrast, the extension to probabilistic cellular automata fails [1].

Theorem 1 (Structure). *Let F be a dynamics from $\mathcal{G}_{\Sigma,\Delta,\pi}$ to $\mathcal{G}_{\Sigma,\Delta,\pi}$. F is causal if and only if it is localizable.*

5   Properties

Causal graph dynamics are closed under composition. This is an important indicator of the robustness of this notion. Such a result holds trivially for classical and reversible cellular automata, but depends on the chosen definition for quantum [11,34,5] and probabilistic cellular automata [2].

Theorem 2 (Composability). *Consider F_1 a causal dynamics induced by the local rule f_1 of radius r_1 (i.e. diameter $d_1 = 2r_1 + 1$). Consider F_2 a causal dynamics induced by the local rule f_2 of radius r_2 (i.e. diameter $d_2 = 2r_2 + 1$). Then $F_2 \circ F_1$ is a causal dynamics induced by the local rule g of radius $r'' = 2r_1r_2 + r_1 + r_2$ (i.e. diameter $d'' = d_1d_2$) from $\mathcal{D}^{r''}$ to $\mathcal{G}_{\Sigma,\Delta,\pi}$ which maps $N_v^{r''}$ to*

$$\bigcup_{v' \in f_1(N_v^{r_1})} f_2\Big(\big(\bigcup_{v \in N_v^{r''}} f_1(N_v^{r_1})\big)_{v'}^{r_2}\Big)$$

We also show that causal dynamics of radius one are universal.

Proposition 2 (Universality of radius one). *Consider F a causal dynamics of radius $r = 2^l$ over $\mathcal{G}_{\Sigma,\Delta,\pi}$. There exists F' a causal dynamics of radius 1 over $\mathcal{G}_{\Sigma^{l+1},\Delta\cup\{*\},\pi^r}$, such that $F'^{l+1} = F$.*

The notion of causality (see Definition 9), is based on the mathematical notion of uniform continuity: the radius r is independent of the vertex v and of the graph G. It is well-known that in general uniform continuity implies continuity, and that on compact spaces continuity implies uniform continuity. Such results have been extended to cellular automata [16]. This is also true of graph dynamics. Uniform continuity always implies continuity, and the converse holds when the state spaces are finite.

   We now develop a notion of continuity, and find out that it is equivalent to limit-preservation.

Definition 15 (Continuous dynamics). *A dynamics F from $\mathcal{G}_{\Sigma,\Delta,\pi}$ to $\mathcal{G}_{\Sigma,\Delta,\pi}$ is continuous if and only if:*

$$\forall r', \forall v', v \in a(v'), \forall G, \exists r, \forall H, \quad [G_v^r = H_v^r \Rightarrow F(G)_{v'}^{r'} = F(H)_{v'}^{r'}]$$

with a(.) the antecedent codynamics of F.

Definition 16 (Limit at A). *Consider a function $r \mapsto (G(r), A)$ from $\mathbb{N}$ to $\mathcal{P}_{\Sigma, \Delta, \pi}$. We say that it converges to (G, A) if and only if for all r there exists s such that for all $s' \geq s$, $G(s')^r_A = G^r_A$.*

Proposition 3 (Continuity as limit preservation). *Consider a dynamics F having antecedent codynamics $a(.)$. F is continuous if and only if it preserves limits, meaning that if the function $s \mapsto (G(s), A)$ converges to (G, A), then the function $s \mapsto (F(G(s)), a^{-1}(A))$ converges to $(F(G), a^{-1}(A))$.*

Corollary 1 (Causality implies limit preservation). *If a dynamics F is causal then it is uniformly continuous, hence it is continuous and it preserves limits.*

The converse is not true in general, and will now be investigated in the finite case. For the next two theorems the graphs may still be infinite in size, but their set of states of vertices and edges are supposed to be finite. Both suppositions are necessary to obtain the compactness property, which works modulo isomorphism. In topology, continuity and compactness entail uniform continuity. We do not have a clear topology for our graphs but a similar result holds.

Theorem 3 (Continuity and causality). *Consider F a dynamics from $\mathcal{G}_{\Sigma, \Delta, \pi}$ to $\mathcal{G}_{\Sigma, \Delta, \pi}$, with Σ and Δ finite. If F is continuous, then it verifies the first causality condition, and conversely.*

A causal dynamics is said to be *invertible* if it has an inverse. It is said to be *reversible* if this inverse is itself a causal dynamics. As for the cellular automata, invertibility implies reversibility: if a causal dynamics has an inverse, then this inverse is also a causal dynamics. The proof of this invertibility theorem is similar to the proof that continuity implies uniform continuity, both proceeding by extracting converging subsequences from arbitrary ones.

Definition 17 (Invertible dynamics). *A dynamics F from $\mathcal{G}_{\Sigma, \Delta, \pi}$ to $\mathcal{G}_{\Sigma, \Delta, \pi}$ is invertible if and only if there exists a dynamics $F^\dagger$ such that $F^\dagger \circ F = Id$ and $F \circ F^\dagger = Id$. This inverse is unique.*

Theorem 4 (Causal, invertible, reversible). *Consider F a causal dynamics from $\mathcal{G}_{\Sigma, \Delta, \pi}$ to $\mathcal{G}_{\Sigma, \Delta, \pi}$, with Σ and Δ finite. If F is invertible with dynamics $F^\dagger$, then $F^\dagger$ is also causal.*

6 Future Work

All of these results reinforce the impression that the notion of causal graph dynamics we have reached is both general and robust; and at the same time concrete enough so that non-trivial facts can be said about them, and useful models be instantiated. Similar specific-purpose models are in fact in use in a variety of contexts ranging from social networks to epidemiology [22,32] or physics [21].

Several of these, however, include an extra ingredient of non-determinism, probabilities, or even quantum theory, which often are key ingredients of the physical situation at hand — whereas the setting of this paper has remained nicely deterministic for now. Hence, studying probabilistic or quantum versions of causal graph dynamics ought to be promising.

There are plenty positive theoretical results about cellular automata that need to be reevaluated in this more general context, in terms of set-theoretical properties, structure, order, dynamical properties, computability etc. More interestingly even, negative results, such as the suspected anisotropy of Cellular automata, need to be reevaluated. Good wills are welcome. More concretely, we leave it as an open question whether condition (*ii*) of Definition 5 can be relaxed to its binary form and the same results be obtained.

Acknowledgement. This research was funded by ANR CausaQ. We thank Rachid Echahed, Renan Fargetton, Miguel Lezama, Mehdi Mhalla, Simon Perdrix, Guillaume Theyssier, Eric Thierry and Nicolas Trotignon for inspiring discussions and pointers.

This work is dedicated to Angelo.

References

1. Arrighi, P., Dowek, G.: On the Completeness of Quantum Computation Models. In: Ferreira, F., Löwe, B., Mayordomo, E., Mendes Gomes, L. (eds.) CiE 2010. LNCS, vol. 6158, pp. 21–30. Springer, Heidelberg (2010)
2. Arrighi, P., Fargetton, R., Nesme, V., Thierry, E.: Applying Causality Principles to the Axiomatization of Probabilistic Cellular Automata. In: Löwe, B., Normann, D., Soskov, I., Soskova, A. (eds.) CiE 2011. LNCS, vol. 6735, pp. 1–10. Springer, Heidelberg (2011)
3. Arrighi, P., Nesme, V.: A simple block representation of Reversible Cellular Automata with time-simmetry. In: 17th International Workshop on Cellular Automata and Discrete Complex Systems, AUTOMATA 2011, Santiago de Chile (November 2011)
4. Arrighi, P., Nesme, V., Werner, R.: Unitarity plus causality implies localizability. J. of Computer and Systems Sciences 77, 372–378 (2010); QIP 2010 (long talk)
5. Arrighi, P., Nesme, V., Werner, R.: One-Dimensional Quantum Cellular Automata over Finite, Unbounded Configurations. In: Martín-Vide, C., Otto, F., Fernau, H. (eds.) LATA 2008. LNCS, vol. 5196, pp. 64–75. Springer, Heidelberg (2008)
6. Boehm, P., Fonio, H.R., Habel, A.: Amalgamation of graph transformations: a synchronization mechanism. Journal of Computer and System Sciences 34(2-3), 377–408 (1987)
7. Cavaliere, M., Csikasz-Nagy, A., Jordan, F.: Graph transformations and game theory: A generative mechanism for network formation. University of Trento, Technical Report CoSBI 25/2008 (2008)
8. Ceccherini-Silberstein, T., Coornaert, M.: Cellular automata and groups. Springer (2010)
9. Derbel, B., Mosbah, M., Gruner, S.: Mobile Agents Implementing Local Computations in Graphs. In: Ehrig, H., Heckel, R., Rozenberg, G., Taentzer, G. (eds.) ICGT 2008. LNCS, vol. 5214, pp. 99–114. Springer, Heidelberg (2008)

10. Durand-Lose, J.: Representing reversible cellular automata with reversible block cellular automata. Discrete Mathematics and Theoretical Computer Science 145, 154 (2001)

11. Durr, C., Santha, M.: A decision procedure for unitary linear quantum cellular automata. In: Proceedings of the 37th IEEE Symposium on Foundations of Computer Science, pp. 38–45. IEEE (1996)

12. Ehrig, H., Ehrig, K., Prange, U., Taentzer, G.: Fundamentals of algebraic graph transformation. Springer-Verlag New York Inc. (2006)

13. Ehrig, H., Lowe, M.: Parallel and distributed derivations in the single-pushout approach. Theoretical Computer Science 109(1-2), 123–143 (1993)

14. Gabbay, M.J., Pitts, A.M.: A new approach to abstract syntax with variable binding. Formal Aspects of Computing 13(3), 341–363 (2002)

15. Giavitto, J.L., Spicher, A.: Topological rewriting and the geometrization of programming. Physica D: Nonlinear Phenomena 237(9), 1302–1314 (2008)

16. Hedlund, G.A.: Endomorphisms and automorphisms of the shift dynamical system. Math. Systems Theory 3, 320–375 (1969)

17. Herrmann, F., Margenstern, M.: A universal cellular automaton in the hyperbolic plane. Theoretical Computer Science 296(2), 327–364 (2003)

18. Kari, J.: Representation of reversible cellular automata with block permutations. Theory of Computing Systems 29(1), 47–61 (1996)

19. Klales, A., Cianci, D., Needell, Z., Meyer, D.A., Love, P.J.: Lattice gas simulations of dynamical geometry in two dimensions. Phys. Rev. E 82(4), 046705 (2010)

20. Kolmogorov, A.N., Uspensky, V.A.: On the definition of an algorithm. Uspekhi Matematicheskikh Nauk 13(4), 3–28 (1958)

21. Konopka, T., Markopoulou, F., Smolin, L.: Quantum graphity. Arxiv preprint hep-th/0611197 (2006)

22. Kozma, B., Barrat, A.: Consensus formation on adaptive networks. Phys. Rev. E 77, 016102 (2008)

23. Kreowski, H.-J., Kuske, S.: Autonomous Units and Their Semantics - The Parallel Case. In: Fiadeiro, J.L., Schobbens, P.-Y. (eds.) WADT 2006. LNCS, vol. 4409, pp. 56–73. Springer, Heidelberg (2007)

24. Kurth, W., Kniemeyer, O., Buck-Sorlin, G.: Relational Growth Grammars – A Graph Rewriting Approach to Dynamical Systems with a Dynamical Structure. In: Banâtre, J.-P., Fradet, P., Giavitto, J.-L., Michel, O. (eds.) UPP 2004. LNCS, vol. 3566, pp. 56–72. Springer, Heidelberg (2005)

25. Lathrop, J.I., Lutz, J.H., Patterson, B.: Multi-Resolution Cellular Automata for Real Computation. In: Löwe, B., Normann, D., Soskov, I., Soskova, A. (eds.) CiE 2011. LNCS, vol. 6735, pp. 181–190. Springer, Heidelberg (2011)

26. Löwe, M.: Algebraic approach to single-pushout graph transformation. Theoretical Computer Science 109(1-2), 181–224 (1993)

27. Murray, J.D.: Mathematical biology. ii: Spatial models and biomedical applications. In: Biomathematics, 3rd edn., vol. 18, Springer (2003)

28. Papazian, C., Rémila, É.: Hyperbolic Recognition by Graph Automata. In: Widmayer, P., Triguero, F., Morales, R., Hennessy, M., Eidenbenz, S., Conejo, R. (eds.) ICALP 2002. LNCS, vol. 2380, pp. 330–342. Springer, Heidelberg (2002)

29. Róka, Z.: Simulations between cellular automata on Cayley graphs. Theoretical Computer Science 225(1-2), 81–111 (1999)

30. Rozenberg, G.: Handbook of graph grammars and computing by graph transformation: Foundations, vol. 1. World Scientific (2003)

31. Sayama, H.: Generative network automata: A generalized framework for modeling complex dynamical systems with autonomously varying topologies. In: IEEE Symposium on Artificial Life, ALIFE 2007, pp. 214–221. IEEE (2007)

32. Scherrer, A., Borgnat, P., Fleury, E., Guillaume, J.-L., Robardet, C.: Description and simulation of dynamic mobility networks. Computer Networks 52(15), 2842–2858 (2008)

33. Schönhage, A.: Storage modification machines. SIAM Journal on Computing 9, 490 (1980)

34. Schumacher, B., Werner, R.: Reversible quantum cellular automata. ArXiv preprint quant-ph/0405174 (2004)

35. Sieg, W.: Church without dogma: Axioms for computability. In: New Computational Paradigms, pp. 139–152 (2008)

36. Sorkin, R.: Time-evolution problem in Regge calculus. Phys. Rev. D 12(2), 385–396 (1975)

37. Taentzer, G.: Parallel and distributed graph transformation: Formal description and application to communication-based systems. PhD thesis, Technische Universitat Berlin (1996)

38. Taentzer, G.: Parallel high-level replacement systems. Theoretical Computer Science 186(1-2), 43–81 (1997)

39. Tomita, K., Murata, S., Kamimura, A., Kurokawa, H.: Self-description for Construction and Execution in Graph Rewriting Automata. In: Capcarrère, M.S., Freitas, A.A., Bentley, P.J., Johnson, C.G., Timmis, J. (eds.) ECAL 2005. LNCS (LNAI), vol. 3630, pp. 705–715. Springer, Heidelberg (2005)

40. Tomita, K., Kurokawa, H., Murata, S.: Graph automata: natural expression of self-reproduction. Physica D: Nonlinear Phenomena 171(4), 197–210 (2002)

41. Tomita, K., Kurokawa, H., Murata, S.: Graph-rewriting automata as a natural extension of cellular automata. In: Gross, T., Sayama, H. (eds.) Adaptive Networks. Understanding Complex Systems, vol. 51, pp. 291–309. Springer, Heidelberg (2009)

42. Tomita, K., Murata, S., Kurokawa, H.: Asynchronous Graph-Rewriting Automata and Simulation of Synchronous Execution. In: Almeida e Costa, F., Rocha, L.M., Costa, E., Harvey, I., Coutinho, A. (eds.) ECAL 2007. LNCS (LNAI), vol. 4648, pp. 865–875. Springer, Heidelberg (2007)

43. von Mammen, S., Phillips, D., Davison, T., Jacob, C.: A Graph-Based Developmental Swarm Representation and Algorithm. In: Dorigo, M., Birattari, M., Di Caro, G.A., Doursat, R., Engelbrecht, A.P., Floreano, D., Gambardella, L.M., Groß, R., Şahin, E., Sayama, H., Stützle, T. (eds.) ANTS 2010. LNCS, vol. 6234, pp. 1–12. Springer, Heidelberg (2010)

Degree Lower Bounds of Tower-Type for Approximating Formulas with Parity Quantifiers

Albert Atserias[1,*] and Anuj Dawar[2,**]

[1] Universitat Politècnica de Catalunya, Barcelona, Spain
[2] University of Cambridge, Cambridge, UK

Abstract. Kolaitis and Kopparty have shown that for any first-order formula with parity quantifiers over the language of graphs there is a family of multi-variate polynomials of constant degree that agree with the formula on all but a $2^{-\Omega(n)}$-fraction of the graphs with n vertices. The proof bounds the degree of the polynomials by a tower of exponentials in the nesting depth of parity quantifiers in the formula. We show that this tower-type dependence is necessary. We build a family of formulas of depth q whose approximating polynomials must have degree bounded from below by a tower of exponentials of height proportional to q. Our proof has two main parts. First, we adapt and extend known results describing the joint distribution of the parity of the number of copies of small subgraphs on a random graph to the setting of graphs of growing size. Secondly, we analyse a variant of Karp's graph canonical labeling algorithm and exploit its massive parallelism to get a formula of low depth that defines an almost canonical pre-order on a random graph.

1 Introduction

Since the 0-1 law for first-order logic was established [1,2], there has been much interest in exploring the asymptotic properties of definable classes of graphs. Many extensions of first-order logic have been shown to have a 0-1 law (see for instance [3,4]) and in many other cases, weaker forms of convergence have been established (see [5]). A recent, remarkable result in this vein is that of Kolaitis and Kopparty [6] who study FO[⊕], the extension of first-order logic with *parity quantifiers*. They show that for every FO[⊕] sentence ϕ, there are two rational numbers a_0, a_1 such that for $i \in \{0, 1\}$, as n approaches infinity, the probability that the random graph $G(2n + i; p)$ (for any constant p) satisfies ϕ approaches a_i. So, ϕ has asymptotic probability a_0 on the graphs of even cardinality and a_1 on those of odd cardinality. The proof of this result brings entirely new methods to the analysis of the asymptotic behaviour of logics on graphs, based on discrete analysis and polynomials over finite fields. In particular, it ties this to the study of approximations of circuits by low-degree polynomials.

* Supported in part by MICIIN project TIN2010 - 20967 - C04 - 04 (TASSAT).
** Supported in part by EPSRC grant EP/H026835.

A. Czumaj et al. (Eds.): ICALP 2012, Part II, LNCS 7392, pp. 67–78, 2012.
© Springer-Verlag Berlin Heidelberg 2012

The 0-1 law for first-order logic, in its general form is a quantifier-elimination result. It states that for any first-order formula ϕ, there is a quantifier-free formula θ such that ϕ is equivalent to θ almost surely. To be precise, ϕ and θ agree on a fraction $1 - 2^{-\epsilon n}$ of the graphs on n vertices. We can say that any first-order formula is well approximated by a quantifier-free formula. This is similar to the phenomenon of depth-reduction for circuits which has a long history in computational complexity theory. For instance, Allender showed that AC^0-circuits have equivalent TC^0-circuits of depth 3 and quasi-polynomial size [7]. The result of Beigel and Tarui that general ACC^0-circuits of polynomial-size have equivalent depth-2 circuits of quasi-polynomial size with a symmetric gate at the root [8] has been exploited to remarkable effect recently in the work of Williams [9]. In the context of approximation, one of the best known examples is the Razborov-Smolensky approximation of $AC^0[\oplus]$-circuits by multi-variate polynomials over $\mathbb{Z}_2$ of polylogarithmic degree [10,11]. The method yields an approximation that agrees on a fraction $1 - 2^{-(\log n)^c}$ of the inputs.

The Kolaitis-Kopparty result mentioned above is proved by a depth-reduction argument of a similar kind that exploits the higher degree of symmetry that $FO[\oplus]$-formulas have over $AC^0[\oplus]$-circuits. They prove that every $FO[\oplus]$ formula ϕ is well-approximated by a formula of a special form, which is a Boolean combination of quantifier-free formulas and polynomials over $\mathbb{Z}_2$. The polynomials have as variables X_{uv} for every potential edge $\{u, v\}$ over the vertex-set $\{1, \ldots, n\}$. For example, the polynomial that gives the parity of the number of triples that extend the vertex u to a triangle is

$$\sum_{\substack{v \\ v \neq u}} \sum_{\substack{w \\ w \neq u \\ w \neq v}} X_{uv} X_{vw} X_{wu}. \tag{1}$$

The construction relies on a quantifier-elimination argument which, intriguingly, incurs an exponential loss in the degree of the resulting polynomials at the elimination of each parity quantifier. The final outcome yields, for a fixed formula ϕ, polynomials of degree d, where the dependence of d on the quantifier rank of ϕ is non-elementary. In this paper we prove that this non-elementary dependence cannot be avoided. To be precise, we construct an explicit family of $FO[\oplus]$ formulas ϕ_q of quantifier rank q and prove that they cannot be approximated by a Boolean combination of polynomials of degree bounded by an elementary function of q. Specifically, we prove the following:

Theorem 1. *There exists a constant $c > 0$ such that for every large enough integer q, every $\epsilon > 0$, and every large enough integer n, there exists a $FO[\oplus]$-formula $\phi(u, v, w)$ of quantifier rank q such that, for every Boolean combination p of quantifier-free formulas and $FO[\oplus]$-polynomials of degree bounded by a tower of exponentials of height at most q/c, the formulas ϕ and p must disagree on a fraction $1 - \epsilon$ of all graphs with n vertices.*

By an $FO[\oplus]$-polynomial we mean a formula that has a direct translation as a bounded-degree polynomial over $\mathbb{Z}_2$: a sequence of parity quantifiers followed by a conjunction of atomic facts.

Theorem 1 should be contrasted with the 0-1 law for first-order logic. In that case the approximating formula is quantifier-free, and such formulas translate into polynomials of degree at most polynomial in the number of free variables.

Proof outline and techniques. Our proof relies on two technical ingredients. On one hand we analyse a canonical labelling algorithm for graphs due to Karp [12] (see [13] for another view on the logical definability of Karp's canonical labeling). We exploit its massive parallelism to build an FO[$\oplus$]-formula $\psi(u,v)$ of depth $O(\log^* n)$. The formula is such that, on almost every n-vertex graph, it defines a linear pre-order of width at most two on the set of vertices. The second ingredient is a refined analysis of one of the key tools from [6]. Using and extending their techniques for estimating the frequencies mod 2 of subgraph copies, we show that every FO[$\oplus$]-polynomial $p(u,v,w)$ of quantifier rank $\log\log\log n$ must be unable to distinguish some triple of distinct vertices (a,b,c) from any of its permutations, with high probability.

From these two ingredients, the lower bound follows by taking the formula $\psi(u,v) \wedge \psi(v,w)$. On one hand this formula distinguishes at least one permutation of the vertices (a,b,c) from some other because by linearity of the pre-order the classes they lie in must be comparable, but by the width-2 condition on the pre-order not all three vertices can sit in the same class. On the other hand no formula of quantifier rank $\log\log\log n$ is able to distinguish any permutation of some triple (a,b,c) from the others. Since the quantifier rank of ϕ is still $O(\log^* n)$, the tower-type lower bound follows. Proofs omitted from this paper may be found in the extended version available at http://eccc.hpi-web.de/report/2012/015/.

2   Preliminaries

We use $[n]$ to denote the set $\{1,\dots,n\}$. For an arithmetic expression E that contains $\pm$ and $\mp$ among its operations, we write $A = B \pm E$ to mean that $|A - B| \leq |E|$, where $|E|$ stands for the largest quantity that can be made by replacing each occurrence of $\pm$ or $\mp$ by $+$ or $-$. We identify the nodes of a complete rooted binary tree with the binary strings that start with the symbol 1: the root is 1, the left child of t is $t0$ and the right child of t is $t1$. The *level-order* of a complete binary tree is $1, 10, 11, 100, 101, 110, 111, 1000, 1001, \dots$, i.e. ordered first by length, and within each length, in lexicographical order. Note that if the strings are interpreted as numbers written in binary, this is the usual order of the natural numbers. For a natural number $n \geq 1$, we write $\mathrm{bin}_2(n)$ for its unique binary encoding with a leading one.

Let G and H be graphs. A homomorphism from G to H is a mapping $h : V(G) \to V(H)$ such that if $\{u,v\} \in E(G)$, then $\{h(u),h(v)\} \in E(H)$. Let $\mathrm{Hom}(G,H)$ denote the collection of all homomorphisms from G to H.

The collection of FO[$\oplus$]-formulas over the language of graphs is the smallest class of formulas that contains the atomic formulas $E(x,y)$ and the equalities $x = y$, and is closed under negation, conjunction and disjunction, universal

and existential quantification, and parity quantification; i.e. quantification of
the form $\oplus x\ \phi(x)$. The meaning of $\oplus x\ \phi(x)$ is that there is an odd number
of vertices that satisfy $\phi(x)$. For a tuple $\mathbf{a} = (a_1, \ldots, a_k)$ and a permutation
$\pi \in S_k$, we write $\mathbf{a} \circ \pi$ for the tuple $(a_{\pi(1)}, \ldots, a_{\pi(k)})$. If $p(x_1, \ldots, x_k)$ is a formula
with free variables $x_1, \ldots, x_k$, and $y_1, \ldots, y_k$ are variables or constants, we write
$p(y_1, \ldots, y_k)$ for the result of replacing each occurrence of x_i by y_i. This applies
also to the case where $y_1, \ldots, y_k$ is a permutation of $x_1, \ldots, x_k$.

3   Fooling Polynomials of Low Degree

In this section we show that every FO[$\oplus$]-formula of a certain general form
corresponding to polynomials of low degree is not able to distinguish some triple
from any of its permutations, with high probability. Roughly speaking, the proof
strategy is as follows. Fix such a formula $p(x, y, z)$. For every fixed $a, b, c \in [n]$,
let $Y(a, b, c)$ be the event that p cannot distinguish any two permutations of
a, b, c. Ideally we would like to show that the event $Y(a, b, c)$ has non-negligible
probability of happening, and that if $a', b', c' \in [n]$ is a triple disjoint from a, b, c,
then the events $Y(a, b, c)$ and $Y(a', b', c')$ are almost independent. The result
would then follow from an application of Chebyshev's inequality. Unfortunately
we are not able to do this directly, so we take a detour.

Formulas and polynomials. In this section, we define the formulas to which our
result applies. In short, they are Boolean combinations of FO[$\oplus$]-*polynomials*.
An FO[$\oplus$]-*polynomial* is a formula of FO[$\oplus$] consisting of a sequence of parity
quantifiers followed by a conjunction of atomic formulas. Thus, in its general
form, an FO[$\oplus$]-polynomial p with free variables $u_1, \ldots, u_k$ is a formula of the
form

$$\oplus u_{k+1} \cdots \oplus u_m \left(\bigwedge_{i \neq j} u_i \neq u_j \wedge \bigwedge_{\ell=1}^{d} E(u_{i_\ell}, u_{j_\ell}) \right), \tag{2}$$

where i and j range over $[m]$ and $i_1, \ldots, i_d$, $j_1, \ldots, j_d$ are indices in $[m]$. The
number d of atomic facts in the conjunction is the *degree* of p.

A conjunction of atomic formulas as in the matrix of the formula (2) corre-
sponds to a graph H on $\{u_1, \ldots, u_m\}$. H has an edge between u_{i_ℓ} and u_{j_ℓ} for
each $\ell \in [d]$. Thus, the formula expresses the parity of the number of extensions
of $u_1, \ldots, u_k$ to a copy of H. We use the notation $\oplus H(u_1, \ldots, u_k)$ to denote this
formula. Note that the degree of $\oplus H(u_1, \ldots, u_k)$ is the number of edges of H.

Example 2. If H is a triangle containing vertex u, then $\oplus H(u)$ is the formula
that expresses the parity of the number of extensions of u to a triangle. Formally,
$\oplus H(u)$ is the formula

$$\oplus v\ \oplus w\ (u \neq v \wedge u \neq w \wedge v \neq w \wedge E(u, v) \wedge E(v, w) \wedge E(w, u)).$$

Note that over undirected graphs, this formula is always false. In general, if
$H(u_1, \ldots, u_k)$ has an even number of automorphisms that fix $u_1, \ldots, u_k$, then
$\oplus H(u_1, \ldots, u_k)$ will always be false.

There is a precise sense in which FO[$\oplus$]-polynomials correspond to polynomials over the Boolean edge-variables X_{uv}. For example, the formula in Example 2 corresponds to the family of degree-3 polynomials as u ranges over $[n]$.

Independence and plan of action. The main obstacle to carrying out the argument sketched at the beginning of this section is that it is not true, in general, that the events $Y(a, b, c)$ and $Y(a', b', c')$ are almost independent, even if a, b, c, a', b', c' are all different. The reason is that the formula $p(x, y, z)$ may include statements about the graph G which do not involve the free variables. These are true or false independently of the choice of a, b, c or a', b', c' and thus create correlations between $Y(a, b, c)$ and $Y(a', b', c')$.

The key observation at this point is that the *full type* of (x, y, z) in terms of its atomic type (the pattern of connections and equalities among x, y and z) and the truth values of its $\oplus H$'s as H ranges over all small graphs that contain x, y and z as vertices is enough to determine the truth value of $p(x, y, z)$. Thus, if we are able to find a full type implying $p(x, y, z)$ that is symmetric in x, y and z, we would have reduced the case of general $p(x, y, z)$ to the case of a $p(x, y, z)$ that consists of a single term. The argument that we use is a bit more delicate than this, but this is the main idea.

Normal forms. In this section we introduce some definitions and discuss two different types of *normal forms* for Boolean combinations of FO[$\oplus$]-polynomials.

An *I-labeled graph* is a graph with some vertices labeled by elements of I in such a way that, for every $i \in I$ there is exactly one vertex labeled i, and the set of labeled vertices induces an independent set. The set of labeled vertices of an I-labeled graph H is denoted by $\mathcal{L}(H)$. The vertex labeled by $i \in I$ is denoted by $H(i)$. An I-labeled graph H is *label-connected* if $H \setminus \mathcal{L}(H)$ is connected. Let Conn_I^t be the set of all I-labeled label-connected graphs with at most t unlabeled vertices. We say that H *depends on* label $i \in I$ if $H(i)$ is not an isolated node. We say that H is *label-dependent* if it depends on all its labels. Let $\mathrm{Conn}_I^{*,t}$ be the set of all labeled graphs in Conn_I^t that are label-dependent.

A *k-labeled graph* is a $[k]$-labeled graph. A $\leq k$-*labeled graph* is an I-labeled graph for some $I \subseteq [k]$. Let H be a $\leq k$-labeled graph with labels $I \subseteq [k]$. A *homomorphism* from H to a pair $(G, \mathbf{a})$, where G is a graph and $\mathbf{a} = (a_1, \ldots, a_k)$ is a tuple in V_G^k, is a map $\chi \in \mathrm{Hom}(H, G)$ such that $\chi(H(i)) = a_i$ for each $i \in I$. It is *injective* if for any distinct $a, b \in V_H$ such that $\{a, b\} \not\subseteq \mathcal{L}(H)$ we have $\chi(a) \neq \chi(b)$. Write $\oplus H(G, \mathbf{a})$ for the parity of the number of injective homomorphisms from H to $(G, \mathbf{a})$. We usually omit G and write $\oplus H(\mathbf{a})$.

A *KK-normal form* of degree c with free-variables $\mathbf{x} = x_1, \ldots, x_k$ is a Boolean combination of the atomic types on the variables $\mathbf{x}$ and formulas $\oplus H(\mathbf{x})$ as H ranges over the k-labeled label-connected graphs with labeled vertices $\mathbf{x}$ and at most $c - \binom{k}{2}$ edges with at least one non-labeled endpoint. A *regular normal form* of degree c with free-variables $\mathbf{x}$ is a Boolean combination of the atomic types on the variables $\mathbf{x}$ and the formulas $\oplus H(\mathbf{x})$ as H ranges over the $\leq k$-labeled label-connected, label-dependent graphs with labeled vertices within $\mathbf{x}$ and at most $c - \binom{k}{2}$ edges with at least one non-labeled endpoint.

Lemma 3. *Let $k \geq 0$ and $c \geq \binom{k}{2}$ be integers and let $\phi(x_1, \ldots, x_k)$ be a formula of $\mathrm{FO}[\oplus]$ that implies $x_i \neq x_j$ when $i \neq j$. The following are equivalent:*

1. *ϕ is equivalent to a Boolean combination of $\mathrm{FO}[\oplus]$-polynomials of degree $\leq c$,*
2. *ϕ is equivalent to a KK-normal form of degree $\leq c$,*
3. *ϕ is equivalent to a regular normal form of degree $\leq c$.*

Distribution of frequency vectors. The frequency vector of degree t in a graph G is the $\{0,1\}$-vector indexed by the set of all connected graphs with at most t vertices where the component indexed by H is $\oplus H(G)$, i.e. the parity of the number of occurrences of H in G. Kolaitis and Kopparty give an analysis of the distribution of frequency vectors in a random graph $G \sim G(n, 1/2)$, for constant t. Our aim in the present section is to extend this analysis to degrees that grow with n and to $\leq k$-labeled graphs.

Let $\mathrm{Conn}_{\leq k}^t$ be the set of all $\leq k$-labeled label-connected graphs with at most t unlabeled vertices. Let $\mathrm{Conn}_{\leq k}^{*,t}$ be the subset of $\mathrm{Conn}_{\leq k}^t$ containing all graphs that are label-dependent. Let G be a graph, let $\mathbf{a}$ be a tuple in V_G^k, and let $t \geq 0$ be an integer. Let $\mathrm{freq}_{\leq k, G}^{*,t}(\mathbf{a})$ be the $\{0,1\}$-vector indexed by the elements $\mathrm{Conn}_{\leq k}^{*,t}$ that has $\oplus H(\mathbf{a})$ component indexed by H. If τ is an atomic type on $x_1, \ldots, x_k$ that forces $x_i \neq x_j$ for $i \neq j$, let $\mathrm{FFreq}^*(\tau, \leq k, t)$ denote the set of all *feasible frequency vectors*. These are all $\{0,1\}$-vectors indexed by $\mathrm{Conn}_{\leq k}^{*,t}$ whose component F belongs to $\mathrm{aut}(F) \cdot \mathbb{Z}/2\mathbb{Z}$. Here $\mathrm{aut}(F)$ denotes the number of automorphisms of F that fix the labels. Let $\mathrm{FFreq}_n^*(\tau, \leq k, t)$ denote the set of $f \in \mathrm{FFreq}^*(\tau, \leq k, t)$ such that $f_{K_1(\emptyset)} = n \mod 2$, where $K_1(\emptyset)$ is the graph with no labels and exactly one unlabeled vertex.

The next lemma describes the distribution of $\mathrm{freq}_{\leq k, G}^{*,t}(\mathbf{a})$ in a random graph. This is analogous to Theorem 2.4 in [6] extended to growing degrees up to $\log \log \log n$, and extended from k-labeled graphs to $\leq k$-labeled graphs.

Lemma 4. *For every $k \geq 0$ there exists $n_0 \geq 0$ such that for every $n \geq n_0$, every atomic type τ on k variables that forces all of them distinct, every $c \leq \log \log \log n$, and every k-tuple $\mathbf{a}$ of distinct elements in $[n]$, the distribution of $\mathrm{freq}_{\leq k, G}^{*,c}(\mathbf{a})$ as $G = G(n, 1/2 \mid \tau(\mathbf{a}))$ is $2^{-\Omega_k(n/\log n)}$-close in statistical distance from the uniform distribution over $\mathrm{FFreq}_n^*(\tau, \leq k, c)$.*

The argument itself. Finally we reached the point where we can execute the plan sketched at the beginning of Section 3. Fix a positive integer k (for the application in Section 5 it suffices to take $k = 3$) and let $p(x_1, \ldots, x_k)$ be a regular normal form of degree $c \leq \log \log \log n$. For every $\mathbf{a} = (a_1, \ldots, a_k) \in [n]^k$, define the following indicator random variables:

$$X(\mathbf{a}) := \mathbb{I}[\, p(\mathbf{a}) \not\leftrightarrow p(\mathbf{a} \circ \pi) \text{ for some } \pi \in S_k \,],$$
$$Y(\mathbf{a}) := \mathbb{I}[\, p(\mathbf{a}) \leftrightarrow p(\mathbf{a} \circ \pi) \text{ for every } \pi \in S_k \,].$$

Obviously, $X(\mathbf{a}) = 1 - Y(\mathbf{a})$, and $Y(\mathbf{a})$ is the indicator random variable for the event that p does not distinguish any two permuted versions of $\mathbf{a}$. Our goal is to show that $Y(\mathbf{a})$ holds for some $\mathbf{a}$ with high probability and for this we will follow the plan sketched in Section 3.

Write $p(\mathbf{x})$ as a DNF on the (Boolean) variables

$$\tau_1(\mathbf{x}), \ldots, \tau_r(\mathbf{x}), \oplus H_1(\mathbf{x}), \ldots, \oplus H_\ell(\mathbf{x}),$$

where $\tau_1, \ldots, \tau_r$ are the atomic types on $x_1, \ldots, x_k$, and $H_1, \ldots, H_\ell$ are the $\leq k$-labeled label-connected, label-dependent graphs with labeled vertices within $x_1, \ldots, x_k$ and at most $c - \binom{k}{2}$ edges with at least one non-labeled endpoint. Let us assume that for $j \in [e]$, $\mathrm{aut}(H_j)$ is odd and for $j \in [\ell] \setminus [e]$, $\mathrm{aut}(H_j)$ is even. Also assume that $H_1, \ldots, H_f$ are the graphs among $H_1, \ldots, H_e$ that have no label. Since exactly one atomic type must hold and each $\oplus H_j(\mathbf{a})$ is false for $j \in [\ell] \setminus [e]$, we may assume that each term in the DNF formula has the form

$$\tau_i(\mathbf{x}) \cdot \prod_{j \in K} \oplus H_j \cdot \prod_{j \in K'} \overline{\oplus H_j} \cdot \prod_{j \in I} \oplus H_j(\mathbf{x}) \cdot \prod_{j \in I'} \overline{\oplus H_j(\mathbf{x})}. \tag{3}$$

for some $i \in [r]$, a partition (K, K') of $[f]$, and a partition (I, I') of $[e] \setminus [f]$.

Next note that for every permutation $\pi \in S_k$, the sequence of Boolean variables $\tau_1(\mathbf{x} \circ \pi), \ldots, \tau_r(\mathbf{x} \circ \pi)$ is equivalent to a permutation of the sequence of Boolean variables $\tau_1(\mathbf{x}), \ldots, \tau_r(\mathbf{x})$. Similarly, the sequence of Boolean variables $\oplus H_{f+1}(\mathbf{x} \circ \pi), \ldots, \oplus H_e(\mathbf{x} \circ \pi)$ is equivalent to a permutation of the sequence $\oplus H_{f+1}(\mathbf{x}), \ldots, \oplus H_e(\mathbf{x})$. Therefore $p(\mathbf{x})$ and $p(\mathbf{x} \circ \pi)$ are functions of the same Boolean variables and we can write $p(\mathbf{x} \circ \pi)$ also as a DNF formula with terms of the type (3).

For every $K \subseteq [f]$, let R_K be the term $R_K := \prod_{j \in K} \oplus H_j \cdot \prod_{j \in K'} \overline{\oplus H_j}$, where $K' = [f] \setminus K$. Recall that $H_1, \ldots, H_f$ are all label-free and therefore R_K does not depend on $\mathbf{x}$. Similarly, for every $I \subseteq [e] \setminus [f]$, let $S_I(\mathbf{x})$ be the term $S_I(\mathbf{x}) := \prod_{j \in I} \oplus H_j(\mathbf{x}) \cdot \prod_{j \in I'} \overline{\oplus H_j(\mathbf{x})}$, where $I' = ([e] \setminus [f]) \setminus I$. For every $K \subseteq [f]$, let $p_K(\mathbf{x})$ denote the disjunction of the terms in $p(\mathbf{x})$ that are consistent with R_K. Therefore $p(\mathbf{x})$ is equivalent to the disjunction $\bigvee_{K \subseteq [f]} p_K(\mathbf{x})$.

Define the "all-positive-term" as follows: $Z_K(\mathbf{x}) := \sigma(\mathbf{x}) \cdot R_K \cdot S_{[e] \setminus [f]}(\mathbf{x})$, where σ is the atomic type that forces $x_i \neq x_j$ for $i \neq j$, and all possible edges among different x_i, x_j. We show that for every $\mathbf{a} \in [n]^k$, the event $Z_K(\mathbf{a}) = 1$ implies $p(\mathbf{a}) \leftrightarrow p(\mathbf{a} \circ \pi)$ for every $\pi \in S_k$.

Lemma 5. $Z_K(\mathbf{a}) \leq Y(\mathbf{a})$.

At this point it suffices to show that for every $K \subseteq [f]$, the event $Z_K(\mathbf{a}) = 1$ holds for some $\mathbf{a} \in [n]^k$ with high probability in the probability space conditioned on R_K. From now on, for every event A, write $\Pr_K[A] := \Pr[A \mid R_K]$.

We first compute the probability of $Z_K(\mathbf{a})$ for $\mathbf{a} \in [n]^k$ with $a_i \neq a_j$ for $i \neq j$ in this space. Let δ be the maximum, over all atomic types $\tau(\mathbf{x})$ that force $x_i \neq x_j$ for $i \neq j$, of the statistical distance between the distribution $\mathrm{freq}_{\leq k, G}^{*, c}(\mathbf{a})$ as $G = G(n, 1/2 \mid \tau(\mathbf{a}))$ and the uniform distribution over $\mathrm{FFreq}^*(\tau, \leq k, c)$. By symmetry, δ does not depend on $\mathbf{a}$ provided $a_i \neq a_j$ for $i \neq j$.

Lemma 6. $\mathrm{Pr}_K[\, Z_K(\mathbf{a})\,] = \left(\frac{2^{-e}}{2^{-f}} \pm \delta\frac{1}{2^{-f}\pm\delta} \mp \delta\frac{2^{-e}}{2^{-f}\cdot(2^{-f}\pm\delta)}\right)\cdot 2^{-\binom{k}{2}}$.

Next we compute, for every $\mathbf{a}, \mathbf{a}' \in [n]^k$ with all $a_1, \ldots, a_k, a'_1, \ldots, a'_k$ distinct, the probability of $Z_K(\mathbf{a}) \cdot Z_K(\mathbf{a}')$ in the space conditioned on R_K. Let γ be the maximum, over all atomic types $\tau(\mathbf{x}, \mathbf{x}')$ that force $x_1, \ldots, x_k, x'_1, \ldots, x'_k$ to be distinct, of the statistical distance between the distribution $\mathrm{freq}^{*,c}_{<2k,G}(\mathbf{a}, \mathbf{a}')$ as $G = G(n, 1/2 \mid \tau(\mathbf{a}, \mathbf{a}'))$ and the uniform distribution over $\mathrm{FFreq}^*(\tau, \leq 2k, c)$. By symmetry, γ does not depend on $\mathbf{a}, \mathbf{a}'$ provided they are all distinct.

Lemma 7. $\mathrm{Pr}_K[\, Z_K(\mathbf{a}) \cdot Z_K(\mathbf{a}')\,] = \left(\frac{2^{-2e}}{2^{-2f}} \pm \gamma\frac{1}{2^{-f}\pm\gamma} \mp \gamma\frac{2^{-2e+f}}{2^{-f}\cdot(2^{-f}\pm\gamma)}\right)\cdot 2^{-2\binom{k}{2}}$.

Let us note at this point that the number of $\leq k$-labeled graphs with at most c non-labeled vertices is bounded by $2^{(c+k)^2}$. Therefore, if ℓ is the number of $\leq k$-labeled label-connected graphs with at most c non-labeled vertices, then $\ell \leq \frac{1}{2}\log n$ for sufficiently large n, and in particular $2^\ell \leq \sqrt{n}$. We use this to prove the main consequence of this analysis up to now:

Lemma 8. *Let* $\mathbf{a}, \mathbf{a}' \in [n]^k$ *be such that* $a_1, \ldots, a_k, a'_1, \ldots, a'_k$ *are all distinct. The following hold:*

1. $\mathrm{Pr}_K[\, Z_K(\mathbf{a})\,] \geq n^{-1/2} \cdot 2^{-\binom{k}{2}} - 2^{-\Omega_k(n/\log n)}$.
2. $|\mathrm{Pr}_K[\, Z_K(\mathbf{a}) \cdot Z_K(\mathbf{a}')\,] - \mathrm{Pr}_K[\, Z_K(\mathbf{a})\,] \cdot \mathrm{Pr}_K[\, Z_K(\mathbf{a}')\,]| \leq 2^{-\Omega_k(n/\log n)}$.

Now we conclude by stating the main result of this section:

Lemma 9. *For every* $k > 0$ *and* $\epsilon > 0$, *there exists* $n_0 \geq 0$ *such that for every* $n \geq n_0$ *and every KK-normal form* $p(x_1, \ldots, x_k)$ *of degree bounded by* $\log \log \log n$, *for* $G \sim G(n, 1/2)$, *the probability that there exists* $\mathbf{a} \in [n]^k$ *with* $a_i \neq a_j$ *for* $i \neq j$ *such that* $p(\mathbf{a}) \leftrightarrow p(\mathbf{a} \circ \pi)$ *for every* $\pi \in S_k$ *is at least* $1 - \epsilon$.

4   Defining a Linear Pre-order of Width Two

In this section we construct the formula of low quantifier rank that defines a linear pre-order of width 2 with high probability. The proof strategy is to analyse a variant of an algorithm for graph canonization due to Karp [12], and to exploit its massive implicit parallelism to get formulas of low depth.

Plan of action. Informally, the graph canonization algorithm works as follows. For a given graph G, split the vertices into two classes: those of even degree and those of odd degree. Inductively, we split the classes further by dividing the vertices according to the parity of the numbers of neighbours they have in each of the existing classes. We continue this process until no more classes are split.

We will need three facts about this process: (1) that for $G \sim G(n, 1/2)$ the process will reach a state where each class has at most two vertices with high probability, (2) that this will happen in fewer than n "generations" of the splitting process with high probability, and (3) that the process is massively parallel: all the classes created between the $\ell/2$-th generation and the ℓ-th generation are definable in terms of the classes created in the $(\log_2 \ell)$-th generation.

Splitting procedure. Let $G = (V, E)$ be an undirected graph. For a vertex x and a set B, we write $p(G, x, B)$ for the parity of the number of neighbours that x has in B. We extend this to sets: $p(G, A, B) = \sum_{x \in A} p(G, x, B) \mod 2$.

A *splitting tree* for G is a rooted binary tree T with each node t carrying a label $L_t \subseteq V$ and a sign $M_t \in \{+, -\}$ denoting whether t is *marked* or *unmarked*, and satisfying the following properties:

1. the label of the root is V,
2. no two siblings are marked,
3. if t is an internal node, then $L_{t0} \cup L_{t1} = L_t$ and $L_{t0} \cap L_{t1} = \emptyset$,
4. if s is a leaf, $x, y \in L_s$ and t is marked, then $p(G, x, L_t) = p(G, y, L_t)$.

Given a splitting tree T for G, let $R(T)$ denote the set of unmarked nodes that are either the root or are a left child. Let $R'(T)$ be the set of nodes in $R(T)$ that are either the root or such that their label and the label of their sibling are both non-empty. One step of the splitting procedure works as follows:

1. let t be the least node in $R(T)$ in level-order and mark it,
2. for every leaf s, let $L_{sa} := \{x \in L_s : p(G, x, L_t) = a\}$ for $a = 0$ and $a = 1$,
3. make $s0$ and $s1$ the left and right children of s and leave them unmarked.

Let $\mathcal{P}(T)$ be the result of applying one step of the splitting procedure to T. If the node t that is chosen in (1) also belongs to $R'(T)$ we say that the step is *proper*, otherwise *improper*. When $R'(T)$ is empty we say that the procedure *stalls* at T. Note that when it stalls it will never make a proper step again. The procedure starts at the splitting tree T_0 that has only an unmarked root labeled by V.

Analysis of the splitting procedure. Let T_0 be the tree that has only an unmarked root labeled by V. For $k \geq 1$, let $T_k := \mathcal{P}(T_{k-1})$. Ideally we would like to show that after a modest number of steps, all leaves of the splitting tree are labeled by singletons or empty sets. Unfortunately the splitting procedure is not able to produce a tree with this property in general, not even with high probability on a random graph. The best we will be able to show is that for a randomly generated graph, with high probability all leaves will have at most two vertices.

We identify three desirable properties of T_k, where the third is our goal:

(A_k): T_k has $L_t \neq \emptyset$ for every node t,
(B_k): T_k has been generated through proper steps only,
(C_k): T_k has $|L_t| \leq 2$ for every leaf t.

In the following we show:

1. property (A_k) holds with high probability for small k,
2. property (A_k) implies (B_{2^k}) for every graph and every $k \geq 0$,
3. conditioned on (B_{2^k}), (C_{2^k}) holds with high probability for small k.

Before we analyse the probability of (A_k) we need to introduce some terminology and a lemma from [12]. Let T be a splitting tree for some graph H on the vertices V. To every node $t \in R'(T)$ we associate a set $S_t \subseteq V$: the set of all $x \in L_t$

for which t is the unique minimal node in $R'(T)$ containing x. Let $\mathcal{S}(T)$ be the collection of all such sets. For every $t \in R'(T)$, let β_t be t together with the set of nodes $s \in R'(T)$ such that $s \neq t$ and L_s is a maximal subset of L_t. Note that $S_t = \triangle_{s \in \beta_t} L_s$, where $\triangle$ denotes symmetric difference. Define $\ell(x, S_t) := \sum_{s \in \beta_t} p(H, x, L_s) \mod 2$. We will say that another graph G on the vertices V is consistent with T if $p(G, x, L_t) = p(H, x, L_t)$ holds for every $x \in V$ and every properly marked node t.

We state a consequence of Lemmas 4 and 5 in [12]

Lemma 10. *Let T be the splitting tree of some graph on the vertices V and let H be chosen uniformly at random among the graphs on the vertices V that are consistent with T. If $t \in R'(T)$, then the distribution of $\{p(H, x, L_t)\}_{x \in V}$ is uniform over the assignments that satisfy the constraints*

$$p(H, S, L_t) = \ell(L_t, S) \quad \text{for every } S \in \mathcal{S}(T_k) \setminus \{Y\},$$

where Y is the unique set in $\mathcal{S}(T)$ of which L_t is a proper subset.

In order to be able to make use of this lemma it is important to notice that if G denotes a random graph drawn from $G(n, 1/2)$ and $T_0, T_1, \ldots$ denotes the sequence of splitting trees produced by G, then the distribution of T_{k+1} conditioned on T_k is equally produced as follows: first choose a graph H uniformly at random among those consistent with T_k, and then run one step of the splitting procedure on T_k with respect to H. This follows from the fact that the restriction of a uniform distribution to a subset of its support is uniformly distributed on that subset.

Now we can analyse the probability of (A_k):

Lemma 11. *Let $n \geq 1$ and $k \geq 1$ be integers such that $4k \leq \log_2 n$, and let $G \sim G(n, 1/2)$. Then, the probability that (A_k) fails is $2^{k+1} \cdot \exp(-n/2^{6k})$.*

Next we observe that (A_k) implies (B_{2^k}).

Lemma 12. *For any graph G and $k \geq 0$. $(A_k) \Rightarrow (B_{2^k})$.*

Finally we note that 3-element sets split with high probability if enough steps are proper. This is similar to Lemma 7 in [12].

Lemma 13. *Let $G \sim G(n, 1/2)$ and let $k \geq 0$. Then, the probability that (B_k) holds and (C_k) fails is at most $\binom{n}{3} \cdot 2^{-2k}$.*

We are ready to synthesize what we learned in a single lemma. In its statement, the choice of parameters is made to minimize the probability of failure.

Lemma 14. *Let $G \sim G(n, 1/2)$. Then, the probability that $T_{\lceil n^{1/5} \rceil}$ does not satisfy $(C_{\lceil n^{1/5} \rceil})$ is at most $2^{-\Omega(n^{1/6})}$.*

Proof. Choose $k = \lceil \frac{1}{5} \log_2 n \rceil$ in Lemma 11 and $k = \lceil n^{1/5} \rceil$ in Lemma 13 and link them through Lemma 12.

Defining the splitting steps. We now show that sets L_t of the splitting trees T_k are definable by formulas $\psi_t(x)$ of very low quantifier rank. Note that the nodes at depth ℓ are generated by the ℓ-th splitting step. For every non-root node t in a splitting tree T, let $v_T(t)$ be the node of T that generated t. In the following let $u(1) := 1$ and $u(\ell) := \mathrm{bin}_2(2(\ell - 1))$ for every $\ell \geq 2$.

Lemma 15. *Let G be a graph and let $k \geq \ell \geq 1$. Then, for every node t at depth ℓ in T_k, we have $v_{T_k}(t) = u(\ell)$.*

Now, for $a_1, \ldots, a_\ell \in \{0, 1\}$, define

$$\psi_{1a_1 \cdots a_\ell}(x) := \bigwedge_{\substack{i=1 \\ a_i=1}}^{\ell} \oplus z \left(\psi_{u(i)}(z) \wedge E(x, z) \right) \wedge \bigwedge_{\substack{i=1 \\ a_i=0}}^{\ell} \neg \oplus z \left(\psi_{u(i)}(z) \wedge E(x, z) \right).$$

Note that $\psi_1(x)$ is true since then the conjunctions are empty. We show that $\psi_t(x)$ are the formulas we are after.

Lemma 16. *Let G be a graph and let $k \geq \ell \geq 0$. Then, for every node t at depth at most ℓ in T_k, the formula $\psi_t(x)$ defines the set L_t in G.*

Note that the quantifier rank of $\psi_t(x)$ depends only on the depth of t. Therefore, let $q(\ell)$ be the quantifier rank of $\psi_t(x)$ for some and hence every t of depth ℓ. Note that $q(\ell)$ is monotone non-decreasing.

Lemma 17. $q(\ell) = O(\log^* \ell)$.

We are ready to prove the main lemma of this section.

Lemma 18. *For every $\epsilon > 0$, there is $n_0 \geq 0$ such that for every $n \geq n_0$ there is a formula $\psi(x, y)$ of quantifier rank $O(\log^* n)$ such that, for $G \sim G(n, 1/2)$, the probability that ψ defines a linear pre-order of width at most 2 is at least $1 - \epsilon$.*

5   The Lower Bound and Final Remarks

Here we put it all together to prove Theorem 1.

Proof of Theorem 1. Fix $\delta > 0$ and choose n_0 large enough and $n \geq n_0$. Let $\psi(x, y)$ be the formula from Lemma 18 for $\epsilon = \delta/2$. Let $\phi(x, y, z) := \psi(x, y) \wedge \psi(y, z)$. Let q be the quantifier rank of ϕ, which is the same as ψ. We have $q \leq c \log^* n$ for some constant $c > 0$. We claim that ϕ witnesses the theorem.

Suppose $p(x, y, z)$ is a Boolean combination of quantifier-free formulas and $FO[\oplus]$-polynomials of degree bounded by a tower of exponentials of height $q/c-3$ that agrees with $\phi(x, y, z)$ on more than a δ-fraction of graphs with n vertices. By Lemma 3 we may assume that $p(x, y, z)$ is a regular normal form of the same degree. The degree of $p(x, y, z)$ is bounded by $\log \log \log n$. If n_0 is large enough, with probability at least $1 - \delta/2$ there exists a triple a, b, c of distinct vertices for which $Y(a, b, c)$ holds. Also if n_0 is large enough, with probability at least $1 - \delta/2$ the formula $\psi(x, y)$ defines a linear pre-order of width at most 2. By the union bound, with positive probability the three conditions below hold:

1. $\phi(x, y, z)$ and $p(x, y, z)$ agree on G,
2. $\psi(x, y)$ defines a linear pre-order $\preceq$ of width at most 2 in G,
3. there exists a triple of distinct vertices a, b, c of G for which $Y(a, b, c)$ holds.

Since no graph can satisfy them simultaneously, this yields a contradiction.

Final remarks. The lower bound is achieved by a formula with free variables. In particular, when we say that $\phi(x, y, z)$ and $p(x, y, z)$ disagree on many graphs, what we mean is that, on such graphs, the ternary relations on the set of vertices that are defined by $\phi(x, y, z)$ and $p(x, y, z)$ are not identical. It would be nice to obtain the same kind of lower bound for *sentences*, i.e. formulas without free variables. A candidate such sentence could be the one saying that the number of edges between the minimum and the maximum classes in the pre-order is odd. However we were not able to prove that this sentence must be uncorrelated to any low degree FO[$\oplus$]-polynomial. We leave this as an interesting open problem.

Acknowledgment. We are grateful to Swastik Kopparty for discussions and comments on a previous version of this paper.

References

1. Fagin, R.: Probabilities on finite models. J. Symb. Log. 41(1), 50–58 (1976)
2. Glebskiĭ, Y.V., Kogan, D.I., Ligon'kiĭ, M., Talanov, V.A.: Range and degree of realizability of formulas in the restricted predicate calculus. Kibernetika 2, 17–28 (1969)
3. Kolaitis, P.G., Vardi, M.Y.: Infinitary logics and 0-1 laws. Information and Computation 98(2), 258–294 (1992)
4. Dawar, A., Grädel, E.: Properties of almost all graphs and generalized quantifiers. Fundam. Inform. 98(4), 351–372 (2010)
5. Compton, K.: 0-1 laws in logic and combinatorics. In: Rival, I. (ed.) NATO Advanced Study Institute on Algorithms and Order, pp. 353–383. Kluwer (1989)
6. Kolaitis, P.G., Kopparty, S.: Random graphs and the parity quantifier. In: STOC 2009: Proc. 41st ACM Symp. Theory of Computing, pp. 705–714 (2009)
7. Allender, E.: A note on the power of threshold circuits. In: FOCS 1989: Proc. 30th Symp. on Foundations of Computer Science, pp. 580–584 (1989)
8. Beigel, R., Tarui, J.: On acc. Computational Complexity 4, 350–366 (1994)
9. Williams, R.: Non-uniform acc circuit lower bounds. In: Proc. 26th IEEE Conf. Computational Complexity, pp. 115–125 (2011)
10. Razborov, A.A.: Lower bounds on the size of bounded depth networks over a complete basis with logical addition. Math. Notes of the Academy of Sciences of the USSR 41, 333–338 (1987)
11. Smolensky, R.: Algebraic methods in the theory of lower bounds for boolean circuit complexity. In: STOC 1987: Proc. 19th ACM Symp. Theory of Computing, pp. 77–82 (1987)
12. Karp, R.M.: Probabilistic analysis of a canonical numbering algorithm for graphs. In: Proceedings of the AMS Symposium in Pure Mathematics, vol. 34, pp. 365–378 (1979)
13. Hella, L., Kolaitis, P.G., Luosto, K.: Almost everywhere equivalence of logics in finite model theory. Bulletin of Symbolic Logic 2, 422–443 (1996)

Monadic Datalog Containment

Michael Benedikt[1], Pierre Bourhis[1], and Pierre Senellart[2]

[1] Oxford University, Oxford, United Kingdom
`firstname.lastname@cs.ox.ac.uk`
[2] Institut Mines–Télécom; Télécom ParisTech; CNRS LTCI, Paris, France
`pierre.senellart@telecom-paristech.fr`

Abstract. We reconsider the problem of containment of monadic data-log (MDL) queries in unions of conjunctive queries (UCQs). Prior work has dealt with special cases, but has left the precise complexity character-ization open. We begin by establishing a 2EXPTIME lower bound on the MDL/UCQ containment problem, resolving an open problem from the early 90's. We then present a general approach for getting tighter bounds on the complexity, based on analysis of the number of mappings of queries into tree-like instances. We use the machinery to present an important case of the MDL/UCQ containment problem that is in co-NEXPTIME, and a case that is in EXPTIME. We then show that the technique can be used to get a new tight upper bound for containment of tree automata in UCQs. We show that the new MDL/UCQ upper bounds are tight.

1 Introduction

Context. Datalog represents a standard model for querying data with recursion. The basic problems of evaluation, equivalence, and containment thus have been the object of study for several decades. Shmueli [1] showed that containment (and equivalence) of datalog is undecidable. Decidable subclasses were subsequently isolated [2,3], focusing on restricting the form of recursion used. Chaudhuri and Vardi [4,5] provide an extensive study of the containment of datalog queries in nonrecursive datalog queries, showing in particular that the problem is decidable. They also show that it is 2EXPTIME-complete to decide containment of a datalog query within a union of conjunctive queries (UCQ).

In this article we focus on *monadic datalog* (MDL) – the fragment in which all intensional predicates are unary. Cosmodakis, Gaifman, Kanellakis, and Vardi [6] showed that containment of monadic datalog queries is in 2EXPTIME, and is EXPTIME-hard, leaving open the question of a tight bound. In another article, Chaudhuri and Vardi [7] prove a co-NEXPTIME upper bound for containment of a unary MDL query (i.e., a query with one output variable) in a union of unary connected conjunctive queries. Their co-NEXPTIME upper bound does not apply to Boolean queries, and, in fact, we will show that the problem is 2EXPTIME-hard even for connected Boolean queries. It also does not apply when either the MDL or the conjunctive queries have constants.

Since the work of Chaudhuri and Vardi, the "fine structure" of the containment problem between recursive and non-recursive queries has remained

A. Czumaj et al. (Eds.): ICALP 2012, Part II, LNCS 7392, pp. 79–91, 2012.
© Springer-Verlag Berlin Heidelberg 2012

mysterious. What computation can one code up in a MDL/UCQ containment problem? What features of recursive queries make the containment problem hard? Put another way: What restriction on the queries or on the instances can make the problem easier? A better understanding of these issues can shed light on a number of other questions concerning containment between recursive and non-recursive queries: e.g., containment under limited access patterns [8,9], and containment of unions of conjunctive two-way regular path queries [10] in non-recursive queries.

Contributions. We start by showing that containment of MDL in UCQs is 2EXPTIME-complete, with the upper bound holding in the presence of constants and the lower bound even in their absence. This resolves the main open question from [7]. Since this is a special case of monadic datalog containment, it also shows that MDL containment is 2EXPTIME-complete, closing the gap of [6].

We then look to understand the factors leading to high complexity. We start by revisiting the 2EXPTIME upper bound for containment of Chaudhuri and Vardi, giving a refinement of the main two tools used there: reduction to tree-shaped models, and counting the number of different types of nodes in the models. We present a property of a class of instances, the Unique Mapping Condition, that suffices to show that the number of types reduces from doubly-exponential to exponential. We then show that whenever this condition holds, containment in UCQs goes down to co-NEXPTIME, and containment of constant-free queries goes down to EXPTIME. We give two settings where the Unique Mapping Condition holds. One is that of "globally extensionally limited" MDL (GEMDL) queries over general relational structures: we limit the number of occurrences of an extensional predicate in the program. A second is when the models are trees. In the second case, we get new upper bounds of containment of tree automata in UCQs. We believe that the upper bound techniques give new insight into implication problems for recursive queries in UCQs, and more generally for the understanding of homomorphisms of conjunctive queries in tree-like structures.

We round off these upper bound results with two complementary lower bounds, showing that bounds on containment of GEMDL queries are tight, and showing that slight generalizations of the GEMDL condition push the complexity of containment in UCQs back to 2EXPTIME. The lower bound arguments require a much more involved coding argument than for general MDL.

Organization. Section 2 contains the basic definitions. Section 3 discusses monadic datalog containment in general, and presents our main lower bound, exploring its consequences. Section 4 deals with the main new subclasses for which better bounds can be provided. Section 5 gives conclusions and discusses related work.

2   Background

Datalog. An *atom* over a relational signature S is an expression $R(x_1 \ldots x_n)$ where R is an n-ary predicate of S and the x_i are either variables (from some countable set) or constants.

We consider the following three simple positive query languages over S: (a) conjunctive queries (CQs); (b) unions of conjunctive queries (UCQs); (c) positive queries (PQs). They are respectively defined as the fragments of first-order logic over S built up from relation atoms and, respectively, (a) $\exists$ and $\wedge$; (b) disjunctions of CQs; (c) $\exists$, $\vee$, and $\wedge$.

A *datalog program* over S consists of: (i) A relational signature S. (ii) A set of rules of the form $A \leftarrow \varphi$, where φ is a conjunction of atoms over S, and A is an atom over S. A is the *head* of the rule and φ is the *body* of the rule. We will often identify φ with the set of atomic formulas in it, writing $A_1(\boldsymbol{x}_1), \ldots, A_n(\boldsymbol{x}_n)$ instead of $A_1(\boldsymbol{x}_1) \wedge \ldots \wedge A_n(\boldsymbol{x}_n)$. A variable that occurs in the head of rule r is a *free variable* of r. Other variables are *bound* in r. We require that every free variable occurs in the body. (iii) A distinguished predicate $Goal$ of S which occurs in the head of a rule, referred to as the *goal predicate*.

The relational symbols that do not occur in the head of any rule are the *input* or *extensional predicates*, while the others are *intensional predicates*. Monadic datalog (MDL) denotes the sublanguage where all intensional predicates are monadic (unary). For a datalog program Q, we denote by $\mathrm{Const}(Q)$ the constants used in Q.

Semantics. For a datalog program, an intensional predicate P, a structure D interpreting the input predicates and constants, we define the output of P in D, denoted $P(D)$, as the fixpoint of the relations $P_i(D)$ defined via the following process, starting with $P_0(D) = \emptyset$:

- Let D^i be the expansion of D with $P_i(D)$ for all intensional P.
- If r is a rule with $P(x_1 \ldots x_l)$ in the head, $\boldsymbol{w}$ the bound variables of r, and $\varphi(\boldsymbol{x}, \boldsymbol{w})$ the body of r, let $P_{i+1}^r(D)$ be defined by: $\{\boldsymbol{c} \in \mathrm{dom}(D)^l : D^i \models \exists \boldsymbol{w}\, \varphi(\boldsymbol{c}, \boldsymbol{w})\}$ where $\mathrm{dom}(D)$ is the active domain of D.

We let $P_{i+1}(D)$ denote the union of $P_{i+1}^r(D)$ over all r with P in the head. Finally, the *query result* of Q on D is the evaluation of the goal predicate of Q on D. We will often assume $Goal()$ is nullary, in which case the result of the query on D is the Boolean true iff $Goal()$ holds in D. We will alternatively refer to a *datalog query* to emphasize that we are only interested in the output on the goal predicate. Note that, unlike in many works on datalog, we allow constants. Under these semantics, it is easy to check that any UCQ can be transformed in linear-time into an equivalent MDL query.

Containment. The main problem we deal with in this work is the classical notion of query containment.

Definition 1. *Let Q and Q' be two queries over a relational structure S. We say Q is* contained *in Q', denoted $Q \sqsubseteq Q'$, if, for any database D over S, $Q(D) \subseteq Q'(D)$.*

We focus on containment for Boolean queries in the remainder of the paper. However, our results apply to the unary case as well, thanks to the following:

Proposition 1. *There are polynomial-time one-to-one reductions between the containment of Boolean MDL queries in Boolean UCQs and that of MDL queries in UCQs. A similar statement holds for positive queries.*

The direction from non-Boolean to Boolean in the proposition above will imply that our lower bounds apply to unary MDL as well. The other direction will be used to transfer upper bounds. Our upper bounds will be proven for several restricted classes of Boolean MDL. The analogous definitions for non-Boolean MDL will be obvious, and all of these classes will easily be seen to be preserved by the transformation from above. Hence the argument above implies that these upper bounds also apply to the corresponding non-Boolean problem.

Note that this simple argument does not apply to the results of Chaudhuri and Vardi [7] on containment of connected unary MDL queries into union of unary connected queries since connectedness is not preserved by the reduction.

Tree-Like Instances. Before discussing the complexity of the containment problem, we introduce the notion of *tree-like instance* that will be used in various proofs throughout the paper.

A *tree-like instance* for an MDL query Q over S is an instance I for the relations of S, plus a tree $T(I)$ whose nodes n are associated with both a value v appearing in I and a collection of atoms of I, called the *bag* of n and denoted $\text{bag}(n)$. We say v is the *output value* of the node n and we require that:

1. all values appearing in atoms of $\text{bag}(n)$ are either in $\text{Const}(Q)$, are output values of n, or are output values of one of n's children in the tree;
2. for every value $v \notin \text{Const}(Q)$ that appears in I, v is the output value of exactly one node n, and any atoms containing v must be in the bag of n or its parent.

We denote a node n by a pair $(v, \text{bag}(n))$ of its output value and bag of facts.

A fundamental fact is that there is always a tree-like witness for non-containment of a monadic datalog query in a UCQ. A similar result appears in various other places in the literature, such as [7–9,11], but we will require a version of this result valid also in the presence of constants. Note that tree-like instances are a special kind of tree-decomposition, with additional structure.

Proposition 2. *If an MDL query Q is not contained in a UCQ Q', then there is a tree-like instance I such that I satisfies $Q \wedge \neg Q'$. Furthermore, the size of the bags of I can be taken to be polynomial in the size of Q and Q'.*

3 Monadic Datalog Containment in General

We first note that containment of monadic datalog queries in positive queries is known to be in 2EXPTIME: in the absence of constants, this follows from results of ten Cate and Segoufin [12], who define a language closed under negation that subsumes both MDL and PQs. This also holds in the presence of constants:

Theorem 1 (Proposition 4.11 of [13]). *The problem of containment of monadic datalog (with constants) in PQs (with constants) is in 2EXPTIME.*

Indeed, [13] shows the same bound for containment of datalog in positive queries with constants. We now show the first main result of this article, a corresponding lower bound of 2EXPTIME-hardness holding even in the absence of constants.

Theorem 2. *Containment of monadic datalog in UCQs is* 2EXPTIME-*hard, even for connected Boolean queries. Hence containment of MDL in MDL is* 2EXPTIME-*hard.*

The proof proceeds by creating a Datalog program Q and UCQ Q' such that models of $Q \wedge \neg Q'$ code the computation of an alternating EXPSPACE Turing Machine. The proof relies on a coding trick developed by Björklund, Martens, and Schwentick in the context of trees [14]. The trick allows one to check, given two paths coding an n-length bit string, whether they code the same string. Under a naive coding one needs a positive query to do this, checking a conjunction of disjunctions; in the Björklund et. al. coding, one can use a UCQ to do the check. The coding of the machine uses a datalog program of a very restricted form: extensional relations have arity at most 2, rule bodies are acyclic, and the program is nonpersistent [7]. Note that the co-NEXPTIME membership result of [7] of nonpersistent datalog programs in unions of connected queries only holds for non-Boolean queries.

4   Reducing the Complexity

To lower the complexity, we consider a restriction on the form of MDL queries that bounds the use of extensional predicates across all rules. As we will show, this restriction is relevant to the problem of containment under limited access patterns [8, 9]. Enforcing this bound will suffice to ensure the existence of witnesses of non-containment in any UCQ of size at most exponential, which easily shows that the complexity reduces to co-NEXPTIME. Furthermore, if constants are forbidden, we will show the problem is in EXPTIME. In addition of giving tight complexity bounds for this restricted language, we also provide lemmas about the structure of witnesses to non-containment, that are of interest in themselves and can serve to prove similar complexity results in other settings.

4.1   Equivalence of Nodes in Tree-Like Instances

Before considering our restricted language, we revisit the argument showing the 2EXPTIME upper bound of containment of monadic datalog in unions of conjunctive queries. This result is proved in [4, 5] in the absence of constants, and is generalized by Theorem 1.

Note that *in all of the discussion of upper bounds in this paper, we usually reason with Q' being a single CQ, but then state results for UCQs.* The extension of the upper bound techniques and results to UCQs will always be straightforward.

By Proposition 2 in considering containment we can restrict ourselves to tree-like instances. Then we introduce a notion of "special subquery" which is used to define an equivalence relation on the nodes of a tree like instance, called *IQ-equivalence*. Finally, we show that the tree-like counterexamples to containment are given by a collection of tree automata whose states are IQ equivalence classes. This allows us to relate the time complexity of finding a counterexample by the number of IQ equivalence classes. We show how this gives a generic recipe for getting bounds on the containment problem.

IQ Classes. How big does a tree-like counterexample to containment have to be? Intuitively, one only needs nodes that represent the different kinds of behavior with respect to Q', since two nodes with the same behavior can be collapsed. A crude notion of "same behavior w.r.t. Q'" would be to identify a node with the collection of subqueries, substituted with constants, that simultaneously hold at that node. Such an abstraction would easily lead to a doubly-exponential bound on the size of a counterexample to containment. Our main contribution is a finer notion of similarity that takes advantage of the restricted structure of tree-like instances. Intuitively, we do not care about all subqueries that map to a node, but only about the way that the subqueries impact what is happening at other nodes. That interaction is captured by the restriction of the mapping to the root, and by the variables that are mapped to constants, since constants cut across the tree structure. We can thus think of fixing both a mapping to constants (derived subquery below) and a root interface. Naïvely, our root interface would require us to specify the mapping to the root bag completely. Instead we allow ourselves to fix only two things: 1) the set of variables that map to the output value of the root; 2) for each single atom at a time, and each connected component in the query obtained by removing this atom, information about whether the corresponding query is *mappable*. We formalize this idea of "mappable interface queries" below.

A *derived query* of Q' is one obtained by substituting variables with constants (where multiple variables can map to the same constant).

Given a query Q', an atom A of Q', and a subset X of the variables of Q', a subquery Q'' of Q' is *fresh-connected for X and A* iff A occurs in Q'' and:

- (closure) Q'' is closed under variables not in X – if it contains one atom with such a variable x, it contains all atoms containing x.
- (connectedness) for every variable x of Q'' not in X there is a path from x to a variable in atom A, where the path is in the graph connecting two variables if they co-occur in an atom of Q'';
- (fresh witnesses) every atom in Q'' other than A must contain a variable outside of X.

A query Q'' and an atom A are *pointed mapped* relative to a subset X of its variables and a node n in a tree-like instance iff there exists a homomorphism h from Q'' to the subtree rooted at n such that the set of variables mapping to the output value of n is equal to X, no variable of Q'' is mapped to a query constant, and $h(A) \in \text{bag}(n)$. Thus variables outside of X map to values that are *fresh* for n – neither a query constant nor the output value of n.

We define an equivalence relation on nodes n in tree-like instances.

Two nodes $n = (v, b)$ and $n' = (v', b')$ are *IQ-equivalent* (short for "Interface Query equivalent") with respect to a set C of constants iff

- either a) v is in C and $v' = v$; or b) neither v nor v' is in C;
- for each atom A and derived subquery Q'' of Q', for each subset X of the variables of Q'', for each fresh-connected subquery Q''' of Q'' for X and A, Q''' is pointed mapped to the subtree rooted at n relative to X iff Q''' is pointed mapped to the subtree rooted at n' relative to X.

Composition of IQ Classes. Let C be the constants occurring in the MDL query Q and in the CQ Q'. Let I be a tree-like instance, n and n' two nodes of $T(I)$ with respective sets of children $\{n_1, \ldots n_k\}$ and $\{n'_1 \ldots n'_{k'}\}$. An isomorphism φ from the values of bag(n) to the values of bag(n') is a *C-isomorphism* iff it preserves constants of C and the output value of n maps to the output value of n'.

We can now state our composition lemma: in the presence of a C-isomorphism, IQ-equivalence of the children of two nodes implies IQ-equivalence of the nodes.

Lemma 1. *Let I be a tree-like instance. Let n, $n_1 \ldots n_k$ be a node of I and its children, and n', $n'_1 \ldots n'_k$ a node of I and its children. Suppose n_i is IQ-equivalent to n'_i w.r.t. constants C for $1 \leqslant i \leqslant k$ and that there exists a C-isomorphism φ from bag(n) to bag(n') such that if the output value of n_i is in domain(φ), then $\varphi(n_i)$ is the output value of n'_i, and vice versa. Then n and n' are IQ-equivalent w.r.t. to C.*

From IQ Classes to Automata. It is easy to use the composition lemma to show that given a counterexample to containment of MDL Q in UCQ Q', another counterexample can be obtained where any non-root nodes have distinct IQ classes – i.e., one can bound the size of a counterexample instance for containment by the number of IQ classes. In fact, IQ classes capture the "state" of a node in the tree skeleton in a literal sense: they can be used as states in a tree automaton that accepts such skeletons. Let $B(C, d, S)$ be the set of C-isomorphism types of bags with d atoms over schema S. Then every tree-like instance I is associated with a $B(C, d, S)$-labeled (ranked) tree Code(I). Note that the C-isomorphism type indicates which value is the output value.

Lemma 2. *For each C, d, S, and CQ Q', there is a non-deterministic bottom-up tree automaton $A = \bigcup_{1 \leqslant i \leqslant k} A_i$ that accepts the set of all Code(I) with I a tree-like instance not satisfying Q', where k is bounded by a polynomial in the number of subsets of connected subqueries of derived queries of Q', and the size of each A_i is bounded by a polynomial in the number of IQ classes of Q'.*

How does this help us? We can intersect the automaton A above with an automaton restricting the shape of tree-like instances: e.g., to tree-like models of an MDL query Q, or special tree-like models. Using the fact that non-empty automata have examples with DAG representations polynomial in the automaton size, we see that whenever containment fails, there is a counterexample of size at most the maximum of $|A_i|$, which is bounded by the number of IQ classes. This gives a non-deterministic algorithm for guessing a counterexample to containment. But the result also gives a deterministic time bound in the size of A. In the general case of MDL containment in UCQs, the number of IQ classes is doubly-exponential, and the number k above is also doubly-exponential, and so the automata-theoretic approach gives the same bound as in Chaudhuri and Vardi [5]. As we have seen in the previous section, this is the best we can do, since the bound is tight. We can do better if we can restrict to a class of instances that satisfies the Unique Mapping Condition:

Definition 2. *A class of tree-like instances satisfies the Unique Mapping Condition (UMC) if for any node n, any conjunctive query Q', and any atom A there* *exists at most one triple Q'', X, and Q''' such that: Q'' is a derived subquery of Q', X is a subset of the variables, Q''' is fresh-connected for X, A, and there is a pointed map of Q''' to n for X.*

The UMC should shed more light on the conditions of fresh-connected queries: it is easy to see that for a disconnected query we will not have a unique mapping even for very restricted structures; and if we do not look at connected queries that are somehow "maximal," we cannot get uniqueness. We will show that the Unique Mapping Condition suffices to get better upper bounds. The following is easy to see:

Proposition 3. *Restricted to a class of instances where the UMC applies, the number of IQ classes is exponential. Thus satisfiability of UCQs restricted to such a class is in co-NEXPTIME.*

Proof. By the UMC, the number of classes is bounded by the number of functions from atoms to triples (Q'', X, Q'''), which is exponential in Q'. The existence of an exponential-sized bound on counterexample size follows using Lemma 2. □

If the queries are required not to have constants, the problem moves from NEXPTIME to EXPTIME.

Theorem 3. *Suppose we have a class C of instances that has the UMC, has no constants, and whose codes can be described by a tree automaton of size at most exponential. Then the problem of determining whether a member of C satisfies the negation of a UCQ is in EXPTIME.*

Proof. The number of IQ classes is exponential in Q'. Due to the absence of constants, there exists only one derived query from Q', which is Q' itself. So the number of connected subqueries of Q' is polynomial in Q'. This in turn implies that the number of subsets is exponential in the size of Q'. The rest follows from Lemma 2 and the fact that emptiness of tree-automata is in PTIME. □

4.2 Global Restrictions and Diversified Instances

We are now ready to get an improved upper bound. We first look at arbitrary instances, but put a strong restriction on queries.

A monadic datalog query, possibly with constants, is *globally extensionally restricted* (GEMDL) if every extensional predicate appears in only one rule, and occurs only once in that rule. An MDL query is *almost globally extensionally restricted* (AGEMDL) if the goal predicate never occurs in the body of a rule, and every extensional predicate has only one occurrence in a rule other than a rule for the goal predicate.

Informally, AGEMDL queries allow UCQs built over intensional predicates, where extensional predicates are partitioned into classes where each rule uses predicates in a particular class.

An interesting example of AGEMDL comes from the area of *limited-access querying* [15]. In limited-access querying, one has a schema where each relation of arity l comes with a set of access methods, each with a subset of $[1 \dots l]$, the *input positions*. In an instance of the database, the accessible data is limited to what can be extracted starting with a set of seed constants, iteratively putting the values in for input positions and getting the resulting output [16]. Restrict for now to access methods having at most one output position. Formally, we can consider the monadic datalog program with rules

$$\begin{cases} \text{Accessible}(c) \leftarrow & \text{where } c \text{ is any constant in the seed set} \\ \text{Accessible}(x_{j_k}) \leftarrow R(\boldsymbol{x}), \text{Accessible}(x_{j_1}), \ldots, \text{Accessible}(x_{j_p}) \end{cases}$$

whenever relation l has an access method with input positions $j_1 \ldots j_p$ and output position j_k. Given an instance I and set of constants C, we let $\text{Accessible}(I, C)$ be the output of Accessible where C is the set of constants.

Given queries Q and Q' that use constants C the problem of query containment of Q in Q' relative to access patterns asks whether Q will always return a subset of Q' when applied to tuples that are accessible. That is, it asks whether the set $\{\boldsymbol{c} \mid \boldsymbol{c} \in Q(I) \wedge \bigwedge_i c_i \in \text{Accessible}(I, C)\}$ is contained in $\{\boldsymbol{c} \mid \boldsymbol{c} \in Q'(I)\}$.

Consider the case where every relation has only one access method. This is the standard assumption in much of the literature [17]. The rules above then satisfy the GEMDL restriction. We can thus express the evaluation of UCQ Q over accessible data using a single additional goal predicate, expressing Q and the restriction that all witness variables are accessible. This will be an AGEMDL query. It follows that the question of *containment of unions of conjunctive queries under limited access patterns* (see [9,17] for the formal definition) can be expressed as the containment of a AGEMDL query in a UCQ. We now claim:

Theorem 4. *The containment problem for AGEMDL queries in union of conjunctive queries (with constants) is in* co-NEXPTIME *(and is* co-NEXPTIME-*complete).*

From this and the observation above we have the following result, also shown in [9]:

Corollary 1. *Containment of UCQ queries with constants under limited access patterns, where every relation has only a single access method and every access method has at most one output position, is complete for* co-NEXPTIME.

The technique can also be used for access methods with arbitrary numbers of positions, but this requires extending the definition of AGEMDL to allow rules with multiple head predicates, with the restriction that these rules need to be nonpersistent [7].

We now begin the proof of the upper bound in Theorem 4. The AGEMDL restriction implies a stronger kind of tree-like instance. A *diversified tree-like instance* is a tree-like instance in which: (i) for each node n which is not the root, for each relation R, there exists at most one fact in bag(n) having the relation name R; (ii) we cannot have a fresh value v (a value that is not a constant or the output value of the root) such that there are two distinct facts in the instance with the same relation name which have v in the same position.

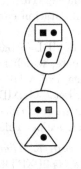

Fig. 1. Diversified tree-like instance

The first condition is a local one, saying we do not have self-joins within a bag, while the second one is global, saying self-joins across bags must not have the joined variable in the same position. A figure showing a diversified tree-like instance is given in Figure 1. The large ellipses represent bags, with different relations represented by different shapes in a bag. The dark disc represents a common value shared across and within bags.

We now have the following refinement of Lemma 2.

Proposition 4. *If a AGEMDL query Q is not contained in UCQ Q', then there is a diversified tree-like instance I such that I satisfies $Q \wedge \neg Q'$ and for each n of I, the size of its bag is polynomial in Q.*

The following links diversified instances to the IQ-class machinery:

Theorem 5. *The class of diversified tree-like instances satisfies the UMC.*

Note that X and Q'' always determine a single Q''', so the key is that, in this restricted setting, we have only one way to map variables to constants, and only one way to select which variables map to the output value of the root. The idea is that if we had a choice of how to map an atom, the choice would be between either atoms in the same bag (forbidden by the first condition of diversified instances) or atoms in neighboring bags (forbidden by the second condition).

From the result above, Lemma 3 and Theorem 3 we get upper bounds for AGEMDL; we also have matching lower bounds:

Corollary 2. *The containment of an AGEMDL query with constants in a UCQ with constants can be decided in co-NEXPTIME. In the absence of constants the problem is in EXPTIME.*

Theorem 6. *The problem of determining whether an AGEMDL query is contained in a UCQ is EXPTIME-hard, and with constants the problem is co-NEXPTIME-hard.*

The co-NEXPTIME argument is by reduction to a prior result in [9]. The EXPTIME-hardness is proven via an encoding of the execution of a PSPACE alternating Turing machine. The coding is simpler than for Theorem 2, since a configuration can be directly coded as a polynomial-sized vertical chain in a tree.

AGEMDL forbids a relation from occurring more than once outside of the goal predicate, which is a strong restriction. A simple generalization is to consider the class of MDL queries where relations can occur in a bounded number of rules. Let k-GEMDL be the class obtained by replacing "in only one rule", with "in at most k rules" in the definition of GEMDL (while still restricting to one occurrence per rule). In this case, the complexity jumps back up to 2EXPTIME.

Theorem 7. *There is a k such that the problem of k-GEMDL containment in a UCQ is 2EXPTIME-hard.*

The proof is quite involved. The main challenge, compared to that of Theorem 2, is that we cannot use a generic *Child* relation to code the tree structure of an alternating EXPSPACE computation, since this would require two occurrences of the same atom in one rule. We also cannot use separate predicates for left and

Table 1. Complexity of containment of monadic datalog queries in UCQs and PQs. In all cases, problems are complete in the class shown.

	UCQ	PQ
MDL	2EXPTIME	2EXPTIME
MDL with constants	2EXPTIME	2EXPTIME
GEMDL	EXPTIME	2EXPTIME
GEMDL with constants	co-NEXPTIME	2EXPTIME
k-**GEMDL with constants**	2EXPTIME	2EXPTIME

right child to code the structure: this would make it impossible to use a UCQ to compare paths of the trees that represent addresses of cells. Thus we develop a new coding relying on a combination of a linear structure and "value joins".

We also note that, for positive queries, the GEMDL restriction does not help:

Theorem 8. *GEMDL containment in a positive query is* 2EXPTIME-*hard.*

The lower bound proof uses a simple coding of an alternating EXPSPACE computation, using the power of positive queries to compare paths.

A diversified tree-like instance still allows joins. What about "pure tree structures" – tree-like instances representing labeled unranked trees with the child relation? These are not a special case of diversified tree-like instances, since one cannot have all child facts of a given vertex in the same bag. Still, it is not surprising that the UMC also applies to pure trees:

Theorem 9. *Tree structures have the UMC.*

From the theorem above and the small model machinery presented earlier, we get the following new bound, which answers a question left open from [14].

Corollary 3. *Containment of a tree automaton in a UCQ is in* EXPTIME.

The above results complements a recent theorem of Björklund et al. [18], which shows that the above problem is hard for EXPTIME.

5 Conclusions and Related Work

In this paper we have revisited the containment problem for recursive queries in UCQs. We started by showing that the problem is hard for doubly-exponential time in general. We then analyzed the phenomenon of tree-like models for recursive queries in more detail, and give a parameter – the number of IQ classes – that controls the size of a minimal counterexample to containment. We have shown that if a logic has models that are "very tree-like", then the number of distinct ways a CQ can map into the model is limited, and thus an exponential bound can be shown on the number of IQ classes. We have applied this analysis to two logics and two collections of instances – GEMDL and tree automata. But we believe that it can be applied to other fragments of MDL.

Our results on containment of MDL are summarized in Table 1.

Special cases of the containment problem of monadic datalog in UCQs have been studied in the past. The original Chaudhuri and Vardi article [7] proved a co-NEXPTIME bound of containment of connected unary MDL queries in UCQs, where the queries were unary. Their paper extends earlier work by Courcelle [11], who noted the connection with graph decompositions.

Segoufin and ten Cate [12] define the language UNFP which can express the conjunction of an MDL query with a negated UCQ; they show that the satisfiability for this language is 2EXPTIME-complete. An EXPTIME bound on satisfiability is shown for a fragment called Simple UNFP; the fragment cannot express UCQs, much less the negation of a UCQ conjoined with an MDL query. Björklund et al. [14] study containment of tree automata in UCQs with child and descendant. We make use of their lower bound technique in our first result, while also refining one of their upper bounds in the absence of descendant.

Our main upper bound technique originates from our work on limited access querying [9] – there we showed a co-NEXPTIME bound for a particular kind of MDL/UCQ containment problem, using a special case of the technique. Our upper bounds here are an abstraction of the idea in [9], relating it to tree-like instances. Our lower bounds can be seen as exploring the limits of this method. Note that [9] also contains errors that are discussed (and in one case, corrected) in [13] – see the end of the related work section of [13].

Acknowledgments. Bourhis is supported by EPSRC EP/G004021/1 (the Engineering and Physical Sciences Research Council, UK). Senellart is partly supported by ERC grant Webdam, agreement 226513.

References

1. Shmueli, O.: Equivalence of datalog queries is undecidable. J. Log. Prog. 15(3) (1993)
2. Bonatti, P.A.: On the decidability of containment of recursive datalog queries. In: PODS (2004)
3. Calvanese, D., De Giacomo, G., Vardi, M.Y.: Decidable containment of recursive queries. Theoretical Computer Science 336(1) (2005)
4. Chaudhuri, S., Vardi, M.Y.: On the equivalence of recursive and nonrecursive Datalog programs. In: PODS (1992)
5. Chaudhuri, S., Vardi, M.Y.: On the equivalence of recursive and nonrecursive Datalog programs. JCSS 54(1) (1997)
6. Cosmadakis, S.S., Gaifman, H., Kanellakis, P.C., Vardi, M.Y.: Decidable optimization problems for database logic programs. In: STOC (1988)
7. Chaudhuri, S., Vardi, M.Y.: On the complexity of equivalence between recursive and nonrecursive Datalog programs. In: PODS (1994)
8. Calì, A., Martinenghi, D.: Conjunctive Query Containment under Access Limitations. In: Li, Q., Spaccapietra, S., Yu, E., Olivé, A. (eds.) ER 2008. LNCS, vol. 5231, pp. 326–340. Springer, Heidelberg (2008)
9. Benedikt, M., Gottlob, G., Senellart, P.: Determining relevance of accesses at runtime. In: PODS (2011)
10. Calvanese, D., Giacomo, G.D., Lenzerini, M., Vardi, M.Y.: Containment of conjunctive regular path queries with inverse. In: KR (2000)

11. Courcelle, B.: Recursive queries and context-free graph grammars. Theoretical Computer Science 78(1) (1991)
12. ten Cate, B., Segoufin, L.: Unary negation. In: STACS (2011)
13. Benedikt, M., Bourhis, P., Ley, C.: Querying schemas with access paths. PVLDB (2012)
14. Björklund, H., Martens, W., Schwentick, T.: Optimizing Conjunctive Queries over Trees Using Schema Information. In: Ochmański, E., Tyszkiewicz, J. (eds.) MFCS 2008. LNCS, vol. 5162, pp. 132–143. Springer, Heidelberg (2008)
15. Rajaraman, A., Sagiv, Y., Ullman, J.D.: Answering queries using templates with binding patterns. In: PODS (1995)
16. Li, C., Chang, E.Y.: Answering queries with useful bindings. ACM TODS 26(3) (2001)
17. Calì, A., Martinenghi, D.: Querying the deep web. In: EDBT (2010)
18. Björklund, H., Martens, W., Schwentick, T.: Validity of tree pattern queries with respect to schema information (2012) (unpublished draft)

A Machine-Independent Characterization of Timed Languages

Mikołaj Bojańczyk* and Sławomir Lasota**

Institute of Informatics, University of Warsaw

Abstract. We use a variant of Fraenkel-Mostowski sets (known also as nominal sets) as a framework suitable for stating and proving the following two results on timed automata. The first result is a machine-independent characterization of languages of deterministic timed automata. As a second result we define a class of automata, called by us timed register automata, that extends timed automata and is effectively closed under minimization.

1 Introduction

This paper studies minimization of deterministic timed automata [2]. Existing approaches to this problem explicitly minimize various resources used by an automaton, such a locations or clocks, see [1,8,14,15,16]. We take a different approach, which abstracts away from the syntax of a timed automaton, and focuses on the recognized language, and specifically its Myhill-Nerode equivalence relation. Our notion of minimality is described by the following definition.

Definition 1. *An automaton for a language L is called* minimal *if for every two words w, w' the following conditions are equivalent:*

- *The words are equivalent with respect to Myhill-Nerode equivalence.*
- *The states reached after reading the words are equal.*

In the case of a deterministic timed automaton, the term "state" refers to the location (or control state) and the valuation of clocks. One of the main contributions of this paper is a minimization algorithm for deterministic timed automata. Of course in the case of timed automata, Myhill-Nerode equivalence has infinitely many equivalence classes, e.g. in the language

$$\{t_1 \cdots t_n \in \mathbb{R}^* : t_i = t_{i-1} + 1 \text{ for all } i \in \{2, \ldots, n\}\},$$

the equivalence class of a word is determined by its last letter.

A New Automaton Model. There is a technical problem with minimizing deterministic timed automata: the minimization process might leave the class of timed automata, as witnessed by the following example.

---

* Supported by the ERC Starting Grant "Sosna".

** Supported by the FET-Open grant agreement FOX, number FP7-ICT-233599.

A. Czumaj et al. (Eds.): ICALP 2012, Part II, LNCS 7392, pp. 92–103, 2012.
© Springer-Verlag Berlin Heidelberg 2012

Example 1. Consider the following language $L \subseteq \mathbb{R}^*$. A word belongs to L if and only if it has exactly three letters $t_1, t_2, t_3 \in \mathbb{R}$, and the following conditions hold.

- The letter t_2 belongs to the open interval $(t_1; t_1 + 2)$;
- The letter t_3 belongs to the open interval $(t_1 + 2; t_1 + 3)$;
- The letters t_2 and t_3 have the same fractional part, i.e. $t_3 - t_2 \in \mathbb{Z}$.

This language is recognized by a deterministic timed automaton. After reading the first two letters t_1 and t_2, the automaton stores t_1 and t_2 in its clocks. This automaton is not minimal in the sense of Definition 1. The reason is that the words $(0, 0.5)$ and $(0, 1.5)$ are equivalent with respect to Myhill-Nerode equivalence, but the automaton reaches two different states. Any other timed automaton would also reach different states, as timed automata may reset clocks only on time-stamps seen in the input word (unless ε-transitions are allowed).

Because of the example above, we need a new definition of automata. We propose a straightforward modification of timed automata, which we call *timed register automata*. Roughly speaking, a timed register automaton works like a timed automaton, but it can modify its clocks, e.g. increment or decrement them by integers[1]. For instance, in language L from Example 1, the minimal automaton stores not the actual letter t_2, but the unique number in the interval $(t_1; t_1 + 1)$ that has the same fractional part as t_2.

We prove that timed register automata can be effectively minimized.

Typically, minimization corresponds to optimization of resources of an automaton. In case of timed automata, the resources seem to be locations and clocks, but maybe also constants used in the guards, anything else? One substantial novelty of our approach is that the kind of resource we optimize is not chosen ad hoc, but derived directly from Myhill-Nerode equivalence. Myhill-Nerode equivalence is an abstract concept; and therefore we need a tool that is well-suited to abstract concepts. The tool we use is *Fraenkel-Mostowski sets*.

Fraenkel-Mostowski Sets. By these we mean a set theory different from the standard one, originating in the work of Fraenkel and Mostowski (see [10] for the references), and thus called by us *Fraenkel-Mostowski sets* (FM sets in short). Much later a special case of this set theory has been rediscovered by Gabbay and Pitts [11,10] in the semantics community, as a convenient way of describing binding of variable names. Motivated by this important application, Gabbay and Pitts use the name *nominal sets* for the special case of FM sets they consider. Finally, FM sets (under the name "nominal G-sets") have been used in [3] to minimize automata over infinite alphabets, such as Francez-Kaminski finite-memory automata [9]. The paper [3] is the direct predecessor of the present paper.

In the setting of [3] (see also the full version [4]), FM sets are parametrized by a *data symmetry*, consisting of a set of data values together with a group G of permutations of this set. For instance, finite-memory automata are suitably represented in the data symmetry of all permutations of data values.

---

[1] A certain restriction to the model is required to avoid capturing Minsky machines.

To model timed automata, and even timed register automata, we choose the *timed symmetry*, based on the group of automorphisms of the structure[2]

$$(\mathbb{R}, <, +1).$$

Despite that this data symmetry presents several technical challenges, we show that FM sets can be used to solve nontrivial algorithmic problems, such as the minimization problem. A more accurate description of this paper is that we study automata in FM sets under the timed symmetry; and these automata happen to capture timed automata, and even timed register automata. In particular, we study languages where the timestamps appearing in a word are not necessarily increasing.

The second principal contribution of this paper is an exact characterization of the languages recognized by deterministic timed automata. The characterization is in the style of the Myhill-Nerode theorem, and is machine-independent, in the sense that it does not refer to any notion of recognizing device.

Summary of Contributions. Below are the main contributions of our paper.

1. We introduce a new class of automata, called timed register automata, which generalize timed automata.
2. We prove that, unlike for deterministic timed automata, deterministic timed register automata are closed under minimization. We also give a minimization algorithm for timed register automata (Theorem 3 in Section 2).
3. We study automata in Fraenkel-Mostowski sets, under the timed symmetry.
4. We prove a kind of Myhill-Nerode theorem, which characterizes exactly the languages of deterministic timed automata (Theorem 5 in Section 4).

Related Research. We only mention here a few related papers we are aware of. Minimization of (nondeterminstic) timed automata has been studied in particular in [1,14,16], with respect to bisimulation equivalence. As we mention later, our approach extends easily to bisimulation. On the negative side, minimization of nondeterministic automata with respect to language equivalence is undecidable, cf. [15,8]. A characterization of deterministic timed languages using finite monoids has been proposed in [6]. Our characterization is of a different nature, being based on *orbit-finiteness* of the set of equivalence classes of Myhill-Nerode equivalence. Another machine-independent characterization of deterministic timed languages has been given in [13].

2   Timed Register Automata

In this paper, we study timed automata as a special case of automata where the alphabet is of the form $A \times \mathbb{R}$, where A is a finite set and $\mathbb{R}$ is the real numbers. In a letter $(a, t) \in A \times \mathbb{R}$, we call a the label and t the timestamp. Timed automata accept only words where the timestamps increase from left to right, call such words *monotonic*. Unlike timed automata, some of the automata we study in this paper can accept non-monotonic words.

---

[2] Studying this group has been suggested to us by James Worrell.

Constraints. A *constraint* over variables $x_1, \ldots, x_n$ is any quantifier free formula that uses the variables, the binary predicate $\leq$, and the unary function $+1$. Examples of constraints include

$$x \leq (y+1)+1 \quad \wedge \quad (y+1)+1 \leq x.$$

When writing constraints, we sometimes use syntactic sugar, for instance writing the above constraint as $x = y + 2$. A constraint over variables $x_1, \ldots, x_n$ defines a subset $X \subseteq \mathbb{R}^n$.

A constraint φ is called *maximal* if every other constraint on the same variables is either implied by φ, or inconsistent with φ. An example of a maximal constraint is

$$x_2 = x_1 + 1 \quad \wedge \quad x_2 < x_3 < x_2 + 1.$$

The constraint $x < y < x + 2$ is not maximal, since it is independent with $y < x + 1$. Not every constraint is equivalent to a finite disjunction of maximal constraints, for instance the constraint $x < y$.

Clearly, maximal constraints describe those *regions* that are bounded.

Timed Register Automata. We now define an automaton model, which can recognize languages over alphabets of the form $A \times \mathbb{R}$. A *(nondeterministic) timed register automaton* $\mathcal{A}$ is given by the following ingredients.

- A finite set A of *labels*.
- A finite set Loc of *locations*, also called *control states*.
- Subsets of the locations for the *initial* and *final* locations.
- For each location $l \in$ Loc, a set X_l of register names[3].
- For every two locations $l, k \in$ Loc, and every label $a \in A$, a constraint (not necessarily maximal) which defines a subset

$$\delta_{l,a,k} \subseteq \mathbb{R}^{X_l} \times \mathbb{R} \times \mathbb{R}^{X_k}.$$

We assume that every initial location has an empty set of register names.

A *state* of the automaton is defined to be a pair (l, η), where l is a location and η is a function, called the *register valuation*, of the form $\eta : X_l \to \mathbb{R}$. We write $Q_{\mathcal{A}}$ for the set of states of an automaton $\mathcal{A}$. This set is infinite if the automaton uses registers.

The semantics of the automaton is defined in the standard way. One defines the *transition relation*

$$\delta_{\mathcal{A}} \subseteq Q_{\mathcal{A}} \times (A \times \mathbb{R}) \times Q_{\mathcal{A}},$$

to be the set of triples $(l, \eta), (a, t), (k, \mu)$ such that $(\eta, t, \mu) \in \delta_{l,a,k}$. A run over an input word from $(A \times \mathbb{R})^*$ is a sequence of states that starts in an initial state and is consistent with the transition relation.

---

[3] A simplified version, where the set of register names does not depend on the location, would not minimize well. The reason is that the number of reals necessary to remember may depend on location. Ignoring some minor differences, the simplified version resembles updatable timed automata of [5].

A timed register automaton is called *deterministic* if there is one initial location and the transition relation $\delta_{\mathcal{A}}$ is a function $\delta_{\mathcal{A}} : Q_{\mathcal{A}} \times (A \times \mathbb{R}) \to Q_{\mathcal{A}}$.

Timed register automata, as defined above, are too powerful (a similar undecidability result is shown in [5]):

Theorem 1. *Emptiness is undecidable for deterministic timed register automata.*

Proof. By simulating a Minsky machine. The automaton has three register names: x, y, z. The idea is that z represents zero, $x - z$ is the value of the first counter and $y - z$ is the value of the second counter. Since the automaton can use the $+1$ in its transition relation, it can increment and decrement the counters. The zero tests are simulated by testing $x = z$ or $y = z$. □

The reason why the undecidability proof above works is that we allow a state to store, at the same time, real numbers which are very far from each other. This motivates a restriction on timed register automata to be defined now.

Constrained Timed Register Automata. In a *constrained timed register automaton*, for each location l there is a maximal constraint φ_l over the register names of l, called the *legality constraint*. In a constrained automaton, the notion of state is changed: a state (l, μ) must be such that the register valuation μ satisfies the constraint φ_l. Despite the different semantics, a constrained timed register automaton can be easily seen to be a special case of a timed register automaton, because legality constraints can be enforced by the transition relation.

The idea of adding legality constraints might seem an ugly fix. As we shall see later, constrained timed register automata have an elegant interpretation in terms of FM sets. Also, they are powerful enough to simulate timed automata.

Theorem 2. *Emptiness is decidable for constrained timed register automata.*

As our first main result, we state:

Theorem 3. *The class of constrained timed register automata is closed under minimization. There is an algorithm that computes, for a given constrained timed register automaton, the minimal automaton.*

Speaking abstractly, the minimal automaton is the syntactic automaton, or, in other words, the quotient of a given automaton by language equivalence; this will become apparent when in Section 3 we will observe that timed register automata are a subclass of automata in FM sets under the timed symmetry. Speaking concretely, we minimize the number of locations, and the number of register variables in each location.

Our minimization algorithm adopts the classical idea of iterative partition refinement, and works equally well for bisimulation of nondeterministic automata.

Timed Automata. Timed automata [2] are defined similarly as timed register automata above. A timed automaton has a number of clock variables, that may be used to store the current timestamp and to compare it against timestamps read later on. The transition relation of a timed automaton is described using a subset of constraints, in the sense of the above definition. With these respects, timed automata seem to be a subclass of constrained timed register automata.

Timed automata have however one additional feature, not reflected in our definitions above: the clock variables are initially set to 0. In consequence, only non-negative timestamps are considered. Intuitively, a timed automaton is aware of the time that has elapsed from some *absolute* moment 0, while our automata are only aware of the *relative* time separating timestamps in the input. In particular, languages recognized by timed register automata are always closed under translations, i.e., for any $d \in \mathbb{R}$, the permutation $x \mapsto x + t$ preserves L:

$$L + t = L.$$

A language $L \subseteq (A \times \mathbb{R}^{\geq 0})^*$ can be encoded as the following language closed under translations, which has essentially the same structure as L:

$$\overrightarrow{L} \;=\; \bigcup_{\substack{t \in \mathbb{R} \\ a \in A}} ((a,0)\, L) + t \;=\; \bigcup_{\substack{t \in \mathbb{R} \\ a \in A}} (a,t)\, (L + t) \;\subseteq\; (A \times \mathbb{R})^*.$$

Thus, in this paper we only consider languages that are closed under translations. On the level of timed automata, this property may be enforced by assuming that all the clock variables are *uninitialized* (that is, initially undefined), similarly like in finite memory automata of Francez and Kaminski [9].

Theorem 4. *For every (deterministic) timed automaton with uninitialized clocks one can compute an equivalent (deterministic) constrained timed register automaton.*

The idea of the proof is to translate regions of a timed automaton to locations of a timed register automaton. Unbounded regions are eliminated by projecting onto bounded coordinates. One additional register checks monotonicity.

Constrained timed register automata are strictly more expressible than timed automata, as shown in the example below.

Example 2. Let A be a singleton, thus $A \times \mathbb{R}$ is essentially $\mathbb{R}$. The language

$$L = \{t_1 \ldots t_n \;:\; n \geq 2,\; t_n - t_1 \in \mathbb{N},\; t_{i+1} - t_i \leq 1 \text{ for } i < n\}$$

is not recognized by a timed automaton, but is recognized by a deterministic constrained timed register automaton with two registers. The automaton stores initially t_1 in its register, and then increments its value, say t, by 1 at every input letter greater than t. It accepts whenever an input letter equals t.

Due to Theorem 4, the minimization algorithm of Theorem 3 works for deterministic timed automata as well. How does the definition of minimality from Definition 1 correspond to resources of a timed automaton? The most appropriate to say is that we minimize the number of regions, and the number of clocks in each region. Indeed, as regions of timed automata are translated to locations of timed register automata, each region may be optimized independently. We however honestly note that the number of locations of the minimal automaton may be greater than the number of locations of an original timed automaton.

3 Fraenkel-Mostowski Sets and Their Automata

The definition of Fraenkel-Mostowski sets (FM sets) is parametrized by a *data symmetry* $(\mathbb{D}, G)$, which consists of a set $\mathbb{D}$ of *data values* and a subgroup G of the group of all bijections of $\mathbb{D}$. Examples of data symmetries include:

- The *classical symmetry*, where the set of data values is empty, and the group has only the identity. FM sets in the classical symmetry are going to be normal sets.
- The *equality symmetry*, where the set of data values is a countably infinite set, and the group contains all bijections. FM sets in the equality symmetry are essentially the same thing as nominal sets in [11] or FM sets in [10].
- The *timed symmetry*, where the set of data values is the real numbers, and the group contains all permutations of real numbers that preserve the order relation $\leq$ and the successor function $x \mapsto x + 1$ (we call such permutations *timed permutations*). This is the data symmetry that we use in this paper.

Intuitively speaking, normal sets are built out of empty sets and brackets { and }. The intuition behind FM sets is that they can in addition use data values as atomic elements. Our presentation below is motivated by [10].

Fix a data symmetry $(\mathbb{D}, G)$. Consider first the *cumulative hierarchy* of sets with data values, which is a hierarchy of sets indexed by ordinal numbers and defined as follows. The empty set is the unique set of rank 0. A set of rank α is any set whose elements are sets of rank smaller than α, or data values. A permutation $\pi \in G$ can be applied to a set X in the hierarchy, by renaming the data values belonging to X, and the data values belonging to elements of X, and so on. The resulting set, which has the same rank, is denoted by $X \cdot \pi$.

A set C of data values is said to be a *support* of a set X in the cumulative hierarchy if $X \cdot \pi = X \cdot \sigma$ holds for every permutations $\pi, \sigma \in G$ which agree on elements of C. A set is called *finitely supported* if it has some finite support. We use the name *FM set* for a set in the cumulative hierarchy which is hereditarily finitely supported, which means that it is finitely supported, the sets belonging to it are finitely supported, and so on.

The support of an FM set is not unique, e.g. supports are closed under adding data values. A set with empty support is called *equivariant*.

Example 3. An example of an equivariant FM set in the timed symmetry is $\mathbb{R}$ itself. Another example is $\mathbb{R}^*$. A tuple $(x_1, \ldots, x_n) \in \mathbb{R}^*$ is supported by the set $\{x_1, \ldots, x_n\}$. The set $\mathbb{R} - \{0\}$ is not equivariant; it is supported by $\{0\}$.

For some data symmetries, including the classical and equality ones, one can show that every FM set has the least support. However, FM sets in the timed symmetry do not have least supports. For instance, the set $\mathbb{R} - \{0\}$ is not supported by the empty set, but it is supported by the sets $\{0\}$ or $\{1\}$. This is because if π is a timed permutation, then $\pi(1) = 1$ is equivalent to $\pi(0) = 0$.

In many respects, FM sets behave like normal sets. For instance, if X, Y are FM sets, then $X \times Y$, $X \cup Y$, X^* and the finite powerset of X are all FM sets.

Another example is the family of subsets of X that have finite supports. The appropriate notion of a function between FM sets X and Y is that of a *finitely supported function*, which is a function from X to Y whose graph is an FM set.

Orbit-finite FM sets. From our perspective, the key property of FM sets is their more relaxed notion of finiteness. Suppose that X is an FM set. For a set of data values C, define the C-orbit of an element $x \in X$ to be the set

$$\{x \cdot \pi : \pi \in G \text{ and } \pi \text{ is the identity on } C\}.$$

If C supports X, then the C-orbits form a partition of X. The set X is called orbit-finite if the partition into C-orbits has finitely many parts, for some C which supports X. Observe that the number of C-orbits increases as the set C grows. Therefore, a set is orbit-finite if it has a finite number of orbits for some minimal set C that supports it. In particular, an equivariant set is orbit-finite if and only if it has finitely many $\emptyset$-orbits.

In many data symmetries orbit-finite sets are closed under product, but not in the timed symmetry as illustrated in Example 4.

Example 4. The set $\mathbb{R}$ is orbit-finite, namely it has one $\emptyset$-orbit. The set $\mathbb{R}^2$ is not orbit-finite. The $\emptyset$-orbits are of the following form:

$$\{(x,y) : x - y = k\} \qquad \{(x,y) : x - y \in (k, k+1)\} \qquad \text{for all } k \in \mathbb{Z}.$$

Observe that two orbits of the first kind, say $\{(x,y) : x - y = k\}$ and $\{(x,y) : x - y = l\}$, are equivariantly isomorphic via the mapping $(x,y) \mapsto (x, y + (k - l))$. Likewise, every two orbits of the second kind are mutually isomorphic. Another example of two isomorphic but distinct orbits in $\mathbb{R}^*$ is $\mathbb{R}$ and $\{(x,x,x) : x \in \mathbb{R}\}$. There are infinitely many equivariant isomorphisms between these two orbits, including $x \mapsto (x,x,x)$ and $x \mapsto (x+1, x+1, x+1)$.

Automata. The definition of automata in FM sets is exactly like the definition of automata in normal sets, except that the notion of finiteness is relaxed to orbit-finiteness. Specifically, a *nondeterministic FM automaton* is a tuple

$$(A, Q, I, F, \delta) \qquad I, F \subseteq Q \qquad \delta \subseteq Q \times A \times Q$$

where the alphabet A, states Q, initial states $I \subseteq Q$, final states $F \subseteq Q$ and transitions $\delta \subseteq Q \times A \times Q$ are FM sets, and all of them except for δ are required to be orbit-finite. (We come back to the orbit-finiteness of δ in Example 5.) The definition of acceptance is as usual for automata. An automaton is called *equivariant* if all of its components are equivariant. From now on, we only study equivariant automata.

Example 5. Consider the language $L \subseteq \mathbb{R}^*$ which contains words where some letter appears twice. This language is recognized by a nondeterministic FM automaton whose states are: an initial state q, one state q_x for each real number x, and a single accepting state $\top$. The transition relation contains triples

$$(q, x, q) \qquad (q, x, q_x) \qquad (q_x, y, q_x) \qquad (q_x, x, \top) \qquad (\top, x, \top)$$

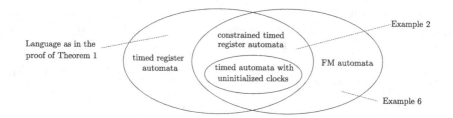

Fig. 1. Timed register automata, and FM automata in the timed symmetry

for every real numbers x, y. The transition relation is not orbit-finite, because the set of transitions (q_x, y, q_x) is isomorphic to $\mathbb{R}^2$. In general, the transition relation will necessarily have infinitely many orbits in any automaton which stores real numbers in its state, and which reads arbitrary input letters.

A *deterministic FM automaton* is the special case of a nondeterministic one, where the transition relation is a function $\delta : Q \times A \to Q$, and where the set of initial states contains only one state. From now on, we only study equivariant deterministic automata and work exclusively in the timed symmetry.

Comparing the Models. So far, we have introduced two kinds of automata. In Section 2, we have introduced timed register automata, and we have identified a subclass of constrained timed register automata. In Section 3, we have introduced automata in FM sets. In this section, we show that in the specific case of FM sets in the timed symmetry, the two kinds of automata are closely related. We only study the deterministic case, but the nondeterministic case is analogous. The results are summed up in Figure 1.

We first show that a deterministic timed register automaton is almost a special case of a deterministic FM automaton. The input alphabet, which is a set of the form $A \times \mathbb{R}$, for a finite set A is an equivariant orbit-finite FM set. The number of orbits is the size of A, because permutations of data values (= timestamps) do not change the labels. Recall that a state of a timed register automaton consists of a location and a valuation of registers. Thus the set of all states is an equivariant FM set, since it is basically a set of tuples of real numbers. In the same way, the initial and accepting states are equivariant subsets, because they are identified by their locations, and locations are not changed by permutations of data values. Finally, transition function of a timed register automaton is equivariant, because it is defined in terms of the order and successor, both preserved by timed permutations.

So why is a deterministic timed register automaton not necessarily an FM automaton? Because the states are not, in general, an orbit-finite FM set. For instance, if an automaton has two registers in some location, then its states will not be orbit-finite for the same reason as $\mathbb{R}^2$. This is where the constraints on the register valuations, as defined in Section 2, come in. The following lemma shows that maximal constraints can be used to enforce an orbit-finite state space.

Lemma 1. *The following conditions are equivalent for a subset $X \subseteq \mathbb{R}^n$:*

- *X is equivariant and has one orbit.*
- *X is defined by a maximal constraint.*

As a conclusion, a constrained timed register automaton is exactly the same thing as a timed register automaton, whose state space is orbit-finite.

So far we have shown that constrained timed register automata are included in FM automata. The inclusion is strict, as the transition function in a timed register automaton is defined by constraints, while in the abstract definition, the transition function is only required to be equivariant. Not all equivariant transition functions are definable by constraints, as shown in the following example.

Example 6. Suppose that $K \subseteq \mathbb{Z}$ is any set of integers, e.g. the prime numbers. Consider the language $\mathrm{diff}(K) = \{t_1 t_2 \in \mathbb{R}^2 : t_2 - t_1 \in K\}$. Regardless of K, this language can be recognized by a deterministic FM automaton. The state space of the automaton has four orbits: an initial state ϵ, an accepting state $\top$, a rejecting sink state $\bot$, and one state q_t for every real number t. The automaton starts in the initial state ϵ. The transition function is:

$$\delta(\epsilon, t) = q_t \qquad \delta(\bot, t) = \bot \qquad \delta(q_s, t) = \begin{cases} \top & \text{if } t - s \in K \\ \bot & \text{otherwise} \end{cases} \qquad \delta(\top, t) = \top$$

The transition function is easily seen to be equivariant. For most K, however, it is not defined by a constraint (one argument is that there are uncountably many choices for K, and only countably many choices for a constraint).

Example 6 implies that the abstract definition of a deterministic FM automaton is too powerful. For instance, arbitrary FM automata cannot be represented in a finite way. Restricting equivariant functions to those definable by constraints makes the automata manageable, but it is not necessarily the only solution to the problem. We do not investigate other solutions in this paper.

4   Characterization of Deterministic Timed Automata

In this section we provide a machine-independent characterization of the class of languages recognized by deterministic timed automata.

Every language recognized by a deterministic timed automaton with uninitialized clocks is equivariant and contains only monotonic words. Finally, the set of equivalence classes of Myhill-Nerode equivalence is orbit-finite. As shown in Example 6, these conditions are not sufficient even to characterize nondeterministic orbit-finite timed register automata. One additionally needs to say, roughly, that only recent timestamps can be remembered, and older timestamps must be forgotten. Our formulation of this condition is as follows.

For two nonempty words $u, w \in (A \times \mathbb{R})^+$ (think of monotonic words) and $M \in \mathbb{N}$ we write $u <_M w$ to mean that the first timestamp in w is larger than the last timestamp in u, by at least M.

Definition 2. *Let $M \in \mathbb{N}$. A language $L \subseteq (A \times \mathbb{R})^*$ is called M-forgetful if for every words $u, w \in (A \times \mathbb{R})^+$, $v \in (A \times \mathbb{R})^*$ and a timed permutation π such that $v \cdot \pi = v$, $u <_M w$ and $u \cdot \pi <_M w$, it holds:*

$$u v w \in L \iff (u \cdot \pi) v w \in L. \tag{1}$$

Observe that M-forgetfulness implies M'-forgetfulness for all $M' > M$. Note that $v \cdot \pi = v$ implies $(u v) \cdot \pi = (u \cdot \pi) v$ and that if L is equivariant then the property (1) may be equivalently written as $u v w \in L \iff u v (w \cdot \pi) \in L$.

Example 7. The language L from Example 2 in Section 2 is not M-forgetful for any $M \geq 0$. Indeed, instantiating Definition 2 with

$$u = 0.4 \qquad v = 1.2 \; 2.2 \; \ldots \; M{+}0.2 \; M{+}1.2 \qquad w = M{+}1.4$$

and any timed permutation π satisfying $\pi(0.4) = 0.3$ and $\pi(0.2) = 0.2$, we get a contradiction, as $0.4 \; v \; w \in L$ while $0.3 \; v \; w \notin L$.

Example 8. The language of all monotonic words is 0-forgetful. The language "for some timestamp t, both t and $t + 3$ appear in the word" is 3-forgetful but not 2-forgetful.

Theorem 5. *Let A be a finite set of labels. For a language $L \subseteq (A \times \mathbb{R})^*$, the following conditions are equivalent:*

- *L is recognized by a deterministic timed automaton with uninitialized clocks.*
- *L satisfies simultaneously the following conditions:*
  1. *L is equivariant;*
  2. *L contains only monotonic words;*
  3. *L is M-forgetful for some threshold $M > 0$; and*
  4. *the set of equivalence classes of the Myhill-Nerode equivalence $\sim_L$ is orbit-finite.*

Note that the set of equivalence classes of $\sim_L$ is an (equivariant) FM set when L is an (equivariant) FM set. Even in presence of condition 3, condition 4 is still necessary, as shown by the following example.

Example 9. Consider the language containing all monotonic timed words of the form $t_1 t_2 \ldots t_n (t_1{+}1) (t_2{+}1) \ldots (t_n{+}1)$, for $n \geq 0$. The language is 1-forgetful, but its syntactic automaton is orbit-infinite.

5   Future Work

Our approach based on Fraenkel-Mostowski sets may be further elaborated.

We consider a subclass of orbit-finite automata where the transition function (or relation) is definable by constraints. These restrictions are sufficient to capture timed automata, but there may be other manageable restrictions that are more liberal. As a natural continuation of this work we plan to pursue automata with semi-linear transition functions. We suppose that one would be able to capture in this framework, among the others, periodic time constraints, cf. [7], or some subclasses of hybrid automata, like linear hybrid automata [12].

Another urgent challenge is to relate our approach to the previous work, in particular to minimization of [1,14,16] and to characterizations of [6] and [13].

Acknowledgments. We kindly thank anonymous reviewers for insightful comments and valuable suggestions.

References

1. Alur, R., Courcoubetis, C., Halbwachs, N., Dill, D.L., Wong-Toi, H.: Minimization of Timed Transition Systems. In: Cleaveland, W.R. (ed.) CONCUR 1992. LNCS, vol. 630, pp. 340–354. Springer, Heidelberg (1992)
2. Alur, R., Dill, D.L.: A theory of timed automata. Theor. Comput. Sci. 126(2), 183–235 (1994)
3. Bojańczyk, M., Klin, B., Lasota, S.: Automata with group actions. In: Proc. LICS 2011, pp. 355–364 (2011)
4. Bojańczyk, M., Klin, B., Lasota, S.: Automata theory in nominal sets (submitted, 2012), http://www.mimuw.edu.pl/~sl/PAPERS/lics11full.pdf
5. Bouyer, P., Dufourd, C., Fleury, E., Petit, A.: Updatable timed automata. Theor. Comput. Sci. 321(2-3), 291–345 (2004)
6. Bouyer, P., Petit, A., Thérien, D.: An algebraic approach to data languages and timed languages. Inf. Comput. 182(2), 137–162 (2003)
7. Choffrut, C., Goldwurm, M.: Timed automata with periodic clock constraints. Journal of Automata, Languages and Combinatorics 5(4), 371–404 (2000)
8. Finkel, O.: Undecidable problems about timed automata. CoRR, abs/0712.1363 (2007)
9. Francez, N., Kaminski, M.: Finite-memory automata. TCS 134(2), 329–363 (1994)
10. Gabbay, M.: Foundations of nominal techniques: logic and semantics of variables in abstract syntax. Bulletin of Symbolic Logic 17(2), 161–229 (2011)
11. Gabbay, M., Pitts, A.M.: A new approach to abstract syntax with variable binding. Formal Asp. Comput. 13(3-5), 341–363 (2002)
12. Henzinger, T.A.: The theory of hybrid automata. In: LICS, pp. 278–292 (1996)
13. Maler, O., Pnueli, A.: On Recognizable Timed Languages. In: Walukiewicz, I. (ed.) FOSSACS 2004. LNCS, vol. 2987, pp. 348–362. Springer, Heidelberg (2004)
14. Springintveld, J., Vaandrager, F.W.: Minimizable Timed Automata. In: Jonsson, B., Parrow, J. (eds.) FTRTFT 1996. LNCS, vol. 1135, pp. 130–147. Springer, Heidelberg (1996)
15. Tripakis, S.: Folk theorems on the determinization and minimization of timed automata. Inf. Process. Lett. 99(6), 222–226 (2006)
16. Yannakakis, M., Lee, D.: An efficient algorithm for minimizing real-time transition systems. Formal Methods in System Design 11(2), 113–136 (1997)

Regular Languages of Infinite Trees That Are Boolean Combinations of Open Sets

Mikołaj Bojańczyk and Thomas Place*

University of Warsaw

Abstract. In this paper, we study boolean (not necessarily positive) combinations of open sets. In other words, we study positive boolean combinations of safety and reachability conditions. We give an algorithm, which inputs a regular language of infinite trees, and decides if the language is a boolean combination of open sets.

1 Introduction

In this paper, we work with infinite binary trees labeled by a finite alphabet. The set of trees can be interpreted as a compact metric space. The distance between two different trees is 2^{-n}, where n is the smallest depth where the two trees are different. In the topology induced by this distance, a set of trees L is open if for every tree $t \in L$, there is a finite prefix of t such that changing nodes outside the prefix does not affect membership in L. In other words, an open set is a reachability language. We are interested in understanding finite boolean combinations, not necessarily positive, of open sets. The main result of this paper is:

Theorem 1.1. *The following problem is EXPTIME complete. Given a nondeterministic parity automaton on infinite trees, decide if the recognized language is a boolean combination of open sets.*

In other words, this paper provides an effective characterization of boolean (not necessarily positive) combinations of open sets, within the class of regular languages of infinite trees.

A similar version of the problem, where one asks if L is simply an open set, and not a finite boolean combination of open sets, is significantly simpler. Here is the solution to this simpler problem which is folklore to the best of our knowledge. The key observation is that the topological closure of a tree language L is the set

$$closure(L) = \{t : \text{every finite prefix of } t \text{ can be extended to some tree in } L\}.$$

A language L is open if and only if its complement L^c satisfies $L^c = closure(L^c)$. Automata for both L^c and $closure(L^c)$ can be computed based on the automaton for L, and then one can test two regular languages for equality.

---

* Both authors supported by ERC Starting Grant "Sosna". A full version of this paper can be found at www.mimuw.edu.pl/~bojan.

A. Czumaj et al. (Eds.): ICALP 2012, Part II, LNCS 7392, pp. 104–115, 2012.
© Springer-Verlag Berlin Heidelberg 2012

The difficulty in Theorem 1.1 is dealing with the boolean combinations.

Our approach to the problem uses forest algebra for infinite trees [4]. We intended to achieve two complementary goals: use the algebra to understand boolean combinations of open sets; and use boolean combinations of open sets to understand the algebra.

Goal 1: Understand Boolean Combinations of Open Sets. We believe that giving an effective characterization for a class $\mathscr{L}$ of regular languages can be the most mathematically rewarding thing that one can do with $\mathscr{L}$. The ostensible goal of an effective characterization – the algorithm deciding membership in $\mathscr{L}$ – is usually less interesting than the insight into the structure of $\mathscr{L}$ that is needed to get the algorithm. A famous example is the theorem of Schützenberger and McNaughton/Papert, which makes a beautiful connection between logic and algebra: a word language is definable in first-order logic if and only if its syntactic monoid is aperiodic [8,6].

We believe that our study of boolean combinations of open sets achieves this goal. We discover that this class of languages has a rich structure, which is much more complex in the case of infinite trees than in the case of infinite words. On our way to Theorem 1.1, we provide three conditions which are equivalent to being a boolean combination of open sets, see Theorem 5.3. Two of these conditions are stated in terms of games, and one is stated in terms of algebraic equations. We believe that each of these conditions are interesting in their own right.

Goal 2: Understand Algebra for Infinite Trees. The algebraic theory of languages of finite words is well studied, using monoids and semigroups, see the book by Straubing [9]. The algebraic theory of languages of infinite words is also well understood, see the book by Perrin and Pin [7]. There has been quite a lot of recent work on algebra for finite trees [2], but the theory is still not mature. Finally, the algebraic theory of infinite trees is very far from being understood, despite some work [4,1].

We believe that our study of boolean combinations of open sets has highlighted the kind of tools that might be important in the algebraic theory of infinite trees. An important theme is the use of games. As mentioned previously, we characterize boolean combinations of opens sets in terms of games, and a set of two identities. Even in the identities, there is a hidden game, which is played in the algebra. We see this as evidence that the algebraic theory of infinite trees will need to take games into account.

Organization of the Paper. In Section 2 we give our first characterization of boolean combination of open sets using games. Note that this characterization is not specific to trees and works for all topological spaces. Unfortunately, it is not effective. In Section 3, we provide basic definitions for trees and algebra. In Section 4, we make a sharper analysis of the non-effective characterization of Section 2 in the setting of trees. Finally, in Section 5 we state our effective characterization using algebraic identities.

2   A Game Characterization

We begin by studying boolean combinations of open sets in arbitrary topological spaces. Fix a topological space. In this paper, we are interested in the topological space of infinite trees, but the discussion in this section works for all spaces.

Let $X_1, \ldots, X_n$ be arbitrary subsets of the topological space. We define a game

$$\mathscr{H}(X_1, \ldots, X_n)$$

which is played by two players, called Alternator and Constrainer. The game is played in n rounds. At the beginning of each round $i \in \{1, \ldots, n\}$, there is an open set U_i. Initially, U_1 is the whole space. Round i of the game is played as follows.

- Alternator chooses a point $x_i \in U_i \cap X_i$. If there is no such point x_i, the game is interrupted and Constrainer wins immediately. Otherwise,
- Constrainer chooses an open set $U_{i+1} \ni x_i$, and the next round is played.

If Alternator manages to survive n rounds, then he wins.

A base for a topology is a set of 'base open sets' such that open sets are obtained as infinite unions of these sets. The following lemma shows that the rules of the game could be changed such that Constrainer can only pick base open sets.

Lemma 2.1. *Choose some base for the topology. If Constrainer has a winning strategy, then he has a winning strategy which uses base open sets for $U_1, \ldots, U_n$.*

Suppose that X is a set and $n \in \mathbb{N}$. We write $\mathscr{H}_{\in\notin}(X, n)$ for the game where Alternator needs to alternate n times between X and its complement, that is:

$$\mathscr{H}(X_1, \ldots, X_n) \qquad \text{where } X_i = \begin{cases} X & \text{when } i \text{ is odd} \\ \text{the complement of } X & \text{when } i \text{ is even} \end{cases}.$$

Example 2.2. Consider the space of real numbers, and let X be the rational numbers. Then for every n, Alternator wins the game $\mathscr{H}_{\in\notin}(X, n)$.

Example 2.3. In the real numbers, let X be the complement of $\{1/n : n \in \mathbb{N}\}$. Alternator wins $\mathscr{H}_{\in\notin}(X, 3)$. In the first round, Alternator plays $0 \in X$. In the second round, Alternator plays $1/n \notin X$ for some large n depending on Constrainer's move. In the third round, Alternator plays $1/n + \epsilon \in X$, for some small ϵ depending on Constrainer's move. Constrainer wins $\mathscr{H}_{\in\notin}(X, n)$ for $n \geq 4$.

Proposition 2.4. *The following conditions are equivalent for a set X:*

- *X is a finite boolean combination of open sets*
- *Constrainer wins the game $\mathscr{H}_{\in\notin}(X, n)$ for all but finitely many n.*

Refinement lemma. We now state a lemma, which shows that if the topological space is a metric space (this is the case of trees) Alternator's winning sets can be refined in an arbitrary finite way.

Lemma 2.5. *Assume the topological space is a metric space and let $X_1, \ldots, X_n$ be sets. For $i \in \{1, \ldots, n\}$, let $\mathcal{Y}_i$ a finite family of sets partitioning X_i. If Alternator wins*

$$\mathcal{H}(X_1, \ldots, X_n)$$

then there exist $Y_1 \in \mathcal{Y}_1, \ldots, Y_n \in \mathcal{Y}_n$ such that Alternator wins

$$\mathcal{H}(Y_1, \ldots, Y_n).$$

3   Preliminaries on Trees

Trees. We use possibly infinite trees where every node has zero or two children. For a finite alphabet A, we denote by H_A the set of infinite binary trees labeled over A. Notions of node, leaf, child, root, descendant are defined as usual. We write '$<$' the descendant relation (the smallest node being the root of the tree). If t is a tree and x a node of t, we write $t(x)$ the label of x in t, and $t|_x$ for the *subtree of t at x*.

Multicontexts. A *multicontext* is a tree with some distinguished unlabeled leaves called its *ports*. The number of ports is called the *arity*, there might be infinitely many ports. Given a multicontext C and a valuation η which maps ports to trees, we write $C[\eta]$ for the tree obtained by replacing each port x by the tree $\eta(x)$. A tree $C[\eta]$ is said to *extend* the multicontext C, conversely the multicontext C is said to be a *prefix* of the tree $C[\eta]$. The set of all trees extending a multicontext C is denoted by $C[*]$. The following picture shows three multicontexts, with arities $0, 1$ and 2, with the ports depicted by squares.

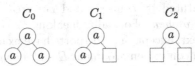

The multicontext C_0 is a tree, and $C_0[*]$ is $\{C_0\}$. The multicontext C_1 is a prefix of every tree where the root label is a, and the left child of the root is a leaf with label a. Finally, C_2 is a prefix of every tree with root label a. We are mostly interested in finite prefixes, which are multicontexts where every path ends in a leaf, which is either a port or a normal leaf.

Contexts. A *context* is a multicontext with exactly one port. We write V_A the set of contexts over A. We write C, D for contexts. Given two contexts C, D, we write $C \cdot D$ for the context obtained by replacing the port of C with D. One can verify that $\cdot$ is associative, therefore, $(V_A, \cdot)$ is a monoid (with the empty context, denoted by $\square$, as neutral element). V_A also acts on H_A, with $C \cdot t$ defined as the tree obtained by replacing the port of C with t. Finally, we write C^∞ for the infinite tree $C \cdot C \cdot C \cdots$.

3.1   Tree Languages and Algebra

We are mainly concerned with regular languages of infinite trees. This is the class of languages of infinite trees that is recognized by nondeterministic parity automata; or equivalently recognized by alternating parity automata; or equivalently can be defined in monadic second-order logic. See [5].

Recall that our goal is to decide if a given regular language L of infinite trees is a boolean combination of open sets. It will be important for us to work with a canonical representation of L. As our canonical representation, we use equivalence classes of trees and contexts with respect to a natural Myhill-Nerode style equivalence, see below.

The equivalence classes form a kind of algebra, which is similar to the forest algebra for infinite trees from [4]. The similarity is that both algebras represent infinite trees. The difference is that the algebra in [4] represents finitely branching unranked trees, while the algebra in this paper represents binary trees. We do not use unranked trees because for unranked finitely branching trees, the topological space is not compact. This is because there is no converging subsequence in a sequence of trees where the n-th tree has n children of the root.

Myhill-Nerode Equivalence. Fix L a language of trees over an alphabet A. We define two Myhill-Nerode equivalence relations: one for trees and one for contexts.

Let C be a multicontext, possibly with infinitely many ports. For a tree t, we write $C[t]$ for the tree obtained by putting t in all ports of C. In the Myhill-Nerode equivalence for trees, we say trees t and t' are L-equivalent if

$$C[t] \in L \Leftrightarrow C[t'] \in L \qquad \text{for every multicontext } C.$$

To give a similar definition for contexts, we use a variant of multicontexts where the ports can be substituted by contexts and not trees. Such a multicontext is called a *context environment*. Formally speaking, a context environment is defined like a multicontext, except that the ports have exactly one child (instead of being leaves). Given a context environment E and a context C, we write $C[E]$ for the tree obtained by substituting C for every port of E, as in the following picture:

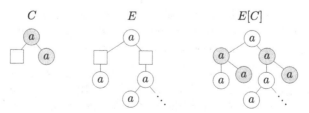

We define two contexts C and C' to be L-equivalent if

$$E[C] \in L \Leftrightarrow E[C'] \in L \qquad \text{for every context environment } E.$$

The Algebra. We write H_L for the equivalence classes of trees with respect to L, and V_L for the equivalence classes of contexts with respect to L. We write:

$$\alpha : (H_A, V_A) \to (H_L, V_L)$$

for the two-sorted function which maps trees and contexts to their L-equivalence classes; this function is called the *syntactic morphism of L*. We use the name *tree type* for elements of H_L and *context type* for elements of V_L. It is not difficult to show that both H_L and V_L are finite when L is regular. The syntactic morphism can be computed based on a nondeterministic tree automaton recognizing L, in exponential time [4]. Finiteness of H_L and V_L is necessary but not sufficient for regularity, for instance both H_L and V_L are finite for any language defined in the logic MSO+U [3].

Lemma 3.1. *The following operations respect L-equivalence.*

1. *For every multicontext D, the operation: $t \mapsto D[t]$.*
2. *For every context environment E, the operation: $C \mapsto E[C]$.*
3. *The composition of contexts $(C_1, C_2) \mapsto C_1 \cdot C_2$.*
4. *Substituting a tree in the port of a context: $(C, t) \mapsto C \cdot t$.*
5. *Infinite iteration of a context: $C \mapsto C^\infty$.*
6. *For every letter a, the operations $t \mapsto a(t, \Box)$ and $t \mapsto a(\Box, t)$.*

It follows that the above operations can be applied to elements of the syntactic algebra.

Idempotents. Given any finite monoid V, there is (folklore) a number $\omega(V)$ (denoted by ω when V is understood from the context) such that for each element v of V, v^ω is an idempotent: $v^\omega = v^\omega v^\omega$.

4   Boolean Combinations of Open Sets of Trees

As mentioned in the introduction, we use prefix topology on trees, which yields a topology identical to that of the Cantor space. In this topology, a *base open set* is defined to be any set $C[*]$, where C is a *finite* multicontext. Open sets are defined to be arbitrary unions of base open sets. This topology is the same as the topology generated by a distance, which says that trees are at distance 2^{-n} where n is the smallest depth where the two trees differ. This paper is about finite boolean combination of open sets. Typical boolean combinations of open sets include

- Trees over alphabet $\{a, b, c\}$ which contain at least one a and no b's.
- Trees over alphabet $\{a, b\}$ which contain two or five a's.

Languages, which are not boolean combinations of open sets include

- Trees over alphabet $\{a, b\}$ with finitely many a's.
- Trees over alphabet $\{a, b\}$ with a finite and even number of a's.

Let us revisit the game from Proposition 2.4 in the case of trees. In this special case, points are trees. By Lemma 2.1, we may assume that Constrainer uses base open sets, which are finite multicontexts. The game begins with the whole space, which corresponds to the empty multicontext. In each round, Alternator chooses a tree that extends the current multicontext, and then Constrainer chooses a finite multicontext that is a prefix of the tree chosen by Alternator.

Example 4.1. Consider an alphabet $\{a, b\}$ and the language $L=$"infinitely many a's". It is not difficult to see that Alternator can win the game $\mathcal{H}_{\in\notin}(L, n)$ for every $n \in \mathbb{N}$. This is because every finite multicontext can be extended to a tree with finitely many a's, or to a tree with infinitely many a's. By Proposition 2.4, L is not a boolean combination of open sets.

Proposition 2.4 helps us understand finite boolean combinations of open sets, but it is not an effective characterization. To be effective, we should be able to decide if Alternator wins $\mathcal{H}_{\in\notin}(L, n)$ for every n. The following simple lemma shows how to decide the winner for a given n.

Lemma 4.2. *Given regular tree languages $L_1, \ldots, L_n$, one can decide who wins $\mathcal{H}(L_1, \ldots, L_n)$. In particular, given L and n, one can decide who wins $\mathcal{H}_{\in\notin}(L, n)$.*

Proof. The statement "Alternator wins the game $\mathcal{H}(L_1, \ldots, L_n)$" can be formalized in monadic second-order logic on the complete binary tree, by a formula which can be computed based on the languages $L_1, \ldots, L_n$. Therefore, the winner can be decided using Rabin's theorem.   □

The above lemma gives a semi-algorithm for deciding if a regular language is a finite boolean combination of open sets. For $n = 1, 2, \ldots$, use Lemma 4.2 to compute the winner of $\mathcal{H}_{\in\notin}(L, n)$. If Constrainer wins the game for some n, then he also wins the game for $n + 1, n + 2, \ldots$ and therefore the algorithm can terminate and declare that the L is a finite boolean combination of open sets. If the language is *not* a finite boolean combination of open sets, then the algorithm does not terminate.

Observe that even when the algorithm *does* terminate, it does multiple calls to Rabin's theorem, which has non-elementary complexity.

The main contribution of this paper is a finer analysis of the problem, which yields an algorithm (not a semi-algorithm) deciding if a tree language is a finite boolean combination of open sets.

4.1   The Infinite Game

Proposition 2.4 can be rephrased as: a language L is *not* a finite boolean combination of open sets if and only if player Alternator can win $\mathcal{H}_{\in\notin}(L, n)$ for arbitrarily large n. One could imagine a variant of the game, more difficult for Alternator, where infinitely many rounds have to be played. Call the infinite variant $\mathcal{H}_{\in\notin}(L, \infty)$. In Example 2.2, which is about rational numbers, Alternator can win the infinite game.

It is clear that if Alternator wins $\mathscr{H}_{\in\notin}(X, \infty)$, then he also wins $\mathscr{H}_{\in\notin}(X, n)$ for every n. We show a counterexample for the converse implication, which is a regular tree language. This counterexample language necessarily uses trees, because the converse implication holds for regular languages of infinite words.

The counterexample language L is the set of trees over $\{a, b\}$ of the form:

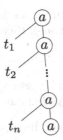

such that n is some natural number, and for each $i \in \{1, \ldots, n\}$, the tree t_i is either finite, or contains no b nodes. Observe that in a tree from L, the rightmost branch is necessarily finite.

Fact 1. *Alternator loses $\mathscr{H}_{\in\notin}(L, \infty)$, but wins $\mathscr{H}_{\in\notin}(L, n)$ for every n.*

5   The Effective Characterization

In this section, we present the main result of the paper.

The Context Game. So far we have worked with a game $\mathscr{H}(L_1, \ldots, L_n)$, for tree languages $L_1, \ldots, L_n$. We define a similar game for languages of contexts. Recall that contexts are defined as a special case of trees, with an additional port label that appears in exactly one leaf. From the distance on trees, we get a distance on contexts. This yields the definition of a game for a sequence $K_1, \ldots, K_n$ of context languages. To avoid confusion between trees and contexts, we denote the context game by $\mathscr{V}(K_1, \ldots, K_n)$.

Games on Types. Consider a language L and its syntactic morphism

$$\alpha_L : (H_A, L_A) \to (H_L, V_L).$$

Recall that by definition of the syntactic morphism, a type $h \in H_L$ is actually equal to the set of trees $\alpha_L^{-1}(h)$. Therefore, it makes sense to talk about the game $\mathscr{H}(h_1, \ldots, h_n)$ for a sequence of tree types. Likewise for context types. Define

$$\mathscr{H}_L \stackrel{\text{def}}{=} \{(h_1, \ldots, h_n) \in (H_L)^n : n \in \mathbb{N} \text{ and Alternator wins } \mathscr{H}(h_1, \ldots, h_n)\}$$

$$\mathscr{V}_L \stackrel{\text{def}}{=} \{(v_1, \ldots, v_n) \in (V_L)^n : n \in \mathbb{N} \text{ and Alternator wins } \mathscr{V}(v_1, \ldots, v_n)\}$$

A comment on notation is in order here. The sets $\mathscr{H}_L$ and $\mathscr{V}_L$ contain words, over alphabets H_L and V_L, respectively. Usually when dealing with words, one omits the brackets and commas, and writes abc instead of (a, b, c). When the alphabet is V_L, this leads to ambiguity, since the expression vwu can be interpreted as: 1) a word with a single letter obtained by multiplying v, w, u in the context monoid V_L; or 2) a three-letter word over the alphabet V_L. These two interpretations should not be confused, so we write $(v_1, \ldots, v_n)$ for n-letter words over the alphabet V_L. For the sake of uniformity, we also write $(h_1, \ldots, h_n)$ for n-letter words in over the alphabet H_L, although there is no risk of ambiguity here.

Fact 2. *Both $\mathscr{H}_L$ and $\mathscr{V}_L$ are regular languages of finite words.*

Proof. Both languages are closed under removing letters. Every language closed under removing letters is regular, by Higman's lemma.   □

The above fact is amusing, but useless, because it does not say how to compute automa for $\mathscr{H}_L$ and $\mathscr{V}_L$ as a function of the language L.[1] If we are not interested in efficiency, membership in $\mathscr{H}_L$ can be decided with Lemma 4.2. The same kind of algorithm works for $\mathscr{V}_L$. Later on, we give a more efficient algorithm.

$(h_1, \ldots, h_n) \in \mathscr{H}_L$	implies	$(C[h_1], \ldots, C[h_n]) \in \mathscr{H}_L$
$(v_1, \ldots, v_n) \in \mathscr{V}_L$	implies	$(E[v_1], \ldots, E[v_n]) \in \mathscr{H}_L$
$(v_1, \ldots, v_n), (w_1, \ldots, w_n) \in \mathscr{V}_L$	implies	$(v_1 w_1, \ldots, v_n w_n) \in \mathscr{V}_L$
$(v_1, \ldots, v_n) \in \mathscr{V}_L, (h_1, \ldots, h_n) \in \mathscr{H}_L$	implies	$(v_1 h_1, \ldots, v_n h_n) \in \mathscr{H}_L$
$(v_1, \ldots, v_n) \in \mathscr{V}_L$	implies	$(v_1^\infty, \ldots, v_n^\infty) \in \mathscr{H}_L$
$(h_1, \ldots, h_n) \in \mathscr{H}_L$	implies	$(a[\square, h_1], \ldots, a[\square, h_n]) \in \mathscr{V}_L$
$(h_1, \ldots, h_n) \in \mathscr{H}_L$	implies	$(a[h_1, \square], \ldots, a[h_n, \square]) \in \mathscr{V}_L$

Table 1. Closure properties of $\mathscr{H}_L$ and $\mathscr{V}_L$. C is a multicontext, E is a context environment, and a is a letter.

Lemma 5.1. *The sets $\mathscr{H}_L$ and $\mathscr{V}_L$ satisfy the closure properties in Table 1.*

Notice the similarity of Table 1 with the operations in Lemma 3.1. Another way of stating Lemma 5.1 is that for every $n \in \mathbb{N}$, $(\mathscr{H}_L, \mathscr{V}_L)$ restricted to sequences of length n is a subalgebra of the the the n-fold power of the syntactic algebra (H_L, V_L).

We define the *alternation* of a word to be its length, after iteratively eliminating letters that are identical to their predecessors. The alternation of $abaabbb$ is 4. We say that a set of words has *unbounded alternation* if it contains words with arbitrarily large alternation.

Proposition 5.2. *For a regular language L of infinite trees, the following conditions are equivalent.*

- *Alternator wins the game $\mathscr{H}_{\in\notin}(L, n)$ for infinitely many n.*
- *The set $\mathscr{H}_L$ has unbounded alternation.*

---

[1] To the best of our knowledge it is possible, although unlikely, that computing $\mathscr{H}_L$ and $\mathscr{V}_L$ is undecidable.

Proof. We begin with the top-down implication. We show that if Alternator wins the game $\mathcal{H}_{\in\notin}(L, n)$, then $\mathcal{H}_L$ contains a word of length n where every two consecutive letters are different. Suppose then that Alternator wins $\mathcal{H}_{\in\notin}(L, n)$, which means that he wins $\mathcal{H}(L_1, \ldots, L_n)$ where L_i is L for odd-numbered rounds and its complement for even-numbered rounds. Both L and its complement can be partitioned into tree types. By Lemma 2.5, Alternator wins $\mathcal{H}(h_1, \ldots, h_n)$ for some sequence of types, such that h_i is included in L or its complement, depending on the parity of i. In particular, the consecutive types are different.

We now do the bottom-up implication. Suppose that $\mathcal{H}_L$ has unbounded alternation. Since $\mathcal{H}_L$ is closed under removing letters, there must be some $g, h \in H_L$ such that $\mathcal{H}_L$ contains all the words

$$(g, h), (g, h, g, h), (g, h, g, h, g, h), \ldots. \tag{1}$$

Since g and h are different elements of the syntactic algebra, it follows that there must be some multicontext C such that the tree type $C[g]$ is contained in L, while the tree type $C[h]$ is disjoint with L. By applying Lemma 5.1 to (1), we conclude $\mathcal{H}_L$ contains all the words

$$(C[g], C[h]), (C[g], C[h], C[g], C[h]), (C[g], C[h], C[g], C[h], C[g], C[h]), \ldots.$$

It follows that Alternator can alternate arbitrarily long between the language L and its complement.   □

The Main Theorem. So far, we have proved Propositions 2.4 and 5.2, which characterize finite boolean combinations of open sets in terms of games. Neither of these game characterizations is effective. We now add a final characterization, which uses identities and is effective.

Theorem 5.3 (Main Theorem). *For a regular language L of infinite trees, the following conditions are equivalent.*

1. *L is a finite boolean combination of open sets.*
2. *Constrainer wins the game $\mathcal{H}_{\in\notin}(L, n)$ for all but finitely many n.*
3. *The set $\mathcal{H}_L$ has bounded alternation.*
4. *The following identities are satisfied.*

$$u^\omega w w^\omega = u^\omega v w^\omega = u^\omega u w^\omega \qquad \text{if } (u, v, w) \in \mathcal{V}_L \text{ or } (w, v, u) \in \mathcal{V}_L \tag{2}$$

$$(u_2 w_2^\omega v)^\omega u_1 w_1^\infty = (u_2 w_2^\omega v)^\infty \qquad \text{if } (u_1, u_2) \in \mathcal{V}_L \text{ and } (w_1, w_2) \in \mathcal{V}_L \tag{3}$$

Theorem 5.3 implies Theorem 1.1 from the introduction, which says that one can decide if the language recognized by a nondeterministic parity automaton on infinite trees, is a boolean combination of open sets. Indeed: there are finitely many sequences of length two and three in $\mathcal{V}_L$. These sequences can be computed using Lemma 4.2. It is then sufficient to test if the identities from item 4 are valid by checking all combinations. A more detailed proof, together with the EXPTIME completeness, can be found in the full version of the paper.

In Propositions 2.4 and 5.2, we have shown that conditions 1, 2 and 3 in Theorem 5.3 are equivalent. It remains to show that conditions 3 and 4 are equivalent. The proof of the implication from 4 to 3 is the technical core of this paper, and is in the full version. Below we prove the much simpler converse implication, which at least can serve to illustrate the identities.

Implication from 3 to 4. We prove the contrapositive: if one of the identities (2) or (3) is violated, then $\mathscr{H}_L$ has unbounded alternation.

Suppose first that (2) is violated. We will show that $\mathscr{V}_L$ has unbounded alternation. This is enough, by the following lemma.

Lemma 5.4. *If $\mathscr{V}_L$ has unbounded alternation, then so does $\mathscr{H}_L$.*

The assumption that (2) is violated says that there are $u, v, w \in V_L$ such that

$$(u, v, w) \in \mathscr{V}_L \qquad \text{or} \qquad (w, v, u) \in \mathscr{V}_L,$$

but the three context types $u^\omega w w^\omega$, $u^\omega v w^\omega$ and $u^\omega u w^\omega$ are not all equal. If the three context types are not equal, then second one must be different from either the first one or the third one. We only do the proof for the case when $(u, v, w) \in \mathscr{V}$ and when $u^\omega w w^\omega \neq u^\omega v w^\omega$. For nonzero $n \in \mathbb{N}$ and $i \in \{1, \ldots, n\}$, define

$$\boldsymbol{w}_{(i,n)} = (\; \overbrace{u, \ldots, u}^{2n - 2i + 1 \text{ times}} \;, v, \; \overbrace{w, \ldots, w}^{2(i-1) \text{ times}} \;) \quad \in (H_L)^{2n}.$$

This word is obtained from (u, v, w) by duplicating some letters, and therefore it belongs to $\mathscr{V}_L$. For some n, consider the words

$$\boldsymbol{w}_{(1,n)}, \ldots, \boldsymbol{w}_{(n,n)} \in \mathscr{V}_L.$$

These are n words of length $2n$. Let us multiply all these words coordinate-wise, yielding a word $\boldsymbol{w}$, also of length $2n$, which is depicted in the following picture:

Choose some k, and take $n = k \cdot \omega + 1$, and $i \in \{\omega, 2\omega, \ldots, (k-1) \cdot \omega\}$. Consider letters $2i + 1$ and $2i + 2$ in the word $\boldsymbol{w}$, which are

$$u^i w^{n-i} = u^\omega w w^\omega \qquad u^i v w^{n-i-1} = u^\omega v w^\omega.$$

By assumption, these letters are different, and therefore the word w has alternation at least k. Because k was chosen arbitrarily, it follows that $\mathscr{V}_L$ has unbounded alternation.

Consider now the case when (3) is violated. This means $\mathscr{V}_L$ contains pairs (u_1, u_2) and (w_1, w_2) such that for some $v \in V_L$,

$$e^\infty \neq e u_1 w_1^\infty \qquad \text{for } e \overset{\text{def}}{=} (u_2 w_2^\omega v)^\omega.$$

Lemma 5.5. *Let* u_1, u_2, w_1, w_2 *and* e *be as above. If* $(h_1, \ldots, h_n) \in \mathscr{H}_L$, *then*

$$(e^\infty, e h_1, \ldots, e h_n) \in \mathscr{H}_L \tag{4}$$

$$(e u_1 w_1^\infty, e h_1, \ldots, e h_n) \in \mathscr{H}_L \tag{5}$$

Using the lemma, one shows by induction that for every n, the sequence which alternates n times between e^∞ and $e^\infty u_1 w_1^\infty$ belongs to $\mathscr{H}_L$. This completes the proof of the implication from 3 to 4, and also of Theorem 5.3.

6   Conclusion

We have proved an effective characterization of boolean combination of open sets. We hope that the characterization sheds some light on the nature of such languages. Also, we hope that the technical tools we developed, involving both algebra and games, will be useful in future work on regular languages of infinite trees. One class of particular interest is Weak-MSO (i.e. MSO where set quantification is restricted to finite sets). This class can be characterized by wreath products of boolean combinations of open sets.

References

1. Blumensath, A.: An algebraic proof of Rabin's theorem (unpublished manuscript)
2. Bojańczyk, M.: Algebra for trees. In: Handbook of Automata Theory. European Mathematical Society Publishing House (to appear)
3. Bojańczyk, M.: A Bounding Quantifier. In: Marcinkowski, J., Tarlecki, A. (eds.) CSL 2004. LNCS, vol. 3210, pp. 41–55. Springer, Heidelberg (2004)
4. Bojańczyk, M., Idziaszek, T.: Algebra for Infinite Forests with an Application to the Temporal Logic EF. In: Bravetti, M., Zavattaro, G. (eds.) CONCUR 2009. LNCS, vol. 5710, pp. 131–145. Springer, Heidelberg (2009)
5. Grädel, E., Thomas, W., Wilke, T. (eds.): Automata, Logics, and Infinite Games. LNCS, vol. 2500. Springer, Heidelberg (2002)
6. McNaughton, R., Papert, S.: Counter-Free Automata. MIT Press (1971)
7. Perrin, D., Pin, J.-É.: Infinite Words. Elsevier (2004)
8. Schützenberger, M.P.: On finite monoids having only trivial subgroups. Information and Control 8 (1965)
9. Straubing, H.: Finite Automata, Formal Languages, and Circuit Complexity. Birkhäuser (1994)

Toward Model Theory with Data Values

Mikołaj Bojańczyk and Thomas Place*

University of Warsaw

Abstract. We define a variant of first-order logic that deals with data words, data trees, data graphs etc. The definition of the logic is based on Fraenkel-Mostowski sets (FM sets, also known as nominal sets). The key idea is that we allow infinite disjunction (and conjunction), as long as the set of disjuncts (conjunct) is finite modulo renaming of data values. We study model theory for this logic; in particular we prove that the infinite disjunction can be eliminated from formulas.

1 Introduction

This paper uses Fraenkel-Mostowski sets (FM sets, also known as nominal sets) to study logics that describe properties of objects such as data words, data trees, or data graphs.

Suppose that $\mathbb{D}$ is an infinite set of *data values*, also called *atoms* or *ur-elements*, whose elements can only be compared for equality. A *data word* is a word (trees and graphs can also be considered, of course) whose positions are labelled by an alphabet that is not necessarily finite, but which can refer to data values in a finite way, such as in the following examples of alphabets.

$$\mathbb{D} \qquad \{0,1\} \times \mathbb{D} \qquad (\mathbb{D}^2 \cup \mathbb{D}) \qquad \{0,1\} \times \{\{c,d\} : c,d \in \mathbb{D}\}.$$

The statement "data values can only be compared for equality" is formalized by saying that properties of data words should be invariant under the action of the group of bijections of $\mathbb{D}$. The statement "refer to data values in a finite way" is formalized by saying that the alphabet contains finitely many elements, modulo bijections of data values. For instance, modulo bijections, the set $\mathbb{D}^2 \cup \mathbb{D}$ has three elements, which look like this: $(d,e), (d,d)$ and d.

Properties that are invariant under the action of the group of bijections include:

1. Data words over the alphabet $\mathbb{D}$ where all positions have different labels.
2. Data words over the alphabet $\mathbb{D}$ with at least six distinct letters.
3. Graphs with edges labelled by $\mathbb{D}^2$ where for each vertex, all outgoing edges have the same data value on the first coordinate.

---

* Both authors supported by ERC Starting Grant "Sosna". A full version of this paper can be found at www.mimuw.edu.pl/~bojan.

A. Czumaj et al. (Eds.): ICALP 2012, Part II, LNCS 7392, pp. 116–127, 2012.
© Springer-Verlag Berlin Heidelberg 2012

There is a more relaxed notion of invariance: a property is called *finitely supported* if there is a finite set of data values $C \subseteq \mathbb{D}$, such that the property is invariant under the action of permutations that preserve C. The set C is called the *support*. For instance, if we choose some two elements $c, d \in \mathbb{D}$, then

4. Data words over alphabet $\mathbb{D}$ which begin with c and end with d.

is a finitely supported property, namely supported by $C = \{c, d\}$.

To give examples of properties that are not finitely supported, one needs additional assumptions on $\mathbb{D}$. For instance, if we assume that $\mathbb{D}$ is the natural numbers, then "words in $\mathbb{D}^*$ which contain only even numbers" is not a finitely supported property of data words.

The notion of finitely supported sets is the cornerstone of "permutation models" of set theory, which were studied by logicians such as Fraenkel and Mostowski starting in the 1920's. Permutation models were rediscovered, under the name "nominal sets", by Gabbay and Pitts in [3], see also [4], as an elegant approach to deal with binding and fresh names in the syntax of programming languages and logical formulas. When dealing with syntax, one thinks of data values as being variable names. Finally, these sets were rediscovered by the automata community, as an approach to describing languages of data words [2].[1]

Logic on data words and data trees. The direct predecessor of this paper is [2], which uses FM sets (under the name nominal sets) to talk about automata on data words. In the present paper, we use FM sets to talk about logics on data words (and more general structures). We define:

- A notion of *FM relational structure*. This notion generalizes data words, data trees, data graphs, etc. One can apply a permutation of data values to an FM relational structure, and get another FM relational structure.
- A notion of *FM first-order logic*. The formulas are evaluated in FM relational structures. The formulas form an FM set. The previously stated examples of properties of data words and data graphs are definable in the logic, including example 4.

Logics for data words have been extensively studied in the special case of data words and data trees with alphabets of the form $A \times \mathbb{D}$, where A is a finite set. In this special case, the approach of [6] is to use: a binary predicate $x \sim y$ which says that two positions carry the same data value; as well as a unary predicate $a(x)$ for each $a \in A$. The satisfiability problem for the logic is undecidable for most variants, see [6]. In the special case of alphabets $A \times \mathbb{D}$, our abstract definition of FM first-order logic coincides with the existing definition.

---

[1] There are two names for the sets that can be used: "FM sets" as in mathematical logic, or "nominal sets" as in the study of name binding. In this paper, we decided to use the name "FM sets". The main reason is that our application of Frankel-Mostowski set theory is not principally concerned with the use of names and their binding. An additional reason is that, like in Fraenkel-Mostwoski set theory, but unlike in the study of name binding, we are often interested in data values with additional structure, such as a linear order.

Even for words, the choice of logic is not obvious for some alphabets. Consider the alphabet "sets of data values of size at most 3". A natural predicate would be $x \subseteq y$, saying that the set in the label of x is a subset of the set in the label of y. Another kind of predicate, not definable in terms of $x \subseteq y$, could be

$$|x_1 \cup \cdots \cup x_n| = k \qquad \text{for } n, k \in \mathbb{N}.$$

Which predicates should be allowed in the logic? Our definition implies that they are all allowed. We do not address the question of a minimal choice of predicates, i.e. which predicates can be defined in terms of others.

Parse trees. On a definitional level, the principal idea in this paper is to allow parse trees of formulas where the branching degree is not finite, but finite modulo bijections of data values (we call this orbit-finite branching). In normal sets, the parse tree of an expression (a formula of first-order logic, a regular expression, an arithmetic expression, etc.) is a finite tree. In FM sets, one can have a more relaxed parse tree: for each node, the set of child subtrees is only required to be finite modulo bijections of data values[2]. For instance, if for each data value $d \in \mathbb{D}$ we have a formula φ_d, and the function $d \mapsto \varphi_d$ is finitely-supported, then it makes sense to consider the infinite disjunction $\bigvee_{d \in \mathbb{D}} \varphi_d$. On a technical level, the main contribution of this paper is Theorem 5.2 which says that the infinite disjunction can be eliminated from formulas.

Related work. A logic for nominal sets, called nominal logic, was studied by Pitts in [5]. Nominal logic and the logic from this paper have different goals: nominal logic is designed to axiomatise nominal sets, while the formulas in this paper are used to define languages of data words and similar objects. Also, the logics are defined differently: the formulas and models for nominal logic are defined in normal set theory; while the formulas and models in this paper are defined inside FM set theory[3]. Finally, the principal technical result of this paper is elimination of infinite disjunction, this result cannot be even stated in the language of [5].

2   Preliminaries

Data symmetry. The notion of FM sets is parametrized by a set of *data values* $\mathbb{D}$, and a group G of bijections on $\mathbb{D}$. The group G need not contain all bijections of $\mathbb{D}$. The idea is that $\mathbb{D}$ has some structure, and G contains the structure-preserving bijections. The pair $(\mathbb{D}, G)$ is called a *data symmetry*. In this paper, we use the following data symmetries:

- The set $\mathbb{D}$ is empty, and G has only the identity element. We call this the *classical symmetry*. FM sets in the classical symmetry are normal sets.

---

[2] This appears already explicitly in [1], where terms of λ-calculus have orbit-finitely branching parse-trees. Implicitly, the idea goes back the work of Gabbay and Pitts, where the whole point of nominal sets was to model the use of binding.

[3] One could say that our logic is an internal logic for FM sets, while the logic of [5] is external.

- The set $\mathbb{D}$ is a countable set, say the natural numbers. The group G consists of all bijections on $\mathbb{D}$. We call this the *equality symmetry*. FM sets in the equality symmetry are the same thing as nominal sets [3,4].
- The set $\mathbb{D}$ is the vertices of the undirected countable homogeneous graph (also called the Rado graph), and the group G is the group of automorphisms of this graph. We call this the *graph symmetry*.

FM set. Consider first the *cumulative hierarchy* of sets with data values, which is a hierarchy of sets indexed by ordinal numbers and defined as follows. The empty set is the unique set of rank 0. A set of rank α is any set whose elements are sets of rank smaller than α, or data values. A permutation π of data values can be applied to a set X in the hierarchy, by renaming the elements of X, and the elements of elements of X, and so on. The resulting set, which has the same rank, is denoted by $X \cdot \pi$.

A set C of data values is said to be a *support* of a set X in the cumulative hierarchy if $X \cdot \pi = X \cdot \sigma$ holds for every permutations π, σ in the group from the data symmetry which agree on elements of C. A set is called *finitely supported* if it has some finite support. We use the name *FM set* for a set in the cumulative hierarchy which is hereditarily finitely supported, which means that it is finitely supported, the sets in it are finitely supported, and so on[4].

The support of an FM set is not unique, e.g. supports are closed under adding data values. A set with empty support is called *equivariant*.

In many respects, FM sets behave like normal sets. For instance, if X, Y are FM sets, then $X \times Y$, $X \cup Y$, X^* and the finite powerset of X are all FM sets. Another example is the family of subsets of X that have finite supports. The appropriate notion of a function between FM sets X and Y is that of a *finitely supported function*, which is a function from X to Y whose graph is an FM set.

Observe that FM sets in the classical symmetry are simply sets (equipped with the only possible action). Therefore the classical symmetry corresponds to classical set theory, without data values.

Orbit-finite FM sets. Suppose that X is an FM set. For a set of data values C, define the C-orbit of an element $x \in X$ to be the set $\{x \cdot \pi : \pi \in G_C\}$. If C supports X, then the C-orbits form a partition of X. The set X is called orbit-finite if it the partition into C-orbits has finitely parts, for some C which supports X. For some data symmetries, including the classical, equality and graph symmetries discussed in this paper, the notion of orbit-finiteness does not depend on the choice of support [1]. In other words, for these data symmetries, if two sets C and D support an FM set X, then X has finitely many C-orbits if and only if it has finitely many D-orbits.

In this paper, we are mostly interested in FM sets that are orbit-finite.

---

[4] The definition here is based on Definition 10.6 in [4], except that we use the name *FM set* for what [4] calls elements of $\mathcal{HFS}$.

3   Relational Structures

The discussion in this section – and the next Section 4 – makes sense in any data symmetry. Fix some data symmetry $(\mathbb{D}, G)$ for this section and the next.

One of the key ideas of finite model theory in computer science is that a combinatorial object, such as a word, tree, or graph, can be treated as a model for a logical formula. For instance, in the case of words over an alphabet $\{a, b\}$, a word with n positions can be interpreted as a relational structure where the domain is the set of positions $\{1, \ldots, n\}$, there are two unary predicates $a(x)$ and $b(x)$ for labels, and there is a binary predicate $x \leq y$ for the order on positions. Using this interpretation, one can define properties of words using first-order logic, e.g. the set of words that end with b is defined by the formula

$$\forall x \exists y \quad x \leq y \wedge b(y).$$

The goal of this paper is to define a similar notion of logic for combinatorial objects that contain data values. In particular, our definition should cover data words and data trees.

In standard sets, not FM sets, a relational structure can be seen as a hyper-edge colored directed hypergraph. For instance, the relational structure corresponding to the word aab is the following hypergraph.

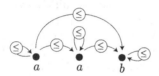

We adapt this definition to FM sets as follows. For instance a binary predicate will not just say yes/no to each directed edge, but it can also color the edge, e.g. by a data value.

An *FM predicate* R consists of an orbit-finite FM set colors(R) and a natural number arity(R). An *FM signature* Σ is a finite set of FM predicates. An *FM relational structure* $\mathfrak{A}$ over Σ consists of:

- A set dom($\mathfrak{A}$), called the *domain* of the structure, which is an FM set.
- For every predicate R in the signature, a finitely supported partial function

$$R^{\mathfrak{A}} : \mathrm{dom}(\mathfrak{A})^{\mathrm{arity}(R)} \rightharpoonup \mathrm{colors}(R),$$

  called the *interpretation of R*.

For a fixed FM signature Σ, the set of FM relational structures over Σ is itself a FM set, because an FM relational structure is nothing other than a domain (which has empty support, since it is equipped with the trivial action) and a finite tuple of finitely-supported partial functions[5]. Therefore, if $\mathfrak{A}$ is a FM relational

---

[5] Formally speaking, this is an FM class, because all the domains do not form a set.

structure and $\pi \in G$, then also $\mathfrak{A} \cdot \pi$ is a FM relational structure. Both $\mathfrak{A}$ and $\mathfrak{A} \cdot \pi$ have the same domains, only different interpretations. If R is a predicate of arity n, and $x_1, \ldots, x_n$ are in the domain of $\mathfrak{A}$, then

$$R^{\mathfrak{A} \cdot \pi}(x_1, \ldots, x_n) = (R^{\mathfrak{A}}(x_1, \ldots, x_n)) \cdot \pi.$$

In particular, either both sides of the equality above are defined, or neither are (recall that interpretations are partial functions.)

An FM relational structure is called finite if its domain is finite. In such a case, an interpretation is a finite tuple of colors. A tuple of objects taken from an FM set necessarily has finite support, and therefore in the case of finite relational structures, the requirement on finitely supported interpretations is redundant.

Example 3.1. Data words can be modeled as FM relational structures. We use the name *FM alphabet* for any orbit-finite set A. To an FM alphabet A, we associate an FM signature Σ_A with two predicates:

- The alphabet predicate R_A, which has arity 1 and colors A.
- The order predicate $R_<$, which has arity 2 and only one color $\{<\}$.

For a data word $w = a_1 \cdots a_n \in A^*$, we define a corresponding FM relational structure $\mathfrak{A}_w$ over the signature Σ_A as follows. The domain $dom(\mathfrak{A}_w)$ is the set $\{1, \ldots, n\}$ of positions. The interpretation of the alphabet predicate maps each position to its label, and the interpretation of the order predicate maps a pair (i, j) to $<$ if and only if $i < j$.

Example 3.2. Edge-labelled directed data graphs can be modeled as FM relational structures. To an FM alphabet A, we associate an FM signature Σ_A with one predicate R_A with arity 2 and colors A, called the edge label predicate. A structure over this FM signature describes a directed graph, where edges are labelled by A. Because the interpretation is a partial function, every ordered pair of nodes is connected by zero or one edge.

4   Logic

In this section, we define a variant of first-order logic which is used to define properties of FM relational structures. Before giving the actual definition, we enumerate the postulates it should satisfy:

1. The set of formulas is itself an FM set, and the satisfaction relation is equivariant. That is

$$\mathfrak{A} \models \varphi \quad \text{iff} \quad \mathfrak{A} \cdot \pi \models \varphi \cdot \pi$$

   holds for every $\pi \in G$, FM relational structure $\mathfrak{A}$ and formula φ.
2. Orbit-finite disjunction is allowed. That is, if Γ is an orbit-finite FM set of formulas, then also $\bigvee \Gamma$ is a formula.

Below, in Section 4.1, we give a definition which satisfies the above postulates.

4.1   Definition of FM First-Order Logic

To choose the predicates for our logic, we use a semantic approach: every isomorphism-closed property of a tuple of elements is going to be a predicate. An isomorphism between two FM relational structures $\mathfrak{A}$ and $\mathfrak{B}$ is a finitely supported bijection

$$f : \mathrm{dom}(\mathfrak{A}) \to \mathrm{dom}(\mathfrak{A})$$

such that for every k-ary predicate R in the signature, we have

$$R^{\mathfrak{A}}(a_1, \ldots, a_k) = R^{\mathfrak{B}}(f(a_1), \ldots, f(a_k)) \qquad \text{for every } a_1, \ldots, a_k \in \mathrm{dom}(\mathfrak{A}).$$

Example 4.1. Consider the equality symmetry, and a signature with one unary predicate P, whose colors are $\mathbb{D}$. Suppose that $1, 2$ are data values. Let $\mathfrak{A}$ be a structure whose domain is $\{1\}$, and where the interpretation is $P^{\mathfrak{A}}(1) = 1$. Let $\mathfrak{B}$ be a structure whose domain is $\{2\}$, and where the interpretation is $P^{\mathfrak{B}}(2) = 1$. Then $\mathfrak{A}$ and $\mathfrak{B}$ are isomorphic.

An *atomic type of arity* n is (the isomorphism type of) a structure $\mathfrak{A}$, together with an n-tuple of elements, such that every element of the domain of $\mathfrak{A}$ appears in the tuple. The domain of the atomic type has at most n elements, but might be smaller if the tuple contains repetitions. We write $\mathrm{atoms}_n(\Sigma)$ for the set of atomic types of arity n. If $\mathfrak{A}$ is a structure, and $\bar{a}$ is a (possibly repeating) tuple of elements in $\mathrm{dom}(\mathfrak{A})$, then we define $\mathfrak{A}|\bar{a}$ to be the atomic type obtained from $\mathfrak{A}$ by only keeping the elements from $\bar{a}$.

Fact 1. *In the classical, equality and graph symmetries, the set* $\mathrm{atoms}_n(\Sigma)$ *is orbit-finite.*

A *basic type of arity* n is defined to be any finitely supported subset of $\mathrm{atoms}_n(\Sigma)$.

Example 4.2. Consider the FM alphabet $\binom{\mathbb{D}}{2}$, which is defined to be the family of two-element subsets of $\mathbb{D}$. Consider data words over this alphabet, as in Example 3.1. A basic type $\mathcal{B}$ of arity 2 could say that that the set in the label of the first distinguished position has non-empty intersection with the set in the label in the second distinguished position. This basic type is not only finitely supported, but also equivariant.

Example 4.3. This example also concerns data words over $\binom{\mathbb{D}}{2}$. In this example, as in subsequent examples, we assume that the data values are natural numbers. A non-equivariant basic type $\mathcal{B}_9$ of arity 2 says that the sets in the label of the first and second distinguished position both contain the data value $9 \in \mathbb{D}$.

We define *FM first-order logic for a relational signature* Σ as follows. As predicates, we use basic types in the following sense: a basic type $\mathcal{B}$ of arity n is a predicate of arity n, with the semantics

$$\mathfrak{A}, \bar{a} \models \mathcal{B}(x_1, \ldots, x_n) \qquad \text{iff} \qquad \mathfrak{A}|\bar{a} \in \mathcal{B}.$$

Furthermore, formulas can use boolean combinations $\{\vee, \wedge, \neg\}$ as well as quantifiers $\{\forall, \exists\}$. We will add one more connective, but to define this connective we need to discuss the action of G on formulas. When applying a permutation $\pi \in G$ to a formula φ, the structure of the formula, the connectives $\vee, \wedge, \neg, \forall, \exists$ as well as the variables are not changed. The only thing that changes is the basic types: a set of atomic types $\mathcal{B}$ is mapped to the set $\mathcal{B} \cdot \pi$.

Example 4.4. Consider the basic type $\mathcal{B}_9$ from Example 4.3, and the formula

$$\exists x \exists y \; x \neq y \wedge \mathcal{B}_9(x, y).$$

This formula, call it φ_9, says that the data value 9 appears in the label of at least two positions. Consider a permutation $\pi \in G$, which maps 9 to 8. Then the formula $\varphi_9 \cdot \pi$ says that the data value 8 appears in the label of at least two positions.

It is not difficult to see that every formula has finite support. The reason is that every formula uses a finite number of basic types, and each basic type has finite support by definition.

We now define the remaining connective, which is called *orbit-finite disjunction*. Consider an orbit-finite FM set of already defined formulas Γ. We allow a disjuction over this set $\bigvee \Gamma$, with the expected semantics. Orbit-finite disjunction is the last connective of the logic, and the definition of FM first-order logic is now complete.

Example 4.5. Consider the formula φ_9 in Example 4.4. This formula can be defined for any data value d, not just 9, and it is easy to see that the set $\{\varphi_d : d \in \mathbb{D}\}$ is an orbit-finite FM set of formulas. Therefore, we can use the orbit-finite disjunction

$$\bigvee \{\varphi_d : d \in \mathbb{D}\} \qquad \text{also written as} \qquad \bigvee_{d \in \mathbb{D}} \varphi_d.$$

The disjunction above says that some data value appears in the label of at least two positions. Observe that the above formula can be expressed, without orbit-finite disjunction, by using the predicate $\mathcal{B}$ from Example 4.2:

$$\exists x \exists y \; x \neq y \wedge \mathcal{B}(x, y).$$

Example 4.6. In the previous case, the set $\{\varphi_d : d \in \mathbb{D}\}$ was equivariant. One can also use non-equivariant sets, such as

$$\bigvee_{d \in \mathbb{D} - \{9\}} \varphi_d.$$

Non-equivariant sets are useful for nesting formulas, e.g.

$$\bigwedge_{e \in \mathbb{D}} \bigvee_{d \in \mathbb{D} - \{d\}} \varphi_d.$$

The formula above says that there are two data values that appear in the label of at least two positions.

5   Elimination of Orbit-Finite Disjunction

Recall that in Example 4.5, we were able to eliminate orbit-finite disjunction. The technique was to push the disjunction into the basic types. This technique can fail, e.g. in the graph symmetry, as shown by the following theorem.

Theorem 5.1. *Consider the graph symmetry. Let $L \subseteq \mathbb{D}^*$ be the set of words $d_1 \cdots d_n$ such that n is even, all letters are distinct, and for every $i, j \in \{1, \ldots, n\}$ there is no graph edge from d_i to d_j. This set is definable in FM first-order logic with orbit-finite disjunction, but not by a formula without orbit-finite disjunction.*

The main technical result of this paper is the following theorem, which says that orbit-finite disjunction can be eliminated in the equality symmetry.

Theorem 5.2. *Consider the equality symmetry. Every formula of FM first-order logic is equivalent to a formula that does not use orbit-finite disjunction.*

The proof can be found in the full version of the paper. It uses a notion of functionality, which we believe to be of independent interest, and which is discussed in Section 6. When the colors used by the predicates are just the set $\mathbb{D}$ of data values, Theorem 5.2 is straightforward. The main difficulty is dealing with non-standard sets of colors. We illustrate this with the following examples which show Theorem 5.2 in action for increasingly complicated sets of colors.

Example 5.3. Consider data words over the alphabet $A = \{a, b\} \times \mathbb{D}$. If a position carries the letter $(\sigma, d) \in A \times \mathbb{D}$, then we say that that it has label σ and data value d. Consider the language: "some data value appears only on positions with label a". This language is expressed by the formula

$$\bigvee_d \exists x \ d(x) \wedge \forall x \ d(x) \Rightarrow a(x),$$

where $d(x)$ is the basic type which holds for positions where the data value is d. The orbit-finite disjunction in this formula can be eliminated by encoding the data value d by a position y:

$$\exists y \forall x \ x \sim y \Rightarrow a(x),$$

where $x \sim y$ says that x and y carry the same data value, also a basic type. The same trick, of encoding a data value by a position, works for every formula over this alphabet.

Example 5.4. For $k \in \mathbb{N}$, consider the alphabet

$$B_k = \{a, b\} \times P_{\leq k}(\mathbb{D}),$$

which is like in the previous example, except that the second coordinate is now not a single data value, but an (unordered) set of at most k data values. Let us first study the case of $k = 2$. Consider the language "some data value appears (in the set) only on positions with label a".

$$\bigvee_d \exists x \ (d \in x) \land \forall x \ (d \in x) \Rightarrow a(x),$$

where $d \in x$ is a unary basic type, which selects positions that contain d. Let us use the name *witness* for a data value d which satisfies the formula $\forall x \ (d \in x) \Rightarrow a(x)$. The trick from Example 5.3 was to encode a witness by a position. This trick does not always work for the alphabet B. Consider for instance the word

$$(a, \{1\})(b, \{2,3\})(a, \{1,3\})(a, \{2,4\})(a, \{5,6\})(a, \{5,6\})$$

The witnesses are the data values $1, 4, 5, 6$. The witness 1 can be defined in terms of the first position: it is the unique data value in the set $\{1\}$, which appears in the first position. The witness 4 can be defined in terms of two positions: it is the unique data value which appears in the set $\{2,4\}$ on the fourth position but not in the set $\{2,3\}$ on the second position. Finally, witnesses 5 and 6 can only be defined as a set of size two; they cannot be distinguished. One can see that these three types of witnesses are the only possible ones for $k = 2$. All of these three types can be captured by the following formula, which does not use orbit-finite disjunction:

$$\exists y_1 \exists y_2 \forall x (\emptyset \subsetneq y_1 \cap y_2 \subseteq x) \Rightarrow a(x),$$

where $\emptyset \subsetneq y_1 \cap y_2 \subseteq x$ is a basic type, which says that the intersection of the sets of data values in y_1 and y_2 is non-empty and included in the set in x.

For $k > 2$, one needs more complicated expressions to define some data values, such as: "the data value that appears in positions five, six and seven, but not eight and two". Also, one can have sets of up to k data values that cannot be distinguished from each other.

5.1 Standard Data Words

Consider the special case where the models are data words as in Example 3.1 and the alphabet is of the form $A_{\text{fin}} \times \mathbb{D}$, where A_{fin} is a finite set. As mentioned in the introduction, there is an established logic for words over this kind of alphabet, which has a predicate $x < y$ for the position order, a predicate $x \sim y$ for equal data values, and a label predicate $a(x)$ for every $a \in A_{\text{fin}}$ (note that this logic does not allow orbit-finite disjunctions). A simple consequence of Theorem 5.2 is the following theorem:

Theorem 5.5. *Let A_{fin} be a finite set. Let L be an equivariant language over the alphabet $A_{\text{fin}} \times \mathbb{D}$. The following conditions are equivalent:*

1. *L is definable by a formula of FM first-order logic, possibly including orbit-finite disjunction.*
2. *L is definable by a formula of the standard first-order logic for data words, which has predicates for the position order $x < y$, equal data value $x \sim y$, and the labels $\{a(x)\}_{a \in A_{\text{fin}}}$.*

6 Functionality and Locality

In this section we define a key concept for the proof of Theorem 5.2. Our proof technique is to encode data values in elements of the domain of the relational structure (which corresponds to positions in the case of data words). As illustrated in Example 5.4:

1. Sometimes, more than one element is needed to define a data value;
2. Sometimes, a data value can only be defined in combination with some indistinguishable other data value;
3. Sometimes, both problems above hold simultaneously.

This section is devoted to a study of how one can define a data value, or more generally an element of some orbit finite set, in terms of a relational structure.

Functionality. In normal sets, without data values and group actions, the expression "f is a function of g" makes sense only when both f and g are functions with a common domain. For instance, one can say "the area of a circle is a function of its radius", which is formalized as two functions on the domain of circles, the area and radius functions. Another example: "a person's taste in football is a function of their sympathy for Real Madrid".

With data values, the notion of functionality makes sense for arbitrary objects. Suppose that x is an FM set or a data value, likewise y. Let C be a finite set of data values. We say that y *is a C-supported function of x* if there is a C-supported function

$$f : X \to Y \qquad \text{such that } x \in X \text{ and } y \in Y$$

which maps x to y. In the spacial case of $C = \emptyset$, we say that y is an equivariant function of x. (In the classical symmetry, which corresponds to normal sets, one can always take $X = \{x\}$, $Y = \{y\}$. In this case, every y is an equivariant function of every x, which is why the definition is not interesting.)

Example 6.1. The data value 2 is an equivariant function of the three-letter data word 123. In this case X is the set of data words of length three, Y is $\mathbb{D}$, and f maps a word to its second letter (there are other choices for X, Y and f). The data value 2 is an equivariant function of $\{2\} \in P(\mathbb{D})$. The data value 2 is not an equivariant function of the set $\{1, 2, 3\} \in P(\mathbb{D})$, or of the empty set $\emptyset \in P(\mathbb{D})$. The data value 2 is a $\{2\}$-supported function, but not a $\{3\}$-supported function, of the data value 1.

We can now state the main theorem of this section which concerns the first issue in the list at the beginning of this section: more than one element might be needed to define a data value.

Theorem 6.2 (Local Functionality Theorem). *Let X be an orbit-finite FM set, and Σ an FM relational signature. Let C be a finite set of data values that supports X and Σ. There is some $k \in \mathbb{N}$ such that for every $x \in X$ and every nominal relational structure $\mathfrak{A}$, the following conditions are equivalent*

– x is a C-supported function of $\mathfrak{A}$;
– x is a C-supported function of $\mathfrak{A}|\bar{a}$, for some tuple $\bar{a} \in (\mathrm{dom}(\mathfrak{A}))^k$.

The point of Theorem 6.2 is that the bound k depends on X, Σ and C, but not on $\mathfrak{A}$. When proving Theorem 5.2, we use this result in the following form: if a parameter $i \in I$ of an orbit-finite disjunction $\bigvee_{i \in I} \varphi_i$ is an equivariant function of a model, then it is an equivariant function of a small tuple $\bar{a}$, and the tuple can be captured using k existential quantifiers.

7   Conclusions

We have defined a notion of first-order logic for models that talk about data values. The main technical result is that orbit-finite disjunction can be eliminated in the equality symmetry. Possibilities of future work include:

– Using orbit-finite disjunction, one gets a natural notion of star-free languages of data words. Is this notion equivalent to FM first-order logic?
– Elimination of orbit-finite disjunction works in the equality symmetry, but not in the graph symmetry. In which symmetries does it work? We conjecture that it also works in the total order symmetry and the forest order symmetry, see [1]. We intend to investigate this issue further.
– Can one use the syntax of FM first-order logic to define new fragments of first-order logic on data words that have decidable satisfiability?

Acknowledgement. We would like to thank Nathanaël Fijalkow, Bartek Klin, and the anonymous referees for their comments and suggestions.

References

1. Bojanczyk, M., Braud, L., Klin, B., Lasota, S.: Towards nominal computation. In: POPL, pp. 401–412 (2012)
2. Bojanczyk, M., Klin, B., Lasota, S.: Automata with group actions. In: LICS, pp. 355–364 (2011)
3. Gabbay, M.J., Pitts, A.M.: A new approach to abstract syntax with variable binding. Formal Asp. Comput. 13(3-5), 341–363 (2002)
4. Gabbay, M.J.: Foundations of nominal techniques: logic and semantics of variables in abstract syntax. Bulletin of Symbolic Logic 17(2), 161–229 (2011)
5. Pitts, A.M.: Nominal logic, a first order theory of names and binding. Inf. Comput. 186(2), 165–193 (2003)
6. Segoufin, L.: Automata and Logics for Words and Trees over an Infinite Alphabet. In: Ésik, Z. (ed.) CSL 2006. LNCS, vol. 4207, pp. 41–57. Springer, Heidelberg (2006)

Robust Reachability in Timed Automata: A Game-Based Approach*

Patricia Bouyer, Nicolas Markey, and Ocan Sankur

LSV, CNRS & ENS Cachan, France
{bouyer,markey,sankur}@lsv.ens-cachan.fr

Abstract. Reachability checking is one of the most basic problems in verification. By solving this problem, one synthesizes a strategy that dictates the actions to be performed for ensuring that the target location is reached. In this work, we are interested in synthesizing "robust" strategies for ensuring reachability of a location in a timed automaton; with "robust", we mean that it must still ensure reachability even when the delays are perturbed by the environment. We model this perturbed semantics as a game between the controller and its environment, and solve the parameterized robust reachability problem: we show that the existence of an upper bound on the perturbations under which there is a strategy reaching a target location is EXPTIME-complete.

1 Introduction

Timed automata [2] are a timed extension of finite-state automata. They come with an automata-theoretic framework to design, model, verify and synthesize systems with timing constraints. One of the most basic problems in timed automata is the reachability problem: given a timed automaton and a target location, is there a path that leads to that location? This can be rephrased in the context of control as follows: is there a *strategy* that dictates how to choose time delays and edges to be taken so that a target location is reached? This problem has been solved long ago [2], and efficient algorithms have then been developed and implemented [13,18].

However, the abstract model of timed automata is an idealization of real timed systems. For instance, we assume in timed automata that strategies can choose the delays with arbitrary precision. In particular, the delays can be arbitrarily close to zero (the system is arbitrarily fast), and clock constraints can enforce exact delays (time can be measured exactly). Although these assumptions are natural in abstract models, they need to be justified after the design phase. Indeed the situation is different in real-world systems: digital systems have response times that may not be negligible, and control software cannot ensure timing constraints exactly, but only up to some error, caused by clock imprecisions, measurement errors, and communication delays. A good control

---

* This work has been partly supported by project ImpRo (ANR-10-BLAN-0317).

A. Czumaj et al. (Eds.): ICALP 2012, Part II, LNCS 7392, pp. 128–140, 2012.
© Springer-Verlag Berlin Heidelberg 2012

software must be *robust, i.e.*, it must ensure good behavior in spite of small imprecisions [11,12].

In this work, we are interested in the synthesis of robust strategies in timed automata for reachability objectives, taking into account response times and imprecisions. We propose to model the problem as a game between a controller (that will guide the system) and its environment. In our semantics, which is parameterized by some $0 < \delta_P \leq \delta_R$, the controller chooses to delay an amount $d \geq \delta_R$, and the system delays d', where d' is chosen by the environment satisfying $|d - d'| \leq \delta_P$. We say that a given location is *robustly reachable* if there exist parameters $0 < \delta_P \leq \delta_R$ such that the controller has a winning strategy ensuring that the location is reached against any strategy of the environment. If δ_P and δ_R are fixed, this can be solved using techniques from control theory [3]. However δ_P, δ_R are better seen as parameters here, representing imprecisions in the implementation of the system (they may depend on the digital platform on which the system is implemented), and whose values may not be available in the design phase. To simplify the presentation, but w.l.o.g., we assume in this paper that $\delta = \delta_P = \delta_R$; our algorithm can easily be adapted to the general case (by adapting the shrink operator in Section 3).

Note that this semantics was studied in [6] for timed games with *fixed* parameters, where the parameterized version was presented as a challenging open problem. We solve this problem for reachability objectives in timed automata: we show that deciding the existence of $\delta > 0$, and of a strategy for the controller so as to ensure reachability of a given location (whatever the imprecision, up to δ), is EXPTIME-complete. Moreover, if there is a strategy, we can compute a *uniform* one, which is parameterized by δ, using *shrunk difference bound matrices* (shrunk DBMs) that we introduced recently [17]. In this case, our algorithm provides a bound $\delta_0 > 0$ such that the strategy is correct for all $\delta \in [0, \delta_0]$. Our strategies also give quantitative information on how perturbations accumulate or can compensate. Technically, our work extends shrunk DBMs by *constraints*, and establishes non-trivial algebraic properties of this data structure (Section 3). The main result is then obtained by transforming the infinite-state game into a finite abstraction, which we prove can be used to symbolically compute a winning strategy, if any (see Section 4).

By lack of space, technical proofs have been omitted; they can be found in [5].

2   Robust Reachability in Timed Automata

2.1   Timed Automata and Robust Reachability

Given a finite set of clocks $\mathcal{C}$, we call *valuations* the elements of $\mathbb{R}_{\geq 0}^{\mathcal{C}}$. For a subset $R \subseteq \mathcal{C}$ and a valuation v, $v[R \leftarrow 0]$ is the valuation defined by $v[R \leftarrow 0](x) = v(x)$ for $x \in \mathcal{C} \setminus R$ and $v[R \leftarrow 0](x) = 0$ for $x \in R$. Given $d \in \mathbb{R}_{\geq 0}$ and a valuation v, the valuation $v + d$ is defined by $(v + d)(x) = v(x) + d$ for all $x \in \mathcal{C}$. We extend these operations to sets of valuations in the obvious way. We write $\mathbf{0}$ for the valuation that assigns 0 to every clock.

An atomic clock constraint is a formula of the form $k \preceq x \preceq' l$ or $k \preceq x - y \preceq' l$ where $x, y \in \mathcal{C}$, $k, l \in \mathbb{Z} \cup \{-\infty, \infty\}$ and $\preceq, \preceq' \in \{<, \leq\}$. A *guard* is a conjunction of atomic clock constraints. A valuation v satisfies a guard g, denoted $v \models g$, if all constraints are satisfied when each $x \in \mathcal{C}$ is replaced with $v(x)$.

Definition 1 ([2]). *A* timed automaton $\mathcal{A}$ *is a tuple* $(\mathcal{L}, \mathcal{C}, \ell_0, E)$, *consisting of finite sets* $\mathcal{L}$ *of locations,* $\mathcal{C}$ *of clocks,* $E \subseteq \mathcal{L} \times \Phi_{\mathcal{C}} \times 2^{\mathcal{C}} \times \mathcal{L}$ *of edges, and where* $\ell_0 \in \mathcal{L}$ *is the initial location. An edge* $e = (\ell, g, R, \ell')$ *is also written as* $\ell \xrightarrow{g, R} \ell'$.

Standard semantics of timed automata is usually given as a timed transition system. To capture robustness, we define the semantics as a game where perturbations in delays are uncontrollable. Given a timed automaton $\mathcal{A} = (\mathcal{L}, \mathcal{C}, \ell_0, E)$ and $\delta > 0$, we define the *perturbation game* of $\mathcal{A}$ w.r.t. δ as a two-player turn-based timed game $\mathcal{G}_\delta(\mathcal{A})$ between players *Controller* and *Perturbator*. The state space of $\mathcal{G}_\delta(\mathcal{A})$ is partitioned into $V_C \cup V_P$ where $V_C = \mathcal{L} \times \mathbb{R}^{\mathcal{C}}_{\geq 0}$ is the set of states that belong to Controller and $V_P = \mathcal{L} \times \mathbb{R}^{\mathcal{C}}_{\geq 0} \times \mathbb{R}_{\geq 0} \times E$ is the set of states that belong to Perturbator. The initial state is $(\ell_0, \mathbf{0})$ and belongs to Controller. The transitions are defined as follows: from any state $(\ell, v) \in V_C$, there is a transition to $(\ell, v, d, e) \in V_P$ whenever $d \geq \delta$, $e = (\ell, g, R, \ell')$ is an edge such that $v + d \models g$. Then, from any such state $(\ell, v, d, e) \in V_P$, there is a transition to $(\ell', (v + d + \epsilon)[R \leftarrow 0]) \in V_C$, for any $\epsilon \in [-\delta, \delta]$.

We assume familiarity with basic notions in game theory, and quickly survey the main definitions. A run in $\mathcal{G}_\delta(\mathcal{A})$ is a finite or infinite sequence of consecutive states starting at $(\ell_0, \mathbf{0})$. It is said maximal if it is infinite or cannot be extended. A strategy for Controller is a function that assigns to every non-maximal run ending in some $(\ell, v) \in V_C$, a pair (d, e) where $d \geq \delta$ and e is an edge enabled at $v + d$ (i.e., there is a transition from (ℓ, v) to (ℓ, v, d, e)). A run ρ is compatible with a strategy f if for every prefix ρ' of ρ ending in V_C, the next transition along ρ after ρ' is given by f. Given a target location ℓ, a strategy f is winning for the reachability objective defined by ℓ whenever all maximal runs that are compatible with f visit ℓ.

Observe that we require at any state (ℓ, v), that Controller should choose a delay $d \geq \delta$ and an edge e that is enabled after the chosen delay d. The edge chosen by Controller is always taken but there is no guarantee that the guard will be satisfied exactly when the transition takes place. In fact, Perturbator can perturb the delay d chosen by Controller by any amount $\epsilon \in [-\delta, \delta]$, including those that do not satisfy the guard. Notice that $\mathcal{G}_0(\mathcal{A})$ corresponds to the standard (non-robust) semantics of $\mathcal{A}$. We are interested in the following problem.

Problem 1 (Parameterized Robust Reachability). Given a timed automaton $\mathcal{A}$ and a target location ℓ, decide whether there exists $\delta > 0$ such that Controller has a winning strategy in $\mathcal{G}_\delta(\mathcal{A})$ for the reachability objective ℓ.

Notice that we are interested in the parameterized problem: δ is not fixed in advance. For fixed parameter, the problem can be formulated as a usual timed game, see [6]. Our main result is the decidability of this parameterized problem. Moreover, if there is a solution, we compute a strategy represented by parameterized difference-bound matrices where δ is the parameter; the strategy is thus *uniform* with respect to δ. In fact, we provide a bound $\delta_0 > 0$ such that the strategy is winning for Controller for *any* $\delta \in [0, \delta_0]$. These strategies also provide a quantitative information on how much the perturbation accumulates (See Fig. 3). The main result of this paper is the following:

Theorem 2. *Parameterized robust reachability is* EXPTIME*-complete.*

Checking parameterized robust reachability is different from usual reachability checking mainly for two reasons. First, in order to reach a given location, Controller has to choose the delays along a run, so that these perturbations do not accumulate and block the run. In particular, it shouldn't play too close to the borders of the guards (see Fig. 3). Second, due to these uncontrollable perturbations, some regions that are not reachable in the absence of perturbation can become reachable (see Fig. 4). So, Controller must also be able to win from these new regions. The regions that become reachable in our semantics are those *neighboring* reachable regions. The characterization of these neighboring regions is one of the main difficulties in this paper (see Section 3.5).

2.2 Motivating Example: Robust Real-Time Scheduling

An application of timed automata is the synthesis of schedulers in various contexts [1]. We show that robust reachability can help providing a better schedulability analysis: we show that schedulers synthesized by standard reachability analysis may not be robust: even the slightest *decrease* in task execution times can result in a large *increase* in the total time. This is a phenomenon known as *timing anomalies*, first identified in [9].

Consider the scheduling problem described in Fig. 1, inspired by [16]. Assume that we look for a *greedy* (*i.e.*, work-conserving) scheduler, that will immediately start executing a task if a machine is free for execution on an available task. What execution time can guarantee a greedy scheduling policy on this instance?

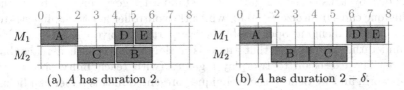

(a) A has duration 2.   (b) A has duration $2 - \delta$.

Fig. 1. Consider tasks A, B, C of duration 2 and D, E of duration 1. Dependences between tasks are as follows: $A \to B$ and $C \to D, E$, meaning *e.g.* that A must be completed before B can start. Task A must be executed on machine M_1 and tasks B, C on machine M_2. Moreover, task C cannot be scheduled before 2 time units (which could be modelled using an extra task). Fig. 1(a) shows the optimal greedy schedule for these tasks under these constraints, while Fig. 1(b) shows the outcome of any greedy scheduler when the duration of task A is less than 2.

One can model this problem as a timed automaton, and prove, by classical reachability analysis, that these tasks can be scheduled using a greedy policy within six time units. However the scheduler obtained this way may not be robust, as illustrated in Fig. 1(b). If the duration of task A unexpectedly drops by a small amount $\delta > 0$, then any greedy scheduler will schedule task B before task C, since the latter is not ready for execution at time $2 - \delta$. This yields a scheduling of tasks in $8 - \delta$ time units.

Our robust reachability algorithm is able to capture such phenomena, and can provide correct and robust schedulers. In fact, it would answer that the tasks are not schedulable in six time units (with a greedy policy), but only in eight time units.

2.3   Related Work: Robustness in Timed Automata and Games

There has been a recent effort to consider imprecisions inherent to real systems in the theory of timed systems. In particular there has been several attempts to define convenient notions of robustness for timed automata, see [14] for a survey.

The approach initiated in [15,8,7] is the closest to our framework/proposition. It consists in *enlarging* all clocks constraints of the automaton by some parameter δ, that is transforming each constraint of the form $x \in [a, b]$ into $x \in [a - \delta, b + \delta]$, and in synthesizing $\delta > 0$ such that all runs of the enlarged automaton satisfy a given property. This can be reformulated as follows: does there exists some $\delta > 0$ such that whatever Controller and Perturbator do in $\mathcal{G}_\delta(\mathcal{A})$, a given property is satisfied. This is therefore the universal counterpart of our formulation of the parameterized robustness problem. It has been shown that this universal parameterized robust model-checking is no more difficult (in terms of complexity) than standard model-checking. This has to be compared with our result, where complexity goes up from PSPACE to EXPTIME.

Another work that is close to ours is that of [6]. The authors consider general two-player (concurrent) games with a fixed lower bound on delays, where chosen delays can be changed by some fixed value δ. It is then shown that winning strategies can be synthesized: In fact, when δ is fixed, the semantics can simply be encoded by a usual timed game, and standard algorithms can be applied. Whether one can synthesize $\delta > 0$ for which the controller has a winning strategy was left as a challenging open problem. We partially solve this open problem here, under the assumption that there is a single player with a reachability objective. The extension to two-player games (with reachability objective) is ongoing work, and we believe the techniques presented in this paper can be used for that purpose.

Finally, [10] studies a topological and language-based approach to robustness, where (roughly) a timed word is accepted by the automaton if, and only if, one of its neighborhoods is accepted. This is not related to our formalization.

3 Shrinking DBMs

3.1 Regions, Zones and DBMs

We assume familiarity with the notions of regions and zones (see [4]). For two regions r and r', we write $r \lessdot r'$ if r' is the immediate (strict) time-successor of r. A *zone* is a set of clock valuations satisfying a guard.

We write $\mathcal{C}_0$ for the set $\mathcal{C} \cup \{0\}$. A *difference-bound matrix (DBM)* is a $|\mathcal{C}_0| \times |\mathcal{C}_0|$-matrix over $(\mathbb{R} \times \{<, \leq\}) \cup \{(\infty, <)\}$. A DBM M naturally represents a zone (which we abusively write M as well), defined as the set of valuations v such that, for all $x, y \in \mathcal{C}_0$, writing $(M_{x,y}, \prec_{x,y})$ for the (x, y)-entry of M, it holds $v(x) - v(y) \prec_{x,y} M_{x,y}$ (where $v(0) = 0$). For any DBM M, let $\mathsf{G}(M)$ denote the graph over nodes $\mathcal{C}_0$, where the weight of the edge $(x, y) \in \mathcal{C}_0^2$ is $(M_{x,y}, \prec_{x,y})$. The normalization of M corresponds to assigning to each edge (x, y) the weight of the shortest path in $\mathsf{G}(M)$. We say that M is *normalized* when it is stable under normalization.

3.2 Shrinking

Consider the automaton $\mathcal{A}$ of Fig. 2, where the goal is to reach ℓ_3. If there is no perturbation or lower bound on the delays between transitions (*i.e.*, $\delta = 0$), then the states from which Controller can reach location ℓ_3 can be computed backwards. One can reach ℓ_3 from location ℓ_2 and any state in the zone $X = (x \leq 2) \wedge (y \leq 1) \wedge (1 \leq x - y)$, shown by (the union of the light and dark) gray areas on Fig. 3 (left); this is the set of time-predecessors of the corresponding guard. The set of winning states from location ℓ_1 is the zone $Y = (x \leq 2)$, shown in Fig. 3 (right), which is simply the set of predecessors of X at ℓ_2. When $\delta > 0$ however, the set of winning states at ℓ_2 is a "shrinking" of X, shown by the dark gray area. If the value of the clock x is too close to 2 upon arrival in ℓ_2, Controller will fail to satisfy the guard $x = 2$ due to the lower bound δ on the delays. Thus, the winning states from ℓ_2 are described by $X \cap (x \leq 2 - \delta)$. Then, this shrinking is backward propagated to ℓ_1: the winning states are $Y \cap (x \leq 2 - 2\delta)$, where we "shrink" Y by 2δ in order to compensate for a possible perturbation.

An important observation here is that when $\delta > 0$ is small enough, so that both $X \cap (x \leq 2 - \delta)$ and $Y \cap (x \leq 2 - 2\delta)$ are non-empty, these sets precisely describe the winning states. Thus, we have a *uniform* description of the winning states for "all small enough $\delta > 0$". We now define *shrunk DBMs*, a data structure we introduced in [17], in order to manipulate "shrinkings" of zones.

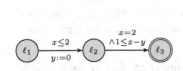

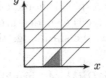

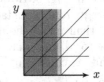

Fig. 2. Automaton $\mathcal{A}$ **Fig. 3.** Winning states in ℓ_2 (left) and in ℓ_1 (right)

3.3   Shrunk DBMs

For any interval $[a, b]$, we define the *shrinking operator* as $\mathsf{shrink}_{[a,b]}(Z) = \{v \mid v + [a, b] \subseteq Z\}$ for any zone Z. We only use operators $\mathsf{shrink}_{[0,\delta]}$ and $\mathsf{shrink}_{[-\delta,\delta]}$ in the sequel. For a zone Z represented as a DBM, $\mathsf{shrink}_{[0,\delta]}(Z)$ is the DBM $Z - \delta \cdot \mathbb{1}_{\mathcal{C} \times \{0\}}$ and $\mathsf{shrink}_{[-\delta,\delta]}(Z)$ is the DBM $Z - \delta \cdot \mathbb{1}_{\mathcal{C} \times \{0\} \cup \{0\} \times \mathcal{C}}$, for any $\delta > 0$.

Our aim is to handle these DBMs symbolically. For this, we define *shrinking matrices (SM)*, which are nonnegative integer square matrices with zeroes on their diagonals. A *shrunk DBM* is then a pair (M, P) where M is a DBM, P is a shrinking matrix [17]. The meaning of this pair is that we consider DBMs $M - \delta P$ where $\delta \in [0, \delta_0]$ for some $\delta_0 > 0$. In the sequel, we abusively use "for all small enough $\delta > 0$" meaning "there exists $\delta_0 > 0$ such that for all $\delta \in [0, \delta_0]$". We also adopt the following notation: when we write a statement involving a shrunk DBM (M, P), we mean that the statement holds for $(M - \delta P)$ for all small enough $\delta > 0$. For instance, $(M, P) = \mathsf{Pre}_{\mathsf{time}}((N, Q))$ means that $M - \delta P = \mathsf{Pre}_{\mathsf{time}}((N - \delta Q))$ for all small enough $\delta > 0$. In the same vein, shrunk DBMs can be re-shrunk, and we write $\mathsf{shrink}((M, P))$ (resp. $\mathsf{shrink}^+((M, P))$) for the shrunk DBM (N, Q) such that $N - \delta Q = \mathsf{shrink}_{[-\delta,\delta]}(M - \delta P)$ (resp. $N - \delta Q = \mathsf{shrink}_{[0,\delta]}(M - \delta P)$) for all small enough $\delta > 0$.

It was shown in [17] that when usual operations are applied on shrunk DBMs, one always obtain shrunk DBMs, whose shrinking matrices can be computed. We refer to [4,17] for the formal definitions of these operations.

Lemma 3 ([17]). *Let $M = f(N_1, \ldots, N_k)$ be an equation between normalized DBMs $M, N_1, \ldots, N_k$, using the operators $\mathsf{Pre}_{\mathsf{time}}$, $\mathsf{Unreset}_R$, $\cap$, shrink and shrink$^+$ and let $P_1, \ldots, P_k$ be SMs. Then, there exists a SM Q such that (M, Q) is normalized and $(M, Q) = f((N_1, P_1), \ldots, (N_k, P_k))$. Moreover, Q and the corresponding upper bound on δ can be computed in polynomial time.*

3.4   Shrinking constraints

Consider a transition of a timed automaton, as depicted on the figure at right. From region r_0, the game can reach regions r_1, r_2, r_3, depending on the move of Perturbator. Therefore, in order to win, Controller needs a winning strategy from all three regions. One can then inductively look for winning strategies from these regions; this will generally require shrinking, as exemplified in Fig. 3. However, not all shrinkings of these regions provide a winning strategy from r_0. In fact, r_1 (resp. r_3) should not shrink from the right (resp. left) side: their union should include the shaded area, thus points that are arbitrarily close to r_2. In order to define the shrinkings that are useful to us, we introduce shrinking constraints.

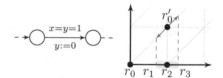

Fig. 4. Perturbing one transition

Definition 4. *Let M be a DBM. A shrinking constraint for M is a $|\mathcal{C}_0| \times |\mathcal{C}_0|$ matrix over $\{0, \infty\}$. A shrinking matrix P is said to* respect *a shrinking*

$$
\begin{array}{c}
\quad\ \ 0 \qquad\qquad x \qquad\qquad y \\
\begin{array}{c} 0 \\ x \\ y \end{array}
\left(
\begin{array}{ccc}
(0,\leq),0 & (0,<),\infty & (0,<),0 \\
(3,<),\infty & (0,\leq),0 & (2,<),0 \\
(3,<),\infty & (0,<),\infty & (0,\leq),0
\end{array}
\right)
\end{array}
$$

(a) A constrained DBM $\langle M, S \rangle$ and its representation   (b) A shrinking of $\langle M, S \rangle$

Fig. 5. Consider a zone defined by $0 < x < 3$, $0 < y < 3$, and $0 < x - y < 2$. Let the shrinking constraint S be defined by $S_{0,y} = 0$, $S_{x,y} = 0$, and $S_{z,z'} = \infty$ for other components. The resulting $\langle M, S \rangle$ is depicted on the left, as a matrix (where, for convenience, we merged both matrices into a single one) and as a constrained zone (where a thick segment is drawn for any boundary that is not "shrinkable", i.e., with $S_{z,z'} = 0$). On the right, the dark gray area represents a shrinking of M that satisfies S.

constraint S if $P \leq S$, where the comparison is component-wise. A pair $\langle M, S \rangle$ of a DBM and a shrinking constraint is called a constrained DBM.

Shrinking constraints specify which facets of a given zone one is (not) allowed to shrink (see Fig. 5). A shrinking constraint S for a DBM M is said to be *well* if for any SM $P \leq S$, (M, P) is non-empty. A *well constrained DBM* is a constrained DBM given with a well shrinking constraint. We say that a shrinking constraint S for a DBM M is *normalized* if it is the minimum among all equivalent shrinking constraints: for any shrinking constraint S' if for all SMs P, $P \leq S \Leftrightarrow P \leq S'$, then $S \leq S'$. One can show that any shrinking constraint can be made normalized, by a procedure similar to the normalization of DBMs. Lemma 5 shows that shrinking constraints can be propagated along operations on DBMs. This is illustrated in Fig. 6.

Lemma 5. *Let M, N, N' be normalized non-empty DBMs.*

1. *Assume that $M = \mathsf{Pre}_{time}(N)$, $M = N \cap N'$, or $M = \mathsf{Unreset}_R(N)$. Then, for any normalized well shrinking constraint S for M, there exists a well shrinking constraint S' for N such that for any SM Q, the following holds: $Q \leq S'$ iff the SM P s.t. $(M, P) = \mathsf{Pre}_{time}((N, Q))$ (respectively, $(M, P) = (N, Q) \cap N'$ or $(M, P) = \mathsf{Unreset}_R((N, Q)))$ satisfies $P \leq S$.*
2. *Assume that $M = N \cap N'$. For any well shrinking constraint S for N, there exists a shrinking constraint S' for M such that for any SM Q, the following holds: $Q \leq S'$ iff a SM $P \leq S$ s.t. $(N, P) \cap N' \subseteq (M, Q)$. Moreover, if $(N, P) \cap N' \neq \emptyset$ for all SMs $P \leq S$, then S' is well.*

Let us comment on Fig. 6(a), and how it can be used for our purpose. Assume there is an edge guarded by N (the whole gray area in the right) without resets. In the non-robust setting, this guard can be reached from any point of M (the whole gray area in the left). If we have a shrinking constraint S on M, and we want to synthesize a winning strategy from a shrinking of M satisfying S, then Lemma 5 gives the shrinking constraint S' for N, with the following property: given any shrinking (N, Q), we can find $P \leq S$ with $(M, P) = \mathsf{Pre}_{time}((N, Q))$

(hence, we can delay into (N, Q)), if, and only if Q satisfies $Q \leq S'$. The problem is now "reduced" to finding a winning strategy from $\langle N, S' \rangle$. However, forward-propagating these shrinking constraints is not always that easy. We also need to deal with resets, with the fact that Controller has to choose a delay greater than $\delta > 0$, and also with the case where there are several edges leaving a location. This is the aim of the following developments.

3.5 Neighborhoods

We now consider constrained regions, which are constrained DBMs in which the DBM represents a region. Fig. 4 shows that if Controller plays to a region, then Perturbator can reach some of the surrounding regions, shown by the arrows. To characterize these, we define the set of *neighboring regions* of $\langle r, S \rangle$ as,

$$\mathcal{N}_{r,S} = \left\{ r' \ \middle| \ r' \lessdot^* r \text{ or } r \lessdot^+ r', \text{ and } \forall Q \leq S. \ r' \cap \mathsf{enlarge}((r, Q)) \neq \emptyset \right\}$$

where $\mathsf{enlarge}((r, Q))$ is the shrunk DBM (M, P) such that $v + [-\delta, \delta] \subseteq M - \delta P$ for every $v \in r - \delta Q$. This is the set of regions that have "distance" at most δ to any shrinking of the constrained region (r, S). We write $\mathsf{neighbor}\langle r, S \rangle = \bigcup_{r' \in \mathcal{N}_{r,S}} r'$.

Lemma 6 (Neighborhood). *Let $\langle r, S \rangle$ be a well constrained region. Then $\mathsf{neighbor}\langle r, S \rangle$ is a zone. If N is the corresponding normalized DBM, there exists a well shrinking constraint S' such that for every SM Q, $Q \leq S'$ iff the SM P defined by $(r, P) = r \cap \mathsf{shrink}((N, Q))$, satisfies $P \leq S$. The pair $\langle N, S' \rangle$ is the constrained neighborhood of $\langle r, S \rangle$, and it can be computed in polynomial time.*

Constrained neighborhoods are illustrated in Fig. 7.

3.6 Two Crucial Properties for the Construction of the Abstraction

The following lemma characterizes, given a constrained region $\langle r, S \rangle$, the set of constrained regions $\langle r', S_{r'} \rangle$ such that any shrunk region satisying $\langle r', S_{r'} \rangle$ can be reached by delaying from some shrunk region satisfying $\langle r, S \rangle$.

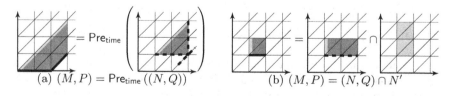

(a) $(M, P) = \mathsf{Pre}_{\mathsf{time}}((N, Q))$ (b) $(M, P) = (N, Q) \cap N'$

Fig. 6. The figures illustrate the first item in Lemma 5. In each case, DBMs M, N and N' are fixed and satisfy the "unshrunk" equation. The thick plain segments represent the fixed shrinking constraint S. The dashed segments represent resulting constraint S'. For any SM Q, we have $Q \leq S'$ iff there is an SM $P \leq S$ that satisfies the equation.

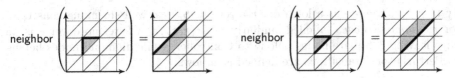

Fig. 7. Constrained neighborhood of two constrained regions. Notice that inside any shrinking of the constrained region, there is always a valuation such that a perturbation of $[-\delta, \delta]$ moves the valuation to any region of the neighborhood.

Lemma 7. *Let $\langle r, S \rangle$ be a well constrained region, and r' be a region such that $r \prec^* r'$. Then the following properties are equivalent:*

1. *there exists a well shrinking constraint S' (which can be computed in polynomial time) such that for every SM Q, $Q \leq S'$ iff the SM P such that $(r, P) = r \cap shrink^+(Pre_{time}((r', Q)))$, satisfies $P \leq S$;*
2. *neighbor$\langle r, S \rangle \subseteq Pre_{time}(r')$;*

Note that this lemma may not hold for all r' with $r \prec r'$. Consider the constrained region $\langle r, S \rangle$ on the right of Fig. 7, and let r' be the first triangle region above r: any valuation arbitrarily close to the thick segments will be in $r - \delta P$ for any $P \leq S$, but it can only reach r' by delaying less than δ time units.

Lemma 8. *Let $\langle r, S \rangle$ be a well constrained region, and let $R \subseteq C$. Let $\mathcal{N}$ be the set of neighboring regions of $\langle r, S \rangle$, and $\mathcal{N}' = \{r'[R \leftarrow 0] \mid r' \in \mathcal{N}\}$. Then, there exist well shrinking constraints $S_{r''}$ for all $r'' \in \mathcal{N}'$ such that for any $(Q_{r''})_{r'' \in \mathcal{N}'}$, we have $Q_{r''} \leq S_{r''}$ for all $r'' \in \mathcal{N}'$ iff there exists $P \leq S$ such that*

$$(r, P) \subseteq r \cap shrink(\bigcup_{r' \in \mathcal{N}} (r' \cap Unreset_R((r'', Q_{r''})))).$$

with $r'' = r'[R \leftarrow 0]$. Moreover, all $\langle r'', S_{r''} \rangle$ can be computed in polynomial time.

This lemma gives for instance the shrinking constraints that should be satisfied in r_1, r_2 and r_3, in Fig. 4, once shrinking constraint in r'_0 is known. In this case, the constraint in r'_0 is 0 everywhere since it is a punctual region. The neighborhood $\mathcal{N}$ of r'_0 is composed of r'_0 and two extra regions (defined by $(0 < x < 1) \wedge (x = y)$ and $(1 < x < 2) \wedge (x = y)$). If there are shrinkings of regions r_1, r_2, r_3 satisfying the corresponding shrinking constraints (given in the lemma), and from which Controller wins, then one can derive a shrinking of r'_0, satisfying its constraint, and from which Controller wins. In the next section, we define the game $\mathcal{RG}(\mathcal{A})$ following this idea, and explain how it captures the game semantics for robustness.

4   A Finite Game Abstraction

Let $\mathcal{A} = (\mathcal{L}, \mathcal{C}, \ell_0, E)$ be a timed automaton. We define a finite turn-based game $\mathcal{RG}(\mathcal{A})$ on a graph whose nodes are of two sorts: *square nodes* labelled by (ℓ, r, S_r), where ℓ is a location, r a region, S_r is a well shrinking constraint for r; *diamond nodes* labelled similarly by (ℓ, r, S_r, e) where moreover e is an edge leaving ℓ. Square nodes belong to Controller, while diamond nodes belong to Perturbator. Transitions are defined as follows:

(a) From each square node (ℓ, r, S_r), for any edge $e = (\ell, g, R, \ell')$ of $\mathcal{A}$, there is a transition to the diamond node $(\ell, r', S_{r'}, e)$ if the following conditions hold:
    (i) $r \lessdot^* r'$ and $r' \subseteq g$;
    (ii) $S_{r'}$ is such that for all SMs Q, $Q \leq S_{r'}$ iff there exists $P \leq S_r$ with

$$(r, P) = r \cap \mathsf{shrink}^+(\mathsf{Pre}_{\mathsf{time}}((r', Q)))$$

(b) From each diamond node (ℓ, r, S_r, e), where $e = (\ell, g, R, \ell')$ is an edge of $\mathcal{A}$, writing $\mathcal{N}$ for the set of regions in the neighborhood of (r, S_r) and $\mathcal{N}' = \{r'[R \leftarrow 0] \mid r' \in \mathcal{N}\}$, there are transitions to all square nodes $(\ell', r'', S_{r''})$ with $r'' \in \mathcal{N}'$, and $(S_{r''})_{r'' \in \mathcal{N}'}$ are such that for all SMs $(Q_{r''})_{r'' \in \mathcal{N}'}$, it holds $Q_{r''} \leq S_{r''}$ for every $r'' \in \mathcal{N}'$ iff there exists $P \leq S_r$ such that

$$(r, P) \subseteq r \cap \mathsf{shrink}(\bigcup_{r' \in \mathcal{N}} (r' \cap \mathsf{Unreset}_R((r'', Q_{r''})))) \qquad (\text{where } r'' = r'[R \leftarrow 0])$$

Intuitively, the transitions from the square nodes are the decisions of Controller. In fact, it has to select a delay and a transition whose guard is satisfied. Then Perturbator can choose any region in the neighborhood of the current region, and, after reset, this determines the next state.

Note that $\mathcal{RG}(\mathcal{A})$ can be computed, thanks to Lemmas 7 and 8, and has exponential-size. Observe also that $\mathcal{RG}(\mathcal{A})$ is constructed in a forward manner: we start by the initial constrained region (*i.e.* the region of valuation $\mathbf{0}$ with the zero matrix as shrinking constraint), and compute its successors in $\mathcal{RG}(\mathcal{A})$. Then, if Controller has a winning strategy in $\mathcal{RG}(\mathcal{A})$, we construct a winning strategy for $\mathcal{G}_\delta(\mathcal{A})$ by a backward traversal of $\mathcal{RG}(\mathcal{A})$, using Lemmas 7 and 8. Thus, we construct $\mathcal{RG}(\mathcal{A})$ by propagating shrinking constraints forward, but later do a backward traversal in it. The correctness of the construction is stated as follows.

Proposition 9. *Controller has a winning strategy in $\mathcal{RG}(\mathcal{A})$ if, and only if there exists $\delta_0 > 0$ such that Controller wins $\mathcal{G}_\delta(\mathcal{A})$ for all $\delta \in [0, \delta_0]$.*

Note that as we compute a winning strategy for Controller (if any) by Proposition 9, we can also compute a corresponding δ_0. One can show, by a rough estimation, that $1/\delta_0$ is at worst doubly exponential in the size of $\mathcal{A}$.

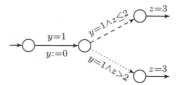

Fig. 8. Conjunction

Let us point out an interesting intermediary result of the proof: given a winning strategy for Perturbator in $\mathcal{RG}(\mathcal{A})$, we show that there is a winning strategy for Perturbator in $\mathcal{G}_\delta(\mathcal{A})$ that keeps the compatible runs close to borders of regions where shrinking constraints are 0.

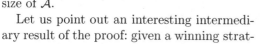

The upper bound of Theorem 2 is a consequence of the above proposition, since $\mathcal{RG}(\mathcal{A})$ has exponential size and finite reachability games can be solved in time polynomial in the size of the game. The EXPTIME lower bound is obtained by simulating an alternating-time linear-bounded Turing machine. Simulation of the transitions is rather standard in timed-automata literature (though we must be careful here as delays can be perturbed). The difficult point is to simulate conjunctions: this is achieved using the module of Fig. 8. From the initial state, Controller has no choice but to play the first transition when $y = 1$. Perturbator can either anticipate or delay this transition, which will determine which of the dashed or dotted transitions is available next. This way, Perturbator decides by which end the module is exited.

5 Conclusion

We considered a game-based approach to robust reachability in timed automata. We proved that robust schedulers for reachability objectives can be synthesized, and that the existence of such a scheduler is EXPTIME-complete (hence harder than classical reachability [2]). We are currently working on a zone-based version of the algorithm, and on extending the techniques of this paper to the synthesis of robust controllers in timed games, which will answer an open problem posed in [6] for reachability objectives. Natural further works also include the synthesis of robust schedulers for safety objectives. This seems really challenging, and the abstraction we have built here is not correct in this case (it requires at least a notion of *profitable cycles à la* [7]). Another interesting direction for future work is to assume imprecisions are probabilistic, that is, once Controller has chosen a delay d, the real delay is chosen in a stochastic way in the interval $[d - \delta, d + \delta]$.

References

1. Adbeddaïm, Y., Asarin, E., Maler, O.: Scheduling with timed automata. TCS 354(2), 272–300 (2006)
2. Alur, R., Dill, D.L.: A theory of timed automata. TCS 126(2), 183–235 (1994)
3. Asarin, E., Maler, O., Pnueli, A., Sifakis, J.: Controller synthesis for timed automata. In: SSSC 1998, pp. 469–474. Elsevier (1998)
4. Bengtsson, J., Yi, W.: Timed Automata: Semantics, Algorithms and Tools. In: Desel, J., Reisig, W., Rozenberg, G. (eds.) ACPN 2003. LNCS, vol. 3098, pp. 87–124. Springer, Heidelberg (2004)
5. Bouyer, P., Markey, N., Sankur, O.: Robust reachability in timed automata: A game-based approach. Technical Report LSV-12-07, Lab. Specification & Verification, ENS Cachan, France (May 2012)
6. Chatterjee, K., Henzinger, T.A., Prabhu, V.S.: Timed parity games: Complexity and robustness. LMCS 7(4) (2010)
7. De Wulf, M., Doyen, L., Markey, N., Raskin, J.-F.: Robust safety of timed automata. FMSD 33(1-3), 45–84 (2008)
8. De Wulf, M., Doyen, L., Raskin, J.-F.: Almost ASAP semantics: From timed models to timed implementations. FAC 17(3), 319–341 (2005)

9. Graham, R.L.: Bounds on multiprocessing timing anomalies. SIAM J. Applied Maths 17(2), 416–429 (1969)
10. Gupta, V., Henzinger, T.A., Jagadeesan, R.: Robust Timed Automata. In: Maler, O. (ed.) HART 1997. LNCS, vol. 1201, pp. 331–345. Springer, Heidelberg (1997)
11. Henzinger, T.A., Sifakis, J.: The Embedded Systems Design Challenge. In: Misra, J., Nipkow, T., Karakostas, G. (eds.) FM 2006. LNCS, vol. 4085, pp. 1–15. Springer, Heidelberg (2006)
12. Kopetz, H.: Real-Time Systems: Design Principles for Distributed Embedded Applications. Springer (2011)
13. Larsen, K.G., Pettersson, P., Yi, W.: UPPAAL in a nutshell. Intl J. STTT 1(1-2), 134–152 (1997)
14. Markey, N.: Robustness in real-time systems. In: SIES 2011, pp. 28–34. IEEE Comp. Soc. Press (2011)
15. Puri, A.: Dynamical properties of timed automata. DEDS 10(1-2), 87–113 (2000)
16. Reineke, J., Wachter, B., Thesing, S., Wilhelm, R., Polian, I., Eisinger, J., Becker, B.: A definition and classification of timing anomalies. In: WCET 2006 (2006)
17. Sankur, O., Bouyer, P., Markey, N.: Shrinking timed automata. In: FSTTCS 2011. LIPIcs, vol. 13, pp. 375–386. LZI (2011)
18. Yovine, S.: Kronos: A verification tool for real-time systems. Intl J. STTT 1(1-2), 123–133 (1997)

Minimizing Expected Termination Time in One-Counter Markov Decision Processes[*]

Tomáš Brázdil[1,**], Antonín Kučera[1,**], Petr Novotný[1,**], and Dominik Wojtczak[2,**]

[1] Faculty of Informatics, Masaryk University
{xbrazdil,kucera,xnovot18}@fi.muni.cz
[2] Department of Computer Science, University of Liverpool
d.wojtczak@liv.ac.uk

Abstract. We consider the problem of computing the value and an optimal strategy for minimizing the expected termination time in one-counter Markov decision processes. Since the value may be irrational and an optimal strategy may be rather complicated, we concentrate on the problems of approximating the value up to a given error $\varepsilon > 0$ and computing a finite representation of an ε-optimal strategy. We show that these problems are solvable in exponential time for a given configuration, and we also show that they are computationally hard in the sense that a polynomial-time approximation algorithm cannot exist unless P=NP.

1 Introduction

In recent years, a lot of research work has been devoted to the study of stochastic extensions of various automata-theoretic models such as pushdown automata, Petri nets, lossy channel systems, and many others. In this paper we study the class of *one-counter Markov decision processes (OC-MDPs)*, which are infinite-state MDPs [20,14] generated by finite-state automata operating over a single unbounded counter. Intuitively, an OC-MDP is specified by a finite directed graph $\mathcal{A}$ where the nodes are control states and the edges correspond to transitions between control states. Each control state is either stochastic or non-deterministic, which means that the next edge is chosen either randomly (according to a fixed probability distribution over the outgoing edges) or by a controller. Further, each edge either increments, decrements, or leaves unchanged the current counter value. A *configuration* $q(i)$ of an OC-MDP $\mathcal{A}$ is given by the current control state q and the current counter value i (for technical convenience, we also allow negative counter values, although we are only interested in runs where the counter stays non-negative). The outgoing transitions of $q(i)$ are determined by the edges of $\mathcal{A}$ in the natural way.

Previous works on OC-MDPs [4,2,3] considered mainly the objective of *maximizing/minimizing termination probability*. We say that a run initiated in a configuration $q(i)$ *terminates* if it visits a configuration with zero counter. The goal of the controller

---

[*] The full version of this paper can be found at http://arxiv.org/abs/1205.1473.
[**] Tomáš Brázdil is supported by the Czech Science Foundation, grant No. P202/12/P612. Antonín Kučera and Petr Novotný are supported by the Czech Science Foundation, grant No. P202/10/1469. Dominik Wojtczak is supported by EPSRC grant EP/G050112/2.

is to play so that the probability of all terminating runs is maximized (or minimized). In this paper, we study a related objective of *minimizing the expected termination time*. Formally, we define a random variable T over the runs of $\mathcal{A}$ such that $T(\omega)$ is equal either to ∞ (if the run ω is non-terminating) or to the number of transitions needed to reach a configuration with zero counter (if ω is terminating). The goal of the controller is to minimize the expectation $\mathbb{E}(T)$. The *value* of $q(i)$ is the infimum of $\mathbb{E}(T)$ over all strategies. It is easy to see that the controller has a memoryless deterministic strategy which is optimal (i.e., achieves the value) in every configuration. However, since OC-MDPs have infinitely many configurations, this does not imply that an optimal strategy is finitely representable and computable. Further, the value itself can be irrational. Therefore, we concentrate on the problem of *approximating* the value of a given configuration up to a given (absolute or relative) error $\varepsilon > 0$, and computing a strategy which is *ε-optimal* (in both absolute and relative sense). Our main results can be summarized as follows:

- *The value and optimal strategy can be effectively approximated up to a given relative/absolute error in exponential time.* More precisely, we show that given an OC-MDP $\mathcal{A}$, a configuration $q(i)$ of $\mathcal{A}$ where $i \geq 0$, and $\varepsilon > 0$, the value of $q(i)$ up to the (relative or absolute) error ε is computable in time exponential in the encoding size of $\mathcal{A}$, i, and ε, where all numerical constants are represented as fractions of binary numbers. Further, there is a history-dependent deterministic strategy σ computable in exponential time such that the absolute/relative difference between the value of $q(i)$ and the outcome of σ in $q(i)$ is bounded by ε.
- *The value is not approximable in polynomial time unless P=NP.* This hardness result holds even if we restrict ourselves to configurations with counter value equal to 1 and to OC-MDPs where every outgoing edge of a stochastic control state has probability $1/2$. The result is valid for absolute as well as relative approximation.

Let us sketch the basic ideas behind these results. The upper bounds are obtained in two steps. In the first step (Section 3.1), we analyze the special case when the underlying graph of $\mathcal{A}$ is strongly connected. We show that minimizing the expected termination time is closely related to minimizing the expected increase of the counter per transition, at least for large counter values. We start by computing the minimal expected increase of the counter per transition (denoted by $\bar{x}$) achievable by the controller, and the associated strategy σ. This is done by standard linear programming techniques developed for optimizing the long-run average reward in finite-state MDPs (see, e.g., [20]) applied to the underlying finite graph of $\mathcal{A}$. Note that σ depends only on the current control state and ignores the current counter value (we say that σ is *counterless*). Further, the encoding size of $\bar{x}$ is *polynomial* in the encoding size of $\mathcal{A}$ (which we denote by $\|\mathcal{A}\|$). Then, we distinguish two cases.

Case (A), $\bar{x} \geq 0$. Then the counter does not have a tendency to decrease *regardless* of the controller's strategy, and the expected termination time value is infinite in all configurations $q(i)$ such that $i \geq |Q|$, where Q is the set of control states of $\mathcal{A}$ (see Proposition 5. A). For the finitely many remaining configurations, we can compute the value and optimal strategy precisely by standard methods for finite-state MDPs.

Case (B), $\bar{x} < 0$. Then, one intuitively expects that applying the strategy σ in an initial configuration $q(i)$ yields the expected termination time about $i/|\bar{x}|$. Actually, this is *almost* correct; we show (Proposition 5. B.2) that this expectation is bounded by $(i + U)/|\bar{x}|$, where $U \geq 0$ is a constant depending only on $\mathcal{A}$ whose size is at most exponential in $\|\mathcal{A}\|$. Further, we show that an *arbitrary* strategy π applied to $q(i)$ yields the expected termination time *at least* $(i - V)/|\bar{x}|$, where $V \geq 0$ is a constant depending only on $\mathcal{A}$ whose size is at most exponential in $\|\mathcal{A}\|$ (Proposition 5. B.1). In particular, this applies to the *optimal* strategy π^* for minimizing the expected termination time. Hence, π^* can be more efficient than σ, but the difference between their outcomes is bounded by a constant which depends only on $\mathcal{A}$ and is at most exponential in $\|\mathcal{A}\|$. We proceed by computing a sufficiently large k so that the probability of increasing the counter to $i + k$ by a run initiated in $q(i)$ is inevitably (i.e., under any optimal strategy) so small that the controller can safely switch to the strategy σ when the counter reaches the value $i + k$. Then, we construct a *finite-state* MDP $\mathcal{M}$ and a reward function f over its transitions such that

- the states are all configurations $p(j)$ where $0 \leq j \leq i + k$;
- all states with counter values less than $i + k$ "inherit" their transitions from $\mathcal{A}$; configurations of the form $p(i + k)$ have only self-loops;
- the self-loops on configurations where the counter equals 0 or $i+k$ have zero reward, transitions leading to configurations where the counter equals $i + k$ have reward $(i + k + U)/|\bar{x}|$, and the other transitions have reward 1.

In this finite-state MDP $\mathcal{M}$, we compute an optimal memoryless deterministic strategy ϱ for the total accumulated reward objective specified by f. Then, we consider another strategy $\hat{\sigma}$ for $q(i)$ which behaves like ϱ until the point when the counter reaches $i + k$, and from that point on it behaves like σ. It turns out that the absolute as well as relative difference between the outcome of $\hat{\sigma}$ in $q(i)$ and the value of $q(i)$ is bounded by ε, and hence $\hat{\sigma}$ is the desired ε-optimal strategy.

In the general case when $\mathcal{A}$ is not necessarily strongly connected (see Section 3.2), we have to solve additional difficulties. Intuitively, we split the graph of $\mathcal{A}$ into maximal end components (MECs), where each MEC can be seen as a strongly connected OC-MDP and analyzed by the techniques discussed above. In particular, for every MEC C we compute the associated $\bar{x}_C$ (see above). Then, we consider a strategy which tries to reach a MEC as quickly as possible so that the expected value of the fraction $1/|\bar{x}_C|$ is *minimal*. After reaching a target MEC, the strategy starts to behave as the strategy σ discussed above. It turns out that this particular strategy cannot be much worse than the optimal strategy (a proof of this claim requires new observations), and the rest of the argument is similar as in the strongly connected case.

The lower bound, i.e., the result saying that the value cannot be efficiently approximated unless P=NP (see Section 4), seems to be the first result of this kind for OC-MDPs. Here we combine the technique of encoding propositional assignments presented in [18] (see also [16]) with some new gadgets constructed specifically for this proof (let us note that we did not manage to improve the presented lower bound to PSPACE by adapting other known techniques [15,21,17]). As a byproduct, our proof also reveals that the optimal strategy for minimizing the expected termination time *cannot* ignore the precise counter value, even if the counter becomes very large. In our example, the (only) optimal strategy is *eventually periodic* in the sense that for a sufficiently large counter value i, it is only "i modulo c" which matters, where c is a fixed (exponentially large) constant. The question whether there *always* exists an optimal eventually periodic strategy is left open. Another open question is whether our results can be extended to stochastic games over one-counter automata.

Due to space constraints, most proofs are omitted and can be found in [7].

Related Work: One-counter automata can also be seen as pushdown automata with one letter stack alphabet. Stochastic games and MDPs generated by pushdown automata and stateless pushdown automata (also known as BPA) with termination and reachability objectives have been studied in [12,13,5,6]. To the best of our knowledge, the only prior work on the expected termination time (or, more generally, total accumulated reward) objective for a class of infinite-state MDPs or stochastic games is [10], where this problem is studied for stochastic BPA games. However, the proof techniques of [10] are not directly applicable to one-counter automata.

The termination objective for one-counter MDPs and games has been examined in [4,2,3], where it was shown (among other things) that the equilibrium termination probability (i.e., the termination value) can be approximated up to a given precision in exponential time, but no lower bound was provided. In this paper, we build on some of the underlying observations presented in [4,2,3]. In particular, we employ the submartingale of [3] to derive certain bounds in Section 3.1.

The games over one-counter automata are also known as "energy games" [8,9]. Intuitively, the counter is used to model the amount of currently available energy, and the aim of the controller is to optimize the energy consumptions.

Finally, let us note that OC-MDPs can be seen as discrete-time Quasi-Birth-Death Processes (QBDs, see, e.g., [19,11]) extended with a control. Hence, the theory of one-counter MDPs and games is closely related to queuing theory, where QBDs are considered as a fundamental model.

2   Preliminaries

Given a set A, we use $|A|$ to denote the cardinality of A. We also write $|x|$ to denote the absolute value of a given $x \in \mathbb{R}$, but this should not cause any confusions. The encoding size of a given object B is denoted by $\|B\|$. The set of integers is denoted by $\mathbb{Z}$, and the set of positive integers by $\mathbb{N}$.

We assume familiarity with basic notions of probability theory. In particular, we call a probability distribution f over a discrete set A *positive* if $f(a) > 0$ for all $a \in A$.

Definition 1 (MDP). *A Markov decision process (MDP) is a tuple* $\mathcal{M} = (S, (S_0, S_1), \rightsquigarrow, Prob)$, *consisting of a countable set of states* S *partitioned into the sets* S_0 *and* S_1 *of stochastic and non-deterministic states, respectively. The edge relation* $\rightsquigarrow \subseteq S \times S$ *is total, i.e., for every* $r \in S$ *there is* $s \in S$ *such that* $r \rightsquigarrow s$. *Finally, Prob assigns to every* $s \in S_0$ *a positive probability distribution over its outgoing edges.*

A *finite path* is a sequence $w = s_0 s_1 \cdots s_n$ of states such that $s_i \rightsquigarrow s_{i+1}$ for all $0 \le i < n$. We write $len(w) = n$ for the length of the path. A *run* is an infinite sequence ω of states such that every finite prefix of ω is a path. For a finite path, w, we denote by $Run(w)$ the set of runs having w as a prefix. These generate the standard σ-algebra on the set of all runs. If w is a finite path or a run, we denote by $w(i)$ the i-th state of sequence w.

Definition 2 (OC-MDP). *A* one-counter MDP (OC-MDP) *is a tuple* $\mathcal{A} = (Q, (Q_0, Q_1), \delta, P)$, *where* Q *is a finite non-empty set of* control states *partitioned into* stochastic and non-deterministic states *(as in the case of MDPs),* $\delta \subseteq Q \times \{+1, 0, -1\} \times Q$ *is a set of* transition rules *such that* $\delta(q) := \{(q, i, r) \in \delta\} \neq \emptyset$ *for all* $q \in Q$, *and* $P = \{P_q\}_{q \in Q_0}$ *where* P_q *is a positive rational probability distribution over* $\delta(q)$ *for all* $q \in Q_0$.

In the rest of this paper we often write $q \xrightarrow{i} r$ to indicate that $(q, i, r) \in \delta$, and $q \xrightarrow{i,x} r$ to indicate that $(q, i, r) \in \delta$, q is stochastic, and $P_q(q, i, r) = x$. Without restrictions, we assume that for each pair $q, r \in Q$ there is at most one i such that $(q, i, r) \in \delta$. The encoding size of $\mathcal{A}$ is denoted by $\|\mathcal{A}\|$, where all numerical constants are encoded as fractions of binary numbers. The set of all *configurations* is $C := \{q(i) \mid q \in Q, i \in \mathbb{Z}\}$.

To $\mathcal{A}$ we associate an infinite-state MDP $M_{\mathcal{A}}^{\infty} = (C, (C_0, C_1), \rightsquigarrow, Prob)$, where the partition of C is defined by $q(i) \in C_0$ iff $q \in Q_0$, and similarly for C_1. The edges are defined by $q(i) \rightsquigarrow r(j)$ iff $(q, j - i, r) \in \delta$. The probability assignment $Prob$ is derived naturally from P and we write $q(i) \overset{x}{\rightsquigarrow} r(j)$ to indicate that $q(i) \in C_0$, and $Prob(q(i))$ assigns probability x to the edge $q(i) \rightsquigarrow r(j)$.

By forgetting the counter values, the OC-MDP $\mathcal{A}$ also defines a finite-state MDP $M_{\mathcal{A}} = (Q, (Q_0, Q_1), \rightsquigarrow, Prob')$. Here $q \rightsquigarrow r$ iff $(q, i, r) \in \delta$ for some i, and $Prob'$ is derived in the obvious way from P by forgetting the counter changes.

Strategies and Probability. Let M be an MDP. A *history* is a finite path in M, and a *strategy* (or *policy*) is a function assigning to each history ending in a state from S_1 a distribution on edges leaving the last state of the history. A strategy σ is *pure* (or *deterministic*) if it always assigns 1 to one edge and 0 to the others, and *memoryless* if $\sigma(ws)$ depends just on the last state s, for every $w \in S^*$.

Now consider some OC-MDP $\mathcal{A}$. A strategy σ over the histories in $M_{\mathcal{A}}^{\infty}$ is *counterless* if it is memoryless and $\sigma(q(i)) = \sigma(q(j))$ for all i, j. Observe that every strategy σ for $M_{\mathcal{A}}$ gives a unique strategy σ' for $M_{\mathcal{A}}^{\infty}$ which just forgets the counter values in the history and plays as σ. This correspondence is bijective when restricted to memoryless strategies in $M_{\mathcal{A}}$ and counterless strategies in $M_{\mathcal{A}}^{\infty}$, and it is used implicitly throughout the paper.

Fixing a strategy σ and an initial state s, we obtain in a standard way a probability measure $\mathbb{P}_s^{\sigma}(\cdot)$ on the subspace of runs starting in s. For MDPs of the form $M_{\mathcal{A}}^{\infty}$ for some OC-MDP $\mathcal{A}$, we consider two sequences of random variables, $\{C^{(i)}\}_{i \geq 0}$ and $\{S^{(i)}\}_{i \geq 0}$, returning the current counter value and the current control state after completing i transitions.

Termination Time in OC-MDPs. Let $\mathcal{A}$ be an OC-MDP. A run ω in $M_{\mathcal{A}}^{\infty}$ *terminates* if $\omega(j) = q(0)$ for some $j \geq 0$ and $q \in Q$. The associated *termination time*, denoted by $T(\omega)$, is the least j such that $\omega(j) = q(0)$ for some $q \in Q$. If there is no such j, we put $T(\omega) = \infty$, where the symbol ∞ denotes the "infinite amount" with the standard conventions, i.e., $c < \infty$ and $\infty + c = \infty + \infty = \infty \cdot d = \infty$ for arbitrary real numbers c, d where $d > 0$.

For every strategy σ and a configuration $q(i)$, we use $\mathbb{E}^{\sigma} q(i)$ to denote the expected value of T in the probability space of all runs initiated in $q(i)$ where $\mathbb{P}^{\sigma}_{q(i)}(\cdot)$ is the underlying probability measure. The *value* of a given configuration $q(i)$ is defined by $\mathrm{Val}(q(i)) := \inf_{\sigma} \mathbb{E}^{\sigma} q(i)$. Let $\varepsilon \geq 0$ and $i \geq 1$. We say that a constant v approximates $\mathrm{Val}(q(i))$ up to the absolute or relative error ε if $|\mathrm{Val}(q(i)) - v| \leq \varepsilon$ or $|\mathrm{Val}(q(i)) - v| / \mathrm{Val}(q(i)) \leq \varepsilon$, respectively. Note that if v approximates $\mathrm{Val}(q(i))$ up to the absolute error ε, then it also approximates $\mathrm{Val}(q(i))$ up to the relative error ε because $\mathrm{Val}(q(i)) \geq 1$. A strategy σ is (absolutely or relatively) ε-*optimal* if $\mathbb{E}^{\sigma} q(i)$ approximates $\mathrm{Val}(q(i))$ up to the (absolute or relative) error ε. A 0-optimal strategy is called *optimal*.

It is easy to see that there is a memoryless deterministic strategy σ in $\mathcal{M}_{\mathcal{A}}^{\infty}$ which is optimal in every configuration of $\mathcal{M}_{\mathcal{A}}^{\infty}$. First, observe that for all $q \in Q_0$, $q' \in Q_1$, and $i \neq 0$ we have that

$$\mathrm{Val}(q(i)) = 1 + \sum_{q(i) \overset{x}{\rightsquigarrow} r(j)} x \cdot \mathrm{Val}(r(j))$$
$$\mathrm{Val}(q'(i)) = 1 + \min\{\mathrm{Val}(r(j)) \mid q'(i) \rightsquigarrow r(j)\}.$$

We put $\sigma(q'(i)) = r(j)$ if $q'(i) \rightsquigarrow r(j)$ and $\mathrm{Val}(q'(i)) = 1 + \mathrm{Val}(r(j))$ (if there are several candidates for $r(j)$, any of them can be chosen). Now we can easily verify that σ is indeed optimal in every configuration.

3   Upper Bounds

The goal of this section is to prove the following:

Theorem 3. *Let $\mathcal{A}$ be an OC-MDP, $q(i)$ a configuration of $\mathcal{A}$ where $i \geq 0$, and $\varepsilon > 0$.*

1. *The problem whether $\mathrm{Val}(q(i)) = \infty$ is decidable in polynomial time.*
2. *There is an algorithm that computes a rational number v such that $|\mathrm{Val}(q(i)) - v| \leq \varepsilon$, and a strategy σ that is absolutely ε-optimal starting in $q(i)$. The algorithm runs in time exponential in $\|\mathcal{A}\|$ and polynomial in i and $1/\varepsilon$. (Note that v then approximates $\mathrm{Val}(q(i))$ also up to the relative error ε, and σ is also relatively ε-optimal in $q(i)$).*

For the rest of this section, we fix an OC-MDP $\mathcal{A} = (Q, (Q_0, Q_1), \delta, P)$. First, we prove Theorem 3 under the assumption that $\mathcal{M}_{\mathcal{A}}$ is *strongly connected* (Section 3.1). A generalization to arbitrary OC-MDPs is then given in Section 3.2.

3.1   Strongly Connected OC-MDP

Let us assume that $\mathcal{M}_{\mathcal{A}}$ is strongly connected, i.e., for all $p, q \in Q$ there is a finite path from p to q in $\mathcal{M}_{\mathcal{A}}$. Consider the linear program of Figure 1. Intuitively, the variable x encodes a lower bound on the long-run trend of the counter value. The program corresponds to the one used for optimizing the long-run average reward in Sections 8.8 and 9.5 of [20], and hence we know it has a solution.

Lemma 4 ([20]). *There is a rational solution $\left(\bar{x}, (\bar{z}_q)_{q \in Q} \right) \in \mathbb{Q}^{|Q|+1}$ to $\mathcal{L}$, and the encoding size[1] of the solution is polynomial in $\|\mathcal{A}\|$*

---

[1] Recall that rational numbers are represented as fractions of binary numbers.

maximize x, subject to

$$z_q \le -x + k + z_r \qquad\qquad \text{for all } q \in Q_1 \text{ and } (q,k,r) \in \delta,$$

$$z_q \le -x + \sum_{(q,k,r)\in\delta} P_q((q,k,r)) \cdot (k + z_r) \qquad\qquad \text{for all } q \in Q_0,$$

Fig. 1. The linear program $\mathcal{L}$ over x and z_q, where $q \in Q$

Note that $\bar{x} \ge -1$, because for any fixed $x \le -1$ the program $\mathcal{L}$ trivially has a feasible solution. Further, we put $V := \max_{q\in Q} \bar{z}_q - \min_{q\in Q} \bar{z}_q$. Observe that $V \in \exp\left(\|\mathcal{A}\|^{O(1)}\right)$ and V is computable in time polynominal in $\|\mathcal{A}\|$.

Proposition 5. *Let $\left(\bar{x}, (\bar{z}_q)_{q\in Q}\right)$ be a solution of $\mathcal{L}$.*

(A) If $\bar{x} \ge 0$, then $\mathrm{Val}(q(i)) = \infty$ for all $q \in Q$ and $i \ge |Q|$.
(B) If $\bar{x} < 0$, then the following holds:
   (B.1) For every strategy π and all $q \in Q$, $i \ge 0$ we have that $\mathbb{E}^\pi q(i) \ge (i - V)/|\bar{x}|$.
   (B.2) There is a counterless strategy σ and a number $U \in \exp\left(\|\mathcal{A}\|^{O(1)}\right)$ such that for all $q \in Q$, $i \ge 0$ we have that $\mathbb{E}^\sigma q(i) \le (i + U)/|\bar{x}|$. Moreover, σ and U are computable in time polynomial in $\|\mathcal{A}\|$.

First, let us realize that Proposition 5 implies Theorem 3. To see this, we consider the cases $\bar{x} \ge 0$ and $\bar{x} < 0$ separately. In both cases, we resort to analyzing a finite-state MDP $\mathcal{G}_K$, where K is a suitable natural number, obtained by restricting $\mathcal{M}_\mathcal{A}^\infty$ to configurations with counter value at most K, and by substituting all transitions leaving each $p(K)$ with a self-loop of the form $p(K) \rightsquigarrow p(K)$.

First, let us assume that $\bar{x} \ge 0$. By Proposition 5 (A), we have that $\mathrm{Val}(q(i)) = \infty$ for all $q \in Q$ and $i \ge |Q|$. Hence, it remains to approximate the value and compute ε-optimal strategy for all configurations $q(i)$ where $i \le |Q|$. Actually, we can even compute these values precisely and construct a strategy $\hat{\sigma}$ which is optimal in each such $q(i)$. This is achieved simply by considering the finite-state MDP $\mathcal{G}_{|Q|}$ and solving the objective of minimizing the expected number of transitions needed to reach a state of the form $p(0)$, which can be done by standard methods in time polynomial in $\|\mathcal{A}\|$.

If $\bar{x} < 0$, we argue as follows. The strategy σ of Proposition 5 (B.2) is not necessarily ε-optimal in $q(i)$, so we cannot use it directly. To overcome this problem, consider an *optimal* strategy π^* in $q(i)$, and let x_ℓ be the probability that a run initiated in $q(i)$ (under the strategy π^*) visits a configuration of the form $r(i+\ell)$. Obviously, $x_\ell \cdot \min_{r\in Q}\{\mathbb{E}^\sigma r(i+\ell)\} \le \mathbb{E}^\sigma q(i)$, because otherwise π^* would not be optimal in $q(i)$. Using the lower/upper bounds for $\mathbb{E}^{\pi^*} r(i+\ell)$ and $\mathbb{E}^\sigma q(i)$ given in Proposition 5 (B), we obtain $x_\ell \le (i + U)/(i + \ell - V)$. Then, we compute $k \in \mathbb{N}$ such that

$$x_k \cdot \left(\max_{r\in Q}\left\{(i + k + U)/|\bar{x}| - \mathbb{E}^{\pi^*} r(i+k)\right\}\right) \le \varepsilon$$

A simple computation reveals that it suffices to choose any k such that

$$k \ge \frac{(i + U) \cdot (U + V)}{\varepsilon \cdot |\bar{x}|} + V - i,$$

so the value of k is exponential in $\|\mathcal{A}\|$ and polynomial in i and $1/\varepsilon$. Now, consider $\mathcal{G}_{i+k}$, and let f be a reward function over the transitions of $\mathcal{G}_{i+k}$ such that the loops on configurations where the counter equals 0 or $i + k$ have zero reward, transitions leading to configurations of the form $r(i+k)$ have reward $(i + k + U)/|\bar{x}|$, and all of the remaining transitions have reward 1. Now we solve the finite-state MDP $\mathcal{G}_{i+k}$ with the objective of minimizing the total accumulated reward. Note that an optimal strategy ϱ in $\mathcal{G}_{i+k}$ is computable in time polynomial in the size of $\mathcal{G}_{i+k}$ [20]. Then, we define the corresponding strategy $\hat{\sigma}$ in $M_{\mathcal{A}}^{\infty}$, which behaves like ϱ until the counter reaches $i + k$, and from that point on it behaves like the counterless strategy σ. It is easy to see that $\hat{\sigma}$ is indeed ε-optimal in $q(i)$.

Proof of Proposition 5. Similarly as in [3], we use the solution $(\bar{x}, (\bar{z}_q)_{q \in Q}) \in \mathbb{Q}^{|Q|+1}$ of $\mathcal{L}$ to define a suitable submartingale, which is then used to derive the required bounds. In [3], Azuma's inequality was applied to the submartingale to prove exponential tail bounds for termination probability. In this paper, we need to use the optional stopping theorem rather than Azuma's inequality, and therefore we need to define the submartingale relative to a suitable filtration so that we can introduce an appropriate stopping time (without the filtration, the stopping time would have to depend just on numerical values returned by the martingale, which does not suit our purposes).

Given the solution $(\bar{x}, (\bar{z}_q)_{q \in Q}) \in \mathbb{Q}^{|Q|+1}$ from Lemma 4, we define a sequence of random variables $\{m^{(i)}\}_{i \geq 0}$ by setting

$$m^{(i)} := \begin{cases} C^{(i)} + \bar{z}_{S^{(i)}} - i \cdot \bar{x} & \text{if } C^{(j)} > 0 \text{ for all } j, \ 0 \leq j < i, \\ m^{(i-1)} & \text{otherwise.} \end{cases}$$

Note that for every history u of length i and every $0 \leq j \leq i$, the random variable $m^{(j)}$ returns the same value for every $\omega \in Run(u)$. The same holds for variables $S^{(j)}$ and $C^{(j)}$. We will denote these common values $m^{(j)}(u)$, $S^{(j)}(u)$ and $C^{(j)}(u)$, respectively. Using the same arguments as in Lemma 3 of [3], one may show that for every history u of length i we have $\mathbb{E}(m^{(i+1)} \mid Run(u)) \geq m^{(i)}(u)$. This shows that $\{m^{(i)}\}_{i \geq 0}$ is a *submartingale relative to the filtration* $\{\mathcal{F}_i\}_{i \geq 0}$, where for each $i \geq 0$ the σ-algebra $\mathcal{F}_i$ is the σ-algebra generated by all $Run(u)$ where $len(u) = i$. Intuitively, this means that value $m^{(i)}(\omega)$ is uniquely determined by prefix of ω of length i and that the process $\{m^{(i)}\}_{i \geq 0}$ has nonnegative average change. For relevant definitions of (sub)martingales see, e.g., [22]. Another important observation is that $|m^{(i+1)} - m^{(i)}| \leq 1 + \bar{x} + V$ for every $i \geq 0$, i.e., the differences of the submartingale are bounded.

Lemma 6. *Under an arbitrary strategy π and with an arbitrary initial configuration $q(j)$ where $j \geq 0$, the process $\{m^{(i)}\}_{i \geq 0}$ is a submartingale (relative to the filtration $\{\mathcal{F}_i\}_{i \geq 0}$) with bounded differences.*

Part (A) of Proposition 5. This part can be proved by a routine application of the optional stopping theorem to the martingale $\{m^{(i)}\}_{i \geq 0}$. Let $\bar{z}_{\max} := \max_{q \in Q} \bar{z}_q$, and consider a configuration $p(\ell)$ where $\ell + \bar{z}_p > \bar{z}_{\max}$. Let σ be a strategy which is optimal in every configuration. Assume, for the sake of contradiction, that $Val(p(\ell)) < \infty$.

Let us fix $k \in \mathbb{N}$ such that $\ell + \bar{z}_p < \bar{z}_{\max} + k$ and define a stopping time τ which returns the first point in time in which either $m^{(\tau)} \geq \bar{z}_{\max} + k$, or $m^{(\tau)} \leq \bar{z}_{\max}$. To apply the optional stopping theorem, we need to show that the expectation of τ is finite.

We argue that every configuration $q(i)$ with $i \geq 1$ satisfies the following: under the optimal strategy σ, a configuration with counter height $i - 1$ is reachable from $q(i)$ in at most $|Q|^2$ steps (i.e., with a bounded probability). To see this, realize that for every configuration $r(j)$ there is a successor, say $r'(j')$, such that $\mathrm{Val}(r(j)) > \mathrm{Val}(r'(j'))$. Now consider a run w initiated in $q(i)$ obtained by subsequently choosing successors with smaller and smaller values. Note that whenever $w(j)$ and $w(j')$ with $j < j'$ have the same control state, the counter height of $w(j')$ must be strictly smaller than the one of $w(j)$ because otherwise the strategy σ could be improved (it suffices to behave in $w(j)$ as in $w(j')$). It follows that there must be $k \leq |Q|^2$ such that the counter height of $w(k)$ is $i-1$. From this we obtain that the expected value of τ is finite because the probability of terminating from any configuration with bounded counter height is bounded from zero. Now we apply the optional stopping theorem and obtain $\mathbb{P}^{\sigma}_{p(\ell)}(m^{(\tau)} \geq \bar{z}_{\max}+k) \geq c/(k+d)$ for suitable constants $c, d > 0$. As $m^{(\tau)} \geq \bar{z}_{\max} + k$ implies $C^{(\tau)} \geq k$, we obtain that

$$\mathbb{P}^{\sigma}_{p(\ell)}(T \geq k) \quad \geq \quad \mathbb{P}^{\sigma}_{p(\ell)}(C^{(\tau)} \geq k) \quad \geq \quad \mathbb{P}^{\sigma}_{p(\ell)}(m^{(\tau)} \geq \bar{z}_{\max} + k) \quad \geq \quad \frac{c}{k+d}$$

and thus

$$\mathbb{E}^{\sigma} p(\ell) \quad = \quad \sum_{k=1}^{\infty} \mathbb{P}^{\sigma}_{p(\ell)}(T \geq k) \quad \geq \quad \sum_{k=1}^{\infty} \frac{c}{k+d} \quad = \quad \infty$$

which contradicts our assumption that σ is optimal and $\mathrm{Val}(p(\ell)) < \infty$.

It remains to show that $\mathrm{Val}(p(\ell)) = \infty$ even for $\ell = |Q|$. This follows from the following simple observation:

Lemma 7. *For all $q \in Q$ and $i \geq |Q|$ we have that* $\mathrm{Val}(q(i)) < \infty$ *iff* $\mathrm{Val}(q(|Q|)) < \infty$.

The "only if" direction of Lemma 7 is trivial. For the other direction, let $\mathcal{B}_k$ denote the set of all $p \in Q$ such that $\mathrm{Val}(p(k)) < \infty$. Clearly, $\mathcal{B}_0 = Q$, $\mathcal{B}_k \subseteq \mathcal{B}_{k-1}$, and one can easily verify that $\mathcal{B}_k = \mathcal{B}_{k+1}$ implies $\mathcal{B}_k = \mathcal{B}_{k+\ell}$ for all $\ell \geq 0$. Hence, $\mathcal{B}_{|Q|} = \mathcal{B}_{|Q|+\ell}$ for all ℓ. Note that Lemma 7 holds for general OC-MDPs (i.e., we do not need to assume that $\mathcal{M}_{\mathcal{A}}$ is strongly connected).

Part (B1) of Proposition 5. Let π be a strategy and $q(i)$ a configuration where $i \geq 0$. If $\mathbb{E}^{\pi} q(i) = \infty$, we are done. Now assume $\mathbb{E}^{\pi} q(i) < \infty$. Observe that for every $k \geq 0$ and every run ω, the membership of ω into $\{T \leq k\}$ depends only on the finite prefix of ω of length k. This means that T is a stopping time relative to filtration $\{\mathcal{F}_n\}_{n \geq 0}$. Since $\mathbb{E}^{\pi} q(i) < \infty$ and the submartingale $\{m^{(n)}\}_{n \geq 0}$ has bounded differences, we can apply the optional stopping theorem and obtain $\mathbb{E}^{\pi}(m^{(0)}) \leq \mathbb{E}^{\pi}(m^{(T)})$. But $\mathbb{E}^{\pi}(m^{(0)}) = i + \bar{z}_q$ and $\mathbb{E}^{\pi}(m^{(T)}) = \mathbb{E}^{\pi} \bar{z}_{S(T)} + \mathbb{E}^{\pi} q(i) \cdot |\bar{x}|$. Thus, we get $\mathbb{E}^{\pi} q(i) \geq (i + \bar{z}_q - \mathbb{E}^{\pi} \bar{z}_{S(T)})/|\bar{x}| \geq (i - V)/|\bar{x}|$.

Part (B2) of Proposition 5. First we show how to construct the desired strategy σ. Recall again the linear program $\mathcal{L}$ of Figure 1. We have already shown that this program has an optimal solution $(\bar{x}, (\bar{z}_q)_{q \in Q}) \in \mathbb{Q}^{|Q|+1}$, and we assume that $\bar{x} < 0$. By the strong duality theorem, this means that the linear program dual to $\mathcal{L}$ also has a feasible solution $((\bar{y}_q)_{q \in Q_0}, (\bar{y}_{(q,i,q')})_{q \in Q_1, (q,i,q') \in \delta})$. Let

$$D = \{q \in Q_0 \mid \bar{y}_q > 0\} \cup \{q \in Q_1 \mid \bar{y}_{(q,i,q')} > 0 \text{ for some } (q, i, q') \in \delta\}.$$

By Corollary 8.8.8 of [20], the solution $\left((\bar{y}_q)_{q \in Q_0}, (\bar{y}_{(q,i,q')})_{q \in Q_1, (q,i,q') \in \delta}\right)$ can be chosen so that for every $q \in Q_1$ there is at most one transition (q, i, q') with $\bar{y}_{(q,i,q')} > 0$. Following the construction given in Section 8.8 of [20], we define a counterless deterministic strategy σ such that

- in a state $q \in D \cap Q_1$, the strategy σ selects the transition (q, i, q') with $\bar{y}_{(q,i,q')} > 0$;
- in the states outside D, the strategy σ behaves like an optimal strategy for the objective of reaching the set D.

Clearly, the strategy σ is computable in time polynomial in $\|\mathcal{A}\|$. In the full version of this paper [7], we show that σ indeed satisfies Part (B.2) of Proposition 5.

3.2 General OC-MDP

In this section we prove Theorem 3 for general OC-MDPs, i.e., we drop the assumption that $\mathcal{M}_{\mathcal{A}}$ is strongly connected. We say that $C \subseteq Q$ is an *end component* of $\mathcal{A}$ if C is strongly connected and for every $p \in C \cap Q_0$ we have that $\{q \in Q \mid p \leadsto q\} \subseteq C$. A *maximal end component (MEC)* of $\mathcal{A}$ is an end component of $\mathcal{A}$ which is maximal w.r.t. set inclusion. The set of all MECs of $\mathcal{A}$ is denoted by $MEC(\mathcal{A})$. Every $C \in MEC(\mathcal{A})$ determines a strongly connected OC-MDP $\mathcal{A}_C = (C, (C \cap Q_0, C \cap Q_1), \delta \cap (C \times \{+1, 0, -1\} \times C), \{P_q\}_{q \in C \cap Q_0})$. Hence, we may apply Proposition 5 to $\mathcal{A}_C$, and we use $\bar{x}_C$ to denote the constant of Proposition 5 computed for $\mathcal{A}_C$.

Part 1. of Theorem 3. We show how to compute, in time polynomial in $\|\mathcal{A}\|$, the set $Q_{fin} = \{p \in Q \mid \mathrm{Val}(p(k)) < \infty \text{ for all } k \geq 0\}$. From this we easily obtain Part 1. of Theorem 3, because for every configuration $q(i)$ where $i \geq 0$ we have the following:

- if $i \geq |Q|$, then $\mathrm{Val}(q(i)) < \infty$ iff $q \in Q_{fin}$ (see Lemma 7);
- if $i < |Q|$, then $\mathrm{Val}(q(i)) < \infty$ iff the set $\{p(0) \mid p \in Q\} \cup \{p(|Q|) \mid p \in Q_{fin}\}$ can be reached from $q(i)$ with probability 1 in the finite-state MDP $\mathcal{G}_{|Q|}$ defined in Section 3.1 (here we again use Lemma 7).

So, it suffices to show how to compute the set Q_{fin} in polynomial time.

Proposition 8. *Let $Q_{<0}$ be the set of all states from which the set $H = \{q \in Q \mid q \text{ belongs to a MEC } C \text{ satisfying } \bar{x}_C < 0\}$ is reachable with probability 1. Then $Q_{fin} = Q_{<0}$. Moreover, the membership to $Q_{<0}$ is decidable in time polynomial in $\|\mathcal{A}\|$.*

Part 2. of Theorem 3. First, we generalize Part (B) of Proposition 5 into the following:

Proposition 9. *For every $q \in Q_{fin}$ there is a number t_q computable in time polynomial in $\|\mathcal{A}\|$ such that $-1 \leq t_q < 0$, $1/|t_q| \in \exp\left(\|\mathcal{A}\|^{O(1)}\right)$, and the following holds:*

(A) *There is a counterless strategy σ and a number $U \in \exp(\|\mathcal{A}\|^{O(1)})$ such that for every configuration $q(i)$ where $q \in Q_{fin}$ and $i \geq 0$ we have that $\mathbb{E}^\sigma q(i) \leq i/|t_q| + U$. Moreover, both σ and U are computable in time polynomial in $\|\mathcal{A}\|$.*

(B) *There is a number $L \in \exp(\|\mathcal{A}\|^{O(1)})$ such that for every strategy π and every config-uration $q(i)$ where $i \geq |Q|$ we have that $\mathbb{B}^\pi \geq i/|t_q| - L$. Moreover, L is computable in time polynomial in $\|\mathcal{A}\|$.*

Once the Proposition 9 is proved, we can compute an ε-optimal strategy for an arbitrary configuration $q(i)$ where $q \in Q_{fin}$ and $i \geq |Q|$ in exactly the same way (and with the same complexity) as in the strongly connected case. Actually, it can also be used to compute the approximate values and ε-optimal strategies for configurations $q(j)$ such that $q \notin Q_{fin}$ or $1 \leq j < |Q|$. Observe that

- if $q \notin Q_{fin}$ and $j \geq |Q|$, the value is infinite by Part 1;
- otherwise, we construct the finite-state MDP $G_{|Q|}$ (see Section 3.1) where the loops on configurations with counter value 0 have reward 0, the loops on configurations of the form $r(|Q|)$ have reward 0 or 1, depending on whether $r \in Q_{fin}$ or not, transitions leading to $r(|Q|)$ where $r \in Q_{fin}$ are rewarded with some ε-approximation of $\mathrm{Val}(r(|Q|))$, and all other transitions have reward 1. The reward function can be computed in time exponential in $\|\mathcal{A}\|$ by Proposition 9, and the minimal total accumulated reward from $q(j)$ in $G_{|Q|}$, which can be computed by standard algorithms, is an ε-approximation of $\mathrm{Val}(q(j))$. The corresponding ε-optimal strategy can be computed in the obvious way.

The missing proofs of Propositions 8 and 9 can be found in [7].

4 Lower Bounds

In this section, we show that approximating $\mathrm{Val}(q(i))$ is computationally hard, even if $i = 1$ and the edge probabilities in the underlying OC-MDP are all equal to $1/2$. More precisely, we prove the following:

Theorem 10. *The value of a given configuration $q(1)$ cannot be approximated in poly-nomial time up to a given absolute/relative error $\varepsilon > 0$ unless P=NP, even if all outgoing edges of all stochastic control states in the underlying OC-MDP have probability $1/2$.*

The proof of Theorem 10 is split into two phases, which are relatively independent. First, we show that given a propositional formula φ, one can efficiently compute an OC-MDP $\mathcal{A}$, a configuration $p(K)$ of $\mathcal{A}$, and a number N such that the value of $p(K)$ is either $N - 1$ or N depending on whether φ is satisfiable or not, respectively. The numbers K and N are exponential in $\|\varphi\|$, which means that their encoding size is polynomial (we represent all numerical constants in binary). Here we use the technique of encoding propositional assignments into counter values presented in [18], but we also need to invent some specific gadgets to deal with our specific objective. The first part already implies that approximating $\mathrm{Val}(q(i))$ is computationally hard. In the second phase, we show that the same holds also for configurations where the counter is initiated to 1. This is achieved by employing another gadget which just increases the counter to an exponentially high value with a sufficiently large probability. The two phases are elaborated in [7].

References

1. Proceedings of FST&TCS 2010. LIPIcs, vol. 8. Schloss Dagstuhl (2010)
2. Brázdil, T., Brožek, V., Etessami, K.: One-counter stochastic games. In: Proceedings of FST&TCS 2010 [1], pp. 108–119
3. Brázdil, T., Brožek, V., Etessami, K., Kučera, A.: Approximating the Termination Value of One-Counter MDPs and Stochastic Games. In: Aceto, L., Henzinger, M., Sgall, J. (eds.) ICALP 2011, Part II. LNCS, vol. 6756, pp. 332–343. Springer, Heidelberg (2011)
4. Brázdil, T., Brožek, V., Etessami, K., Kučera, A., Wojtczak, D.: One-counter Markov decision processes. In: Proceedings of SODA 2010, pp. 863–874. SIAM (2010)
5. Brázdil, T., Brožek, V., Forejt, V., Kučera, A.: Reachability in recursive Markov decision processes. I&C 206(5), 520–537 (2008)
6. Brázdil, T., Brožek, V., Kučera, A., Obdržálek, J.: Qualitative reachability in stochastic BPA games. I&C 208(7), 772–796 (2010)
7. Brázdil, T., Kučera, A., Novotný, P., Wojtczak, D.: Minimizing expected termination time in one-counter Markov decision processes. CoRR abs/1205.1473 (2012)
8. Chatterjee, K., Doyen, L.: Energy Parity Games. In: Abramsky, S., Gavoille, C., Kirchner, C., Meyer auf der Heide, F., Spirakis, P.G. (eds.) ICALP 2010, Part II. LNCS, vol. 6199, pp. 599–610. Springer, Heidelberg (2010)
9. Chatterjee, K., Doyen, L., Henzinger, T., Raskin, J.F.: Generalized mean-payoff and energy games. In: Proceedings of FST&TCS 2010 [1], pp. 505–516
10. Etessami, K., Wojtczak, D., Yannakakis, M.: Recursive Stochastic Games with Positive Rewards. In: Aceto, L., Damgård, I., Goldberg, L.A., Halldórsson, M.M., Ingólfsdóttir, A., Walukiewicz, I. (eds.) ICALP 2008, Part I. LNCS, vol. 5125, pp. 711–723. Springer, Heidelberg (2008)
11. Etessami, K., Wojtczak, D., Yannakakis, M.: Quasi-birth-death processes, tree-like QBDs, probabilistic 1-counter automata, and pushdown systems. Performance Evaluation 67(9), 837–857 (2010)
12. Etessami, K., Yannakakis, M.: Recursive Markov Decision Processes and Recursive Stochastic Games. In: Caires, L., Italiano, G.F., Monteiro, L., Palamidessi, C., Yung, M. (eds.) ICALP 2005. LNCS, vol. 3580, pp. 891–903. Springer, Heidelberg (2005)
13. Etessami, K., Yannakakis, M.: Efficient Qualitative Analysis of Classes of Recursive Markov Decision Processes and Simple Stochastic Games. In: Durand, B., Thomas, W. (eds.) STACS 2006. LNCS, vol. 3884, pp. 634–645. Springer, Heidelberg (2006)
14. Filar, J., Vrieze, K.: Competitive Markov Decision Processes. Springer (1996)
15. Göller, S., Lohrey, M.: Branching-time model checking of one-counter processes. In: Proceedings of STACS 2010. LIPIcs, vol. 5, pp. 405–416. Schloss Dagstuhl (2010)
16. Jančar, P., Kučera, A., Moller, F., Sawa, Z.: DP lower bounds for equivalence-checking and model-checking of one-counter automata. I&C 188(1), 1–19 (2004)
17. Jančar, P., Sawa, Z.: A note on emptiness for alternating finite automata with a one-letter alphabet. IPL 104(5), 164–167 (2007)
18. Kučera, A.: The complexity of bisimilarity-checking for one-counter processes. TCS 304(1-3), 157–183 (2003)
19. Latouche, G., Ramaswami, V.: Introduction to Matrix Analytic Methods in Stochastic Modeling. ASA-SIAM series on statistics and applied probability (1999)
20. Puterman, M.: Markov Decision Processes. Wiley (1994)
21. Serre, O.: Parity Games Played on Transition Graphs of One-Counter Processes. In: Aceto, L., Ingólfsdóttir, A. (eds.) FOSSACS 2006. LNCS, vol. 3921, pp. 337–351. Springer, Heidelberg (2006)
22. Williams, D.: Probability with Martingales. Cambridge University Press (1991)

Prefix Rewriting for Nested-Words and Collapsible Pushdown Automata*

Christopher Broadbent

LIAFA (CNRS and Paris 7), Paris, France

Abstract. We introduce two natural variants of prefix rewriting on nested-words. One captures precisely the transition graphs of order-2 pushdown automata and the other precisely those of order-2 *collapsible* pushdown automata (2-CPDA). To our knowledge this is the first precise 'external' characterisation of 2-CPDA graphs and demonstrates that the class is robust and hence interesting in its own right. The comparison with our characterisation for 2-PDA graphs also gives an idea of what 'collapse means' in terms outside of higher-order automata theory. Additionally, a related construction gives us a decidability result for first-order logic on a natural subclass of 3-CPDA graphs, which in some sense is optimal.

1 Introduction

Higher-order recursion schemes are systems of rewrite rules on typed non-terminal symbols. They provide an excellent basis for modelling higher-order functional programs and so have enticing prospects for software verification.

Whilst a conventional *'order-1'* pushdown automaton (1-PDA) has access to a stack of atomic symbols, an order-n pushdown automaton (n-PDA) employs a stack that itself contains other stacks—an order-$(n+1)$ stack is a stack of order-n stacks. Unfortunately such devices share expressivity with only those recursion schemes satisfying a constraint called *safety* [17]. For expressive parity with unrestricted schemes, the structure of higher-order stacks must be enriched with 'links' and an associated *'collapse'* operation, giving the order-n *collapsible pushdown automata* (n-CPDA) [18,14].

This paper focuses on the *configuration graphs* of these automata rather than the trees that they generate. Roughly speaking, such graphs consist of a set of stack configurations with a directed edge connecting one configuration to another whenever the underlying automaton would be able to transition between them. The ϵ-*closure* of such a graph 'glues together' nodes linked by an arbitrarily large number of ϵ-transitions. The ϵ-closures of configuration graphs of n-PDA are well understood, coinciding precisely [9] with the nth level of the *Caucal Hierarchy* [11], a robust class of graphs with decidable MSO theories. By contrast there exists a 2-CPDA yielding a transition graph with undecidable MSO theory [14]. Progress on the question of *first-order logic* and CPDA graphs remained open until recently when Kartzow demonstrated that first-order logic is decidable on 2-CPDA graphs [15].

---

* Supported by La Fondation Sciences Mathématiques de Paris and by the project AMIS (ANR 2010 JCJC 0203 01 AMIS)

Full version: `mjolnir.cs.ox.ac.uk/~chrb/isophilic.pdf`

A. Czumaj et al. (Eds.): ICALP 2012, Part II, LNCS 7392, pp. 153–164, 2012.
© Springer-Verlag Berlin Heidelberg 2012

Kartzow's proof shows that 2-CPDA graphs are *tree automatic*. A relational structure is called *automatic* [16,5] with respect to some class of objects (such as finite words, ω-words or trees) if its domain can be represented as a set of such objects recognised by a finite automaton, together with finite automata recognising relations by acting synchronously on tuples of objects. Since such automata on words or trees have good closure properties, automatic structures enjoy decidable first-order theories.

Other than decidability, Kartzow's work is interesting in that it provides an *external* description of the relations between nodes in the CPDA graph using a natural presentation that is *prima facie* quite different to the internal workings of a CPDA. However, whilst all 2-CPDA graphs are tree automatic, not all tree automatic graphs are 2-CPDA. We build on Kartzow's work and give a *precise* characterisation of 2-CPDA graphs.

First Contribution: Seeking inspiration from conventional pushdown automata (order-1), we recall that their transition graphs can be precisely characterised by *prefix rewriting* on words [10,19]. We generalise prefix rewriting to *nested-words* [2,3], which are strings endowed with pointers between positions arranged in a well-nested manner. In fact for nested-words there are two ways of defining a set of prefixes; one which ignores the pointer structure (a *'rational prefix set'*) and another which makes use of it (a *'summary prefix set'*). Rewrite systems mapping rational prefix sets to rational prefix sets characterise *precisely* the ϵ-closures of 2-PDA graphs, whilst those mapping summary prefix sets to rational sets characterise *precisely* the ϵ-closures of 2-CPDA graphs.

This gives an account of the difference between 2-PDA and 2-CPDA and also shows that the 2-CPDA graphs form a robust class of interest in its own right. (The Caucal hierarchy already told us this for 2-PDA). We also hope that the rewrite systems (and associated notions of automaticity) will turn out to be of more general use.

Second Contribution: Representing stacks of 2-CPDA using nested-words allows us to represent stacks of a restriction of 3-CPDA, called $3_{\{2\}}$-*CPDA*, with *nested-trees*. Via Gaifman's Locality Theorem [13] and the Boolean closure of nested-tree automata acting on trees with a bounded number of branches, one can show that first-order logic is decidable on $3_{\{2\}}$-CPDA graphs *without ϵ-closure*. This generalisation of Kartzow's decidability result is *optimal* in that adding *any one of* order-3 links, ϵ-closure *or* going to order-4, generates graphs with undecidable first-order theory [7].

Related Work

Our work is heavily indebted to that of Kartzow [15]. Indeed there are close connections between nested-words and trees; each type of structure can be encoded as the other. However, whilst every structure generated by our prefix rewrite systems is tree automatic, the converse is not true; tree automaticity is too strong to give the requisite precise characterisation. Nevertheless, in order to convert 2-(C)PDA into prefix rewrite systems, we use a construction that is very similar to that of Kartzow [15]. However, we differ from Kartzow in the way that we show reachability to be definable; our method exploits the fact we are working with nested-words rather than trees.

Our proof that a Kartzow-style construction works for nested-words is most neatly formulated using a notion of 'automaticity' that turns out to be equivalent to our prefix-rewrite systems. One challenge is developing a way in which nested-word automata can act synchronously on nested-words, even if their nesting structure differs. Arenas *et al.* provided a comprehensive solution to this problem [4]; we differ by dramatically reducing the power of synchronisation to the point where synchronisation may only occur on the initial segments that the nested-words have in common.

Carayol [8] also defined a generalisation of prefix rewriting for n-PDA (without collapse) which is quite different from our own in that it is described in terms of stack operations. Carayol's advantages include being able to deal with all orders (rather than just level-2) and obtaining some good logical characterisations for these graphs. Our strong points are being able to say something about collapse, as well as formulating our rewrite systems in terms naturally understood without referring to higher-order stacks.

2   Prefix Rewriting on Nested-Words

2.1   Nested-Words

Recall that a Σ-labelled *word* is a map $w\ :\ \mathbf{dom}(w)\longrightarrow \Sigma$ where $\mathbf{dom}(w)$ is a downward closed subset of $\mathbb{N}$. A *(semi-)nested-word* is a word endowed with 'back pointers' that are arranged in a well-nested manner.

Definition 1. *A semi-nested-word over Σ is a pair $w^{\frown E}$ where w is a Σ-labelled word and $E : \mathbf{dom}(w) \rightharpoonup \mathbf{dom}(w)$ is a partial map such that $\mathbf{dom}(E) \cap \mathbf{img}(E) = \emptyset$ and for all $x \in \mathbf{dom}(E)$, $E(x) < x$ and $E(x) \le E(y)$ for every $y \in \mathbf{dom}(E)$ such that $E(x) < y < x$. We call it a nested-word if the last requirement is strengthened to $E(x) < y < x$ implying $E(x) \lneq E(y)$.*

Graphically we represent E using pointers as in Figure 1. This set of nested-words is denoted $\mathbf{NWord}(\Sigma)$.

The definition is very similar to that given by Alur *et al.* [3] with the main difference that we disallow 'unmatched calls'—we cannot have a position in the word being the target of a pointer without the pointer having a corresponding source. This is important when considering a

Fig. 1. Semi-nested and nested-word

prefix of a nested-word $w^{\frown E}$ which we define to be a nested-word $w\restriction_S{}^{\frown E\restriction_S}$ for some downward closed subset $S \subseteq \mathbf{dom}(w)$. (We often abbreviate this to $u^{\frown E\restriction_u}$ where u is a prefix of w.) The requirement that a prefix of a nested-word be itself a nested-word removes information about pointers sourced in the suffix and targeted at the prefix. It is useful to have a notion of prefix for which such information is retained. For this reason we make a distinction with *visibly pushdown words* [2] which coincide with Alur *et al.*'s notion of nested-words but differ from our own. *Left* of Figure 2 illustrates this.

Definition 2. *Given an alphabet Σ, the associated* visibly pushdown alphabet $\mathfrak{V}(\Sigma)$ *is given by $\mathfrak{V}(\Sigma) := \Sigma \cup \acute{\Sigma} \cup \dot{\Sigma}$, where $\acute{\Sigma} = \{\, \mathring{a}\ :\ a \in \Sigma \,\}$ and $\dot{\Sigma} = \{\, \dot{a}\ :\ a \in \Sigma \,\}$. Given $w^{\frown E} \in \mathbf{NWord}(\Sigma)$ we define the* visibly pushdown word $\mathfrak{V}(w^{\frown E})$ *to be the $\mathfrak{V}(\Sigma)$-word w' where $\mathbf{dom}(w') = \mathbf{dom}(w)$ and for each $x \in \mathbf{dom}(w')$ with $w(x) = a$:*

$$w'(x) := \begin{cases} \dot{a} & \text{if } x \in \mathbf{img}(E) \\ a & \text{if } x \notin \mathbf{img}(E) \cup \mathbf{dom}(E) \\ \mathring{a} & \text{if } x \in \mathbf{dom}(E) \end{cases}$$

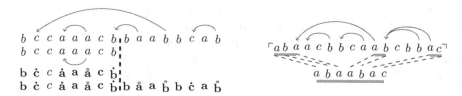

Fig. 2. *Left*: Prefix of a nested-word *vs.* visibly pushdown word. *Right*: Summary

Note that $\mathfrak{V}$ does not make sense for semi-nested words as the pointer structure could not be uniquely recovered from the image of the map.

Another concept that we borrow from Alur *et al.* is the idea of the *summary* $\ulcorner w^{\frown E} \urcorner$ of a nested-word $w^{\frown E}$. This is the string that is obtained by reading the nested-word from left to right 'skipping over' the back edges as they are found. This concept applies equally well to semi-nested words.

Definition 3. *The* summary $\ulcorner w \urcorner$ *of a (semi-)nested-word w is defined recursively as follows:*

$$\ulcorner w\, a \urcorner := \ulcorner w \urcorner\, a \;\text{ if } a \text{ sources no pointer}$$

$$\ulcorner w_0\, \overset{\frown}{a_0\, w_1}\, a_1 \urcorner := \ulcorner w_0 \urcorner\, a_0\, a_1$$

A *(semi-)nested-word automaton* $\mathcal{A}$ is a finite state machine that reads (semi-)nested-words $w^{\frown E}$ from left to right. It differs from a standard finite word automaton in that it may be sensitive to whether the node it is reading is in $\mathbf{dom}(E)$, $\mathbf{img}(E)$ or neither, and when reading a position $u \in \mathbf{dom}(E)$ its next transition may depend on the label $w(u)$ of u, its current state *and also* the state that it was in when reading $E(u)$. The *language* $\mathcal{L}(\mathcal{A})$ recognised by $\mathcal{A}$ consists of those (semi)-nested-words on which $\mathcal{A}$ has a run ending in one of a select number of 'accepting states'. Nested-word automata can be *determinised* and are hence closed under Boolean operations [2,3]. Languages so recognised are called *regular nested-word languages*.

2.2   Prefix and Suffix Rewriting

Whilst conventionally prefix rewriting [10] is employed, we find it technically convenient to work with *suffix* rewrite systems. We only define the latter, but they are interconvertable with a 'prefix version' due to the fact that determinised nested-word automata are reversible.

A *suffix-rewrite system* would traditionally consist of a regular language $\mathcal{L}$ of standard (non-nested) words together with pairs of regular languages $(\mathcal{L}_i, \mathcal{L}'_i)$ each assigned a label e_i. The *suffix recognisable graph* generated by such a system has $\mathcal{L}$ as its node-set and an e_i labelled edge from a word of the form uv to a word of the form uv' whenever $v \in \mathcal{L}_i$ and $v' \in \mathcal{L}'_i$. Call the resulting class of structures $\mathcal{RW}$.

All of our generalisations have the form:

$$\mathcal{RW} = \left\langle\, \mathcal{L}, \mathcal{L}_1 \xrightarrow{e_1} \mathcal{L}'_1, \ldots, \mathcal{L}_k \xrightarrow{e_k} \mathcal{L}'_k \,\right\rangle$$

where $\mathcal{L}$ is a regular *nested*-word language over some alphabet Σ and $\mathcal{L}_i$, $\mathcal{L}'_i$ are regular languages of *non-nested words* over that same alphabet. The nodes of the graph generated are always the elements of $\mathcal{L}$.

The 'dangling pointers' targeting the suffix whose sources disappear during the rewriting of a suffix must be 'preserved'. The language $\mathcal{L}$ will constrain where their new sources in the new suffix may be located. In order to express this, we use $\mathfrak{V}$ to mark the prefix in order to track the targets of pointers that originate there. Given a nested-word of the form $(uv)^{\frown E}$, we abuse notation and write $\mathfrak{V}(u^{\frown E})$ to mean the prefix of $\mathfrak{V}((uv)^{\frown E})$ corresponding to u (noting E defines all positions in u that were originally sources.) A

- *rat-rat* system has an e_i-labelled edge from $(uv)^{\frown E}$ to $(uv')^{\frown E'}$ if $v \in \mathcal{L}_i$, $v' \in \mathcal{L}'_i$ and $\mathfrak{V}(u^{\frown E}) = \mathfrak{V}(u^{\frown E'})$. Denote this class of graphs by $\mathcal{RW}_{rr}$.
- *sum-rat* system has an e_i-labelled edge from $(uv)^{\frown E}$ to $(uv')^{\frown E'}$ if $\ulcorner v^{\frown E}{}_v \urcorner \in \mathcal{L}_i$, $v' \in \mathcal{L}'_i$ and $\mathfrak{V}(u^{\frown E}) = \mathfrak{V}(u^{\frown E'})$. Denote this class of graphs by $\mathcal{RW}_{sr}$.

Note that sum-rat rewrite systems subsume their rat-rat counterparts. It is always possible to enrich the words in the domain so that each node $u \in \mathbf{dom}(E)$ with corresponding $u' \in \mathbf{img}(E)$ is decorated with a function mapping each state q of a given regular automaton to the sets of states that it could be in at node u had it started reading the word from u' in state q.

Example 1. The following is an example of a traditional rational suffix rewrite system (the suffix analogue of [10]). Consider $\mathcal{L} = (a + b)^*$ and a rule: $e : a^*b \longrightarrow b^*a$. Then we have the following edges in the graph:

$$abaa\underline{aaa}b \xrightarrow{\quad e \quad} abaa\underline{abba} \quad \text{and} \quad aaa\underline{aaaaaa}b \xrightarrow{\quad e \quad} aaa\underline{bbbbba}$$

Example 2. We now give an example of a rat-rat system. Consider an $\mathcal{L}$ based on $(a + b)^*$ that allows arbitrary well-nested pointers. The rule: $e : a^*b \longrightarrow b^*a$ interpreted as a rat-rat rule would now give the following edges in the graph:

$$aabaaa\underline{aaab} \xrightarrow{\quad e \quad} aabaa\underline{abba} \quad \text{and} \quad aaa\underline{aaaaaaab} \xrightarrow{\quad e \quad} aaa\underline{bbbbba}$$

To give an example of a sum-rat system, let us interpret the rule as a sum-rat rule. This now gives the following edge (which would not belong to the rat-rat interpretation).

$$aaa\underline{aaabab} \xrightarrow{\quad e \quad} aaa\underline{bbbba}$$

3   (C)PDA Graphs

3.1   Higher-Order Stacks

Let us fix a stack-alphabet Γ. For higher-order automata this alphabet must be finite, but it is convenient for definitions to allow it to be infinite. An *order*-1 stack over Γ is just a string of the form $[\, \gamma \,]$ where $\gamma \in \Gamma^*$. Let us refer to the set of order-1 stacks over Γ as $stack_1(\Gamma)$. For $n \in \mathbb{N}$ the set of *order-$(n+1)$* stacks is: $stack_{n+1}(\Gamma) := stack_1(stack_n(\Gamma))$. We allow the following operations on an order-1 stack s for every $a \in \Gamma$: $push_1^a([\ a_1 \cdots a_m\ ]) := [\ a_1 \cdots a_m a\ ]$, $pop_1([\ a_1 \cdots a_m a_{m+1}\ ]) := [\ a_1 \cdots a_m\ ]$, $nop(s) := s$. We allow the following *order-$(n+1)$* operations on an order-$(n+1)$ stack s, where θ is any *order-n* operation: $push_{n+1}([\ s_1 \cdots s_m\ ]) := [\ s_1 \cdots s_m s_m\ ]$, $pop_{n+1}([\ s_1 \cdots s_m s_{m+1}\ ]) := [\ s_1 \cdots s_m\ ]$, $\theta([\ s_1 \cdots s_{m-1} s_m\ ]) := [\ s_1 \cdots s_{m-1} \theta(s_m)\ ]$. Where s is an $(n+1)$-stack we write $top_{n+1}(s)$ to denote the top-most (right-most) order-n stack (abusing notation with $top_{n+1}(s) := s$ if s is an n-stack) and $top_1(s)$ as the top atomic element. Define $|s| := m$ for $s = [\ s_1 \cdots s_m\ ]$.

3.2   Collapsible Pushdown Stacks

CPDA include a new $push_1^{a,k}$ operation, which attaches a k-*link* from the newly created a element that points to the k-stack below. The targets of these links are preserved during higher-order *push* operations in the manner illustrated by the following example:

Example 3.

$$[[[abca]\ [ab]]] \xrightarrow{push_1^{a,2};push_2;push_3;push_1^{c,3};push_3}$$

$$[[[abca]\ [aba]\ [aba]]\ [[abca]\ [aba]\ [abac]]\ [[abca]\ [aba]\ [abac]]]$$

We offer fine control over the orders of links that a collapsible stack may contain. We include the orders of links as a subscript, so an order-n_S stack is an order-n stack equipped with order-i links for each $i \in S$. Formally the *S-collapsible pushdown alphabet* $\Gamma^{[S]}$ (for $S \subseteq \mathbb{N}$) induced by an alphabet Γ is the set $\Gamma \times S \times \mathbb{N}$. The set of order-$n_S$ collapsible stacks $stack_{n_S}^C(\Gamma)$ is defined by: $stack_{n_S}^C(\Gamma) := stack_n(\Gamma^{[S]})$. An atomic element $(a, l, p) \in \Gamma^{[S]}$ has label a with a link of order l. If $l < n$ the target of a link is the pth $(l-1)$-stack of the l-stack in which the element resides.

We thus introduce the operations $push_1^{a,l}(s) := push_1^{(a,l,|top_{l+1}(s)|-1)}(s)$ and $collapse(s) := pop_l^{|top_{l+1}(s)|-p}$ where $top_1(s) = (a, l, p)$ and θ^m represents the operation θ iterated m times. From now on we will abuse notation and consider $top_1(s) := a$. Write Θ_{n_S} to denote the set of order-n_S collapsible stack operations.

3.3   The Automata and Their Graphs

Let $n \in \mathbb{N}$ and let $S \subseteq [2..n]$. An n_S-*CPDA* (order-n_S collapsible pushdown automaton) $\mathcal{A}$ is a tuple: $\langle\ \Sigma, \Pi, Q, q_0, \Gamma, R_{a_1}, R_{a_2}, \ldots, R_{a_r}, P_{b_1}, P_{b_2}, \ldots, P_{b_{r'}}\ \rangle$ where Σ is a finite set of transition labels $\{a_1, a_2, \ldots, a_r\}$; Π is a finite set of configuration labels $\{b_1, b_2, \ldots, b_{r'}\}$; Q is a finite set of control-states; $q_0 \in Q$ is an initial control-state; Γ is a finite stack alphabet; each R_{a_i} is the a_i-labelled transition relation with $R_{a_i} \subseteq Q \times \Gamma \times \Theta_{n_S} \times Q$; each P_{b_i} is the b_i-labelled unary predicate specified by $P_{b_i} \subseteq Q \times \Gamma$. Remaining consistent with the definitions in the literature, an n-CPDA is an $n_{[n..2]}$-CPDA and an n-PDA is an $n_\emptyset$-CPDA.

A *configuration* of an n_S-CPDA $\mathcal{A}$ is a pair (q, s) where q is a control-state and s is an n_S-stack. Such a configuration *satisfies* the predicate $b_i \in \Pi$ just in case $(q, top_1(s)) \in P_{b_i}$. We say $\mathcal{A}$ can a_i-*transition from* (q, s) *to* $(q', \theta(s))$, written $(q, s) \xrightarrow{a_i} (q', \theta(a))$, iff $(q, top_1(s), \theta, q') \in R_{a_i}$. Let us further say that (q', s') *can be reached from* (q, s) *in* $\mathcal{A}$ *with path labeled in* $\mathcal{L}$ for some $\mathcal{L} \subseteq \Sigma^*$ iff $(q, s) \xrightarrow{a_{i_1}} (q_1, s_1) \xrightarrow{a_{i_2}} (q_2, s_2) \xrightarrow{a_{i_3}} \cdots (q_{m-1}, s_{m-1}) \xrightarrow{a_{i_m}} (q', s')$ for some configurations $(q_1, s_1), \ldots, (q_{m-1}, s_{m-1})$ where $a_{i_1} a_{i_2} a_{i_3} \cdots a_{i_m} \in \mathcal{L}$. We write $(q, s) \xrightarrow{\mathcal{L}} (q', s')$ to mean this. The set of *reachable configurations* of $\mathcal{A}$ is given by:

$$R(\mathcal{A}) := \{ (q, s) : (q_0, \bot_n) \xrightarrow{\Sigma^*} (q, s) \}$$

where $\bot_n$ is the empty n-stack.

Definition 4. *The configuration graph of (graph generated by) $\mathcal{A}$ has domain $R(\mathcal{A})$, unary predicates Π and directed edges Σ between configurations. We write $\mathcal{G}(\mathcal{A})$ to denote this graph. The ϵ-closure $\mathcal{G}^\epsilon(\mathcal{A})$ is induced from $\mathcal{G}(\mathcal{A})$ by restricting the domain to nodes reachable from the initial configuration by a path of the form $\xrightarrow{\epsilon^* b}$ for $b \neq \epsilon$, and defining a-labelled edges between vertices related by $\xrightarrow{\epsilon^* a}$ in $\mathcal{G}(\mathcal{A})$.*

First-Order Logic on such a graph has the unary and binary relation labels as predicates, together with existential and universal quantification over nodes. A *sentence* is a formula with no free variables and the *first-order theory* of a structure is the set of first-order sentences that are true of it.

4   Characterising 2-(C)PDA Graphs as Prefix Rewrite Systems

The following relationship holds:

Theorem 1. *The class of graphs $\mathcal{RW}_{sr}$ contains precisely the ϵ-closures of 2-CPDA graphs. The class of graphs $\mathcal{RW}_{rr}$ contains precisely the ϵ-closures of 2-PDA graphs.*

We thus see that the power of the 'collapse' operation corresponds precisely to the ability of a rewrite system to look at summaries. Note that this characterisation also makes clear the inherent *asymmetry* of the collapse operation; whilst 2-PDA are characterised using 'symmetric rules' ('rat' on both sides) this is not the case for 2-CPDA.

We now sketch how the above theorem is proved.

4.1   From Rewrite Systems to 2-(C)PDA

We define maps encoding nested-words as order-2 stacks:

$$\mathfrak{S} : \mathbf{NWord}(\Sigma) \longrightarrow stack_2(\mathfrak{V}(\Sigma)) \quad \text{and} \quad \mathfrak{S}^{\mathcal{C}} : \mathbf{NWord}(\Sigma) \longrightarrow stack_{2_2}^{\mathcal{C}}(\mathfrak{V}(\Sigma))$$

Fig. 3. The 2-CPDA stack corresponding to a nested-word

Consider $w^{\frown E} \in \mathbf{NWord}(\Sigma)$. We define $\mathfrak{S}^{\mathcal{C}}(w^{\frown E}) := \mathfrak{S}^{\mathcal{C}-}(\mathfrak{V}(w^{\frown E}))$ where:

$$\mathfrak{S}^{\mathcal{C}-}(\epsilon) \quad := \quad \bot_2$$
$$\mathfrak{S}^{\mathcal{C}-}(v\,\mathring{a}) \quad := \quad (push_2; push_1^{\mathring{a},2})(\mathfrak{S}^{\mathcal{C}}(v))$$
$$\mathfrak{S}^{\mathcal{C}-}(v\,a) \quad := \quad push_1^a(\mathfrak{S}^{\mathcal{C}}(v))$$
$$\mathfrak{S}^{\mathcal{C}-}(v\,\overset{\frown}{\mathring{a}\cdots c\,\mathring{b}}) \quad := \quad push_1^{\mathring{b}}(\mathfrak{S}^{\mathcal{C}}(v\,\mathring{a}\cdots c) :: top_2(\mathfrak{S}^{\mathcal{C}}(v\,\mathring{a})))$$

where we define $[\, s_1 \cdots s_m \,] :: s_{m+1} := [\, s_1 \cdots s_m\ s_{m+1} \,]$. We define $\mathfrak{S}(w^{\frown E})$ in the same way as $\mathfrak{S}^{\mathcal{C}}(w^{\frown E})$, ignoring stack pointers.

Example 4. Consider the nested-word:

$$w^{\frown E} := a\,b\,a\,\overset{\frown}{b}\,b\,\overset{\frown}{b}\,c\,a\,c\,a\,b\,a\,a\,b\,a\,c\,b\,a$$

We illustrate $\mathfrak{S}^{\mathcal{C}}(w^{\frown E})$ in Figure 3.

We construct a 2-PDA (resp. 2-CPDA) that performs a series of transitions with labels of the form $\epsilon^* e$ from stack $\mathfrak{S}(w^{\frown E})$ to $\mathfrak{S}(w'^{\frown E'})$ (*resp.* $\mathfrak{S}^{\mathcal{C}}(w^{\frown E})$ to $\mathfrak{S}^{\mathcal{C}}(w'^{\frown E'})$) iff $w^{\frown E}$ suffix rewrites to $w'^{\frown E'}$ via an e-labelled rule.

In actual fact we extend the stack alphabet so that an atom a in the stack, corresponding to a letter a in a nested-word $w^{\frown E}$, is annotated with state information concerning a run on $w^{\frown E}$ by the deterministic nested-word automaton $\mathcal{A}$ recognising $\mathcal{L}$. The definition of $\mathfrak{S}(w^{\frown E})$ allows the (C)PDA to 'lookup' the state of $\mathcal{A}$ at the target $\mathring{a}$ of a pointer when reading its source $\mathring{b}$. This is because $\mathring{b}$ will always be recorded immediately above $\mathring{a}$ in some 1-stack. In order to skip over a pointer sourced at $\mathring{b}$ one may simply pop_1 and then *collapse* from its target $\mathring{a}$. This is useful when 'reading a summary backwards' for the left-hand side of a sum-rat rule.

If one wishes to ignore pointers (as needed for the left-hand-side of a rat-rat rule) one may use pop_2 at $\mathring{a}$ instead of pop_1; *collapse*. For this a 2-PDA suffices.

4.2   (Tree) Isophilic and Dendrisophilic Structures

In order to establish the converse direction, from 2-CPDA to rewrite systems, we use an intermediate representation of the graphs; this is a kind of 'automaticity' for nested-words. There are two main variants which we call *isophilic* and *dendrisophilic*. We also introduce a notion of automaticity with respect to '*nested-trees*', which will be able to describe $3_{\{2\}}$-CPDA graphs and lead to our new decidability result for first-order logic.

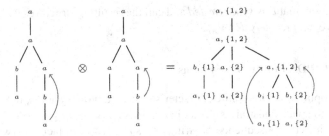

Fig. 4. Convolution of Two Trees

A *quasi-nested-tree* is a tree whose branches are nested-words and a *nested-tree* [1] is one insisting that every target of a pointer has a corresponding source on each branch. Nested-tree automata act top down on a nested-tree, and like their nested-word counterparts are allowed to look up the state they were in at the target of a pointer when reading its source. We can represent pairs (and indeed k-tuples) of nested-trees using '*convolutions*' as illustrated in Figure 1. The idea is that the convolution 'splits' whenever there is a difference in label or pointer.

Convolutions enable us to define a notion of 'relations recognised by a nested-tree automaton'. Unfortunately these automata have poor closure properties and so 'nested-tree automatic structures' do not necessarily have decidable first-order theories. In order to address this, we consider the '*wreath product*' $\mathcal{B}^{\mathcal{C}}$ of a restricted nested-tree automaton $\mathcal{B}$ and a standard tree automaton $\mathcal{C}$ (that is oblivious to pointers). $\mathcal{C}$ is allowed to inspect the state of $\mathcal{B}$ whilst $\mathcal{B}$ is restricted in various different ways so as to give Boolean closure. Each kind of restriction yields a different class of automatic structures.

Path-Nested   Spine Nested

Fig. 5. The black lines indicate the branches along which $\mathcal{B}$ must be deterministic and ignore branching, whilst along the grey it may be unrestricted

One possible restriction is to insist that $\mathcal{B}$ is *path-nested*, which means that it must act as a deterministic nested-word automaton along each branch. Structures whose relations (as convolutions) are recognised by path-nested automata are called *tree isophilic*.

Another way to recover Boolean closure is to restrict the domain to nested-*words*. In this case a k-ary relation will be represented by convolutions having exactly k branches. Since it can be seen that *unrestricted* nested-tree automata acting on trees with a bounded number of branches *are Boolean closed*, we get sensible notions of automaticity for nested-words, even when $\mathcal{B}$ is unrestricted.

For unrestricted nested-word automata it turns out that we subsume the word automatic structures. However, there is still good reason to restrict $\mathcal{B}$ even for nested-words. Indeed let us restrict our attention to structures with only binary relations. If $\mathcal{B}$ is path-nested we describe the structures as *isophilic* and these coincide *precisely* with $\mathcal{RW}_{rr}$. If $\mathcal{B}$ is deterministic along the right-most branch (spine) of a tree (but unrestricted elsewhere) we say that $\mathcal{B}$ is *spine-nested* and call the resulting structures *dendrisophilic*. In this case we get *precisely* $\mathcal{RW}_{sr}$.

4.3   From (C)PDA to Rewrite Systems

We use the isophilic/dendrisophilc characterisation of rewrite systems, but only sketch the ideas due to space constraints. Note that an order-2 stack $[\, s_1\ s_2\ \cdots\ s_m\,]$ without links can be viewed as a sequence of strings $s_1, s_2, \ldots, s_m$. In fact even when 2-links are added, it is possible to represent a stack as such a sequence. Roughly speaking one can 'colour' the atoms in the stack so that the link of the top element of s_m always points to s_j, where j is the least number such that $s_m \sqsubseteq s_i$ for every i with $j < i \le m$. Intuitively this works because 2-links track the point at which an atom was first created.

Given a sequence of strings $abccba, abcbba, abcbaab, abccbbb, abccbbc, bba$ we can *uniquely* encode it as a semi-nested-word:

where the elements of the sequence are recovered by considering the summaries at the element preceding each $\bullet$. Unicity is achieved by adding an atom to the right of a $\bullet$ only if 'really necessary'—*i.e.* only if the pointer from the $\bullet$ cannot be made to point to another instance of that atom by moving it right one space along the previous summary.

Whilst this is the intuitive representation of a sequence of strings (and hence of an order-2 stack), it is necessary to encode it as a nested-word with pointers such that the convolution does not split too early to make the requisite comparisons. Once a suitable encoding has been made, individual stack operations can be represented in the same manner as by Kartzow; for example a $push_2$ can be represented by 'adding a $\bullet$' with the 'same' (modulo the encoding) target as its preceding $\bullet$. Dendrisophilicity is necessary for *collapse* as a small dose of non-determinism is needed to guess the j (as described above) characterising the target of the top link. All other stack operations are isophilic.

However, we differ from Kartzow in that our nested-word setting allows us to show the definability of ϵ-closure as the result of a more general construction witnessing the definability of reachability in all isophilic and dendrisophilic structures. Observe that sequences of nested-words can be represented as quasi-nested-trees, as illustrated

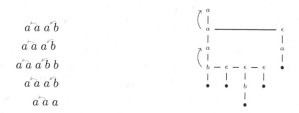

Fig. 6. Encoding a sequence of nested-words

in Figure 6. In fact this quasi-nested tree will be a genuine nested-tree after running the branches through our 'encoding'. Moreover pairs of branches having adjacent leaves can be viewed as forming a binary convolution. This means that a nested-tree automaton $\mathcal{A}$ can be constructed that verifies every 'convolution' with respect to an isophilic/dendrisophilic relation. If this verification is successful, the tree must represent a *path* in the structure. By 'projecting away' all but the left and right most branches of the tree, plus a bit of additional manipulation, it turns out that $\mathcal{A}$ becomes a path-nested/spine-nested automaton recognising the reachability relation.

5 First-Order Logic on $3_{\{2\}}$-CPDA Graphs

Since the stack of a $3_{\{2\}}$-CPDA can be viewed as simply a sequence of 2-CPDA stacks, we can use the nested-word encoding of order-2 stacks combined with Figure 6 to encode $3_{\{2\}}$-stacks as a nested-tree. Unfortunately, in order to recognise the representations of $3_{\{2\}}$-stacks *reachable* by a given $3_{\{2\}}$-CPDA, we require an unrestricted nested-tree automaton. (By contrast, 3-PDA graphs are tree isophilic.)

However, we can still establish that $3_{\{2\}}$-CPDA graphs have decidable first-order theories if there is no ϵ-closure. This is because without ϵ-closure the number of 2-stacks in a $3_{\{2\}}$-stack will change by at most one with a single transition, and hence by at most k after k-transitions. Thus if we restrict our attention to comparing encodings of configurations that are at most k transitions apart, we can do so by comparing just the right-most $(2k + 1)$ branches of their convolutions. Automata acting on a bounded number of branches are Boolean closed. Moreover by Gaifman's Locality Theorem [13] it turns out that one only needs Boolean closure on these local comparisons to decide whether any given first-order sentence holds of the graph. Thus we get:

Theorem 2. *Every $3_{\{2\}}$-CPDA graph (w/o ϵ-closure) has decidable first-order theory.*

6 Further Directions

The progression from words to nested-words to nested-trees (each encoding sequences of the previous) can be used to encode all ordinals strictly less than ω^ω, ω^{ω^ω} (isophilic) and $\omega^{\omega^{\omega^\omega}}$ (tree isophilic) in a similar way to the encoding of stacks of increasing order. An upper bound on the well-orders definable by *tree* isophilic structures would be good; since isophilic structures are tree automatic, ω^{ω^ω} is tight [12] (also implied by [6]).

The way in which we encode nested-words as stacks reminds us a little of the way in which 'traversals' (semantics of a recursion scheme) are 'computed' by CPDA [14]. It would be interesting to explore any connections further.

Acknowledgement. The author thanks the anonymous reviewers for their suggestions.

References

1. Alur, R., Chaudhuri, S., Madhusudan, P.: Languages of Nested Trees. In: Ball, T., Jones, R.B. (eds.) CAV 2006. LNCS, vol. 4144, pp. 329–342. Springer, Heidelberg (2006)

2. Alur, R., Madhusudan, P.: Visibly Pushdown Languages. In: STOC, pp. 202–211. ACM (2004)
3. Alur, R., Madhusudan, P.: Adding nesting structure to words. Journal of the ACM (JACM) 56(3), 1–43 (2009)
4. Arenas, M., Barcelo, P., Libkin, L.: Regular Languages of Nested Words: Fixed Points, Automata, and Synchronization. Theory Comput. Syst. 49(3), 639–670 (2011)
5. Blumensath, A., Gradel, E.: Automatic structures. In: LICS, pp. 51–62. IEEE Comput. Soc. (2000)
6. Braud, L., Carayol, A.: Linear Orders in the Pushdown Hierarchy. In: Abramsky, S., Gavoille, C., Kirchner, C., Meyer auf der Heide, F., Spirakis, P.G. (eds.) ICALP 2010, Part II. LNCS, vol. 6199, pp. 88–99. Springer, Heidelberg (2010)
7. Broadbent, C.: The Limits of Decidability for First-Order Logic on CPDA Graphs. In: STACS (2012)
8. Carayol, A.: Regular Sets of Higher-Order Pushdown Stacks. In: Jedrzejowicz, J., Szepietowski, A. (eds.) MFCS 2005. LNCS, vol. 3618, pp. 168–179. Springer, Heidelberg (2005)
9. Carayol, A., Wöhrle, S.: The Caucal Hierarchy of Infinite Graphs in Terms of Logic and Higher-Order Pushdown Automata. In: Pandya, P.K., Radhakrishnan, J. (eds.) FSTTCS 2003. LNCS, vol. 2914, pp. 112–123. Springer, Heidelberg (2003)
10. Caucal, D.: On Infinite Transition Graphs Having a Decidable Monadic Theory. In: Meyer auf der Heide, F., Monien, B. (eds.) ICALP 1996. LNCS, vol. 1099, pp. 194–205. Springer, Heidelberg (1996)
11. Caucal, D.: On Infinite Terms Having a Decidable Monadic Theory. In: Diks, K., Rytter, W. (eds.) MFCS 2002. LNCS, vol. 2420, pp. 165–176. Springer, Heidelberg (2002)
12. Delhommé, C.: Automaticité des ordinaux et des graphes homogènes. Comptes Rendus Mathematique 339(1), 5–10 (2004)
13. Gaifman, H.: On Local and Non-Local Properties. In: Stern, J. (ed.) Proceedings of the Herbrand Symposium. Studies in Logic and the Foundations of Mathematics, vol. 107, pp. 105–135. Elsevier (1982)
14. Hague, M., Murawski, A.S., Ong, C.-H.L., Serre, O.: Collapsible Pushdown Automata and Recursion Schemes. In: LICS. IEEE Computer Society (2008)
15. Kartzow, A.: Collapsible Pushdown Graphs of Level 2 are Tree-Automatic. In: STACS, pp. 501–512 (2010)
16. Khoussainov, B., Nerode, A.: Automatic Presentations of Structures. In: Leivant, D. (ed.) LCC 1994. LNCS, vol. 960, pp. 367–392. Springer, Heidelberg (1995)
17. Knapik, T., Niwiński, D., Urzyczyn, P.: Higher-Order Pushdown Trees Are Easy. In: Nielsen, M., Engberg, U. (eds.) FOSSACS 2002. LNCS, vol. 2303, pp. 205–222. Springer, Heidelberg (2002)
18. Knapik, T., Niwiński, D., Urzyczyn, P., Walukiewicz, I.: Unsafe Grammars and Panic Automata. In: Caires, L., Italiano, G.F., Monteiro, L., Palamidessi, C., Yung, M. (eds.) ICALP 2005. LNCS, vol. 3580, pp. 1450–1461. Springer, Heidelberg (2005)
19. Stirling, C.: Decidability of Bisimulation Equivalence for Pushdown processes. Tech. Report, Division of Informatics, University of Edinburgh (2000)

A Saturation Method for Collapsible Pushdown Systems*

Chris Broadbent[1], Arnaud Carayol[2], Matthew Hague[1,2], and Olivier Serre[1]

[1] LIAFA, Université Paris Diderot – Paris 7 & CNRS
[2] LIGM, Université Paris-Est & CNRS

Abstract. We introduce a natural extension of collapsible pushdown systems called annotated pushdown systems that replaces collapse links with stack annotations. We believe this new model has many advantages. We present a saturation method for global backwards reachability analysis of these models that can also be used to analyse collapsible pushdown systems. Beginning with an automaton representing a set of configurations, we build an automaton accepting all configurations that can reach this set. We also improve upon previous saturation techniques for higher-order pushdown systems by significantly reducing the size of the automaton constructed and simplifying the algorithm and proofs.

1 Introduction

Via languages such as C++, Haskell, Javascript, Python, or Scala, modern day programming increasingly embraces higher-order procedure calls. This is a challenge for software verification, which usually does not model recursion accurately, or models only first-order calls (e.g. SLAM [2] and Moped [29]). Collapsible pushdown systems (collapsible PDS) are an automaton model of (higher-order recursion) schemes [11,24], which allow reasoning about higher-order recursion.

Collapsible pushdown systems are a generalisation of higher-order pushdown systems (higher-order PDS). Higher-order PDS provide a model of schemes subject to a technical constraint called *safety* [23,19] and are closely related to the Caucal hierarchy [9]. These systems extend the stack of a pushdown system to allow a nested "stack-of-stacks" structure. Recently it has been shown by Parys that safety is a genuine constraint on definable traces [26]. Hence, to model higher-order recursion fully, we require collapsible PDS, which — using an idea from *panic automata* [20] — add additional *collapse* links to the stack structure. These links allow the automaton to return to the context in which a character was added to the stack.

These formalisms are known to have good model-checking properties. For example, it is decidable whether a given μ-calculus formula holds on the execution graph of a scheme [24] (or collapsible PDS [14]). Although, the complexity of

---

* Supported by Fond. Sci. Math. Paris, AMIS (ANR 2010 JCJC 0203 01 AMIS), FREC (ANR 2010 BLAN 0202 02 FREC) and VAPF (Région IdF). The full version of this paper is available from http://hal.archives-ouvertes.fr/hal-00694991.

A. Czumaj et al. (Eds.): ICALP 2012, Part II, LNCS 7392, pp. 165–176, 2012.
© Springer-Verlag Berlin Heidelberg 2012

such analyses is high — for an order-n collapsible PDS, reachability checking is complete for $(n-1)$-EXPTIME, while μ-calculus is complete for n-EXPTIME — the problem becomes PTIME if the arity of the recursion scheme, and the number of alternations in the formula, is bounded. The same holds true for collapsible PDS when the number of control states is bounded. Furthermore, when translating from a scheme to a collapsible PDS, it is the arity that determines the number of control states [14]. It has been shown by Kobayashi [21] that these analyses can be performed in practice. For example, resource usage properties of programs of orders up to five can be verified in a matter of seconds.

Kobayashi's approach uses *intersection types*. In the order-1 case, an alternative approach called *saturation* has been successfully implemented by tools such as Moped [29] and PDSolver [16]. Saturation techniques begin with a small automaton — representing a set of configurations — and add new transitions as they become necessary until a fixed point is reached. These algorithms, then, naturally do not pay the worst case complexity immediately, and hence, represent ideal algorithms for efficient verification. Furthermore, they also provide a solution to the *global* model checking problem: that is, determining the set of all system states that satisfy a property. This is particularly useful when, for example, composing analyses. Furthermore, when testing reachability from a given initial state, we may terminate the analysis as soon as this state is found. That is, we do not need to compute the whole fixed point.

Our first contribution is a new model of higher-order execution called *annotated pushdown systems* (annotated PDS)[1], which replace the collapse links of a collapsible PDS with annotations containing the stack the link pointed to. In addition to allowing a more straightforward definition of regularity and greatly simplifying the proofs of the paper, this model provides a more natural handling of collapse links, highlighting their connection with closures. In addition, configuration graphs of this model are isomorphic to those of collapsible PDS when restricted to configurations reachable from the initial configuration.

Our second contribution is a saturation method for backwards reachability analysis of annotated pushdown systems that can also be applied *as-is* to collapsible PDS. This is a global model-checking algorithm that is based on saturation techniques for higher-order pushdown automata [5,15,30]. Our algorithm handles alternating (or "two-player") as well as non-alternating systems.

In addition to the extension to annotated pushdown systems, the algorithm improves on Hague and Ong's construction for higher-order PDS [15] since the number of states introduced by the construction is no longer multiplied by the number of iterations it takes to reach a fixed point, potentially leading to a large reduction in the size of the automata constructed. In addition, both the presentation and the proofs of correctness are much less involved.

Related Work. In addition to the works mentioned above, solutions to global model checking problems have been proposed by Broadbent *et al.* [6]. Additionally, Salvati and Walukiewicz provide a global analysis technique for μ-calculus

---

[1] Kartzow and Parys have independently introduced a similar model [18].

properties using a Krivine machine model of schemes [28]. However, there are currently no versions of these algorithms available that do not pay immediately the exponential blow up.

Extensions of schemes with pattern matching have also been considered by Ong and Ramsay [25]. A recent algorithm by Kobayashi speeds up his techniques using an over-approximating least fixed point computation to give an initial input to a greatest fixed point computation [22]. Like saturation this is a 'bottom-up' approach and it would be interesting to see whether there are connections. Extensions of higher-order PDS to concurrent settings have also been considered by Seth [31].

The saturation technique has proved popular in the literature. It was introduced by Bouajjani *et al.* [4] and Finkel *et al.* [13] and based on a string rewriting algorithm by Benois [3]. It has since been extended to Büchi games [7], parity and μ-calculus conditions [16], and concurrent systems [32,1], as well as weighted pushdown systems [27]. In addition to various implementations, efficient versions of these algorithms have also been developed [12,33].

2   Preliminaries

2.1   Annotated Pushdown Systems

We define annotated stacks, their operations, and annotated pushdown systems.

Annotated Stacks. Let Σ be a set of stack symbols. We define a notion of annotated higher-order stack. Intuitively, an annotated stack of order-n is an order-n stack in which stack symbols have attached annotated stacks of order at most n. For the rest of the formal definitions, we fix the maximal order to n, and use k to range between n and 1. We simultaneously define for all $1 \leq k \leq n$, the set Stacks_k^n of stacks of order-k whose symbols are annotated by stacks of order at most n. Note, we use subscripts to indicate the order of a stack.

Definition 1 (Annotated Stacks). *The family of sets* $(\mathrm{Stacks}_k^n)_{1 \leq k \leq n}$ *is the smallest family (for point-wise inclusion) such that:*

- *for all* $2 \leq k \leq n$, Stacks_k^n *is the set of all (possibly empty) sequences* $[w_1 \ldots w_\ell]_k$ *with* $w_1, \ldots, w_\ell \in \mathrm{Stacks}_{k-1}^n$.
- Stacks_1^n *is all sequences* $[a_1{}^{w_1} \ldots a_\ell{}^{w_\ell}]_1$ *with* $\ell \geq 0$ *and for all* $1 \leq i \leq \ell$, a_i *is a stack symbol in* Σ *and* w_i *is an annotated stack in* $\bigcup_{1 \leq k \leq n} \mathrm{Stacks}_k^n$.

Observe that the above definition uses a least fixed-point. This ensures that all stacks are finite; in particular a stack cannot contain itself as an annotation. When the maximal order n is clear, we simply write Stacks_k instead of Stacks_k^n. We also write order-k stack to designate an annotated stack in Stacks_k^n.

An order-n stack can be represented naturally as an edge-labelled tree over the alphabet $\{[_{n-1}, \ldots, [_1,]_1, \ldots,]_{n-1}\} \uplus \Sigma$, with Σ-labelled edges having a second target to the tree representing the annotation. For technical convenience, a tree representing an order-k stack does not use $[_k$ or $]_k$ symbols (these appear uniquely at the beginning and end of the stack). An example order-3 stack is given below, with only a few annotations shown. The annotations are order-3 and order-2 respectively.

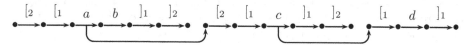

Given an order-n stack $w = [w_1 \ldots w_\ell]_n$, we define $top_{n+1}(w) = w$ and

$$top_n([w_1 \ldots w_\ell]_n) = w_1 \qquad \text{when } \ell > 0$$
$$top_n([]_n) = []_{n-1} \qquad \text{otherwise}$$
$$top_k([w_1 \ldots w_\ell]_n) = top_k(w_1) \text{ when } k < n \text{ and } \ell > 0$$

noting that $top_k(w)$ is undefined if $top_{k'}(w)$ is empty for any $k' > k$.

We write $u :_k v$ — where u is order-$(k-1)$ — to denote the stack obtained by placing u on top of the top_k stack of v. That is, if $v = [v_1 \ldots v_\ell]_k$ then $u :_k v = [uv_1 \ldots v_\ell]_k$, and if $v = [v_1 \ldots v_\ell]_{k'}$ with $k' > k$, $u :_k v = [u :_k v_1, \ldots, v_\ell]_{k'}$. This composition associates to the right. For example, the order-3 stack above can be written $[[[a^w b]_1]_2]_3$ and also $u :_3 v$ where u is the order-2 stack $[[a^w b]_1]_2$ and v is the empty order-3 stack $[]_3$. Then $u :_3 u :_3 v$ is $[[[a^w b]_1]_2[[a^w b]_1]_2]_3$.

Operations on Order-n Annotated Stacks. The following operations can be performed on an order-n stack. We say $o \in \mathcal{O}_n$ is of order-k when k is minimal such that $o \in \mathcal{O}_k$. For example, $push_k$ is of order k.

$$\mathcal{O}_n = \{pop_1, \ldots, pop_n\} \cup \{push_2, \ldots, push_n\} \cup \{collapse_2, \ldots, collapse_n\} \cup$$
$$\{ push_a^1, \ldots, push_a^n, rew_a \mid a \in \Sigma \}$$

We define each stack operation for an order-n stack w. Annotations are created by $push_a^k$, which add a character to the top of a given stack w annotated by $top_{k+1}(pop_k(w))$. This gives a access to the context in which it was created. In Section 3.2 we give several examples of these operations.

1. We set $pop_k(u :_k v) = v$.
2. We set $push_k(u :_k v) = u :_k u :_k v$.
3. We set $collapse_k\left(a^{u'} :_1 u :_{(k+1)} v\right) = u' :_{(k+1)} v$ when u is order-k and $n > k \geq 1$; and $collapse_n(a^u :_1 v) = u$ when u is order-n.
4. We set $push_b^k(w) = b^u :_1 w$ where $u = top_{k+1}(pop_k(w))$.
5. We set $rew_b(a^u :_1 v) = b^u :_1 v$.

Annotated Pushdown Systems. We are now ready to define annotated PDS.

Definition 2 (Annotated Pushdown Systems). *An order-n alternating annotated pushdown system (annotated PDS) is a tuple $\mathcal{C} = (\mathcal{P}, \Sigma, \mathcal{R})$ where $\mathcal{P}$ is a finite set of control states, Σ is a finite stack alphabet, and the set $\mathcal{R} \subseteq (\mathcal{P} \times \Sigma \times \mathcal{O}_n \times \mathcal{P}) \cup (\mathcal{P} \times 2^\mathcal{P})$ is a set of rules.*

We write *configurations* of an annotated PDS as a pair $\langle p, w \rangle$ where $p \in \mathcal{P}$ and $w \in Stacks_n$. We write $\langle p, w \rangle \longrightarrow \langle p', w' \rangle$ to denote a transition from a rule (p, a, o, p') with $top_1(w) = a$ and $w' = o(w)$. Furthermore, we have a transition $\langle p, w \rangle \longrightarrow \{ \langle p', w \rangle \mid p' \in P \}$ whenever we have a rule $p \to P$. A non-alternating annotated PDS has no rules of this second form. We write C to denote a set of configurations.

Collapsible Pushdown Systems. Annotated pushdown systems are based on collapsible PDS. In this model, stacks do not contain order-k annotations, rather they have order-k links to an order-k stack occurring lower down in the top-most order-$(k + 1)$ stack. We define the model formally in the full version. We give an example below, where links are marked with their order.

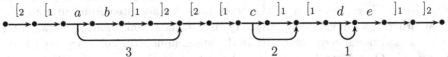

The set $\mathcal{O}_n$ is the same as in the annotated version. Collapse links are created by the $push_a^k$ operation, which augments a with a link to pop_k of the stack being pushed onto. A $collapse_k$ returns to the stack that is the target of the link.

Collapsible vs. Annotated. To an order-n stack w with links, we associate a canonical annotated stack $[\![w]\!]$ where each link is replaced by the annotated version of the link's target. We inductively and simultaneously define $[\![w]\!]_k$ which is the annotated stack representing $top_k(w)$.

$$
\begin{cases}
[\![\, [\,]_k :_{k+1} v \,]\!]_k = [\,]_k \\
[\![\, u :_{k'+1} v \,]\!]_k = [\![u]\!]_{k'} :_{k'+1} [\![v]\!]_k & \text{where } 0 < k' < k \\
[\![\, a^* :_1 v \,]\!]_k = a^{[\![collapse_{k'} (a^* :_1 v)]\!]_{k'}} :_1 [\![v]\!]_k & \text{where } * \text{ is an order-}k' \text{ link}
\end{cases}
$$

For example, the order-3 stack above becomes $[[[a^{w_1} b]_1]_2 [[c^{w_2}]_1 [d^{w_3} e]_1]_2]_3$ where $w_1 = [[[c^{w_2}]_1 [d^{w_3} e]_1]_2]_3$, $w_2 = [[d^{w_3} e]_1]_2$ and $w_3 = [e]_1$.

Note that some annotated stacks such as $[[a^w]_1]_2$ with $w = [[b^{[]_1}]_1]_2$ do not correspond to any stacks with links. However for all order-n stacks with links w and for any operation o of order at most n, we have $[\![o(w)]\!] = o([\![w]\!])$.

Remark 1. The configuration graphs of annotated pushdown systems of order-n are isomorphic to their collapsible counter-part when restricted to configurations reachable from the initial configuration. This implies annotated pushdown automata generate the same trees as higher-order recursion schemes, as in [6].

2.2 Regularity of Annotated Stacks

We will present an algorithm that operates on sets of configurations. For this we use order-n stack automata, thus defining a notion of regular sets of stacks. These have a nested structure based on a similar automata model by Bouajjani and Meyer [5]. The handling of annotations is similar to automata introduced by Broadbent *et al.* [6], except we read stacks top-down rather than bottom-up.

Definition 3 (Order-n Stack Automata). *An order-n stack automaton*

$$A = (\mathcal{Q}_n, \ldots, \mathcal{Q}_1, \Sigma, \Delta_n, \ldots, \Delta_1, \mathcal{F}_n, \ldots, \mathcal{F}_1)$$

is a tuple where Σ is a finite stack alphabet, and

1. *for all $n \geq k \geq 2$, we have $\mathcal{Q}_k$ is a finite set of states, $\Delta_k \subseteq \mathcal{Q}_k \times \mathcal{Q}_{k-1} \times 2^{\mathcal{Q}_k}$ is a transition relation, and $\mathcal{F}_k \subseteq \mathcal{Q}_k$ is a set of accepting states, and*
2. *$\mathcal{Q}_1$ is a finite set of states, $\Delta_1 \subseteq \bigcup_{2 \leq k \leq n} (\mathcal{Q}_1 \times \Sigma \times 2^{\mathcal{Q}_k} \times 2^{\mathcal{Q}_1})$ a transition relation, and $\mathcal{F}_1 \subseteq \mathcal{Q}_1$ a set of accepting states.*

Stack automata are alternating automata that read the stack in a nested fashion. Order-k stacks are recognised from states in $\mathcal{Q}_k$. A transition $(q, q', Q) \in \Delta_k$ from q to Q for some $k > 1$ can be fired when the top_{k-1} stack is accepted from $q' \in \mathcal{Q}_{(k-1)}$. The remainder of the stack must be accepted from all states in Q. At order-1, a transition (q, a, Q_{br}, Q) is a standard alternating a-transition with the additional requirement that the stack annotating a is accepted from all states in Q_{br}. A stack is accepted if a subset of $\mathcal{F}_k$ is reached at the end of each order-k stack. In the full version, we formally define the runs of a stack automaton. We write $w \in \mathcal{L}_q(A)$ whenever w is accepted from a state q.

A (partial) run is pictured below, using $q_3 \xrightarrow{q_2} Q_3 \in \Delta_3, q_2 \xrightarrow{q_1} Q_2 \in \Delta_2$ and $q_1 \xrightarrow[Q_{br}]{a} Q_1 \in \Delta_1$. The node labelled Q_{br} begins a run on the stack annotating a.

Remark 2. In the full version, we show several results on stack automata: membership testing is linear time; emptiness is PSPACE-complete; the sets of stacks accepted by these automata form an effective Boolean algebra (note that complementation causes a blow-up in the size of the automaton); and they accept the same family of collapsible stacks as the automata used by Broadbent et al. [6].

3   Algorithm

Given an annotated PDS $\mathcal{C}$ and a stack automaton A_0 with a state $q_p \in \mathcal{Q}_n$ for each control state p in $\mathcal{C}$, we define $Pre^*_{\mathcal{C}}(A_0)$ as the smallest set such that $Pre^*_{\mathcal{C}}(A_0) \supseteq \{ \langle p, w \rangle \mid w \in \mathcal{L}_{q_p}(A_0) \}$, and

$$Pre^*_{\mathcal{C}}(A_0) \supseteq \left\{ \langle p, w \rangle \;\middle|\; \begin{array}{l} \exists \langle p, w \rangle \longrightarrow \langle p', w' \rangle \text{ with } \langle p', w' \rangle \in Pre^*_{\mathcal{C}}(A_0) \vee \\ \exists \langle p, w \rangle \longrightarrow C \text{ and } C \subseteq Pre^*_{\mathcal{C}}(A_0) \end{array} \right\}$$

recalling that C denotes a set of configurations. We build a stack automaton recognising $Pre^*_{\mathcal{C}}(A_0)$. We begin with A_0 and iterate a saturation function denoted Γ — which adds new transitions to A_0 — until a 'fixed point' has been reached. That is, we iterate $A_{i+1} = \Gamma(A_i)$ until $A_{i+1} = A_i$. As the number of states is bounded, we eventually obtain this, giving us the following theorem.

Theorem 1. *Given an alternating annotated pushdown system $\mathcal{C}$ and a stack automaton A_0, we can construct an automaton A accepting $Pre^*_{\mathcal{C}}(A_0)$.*

The construction runs in n-EXPTIME for both alternating annotated PDS and collapsible PDS — which is optimal — and can be improved to $(n-1)$-EXPTIME for non-alternating collapsible PDS when the initial automaton satisfies a certain notion of *non-alternation*, again optimal. Correctness and complexity are discussed in subsequent sections.

3.1   Notation and Conventions

Number of Transitions. We assume for all $q \in \mathcal{Q}_k$ and $Q \subseteq \mathcal{Q}_k$ that there is at most one transition of the form $q \xrightarrow{q'} Q \in \Delta_k$. This condition can easily be ensured on A_0 by replacing pairs of transitions $q \xrightarrow{q_1} Q$ and $q \xrightarrow{q_2} Q$ with a single transition $q \xrightarrow{q'} Q$, where q' accepts the union of the languages of stacks accepted from q_1 and q_2. The construction maintains this condition.

Short-Form Notation. We introduce some short-form notation for runs. Consider the example run in Section 2.2. In this case, we write $q_3 \xrightarrow[Q_{br}]{a} (Q_1, Q_2, Q_3)$, $q_3 \xrightarrow{q_1} (Q_2, Q_3)$, and $q_3 \xrightarrow{q_2} (Q_3)$. In general, we write

$$q \xrightarrow[Q_{br}]{a} (Q_1, \ldots, Q_k) \text{ and } q \xrightarrow{q'} (Q_{k'+1}, \ldots, Q_k).$$

In the first case, $q \in \mathcal{Q}_k$ and there exist $q_{k-1}, \ldots, q_1$ such that $q \xrightarrow{q_{k-1}} Q_k \in \Delta_k, q_{k-1} \xrightarrow{q_{k-2}} Q_{k-1} \in \Delta_{k-1}, \ldots, q_1 \xrightarrow[Q_{br}]{a} Q_1 \in \Delta_1$. Thus, we capture nested sequences of initial transitions from q. Since we assume at most one transition between any state and set of states, the intermediate states $q_{k-1}, \ldots, q_1$ are uniquely determined by q, a, Q_{br} and $Q_1, \ldots, Q_k$.

In the second case $q \in \mathcal{Q}_k$, $q' \in \mathcal{Q}_{k'}$, and there exist $q_{k-1}, \ldots, q_{k'+1}$ with $q \xrightarrow{q_{k-1}} Q_k \in \Delta_k, q_{k-1} \xrightarrow{q_{k-2}} Q_{k-1} \in \Delta_{k-1}, \ldots, q_{k'+2} \xrightarrow{q_{k'+1}} Q_{k'+2} \in \Delta_{k'+2}$ and $q_{k'+1} \xrightarrow{q'} Q_{k'+1} \in \Delta_{k'+1}$.

We lift the short-form transition notation to transitions from sets of states. We assume that state-sets $\mathcal{Q}_n, \ldots, \mathcal{Q}_1$ are disjoint. Suppose $Q = \{q_1, \ldots, q_\ell\}$ and for all $1 \le i \le \ell$ we have $q_i \xrightarrow[Q_{br}^i]{a} (Q_1^i, \ldots, Q_k^i)$. Then we have $Q \xrightarrow[Q_{br}]{a} (Q_1, \ldots, Q_k)$ where $Q_{br} = \bigcup_{1 \le i \le \ell} Q_{br}^i$ and for all k, $Q_k = \bigcup_{1 \le i \le \ell} Q_k^i$. Because an annotation can only be of one order, we insist that $Q_{br} \subseteq \mathcal{Q}_k$ for some k.

Finally, we remark that a transition to the empty set is distinct from having no transition.

Initial States. We say a state is *initial* if it is of the form $q_p \in Q_n$ for some control state p or if it is a state $q_k \in Q_k$ for $k < n$ such that there exists a transition $q_{k+1} \xrightarrow{q_k} Q_{k+1}$ in Δ_{k+1}. We make the assumption that all initial states do not have any incoming transitions and that they are not final[2]

---

[2] Hence automata cannot accept empty stacks from initial states. This can be overcome by introducing a bottom-of-stack symbol.

Adding Transitions. Finally, when we add a transition $q_n \xrightarrow[Q_{br}]{a} (Q_1, \ldots, Q_n)$ to the automaton, then for each $n \geq k > 1$, we add $q_k \xrightarrow{q_{k-1}} Q_k$ to Δ_k (if a transition between q_k and Q_k does not already exist, otherwise we use the existing transition and state q_{k-1}) and add $q_1 \xrightarrow[Q_{br}]{a} Q_1$ to Δ_1.

3.2 The Saturation Function

Given an annotated PDS $\mathcal{C} = (\mathcal{P}, \Sigma, \mathcal{R})$, we define the saturation function. Examples can be found below.

Definition 4 (The Saturation Function Γ). *Given an order-n stack automaton A we define $A' = \Gamma(A)$ such that A' is A plus, for each $(p, a, o, p') \in \mathcal{R}$,*

1. *when $o = pop_k$, for each $q_{p'} \xrightarrow{q_k} (Q_{k+1}, \ldots, Q_n)$ in A, add to A'*

$$q_p \xrightarrow[\emptyset]{a} (\emptyset, \ldots, \emptyset, \{q_k\}, Q_{k+1}, \ldots, Q_n) ,$$

2. *when $o = push_k$, for each $q_{p'} \xrightarrow[Q_{br}]{a} (Q_1, \ldots, Q_k, \ldots, Q_n)$ and $Q_k \xrightarrow[Q'_{br}]{a} (Q'_1, \ldots, Q'_k)$ in A, add to A' the transition*

$$q_p \xrightarrow[Q_{br} \cup Q'_{br}]{a} (Q_1 \cup Q'_1, \ldots, Q_{k-1} \cup Q'_{k-1}, Q'_k, Q_{k+1}, \ldots, Q_n) ,$$

3. *when $o = collapse_k$, when $k = n$, add $q_p \xrightarrow[\{q_{p'}\}]{a} (\emptyset, \ldots, \emptyset)$, and when $k < n$, for each transition $q_{p'} \xrightarrow{q_k} (Q_{k+1}, \ldots, Q_n)$ in A, add to A' the transition*

$$q_p \xrightarrow[\{q_k\}]{a} (\emptyset, \ldots, \emptyset, Q_{k+1}, \ldots, Q_n) ,$$

4. *when $o = push_b^k$ for all transitions $q_{p'} \xrightarrow[Q_{br}]{b} (Q_1, \ldots, Q_n)$ and $Q_1 \xrightarrow[Q'_{br}]{a} Q'_1$ in A with $Q_{br} \subseteq Q_k$, add to A' the transition*

$$q_p \xrightarrow[Q'_{br}]{a} (Q'_1, Q_2, \ldots, Q_k \cup Q_{br}, \ldots, Q_n) ,$$

5. *when $o = rew_b$ for each transition $q_{p'} \xrightarrow[Q_{br}]{b} (Q_1, \ldots, Q_n)$ in A, add to A' the transition $q_p \xrightarrow[Q_{br}]{a} (Q_1, \ldots, Q_n)$.*

Finally, for every rule $p \to P$, let $Q = \{ q_{p'} \mid p' \in P \}$, then, for each $Q \xrightarrow[Q_{br}]{a} (Q_1, \ldots, Q_n)$, add a transition $q_p \xrightarrow[Q_{br}]{a} (Q_1, \ldots, Q_n)$. For convenience, the state-sets of A' are defined implicitly from the states used in the transition relations.

Examples. All examples except one use the order-2 stack w', labelled by a run of a stack automaton, pictured below, where the sub-script indicates states in $\mathcal{Q}_1$ or $\mathcal{Q}_2$. Recall that the first transition of the run can be written $q_{p'} \xrightarrow[Q_{br}^2]{a} (Q_1^1, Q_2^1)$.

Example of (p, a, pop_2, p') Consider the stack w pictured below with $pop_2(w) = w'$. By the construction, we add a transition $q_p \xrightarrow{a}_{\emptyset} (\emptyset, \{q_{p'}\})$ giving the run below labelling w, where q_1' is the state labelling the new transition $q_{p'} \xrightarrow{q_1'} \{q_{p'}\}$.

$$q_p \xrightarrow{[_1} q_1' \xrightarrow{a} \emptyset \xrightarrow{c} \emptyset \xrightarrow{]_1} \emptyset \xrightarrow{[_1} q_{p'} \xrightarrow{a} q_1 \underset{}{\overset{[_1}{\rightharpoonup}} Q_1^1 \xrightarrow{]_1} Q_2^1 \xrightarrow{} \cdots \quad Q_{br}^1 \xrightarrow{[_1} \cdots$$

Example of $(p, a, push_2, p')$ Consider the stack w below with $push_2(w) = w'$. Take $Q_2^1 \xrightarrow{a}_{Q_{br}^1} (Q_1^4, Q_2^2)$ from the node labelled Q_2^1 in the run over w'. By the construction, we add $q_p \xrightarrow[Q_{br}^1 \cup Q_{br}^2]{a} (Q_1^1 \cup Q_1^4, Q_2^2)$ and obtain a run over w

$$q_p \xrightarrow{[_1} q_1' \underset{}{\overset{a}{\rightharpoonup}} Q_1^1 \cup Q_1^4 \xrightarrow{]_1} Q_2^2 \qquad Q_{br}^1 \cup Q_{br}^2 \xrightarrow{[_1} \cdots \ .$$

where q_1' is the state used by the new transition. This run combines the runs over the top two order-1 stacks of w', ensuring any stack accepted could appear twice on top of a stack already accepted. That is, $push_2(w) = w'$ is in $Pre_C^*(A_0)$.

Example of $(p, a, collapse_2, p')$ Consider the stack w below with $collapse_2(w) = w'$. By the construction, we add a transition $q_p \xrightarrow[\{q_{p'}\}]{a} (\emptyset, \emptyset)$; hence, we have the run below, where q_1' is the state labelling the new transition $q_{p'} \xrightarrow{q_1'} \{\emptyset\}$.

$$q_p \xrightarrow{[_1} q_1' \underset{}{\overset{a}{\rightharpoonup}} \emptyset \xrightarrow{c} \emptyset \xrightarrow{]_1} \emptyset \xrightarrow{[_1} \emptyset \xrightarrow{]_1} \emptyset \quad q_{p'} \xrightarrow{a} q_1 \underset{}{\overset{[_1}{\rightharpoonup}} Q_1^1 \xrightarrow{]_1} Q_2^1 \xrightarrow{} \cdots \quad Q_{br}^1 \xrightarrow{[_1} \cdots$$

Example of $(p, a, push_b^2, p')$ The stack of our running example cannot be constructed via a $push_b^2$ operation. Hence, we use the following stack and run for w'

$$q_{p'} \xrightarrow{[_1} q_1 \underset{}{\overset{b}{\rightharpoonup}} Q_1^1 \xrightarrow{a} Q_1^2 \xrightarrow{]_1} Q_2^1 \xrightarrow{[_1} Q_1^3 \xrightarrow{a} Q_1^4 \xrightarrow{]_1} Q_2^2 \quad Q_{br} \xrightarrow{[_1} Q_1^5 \xrightarrow{a} Q_1^6 \xrightarrow{]_1} Q_2^3$$

with $q_{p'} \xrightarrow[Q_{br}]{b} (Q_1^1, Q_2^1)$ and $Q_1^1 \xrightarrow{a}_{\emptyset} Q_1^2$. The algorithm adds $q_p \xrightarrow{a}_{\emptyset} (Q_1^2, Q_2^1 \cup Q_{br})$. This gives us a run on the stack w such that $push_b^2(w) = w'$, where q_1' is the order-1 state labelling the new order-2 transition.

$$q_{p'} \xrightarrow{[_1} q_1' \xrightarrow{a} Q_1^2 \xrightarrow{]_1} Q_2^1 \cup Q_{br} \xrightarrow{[_1} Q_1^3 \cup Q_1^5 \xrightarrow{a} Q_1^4 \cup Q_1^6 \xrightarrow{]_1} Q_2^2 \cup Q_2^3$$

4 Correctness and Complexity

Theorem 2. *For a given* C *and* A_0, *let* $A = A_i$ *where* i *is the least index such that* $A_{i+1} = \Gamma(A_i)$. *We have* $w \in \mathcal{L}_{q_p}(A)$ *iff* $\langle p, w \rangle \in Pre_C^*(A_0)$.

The proof is in the full version. Completeness is by a straightforward induction over the "distance" to A_0. Soundness is the key technical challenge. The idea is to assign a "meaning" to each state of the automaton. For this, we define what it means for an order-k stack w to satisfy a state $q \in \mathcal{Q}_k$, which is denoted $w \models q$.

Definition 5 ($w \models q$). *For any $Q \subseteq \mathcal{Q}_k$ and any order-k stack w, we write $w \models Q$ if $w \models q$ for all $q \in Q$, and we define $w \models q$ by a case distinction on q.*

1. *q is an initial state in $\mathcal{Q}_n$. Then for any order-n stack w, we say that $w \models q$ if $\langle q, w \rangle \in Pre_{\mathcal{C}}^*(A_0)$.*
2. *q is an initial state in $\mathcal{Q}_k$, labeling a transition $q_{k+1} \xrightarrow{q} Q_{k+1} \in \Delta_{k+1}$. Then for any order-$k$ stack w, we say that $w \models q$ if for all order-$(k+1)$ stacks v s.t. $v \models Q_{k+1}$, then $w :_{(k+1)} v \models q_{k+1}$.*
3. *q is a non-initial state in $\mathcal{Q}_k$. Then for any order-k stack w, we say that $w \models q$ if A_0 accepts w from q.*

We show the automaton constructed is sound with respect to this meaning. That is, for all $q_k \xrightarrow[Q_{br}]{a} (Q_1, \dots, Q_k)$, we can place a^u, for any $u \models Q_{br}$, on top of any stack satisfying $Q_1, \dots, Q_k$ and obtain a stack that satisfies q_k. By induction over the length of the stack, this property extends to complete stacks. That is, a stack is accepted from a state only if it is in its meaning. Since states q_p are assigned their meaning in $Pre_{\mathcal{C}}^*(A_0)$, we obtain soundness of the construction.

The construction is also sound for collapsible stacks. That is, $\langle p, w \rangle$ belongs to $Pre_{\mathcal{C}}^*(A_0)$ where $\mathcal{C}$ is a collapsible PDS and A_0 accepts collapsible stacks iff $\langle p, [\![w]\!]_n \rangle$ belongs to $Pre_{\mathcal{C}}^*(A_0)$ where $\mathcal{C}$ and A_0 are interpreted over annotated stacks. This is due to the commutativity of $[\![o(w)]\!] = o([\![w]\!])$.

Proposition 1. *The saturation construction for an alternating order-n annotated PDS $\mathcal{C}$ and an order-n stack automaton A_0 runs in n-EXPTIME, which is optimal.*

Proof. Let $2 \uparrow_0 \ell = \ell$ and $2 \uparrow_{i+1} \ell = 2^{2\uparrow_i \ell}$. The number of states of A is bounded by $2 \uparrow_{(n-1)} \ell$ where ℓ is the size of $\mathcal{C}$ and A_0: each state in $\mathcal{Q}_k$ was either in A_0 or comes from a transition in Δ_{k+1}. Since the automata are alternating, there is an exponential blow up at each order except at order-n. Each iteration of the algorithm adds at least one new transition. Only $2 \uparrow_n \ell$ transitions can be added. Since the reachability problem for alternating higher-order pushdown systems is complete for n-EXPTIME [15], our algorithm is optimal.

It is known that the complexity of reachability for non-alternating collapsible PDS is in $(n-1)$-EXPTIME. The cause of the additional exponential blow up is in the alternation of the stack automata. In the full version we show that, for a suitable notion of *non-alternating* stack automata, our algorithm can be adapted to run in $(n-1)$-EXPTIME, when the collapsible PDS is also non-alternating.

Furthermore, the algorithm is PTIME for a fixed order and number of control states. If we obtained $\mathcal{C}$ from a scheme, the number of control states is given by the arity of the scheme [14]. Since the arity and order are expected to be small, we are hopeful that our algorithm will perform well in practice.

5 Perspectives

There are several avenues of future work. First, we intend to generalise our saturation technique to computing winning regions of *parity* conditions, based on the order-1 case [17]. This will permit verification of more general specifications. We also plan to design a prototype tool to test the algorithm in practice

An important direction is that of counter example generation. When checking safety property, it is desirable to provide a trace witnessing a violation of the property. This can be used to repair the bug and as part of a *counter-example guided abstraction refinement (CEGAR)* loop enabling efficient verification algorithms. However, finding *shortest* counter examples — due to its tight connection with pumping lemmas — will present a challenging and interesting problem.

Saturation techniques have been extended to concurrent order-1 pushdown systems [32,1]; concurrency at higher-orders would be interesting.

It will also be interesting to study notions of regularity of annotated stacks. In our notion of regularity, the forwards reachability set is not regular, due to the copy operation $push_k$. This problem was addressed by Carayol for higher-order stacks [8]; adapting these techniques to annotated PDS is a challenging problem.

References

1. Atig, M.F.: Global model checking of ordered multi-pushdown systems. In: FSTTCS, pp. 216–227 (2010)
2. Ball, T., Rajamani, S.K.: The SLAM project: Debugging system software via static analysis. In: POPL, pp. 1–3 (2002)
3. Benois, M.: Parties rationnelles du groupe libre. Comptes-Rendus de l'Acamdémie des Sciences de Paris, Série A, 1188–1190 (1969)
4. Bouajjani, A., Esparza, J., Maler, O.: Reachability Analysis of Pushdown Automata: Application to Model-Checking. In: Mazurkiewicz, A., Winkowski, J. (eds.) CONCUR 1997. LNCS, vol. 1243, pp. 135–150. Springer, Heidelberg (1997)
5. Bouajjani, A., Meyer, A.: Symbolic Reachability Analysis of Higher-Order Context-Free Processes. In: Lodaya, K., Mahajan, M. (eds.) FSTTCS 2004. LNCS, vol. 3328, pp. 135–147. Springer, Heidelberg (2004)
6. Broadbent, C.H., Carayol, A., Ong, C.-H.L., Serre, O.: Recursion schemes and logical reflection. In: LiCS, pp. 120–129 (2010)
7. Cachat, T.: Games on Pushdown Graphs and Extensions. PhD thesis, RWTH Aachen (2003)
8. Carayol, A.: Regular Sets of Higher-Order Pushdown Stacks. In: Jedrzejowicz, J., Szepietowski, A. (eds.) MFCS 2005. LNCS, vol. 3618, pp. 168–179. Springer, Heidelberg (2005)
9. Carayol, A., Wöhrle, S.: The Caucal Hierarchy of Infinite Graphs in Terms of Logic and Higher-Order Pushdown Automata. In: Pandya, P.K., Radhakrishnan, J. (eds.) FSTTCS 2003. LNCS, vol. 2914, pp. 112–123. Springer, Heidelberg (2003)
10. Chandra, A.K., Kozen, D., Stockmeyer, L.J.: Alternation. J. ACM 28(1), 114–133 (1981)
11. Damm, W.: The IO- and OI-hierarchies. Theor. Comput. Sci. 20, 95–207 (1982)
12. Esparza, J., Hansel, D., Rossmanith, P., Schwoon, S.: Efficient Algorithms for Model Checking Pushdown Systems. In: Emerson, E.A., Sistla, A.P. (eds.) CAV 2000. LNCS, vol. 1855, pp. 232–247. Springer, Heidelberg (2000)

13. Finkel, A., Willems, B., Wolper, P.: A direct symbolic approach to model checking pushdown systems. Electr. Notes Theor. Comput. Sci. 9, 27–37 (1997)
14. Hague, M., Murawski, A.S., Ong, C.-H.L., Serre, O.: Collapsible pushdown automata and recursion schemes. In: LiCS, pp. 452–461 (2008)
15. Hague, M., Ong, C.-H.L.: Symbolic backwards-reachability analysis for higher-order pushdown systems. Logical Methods in Computer Science 4(4) (2008)
16. Hague, M., Ong, C.-H.L.: Analysing Mu-Calculus Properties of Pushdown Systems. In: van de Pol, J., Weber, M. (eds.) SPIN 2010. LNCS, vol. 6349, pp. 187–192. Springer, Heidelberg (2010)
17. Hague, M., Ong, C.-H.L.: A saturation method for the modal μ-calculus over pushdown systems. Inf. Comput. 209(5), 799–821 (2010)
18. Kartzow, A., Parys, P.: Strictness of the Collapsible Pushdown Hierarchy. arXiv:1201.3250v1 [cs.FL] (2012)
19. Knapik, T., Niwiński, D., Urzyczyn, P.: Higher-Order Pushdown Trees Are Easy. In: Nielsen, M., Engberg, U. (eds.) FOSSACS 2002. LNCS, vol. 2303, pp. 205–222. Springer, Heidelberg (2002)
20. Knapik, T., Niwiński, D., Urzyczyn, P., Walukiewicz, I.: Unsafe Grammars and Panic Automata. In: Caires, L., Italiano, G.F., Monteiro, L., Palamidessi, C., Yung, M. (eds.) ICALP 2005. LNCS, vol. 3580, pp. 1450–1461. Springer, Heidelberg (2005)
21. Kobayashi, N.: Higher-order model checking: From theory to practice. In: LiCS, pp. 219–224 (2011)
22. Kobayashi, N.: A Practical Linear Time Algorithm for Trivial Automata Model Checking of Higher-Order Recursion Schemes. In: Hofmann, M. (ed.) FOSSACS 2011. LNCS, vol. 6604, pp. 260–274. Springer, Heidelberg (2011)
23. Maslov, A.N.: Multilevel stack automata. Problems of Information Transmission 15, 1170–1174 (1976)
24. Ong, C.-H.L.: On model-checking trees generated by higher-order recursion schemes. In: LiCS, pp. 81–90 (2006)
25. Ong, C.-H.L., Ramsay, S.J.: Verifying higher-order functional programs with pattern-matching algebraic data types. In: POPL, pp. 587–598 (2011)
26. Parys, P.: Collapse operation increases expressive power of deterministic higher order pushdown automata. In: STACS, pp. 603–614 (2011)
27. Reps, T.W., Schwoon, S., Jha, S., Melski, D.: Weighted pushdown systems and their application to interprocedural dataflow analysis. Sci. Comput. Program. 58(1-2), 206–263 (2005)
28. Salvati, S., Walukiewicz, I.: Krivine Machines and Higher-Order Schemes. In: Aceto, L., Henzinger, M., Sgall, J. (eds.) ICALP 2011, Part II. LNCS, vol. 6756, pp. 162–173. Springer, Heidelberg (2011)
29. Schwoon, S.: Model-checking Pushdown Systems. PhD thesis, Technical University of Munich (2002)
30. Seth, A.: An alternative construction in symbolic reachability analysis of second order pushdown systems. In: RP, pp. 80–95 (2007)
31. Seth, A.: Games on Higher Order Multi-stack Pushdown Systems. In: Bournez, O., Potapov, I. (eds.) RP 2009. LNCS, vol. 5797, pp. 203–216. Springer, Heidelberg (2009)
32. Suwimonteerabuth, D., Esparza, J., Schwoon, S.: Symbolic Context-Bounded Analysis of Multithreaded Java Programs. In: Havelund, K., Majumdar, R. (eds.) SPIN 2008. LNCS, vol. 5156, pp. 270–287. Springer, Heidelberg (2008)
33. Suwimonteerabuth, D., Schwoon, S., Esparza, J.: Efficient Algorithms for Alternating Pushdown Systems with an Application to the Computation of Certificate Chains. In: Graf, S., Zhang, W. (eds.) ATVA 2006. LNCS, vol. 4218, pp. 141–153. Springer, Heidelberg (2006)

Regular Languages Are
Church-Rosser Congruential*

Volker Diekert[1], Manfred Kufleitner[1,**], Klaus Reinhardt[2], and Tobias Walter[1]

[1] Institut für Formale Methoden der Informatik
University of Stuttgart, Germany
[2] Wilhelm-Schickard-Institut für Informatik
University of Tübingen, Germany

Abstract. This paper proves a long standing conjecture in formal language theory. It shows that all regular languages are Church-Rosser congruential. The class of Church-Rosser congruential languages was introduced by McNaughton, Narendran, and Otto in 1988. A language L is Church-Rosser congruential if there exists a finite, confluent, and length-reducing semi-Thue system S such that L is a finite union of congruence classes modulo S. It was known that there are deterministic linear context-free languages which are not Church-Rosser congruential, but on the other hand it was strongly believed that all regular languages are of this form. This paper solves the conjecture affirmatively by actually proving a more general result.

Keywords. String rewriting, Church-Rosser system, regular language, finite monoid, finite semigroup, local divisor.

1 Introduction

It has been a long standing conjecture in formal language theory that all regular languages are Church-Rosser congruential. The class of Church-Rosser congruential languages was introduced in 1988 by McNaughton, Narendran, and Otto [9], see also [10]. A language L is Church-Rosser congruential, if there exists a finite confluent, and length-reducing semi-Thue system S such that L is a finite union of congruence classes modulo S. One of the main motivations to consider this class of languages is that the membership problem for L can be solved in linear time; this is done by computing normal forms using the system S, followed by a table look-up. For this it is not necessary that the quotient monoid A^*/S is finite, it is enough that L is a finite union of congruence classes modulo S. It is not hard to see that $\{a^n b^n \mid n \in \mathbb{N}\}$ is Church-Rosser congruential, but $\{a^m b^n \mid m, n \in \mathbb{N} \text{ and } m \geq n\}$ is not. This led the authors of [9] to the more technical notion of Church-Rosser languages; this class of languages captures

---

* A version with full proofs can be found on arXiv:1202.1148 [4].
** The second author was supported by the German Research Foundation (DFG) under grant DI 435/5-1.

A. Czumaj et al. (Eds.): ICALP 2012, Part II, LNCS 7392, pp. 177–188, 2012.
© Springer-Verlag Berlin Heidelberg 2012

all deterministic context-free languages. For more results about Church-Rosser languages see e.g. [2, 10, 15, 16].

From the very beginning it was strongly believed that all regular languages are Church-Rosser congruential in the pure sense. However, after some significant initial progress [10, 11, 12, 13, 14] there was stagnation.

Before 2011 the most advanced result was the one announced in 2003 by Reinhardt and Thérien [14]. According to this manuscript the conjecture is true for all regular languages where the syntactic monoid is a group. However, the manuscript has never been published as a refereed paper and there are some flaws in its presentation. The main problem with [14] has however been quite different for us. The statement is too weak to be useful in the induction for the general case. So, instead of being able to use [14] as a black box, we shall prove a more general result in the setting of weight-reducing systems. This part about group languages is a cornerstone in our approach.

The other ingredient to our paper has been established only very recently. Knowing that the result is true if the syntactic monoid is a group, we started looking at aperiodic monoids. Aperiodic monoids correspond to star-free languages and the first two authors together with Weil proved that all star-free languages are Church-Rosser congruential [6]. Our proof became possible by *loading the induction hypothesis*. This means we proved a much stronger statement. We showed that for every star-free language $L \subseteq A^*$ there exists a finite confluent semi-Thue system $S \subseteq A^* \times A^*$ such that the quotient monoid A^*/S is finite (and aperiodic), L is a union of congruence classes modulo S, and moreover all right-hand sides of rules appear as scattered subwords in the corresponding left-hand side. We called the last property *subword-reducing*, and it is obvious that every subword-reducing system is length-reducing. We have little hope that such a strong result could be true in general. Indeed here we step back from subword-reducing to weight-reducing systems.

We prove in Theorem 2 the following result: Let $L \subseteq A^*$ be a regular language and $\|a\| \in \mathbb{N} \setminus \{0\}$ be a positive weight for every letter $a \in A$ (e.g., $\|a\| = |a| = 1$). Then we can construct for the given weight a finite, confluent and weight-reducing semi-Thue system $S \subseteq A^* \times A^*$ such that the quotient monoid A^*/S is finite and recognizes L. In particular, L is a finite union of congruence classes modulo S. Using the notation of Niemann [11], this implies that regular languages are *strongly* Church-Rosser congruential.

Note that this gives us another characterization for the class of regular languages. By Corollary 1 we see that a language $L \subseteq A^*$ is regular if and only if L is recognized by a finite Church-Rosser system S with finite index. As a consequence, a long standing conjecture about regular languages has been solved positively.

2   Preliminaries

Throughout this paper, A is a finite alphabet. An element of A is called a *letter*. The set A^* is the free monoid generated by A. It consists of all finite sequences of letters from A. The elements of A^* are called *words*. The empty word is denoted by 1. The *length* of a word u is denoted by $|u|$. We have $|u| = n$ for $u = a_1 \cdots a_n$ where $a_i \in A$. The empty word has length 0, and it is the only word with this property. The set of words of length at most n is denoted by $A^{\leq n}$, and the set of all nonempty words is A^+. We generalize the length of a word by introducing weights. A *weighted alphabet* $(A, \|\cdot\|)$ consists of an alphabet A equipped with a weight function $\|\cdot\| : A \to \mathbb{N} \setminus \{0\}$. The *weight* of a letter $a \in A$ is $\|a\|$ and the *weight* $\|u\|$ of a word $u = a_1 \cdots a_n$ with $a_i \in A$ is $\|a_1\| + \cdots + \|a_n\|$. The weight of the empty word is 0. The length is the special weight with $\|a\| = 1$ for all $a \in A$. A word u is a *factor* of a word v if there exist $p, q \in A^*$ such that $puq = v$, and u is a *proper factor* of v if $pq \neq 1$. The word u is a *prefix* of v if $uq = v$ for some $q \in A^*$, and it is a *suffix* of v if $pu = v$ for some $p \in A^*$. We say that u is a factor (resp. prefix, resp. suffix) of v^+ if there exists $n \in \mathbb{N}$ such that u is a factor (resp. prefix, resp. suffix) of v^n. Two words $u, v \in A^*$ are *conjugate* if there exist $p, q \in A^*$ such that $u = pq$ and $v = qp$. An integer $m > 0$ is a *period* of a word $u = a_1 \cdots a_n$ with $a_i \in A$ if $a_i = a_{i+m}$ for all $1 \leq i \leq n - m$. A word $u \in A^+$ is *primitive* if there exists no $v \in A^+$ such that $u = v^n$ for some integer $n > 1$. It is a standard fact that a word u is not primitive if and only if $u^2 = puq$ for some $p, q \in A^+$. This follows immediately from the result from combinatorics on words that $xy = yx$ if and only if x and y are powers of a common root; see e.g. [8, Section 1.3].

A monoid M *recognizes* a language $L \subseteq A^*$ if there exists a homomorphism $\varphi : A^* \to M$ such that $L = \varphi^{-1}\varphi(L)$. A language $L \subseteq A^*$ is *regular* if it is recognized by a finite monoid. There are various other well-known characterizations of regular languages; e.g., regular expressions, finite automata or monadic second order logic. Regular languages L can be classified in terms of structural properties of the monoids recognizing L. In particular, we consider group languages; these are languages recognized by finite groups.

A *semi-Thue system* over A is a subset $S \subseteq A^* \times A^*$. In this paper, all semi-Thue systems are finite. The elements of S are called *rules*. We frequently write $\ell \to r$ for rules (ℓ, r). A system S is called *length-reducing* if we have $|\ell| > |r|$ for all rules $\ell \to r$ in S. It is called *weight-reducing* with respect to some weighted alphabet $(A, \|\cdot\|)$, if $\|\ell\| > \|r\|$ for all rules $\ell \to r$ in S. Every system S defines the rewriting relation $\underset{S}{\Longrightarrow} \subseteq A^* \times A^*$ by setting $u \underset{S}{\Longrightarrow} v$ if there exist $p, q, \ell, r \in A^*$ such that $u = p\ell q$, $v = prq$, and $\ell \to r$ is in S.

By $\underset{S}{\overset{*}{\Longrightarrow}}$ we mean the reflexive and transitive closure of $\underset{S}{\Longrightarrow}$. By $\underset{S}{\overset{*}{\Longleftrightarrow}}$ we mean the symmetric, reflexive, and transitive closure of $\underset{S}{\Longrightarrow}$. We also write $u \underset{S}{\overset{*}{\Longleftarrow}} v$ whenever $v \underset{S}{\overset{*}{\Longrightarrow}} u$. The system S is *confluent* if for all $u \underset{S}{\overset{*}{\Longleftrightarrow}} v$ there is some w such that $u \underset{S}{\overset{*}{\Longrightarrow}} w \underset{S}{\overset{*}{\Longleftarrow}} v$. It is *locally confluent* if for all $v \underset{S}{\Longleftarrow} u \underset{S}{\Longrightarrow} v'$ there exists w such that $v \underset{S}{\overset{*}{\Longrightarrow}} w \underset{S}{\overset{*}{\Longleftarrow}} v'$. If S is locally confluent and weight-reducing for some weight, then S is confluent; see e.g. [1, 7]. Note that $u \underset{S}{\Longrightarrow} v$ implies that $\|u\| > \|v\|$ for weight-reducing systems. The relation $\underset{S}{\overset{*}{\Longleftrightarrow}} \subseteq A^* \times A^*$ is a congruence, hence the congruence classes $[u]_S = \{v \in A^* \mid u \underset{S}{\overset{*}{\Longleftrightarrow}} v\}$ form a monoid which is denoted by A^*/S. The size of A^*/S is called the *index of S*. A finite semi-Thue system S can be viewed as a finite set of defining relations. Hence, A^*/S becomes a finitely presented monoid $A^*/\{\ell = r \mid (\ell, r) \in S\}$. By $\mathrm{IRR}_S(A^*)$ we denote the set of irreducible words in A^*, i.e., the set of words where no left-hand side occurs as a factor.

Whenever the weighted alphabet $(A, \|\cdot\|)$ is fixed, a finite semi-Thue system $S \subseteq A^* \times A^*$ is called a *weighted Church-Rosser system* if it is finite, weight-reducing for $(A, \|\cdot\|)$, and confluent. Hence, a finite semi-Thue system S is a weighted Church-Rosser system if and only if (1) we have $\|\ell\| > \|r\|$ for all rules $\ell \to r$ in S and (2) every congruence class has exactly one irreducible element. In particular, for weighted Church-Rosser systems S, there is a one-to-one correspondence between A^*/S and $\mathrm{IRR}_S(A^*)$. A *Church-Rosser system* is a finite, length-reducing, and confluent semi-Thue system. In particular, every Church-Rosser system is a weighted Church-Rosser system. A language $L \subseteq A^*$ is called a *Church-Rosser congruential language* if there is a finite Church-Rosser system S such that L can be written as a finite union of congruence classes $[u]_S$.

Definition 1. *Let $\varphi : A^* \to M$ be a homomorphism and let S be a semi-Thue system. We say that φ factorizes through S if for all $u, v \in A^*$ we have:*

$$u \underset{S}{\Longrightarrow} v \quad implies \quad \varphi(u) = \varphi(v).$$

Note that if S is a semi-Thue system and $\varphi : A^* \to M$ factorizes through S, then $\psi([u]_S) = \varphi(u)$ is well-defined and the following diagram commutes (here, $\pi(u) = [u]_S$ is the canonical homomorphism).

$$
\begin{array}{ccc}
 & A^*/S & \\
\overset{\pi}{\nearrow} & & \downarrow \psi \\
A^* \xrightarrow[\varphi]{} & & M
\end{array}
$$

3 Finite Groups

Our main result is that every homomorphism $\varphi : A^* \to M$ to a finite monoid M factorizes through a Church-Rosser system S. Our proof of this theorem distinguishes whether or not M is a group. Thus, we first prove this result for groups. Before we turn to the general group case, we show that for some particular groups, proving the claim is easy. The techniques developed here will also be used when proving the result for arbitrary finite groups.

3.1 Groups without Proper Cyclic Quotient Groups

The aim of this section is to show that finding a Church-Rosser system is very easy for many cases. This list includes presentations of finite (non-cyclic) simple groups, but it goes far beyond this. Let $\varphi : A^* \to G$ be a homomorphism to a finite group, where $(A, \|\cdot\|)$ is a weighted alphabet. This defines a regular language $L_G = \{w \in A^* \mid \varphi(w) = 1\}$. Let us assume that the greatest common divisor $\gcd\{\|w\| \mid w \in L_G\}$ is equal to one; e.g. $\{6, 10, 15\} \subseteq \{\|w\| \mid w \in L_G\}$. Then there are two words $u, v \in L_G$ such that $\|u\| - \|v\| = 1$. Now we can use these words to find a constant d such that all $g \in G$ have a representing word v_g with the exact weight $\|v_g\| = d$. To see this, start with some arbitrary set of representing words v_g. We multiply words v_g with smaller weight with u and words v_g with larger weights with v until all weights are equal.

The final step is to define the following weight-reducing system

$$S_G = \{w \to v_{\varphi(w)} \mid w \in A^* \text{ and } d < \|w\| \leq d + \max\{\|a\| \mid a \in A\}\}.$$

Confluence of S_G is trivial; and every language recognized by φ is also recognized by the canonical homomorphism $A^* \to A^*/S_G$.

Now assume that we are not so lucky, i.e., $\gcd\{\|w\| \mid w \in L_G\} > 1$. This means there is a prime number p such that p divides $\|w\|$ for all $w \in L_G$. Then, the homomorphism of A^* to $\mathbb{Z}/p\mathbb{Z}$ defined by $a \mapsto \|a\|$ mod p factorizes through φ and $\mathbb{Z}/p\mathbb{Z}$ becomes a quotient group of G. This can never happen if $\varphi(A^*)$ is a simple and non-cyclic subgroup of G, because a simple group does not have any proper quotient group. But there are many other cases where a natural homomorphism $A^* \to G$ for some weighted alphabet $(A, \|\cdot\|)$ satisfies the property $\gcd\{\|w\| \mid w \in L_G\} = 1$ although G has a non-trivial cyclic quotient group. Just consider the length function and a presentation by standard generators for dihedral groups D_{2n} or the permutation groups S_n where n is odd.

For example, let $G = D_6 = S_3$ be the permutation group of a triangle. Then G is generated by elements τ and ρ with defining relations $\tau^2 = \rho^3 = 1$ and $\tau\rho\tau = \rho^2$. The following six words of length 3 represent all six group elements:

$$1 = \rho^3, \ \rho = \rho\tau^2, \ \rho^2 = \tau\rho\tau, \ \tau = \tau^3, \ \tau\rho = \rho^2\tau, \ \tau\rho^2.$$

The corresponding monoid $\{\rho, \tau\}^*/S_G$ has 15 elements. More systematically, one could obtain a normal form of length 5 for each of the group elements in $\{1, \rho, \rho^2, \tau, \tau\rho, \tau\rho^2\}$ by adding factors ρ^3 and τ^2. For example, this could lead to the set of normal forms $\{\tau^2\rho^3, \tau^4\rho, \rho^5, \tau^5, \tau\rho^4, \tau^3\rho^2\}$. We will use this pumping idea in our proof of the general case for finding normal forms of approximately the same size.

It is much harder to find a Church-Rosser system for the homomorphism $\varphi : \{a, b, c\}^* \to \mathbb{Z}/3\mathbb{Z}$ where $\varphi(a) = \varphi(b) = \varphi(c) = 1$ mod 3. Restricting φ to the submonoid $\{a, b\}^*$ makes the situation simpler. Still it is surprisingly complicated. A possible Church-Rosser system $S \subseteq \{a, b\}^* \times \{a, b\}^*$ of finite index such that the restriction of φ factorizes through S is given by:

$$S = \left\{ aaa \to 1, \ baab \to b, \ (ba)^3 b \to b, \ bb\,u\,bb \to b^{|u|+1} \ \middle| \ 1 \leq |u| \leq 3 \right\}.$$

There are 272 irreducible elements and the longest irreducible word has length 16. Note that the last set of rules has bb as a prefix and as a suffix on both sides of every rule. The idea of preserving end markers such as $\omega = bb$ in the above example is essential for the solution of the general case, too.

In some sense this phenomenon suggests that finite cyclic groups or more general commutative groups are obstacles to find a simple construction for Church-Rosser systems.

3.2   The General Case for Group Languages

In this section, we consider arbitrary groups. We start with some simple properties of Church-Rosser systems. Then, in Theorem 1, we state and prove that group languages are Church-Rosser congruential.

Lemma 1. *Let $(A, \|\cdot\|)$ be a weighted alphabet, let $d \in \mathbb{N}$, and let $S \subseteq A^* \times A^*$ be a weighted Church-Rosser system such that $\mathrm{IRR}_S(A^*)$ is finite. Then*

$$S_d = \left\{ u\ell v \to urv \mid u, v \in A^d \text{ and } \ell \to r \in S \right\}$$

is a weighted Church-Rosser system satisfying: (1) $\mathrm{IRR}_{S_d}(A^*)$ is finite, (2) all words of length at most $2d$ are irreducible with respect to S_d, and (3) the mapping $[u]_{S_d} \mapsto [u]_S$ for $u \in A^*$ is well-defined and yields a surjective homomorphism from A^*/S_d onto A^*/S.

Proof. First, one shows that local confluence of S transfers to local confluence of S_d. The remaining proof is straightforward and therefore left to the reader.   □

Lemma 2. *Let $\Delta \subseteq A^+$ be a set of words such that all words in Δ have length at most n. If $u \in A^{>2n}$ is not a factor of some δ^+ for $\delta \in \Delta$, then there is a proper factor v of u which is also not a factor of some δ^+ for $\delta \in \Delta$.*

Proof. Assume that such a factor v of u does not exist. Let $u = awb$ for $a, b \in A$. Then aw is a factor of δ^+ and wb is a factor of δ'^+ for some $\delta, \delta' \in \Delta$. Let $p = |\delta|$ and $q = |\delta'|$. Now, p is a period of aw and q is a period of wb. Thus p and q are both periods of w. Since $|w| \geq 2n - 1 \geq p + q - \gcd(p, q)$, we see that $\gcd(p, q)$ is also a period of w by the Periodicity Lemma of Fine and Wilf [8, Section 1.3]. The $(p + 1)$-th letter in aw is a. Going in steps $\gcd(p, q)$ to the left or to the right in w, we see that the $(q + 1)$-th letter in aw is a. Thus awb is a factor of δ'^+, which is a contradiction.   □

We are now ready to prove the main result of this section: Group languages are Church-Rosser congruential. An outline of the proof is as follows. By induction on the size of the alphabet, we show that every homomorphism $\varphi : A^* \to G$ factorizes through a weighted Church-Rosser system S with finite index. Remove some letter c from the alphabet A. This leads to a system R for the remaining letters B. Lemma 1 allows us to assume that certain words are irreducible. Then we set $K = \mathrm{IRR}_R(B^*)c$ which is a prefix code in A^*. We consider K as a new alphabet. Essentially, it is this situation where weighted alphabets come into play because we can choose the weight of K such that it is compatible with the weight over the alphabet A. Over K, we introduce two sets of rules T_Δ and T_Ω.

The T_Δ-rules reduce long repetitions of short words Δ, and the T_Ω-rules have the form $\omega\,u\,\omega \to \omega\,v_g\,\omega$. Here, Ω is some finite set of markers and $\omega \in \Omega$ is such a marker. The word v_g is a normal form for the group element g. The T_Ω-rules reduce long words without long repetitions of short words. We show that T_Δ and T_Ω are confluent and that their union has finite index over K^*. Since by construction all rules in $T = T_\Delta \cup T_\Omega$ are weight-reducing, the system T is a weighted Church-Rosser system over K^* with finite index such that $\varphi : K^* \to G$ factorizes through T. Since $K \subseteq A^*$, we can translate the rules $\ell \to r$ in T over K^* to rules $c\ell \to cr$ over A^*. This leads to the set of T'-rules over A^*. The letter c at the beginning of the T'-rules is required to shield the T'-rules from R-rules. Finally, we show that $S = R \cup T'$ is the desired system over A^*.

Theorem 1. *Let $(A, \|\cdot\|)$ be a weighted alphabet and let $\varphi : A^* \to G$ be a homomorphism to a finite group G. Then there exists a weighted Church-Rosser system S with finite index such that φ factorizes through S.*

Proof. We may assume that φ is surjective. In the following, n denotes the exponent of G; this is the least positive integer n such that $g^n = 1$ for all $g \in G$. The proof is by induction on the size of the alphabet A. If $A = \{c\}$, then we set $S = \{c^n \to 1\}$. Let now $A = \{a_0, \dots, a_{s-1}, c\}$ and let a_0 have minimal weight. We set $B = A \setminus \{c\}$. Let

$$\gamma_i = a_{i \bmod s}^{n + \lfloor i/s \rfloor} c.$$

Since A and $\{a_0 c, \dots, a_{s-1} c, c\}$ generate the same subgroups of G and since every element $a_j c \in G$ occurs infinitely often as some γ_i, there exists $m > 0$ such that for every $g \in G$ there exists a word

$$v_g = \gamma_0^{n_0} \cdots \gamma_m^{n_m} \gamma_0$$

with $n_i > 0$ satisfying $\varphi(v_g) = g$ and $\|v_g\| - \|v_h\| < n \|a_0\|$ for all $g, h \in G$. The latter property relies on $\|\gamma_0\| + \|a_0\| = \|\gamma_s\|$ and pumping with γ_0^n and γ_s^n which both map to the neutral element of G: Assume $\|v_g\| - \|v_h\| \geq n \|a_0\|$ for some $g, h \in G$. Then we do the following. All v_g with maximal weight are multiplied by γ_0^n on the left, and for all other words v_h the exponent n_s of γ_s is replaced by $n_s + n$. After that, the maximal difference $\|v_g\| - \|v_h\|$ has decreased at least by 1 (and at most by $n \|a_0\|$). We can iterate this procedure until the weights of all v_g differ less than $n \|a_0\|$. Let

$$\Gamma = \{\gamma_0, \dots, \gamma_m\}$$

be the generators of the v_g. By induction there exists a weighted Church-Rosser system R for the restriction $\varphi : B^* \to G$ satisfying the statement of the theorem. By Lemma 1, we can assume $\Gamma \subseteq \mathrm{IRR}_R(B^*)\, c$. Thus $v_g \in \mathrm{IRR}_R(A^*)$ for all $g \in G$. Let

$$K = \mathrm{IRR}_R(B^*)\, c.$$

The set K is a prefix code in A^*. We consider K as an extended alphabet and its elements as extended letters. The weight $\|u\|$ of $u \in K$ is its weight as a word over A. Each γ_i is a letter in K. The homomorphism $\varphi : A^* \to G$ can be interpreted as a homomorphism $\varphi : K^* \to G$; it is induced by $u \mapsto \varphi(u)$ for $u \in K$. The length lexicographic order on B^* induces a linear order $\leq$ on $\mathrm{IRR}_R(B^*)$ and hence also on K. Here, we define $a_0 < \cdots < a_{s-1}$. The words v_g can be read as words over the weighted alphabet $(K, \|\cdot\|)$ satisfying the following five properties: First, v_g starts with the extended letter γ_0. Second, the last two extended letters of v_g are $\gamma_m \gamma_0$. Third, all extended letters in v_g are in non-decreasing order from left to right with respect to $\leq$, with the sole exception of the last letter γ_0 which is smaller than its predecessor γ_m. The fourth property is that all extended letters in v_g have a weight greater than $n \|a_0\|$. And the last important property is that all differences $\|v_g\| - \|v_h\|$ are smaller than $n \|a_0\|$. Let

$$\Delta = \left\{ \delta \in K^+ \mid \delta \in K \text{ or } \|\delta\| \leq n \|a_0\| \right\}.$$

Note that Δ is closed under conjugation, i.e., if $uv \in \Delta$ for $u, v \in K^*$, then $vu \in \Delta$. We can think of Δ as the set of all "short" words. Choose $t \geq n$ such that all normal forms v_g have no factor δ^{t+n} for $\delta \in \Delta$ and such that $\|c^t\| \geq \|u\|$ for all $u \in K^{2n}$. Note that $c \in \Delta$ has the smallest weight among all words in Δ.

The first set of rules over the extended alphabet K deals with long repetitions of short words: The Δ-rules are

$$T_\Delta = \left\{ \delta^{t+n} \to \delta^t \mid \delta \in \Delta \text{ and } \delta \text{ is primitive} \right\}.$$

Let $F \subseteq K^*$ contain all words which are a factor of some δ^+ for $\delta \in \Delta$ and let $J \subseteq K^+$ be minimal such that $K^* J K^* = K^* \setminus F$. By Lemma 2, we have $J \subseteq K^{\leq 2n}$. In particular, J is finite. Since J and Δ are disjoint, all words in J have a weight greater than $n \|a_0\|$. Let Ω contain all $\omega \in J$ such that $\omega \in \Gamma K^*$ implies $\omega = \gamma \gamma'$ for some $\gamma, \gamma' \in \Gamma$ with $\gamma > \gamma'$, i.e.,

$$\Omega = J \cap \left\{ \omega \in K^* \mid \omega \notin \Gamma K^* \text{ or } \omega = \gamma \gamma' \text{ for } \gamma, \gamma' \in \Gamma \text{ with } \gamma > \gamma' \right\}.$$

As we will see below, every sufficiently long word without long Δ-repetitions contains a factor $\omega \in \Omega$.

Claim 1. *There exists a bound $t' \in \mathbb{N}$ such that every word $u \in K^*$ with $\|u\| \geq t'$ contains a factor $\omega \in \Omega$ or a factor of the form δ^{t+n} for $\delta \in \Delta$.*

The proof of Claim 1 can be found on arXiv [4].

Since Δ is closed under factors, u contains no factor of the form δ^{t+n} for $\delta \in \Delta$ if and only if $u \in \mathrm{IRR}_{T_\Delta}(K^*)$. In particular, it is no restriction to only allow primitive words from Δ in the rules T_Δ. Every sufficiently long word u' can be written as $u' = u_1 \cdots u_k$ with $\|u_i\| \geq t'$ and k sufficiently large. Thus, by repeatedly applying Claim 1, there exists a non-negative integer t_Ω such that every word $u' \in \mathrm{IRR}_{T_\Delta}(K^*)$ with $\|u'\| \geq t_\Omega$ contains two occurrences of the same $\omega \in \Omega$ which are far apart. More precisely, u' has a factor $\omega u \omega$ with $\|u\| > \|v_g\|$ for all $g \in G$.

This suggests rules of the form $\omega u \omega \to \omega v_{\varphi(u)} \omega$; but in order to ensure confluence we have to limit their use. For this purpose, we equip Ω with a linear order $\preceq$ such that $\gamma_m \gamma_0$ is the smallest element, and every element in $\Omega \cap K^+\gamma_0$ is smaller than all elements in $\Omega \setminus K^+\gamma_0$. By making t_Ω bigger, we can assume that every word u' with $\|u'\| \geq t_\Omega$ contains a factor $\omega u \omega$ such that

- $\|u\| > \|v_g\|$ for all $g \in G$, and

- for every factor $\omega' \in \Omega$ of $\omega u \omega$ we have $\omega' \preceq \omega$.

The following claim is one of the main reasons for using the above definition of the normal forms v_g, and also for excluding all words $\omega \in \Gamma K^*$ in the definition of Ω except for $\omega = \gamma\gamma' \in \Gamma^2$ with $\gamma > \gamma'$.

Claim 2. *Let* $\omega, \omega' \in \Omega$ *and* $g \in G$. *If* $\omega v_g \omega \in K^*\omega' K^*$, *then* $\omega' \preceq \omega$.

The proof of Claim 2 can be found on arXiv [4].

We are now ready to define the second set of rules over the extended alphabet K. They are reducing long words without long repetitions of words in Δ. We set

$$T'_\Omega = \left\{ \omega u \omega \to \omega v_{\varphi(u)} \omega \mid \begin{array}{l} \|v_{\varphi(u)}\| < \|u\| \leq t_\Omega \text{ and} \\ \omega u \omega \text{ has no factor } \omega' \in \Omega \text{ with } \omega \prec \omega' \end{array} \right\}.$$

Whenever there is a shorter rule in $T'_\Omega \cup T_\Delta$ then we want to give preference to this shorter rule. Thus the Ω-rules are

$$T_\Omega = \left\{ \ell \to r \in T'_\Omega \mid \begin{array}{l} \text{there is no rule } \ell' \to r' \in T'_\Omega \cup T_\Delta \\ \text{such that } \ell' \text{ is a proper factor of } \ell \end{array} \right\}.$$

Let now $T = T_\Delta \cup T_\Omega$.

Claim 3. *The system* T *is locally confluent over* K^*.

The proof of Claim 3 can be found on arXiv [4].

Since all rules in T are weight-reducing, it follows from Claim 3 that T is confluent. Moreover, all rules $\ell \to r$ in T satisfy $\varphi(\ell) = \varphi(r)$. We conclude that T is a weighted Church-Rosser system such that K^*/T is finite and $\varphi : K^* \to G$ factorizes through T. Remember that every element in K^* can be read as a sequence of elements in A^*. Thus every $u \in K^*$ can be interpreted as a word $u \in A^*$. We use this interpretation in order to apply the rules in T to words in A^*; but in order to not destroy K-letters when applying rules in R, we have to guard the first K-letter of every T-rule by appending the letter c. This leads to the system

$$T' = \{c\ell \to cr \in A^* \times A^* \mid \ell \to r \in T\}.$$

Combining the rules R over the alphabet B with the T'-rules yields

$$S = R \cup T'.$$

Since left sides of R-rules and of T'-rules can not overlap, the system S is confluent. By definition, each S-rule is weight-reducing. This means that S is a weighted Church-Rosser system. The sets $\mathrm{IRR}_S(A^*)$ and A^*/S are finite. Since $\ell \to r$ in S satisfies $\varphi(\ell) = \varphi(r)$, the homomorphism φ factorizes through S. $\square$

4   Arbitrary Finite Monoids

This section contains the main result of this paper. We show that every homomorphism $\varphi : A^* \to M$ to a finite monoid M factorizes through a weighted Church-Rosser system S with finite index. The proof relies on Theorem 1 and on a construction called local divisors. The notion of *local divisor* has turned out to be a rather powerful tool when using inductive proofs for finite monoids, see e.g. [3, 5, 6]. The same is true in this paper. The definition of a local divisor is as follows: Let M be a monoid and let $c \in M$. We equip $cM \cap Mc$ with a monoid structure by introducing a new multiplication $\circ$ as follows:

$$xc \circ cy = xcy.$$

It is straightforward to see that $\circ$ is well-defined and $(cM \cap Mc, \circ)$ is a monoid with neutral element c.

The following observation is crucial. If $1 \in cM \cap Mc$, then c is a unit. Thus if the monoid M is finite and c is not a unit, then $|cM \cap Mc| < |M|$. The set $M' = \{x \mid cx \in Mc\}$ is a submonoid of M, and $c \cdot : M' \to cM \cap Mc : x \mapsto cx$ is a surjective homomorphism. Since $(cM \cap Mc, \circ)$ is the homomorphic image of a submonoid, it is a divisor of M. We therefore call $(cM \cap Mc, \circ)$ the *local divisor* of M at c.

4.1   The Main Result

We are now ready to prove our main result. Let $(A, \|\cdot\|)$ be a weighted alphabet. Then every homomorphism $\varphi : A^* \to M$ to a finite monoid M factorizes through a weighted Church-Rosser system S with finite index. The proof uses induction on the size of M and the size of A. If $\varphi(A^*)$ is a group, then we apply Theorem 1; and if $\varphi(A^*)$ is not a group, then we find a letter $c \in A$ such that c is not a unit. Thus in this case we can use local divisors.

Theorem 2. *Let $(A, \|\cdot\|)$ be a weighted alphabet and let $\varphi : A^* \to M$ be a homomorphism to a finite monoid M. Then there exists a weighted Church-Rosser system S of finite index such that φ factorizes through S.*

Proof. The proof is by induction on $(|M|, |A|)$ with lexicographic order. If $\varphi(A^*)$ is a group, then the claim follows by Theorem 1. If $\varphi(A^*)$ is not a group, then there exists $c \in A$ such that $\varphi(c)$ is not a unit. Let $B = A \setminus \{c\}$. By induction on the size of the alphabet there exists a weighted Church-Rosser system R for the restriction $\varphi : B^* \to M$ satisfying the statement of the theorem. Let

$$K = \mathrm{IRR}_R(B^*)c.$$

We consider the prefix code K as a weighted alphabet. The weight of a letter $uc \in K$ is the weight $\|uc\|$ when read as a word over the weighted alphabet $(A, \|\cdot\|)$. Let $M_c = \varphi(c)M \cap M\varphi(c)$ be the local divisor of M at $\varphi(c)$. We let $\psi : K^* \to M_c$ be the homomorphism induced by $\psi(uc) = \varphi(cuc)$ for $uc \in K$.

By induction on the size of the monoid there exists a weighted Church-Rosser system $T \subseteq K^* \times K^*$ for ψ satisfying the statement of the theorem. Suppose $\psi(\ell) = \psi(r)$ for $\ell, r \in K^*$ and let $\ell = u_1 c \cdots u_j c$ and $r = v_1 c \cdots v_k c$ with $u_i, v_i \in \mathrm{IRR}_R(B^*)$. Then

$$\varphi(c\ell) = \varphi(cu_1 c) \circ \cdots \circ \varphi(cu_j c)$$
$$= \psi(u_1 c) \circ \cdots \circ \psi(u_j c)$$
$$= \psi(\ell) = \psi(r) = \varphi(cr).$$

This means that every T-rule $\ell \to r$ yields a φ-invariant rule $c\ell \to cr$. We can transform the system $T \subseteq K^* \times K^*$ for ψ into a system $T' \subseteq A^* \times A^*$ for φ by

$$T' = \{c\ell \to cr \in A^* \times A^* \mid \ell \to r \in T\}.$$

Since T is confluent and weight-reducing over K^*, the system T' is confluent and weight-reducing over A^*. Combining R and T' leads to $S = R \cup T'$. The left sides of a rule in R and a rule in T' cannot overlap. Therefore, S is a weighted Church-Rosser system such that φ factorizes through A^*/S. Suppose that every word in $\mathrm{IRR}_T(K^*)$ has length at most k. Here, the length is over the extended alphabet K. Similarly, let every word in $\mathrm{IRR}_R(B^*)$ have length at most m. Then

$$\mathrm{IRR}_S(A^*) \subseteq \{u_0 c u_1 \cdots c u_{k'+1} \mid u_i \in \mathrm{IRR}_R(B^*),\ k' \leq k\}$$

and every word in $\mathrm{IRR}_S(A^*)$ has length at most $(k+2)m$. In particular $\mathrm{IRR}_S(A^*)$ and A^*/S are finite. $\qquad\square$

The following corollary is a straightforward translation of the result in Theorem 2 about homomorphisms to a statement about regular languages.

Corollary 1. *A language $L \subseteq A^*$ is regular if and only if there exists a Church-Rosser system S of finite index such that $L = \bigcup_{u \in L}[u]_S$. In particular, all regular languages are Church-Rosser congruential.*

5   Conclusion

We have shown that all regular languages are Church-Rosser congruential. The proof has been done by loading the induction hypothesis. Our result says that for all $\varphi : A^* \to M$ to a finite monoid M and all weights $\|\cdot\| : A \to \mathbb{N} \setminus \{0\}$ there exists a weighted Church-Rosser system S of finite index such that φ factorizes through S. A very interesting question is whether we can change quantifiers. Is it true that for all such φ there exists a finite confluent system S of finite index such that φ factorizes through S and which is weight-reducing for all weights? Note that whether a system is weight-reducing for all weights is a natural condition on the number of letters in the Parikh image. Thus, we can call such a system *Parikh-reducing*. This result is true for aperiodic monoids [6], because every subword-reducing system is Parikh-reducing.

Another problem for future research is which algebraic invariants of M can be maintained in A^*/S. For example, if M satisfies the equation $x^{t+p} = x^t$, then our construction yields that A^*/S satisfies an equation $x^{s+p} = x^s$ for some s large enough. We conjecture that we must choose $s > t$, in general. In particular, we doubt that we can choose A^*/S to be a group, even if M is a (cyclic) finite group. However proving such a *lower bound result* seems to be a hard task.

The consideration of weights makes perfect sense in the setting of [9], too. If a language is a finite union of congruence classes w.r.t. to some finite confluent and weight-reducing system, then it has essentially the very same nice properties as a Church-Rosser congruential language. This program goes beyond regular languages and might lead to interesting new results in the interplay of string rewriting, formal languages, and algebra.

References

[1] Book, R., Otto, F.: String-Rewriting Systems. Springer (1993)
[2] Buntrock, G., Otto, F.: Growing context-sensitive languages and Church-Rosser languages. Information and Computation 141, 1–36 (1998)
[3] Diekert, V., Gastin, P.: First-order definable languages. In: Logic and Automata: History and Perspectives. Texts in Logic and Games, pp. 261–306. Amsterdam University Press (2008)
[4] Diekert, V., Kufleitner, M., Reinhardt, K., Walter, T.: Regular languages are Church-Rosser congruential. ArXiv e-prints, arXiv:1202.1148 (February 2012)
[5] Diekert, V., Kufleitner, M., Steinberg, B.: The Krohn-Rhodes Theorem and Local Divisors. ArXiv e-prints, arXiv:1111.1585 (November 2011)
[6] Diekert, V., Kufleitner, M., Weil, P.: Star-free languages are Church-Rosser congruential. Theoretical Computer Science (2012), doi:10.1016/j.tcs.2012.01.028
[7] Jantzen, M.: Confluent String Rewriting. EATCS Monographs on Theoretical Computer Science, vol. 14. Springer (1988)
[8] Lothaire, M.: Combinatorics on Words. Encyclopedia of Mathematics and its Applications, vol. 17. Addison-Wesley (1983)
[9] McNaughton, R., Narendran, P., Otto, F.: Church-Rosser Thue systems and formal languages. J. ACM 35(2), 324–344 (1988)
[10] Narendran, P.: Church-Rosser and related Thue systems. Doctoral dissertation, Dept. of Mathematical Sciences, Rensselaer Polytechnic Institute, Troy, NY, USA (1984)
[11] Niemann, G.: Church-Rosser Languages and Related Classes. Kassel University Press, PhD thesis (2002)
[12] Niemann, G., Otto, F.: The Church-Rosser languages are the deterministic variants of the growing context-sensitive languages. Inf. Comput. 197, 1–21 (2005)
[13] Niemann, G., Waldmann, J.: Some Regular Languages That Are Church-Rosser Congruential. In: Kuich, W., Rozenberg, G., Salomaa, A. (eds.) DLT 2001. LNCS, vol. 2295, pp. 330–339. Springer, Heidelberg (2002)
[14] Reinhardt, K., Thérien, D.: Some more regular languages that are Church Rosser congruential. In: 13. Theorietag, Automaten und Formale Sprachen, Herrsching, Germany, pp. 97–103 (2003)
[15] Woinowski, J.R.: Church-Rosser Languages and Their Application to Parsing Problems. PhD Thesis, TU Darmstadt (2001)
[16] Woinowski, J.R.: The context-splittable normal form for Church-Rosser language systems. Inf. Comput. 183, 245–274 (2003)

Time and Parallelizability Results for Parity Games with Bounded Treewidth

John Fearnley and Sven Schewe

Department of Computer Science, University of Liverpool, Liverpool, UK

Abstract. Parity games are a much researched class of games in NP ∩ CoNP that are not known to be in P. Consequently, researchers have considered specialised algorithms for the case where certain graph parameters are small. In this paper, we show that, if a tree decomposition is provided, then parity games with bounded treewidth can be solved in $O(k^{3k+2} \cdot n^2 \cdot (d+1)^{3k})$ time, where n, k, and d are the size, treewidth, and number of priorities in the parity game. This significantly improves over previously best algorithm, given by Obdržálek, which runs in $O(n \cdot d^{2(k+1)^2})$ time. Our techniques can also be adapted to show that the problem lies in the complexity class NC^2, which is the class of problems that can be efficiently parallelized. This is in stark contrast to the general parity game problem, which is known to be P-hard, and thus unlikely to be contained in NC.

1 Introduction

A parity game is a two player game that is played on a finite directed graph. The problem of solving a parity game is known to lie in NP ∩ CoNP [9]. However, despite much effort, this problem is not known to be in P. Due to the apparent difficulty of solving parity games, recent work has considered special cases, where the input is restricted in some way. In particular, people have studied parity games where the input graph is restricted by a graph parameter. For example, parity games have been shown to admit polynomial time algorithms whenever the input graph has bounded treewidth [11], DAG-width [3], or entanglement [4]. In this paper we study the parity game problem for graphs of bounded treewidth.

Parity games are motivated by applications in model checking. The problem of solving a parity game is polynomial-time equivalent to the modal μ-calculus model checking problem [7,12]. In model checking, we typically want to check whether a large system satisfies a much smaller formula. It has been shown that many practical systems have bounded treewidth. For example, it has been shown that the control flow graphs of goto free Pascal programs have treewidth at most 3, and that the control flow graphs of goto free C programs have treewidth at most 6 [13]. The same paper also shows that tree decompositions, which are very costly to compute in general, can be generated in linear time with small constants for these control flow graphs. Moreover, Obdržálek has shown that, if the input system has treewidth k, and if the μ-calculus formula has m sub-formulas, then the modal μ-calculus model checking problem can be solved by

A. Czumaj et al. (Eds.): ICALP 2012, Part II, LNCS 7392, pp. 189–200, 2012.
© Springer-Verlag Berlin Heidelberg 2012

determining the winner of a parity game that has treewidth at most $k \cdot m$ [11]. Since m is usually much smaller than the size of the system, we have a strong motivation for solving parity games with bounded treewidth.

The first algorithm for parity games with bounded treewidth was given by Obdržálek [11], and it runs in $O(n \cdot k^2 \cdot d^{2(k+1)^2})$ time, while using $d^{O(k^2)}$ space, where d is the number of priorities in the parity game, and k is the treewidth of the game. This result shows that, if the treewidth of a parity game is bounded, then there is a polynomial time algorithm for solving the game. However, since the degree of this polynomial depends on k^2, this algorithm is not at all practical. For example, if we wanted to solve the model checking problem for the control flow graph of a C program with treewidth 6, and a μ-calculus formula with a single sub-formula, the running time of the algorithm will already be $O(n \cdot d^{98})$!

Obdržálek's result was generalised by Berwanger et al. to the problem of solving a parity game with small DAG-width [3]. In order to obtain their result, Berwanger et al. introduced a more efficient data structure, and it is possible to modify Obdržálek's algorithm to make use of this data structure when solving parity games with small treewidth. This hybrid algorithm obtains a better space complexity of $d^{O(k)}$. However, some operations used by the algorithm of Berwanger et al. still require $d^{O(k^2)}$ time, and thus the hybrid algorithm does not run faster than Obdržálek's original algorithm.

More recently, an algorithm has been proposed for graphs with "medium" tree width [8]. This algorithm runs in time $n^{O(k \cdot \log n)}$, and it is therefore better than Obdržálek's algorithm whenever $k \in \omega(\log n)$. On the other hand, this algorithm does not provide an improvement for parity games with small treewidth.

Our Contribution. In this paper, we take a step towards providing a practical algorithm for this problem. We present a method that allows the problem to be solved in logarithmic space on an alternating Turing machine. From this, we are able to derive an algorithm that runs in $O(k^{3k+2} \cdot n^2 \cdot (d+1)^{3k})$ time on a deterministic Turing machine, if a tree decomposition is provided. Since the degree of the polynomials here is $O(k)$, rather than $O(k^2)$, this algorithm is dramatically faster than the algorithm of Obdržálek, for all values of k.

If a tree decomposition is not provided, then the approximation algorithm of Amir [2] can be applied to find a 4.5-approximate tree decomposition, which causes our overall running time to be $O((4.5k)^{13.5k+2} \cdot n^2 \cdot (d+1)^{13.5k})$. Precise algorithms for computing a tree decomposition, such as the algorithm of Bodlaender [5], are unsuitable here, as their complexity includes a factor of $2^{O(k^3)}$, which dwarfs any potential gains made by our algorithm. Note that Obdržálek's algorithm also requires a tree decomposition as input. This causes the gap between the two algorithms is even wider, as using a 4.5-approximate tree decomposition causes Obdržálek's algorithm to run in $O(n \cdot d^{2(4.5k+1)^2})$ time.

We are also able to adapt these techniques to show a parallelizability result for parity games with bounded treewidth. We are able to provide an alternating Turing machine that solves the problem in $O(k^2 \cdot (\log |V|)^2)$ time while using $O(k \cdot \log |V|)$ space. This version of the algorithm does not require a precomputed tree decomposition. Hence, using standard results in complexity theory [1], we have that the problem lies in the complexity class $NC^2 \subseteq NC$, which is the class of problems that can be efficiently parallelized.

This result can be seen in stark contrast to the complexity of parity games on general graphs: parity games are known to be P-hard by a reduction from reachability games, and P-hardness is considered to be strong evidence that an efficient parallel algorithm does not exist. Our result here shows that, while we may be unable to efficiently parallelize the μ-calculus model checking problem itself, we can expect to find efficient parallel algorithms for the model checking problems that appear in practice.

2   Preliminaries

A parity game is a tuple $(V, V_0, V_1, E, \text{pri})$, where V is a set of vertices and E is a set of edges, which together form a finite directed graph. We assume that every vertex has at least one outgoing edge. The sets V_0 and V_1 partition V into vertices belonging to player Even and player Odd, respectively. The function $\text{pri} : V \to D$ assigns a *priority* to each vertex from the set of priorities $D \subseteq \mathbb{N}$.

We define the *significance* ordering $\prec$ over D. This ordering represents how attractive each priority is to player Even. For two priorities $a, b \in \mathbb{N}$, we have $a \prec b$ if one of the following conditions holds: (1) a is odd and b is even, (2) a and b are both even and $a < b$, or (3) a and b are both odd and $a > b$. We say that $a \preceq b$ if either $a \prec b$ or $a = b$.

At the beginning of the game, a token is placed on a starting vertex v_0. In each step, the owner of the vertex that holds the token must choose one outgoing edge from that vertex and move the token along it. In this fashion, the two players form an infinite path $\pi = \langle v_0, v_1, v_2, \ldots \rangle$, where $(v_i, v_{i+1}) \in E$ for every $i \in \mathbb{N}$. To determine the winner of the game, we consider the set of priorities that occur *infinitely often* along the path. This is defined to be: $\text{Inf}(\pi) = \{d \in \mathbb{N} :$ For all $j \in \mathbb{N}$ there is an $i > j$ such that $\text{pri}(v_i) = d\}$. Player Even wins the game if the highest priority occurring infinitely often is even, and player Odd wins the game if it is odd. In other words, player Even wins the game if and only if $\max(\text{Inf}(\pi))$ is even.

A positional strategy for Even is a function that chooses one outgoing edge for every vertex in V_0. A strategy is denoted by $\sigma : V_0 \to V$, with the condition that $(v, \sigma(v)) \in E$ for every Even vertex v. Positional strategies for player Odd are defined analogously. The sets of positional strategies for Even and Odd are denoted by Σ_0 and Σ_1, respectively. Given two positional strategies σ and τ, for Even and Odd, respectively, and a starting vertex v_0, there is a unique path $\langle v_0, v_1, v_2 \ldots \rangle$, where $v_{i+1} = \sigma(v_i)$ if v_i is owned by Even, and $v_{i+1} = \tau(v_i)$ if v_i is owned by Odd. This path is known as the *play* induced by the two strategies σ and τ, and will be denoted by $\text{Play}(v_0, \sigma, \tau)$.

For each $\sigma \in \Sigma_0$, we define $G \restriction \sigma$ to be the modification of G where Even is forced to play σ. That is, an edge $(v, u) \in E$ is included in $G \restriction \sigma$ if either $v \in V_1$, or $v \in V_0$ and $\sigma(v) = u$. We define $G \restriction \tau$ for all $\tau \in \Sigma_1$ analogously.

An infinite path $\langle v_0, v_1, \ldots \rangle$ is said to be *consistent* with an Even strategy $\sigma \in \Sigma_0$ if $v_{i+1} = \sigma(v_i)$ for every i such that $v_i \in V_0$. If $\sigma \in \Sigma_0$ is strategy for Even, and v_0 is a starting vertex then we define $\text{Paths}(v_0, \sigma)$ to give every path starting at v_0 that is consistent with σ. An Even strategy $\sigma \in \Sigma_0$ is called a *winning strategy* for a vertex $v_0 \in V$ if $\max(\text{Inf}(\pi))$ is even for all $\pi \in \text{Paths}_0(v_0, \sigma)$. The strategy σ is said to be winning for a set of vertices $W \subseteq V$ if it is winning for all $v \in W$. Winning strategies for player Odd are defined analogously.

A game is said to be *positionally determined* if one of the two players always has a positional winning strategy. We now give a fundamental theorem, which states that parity games are positionally determined.

Theorem 1 ([6,10]). *In every parity game, the set of vertices V can be partitioned into winning sets (W_0, W_1), where Even has a positional winning strategy for W_0, and Odd has a positional winning strategy for W_1.*

To define the treewidth of a parity game, we will use the treewidth of the undirected graph that is obtained when the orientation of the edges in the game is ignored. Therefore, we will use the following definition of a tree decomposition.

Definition 2 (Tree Decomposition). *For each game G, the pair (T, X), where $T = (I, J)$ is an undirected tree and $X = \{X_i : i \in I\}$ is a family of subsets of V, is a tree decomposition of G if all of the following hold:*

1. $\bigcup_{i \in I} X_i = V$,
2. *for every $(v, u) \in E$ there is an $i \in I$ such that $v \in X_i$ and $u \in X_i$, and*
3. *for every $i, j \in I$, if $k \in I$ is on the unique path from i to j in T, then $X_i \cap X_j \subseteq X_k$.*

The *width* of a tree decomposition (T, X) is $\max\{|X_i| - 1 : i \in I\}$. The *treewidth* of a game G is the smallest width that can be obtained by a tree decomposition of G. We can apply the algorithm of Amir [2] to approximate the treewidth of the graph. This algorithm takes a graph G and an integer k, and in $O(2^{3k} n^2 k^{3/2})$ time either finds a tree decomposition of width at most $4.5k$ for G, or reports that the tree-width of G is larger than k.

3   Strategy Profiles

In this section we define strategy profiles, which are data structures that allow a player to give a compact representation of their strategy. Our algorithms in Sections 4 and 5 will use strategy profiles to allow the players to declare their strategies in a small amount of space. Throughout this section we will assume that there is a *starting vertex* $s \in V$ and a set of *final vertices* $F \subseteq V$. Let $\sigma \in \Sigma_0$ be a strategy for Even. The strategy profile of σ describes the outcome of a modified parity game that starts at s, terminates whenever a vertex $u \in F$ is encountered, and in which Even is restricted to only play σ.

For each $u \in F$, we define $\text{Paths}(\sigma, s, F, u)$ to be the set of paths from s to u that are consistent with σ and that do not visit a vertex in F. More formally,

Paths(σ, s, F, u) contains every path of the form $\langle v_0, v_1, v_2, \ldots v_k \rangle$ in $G \restriction \sigma$, for which the following conditions hold: (1) the vertex $v_0 = s$ and the vertex $v_k = u$, and (2) for all i in the range $0 \leq i \leq k-1$ we have $v_i \notin F$. For each strategy $\tau \in \Sigma_1$, we define the functions Paths(τ, s, F, u) for each $u \in F$ analogously.

For each $u \in F$, the function Exit(σ, s, F, u), gives the best possible priority (according to $\prec$) that Odd can visit when Even plays σ and Odd chooses to move to u. This function either gives a priority $p \in D$, or, if Odd can never move to u when Even plays σ, the function gives a special symbol $-$, which stands for "unreachable". We will also define this function for Odd strategies $\tau \in \Sigma_1$. Formally, for every finite path $\pi = \langle v_1, v_2, \ldots, v_k \rangle$, we define MaxPri$(\pi) =$ $\max\{\text{pri}(v_i) : 1 \leq i \leq k\}$. Furthermore, we define, for $\sigma \in \Sigma_0$ and $\tau \in \Sigma_1$:

$$\text{MinPath}(\sigma, s, F, u) = \min_{\preceq}\{\text{MaxPri}(\pi) : \pi \in \text{Paths}(\sigma, s, F, u)\},$$

$$\text{MaxPath}(\tau, s, F, u) = \max_{\preceq}\{\text{MaxPri}(\pi) : \pi \in \text{Paths}(\tau, s, F, u)\}.$$

For every strategy $\chi \in \Sigma_0 \cup \Sigma_1$ and every $u \in F$ we define:

$$\text{Exit}(\chi, s, F, u) = \begin{cases} - & \text{if Paths}(\chi, s, F, u) = \emptyset, \\ \text{MinPath}(\chi, s, F, u) & \text{if Paths}(\chi, s, F, u) \neq \emptyset \text{ and } \chi \in \Sigma_0, \\ \text{MaxPath}(\chi, s, F, u) & \text{if Paths}(\chi, s, F, u) \neq \emptyset \text{ and } \chi \in \Sigma_1. \end{cases}$$

We can now define the strategy profile for each strategy $\chi \in \Sigma_0 \cup \Sigma_1$. We define Profile$(\chi, s, F)$ to be a function $F \to D \cup \{-\}$ such that Profile$(\chi, s, F)(u) =$ Exit(χ, s, F, u) for each $u \in F$.

4   Time Complexity Results for Parity Games

Let G be a parity game with a tree decomposition $(T = (I, J), X)$ of width k. In this section we give a game that allows us to determine the winner of G for some starting vertex s. We will show that this game can be solved in $2 \cdot \log_2 |V| + (3k + 2) \cdot \log_2 k + 3k \cdot \log_2(|D| + 1) + O(1)$ space on an alternating Turing machine, and hence $O(k^{3k+2} \cdot |V|^2 \cdot (|D| + 1)^{3k})$ time on a deterministic Turing machine.

A basic property of tree decompositions is that each set $X_i \in X$ is a *separator* in the game graph. This means that, if $v \in X_j \setminus X_i$ and $u \in X_k \setminus X_i$, and if the path from j to k in T passes through X_i, then all paths from v to u in G must pass through a vertex in X_i. Our approach is to choose some separator $S \in X$, such that our starting vertex $s \in S$, and to simulate a run of the parity game using only the vertices in S. We will call this the *simulation game*.

One possible run of the simulation game is shown in Figure 1a. The large circle depicts the separator S, the boxes represent Even vertices, and the triangles represent Odd vertices. Since no two vertices share the same priority in this example, we will use the priorities to identify the vertices. As long as both players choose to remain in S, the parity game is played as normal. However, whenever one of the two players chooses to move to a vertex v with $v \notin S$ we

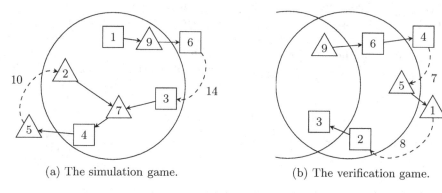

(a) The simulation game.   (b) The verification game.

Fig. 1. Example runs for the two types of game

simulate the parity game using a strategy profile: Even is required to declare a strategy in the form of a strategy profile P for v and S. Odd then picks some vertex $u \in S$, and moves there with priority $P(u)$. In the diagram, the dashed edges represent these simulated decisions. For example, when the path moved to the vertex 6, Even gave a strategy profile P with $P(3) = 14$, and Odd decided to move to the vertex 3. Together, the simulated and real edges will eventually form a cycle, and the winner of the game is the winner of this cycle. In our example, Even wins the game because the largest priority on the cycle is 10.

If Even always gives strategy profiles that correspond to some strategy $\sigma \in \Sigma_0$, then the outcome of our simulated game will match the outcome that would occur in the real parity game. On the other hand, it is possible that Even could lie by giving a strategy profile P for which he has no strategy. To deal with this, Odd is allowed to reject P, which causes the two players to move to a *verification game*, where Even is required to prove that he does have a strategy for P.

If the simulation game is played on a separator X_i, then the verification game will be played on a separator X_j where (i, j) is an edge in the tree decomposition. Suppose that the separator game ended when Odd rejected a strategy profile P for a vertex v, and note that we must have $v \notin S$. By the properties of a tree decomposition, there is a unique separator X_k such that $v \in X_k$, and k is the closest node to i in T. We define $\text{Next}(i, v)$ to be the unique edge (i, j) on the path from i to k. If the separator game on $S = X_i$ ends when the strategy profile P for a vertex v is rejected, then the verification game will be played on $X_j \cup \{v\}$, where $j = \text{Next}(i, v)$.

Suppose that Odd rejected Even's first strategy profile in the game shown in Figure 1a. Hence, Even must show that his strategy profile P, that contains $P(3) = 14$, is correct in the verification game. An example run of this verification game is shown in Figure 1b. The leftmost circle represents S, the original separator, and the other circle represents S', the separator chosen by Next. The verification game proceeds as before, by simulating a parity game on S'. However, we add the additional constraint that, if a vertex $u \in S$ is visited, then the game ends: Even wins the game if the largest priority p seen on the path to u has

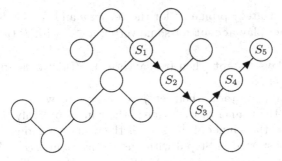

Fig. 2. The outcome when Odd repeatedly rejects Even's strategy profile

$p \succeq P(u)$, and Odd wins otherwise. In this example, Even loses the verification game because the largest priority seen during the path is 8, and Even's strategy profile in Figure 1a claimed that the largest priority p should satisfy $p \succeq 14$.

Note that the verification game also uses strategy profiles to simulate the moves made outside S'. Since Even could also give a false strategy profile in the verification game, Odd is still allowed to reject a strategy profile, and create a second verification game. Figure 2 shows the outcome when Odd repeatedly rejects Even's strategy profiles. The diagram depicts the tree decomposition, and each circle is a separator in the parity game. The game started with the simulation game on S_1, moved to the verification game on S_2, and so on.

An important observation is that we only need to remember the strategy profile for the game immediately before the one we are playing. This is because we always play the game on a separator. For example, when we play the game on S_3, it does not matter what strategy profile was rejected for S_1, because we must pass through a vertex in S_2 to reach S_1. This will be crucial in showing that the game can be solved in logarithmic space by an alternating Turing machine.

We now formally describe the verification game. Given two separators $S, F \in X$, a starting vertex $s \in S$, and a strategy profile P for s and F, we define $\mathsf{Verify}_G(S, F, s, P)$ as follows. The game maintains a variable c to store the current vertex, which is initially set so that $c = s$. It also maintains a sequence of triples Π, where each entry is of the form (v, p, u) with $v, u \in S \cup F$ and $p \in D$. The sequence Π represents the path that the two players form through S.

To define our game, we will require notation to handle the path Π. We extend MaxPri for paths of the form $\Pi = \langle(v_1, p_1, v_2), (v_2, p_2, v_3), \ldots, (v_{j-1}, p_{j-1}, v_j)\rangle$ by defining $\mathsf{MaxPri}(\Pi) = \{\max(p_i) : 1 \leq i \leq j - 1\}$. Furthermore, if a path Π consists of an initial path $\langle(v_1, p_1, v_2), \ldots, (v_{j-1}, p_{j-1}, v_j)\rangle$ followed by a cycle $\langle(v_j, p_j, v_{j+1}), \ldots, (v_{c-1}, p_{c-1}, v_j)\rangle$, then we define $\mathsf{Winner}(\Pi)$ to be Even if $\max(\{p_i : j \leq i \leq c - 1\})$ is even, and Odd otherwise.

We are now able to define $\mathsf{Verify}_G(S, F, s, P)$. The game is played in rounds, and each round proceeds as follows.

1. An edge (c, v) is selected by Even if $c \in V_0$ or by Odd if $c \in V_1$.
2. If $v \in F \cup S$, then the tuple $(c, \mathrm{pri}(v), v)$ is added to Π, the vertex c is set to v, and the game moves to Step 5.

3. Even gives a strategy profile P' for the vertex v and the set S.
4. Odd can either play `accept` for some vertex $u \in S$ with $P'(u) \neq -$, or play `reject`.
   - If Odd plays `accept`, then $(c, \max(\mathrm{pri}(v), P'(u)), u)$ is appended to Π, and c is set to u.
   - If Odd plays `reject`, then let $j = \mathrm{Next}(i, v)$, where $S = X_i$. If $v \in X_j$, then the winner of the game is the winner of $\mathsf{Verify}_G(X_j, X_i, v, P')$. Otherwise, the winner of the game is the winner of $\mathsf{Verify}_G(X_j, X_i, c, P')$, when it is started in Step 3 with an identical v and $\Pi = \emptyset$.
5. If $c \in F$, then the game stops. If $P(c) = -$, then Odd wins the game. Otherwise, Even wins the game if $\mathrm{MaxPri}(\Pi) \succeq P(c)$ and Odd wins if $\mathrm{MaxPri}(\Pi) \prec P(c)$.
6. If Π ends in a cycle, then the winner of the game is $\mathrm{Winner}(\Pi)$.

The simulation game is very similar to the verification game, but we do not need the parameters F and P. In fact, we define the simulation game for a separator $S \in X$ and starting vertex $s \in S$ to be $\mathsf{Verify}(S, \emptyset, s, \emptyset)$. Note that the game continues to be well defined with these parameters. The only difference is that Step 5 can never be triggered. Hence the game only ends when the two players construct a cycle, or when Odd rejects a strategy profile given by Even.

We now argue that the simulation game correctly determines the winner of the parity game. Suppose that Even has a winning strategy $\sigma \in \Sigma_0$ for some vertex $s \in W_0$. Even can win $\mathsf{Verify}_G(S, \emptyset, s, \emptyset)$ by following the moves prescribed by σ in Step 1, and by giving $\mathrm{Profile}(\sigma, v, S)$ for P' in Step 3. On the other hand, suppose that Odd has a winning strategy $\tau \in \Sigma_1$ for some vertex $s \in W_1$. We say that τ *refutes* a strategy profile P if one of the following conditions holds: (1) for all $u \in S$, we have $P(u) = -$, or (2) for all $u \in S$, we have either $\mathrm{Profile}(\tau, s, S)(u) \neq -$ and $P(u) = -$, or $\mathrm{Profile}(\tau, s, S)(u) \prec P(u)$. Odd can win $\mathsf{Verify}_G(S, \emptyset, s, \emptyset)$ by following τ in Step 1, and playing `reject` in Step 4 only in the case where τ refutes P'. If τ does not refute P', then there must exist a vertex $u \in S$ with $\mathrm{Profile}(\tau, v, S)(u) \succeq P'(u)$. Odd finds such a vertex u, and plays `accept`. We can show that these strategies are correct.

Lemma 3. *If $s \in V$ and $S \in X$ with $s \in S$, then Even can win $\mathsf{Verify}_G(S, \emptyset, s, \emptyset)$ if and only if $s \in W_0$.*

We now give the main result of this section, which is that $\mathsf{Verify}(S, \emptyset, s, \emptyset)$ can be solved in $2 \cdot \log_2 |V| + (3k + 2) \cdot \log_2 k + 3k \cdot \log_2(|D| + 1) + O(1)$ space on an alternating Turing machine, and hence $O(k^{3k+2} \cdot |V|^2 \cdot (|D| + 1)^{3k})$ time on a deterministic Turing machine. To solve the game with an alternating Turing machine, we use the existential states to simulate the moves of Even, and the universal states to simulate the moves of Odd. To implement the game, we must allocate space to store the sets S and F, the vertices c and v, the strategy profiles P and P', and the path Π. It can be verified that all of these can be stored in the required space bound. Hence, we obtain the main result of this section.

Theorem 4. *There is an algorithm that, when given a parity game G and a tree decomposition of width k, terminates in $O(k^{3k+2} \cdot |V|^2 \cdot (|D|+1)^{3k})$ time and outputs the partition (W_0, W_1).*

5   Parity Games with Bounded Treewidth Are In NC²

In this section we show that parity games with bounded tree width lie in the complexity class NC². In order to do this, we will show that there is an alternating Turing machine that takes $O(k^2 \cdot (\log |V|)^2)$ time, and uses $O(k \cdot \log |V|)$ space to solve a parity game whose treewidth is k. Note that the alternating Turing machine considered in Section 4 could take $O(|V|)$ time. Hence, in this section we show a weaker space bound in exchange for also bounding the running time of the alternating Turing machine.

We will define the game VerifyNC, which is identical to the game Verify, except for the method used to choose the next separator when odd plays reject. The game Verify can take up to $O(|V|)$ steps to terminate, because it always chooses an adjacent separator when Odd plays reject (recall Figure 2). In contrast to this, the game VerifyNC will give Odd the freedom to choose the set of vertices that the next game will be played on.

However, giving Odd this freedom comes at a price. Whereas, in the game Verify we only had to remember one previous strategy profile, in the game VerifyNC we may have to remember several previous strategy profiles. Formally, a *record* is a triple (F, p, P), where $F \subseteq V$ is a set of vertices, P is a strategy profile for F, and $p \in D$ is the largest priority that has been seen since P was rejected. When a record is created, we will use $(F, -, P)$ to indicate that no priority has been seen since P was rejected. Given a record (F, p, P), and a priority p', we define $\text{Update}((F, p, P), p') = (F, \max(p, p'), P)$, where we have $\max(-, p') = p'$. A *history* is a set of records. Given a history $\mathcal{F}$ and a priority p', we define: $\text{Update}(\mathcal{F}, p') = \{\text{Update}((F, p, P), p') : (F, p, P) \in \mathcal{F}\}$. Note that we cannot allow our history to contain a large number of records, because we require our alternating Turing machine to use at most $O(k \cdot \log |V|)$ space. Hence, we will insist that the history may contain at most 3 records, and we will force Odd to delete some records whenever the history contains more than 3 records. This restriction will be vital for showing that our alternating Turing machine meets the required space bounds.

We now formally define VerifyNC$(S, \mathcal{F}, s)$, where $S \subseteq V$ is a set of vertices, where $\mathcal{F}$ is a history, and where $s \in V$ is a starting vertex. These definitions mostly follow the game Verify, except where modifications are needed to deal with the history $\mathcal{F}$. The game is played in rounds. As before, we maintain a current vertex c, which is initially set to s, and we maintain a sequence of triples Π, where each entry is of the form (v, p, u) with $v, u \in S$, and $p \in D$. The sequence Π is initially empty. Each round of VerifyNC$(S, \mathcal{F}, s)$ proceeds as follows:

1. An edge (c, v) is selected by Even if $c \in V_0$ or by Odd if $c \in V_1$.
2. If $v \in S$ or if $v \in F$ for some $(F, p, P) \in \mathcal{F}$, then the tuple $(c, \text{pri}(v), v)$ is added to Π, the vertex c is set to v, and the game moves to Step 5.

3. Even gives a strategy profile P' for the vertex v and the set S.
4. Odd can either play accept for some vertex $u \in S$ with $P'(u) \neq -$, or play reject.
   - If Odd plays accept, then $(c, \max(\text{pri}(v), P'(u)), u)$ is appended to Π, and c is set to u.
   - If Odd plays reject, then:
     • The history $\mathcal{F}'$ is obtained by setting $\mathcal{F}' = \text{Update}(\mathcal{F}, \text{MaxPri}(\Pi))$.
     • The record $(F, -, P')$ is added to $\mathcal{F}'$. Odd may choose to delete one or more records in $\mathcal{F}'$. If $|\mathcal{F}'| > 3$, then Odd must delete enough records so that $|\mathcal{F}'| \leq 3$.
     • Odd selects a set of vertices $S' \subseteq V$ such that $|S'| \leq k+1$ and $v \in S'$.
     • The winner of the game is the winner of $\text{VerifyNC}(S', \mathcal{F}', v)$.
5. If $c \in F$ for some $(F, p, P) \in \mathcal{F}$, then the game stops. If $P(c) = -$ then Odd wins the game. Otherwise, let $p' = \max(\text{MaxPri}(\Pi), p)$. Even wins the game if $p' \succeq P(c)$ and Odd wins if $p' \prec P(c)$.
6. If Π ends in a cycle, then the winner of the game is $\text{Winner}(\Pi)$.

Note that this game may not end, since Odd could, for example, always play reject, and remove all records from $\mathcal{F}$. It is for this reason that we define a limited-round version of the game, which will be denoted by $\text{VerifyNC}^r(S, \mathcal{F}, s)$, where $r \in \mathbb{N}$. This game is identical to $\text{VerifyNC}(S, \mathcal{F}, s)$ for the first r rounds. However, if the game has not ended before the start of round $r + 1$, then Even is declared to be the winner.

If we ignore, for the moment, the issue of selecting the set S', then the strategies that were presented for Verify continue to work for VerifyNC. That is, if Even has a winning strategy $\sigma \in \Sigma_0$ for a vertex s, then Even has a winning strategy for $\text{VerifyNC}^r(\{s\}, \emptyset, s)$ for all r. Moreover, if Odd has a winning strategy $\tau \in \Sigma_1$ for s, and if the game ends before round $r + 1$, then Odd has a winning strategy for $\text{VerifyNC}^r(\{s\}, \emptyset, s)$. Both of these properties hold regardless of the set S' that happens to be chosen by Odd in Step 4.

Lemma 5. *No matter what sets S' are chosen in Step 4, if $s \in W_0$, then Even can force a win in $\text{VerifyNC}^r(\{s\}, \emptyset, s)$. Moreover, if $s \in W_1$, and if the game ends before round $r + 1$, then Odd can force a win in $\text{VerifyNC}^r(\{s\}, \emptyset, s)$.*

We now prove the key result of this section: if the input graph has treewidth bounded by k, then Odd has a strategy for choosing S' in Step 4 that forces the game to terminate after at most $O(k \cdot \log |V|)$ verification games have been played. Although our strategy will make use of a tree decomposition, note that we do not need to be concerned with actually computing this tree decomposition. The purpose of presenting this strategy is to show that there exists a sequence of sets S' that Odd can choose in order to make the game terminate quickly. The existence of an Odd strategy with this property is sufficient to imply that an implementation of VerifyNC on an alternating Turing machine will terminate quickly whenever the input graph has bounded treewidth.

We present two strategies that Odd can use to select S'. The first strategy is called Slice, and is shown in Figure 3a. The figure shows a tree decomposition,

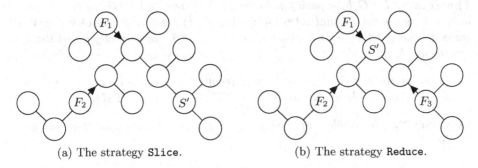

(a) The strategy Slice.   (b) The strategy Reduce.

Fig. 3. A method that Odd can use to win using $\log |V|$ separators in VerifyNC

and two records F_1 and F_2. Our task is to select a separator in subtree between F_1 and F_2. We do so using the following well known lemma about trees.

Lemma 6. *For every tree $T = (I, J)$ with $|I| \geq 3$, there is an $i \in I$ such that removing i from T splits T into parts, where each part has at most $\frac{2}{3}|I|$ vertices.*

We define the strategy Slice to select a separator that satisfies Lemma 6. For example, in Figure 6 the separator S' splits the tree into 3 parts, having 6, 1, and 1 vertices, respectively. Since there were originally 9 vertices in the subtree between F_1 and F_2, we have that S' satisfies Lemma 6.

The second strategy is called Reduce, and it is shown in Figure 3b. It is used whenever there are three records in $\mathcal{F}$. It selects the unique vertex that lies on the paths between these three records. It can be seen in Figure 3b that the set S' lies on the unique vertex that connects F_1, F_2, and F_3. The purpose of this strategy is to reduce the number of records in $\mathcal{F}$. It can be seen that, no matter how the game on S' ends, we will be able to forget at least two of the three records F_1, F_2, and F_3, while adding only one record for S'.

Therefore, Odd's strategy for picking S' is to use Slice until $|\mathcal{F}| = 3$, and then to alternate using Slice and using Reduce. It can be shown that this strategy always forces the game to end after at most $O(\log |V|)$ verification games, and hence $O(k \cdot \log |V|)$ rounds, have been played. This strategy also ensures that we never have to remember more than three records.

Lemma 7. *If G has a tree decomposition of width k, then Odd has a strategy for chosing the set S' in Step 4 that forces VerifyNC$(\{s\}, \emptyset, s)$ to terminate after at most $O(k \cdot \log |V|)$ rounds.*

Thus, if we solve VerifyNC$_G^{O(k \log |V|)}(\{s\}, \emptyset, s)$ for some vertex $s \in V$, Lemmas 5 and 7 imply that, if the input graph has treewidth bounded by k, and if Odd has a winning strategy for s, then Odd can win VerifyNC$_G^{O(k \log |V|)}(\{s\}, \emptyset, s)$. On the other hand, if Even wins VerifyNC$_G^{O(k \log |V|)}(\{s\}, \emptyset, s)$, then either the treewidth of the graph must be greater than k, or Even has a winning strategy for s. We can use these two properties to prove the following theorem, which is the main result of this section.

Theorem 8. *Let G be a parity game and k be a parameter. There is an alternating Turing machine that takes $O(k^2 \cdot (\log |V|)^2)$ time, and uses $O(k \cdot \log |V|)$ space, to either determine the winner of a vertex $s \in V$, or correctly report that G has treewidth larger than k.*

A simple corollary of the time and space results for our alternating Turing machine is that our problem lies in the class NC^2 [1, Theorem 22.15].

Corollary 9. *The problem of solving a parity game with bounded treewidth lies in NC^2.*

References

1. Allender, E., Loui, M.C., Regan, K.W.: Complexity classes. In: Atallah, M.J., Blanton, M. (eds.) Algorithms and Theory of Computation Handbook, p. 22. Chapman & Hall/CRC (2010)
2. Amir, E.: Efficient approximation for triangulation of minimum treewidth. In: Proc. of UAI, pp. 7–15. Morgan Kaufmann Publishers Inc., San Francisco (2001)
3. Berwanger, D., Dawar, A., Hunter, P., Kreutzer, S.: DAG-Width and Parity Games. In: Durand, B., Thomas, W. (eds.) STACS 2006. LNCS, vol. 3884, pp. 524–536. Springer, Heidelberg (2006)
4. Berwanger, D., Grädel, E.: Entanglement – A Measure for the Complexity of Directed Graphs with Applications to Logic and Games. In: Baader, F., Voronkov, A. (eds.) LPAR 2004. LNCS (LNAI), vol. 3452, pp. 209–223. Springer, Heidelberg (2005)
5. Bodlaender, H.L.: A linear-time algorithm for finding tree-decompositions of small treewidth. SIAM Journal on Computing 25(6), 1305–1317 (1996)
6. Emerson, E.A., Jutla, C.S.: Tree automata, μ-calculus and determinacy. In: Proc. of FOCS, pp. 368–377. IEEE Computer Society Press (October 1991)
7. Emerson, E.A., Jutla, C.S., Sistla, A.P.: On Model-Checking for Fragments of μ-Calculus. In: Courcoubetis, C. (ed.) CAV 1993. LNCS, vol. 697, pp. 385–396. Springer, Heidelberg (1993)
8. Fearnley, J., Lachish, O.: Parity Games on Graphs with Medium Tree-Width. In: Murlak, F., Sankowski, P. (eds.) MFCS 2011. LNCS, vol. 6907, pp. 303–314. Springer, Heidelberg (2011)
9. McNaughton, R.: Infinite games played on finite graphs. Ann. Pure Appl. Logic 65(2), 149–184 (1993)
10. Mostowski, A.W.: Games with forbidden positions. Technical Report 78, University of Gdańsk (1991)
11. Obdržálek, J.: Fast Mu-Calculus Model Checking when Tree-Width Is Bounded. In: Hunt Jr., W.A., Somenzi, F. (eds.) CAV 2003. LNCS, vol. 2725, pp. 80–92. Springer, Heidelberg (2003)
12. Stirling, C.: Local Model Checking Games (Extended Abstract). In: Lee, I., Smolka, S.A. (eds.) CONCUR 1995. LNCS, vol. 962, pp. 1–11. Springer, Heidelberg (1995)
13. Thorup, M.: All structured programs have small tree width and good register allocation. Information and Computation 142(2), 159–181 (1998)

Nominal Completion
for Rewrite Systems with Binders[*]

Maribel Fernández[1] and Albert Rubio[2]

[1] Dept. of Informatics, King's College London, UK
Maribel.Fernandez@kcl.ac.uk
[2] Technical University of Catalonia, LSI, Barcelona, Spain
rubio@lsi.upc.edu

Abstract. We design a completion procedure for nominal rewriting systems, based on a generalisation of the recursive path ordering to take into account alpha equivalence. Nominal rewriting generalises first-order rewriting by providing support for the specification of binding operators. Completion of rewriting systems with binders is a notably difficult problem; the completion procedure presented in this paper is the first to deal with binders in rewrite rules.

Keywords: nominal syntax, rewriting, orderings, completion.

1   Introduction

Term rewriting systems [2,31] are used to model the dynamics (e.g., deduction and evaluation) of formal systems. In particular, rewriting can be used to decide equality of terms modulo an equational theory, if the set of equational axioms can be oriented to form a *confluent and terminating* set of rules: To decide if two terms are equal we rewrite them as much as possible and compare their irreducible forms. The problem is that the system obtained by orienting the axioms of the theory is not necessarily confluent and terminating. For this reason, *completion procedures* [23] have been designed to attempt to transform the system into an equivalent confluent and terminating one.

Completion procedures rely on *well founded orderings* on terms to check termination of the rewrite system. This is because in a terminating system, local confluence implies confluence [26] and local confluence can be checked by computing overlaps of rules. The idea is that if a given system is terminating but not confluent, then to obtain a confluent one it is sufficient to compute its critical pairs (using a unification algorithm to check if rules overlap) and add rules to join them; the new rules must also be oriented according to a well founded ordering to ensure termination.

Many formal systems include binding operators, therefore, α-equivalence, that is, equivalence modulo renaming of bound variables, must be considered when

---

[*] Partially supported by the Spanish MICINN under grant TIN 2010-21062-C02-01

A. Czumaj et al. (Eds.): ICALP 2012, Part II, LNCS 7392, pp. 201–213, 2012.
© Springer-Verlag Berlin Heidelberg 2012

defining rewriting, reduction orderings and completion procedures. One alternative is to define binders through functional abstraction, as done in higher-order rewriting systems (e.g., CRSs [22], HRSs [25] and ERSs [21]), using the λ-calculus as a metalanguage. However, since not only α-equivalence but also some form of β-reduction is used, we can no longer rely on simple notions such as syntactic unification. The syntax/type restrictions imposed on rules in these systems has prevented the design of completion procedures, as explained in [27].

Alternatively, the nominal approach [17,28], which we follow in this paper, does not rely on the λ-calculus as metalanguage. It distinguishes between object-level variables (called *atoms*), which can be abstracted but behave similarly to constants, and metalevel variables or just *variables*, which are first-order in that they cannot be abstracted and substitution does not avoid capture of free atoms. On nominal terms [32,13] α-equivalence is axiomatised using a *freshness relation* between atoms and terms, and unification modulo α-equivalence is decidable [32].

Nominal rewriting systems (NRSs) [13] are rewriting systems on nominal terms. A step of nominal rewriting involves matching modulo α, which, if we only use *closed* rules, can be implemented in linear time [9]. Closed rules are, roughly speaking, rules which preserve abstracted atoms during reductions — a natural restriction which is also imposed on CRSs, HRSs and ERSs by definition.

Nominal rewriting can be used to automate equational reasoning on nominal terms (see [16,11] for related notions of nominal equational reasoning), if the equational theory is represented by a confluent and terminating NRS [14]. We also have a critical pair lemma (see [13]) which can be used to derive confluence of terminating systems. The only ingredient missing to design a completion procedure for NRSs is a tool to check termination.

In this paper, we generalise the well-known recursive path ordering (rpo) [12] to deal with nominal terms and α-equivalence; we call it nominal recursive path ordering (nrpo). Our main contribution is a proof that the nrpo inherits the properties of the rpo, and can therefore be used to check termination of NRSs. Armed with a method to check termination, we then focus on obtaining confluence via completion. Our second contribution is the design of a completion procedure à la Knuth and Bendix [23] for closed NRSs using the nrpo. An implementation is currently being developed; it is available from http://www.lsi.upc.edu/~albert/normal-completion.tar.gz

Related work. Algebraic λ-calculi [8,19] combine the λ-calculus with a set of term rewriting rules. A powerful generalisation of the rpo for algebraic λ-terms was presented by Jouannaud and Rubio [20,6]: the *higher-order rpo* (HORPO).

The HORPO can also be applied to higher-order rewriting systems (other orderings for these systems were proposed, see e.g., in [18,4]). The nrpo, in contrast, does not include β-reduction and cannot be directly applied to algebraic λ-calculi or higher-order rewriting systems, but, thanks to the fact that it does not need to be compatible with β-reduction (only with α-equivalence), it can handle systems that the HORPO cannot handle (we give examples in Section 3).

Completion has been generalised to systems that use higher-order functions but no binders, i.e. with a first-order syntax [24]. No completion procedures are available for higher-order rewriting systems.

Overview: Section 2 recalls nominal syntax and rewriting. In Section 3 we define orderings for nominal terms, and in Section 4 completion for NRSs. Section 5 concludes.

2   Preliminaries

A *nominal signature* Σ is a set of term-formers, or *function symbols*, $f, g, \ldots$, each with a fixed arity. In the following, we fix a countably infinite set $\mathcal{X}$ of *variables* ranged over by $X, Y, Z, \ldots$, and a countably infinite set $\mathcal{A}$ of *atoms* ranged over by $a, b, c, \ldots$, and assume that Σ, $\mathcal{X}$ and $\mathcal{A}$ are pairwise disjoint.

A *swapping* is a pair of atoms, written $(a\ b)$. *Permutations* π are bijections on atoms, represented by lists of swappings (Id denotes the *identity permutation*). We write π^{-1} for the inverse of π and $\pi \circ \pi'$ for the composition of π and π'. For example, if $\pi = (a\ b)(b\ c)$ then $\pi^{-1} = (b\ c)(a\ b)$.

Nominal terms, or just *terms*, are generated by the grammar

$$s, t ::= a \mid \pi \cdot X \mid [a]s \mid f(s_1, \ldots, s_n)$$

and called, respectively, atoms, moderated variables or simply variables, abstractions and function applications (if $n = 0$ or $n = 1$ we may omit the brackets). We abbreviate $\text{Id} \cdot X$ as X if there is no ambiguity.

We write $V(t)$ (resp. $A(t)$) for the set of variables (resp. atoms) occurring in t (we use the same notation for pairs of terms, contexts, etc.) and $root(t)$ for the symbol at the root of the term t. *Ground terms* have no variables: $V(t) = \emptyset$.

An abstraction $[a]t$ is intended to represent t with a bound. We call occurrences of a *abstracted* if they are under an abstraction, and *unabstracted* (or free) otherwise. For examples of nominal terms, we refer the reader to [32,13].

The *action of a permutation* π on a term t is defined by induction: $\text{Id} \cdot t = t$ and $(a\ b)\pi \cdot t = (a\ b) \cdot (\pi \cdot t)$, where a swapping acts on terms as follows:

$$(a\ b) \cdot a = b \quad (a\ b) \cdot b = a \quad (a\ b) \cdot c = c \quad (c \notin \{a, b\})$$
$$(a\ b) \cdot (\pi \cdot X) = ((a\ b) \circ \pi) \cdot X \qquad (a\ b) \cdot [c]t = [(a\ b) \cdot c](a\ b) \cdot t$$
$$(a\ b) \cdot f(t_1, \ldots, t_n) = f((a\ b) \cdot t_1, \ldots, (a\ b) \cdot t_n)$$

Substitutions are generated by the grammar: $\sigma ::= \text{Id} \mid [X \mapsto s]\sigma$. We use the same notation for the identity substitution and permutation, and also for composition, since there will be no ambiguity. Substitutions act on variables, without avoiding capture of atoms. We write $t\sigma$ for the application of σ on t, defined as follows: $t\text{Id} = t$, $t[X \mapsto s]\sigma = (t[X \mapsto s])\sigma$, and

$$a[X \mapsto s] = a \quad (\pi \cdot X)[X \mapsto s] = \pi \cdot s \quad (\pi \cdot Y)[X \mapsto s] = \pi \cdot Y \quad (X \neq Y)$$
$$([a]t)[X \mapsto s] = [a](t[X \mapsto s]) \quad f(t_1, \ldots, t_n)[X \mapsto s] = f(t_1[X \mapsto s], \ldots, t_n[X \mapsto s])$$

Definition 1. *A* freshness constraint *is a pair* $a\#t$ *of an atom and a term. A* freshness context *(ranged over by* Δ, ∇, Γ), *is a set of constraints of the form* $a\#X$. Freshness *and* α-equivalence *judgements* $\Delta \vdash a\#t$ *and* $\Delta \vdash s \approx_\alpha t$ *are derived using the rules below[1], where* a, b, c *are distinct atoms and* $ds(\pi, \pi') = \{a \mid \pi{\cdot}a \neq \pi'{\cdot}a\}$ *(difference set).*

$$\frac{}{\Delta \vdash a\#b}\ (\#\mathbf{ab}) \qquad \frac{\pi^{-1}{\cdot}a\#X \in \Delta}{\Delta \vdash a\#\pi{\cdot}X}\ (\#\mathbf{X}) \qquad \frac{\Delta \vdash a\#s_1 \ \cdots \ \Delta \vdash a\#s_n}{\Delta \vdash a\#f(s_1, \ldots, s_n)}\ (\#\mathbf{f})$$

$$\frac{}{\Delta \vdash a\#[a]s}\ (\#[\mathbf{a}]) \qquad \frac{\Delta \vdash a\#s}{\Delta \vdash a\#[b]s}\ (\#[\mathbf{b}]) \qquad \frac{}{\Delta \vdash a \approx_\alpha a}\ (\approx_\alpha\mathbf{a})$$

$$\frac{\forall a \in ds(\pi, \pi') : a\#X \in \Delta}{\Delta \vdash \pi{\cdot}X \approx_\alpha \pi'{\cdot}X}\ (\approx_\alpha\mathbf{X}) \qquad \frac{\Delta \vdash s_1 \approx_\alpha t_1 \ \cdots \ \Delta \vdash s_n \approx_\alpha t_n}{\Delta \vdash f(s_1, \ldots, s_n) \approx_\alpha f(t_1, \ldots, t_n)}\ (\approx_\alpha\mathbf{f})$$

$$\frac{\Delta \vdash s \approx_\alpha t}{\Delta \vdash [a]s \approx_\alpha [a]t}\ (\approx_\alpha[\mathbf{a}]) \qquad \frac{\Delta \cup \Delta_{c\#s,t} \vdash (c\ a){\cdot}s \approx_\alpha (c\ b){\cdot}t \quad c \notin A([a]s, [b]t)}{\Delta \vdash [a]s \approx_\alpha [b]t}\ (\approx_\alpha[\mathbf{b}])$$

where $\Delta_{c\#s,t}$ *is a freshness context such that* $\Delta_{c\#s,t} \vdash c\#s$ *and* $\Delta_{c\#s,t} \vdash c\#t$; *such a context always exists if* $c \notin A(s,t)$.

Definition 2. *A* nominal rewrite rule $R = \nabla \vdash l \to r$ *is a tuple of a freshness context* ∇ *and terms* l *and* r *such that* $V(r, \nabla) \subseteq V(l)$.

A nominal rewrite system *(NRS) is an equivariant set* $\mathcal{R}$ *of nominal rewrite rules, that is, a set of nominal rules that is closed under permutations. We shall generally equate a set of rewrite rules with its equivariant closure.*

Example 1. The following rules compute prenex normal forms in first-order logic; the signature has term-formers $\forall$, $\exists$, $\neg$, $\wedge$, $\vee$ ($\wedge$ and $\vee$ are infix). Intuitively, equivariance means that the choice of atoms in rules is not important (for more explanations and examples, we refer the reader to [13]).

$$a\#P \vdash P \wedge \forall[a]Q \to \forall[a](P \wedge Q) \qquad a\#P \vdash (\forall[a]Q) \wedge P \to \forall[a](Q \wedge P)$$
$$a\#P \vdash P \vee \forall[a]Q \to \forall[a](P \vee Q) \qquad a\#P \vdash (\forall[a]Q) \vee P \to \forall[a](Q \vee P)$$
$$a\#P \vdash P \wedge \exists[a]Q \to \exists[a](P \wedge Q) \qquad a\#P \vdash (\exists[a]Q) \wedge P \to \exists[a](Q \wedge P)$$
$$a\#P \vdash P \vee \exists[a]Q \to \exists[a](P \vee Q) \qquad a\#P \vdash (\exists[a]Q) \vee P \to \exists[a](Q \vee P)$$
$$\vdash \neg(\exists[a]Q) \to \forall[a]\neg Q \qquad\qquad \vdash \neg(\forall[a]Q) \to \exists[a]\neg Q$$

Nominal rewriting [13] operates on 'terms-in-contexts', written $\Delta \vdash s$. Below C varies over terms with exactly one occurrence of a distinguished variable Id-$\cdot$-, or just -. We write $C[s]$ for $C[\text{-}{\mapsto}s]$, and $\Delta \vdash \nabla\theta$ for $\{\Delta \vdash a\#X\theta \mid a\#X \in \nabla\}$.

Definition 3. *A term* s rewrites with $R = \nabla \vdash l \to r$ *to* t *in* Δ, *written* $\Delta \vdash s \to_R t$ *(as usual, we assume* $V(R) \cap V(\Delta, s) = \emptyset$), *if* $s = C[s']$ *and there exists* θ *such that* $\Delta \vdash \nabla\theta$, $\Delta \vdash l\theta \approx_\alpha s'$ *and* $\Delta \vdash C[r\theta] \approx_\alpha t$. *The rewrite relation* $\to^*$ *is the reflexive and transitive closure of this relation. If* t *does not rewrite in the context* Δ *we say that it is in* normal form *in* Δ.

---

[1] The version of $(\approx_\alpha[\mathbf{b}])$ used here is equivalent to the standard presentation [32], and it is easier to use in orderings in the next sections.

An NRS is *terminating* if all the rewrite sequences are finite. It is *confluent* if, for all Δ, s, t, t' such that $\Delta \vdash s \to^* t$ and $\Delta \vdash s \to^* t'$, there exists u such that $\Delta \vdash t \to^* u$ and $\Delta \vdash t' \to^* u$. Local confluence (a weaker property) can be checked by computing critical pairs [13], using nominal unification [32].

Definition 4. *Suppose* $R_i = \nabla_i \vdash l_i \to r_i$ *for* $i = 1, 2$ *are copies of two rules in* $\mathcal{R}$ *such that* $V(R_1) \cap V(R_2) = \emptyset$ *(R_1 and R_2 could be copies of the same rule), $l_1 = C[l_1']$ and there is a most general unifier* (Γ, θ) *such that* $\Gamma \vdash l_1'\theta \approx_\alpha l_2\theta$ *and* $\Gamma \vdash \nabla_i \theta$ *for* $i = 1, 2$. *Then* $\Gamma \vdash (r_1\theta, C\theta[r_2\theta])$ *is a* critical pair. *If* $C = $ - *and* R_1, R_2 *are copies of the same rule, or if* l_1' *is a variable, then the critical pair is* trivial.

We are interested in closed rules (roughly, rules without unabstracted atoms); this notion is defined below. All the systems that can be specified in standard higher-order rewriting formalisms are closed [15].

Given a term in context $\nabla \vdash t$, or more generally, a pair $P = \nabla \vdash (l, r)$ (this could be a rule $R = \nabla \vdash l \to r$), we write $P^n = \nabla^n \vdash (l^n, r^n)$ to denote a *freshened variant* of P, i.e., a version where the atoms and variables have been replaced by 'fresh' ones. We shall always explicitly say what P^n is freshened for when this is not obvious. For example, a freshened version of $(a\#X \vdash X \to X)$ with respect to itself and to $a'\#X \vdash a'$ is $(a''\#X' \vdash X' \to X')$, where $a'' \neq a, a'$ and $X' \neq X$. We will write $A(P')\#V(P)$ to mean that all atoms occurring in P' are fresh for each of the variables occurring in P.

Definition 5. *Let* $\nabla^n \vdash t^n$ *be a freshened version of* $\nabla \vdash t$. *We say that* $\nabla \vdash t$ *is* closed *if there exists a substitution* σ *such that* $\nabla, A(\nabla^n \vdash t^n)\#V(\nabla \vdash t) \vdash \nabla^n\sigma$ *and* $\nabla, A(\nabla^n \vdash t^n)\#V(\nabla \vdash t) \vdash t^n\sigma \approx_\alpha t$. *This can be checked using the nominal matching algorithm [9]. This definition applies to nominal rewrite rules, and to pairs* $\nabla \vdash (l, r)$ *in general, using* $(,)$ *as a term-former.*

Let R^n *be a freshened version of the rule R with respect to Δ, s, t (as shown in [13], it does not matter which particular freshened R^n we choose). We write* $\Delta \vdash s \to^c_R t$ *if* $\Delta, A(R^n)\#V(\Delta, s) \vdash s \to_{R^n} t$ *and call this a* closed *rewriting step.*

Closed NRSs inherit properties of first-order rewriting systems such as the Critical Pair Lemma: If all non-trivial critical pairs of a closed nominal rewrite system are joinable, then it is locally confluent [13]. In the rest of the paper we will always work with closed NRSs, using closed rewriting.

3   Reduction Orderings for Nominal Terms

In this section we define two relations on nominal terms, inspired by the recursive path ordering (rpo) [12]. The rpo is a well-founded ordering on first-order terms based on simple syntactic comparisons, using a precedence on function symbols. It has been widely used to prove termination of first-order rewriting systems.

The first ordering that we define equates all the atoms (by collapsing them into a unique atom a) and uses a precedence on the signature Σ of a nominal rewriting system extended with a and an abstraction operator for a.

To define this ordering, we use a translation function $\hat{\cdot}$ such that $\hat{t}$ coincides with t except that all the atoms have been replaced by a distinguished atom a. For example, the translation of $f([b]b, c)$ is $f([a]a, a)$ (we omit the inductive definition of this translation function). Then, to define $\Delta \vdash s > t$, we compare $\hat{s}$ and $\hat{t}$ (seen as first-order terms), using the standard rpo, with a precedence $>_\Sigma$ on $\Sigma \cup \{a, [a]\cdot\}$, such that $f >_\Sigma [a]\cdot >_\Sigma a$ for all $f \in \Sigma$. We give a direct definition of $>$ below.

Definition 6. Let Δ be a freshness context, s, t nominal terms over Σ, and $>_\Sigma$ a precedence on Σ. We write $\Delta \vdash s > t$, if $\hat{s} > \hat{t}$ as defined below, where $\geq$ is the reflexive closure of $>$, and $>_{mul}$ is the multiset extension of $>$.

1. $[a]s > t$ if $s \geq t$
2. $s > [a]t$ if $root(s) \in \Sigma$ and $s > t$
3. $[a]s > [a]t$ if $s > t$
4. $[a]s > a$
5. $f(s_1, \ldots, s_n) > t$ if $s_i \geq t$ for some i, $f \in \Sigma$
6. $f(s_1, \ldots, s_n) > g(t_1, \ldots, t_m)$ if $f >_\Sigma g$ and $f(s_1, \ldots, s_n) > t_i$ for all i
7. $f(s_1, \ldots, s_n) > a$
8. $f(s_1, \ldots, s_n) > g(t_1, \ldots, t_m)$ if $f =_\Sigma g$ and $\{s_1, \ldots, s_n\} >_{mul} \{t_1, \ldots, t_m\}$.

In the definition above, we have not included a and the abstraction operator in the precedence, but we have given explicitly the cases for comparison with abstraction and atoms (which correspond to $f > [a]\cdot > a$ in the precedence).

For example, the rules in Example 1 can be ordered using a precedence $>_\Sigma$ such that $\wedge >_\Sigma \forall$, $\wedge >_\Sigma \exists$, $\vee >_\Sigma \forall$, $\vee >_\Sigma \exists$, $\neg >_\Sigma \forall$, $\neg >_\Sigma \exists$.

It is easy to see that the relation specified in Definition 6 satisfies all the required conditions to be used as a reduction ordering: it is transitive, irreflexive, well-founded, and preserved by context and substitution, because we have $\Delta \vdash s > t$ if and only if $\hat{s} >_{rpo} \hat{t}$. It is also preserved by $\approx_\alpha$, that is, it works uniformly in α-equivalence classes, because all the terms in the α-equivalence class of t are mapped to the same term $\hat{t}$.

Example 2. This example is taken from [6], where terms in a λ-calculus extended with constants are compared using the parametric HORPO (an extension of the HORPO). First, note that $\vdash f(g([x]f(x, x)), g([x]f(x, x))) > [y]f(y, y)$ simply because $g([a]f(a, a)) > a$ since a is smaller than terms rooted by function symbols. More interestingly, the following terms can be compared using $>$ but are not comparable under the parametric HORPO [6]:

$$\vdash f(g(X, Y))) > g(X, [a]f(h(Y, a)))$$

To show that this holds, assume $f >_\Sigma g >_\Sigma h$. We show $\vdash f(g(X, Y))) > X$ and $\vdash f(g(X, Y))) > [a]f(h(Y, a)))$. The first one is trivial since it is included in the subterm relation. For the second, we show $\vdash f(g(X, Y))) > f(h(Y, a)))$, which requires $g(X, Y) > h(Y, a)$. Since $g(X, Y) > Y$, and $g(X, Y) > a$, we are done.

In contrast with higher-order versions of the rpo, our ordering does not deal with β-reduction, but it can compare terms that are not comparable in the HORPO. This is because it takes advantage of the fact that the set of atoms and the set of variables are disjoint, and substitutions only affect variables.

It is easy to see that this ordering is not stable under the usual notion of capture-avoiding substitution for atoms (since we are equating all the atoms!). This is not necessarily a problem: substitution for atoms is not a primitive operation for nominal terms; it can be defined using closed nominal rewriting rules (see Example 43 in [13]), which the ordering $>$ does orient. However, for certain applications (e.g., to define type theories with dependent types), the nominal framework has been extended to include capture-avoiding atom-substitution as a primitive operation [29,10]. To address this point, instead of replacing all atoms by the same one as in Definition 6, we replace and equate abstracted atoms only. For this, we first define $\bowtie$, a generalisation of $\approx_\alpha$.

Definition 7. *Let* $\mathbb{C} = \{c_1, c_2, \ldots\}$ *be an infinite set of distinguished atoms. Judgements* $\Delta \vdash s \bowtie_{\mathbb{C}} t$, *or just* $\Delta \vdash s \bowtie t$ *if* $\mathbb{C}$ *is clear, are derived using:*

$$\frac{}{\Delta \vdash a \bowtie a}\,(\bowtie\mathsf{a}) \qquad \frac{c_i, c_j \in \mathbb{C}}{\Delta \vdash c_i \bowtie c_j}\,(\bowtie\mathsf{c}) \qquad \frac{\forall a \in ds(\pi, \pi') : (a \# X \in \Delta \ or \ \pi{\cdot}a, \pi'{\cdot}a \in \mathbb{C})}{\Delta \vdash \pi{\cdot}X \bowtie \pi'{\cdot}X}\,(\bowtie\mathbf{X})$$

$$\frac{\Delta \vdash s_1 \bowtie t_1 : \cdots \ \Delta \vdash s_n \bowtie t_n}{\Delta \vdash f(s_1, \ldots, s_n) \bowtie f(t_1, \ldots, t_n)}\,(\bowtie\mathsf{f}) \qquad \frac{\Delta \vdash s \bowtie t}{\Delta \vdash [a]s \bowtie [a]t}\,(\bowtie[\mathsf{a}])$$

$$\frac{\Delta \cup \Delta_{c\#s,t} \vdash (c\ a){\cdot}s \bowtie (c\ b){\cdot}t \quad c \in \mathbb{C}, c \notin A([a]s, [b]t)}{\Delta \vdash [a]s \bowtie [b]t}\,(\bowtie[\mathsf{b}])$$

The relation $\bowtie$ equates all the atoms in $\mathbb{C}$ (see rules $(\bowtie\mathsf{c})$ and $(\bowtie\mathbf{X})$). It has similar properties to $\approx_\alpha$ regarding freshness constraints and permutations.

Property 1. 1. If $\Delta \vdash s \approx_\alpha t$ then $\Delta \vdash s \bowtie t$.
  2. If $\Delta \vdash s \bowtie t$ and $\Delta \vdash a\#s$ then $\Delta \vdash a\#t$ or $(a \in A(s,t), a \in \mathbb{C})$.
  3. If $\Delta \vdash s \bowtie t$ then $\Delta \vdash \pi{\cdot}s \bowtie \pi{\cdot}t$, for any π such that
    $\forall c \in \mathbb{C}, \pi{\cdot}c \in \mathbb{C}$ or $(c \notin A(s,t)$ and $\Delta \vdash c\#s,t)$.

Using the properties above, we can prove that $\bowtie$ is an equivalence relation.

Property 2. The relation $\bowtie$ is reflexive, symmetric and transitive.

In order to define the *nominal rpo*, $>_{nrpo}$, we consider a precedence $>_\Sigma$ on the signature Σ of an NRS, as in Definition 6. The main difference is that now we define $>_{nrpo}$ using the auxiliary relation $\bowtie$ since abstracted atoms will be replaced by fresh atoms from $\mathbb{C}$. Also, $\geq_{nrpo}$ is now the union of $>_{nrpo}$ and $\approx_\alpha$, that is, syntactic equality is replaced by the more general $\approx_\alpha$. We use permutations in Definition 8 to ensure that substitutions (of terms for variables) are dealt with correctly, and we use the atoms in $\mathbb{C}$ to ensure that atom-substitutions do not affect the ordering. Freshness contexts are needed to deal with $\approx_\alpha$.

Definition 8 (nrpo). *Let $>_\Sigma$ be a precedence on the signature Σ of an NRS, Δ a freshness context, and s, t terms. Let $\mathbb{C} = \{c_1, c_2, \ldots\}$ be an infinite set of atoms such that $\mathbb{C} \cap A(\Delta, s, t) = \emptyset,$[2] and $\Delta_{c\#s,t}$ a freshness context such that $\Delta_{c\#s,t} \vdash c\#s, c\#t$; such a context always exists if $c \notin A(s,t)$.*

We define $\Delta \vdash s >_{nrpo} t$, or just $\Delta \vdash s > t$, by cases, where $\geq$ is $> \cup \bowtie$.

1. *$\Delta \vdash [a]s > t$ if $\Delta \cup \Delta_{c\#s,t} \vdash (a\ c) \cdot s \geq t$, for an arbitrary $c \in \mathbb{C} - A(\Delta, [a]s, t)$.*
2. *$\Delta \vdash s > [a]t$ if $root(s) \in \Sigma$ and $\Delta \cup \Delta_{c\#s,t} \vdash s > (a\ c) \cdot t$, for an arbitrary $c \in \mathbb{C} - A(\Delta, s, [a]t)$.*
3. *$\Delta \vdash [a]s > [b]t$ if $\Delta \cup \Delta_{c\#s,t} \vdash (a\ c) \cdot s > (b\ c) \cdot t$, for an arbitrary $c \in \mathbb{C} - A(\Delta, [a]s, [b]t)$.*
4. *$\Delta \vdash [a]s > [a]t$ if $\Delta \vdash s > t$.*
5. *$\Delta \vdash [a]s > c$ if $c \in \mathbb{C}$.*
6. *$\Delta \vdash f(s_1, \ldots, s_n) > t$ if $\Delta \vdash s_i \geq t$ for some i.*
7. *$\Delta \vdash f(s_1, \ldots, s_n) > g(t_1, \ldots, t_m)$ if $f >_\Sigma g$ and $\Delta \vdash f(s_1, \ldots, s_n) > t_i$ for all i.*
8. *$\Delta \vdash f(s_1, \ldots, s_n) > c$ if $c \in \mathbb{C}$.*
9. *$\Delta \vdash f(s_1, \ldots, s_n) > g(t_1, \ldots, t_m)$ if $f =_\Sigma g$ and $\Delta \vdash \{s_1, \ldots, s_n\} >_{mul} \{t_1, \ldots, t_m\}$.*

Before showing that $>_{nrpo}$ is indeed a reduction ordering that can be used to check termination of nominal rewriting (Property 4), we give an example of application.

Example 3. For all the rules $\Delta \vdash l \rightarrow r$ in Example 1, $\Delta \vdash l >_{nrpo} r$ if we consider a precedence $>_\Sigma$ such that $\wedge >_\Sigma \forall$, $\wedge >_\Sigma \exists$, $\vee >_\Sigma \forall$, $\vee >_\Sigma \exists$, $\neg >_\Sigma \forall$, $\neg >_\Sigma \exists$. We show the derivation for the first rule only:

Since $\wedge >_\Sigma \forall$, it is sufficient to show $a\#P \vdash P \wedge \forall[a]Q > [a](P \wedge Q)$ and since we now have an abstraction in the right-hand side, it is sufficient to show

$$a\#P, c\#P, c\#Q \vdash P \wedge \forall[a]Q > (a\ c) \cdot P \wedge (a\ c) \cdot Q.$$

We now use case (9), and compare their subterms. Since $a\#P, c\#P, c\#Q \vdash P \approx_\alpha (a\ c) \cdot P$, it boils down to proving $a\#P, c\#P, c\#Q \vdash \forall[a]Q > (a\ c) \cdot Q$, for which it is sufficient to show $a\#P, c\#P, c\#Q \vdash [a]Q \geq (a\ c) \cdot Q$ (by case 6). This is a consequence of $a\#P, c\#P, c\#Q, c'\#Q \vdash (a\ c') \cdot Q \geq (a\ c) \cdot Q$, which holds since $ds((a\ c'), (a\ c)) = \{a, c, c'\}$ and we can use $(\bowtie\mathbf{X})$ to derive

$$a\#P, c\#P, c\#Q, c'\#Q \vdash (a\ c') \cdot Q \bowtie (a\ c) \cdot Q$$

We can also orient the (explicit) substitution rules for the λ-calculus (see [13], Example 43) using a precedence where $subst >_\Sigma app$ and $subst >_\Sigma \lambda$. The β rule can be oriented with a precedence such that $app >_\Sigma subst$, but, as expected, we cannot orient both β and the substitution rules since the system specifies the untyped λ-calculus, which is non-terminating. We give more examples below.

---

[2] Any such set $\mathbb{C}$ can be used due to equivariance.

Property 3. Assume $\Delta \vdash s \bowtie s'$ and $\Delta \vdash t \bowtie t'$. If $\Delta \vdash s >_{nrpo} t$ then also $\Delta \vdash s' >_{nrpo} t$ and $\Delta \vdash s >_{nrpo} t'$. Hence $>_{nrpo}$ is α-compatible.

Property 4 (Reduction Ordering). The nominal rpo has the following properties:

1. If $\Delta \vdash s >_{nrpo} t$ then:
   (a) For all $C[\text{-}]$, $\Delta \vdash C[s] >_{nrpo} C[t]$.
   (b) For all π, $\Delta \vdash \pi{\cdot}s >_{nrpo} \pi{\cdot}t$.
   (c) For all Γ such that $\Gamma \vdash \Delta\sigma$, $\Gamma \vdash s\sigma >_{nrpo} t\sigma$.
2. It is a decidable, irreflexive and transitive relation.
3. It is well founded when the precedence is well-founded: there are no infinite descending chains $\Delta \vdash s_1 >_{nrpo} s_2 >_{nrpo} \cdots$

The ordering $>_{nrpo}$ is also stable under capture-avoiding substitution for atoms. Due to space constraints we omit the definition of capture-avoiding substitution for atoms, and simply state the result for ground terms: If $\vdash s >_{nrpo} t$ then also $\vdash s\tau >_{nrpo} t\tau$, where τ is a capture-avoiding atom substitution.

Example 4. Assume $f >_\Sigma g >_\Sigma h$. We can use $>_{nrpo}$ to compare the terms $f(g([x]f(x,x)), g([x]f(x,x)))$ and $[y]f(y,y)$, as in Example 2:

$$\vdash f(g([x]f(x,x)), g([x]f(x,x))) >_{nrpo} [y]f(y,y)$$

Unlike $>$ in Example 2, $>_{nrpo}$ cannot compare $f(g(X,Y))$ and $g(X, [a]f(h(Y,a)))$. This is not surprising: $>_{nrpo}$ would not be stable by atom substitution if it ordered these terms (Y can be instantiated with a term containing a, then a is free in the left and bound in the right). HORPO cannot compare these terms either. However, unlike HORPO, $>_{nrpo}$ does use freshness information, and can compare these terms if a is fresh for Y: $a\#Y \vdash f(g(X,Y)) >_{nrpo} g(X, [a]f(h(Y,a)))$. For this, we show $a\#Y \vdash f(g(X,Y)) > X$ and $a\#Y \vdash f(g(X,Y)) > [a]f(h(Y,a)))$. The first one is trivial. For the second, we show $a\#Y, c\#Y, c\#X \vdash f(g(X,Y))) > f(h((a\ c){\cdot}Y, c)))$, which requires $a\#Y, c\#Y, c\#X \vdash g(X,Y) > (a\ c){\cdot}Y$ and $a\#Y, c\#Y, c\#X \vdash g(X,Y) > c$. The first holds because $a\#Y, c\#Y, c\#X \vdash Y \approx_\alpha (a\ c){\cdot}Y$, whereas the second holds directly by Definition 8, case 8.

Theorem 1. *If for all $\nabla \vdash l \to r$ in an NRS $\mathcal{R}$ we have $\nabla \vdash l > r$, where $>$ is one of the orderings defined in this section, then $\mathcal{R}$ is terminating.*

This theorem requires all the rules to be oriented. An NRS may have an infinite number of variants of each rule due to equivariance, however, since the ordering is itself equivariant, it is sufficient to orient one variant of each rule.

4 Completion for Nominal Rewriting Systems

Let E be a nominal equational theory defined by a set of axioms $\Delta_i \vdash s_i = t_i$. If the axioms can be oriented to form a confluent and terminating closed rewrite system, then closed nominal rewriting can be used to efficiently decide equality modulo E and $\approx_\alpha$, as shown in [14]. In this section we describe techniques to obtain a confluent and terminating set of rules equivalent to E.

Completion procedures were first described by Knuth and Bendix [23] to attempt to complete first-order systems that are terminating. The idea is to add rules to ensure that all the critical pairs are joinable (i.e., both terms have the same normal form), while preserving termination. Completion may fail (for instance, we might not be able to orient a critical pair) or may not terminate (generating new critical pairs for the added rules). For higher-order rewriting systems, the difficulty lies not only in the definition of a suitable ordering, but also in that after computing an orientable critical pair, we might not be able to add the corresponding rule due to syntactic or type restrictions, as mentioned in [27]. In the case of closed NRSs, critical pairs are also closed (Lemma 1). Thus, we can design a completion procedure à la Knuth and Bendix for nominal systems.

Lemma 1.   1. If $\Delta \vdash t$ is closed and $\Delta \vdash t \approx_\alpha t'$ then $\Delta \vdash t'$ is closed.
2. If $\Delta \vdash (s,t)$ is a critical pair from a closed NRS, then $\Delta \vdash (s,t)$ is closed.
3. If $\Delta \vdash s$ is closed and $\Delta \vdash s \to_{\mathcal{R}}^c t$ using a closed rule, $\Delta \vdash (s,t)$ is closed.
4. If $\Delta \vdash (s,t)$ is closed and $\Delta \vdash t \to_{\mathcal{R}}^c u$ using a closed rule, then $\Delta \vdash (s,u)$ is closed.

The completion procedure for NRSs is specified as a set of transformation rules on pairs $(E, \mathcal{R})$ consisting of a set of closed nominal equations and a set of nominal rewriting rules, in the style of Bachmair et al. [3] (see also [5]).

Input: $(E, \emptyset)$ and a (well-founded) reduction ordering $>$, such as $>_{nrpo}$.

Transformation rules:

$$
\begin{array}{lll}
(E, \mathcal{R}) & \Rightarrow (E \cup \Delta \vdash s = t, \mathcal{R}) & \text{if } \Delta \vdash (s,t) \text{ is a critical pair of } \mathcal{R} \\
(E \cup \Delta \vdash s = t, \mathcal{R}) & \Rightarrow (E, \mathcal{R} \cup \Delta \vdash s \to t) & \text{if } \Delta \vdash s > t \\
(E \cup \Delta \vdash s = t, \mathcal{R}) & \Rightarrow (E, \mathcal{R} \cup \Delta \vdash t \to s) & \text{if } \Delta \vdash t > s \\
(E \cup \Delta \vdash s = t, \mathcal{R}) & \Rightarrow (E, \mathcal{R}) & \text{if } \Delta \vdash s \approx_\alpha t \\
(E \cup \Delta \vdash s = t, \mathcal{R}) & \Rightarrow (E \cup \Delta \vdash s' = t, \mathcal{R}) & \text{if } \Delta \vdash s \to_{\mathcal{R}}^c s' \\
(E \cup \Delta \vdash s = t, \mathcal{R}) & \Rightarrow (E \cup \Delta \vdash s = t', \mathcal{R}) & \text{if } \Delta \vdash t \to_{\mathcal{R}}^c t' \\
(E, \mathcal{R} \cup \Delta \vdash s \to t) & \Rightarrow (E, \mathcal{R} \cup \Delta \vdash s \to t') & \text{if } \Delta \vdash t \to_{\mathcal{R}}^c t' \\
(E, \mathcal{R} \cup \Delta \vdash s \to t) & \Rightarrow (E \cup \Delta \vdash s' = t, \mathcal{R}) & \text{if } \Delta \vdash s \to_{\mathcal{R}}^c s' \\
& & \text{using } \nabla \vdash l \to r \in \mathcal{R} \text{ and} \\
& & l \text{ does not reduce with } \Delta \vdash s \to t
\end{array}
$$

As in the case of first-order rewriting, the rule that computes critical pairs should be used in a controlled way, to avoid computing the same pair repeatedly. Different versions of completion use different strategies to apply the rules above. In general, the completion procedure may loop, fail if it reaches a pair $(E, \mathcal{R})$ where the equations in E cannot be oriented, and succeed if it reaches an irreducible state $(\emptyset, \mathcal{R})$. In the latter case, $\mathcal{R}$ is the *result* of the procedure.

Theorem 2. *If completion succeeds for a set E of closed nominal equations, then the resulting $\mathcal{R}$ is confluent and terminating, and $\leftrightarrow_{\mathcal{R}}^{c*}$ coincides with $=_E$.*

Example 5. The following rules are inspired by an example in [27]. As ordering take nrpo with any precedence.

$$(\eta) \quad a\#X \vdash \lambda([a]app(X,a)) \to X$$
$$(\bot) \qquad app(\bot, Y) \to \bot$$

There is a critical pair since the unification problem $a\#X, app(X,a)$?$\approx$? $app(\bot, Y)$ has solution $\{X \mapsto \bot, Y \mapsto a\}$: $\lambda([a]\bot) = \bot$ can be oriented into $\lambda([a]\bot) \to \bot$.

Example 6. The following rules are inspired by the summation example in [30]. As ordering take nrpo with precedence $\Sigma > +$ and $\Sigma > app > id$.

$$a\#F \vdash \Sigma(0, [a]app(F,a)) \to 0$$
$$a\#F \vdash \Sigma(s(N), [a]app(F,a)) \to app(F, s(N)) + \Sigma(N, [a]app(F,a))$$
$$id(X) \to X$$
$$app(id0, X) \to id(X)$$

With the fourth rule applied on the first and second we obtain two critical pairs: $\Sigma(0, [a]id(a)) = 0$ and $\Sigma(s(N), [a]id(a)) = app(id0, s(N)) + \Sigma(N, [a]app(id0, a))$ which can be simplified and oriented as follows: $\Sigma(0, [a]a) \to 0$ and $\Sigma(s(N), [a]a) \to s(N) + \Sigma(N, [a]a)$.

5   Conclusions

We have defined two extensions of the rpo to deal with α-equivalence classes of terms using the nominal approach. The first extension is simple and remains close to the first-order rpo. The second is also stable under atom-substitution.

Using the orderings we have designed a completion procedure for NRSs. It would be interesting to see if the technique can be adapted to some restricted classes of higher-order rewriting systems, and to other term orderings (e.g., using polynomial interpretations [2]) or more sophisticated methods, such as dependency pairs [1] or the monotonic semantic path ordering [7].

Acknowledgements: The first author thanks the participants at HOR 2010 for useful comments during her talk, which covered some of the topics in this paper. We thank Christian Urban for comments on a previous version of the paper, and Jamie Gabbay for providing the LaTeX command for t^u.

References

1. Arts, T., Giesl, J.: Termination of Term Rewriting Using Dependency Pairs. Theoretical Computer Science 236, 133–178 (2000)
2. Baader, F., Nipkow, T.: Term rewriting and all that. Cambridge University Press, Great Britain (1998)
3. Bachmair, L., Dershowitz, N., Hsiang, J.: Orderings for equational proofs. In: Proc. Symp. Logic in Computer Science, Boston, pp. 346–357 (1986)

4. Blanqui, F.: Termination and Confluence of Higher-Order Rewrite Systems. In: Bachmair, L. (ed.) RTA 2000. LNCS, vol. 1833, pp. 47–61. Springer, Heidelberg (2000)

5. Bachmair, L., Dershowitz, N., Plaisted, D.A.: Completion Without Failure. In: Ait-Kaci, H., Nivat, M. (eds.) Resolution of Equations in Algebraic Structures. Rewriting Techniques, vol. 2, ch. 1. Academic Press, New York (1989)

6. Blanqui, F., Jouannaud, J.-P., Rubio, A.: The Computability Path Ordering: The End of a Quest. In: Kaminski, M., Martini, S. (eds.) CSL 2008. LNCS, vol. 5213, pp. 1–14. Springer, Heidelberg (2008)

7. Borralleras, C., Ferreira, M., Rubio, A.: Complete Monotonic Semantic Path Orderings. In: McAllester, D. (ed.) CADE 2000. LNCS (LNAI), vol. 1831, pp. 346–364. Springer, Heidelberg (2000)

8. Breazu-Tannen, V., Gallier, J.: Polymorphic rewriting conserves algebraic strong normalization. Theoretical Computer Science 83(1) (1991)

9. Calvès, C., Fernández, M.: Matching and Alpha-Equivalence Check for Nominal Terms. Journal of Computer and System Sciences, Special issue: Selected papers from WOLLIC 2008 (2009)

10. Cheney, J.: A Dependent Nominal Type Theory. Logical Methods in Computer Science 8(1) (2012)

11. Clouston, R.A., Pitts, A.M.: Nominal Equational Logic. In: Cardelli, L., Fiore, M., Winskel, G. (eds.) Computation, Meaning and Logic. Articles dedicated to Gordon Plotkin. Electronic Notes in Theoretical Computer Science, vol. 1496. Elsevier (2007)

12. Dershowitz, N.: Orderings for Term-Rewriting Systems. Theoretical Computer Science 17(3), 279–301 (1982)

13. Fernández, M., Gabbay, M.J.: Nominal rewriting. Information and Computation 205(6) (2007)

14. Fernández, M., Gabbay, M.J.: Closed nominal rewriting and efficiently computable nominal algebra equality. In: Proc. 5th Int. Workshop on Logical Frameworks and Metalanguages: Theory and Practice (LFMTP 2010). EPTCS, vol. 34 (2010)

15. Fernández, M., Gabbay, M.J., Mackie, I.: Nominal rewriting systems. In: Proceedings of the 6th ACM-SIGPLAN Symposium on Principles and Practice of Declarative Programming (PPDP 2004), Verona, Italy. ACM Press (2004)

16. Gabbay, M.J., Mathijssen, A.: Nominal universal algebra: equational logic with names and binding. Journal of Logic and Computation 19(6), 1455–1508 (2009)

17. Gabbay, M.J., Pitts, A.M.: A new approach to abstract syntax with variable binding. Formal Aspects of Computing 13, 341–363 (2001)

18. Hamana, M.: Semantic Labelling for Proving Termination of Combinatory Reduction Systems. In: Escobar, S. (ed.) WFLP 2009. LNCS, vol. 5979, pp. 62–78. Springer, Heidelberg (2010)

19. Jouannaud, J.-P., Okada, M.: Executable higher-order algebraic specification languages. In: Proceedings, Sixth Annual IEEE Symposium on Logic in Computer Science, pp. 350–361. IEEE Computer Society Press (1991)

20. Jouannaud, J.-P., Rubio, A.: Polymorphic Higher-Order Recursive Path Orderings. Journal of ACM 54(1) (2007)

21. Khasidashvili, Z.: Expression reduction systems. In: Proceedings of I. Vekua Institute of Applied Mathematics, Tbilisi, vol. 36, pp. 200–220 (1990)

22. Klop, J.-W., van Oostrom, V., van Raamsdonk, F.: Combinatory reduction systems, introduction and survey. Theoretical Computer Science 121, 279–308 (1993)

23. Knuth, D., Bendix, P.: Simple word problems in universal algebras. In: Leech, J. (ed.) Computational Problems in Abstract Algebra, pp. 263–297. Pergamon Press, Oxford (1970)

24. Kusakari, K., Chiba, Y.: A higher-order Knuth-Bendix procedure and its applications. IEICE Transactions on Information and Systems E90-D(4), 707–715 (2007)

25. Mayr, R., Nipkow, T.: Higher-order rewrite systems and their confluence. Theoretical Computer Science 192, 3–29 (1998)

26. Newman, M.H.A.: On theories with a combinatorial definition of equivalence. Annals of Mathematics 43(2), 223–243 (1942)

27. Nipkow, T., Prehofer, C.: Higher-Order Rewriting and Equational Reasoning. In: Bibel, W., Schmitt, P. (eds.) Automated Deduction — A Basis for Applications. Volume I: Foundations. Applied Logic Series, vol. 8, pp. 399–430. Kluwer (1998)

28. Pitts, A.M.: Nominal logic, a first order theory of names and binding. Information and Computation 186, 165–193 (2003)

29. Pitts, A.M.: Structural Recursion with Locally Scoped Names. Journal of Functional Programming 21(3), 235–286 (2011)

30. van de Pol, J.C.: Termination of higher-order rewrite systems. PhD thesis, Utrecht University, Utrecht, The Netherlands (December 1996)

31. Terese: Term Rewriting Systems. Cambridge Tracts in Theoretical Computer Science, vol. 55. Cambridge University Press (2003)

32. Urban, C., Pitts, A.M., Gabbay, M.J.: Nominal unification. Theoretical Computer Science 323, 473–497 (2004)

Discrete Generalised Polynomial Functors

(Extended Abstract)

Marcelo Fiore[*]

Computer Laboratory, University of Cambridge
Marcelo.Fiore@cl.cam.ac.uk

Abstract. We study generalised polynomial functors between presheaf categories, developing their mathematical theory together with computational applications. The main theoretical contribution is the introduction of discrete generalised polynomial functors, a class that lies in between the classes of cocontinuous and finitary functors, and is closed under composition, sums, finite products, and differentiation. A variety of applications are given: to the theory of nominal algebraic effects; to the algebraic modelling of languages, and equational theories there of, with variable binding and polymorphism; and to the synthesis of dependent zippers.

Keywords: presheaf categories, Kan extensions, generalised logic, polynomial functors, equational systems, algebraic effects, abstract syntax, dependent programming.

1 Introduction

The recurrent appearance of a structure in mathematical practice, as substantiated by interesting examples and applications, is a strong indicator of a worthwhile theory lurking in the background that merits development. This work is a direct outgrowth of this viewpoint as it arises from the interaction of two ubiquitous mathematical structures —presheaf categories and polynomial functors— in the context of computational applications. The paper thus contributes to the continued search for the foundational mathematical structures that underlie computer science.

Presheaf categories are functor categories of the form $Set^{\mathbb{C}}$ for $\mathbb{C}$ a small category and Set the category of sets and functions. They enrich the universe of constant sets to one of variable sets [28]. The crucial import of this being that the mode of variation, as given by the parameter small category, translates to new, often surprising, internal structure in the presheaf category. As such, the use of presheaf categories in computer science applications has been prominent: *e.g.*, in programming language theory [33,31,15], lambda calculus [34,25], domain theory [14,16], concurrency theory [13,35,5], and type theory [23,7].

Polynomial functors, *i.e.* polynomial constructions in or between categories, have also featured extensively; especially in the semantic theory of ADTs (see *e.g.* [37]). In modern theories of data structure there has been a need for the

---

[*] Partially supported by ERC ECSYM.

A. Czumaj et al. (Eds.): ICALP 2012, Part II, LNCS 7392, pp. 214–226, 2012.
© Springer-Verlag Berlin Heidelberg 2012

naive notion of polynomial functor as a sum-of-products construction to evolve to more sophisticated ones. In connection with type-theoretic investigations, this arose in the work of Gambino and Hyland [17] on the categorical study of W-types in Martin Löf Type Theory, and in the work on (indexed) containers of Abbott, Altenkirch, Ghani, and Morris [1,4] on data structure in dependent programming. Type theoretically, the extension can be roughly understood as generalising from sums and products to Σ and Π types.

It is with the above level of generality that we are concerned here, in the particular context of presheaf categories. Our motivation stems from a variety of applications that require these new generalised polynomial functors. A case in point, treated in some detail in the paper, is the use of a class of discrete generalised polynomial functors as formal semantic counterparts of the informal vernacular rules that one encounters in presentations of syntactic structure. A main contribution of the paper shows the flexibility and expressiveness of our theory to the extent of being able to encompass languages, and equational theories thereof, with variable binding and polymorphism.

The work is presented in three parts, with Secs. 2 and 3 providing the necessary background, Secs. 4 and 5 developing the mathematical theory, and Secs. 6 and 7 dwelling on applications.

2   Generalised Logic

This section recalls the basics of Lawvere's generalised logic [27]. Emphasis is placed on the categorical view of quantifiers as adjoints [26] that plays a central role in our development.

The basic categorical modelling of quantifiers as adjoints arises from the consideration of the contravariant powerset construction on sets, for which we will write $\wp$. Indeed, for all functions $f : X \to Y$, the monotone function $\wp f : \wp Y \to \wp X$ (where, for $T \in \wp Y$ and $x \in X$, $x \in \wp f(T) \Leftrightarrow fx \in T$) has both a left and a right adjoint, that roughly correspond to existential and universal quantification. More precisely, there are Galois connections

$$\exists_f \dashv \wp f \dashv \forall_f : \wp X \to \wp Y$$

given, for $S \in \wp X$ and $y \in Y$, by

$$y \in \exists_f(S) \iff \exists x \in X.\ fx = y \wedge x \in S \ , \tag{1}$$

$$y \in \forall_f(S) \iff \forall x \in X.\ y = fx \Rightarrow x \in S \ . \tag{2}$$

The categorical viewpoint of quantifiers generalises from sets to categories by considering the contravariant presheaf construction on small categories. For a small category $\mathbb{C}$, let $\mathcal{P}\mathbb{C}$ be the functor category $\mathbf{Set}^{\mathbb{C}}$ of covariant presheaves and, for a functor $f : \mathbb{X} \to \mathbb{Y}$ between small categories, let $f^* : \mathcal{P}\mathbb{Y} \to \mathcal{P}\mathbb{X}$ be given by $P \mapsto Pf$. A fundamental result states that there are adjunctions

$$f_! \dashv f^* \dashv f_* : \mathcal{P}\mathbb{X} \to \mathcal{P}\mathbb{Y} \ . \tag{3}$$

For $P : \mathbb{X} \to \mathbf{Set}$, $f_!(P)$ is a left Kan extension of P along f; whilst f_*P is a right Kan extension of P along f. Importantly for our development, these can be expressed by the following coend and end formulas

$$f_!P(y) = \int^{x\in\mathbb{X}} \mathbb{Y}(fx,y) \times Px \ , \tag{4}$$

$$f_*P(y) = \int_{x\in\mathbb{X}} [\mathbb{Y}(y,fx) \Rightarrow Px] \tag{5}$$

for $P \in \mathcal{P}\mathbb{X}$ and $y \in \mathbb{Y}$. (See *e.g.* [29].)

Example 2.1. Whenever necessary, we will identify a set (resp. a function) with its induced discrete category (resp. functor). For a function $f : X \to Y$, the formulas (4) and (5) for $f_!, f_* : \mathcal{P}X \to \mathcal{P}Y$ simplify to give, for $P \in \mathcal{P}X$ and $y \in Y$,

$$f_!P(y) \cong \coprod_{\{x\in X|fx=y\}} P(x) \ , \tag{6}$$

$$f_*P(y) \cong \prod_{\{x\in X|y=fx\}} P(x) \ . \tag{7}$$

Coends are quotients of sums under a compatibility equivalence relation and, as such, correspond to a generalised form of existential quantification; ends are restrictions of products under a parametricity condition and, dually, correspond to a generalised form of universal quantification (see *e.g.* [27]). In this respect, the formulas (6) and (7) are intensional generalisations of the formulas (1) and (2). Further in this vein, the reader is henceforth encouraged to use the following translation table

sum	disjunction
coend	existential quantification
product	conjunction
end	universal quantification
exponential	implication
hom	equality
presheaf application	predicate membership

to provide an intuitive logical reading of generalised categorical structures. In particular, note for instance that the categorical formulas (4) and (5) translate into the logical formulas (1) and (2).

3   Polynomial Functors

The main objects of study and application in the paper are given by a general notion of polynomial functor between presheaf categories. This will be introduced in the next section. Here, as preliminary motivating background, we briefly recall an analogous notion of polynomial functor between slices of locally cartesian closed categories as it arose in the work of Gambino and Hyland [17] and of Abbott, Altenkirch, Ghani, and Morris [1,4]. See [18] for further details.

In a type-theoretic setting, starting with the informal idea that a polynomial is a sum-of-products construction, one can consider them as constructions

$$X \mapsto \sum_{j\in J} \prod_{i\in I(j)} X \tag{8}$$

arising from dependent pairs $(J : \text{Set}, I : J \to \text{Set})$, where each $j \in J$ can be naturally understood as a constructor with arities in $I(j)$. The categorical formalisation of this idea is founded on the view of Σ and Π types as adjoints. Recall that these arise as follows

$$\Sigma_f \dashv R_f \dashv \Pi_f : \mathscr{C}/A \to \mathscr{C}/B \tag{9}$$

where, for an object C of a category $\mathscr{C}$, we write $\mathscr{C}/C$ for the slice category of $\mathscr{C}$ over C and, for $f : A \to B$ in $\mathscr{C}$, we set $\Sigma_f : \mathscr{C}/A \to \mathscr{C}/B : (a : C \to A) \mapsto (f a : C \to B)$. Thus, (8) corresponds to

$$X \mapsto \Sigma_{J \to 1} \Pi_{I \to J} X \tag{10}$$

for a specification of constructors J and arities $I \to J$. More generally, we have the following definitions extending the construction (10) to slices.

Definition 3.1. *1. A polynomial in a category is a diagram $A \leftarrow I \to J \to B$.*
2. The polynomial functor induced by a polynomial $P = (A \xleftarrow{s} I \xrightarrow{f} J \xrightarrow{t} B)$ in $\mathscr{C}$ is the composite

$$F_P = \Sigma_t \, \Pi_f \, R_s : \mathscr{C}/A \to \mathscr{C}/B \ . \tag{11}$$

Example 3.1. To grasp the definition, it might be convenient to instantiate it in the category of sets. There, one sees that modulo the equivalence $\textbf{Set}/S \simeq \textbf{Set}^S$,

$$F_{(A \xleftarrow{s} I \xrightarrow{f} J \xrightarrow{t} B)} \langle X_a \rangle_{a \in A} = \Big\langle \coprod_{j \in J_b} \prod_{i \in I_j} X_{si} \Big\rangle_{b \in B} \tag{12}$$

for $J_b = \{j \in J \mid tj = b\}$ and $I_j = \{i \in I \mid fi = j\}$. These are the normal functors of Girard [19].

4 Generalised Polynomial Functors

The notion, though not the terminology, of polynomial in a category appeared in the work of Tambara [36] as an abstract setting for inducing polynomial constructions in the presence of additive and multiplicative transfer structure. The transfer structure provided by Σ and Π types (9) gives rise to polynomial functors between slice categories (11). Our interest here is in the transfer structure provided by existential and universal quantification in generalised logic (3) as giving rise to generalised polynomial functors between presheaf categories (13).

Definition 4.1. *The generalised polynomial functor induced by a polynomial $P = (\mathbb{A} \xleftarrow{s} \mathbb{I} \xrightarrow{f} \mathbb{J} \xrightarrow{t} \mathbb{B})$ in the category of small categories and functors is the composite*

$$F_P = t_! \, f_* \, s^* : \mathcal{P}\mathbb{A} \to \mathcal{P}\mathbb{B} \ . \tag{13}$$

That is,

$$F_P X b = \int^{j \in \mathbb{J}} \mathbb{B}(tj, b) \times \int_{i \in \mathbb{I}} \left[\mathbb{J}(j, fi) \Rightarrow X(si) \right] .$$

The closure under natural isomorphism of these functors yields the class of generalised polynomial functors.

A polynomial is said to represent a functor between presheaf categories whenever the latter is naturally isomorphic to the generalised polynomial functor induced by the former.

Example 4.1. 1. Write $\oint P$ for the category of elements of a presheaf $P \in \mathcal{P}\mathbb{C}$ and let π be the projection functor $\oint P \to \mathbb{C} : (c \in \mathbb{C}, p \in Pc) \mapsto c$.
Modulo the equivalence $\mathcal{P}(\mathbb{C})/P \simeq \mathcal{P}(\oint P)$, polynomial functors $\mathcal{P}(\mathbb{C})/P \to \mathcal{P}(\mathbb{C})/Q$ are subsumed by generalised polynomial functors $\mathcal{P}(\oint P) \to \mathcal{P}(\oint Q)$. The two notions coinciding for $\mathbb{C} = \mathbf{1}$.
It follows, for instance, that for every presheaf P, (i) the product endofunctor $(-) \times P$ and (ii) the exponential endofunctor $(-)^P$ are generalised polynomial.
2. Constant functors between presheaf categories are generalised polynomial.
3. Every cocontinuous functor between presheaf categories is generalised polynomial.

The definition of generalised polynomial functor extends to the multi-ary case as follows.

Definition 4.2. *For indexed families of small categories $\{\mathbb{A}_i\}_{i \in I}$ and $\{\mathbb{B}_j\}_{j \in J}$, a functor $\prod_{i \in I} \mathcal{P}\mathbb{A}_i \to \prod_{j \in J} \mathcal{P}\mathbb{B}_j$ is generalised polynomial iff so is the composite functor $\mathcal{P}(\coprod_{i \in I} \mathbb{A}_i) \cong \prod_{i \in I} \mathcal{P}\mathbb{A}_i \to \prod_{j \in J} \mathcal{P}\mathbb{B}_j \cong \mathcal{P}(\coprod_{j \in J} \mathbb{B}_j)$.*

The examples below play an important role in applications.

Example 4.2. For a small monoidal category $\mathbb{C}$, Day's monoidal-convolution tensor product [8] on $\mathcal{P}\mathbb{C}$ is a generalised polynomial functor $\mathcal{P}(\mathbb{C})^2 \to \mathcal{P}(\mathbb{C})$. Furthermore, for every $c \in \mathbb{C}$, monoidal-convolution exponentiation to the representable $yc \in \mathcal{P}\mathbb{C}$ is a generalised polynomial endofunctor on $\mathcal{P}\mathbb{C}$.

Proposition 4.1. *The class of generalised polynomial functors is closed under sums and finite products.*

5   Discrete Generalised Polynomial Functors

We introduce a simple subclass of generalised polynomial functors: the discrete ones. The results of this section show that they have a rich theory; those of Section 7 provide a range of sample applications.

Notation. For a set L and a category $\mathscr{C}$, let $L \cdot \mathscr{C} = \coprod_{\ell \in L} \mathscr{C}$ with ∇_L the codiagonal functor $[\mathrm{Id}]_{\ell \in L} : L \cdot \mathscr{C} \to \mathscr{C}$.

Definition 5.1. *The* class of *discrete generalised polynomial functors is induced by* discrete polynomials, *defined to be those of the form*

$$P = \left(\mathbb{A} \xleftarrow{\ s\ } \coprod_{k \in K} L_k \cdot \mathbb{J}_k \xrightarrow{\ \coprod_{k \in K} \nabla_{L_k}\ } \coprod_{k \in K} \mathbb{J}_k \xrightarrow{\ t\ } \mathbb{B} \right)$$

where L_k is finite for all $k \in K$.

One then has that: $F_P X\, b \cong \coprod_{k \in K} \int^{j \in \mathbb{J}_k} \mathbb{B}(t\iota_k j, b) \times \prod_{\ell \in L_k} X(s\iota_k \iota_\ell j)$.

Remark. The class of discrete generalised polynomial functors is the closure under sums of the class of functors induced by *uninomials*, defined as diagrams

$$\mathbb{A} \longleftarrow L \cdot \mathbb{J} \xrightarrow{\nabla_L} \mathbb{J} \longrightarrow \mathbb{B}$$

with L finite.

Example 5.1. 1. For sets A and B, the discrete generalised polynomial functors $\mathcal{P}A \to \mathcal{P}B$ are as in (12) with I_j finite for all $j \in J$.
 2. The generalised polynomial functors of Examples 4.1 (1(i)), (2), (3) and 4.2 are discrete.

The theory of discrete generalised polynomial functors will be developed elsewhere. An outline of main results and constructions on them follows.

Definition 5.2. *A functor is said to be* inductive *whenever it is finitary and preserves epimorphisms.*

Proposition 5.1. *Discrete generalised polynomial functors are inductive. Thus, they admit inductively constructed free algebras.*

Theorem 5.1. *The class of discrete generalised polynomial functors is closed under sums, finite products, and composition.*

Proposition 5.2. *The 2-category of small categories, discrete generalised polynomial functors, and natural transformations is cartesian. As such, and up to biequivalence, it subsumes the cartesian bicategory of small categories, profunctors, and natural transformations.*

Definition 5.3. *The* differential *of a uninomial $M = (\mathbb{A} \xleftarrow{\ s\ } L \cdot \mathbb{J} \xrightarrow{\nabla_L} \mathbb{J} \xrightarrow{\ t\ } \mathbb{B})$ is the discrete polynomial given by*

$$\partial M = \left(\mathbb{A} \xleftarrow{\ s'\ } \coprod_{(L_0, \ell_0) \in L'} L_0 \cdot \widetilde{\mathbb{J}} \xrightarrow{\ \coprod_{(L_0,\ell_0) \in L'} \nabla_{L_0}\ } L' \cdot \widetilde{\mathbb{J}} \xrightarrow{\ t'\ } \mathbb{A}^\circ \times \mathbb{B} \right)$$

for $L' = \{(L_0, \ell_0) \in \wp(L) \times L \mid L_0 \cap \{\ell_0\} = \emptyset, L_0 \cup \{\ell_0\} = L\}$; $\widetilde{\mathbb{J}} = \oint \mathrm{hom}_{\mathbb{J}}$ (the twisted arrow category of $\mathbb{J}$); $s' = [s\,(\iota_0 \cdot \pi_2)]_{(L_0,\ell_0) \in L'}$ for $\iota_0 : L_0 \hookrightarrow L$; and $t' = [\langle (s\,\iota_{\ell_0})^\circ \pi_1, t\,\pi_2 \rangle]_{(L_0,\ell_0) \in L'}.$

For a discrete polynomial P expressed as the sum of uninomials $\coprod_{i \in I} M_i$, we set $\partial P = \coprod_{i \in I} \partial M_i$.

We have that:

$$F_{\partial M}\, X\,(a, b) \cong \int^{j \in \mathbb{J}} \mathbb{B}(tj, b) \times \coprod_{(L_0, \ell_0) \in L'} \mathbb{A}(a, s\iota_{\ell_0} j) \times \prod_{\ell \in L_0} X(s\iota_\ell j)\ .$$

6   Equational Systems

We reformulate and extend the abstract theory of equational systems of Fiore and Hur [10] to more directly provide a framework for the specification of equational presentations. This we apply to generalised polynomial functors in the next section.

Notation. Writing Σ-Alg for the category of Σ-algebras of an endofunctor Σ on a category $\mathscr{C}$, let U denote the forgetful functor Σ-Alg $\to \mathscr{C} : (C, \Sigma C \overset{\gamma}{\to} C) \mapsto C$ and σ the natural transformation $\Sigma \mathrm{U} \Rightarrow \mathrm{U}$ with $\sigma_{(C,\gamma)} = \gamma$.

Definition 6.1.   *1. An* inductive equational system *is a structure* $(\mathscr{C} : \Sigma \triangleright \Gamma \vdash \lambda \equiv \rho)$ *with* $\mathscr{C}$ *a cocomplete category,* Σ, Γ *inductive endofunctors on* $\mathscr{C}$, *and* λ, ρ *natural transformations* $\Gamma \mathrm{U} \Rightarrow \mathrm{U} : \Sigma$-Alg $\to \mathscr{C}$.

2. *For an inductive equational system* $S = (\mathscr{C} : \Sigma \triangleright \Gamma \vdash \lambda \equiv \rho)$, *the* category of S-algebras, S-Alg, *is the full subcategory of* Σ-Alg *determined by the* Σ-algebras that equalise λ and ρ.

Categories of algebras for inductive equational systems have the expected properties.

Theorem 6.1 ([10]). *For an inductive equational system* $S = (\mathscr{C} : \Sigma \triangleright \Gamma \vdash \lambda \equiv \rho)$, S-Alg *is reflective in* Σ-Alg *(with inductively constructed free S-algebras over Σ-algebras), the forgetful functor S-Alg $\to \mathscr{C}$ is monadic (with inductively constructed free algebras), and S-Alg is complete and cocomplete.*

It is important in applications to have a systematic framework for specifying inductive equational systems. To this end, we introduce rules of a somewhat syntactic character for deriving *natural terms*. These are judgements $F \vdash_n \varphi$ with $n \in \mathbb{N}$, F an inductive functor $\mathscr{C}^n \to \mathscr{C}$, and φ a natural transformation $F \mathrm{U}_n \Rightarrow \mathrm{U} : \Sigma$-Alg $\to \mathscr{C}$ for $\mathrm{U}_n = \langle \mathrm{U}, \dots, \mathrm{U} \rangle : \Sigma$-Alg $\to \mathscr{C}^n$.

$$\frac{}{\Sigma \vdash_1 \sigma} \qquad\qquad \frac{}{\Pi_i \vdash_n \mathrm{id}} \ (1 \leq i \leq n)$$

$$\frac{F \vdash_n \varphi \quad F_i \vdash_m \varphi_i \ (1 \leq i \leq n)}{F \circ (F_1, \dots, F_n) \vdash_m \varphi \circ F(\varphi_1, \dots, \varphi_n)} \qquad \frac{\gamma : G \Rightarrow F \quad F \vdash_n \varphi}{G \vdash_n \varphi \circ \gamma_{\mathrm{U}_n}} \ (G \text{ inductive})$$

Within this new framework, the specification of an equational presentation $\Gamma \vdash \lambda \equiv \rho$ is done by setting $\Gamma = \coprod_{i \in I} \Gamma_i$ and $\lambda = [\lambda_i]_{i \in I}$, $\rho = [\rho_i]_{i \in I}$ for natural terms $\Gamma_i \vdash_1 \lambda_i$ and $\Gamma_i \vdash_1 \rho_i$ for $i \in I$.

7   Applications

We give applications of generalised polynomial functors to nominal effects, abstract syntax, and dependent programming.

7.1   Nominal Effects

The algebraic approach to computational effects of Plotkin and Power [32] regards the view of notions of computation as monads of Moggi [31] as derived from algebraic structure. This section illustrates the use of the theories of generalised polynomial functors and of equational systems in this context. We focus on nominal effects, explaining that the algebraic theory of the π-calculus of Stark [35] is an inductive equational system.

The algebraic theory of the π-calculus is built from the combination of three sub-theories: for non-determinism, communication, and name creation. Whilst each of the sub-theories is subject to algebraic laws of their own, the overall theory is obtained by further laws of interaction between them; see [35, Sec. 3] for details. The signature of π-algebras gives rise to an inductive generalised polynomial endofunctor on $\mathcal{P}I$, for I a skeleton of the category of finite sets and injections. All the laws, that in [35] are either informally presented by syntactic equations or formally presented by commuting diagrams, can be seen to arise as equational presentations of natural terms. The monadicity of π-calculus algebras is thus a consequence of the abstract theory of inductive equational systems.

7.2   Abstract Syntax

An algebraic theory for polymorphic languages is developed within the framework of discrete generalised polynomial functors. This is done in three successive steps by considering first variable binding and capture-avoiding substitution, and then polymorphism.

Variable Binding. We start by recasting the algebraic approach to variable binding of Fiore, Plotkin and Turi [15] in the setting of discrete polynomial functors. This we exemplify by means of the syntax of types in the polymorphic lambda calculus. Recall that this is given by the following informal vernacular rules (see *e.g.* [6]):

$$(\text{Var}) \ \frac{}{\Delta, X : * \vdash X : *} \qquad (\Rightarrow) \ \frac{\Delta \vdash A : * \quad \Delta \vdash B : *}{\Delta \vdash A \Rightarrow B : *} \qquad (\forall) \ \frac{\Delta, X : * \vdash A : *}{\Delta \vdash \forall X : *. A : *} \quad (14)$$

Our first task here will be to explain how these, and thereby the syntax they induce, are formalised as discrete generalised polynomial functors.

We need first to describe the mathematical structure of variable contexts. To this end, define a *scalar* in a category $\mathscr{C}$ to be an object $C \in \mathscr{C}$ equipped with a coproduct structure $\imath : Z \to (Z \bullet C) \leftarrow C : \jmath$ for all objects $Z \in \mathscr{C}$. In the vein of [9], for a set of sorts S, define the category of S-sorted variable contexts $\boldsymbol{C}[S]$ to be the free category with scalars $\langle s \rangle \in \boldsymbol{C}[S]$ for all sorts $s \in S$. The mathematical development does not depend on explicit descriptions of $\boldsymbol{C}[S]$, rather these provide different implementations. To fix ideas, however, the reader may take $\boldsymbol{C}[S]$ to be a skeleton of the comma category $\boldsymbol{\mathcal{F}inSet} \downarrow S$ of S-indexed finite sets.

Writing C for the category of mono-sorted contexts $C[\{*\}]$, which is a skeleton of $\mathcal{F}in\mathcal{S}et$, each of the rules (14) directly translates as a uninomial as follows:

$$(\text{Var}) \quad C \longleftarrow 0 \cdot C \xrightarrow{\nabla_0} C \xrightarrow{-\bullet\langle*\rangle} C$$

$$(\Rightarrow) \quad C \xleftarrow{[\text{id},\text{id}]} 2 \cdot C \xrightarrow{\nabla_2} C \xrightarrow{\text{id}} C \qquad (15)$$

$$(\forall) \quad C \xleftarrow{-\bullet\langle*\rangle} 1 \cdot C \xrightarrow{\nabla_1} C \xrightarrow{\text{id}} C$$

Note that the maps $\nabla_n : n \cdot C \to C$ correspond to the number of premises in the rule and that the associated maps $C \leftarrow n \cdot C$ describe the type of context needed for each of the premises. On the other hand, the maps $C \to C$ describe the type of context needed in the conclusion. Indeed, note that an algebra structure for the functor induced by a uninomial $C \xleftarrow{s} n \cdot C \xrightarrow{\nabla_n} C \xrightarrow{t} C$ on a presheaf $P \in \mathcal{P}C$ corresponds to a family of functions $\prod_{i=1}^{n} P(s_i(\Delta)) \to P(t(\Delta))$ natural for $\Delta \in C$.

The sum of the uninomials (15) yields the discrete generalised polynomial endofunctor Σ on $\mathcal{P}C$ given by

$$\Sigma(P)(\Delta) = C(\langle*\rangle, \Delta) + P(\Delta) \times P(\Delta) + P(\Delta \bullet \langle*\rangle) \ ,$$

whose initial algebra universally describes the syntax of polymorphic types in context up to α-equivalence.

Capture-Avoiding Substitution. Within this framework, one can algebraically account for the operation of capture-avoiding substitution. This is achieved by introducing a suitably axiomatised operator for substitution (see [15, Secs. 3 & 4]), that specifies its basic properties and renders it a derived operation. An outline follows.

The uninomial for the substitution operator is on the left below

$$(\varsigma) \quad C \xleftarrow{[-\bullet\langle*\rangle,\text{id}]} 2 \cdot C \xrightarrow{\nabla_2} C \xrightarrow{\text{id}} C \qquad \frac{\Delta, X : * \vdash A : * \quad \Delta \vdash B : *}{\Delta \vdash \varsigma(X :*.\, A, B) : *}$$

and, as such, arises from the informal vernacular rule on the right above. Consequently, one extends (Var), $(\Rightarrow)$, $(\forall)$ with (ς) to obtain the discrete generalised polynomial endofunctor Σ' on $\mathcal{P}C$ given by

$$\Sigma'(P)(\Delta) = \Sigma(P)(\Delta) + P(\Delta \bullet \langle*\rangle) \times P(\Delta) \ ,$$

and then equips it with an equational presentation axiomatising the Σ-substitution algebras of [15, Sec. 4]. This axiomatisation can be equationally presented by means of natural terms as explained at the end of Sec. 6. Thereby, it determines an inductive generalised equational system by construction. Its initial algebra universally describes the syntax of polymorphic types in context up to α-equivalence equipped with the operation of single-variable capture-avoiding substitution. (This development applies to general second-order algebraic presentations, for which see [11,12].)

Polymorphism. As we proceed to show, our framework is expressive enough to also allow for the modelling of polymorphic languages. This we exemplify with the term syntax of the polymorphic lambda calculus, for which the informal vernacular rules follow (see *e.g.* [6]):

$$(\text{var}) \quad \frac{\Delta \vdash A : *}{\Delta; \Gamma, x : A \vdash x : A}$$

$$(\text{app}) \quad \frac{\Delta; \Gamma \vdash t : A \Rightarrow B \quad \Delta; \Gamma \vdash u : A}{\Delta; \Gamma \vdash t(u) : B} \qquad (\text{abs}) \quad \frac{\Delta; \Gamma, x : A \vdash t : B}{\Delta; \Gamma \vdash \lambda x : A. t : A \Rightarrow B} \quad (16)$$

$$(\text{App}) \quad \frac{\Delta; \Gamma \vdash t : \forall X {:} {*}. A \quad \Delta \vdash B}{\Delta; \Gamma \vdash t[B] : A[B/X]} \qquad (\text{Abs}) \quad \frac{\Delta, X : *; \Gamma \vdash t : A}{\Delta; \Gamma \vdash \Lambda X {:} {*}. t : \forall X {:} {*}. A}$$

(In the rule (Abs) the type variable X is not free in any type of the context Γ.)

Following Hamana [21], we set up the algebraic framework on presheaves over type and term variable contexts obtained from indexed categories by means of the Grothendieck construction [20]. Recall that from $\mathcal{K} : \mathcal{H} \to \mathbf{CAT}$ this yields the category, which we will denote $\oint^{K \in \mathcal{H}} \mathcal{K}(K)$, with objects (K, k) for $K \in \mathcal{H}$ and $k \in \mathcal{K}(K)$, and morphisms $(f, g) : (K, k) \to (K', k')$ for $f : K \to K'$ in $\mathcal{H}$ and $g : \mathcal{K}(f)(k) \to k'$ in $\mathcal{K}(K')$.

Let $[\nu, \Rightarrow, \forall, \varsigma] : \Sigma'(T) \to T$ in $\mathcal{PC}$ be an initial algebra for the equational system of polymorphic types with substitution. For $G = \oint^{\Delta \in C} C[T\Delta] \times T(\Delta)$, the rules (16) directly translate into uninomials as follows:

(var) $\quad G \longleftarrow 0 \cdot G \xrightarrow{\nabla_0} G \xrightarrow{\oint \langle \bullet, \mathrm{id} \rangle} G$

$\qquad$ where $\bullet_S : C[S] \times S \to C[S] : \Gamma, s \mapsto \Gamma \bullet \langle s \rangle$

(app) $\quad G \xleftarrow{[\oint(\mathrm{id} \times \Rightarrow), \oint(\mathrm{id} \times \pi_1)]} 2 \cdot G_\mathrm{a} \xrightarrow{\nabla_2} G_\mathrm{a} \xrightarrow{\oint(\mathrm{id} \times \pi_2)} G$

$\qquad$ where $G_\mathrm{a} = \oint^{\Delta \in C} C[T\Delta] \times T(\Delta) \times T(\Delta)$

(abs) $\quad G \xleftarrow{\oint \langle \bullet, \mathrm{id} \rangle} 1 \cdot G_\mathrm{a} \xrightarrow{\nabla_1} G_\mathrm{a} \xrightarrow{\oint(\mathrm{id} \times \Rightarrow)} G$

(App) $\quad G \xleftarrow{[\oint(\mathrm{id} \times (\forall \pi_1)), \oint(\mathrm{id} \times \pi_2)]} 2 \cdot G_\mathrm{App} \xrightarrow{\nabla_2} G_\mathrm{App} \xrightarrow{\oint(\mathrm{id} \times \varsigma)} G$

$\qquad$ where $G_\mathrm{App} = \oint^{\Delta \in C} C[T\Delta] \times T(\Delta \bullet \langle * \rangle) \times T(\Delta)$

(Abs) $\quad G \xleftarrow{\imath^\#} 1 \cdot G_\mathrm{Abs} \xrightarrow{\nabla_1} G_\mathrm{Abs} \xrightarrow{\oint(\mathrm{id} \times \forall)} G$

$\qquad$ where $G_\mathrm{Abs} = \oint^{\Delta \in C} C[T\Delta] \times T(\Delta \bullet \langle * \rangle)$

$\qquad$ and $\imath^\#(\Delta; \Gamma, A) = (\Delta \bullet \langle * \rangle; C[T\imath](\Gamma), A)$

Note again that the maps $\nabla_n : n \cdot G \to G$ correspond to the number of premises in the rules and that the associated maps $G \leftarrow n \cdot G$ describe the type of context and types needed for each premise; whilst the maps $\mathbb{G} \to G$ describe the context and type needed in the conclusion. One can therefore see rules as syntactic specifications of uninomials.

The sum of the above uninomials yields a discrete generalised polynomial endofunctor on $\mathcal{P}G$ whose initial algebra universally describes the syntax of typed polymorphic terms in context up to α-equivalence. One can even go further and equationally axiomatise the operations of type-in-term and term-in-term substitution leading to a purely algebraic notion of model for polymorphic simple type theories. Details will appear elsewhere.

7.3   Dependent Programming

McBride [30] observed that the formal derivative of the type constructor of an ADT yields the type constructor for the one-hole contexts of the ADT, and use this as the basis for a generic framework for developing *zippers* for data-structure navigation as originally conceived by Huet [24]. In dependent programming, these ideas were revisited by Abbott, Altenkirch, Ghani, and McBride [2,3] for containers [1]. Following this line, Hamana and Fiore [22, Sec. 5] considered partial derivation for set-theoretic indexed containers (*i.e.* polynomial functors between slice categories of *Set*) and applied the construction to synthesise dependent zippers for GADTs. This section shows that the more general notion of differential for discrete generalised polynomial functors introduced in this paper (Definition 5.3) leads to more refined zippers, with staging information. For brevity, we only consider an example.

Dependent Zippers. The differentials of the uninomials (15) are as follows

$$(\mathrm{Var}')\quad \boldsymbol{C} \leftarrow \boldsymbol{0} \rightarrow \boldsymbol{0} \rightarrow C^\circ \times \boldsymbol{C}$$

$$(\Rightarrow')\quad \boldsymbol{C} \xleftarrow{\;[\pi_2,\pi_2]\;} 2 \cdot \widetilde{\boldsymbol{C}} \xrightarrow{\;\mathrm{id}\;} 2 \cdot \widetilde{\boldsymbol{C}} \xrightarrow{\;[\pi,\pi]\;} C^\circ \times \boldsymbol{C}$$

$$(\forall')\quad \boldsymbol{C} \leftarrow \boldsymbol{0} \rightarrow \widetilde{\boldsymbol{C}} \xrightarrow{\;\langle(-\bullet\langle*\rangle)\,\pi_1,\,\pi_2\rangle\;} C^\circ \times \boldsymbol{C}$$

from which it follows that the induced discrete generalised polynomial functor $\partial\Sigma : \mathcal{P}C \rightarrow \mathcal{P}(C^\circ \times C)$ is given by

$$\partial\Sigma(P)(\Delta',\Delta) = 2 \times C(\Delta',\Delta) \times P(\Delta) + C\big(\Delta',\Delta \bullet \langle*\rangle\big)\ .$$

For an algebra $[\nu, \Rightarrow, \forall] : \Sigma S \rightarrow S$, the construction $\partial\Sigma(S)(\Delta',\Delta)$ is that of the one-hole contexts at stage Δ for elements of S at stage Δ'. Indeed, we have a staged map for plugging components

$$\mathsf{plug} : S(\Delta') \times \partial\Sigma(S)(\Delta',\Delta) \rightarrow S(\Delta)$$

with $\mathsf{plug}\big(A, (0, f, B)\big) = \big(B \Rightarrow Sf(A)\big)$, $\mathsf{plug}\big(A, (1, f, B)\big) = \big(Sf(A) \Rightarrow B\big)$, and $\mathsf{plug}(A, f) = \forall\big(Sf(A)\big)$; and staged maps for focussing on sub-components

$$\mathsf{focus}_0, \mathsf{focus}_1 : S(\Delta') \times S(\Delta') \times C(\Delta',\Delta) \longrightarrow S(\Delta') \times \partial\Sigma(S)(\Delta',\Delta)$$

$$\mathsf{focus} : \quad S(\Delta' \bullet \langle*\rangle) \times C(\Delta',\Delta) \longrightarrow S(\Delta' \bullet \langle*\rangle) \times \partial\Sigma(S)(\Delta' \bullet \langle*\rangle, \Delta)$$

with $\mathsf{focus}_0\big((A, B), f\big) = \big(A, \big(1, f, Sf(B)\big)\big)$, $\mathsf{focus}_1\big((A, B), f\big) = \big(B, \big(0, f, Sf(A)\big)\big)$, and $\mathsf{focus}(A, f) = (A, f \bullet \langle*\rangle)$.

References

1. Abbott, M., Altenkirch, T., Ghani, N.: Categories of Containers. In: Gordon, A.D. (ed.) FOSSACS 2003. LNCS, vol. 2620, pp. 23–38. Springer, Heidelberg (2003)
2. Abbott, M., Altenkirch, T., Ghani, N., McBride, C.: Derivatives of Containers. In: Hofmann, M.O. (ed.) TLCA 2003. LNCS, vol. 2701, pp. 16–30. Springer, Heidelberg (2003)
3. Abbott, M., Altenkirch, T., Ghani, N., McBride, C.: ∂ is for data - differentiating data structures. Fundamenta Informaticae 65, 1–28 (2005)
4. Altenkirch, T., Morris, P.: Indexed containers. In: LICS 2009, pp. 277–285 (2009)
5. Cattani, G.L., Winskel, G.: Profunctors, open maps, and bisimulation. Mathematical Structures in Computer Science 15, 553–614 (2005)
6. Crole, R.: Categories for Types. Cambridge University Press (1994)
7. Danvy, O., Dybjer, P. (eds.): Preliminary Proceedings of the APPSEM Workshop on Normalisation by Evaluation (1998)
8. Day, B.: On closed categories of functors. In: Reports of the Midwest Category Seminar IV. LNM, vol. 137, pp. 1–38. Springer (1970)
9. Fiore, M.: Semantic analysis of normalisation by evaluation for typed lambda calculus. In: PPDP 2002, pp. 26–37 (2002)
10. Fiore, M., Hur, C.-K.: On the construction of free algebras for equational systems. Theoretical Computer Science 410, 1704–1729 (2008)
11. Fiore, M., Hur, C.-K.: Second-Order Equational Logic (Extended Abstract). In: Dawar, A., Veith, H. (eds.) CSL 2010. LNCS, vol. 6247, pp. 320–335. Springer, Heidelberg (2010)
12. Fiore, M., Mahmoud, O.: Second-Order Algebraic Theories. In: Hliněný, P., Kučera, A. (eds.) MFCS 2010. LNCS, vol. 6281, pp. 368–380. Springer, Heidelberg (2010)
13. Fiore, M., Moggi, E., Sangiorgi, D.: A fully-abstract model for the pi-calculus. Information and Computation 179, 76–117 (2002)
14. Fiore, M., Plotkin, G., Power, A.J.: Complete cuboidal sets in Axiomatic Domain Theory. In: LICS 1997, pp. 268–279 (1997)
15. Fiore, M., Plotkin, G., Turi, D.: Abstract syntax and variable binding. In: LICS 1999, pp. 193–202 (1999)
16. Fiore, M., Rosolini, G.: Domains in $\mathcal{H}$. Theoretical Computer Science 264, 171–193 (2001)
17. Gambino, N., Hyland, M.: Wellfounded Trees and Dependent Polynomial Functors. In: Berardi, S., Coppo, M., Damiani, F. (eds.) TYPES 2003. LNCS, vol. 3085, pp. 210–225. Springer, Heidelberg (2004)
18. Gambino, N., Kock, J.: Polynomial functors and polynomial monads. ArXiv:0906:4931 (2010)
19. Girard, J.-Y.: Normal functors, power series and λ-calculus. Annals of Pure and Applied Logic 37, 129–177 (1988)
20. Grothendieck, A.: Catégories fibrées et descente. In: SGA1. LNM, vol. 224. Springer (1971)
21. Hamana, M.: Polymorphic Abstract Syntax via Grothendieck Construction. In: Hofmann, M. (ed.) FOSSACS 2011. LNCS, vol. 6604, pp. 381–395. Springer, Heidelberg (2011)
22. Hamana, M., Fiore, M.: A foundation for GADTs and Inductive Families: Dependent polynomial functor approach. In: WGP 2011, pp. 59–70 (2011)

23. Hofmann, M.: Syntax and semantics of dependent types. In: Semantics and Logics of Computation, pp. 79–130. Cambridge University Press (1997)
24. Huet, G.: The zipper. Journal of Functional Programming 7, 549–554 (1997)
25. Jung, A., Tiuryn, J.: A New Characterization of Lambda Definability. In: Bezem, M., Groote, J.F. (eds.) TLCA 1993. LNCS, vol. 664, pp. 245–257. Springer, Heidelberg (1993)
26. Lawvere, F.W.: Adjointness in foundations. Dialectica 23 (1969)
27. Lawvere, F.W.: Metric spaces, generalized logic, and closed categories. Rend. del Sem. Mat. e Fis. di Milano 43, 135–166 (1973)
28. Lawvere, F.W.: Continuously variable sets: algebraic geometry = geometric logic. In: Proc. Logic Colloq. 1973, pp. 135–156 (1975)
29. Mac Lane, S.: Categories for the working mathematician. Springer (1971)
30. McBride, C.: The derivative of a regular type is its type of one-hole contexts (2001) (unpublished)
31. Moggi, E.: Notions of computation and monads. Information and Computation 93, 55–92 (1991)
32. Plotkin, G., Power, A.J.: Notions of Computation Determine Monads. In: Nielsen, M., Engberg, U. (eds.) FOSSACS 2002. LNCS, vol. 2303, pp. 342–356. Springer, Heidelberg (2002)
33. Reynolds, J.: Using functor categories to generate intermediate code. In: POPL 1995, pp. 25–36 (1995)
34. Scott, D.: Relating theories of the λ-calculus. In: To H.B. Curry: Essays in Combinatory Logic, Lambda Calculus and Formalisms. Academic Press (1980)
35. Stark, I.: Free-Algebra Models for the π-Calculus. In: Sassone, V. (ed.) FOSSACS 2005. LNCS, vol. 3441, pp. 155–169. Springer, Heidelberg (2005)
36. Tambara, D.: On multiplicative transfer. Comm. Alg. 21, 1393–1420 (1993)
37. Wagner, E.: Algebraic specifications: some old history and new thoughts. Nordic J. of Computing 9, 373–404 (2002)

Computing Game Metrics
on Markov Decision Processes[*]

Hongfei Fu[**]

Lehrstuhl für Informatik II, RWTH Aachen, Germany
hongfeifu@informatik.rwth-aachen.de

Abstract. In this paper we study the complexity of computing the game bisimulation metric defined by de Alfaro *et al.* on Markov Decision Processes. It is proved by de Alfaro *et al.* that the undiscounted version of the metric is characterized by a quantitative game μ-calculus defined by de Alfaro and Majumdar, which can express reachability and ω-regular specifications. And by Chatterjee *et al.* that the discounted version of the metric is characterized by the discounted quantitative game μ-calculus. In the discounted case, we show that the metric can be computed exactly by extending the method for Labelled Markov Chains by Chen *et al.* And in the undiscounted case, we prove that the problem whether the metric between two states is under a given threshold can be decided in NP ∩ coNP, which improves the previous PSPACE upperbound by Chatterjee *et al.*

Keywords: Markov Decision Processes, Game Metrics, Approximate Bisimulation.

1 Introduction

In recent years, probabilistic behavioral equivalences have been extensively studied. Many equivalence notions for probabilistic systems such as probabilistic (bi)simulation [18,22,16] have been established. And many efficient algorithms have been proposed for these notions [5,7]. Generally, probabilistic (bi)simulation is a class of formal notions judging whether two probabilistic systems are equivalent. In practical situations, they are often used to compare if the implemented system is semantically equivalent to the specification. One can also tackle the state explosion problem in model checking by reducing a large probabilistic system into its (possibly much smaller) quotient system w.r.t probabilistic (bi)simulation. This is because the quotient system is equivalent to the original one in the sense that they satisfy the same set of logical formulae [4,22].

However, the definition of probabilistic bisimulation relies on exact probability values and a slight variation of probability values will differentiate two originally equivalent systems. In this sense, probabilistic bisimulation is too restrictive and

---

[*] Full version available at [13]

[**] Supported by a CSC scholarship.

A. Czumaj et al. (Eds.): ICALP 2012, Part II, LNCS 7392, pp. 227–238, 2012.
© Springer-Verlag Berlin Heidelberg 2012

not robust, and a notion of *approximate bisimulation* or *bisimilarity metric* is needed. In the context of approximate bisimulation, the difference of two states is measured by a value in $[0, 1]$, rather than by a boolean argument stating they are either "equal" or "different". This yields a smooth, quantitative notion of probabilistic bisimulation. The smaller the value, the more alike the behaviours of the two states are. In particular, the value is zero if they are probabilistic bisimilar. In practical situations, the value for the difference would suggest if one component can substitute another: if the value between them is small enough, then one may choose the cheaper component.

The notion of approximate bisimulation is first considered by Giacalone *et al* [14]. They defined a notion of "ϵ-bisimilarity" to measure the distance between two probabilistic processes encoded in the PCCS calculus, which extends Milner's CCS [19] with probabilities. Then various notions of approximate bisimulation are defined on discrete-time probabilistic systems such as Labelled Markov Chains (LMC) [6], Markov Decision Processes (MDP) [12,3,10,23] and Concurrent Games [3], and continuous-time probabilistic systems such as Labelled Markov Processes (LMP) [20], Continuous-Time Markov Chains (CTMC) [15] and Stochastic Hybrid Systems [17].

Here we focus on the bisimilarity metric on concurrent games by de Alfaro *et al.* [3], called *game bisimulation metric*. It is proved that this metric is characterized by a quantitative game μ-calculus [2], where various properties such as the maximum reachability (to reach some set) and the maximum recurrent reachability (to reach some set infinitely often) can be expressed. This means that this metric serves as the exact bound for the differences of these properties across states. Furthermore, Chatterjee *et al.* [8] proved that this metric is a tight bound for the difference in long-run average and discounted average behaviors. In this paper, we will also study a discounted version of this game bisimulation metric [1,8]. In the discounted version, future difference is discounted by a factor and does not contribute fully to the metric, which is in contrast to the case of the original undiscounted metric. Analogous to the undiscounted version, the discounted metric is characterized by a discounted quantitative μ-calculus [1,8].

If one restricts the game bisimulation metric to MDPs (a turn-based degenerate class of concurrent games) and LMCs (MDPs without nondeterminism), one can obtain a metric on MDPs and LMCs, respectively. In this paper we consider the game bisimulation metric [3] on MDPs. We briefly compare this metric to another two metrics on Markov Decision Processes which are related with strong probabilistic bisimulation, namely the metrics by Ferns *et al.* [12] and Desharnais *et al.* [10,23]. The three metrics are different. Both the game bisimulation metric and the metric by Ferns *et al.* are defined as a least fixpoint on the lattice of pseudometrics, however the latter focuses on the difference in accumulated rewards. The one by Desharnais is defined directly as a binary relation and focuses on one-step difference. It is shown that the metric by Desharnais *et al.* is PTIME-decidable [23]. However, this metric does not serve as a bound for properties such as the reachability probability to some state labelled with a. This can be observed in [10, Example 7], where we may label a on states s_n and t_n. In this example, $d(s,t) \leq 0.1$. However the difference between s,t in the probability to reach $\{s_n, t_n\}$ is $1 - 0.95^n$, which approaches 1 when n goes to infinity. On the other hand, the game bisimulation metric serves as a bound for this property, since this property can be encoded in the quantitative game μ-calculus [2]. It is also worth noting that the game bisimulation metric on LMCs coincides with the metric by van Breugel $et\ al.$ [6].

In this paper, we study the complexity of computing the discounted and undiscounted game bisimulation metric [3,1,8] on Markov Decision Processes. It is shown by Chatterjee $et\ al.$ [8] that the undiscounted metric can be decided in PSPACE. In other words, one can decide in PSPACE whether the undiscounted metric between two states in an MDP is under a given threshold. And very recently, Chen $et\ al.$ [9] proved that the undiscounted metric is PTIME-decidable on LMCs. Here we prove that the undiscounted metric (on MDPs) can be decided in NP ∩ coNP, which is one-step closer to obtain the PTIME-decidability of the problem. We prove this result by establishing a notion of "self-closed" sets, which in some sense characterizes this metric. We remark that the method devised by Chen $et\ al.$ [9] cannot be (at least directly) extended to Markov Decision Processes. This is because their method heavily relies on the fact that the metric is the unique fixpoint of a bisimulation-minimal LMC, which generally is not the case on MDPs. For the discounted case, we show that the discounted metric can be computed exactly in polynomial time by simply extending the method by Chen $et\ al.$ [9] for LMCs.

The organization of this paper is as follows: Section 2 introduces Markov Decision Processes. Section 3 introduces the discounted and undiscounted game bisimulation metric on MDPs. In Section 4 we discuss approximations of the game metrics, where we derive that the discounted metric on MDPs can be computed exactly in polynomial time. Then in Section 5 we prove that the undiscounted metric on MDPs can be decided in NP ∩ coNP. Section 6 concludes the paper.

2   Markov Decision Processes

We define Markov Decision Processes (MDP) in the context of game structures, following the definitions in [3].

Definition 1. *Let S be a finite set. A function $\mu : S \to [0,1] \cap \mathbb{Q}$ is a probability distribution over S if $\sum_{s \in S} \mu(s) = 1$. We denote the set of probability distributions over S by $\mathrm{Dist}(S)$.*

Definition 2. *A Markov Decision Process is a tuple $(S, \mathcal{V}, [\cdot], Moves, \Gamma, \delta)$ which consists of the following components:*

- *A finite set $S = \{s, t \ldots\}$ of states;*
- *A finite set $\mathcal{V}$ of observational variables;*
- *A variable interpretation $[\cdot] : \mathcal{V} \times S \mapsto [0,1] \cap \mathbb{Q}$, which associates with each variable $\mathrm{v} \in \mathcal{V}$ a valuation $[\mathrm{v}]$;*

- A *finite set Moves = {a, b . . . } of* moves;
- A *move assignments* $\Gamma : S \mapsto 2^{Moves}\backslash\emptyset$, *which associates with each state* $s \in S$ *the nonempty set* $\Gamma(s) \subseteq Moves$ *of moves available at state s.*
- A *probabilistic transition function* $\delta : S \times Moves \mapsto \text{Dist}(S)$, *which gives the probability* $\delta(s, a)(t)$ *of a transition from s to t through the move* $a \in \Gamma(s)$.

Intuitively, Γ is the set of moves available at each state which can be controlled by a (sole) player that tries to maximize or minimize certain property.

Below we define *mixed moves* [3] on a Markov Decision Process. Intuitively, a mixed move is a probabilistic combination of single moves. This notion corresponds to randomized strategies on Markov Decision Processes, which coincides with combined transitions defined in [22].

Definition 3. *Let* $(S, \mathcal{V}, [\cdot], Moves, \Gamma, \delta)$ *be an MDP. A* mixed move *at state* $s \in S$ *is a probability distribution over* $\Gamma(s)$. *We denote by* $\mathcal{D}(s) = \text{Dist}(\Gamma(s))$ *the set of mixed moves at state s. We extend the probability transition function* δ *to mixed moves as follows: for* $s \in S$ *and* $x \in \mathcal{D}(s)$, *we define* $\delta(s, x)$ *by* $\delta(s, x)(t) := \sum_{a \in \Gamma(s)} x(a) \cdot \delta(s, a)(t)$, *for* $t \in S$.

3   Game Metrics on Markov Decision Processes

In this section we define discounted and undiscounted game bisimulation metrics on MDPs [3,8]. Both of them are defined as a least fixpoint on the complete lattice of pseudometrics w.r.t an MDP. For technical reasons we also extend these definitions to premetrics, a wider class of pseudometrics.

Below we fix an MDP $(S, \mathcal{V}, [\cdot], Moves, \Gamma, \delta)$. The following definition illustrates the concepts of premetrics and pseudometrics.

Definition 4. *A function* $d : S \times S \rightarrow [0, 1]$ *is a* premetric *iff* $d(s, s) = 0$ *for all* $s \in S$. *A premetric d is further a* pseudometric *iff for all* $r, s, t \in S$, $d(s, t) = d(t, s)$ *(symmetry) and* $d(r, t) \leq d(r, s) + d(s, t)$ *(triangle inequality). We denote the set of premetrics (resp. pseudometrics) by* $\mathcal{M}_r$ *(resp.* $\mathcal{M}_p$).

Given $d_1, d_2 \in \mathcal{M}_\kappa$ (where $\kappa \in \{r, p\}$), we define the partial order $d_1 \leq d_2$ in the pointwise fashion, i.e., $d_1 \leq d_2$ iff $d_1(s, t) \leq d_2(s, t)$ for all $s, t \in S$. It is not hard to prove the following lemma [20,6,3].

Lemma 1. *For* $\kappa \in \{r, p\}$, *the structure* $(\mathcal{M}_\kappa, \leq)$ *is a complete lattice.*

We concern the least fixpoint of $(\mathcal{M}_p, \leq)$ w.r.t a monotone function H^α, where $\alpha \in [0, 1]$ is a discount factor. The function H^α is defined as follows.

Definition 5. *Given* $\mu, \nu \in \text{Dist}(S)$, *we define* $\mu \otimes \nu$ *as the following set*

$$\{\lambda : S \times S \mapsto [0, 1] \mid (\forall u \in S. \sum_{v \in S} \lambda(u, v) = \nu(u)) \wedge (\forall v \in S. \sum_{u \in S} \lambda(u, v) = \mu(v))\}$$

Further we lift $d \in \mathcal{M}_r$ to $\mu, \nu \in \mathrm{Dist}(S)$ as follows,

$$d(\mu, \nu) := \inf_{\lambda \in \mu \otimes \nu} \left(\sum_{u,v \in S} d(u,v) \cdot \lambda(u,v) \right)$$

Then the function $H_\kappa^\alpha : \mathcal{M}_\kappa \mapsto \mathcal{M}_\kappa$ is defined as follows: given $d \in \mathcal{M}_\kappa$, $H_\kappa^\alpha(d)(s,t) = \max\{p(s,t), \alpha \cdot H_1(d)(s,t), \alpha \cdot H_2(d)(s,t)\}$ for $s,t \in S$, for which:

- $p(s,t) = \max_{v \in \mathcal{V}} |[v](s) - [v](t)|$;
- $H_1(d)(s,t) = \sup_{a \in \Gamma(s)} \inf_{y \in \mathcal{D}(t)} d(\delta(s,a), \delta(t,y))$;
- $H_2(d)(s,t) = H_1(d)(t,s)$.

We denote by d_κ^α the least fixpoint of H_κ^α, i.e., $d_\kappa^\alpha = \bigsqcap\{d \in \mathcal{M}_\kappa \mid H_\kappa^\alpha(d) \leq d\}$.

One can verify that H_κ^α is indeed a monotone function on $(\mathcal{M}_\kappa, \leq)$ [8,3]. The pseudometric d_p^1 corresponds to the undiscounted game bisimulation metric [3], and the pseudometric d_p^α with $\alpha \in [0,1)$ corresponds to the discounted metric with discount factor α [8]. Note that the definitions of H_κ^α and d_κ^α take a different form from the original ones [3,8] which cover concurrent games. However by [8, Lemma 1 and Lemma 2], these two definitions are equivalent on Markov Decision Processes.

Note that the set $\mu \otimes \nu$ is a bounded polyhedron on the vector space $S \times S \mapsto \mathbb{R}$. Thus $d(\mu, \nu)$ equals the optimal value of the linear programming (LP) problem with feasible region $\mu \otimes \nu$ and objective function $\min \sum_{u,v \in S} d(u,v) \cdot \lambda(u,v)$. We denote by $\mathrm{OP}[d](\mu, \nu)$ the set of optimum solutions that reach the optimal value $d(\mu, \nu)$.

Remark 1. It is proved in [8, Page 16] that for $d \in \mathcal{M}_r$, $s,t \in S$ and $a \in \Gamma(s)$, the value $h[d](s,a,t) := \inf_{y \in \mathcal{D}(t)} d(\delta(s,a), \delta(t,y))$ equals the optimal value of the LP problem $\mathrm{LP}[d](s,a,t)$ whose variables are $\{\lambda_{u,v}\}_{u,v \in S}$ and $\{y_b\}_{b \in \Gamma(t)}$, whose objective function is $\min \sum_{u,v \in S} d(u,v) \cdot \lambda_{u,v}$, and whose feasible region is the bounded polyhedron $P(s,a,t)$ specified by:

$$\sum_{v \in S} \lambda_{u,v} = \sum_{b \in \Gamma(t)} \delta(t,b)(u) \cdot y_b \qquad \sum_{u \in S} \lambda_{u,v} = \delta(s,a)(v)$$
$$\sum_{b \in \Gamma(t)} y_b = 1 \qquad \lambda_{u,v} \geq 0 \qquad y_b \geq 0$$

Thus $h[d](s,a,t)$ (and $H_\kappa^\alpha(d)$) can be computed in polynomial time [8, Theorem 4.3]. Further there is $y^* \in \mathcal{D}(t)$ such that $h[d](s,a,t) = d(\delta(s,a), \delta(t,y^*))$, where y^* can be $\{y_b^*\}_{b \in \Gamma(t)}$ of some optimum solution $\{\lambda_{u,v}^*\}_{u,v \in S}, \{y_b^*\}_{b \in \Gamma(t)}$.

Below we give an example. Consider the MDP with states $\{s_1, s_2, t_1, t_2\}$, moves $\{a, b\}$ and variables $\{u_1, u_2, v\}$. The variable evaluation is specified by: $u_i(t_i) = 1$ and $u_i(s) = 0$ for $s \neq t_i$; $v(s_1) = v(s_2) = 1$ and $v(s) = 0$ for $s \in S \setminus \{s_1, s_2\}$. The probability transition function is depicted in Fig. 1. One can verify that the set of undiscounted fixpoints on this MDP is

$$\{d \in \mathcal{M}_p \mid d(s_1, s_2) \in [\epsilon, 1], d(s,t) = 1 \text{ if } s \neq t \text{ and } \{s,t\} \neq \{s_1, s_2\}\}.$$

Note that even in this simple example which is bisimulation minimal, there exists no unique fixpoint. This is in contrast to the case on Labelled Markov Chains [9].

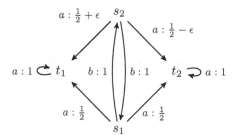

Fig. 1.

4   Approximations of the Game Metrics

In this section we describe the approximation of the discounted and undiscounted game metric by Picard's Iteration. Then we prove that the discounted metric can be computed exactly in polynomial time by simply extending the method by Chen *et al.* [9].

We fix an MDP $(S, \mathcal{V}, [\cdot], Moves, \Gamma, \delta)$. The size of the MDP, denoted by M, is the space needed to store the MDP, where all numerical values appearing in $\mathcal{V}$ and δ are represented in binary. We denote by $\|\alpha\|$ the space needed to represent the rational number α in binary. Below we define approximations of d_κ^α.

Definition 6. *The family $\{d_i^{\kappa,\alpha}\}_{i \in \mathbb{N}_0}$ of approximants of d_κ^α is inductively defined as follows: $d_0^{\kappa,\alpha} := \mathbf{0}$ (i.e., $d_0^{\kappa,\alpha}(s,t) = 0$ for all $s,t \in S$); $d_{i+1}^{\kappa,\alpha} = H_\kappa^\alpha(d_i^{\kappa,\alpha})$. We denote by $d_\omega^{\kappa,\alpha}$ the limit of $\{d_i^{\kappa,\alpha}\}_{i \in \mathbb{N}_0}$, i.e., $d_\omega^{\kappa,\alpha}(s,t) = \lim_{i \to \infty} d_i^{\kappa,\alpha}(s,t)$ for all $s,t \in S$. (By monotonicity of H_κ^α, one can prove inductively that $d_i^{\kappa,\alpha} \leq d_{i+1}^{\kappa,\alpha}$ for $i \in \mathbb{N}_0$. Thus $d_\omega^{\kappa,\alpha}$ exists.)*

For $\alpha < 1$, it is not hard to prove that $d_\omega^{\kappa,\alpha} = d_\kappa^\alpha$ since H_κ^α is a contraction mapping. Below we define $\|d_1 - d_2\| = \max_{s,t \in S} |d_1(s,t) - d_2(s,t)|$.

Lemma 2. *Suppose $\alpha \in [0,1)$. Then $\|d_i^{\kappa,\alpha} - d_\kappa^\alpha\| \leq \alpha^i$ and hence $d_\omega^{\kappa,\alpha} = d_\kappa^\alpha$.*

The situation when $\alpha = 1$ still holds [3], although H_κ^1 is not necessarily a contraction mapping.

Lemma 3. [3] $d_\omega^{\kappa,1} = d_\kappa^1$.

Here we derive a corollary from Lemma 3 which states that $d_r^1 = d_p^1$. This allows us to reason about d_p^1 on the lattice of premetrics.

Corollary 1. $d_r^1 = d_p^1 = \bigcap \{d \in \mathcal{M}_r \mid H_r^1(d) \leq d\}$.

Proof. This follows directly from Lemma 3 and the fact that $d_0^{r,1} = d_0^{p,1}$ (which implies $d_\omega^{r,1} = d_\omega^{p,1}$). □

Below we follow Chen *et al.* [9] to prove that for a fixed $\alpha \in [0,1)$, d_κ^α can be computed exactly in polynomial time. To this purpose, we prove two technical lemmas as follows.

Lemma 4. *For* $\alpha \in [0,1] \cap \mathbb{Q}$, d_κ^α *is a rational vector of size polynomial in* M *and* $\|\alpha\|$.

Lemma 5. *For a fixed* $\alpha \in [0,1) \cap \mathbb{Q}$ *and any* $\epsilon > 0$, *we can compute, in polynomial time in* M *and* $\|\epsilon\|$, *a* $d \in \mathcal{M}_p$ *such that* $\|d - d_\kappa^\alpha\| \leq \epsilon$.

Then we apply the method by Chen *et al.* [9].

Theorem 1. *For a fixed* $\alpha \in [0,1) \cap \mathbb{Q}$, d_κ^α *can be computed exactly in polynomial time in* M.

Proof. By Lemma 5, we can find in polynomial time in M and $\|\epsilon\|$ a vector that is ϵ-close to d_κ^α. And by Lemma 4, d_κ^α is a rational vector of size polynomial in M. So we can use the continued fraction algorithm [21, Section 6] to compute d_κ^α in polynomial time, as is illustrated in [9] and [11, Page 2540]. □

5   Complexity for the Undiscounted Metric

In this section we prove that the undiscounted metric d_p^1 can be decided in NP ∩ coNP. More formally, we prove that the problem MDPMetric:

- **Input:** a MDP $(S, \mathcal{V}, [\cdot], Moves, \Gamma, \delta)$, $s_{\text{in}}, t_{\text{in}} \in S$ and a number $\epsilon \in \mathbb{Q}_{\geq 0}$
- **Output:** whether $d_p^1(s_{\text{in}}, t_{\text{in}}) \leq \epsilon$ or not

lies in NP and coNP, where numerical values in $\mathcal{V}$, δ and ϵ are represented in binary. Recall that $d_p^1 = d_r^1$ (Corollary 1), so we can work on the lattice $(\mathcal{M}_r, \leq)$ instead of $(\mathcal{M}_p, \leq)$. For convenience we shall abbreviate d_r^1 as d_r.

Our proof method is divided into three steps: First we establish a characterization of the least fixpoint d_r called "self-closed" sets; Then we show that whether a given $d \in \mathcal{M}_r$ equals d_r is polynomial-time decidable; Finally, we complete the proof by showing how we can guess a premetric $d \in \mathcal{M}_r$ which is also a fixpoint of H_r^1.

Below we fix an MDP $(S, \mathcal{V}, [\cdot], Moves, \Gamma, \delta)$. First we introduce the characterization of d_r, called "self-closed" sets, as follows:

Definition 7. *Let* $d \in \mathcal{M}_r$ *satisfying* $d = H_r^1(d)$. *A subset* $X \subseteq S \times S$ *is self-closed w.r.t* d *iff for all* $(s, t) \in X$, *the following conditions hold:*

1. $d(s, t) > p(s, t)$ *(i.e.,* $d(s, t) \neq p(s, t)$ *from Definition 5);*
2. *for all* $a \in \Gamma(s)$ *such that* $d(s, t) = \inf_{y \in \mathcal{D}(t)} d(\delta(s, a), \delta(t, y))$, *there is* $y^* \in \mathcal{D}(t)$ *and* $\lambda^* \in \mathrm{OP}[d](\delta(s, a), \delta(t, y^*))$ *such that* $d(s, t) = d(\delta(s, a), \delta(t, y^*))$ *and* $\lfloor \lambda^* \rfloor \subseteq X$.
3. *for all* $b \in \Gamma(t)$ *such that* $d(s, t) = \inf_{x \in \mathcal{D}(s)} d(\delta(t, b), \delta(s, x))$, *there is* $x^* \in \mathcal{D}(s)$ *and* $\lambda^* \in \mathrm{OP}[d](\delta(t, b), \delta(s, x^*))$ *such that* $d(s, t) = d(\delta(t, b), \delta(s, x^*))$ *and* $\lfloor \lambda^* \rfloor \subseteq X$.

where $\lfloor \lambda \rfloor := \{(u, v) \in S \times S \mid \lambda(u, v) > 0\}$ *for a given* λ.

Intuitively, a self-closed set X w.r.t d is a set such that for all $(s,t) \in X$, the value $d(s,t)$ can be reached by some λ with $\lfloor \lambda \rfloor \subseteq X$. This allows us to reduce all $\{d(u,v)\}_{(u,v) \in X}$ simultaneously by a small amount so that d still is a pre-fixpoint of H_r^1. Thus if d has a nonempty self-closed set, then d is not the least fixpoint d_r. Below we show that nonempty self-closed sets in some sense characterize d_r.

Theorem 2. *Let $d \in \mathcal{M}_r$ satisfying $d = H_r^1(d)$. If $d \neq d_r$, then there exists a nonempty self-closed set X w.r.t d.*

Proof. Suppose $d \neq d_r$, we construct a nonempty self-closed set X as described below. Define $\theta(s,t) = d(s,t) - d_r(s,t)$. Then $\theta(s,t) \geq 0$ for all $s,t \in S$, and there is (s,t) such that $\theta(s,t) > 0$. Define X to be the following set:

$$X := \{(s,t) \in S \times S \mid \theta(s,t) = \max\{\theta(u,v) \mid (u,v) \in S \times S\}\}$$

We prove that X is a nonempty self-closed set. The non-emptiness of X is clear. We further prove that all $(s,t) \in X$ satisfy the conditions specified in Definition 7. Note that $\theta(s,t) > 0$ for all $(s,t) \in X$. Consider an arbitrary $(s,t) \in X$:

1. It is clear that $d(s,t) > p(s,t)$, otherwise $d(s,t) = d_r(s,t) = p(s,t)$ by Definition 5 and $\theta(s,t) = 0$. So (s,t) satisfies the first condition in Definition 7.
2. Let $a \in \Gamma(s)$ be a move such that $d(s,t) = \inf_{y \in \mathcal{D}(t)} d(\delta(s,a), \delta(t,y))$. Since d_r is a fixpoint of H_r^1, $d_r(s,t) \geq \inf_{y \in \mathcal{D}(t)} d_r(\delta(s,a), \delta(t,y))$. Choose a $y^* \in \mathcal{D}(t)$ that reaches the value $\inf_{y \in \mathcal{D}(t)} d_r(\delta(s,a), \delta(t,y))$. By the definition of X, $\theta(u,v) \leq \theta(s,t)$ for all $(u,v) \in S \times S$. Thus for all $\lambda \in \delta(s,a) \otimes \delta(t,y^*)$,

$$\sum_{u,v \in S} d_r(u,v) \cdot \lambda(u,v) \geq \sum_{u,v \in S}(d(u,v) - \theta(s,t)) \cdot \lambda(u,v)$$
$$= \left(\sum_{u,v \in S} d(u,v) \cdot \lambda(u,v)\right) - \theta(s,t)$$

The last equality is obtained by $\sum_{u,v \in S} \lambda(u,v) = 1$. By taking the infimum at the both sides, we obtain $d(\delta(s,a), \delta(t,y^*)) \leq d_r(\delta(s,a), \delta(t,y^*)) + \theta(s,t)$. Thus, we have:

$$d(s,t) \leq d(\delta(s,a), \delta(t,y^*)) \leq d_r(\delta(s,a), \delta(t,y^*)) + \theta(s,t) \leq d_r(s,t) + \theta(s,t)$$

where the last one equals $d(s,t)$. This means that $d(s,t) = d(\delta(s,a), \delta(t,y^*))$ and $d_r(s,t) = d_r(\delta(s,a), \delta(t,y^*))$. Let $\lambda^* \in \mathrm{OP}[d_r](\delta(s,a), \delta(t,y^*))$ be an optimum solution. We prove that $\lambda^* \in \mathrm{OP}[d](\delta(s,a), \delta(t,y^*))$ and $\lfloor \lambda^* \rfloor \subseteq X$. This can be observed as follows:

$$\sum_{u,v \in S} d(u,v) \cdot \lambda^*(u,v)$$
$$\geq d(s,t) \qquad\qquad\qquad (\text{by } d(s,t) = d(\delta(s,a), \delta(t,y^*)))$$
$$= d_r(s,t) + \theta(s,t)$$
$$= \sum_{u,v \in S} d_r(u,v) \cdot \lambda^*(u,v) + \theta(s,t) \quad (\text{by } d_r(s,t) = d_r(\delta(s,a), \delta(t,y^*)))$$
$$= \sum_{u,v \in S}(d_r(u,v) + \theta(s,t)) \cdot \lambda^*(u,v)$$
$$\geq \sum_{u,v \in S} d(u,v) \cdot \lambda^*(u,v)$$

Then it must be the case that $\lambda^* \in \mathrm{OP}[d](\delta(s,a), \delta(t,y^*))$ and $\theta(u,v) = \theta(s,t)$ whenever $\lambda^*(u,v) > 0$. The latter implies $\lfloor \lambda^* \rfloor \subseteq X$. So (s,t) satisfies the second condition in Definition 7.

3. It can be argued symmetrically to the second condition that the third condition is also satisfied.

Hence in conclusion, X is a nonempty self-closed set.   □

Theorem 3. *Let $d \in \mathcal{M}_r$ such that $d = H_r^1(d)$. If there exists a nonempty self-closed set $X \subseteq S \times S$ w.r.t d, then $d \neq d_r$.*

Proof. Suppose X is a nonempty self-closed set w.r.t d. We construct a premetric $d' \lneq d$ such that $H_r^1(d') \leq d'$. For each $s, t \in S$, $a \in \Gamma(s)$ and $b \in \Gamma(t)$, define

- $\theta[s, a, t] := d(s, t) - \inf_{y \in \mathcal{D}(t)} d(\delta(s, a), \delta(t, y))$.
- $\theta[s, t, b] := d(s, t) - \inf_{x \in \mathcal{D}(s)} d(\delta(t, b), \delta(s, x))$.

Note that $\theta[s, a, t]$ and $\theta[s, t, b]$ are always non-negative since d is a fixpoint of H_r^1. Further we define

- $\theta_1 := \min\{\theta[s, a, t] \mid (s, t) \in X, a \in \Gamma(s) \text{ and } \theta[s, a, t] > 0\}$;
- $\theta_2 := \min\{\theta[s, t, b] \mid (s, t) \in X, b \in \Gamma(t) \text{ and } \theta[s, t, b] > 0\}$.
- $\theta_3 := \min\{d(s, t) - p(s, t) \mid (s, t) \in X\}$

where $\min \emptyset := 0$. Finally we define $\theta := \min\{\theta' \mid \theta' \in \{\theta_1, \theta_2, \theta_3\} \text{ and } \theta' > 0\}$. Note that $\theta > 0$ since $\theta_3 > 0$. Then we construct $d' \in \mathcal{M}_r$ by:

$$d'(s, t) := \begin{cases} d(s, t) - \frac{1}{2}\theta & \text{if } (s, t) \in X \\ d(s, t) & \text{if } (s, t) \notin X \end{cases}$$

It is clear that $d' \lneq d$ since X is non-empty and $\theta > 0$. We prove that $H_r^1(d') \leq d'$.

Let $(s, t) \in S \times S$. Suppose first that $(s, t) \notin X$. Then by $d' \leq d$ we have $H_r^1(d') \leq H_r^1(d)$. Thus $H_r^1(d')(s, t) \leq d'(s, t)$ by $d'(s, t) = d(s, t)$ and $d(s, t) = H_r^1(d)(s, t)$. Suppose now that $(s, t) \in X$. Consider an arbitrary move $a \in \Gamma(s)$. We clarify two cases below:

(i) $\theta[s, a, t] > 0$. Then $\theta_1 > 0$ and $\theta \leq \theta_1 \leq \theta[s, a, t]$. So we have

$$d'(s, t) > d(s, t) - \theta[s, a, t] = \inf_{y \in \mathcal{D}(t)} d(\delta(s, a), \delta(t, y)) \geq \inf_{y \in \mathcal{D}(t)} d'(\delta(s, a), \delta(t, y))$$

(ii) $\theta[s, a, t] = 0$. Since X is self-closed, there is $y^* \in \mathcal{D}(t)$ such that $d(s, t) = d(\delta(s, a), \delta(t, y^*))$ and $\lambda^* \in \mathrm{OP}[d](\delta(s, a), \delta(t, y^*))$ such that $\lfloor \lambda^* \rfloor \subseteq X$. By $\lfloor \lambda^* \rfloor \subseteq X$ we obtain

$$\sum_{u, v \in S} d'(u, v) \cdot \lambda^*(u, v) = \sum_{u, v \in S} d(u, v) \cdot \lambda^*(u, v) - \frac{1}{2}\theta = d(\delta(s, a), \delta(t, y^*)) - \frac{1}{2}\theta$$

Then:

$$d'(\delta(s, a), \delta(t, y^*)) \leq \sum_{u, v \in S} d'(u, v) \cdot \lambda^*(u, v) = d(\delta(s, a), \delta(t, y^*)) - \frac{1}{2}\theta = d'(s, t)$$

Thus we have $\inf_{y \in \mathcal{D}(t)} d'(\delta(s, a), \delta(t, y)) \leq d'(\delta(s, a), \delta(t, y^*)) \leq d'(s, t)$. Symmetrically, we can prove that $\inf_{x \in \mathcal{D}(s)} d'(\delta(t, b), \delta(s, x)) \leq d'(s, t)$ for all $b \in \Gamma(t)$. Also by the definition of θ, we have $d'(s, t) > p(s, t)$. So we also obtain $H_r^1(d')(s, t) \leq d'(s, t)$. Thus $H_r^1(d') \leq d'$ and hence $d_r \leq d' \lneq d$ by Corollary 1.

□

Note that in the proof d' may not be a pseudometric, especially the triangle inequality may not hold. This is the reason why we need Corollary 1.

Thus for each fixpoint d, $d \neq d_r$ iff there exists a nonempty self-closed set w.r.t d. This characterization means that to check whether $d \neq d_r$, we can equivalently check whether there exists a nonempty self-closed set. The intuition to check the latter is that for self-closed sets X, Y w.r.t d, $X \cup Y$ is still a self-closed set w.r.t d. Thus there exists a largest self-closed set Z w.r.t d.

Theorem 4. *Denote by* $FP := \{d \in \mathcal{M}_r \mid d = H_r^1(d) \text{ and } \forall s, t \in S.d(s,t) \in \mathbb{Q}\}$ *the set of rational fixpoints of* H_r^1. *The problem whether a given* $d \in FP$ *equals* d_r *is decidable in polynomial time.*

Proof. (Sketch) Let Z be the largest self-closed set w.r.t d. By Theorem 2 and Theorem 3, $d \neq d_r$ iff Z is nonempty. In general, a refinement algorithm is devised to compute Z in polynomial time. Using this algorithm, we can check whether $d \neq d_r$ by checking whether Z is nonempty. □

Below we present the last step to complete the proof for the membership in NP ∩ coNP, where we illustrate how we can guess an element from FP.

Theorem 5. *The problem* MDPMetric *can be decided in* NP ∩ coNP.

Proof. (Sketch) We only prove the membership of coNP, since the membership of NP is similar. Let $(S, \mathcal{V}, [\cdot], Moves, \Gamma, \delta)$, (s_{in}, t_{in}) and ϵ be the input. Our strategy to obtain an NP algorithm to decide whether $d_r(s_{in}, t_{in}) > \epsilon$ is as follows:

1. We guess a $d \in FP$ by guessing vertices of some polyhedra.
2. We check whether $d = d_r$ by Theorem 4. If $d = d_r$ then we compare $d(s,t)$ and ϵ and return the result, otherwise we abort.

Below we show that how we can guess a $d \in FP$ using polynomial bits. The guessing procedure is illustrated step by step as follows.

1. For each $s, t \in S$ and $a \in \Gamma(s)$, we guess a vertex of $P(s, a, t)$ (cf. Remark 1) as follows:
   - We guess $|S|^2 + |\Gamma(t)|$ constraints (denoted by $Con(s, a, t)$) specified in each polyhedron $P(s, a, t)$.
   - We check whether for all s, a, t, the linear equality system obtained by modifying all comparison operators (e.g., $\geq$, $\leq$) in $Con(s, a, t)$ to equality (i.e. $=$) has a unique solution. If not, we abort the guessing.
   - We compute the unique solutions. We denote these unique solutions by $\{\lambda(s, a, t)_{u,v}\}_{u,v \in S}$, $\{y(s, a, t)_b\}_{b \in \Gamma(t)}$ for s, a, t.
   - We check if $\{\{\lambda(s, a, t)_{u,v}\}_{u,v \in S}, \{y(s, a, t)_b\}_{b \in \Gamma(t)}\} \in P(s, a, t)$ for all s, a, t. If not, then we abort the guessing.
2. We find in polynomial time [21] a premetric $d \in \mathcal{M}_r$ where $\{d(s,t)\}_{s,t \in S}$ is an (arbitrary) optimum solution of the LP problem with variables $\{\mathfrak{d}_{s,t}\}_{s,t \in S}$, objective function $\min \sum_{s,t \in S} \mathfrak{d}_{s,t}$, and the feasible region specified by:
   (a) $\mathfrak{d}_{s,s} = 0$ and $0 \leq \mathfrak{d}_{s,t} \leq 1$ for $s, t \in S$

(b) $\mathfrak{d}_{s,t} \geq p(s,t)$ for $s,t \in S$

(c) $\mathfrak{d}_{s,t} \geq \sum_{u,v \in S} \lambda(s,a,t)_{u,v} \cdot \mathfrak{d}_{u,v}$ for $(s,t) \in S \times S$ and $a \in \Gamma(s)$.

(d) $\mathfrak{d}_{s,t} \geq \sum_{u,v \in S} \lambda(t,b,s)_{u,v} \cdot \mathfrak{d}_{u,v}$ for $(s,t) \in S \times S$ and $b \in \Gamma(t)$.

If the linear programming system above has no feasible solution, we abort the guessing. Otherwise we proceed to the next step.

3. We check whether $d \in FP$, which can be done in polynomial time. We abort if the checking is unsuccessful. Otherwise we have guessed a $d \in \mathrm{FP}$.

Then we check whether $d = d_r$ and (if yes) return the comparison result of $d(s,t) > \epsilon$. It is clear that if we can guess a $d \in \mathrm{FP}$ and check successfully that $d = d_r$, then the returned result is correct. Below we prove that d_r can be guessed through the guessing procedure.

Consider Step 1. For all $s,t \in S$ and $a \in \Gamma(s)$, we can guess $Con(s,a,t)$ whose unique solution $\{\{\lambda(s,a,t)_{u,v}\}_{u,v \in S}, \{y(s,a,t)_b\}_{b \in \Gamma(t)}\}$ is a vertex of $P(s,a,t)$ that reaches the optimal value $h[d_r](s,a,t)$ of $\mathrm{LP}[d_r](s,a,t)$ (cf. Remark 1) [21, Section 8]. Then in Step 2, we prove that $\{d_r(s,t)\}_{s,t \in S}$ is the unique optimum solution of the LP problem specified in Step 2. Thus d_r can be computed polynomially in Step 2.   □

6   Conclusion and Related Work

We have shown that for Markov Decision Processes, the discounted game bisimulation metric [1,8] can be computed exactly in polynomial time, and the undiscounted game bisimulation metric [3] can be decided in NP ∩ coNP. Our results extend the one for the discounted metric on Labelled Markov Chains [9], and improves the PSPACE upperbound for the undiscounted metric on Markov Decision Processes [8]. It is proved by Chen et al. [9] that the undiscounted metric on Labelled Markov Chains can be decided in polynomial time, however their result cannot be directly applied to Markov Decision Processes. The exact complexity for the undiscounted metric could be of theoretical interest. It is also worth noting that deciding the undiscounted metric on concurrent games is at least as hard as the square-root sum problem [8], which is in PSPACE but whose inclusion in NP is a long-standing open problem [11].

Acknowledgement. Thanks to anonymous referees for valuable comments. The author is supported by a CSC scholarship.

References

1. de Alfaro, L., Henzinger, T.A., Majumdar, R.: Discounting the Future in Systems Theory. In: Baeten, J.C.M., Lenstra, J.K., Parrow, J., Woeginger, G.J. (eds.) ICALP 2003. LNCS, vol. 2719, pp. 1022–1037. Springer, Heidelberg (2003)

2. de Alfaro, L., Majumdar, R.: Quantitative solution of omega-regular games. J. Comput. Syst. Sci. 68(2), 374–397 (2004)

3. de Alfaro, L., Majumdar, R., Raman, V., Stoelinga, M.: Game relations and metrics. In: LICS, pp. 99–108. IEEE Computer Society (2007)

4. Aziz, A., Singhal, V., Balarin, F.: It Usually Works: The Temporal Logic of Stochastic Systems. In: Wolper, P. (ed.) CAV 1995. LNCS, vol. 939, pp. 155–165. Springer, Heidelberg (1995)
5. Baier, C., Engelen, B., Majster-Cederbaum, M.E.: Deciding bisimilarity and similarity for probabilistic processes. J. Comput. Syst. Sci. 60(1), 187–231 (2000)
6. van Breugel, F., Sharma, B., Worrell, J.: Approximating a behavioural pseudometric without discount for probabilistic systems. Logical Methods in Computer Science 4(2) (2008)
7. Cattani, S., Segala, R.: Decision Algorithms for Probabilistic Bisimulation. In: Brim, L., Jančar, P., Křetínský, M., Kučera, A. (eds.) CONCUR 2002. LNCS, vol. 2421, pp. 371–385. Springer, Heidelberg (2002)
8. Chatterjee, K., de Alfaro, L., Majumdar, R., Raman, V.: Algorithms for game metrics (full version). Logical Methods in Computer Science 6(3) (2010)
9. Chen, D., van Breugel, F., Worrell, J.: On the Complexity of Computing Probabilistic Bisimilarity. In: Birkedal, L. (ed.) FOSSACS 2012. LNCS, vol. 7213, pp. 437–451. Springer, Heidelberg (2012)
10. Desharnais, J., Laviolette, F., Tracol, M.: Approximate analysis of probabilistic processes: Logic, simulation and games. In: QEST, pp. 264–273. IEEE Computer Society (2008)
11. Etessami, K., Yannakakis, M.: On the complexity of Nash equilibria and other fixed points. SIAM J. Comput. 39(6), 2531–2597 (2010)
12. Ferns, N., Panangaden, P., Precup, D.: Metrics for finite Markov decision processes. In: McGuinness, D.L., Ferguson, G. (eds.) AAAI, pp. 950–951. AAAI Press/The MIT Press (2004)
13. Fu, H.: Computing game metrics on Markov decision processes. Tech. Rep. AIB-2012-08, RWTH Aachen (May 2012), http://aib.informatik.rwth-aachen.de/
14. Giacalone, A., Jou, C.C., Smolka, S.A.: Algebraic reasoning for probabilistic concurrent systems. In: Proc. IFIP TC2 Working Conference on Programming Concepts and Methods, pp. 443–458. North-Holland (1990)
15. Gupta, V., Jagadeesan, R., Panangaden, P.: Approximate reasoning for real-time probabilistic processes. In: QEST, pp. 304–313. IEEE Computer Society (2004)
16. Jonsson, B., Larsen, K.G.: Specification and refinement of probabilistic processes. In: LICS, pp. 266–277. IEEE Computer Society (1991)
17. Julius, A.A., Girard, A., Pappas, G.J.: Approximate bisimulation for a class of stochastic hybrid systems. In: American Control Conference, pp. 4724–4729. IEEE, Portland (2006)
18. Larsen, K.G., Skou, A.: Bisimulation through probabilistic testing. Inf. Comput. 94(1), 1–28 (1991)
19. Milner, R.: Communication and concurrency. Prentice-Hall, Inc., Upper Saddle River (1989)
20. Panangaden, P.: Labelled Markov Processes. Imperial College Press (2009)
21. Schrijver, A.: Theory of Linear and Integer Programming. John Wiley & Sons, Inc., New York (1986)
22. Segala, R., Lynch, N.A.: Probabilistic simulations for probabilistic processes. Nord. J. Comput. 2(2), 250–273 (1995)
23. Tracol, M., Desharnais, J., Zhioua, A.: Computing distances between probabilistic automata. In: Massink, M., Norman, G. (eds.) QAPL. EPTCS, vol. 57, pp. 148–162 (2011)

Deciding First Order Properties of Matroids[*]

Tomáš Gavenčiak[1,**], Daniel Král'[2,3,***], and Sang-il Oum[4,†]

[1] Department of Applied Mathematics, Faculty of Mathematics and Physics,
Charles University, Malostranské náměstí 25, 118 00 Prague, Czech Republic
gavento@kam.mff.cuni.cz
[2] Department of Mathematics, University of West Bohemia,
Univerzitní 8, 306 14 Pilsen, Czech Republic
dankral@kma.zcu.cz
[3] Computer Science Institute, Faculty of Mathematics and Physics,
Charles University, Malostranské náměstí 25, 118 00 Prague, Czech Republic
[4] Department of Mathematical Sciences, KAIST,
291 Daehak-ro Yuseong-gu, Daejeon 305-701, South Korea
sangil@kaist.edu

Abstract. Frick and Grohe [J. ACM 48 (2006), 1184–1206] introduced
a notion of graph classes with locally bounded tree-width and established
that every first order property can be decided in almost linear time in
such a graph class. Here, we introduce an analogous notion for matroids
(locally bounded branch-width) and show the existence of a fixed parameter algorithm for first order properties in classes of regular matroids
with locally bounded branch-width. To obtain this result, we show that
the problem of deciding the existence of a circuit of length at most k
containing two given elements is fixed parameter tractable for regular
matroids.

1 Introduction

Classes of graphs with bounded tree-width play an important role both in structural and algorithmic graph theory. Their importance from the structural point
of view stems from their close relation to graph minors. With respect to algorithms, there have been an enormous amount of reports on designing efficient algorithms for classes of graphs with bounded tree-width for various NP-complete
problems. Most of such algorithms have been put into a uniform framework by
Courcelle [2] who showed that every monadic second order can be decided in
linear-time for a class of graphs with bounded tree-width. The result can also be
extended to computing functions described in the monadic second order logic [3].

---

[*] A full version of this contribution is available as arXiv:1108.5457.
[**] This author was supported by the grant GAUK 64110.
[***] This author was supported by the grant GAČR P202/11/0196.
[†] This author was supported by Basic Science Research Program through the Na-
tional Research Foundation of Korea(NRF) funded by the Ministry of Education,
Science and Technology(2011-0011653) and also by TJ Park Junior Faculty Fel-
lowship.

A. Czumaj et al. (Eds.): ICALP 2012, Part II, LNCS 7392, pp. 239–250, 2012.
© Springer-Verlag Berlin Heidelberg 2012

Not all minor-closed graph classes have bounded tree-width and though some important graph properties can be tested in linear or almost linear time for these classes. One example of such property is whether a graph has an independent of size at most k. Clearly, such a property can be tested in polynomial time for a fixed k. However, for the class of planar graphs, this can be tested in linear time for any fixed k. More generally, if a class of graphs have locally bounded tree-width (e.g., this holds the class of planar graphs), the presence of a subgraph isomorphic to a given fixed subgraph H can be tested in linear time [7,8]. Frick and Grohe [9] developed a framework unifying such algorithms. They showed that every first order property can be tested in almost linear time for graphs in a fixed class of graphs with locally bounded tree-width. Here, almost linear time stands for the existence of an algorithm A_ε running in time $O(n^{1+\varepsilon})$ for every $\varepsilon > 0$. More general results in this direction apply to classes of graphs locally excluding a minor [4] or classes of graphs with bounded expansion [6].

Matroids are combinatorial structures generalizing graphs and linear independance. The reader is referred to Section 2 if (s)he is not familiar with this concept. For the exposition now, let us say that every graph G can be associated with a matroid $M(G)$ that captures the structure of cycles and cuts of G. Though tree-width can be generalized to matroids [16,17], a more natural decomposition parameter for matroids is branch-width which also exists for graphs [22]. The two notions are closely related: if G is a graph with tree-width t and its matroid $M(G)$ has branch-width b, then it holds that $b \leq t + 1 \leq 3b/2$ [22].

Hliněný [11,13] established an analogue of the result of Courcelle [2] for classes of matroids with bounded branch-width that are represented over a fixed finite field. He showed that every monadic second order property can be decided in cubic time for matroids from such a class which are given by their representation over the field (the decomposition of a matroid need not be given as it can be efficiently computed, see Subsection 2.4 for more details). Since any graphic matroid, i.e., a matroid corresponding to a graph, can be represented over any (finite) field and its representation can be easily computed from the corresponding graph, Hliněný's result generalizes Courcelle's result to matroids.

The condition that a matroid is given by its representation seems necessary as there are several negative results on deciding representability of matroids: Seymour [24] argued that it is not possible to test representability of a matroid given by the independance oracle in subexponential time. His construction readily generalizes to every finite field and the matroids appearing in his argument have branch-width at most three. If the input matroid is represented over rationals, it can be tested in polynomial-time whether it is binary [23] even without constraining its branch-width, but for every finite field $\mathbb{F}$ of order four or more and every $k \geq 3$, deciding representability over $\mathbb{F}$ is NP-hard [14] for matroids with branch-width at most k given by their representation over rationals.

In this paper, we propose a notion of locally bounded branch-width and establish structural and algorithmic results for such classes of matroids. In particular, we aim to introduce an analogue of the result of Frick and Grohe [9] for matroids in the spirit how the result of Hliněný [13] is an analogue of that of Courcelle [2].

Table 1. Fixed parameter algorithms for testing first order and monadic second order properties for classes of graphs and matroids. Our result is printed in bold

Properties	Graphs	Matroids
Monadic second order	bounded tree-width [2]	bounded branch-width and represented over a finite field [13]
First order	locally bounded tree-width [9]	**locally bounded branch-width and regular**

Our main result (Theorem 7) asserts that testing first order properties in classes of regular matroids with locally bounded branch-width that are given by the independance oracle is fixed parameter tractable. A formal definition of matroid FO properties including some examples and discussion of the relation of graph and matroid properties can be found in Subsection 2.5.

The definition of graph classes with locally bounded tree-width is based on distances between vertices, a concept missing (and impossible to be analogously defined) in matroids. Hence, we introduce a sense of locality based on circuits of matroids and we show in Section 3 that if a class of graphs has locally bounded tree-width, then the corresponding class of matroids has locally bounded branch-width. This justifies the notion to be appropriately defined. We then complete the algorithmic picture (see Table 1) by showing that every first order property is fixed parameter tractable for classes of regular matroids with locally bounded branch-width. Recall that all graphic matroids are regular.

In one step of our algorithm, we need to decide the existence of a circuit containing two given elements with length at most k in a fixed parameter way (when parametrized by k). However, the Maximum likelihood decoding problem for binary inputs, which is known to be W[1]-hard [5], can be reduced to this problem (in the Maximum likelihood decoding problem, one is given a target vector v and a matrix M and the task is to decide whether there exists a vector v' with at most k non-zero coordinates such that $v = Mv'$). Hence, it is unlikely that the existence of a circuit with length at most k containing two given elements is fixed parameter tractable even for binary matroids. So, we had to restrict to a subclass of binary matroids. We have chosen regular matroids and established the fixed parameter tractability of deciding the existence of a circuit containing two given elements with length at most k in this class of matroids (Theorem 6). We think that this result might be of independent interest.

2 Notation

2.1 Tree-Width and Local Tree-Width

A *tree-decomposition* of a graph G with vertex set V and edge set E is a tree T such that every node x of T is assigned a subset V_x of V and such that

- for every edge $vw \in E$, there exists a node x such that $\{v, w\} \subseteq V_x$, and
- for every vertex $v \in V$, the nodes x with $v \in V_x$ induce a subtree of T.

The *width* of a tree-decomposition is $\max_{x \in V(T)} |V_x| - 1$ and the *tree-width* of a graph G is the minimum width of its tree-decomposition. A graph has tree-width at most one if and only if it is a forest. A class $\mathcal{G}$ of graphs has *bounded tree-width* if there exists k such that the tree-width of any graph in $\mathcal{G}$ is at most k.

The set of vertices at distance at most d from a vertex v of G is denoted by $N_d^G(v)$ and it is called the *d-neighborhood* of v. A class $\mathcal{G}$ of graphs has *locally bounded tree-width* if there exists a function $f : \mathbb{N} \to \mathbb{N}$ such that the tree-width of the subgraph induced by the vertices of $N_d^G(v)$ is at most $f(d)$ for every graph $G \in \mathcal{G}$, every vertex v of G and every non-negative integer d.

2.2 Matroids

We briefly introduce basic definitions related to matroids; further exposition on matroids can be found in several monographs on the subject, e.g., in [20, 26]. A *matroid* is a pair $(E, \mathcal{I})$ where E are the elements of M and $\mathcal{I}$ is a family of subsets of E, whose elements are referred as *independent*, such that

1. the empty set is contained in $\mathcal{I}$,
2. if X is contained in $\mathcal{I}$, then every subset of X is also contained in $\mathcal{I}$, and
3. if X and Y are contained in $\mathcal{I}$ and $|X| < |Y|$, then there exists $y \in Y$ such that $X \cup \{y\}$ is also contained in $\mathcal{I}$.

The set E is the *ground set* of M. If $M = (E, \mathcal{I})$ is a matroid and $E' \subseteq E$, then $(E', \mathcal{I} \cap 2^{E'})$ is a matroid which we refer to as the matroid M *restricted* to E'.

The maximum size of an independent subset of a set X is called the *rank* of X. All inclusion-wise maximum subsets of X have the same size. The rank function of a matroid, i.e., the function $r : 2^E \to \mathbb{N} \cup \{0\}$ assigning a subset X its rank, is submodular which means that $r(A \cap B) + r(A \cup B) \le r(A) + r(B)$ for any two subsets A and B of E. The inclusion-wise maximal subsets contained in $\mathcal{I}$ are called *bases*. An element e of M such that $r(\{e\}) = 0$ is called a *loop*; on the other hand, if $r(E \setminus \{e\}) = r(E) - 1$, then e is a *coloop*.

Two most classical examples of matroids are vector matroids and graphic matroids. In *vector matroids*, the elements are vectors and a subset X is independent if the vectors in X are linearly independent. The dimension of the linear hull of the vectors in X is equal to the rank of X. A matroid M is *representable* over a field $\mathbb{F}$ if it is isomorphic to a vector matroid over $\mathbb{F}$; the corresponding assignment of vectors to the elements of M is a *representation* of M over $\mathbb{F}$. A matroid representable over the two-element field is called *binary*.

If G is a graph, its *cycle matroid* $M(G)$ is the matroid with ground set formed by the edges of G and a subset X is independent if G does not contain a cycle formed by some edges of X. A matroid is *graphic* if it is a cycle matroid of a graph. Dually, the *cut matroid* $M^*(G)$ is the matroid with the same ground set as $M(G)$ but a subset X is independent if G and $G \setminus X$ have the same number of components. Cut matroids are called *cographic*.

A *circuit* of a matroid M is a subset X of the elements that is an inclusion-wise minimal subset that is not independent. The following is well-known.

Proposition 1. *Let C_1 and C_2 be two circuits of a matroid with $C_1 \cap C_2 \neq \emptyset$ and e and f two distinct elements in $C_1 \cup C_2$. There exists a circuit $C \subseteq C_1 \cup C_2$ such that $\{e, f\} \subseteq C$.*

The dual matroid M^* of a matroid $M = (E, \mathcal{I})$ is the matroid with ground set E such that a set X is independent in M^* if and only if the complement of X contains a base of M. A circuit of M^* is referred to as a *cocircuit* of M.

A *k-separation* of a matroid M is a partition of its ground set into two sets A and B such that $r(A) + r(B) - r(M) = k - 1$. A matroid M is *connected* if it has no 1-separation. This is equivalent to the statement that every pair e_1 and e_2 of elements of M is contained in a circuit. The notions of connectivity for matroids and graphs are different. A matroid $M(G)$ corresponding to a graph G is connected if and only if G is 2-connected.

We now explain how a binary matroid can be obtained from two smaller binary matroids by gluing along a small separation. Before doing so, we need one more definition: a *cycle* is a disjoint union of circuits in a binary matroid. Let M_1 and M_2 be two binary matroids with ground sets E_1 and E_2, respectively. Suppose that $\min\{|E_1|, |E_2|\} \leq |E_1 \triangle E_2|$ and that one of the following holds.

- E_1 and E_2 are disjoint.
- $E_1 \cap E_2$ contains a single element which is neither a loop nor a coloop in M_1 or M_2.
- $E_1 \cap E_2$ is a 3-element circuit in both M_1 and M_2 and $E_1 \cap E_2$ contains no cocircuit in M_1 or M_2.

We define $M = M_1 \triangle M_2$ to be the binary matroid with ground set $E_1 \triangle E_2$ such that a set $C \subseteq E_1 \triangle E_2$ is a cycle in M if and only if there exist cycles C_i in M_i, $i = 1, 2$, such that $C = C_1 \triangle C_2$. Based on which of the three cases apply, we say that M is a *1-sum*, a *2-sum* or a *3-sum* of M_1 and M_2. If M is a k-sum of M_1 and M_2, then the partition $(E_1 \setminus E_2, E_2 \setminus E_1)$ is a k-separation of M.

A matroid M is *regular* if it is representable over any field. Seymour's Regular Matroid Decomposition Theorem [23] asserts that every regular matroid can be obtained by a series of 1-, 2- and 3-sums of graphic matroids, cographic matroids and copies of a particular 10-element matroid referred as R_{10}.

2.3 Matroid Algorithms

The area of matroid algorithms involves several surprising aspects. So, we provide a brief introduction. The running time of algorithms is measured in the number of elements of an input matroid. Since the number of non-isomorphic n-element matroids is double exponential in n, an input matroid cannot be presented to an algorithm using subexponential space. So, one has to resolve how an input matroid is given. One possibility is that the input matroid is given by an oracle (a black-box function) that returns whether a given subset of the ground set is independent. Other possibilities include giving the matroid by its representation over a field. Since algorithms working with oracle-given matroids are the most general, we assume that an input matroid is oracle-given unless stated otherwise.

Though it is easy to construct a representation over $\mathbb{GF}(2)$ for an oracle-given binary matroid in quadratic time (under the promise that the input matroid is binary), deciding whether an oracle-given matroid is binary cannot be solved in subexponential time [24]. However, it can be decided in polynomial time whether an oracle-given matroid is graphic [24]. Moreover, if the matroid is graphic, the corresponding graph can be found efficiently, see, e.g., [26, Chapter 10].

Theorem 1. *There exists a polynomial-time algorithm that decides whether an oracle-given matroid M is graphic and, if so, it finds a graph G such that $M = M(G)$.*

Similarly, one can test whether a given matroid is regular [25].

Theorem 2 (Seymour [23]). *There exists a polynomial-time algorithm that decides whether an oracle-given matroid M is regular and, if so, it outputs a sequence containing graphic matroids, cographic matroids and the matroid R_{10} along with the sequence 1-, 2- and 3-sums yielding M.*

2.4   Branch-Width and Local Branch-Width

A *branch-decomposition* of a matroid $M = (E, \mathcal{I})$ is a subcubic tree T with leaves one-to-one corresponding to the elements of M. Each edge e of T splits T into two subtrees and thus it naturally partitions the elements of M into two subsets A and B, those corresponding to the leaves of the two subtrees. The *width of the edge e* is $r(A) + r(B) - r(M) + 1$ where r is the rank function of M, and the *width of the branch-decomposition* is the maximum width of an edge. The *branch-width* of M is the minimum width of its branch-decomposition. The branch-width of the cycle matroid $M(G)$ of a graph G is related to the tree-width of G.

Proposition 2 ([22]). *Let G be a graph. If b is the branch-width of the cycle matroid $M(G)$ of G and t is the tree-width of G, then $b - 1 \le t \le 3b/2$.*

Though the branch-width of a matroid cannot be efficiently computed, testing whether a matroid has branch-width at most k (when k is fixed) can be done in polynomial time [19]. For matroids represented over a finite field, a cubic-time approximation algorithm for every fixed branch-width exists [12] and the result has been extended to a cubic-time algorithm deciding fixed branch-width:

Theorem 3 (Hliněný and Oum [15]). *Let k be a fixed integer and $\mathbb{F}$ a fixed finite field. There is a cubic-time algorithm that decides whether the branch-width of a matroid M represented over $\mathbb{F}$ is at most k and, if so, it constructs a branch-decomposition of M with width at most k.*

One obstacle to be overcome when defining local branch-width is the absence of the natural notion of distance in matroids. As in the definition of matroid connectivity, we utilize containment in a common circuit. Let M be a matroid. The *distance* of two elements e and f of M is the minimum size of a circuit containing both e and f; if $e = f$, the distance is equal to zero, and if there is no such circuit, then it is set to infinity. By Proposition 1, the distance function defined satisfies the triangle-inequality. For an element e of M and a positive integer d, $N_d^M(e)$ denotes the set containing all elements of M at distance at most d from e. A class $\mathcal{M}$ of matroids has *locally bounded branch-width* if there exists a function $f : \mathbb{N} \to \mathbb{N}$ such that for every matroid $M \in \mathcal{M}$, every element e of M and every positive integer d, the branch-width of the matroid M restricted to the elements of $N_d^M(e)$ is at most $f(d)$.

2.5   FO and MSO Properties

A *monadic second order formula* ψ for a graph property is built up in the usual way from variables for vertices, edges, vertex and edge sets, the equality and element-set signs $=$ and $\in$, the logic connectives $\wedge$, $\vee$ and $\neg$, the quantification over variables and the predicate indicating whether a vertex is incident with an edge; this predicate encodes the input graph. A *first order formula* is a monadic second order formula with no variables for vertex and edge sets. A formula with no free variables is called a *sentence*. A graph property is *first order (monadic second order) property* if there exists a first order (monadic second order) sentence describing the property. We further abbreviate "first order" to FO and "monadic second order" to MSO.

Similarly, we define FO and MSO formulas for matroids. Such formulas contain the unary predicate Indep_M defined on subsets of the elements indicating whether a set of elements is independent; this predicate encodes the input matroid. MSO formulas contain variables for elements and their sets and the predicate describing the independent set of matroids. FO formulas do not contain variables for sets of elements but we allow variables for elements to be used to form sets (otherwise, we could not use the predicate Indep_M). For example, if x_1, x_2 and x_3 are variables for elements, the set $\{x_1, x_2, x_3\}$ may appear in an FO formula and, in particular, an FO formula may contain the predicate $\mathrm{Indep}_M(\{x_1, x_2, x_3\})$. Finally, FO and MSO matroid properties are those that can be described by FO and MSO sentences, respectively.

An example of an FO matroid property is the presence of a fixed matroid M_0 in an input matroid. For instance, if M_0 is the uniform matroid $U_{2,3}$, then the corresponding FO sentence is

$$\exists a, b, c; \mathrm{Indep}_M(\{a, b\}) \wedge \mathrm{Indep}_M(\{a, c\}) \wedge \mathrm{Indep}_M(\{b, c\}) \wedge \neg \mathrm{Indep}_M(\{a, b, c\}) .$$

Another example of an FO matroid property is the existence of k disjoint triangles in a matroid.

As indicated in Table 1, MSO graph properties can be decided in linear time in classes of graphs with bounded tree-width [2], MSO matroid properties can be decided in cubic time in classes of matroids with bounded branch-width that are given by a representation over a fixed finite field [13] and FO graph properties can be decided in almost linear time in classes of graphs with locally bounded tree-width [9]. Our Theorem 7 completes the picture by establishing that FO matroid properties are fixed parameter tractable in classes of oracle-given regular matroids with locally bounded branch-width. Since the matroids associated to n-vertex trees are isomorphic, not all FO graph properties can be expressed as FO matroid properties. In particular, our Theorem 7 extends the result of Frick and Grohe to matroids in the same spirit as the result of Hliněný extends Courcelle's theorem where the same obstacle also appears.

One of the aspects exploited in [9] is the locality of FO properties. Let $P_1, \ldots, P_k$ be relations on a set E and let r_i be the arity of P_i, $i = 1, \ldots, k$. The *Gaifman graph* for the relations $P_1, \ldots, P_k$ is the graph with vertex set E such that two elements y_1 and y_2 of E are adjacent if there exist an index $i \in \{1, \ldots, k\}$ and elements $x_1, \ldots, x_{r_k}$ of E such that $(x_1, \ldots, x_{r_k}) \in P_i$ and $y_1, y_2 \in \{x_1, \ldots, x_{r_k}\}$. A formula $\psi[x]$ is r-*local* if all quantifiers in ψ are restricted to the elements at distance at most r from x in the Gaifman graph. Gaifman established the following theorem which captures locality of FO properties.

Theorem 4 (Gaifman [10]). *If ψ_0 is a first order sentence with predicates $P_1, \ldots, P_k$, then the sentence ψ_0 is equivalent to a Boolean combination of sentences of the form*

$$\exists x_1, \ldots, x_k \left(\bigwedge_{1 \leq i < j \leq k} d(x_i, x_j) > 2r \wedge \bigwedge_{1 \leq i \leq k} \psi[x_i] \right) \tag{1}$$

where $d(x_i, x_j)$ is the distance between x_i and x_j in the Gaifman graph and ψ is an r-local first order formula for some integer r.

In particular, Theorem 4 reduces deciding FO sentences to r-local formulas.

3   Local Tree-Width vs. Local Branch-Width

We now relate graph classes with locally bounded tree-width and their matroid counterparts. The proof of the next theorem is omitted due to space constraints.

Theorem 5. *Let $\mathcal{G}$ be a class of graphs and $\mathcal{M}$ the class of cycle matroids $M(G)$ for graphs $G \in \mathcal{G}$. If the class $\mathcal{G}$ has locally bounded tree-width, then $\mathcal{M}$ has locally bounded branch-width.*

The converse of Theorem 5 is not true; this does however not harm the view of our results as a generalization of the result of Frick and Grohe since our result apply to classes of graphic matroids corresponding to graph classes with locally bounded tree-width (and, additionally, to several other matroid classes).

Let us give an example witnessing that the converse implication does not hold. Consider a graph G_k obtained from the complete graph K_k of order k by replacing each edge with a path with k edges and by adding a vertex v joined to all the vertices of the subdivision of K_k. The class of graphs $\mathcal{G} = \{G_k, k \in \mathbb{N}\}$ does not have locally bounded tree-width since the subgraph induced by $N_1^G(v)$ in G_k contains K_k as a minor and thus its tree-width is at least $k - 1$.

Let $\mathcal{M}$ be the class of graphic matroids of graphs G_k. We claim that the function $f_M(d) := \max\{3, d+1\}$ witnesses that the class $\mathcal{M}$ has locally bounded branch-width. Let $M = M(G_k)$ and fix an element e of M and an integer $d \geq 3$. If $k < d$, then the tree-width of G_k is at most $k + 1 \leq d$ and thus the branch-width of M is at most $d + 1$ by Proposition 2. In particular, the branch-width of M restricted to any subset of its elements is at most $d + 1$.

We now assume that $k \geq d$. Let w be a vertex incident with e that is different from the vertex v. Any edge contained in $N_d^M(e)$ is incident with a vertex of the subdivision of K_k at distance at most $d - 2 < k$ from w. Since $k \geq d$, the edges in $N_d^M(e)$ contained in the subdivision of K_k form no cycles in G_k and thus they yield a subgraph of tree-width one. Adding the extra vertex v increases the tree-width by at most one and thus the tree-width of the subgraph of G_k formed by the edges of $N_d^M(e)$ is at most two; consequently, the matroid M restricted to $N_d^M(e)$ has branch-width at most three.

4   Constructing Gaifman Graph

Let φ be a first order sentence which contains the predicate Indep_M and let d be the depth of quantification in φ. Introduce new predicates $C_M^1, \ldots, C_M^d$ such that the arity of C_M^k is k and $C_M^k(x_1, \ldots, x_k)$ is true if and only if $\{x_1, \ldots, x_k\}$ is a circuit of M. For instance, $C^1(x_1)$ is true if and only if x_1 is a loop. For an oracle-given matroid M, each predicate $C_M^k(x_1, \ldots, x_k)$, $k = 1, \ldots, d$, can be evaluated in linear time. The *circuit reduction* φ^C of φ is then the sentence obtained from φ by replacing each $\mathrm{Indep}_M(\{x_1, \ldots, x_\ell\})$ by the conjunction:

$$\bigwedge_{1 \leq k \leq \ell} \bigwedge_{1 \leq j_1 < \cdots < j_k \leq \ell} \neg C_M^k(x_{j_1}, \ldots, x_{j_k}) .$$

Observe that a sentence φ is satisfied for M if and only if its circuit reduction φ^C is. Let $G_{M,d}^C$ be the graph with vertex set being the elements of M and two elements of M are adjacent if they are contained in a circuit of length at most d, i.e., their distance in M is at most d. Clearly, the graph $G_{M,d}^C$ is a supergraph of the Gaifman graph for the circuit reduction of φ when d is the depth of quantification of φ. Proposition 1 yields the following (the bound $d\ell$ is not tight).

Lemma 1. *Let φ be a first order sentence and d an integer. If two elements x and y in M are at distance ℓ in the graph $G_{M,d}^C$, then there exists a circuit of M containing both x and y with length at most $d\ell$. In particular, the distance of x and y in M is at most $d\ell$.*

We now focus on constructing the graph $G_{M,d}^C(\varphi)$ efficiently for an oracle-given regular matroid M. We utilize Seymour's Regular Matroid Decomposition Theorem. To use this result, we show that the problem, which we call Minimum dependency weight circuit, is fixed-parameter tractable for graphic and cographic matroids. In this problem, elements of a matroid are assigned weights and we seek a circuit containing given elements with minimum weight; however, if the circuit contains two elements from one of special 3-element circuits, then the weight of the sought circuit gets adjusted (this we refer to as dependency).

Problem: **Minimum dependency weight circuit (MDWC)**

Parameter: an integer ℓ

Input: a matroid M with a set F of one, two or three elements of M
           a collection $\mathcal{T}$ of disjoint 3-element circuits of M
           a positive integral function w on $E(M) \setminus \bigcup_{T \in \mathcal{T}} T$
           non-negative integral functions w_T on 2^T, $T \in \mathcal{T}$, such that
           $w_T(\emptyset) = 0$ and $w_T(A) \geq |A|$ for any non-empty $A \subseteq T$

Output: the minimum weight $w(C)$ of a circuit C with $F \subseteq C$
           where $w(C) = \sum_{\substack{e \in C \setminus \bigcup_{T \in \mathcal{T}} T}} w(e) + \sum_{T \in \mathcal{T}} w_T(C \cap T)$
           if such minimum weight is at most ℓ

To establish the fixed parameter tractability of MDWC problem for regular matroids, we start with graphic matroids. The proof of the next lemma is based on a simple application of the color coding technique introduced in [1] and it is left due to space constraints.

Lemma 2. *Minimum dependency weight circuit problem is fixed parameter tractable in the class of graphic matroids.*

Before we handle the case of cographic matroids, we need one more definition. If M is a cographic matroid, then a family $\mathcal{C}$ of its circuits is *simple* if there exists a graph G corresponding to M such that every circuit of $\mathcal{C}$ corresponds to a cut around a vertex of G (in particular, if $\mathcal{C}$ is empty, it is simple).

The proof of the next lemma is complicated. We adopt the method of Kawarabayashi and Thorup [18] who proved that finding an edge-cut with at most ℓ edges and at least k components is fixed parameter tractable. However, the cut we seek, besides weight computing issues, is required to include the edges of F and, more importantly, to be inclusion-wise minimal. Further details are ommited due to space constraints.

Lemma 3. *Minimum dependency weight circuit problem is fixed parameter tractable for cographic matroids providing that $\mathcal{T}$ is simple.*

Lemmas 2 and 3 imply the main theorem of this section. Theorem 1 gives a recursive decomposition of regular matroids to graphic and cographic matroids and a simple dynamic programming algorithm then yields the result.

Theorem 6. *The problem of deciding the existence of a circuit of length at most d containing two given elements of a regular matroid is fixed parameter tractable when parameterized by d.*

Since two elements of a matroid M are adjacent in $G_{M,d}^C$ if and only if they are contained in a common circuit of size at most d, Theorem 6 yields the following.

Corollary 1. *The graph $G_{M,d}^C$ can be constructed in polynomial time for an oracle-given regular matroid M and an integer d and the degree of the polynomial in the estimate on its running time is independent of d.*

5 Deciding First Order Properties

Since first order formulas are special cases of monadic second order formulas, the results of Hliněný [11, 13] imply the following.

Lemma 4. *Let $\mathcal{M}$ be a class of regular matroids with locally bounded branch-width. There exists a cubic-time algorithm that given an oracle-given $M \in \mathcal{M}$, an element x of M, the graph $G_{M,d}^C$ and an r-local formula $\psi[x]$ with predicates $C_M^1, \ldots, C_M^d$ (the locality of ψ is measured in $G_{M,d}^C$) decides whether $\psi[x]$ is satisfied in M.*

We have now introduced all tools we need. Due to space limitations, we can only briefly sketch the main steps. By Theorem 4, it is enough to present how to decide sentences of the form (1). For the r-local formula ψ appearing in such a sentence, we compute for every element x of an input matroid whether $\psi[x]$ holds; this can be done in a fixed parameter way by Lemma 4. Then, using an argument analogous to that used by Frick and Grohe in [9], we decide whether there exist at least k such elements at mutual distance at least $2r$ in the Gaifman graph.

Theorem 7. *Let ψ_0 be a first order sentence and $\mathcal{M}$ a class of regular matroids with locally bounded branch-width. There exists a polynomial-time algorithm that decides whether an oracle-given n-element matroid $M \in \mathcal{M}$ satisfies ψ_0. Moreover, the degree of the polynomial in the estimate on its running time is independent of ψ_0 and $\mathcal{M}$, i.e., testing first order properties in classes of regular matroids with locally bounded branch-width is fixed parameter tractable.*

Acknowledgement. The authors would like to thank Ken-ichi Kawarabayashi for discussing the details of his fixed parameter algorithm for computing multi-way cuts given in [18] during the NII workshop—Graph Algorithms and Combinatorial Optimization.

References

1. Alon, N., Yuster, R., Zwick, U.: Color-coding: a new method for finding simple paths, cycles and other small subgraphs within large graphs. In: Proc. STOC 1994, pp. 326–335. ACM (1994)
2. Courcelle, B.: The monadic second-order logic of graph I. Recognizable sets of finite graphs. Inform. and Comput. 85, 12–75 (1990)

3. Courcelle, B.: The expression of graph properties and graph transformations in monadic second-order logic. In: Rozenberg, G. (ed.) Handbook of Graph Grammars and Computing by Graph Transformations. Foundations, vol. 1, pp. 313–400. World Scientific (1997)
4. Dawar, A., Grohe, M., Kreutzer, S.: Locally excluding a minor. In: Proc. LICS 2007, pp. 270–279. IEEE Computer Society Press (2007)
5. Downey, R.G., Fellows, M.R., Vardy, A., Whittle, G.: The parametrized complexity of some fundamental problems in coding theory. SIAM J. Comput. 29, 545–570 (1999)
6. Dvořák, Z., Král', D., Thomas, R.: Deciding first-order properties for sparse graphs. In: Proc. FOCS 2010, pp. 133–142 (2010)
7. Eppstein, D.: Subgraph isomorphism in planar graphs and related problems. J. Graph Algor. Appli. 3, 1–27 (1999)
8. Eppstein, D.: Diameter and treewidth in minor-closed graph families. Algorithmica 27, 275–291 (2000)
9. Frick, M., Grohe, M.: Deciding first-order properties of locally tree-decomposable structures. Journal of the ACM 48, 1184–1206 (2001)
10. Gaifman, H.: On local and non-local properties. In: Proc. of Herbrand Symposium, Logic Colloquium 1981. North-Holland, Amsterdam (1982)
11. Hliněný, P.: On Matroid Properties Definable in the MSO Logic. In: Rovan, B., Vojtáš, P. (eds.) MFCS 2003. LNCS, vol. 2747, pp. 470–479. Springer, Heidelberg (2003)
12. Hliněný, P.: A parametrized algorithm for matroid branch-width. SIAM J. Computing 35(2), 259–277 (2005)
13. Hliněný, P.: Branch-width, parse trees and monadic second-order logic for matroids. J. Combin. Theory Ser. B 96, 325–351 (2006)
14. Hliněný, P.: On Matroid Representability and Minor Problems. In: Královič, R., Urzyczyn, P. (eds.) MFCS 2006. LNCS, vol. 4162, pp. 505–516. Springer, Heidelberg (2006)
15. Hliněný, P., Oum, S.: Finding branch-decomposition and rank-decomposition. SIAM J. Computing 38, 1012–1032 (2008)
16. Hliněný, P., Whittle, G.: Matroid tree-width. European J. Combin. 27, 1117–1128 (2006)
17. Hliněný, P., Whittle, G.: Addendum to matroid tree-width. European J. Combin. 30, 1036–1044 (2009)
18. Kawarabayashi, K., Thorup, M.: Minimum k-way cut of bounded size is fixed-parameter tractable. In: Proc. FOCS 2011, pp. 160–169. IEEE (2011)
19. Oum, S., Seymour, P.: Approximating clique-width and branch-width. J. Combin. Theory, Ser. B 96, 514–528 (2006)
20. Oxley, J.G.: Matroid theory. Oxford University Press (1992)
21. Peleg, D.: Distance-dependent distributed directories. Info. Computa. 103, 270–298 (1993)
22. Robertson, N., Seymour, P.D.: Graph minors X. Obstructions to tree-decompositions. J. Combin. Theory Ser. B 52, 153–190 (1991)
23. Seymour, P.: Decomposition of regular matroids. J. Combin. Theory Ser. B 28, 305–359 (1980)
24. Seymour, P.: Recognizing graphic matroids. Combinatorica 1, 75–78 (1981)
25. Truemper, K.: A decomposition theory for matroids V. Testing of matrix total unimodularity. J. Combin. Theory Ser. B 49, 241–281 (1990)
26. Truemper, K.: Matroid decomposition. Academic Press (1992)

Pebble Games with Algebraic Rules[*]

Anuj Dawar and Bjarki Holm

University of Cambridge Computer Laboratory
firstName.lastName@cl.cam.ac.uk

Abstract. We define a general framework of *partition games* for formulating two-player pebble games over finite structures. We show that one particular such game, which we call the *invertible-map game*, yields a family of polynomial-time approximations of graph isomorphism that is strictly stronger than the well-known Weisfeiler-Lehman method. The general framework we introduce includes as special cases the pebble games for finite-variable logics with and without counting. It also includes a *matrix-equivalence game*, introduced here, which characterises equivalence in the finite-variable fragments of matrix-rank logic. We show that the equivalence defined by the invertible-map game is a refinement of the equivalence defined by each of these three other games.

1 Introduction

An important open problem in finite model theory is that of finding a logical characterisation of polynomial-time computability. That is to say, to find a logic in which a class of finite structures is expressible if, and only if, membership in the class is decidable in deterministic polynomial time (PTIME). The exact formulation of the problem (see [8]) requires additional effectivity conditions. It was at one time conjectured by Immerman that the extension of inflationary fixed-point logic with a mechanism for counting (IFPC) suffices for expressing all of PTIME. However, this turns out not to be the case and a counter-example was constructed by Cai, Fürer and Immerman [2]. Noting that this construction and various other examples of properties in PTIME that are not definable in IFPC can be reduced to testing the solvability of systems of linear equations, we introduced in [4] the extension of inflationary fixed-point logic with matrix-rank operators (IFPR). This logic strictly extends the expressive power of IFPC while still being contained in PTIME and it remains an open question whether there are polynomial-time properties that are not definable in IFPR.

In this context, the study of ever more expressive logics has gone hand in hand with the development of tools for proving limitations on those logics. An important class of such tools are the so-called pebble games, which are variations and extensions of the Ehrenfeucht-Fraïssé game for first-order logic. In particular, the *k-pebble game* characterises the relation $\equiv_k^L$ of equivalence in first-order logic

---

[*] Research supported by EPSRC grant EP/H026835/1. An extended version of this paper is available at http://arxiv.org/abs/1205.0913.

with k variables. Since it can be shown that any formula of IFP is invariant under $\equiv_k^L$ for some k, this becomes useful in proving inexpressibility results for IFP. Similarly, inexpressibility results for IFPC are established by showing that a property is not invariant under $\equiv_k^C$ for any k, where $\equiv_k^C$ denotes the relation of equivalence in first-order logic with counting quantifiers and at most k variables. The relation $\equiv_k^C$ has been characterised by two different pebble games: the Immerman-Lander game [12] and the *bijection game* of Hella [9].

In addition to providing a tool for the analysis of logics, these games also provide interesting approximations of the graph isomorphism relation. In particular, it can be shown that the equivalence relation $\equiv_{k+1}^C$ is exactly the relation decided by the k-dimensional Weisfeiler-Lehman method (see [2] for a description of the method and its relationship with $\equiv_k^C$). This is a family of polynomial-time algorithms which approach graph isomorphism in the limit by ever finer approximations. The Cai-Fürer-Immerman construction of a property in PTIME that is not definable in IFPC shows that there is no fixed k such that the k-dimensional Weisfeiler-Lehman algorithm decides graph isomorphism.

In a similar way, the logic of matrix-rank operators defined in [4] yields a family of equivalence relations $\equiv_{k,m,\Omega}^R$ which provide a stratification of graph isomorphism and which can be used to analyse definability in IFPR. Here, $\equiv_{k,m,\Omega}^R$ refers to equivalence in the logic that extends k-variable first-order logic with matrix-rank operators of arity at most m for matrices over GF_p for any p in the finite set of primes Ω. In this paper we give a game that characterises this logical equivalence. This game, which we call the *matrix-equivalence game*, is difficult to use and it remains a challenge to deploy it to establish that there is a PTIME property not closed under $\equiv_{k,m,\Omega}^R$ for any k, m and Ω.

The matrix-equivalence game and the relations $\equiv_{k,m,\Omega}^R$ that it characterises suffer from another limitation as approximations of graph isomorphism. It is not clear whether $\equiv_{k,m,\Omega}^R$ can be decided in polynomial time as the natural algorithm that is obtained from the definition of the matrix-equivalence game runs in exponential time. This leads us to consider an alternative that we call the *invertible-map game*. This is obtained by replacing the algebraic matrix-equivalence condition with a condition of simultaneous similarity of tuples of matrices. As a result we obtain a family of equivalence relations $\approx_{m,\Omega}^k$ which *refine* $\equiv_{k,m,\Omega}^R$. Even though the relations are refinements of $\equiv_{k,m,\Omega}^R$, they seem easier to decide. Using a result of Chistov et al. [3] we are able to show that each of the relations $\approx_{m,\Omega}^k$ is decidable in polynomial time. Therefore, this gives us a family of *polynomial-time* algorithms which, like the Weisfeiler-Lehman method, approximates isomorphism in the limit. This family is strictly stronger than the Weisfeiler-Lehman method in the sense that it can also distinguish the Cai-Fürer-Immerman graphs at some fixed level.

The games we introduce in this paper are formulated as *partition games*, so called because Duplicator is required at each move to give a suitable partition of the game board. This partition has to satisfy certain algebraic conditions which vary according to the game we are considering. It turns out that the games for $\equiv_k^L$ and $\equiv_k^C$ can be formulated as partition games, by replacing the algebraic rules of the matrix-equivalence game with weaker conditions. This provides a general framework for exploring other games and, indeed, other equivalence relations on structures. So far, model-comparison games have been formulated for specific logics. Perhaps we can reverse this and extract suitable logics from well-behaved games? One such challenge is to formulate a logic that corresponds to the invertible map game that we define here.

2 Preliminaries

We assume that all structures are finite and that all vocabularies are finite and relational. Throughout, we commonly write τ to denote a vocabulary. We write $U(\mathbf{A})$ for the universe of a structure $\mathbf{A}$ and write $\|\mathbf{A}\|$ for the cardinality of $U(\mathbf{A})$. We denote the class of all finite τ-structures with fixed tuples of $r \in \mathbb{N}$ parameters by $\text{fin}[\tau; r] := \{(\mathbf{A}, \boldsymbol{a}) \mid \mathbf{A} \in \text{fin}[\tau], \ \boldsymbol{a} \in U(\mathbf{A})^r\}$. We denote tuples $(v_1, \ldots, v_k)$ by $\boldsymbol{v}$ and their length by $\|\boldsymbol{v}\|$. If $\boldsymbol{v}$ is a k-tuple of elements from a set X, $i \in [k]$ and $w \in X$, then we write $\boldsymbol{v}\frac{w}{i}$ for the tuple obtained from $\boldsymbol{v}$ by replacing the i-th component with w; that is, $\boldsymbol{v}\frac{w}{i} = (v_1, \ldots, v_{i-1}, w, v_{i+1}, \ldots, v_k)$. If $m \leq k$, $\boldsymbol{i} = (i_1, \ldots, i_m) \in [k]^m$ is a tuple of distinct integers (an 'index pattern') and $\boldsymbol{w}$ is an m-tuple of elements from X, then we write $\boldsymbol{v}\frac{\boldsymbol{w}}{\boldsymbol{i}} := \boldsymbol{v}\frac{w_1}{i_1} \cdots \frac{w_m}{i_m}$.

Matrices indexed by unordered sets. If F is a field and I, J are finite and non-empty sets then an $I \times J$ *matrix* over F is a function $M : I \times J \to F$. If $\|I\| = \|J\|$ then we say that M is *invertible* if there is a $J \times I$ matrix N such that the product MN is the $I \times I$ identity matrix (equivalently, if there is an $J \times I$ matrix N such that NM is the $J \times J$ identity matrix). In this case we refer to N as the inverse of M, denoted by M^{-1}. A matrix whose rows and columns are indexed by the same set is said to be *square*. Recall that if M is a square $I \times I$ matrix, N is a square $J \times J$ matrix and $\|I\| = \|J\|$, then we say that M and N are *equivalent* if there is an invertible $J \times I$ matrix P and an invertible $I \times J$ matrix Q such that $PMQ = N$. The two matrices M and N are said to be *similar* if there is an invertible $J \times I$ matrix S such that $SMS^{-1} = N$.

We focus on square $\{0,1\}$-matrices whose rows and columns are indexed by tuples of elements from some finite and non-empty set A. Specifically, if $B \subseteq A^{2m}$ for some $m \geq 1$, then we write χ_B for the characteristic function of B, seen as a $\{0,1\}$-matrix indexed by $A^m \times A^m$. That is, χ_B is defined by $(\boldsymbol{a}, \boldsymbol{b}) \mapsto 1$ if $(\boldsymbol{a}, \boldsymbol{b}) \in B$ and $(\boldsymbol{a}, \boldsymbol{b}) \mapsto 0$ otherwise. We refer to $\chi_B(\boldsymbol{a}, \boldsymbol{b})$ as the *characteristic matrix of B*; the underlying field and m are usually clear from the context.

We also consider linear combinations of characteristic matrices. Let $\mathbf{P} \subseteq \wp(A^{2m})$ be a collection of subsets of A^{2m} and $\gamma : \mathbf{P} \to F$ be a function. Then we write $M_\gamma^{\mathbf{P}}$ to denote the $A^m \times A^m$ matrix over F defined by $M_\gamma^{\mathbf{P}} := \sum_{P \in \mathbf{P}} \gamma(P) \cdot \chi_P$. Typically, $\mathbf{P}$ will be a *partition* of A^{2m}; that is, a collection of non-empty and mutually disjoint subsets of A^{2m} (called *blocks*) whose union is A^{2m}.

Finite-variable logics. We write L^k to denote the fragment of first-order logic using only the variables $x_1, \ldots, x_k$ and we write C^k for the extension of L^k with rules for defining counting formulas of the kind $\exists^{\geq i} x \cdot \varphi(x)$, for $i > 0$ (for further details, see [7, 14]). For $(\mathbf{A}, \boldsymbol{a})$ and $(\mathbf{B}, \boldsymbol{b})$ in $\mathrm{fin}[\tau; r]$, we write $(\mathbf{A}, \boldsymbol{a}) \equiv_k^L (\mathbf{B}, \boldsymbol{b})$ to indicate that for any L^k-formula φ, it holds that $(\mathbf{A}, \boldsymbol{a}) \models \varphi$ if, and only if, $(\mathbf{B}, \boldsymbol{b}) \models \varphi$; the relation $\equiv_k^C$ is defined similarly for C^k.

For each integer $i > 0$ and prime p, we define a quantifier $\mathrm{rk}_p^{\geq i}$ which binds exactly $2m$ variables. If $\varphi_1, \ldots, \varphi_{p-1}$ are formulas, $\boldsymbol{x}$ and $\boldsymbol{y}$ are m-tuples of pairwise distinct variables, and $\mathbf{A}$ a structure, then we let

$$\mathbf{A} \models \mathrm{rk}_p^{\geq i}(\boldsymbol{x}, \boldsymbol{y}) \cdot (\varphi_1, \ldots, \varphi_{p-1})$$

if, and only if, the rank of the square $U(\mathbf{A})^m \times U(\mathbf{A})^m$ matrix $\sum_{j=1}^{p-1} j \cdot \chi_{\varphi_j^{\mathbf{A}}}$ (mod p) is at least i over GF_p, the finite field with p elements. If Ω is a finite and non-empty set of primes, then we write $R_{m;\Omega}^k$ to denote the logic built up in the same way as k-variable first-order logic, except that we have rules for constructing formulas with $2m$-ary rank quantifiers over GF_p ($p \in \Omega$) instead of the rules for first-order existential and universal quantifiers. Every formula in L^k or C^k is equivalent to one of $R_{\{p\};2}^{k+1}$ (where p is any prime), for we can simulate existential, universal and unary counting quantifiers by expressing the rank of diagonal matrices (see [4, 10, 13] for details). We write $(\mathbf{A}, \boldsymbol{a}) \equiv_{k,m,\Omega}^R (\mathbf{B}, \boldsymbol{b})$ to indicate that $\mathbf{A}$ and $\mathbf{B}$ agree on all $R_{m;\Omega}^k$-formulas under the assignments $\boldsymbol{a}$ and $\boldsymbol{b}$, respectively. It can be shown that any formula of IFPR_Ω is invariant under $\equiv_{k,m,\Omega}^R$ for some k and m [4, 10, 13], where we write IFPR_Ω for the fragment of fixed-point logic with rank operators restricted to matrices over GF_p for $p \in \Omega$.

Pebble games. A pebble game is a two-player model-comparison game where each of the two players (Spoiler and Duplicator) has a finite number of tokens ('pebbles') for placing on the game board. It can be shown that equivalence in L^k is completely characterised by a pebble game where each player has k pebbles [1, 15, 11]. This correspondence gives a purely combinatorial method for proving inexpressibility results for k-variable logic in general and IFP in particular, since it can be shown that any formula of IFP is invariant under $\equiv_k^L$ for some k. Immerman and Lander [12] and Hella [9] later introduced separate versions of the k-pebble game for analysing the expressiveness of C^k over finite models, which can be used to establish lower bounds for IFPC over finite structures.

Class extensions and extension matrices. We frequently consider relations that arise by extending a fixed tuple of elements in a structure according to some criteria. Consider a formula φ and let $\boldsymbol{a}$ be an assignment of elements of a structure $\mathbf{A}$ to the free variables of φ. Then the set of all pairs (c, d) from $\mathbf{A}$ which, when used to replace the first two elements of $\boldsymbol{a}$ give a satisfying assignment to φ, can be seen as a binary "extension" of $\boldsymbol{a}$ in $\mathbf{A}$, defined by the formula φ. This relation can be viewed as a $\{0, 1\}$-matrix over $\mathbf{A}$ in the usual way, giving us a way to associate a pair $(\mathbf{A}, \boldsymbol{a})$ with a family of matrices over $\mathbf{A}$.

More formally, consider a class $\alpha \subseteq \mathrm{fin}[\tau; k]$ and let $\boldsymbol{i} = (i_1, \ldots, i_n) \in [k]^n$ be a tuple of distinct integers, $n \leq k$. Then we write $\mathrm{ext}_{\boldsymbol{i}}^\alpha$ to denote the functor on $\mathrm{fin}[\tau; k]$ defined by $\mathrm{ext}_{\boldsymbol{i}}^\alpha(\mathbf{A}, \boldsymbol{a}) := \{\boldsymbol{b} \in U(\mathbf{A})^n \mid (\mathbf{A}, \boldsymbol{a}\frac{\boldsymbol{b}}{\boldsymbol{i}}) \in \alpha\}$. We refer to

$\mathrm{ext}_i^\alpha(\mathbf{A}, \boldsymbol{a})$ as the $\boldsymbol{i}$-extension of $(\mathbf{A}, \boldsymbol{a})$ into α. Abusing notation, if φ is a formula whose free variables are all amongst $\boldsymbol{x} = (x_1, \ldots, x_k)$, then we let $\mathrm{ext}_i^\varphi := \mathrm{ext}_i^{\alpha_\varphi}$, where $\alpha_\varphi := \{(\mathbf{A}, \boldsymbol{a}) \in \mathrm{fin}[\tau; k] \mid (\mathbf{A}, \boldsymbol{a}) \models \varphi\}$. That is,

$$\mathrm{ext}_i^\varphi = \{\boldsymbol{b} \in U(\mathbf{A})^n \mid \mathbf{A} \models \varphi[\boldsymbol{a}\tfrac{\boldsymbol{b}}{i}]\} \subseteq U(\mathbf{A})^n.$$

If $n = 2m$, then we write $\mathrm{extmat}_i^\alpha(\mathbf{A}, \boldsymbol{a})$ and $\mathrm{extmat}_i^\varphi(\mathbf{A}, \boldsymbol{a})$ to denote the $U(\mathbf{A})^m \times U(\mathbf{A})^m$ characteristic matrices of $\mathrm{ext}_i^\alpha(\mathbf{A}, \boldsymbol{a})$ and $\mathrm{ext}_i^\varphi(\mathbf{A}, \boldsymbol{a})$, respectively. We refer to such matrices as *extension matrices*.

3   A Game Characterisation of Rank Logics

In this section we give a game characterisation of finite-variable logic with quantifiers for matrix rank. This gives us a combinatorial method for proving lower bounds (inexpressibility results) for fixed-point logic with rank operators.

To motivate this game, we first describe a simple "partition game" that is based on the same game protocol. This is played by two players, Spoiler and Duplicator, on a pair of relational structures $\mathbf{A}$ and $\mathbf{B}$, each with k pebbles labelled $1, \ldots, k$. At each round, Spoiler removes a pebble from $\mathbf{A}$ and the corresponding pebble from $\mathbf{B}$. Unlike the classical pebble game, Duplicator is not allowed to move any pebbles herself (by convention, Spoiler is male and Duplicator female). However, in response to the challenge of the Spoiler, she is allowed to divide the game board into disjoint regions in order to restrict the possible moves that Spoiler is subsequently allowed to make. Specifically, in response to Spoiler's challenge, Duplicator partitions each of $U(\mathbf{A})$ and $U(\mathbf{B})$ into the same number of regions and gives a matching between the regions in $U(\mathbf{A})$ and the regions in $U(\mathbf{B})$. Intuitively, Duplicator's strategy is to gather in each region all those elements that lead to game positions that are sufficiently alike. In turn, Spoiler is allowed to place each of the chosen pebbles on some element of the corresponding structure, with the *restriction* that the two newly pebbled elements have to be within matching regions. That completes a round of the game. Though it may seem that the partition game is biased against the Duplicator, since she is not allowed to place her own pebbles after seeing where Spoiler places his, it can be shown that it is actually equivalent to the standard pebble game.

The idea of dividing the game board into disjoint regions leads to a generic template for designing pebble games. For instance, if we adapt the rules so that any two matching regions must have the same cardinality, we essentially get the bijection game. The "matrix-equivalence game" we describe next is obtained by putting additional linear-algebraic constraints on the matching game regions.

Matrix-equivalence Game. Let k and m be positive integers with $2m \leq k$ and let Ω be a finite and non-empty set of primes. The game board of the k-pebble m-ary *matrix-equivalence game* over Ω (or (k, m, Ω)-matrix-equivalence game for short) consists of two structures $\mathbf{A}$ and $\mathbf{B}$ of the same vocabulary, each with k pebbles labelled $1, \ldots, k$. The first $r \leq k$ pebbles of $\mathbf{A}$ may be initially placed on the elements of an r-tuple $\boldsymbol{a}$ of elements in $\mathbf{A}$ and the corresponding r pebbles in $\mathbf{B}$ on an r-tuple $\boldsymbol{b}$ of elements in $\mathbf{B}$. If $\|\mathbf{A}\| \neq \|\mathbf{B}\|$ or the mapping defined by the initial pebble positions is not a partial isomorphism then Spoiler wins the game immediately. Otherwise, each round of the game proceeds as follows.

1. Spoiler chooses a prime $p \in \Omega$ and picks up $2m$ pebbles in some order from $\mathbf{A}$ and the $2m$ corresponding pebbles in the same order from $\mathbf{B}$.
2. Duplicator has to respond by choosing
   - a partition $\mathbf{P}$ of $U(\mathbf{A})^m \times U(\mathbf{A})^m$,
   - a partition $\mathbf{Q}$ of $U(\mathbf{B})^m \times U(\mathbf{B})^m$, with $\|\mathbf{P}\| = \|\mathbf{Q}\|$, and
   - a bijection $f : \mathbf{P} \to \mathbf{Q}$,
   for which it holds that for all labellings $\gamma : \mathbf{P} \to \mathsf{GF}_p$,

   $$\mathrm{rk}(M_\gamma^{\mathbf{P}}) = \mathrm{rk}(M_{\gamma \circ f^{-1}}^{\mathbf{Q}}). \qquad (\star)$$

   Here the composite map $\gamma \circ f^{-1} : \mathbf{Q} \to \mathsf{GF}_p$ is seen as a labelling of $\mathbf{Q}$.
3. Spoiler next picks a block $P \in \mathbf{P}$ and places the $2m$ chosen pebbles from $\mathbf{A}$ on the elements of some tuple in P (in the order they were chosen earlier) and the corresponding $2m$ pebbles from $\mathbf{B}$ on the elements of some tuple in $f(P)$ (in the same order).

This completes one round in the game. If, after this exchange, the partial map from $\mathbf{A}$ to $\mathbf{B}$ defined by the pebbled positions (in addition to constants) is not a partial isomorphism, or if Duplicator is unable to produce the required partitions, then Spoiler wins the game; otherwise it can continue for another round. Observe that the condition "$\mathrm{rk}(M_\gamma^{\mathbf{P}}) = \mathrm{rk}(M_{\gamma \circ f^{-1}}^{\mathbf{Q}})$" is the same as saying that the two matrices should be *equivalent*, since rank is a complete invariant for matrix equivalence. This explains the name of the game.

Theorem 1. *Duplicator has a winning strategy in the (k, m, Ω)-matrix-equivalence game starting with positions $(\mathbf{A}, \boldsymbol{a})$ and $(\mathbf{B}, \boldsymbol{b})$ if, and only if, $(\mathbf{A}, \boldsymbol{a}) \equiv_{k,m,\Omega}^R (\mathbf{B}, \boldsymbol{b})$.*

We give the proof of Theorem 1 by two separate lemmas, one for each implication. To simplify the proof, we consider only positions $(\mathbf{A}, \boldsymbol{a})$ and $(\mathbf{B}, \boldsymbol{b})$ with $\|\boldsymbol{a}\| = \|\boldsymbol{b}\| = k$; that is, positions where all the pebbles are initially placed on the board.

Lemma 2. *If $(\mathbf{A}, \boldsymbol{a}) \not\equiv_{k,m,\Omega}^R (\mathbf{B}, \boldsymbol{b})$ then Spoiler has a winning strategy in the (k, m, Ω)-matrix-equivalence game starting with positions $(\mathbf{A}, \boldsymbol{a})$ and $(\mathbf{B}, \boldsymbol{b})$.*

Sketch proof. We show that if $(\mathbf{A}, \boldsymbol{a}) \not\equiv_{k,m,\Omega}^R (\mathbf{B}, \boldsymbol{b})$ then Spoiler has a strategy to force the game, in a finite number of rounds, into positions that are not partially isomorphic. Spoiler's strategy is obtained by structural induction on some formula $\varphi \in R_{m;\Omega}^k$ on which the two game positions disagree; this argument broadly resembles similar proofs for the standard pebble games (see e.g. [14]). The main difficulty of the proof is to show that if Duplicator produces partitions $\mathbf{P}$ and $\mathbf{Q}$, then Spoiler can always find a block in one of the partitions that contains both tuples that satisfy φ and tuples that satisfy $\neg\varphi$. Once he has identified such a block, Spoiler can place his pebbles in a way that ensures that the resulting game positions disagree on a formula of quantifier rank less than φ. This gives him a strategy to win the game in a finite number of moves. $\square$

In the proof of the next lemma, we show that if $(\mathbf{A}, \mathbf{a}) \equiv^R_{k,m,\Omega} (\mathbf{B}, \mathbf{b})$, then Duplicator can play one round of the (k, m, Ω)-matrix-equivalence game in a way that ensures that the resulting positions will also be $\equiv^R_{k,m,\Omega}$-equivalent. This gives her a strategy to play the game indefinitely. The idea here is to let Duplicator respond to a challenge of the Spoiler with partitions $\mathbf{P}$ and $\mathbf{Q}$ that are obtained by grouping together in each partition block all the elements realising the same $R^k_{m;\Omega}$-type (with respect to the current game positions). The bijection $f : \mathbf{P} \to \mathbf{Q}$ is similarly defined by pairing together blocks whose elements all realise the same $R^k_{m;\Omega}$-type. We show that if Duplicator plays in this manner, then she can ensure both that condition $(\star)$ is met and that Spoiler is restricted to placing his pebbles in blocks which do not distinguish the two structures.

Lemma 3. *If $(\mathbf{A}, \mathbf{a}) \equiv^R_{k,m,\Omega} (\mathbf{B}, \mathbf{b})$ then Duplicator has a winning strategy in the (k, m, Ω)-matrix-equivalence game starting with positions $(\mathbf{A}, \mathbf{a})$ and $(\mathbf{B}, \mathbf{b})$.*

The next lemma shows that for all m and all finite sets of primes Ω, the equivalence $\equiv^R_{k+2m-1,m,\Omega}$ refines $\equiv^C_k$ on the class of finite structures.

Lemma 4. *Duplicator has a winning strategy in the k-pebble bijection game on $(\mathbf{A}, \mathbf{a})$ and $(\mathbf{B}, \mathbf{b})$ if she has a winning strategy in the $(k+2m-1, m, \Omega)$-matrix-equivalence game starting on $(\mathbf{A}, \mathbf{a})$ and $(\mathbf{B}, \mathbf{b})$, for any Ω and $m \in \mathbb{N}$.*

4   Playing with Invertible Linear Maps

It follows from the pebble-game characterisation of finite-variable counting logics that the equivalence $\equiv^C_k$ on finite structures is decidable in polynomial time. Essentially, this is because the number of possible moves for Duplicator at any particular stage of the game can be inductively combined into a structural invariant that completely characterises the game equivalence, and this invariant can be constructed in polynomial time (see Otto [14] for details). In light of Theorem 1, we may therefore ask whether the matrix-equivalence game for $R^k_{m;\Omega}$ can be used to give a similar result for equivalence in finite-variable rank logic; that is, whether we can decide $\equiv^R_{k,m,\Omega}$ on finite structures in polynomial time. Unfortunately, unlike the case with the counting game, there does not seem to be an effective way to encode complete information about a winning strategy in the matrix-equivalence game into a polynomial-size invariant. The main problem is the game condition $(\star)$; this requires Duplicator to show that each pair of matrices $M^{\mathbf{P}}_\gamma$ and $M^{\mathbf{Q}}_{\gamma \circ f^{-1}}$ are equivalent (that is, have the same rank) but the number of these matrices is *exponential* in the size of the partition.

In an attempt to avoid this exponential number of matrix combinations, we define a modification of the matrix-equivalence game which is based on invertible linear maps. In this game, Duplicator is required to specify a bijection between the partitions of the two game structures as the conjugacy action of a single invertible matrix. In that sense, the invertible-map game can be seen as the natural extension of the bijection game for counting logics, where we replace bijections with invertible maps. In [5] it had been asked whether such a game might characterise definability in finite-variable rank logic. We show that equivalence in the invertible-map game does in fact *refine* the relations $\equiv^R_{k,m,\Omega}$ but we do not know if the refinement is proper. We also establish that equivalence in the invertible-map game can be decided in polynomial time, which is not known to be true for the $\equiv^R_{k,m,\Omega}$ as we discussed above. We see one application of this new game equivalence in the next section, where we define algorithms for testing graph isomorphism by playing the invertible-map game on finite graphs.

4.1   Invertible-Map Game

Let k and m be positive integers with $2m \leq k$ and let Ω be a finite and non-empty set of primes. The game board of the k-pebble m-ary *invertible-map game* over Ω (or (k, m, Ω)-invertible-map game for short) consists of two structures $\mathbf{A}$ and $\mathbf{B}$, each with k pebbles labelled $1, \ldots, k$ (and initial placement of pebbles a over $\mathbf{A}$ and b over $\mathbf{B}$, as before). If $\|\mathbf{A}\| \neq \|\mathbf{B}\|$ or the mapping defined by the initial pebble positions is not a partial isomorphism, then Spoiler wins the game. Otherwise, each round of the game is played as follows.

1. Spoiler chooses a prime $p \in \Omega$ and picks up $2m$ pebbles in some order from $\mathbf{A}$ and the $2m$ corresponding pebbles in the same order from $\mathbf{B}$.
2. Duplicator has to respond by choosing
   - a partition $\mathbf{P}$ of $U(\mathbf{A})^m \times U(\mathbf{A})^m$,
   - a partition $\mathbf{Q}$ of $U(\mathbf{B})^m \times U(\mathbf{B})^m$, with $\|\mathbf{P}\| = \|\mathbf{Q}\|$, and
   - a non-singular $U(\mathbf{B})^m \times U(\mathbf{A})^m$ matrix S over GF_p,
   for which it holds that the map $f : \mathbf{P} \to \mathbf{Q}$ defined by

   $$P \mapsto Q \quad \text{iff} \quad S \cdot \chi_P \cdot S^{-1} = \chi_Q \qquad (\ast)$$

   is *total* and *bijective*, where we view χ_P and χ_Q as matrices over GF_p.
3. Spoiler next picks $P \in \mathbf{P}$ and places the $2m$ chosen pebbles from $\mathbf{A}$ on the elements of some tuple in P (in the order they were chosen earlier) and the corresponding $2m$ pebbles from $\mathbf{B}$ on the elements of some tuple in $f(P)$.

This completes one round in the game. If, after this exchange, the partial map from $\mathbf{A}$ to $\mathbf{B}$ defined by the pebbled positions is not a partial isomorphism, or if Duplicator is unable to produce the necessary triple $(\mathbf{P}, \mathbf{Q}, S)$, then Spoiler has won the game; otherwise it can continue for another round.

We write $(\mathbf{A}, b) \approx^k_{m,\Omega} (\mathbf{B}, b)$ to denote that Duplicator has a strategy to play forever in the (k, m, Ω)-invertible-map game with starting positions $(\mathbf{A}, a)$ and $(\mathbf{B}, b)$. The following lemma shows that increasing k, m or Ω, refines the equivalence relation on $\mathrm{fin}[\tau; k]$. The first two inclusions of the lemma are trivial but the last inclusion requires results on the block structure of those invertible maps that can be played as a part of a winning strategy.

Lemma 5. *For all $k, m \in \mathbb{N}$ with $2m \leq k$ and all finite and non-empty sets of primes $\Omega \subseteq \Gamma$, it holds that $\approx^{k+1}_{m,\Omega} \subseteq \approx^k_{m,\Omega}$, $\approx^k_{m,\Gamma} \subseteq \approx^k_{m,\Omega}$ and $\approx^k_{m+1,\Omega} \subseteq \approx^k_{m,\Omega}$.*

In the matrix-equivalence game, it is clearly sufficient for Duplicator to demonstrate the existence of a single similarity transformation that relates all linear combinations of partition matrices, since similar matrices have the same rank.

Lemma 6. *Duplicator has a winning strategy in the (k, m, Ω)-matrix-equivalence game starting on $(\mathbf{A}, \boldsymbol{a})$ and $(\mathbf{B}, \boldsymbol{b})$ if she has a winning strategy in the (k, m, Ω)-invertible-map game starting on $(\mathbf{A}, \boldsymbol{a})$ and $(\mathbf{B}, \boldsymbol{b})$.*

4.2 Complexity of the Game Equivalence

In this section we show that for each k there is an algorithm that decides whether $(\mathbf{A}, \boldsymbol{a}) \approx^k_{m,\Omega} (\mathbf{B}, \boldsymbol{b})$ in time polynomial in $n p_{\max}$, where n is the size of both $\mathbf{A}$ and $\mathbf{B}$ and $p_{\max}$ is the largest prime in Ω. To simplify our notation, fix k, m, Ω and τ. In order to analyse the structure of the game equivalence, we consider a stratification of $\approx^k_{m,\Omega}$ by the number of rounds in the game. Specifically, we let $\sim_i$ be the binary relation on $\text{fin}[\tau; k]$ defined by $(\mathbf{A}, \boldsymbol{a}) \sim_i (\mathbf{B}, \boldsymbol{b})$ if Duplicator has a strategy to play for up to i rounds in the (k, m, Ω)-invertible-map game on $(\mathbf{A}, \boldsymbol{a})$ and $(\mathbf{B}, \boldsymbol{b})$. This relation can be characterised inductively as follows, where we write $\text{atp}(\mathbf{A}, \boldsymbol{a})$ to denote the atomic type of $\boldsymbol{a}$ in $\mathbf{A}$.

Lemma 7. *For all $(\mathbf{A}, \boldsymbol{a}), (\mathbf{B}, \boldsymbol{b}) \in \text{fin}[\tau; k]$ we have*

$$(\mathbf{A}, \boldsymbol{a}) \sim_0 (\mathbf{B}, \boldsymbol{b}) \quad \textit{iff } \text{atp}(\mathbf{A}, \boldsymbol{a}) = \text{atp}(\mathbf{B}, \boldsymbol{b})$$

$(\mathbf{A}, \boldsymbol{a}) \sim_{i+1} (\mathbf{B}, \boldsymbol{b})$ *iff* $(\mathbf{A}, \boldsymbol{a}) \sim_i (\mathbf{B}, \boldsymbol{b})$ *and for all $p \in \Omega$ and all $\boldsymbol{j} \in [k]^{2m}$
with distinct values there is an invertible matrix
S over GF_p such that for all $\alpha \in \text{fin}[\tau; k]/\sim_i$:*
$$\underbrace{S \cdot \text{extmat}^\alpha_{\boldsymbol{j}}(\mathbf{A}, \boldsymbol{a}) \cdot S^{-1} = \text{extmat}^\alpha_{\boldsymbol{j}}(\mathbf{B}, \boldsymbol{b}).}_{(\star\star)}$$

For the proof of the inductive step of Lemma 7, the "if" direction is fairly straightforward (it essentially specifies a sufficient response for Duplicator in one round of the game). For the converse, it needs to be shown that any partition played by Duplicator as a part of an $(i + 1)$-round winning strategy has to be a refinement of the partition of the corresponding structure into $\sim_i$-equivalence classes. Combining Lemma 7 with the fact that the matrix-similarity relation is transitive, we get the following result.

Corollary 8. $\sim_i$ *is an equivalence relation on $\text{fin}[\tau; k]$ for each $i \in \mathbb{N}_0$.*

In particular, since $\approx^k_{m,\Omega}$ coincides with the intersection of the $\sim_i$ over all i, it follows that $\approx^k_{m,\Omega}$ is an equivalence relation on $\text{fin}[\tau; k]$.

Now consider some $(\mathbf{A}, \boldsymbol{a})$ and $(\mathbf{B}, \boldsymbol{b})$ in $\text{fin}[\tau; k]$ and assume that $\|\mathbf{A}\| = \|\mathbf{B}\| = n$. Since the number of distinct positions in the game starting with $(\mathbf{A}, \boldsymbol{a})$ and $(\mathbf{B}, \boldsymbol{b})$ is bounded by a polynomial in n, it follows that there is some polynomial $q : \mathbb{N} \to \mathbb{N}$ (depending only on k and τ) such that if Duplicator can play the game for at least $q(n)$ rounds, then she has a strategy to play forever.

In other words, $(\mathbf{A}, \boldsymbol{a}) \approx_{m,\Omega}^{k} (\mathbf{B}, \boldsymbol{b})$ if, and only if, $(\mathbf{A}, \boldsymbol{a}) \sim_{q(n)} (\mathbf{B}, \boldsymbol{b})$. To decide $(\mathbf{A}, \boldsymbol{a}) \approx_{m,\Omega}^{k} (\mathbf{B}, \boldsymbol{b})$, we inductively construct the graph of $\approx_{m,\Omega}^{k}$, restricted to $\mathbf{A}$ and $\mathbf{B}$, as follows. Initially, we partition the elements of $U(\mathbf{A})^k \cup U(\mathbf{B})^k$ by their atomic equivalance, which is just $\sim_0$. For the induction step, suppose we have constructed $\sim_i$. Then to compute the refinement $\sim_{i+1}$, we consider each $\sim_i$-equivalent pair $(\boldsymbol{c}, \boldsymbol{d})$ and check whether condition $(\star\star)$ of Lemma 7 is satisfied. That is, for each $p \in \Omega$ and $\boldsymbol{j} \in [k]^{2m}$, we let $\mathcal{C} = (C_\alpha)$ and $\mathcal{D} = (D_\alpha)$ be the families of extension matrices defined by $\boldsymbol{j}$ over $\boldsymbol{c}$ and $\boldsymbol{d}$, respectively, indexed by all equivalence classes α of $\sim_i$ (where $C_\alpha = \mathrm{extmat}_{\boldsymbol{j}}^\alpha(\mathbf{A}, \boldsymbol{c})$ if $\boldsymbol{c}$ is defined over $\mathbf{A}$, and similarly for D_α). Here it is important to note that it suffices to consider only equivalence classes of $\sim_i$ restricted to $\mathbf{A}$ and $\mathbf{B}$. Therefore, the number of extension matrices that we need to consider is bounded by a polynomial in n.

At this stage it remains to determine whether the pair of matrix tuples $\mathcal{C}$ and $\mathcal{D}$ are *simultaneously similar*: that is, whether there is a non-singular matrix S such that $S \cdot C_\alpha \cdot S^{-1} = D_\alpha$ for all $C_\alpha \in \mathcal{C}$. By a result of Chistov et al. [3], this problem is in polynomial time over all finite fields.

Proposition 9 (Chistov, Karpinsky and Ivanyov). *There is a deterministic algorithm that, given two families of $N \times N$ matrices $\mathcal{C} = (C_1, \ldots, C_l)$ and $\mathcal{D} = (D_1, \ldots, D_l)$ over a finite field GF_q, determines in time $\mathrm{poly}(N, l, q)$ whether $\mathcal{C}$ and $\mathcal{D}$ are simultaneously similar.*

By our discussion above, it follows that we can construct the graph of $\approx_{m,\Omega}^{k}$ restricted to $\mathbf{A}$ and $\mathbf{B}$ in a polynomial number of steps. At each step, we need to check a polynomial number of matrix tuples for simultaneous similarity, where each tuple has polynomial length. This gives us a proof of the following theorem.

Theorem 10. *For each vocabulary τ there is a deterministic algorithm that, given $(\mathbf{A}, \boldsymbol{a}), (\mathbf{B}, \boldsymbol{b}) \in \mathrm{fin}[\tau; k]$ (with $\|\mathbf{A}\| = \|\mathbf{B}\| = n$), $m \in \mathbb{N}$ with $2m \le k$ and a finite set of primes Ω, decides whether $(\mathbf{A}, \boldsymbol{a}) \approx_{m,\Omega}^{k} (\mathbf{B}, \boldsymbol{b})$ in time $(np)^{\mathcal{O}(k)}$ where p is the largest prime in Ω.*

Observe that this implies that for each fixed k, we can decide $\approx_{m,\Omega}^{k}$ in polynomial time, where Ω can be a part of the input and $m \le k$.

5   Application to the Graph Isomorphism Problem

By considering the invertible-map game equivalence $\approx_{m,\Omega}^{k}$ on the class of all finite graphs, we get a family of polynomial-time algorithms for stratifying the graph isomorphism relation. More specifically, for each k, m and Ω, we write $\mathrm{IM}_{m,\Omega}^{k}$ to denote the following algorithm on a pair of finite graphs $\mathbf{G}$ and $\mathbf{H}$:

If $\|\mathbf{G}\| \ne \|\mathbf{H}\|$ then output "not isomorphic". Otherwise, compute $\approx_{m,\Omega}^{k}$ (restricted to $\mathbf{G}$ and $\mathbf{H}$) on the set $U(\mathbf{G})^k \cup U(\mathbf{H})^k$ by applying the algorithm of Theorem 10 for all tuples in $U(\mathbf{G})^k \cup U(\mathbf{H})^k$. If the result is that there is some equivalence class α of $\approx_{m,\Omega}^{k}$ such that $\|\alpha \cap U(\mathbf{G})^k\| \ne \|\alpha \cap U(\mathbf{H})^k\|$ then output "not isomorphic"; else output "isomorphic".

It follows from Theorem 10 that $\text{IM}_{m,\Omega}^k$ runs in polynomial time for a fixed k. While the algorithm will always correctly identify isomorphic graphs, it may fail to distinguish between non-isomorphic instances. Furthermore, it can be seen that for each pair of graphs, there is always some value of k for which $\text{IM}_{m,\Omega}^k$ correctly determines isomorphism for all m and Ω.

The procedure for $\text{IM}_{m,\Omega}^k$ outlined above bears a strong resemblance to the Weisfeiler-Lehman method for graph isomorphism (see [2] for a description of the method). It was shown by Cai, Fürer and Immerman [2] that Duplicator has a winning strategy in the $(k+1)$-pebble bijection game on $\mathbf{G}$ and $\mathbf{H}$ if, and only if, $\mathbf{G}$ and $\mathbf{H}$ are not distinguished by the k-dimensional Weisfeiler-Lehman algorithm (WL^k). Combining this with lemmas 4 and 6, we have that

$$\mathbf{G} \text{ and } \mathbf{H} \text{ are distinguished by } \text{WL}^k$$
$$\Rightarrow \mathbf{G} \text{ and } \mathbf{H} \text{ are distinguished by } \text{IM}_{m,\Omega}^{k+2m} \text{ for } \textit{all } m \text{ and prime sets } \Omega.$$

In [2], Cai et al. showed how to construct for each $k \in \mathbb{N}$ a pair of non-isomorphic graphs (named "CFI graphs") that are indistinguishable in WL^k. Later, it was shown by Dawar et al. [4] that there is a fixed sentence of first-order logic with rank operators over GF_2 that can distinguish between any pair of these CFI graphs. This construction was extended by Holm [10], who showed that for any prime p, there are families of non-isomorphic graphs that can be distinguished by first-order logic with rank operators over GF_p but not by any fixed dimension of Weisfeiler-Lehman. Hence, it follows that the family of $\text{IM}_{m,\Omega}^k$ algorithms provide a way of stratifying the graph isomorphism relation which goes beyond that given by the Weisfeiler-Lehman algorithms.

Proposition 11. *For each prime p and $k \geq 1$, there is a pair of non-isomorphic graphs $\mathbf{G}$ and $\mathbf{H}$ that can be distinguished by $\text{IM}_{\{p\},1}^3$ but not by WL^k.*

Finally, we remark that Derksen [6] has described a family of polynomial-time algorithms that also give an approximation to graph isomorphism that goes beyond that of the Weisfeiler-Lehman method. While Derksen's method partly builds on the simultaneous-similarity algorithm of Chistov et al. [3] (Proposition 9), it also draws heavily on techniques from algebraic geometry and category theory and seems different to the game-based approach that we describe. Nevertheless, it is an open problem whether these two approaches can be related.

6 Discussion

A natural question that is raised by the definitions of the games we have presented in this paper, is how to use them to establish inexpressibility results. A step in this direction is presented in [10] where it is shown that for any prime p, there is a property definable in first-order logic with rank operators over GF_p which is not closed under $\equiv_{k,1,\{q\}}^R$ for any k and primes $q \neq p$. This method can be further extended to work for finite sets of primes rather than just single primes. It would be interesting to lift this up to arities higher than 1, but playing the game poses combinatorial difficulties.

Another direction is to establish the precise relationship between the two games we consider. While we show that the invertible-map game gives a refinement of the matrix-equivalence game, it is not known whether this refinement is strict. Might it be the case that for any k and m one can find a k' and m' so that $\equiv^{R}_{k',m',\Omega}$ is a refinement of $\approx^{k}_{m,\Omega}$? One way this might be established is by showing that the relations $\approx^{k}_{m,\Omega}$ are themselves definable in IFPR. If it turns out that this is not the case, then we would have established that there is a PTIME property not in in IFPR. A natural line of investigation would then be to extract from the invertible-map game a suitable logical operator, stronger than the matrix-rank operator, that is characterised by this game.

A more general direction of research is to explore other partition games which can be defined by suitable equivalence conditions on the partition matrices. There is space here for defining new logics and also new isomorphism tests.

References

[1] Barwise, J.: On Moschovakis closure ordinals. Journal of Symbolic Logic 42, 292–296 (1977)
[2] Cai, J.-Y., Fürer, M., Immerman, N.: An optimal lower bound on the number of variables for graph identification. Combinatorica 12(4), 389–410 (1992)
[3] Chistov, A., Ivanyos, G., Karpinski, M.: Polynomial time algorithms for modules over finite dimensional algebras. In: Proc. 1997 International Symposium on Symbolic and Algebraic Computation, pp. 68–74. ACM (1997)
[4] Dawar, A., Grohe, M., Holm, B., Laubner, B.: Logics with rank operators. In: Proc. 24th IEEE Symp. on Logic in Computer Science, pp. 113–122 (2009)
[5] Dawar, A., Holm, B.: Pebble games for logics with counting and rank. In: Cégielski, P. (ed.) Studies in Weak Arithmetics, pp. 99–120. CSLI Publications (2010)
[6] Derksen, H.: The graph isomorphism problem and approximate categories, arXiv:1012.2081v1 (2010)
[7] Ebbinghaus, H.D., Flum, J.: Finite Model Theory. Springer (1999)
[8] Gurevich, Y.: Logic and the challenge of computer science. In: Börger, E. (ed.) Current Trends in Theoretical Computer Science, pp. 1–57. Computer Science Press (1988)
[9] Hella, L.: Logical hierarchies in PTIME. Information and Computation 129, 1–19 (1996)
[10] Holm, B.: Descriptive complexity of linear algebra. PhD thesis. Univ. of Cambridge (2010)
[11] Immerman, N.: Relational queries computable in polynomial time. Information and Control 68, 86–104 (1986)
[12] Immerman, N., Lander, E.: Describing graphs: A first-order approach to graph canonization. In: Complexity Theory Retrospective: In Honor of Juris Hartmanis on the Occasion of His Sixtieth Birthday, July 5, 1998, pp. 59–81 (1990)
[13] Laubner, B.: The Structure of Graphs and New Logics for the Characterization of Polynomial Time. PhD thesis, Humboldt-Universität zu Berlin (2011)
[14] Otto, M.: Bounded Variable Logics and Counting — A Study in Finite Models. Lecture Notes in Logic, vol. 9. Springer (1997)
[15] Poizat, B.: Deux ou trois choses que je sais de L_n. Journal of Symbolic Logic 47(3), 641–658 (1982)

Exponential Lower Bounds and Separation for Query Rewriting*

Stanislav Kikot[1], Roman Kontchakov[1], Vladimir Podolskii[2],
and Michael Zakharyaschev[1]

[1] Department of Computer Science and Information Systems
Birkbeck, University of London, UK
{kikot,roman,michael}@dcs.bbk.ac.uk
[2] Steklov Mathematical Institute, Moscow, Russia
podolskii@mi.ras.ru

Abstract. We establish connections between the size of circuits and formulas computing monotone Boolean functions and the size of first-order and nonrecursive Datalog rewritings for conjunctive queries over *OWL 2 QL* ontologies. We use known lower bounds and separation results from circuit complexity to prove similar results for the size of rewritings that do not use non-signature constants. For example, we show that, in the worst case, positive existential and nonrecursive Datalog rewritings are exponentially longer than the original queries; nonrecursive Datalog rewritings are in general exponentially more succinct than positive existential rewritings; while first-order rewritings can be superpolynomially more succinct than positive existential rewritings.

1 Introduction

First-order (FO) rewritability is the key concept of ontology-based data access (OBDA), which is believed to lie at the foundations of the next generation of information systems. A language $\mathcal{L}$ enjoys *FO-rewritability* if any conjunctive query q over an ontology $\mathcal{T}$, formulated in $\mathcal{L}$, can be transformed into an FO-formula q' such that, for any data $\mathcal{A}$, the certain answers to q over the knowledge base $(\mathcal{T}, \mathcal{A})$ can be found by querying q' over $\mathcal{A}$ using a standard relational database management system (RDBMS). Ontology languages with this property include the *OWL 2 QL* profile of the Web Ontology Language *OWL 2*, which is based on the *DL-Lite* family of description logics [11,4], and fragments of Datalog$^{\pm}$ such as linear or sticky sets of TGDs [9,10]. Various rewriting techniques have been implemented in the systems QuOnto [1], REQUIEM [19], Presto [26], Nyaya [12] and Quest [25].

OBDA via FO-rewritability relies on the empirical fact that RDBMSs are usually very efficient in practice. However, this does not mean that they can efficiently evaluate any given query: after all, for expression complexity, database query answering is PSPACE-complete for FO-queries and NP-complete for conjunctive queries (CQs). Indeed, the first 'naïve' rewritings of CQs over *OWL 2 QL*

---

* A full version of this paper is available at http://arxiv.org/abs/1202.4193.

A. Czumaj et al. (Eds.): ICALP 2012, Part II, LNCS 7392, pp. 263–274, 2012.
© Springer-Verlag Berlin Heidelberg 2012

ontologies turned out to be too lengthy even for modern RDBMSs [11,19]. The next step was to develop various rewriting optimisation techniques [26,12,23,24]; however, they still produced exponential-size — $O((|\mathcal{T}| \cdot |\boldsymbol{q}|)^{|\boldsymbol{q}|})$ — rewritings in the worst case. An alternative two-step combined approach to OBDA with *OWL 2 EL* [18] and *OWL 2 QL* [17] first expands the data by applying the ontology axioms and introducing new individuals required by the ontology, and only then rewrites the query over the expanded data. Yet, even with these extra resources a simple polynomial rewriting was constructed only for the fragment of *OWL 2 QL* without role inclusions; the rewriting for the full language remained exponential. A breakthrough seemed to come in [13], which showed that one can construct, in polynomial time, a nonrecursive Datalog rewriting for some fragments of Datalog$^\pm$ containing *OWL 2 QL*. However, this rewriting uses the built-in predicate $\neq$ and numerical constants that are not present in the original query and ontology. Without additional constants, no FO-rewriting for *OWL 2 QL* can be *constructed* in polynomial time [15] (it remained unclear, however, whether such an FO-rewriting of polynomial size *exists*).

These developments bring forward a spectrum of theoretical and practical questions that could influence the future of OBDA. What is the worst-case size of FO- and nonrecursive Datalog rewritings for CQs over *OWL 2 QL* ontologies? What is the type/shape/size of rewritings we should aim at to make OBDA with *OWL 2 QL* efficient? What extra means (e.g., built-in predicates and constants) can be used in the rewritings? In this paper, we investigate the worst-case size of FO- and nonrecursive Datalog rewritings for CQs over *OWL 2 QL* ontologies depending on the available means. We distinguish between 'pure' rewritings, which cannot use constants that do not occur in the original query, and 'impure' ones, where such constants are allowed. Our results can be summarised as follows:

- An exponential blow-up is unavoidable for pure positive existential rewritings and pure nonrecursive Datalog rewritings. Even pure FO-rewritings with $=$ can blow-up superpolynomially unless $\mathrm{NP} \subseteq \mathrm{P/poly}$.
- Pure nonrecursive Datalog rewritings are in general exponentially more succinct than pure positive existential rewritings.
- Pure FO-rewritings can be superpolynomially more succinct than pure positive existential rewritings.
- Impure positive existential rewritings can always be made polynomial, and so they are exponentially more succinct than pure rewritings.

We obtain these results by first establishing connections between pure rewritings for CQs over *OWL 2 QL* ontologies and circuits for monotone Boolean functions, and then using known lower bounds and separation results for the circuit complexity of such functions as $\mathrm{CLIQUE}_{n,k}$ 'a graph with n nodes contains a k-clique' and $\mathrm{MATCHING}_{2n}$ 'a bipartite graph with n vertices in each part has a perfect matching.'

2   Queries over *OWL 2 QL* Ontologies

By a *signature*, Σ, we understand in this paper any set of constant symbols and predicate symbols (with their arity). Unless explicitly stated otherwise, Σ does *not* contain any predicates with fixed semantics, such as $=$ or $\neq$. In the description logic (or *OWL 2 QL*) setting, constant symbols are called *individual names*, a_i, while unary and binary predicate symbols are called *concept names*, A_i, and *role names*, P_i, respectively, where $i \geq 1$. The language of *OWL 2 QL* is built using these names in the following way. The *roles R*, *basic concepts B* and *concepts C* of *OWL 2 QL* are defined by the grammar:

$$
\begin{array}{llll}
R & ::= & P_i \mid P_i^-, & (P_i(x,y) \mid P_i(y,x)) \\
B & ::= & \bot \mid A_i \mid \exists R, & (\bot \mid A_i(x) \mid \exists y\, R(x,y)) \\
C & ::= & B \mid \exists R.B, & (B(x) \mid \exists y\,(R(x,y) \wedge B(y)))
\end{array}
$$

where the formulas on the right give a first-order translation of the *OWL 2 QL* constructs. An *OWL 2 QL TBox*, $\mathcal{T}$, is a finite set of *inclusions* of the form

$$
\begin{array}{ll}
B \sqsubseteq C, & (\forall x\,(B(x) \rightarrow C(x))) \\
R_1 \sqsubseteq R_2, & (\forall x, y\,(R_1(x,y) \rightarrow R_2(x,y))) \\
B_1 \sqcap B_2 \sqsubseteq \bot, & (\forall x\,(B_1(x) \wedge B_2(x) \rightarrow \bot)) \\
R_1 \sqcap R_2 \sqsubseteq \bot. & (\forall x, y\,(R_1(x,y) \wedge R_2(x,y) \rightarrow \bot))
\end{array}
$$

Note that concepts of the form $\exists R.B$ can only occur in the right-hand side of concept inclusions in *OWL 2 QL*. An *ABox*, $\mathcal{A}$, is a finite set of *assertions* of the form $A_k(a_i)$ and $P_k(a_i, a_j)$. $\mathcal{T}$ and $\mathcal{A}$ together form the *knowledge base* (KB) $\mathcal{K} = (\mathcal{T}, \mathcal{A})$. The semantics for *OWL 2 QL* is defined in the usual way [6], based on interpretations $\mathcal{I} = (\Delta^{\mathcal{I}}, \cdot^{\mathcal{I}})$ with domain $\Delta^{\mathcal{I}}$ and interpretation function $\cdot^{\mathcal{I}}$. The set of individuals in $\mathcal{A}$ is denoted by $\mathrm{ind}(\mathcal{A})$. For concepts or roles E_1, E_2, we write $E_1 \sqsubseteq_{\mathcal{T}} E_2$ if $\mathcal{T} \models E_1 \sqsubseteq E_2$; and we set $[E] = \{E' \mid E \sqsubseteq_{\mathcal{T}} E' \text{ and } E' \sqsubseteq_{\mathcal{T}} E\}$.

A *conjunctive query* (CQ) $q(x)$ is an FO-formula $\exists y\, \varphi(x, y)$, where φ is a conjunction of atoms of the form $A_k(t_1)$ and $P_k(t_1, t_2)$, and each t_i is a *term* (an individual or a variable from x or y). A tuple $a \subseteq \mathrm{ind}(\mathcal{A})$ is a *certain answer* to $q(x)$ over $\mathcal{K} = (\mathcal{T}, \mathcal{A})$ if $\mathcal{I} \models q(a)$ for all $\mathcal{I} \models \mathcal{K}$; in this case we write $\mathcal{K} \models q(a)$.

Query answering over *OWL 2 QL* KBs is based on the fact that, for any consistent KB $\mathcal{K} = (\mathcal{T}, \mathcal{A})$, there is an interpretation $\mathcal{C}_{\mathcal{K}}$ such that, for all CQs $q(x)$ and $a \subseteq \mathrm{ind}(\mathcal{A})$, we have $\mathcal{K} \models q(a)$ iff $\mathcal{C}_{\mathcal{K}} \models q(a)$. The interpretation $\mathcal{C}_{\mathcal{K}}$, called the *canonical model* of $\mathcal{K}$, can be constructed as follows. For each pair $[R], [B]$ with $\exists R.B$ in $\mathcal{T}$ (we assume $\exists R$ is just a shorthand for $\exists R.\top$), we introduce a fresh symbol $w_{[RB]}$ and call it the *witness for* $\exists R.B$. We write $\mathcal{K} \models C(w_{[RB]})$ if $\exists R^- \sqsubseteq_{\mathcal{T}} C$ or $B \sqsubseteq_{\mathcal{T}} C$. Define a *generating relation*, $\rightsquigarrow$, on the set of these witnesses together with $\mathrm{ind}(\mathcal{A})$ by taking:

- $a \rightsquigarrow w_{[RB]}$ if $a \in \mathrm{ind}(\mathcal{A})$, $[R]$ and $[B]$ are $\sqsubseteq_{\mathcal{T}}$-minimal with $\mathcal{K} \models \exists R.B(a)$ and there is no $b \in \mathrm{ind}(\mathcal{A})$ with $\mathcal{K} \models R(a,b) \wedge B(b)$;
- $w_{[R'B']} \rightsquigarrow w_{[RB]}$ if, for some u, $u \rightsquigarrow w_{[R'B']}$, $[R]$, $[B]$ are $\sqsubseteq_{\mathcal{T}}$-minimal with $\mathcal{K} \models \exists R.B(w_{[R'B']})$ and it is not the case that $R' \sqsubseteq_{\mathcal{T}} R^-$ and $\mathcal{K} \models B'(u)$.

If $a \rightsquigarrow w_{[R_1 B_1]} \rightsquigarrow \cdots \rightsquigarrow w_{[R_n B_n]}$, $n \geq 0$, then we say that a *generates the path* $aw_{[R_1 B_1]} \cdots w_{[R_n B_n]}$. Denote by $\mathsf{path}_{\mathcal{K}}(a)$ the set of paths generated by a, and by $\mathsf{tail}(\pi)$ the last element in $\pi \in \mathsf{path}_{\mathcal{K}}(a)$. Then $\mathcal{C}_{\mathcal{K}}$ is defined by taking:

$$\Delta^{\mathcal{C}_{\mathcal{K}}} = \bigcup_{a \in \mathsf{ind}(\mathcal{A})} \mathsf{path}_{\mathcal{K}}(a), \quad a^{\mathcal{C}_{\mathcal{K}}} = a, \text{ for } a \in \mathsf{ind}(\mathcal{A}),$$

$$A^{\mathcal{C}_{\mathcal{K}}} = \{\pi \in \Delta^{\mathcal{C}_{\mathcal{K}}} \mid \mathcal{K} \models A(\mathsf{tail}(\pi))\},$$

$$P^{\mathcal{C}_{\mathcal{K}}} = \{(a, b) \in \mathsf{ind}(\mathcal{A}) \times \mathsf{ind}(\mathcal{A}) \mid \mathcal{K} \models P(a, b)\} \cup$$
$$\{(\pi, \pi \cdot w_{[RB]}) \mid \mathsf{tail}(\pi) \rightsquigarrow w_{[RB]}, \ R \sqsubseteq_{\mathcal{T}} P\} \cup$$
$$\{(\pi \cdot w_{[RB]}, \pi) \mid \mathsf{tail}(\pi) \rightsquigarrow w_{[RB]}, \ R \sqsubseteq_{\mathcal{T}} P^-\}.$$

Theorem 1 ([11,17]). *For every OWL 2 QL KB $\mathcal{K} = (\mathcal{T}, \mathcal{A})$, every CQ $q(x)$ and every $a \subseteq \mathsf{ind}(\mathcal{A})$, $\mathcal{K} \models q(a)$ iff $\mathcal{C}_{\mathcal{K}} \models q(a)$.*

Let Σ be a signature that can be used to formulate queries and ABoxes (remember that Σ does not contain any built-in predicates). Given an ABox $\mathcal{A}$ over Σ, define $\mathcal{I}_{\mathcal{A}}$ to be the interpretation whose domain consists of all individuals in Σ (even if they do not occur in $\mathsf{ind}(\mathcal{A})$) and $\mathcal{I}_{\mathcal{A}} \models E(a)$ iff $E(a) \in \mathcal{A}$, for all predicates $E(x)$.

Given a CQ $q(x)$ and a TBox $\mathcal{T}$, a first-order formula $q'(x)$ over Σ is called an *FO-rewriting for $q(x)$ and $\mathcal{T}$ over Σ* if, for any ABox $\mathcal{A}$ over Σ and any $a \subseteq \mathsf{ind}(\mathcal{A})$, we have $(\mathcal{T}, \mathcal{A}) \models q(a)$ iff $\mathcal{I}_{\mathcal{A}} \models q'(a)$. If q' is an FO-rewriting of the form $\exists y \, \varphi(x, y)$, where φ is built from atoms using only $\land$ and $\lor$, then we call $q'(x)$ a *positive existential rewriting for $q(x)$ and $\mathcal{T}$ over Σ* (or a *PE-rewriting*, for short). The *size* $|q'|$ of q' is the number of symbols in q'.

All known FO-rewritings for CQs and OWL 2 QL ontologies are of exponential size in the worst case. More precisely, for any CQ q and OWL 2 QL TBox $\mathcal{T}$, there exists a PE-rewriting of size $O((|\mathcal{T}| \cdot |q|)^{|q|})$ [11,19,12,17]. One of the main results of this paper is that this bound cannot be substantially improved in general, even for FO-rewritings. On the other hand, we also show that FO-rewritings can be superpolynomially more succinct than PE-rewritings.

We also consider rewritings in the form of nonrecursive Datalog queries. We remind the reader that a *Datalog program*, Π, is a finite set of Horn clauses $\forall x \, (A_1 \land \cdots \land A_m \rightarrow A_0)$, where each A_i is an atom of the form $P(t_1, \ldots, t_l)$ and each t_j is either a variable from x or a constant. A_0 is called the *head* of the clause, and $A_1, \ldots, A_m$ its *body*; all variables occurring in the head must also occur in the body. A predicate P *depends* on a predicate Q in Π if Π contains a clause whose head is P and whose body contains Q. Π is called *nonrecursive* if this dependence relation for Π is acyclic. A *nonrecursive Datalog query* consists of a nonrecursive Datalog program Π and a *goal* G, which is just a predicate. Given an ABox $\mathcal{A}$, a tuple $a \subseteq \mathsf{ind}(\mathcal{A})$ is a *certain answer to (Π, G) over $\mathcal{A}$* if $\Pi, \mathcal{A} \models G(a)$. The *size* $|\Pi|$ of Π is the number of symbols in Π.

We distinguish between *pure* and *impure* Datalog queries [7]. In a *pure query* (Π, G), the clauses in Π do not contain constant symbols in their heads. One reason for considering only pure queries in OBDA is that impure ones can add new facts to the database that do not follow from the intensional knowledge in the background ontology. Impure nonrecursive Datalog queries are known to be more succinct than pure ones.

Given a CQ $q(x)$ and a TBox $\mathcal{T}$, a pure nonrecursive Datalog query (Π, G) is called a *nonrecursive Datalog rewriting for* $q(x)$ *and* $\mathcal{T}$ *over* Σ (or an *NDL-rewriting*, for short) if, for any ABox $\mathcal{A}$ over Σ and any $a \subseteq \text{ind}(\mathcal{A})$, we have $(\mathcal{T}, \mathcal{A}) \models q(a)$ iff $\Pi, \mathcal{A} \models G(a)$ (note that Π may define predicates that are not in Σ, but may not use non-signature constants). Similarly to FO-rewritings, known NDL-rewritings for $OWL\,2\,QL$ are of exponential size [26,12]. Here we show that, in general, one cannot make NDL-rewritings shorter. On the other hand, they can be exponentially more succinct than PE-rewritings.

The rewritings can be much shorter if non-signature predicates and constants become available. As follows from [13], every CQ over an $OWL\,2\,QL$ ontology can be rewritten as a polynomial-size nonrecursive Datalog query if we can use the inequality predicate and at least two distinct constants (cf. also [5], which shows how two constants and $=$ can be used to eliminate definitions from first-order theories without an exponential blow-up). In fact, we observe that, using equality and two distinct constants, any CQ over an $OWL\,2\,QL$ ontology can be rewritten into a PE-query of polynomial size.

3   Boolean Circuits, CNFs and OBDA

To establish the lower and upper bounds for the size of rewritings mentioned above, we show first how the problem of constructing formulas and circuits that compute monotone Boolean functions can be reduced to the problem of finding FO- and NDL-rewritings for CQs over $OWL\,2\,QL$ ontologies.

By an *n-ary Boolean function*, for $n \geq 1$, we mean a function from $\{0,1\}^n$ to $\{0,1\}$. A Boolean function f is *monotone* if $f(\alpha) \leq f(\alpha')$, for all $\alpha \leq \alpha'$, where $\leq$ is the component-wise relation $\leq$ on vectors of $\{0,1\}$.

We remind the reader (for more details see, e.g., [3,14]) that an *n-input Boolean circuit*, $\mathbf{C}$, is a directed acyclic graph with n sources, *inputs*, and one sink, *output*. Every non-source node of $\mathbf{C}$ is called a *gate* and is labelled with either $\wedge$ or $\vee$, in which case it has two incoming edges, or with $\neg$, in which case it has one incoming edge. A circuit is *monotone* if it contains only $\wedge$ and $\vee$ gates. *Boolean formulas* can be thought of as circuits in which every gate has at most one outgoing edge. For an input $\alpha \in \{0,1\}^n$, the *output* of $\mathbf{C}$ on α is denoted by $\mathbf{C}(\alpha)$, and $\mathbf{C}$ is said to *compute* an n-ary Boolean function f if $\mathbf{C}(\alpha) = f(\alpha)$, for every $\alpha \in \{0,1\}^n$. The number of nodes in $\mathbf{C}$ is the *size* of $\mathbf{C}$, denoted $|\mathbf{C}|$.

A *family of Boolean functions* is a sequence $f^1, f^2, \ldots$, where each f^n is an n-ary Boolean function. We say that a family $f^1, f^2, \ldots$ is in the complexity class NP if there exist polynomials p and T and, for each $n \geq 1$, a Boolean circuit $\mathbf{C}^n$ with $n + p(n)$ inputs such that $|\mathbf{C}^n| \leq T(n)$ and, for each $\alpha \in \{0,1\}^n$, we have

$$f^n(\alpha) = 1 \qquad \text{iff} \qquad \mathbf{C}^n(\alpha, \beta) = 1, \quad \text{for some } \beta \in \{0,1\}^{p(n)}.$$

The additional $p(n)$ inputs for β in the $\mathbf{C}^n$ are called *advice inputs*.

Given a family $f^1, f^2, \ldots$ of monotone Boolean functions in NP, we construct a sequence of $OWL\,2\,QL$ TBoxes $\mathcal{T}_{f^n}$ and CQs $\boldsymbol{q}_{f^n}$ without answer variables, as well as ABoxes $\mathcal{A}_{\boldsymbol{\alpha}}$, $\boldsymbol{\alpha} \in \{0,1\}^n$, with a *single* individual such that

$$(\mathcal{T}_{f^n}, \mathcal{A}_{\boldsymbol{\alpha}}) \models \boldsymbol{q}_{f^n} \qquad \text{iff} \qquad f^n(\boldsymbol{\alpha}) = 1, \qquad \text{for all } \boldsymbol{\alpha} \in \{0,1\}^n.$$

Then we show that rewritings for $\boldsymbol{q}_{f^n}$ and $\mathcal{T}_{f^n}$ correspond to Boolean circuits computing f^n. The construction proceeds in two steps: first, we represent the f^n by polynomial-size CNFs (in a way similar to the Tseitin transformation [27]), and then encode those CNFs in terms of $OWL\,2\,QL$ query answering.

Let $f^1, f^2, \ldots$ be a family of Boolean functions in NP and $\mathbf{C}^1, \mathbf{C}^2, \ldots$ be a family of circuits computing the f^n (according to the definition above). We consider the inputs $\boldsymbol{x}$ and the advice inputs $\boldsymbol{y}$ of $\mathbf{C}^n$ as Boolean variables; each of the gates $g_1, \ldots, g_{|\mathbf{C}^n|}$ of $\mathbf{C}^n$ is also thought of as a Boolean variable whose value coincides with the output of the gate on a given input. We assume that $\mathbf{C}^n$ only contains $\neg$- and $\wedge$-gates, and so can be regarded as a set of equations of the form

$$g_i = \neg h_i \qquad \text{or} \qquad g_i = h_i \wedge h_i',$$

where h_i and h_i' are the inputs of the gate g_i, that is, either input variables $\boldsymbol{x}$, advice variables $\boldsymbol{y}$ or other gates $\boldsymbol{g} = (g_1, \ldots, g_{|\mathbf{C}^n|})$. We assume $g_{|\mathbf{C}^n|}$ to be the output of $\mathbf{C}^n$. Now, with each f^n and each $\boldsymbol{\alpha} = (\alpha_1, \ldots, \alpha_n) \in \{0,1\}^n$, we associate the following formula in CNF:

$$\varphi_{f^n}^{\boldsymbol{\alpha}}(\boldsymbol{x}, \boldsymbol{y}, \boldsymbol{g}) \;=\; \bigwedge_{\alpha_j = 0} \neg x_j \;\wedge\; g_{|\mathbf{C}^n|} \;\wedge\; \bigwedge_{g_i = \neg h_i \text{ in } \mathbf{C}^n} \left[(h_i \vee \neg g_i) \wedge (\neg h_i \vee g_i) \right] \;\wedge$$
$$\bigwedge_{g_i = h_i \wedge h_i' \text{ in } \mathbf{C}^n} \left[(h_i \vee \neg g_i) \wedge (h_i' \vee \neg g_i) \wedge (\neg h_i \vee \neg h_i' \vee g_i) \right].$$

The clauses of the last two conjuncts encode the correct computation of the circuit: they are equivalent to $g_i \leftrightarrow \neg h_i$ and $g_i \leftrightarrow h_i \wedge h_i'$, respectively.

Lemma 1. *If f^n is a monotone Boolean function then $f^n(\boldsymbol{\alpha}) = 1$ iff $\varphi_{f^n}^{\boldsymbol{\alpha}}$ is satisfiable, for each $\boldsymbol{\alpha} \in \{0,1\}^n$.*

The second step of the reduction is to encode satisfiability of $\varphi_{f^n}^{\boldsymbol{\alpha}}$ by means of the CQ answering problem in $OWL\,2\,QL$. Denote $\varphi_{f^n}^{\boldsymbol{\alpha}}$ for $\boldsymbol{\alpha} = (0, \ldots, 0)$ by φ_{f^n}. It is immediate from the definitions that, for each $\boldsymbol{\alpha} \in \{0,1\}^n$, the CNF $\varphi_{f^n}^{\boldsymbol{\alpha}}$ can be obtained from φ_{f^n} by removing the clauses $\neg x_j$ for which $\alpha_j = 1$, $1 \leq j \leq n$. The CNF φ_{f^n} contains $d \leq 3|\mathbf{C}^n|$ clauses $C_1, \ldots, C_d$ with $N = |\mathbf{C}^n|$ Boolean variables, which will be denoted by $p_1, \ldots, p_N$.

Let P be a role name and let A_i, X_i^0, X_i^1 and $Z_{i,j}$ be concept names. Consider the TBox $\mathcal{T}_{f^n}$ containing the following inclusions, for $1 \leq i \leq N$, $1 \leq j \leq d$:

$$A_{i-1} \sqsubseteq \exists P^-.X_i^\ell, \qquad\qquad X_i^\ell \sqsubseteq A_i, \quad \text{for } \ell = 0,1,$$

$$X_i^0 \sqsubseteq Z_{i,j} \quad \text{if } \neg p_i \in C_j,$$

$$X_i^1 \sqsubseteq Z_{i,j} \quad \text{if } p_i \in C_j,$$

$$Z_{i,j} \sqsubseteq \exists P.Z_{i-1,j},$$

$$A_0 \sqcap A_i \sqsubseteq \bot, \qquad\qquad A_0 \sqcap \exists P \sqsubseteq \bot,$$

$$A_0 \sqcap Z_{i,j} \sqsubseteq \bot, \quad \text{for } (i,j) \notin \{(0,1), \dots, (0,n)\}.$$

It is not hard to check that $|\mathcal{T}_{f^n}| = O(|\mathbf{C}^n|^2)$. Consider also the CQ

$$\boldsymbol{q}_{f^n} = \exists \boldsymbol{y}\, \exists \boldsymbol{z} \left[A_0(y_0) \wedge \bigwedge_{i=1}^{N} P(y_i, y_{i-1}) \wedge \right.$$

$$\left. \bigwedge_{j=1}^{d} \left(P(y_N, z_{N-1,j}) \wedge \bigwedge_{i=1}^{N-1} P(z_{i,j}, z_{i-1,j}) \wedge Z_{0,j}(z_{0,j}) \right) \right],$$

where $\boldsymbol{y} = (y_0, \dots, y_N)$ and $\boldsymbol{z} = (z_{0,1}, \dots, z_{N-1,1}, \dots, z_{0,d}, \dots, z_{N-1,d})$. Clearly, $|\boldsymbol{q}_{f^n}| = O(|\mathbf{C}^n|^2)$. Note that $\mathcal{T}_{f^n}$ is acyclic and $\boldsymbol{q}_{f^n}$ is tree-shaped and has no answer variables. For each $\boldsymbol{\alpha} = (\alpha_1, \dots, \alpha_n) \in \{0,1\}^n$, we set

$$\mathcal{A}_{\boldsymbol{\alpha}} = \{A_0(a)\} \cup \{Z_{0,j}(a) \mid 1 \le j \le n \text{ and } \alpha_j = 1\}.$$

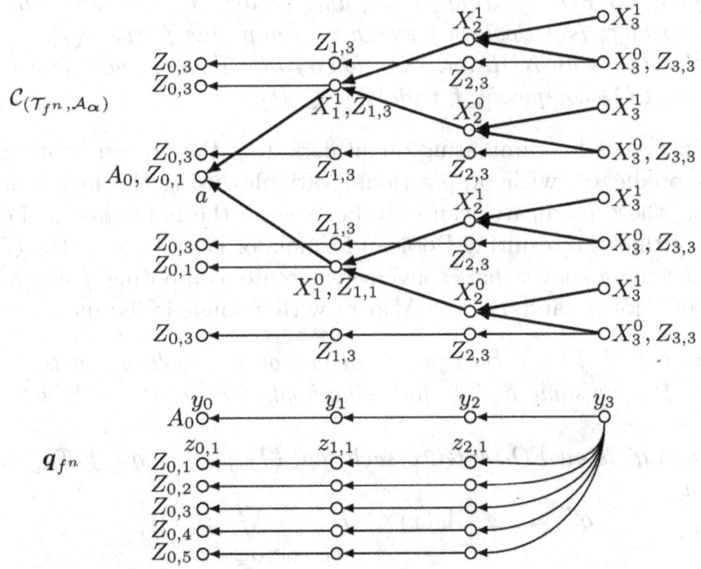

Fig. 1. Canonical model $\mathcal{C}_{(\mathcal{T}_{f^n}, \mathcal{A}_{\boldsymbol{\alpha}})}$ and query $\boldsymbol{q}_{f^n}$ for the Boolean function f^n, $n = 1$, computed by the circuit with one input x, one advice input y and a single $\wedge$-gate. Thus, $N = 3$, $d = 5$ and $\varphi_{f^n}(x, y, g) = \neg x \wedge g \wedge (x \vee \neg g) \wedge (y \vee \neg g) \wedge (\neg x \vee \neg y \vee g)$. Points in X_i^ℓ are also in A_i, for all $1 \le i \le N$; the arrows denote role P and the $Z_{i,j}$ branches in the canonical model are shown only for $j = 1, 3$, i.e., for $\neg x$ and $(x \vee \neg g)$.

We explain the intuition behind the $\mathcal{T}_{f^n}$, $\boldsymbol{q}_{f^n}$ and $\mathcal{A}_{\alpha}$ using the example of Fig. 1, where the query $\boldsymbol{q}_{f^n}$ and the canonical model of $(\mathcal{T}_{f^n}, \mathcal{A}_{\alpha})$, with $\mathcal{A}_{\alpha} = \{A_0(a), Z_{0,1}(a)\}$, are illustrated for some Boolean function. To answer $\boldsymbol{q}_{f^n}$ in the canonical model, we have to check whether $\boldsymbol{q}_{f^n}$ can be homomorphically mapped into it. The variables y_i are clearly mapped to one of the branches of the canonical model from a to a point in A_3, say the lowest one, which corresponds to the valuation for the variables in $\varphi_{f^n}^{\alpha}$ making all of them false. Now, there are two possible ways to map variables $z_{2,1}, z_{1,1}, z_{0,1}$ that correspond to the clause $C_1 = \neg x_1$ in φ_{f^n}. If they are sent to the same branch so that $z_{0,1} \mapsto a$ then $Z_{0,1}(a) \in \mathcal{A}_{\alpha}$, whence the clause C_1 cannot be in $\varphi_{f^n}^{\alpha}$. Otherwise, they are mapped to the points in a side-branch so that $z_{0,1} \not\mapsto a$, in which case $\neg x_1$ must be true under our valuation. Thus, we arrive at the following:

Lemma 2. $(\mathcal{T}_{f^n}, \mathcal{A}_{\alpha}) \models \boldsymbol{q}_{f^n}$ iff $\varphi_{f^n}^{\alpha}$ is satisfiable, for all $\alpha \in \{0,1\}^n$.

We now use this result to reveal a close correspondence between PE-rewritings and monotone Boolean formulas, FO-rewritings and Boolean formulas, and between NDL-rewritings and monotone Boolean circuits.

Lemma 3. Let $f^1, f^2, \ldots$ be a family of monotone Boolean functions in NP, and let $f = f^n$, for some n.

(i) If $\boldsymbol{q}_f'$ is a PE-rewriting for $\boldsymbol{q}_f$ and $\mathcal{T}_f$ then there is a monotone Boolean formula ψ_f computing f with $|\psi_f| \leq |\boldsymbol{q}_f'|$.

(ii) If $\boldsymbol{q}_f'$ is an FO-rewriting for $\boldsymbol{q}_f$ and $\mathcal{T}_f$ over a signature with a single constant then there is a Boolean formula χ_f computing f with $|\chi_f| \leq |\boldsymbol{q}_f'|$.

(iii) If (Π_f, G) is an NDL-rewriting for $\boldsymbol{q}_f$ and $\mathcal{T}_f$ then there is a monotone Boolean circuit $\mathbf{C}_f$ computing f with $|\mathbf{C}_f| \leq |\Pi_f|$.

The proof proceeds by eliminating quantifiers from the given rewriting and replacing its predicates with propositional variables using the fact that, in the ABoxes $\mathcal{A}_{\alpha}$, these predicates can only be true on the individual a. Lemmas 1 and 2 ensure that the resulting Boolean formula or circuit computes f.

The next lemma shows that, conversely, circuits computing f can be turned into rewritings for $\boldsymbol{q}_f$ and $\mathcal{T}_f$ over ABoxes with a single individual.

Lemma 4. Let $f^1, f^2, \ldots$ be a family of monotone Boolean functions in NP, and let $f = f^n$, for some n. The following holds for signatures Σ with a single constant:

(i) Suppose $\boldsymbol{q}'$ is an FO-sentence such that $(\mathcal{T}_f, \mathcal{A}_{\alpha}) \models \boldsymbol{q}_f$ iff $\mathcal{I}_{\mathcal{A}_{\alpha}} \models \boldsymbol{q}'$, for all α. Then

$$q'' = \exists x \left[A_0(x) \wedge \left(q' \vee \bigvee_{A_0 \sqcap B \sqsubseteq_{\mathcal{T}_f} \perp} B(x) \right) \right]$$

is an FO-rewriting for $\boldsymbol{q}_f$ and $\mathcal{T}_f$ with $|q''| = |q'| + O(|\mathbf{C}^n|^2)$.

(ii) Suppose (Π, G) is a pure NDL query with a propositional goal G such that $(\mathcal{T}_f, \mathcal{A}_{\alpha}) \models \boldsymbol{q}_f$ iff $\Pi, \mathcal{A}_{\alpha} \models G$, for all α. Then (Π', G') is an NDL-rewriting for $\mathcal{T}_f$ and $\boldsymbol{q}_f$ with $|\Pi'| = |\Pi| + O(|\mathbf{C}^n|^2)$, where G' is a fresh propositional variable and Π' is obtained by extending Π with the following rules:

- $\forall x \, (A_0(x) \wedge G \to G')$,
- $\forall x \, (A_0(x) \wedge B(x) \to G')$, for all concepts B such that $A_0 \sqcap B \sqsubseteq_{\mathcal{T}_f} \bot$.

(In both statements above, $B(x)$ denotes $\exists y \, P(x,y)$ in the case of $B = \exists P$.)

We are in a position now to formulate our main theorem that connects the size of circuits computing monotone Boolean functions with the size of rewritings for the corresponding queries and ontologies. It follows from Lemmas 1–4.

Theorem 2. *For any family $f^1, f^2, \ldots$ of monotone Boolean functions in* NP, *there exist polynomial-size CQs $\boldsymbol{q}_n$ and OWL2QL TBoxes $\mathcal{T}_n$ such that the following holds:*

(1) *Let $L(n)$ be a lower bound for the size of monotone Boolean formulas computing f^n. Then $|\boldsymbol{q}_n'| \geq L(n)$, for any PE-rewriting $\boldsymbol{q}_n'$ for $\boldsymbol{q}_n$ and $\mathcal{T}_n$.*

(2) *Let $L(n)$ and $U(n)$ be a lower and an upper bound for the size of monotone Boolean circuits computing f^n. Then*
  - *$|\Pi_n| \geq L(n)$, for any NDL-rewriting (Π_n, G) for $\boldsymbol{q}_n$ and $\mathcal{T}_n$;*
  - *there exist a polynomial p and an NDL-rewriting (Π_n, G) for $\boldsymbol{q}_n$ and $\mathcal{T}_n$ over a signature with a single constant such that $|\Pi_n| \leq U(n) + p(n)$.*

(3) *Let $L(n)$ and $U(n)$ be a lower and an upper bound for the size of Boolean formulas computing f^n. Then*
  - *$|\boldsymbol{q}_n'| \geq L(n)$, for any FO-rewriting $\boldsymbol{q}_n'$ for $\boldsymbol{q}_n$ and $\mathcal{T}_n$ over any signature with a single constant;*
  - *there exist a polynomial p and an FO-rewriting $\boldsymbol{q}_n'$ for $\boldsymbol{q}_n$ and $\mathcal{T}_n$ over a signature with a single constant such that $|\boldsymbol{q}_n'| \leq U(n) + p(n)$.*

4   Rewritings Long and Short

We apply Theorem 2 to three concrete families of Boolean functions and show that some queries and ontologies may only have very long rewritings, and some rewritings can be exponentially or superpolynomially more succinct than others.

First we prove that one cannot avoid an exponential blow-up for PE- and NDL-rewritings; moreover, even FO-rewritings can blow-up superpolynomially for signatures with a single constant under the assumption that NP $\not\subseteq$ P/poly (i.e., that NP-complete problems cannot be solved by polynomial-size circuits, which is an open problem; see, e.g., [3]). This can be done using the function CLIQUE$_{n,k}$ of $n(n-1)/2$ variables e_{ij}, $1 \leq i < j \leq n$, which returns 1 iff the graph with vertices $\{1, \ldots, n\}$ and edges $\{\{i,j\} \mid e_{ij} = 1\}$ contains a k-clique. A series of papers, started by Razborov's [22], gave an exponential lower bound for the size of monotone circuits computing CLIQUE$_{n,k}$: $2^{\Omega(\sqrt{k})}$ for $k \leq \frac{1}{4}(n/\log n)^{2/3}$ [2]. For monotone formulas, an even better lower bound is known: $2^{\Omega(k)}$ for $k = 2n/3$ [21]. One can show that there is a nondeterministic circuit with n advice inputs and $O(n^2)$ gates that computes CLIQUE$_{n,k}$. As CLIQUE$_{n,k}$ is NP-complete, the question whether CLIQUE$_{n,k}$ can be computed by a polynomial-size deterministic circuit is equivalent to NP $\subseteq$ P/poly.

Theorem 3. *There is a sequence of CQs q_n of size $O(n)$ and OWL 2 QL TBoxes $\mathcal{T}_n$ of size $O(n)$ such that:*

- *any PE-rewriting for q_n and $\mathcal{T}_n$ is of size $\geq 2^{\Omega(n^{1/4})}$;*
- *any NDL-rewriting for q_n and $\mathcal{T}_n$ is of size $\geq 2^{\Omega((n/\log n)^{1/12})}$;*
- *there does not exist a polynomial-size FO-rewriting for q_n and $\mathcal{T}_n$ over a signature with a single constant unless* NP $\subseteq$ P/poly.

By the Karp-Lipton theorem (see, e.g., [3]), NP $\subseteq$ P/poly implies PH $= \Sigma_2^p$. So we can replace the assumption NP $\not\subseteq$ P/poly with PH $\neq \Sigma_2^p$.

The next result shows that NDL-rewritings can be exponentially more succinct than PE-rewritings. Here we use the function GEN_{n^3} of n^3 variables x_{ijk}, $1 \leq i, j, k \leq n$, defined as follows. We say that 1 *generates* $k \leq n$ if either $k = 1$ or $x_{ijk} = 1$ and 1 generates both i and j. $\mathrm{GEN}_{n^3}(x_{111}, \ldots, x_{nnn})$ returns 1 iff 1 generates n. GEN_{n^3} is clearly a monotone Boolean function computable by polynomial-size monotone circuits. On the other hand, any monotone formula computing GEN_{n^3} is of size 2^{n^ε}, for some $\varepsilon > 0$ [20].

Theorem 4. *There is a sequence of CQs q_n of size $O(n)$ and OWL 2 QL TBoxes $\mathcal{T}_n$ of size $O(n)$ for which there exists a polynomial-size NDL-rewriting over a signature with a single constant, but any PE-rewriting over this signature is of size $\geq 2^{n^\varepsilon}$, for some $\varepsilon > 0$.*

Finally, we show that FO-rewritings can be superpolynomially more succinct than PE-rewritings. We use the function $\mathrm{MATCHING}_{2n}$ with n^2 variables e_{ij}, $1 \leq i, j \leq n$, which returns 1 iff there is a *perfect matching* in the bipartite graph G with vertices $\{v_1^1, \ldots, v_n^1, v_1^2, \ldots, v_n^2\}$ and edges $\{\{v_i^1, v_j^2\} \mid e_{ij} = 1\}$, i.e., a subset E of edges in G such that every node in G occurs exactly once in E. An exponential lower bound $2^{\Omega(n)}$ for the size of monotone formulas computing $\mathrm{MATCHING}_{2n}$ was obtained in [21]. However, there are non-monotone formulas of size $n^{O(\log n)}$ computing this function [8]. On the other hand, it can also be computed by a nondeterministic circuit with n^2 advice inputs and $O(n^2)$ gates.

Theorem 5. *There is a sequence of CQs q_n of size $O(n)$ and OWL 2 QL TBoxes $\mathcal{T}_n$ of size $O(n)$ which has a polynomial-size FO-rewriting over a signature with a single constant, but any PE-rewriting over this signature is of size $\geq 2^{\Omega(2^{\log^{1/2} n})}$.*

In the proof of Theorem 3, we used the CQs $q_n = q_{\mathrm{CLIQUE}_{m,k}}$ containing no constant symbols. It follows that the theorem will still hold if we allow the built-in predicates $=$ and $\neq$ in the rewritings, but disallow the use of constants that *do not occur in the original query*. The situation changes drastically if $=$, $\neq$ and two additional constants, say 0 and 1, are allowed in the rewritings. As shown by Gottlob and Schwentick [13], in this case there is a polynomial-size NDL-rewriting for any CQ and OWL 2 QL TBox. Roughly, the rewriting uses the extra resources to encode in a succinct way the part of the canonical model that is relevant to answering the given query. We call rewritings of this kind *impure* (indicating thereby that they use predicates and constants that do not occur in the original query and ontology). In fact, using the ideas of [5] and [13], one can construct an impure polynomial-size PE-rewriting for any CQ and $OWL\,2\,QL$ TBox. Thus, we obtain the following:

Theorem 6. *Impure PE- and NDL-rewritings for CQs and OWL 2 QL ontologies are exponentially more succinct than pure PE- and NDL-rewritings.*

The difference between short impure and long pure rewritings appears to be of the same kind as the difference between deterministic and nondeterministic Boolean circuits: the impure rewritings can guess (using $=$, 0 and 1) what the pure ones must specify explicitly. It is not clear, however, how the RDBMSs are going to cope with such guesses in practice.

5   Conclusion

The exponential lower bounds for the size of 'pure' rewritings above may look discouraging in the OBDA context. It is to be noted, however, that the ontologies and queries used in their proofs are extremely 'artificial' and never occur in practice (see the analysis in [16]). As demonstrated by the existing description logic reasoners (such as FaCT++, HermiT, Pellet, Racer), *real-world* ontologies can be classified efficiently despite the high worst-case complexity of the classification problem. We believe that practical query answering over $OWL\,2\,QL$ ontologies can be feasible if supported by suitable optimisation and indexing techniques. It also remains to be seen whether polynomial impure rewritings can be used in practice. We conclude the paper by mentioning two open problems. Our exponential lower bounds were proved for a sequence of pairs $(q_n, \mathcal{T}_n)$. It is unclear whether these bounds hold uniformly for all q_n over the same $\mathcal{T}$:

Question 1. Do there exist an $OWL\,2\,QL$ TBox $\mathcal{T}$ and CQs q_n such that any pure PE- or NDL-rewritings for q_n and $\mathcal{T}$ are of exponential size?

As we saw, both FO- and NDL-rewritings are more succinct than PE-rewritings.

Question 2. What is the relation between the size of FO- and NDL-rewritings?

Acknowledgments. We thank the anonymous referees for their constructive feedback and suggestions. This paper was supported by the U.K. EPSRC grant EP/H05099X.

References

1. Acciarri, A., Calvanese, D., De Giacomo, G., Lembo, D., Lenzerini, M., Palmieri, M., Rosati, R.: QuOnto: QUerying ONTOlogies. In: Proc. of the 20th Nat. Conf. on Artificial Intelligence (AAAI 2005), pp. 1670–1671 (2005)
2. Alon, N., Boppana, R.: The monotone circuit complexity of Boolean functions. Combinatorica 7(1), 1–22 (1987)
3. Arora, S., Barak, B.: Computational Complexity: A Modern Approach, 1st edn. Cambridge University Press, New York (2009)

4. Artale, A., Calvanese, D., Kontchakov, R., Zakharyaschev, M.: The DL-Lite family and relations. J. of Artificial Intelligence Research (JAIR) 36, 1–69 (2009)
5. Avigad, J.: Eliminating definitions and Skolem functions in first-order logic. In: Proc. of LICS, pp. 139–146 (2001)
6. Baader, F., Calvanese, D., McGuinness, D., Nardi, D., Patel-Schneider, P.(eds.): The Description Logic Handbook: Theory, Implementation and Applications. Cambridge University Press (2003)
7. Benedikt, M., Gottlob, G.: The impact of virtual views on containment. PVLDB 3(1), 297–308 (2010)
8. Borodin, A., von zur Gathen, J., Hopcroft, J.E.: Fast parallel matrix and gcd computations. In: Proc. of FOCS, pp. 65–71 (1982)
9. Calì, A., Gottlob, G., Lukasiewicz, T.: A general Datalog-based framework for tractable query answering over ontologies. In: Proc. of PODS, pp. 77–86 (2009)
10. Calì, A., Gottlob, G., Pieris, A.: Advanced processing for ontological queries. PVLDB 3(1), 554–565 (2010)
11. Calvanese, D., De Giacomo, G., Lembo, D., Lenzerini, M., Rosati, R.: Tractable reasoning and efficient query answering in description logics: The DL-Lite family. J. of Automated Reasoning 39(3), 385–429 (2007)
12. Gottlob, G., Orsi, G., Pieris, A.: Ontological queries: Rewriting and optimization. In: Proc. of the IEEE Int. Conf. on Data Engineering, ICDE (2011)
13. Gottlob, G., Schwentick, T.: Rewriting ontological queries into small nonrecursive Datalog programs. In: Proc. of DL 2011, vol. 745. CEUR-WS.org (2011)
14. Jukna, S.: Boolean Function Complexity: Advances and Frontiers. Springer (2012)
15. Kikot, S., Kontchakov, R., Zakharyaschev, M.: On (in)tractability of OBDA with OWL 2 QL. In: Proc. of DL 2011, vol. 745. CEUR-WS.org (2011)
16. Kikot, S., Kontchakov, R., Zakharyaschev, M.: Conjunctive query answering with OWL 2 QL. In: Proc. of the 13th Int. Conf. KR 2012. AAAI Press (2012)
17. Kontchakov, R., Lutz, C., Toman, D., Wolter, F., Zakharyaschev, M.: The combined approach to query answering in DL-Lite. In: Proc. of the 12th Int. Conf. KR 2010. AAAI Press (2010)
18. Lutz, C., Toman, D., Wolter, F.: Conjunctive query answering in the description logic EL using a relational database system. In: Proc. of the 21st Int. Joint Conf. on Artificial Intelligence, IJCAI 2009, pp. 2070–2075 (2009)
19. Pérez-Urbina, H., Motik, B., Horrocks, I.: A comparison of query rewriting techniques for DL-Lite. In: Proc. of DL 2009, vol. 477. CEUR-WS.org (2009)
20. Raz, R., McKenzie, P.: Separation of the monotone NC hierarchy. In: Proc. of FOCS, pp. 234–243 (1997)
21. Raz, R., Wigderson, A.: Monotone circuits for matching require linear depth. J. ACM 39(3), 736–744 (1992)
22. Razborov, A.: Lower bounds for the monotone complexity of some Boolean functions. Dokl. Akad. Nauk SSSR 281(4), 798–801 (1985)
23. Rodríguez-Muro, M., Calvanese, D.: Dependencies to optimize ontology based data access. In: Proc. of DL 2011, vol. 745. CEUR-WS.org (2011)
24. Rodríguez-Muro, M., Calvanese, D.: Semantic index: Scalable query answering without forward chaining or exponential rewritings. In: Proc. of the 10th Int. Semantic Web Conf., ISWC (2011)
25. Rodríguez-Muro, M., Calvanese, D.: High performance query answering over DL-Lite ontologies. In: Proc. of the 13th Int. Conf. KR 2012. AAAI Press (2012)
26. Rosati, R., Almatelli, A.: Improving query answering over DL-Lite ontologies. In: Proc. of the 12th Int. Conf. KR 2010. AAAI Press (2010)
27. Tseitin, G.: On the complexity of derivation in propositional calculus. In: Automation of Reasoning 2: Classical Papers on Computational Logic 1967-1970 (1983)

Lattices of Logical Fragments over Words

(Extended Abstract)

Manfred Kufleitner[*] and Alexander Lauser[*]

FMI, University of Stuttgart, Germany
{kufleitner,lauser}@fmi.uni-stuttgart.de

Abstract. This paper introduces an abstract notion of fragments of monadic second-order logic. This concept is based on purely syntactic closure properties. We show that over finite words, every logical fragment defines a lattice of languages with certain closure properties. Among these closure properties are residuals and inverse C-morphisms. Here, depending on certain closure properties of the fragment, C is the family of arbitrary, non-erasing, length-preserving, length-multiplying, or length-reducing morphisms. In particular, definability in a certain fragment can often be characterized in terms of the syntactic morphism. This work extends a result of Straubing in which he investigated certain restrictions of first-order formulae.

As motivating examples, we present *(1)* a fragment which captures the stutter-invariant part of piecewise-testable languages and *(2)* an acyclic fragment of Σ_2. As it turns out, the latter has the same expressive power as two-variable first-order logic FO^2.

1 Introduction

A famous result of Büchi, Elgot, and Trakhtenbrot states that a language of finite words is regular if and only if it is definable in monadic second-order logic [1,7,25]. Later McNaughton and Papert considered first-order logic. They showed that a language is definable in first-order logic if and only if it is star-free [12]. It turned out that the class of first-order definable languages has a huge number of other characterizations; *cf.* [4]. Intuitively, a first-order definable language is easier to describe than a language which is not first-order definable. This leads to a natural notion of descriptive complexity inside the class of regular languages: The simpler the formula to describe a language, the simpler the language. Pursuing this approach, there are several possible restrictions for formulae which come to mind. For example, one can restrict the quantifier depth, the alternation depth, the number of variables, the set of atomic predicates, or the set of quantifiers, just to name a few. There are several problems connected with this approach towards descriptive complexity inside the class of regular languages. First, simplicity of logical formulae is of course not a linear measure. And secondly, how do we test whether some language is definable in a given (infinite) class of formulae?

[*] The authors acknowledge the support by the German Research Foundation (DFG) under grant DI 435/5-1. A full version of this paper is available on-line [10].

A. Czumaj et al. (Eds.): ICALP 2012, Part II, LNCS 7392, pp. 275–286, 2012.
© Springer-Verlag Berlin Heidelberg 2012

In view of the first problem, in some cases one can nonetheless compare the expressive power of classes of formulae. Trivially, if a class of formulae is contained in another class of formulae, then we also have containment for the respective classes of languages. In some cases however, surprising inclusions and equivalences between syntactically incomparable fragments are known. For example, Thérien and Wilke have shown that a language is definable in first-order logic FO^2 with only two different names for the variables if and only if it is definable with one quantifier alternation [23]. Note that FO^2 is a natural restriction since first-order logic with three variables already has the full expressive power of first-order logic [9].

The situation for the second problem is entirely different. Even though there cannot be an algorithm which works for all classes of formulae, there is a very general approach to decide definability in some classes. This technique relies on effective algebraic characterizations. Schützenberger has shown that a language is star-free if and only if its syntactic monoid is aperiodic [16]. Together with the result of McNaughton and Papert, this yields an effective characterization of definability in first-order logic, *i.e.*, for a given regular language one can check whether this language is definable in first-order logic. This kind of correspondence between languages and finite monoids is formalized in Eilenberg's Variety Theorem [6], *e.g.*, star-free languages correspond to finite aperiodic monoids. The main idea is the following. If a class of languages $\mathcal{V}$ has certain closure properties, then there exists a class of finite monoids $\mathbf{V}$ such that a language is in $\mathcal{V}$ if and only if its syntactic monoid is in $\mathbf{V}$. Now, if membership in $\mathbf{V}$ is decidable, then membership in $\mathcal{V}$ becomes decidable because the syntactic monoid can be computed effectively. The closure properties required by Eilenberg's Variety Theorem are Boolean operations, residuals, and inverse morphisms. A class of languages with these closure properties is called a *variety*. There are several variants and extensions of this approach. Pin has shown that there is an Eilenberg correspondence between positive varieties and ordered monoids [13]. A *positive variety* is a class of languages closed under positive Boolean operations, residuals, and inverse morphisms. A $\mathcal{C}$-variety (for some class of morphisms $\mathcal{C}$) is a class of languages closed under Boolean operations, residuals, and inverse $\mathcal{C}$-morphisms. Straubing has given an Eilenberg correspondence between $\mathcal{C}$-varieties and so-called *stamps* [19]. Here, decidability results usually rely on the syntactic morphism and not solely on the syntactic monoid. The work of Straubing was later extended by equational theories, positive $\mathcal{C}$-varieties and a wreath product for $\mathcal{C}$-varieties [2,11,15]. The most extensive generalization of Eilenberg's Variety Theorem is due to Gehrke, Grigorieff, and Pin [8]. They have shown that so-called lattices of languages admit an equational description. A *lattice* is a class of languages closed under positive Boolean operations. Depending on other closure properties such as residuals, the equational description of a lattice can be tightened.

So, in order to apply the existing algebraic framework to a class of formulae, it is important that the resulting class of languages have closure properties like (positive) Boolean operations, residuals, and inverse ($\mathcal{C}$-)morphisms.

In this paper, we introduce a formal notion of *logical fragment* such that language classes defined by fragments admit such closure properties. In addition, almost all logical fragments found in the literature also form fragments in the sense of this paper. Our framework for formal logic is monadic second-order logic over words on a broad base of atoms; it exhausts most variants of first-order logic and monadic second-order logic over words found in the literature, *cf.* [3,5,14,18,19,20,21,22,24]; it includes atomic predicates for order, successor, and modular predicates as well as quantifiers for first-order, second-order and modular counting quantification.

A very general result giving closure properties of logical fragments is due to Straubing [19, Theorem 3]. For several combinations of restrictions within first-order logic (quantifier depth, number of variables, numerical predicates, and set of quantifiers), he showed closure under residuals and inverse $\mathcal{C}$-morphisms. This result can also be derived from our main Theorem 1. Closure properties of logical fragments are often either obtained indirectly (*i.e.*, by showing equivalence with some class of languages for which the closure properties are already known), or rely on methods such as Ehrenfeucht-Fraïssé games; see *e.g.* [18]. However, sometimes Ehrenfeucht-Fraïssé games are difficult to apply, for example if modular quantifiers are involved. In contrast, our syntactic transformation within fragments, directly give the desired closure properties without reverting to Ehrenfeucht-Fraïssé games. In particular, modular quantifiers are handled uniformly as one of many cases of how to compose formulae. Moreover, the proof, which can be found in the full version of this paper [10], is formulated so as to be easily extensible to fragments not considered in this paper.

Finally, we consider two examples which illustrate the formal notion of *fragments* introduced in this paper. Neither of these examples can be easily described using traditional Ehrenfeucht-Fraïssé games. The first example is $\mathbb{B}\Sigma_1[\leq]$, *i.e.*, Boolean combinations of existential negation-free first-order formulae using $\leq$ as the only binary predicate. This leads to stutter-invariant piecewise testable languages. The second example is the fragment of Σ_2-formulae with an acyclic comparison graph. This fragment is expressively complete for two-variable first-order logic FO^2. The vertices of the comparison graph are the variables and the edges reflect the comparisons. The resulting characterization of FO^2 is complementary to the result of Thérien and Wilke who showed that FO^2 and the "semantic" fragment Δ_2 of Σ_2 have the same expressive power [23].

2   Preliminaries

A *language* over an alphabet A is a subset of finite words in A^*. The empty word is 1 and $A^+ = A^* \setminus \{1\}$ is the set of finite nonempty words over A. The set A^* of finite words over A is the free monoid generated by A. It is *finitely generated* if A is a finite set. A *residual* of a language $L \subseteq A^*$ is a language of the form $u^{-1}Lv^{-1} = \{w \in A^* \mid uwv \in L\}$ where $u, v \in A^*$. It is a *left residual* if $v = 1$ and it is a *right residual* if $u = 1$. Let $h : B^* \to A^*$ be a morphism between free monoids. The *inverse image* $h^{-1}(L)$ of L under h is the language

$h^{-1}(L) = \{w \in B^* \mid h(w) \in L\}$ over B. The morphism h is *non-erasing* (respectively, *length-reducing, length-preserving*) if for all $b \in B$ we have $h(b) \in A^+$ (respectively, $h(b) \in A \cup \{1\}$, $h(b) \in A$). The morphism h is *length-multiplying* if there exists $m \in \mathbb{N}$ such that $h(b) \in A^m$ for all $b \in B$. Note that by the universal property of free monoids, a morphism between free monoids is completely determined by its images of the letters. If $\mathcal{C}$ is a family of morphisms, then h is a $\mathcal{C}$-*morphism* if $h \in \mathcal{C}$. We introduce the following families of morphisms: all morphisms $\mathcal{C}_{all}$ between finitely generated free monoids, non-erasing morphisms $\mathcal{C}_{ne}$, length-multiplying morphisms $\mathcal{C}_{lm}$, length-reducing morphisms $\mathcal{C}_{lr}$, length-preserving morphisms $\mathcal{C}_{lp}$.

Logic over Words. We consider *monadic second-order logic* interpreted over finite words. In the context of logic, words are viewed as labeled linear orders. Positions are positive integers with 1 being the first position. Labels come from a fixed countable universe of letters Λ. The set of variables is $\mathbb{V}_1 \dot\cup \mathbb{V}_2$ where $\mathbb{V}_1$ is an infinite set of first-order variables and $\mathbb{V}_2$ is an infinite set of second-order variables. First-order variables range over positions of the word and are denoted by lowercase letters (*e.g.*, $x, y, x_i \in \mathbb{V}_1$) whereas second-order variables range over subsets of positions and are denoted by uppercase letters (*e.g.*, $X, Y, X_i \in \mathbb{V}_2$). Atomic formulae include

- the constants $\top$ (for true) and $\bot$ (for false),
- the 0-ary predicate "empty" which is only true for the empty model,
- the label predicate $\lambda(x) = a$ holds if the position of x is labeled with $a \in \Lambda$,
- the second-order predicate $x \in X$ which is true if x is contained in X

and the following numerical predicates:

- the first-order equality predicate $x = y$,
- the (strict and non-strict) order predicates $x < y$ and $x \leq y$,
- the successor predicate $\mathrm{suc}(x, y)$ with the interpretation $x + 1 = y$,
- the minimum and maximum predicate $\min(x)$ and $\max(x)$ which identify the first and the last position, respectively,
- the modular predicate $x \equiv r \pmod{q}$ which is true if the position of x is congruent to r modulo q.

Formulae φ and ψ can be composed by the Boolean connectives (*i.e.*, by negation $\neg\varphi$, disjunction $\varphi \vee \psi$, and conjunction $\varphi \wedge \psi$), by existential and universal first-order quantification $\exists x\, \varphi$ and $\forall x\, \varphi$, by existential and universal second-order quantification $\exists X\, \varphi$ and $\forall X\, \varphi$, and by modular counting quantification $\exists^{r \bmod q} x\, \varphi$. The latter is true if, modulo q, there are r positions for x which make φ true. Parentheses may be used for disambiguation and to increase readability. The set $\mathrm{FV}(\varphi) \subseteq \mathbb{V}_1 \cup \mathbb{V}_2$ of *free variables* of φ is defined as usual. A *sentence* is a formula without free variables.

We only give a sketch of the *formal semantics* of formulae. A precise definition can be found in the full version of the paper [10]. In the course of the evaluation of a formula, it is necessary to handle formulae with free variables. The idea is to encode their interpretation by enlarging the alphabet to include sets of variables.

A first-order variable evaluates to a position i if the label of i contains the name of this variable. Similarly, a position i is contained in the evaluation of a second-order variable if the variable name is contained in the label of i. Specifically, if φ is a formula and V is a set of variables such that $FV(\varphi) \subseteq V$, then the semantics $[\![\varphi]\!]_V$ is a set of words $w = (a_1, J_1) \cdots (a_n, J_n)$ with $a_i \in \Lambda$ and $J_i \subseteq V$ such that for every first-order variable x there exists exactly one position i such that $x \in J_i$. The *interpretation* of a free first-order variable x is then given by $x(w) = i$ for this unique index. For a second-order variable X the interpretation $X(w) = \{i \in \{1, \ldots, n\} \mid X \in J_i\}$ is the set of positions containing X. With this, it is straightforward to define the semantics so as to coincide with the intuition given above.

We define the following particular classes of formulae:

- MSO is the class of all formulae without the quantifier $\exists^{r \bmod q}$.

- FOMOD is the class of all first-order formulae including modular quantifiers (*i.e.*, without second-order variables).

- FO is the class of all formulae in FOMOD without the quantifier $\exists^{r \bmod q}$.

Let $\mathcal{F}$ be a class of formulae. For a set $\mathcal{P} \subseteq \{\text{empty}, <, \leq, =, \text{suc}, \min, \max, \equiv\}$ of predicates denote by $\mathcal{F}[\mathcal{P}]$ the class of formulae in $\mathcal{F}$ which (apart from $\top$, $\bot$, labels, and atomic formulae of the form $x \in X$) only use predicates in $\mathcal{P}$. We also say that $\mathcal{F}$ *contains* some specific predicate (like the successor predicate) if there is a formula in $\mathcal{F}$ which uses this predicate.

Logic and Languages. For a sentence φ and an alphabet $A \subseteq \Lambda$ the *language (over A) defined by* φ is $L_A(\varphi) = \{a_1 \cdots a_n \mid a_i \in A, (a_1, \emptyset) \cdots (a_n, \emptyset) \in [\![\varphi]\!]_\emptyset\}$. It is the projection onto the letters A of the elements in $[\![\varphi]\!]_\emptyset$. If the alphabet is clear from the context, then we drop it from the subscript and write $L(\varphi)$. For a class of formulae $\mathcal{F}$ the *class of languages $\mathcal{L}(\mathcal{F})$ defined by $\mathcal{F}$* maps every finite alphabet $A \subseteq \Lambda$ to the set

$$\mathcal{L}_A(\mathcal{F}) = \{L_A(\varphi) \mid \varphi \in \mathcal{F} \text{ is a sentence}\}$$

of languages over A. For a class of languages $\mathcal{G}$ the *class of languages defined by $\mathcal{F}$ over $\mathcal{G}$* is the class of languages mapping A to $\mathcal{L}_A(\mathcal{F}) \cap \mathcal{G}_A$. Specifically, the *class of languages defined by $\mathcal{F}$ over nonempty words* maps A to $\mathcal{L}_A(\mathcal{F}) \cap A^+$. Note that in $L_A(\varphi)$, the alphabet A and the set of labels used in a formula φ may well be incomparable; a label predicate $\lambda(x) = a$ with $a \notin A$ will always be false when considering the semantics over A. On the other hand, a formula need of course not use all labels of the alphabet over which structures are built. For example, consider the formula $\exists x \colon \lambda(x) = a$ requiring that there be an a-position. If $a \notin A$, then $L_A(\varphi) = \emptyset$ because all positions of a word w over A are non-a-positions; interpreted over the alphabet $A = \{a, b\}$ this formula defines the language $A^* a A^*$. This might seem unintuitive at first glance but allows a more uniform handling of languages over different alphabets and avoids tedious notation and many case-distinctions.

Fragments. In this section, fragments are introduced as classes of formulae with natural closure properties on the syntactic level. As we shall see in Section 3, these syntactic properties transfer to closure under natural semantic operations.

A *context* is a formula with a unique occurrence of an additional constant predicate ∘ (to be read as "hole"). It is *primitive* if it does not use any label predicate. We shall denote primitive contexts by μ and contexts that *a priori* need not be primitive by ν. The intuition is that ∘ is a place-holder where a formula can be plugged in. Let $\nu(\varphi)$ be the result of substituting φ for the unique occurrence of ∘ in ν. Contexts allow to elegantly describe subformulae as φ is a subformula of ψ if and only if there exists a context ν such that $\psi = \nu(\varphi)$.

Definition 1. *A* fragment $\mathcal{F}$ *is a nonempty class of formulae such that for all primitive contexts μ, all formulae φ, ψ, all $a \in \Lambda$ and all $x, y \in \mathbb{V}_1$:*

1. *If $\mu(\varphi) \in \mathcal{F}$, then $\mu(\top) \in \mathcal{F}$ and $\mu(\bot) \in \mathcal{F}$ and $\mu(\lambda(x) = a) \in \mathcal{F}$,*
2. *$\mu(\varphi \vee \psi) \in \mathcal{F}$ if and only if $\mu(\varphi) \in \mathcal{F}$ and $\mu(\psi) \in \mathcal{F}$,*
3. *$\mu(\varphi \wedge \psi) \in \mathcal{F}$ if and only if $\mu(\varphi) \in \mathcal{F}$ and $\mu(\psi) \in \mathcal{F}$,*
4. *if $\mu(\exists x\, \varphi) \in \mathcal{F}$ and $x \notin \mathrm{FV}(\varphi)$, then $\mu(\varphi) \in \mathcal{F}$.*

A class of formulae $\mathcal{F}$ is closed under negation *if $\varphi \in \mathcal{F}$ implies $\neg\varphi \in \mathcal{F}$.*   ◇

Easy examples of fragments include $\mathrm{FO}^r_{m,n}[\mathcal{P}]$ for all r, m, n and all subsets of predicates $\mathcal{P}$, i.e., the class of all first-order formulae using only r variables, having m blocks of quantifiers and quantifier depth n.

Next, we give an intuition for fragments in terms of local substitution operations. Let $\mathcal{F}$ be a class of formulae and let φ and ψ be formulae. The *syntactic preorder* of $\mathcal{F}$ is defined by $\varphi \leq_{\mathcal{F}} \psi$ if $\mu(\psi) \in \mathcal{F}$ implies $\mu(\varphi) \in \mathcal{F}$ for every primitive context μ. Intuitively $\varphi \leq_{\mathcal{F}} \psi$ means that, with respect to $\mathcal{F}$, the formula φ is syntactically not more "complicated" than ψ. Similarly, we let $\varphi \preceq_{\mathcal{F}} \psi$ if $\nu(\psi) \in \mathcal{F}$ implies $\nu(\varphi) \in \mathcal{F}$ for all contexts ν. The syntactic preorder allows to reformulate some of the axioms of a fragment. For example, property (1) in Definition 1 is equivalent to $\top \leq_{\mathcal{F}} \varphi$ and $\bot \leq_{\mathcal{F}} \varphi$ and $(\lambda(x) = a) \leq_{\mathcal{F}} \varphi$ for all formulae φ. Note that $\varphi \preceq_{\mathcal{F}} \psi$ implies $\varphi \leq_{\mathcal{F}} \psi$. The reverse is however not true for arbitrary classes of formulae. Let for example $\mathcal{F}$ consist of all formulae containing at most one label predicate. In this case, we have $(\lambda(x) = a) \leq_{\mathcal{F}} \top$. If ν is the context $\circ \wedge \lambda(x) = a$, then $\nu(\top) \in \mathcal{F}$ and $\nu(\lambda(x) = a) \notin \mathcal{F}$. Hence $(\lambda(x) = a) \npreceq_{\mathcal{F}} \top$. For fragments on the other hand, this cannot happen because here, $\leq_{\mathcal{F}}$ and $\preceq_{\mathcal{F}}$ are equivalent by the following lemma.

Lemma 1. *If $\mathcal{F}$ is a fragment, then $\varphi \leq_{\mathcal{F}} \psi$ if and only if $\varphi \preceq_{\mathcal{F}} \psi$.*

This provides an intuition for fragments: In a formula from a fragment $\mathcal{F}$, one may replace arbitrary subformulae by $\leq_{\mathcal{F}}$-smaller formulae without leaving $\mathcal{F}$. Note that this is not immediate from the definition of a fragment because in general, primitive contexts are not sufficient to formalize subformulae (as the "rest" of the formula may contain label predicates). On the other hand with finite words as structures in mind, it is also natural to attach an alphabet to a formula, namely the alphabet over which the word models are built; and in this case, primitive contexts do not interfere with the alphabet of the formula.

3   Fragments and C-Varieties

This section summarizes semantic closure properties of fragments. In Proposition 1 and Proposition 2 we give conditions for a fragment to be closed under residuals and inverse morphisms, respectively. The combination of these two propositions gives our main result Theorem 1 which formulates closure properties of languages defined by fragments in terms of C-varieties.

For closure under residuals we need some more assumptions. A fragment $\mathcal{F}$ is *suc-stable* if (1) $(x = y) \leq_{\mathcal{F}} \mathrm{suc}(x, y)$, if (2) empty $\leq_{\mathcal{F}} \max(x) \leq_{\mathcal{F}} \mathrm{suc}(x, y)$, and if (3) empty $\leq_{\mathcal{F}} \min(y) \leq_{\mathcal{F}} \mathrm{suc}(x, y)$.

It is *mod-stable* if (1) $\left(x \equiv s \pmod q \right) \leq_{\mathcal{F}} \left(x \equiv r \pmod q \right)$ for all $r, s \in \mathbb{Z}$, if (2) $\exists^{s \bmod q} x\, \varphi \leq_{\mathcal{F}} \exists^{r \bmod q} x\, \varphi$ for all $r, s \in \mathbb{Z}$, and if (3) $\varphi \leq_{\mathcal{F}} \exists^{r \bmod q} x\, \varphi$ and $\neg \varphi \leq_{\mathcal{F}} \exists^{r \bmod q} x\, \varphi$ whenever $x \notin \mathrm{FV}(\varphi)$.

Consider the left residual, *i.e.*, given a formula φ and a word w, we want to determine the truth value of φ on aw. Conceptually, setting variables to the "phantom" a-position in front of the word is handled syntactically resulting in a formula $a^{-1}\varphi$ defining the residual. With the above stability properties one can maintain the invariant $a^{-1}\varphi \leq_{\mathcal{F}} \varphi$. The actual construction is rather lengthy and can be found in the full version of the paper [10].

Proposition 1. *Let $\mathcal{F}$ be a fragment and suppose that $\mathcal{F}$ is suc-stable and mod-stable. Then the class of languages defined by $\mathcal{F}$ is closed under residuals.*

Note that if $\mathcal{F}$ does not contain suc, max or min, then $\mathcal{F}$ trivially is suc-stable. Similarly, $\mathcal{F}$ is mod-stable if it does not contain any modular predicate.

We now turn to closure under inverse morphisms. A fragment $\mathcal{F}$ is *order-stable* if $(x < y) \leq_{\mathcal{F}} (x \leq y)$ and $(x \leq y) \leq_{\mathcal{F}} (x < y)$. It is *MSO-stable* if (1) $(x \in Y) \leq_{\mathcal{F}} (x \in X)$, if (2) $\exists Y \exists X\, \varphi \leq_{\mathcal{F}} \exists X\, \varphi$, and if (3) $\forall Y \forall X\, \varphi \leq_{\mathcal{F}} \forall X\, \varphi$.

We obtain closure under inverse morphisms as follows. For every morphism $h : B^* \to A^*$ and every formula φ we construct a formula $h^{-1}(\varphi)$ defining the inverse morphic image of $L_A(\varphi)$ with $h^{-1}(\varphi) \leq_{\mathcal{F}} \varphi$. Basically, a position i on $h(w)$ can be represented by its corresponding position on w (called the origin of i) combined with some offset (bounded by the maximal length $|h(b)|$ for letters $b \in B$). For first-order variables the offset is stored syntactically and second-order variables are distributed over several variables, depending on the offset. As for residuals, the actual construction is technically involved, see [10].

Proposition 2. *Let $\mathcal{F}$ be a fragment and let C be a family of morphisms between finitely generated free monoids. Suppose the following:*

1. *If $\mathcal{F}$ contains a second-order quantifier, then $\mathcal{F}$ is MSO-stable or $C \subseteq C_{lr}$.*

2. *If $\mathcal{F}$ contains the predicate $\leq$ or $<$, then $\mathcal{F}$ is order-stable or $C \subseteq C_{lr}$.*

3. *If $\mathcal{F}$ contains the predicate suc, min, max or empty, then $C \subseteq C_{ne}$.*

4. *If $\mathcal{F}$ contains a modular predicate, then $C \subseteq C_{lm}$ and either $\mathcal{F}$ is mod-stable or $C \subseteq C_{lp}$.*

5. *If $\mathcal{F}$ contains a modular quantifier, then $\mathcal{F}$ is mod-stable or $C \subseteq C_{lr}$.*

Then the class of languages defined by $\mathcal{F}$ is closed under inverse C-morphisms.

Typically, if a fragment $\mathcal{F}$ contains more modalities, then either $\mathcal{F}$ has to satisfy more closure properties, or it is closed under fewer inverse morphisms. This trade-off between closure properties and inverse morphisms is given by the implications in Proposition 2, each implication covering certain modalities in $\mathcal{F}$. In particular, Proposition 2 states that every fragment is closed under length-preserving morphisms.

We now turn to $\mathcal{C}$-varieties of which we only give the definition; for details see [15,19]. A *category* $\mathcal{C}$ of morphisms between finitely generated free monoids is a family of morphisms between finitely generated free monoids which contains the identity morphisms and which is closed under composition. A *positive $\mathcal{C}$-variety* is a class of languages which is closed under positive Boolean combinations, residuals and inverse $\mathcal{C}$-morphisms. It is a *$\mathcal{C}$-variety* if it is closed under complement. Examples for categories of morphisms include $\mathcal{C}_{all}$, $\mathcal{C}_{ne}$, $\mathcal{C}_{lm}$, $\mathcal{C}_{lr}$, and $\mathcal{C}_{lp}$.

Our main result is the next theorem from which in particular the main results of a paper by Straubing can be obtained [19, Theorem 3]. Intuitively, the more closure properties some fragment $\mathcal{F}$ has, the larger is the class of inverse morphisms under which $\mathcal{L}(\mathcal{F})$ is closed. In Theorem 1 below this is formalized by a sequence of implications.

Theorem 1. *Let $\mathcal{F}$ be a mod-stable and suc-stable fragment. Let $\mathcal{C}$ be a category of morphisms between finitely generated free monoids. Suppose the following:*

1. *If $\mathcal{F}$ contains a second-order quantifier, then $\mathcal{F}$ is MSO-stable or $\mathcal{C} \subseteq \mathcal{C}_{lr}$.*
2. *If $\mathcal{F}$ contains the predicate $\leq$ or $<$, then $\mathcal{F}$ is order-stable or $\mathcal{C} \subseteq \mathcal{C}_{lr}$.*
3. *If $\mathcal{F}$ contains the predicate suc, min, max or empty, then $\mathcal{C} \subseteq \mathcal{C}_{ne}$.*
4. *If $\mathcal{F}$ contains a modular predicate, then $\mathcal{C} \subseteq \mathcal{C}_{lm}$.*

Then the class of languages defined by $\mathcal{F}$ is a positive $\mathcal{C}$-variety.

Proof. We have to show that $\mathcal{L}(\mathcal{F})$ is closed under union, intersection, residuals and inverse $\mathcal{C}$-morphisms. Using the primitive context $\circ$ it is easy to see that $\mathcal{F}$ is closed under disjunction and conjunction and, consequently, $\mathcal{L}(\mathcal{F})$ is closed under union and intersection. It remains to show that $\mathcal{L}(\mathcal{F})$ is closed under residuals and inverse $\mathcal{C}$-morphisms. Closure under residuals is Proposition 1 and closure under inverse $\mathcal{C}$-morphisms is Proposition 2. $\square$

A *(positive) $*$-variety* is a (positive) $\mathcal{C}_{all}$-variety and a *(positive) $+$-variety* is a (positive) $\mathcal{C}_{ne}$-variety of languages of nonempty words. We get the following corollaries for fragments using equality, order and successor. Note in particular that the predicate "empty" is void over nonempty words and that every first-order fragment trivially is MSO-stable.

Corollary 1. *Let $\mathcal{F} \subseteq \mathrm{MSO}[<, \leq, =]$ be a fragment which is MSO-stable and order-stable. Then $\mathcal{F}$ defines a positive $*$-variety.*

Corollary 2. *Let $\mathcal{F} \subseteq \mathrm{MSO}[<, \leq, =, \mathrm{suc}, \mathrm{min}, \mathrm{max}]$ be an MSO-stable and order-stable fragment. Suppose $\mathrm{min}(y) \leq_{\mathcal{F}} \mathrm{suc}(x, y)$ and $\mathrm{max}(x) \leq_{\mathcal{F}} \mathrm{suc}(x, y)$ for all first-order variables x and y. Then the class of languages defined by $\mathcal{F}$ over nonempty words forms a positive $+$-variety.*

4 Stutter-Invariant Piecewise Testable Languages

A language is a *simple monomial* if it is of the form $A^* a_1 \cdots A^* a_n A^*$. A language $L \subseteq A^*$ is *piecewise testable* if it is a finite Boolean combination of simple monomials. It is *stutter-invariant* if for all words $p, q \in A^*$ and all letters $a \in A$ we have $paq \in L$ if and only if $paaq \in L$.

Let Σ_1 consist of all FO-formulae without negations and without any universal quantifier. Let $\mathbb{B}\Sigma_1$ be the fragment which consists of all Boolean combinations of formulae in Σ_1. By Theorem 1, the class of languages definable in $\mathbb{B}\Sigma_1[\leq]$ forms a $\mathcal{C}_{lr}$-variety. The following proposition describes the class of languages definable in $\mathbb{B}\Sigma_1[\leq]$ in terms of stutter-invariant piecewise testable languages.

Theorem 2. *Let $L \subseteq A^*$ be a language. The following are equivalent:*

1. *L is definable in $\mathbb{B}\Sigma_1[\leq]$.*

2. *L is piecewise testable and stutter-invariant.*

3. *L is a Boolean combination of simple monomials of the form $A^* a_1 \cdots A^* a_n A^*$ with $a_i \neq a_{i+1}$ for all i.*

Proof. We first show "(1) $\Rightarrow$ (2)". If L is $\mathbb{B}\Sigma_1[\leq]$-definable, then of course L is $\mathbb{B}\Sigma_1[<, =]$-definable. The latter is equivalent to L being piecewise testable, see *e.g.* [5]. It is easy to see that the class of languages defined by $\Sigma_1[\leq]$ is stutter-invariant. The claim follows since stutter-invariant languages are closed under Boolean operations.

"(2) $\Rightarrow$ (3)": Let $L \subseteq A^*$ be piecewise testable and stutter-invariant. The intersection of two simple monomials is a finite union of simple monomials. Therefore, we can write $L = \bigcup_{i=1}^{s} (P_i \setminus \bigcup_{j=1}^{t} Q_{i,j})$ where the languages P_i and $Q_{i,j}$ are simple monomials. Suppose $P = (A^* a_1)^{e_1} \cdots (A^* a_n)^{e_n} A^*$ for positive integers e_i and letters $a_i \in A$ with $a_i \neq a_{i+1}$. Then $\mathrm{red}(P) = A^* a_1 \cdots A^* a_n A^*$ is the monomial obtained by discarding successive a_i's with the same label. Note that $\mathrm{red}(P)$ is stutter-invariant and $P \subseteq \mathrm{red}(P)$. It suffices to show $L = \bigcup_i (\mathrm{red}(P_i) \setminus \bigcup_j \mathrm{red}(Q_{i,j}))$. Note that $\bigcup_i (\mathrm{red}(P_i) \setminus \bigcup_j \mathrm{red}(Q_{i,j}))$ is stutter-invariant.

Assume $u_0 \in L$ and $u_0 \notin \bigcup_i (\mathrm{red}(P_i) \setminus \bigcup_j \mathrm{red}(Q_{i,j}))$. Let $I_0 = \emptyset$. We will construct a sequence of words $u_1, u_2, \ldots$ and subsets $I_0 \subsetneq I_1 \subsetneq I_2 \subsetneq \cdots$ of $\{1, \ldots, s\}$ such that each u_k is obtained from u_{k-1} by replacing factors $a \in A$ in u_{k-1} by aa. By stutter-invariance, we have $u_k \in L$ and $u_k \notin \bigcup_i (\mathrm{red}(P_i) \setminus \bigcup_j \mathrm{red}(Q_{i,j}))$ for all $k \geq 0$. Moreover, the index sets I_k will satisfy the invariant

$$u_k \in \bigcup_j Q_{i,j} \quad \text{for all } i \in I_k. \tag{$*$}$$

For $k = 0$, the word u_0 satisfies the invariant $(*)$ for I_0. Let now $k \geq 1$ and let u_{k-1} and I_{k-1} be already defined. Since $u_{k-1} \in L$ there exists $i \in \{1, \ldots, s\} \setminus I_{k-1}$ such that $u_{k-1} \in P_i$. Since $u_{k-1} \notin \mathrm{red}(P_i) \setminus \bigcup_j \mathrm{red}(Q_{i,j})$, we have $u_{k-1} \in \mathrm{red}(Q_{i,j})$ for some j. Let $\mathrm{red}(Q_{i,j}) = A^* a_1 \cdots A^* a_n A^*$ and let $u_{k-1} = v_0 a_1 \cdots v_{n-1} a_n v_n$. Then there exist positive integers e_i such that $u_k = v_0 a_1^{e_1} \cdots v_{n-1} a_n^{e_n} v_n \in Q_{i,j}$. We set $I_k = I_{k-1} \cup \{i\}$. Since u_{k-1} is a scattered subword of u_k we have $u_k \in \bigcup_j Q_{i',j}$ for all $i' \in I_{k-1}$. This establishes the invariant $(*)$ for u_k and I_k. This leads to a contradiction because the sets I_k are strictly increasing but their cardinality is bounded by s. We conclude $L \subseteq \bigcup_i (\mathrm{red}(P_i) \setminus \bigcup_j \mathrm{red}(Q_{i,j}))$.

For the remaining inclusion let $u \in \mathrm{red}(P_i) \setminus \bigcup_j \mathrm{red}(Q_{i,j})$. Let $\mathrm{red}(P_i) = A^* a_1 \cdots A^* a_n A^*$ and $u = v_0 a_1 \cdots v_{n-1} a_n v_n$. There exist positive integers e_i such that $u' = v_0 a_1^{e_1} \cdots v_{n-1} a_n^{e_n} v_n \in P_i$ and stutter-invariance of the languages $\mathrm{red}(Q_{i,j})$ yields $u' \notin \bigcup_j \mathrm{red}(Q_{i,j})$. In particular $u' \notin \bigcup_j Q_{i,j}$ and thus $u' \in L$. By stutter-invariance of L we obtain $u \in L$.

"$(3) \Rightarrow (1)$": Let $P = A^* a_1 \cdots A^* a_n A^*$ with $a_i \neq a_{i+1}$ for all i. Then P is defined by the formula $\exists x_1 \cdots \exists x_n : \bigwedge_{i=1}^n \lambda(x_i) = a_i \wedge \bigwedge_{i=1}^{n-1} x_i \leq x_{i+1}$. Note that in this formula, $x_i \leq x_{i+1}$ implies $x_i < x_{i+1}$ since $a_i \neq a_{i+1}$.   □

A famous result of Simon says that a language L is piecewise testable if and only if the syntactic monoid of L is finite and $\mathcal{J}$-trivial [17]. The latter property is decidable for finite monoids. Moreover, L is stutter-invariant if and only if the image of every letter under the syntactic morphism of L is idempotent. Combining these observations, (2) shows that it is decidable whether a given regular language is definable in $\mathbb{B}\Sigma_1[\leq]$.

5   The Acyclic Fragment of Σ_2

Let Σ_2 consist of all FO-formulae without negations such that no existential quantifier appears in the scope of a universal quantifier, *i.e.*, on every path in the parse-tree all existential quantifiers occur before all universal quantifiers. Prohibiting negations in propositional formulae is not a big restriction since, by De Morgan's laws, one can always assume that negations only occur in front of atomic formulae. However, this would complicate the definition of comparison graphs. The *comparison graph* of a formula φ is the directed graph $G(\varphi) = (V, E)$ with V being the set of variables occurring in φ and $(x, y) \in E$ if and only if one of the atomic formulae $x < y$, $x \leq y$, $x = y$ or $y = x$ occurs in φ. It is *acyclic* if there exist no $x_1, \ldots, x_n \in V$ such that $(x_i, x_{i+1}) \in E$ and $x_n = x_1$. It might seem unnatural to completely disallow negations in the definition of Σ_2. This is however not essential but eases the definition of the comparison graph; actually, negations would reverse the edges of the comparison graph.

Note that the class of formulae in $\Sigma_2[<, \leq]$ with an acyclic comparison graph forms an order-stable fragment thus defining a positive $*$-variety. In fact, the following proposition implies that it defines a $*$-variety even though, syntactically, it is not closed under negation.

Theorem 3. *A language is definable in* $\mathrm{FO}^2[<]$ *if and only if it is definable by a formula in* $\Sigma_2[<, \leq]$ *with an acyclic comparison graph.*

Proof. We only give an outline. The complete proof can be found in the full version of the paper [10]. The proof relies on two famous characterizations of the class of languages definable in $\mathrm{FO}^2[<]$. The first characterization is in terms of unions of unambiguous monomials and the second one is the variety **DA** of finite monoids; see [21,5].

A language of the form $P = A_1^* a_1 \cdots A_n^* a_n A_{n+1}^*$ with $a_i \in A$ and $A_i \subseteq A$ is called a *monomial*. It is *unambiguous* if every word $u \in P$ has a unique factorization $u = u_1 a_1 \cdots u_n a_n u_{n+1}$ with $u_i \in A_i^*$. For the direction from left to right, it suffices to show that every unambiguous monomial $P = A_1^* a_1 \cdots A_n^* a_n A_{n+1}^*$ is definable by a formula in $\Sigma_2[<, \leq]$ with an acyclic comparison graph. There exists some $a_i \notin A_1 \cap A_{n+1}$ since otherwise $(a_1 \cdots a_n)^2$ would admit two different factorizations. By symmetry, we can assume $a_i \notin A_1$. For every word $u \in P$ we consider the factorization $u = q a_i r$ such that a_i does not occur in the prefix q. Then q and r are contained in smaller unambiguous monomials $Q \subseteq (A \setminus \{a_i\})^*$ and $R \subseteq A^*$, respectively, such that $Q a_i R \subseteq P$. By induction, there exist formulae for Q and R. These formulae can be combined into a formula for P using so-called relativizations. The main idea here is that the position of the first a_i is unique and that several variables can be used to identify this position. This allows to maintain an acyclic comparison graph.

For the converse, we show that the syntactic monoid of $L(\varphi)$ is in **DA** if φ is in $\Sigma_2[<, \leq]$ with an acyclic comparison graph. For this, it suffices to show that for some sufficiently large integer $n \geq 1$, we have $p(uv)^n u(uv)^n q \in L(\varphi)$ if and only if $p(uv)^{3n} q \in L(\varphi)$ for all $p, q, u, v \in A^*$. It is easier to describe the outline of the proof using the terminology of Ehrenfeucht-Fraïssé games. We note that in this game, the winning condition is not defined in terms of isomorphisms of game situations and thus, it is not an Ehrenfeucht-Fraïssé game in the usual sense. Since every language definable in $FO^2[<]$ is also definable in $\Sigma_2[<, \leq]$, it follows that if Spoiler starts on the word $p(uv)^{3n} q$, then Duplicator wins for arbitrary comparison graphs [23]. Hence, Spoiler starts on $p(uv)^n u(uv)^n q$. Choosing n large enough, we know that after Spoiler placed his pebbles on $p(uv)^n u(uv)^n q$, there are large gaps to the left and to the right of the central factor u. Duplicator plays as follows: Pebbles outside the center are placed on the respective position on $p(uv)^{3n} q$. For the pebbles in the central part, Duplicators strategy basically is to make as many atomic formulae true on $p(uv)^{3n} q$ as possible. He can do this because the comparison graph is acyclic. In the second round Spoiler places his pebbles on $p(uv)^{3n} q$. Exploiting acyclicity again, Duplicator can use the gaps on $p(uv)^n u(uv)^n q$ to obtain a situation where as many atomic formulae as possible are false on $p(uv)^n u(uv)^n q$. The result is a situation such that if $x_i < x_j$ (respectively, $x_i \leq x_j$) on $p(uv)^n u(uv)^n q$ implies $x_i < x_j$ (respectively, $x_i \leq x_j$) on $p(uv)^{3n} q$. Hence, $p(uv)^n u(uv)^n q \in L$ implies $p(uv)^{3n} q \in L$. $\qquad\square$

Acknowledgments. We would like to thank the anonymous referees for many useful suggestions which helped to improve the presentation of the paper.

References

1. Büchi, J.R.: Weak second-order arithmetic and finite automata. Z. Math. Logik Grundlagen Math. 6, 66–92 (1960)
2. Chaubard, L., Pin, J.-É., Straubing, H.: Actions, wreath products of $\mathcal{C}$-varieties and concatenation product. Theor. Comput. Sci. 356, 73–89 (2006)

3. Chaubard, L., Pin, J.-É., Straubing, H.: First order formulas with modular predicates. In: LICS 2006, pp. 211–220. IEEE Computer Society (2006)
4. Diekert, V., Gastin, P.: First-order definable languages. In: Flum, J., Grädel, E., Wilke, T. (eds.) Logic and Automata: History and Perspectives. Texts in Logic and Games, pp. 261–306. Amsterdam University Press (2008)
5. Diekert, V., Gastin, P., Kufleitner, M.: A survey on small fragments of first-order logic over finite words. Int. J. Found. Comput. Sci. 19(3), 513–548 (2008)
6. Eilenberg, S.: Automata, Languages, and Machines, vol. B. Academic Press (1976)
7. Elgot, C.C.: Decision problems of finite automata design and related arithmetics. Trans. Amer. Math. Soc. 98, 21–51 (1961)
8. Gehrke, M., Grigorieff, S., Pin, J.-É.: Duality and Equational Theory of Regular Languages. In: Aceto, L., Damgård, I., Goldberg, L.A., Halldórsson, M.M., Ingólfsdóttir, A., Walukiewicz, I. (eds.) ICALP 2008, Part II. LNCS, vol. 5126, pp. 246–257. Springer, Heidelberg (2008)
9. Kamp, J.A.W.: Tense Logic and the Theory of Linear Order. PhD thesis, University of California (1968)
10. Kufleitner, M., Lauser, A.: Lattices of logical fragments over words. CoRR, abs/1202.3355 (2012), http://arxiv.org/abs/1202.3355
11. Kunc, M.: Equational description of pseudovarieties of homomorphisms. Theor. Inform. Appl. 37, 243–254 (2003)
12. McNaughton, R., Papert, S.: Counter-Free Automata. The MIT Press (1971)
13. Pin, J.-É.: A variety theorem without complementation. Russian Mathematics (Iz. VUZ) 39, 80–90 (1995)
14. Pin, J.-É.: Expressive power of existential first-order sentences of Büchi's sequential calculus. Discrete Math. 291(1-3), 155–174 (2005)
15. Pin, J.-É., Straubing, H.: Some results on C-varieties. Theor. Inform. Appl. 39, 239–262 (2005)
16. Schützenberger, M.P.: On finite monoids having only trivial subgroups. Inf. Control 8, 190–194 (1965)
17. Simon, I.: Piecewise Testable Events. In: Brakhage, H. (ed.) GI-Fachtagung 1975. LNCS, vol. 33, pp. 214–222. Springer, Heidelberg (1975)
18. Straubing, H.: Finite Automata, Formal Logic, and Circuit Complexity. Birkhäuser (1994)
19. Straubing, H.: On Logical Descriptions of Regular Languages. In: Rajsbaum, S. (ed.) LATIN 2002. LNCS, vol. 2286, pp. 528–538. Springer, Heidelberg (2002)
20. Straubing, H., Thérien, D., Thomas, W.: Regular languages defined with generalized quantifiers. Inf. Comput. 118(2), 289–301 (1995)
21. Tesson, P., Thérien, D.: Diamonds are forever: The variety DA. In: Gomes, G., et al. (eds.) Semigroups, Algorithms, Automata and Languages 2001, pp. 475–500. World Scientific (2002)
22. Tesson, P., Thérien, D.: Logic meets algebra: The case of regular languages. Log. Methods Comput. Sci. 3(1), 1–37 (2007)
23. Thérien, D., Wilke, T.: Over words, two variables are as powerful as one quantifier alternation. In: STOC 1998, pp. 234–240. ACM Press (1998)
24. Thomas, W.: Classifying regular events in symbolic logic. J. Comput. Syst. Sci. 25, 360–376 (1982)
25. Trakhtenbrot, B.A.: Finite automata and logic of monadic predicates. Dokl. Akad. Nauk SSSR 140, 326–329 (1961) (in Russian)

On the Expressive Power of Cost Logics over Infinite Words[*]

Denis Kuperberg[1] and Michael Vanden Boom[2]

[1] LIAFA/CNRS/Université Paris 7, Denis Diderot, France
denis.kuperberg@liafa.jussieu.fr
[2] Department of Computer Science, University of Oxford, England
michael.vandenboom@cs.ox.ac.uk

Abstract. Cost functions are defined as mappings from a domain like words or trees to $\mathbb{N} \cup \{\infty\}$, modulo an equivalence relation $\approx$ which ignores exact values but preserves boundedness properties. Cost logics, in particular cost monadic second-order logic, and cost automata, are different ways to define such functions. These logics and automata have been studied by Colcombet et al. as part of a "theory of regular cost functions", an extension of the theory of regular languages which retains robust equivalences, closure properties, and decidability. We develop this theory over infinite words, and show that the classical results FO = LTL and MSO = WMSO also hold in this cost setting (where the equivalence is now up to $\approx$). We also describe connections with forms of weak alternating automata with counters.

1 Introduction

The theory of regular cost functions is a quantitative extension to the theory of regular languages introduced by Colcombet [4]. Instead of languages being the centrepiece, functions from some set of structures (words or trees over some finite alphabet) to $\mathbb{N} \cup \{\infty\}$ are considered, modulo an equivalence relation $\approx$ which allows distortions but preserves boundedness over all subsets of the domain. Such functions are known as cost functions. This theory subsumes the classical theory of regular languages since a language can be associated with its characteristic function which maps every word (or tree) in the language to 0 and everything else to ∞; it is a strict extension since cost functions can count some behaviour within the input structure.

This theory grew out of two main lines of work: research by Hashiguchi [9], Kirsten [12], and others who were studying problems which could be reduced to whether or not some function was bounded over its domain (the most famous being the star height problem); and research by Bojańczyk and Colcombet [1,2]

---

[*] The full version of the paper can be found at
http://www.liafa.jussieu.fr/~dkuperbe/. The research leading to these results has received funding from the European Union's Seventh Framework Programme (FP7/2007-2013) under grant agreement 259454.

A. Czumaj et al. (Eds.): ICALP 2012, Part II, LNCS 7392, pp. 287–298, 2012.
© Springer-Verlag Berlin Heidelberg 2012

on extensions of monadic second-order logic (MSO) with a quantifier U which can assert properties related to boundedness.

Building on this work, this theory provides a framework which retains the robust closure properties, equivalences, and decidability results that regular languages enjoy. For instance, over finite words regular cost functions can be defined in terms of a cost logic (called cost monadic second-order logic, or CMSO), non-deterministic automata with counters (called B/S-automata), and algebra (called stabilisation semigroups). These relationships can be used to prove that it is decidable whether regular cost functions over finite words are equivalent, up to $\approx$ [4,2]. This decidability is also known over infinite words [3] and finite trees [6]. It is an important open problem on infinite trees: decidability of regular cost functions would imply decidability of the non-deterministic parity index [5].

In this paper, we develop this theory further by studying the expressivity of various cost logics over infinite words, namely the cost versions of linear temporal logic, first-order logic, monadic second-order logic, and weak monadic second-order logic, abbreviated CLTL, CFO, CMSO, and CWMSO, respectively. We also show connections with forms of weak alternating automata with counters.

1.1   Related Work and Motivation

Understanding the relationship between these cost logics and automata is desirable for application in model checking and other verification purposes. For instance, LTL can express "eventually some good condition holds (and this is true globally)". Unfortunately, it is also natural to want to bound the wait time before this good event occurs, but LTL provides no way to express this. Prompt LTL (introduced in [15]) can express this bounded wait time, and already gave rise to interesting decidability and complexity results. CLTL introduced in [13], is a strictly more expressive logic which can also count other types of events (like the number of occurences of a letter), while still retaining nice properties.

This research was also motivated by recent work which cast doubt as to whether the classical equivalences between logics would hold. For instance, the standard method for proving that MSO = WMSO on infinite words relies on McNaughton's Theorem, which states that deterministic Muller automata capture all regular languages of infinite words (WMSO can describe acceptance in terms of partial runs of the deterministic Muller automaton). However, no deterministic model is known for regular cost functions (even over finite words) [4], so this route for proving CMSO = CWMSO was closed to us.

In [18,14], similar logics were explored in the context of infinite trees rather than infinite words. There it was shown that CMSO is strictly more expressive than CWMSO, and that Rabin's famous characterization of WMSO (in terms of definability of the language and its complement using non-deterministic Büchi automata) fails in the cost setting. Based on this previous work, the relationship between these various cost logics over infinite words was not clear.

1.2 Notation

We fix a finite alphabet $\mathbb{A}$, writing $\mathbb{A}^*$ (respectively, $\mathbb{A}^\omega$) for the set of finite (respectively, infinite) words over $\mathbb{A}$. For $a \in \mathbb{A}$, $|u|_a$ denotes the number of a-labelled positions in u, and $|u|$ denotes the length of the word (the length function is noted $|\cdot|$). We write $\mathbb{N}_\infty$ for $\mathbb{N} \cup \{\infty\}$. By convention, $\inf \emptyset = \infty$ and $\sup \emptyset = 0$.

We briefly define $\approx$, see [4] for details. Let E be a set (usually $\mathbb{A}^\omega$) and let $f, g : E \to \mathbb{N}_\infty$. For $X \subseteq E$, $f(X) := \{f(e) : e \in E\}$. We say $f(X)$ is *bounded* if there is $n \in \mathbb{N}$ such that $\inf f(X) \leq n$ (in particular the set $\{\infty\}$ is unbounded). We say $f \approx g$ if for all $X \subseteq E$, $f(X)$ is bounded if and only if $g(X)$ is bounded. For example, $|\cdot|_a \approx 2|\cdot|_a$ but $|\cdot|_a \not\approx |\cdot|_b$. A *cost function* F is an equivalence class of $\approx$, but in practice will be represented by one of its elements $f : E \to \mathbb{N}_\infty$.

1.3 Contributions

In this paper, we show that the classical equivalences of FO = LTL and MSO = WMSO hold in this cost setting. This supports the idea that the cost function theory is a coherent quantitative extension of language theory. We state the full theorems now, and will introduce the precise definitions in later sections.

The first set of results shows that CFO and CLTL are equivalent, up to $\approx$:

Theorem 1. *For a cost function f over infinite words, it is effectively equivalent for f to be recognized by a:*

- *CFO sentence;*
- *very-weak B-automaton;*
- *very-weak B-automaton with one counter;*
- *CLTL formula.*

The second set of results shows that CMSO (which is strictly more expressive than CFO) is equivalent to CWMSO (again, up to $\approx$):

Theorem 2. *For a cost function f over infinite words, it is effectively equivalent for f to be recognized by a:*

- *CMSO sentence;*
- *non-deterministic B/S-Büchi automaton;*
- *quasi-weak B-automaton;*
- *weak B-automaton;*
- *CWMSO sentence.*

2 Cost Logics

2.1 Cost First-Order Logic

We extend first-order logic in order to define cost functions instead of languages. It is called *cost first-order logic*, or CFO. Formulas are defined by the grammar

$$\varphi := a(x) \mid x = y \mid x < y \mid \varphi \wedge \varphi \mid \varphi \vee \varphi \mid \exists x.\varphi \mid \forall x.\varphi \mid \forall^{\leq N} x.\varphi$$

where $a \in \mathbb{A}$, and N is a unique free variable ranging over $\mathbb{N}$. The new predicate $\forall^{\leq N} x.\varphi$ means that φ has to be true for all x, except for at most N "mistakes".

In all of the cost logics, we want to preserve the intuition that increasing the value for N makes it easier to satisfy the formula. In order to make this clear, we will define the logics without negation. An equivalent definition with negation could be given, with the restriction that quantifiers $\forall^{\leq N}.\varphi$ always appear positively (within the scope of an even number of negations); the grammar above could then be viewed as the result of pushing these negations to the leaves.

Given a word $u \in \mathbb{A}^\omega$, an integer $n \in \mathbb{N}$, and a closed formula φ, we say that (u, n) is a model of φ (noted $(u, n) \models \varphi$) if φ is true on the structure u, with n as value for N. If x is a free variable in φ, then we also need to provide a value for x, and we can write $(u, n, i) \models \varphi$, where $i \in \mathbb{N}$ is the valuation for x. Note that $\forall^{\leq N} x.\varphi(x)$ is true with n for N if and only if there exists $X \subseteq \mathbb{N}$ such that the cardinality of X is at most n and for all $i \in \mathbb{N} \setminus X$, $(u, n, i) \models \varphi(x)$. We then associate a cost function $\llbracket \varphi \rrbracket : \mathbb{A}^\omega \to \mathbb{N}_\infty$ to a closed CFO-formula φ by

$$\llbracket \varphi \rrbracket(u) := \inf \{n \in \mathbb{N} : (u, n) \models \varphi\}.$$

We say $\llbracket \varphi \rrbracket$ is the cost function *recognized* by φ.

For instance for $\varphi = \forall^{\leq N} x. \bigvee_{b \neq a} b(x)$, we have $\llbracket \varphi \rrbracket(u) = |u|_a$.

2.2 Cost Monadic Second-Order Logic

We define *cost monadic second-order logic* (CMSO) as an extension of CFO, where we can quantify over sets of positions. The syntax of CMSO is therefore

$$\varphi := a(x) \mid x = y \mid x < y \mid \varphi \wedge \varphi \mid \varphi \vee \varphi \mid \exists x.\varphi \mid \forall x.\varphi \mid \forall^{\leq N} x.\varphi \mid$$
$$\exists X.\varphi \mid \forall X.\varphi \mid x \in X \mid x \notin X.$$

The semantic of CMSO-formulas is defined in the same way as for CFO: if $u \in \mathbb{A}^\omega$ and $n \in \mathbb{N}$, we say that $(u, n) \models \varphi$ if φ is true on the structure u, with n as value for N. We then define $\llbracket \varphi \rrbracket(u) := \inf \{n \in \mathbb{N} : (u, n) \models \varphi\}$.

CMSO was introduced in [4,18] in a slightly different way, as an extension of MSO with predicates $|X| \leq N$ (for a second-order variable X) which appeared positively. The two definitions are equivalent in terms of expressive power.

2.3 Cost Weak Monadic Second-Order Logic

Cost weak monadic second-order logic (CWMSO) was introduced in [18] over infinite trees. The syntax of CWMSO is defined as in CMSO, but the semantics are changed so second-order quantifications range only over finite sets.

CWMSO retains nice properties of WMSO, such as easy dualization, translation to non-deterministic automata models, and equivalence with a form of weak alternating automata with counters. We will only be interested here in cost functions on infinite words, which are a particular case of infinite trees.

2.4   Link with Languages

We can remark that in particular, any FO (resp. MSO) formula φ can be considered as a CFO (resp. CMSO) formula. In this case if L was the language defined by φ, then as a cost formula φ recognizes the cost function $[\![\varphi]\!] = \chi_L$, defined by $\chi_L(u) = \begin{cases} 0 & \text{if } u \in L \\ \infty & \text{if } u \notin L \end{cases}$.

Lemma 1. *If φ is a CFO (resp. CMSO) formula such that $[\![\varphi]\!] \approx \chi_L$ for some language L, then L is definable in FO (resp. MSO).*

Proof. $[\![\varphi]\!] \approx \chi_L$ means that there is a $n \in \mathbb{N}$ such that for all $u \in L$, $[\![\varphi]\!](u) \leq n$, and for all $u \notin L$, $[\![\varphi]\!](u) = \infty$. In particular, all predicates $\forall^{\leq N} x.\psi$ in φ must be verified with $N = n$, when the word is in the language. So we can replace these predicates by the formula $\exists x_1, x_2, \ldots, x_n.\forall x.(\psi \vee \bigvee_{i \in [1,n]} x = x_i)$, expressing that we allow n errors, marked by the x_i's. The resulting formula will be pure FO (resp. MSO), and will recognize L.

Corollary 1. *CMSO is strictly more expressive than CFO.*

Proof. Choose L which is MSO-definable but not FO-definable, like $(aa)^* b^\omega$. By Lemma 1, χ_L is not CFO-definable, but by the first remark it is CMSO-definable.

2.5   Cost Linear Temporal Logic

We now define a cost version of a linear temporal logic, CLTL. This was first introduced in [13] over finite words (and with a slightly different syntax). Formulas are defined by the grammar

$$\varphi := a \mid \varphi \wedge \varphi \mid \varphi \vee \varphi \mid \varphi \mathbf{R} \varphi \mid \varphi \mathbf{U} \varphi \mid \varphi \mathbf{U}^{\leq N} \varphi$$

where a ranges over $\mathbb{A}$, and N is a unique free variable ranging over $\mathbb{N}$. We say that $(u, n, i) \models \varphi$ if u is a model for φ, where $n \in \mathbb{N}$ is the value for N and $i \in \mathbb{N}$ is the position of u from which the formula is evaluated. If i is not specified, the default value is 0. The semantics are defined as usual, with $(u, n, i) \models \psi_1 \mathbf{U} \psi_2$ if there exists $j > i$ such that $(u, n, j) \models \psi_2$ and for all $i < k < j$, $(u, n, k) \models \psi_1$. Similarly, $(u, n, i) \models \psi_1 \mathbf{U}^{\leq N} \psi_2$ if there exists $j > i$ such that $(u, n, j) \models \psi_2$ and $(u, n, k) \models \psi_1$ for all but n positions k in $[i+1, j-1]$. Likewise, $(u, n, i) \models \psi_1 \mathbf{R} \psi_2$ if either $(u, n, j) \models \psi_1$ for all $j > i$ or $(u, n, i) \models \psi_2 \mathbf{U}(\psi_1 \wedge \psi_2)$. We define $[\![\varphi]\!](u) := \inf \{n \in \mathbb{N} : (u, n) \models \varphi\}$.

Notice that we write $\mathbf{U}$ for the next-until operator. From this, the next operator $\mathbf{X}$ can be defined, and the "large" variants of $\mathbf{U}, \mathbf{R}, \mathbf{U}^{\leq N}$ operators, which take into account the current position, can be defined as well (see [7] for more information), and will be noted $\overline{\mathbf{U}}, \overline{\mathbf{R}}, \overline{\mathbf{U}^{\leq N}}$. We also use the standard abbreviations $\mathbf{F}, \mathbf{G}$ for "Eventually" and "Always". We can define the quantitative release $\mathbf{R}^{\leq N}$ by $\psi \mathbf{R}^{\leq N} \varphi \equiv \varphi \mathbf{U}^{\leq N}(\psi \vee \mathbf{G}\varphi)$, and the quantitative always $\mathbf{G}^{\leq N} \varphi \equiv \text{false} \mathbf{R}^{\leq N} \varphi$.

In Sect. 4, we will also use the past variants $\mathbf{S}, \mathbf{Q}, \mathbf{Y}, \mathbf{P}, \mathbf{H}$ of $\mathbf{U}, \mathbf{R}, \mathbf{X}, \mathbf{F}, \mathbf{G}$, respectively (and their quantitative extensions).

3   Cost Automata on Infinite Words

3.1 B-Valuation

Similar to the logic, the automata considered in this paper define functions from $\mathbb{A}^\omega$ to $\mathbb{N}_\infty$. The valuation is based on the classical Büchi acceptance condition and a finite set of counters Γ.

A counter γ is initially assigned value 0 and can be *incremented and checked* $\mathbf{ic}$, left unchanged ε, or *reset* $\mathbf{r}$ to 0. Given an infinite word u_γ over the alphabet $\mathbb{B} := \{\mathbf{ic}, \varepsilon, \mathbf{r}\}$, we define $val_B(u_\gamma) \in \mathbb{N}_\infty$, which is the supremum of all checked values of γ. For instance $val_B((\mathbf{ic}\varepsilon\mathbf{icr})^\omega) = 2$ and $val_B(\mathbf{icric}^2\mathbf{ric}^3\mathbf{r}\ldots) = \infty$. In the case of a finite set of counters Γ and a word u over the alphabet $\{\mathbf{ic}, \varepsilon, \mathbf{r}\}^\Gamma$, $val_B(u) := \max_{\gamma \in \Gamma} val_B(pr_\gamma(u))$ ($pr_\gamma(u)$ is the γ-projection of u).

The set $\mathbb{C} := \mathbb{B}^\Gamma$ is the alphabet of counter actions, that describe the actions on every counter $\gamma \in \Gamma$.

3.2 B- and S-Automata

An *alternating B-Büchi automaton* $\mathcal{A} = \langle Q, \mathbb{A}, F, q_0, \Gamma, \delta \rangle$ on infinite words has a finite set of states Q, alphabet $\mathbb{A}$, initial state $q_0 \in Q$, a set of Büchi states F, a finite set Γ of B-counters, and a transition function $\delta : Q \times \mathbb{A} \to \mathcal{B}^+(\mathbb{C} \times Q)$ (where $\mathcal{B}^+(\mathbb{C} \times Q)$ is the set of positive boolean combinations, written as a disjunction of conjunctions, of elements $(c_i, q_i) \in \mathbb{C} \times Q$).

A run of $\mathcal{A}$ on $u = a_0 a_1 \cdots \in \mathbb{A}^\omega$ is an infinite labelled tree $R = (r, c)$ with $dom(R) \subseteq \mathbb{N}^*$ (words over $\mathbb{N}$), $r : dom(R) \to Q$ and $c : (dom(R) \setminus \{\epsilon\}) \to \mathbb{C}$ with

- $r(\epsilon) = q_0$;
- if $x \in dom(R)$ and $\delta(r(x), a_{|x|}) = \varphi$, then there is some disjunct $(c_1, q_1) \wedge \ldots \wedge (c_k, q_k)$ in φ such that for all $1 \leq j \leq k$, $x \cdot j \in dom(R)$, $r(x \cdot j) = q_j$ and $c(x \cdot j) = c_j$, and for all $j > k$, $x \cdot j \notin dom(R)$.

We say a run is accepting if for every branch π in $R = (r, c)$, there are infinitely many positions x such that $r(x) \in F$.

The behaviour of an alternating automaton can be viewed as a game between two players: Min and Max. Min is in charge of the disjunctive choices and Max chooses a conjunct in the clause picked by Min. Therefore, a run tree fixes all of Min's choices and the branching corresponds to Max's choices.

A *play* is a particular branch of the run: it is the sequence of states and transitions taken according to both players' choices. A *strategy* for some player is a function that gives the next choice of this player, given the history of the play. Notice that a run describes exactly a strategy for Min.

We assign values to runs as follows. Given a branch $\pi = x_0 x_1 \ldots$ in a run R, the value of π is $val_B(\pi) := val_B(c(x_1)c(x_2)\ldots)$. Then $val_B(R)$ is the supremum over all branch values. We call a run of value at most n an n-*run*.

The B-*semantic* of a B-automaton $\mathcal{A}$ is $\llbracket \mathcal{A} \rrbracket_B : \mathbb{A}^\omega \to \mathbb{N}_\infty$ defined by

$$\llbracket \mathcal{A} \rrbracket_B(u) := \inf \{val_B(R) : R \text{ is an accepting run of } \mathcal{A} \text{ on } u\}.$$

The B-semantic minimizes the value over B-accepting runs.

If δ only uses disjunctions, then we say the automaton is *non-deterministic*. In this case, a run tree is just a single branch, and only Min has choices to make. The run is accepting if there are infinitely many Büchi states on its unique branch π, and its value is the value of π. In this case, it can be useful to look at the labelled run-DAG (for Directed Acyclic Graph) G of the automaton over some word u: the set of nodes is $Q \times \mathbb{N}$, and the edges are of the form $(p, i) \xrightarrow{c} (q, i+1)$, where (c, q) is a disjunct in $\delta(p, a_i)$. Runs of $\mathcal{A}$ over u are in bijection with paths in G.

There exists a dual version of B-automata, namely S-automata, where the semantic is reversed: a run remembers the lowest checked value, and this is maximized over all runs. Switching between non-deterministic B- and S-automata corresponds to a complementation procedure on languages. See [4] for details.

Intuitively, non-deterministic B-automata are used to formulate boundedness problems (like star-height), while non-deterministic S-automata are used to answer these problems because they can more easily witness unboundedness.

We will be particularly interested here in non-deterministic B-Büchi (resp. S-Büchi) automata, abbreviated B-NBA (resp. S-NBA).

Weak Automata. We will say that a B-automaton $\mathcal{A} = \langle Q, \mathbb{A}, F, q_0, \Gamma, \delta \rangle$ is a *weak alternating B-automaton* (B-WAA) if it is an alternating B-Büchi automaton such that the state-set Q can be partitioned into $Q_1, \ldots, Q_k$ and there is a partial order $<$ on these partitions satisfying:

- for all $q, q' \in Q_i$, $q \in F$ if and only if $q' \in F$;
- if $q_j \in Q_j$ is reachable from $q_i \in Q_i$ via δ, then $Q_j \leq Q_i$.

This means no cycle visits both accepting and rejecting partitions, so an accepting run must stabilize in an accepting partition on each path in the run tree.

Theorem 3 ([18]). *On infinite trees, B-WAA and CWMSO recognize the same class of cost functions, namely weak cost functions.*

This theorem holds in particular on infinite words. Notice that unlike in the classical case, WCMSO does not characterize the cost functions recognized by both non-deterministic B-Büchi and non-deterministic S-Büchi automata. The class that enjoy this Rabin-style characterization is the quasi-weak class, which strictly contains the weak class, see [14] for more details.

We will show that as in the case of languages, CMSO and CWMSO have the same expressive power on infinite words. It means that the regular class, the quasi-weak class, and the weak class collapse on infinite words. The cost functions definable by any of the automata or logics in Theorem 2 are called *regular cost functions over infinite words*.

Very-Weak Automata. A *very-weak alternating B-automaton* (B-VWAA) is a B-WAA with the additional requirement that each partition is a singleton. That is, there can be no cycle containing 2 or more states. The name follows [7], but these automata are also sometimes known as linear alternating automata, since the condition corresponds to the existence of a linear ordering on states.

4   First-Order Fragment

In this section, we aim to prove Theorem 1.

The classical equivalence of FO and LTL is known as Kamp's Theorem [11]. Converting from CLTL to CFO is standard, since we can describe the meaning of the CLTL operators in CFO, so we omit this part. However, a number of new issues arise in the translation from CFO to CLTL, so we concentrate on this translation in Sect. 4.1.

We then show the connection with B-VWAA. Again, one direction (from CLTL to B-VWAA) is straightforward and only requires one counter (this is adapted from [17]). Moving from B-VWAA (potentially with multiple counters) to CLTL is more interesting, so we describe some of the ideas behind that construction in Sect. 4.2. It uses ideas from [7] but requires some additional work to structure the counter operations in a form that is easily expressible using CLTL.

Example 1. We give an example of a cost function recognizable by a CLTL formula, CFO sentence, and B-VWAA.

Let $\mathbb{A} = \{a, b, c\}$ and $f(u) = \begin{cases} |u|_a & \text{if } |u|_b = \infty \\ \infty & \text{if } |u|_b < \infty \end{cases}$.

Then f is recognized by the CLTL-formula $\varphi = (\mathbf{G}^{\leq N}(b \vee c)) \wedge (\mathbf{GF}b)$ and by the CFO-sentence $\psi = [\forall^{\leq N} x.(b(x) \vee c(x))] \wedge [\forall x.\exists y.(x < y \wedge b(x))]$. f is also recognized by a 3-state B-VWAA: it deterministically counts the number of a's, while Player Max has to guess a point where there is no more b in the future, in order to reject the input word. If the guess is wrong and there is one more b, or if the guess is never made, then the automaton stabilizes in an accepting state.

4.1   CFO to CLTL

Instead of trying to translate CFO directly into CLTL, we first translate a CFO formula into a CLTL formula extended with past operators $\mathbf{Q}, \mathbf{S}, \mathbf{S}^{\leq N}$ (the past versions of $\mathbf{R}, \mathbf{U}, \mathbf{U}^{\leq N}$) and then show how to eliminate these past operators. Let CLTL$_\mathrm{P}$ be CLTL extended with these past operators.

A CLTL$_\mathrm{P}$-formula is *pure past* (resp. *pure future*) if it uses only temporal operators $\mathbf{Q}, \mathbf{S}, \mathbf{S}^{\leq N}$ (resp. $\mathbf{R}, \mathbf{U}, \mathbf{U}^{\leq N}$) and all atoms are under the scope of at least one of these operators. A formula is *pure present* if it does not contain temporal operators. Hence, a pure past (resp. present, future) formula depends only on positions before (resp. equal to, after) the current position in the word. A formula is *pure* if it is pure past, pure present, or pure future. It turns out any CLTL$_\mathrm{P}$ formula can be translated into a boolean combination of pure formulas.

Theorem 4 (Separation Theorem). *CLTL$_P$ has the separation property, i.e. every formula is equivalent to a boolean combination of pure formulas.*

The proof is technical and requires an analysis of a number of different cases of past operators nested inside of future operators (and vice versa). It uses ideas from [8], but new behaviours arise in the cost setting, and have to be treated carefully. The proof proceeds by induction on the *junction depth*, the maximal number of alternations of nested past and future operators, and on the quantifier rank. Each induction step introduces some distortion of the value (the number of mistakes can be squared), but because the junction depth and quantifier rank are bounded in the formula, we end up with an equivalent formula.

We illustrate the idea with an example.

Example 2. Let $\mathbb{A} := \{a, b, c, d\}$. Consider the CLTL$_P$ formula $\varphi = (b\mathbf{U}^{\leq N}c)\mathbf{S}a$. Then φ is equivalent to

$$[(b\mathbf{S}^{\leq N}(c \vee a))\mathbf{S}a] \wedge [(b\overline{\mathbf{U}^{\leq N}}c) \vee (\mathbf{Y}a)]$$

which is a boolean combination of pure formulas. This formula factorizes the input word into blocks separated by c's, since the last a in the past. The first conjunct checks that each block is missing at most N b's. The second conjunct checks that at the previous position, we had either a or $b\mathbf{U}^{\leq N}c$.

We can now prove the desired translation from CFO to CLTL. The proof is adapted from [10]. It proceeds by induction on the quantifier rank of the formula, and makes use of the Separation Theorem. We have to take care of the new quantitative quantifiers, but no problem specific to cost functions arises here.

Proposition 1. *Every CFO-formula can be effectively translated into an equivalent CLTL-formula.*

4.2   *B*-VWAA to CLTL

We use ideas from [17, Theorem 6]. Unlike the classical setting, we must first convert the *B*-VWAA into a more structured form. In this section, we write $\mathbb{C}$ for the set of *hierarchical counter actions* on counters $[1, k]$ such that $\mathtt{ic}_j$ (resp. $\mathtt{r}_j$) performs $\mathtt{ic}$ (resp. $\mathtt{r}$) on counter j, resets counters $j' < j$, and leaves $j' > j$ unchanged. We say a *B*-VWAA is *CLTL-like* if the counters are hierarchical, $\delta(q, a)$ is in disjunctive normal form, each disjunct has at most one conjunct with state q, and all conjuncts with state $q' \neq q$ have counter action $\mathtt{r}_k$.

Lemma 2. *Let $\mathcal{A}$ be a B-VWAA. Then there is a B-VWAA $\mathcal{A}'$ which is CLTL-like and satisfies $[\![\mathcal{A}]\!] \approx [\![\mathcal{A}']\!]$.*

We can then describe a low value run using a CLTL-formula.

Proposition 2. *Let $\mathcal{A}$ be a B-VWAA with k counters which is CLTL-like. For all $\varphi \in \mathcal{B}^+(Q)$, there is a CLTL formula $\theta(\varphi)$ such that $[\![\mathcal{A}_\varphi]\!] \approx [\![\theta(\varphi)]\!]$ where $\mathcal{A}_\varphi$ is the automaton $\mathcal{A}$ starting from the states described in φ.*

Proof. The proof is by induction on $|Q|$. The case $|Q| = 0$ is trivial.

Let $|Q| > 0$ and let q be the highest state in the very-weak ordering. Given some $\varphi \in \mathcal{B}^+(\mathbb{C} \times Q)$, we can treat each element separately and then combine using the original boolean connectives. For elements $q' \neq q$, we can immediately apply the inductive hypothesis.

For an element q, we first write formulas $\theta_{q,c}$ for $c \in \mathbb{C}$ which express the requirements when the automaton selects a disjunct which has one copy which stays in state q and performs operation c (there is only one such operation since $\mathcal{A}$ is CLTL-like). Likewise, we write a formula $\theta_{q,\text{exit}}$ which describes the requirements when $\mathcal{A}$ chooses a disjunct which leaves q. These formulas do not involve q so can be obtained by applying the inductive hypothesis.

While the play stays in state q, we must ensure that transitions with increments are only taken a bounded number of times before resets. For a particular counter γ, this behaviour is approximated by

$$\theta_{q,\text{cycle},\gamma} := (\bigvee_{\gamma' \geq \gamma} \theta_{q,\mathbf{r}_{\gamma'}} \vee \theta_{q,\text{exit}}) \; \overline{\mathbf{R}^{\leq N}} \; (\bigvee_{c < \mathrm{ic}_\gamma} \theta_{q,c}).$$

Putting this together for all $\gamma \in [1, k]$, $\theta_{q,\text{cycle}} := \bigvee_{c \in \mathbb{C}} \theta_{q,c} \wedge \bigwedge_{\gamma \in [1,k]} \theta_{q,\text{cycle},\gamma}$. Finally, this gets wrapped into a statement which ensures correct behaviour in terms of accepting states (i.e. if $q \notin F$ then the play cannot stay forever in q):

$$\theta(q) := \begin{cases} \theta_{q,\text{cycle}} \; \overline{\mathbf{U}} \; \theta_{q,\text{exit}} & \text{if } q \notin F \\ \theta_{q,\text{exit}} \; \overline{\mathbf{R}} \; \theta_{q,\text{cycle}} & \text{if } q \in F \end{cases}.$$

By combining the translation from a B-VWAA with multiple counters to CLTL, and then the translation to a B-VWAA (which uses only one counter) we see that adding counters to B-VWAA does not increase expressivity.

Corollary 2. *Every B-VWAA with k counters is equivalent to a B-VWAA with one counter.*

5   Expressive Completeness of CWMSO

We aim in this part at proving Theorem 2.

The translation from CWMSO to CMSO is standard (since finiteness is expressible in MSO). Likewise, the connection between CMSO and B- and S-automata was proven in [4] for finite words, and its extension to infinite words (and B-NBA and S-NBA automata) is known [3].

As a result, we concentrate on the remaining translations. As mentioned in the introduction, because there is no deterministic model for cost automata, we could not prove that CWMSO = CMSO using the standard method. In this section, we describe an alternative route, which goes via weak automata. Using ideas from [16], we show how to move from B-NBA to B-WAA. This gives an idea about the issues involved in analyzing alternating cost automata over infinite words. We can then use Theorem 3 to move from B-WAA to CWMSO.

5.1 B-NBA to B-WAA

Theorem 5. *For all B-NBA $\mathcal{A}$ with n states and k counters, we can construct an equivalent B-WAA W with $O(n^2 4^k)$ states and k counters.*

Proof. We first transform the B-NBA $\mathcal{A}$ into an equivalent B-NBA $\mathcal{B}$ in the following normal form: every transition leaving a Büchi state resets all counters of $\mathcal{B}$. The principle is the following: because we work on infinite words, $\mathcal{B}$ can choose an n-run of $\mathcal{A}$ and guess for each counter if there will be infinitely many resets, or finitely many increments. In the first case, $\mathcal{B}$ always delays its Büchi states until the next reset. In the second case, $\mathcal{B}$ waits until the last increment, and then add resets on Büchi states. This results in a slightly different function, but still equivalent up to $\approx$. Notice that this transformation cannot be achieved on infinite trees, which is why this result does not hold for trees (see [14]).

Then we use ideas from [16]: we analyze the run DAG of $\mathcal{B}$ and assign ranks to its nodes. Intuitively, these ranks describe how far each node is from a Büchi run. More precisely, for any $n \in \mathbb{N}$ and $u \in \mathbb{A}^\omega$, it is possible to assign a finite rank to every node if and only if there is no n-run of $\mathcal{B}$ on u.

Therefore, we can design a B-WAA that allows Player Min to play transitions of $\mathcal{B}$, and the opponent to guess ranks in order to prove that there is no low value run. This way, if there is an n-run of $\mathcal{B}$ on u, playing this run is a strategy of value n for Player Min in W. Conversely, if there is no n-run of $\mathcal{B}$ on u, then we can prove that Player Min cannot have a strategy of value n in $\mathcal{B}$: if he plays in a way that counters stay below n, then the run will not be Büchi, and Player Max can guess ranks to prove this. The automaton W is defined so that such a play stabilizes in a rejecting partition of W. This shows that $[\![W]\!] = [\![\mathcal{B}]\!] \approx [\![\mathcal{A}]\!]$.

A normal form is needed to make it possible to look for runs of $\mathcal{B}$ that are simultaneously Büchi and low valued. If the automaton is not in normal form, the independence of these two conditions prevents us from defining ranks properly.

6   Conclusion

We lifted various equivalence results on infinite words from languages to cost functions. The proofs needed to take care of new behaviours specific to this quantitative setting. These results show that the classical definitions of logics and automata have been extended in a coherent way to the cost setting, and provide further evidence that the theory of regular cost functions is robust.

We showed that the weak cost functions on infinite words enjoy the same nice properties as in the case of languages. This is in contrast to the case of trees (see [14]), where some classical properties of weak languages only held for the larger class of quasi-weak cost functions.

We also studied the first-order fragment which gave rise to an unexpected result: very-weak B-automata need only one counter to reach their full expressivity. We did not develop here the algebra side of the first-order fragment as it was done in [13], but if this result can be lifted to infinite words (which we think is the case), it would imply algebraic characterization by aperiodic stabilization semigroups, and hence decidablity of membership for the first-order fragment.

Acknowledgments. We would like to thank the referees for their comments, and Thomas Colcombet for making this joint work possible.

References

1. Bojańczyk, M.: A Bounding Quantifier. In: Marcinkowski, J., Tarlecki, A. (eds.) CSL 2004. LNCS, vol. 3210, pp. 41–55. Springer, Heidelberg (2004)
2. Bojańczyk, M., Colcombet, T.: Bounds in w-regularity. In: LICS, pp. 285–296. IEEE Computer Society (2006)
3. Colcombet, T.: Personal communication
4. Colcombet, T.: The Theory of Stabilisation Monoids and Regular Cost Functions. In: Albers, S., Marchetti-Spaccamela, A., Matias, Y., Nikoletseas, S., Thomas, W. (eds.) ICALP 2009, Part II. LNCS, vol. 5556, pp. 139–150. Springer, Heidelberg (2009)
5. Colcombet, T., Löding, C.: The Non-deterministic Mostowski Hierarchy and Distance-Parity Automata. In: Aceto, L., Damgaard, I., Goldberg, L.A., Halldórsson, M.M., Ingólfsdóttir, A., Walukiewicz, I. (eds.) ICALP 2008, Part II. LNCS, vol. 5126, pp. 398–409. Springer, Heidelberg (2008)
6. Colcombet, T., Löding, C.: Regular cost functions over finite trees. In: LICS, pp. 70–79. IEEE Computer Society (2010)
7. Diekert, V., Gastin, P.: First-order definable languages. In: Flum, J., Grädel, E., Wilke, T. (eds.) Logic and Automata. Texts in Logic and Games, vol. 2, pp. 261–306. Amsterdam University Press (2008)
8. Gabbay, D.M., Hodkinson, I., Reynolds, M.: Temporal logic: mathematical foundations and computational aspects. Oxford Logic Guides, Clarendon Press (1994)
9. Hashiguchi, K.: Limitedness theorem on finite automata with distance functions. J. Comput. Syst. Sci. 24(2), 233–244 (1982)
10. Hodkinson, I.M., Reynolds, M.: Separation - past, present, and future. In: We Will Show Them!, vol. 2, pp. 117–142 (2005)
11. Kamp, H.W.: Tense Logic and the Theory of Linear Order. PhD thesis, Computer Science Department, University of California at Los Angeles, USA (1968)
12. Kirsten, D.: Distance desert automata and the star height problem. RAIRO - Theoretical Informatics and Applications 3(39), 455–509 (2005)
13. Kuperberg, D.: Linear temporal logic for regular cost functions. In: Schwentick, T., Dürr, C. (eds.) STACS. LIPIcs, vol. 9, pp. 627–636. Schloss Dagstuhl - Leibniz-Zentrum fuer Informatik (2011)
14. Kuperberg, D., Vanden Boom, M.: Quasi-weak cost automata: A new variant of weakness. In: Chakraborty, S., Kumar, A. (eds.) FSTTCS. LIPIcs, vol. 13, pp. 66–77. Schloss Dagstuhl - Leibniz-Zentrum fuer Informatik (2011)
15. Kupferman, O., Piterman, N., Vardi, M.Y.: From liveness to promptness. Formal Methods in System Design 34(2), 83–103 (2009)
16. Kupferman, O., Vardi, M.Y.: Weak alternating automata are not that weak. ACM Trans. Comput. Log. 2(3), 408–429 (2001)
17. Löding, C., Thomas, W.: Alternating Automata and Logics over Infinite Words (Extended Abstract). In: Watanabe, O., Hagiya, M., Ito, T., van Leeuwen, J., Mosses, P.D. (eds.) IFIP TCS 2000. LNCS, vol. 1872, pp. 521–535. Springer, Heidelberg (2000)
18. Vanden Boom, M.: Weak Cost Monadic Logic over Infinite Trees. In: Murlak, F., Sankowski, P. (eds.) MFCS 2011. LNCS, vol. 6907, pp. 580–591. Springer, Heidelberg (2011)

Coalgebraic Predicate Logic

Tadeusz Litak[1,*], Dirk Pattinson[2], Katsuhiko Sano[3], and Lutz Schröder[4]

[1] Department of Computer Science, University of Leicester
tadeusz.litak@gmail.com
[2] Department of Computing, Imperial College London
dirk@doc.ic.ac.uk
[3] School of Information Science, Japan Advanced Institute of Science and Technology
v-sano@jaist.ac.jp
[4] Department of Computer Science, Friedrich-Alexander-Universität Erlangen-Nürnberg
lutz.schroeder@cs.fau.de

Abstract. We propose a generalization of first-order logic originating in a ne-glected work by C.C. Chang: a natural and generic correspondence language for any types of structures which can be recast as Set-coalgebras. We discuss axioma-tization and completeness results for two natural classes of such logics. Moreover, we show that an entirely general completeness result is not possible. We study the expressive power of our language, contrasting it with both coalgebraic modal logic and existing first-order proposals for special classes of Set-coalgebras (apart for relational structures, also neighbourhood frames and topological spaces). The semantic characterization of expressivity is based on the fact that our language inherits a coalgebraic variant of the Van Benthem-Rosen Theorem. Basic model-theoretic constructions and results, in particular ultraproducts, obtain for the two classes which allow for completeness—and in some cases beyond that.

1 Introduction

Non-relational semantics play an important and ever-increasing role in computer sci-ence, e.g. in concurrency, reasoning about knowledge and agency, description logics and ontologies (see e.g. [1,6,20,16]). Nevertheless, the expressivity of ordinary modal logic is somewhat limited. Just as reasoning about relational structures, reasoning about probabilities, agency, social interactions, or conditionals may require variable binding, interaction of local and global information, or reference to individual states. Moreover, a natural and well-tailored predicate language would allow a transfer of (or at least a comparison with) methods, tools and results of classical and finite model theory.

Thus motivated, we propose *coalgebraic predicate logic (CPL)*: a generic and natural first-order language to reason about such diverse structures as neighbourhood frames, discrete Markov chains, conditional frames, multigraphs and indeed any type of struc-ture that can be understood in terms of Set-coalgebras. In particular, the interpretation

* The first author gratefully acknowledges the support of EPSRC grant EP/G041296/1. Thanks are also due to all the people who kindly discussed these ideas with us, in particular to Alexan-der Kurz, Drew Moshier and referees of all versions of this paper. Work by the fourth author forms part of the DFG project GenMod (SCHR 1118/5-2).

A. Czumaj et al. (Eds.): ICALP 2012, Part II, LNCS 7392, pp. 299–311, 2012.
© Springer-Verlag Berlin Heidelberg 2012

of CPL over Kripke frames (sets with a binary relation) recovers the standard semantics of first-order logic.

Our proposal originates in a largely forgotten paper by C.C. Chang [4] which in contemporary terms can be described as an early contribution to the model theory of *Scott-Montague neighbourhood frames*, i.e., coalgebras for the doubly contravariant powerset functor $\mathcal{N} = \mathcal{Q}\mathcal{Q}$. Chang's original motivation was to simplify model theory for what Montague called *pragmatics* and to replace Montague's many-sorted setting by a single-sorted one. Chang's contributions were primarily of model-theoretic nature. He provided adaptations of (elementary) submodel/extension, elementary chain of models and ultraproduct and established Tarski-Vaught, downward and upward Löwenheim-Skolem theorems. One of the main notable points of [4] are its lucid motivation, natural examples and concise syntax, with only one sort of variables and no need for explicit quantification over neighbourhoods or successors. Here we are going to work with a notational variant of Chang's original syntax which we find even more readable.

The semantics uses the fact that coalgebraic structures can be naturally described in terms of *modal operators*. For example, relational semantics yield an operator $\Diamond$: *there exists a successor* ..., and probability distributions an operator L_p: *with probability* $\geq p$.... More abstractly, (n-ary) modal operators $\heartsuit$ come equipped with a coalgebraic interpretation taking an n-tuple of predicates as arguments. Each operator induces (in the unary case) an atomic formula $t\heartsuit\lceil z : \phi\rceil$ where t is a term, ϕ is a formula of coalgebraic predicate logic and z is a (comprehension) variable. Intuitively, the above formula stipulates that (the denotation of the term) t satisfies property $\heartsuit$, which may parametrically depend on the set of all z that satisfy ϕ. For example, standard modal logic over relational semantics provides a formula $x\Diamond\lceil z : z = y\rceil$ which is semantically equivalent to stipulating that x has y as a successor, i.e., $y \in R(x)$. In the probabilistic setting, validity of $xL_p\lceil y : y \neq x\rceil$ forces that the probability of moving from x to a different state is $\geq p$.

Our aim is to convince the reader that CPL is a fruitful common generalization of both first-order logic and coalgebraic modal logic. Section 2 introduces syntax, semantics and a number of intuitive examples. Section 3 discusses axiomatization and completeness results for two natural classes of structures, including neighbourhood and Kripke frames as extremal cases. Moreover, we show that a fully general completeness result must necessarily fail even for rather natural classes of structures (e.g., Markov chains with non-standard probabilities). Section 4 gives both syntactic and semantic characterizations of coalgebraic modal logic as a fragment of CPL. The semantic characterisation naturally generalizes the van Benthem-Rosen characterization of ordinary modal logic. Section 5 takes first steps in the model theory of CPL and Section 6 concludes.

Related Work. We have already discussed Chang's paper [4] not only in terms of the inspiration of the approach presented here, but also in terms of concrete results on the first-order logic of neighbourhood frames. An alternative, two-sorted language for neighbourhood frames has been proposed in [12, Section 5]. Over neighbourhood frames, the language studied in the present work is a fragment of that of [12]. Without giving full syntactic details, our $x\heartsuit\lceil y : \phi(y)\rceil$ (we restrict the attention to the unary case to keep things simple) can be translated as $\exists u.(x\mathsf{N}u \wedge \forall y.(u\mathsf{E}y \leftrightarrow \phi(y)))$.

First-order formalisms have also been considered for topological spaces, which happen to be particular instances of neighbourhood frames when defined in terms of local neighbourhood bases. In particular, Sgro [28] studies interior operator logic in topology together with interior modalities also for all finite topological powers of the space, which do not seem meaningful in the topological context. This language is the weakest one in the hierarchy of topological languages considered in the classical overview paper [30]. However, the closest reference in this line of work seems to be [17], which does in fact provide a completeness result for the Chang language itself, i.e., a special version of Theorem 7 below. See also [3] for a more contemporary reference.

The relationship between coalgebraic logic and first order logic is the subject of [24], albeit using involved three-sorted syntax and not giving an axiomatization. The technical results of [24] remain valid—indeed, we use them below in Section 4 to establish a van Benthem-Rosen theorem for our language. An explicit embedding of our language into that of *op.cit* is given in the proof of Theorem 15 below. However, one-sorted coalgebraic predicate logic as presented in this paper seems a more natural common generalization of first-order logic and coalgebraic modal logic. It can be shown that our language is a *proper* fragment of that of [24] using, e.g., Example 27 in *op.cit*.

Finally, a different generic first-order logic largely concerned with the Kleisli category of a monad rather than with coalgebras for a functor is introduced and studied in [14]. Of all the languages discussed above, this one seems least related to the present one; indeed, the study of connections with languages like that of [24] is mentioned in [14] as a subject for future research. We also believe the study of possible connections could be of interest.

2 Syntax, Semantics and Examples

We fix a modal similarity type Λ consisting of modal operators $\heartsuit$ and a set Σ of predicate symbols; every $\heartsuit \in \Lambda$ and $P \in \Sigma$ comes with a fixed arity, but instead of writing $\mathrm{ar}P$ or $\mathrm{ar}\heartsuit$, we will just use natural numbers for readability (typically n for $\mathrm{ar}\heartsuit$ and k for $\mathrm{ar}P$). Formulas of *coalgebraic predicate logic (CPL)* over Λ and Σ (denoted as $\mathcal{CPL}_\Lambda\Sigma$, but we will drop Σ wherever possible) are given by the grammar

$$\phi, \psi \quad ::= \quad y_1 = y_2 \mid P(\boldsymbol{x}) \mid \bot \mid \phi \to \psi \mid \forall x.\phi \mid x\heartsuit\lceil y_1 : \phi_1\rceil \ldots \lceil y_n : \phi_n\rceil$$

where $\heartsuit \in \Lambda$ is an n-ary modal operator and $P \in \Sigma$ a k-ary predicate symbol, x, y_i are variables from a fixed set iVar we keep implicit. Booleans and the existential quantifier are defined in the standard way. We do not include function symbols which can be added at no extra cost [4]. In the $\lceil y_i : \phi_i\rceil$ component, y_i is used as a comprehension variable, i.e., $\lceil y_i : \phi_i\rceil$ denotes a subset of the carrier of the model, to which modal operators can be applied in the usual way. In $x\heartsuit\lceil y_1 : \phi_1\rceil \ldots \lceil y_n : \phi_n\rceil$, x is free and y_i is bound in ϕ_i (not elsewhere though!), otherwise the notions of freeness and boundedness are standard. A variable is *fresh* for a formula if it does not have free occurrences in it. A *sentence*, as usual, is a formula without free variables. The notion of a (capture-avoiding) substitution is defined in the expected way: all the usual caveats for quantified variables have to apply now to comprehension variables as well.

Formally, elements of $\mathcal{CPL}_\Lambda$ are interpreted over *coalgebras*, that is, pairs $(C, \gamma : C \to TC)$ consisting of a carrier set C and a transition function γ that maps every world into a set TC of *structured successors*, where $T : \mathbf{Set} \to \mathbf{Set}$ is an endofunctor extending to a Λ-*structure*, i.e. equipped with a set-indexed family of mappings $[\![\heartsuit]\!]_C : (\mathcal{Q}C)^n \to \mathcal{Q}TC$ for every n-ary modal operator $\heartsuit \in \Lambda$ ($\mathcal{Q}$ is the contravariant powerset functor) subject to *naturality*, i.e. $(Tf)^{-1} \circ [\![\heartsuit]\!]_C = [\![\heartsuit]\!]_D \circ (f^{-1})^n$ for every set-theoretic function $f : C \to D$.

A pair $\mathfrak{M} = (C, \gamma, I)$ consisting of a coalgebra $\gamma : C \to TC$ and a predicate interpretation $I : \Sigma \to \bigcup_{n \in \omega} \mathcal{Q}(C^n)$ respecting arities of symbols will be called a *(coalgebraic) model*. In other words, a coalgebraic model consists simply of a **Set**-coalgebra and an ordinary first-order model whose universe coincides with the carrier of the coalgebra. Given a model $\mathfrak{M} = (C, \gamma, I)$ and a valuation $v : \text{iVar} \to C$, we define satisfaction $\mathfrak{M}, v \models \phi$ in the standard way for first-order connectives and for $\heartsuit$ by the clause

$$\mathfrak{M}, v \models x\heartsuit\lceil y_1 : \phi_1 \rceil \ldots \lceil y_n : \phi_n \rceil \iff \gamma(v(x)) \in [\![\heartsuit]\!]_C([\![\phi_1]\!]_C^{y_1}, \ldots, [\![\phi_n]\!]_C^{y_n})$$

where $[\![\phi]\!]_C^y := \{c \in C \mid \mathfrak{M}, v[c/y] \models \phi\}$ and $v[c/y]$ is v modified by mapping y to c. We have the following examples of our setting.

Social Situations and Neighbourhood Frames. The modifications proposed in [4] probably would not be accepted in Montague's account of *pragmatics*, but as noted by Chang himself, the resulting language is particularly well-tailored for reasoning about social situations and relationships between an individual and sets of individuals. The semantics is given in terms of neighbourhood frames, which we capture coalgebraically using $\Lambda := \{\Box\}$ and by putting $TC := \mathcal{Q}\mathcal{Q}C$ (the doubly contravariant powerset functor) which extends to a Λ-structure by $[\![\Box]\!]_C(A) := \{\sigma \in TC \mid A \in \sigma\}$. In the presence of a binary relation $S(x, y)$ that we read as 'x speaks to y' and interpreting $\Box$ as 'enjoyable', the formula $\exists y_1.\exists y_2.(x\Box\lceil z : S(z, y_1)\rceil \wedge x\Box\lceil z : S(z, y_2)\rceil \wedge y_1 \neq y_2)$ reads as 'there are at least two people such that x finds it enjoyable to speak to them' where x determines the truth of this sentence by inspecting the set $\{z : S(z, y_i)\}$ of people speaking to y_i.

Relational First-Order Logic. As already discussed, for $TC := \mathcal{P}C$, i.e., covariant powerset endofunctor, we get a notational variant of ordinary FOL over relational structures.

Facebook Friends and Graded Modal Logic. We obtain a variant of graded modal logic [9] if we consider the similarity type $\Lambda = \{\langle k \rangle \mid k \geq 0\}$ where $\langle k \rangle$ reads as 'more than k successors satisfy ...'. We interpret the ensuing logic over multigraphs: coalgebras for $\mathcal{B}C := \{f : C \to \mathbb{N} \mid f(c) \neq 0 \text{ only finitely often}\}$, extending $\mathcal{B}$ to a Λ-structure by stipulating $[\![\langle k \rangle]\!]_X(A) = \{f \in \mathcal{B}X \mid \sum_{x \in A} f(x) > k\}$ to express that more than k successors (counted with multiplicities) have property A. Given a $\mathcal{B}$-coalgebra $C \xrightarrow{\gamma} \mathcal{B}C$, we can think of elements of C as individuals, and of $\gamma(c)(c')$ as the number of 'likes' (in the sense of Facebook) that c' has received from c. In other words, $\gamma(c)(c') = n$ models the fact that c has pressed the 'like'-button on c''s page n times. In the presence of a binary relation $F(x, y)$ expressing that y is a Facebook-friend of x, the formula $x\langle k\rangle\lceil z : \exists y.F(x,y) \wedge F(y,z)\rceil$ expresses that x likes more than k activities of friends of his/her friends.

Presburger Modal Logic and Arithmetic. A more general set of operators than graded modal logic is that of positive Presburger modal logic [7], which admits integer linear inequalities $\sum a_i \cdot \#(\phi_i) > k$ among formulas (we assume that $a_i \geq 0$). By keeping the same functor $\mathcal{B}$, we can also give the corresponding predicate lifting in a natural way. As before, let C be the supply of individuals but $\gamma(c)(c')$ will be now the number of *posts* of c to c''s wall. In addition to $F(x,y)$ as above, we introduce $T(x,y)$ expressing that y is a follower of x in Twitter and $I(x)$ expressing that x is influential. Then, the formula $\forall x.(x(3 \cdot \#\lceil y : F(x,y)\rceil + 1 \cdot \#\lceil y : T(x,y)\rceil > 10000) \to I(x))$ means that, if x's weighted number of wall posts to his/her Facebook friends and Twitter-followers is greater than ten thousands, then x is influential, provided that Facebook is three times as influential as Twitter.

Combination of Frame Classes. Frame classes can be combined: instead of using the relation symbol R in the previous example, we could consider coalgebras $(C, \gamma : C \to TC)$ where $TC := \mathcal{B}C \times \mathcal{P}C$ gives a multigraph structure and a relational structure, and interpret the operators $\langle k\rangle$ and $\square$ by projecting out the components. We leave it to the reader to express 'x likes more than k activities of friends of his/her friends' in this setting. Alternatively, we can take $T := \mathcal{B} \times \mathcal{Q}\mathcal{Q}$ and combine operators for the Facebook sense of 'like' and Chang's modalities for social situations. A formula $\neg x\heartsuit\lceil y : y\langle 3\rangle\lceil y : y = z\rceil\rceil$ expresses then that x does not fancy the perspective of liking strictly more than 2 of Facebook activities of z (or, to be more precise, the general company of people who do so). The reader may find it entertaining to compare our Facebook examples with these of [27].

Agents and Coalition Logic. Coalgebraically, the semantics of coalition logic [20] or, equivalently, alternating time temporal logic [1] is formulated over game frames $\mathcal{G}(X) = \{(S_i)_{i\in P}, f : \prod_{i\in P} S_i \to X \mid \emptyset \neq S_i \subseteq \mathbb{N}\}$ where P is a (fixed) set of players, S_i is the set of strategies available to player $i \in A$ and f is an *outcome function* that determines the next move of the game, depending on the strategy chosen by each player. We use the modalities $\Lambda = \{[Q] \mid Q \subseteq A\}$ where $[Q]$ reads 'the coalition Q of players can achieve ...'. The functor T extends to a Λ-structure via $[\![Q]\!]_X(A) = \{(f,(S_i)) \in \mathcal{G}X \mid \exists(s_i)_{i\in P}\forall(s_j)_{j\in P\backslash Q}(f(s_i)_{i\in P} \in A)\}$ which gives the standard semantics of coalition logic and alternating time temporal logic. Given a coalgebra $(C, \gamma : C \to \mathcal{G}C)$, we think of C as being the positions of a strategic game, and $\gamma(c)$ as describing the different strategies available to the agents, and their ramifications. In this context, the formula $x[\emptyset]\lceil y : y = x\rceil$ describes that the state x is a dead end: independent of the choice of strategies of the players, the next position will be x itself. The formula $\forall y(x[Q]\lceil z : z = y\rceil \to y[Q]\lceil z : z = x\rceil)$ expresses that— given position x on the board—whenever coalition Q can force a position y on the game board, they also have a (collective) strategy to revert back to x. Universal quantification over x would then ensure that coalition Q enjoys this power, irrespective of the state of the game.

Ludo and Probabilistic Modal Logic. Taking the similarity type $\Lambda = \{\langle p\rangle \mid p \in [0, 1]\cap\mathbb{Q}$ and reading $\langle p\rangle$ as 'with probability at least p', we obtain a localised version of

Halpern's probabilistic first-order logic [11] and $\Lambda_k = \{\langle n/k \rangle \mid n = 0, \ldots, k\}$ restricts to probabilities in the set of multiples of $1/k$. Both logics are interpreted over (local) probability distributions, that is, the Λ-structure given by $\mathcal{D}X = \{\mu : X \to [0, 1] \mid \mu$ has finite support and $\sum_x \mu(x) = 1\}$ where $[\![\langle p \rangle]\!]_X = \{\mu \in \mathcal{D}X \mid \sum_{x \in A} \mu(x) \geq p\}$. If all possible probabilities are contained in some finite set (such as when rolling a die) we consider the sub-structure $\mathcal{D}_k X = \{\mu : X \to \{0, 1/k, \ldots, k/k\} \mid \sum_x \mu(x) = 1\}$ with the same interpretation of the modal operator. Taking the carrier of a model to consist of the positions of a ludo board, the (true) formula $x\langle 1/2 \rangle \lceil y : \langle 1/2 \rangle \lceil z : z = x \rceil \rceil$ expresses the fact that x can capture, with probability $\geq 1/2$ all pieces that could capture x (with the same probability).

Party Invitations and Non-monotonic Conditionals. An example of a binary modality is provided by (conditional) implication $\Rightarrow$, written in infix notation. We interpret $\Rightarrow$ on selection function frames $\mathcal{S}X = \{f : \mathcal{P}(X) \to \mathcal{P}(X)\}$ using $[\![\Rightarrow]\!]_X(A, B) = \{f \in \mathcal{S}X \mid f(A) \cap f(B) \neq \emptyset\}$. The formula $\phi \Rightarrow \psi$ expresses that ψ is possible under condition ϕ. This presentation of conditional logic is equivalent to (but not identical) to the standard presentation [5] and has the technical advantage of boundedness in the second argument (that we will use in Section 3). In the spirit of Chang's original examples concerning social situations, we may read the antecedent of the conditional as 'invited' and the consequent as 'happy'. Given a binary relation ff ('facebook friend') the formula $\exists y(x(\lceil y : \mathsf{ff}(x, y) \rceil \Rightarrow \lceil z : z = y \rceil)$ describes that there is a person (y – possibly Mark Zuckerberg) who is happy if x invites *precisely* her facebook friends to her birthday party. If x also invites non-facebook friends, then the non-monotonicity of the conditional does not allow us to infer anything about y's emotional state.

3 Completeness

In order to state our axiomatization and completeness results, we need an auxiliary notion of *one-step satisfiability*.

Definition 1. Given any supply of primitive symbols D (which can be any set), define $\mathcal{M}^0(D)$ as $A, B ::= d \mid A \to B \mid \bot$ where $d \in D$, $\mathcal{M}_\Lambda^\wedge(D)$ as $W, V ::= \heartsuit d_1 \ldots d_n \mid W \to V \mid \bot$ and $\mathcal{M}_\Lambda^1(D)$ as $X, Y ::= \heartsuit A_1 \ldots A_n \mid X \to Y \mid \bot$; in other words, $\mathcal{M}_\Lambda^1(D) = \mathcal{M}_\Lambda^\wedge(\mathcal{M}^0(D))$. For any $C \in \mathbf{Set}$, given a valuation $\tau : D \to \mathcal{P}(C)$, we write $C, \tau \models A$ if $\tau(A) = \top$. We also set $[\![X]\!]_{TC,\tau}$, i.e., the interpretation of X in the boolean algebra $\mathcal{P}(TC)$ under τ, to be the inductive extension of the assignment $[\![\heartsuit A_1 \ldots A_n]\!]_{TX,\tau} = [\![\heartsuit]\!]_C(\tau(A_1), \ldots, \tau(A_n))$. We write $TC, \tau \models X$ if $[\![X]\!]_{TC,\tau} = TC$, and $t \models_{TC,\tau} X$ if $t \in [\![X]\!]_{TC,\tau}$. A set $\Xi \subseteq \mathcal{M}_\Lambda^1$ is *one-step satisfiable* w.r.t. τ if $\bigcap_{X \in \Xi} [\![X]\!]_{TC,\tau} \neq \emptyset$. If $D \subseteq \mathcal{P}(C)$ and τ is just the inclusion, we will usually drop it from the notation.

Just like in case of coalgebraic modal logic (see Section 4 below), proof systems for CPL are best described in terms of rank-1 rules—or, more precisely, rule schemes.

Definition 2. Fix a collection sVar of schematic variables $a, b, c \ldots$. A *one-step rule* is of the form A/X, $A \in \mathcal{M}^0(\text{sVar})$ and $X \in \mathcal{M}_\Lambda^1(\text{sVar})$. A one-step rule will be called a *one-step axiom scheme* if its premise is empty. A rule is *one-step sound* if $TC, \tau \models X$

whenever $C, \tau \models A$ for a valuation $\tau : \mathsf{sVar} \rightarrow \mathcal{P}(C)$. Given a set $\mathcal{R}$ of one-step rules and a valuation $\tau : \mathsf{sVar} \rightarrow \mathcal{P}(C)$, a set $\Xi \subseteq \mathcal{M}_\Lambda^1(\mathsf{sVar})$ is *one-step consistent (with respect to τ)* [26] if the set $\Xi \cup \{X\sigma \mid \sigma : \mathsf{sVar} \rightarrow \mathcal{M}^0; A/X$ a rule in $\mathcal{R}; C, \tau \models A\sigma\}$ is propositionally consistent.

From now on, we will only consider rule sets one-step sound relatively to a given Λ-structure, so the assumption of one-step soundness will not be mentioned explicitly.

Definition 3. A rule set $\mathcal{R}$ is *strongly 1-step complete (S1SC)* for a Λ-structure if for every $C \in \mathbf{Set}$, any $\Xi \subseteq \mathcal{M}_\Lambda^1 x$ and any $\tau : \mathsf{sVar} \rightarrow \mathcal{P}(C)$, Ξ is one-step satisfiable wrt τ whenever it is one-step consistent wrt τ. We say that a set of rules is *finitary S1SC* if the above holds whenever $\tau : \mathsf{sVar} \rightarrow \mathcal{P}_{fin}(C)$ (but not necessarily for arbitrary τ).

Full S1SC is a somewhat restrictive condition; of all examples in Section 2, it is satisfied by neighbourhood and coalition logic modalities, but not by the remaining ones, which only enjoy finitary S1SC. However, the latter property in itself is too weak to ensure completeness results; we need an additional property of associated predicate liftings.

Definition 4. A modal operator $\heartsuit$ is *k-bounded* in *i*-th argument for $k \in \mathbb{N}$ and with respect to a Λ-structure T if for every $C \in \mathbf{Set}$ and every $\overline{A} \subseteq C$,

$$[\![\heartsuit]\!]_C(A_1, \ldots, A_n) = \bigcup_{B \subseteq A_i, \#B \leq k} [\![\heartsuit]\!]_C(A_1, \ldots, A_{i-1}, B, A_{i+1}, \ldots, A_n).$$

(This implies in particular that $\heartsuit$ is monotonic in the *i*-th argument.) We say that Λ is *bounded* w.r.t. T if every modal operator $\heartsuit \in \Lambda$ for every i smaller than its arity is $k_{\heartsuit,i}$-bounded in i for some $k_{\heartsuit,i}$.

Examples of such operators include—apart from Kripke frames (1-bounded)—graded operators over multigraphs and **positive** Presburger logic. See [25] for details. Note that, e.g., the neighbourhood modality clearly fails to be k-bounded; boundedness is a "Kripke-like" property. The notions of strong one-step completeness and boundedness can be combined for n-ary operators. For example, the binary operator $\Rightarrow$ of conditional logic is strongly one-step complete in the first argument and finitary one-step complete in the second which is expressed by restricting valuations of the second argument to finite sets.

In our axiomatization, we will have to translate one-step rules into predicate axioms. Here is an auxiliary notion:

Definition 5. Let $\sigma : \mathsf{sVar} \rightarrow \mathcal{CPL}_\Lambda$ be a substitution. Then for any $x, y \in \mathsf{iVar}$, let $[\sigma, y, x]$ denote the mapping $\mathcal{M}_\Lambda^1(\mathsf{sVar}) \rightarrow \mathcal{CPL}_\Lambda$ defined as the inductive extension of the mapping sending each $\heartsuit(A_1 \ldots A_n)$ to $x\heartsuit\lceil y : \hat{\sigma}(A_1) \rceil \ldots \lceil y : \hat{\sigma}(A_n) \rceil$, where $\hat{\sigma}$ is the inductive extension of σ to $\mathcal{M}^0$.

Let $\Gamma, \Delta \subseteq \mathcal{CPL}_\Lambda$, let $\mathcal{R}$ be a set of one-step rules and $\phi \in \mathcal{CPL}_\Lambda$. Write $\Gamma \vdash_{\Delta, \mathcal{R}} \phi$ if there are $\gamma_1, \ldots, \gamma_n \in \Gamma$ s.t. $\gamma_1 \rightarrow \ldots \rightarrow \gamma_n \rightarrow \phi$ can be deduced from Δ, EG1–EG6, CONG and ONESTEP in Table 1 using **only Modus Ponens**. This clearly defines *a finitary deducibility relation* in the sense of Goldblatt [10, Sec. 8.1] and being $\vdash_{\Delta, \mathcal{R}}$-*consistent* is equivalent with being *finitely* $\vdash_{\Delta, \mathcal{R}}$-*consistent* in his sense, that is, $\Gamma \vdash_{\Delta, \mathcal{R}} \bot$ iff there is $\Gamma_0 \subseteq_{fin} \Gamma$ s.t. $\Gamma_0 \vdash_{\Delta, \mathcal{R}} \bot$. Note that the axiom CONG is in fact (a syntactic variant of) an axiom already introduced by Chang in [4].

Table 1. Enderton-style [8] Axioms for CPL

Everywhere below, $\forall \overline{y}.$ denotes a sequence of universal quantifiers of arbitrary length, possibly empty.

Axiom schemes valid for arbitrary structures

tautologies of propositional logic, axiomatized for example by:

EG1 $\begin{cases} \forall \overline{y}. ((\phi \to \psi) \to ((\psi \to \chi) \to (\phi \to \chi))) \\ \forall \overline{y}. (((\phi \to \bot) \to \phi) \to \phi) \\ \forall \overline{y}. (\phi \to ((\phi \to \bot) \to \psi)) \end{cases}$

EG2 $\forall \overline{y}. (\forall x.\phi \to \phi)$

EG3 $\forall \overline{y}. (\forall x. (\phi \to \psi) \to (\forall x.\phi \to \forall x.\psi))$

EG4 $\forall \overline{y}. (\phi \to \forall x.\phi)$ (x fresh for ϕ)

EG5 $\forall \overline{y}. (x = x)$

EG6 $\begin{cases} \forall \overline{y}. (x = z \to (P(\overline{u}, x, \overline{v}) \to P(\overline{u}, z, \overline{v}))) \quad (P \in \Sigma \cup \{=\}) \\ \forall \overline{y}. (x = z \to (x \heartsuit \lceil y_1 : \phi_1 \rceil \ldots \lceil y_n : \phi_n \rceil \to z \heartsuit \lceil y_1 : \phi_1 \rceil \ldots \lceil y_n : \phi_n \rceil)) \end{cases}$

CONG $\forall \overline{y}. (\forall x. ((\phi_1 \leftrightarrow \psi_1) \wedge \ldots \wedge (\phi_n \leftrightarrow \psi_n))$
$\to \forall x. (x \heartsuit \lceil x : \phi_1 \rceil \ldots \lceil x : \phi_n \rceil \leftrightarrow x \heartsuit \lceil x : \psi_1 \rceil \ldots \lceil x : \psi_n \rceil))$

ONESTEP $\forall \overline{y}. \forall x. ((\forall z. (A\sigma)) \to [\sigma, z, x](X))$
     (A/X a rule in $\mathcal{R}$, $\sigma : \text{sVar} \to \mathcal{CPL}_\Lambda$ and $[\sigma, x, z] : \mathcal{M}_\Lambda^1(\text{sVar}) \to \mathcal{CPL}_\Lambda$ as in Def. 5)

An additional axiom scheme for predicate liftings k-bounded in argument i

BDPL$_{k,i}$ $\forall \overline{y}. (x \heartsuit \lceil y_1 : \phi_1 \rceil \ldots \lceil y_n : \phi_n \rceil \leftrightarrow \exists z_1 \ldots z_k. (x \heartsuit \lceil y_1 : \phi_1 \rceil \ldots \lceil y_{i-1} : \phi_{i-1} \rceil$
$\lceil y_i : y_i = z_1 \vee \ldots \vee y_i = z_k \rceil \lceil y_{i+1} : \phi_{i+1} \rceil \ldots \lceil y_n : \phi_n \rceil \wedge \bigwedge_{j \leq k} \phi_i[y_i/z_j])$ ($\overline{z}$ fresh for $y_i, \overline{\phi}$)

Definition 6. For any set of additional axioms $\Delta \subseteq \mathcal{CPL}_\Lambda$ and any rule set $\mathcal{R}$, we say that a logic given by Δ and $\mathcal{R}$ is *strongly complete* wrt a given Λ-structure if for any set of sentences $\Gamma \in \mathcal{CPL}_\Lambda$, $\Gamma \nvdash_{\Delta, \mathcal{R}} \bot$ holds **if and only if** there is a coalgebraic Λ-model for Γ where axioms given by Δ hold (and, obviously, the rules in $\mathcal{R}$ are sound) under the reading of all $\heartsuit \in \Lambda$ given by the structure.

Theorem 7 (Completeness). *The set of axioms given in Table 1 is a strongly complete axiomatization of $\mathcal{CPL}_\Lambda$ whenever the Λ-structure satisfies either of the following conditions:*

- *there exists a S1SC rank-1 rule set.*
- *there exists a finitary S1SC rank-1 rule set and each $\heartsuit \in \Lambda$ is bounded.*

Example 8. For the examples discussed in Section 2 the situation is as follows. Completeness holds for neighbourhood models as they have a strongly one-step complete axiomatisation. For all others, but excluding non-monotonic conditionals, finitary one-step complete axiomatisations exist. Boundedness holds for relational models, graded modal logic and the logic of finite probabilities (interpreted over $\mathcal{D}_k$-coalgebras) which gives completeness using Theorem 7. The binary operator $\Rightarrow$ of conditional logic is strongly one-step complete in the first argument and 1-bounded in the second, and, as a consequence, the first-order logic of non-monotonic conditionals is also complete, see [25, Section 2.3] for more details.

Remark 9. The Omitting Types Theorem is a standard result of model theory. Goldblatt [10, Section 8.2] shows how to establish it wherever a Henkin-style completeness proof is available. This covers the two classes of structures in the statement of Theorem 7. Since both formulation and proof are entirely analogous to the standard relational case, we omit the details and refer the reader to [10, Section 8.2]; let us only note that the fact we used variables instead of Henkin constants (making use of advantages of an Enderton-style axiomatization) does not lead to any complications in the proof, in fact making it even simpler in some cases.

We briefly consider those cases where boundedness does not apply. In order to show both how completeness fails and what are possible alternative means to handle such a situation, we introduce a new class of functors/Λ-structures. We believe it to be of independent interest in coalgebraic logic. In the whole subsection, to keep things simple we work with unary $\heartsuit \in \Lambda$.

Definition 10 (ω**-Bounded operators**). A modal operator $\heartsuit$ is ω-*bounded* if for each set X and each $A \subseteq X$, $[\![\heartsuit]\!]_X(A) = \bigcup_{B \subseteq_{fin} A} [\![\heartsuit]\!]_X(B)$.

Example 11. Let D^h be the discrete distributions functor with probabilities taken from hyperreal fields. Explicitly: we intend to model Markov chains with non-standard probabilities; these consist of a set X of states, and at each state x an R_x-valued transition distribution μ_x, where R_x is a hyperreal field (we take this to mean a model of the first-order theory of the reals).These structures are coalgebras for the functor T which maps a set X to the set of pairs (R, μ) where R is a hyperreal field and μ is an R-valued probability measure. This functor is in fact class-valued, which however does not affect the applicability of our coalgebraic analysis (which never requires iterated application of the coalgebraic type functor). We take the modal signature Λ to consist of the operators M_p ('with probability more than p') for $p \in [0, 1] \cap \mathbb{Q}$.

Theorem 12. *Whenever a Λ-structure makes some $\heartsuit \in \Lambda$ ω-bounded without being k-bounded for any $k \in \omega$, strong completeness fails for any non-empty supply of predicate symbols Σ.*

Completeness for the specific case of ω-bounded operators (possibly with some additional assumptions, like a variant of S1SC property) could be restored by means of a deduction system equipped with an explicit ω-rule. A natural candidate is

$$\{\forall y_1, ..., y_k.(\phi[y/y_1] \wedge \cdots \wedge \phi[y/y_k] \rightarrow$$
$$\neg x \heartsuit [y : y = y_1 \vee ... \vee y = y_k]) \mid k \in \omega\}/\neg x \heartsuit [y : \phi].$$

In fact, Henkin-style completeness proofs for logics with infinitary rules work quite naturally in the framework of [10]. We are not pursuing this option here. As we will see below, there are other positive results which can be proved about ω-bounded operators.

4   Correspondence with Coalgebraic Modal Logic

The formulas $\mathcal{CML}_\Lambda(\Sigma)$ of pure (coalgebraic) modal logic in the modal signature Λ over Σ (now all elements of Σ are assumed to be of arity 1) are given by the grammar:
$\phi, \psi ::= P \mid \bot \mid \phi \rightarrow \psi \mid \heartsuit \phi_1 ... \phi_n$. Satisfaction is defined wrt $\mathfrak{M} = (\gamma, I)$ and a specific point $c \in C$ in a standard way, see e.g. [24,25].

Definition and Proposition 13. Define the *coalgebraic standard translation* as $ST_x(P) := P(x)$, $ST_x(\heartsuit \phi_1 \ldots \phi_n) := x \heartsuit \lceil x : ST_x(\phi_1) \rceil \ldots \lceil x : ST_x(\phi_n) \rceil$, $ST_x(\bot) = \bot$, $ST_x(\phi \to \psi) = ST_x(\phi) \to ST_x(\psi)$. Then for any $\phi \in \mathcal{CML}_\Lambda(\Sigma)$, and any $\mathfrak{M} = (\gamma, I), v, c$, we have $\mathfrak{M}, c \vDash \phi$ iff $\mathfrak{M}, v[c/x] \vDash ST_x(\phi)$.

For example, $ST_x(\heartsuit \heartsuit P) = x \heartsuit \lceil x : x \heartsuit \lceil x : P(x) \rceil \rceil$. This definition is more straight-forward than the standard translation into FOL of modal logic over ordinary Kripke frames. Moreover, ST_x uses only one variable from iVar, namely x itself. In the context of standard Kripke models, expressiveness of modal logic is characterized by van Benthem's theorem: modal logic is the bisimulation invariant fragment of first-order logic in the corresponding signature. The finitary analogue of this theorem [21] states that every formula that is bisimulation invariant *over finite models* is equivalent *over finite models* to a modal formula. In the coalgebraic context, replace bisimilarity with behavioural equivalence [29]. Moreover, we need to assume that the language has 'enough' expressive power; e.g., we cannot expect that bisimulation invariant formulas are equivalent to CML formulas over the empty similarity type. This is made precise as follows:

Definition 14. The Λ-structure T is *separating* if, for every set X, every element $t \in TX$ is uniquely determined by the set $\{(\heartsuit, A) \mid \heartsuit \in \Lambda \; n\text{-ary}, A \in \mathcal{P}(X)^n, t \in [\![\heartsuit]\!]_X(A)\}$.

Separation is in general a less restrictive condition than those we needed for completeness proofs. Of all examples introduced in Section 2, the only one which fails it is coalition logic. It was first used to establish the Hennessy-Milner property for coalgebraic logics [18,23] and it is easy to see that all our examples are indeed separating. In particular, separation automatically obtains for Kripke semantics.

Theorem 15. *Suppose that T is separating and $\phi(x)$ is a CPL formula with one free variable. Then ϕ is invariant under behavioural equivalence (over finite models) iff it is equivalent to an infinitary CML formula with finite modal rank (over finite models).*

If we deal with finite similarity types only, the conclusion can be strengthened:

Theorem 16. *Suppose that T is separating, Λ is finite and $\phi(x)$ is a CPL formula with one free variable. Then ϕ is invariant under behavioural equivalence (over finite models) iff ϕ is equivalent to a **finite** CML formula (over finite models).*

The proof uses [24, Theorem 24], which in turn relies on a somewhat less natural three-sorted language. Instantiated to the case of Kripke models, we recover the classical results of [2,21].

5   Beginning Model Theory

We proceed to develop some basic notions of coalgebraic model theory: we introduce an ultraproduct construction on coalgebras, and we prove a downward Lowenheim-Skolem theorem. As is often the case in coalgebraizations of classical model constructions, the structure on the ultraproduct is not uniquely determined, so we refer to the candidate structures as quasi-ultraproducts. Since ultraproducts imply compactness, they will exist only under restrictive conditions, specifically a semantic version of the alternative conditions needed for the completeness theorem. The assumptions needed for the Lowenheim-Skolem theorem are slightly more relaxed.

Observe that if $X = \prod_{\mathfrak{U}} X_i$ is an ultraproduct of sets and (A_i) is a family of subsets $A_i \subseteq X_i$, then $A = \prod_{\mathfrak{U}} A_i := \{x \mid \{i \mid x_i \in A_i\} \in \mathfrak{U}\}$ is a well-defined subset of X.

Definition 17 (Quasi-Ultraproducts of Coalgebras). Let $(C_i) = (X_i, \xi_i)_{i \in I}$ be a family of T-coalgebras, and let $\mathfrak{U}$ be an ultrafilter on I. A coalgebra ξ on the set-ultraproduct $X = \prod_{\mathfrak{U}} X_i$ is called a *quasi-ultraproduct* of the C_i if for every family (A_i) of subsets $A_i \subseteq X_i$, every $x \in \prod_{\mathfrak{U}} X_i$, and every $\heartsuit \in \Lambda$,

$$\xi(x) \in [\![\heartsuit]\!]_X \prod\nolimits_{\mathfrak{U}} A_i \iff \{i \in I \mid \xi_i(x_i) \in [\![\heartsuit]\!]_{C_i}(A_i)\} \in \mathfrak{U}. \tag{1}$$

The notion of quasi-ultraproduct extends naturally to coalgebraic models.

Theorem 18 (Coalgebraic Łoś's Theorem). *If* $\mathfrak{M} = (C, \gamma, V)$ *is a quasi-ultraproduct of* $\mathfrak{M}_i = (C_i, \gamma_i, V_i)$ *for the ultrafilter* $\mathfrak{U}$*, then for every tuple* $(a^1, \ldots, a^n)$ *of states in* C*, where* $a^k = (a_i^k)_{i \in I}$*, and for every CPL formula* $\phi(x_1, \ldots, x_n)$*,* $C \models \phi(a^1, \ldots, a^n) \iff \{i \mid C_i \models \phi(a_i^1, \ldots, a_i^k)\} \in \mathfrak{U}.$

From this theorem, we obtain the usual applications, in particular compactness. The question is, of course, when quasi-ultraproducts exist. A core observation is

Lemma 19. *In the notation of Definition 17, the demands placed on* $\xi(x)$ *by (1) constitute a finitely satisfiable set of one-step formulas.*

The lemma immediately implies that the quasi-ultraproducts exist if the logic is one-step compact, e.g. for neighbourhood logic and coalition logic. This is a mild generalization of the corresponding construction in [4]. Alternatively, we can use bounded operators, along with a semantic version of finitary S1SC axiomatizability:

Definition 20. A Λ-structure is *finitary one-step compact* if for every set X, every finitely satisfiable set $\Phi \subseteq \mathcal{M}_\Lambda^\wedge(\mathcal{P}_{fin}(X))$ of one-step formulas is satisfiable.

Remark 21. Finitary one-step compactness is clearly a consequence of finitary S1SC, hence all our "Kripke-like" cases enjoy this property. Interestingly enough, Example 11 also happens to be finitary one-step compact although its operators are only ω-bounded but not k-bounded. While the ultraproduct construction cannot be available in such cases (cf. Theorem 12), counterparts of some other standard results fare better, notably Lowenheim-Skolem.

Theorem 22. *If a Λ-structure is finitary one-step compact and all its operators are bounded, then it has quasi-ultraproducts.*

Theorem 23 (The Downward Löwenheim-Skolem Theorem). *If a Λ-structure is ω-bounded and finitary one-step compact, then* $\mathcal{CPL}_\Lambda$ *satisfies the downward Löwenheim-Skolem theorem.*

Theorem 24. *If a Λ-structure is one-step compact, then* $\mathcal{CPL}_\Lambda$ *satisfies the downward Löwenheim-Skolem.*

A special case of Theorem 24 for neighbourhood logic has been proved in [4].

6 Conclusions and Further Work

We believe this work opens up several new research avenues. The route towards coalgebraic finite model theory has already been paved in [24], and our Van Benthem-Rosen result is based on the spadework done therein. It is worth observing that Van Benthem-Rosen is a rare instance of a model-theoretic characterization of a fragment of FOL which remains valid over finite models. The only other major one we are aware of is the characterization of existential-positive formulas as exactly those preserved under homomorphisms [22]. The result is relevant to constraint satisfaction problems and to database theory, as existential-positive formulas correspond to unions of conjunctive queries. Interestingly, the proof of Rossman's result relies on Gaifman graphs, which also play a central role in the proof of the Rosen theorem used in [24]. A general CPL variant of Rossman result and development of non-relational database theory seem thus natural research directions.

Generalizations of standard results of *classical* model theory like Beth definability or interpolation and the Keisler-Shelah characterization theorem also seem an interesting research problem. A Herbrand theorem could lead towards an investigation of logic programming in a general coalgebraic setting.

While we are rather satisfied with the shape of our Hilbert-style axiomatization, it would certainly be of interest to study Gentzen-style proof systems. A natural route to explore would be to marry ordinary proof systems for first-order logic with one-step Gentzen systems for CML [19]. This will be in fact the subject of our forthcoming paper.

It remains to be seen which results of *modal model theory* building upon the interplay between modal and predicate languages can be generalized. Specific potential examples include Sahlqvist-type results for suitably well-behaved structures and analogues of results by Fine (does elementary generation imply canonicity, at least wherever the coalgebraic Jónsson-Tarski theorem [15] obtains?) or Hodkinson [13] (is there an algorithm generating a CML axiomatization for CPL-definable classes of coalgebras?).

Finally, a very natural future work from the point of view of the coalgebraic community would be to study models based on coalgebras for endofunctors on other categories than **Set** and variants of CPL with non-boolean propositional bases.

References

1. Alur, R., Henzinger, T.A., Kupferman, O.: Alternating-time temporal logic. J. ACM 49, 672–713 (2002)
2. van Benthem, J.: Modal Correspondence Theory. Ph.D. thesis, Department of Mathematics, University of Amsterdam (1976)
3. ten Cate, B., Gabelaia, D., Sustretov, D.: Modal languages for topology: Expressivity and definability. Ann. Pure Appl. Logic 159(1-2), 146–170 (2009)
4. Chang, C.: Modal model theory. In: Cambridge Summer School in Mathematical Logic. LNM, vol. 337, pp. 599–617. Springer (1973)
5. Chellas, B.: Modal Logic. Cambridge University Press (1980)
6. Cirstea, C., Kurz, A., Pattinson, D., Schröder, L., Venema, Y.: Modal logics are coalgebraic. The Computer J. 54, 31–41 (2011)

7. Demri, S., Lugiez, D.: Presburger Modal Logic Is PSPACE-Complete. In: Furbach, U., Shankar, N. (eds.) IJCAR 2006. LNCS (LNAI), vol. 4130, pp. 541–556. Springer, Heidelberg (2006)

8. Enderton, H.B.: A mathematical introduction to logic. Academic Press (1972)

9. Fine, K.: In so many possible worlds. Notre Dame J. Formal Logic 13, 516–520 (1972)

10. Goldblatt, R.: An abstract setting for Henkin proofs. CSLI Lecture Notes, pp. 191–212. CSLI Publications (1993)

11. Halpern, J.Y.: An analysis of first-order logics of probability. Artif. Intell. 46(3), 311–350 (1990)

12. Hansen, H.H., Kupke, C., Pacuit, E.: Neighbourhood structures: Bisimilarity and basic model theory. Log. Methods Comput. Sci. 5 (2009)

13. Hodkinson, I.: Hybrid formulas and elementarily generated modal logics. Notre Dame J. Formal Logic 47, 443–478 (2006)

14. Jacobs, B.: Predicate logic for functors and monads (2010)

15. Kupke, C., Kurz, A., Pattinson, D.: Ultrafilter Extensions for Coalgebras. In: Fiadeiro, J.L., Harman, N.A., Roggenbach, M., Rutten, J. (eds.) CALCO 2005. LNCS, vol. 3629, pp. 263–277. Springer, Heidelberg (2005)

16. Larsen, K., Skou, A.: Bisimulation through probabilistic testing. Inf. Comput. 94, 1–28 (1991)

17. Makowsky, J.A., Marcja, A.: Completeness theorems for modal model theory with the Montague-Chang semantics I. Math. Logic Quarterly 23, 97–104 (1977)

18. Pattinson, D.: Expressive logics for coalgebras via terminal sequence induction. Notre Dame J. Formal Logic 45, 19–33 (2004)

19. Pattinson, D., Schröder, L.: Cut elimination in coalgebraic logics. Inf. Comput. 208, 1447–1468 (2010)

20. Pauly, M.: A modal logic for coalitional power in games. J. Log. Comput. 12, 149–166 (2002)

21. Rosen, E.: Modal logic over finite structures. J. Logic, Language and Information 6(4), 427–439 (1997)

22. Rossman, B.: Homomorphism preservation theorems. J. ACM 55, 15:1–15:53 (2008)

23. Schröder, L.: A finite model construction for coalgebraic modal logic. J. Log. Algebr. Prog. 73, 97–110 (2007)

24. Schröder, L., Pattinson, D.: Coalgebraic Correspondence Theory. In: Ong, L. (ed.) FOSSACS 2010. LNCS, vol. 6014, pp. 328–342. Springer, Heidelberg (2010)

25. Schröder, L., Pattinson, D.: Named models in coalgebraic hybrid logic. In: Symposium on Theoretical Aspects of Computer Science, STACS 2010. Leibniz Int. Proceedings in Informatics, vol. 5, pp. 645–656. Schloss Dagstuhl – Leibniz-Center of Informatics (2010)

26. Schröder, L., Pattinson, D.: Rank-1 modal logics are coalgebraic. J. Log. Comput. 20, 1113–1147 (2010)

27. Seligman, J., Liu, F., Girard, P.: Logic in the Community. In: Banerjee, M., Seth, A. (eds.) ICLA 2011. LNCS, vol. 6521, pp. 178–188. Springer, Heidelberg (2011)

28. Sgro, J.: The interior operator logic and product topologies. Trans. AMS 258, 99–112 (1980)

29. Staton, S.: Relating coalgebraic notions of bisimulation. Log. Methods Comput. Sci. 7 (2011)

30. Ziegler, A.: Topological model theory. In: Barwise, J., Feferman, S. (eds.) Model-Theoretic Logics. Springer (1985)

Algorithmic Games for Full Ground References

Andrzej S. Murawski[1] and Nikos Tzevelekos[2],[*]

[1] University of Leicester
[2] Queen Mary, University of London

Abstract. We present a full classification of decidable and undecidable cases for contextual equivalence in a finitary ML-like language equipped with full ground storage (both integers and reference names can be stored). The simplest undecidable type is unit $\to$ unit $\to$ unit. At the technical level, our results marry game semantics with automata-theoretic techniques developed to handle infinite alphabets. On the automata-theoretic front, we show decidability of the emptiness problem for register pushdown automata extended with fresh-symbol generation.

1 Introduction

Mutable variables in which numerical values can be stored for future access and update are the pillar of imperative programming. The memory in which the values are deposited can be allocated statically, typically to coincide with the lifetime of the defining block, or dynamically, on demand, with the potential to persist forever. In order to support memory management, modern programming languages feature mechanisms such as *pointers* or *references*, which allow programmers to access memory via addresses. Languages like C (through int*) or ML (via int ref ref) make it possible to store the addresses themselves, which creates the need for storing references to references etc. We refer to this scenario as *full ground storage*. In this paper we study an ML-like language GRef with full ground storage, which permits the creation of references to integers as well as references to integer references, and so on.

We concentrate on contextual equivalence[1] in that setting. Reasoning about program equivalence has been a central topic in programming language semantics since its inception. This is in no small part due to important applications, such as verification problems (equivalence between a given implementation and a model implementation) and compiler optimization (equivalence between the original program and its transform). Specifically, we attack the problem of automated reasoning about our language in a finitary setting, with finite datatypes and with looping instead of recursion, where decidability questions become interesting and the decidability/undecidability frontier can be identified. In particular, it is possible to quantify the impact of higher-order types on decidability, which goes unnoticed in Turing-complete frameworks.

The paper presents a complete classification of cases in which GRef program equivalence is decidable. The result is phrased in terms of the syntactic shape of types.

---

[*] Supported by a Royal Academy of Engineering Research Fellowship.
[1] Two program phrases are regarded as *contextually equivalent*, or simply *equivalent*, if they can be used interchangeably in any context without affecting the observable outcome.

A. Czumaj et al. (Eds.): ICALP 2012, Part II, LNCS 7392, pp. 312–324, 2012.
© Springer-Verlag Berlin Heidelberg 2012

We write $\theta_1, \cdots, \theta_k \vdash \theta$ to refer to the problem of deciding contextual equivalence between two terms M_1, M_2 such that $x_1 : \theta_1, \cdots, x_m : \theta_m \vdash M_i : \theta$ $(i = 1, 2)$. We investigate the problem using a fully abstract game model of GRef.[2] Such a model can be easily obtained by modifying existing models of more general languages, e.g. by either adding type information to Laird's model of untyped references [14] or trimming down our own model for general references [18]. The models are *nominal* [1,14] in that moves may involve elements from an infinite set of *names* to account for reference names. Additionally, each move is equipped with a store whose domain consists of all names that have been revealed (played) thus far and the corresponding values. Note that values of reference types also become part of the domain of the store. This representation grows as the play unfolds and new names are encountered. We shall rely on the model both for decidability and undecidability results. Our work identifies the following undecidable cases as minimal.

$\vdash$ unit $\to$ unit $\to$ unit   (unit $\to$ unit $\to$ unit) $\to$ unit $\vdash$ unit

$\vdash$ ((unit $\to$ unit) $\to$ unit) $\to$ unit   (((unit $\to$ unit) $\to$ unit) $\to$ unit) $\to$ unit $\vdash$ unit

Obviously, undecidability extends to typing judgments featuring syntactic supertypes of those listed above (for instance, when fourth-order types appear on the left-hand side of the turnstile or types of the shape $\theta_1 \to \theta_2 \to \theta_3$ occur on the right). The remaining cases are summarized by typing judgements in which each of $\theta_1, \cdots, \theta_m$ is generated by the grammar given on the left below, and θ by the grammar on the right,

$$\Theta_L ::= \beta \mid \Theta_R \to \Theta_L \qquad\qquad \Theta_R ::= \beta \mid \Theta_1 \to \beta$$

where β stands any ground type and Θ_1 is a first-order type, i.e. $\beta ::= \text{unit} \mid \text{int} \mid \text{ref}^i \text{ int}$ and $\Theta_1 ::= \beta \mid \beta \to \Theta_1$. We shall show that all these cases are in fact decidable. In order to arrive at a decision procedure we rely on effective reducibility to a canonical (β-normal) form. These forms are then inductively translated into a class of automata over infinite alphabets that represent the associated game semantics. Finally, we show that the representations can be effectively compared for equivalence.

The automata we use are especially designed to read moves-with-stores in a single computational step. They are equipped with a finite set of registers for storing elements from the infinite alphabet (names). Moreover, in a single transition step, the content of a subset of registers can be pushed onto the stack (along with a symbol from the stack alphabet), to be popped back at a later stage. We use visibly pushdown stacks [4], i.e. the alphabet can be partitioned into letters that consistently trigger the same stack actions (push, pop or no-op). Conceptually, the automata extend register pushdown automata [6] with the ability to generate fresh names, as opposed to their existing capability to generate names not currently present in registers. Crucially, we can show that the emptiness problem for the extended machine model remains decidable.

Because the stores used in game-semantic plays can grow unboundedly, one cannot hope to construct the automata in such a way that they will accept the full game semantics of terms. Instead we construct automata that, without loss of generality, will accept plays in which the domains of stores are bounded in size. Each such restricted

---

[2] A model is *fully abstract* if it captures contextual equivalence denotationally, i.e. equivalence can be confirmed/disproved by reference to the interpretations of terms.

play can be taken to represent a *set* of real plays compatible with the representation. Compatibility means that values of names omitted in environment-moves (*O*-moves) can be filled in arbitrarily, but values of names omitted in program-moves (*P*-moves) must be the same as in preceding *O*-moves. That is to say, the omissions leading to bounded representation correspond to copy-cat behaviour.

Because we work with representations of plays, we cannot simply use off-the-shelf procedures for checking program equivalence, as the same plays can be represented in different ways: copy-cat behaviour can be modelled explicitly or implicitly via the convention. However, taking advantage of the fact that stacks of two visibly pushdown automata over the same partitioning of the alphabet can be synchronized, we show how to devise another automaton that can run automata corresponding to two terms in parallel and detect inconsistencies in the representations of plays. Exploiting decidability of the associated emptiness problem, we can conclude that GRef program equivalence in the above-mentioned cases is decidable.

Related Work. The investigations into models and reasoning principles for storage have a long history. In this quest, storage of names was regarded by researchers as an indispensable intermediate step towards capturing realistic languages with dynamic-allocated storage, such as ML or Java. Relational methods and environmental bisimulations for reasoning about program equivalence in settings similar to ours were studied in [20,5,13,3,7,21], albeit without decidability results. More foundational work included labelled transition system semantics [11] and game semantics [14,18]. In both cases, it turned out that the addition of name storage simplified reasoning, be it bisimulation-based or game-semantic. In the former case, bisimulation was even un-sound without full ground storage. In the latter case, the game model of integer storage [16] turned out more intricate (complicated store abstractions) than that for full ground or general storage [14,18]. As for decidability results, finitary Reduced ML (integer storage only) was studied by us in [17], yet only judgements of the form $\cdots, \beta \to \beta, \cdots \vdash \beta$ were tackled due to intricacies related to store abstractions (in absence of full ground storage, names cannot be remembered by programs). A closely related language, called RML [2] (integer storage but with *bad references*, that is, constructs of reference type which do not correspond to valid reference cells) was studied in [15,10], but no full classification has emerged yet. In particular, although the class of types shown decidable is common in both languages, the status of the types that we list as undecidable above remains open in the case of RML.

2   GRef

We work with a finitary ML-like language GRef whose types θ are generated according to the following grammar.

$$\theta ::= \beta \mid \theta \to \theta \qquad \beta ::= \text{unit} \mid \gamma \qquad \gamma ::= \text{int} \mid \text{ref}\,\gamma$$

Note that reference types are available for each type of the shape γ (full ground storage). The language is best described as the call-by-value λ-calculus over the ground types β augmented with finitely many constants, do-nothing command, case distinction, looping, and reference manipulation (allocation, dereferencing, assignment). The typing

$$\frac{}{\Gamma \vdash () : \text{unit}} \qquad \frac{i \in \{0, \cdots, max\}}{\Gamma \vdash i : \text{int}} \qquad \frac{(x : \theta) \in \Gamma}{\Gamma \vdash x : \theta} \qquad \frac{\Gamma \vdash M : \text{int} \quad \Gamma \vdash N : \text{unit}}{\Gamma \vdash \text{while } M \text{ do } N : \text{unit}}$$

$$\frac{\Gamma \vdash M : \text{int} \quad \Gamma \vdash N_0 : \theta \quad \cdots \quad \Gamma \vdash N_{max} : \theta}{\Gamma \vdash \text{case}(M)[N_0, \cdots, N_{max}] : \theta} \qquad \frac{\Gamma \vdash M : \theta \to \theta' \quad \Gamma \vdash N : \theta}{\Gamma \vdash MN : \theta'}$$

$$\frac{\Gamma \cup \{x : \theta\} \vdash M : \theta'}{\Gamma \vdash \lambda x^\theta . M : \theta \to \theta'} \qquad \frac{\Gamma \vdash M : \gamma}{\Gamma \vdash \text{ref}_\gamma(M) : \text{ref } \gamma} \qquad \frac{\Gamma \vdash M : \text{ref } \gamma}{\Gamma \vdash !M : \gamma} \qquad \frac{\Gamma \vdash M : \text{ref } \gamma \quad \Gamma \vdash N : \gamma}{\Gamma \vdash M := N : \text{unit}}$$

Fig. 1. Syntax of GRef

rules are given in Figure 1. In what follows, we write $M; N$ for the term $(\lambda z^\theta . N)M$, where z does not occur in N and θ matches the type of M. let $x = M$ in N will stand for $(\lambda x^\theta . N)M$ in general. The operational semantics of the language can be found, e.g. in [18]. Note that, if $max > 0$, reference equality is expressible in our syntax [19].

Definition 1. *We say that the term-in-context $\Gamma \vdash M_1 : \theta$ **approximates** $\Gamma \vdash M_2 : \theta$ (written $\Gamma \vdash M_1 \lesssim M_2$) if $C[M_1] \Downarrow$ implies $C[M_2] \Downarrow$ for any context $C[-]$ such that $\vdash C[M_1], C[M_2] : \text{unit}$. Two terms-in-context are **equivalent** if one approximates the other (written $\Gamma \vdash M_1 \cong M_2$).*

3   Game Semantics

Game semantics views computation as a dialogue between the environment (Opponent, O) and the program (Proponent, P). We give an overview of the fully abstract game model of GRef [14,18]. Let $\mathbb{A} = \biguplus_\gamma \mathbb{A}_\gamma$ be a collection of countably infinite sets of *reference names*, or just *names*. The model is constructed using mathematical objects (moves, plays, strategies) that will feature names drawn from $\mathbb{A}$. Although names underpin various elements of our model, their precise nature is irrelevant. Hence, all of our definitions preserve name-invariance, i.e. our objects are (strong) *nominal sets* [8,22]. Note that we do not need the full power of the theory but mainly the basic notion of name-permutation. For an element x belonging to a (nominal) set X, we write $\nu(x)$ for its name-support, i.e. the set of names occurring in x. Moreover, for any $x, y \in X$, we write $x \sim y$ if x and y are the same up to a permutation of $\mathbb{A}$. Our model is couched in the Honda-Yoshida style of modelling call-by-value computation [9]. Before we define what it means to play our games, let us introduce the auxiliary concept of an arena.

Definition 2. *An arena $A = \langle M_A, I_A, \lambda_A, \vdash_A \rangle$ is given by a set M_A of moves, its subset I_A of initial ones, a labelling function $\lambda_A : M_A \to \{O, P\} \times \{Q, A\}$ and a justification relation $\vdash_A \subseteq M_A \times (M_A \setminus I_A)$.*
In addition, for all $m, m' \in M_A$, we stipulate: $m \in I_A \implies \lambda_A(m) = (P, A)$, $m \vdash_A m' \wedge \lambda_A^{QA}(m) = A \implies \lambda_A^{QA}(m') = Q$, $m \vdash_A m' \implies \lambda_A^{OP}(m) \neq \lambda_A^{OP}(m')$. We write λ_A^{OP} (resp. λ_A^{QA}) for λ_A post-composed with the first (second) projection.

We shall use ι to range over initial moves. Let $\overline{\lambda}_A$ be the OP-complement of λ_A. Given arenas A, B, the arenas $A \otimes B$ and $A \Rightarrow B$ are constructed as in the following figure,

$$M_{A\Rightarrow B} = \{\star\} \uplus M_A \uplus M_B, \ I_{A\Rightarrow B} = \{\star\} \quad M_{A\otimes B} = (I_A \times I_B) \uplus \bar{I}_A \uplus \bar{I}_B, \ I_{A\otimes B} = I_A \times I_B$$

$$\lambda_{A\Rightarrow B} = [\star \mapsto PA, \bar{\lambda}_A[\iota_A \mapsto OQ], \lambda_B] \quad \lambda_{A\otimes B} = [(\iota_A, \iota_B) \mapsto PA, \lambda_A \upharpoonright \bar{I}_A, \lambda_B \upharpoonright \bar{I}_B]$$

$$\vdash_{A\Rightarrow B} = \{(\star, \iota_A), (\iota_A, \iota_B)\} \cup \vdash_A \cup \vdash_B \quad \vdash_{A\otimes B} = \{((\iota_A, \iota_B), m) \mid \iota_A \vdash_A m \vee \iota_B \vdash_B m\}$$
$$\cup \bar{\vdash}_A \cup \bar{\vdash}_B$$

where $\bar{I}_A = M_A \setminus I_A, \bar{\vdash}_A = (\vdash_A \upharpoonright \bar{I}_A \times \bar{I}_A)$ (and similarly for B). Let us write $[i, j]$ for the set $\{i, i+1, \cdots, j\}$. For each type θ we can define the corresponding arena $[\![\theta]\!]$.

$$[\![\text{unit}]\!] = \langle \{\star\}, \{\star\}, \{(\star, PA)\}, \emptyset \rangle \qquad [\![\text{ref } \gamma]\!] = \langle \mathbb{A}_\gamma, \mathbb{A}_\gamma, \{(a, PA) \mid a \in \mathbb{A}_\gamma\}, \emptyset \rangle$$

$$[\![\text{int}]\!] = \langle [0, max], [0, max], \{(i, PA) \mid i \in [0, max]\}, \emptyset \rangle \qquad [\![\theta \to \theta']\!] = [\![\theta]\!] \Rightarrow [\![\theta']\!]$$

Although types are interpreted by arenas, the actual games will be played in *prearenas*, which are defined in the same way as arenas with the exception that initial moves are O-questions. Given arenas A, B we define the prearena $A \to B$ as follows.

$$M_{A\to B} = M_A \uplus M_B \qquad \lambda_{A\to B} = [\bar{\lambda}_A[\iota_A \mapsto OQ], \lambda_B]$$

$$I_{A\to B} = I_A \qquad \vdash_{A\to B} = \{(\iota_A, \iota_B)\} \cup \vdash_A \cup \vdash_B$$

A *store* is a type-sensitive finite partial function $\Sigma : \mathbb{A} \rightharpoonup [0, max] \cup \mathbb{A}$ such that $a \in \text{dom}(\Sigma) \cap \mathbb{A}_{\text{int}}$ implies $\Sigma(a) \in [0, max]$, and $a \in \text{dom}(\Sigma) \cap \mathbb{A}_{\text{ref } \gamma}$ implies $\Sigma(a) \in \text{dom}(\Sigma) \cap \mathbb{A}_\gamma$. We write Sto for the set of all stores. A move-with-store on a (pre)arena A is a pair m^Σ with $m \in M_A$ and $\Sigma \in$ Sto.

Definition 3. *A* justified sequence *on a prearena A is a sequence of moves-with-store on A such that, apart from the first move, which must be of the form ι^Σ with $\iota \in I_A$, every move $n^{\Sigma'}$ in s is equipped with a pointer to an earlier move m^Σ such that $m \vdash_A n$. m is then called the justifier of n.*

For each $S \subseteq \mathbb{A}$ and Σ we define $\Sigma^0(S) = S$ and $\Sigma^{i+1}(S) = \Sigma(\Sigma^i(S)) \cap \mathbb{A} \ (i \geq 0)$. Let $\Sigma^*(S) = \bigcup_i \Sigma^i(S)$. The set of *available names* of a justified sequence is defined inductively by $\text{Av}(\epsilon) = \emptyset$ and $\text{Av}(sm^\Sigma) = \Sigma^*(\text{Av}(s) \cup \nu(m))$. The view of a justified sequence is defined by: $view(\epsilon) = \epsilon$, $view(m^\Sigma) = m^\Sigma$ and $view(s \, \widehat{m^\Sigma t} \, n^{\Sigma'}) = view(s) \, m^\Sigma n^{\Sigma'}$. We shall write $s \sqsubseteq s'$ to mean that s is a prefix of s'.

Definition 4. *Let A be a prearena. A justified sequence s on A is called a* **play***, if it satisfies the conditions below.*

- *No adjacent moves belong to the same player (Alternation).*
- *The justifier of each answer is the most recent unanswered question (Bracketing).*
- *For any $s'm^\Sigma \sqsubseteq s$ with non-empty s', the justifier of m occurs in $view(s')$ (Visibility).*
- *For any $s'm^\Sigma \sqsubseteq s$, $\text{dom}(\Sigma) = \text{Av}(s'm^\Sigma)$ (Frugality).*

Definition 5. *A* **strategy** *σ on a prearena A, written $\sigma : A$, is a set of even-length plays of A satisfying:*

- *If $so^\Sigma p^{\Sigma'} \in \sigma$ then $s \in \sigma$ (Even-prefix closure).*
- *If $s \in \sigma$ and $s \sim t$ then $t \in \sigma$ (Equivariance).*

– *If* $s_1 p_1^{\Sigma_1}, s_2 p_2^{\Sigma_2} \in \sigma$ *and* $s_1 \sim s_2$ *then* $s_1 p_1^{\Sigma_1} \sim s_2 p_2^{\Sigma_2}$ *(Nominal determinacy).*

Following [14,18], GRef-terms $\Gamma \vdash M : \theta$, where $\Gamma = \{x_1 : \theta_1, \cdots, x_n : \theta_n\}$ can be interpreted by strategies for the prearena $[\![\theta_1]\!] \otimes \cdots \otimes [\![\theta_n]\!] \rightarrow [\![\theta]\!]$, which we shall denote by $[\![\Gamma \vdash \theta]\!]$. Given a set of plays X, let us write $\mathsf{comp}(X)$ for the set of complete plays in X, i.e. those in which each occurrence of a question justifies an answer. The interpretation given in [14,18] is then fully abstract in the following sense.

Proposition 6 ([14,18]). *Let* $\Gamma \vdash M_1, M_2 : \theta$ *be* GRef-*terms.* $\Gamma \vdash M_1 \precsim M_2$ *if, and only if,* $\mathsf{comp}([\![\Gamma \vdash M_1 : \theta]\!]) \subseteq \mathsf{comp}([\![\Gamma \vdash M_2 : \theta]\!])$. *Hence,* $\Gamma \vdash M_1 \cong M_2$ *if, and only if,* $\mathsf{comp}([\![\Gamma \vdash M_1 : \theta]\!]) = \mathsf{comp}([\![\Gamma \vdash M_2 : \theta]\!])$.

We shall rely on the result for proving both undecidability and decidability results, by referring to complete plays generated by terms.

Example 7. The name-generating term $\vdash \lambda x^{\mathsf{unit}}.\mathsf{ref}(0) : \mathsf{unit} \rightarrow \mathsf{ref\ int}$ yields complete plays of the shape given below (the corresponding prearena is given on the right).

$$q^\emptyset \quad \star^\emptyset \quad q_0^{\Sigma_0'} \quad a_0^{\Sigma_1} \quad \cdots \quad q_0^{\Sigma_{i-1}'} \quad a_i^{\Sigma_i} \quad q_0^{\Sigma_i'} \quad \cdots$$

$$\begin{array}{cc} q & O \\ | & \\ \star & P \\ | & \\ q_0 & O \\ | & \\ a & P \end{array}$$

where $\Sigma_0' = \emptyset$ and, for all $i > 0$, $\Sigma_i = \Sigma_{i-1}' \cup \{(a_i, 0)\}$, $\mathsf{dom}(\Sigma_i') = \mathsf{dom}(\Sigma_i)$. Moreover, for any $i \neq j$ we have $a_i \neq a_j$. Note that Σ_i' can be different from Σ_i, i.e. the environment is free to change the values stored at all of the locations that have been revealed to it.

Note that in the above example the sizes of stores keep on growing indefinitely. However, the essence of the strategy is already captured by plays of the shape $q \star q_0 a_0^{(a_0,0)} \cdots q_0 a_i^{(a_i,0)} q_0 \cdots$ under the assumption that, whenever a value is missing from the store of an O-move, it is arbitrary and, for P-moves, it is the same as in the preceding O-move. Next we spell out how a sequence of moves-with-store, not containing enough information to qualify as a play, can be taken to represent proper plays.

Definition 8. *Let* $s = m_1^{\Sigma_1} \cdots m_k^{\Sigma_k}$ *be a play over* $\Gamma \vdash \theta$ *and* $t = m_1^{\Theta_1} \cdots m_k^{\Theta_k}$ *be a sequence of moves-with-store. We say that s is an* extension *of t if* $\Theta_i \subseteq \Sigma_i$ $(1 \leq i \leq k)$ *and, for any* $1 \leq i \leq \lfloor k/2 \rfloor$, *if* $a \in \mathsf{dom}(\Sigma_{2i}) \setminus \mathsf{dom}(\Theta_{2i})$ *then* $\Sigma_{2i}(a) = \Sigma_{2i-1}(a)$. *We write* $\mathsf{ext}(t)$ *for the set of all extensions of t.*

Because we cannot hope to encode plays with unbounded stores through automata, our decidability results will be based on representations of plays that capture strategies via extensions.

4   Undecidability Arguments

We begin with undecidable cases. Our argument will rely on queue machines, which are finite-state devices equipped with a queue.

Definition 9. *Let $\mathcal{A}$ be a finite alphabet. A queue machine over $\mathcal{A}$ is specified by* $\langle Q, Q_E, Q_D, init, \delta_E, \delta_D \rangle$, *where Q is a finite set of states such that $Q = Q_E \uplus Q_D$, $init \in Q_E$ is the initial state, $\delta_E : Q_E \to Q \times \mathcal{A}$ is the enqueuing function, whereas $\delta_D : Q_D \times \mathcal{A} \to Q$ is the dequeuing function.*

A queue machine starts at state *init* with an empty queue. Whenever it reaches a state $q \in Q_E$, it will progress to the state $\pi_1 \delta_E(q)$ and $\pi_2 \delta_E(q)$ will be added to the associated queue. If the machine reaches a state $q \in Q_D$ and its queue is empty, the machine is said to *halt*. Otherwise, it moves to the state $\delta_D(q, x)$, where x is the symbol at the head of the associated queue, which is then removed from the queue. The halting problem for queue machines is well known to be undecidable (e.g. [12]). By encoding computation histories of queue machines as plays generated by GRef terms we next show that the equivalence problem for GRef terms must be undecidable. Note that this entails undecidability of the associated notion of term approximation.

Theorem 10. *The contextual equivalence problem is undecidable in the following cases.*
- $\vdash$ unit $\to$ unit $\to$ unit
- (unit $\to$ unit $\to$ unit) $\to$ unit $\vdash$ unit
- (((unit $\to$ unit) $\to$ unit) $\to$ unit) $\to$ unit $\vdash$ unit
- $\vdash$ ((unit $\to$ unit) $\to$ unit) $\to$ unit

We sketch the argument in the first case. The arena used to interpret closed terms of type unit $\to$ unit $\to$ unit has the shape given on the right. We are going to use plays from the arena to represent sequences of queue operations. Enqueuing will be represented by segments of the form $q_0 \star_0$, whereas $q_1 \star_1$ will be used to represent dequeuing. Additionally, in the latter case q_1 will be justified by $\star_0$ belonging to the segment representing the enqueuing of the element that is now being dequeued. For instance, the sequence $EEDEDE$, in which E, D stand for enqueuing and dequeuing respectively, will be represented as follows.

$$q \quad \\ | \\ \star \\ | \\ q_0 \\ | \\ \star_0 \\ | \\ q_1 \\ | \\ \star_1$$

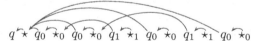

Observe that all such plays are complete. Given a queue machine $\mathbb{Q}$, let us write $\mathrm{hist}(\mathbb{Q})$ for the (prefix-closed) subset of $(E \uplus D)^*$ corresponding to all sequences of queue operations performed by $\mathbb{Q}$. Note that $\mathrm{hist}(\mathbb{Q})$ is finite if and only if $\mathbb{Q}$ halts. Additionally, define $\mathrm{hist}^-(\mathbb{Q})$ to be $\mathrm{hist}(\mathbb{Q})$ from which the longest sequence is removed (if $\mathrm{hist}(\mathbb{Q})$ is infinite and the sequence in question does not exist we set $\mathrm{hist}^-(\mathbb{Q}) = \mathrm{hist}(\mathbb{Q})$). Note that the sequence corresponds to a terminating run and necessarily ends in D.

Lemma 11. *Let $\mathbb{Q}$ be a queue machine. There exist terms $\vdash M, M^- :$ unit$\to$unit$\to$unit of GRef such that* $\mathrm{comp}(\llbracket M \rrbracket)$, $\mathrm{comp}(\llbracket M^- \rrbracket)$ *represent* $\mathrm{hist}(\mathbb{Q})$, $\mathrm{hist}^-(\mathbb{Q})$ *respectively.*

Proof. W.l.o.g. we shall assume that Q can be fitted into int (otherwise, we could use a fixed number of variables to achieve the desired storage capacity). Let $D[-] \equiv C[\lambda x.C_0[\lambda y.C_1[-]]]$, where $C[-], C_0[-], C_1[-]$ are given in Figure 2 ($*$ is a special symbol not in the queue alphabet and Ω is a canonical divergent term). $C_0[-]$ and $C_1[-]$ handle enqueuing and dequeuing respectively. We take M, M^- to be $D[()]$, $D[\text{if } (!STATE \in Q_D \wedge !!LAST = *) \text{ then } \Omega]$ respectively. $\qquad \square$

$C[-] =$ let $STATE = \text{ref}(init)$ in

    let $LAST = \text{ref}(\text{ref}(*))$ in $[-]$

$C_1[-] =$ if $(!STATE \notin Q_D)$ then Ω;

    if $(!!PREV \neq * \vee !SYM = *)$ then Ω;

    $STATE := \delta_D(!STATE, !SYM)$;

    $SYM := *; [-]$

$C_0[-] =$ if $(!STATE \notin Q_E)$ then Ω;

    let $SYM = \text{ref}(\pi_2 \delta_E(!STATE))$ in

    let $PREV = \text{ref}(!LAST)$ in

      $STATE := \pi_1 \delta_E(!STATE)$;

      $LAST := SYM; [-]$

Fig. 2. Simulating a queue machine in $\vdash$ unit $\to$ unit $\to$ unit. The variable $STATE$: ref int contains the current state of the machine. The queue is encoded as a backwards-connected list with elements $(PREV, SYM)$: ref^2 int $\times$ ref int, with last-element pointer $LAST$: ref^2 int. Enqueuing adds a new last element while dequeuing sets the first non-$*$ symbol of the list to $*$.

Observe that $\text{hist}(\mathbb{Q}) = \text{hist}^-(\mathbb{Q})$ exactly when $\mathbb{Q}$ does not halt. Consequently, the problem of deciding $\text{hist}(\mathbb{Q}) = \text{hist}^-(\mathbb{Q})$ is undecidable. Thus, via Proposition 6, we can conclude that program equivalence is undecidable for closed terms of type unit $\to$ unit $\to$ unit. The remaining cases are treated in a similar manner.

5 Decidability

We now focus on a fragment of GRef, called GRef☺, that comprises all types that do *not* fall under the undecidable cases identified earlier.

Definition 12. *Suppose* $\Gamma = x_1 : \theta_1, \cdots, x_m : \theta_m$. *The term-in-context* $\Gamma \vdash M : \theta$ *belongs to* GRef☺ *provided* $\theta_1, \cdots, \theta_m$ *can be generated from* Θ_L *and* θ *is generated from* Θ_R, *where* $\Theta_L ::= \beta \mid \Theta_R \to \Theta_L$, $\Theta_R ::= \beta \mid \Theta_1 \to \beta$ *and* $\Theta_1 ::= \beta \mid \beta \to \Theta_1$.

Put otherwise, we focus on sequents of the form: $\Theta_R \to \cdots \to \Theta_R \to \beta \vdash \Theta_R$, where $\Theta_R = (\beta \to \cdots \to \beta) \to \beta$. In order to show decidability we first translate GRef☺ terms into automata that represent their game semantics. Any GRef☺ term can be effectively converted to an equivalent term in canonical shape, which is captured by the grammar below. Consequently, it suffices to show that program equivalence between terms in canonical form is decidable. Accordingly, in what follows, we focus exclusively on translating terms in canonical shape.

$$C ::= () \mid i \mid x^{\text{ref } \gamma} \mid \lambda x^{\Theta_1}.C \mid \text{case}(x^{\text{int}})[C, \cdots, C] \mid (\text{while } (!x^{\text{ref int}}) \text{ do } C); C$$
$$\mid \text{let } y^\gamma = !x^{\text{ref } \gamma} \text{ in } C \mid (x^{\text{ref int}} := i); C \mid (x^{\text{ref}^2 \gamma} := y^{\text{ref } \gamma}); C$$
$$\mid \text{let } x^{\text{ref int}} = \text{ref}(0) \text{ in } C \mid \text{let } x^{\text{ref}^2 \gamma} = \text{ref}(y^{\text{ref } \gamma}) \text{ in } C \mid \text{let } y^{\Theta_L} = z \, () \text{ in } C$$
$$\mid \text{let } y^{\Theta_L} = z \, i \text{ in } C \mid \text{let } y^{\Theta_L} = z \, x^{\text{ref } \gamma} \text{ in } C \mid \text{let } y^{\Theta_L} = z \, (\lambda x^{\Theta_1}.C) \text{ in } C$$

Each type θ can be written in the form $\theta = \theta_n \to \ldots \to \theta_1 \to \beta$, for types $\theta_1, \ldots, \theta_n$ and base type β. For brevity, we shall write $\theta = (\theta_n, \ldots, \theta_1, \beta)$. We call n the *arity* of θ and denote it by $ar(\theta)$. Next we fix notation for referring to moves that are available in arenas corresponding to GRef☺ typing judgments: each move can be viewed as a pair (l, t) subject to consistency constraints induced by the subtypes which contribute them, e.g. the label corresponding to a tag related to int must be a number from $[0, max]$.

Definition 13. *For every type θ let us define the associated set of labels $\mathcal{L}_\theta$ as follows:*
$\mathcal{L}_{\text{unit}} = \{\star\}$, $\mathcal{L}_{\text{int}} = \{0, \cdots, max\}$, $\mathcal{L}_{\text{ref } \gamma} = \mathbb{A}_\gamma$, $\mathcal{L}_{\theta \to \theta'} = \{\star\}$.
We shall write $\mathcal{L}$ for the set of all labels. Given a $\mathsf{GRef}\odot$ *typing judgement* $\Gamma \vdash M : \theta$
we write $\mathbb{T}$ *for the set of associated tags:*

$$\mathbb{T} = \{c_i, r_i \mid \theta \equiv \theta' \to \beta,\, 0 \le i \le ar(\theta')\} \cup \{c_i^x, r_i^x \mid (x : (\theta_m, \cdots, \theta_1, \beta)) \in \Gamma,\, 0 < i \le m\}$$
$$\cup \{c_{j,i}^x, r_{j,i}^x \mid (x : (\theta_m, \cdots, \theta_1, \beta)) \in \Gamma,\, 0 < j \le m,\, \theta_j \equiv \theta' \to \beta,\, 0 \le i \le ar(\theta')\} \cup \{r_\downarrow\}$$

partitioned as $\mathbb{T} = \mathbb{T}_{\text{push}} \uplus \mathbb{T}_{\text{pop}} \uplus \mathbb{T}_{\text{noop}}$, *where* $\mathbb{T}_{\text{push}} = \{c_i, c_i^x, c_{j,i}^x \mid i > 0\}$, $\mathbb{T}_{\text{pop}} = \{r_i, r_i^x, r_{j,i}^x \mid i > 0\}$ *and* $\mathbb{T}_{\text{noop}} = \{c_0, r_0, c_{j,0}^x, r_{j,0}^x\}$.

The automata we shall rely on are equipped with finitely many, say n, registers for storing elements of $\mathbb{A}$, the first n_r of which are read-only. The content of registers, called *register assignment*, will be described by an injective partial function $\rho : [1, n] \rightharpoonup \mathbb{A}$. The set of all register assignments will be denoted by Reg and we shall use ρ to range over them. The automata will read elements of $\mathcal{L} \times \mathbb{T} \times \mathsf{Sto}$ (corresponding to moves-with-store) in a single transition step. In order to specify what label is to be read in a given step we use *symbolic labels* from the set $\mathbb{L} = \{\star\} \cup [0, max] \cup \{\mathsf{R}_i \mid 1 \le i \le n\}$ (R_i stands for the name stored in the ith register). To designate which tag is to be processed we simply use elements of (the finite set) $\mathbb{T}$. To describe stores, we shall use *symbolic stores* from the set $\mathsf{SSto} = \{S : [1, n] \rightharpoonup [0, max] \cup \{\mathsf{R}_1, \cdots, \mathsf{R}_n\} \mid [1, n_r] \subseteq \text{dom}(S)\}$. Symbolic stores represent stores by use of indices instead of actual names (for example, $S(i) = \mathsf{R}_j$ means that in S the i-th name stores the j-th name). Altogether, in order to define our automata, we will use transition labels from the set

$$\mathsf{TL} = \mathcal{P}([n_r+1, n]) \times \mathbb{L} \times (((\mathbb{T}_{\text{push}} \uplus \mathbb{T}_{\text{pop}}) \times \mathbb{C}_{\text{stack}} \times \mathsf{Mix}) \uplus \mathbb{T}_{\text{noop}}) \times \mathsf{SSto}$$

where Mix is the set of partial injections $\pi : [n_r+1, n] \to [n_r+1, n]$ and $\mathcal{P}$ is powerset. Depending on tags involved, the above set can be partitioned into $\mathsf{TL}_{\text{push}}, \mathsf{TL}_{\text{pop}}, \mathsf{TL}_{\text{noop}}$ respectively. The first component $X \in \mathcal{P}([n_r+1, n])$ of each transition label is responsible for name generation: $|X|$ fresh names are to be generated and placed in registers given in X. When the tag corresponds to an O-move (tags $c_0, r_i, r_i^x, r_{j,i}^x, c_{j,0}^x$ $(i > 0)$) freshness is meant to be interpreted locally, i.e. none of the new names can be present in the current register. For P-moves (tags $r_\downarrow, r_0, c_i, c_i^x, c_{j,i}^x, r_{j,0}^x$ $(i > 0)$), we require *global* freshness, i.e. that the names have not been encountered before by the automaton.

The $((\mathbb{T}_{\text{push}} \uplus \mathbb{T}_{\text{pop}}) \times \mathbb{C}_{\text{stack}} \times \mathsf{Mix}) \uplus \mathbb{T}_{\text{noop}}$ part corresponds to stack actions. Our automata will use a visibly pushdown stack [4], where the tags determine stack actions according to the partition into $\mathbb{T}_{\text{push}}$, $\mathbb{T}_{\text{pop}}$ and $\mathbb{T}_{\text{noop}}$. The stack will be used to store elements from $\mathbb{C}_{\text{stack}} \times \mathsf{Reg}$, where $\mathbb{C}_{\text{stack}}$ is a finite set of stack symbols (thus, on the stack we will store stack symbols together with register assignments). Note that in transition labels the tags from $\mathbb{T}_{\text{push}} \uplus \mathbb{T}_{\text{pop}}$ come with $(s, \pi) \in \mathbb{C}_{\text{stack}} \times \mathsf{Mix}$. For push-tags, s, π indicate that s should be pushed along with register assignment $\rho \circ \pi$, where ρ is the present content of the registers, i.e. we only push the (content of) registers from $\text{cod}(\pi)$ reindexed according to π. For pop-tags, s, π indicate what stack symbol should occur on top of the stack and π spells out the expected relationship between the present register assignment ρ and the assignment ρ' on top of the stack: for popping to take place we require $\rho(i) = \rho'(j)$ iff $(i, j) \in \pi$. The content of registers in $\text{dom}(\rho') \setminus \text{cod}(\pi)$ will then be popped directly into the machine registers without reindexing.

Name-generation is meant to occur before pushing (so the new names can end up on the stack as soon after being generated), but after popping. The symbolic store $S \in$ SSto in a transition label is also interpreted after name generation (so that fresh names can occur in stores). Assuming ρ is the register assignment obtained after popping and name generation, S stipulates that the move which is being read must come with the store $\Sigma = \{ (\rho(i), S(i)) \mid S(i) \in [0, max] \} \cup \{ (\rho(i), \rho(j)) \mid S(i) = R_j \}$ (this definition will be valid because we shall always have $\mathrm{dom}(\rho) = \mathrm{dom}(S)$). Formally, the automata we use are defined as follows.

Definition 14. *An (n_r, n)-**automaton** of type θ is given as $\mathcal{A} = \langle Q, q_0, \rho_0, \delta, F \rangle$ where:*

- *Q is a finite set of states, partitioned into Q_O (O-states) and Q_P (P-states);*
- *$q_0 \in Q_P$ is the initial state; $F \subseteq Q_O$ is the set of final states;*
- *$\rho_0 \in$ Reg is the initial register assignment such that $[1, n_r] \subseteq \mathrm{dom}(\rho_0)$;*
- *$\delta \subseteq (Q_P \times (\mathsf{TL}_{\mathsf{push}} \cup \mathsf{TL}_{\mathsf{noop}}) \times Q_O) \cup (Q_O \times (\mathsf{TL}_{\mathsf{pop}} \cup \mathsf{TL}_{\mathsf{noop}}) \times Q_P) \cup (Q_O \times \mathsf{Mix} \times Q_O) \cup (Q_P \times \mathsf{Mix} \times Q_P)$ is the transition relation.*

Additionally, if θ is a base type then there is a unique final state q_F, while $\delta \upharpoonright \{q_F\} = \emptyset$ (no outgoing transitions) and $\delta^{-1} \upharpoonright \{q_F\} \subseteq \{q_F\} \times \mathsf{TL}_{\mathsf{noop}} \times Q_P$ (reach only by no-op).

Note that in addition to the labelled transitions discussed earlier, we allow ϵ-transitions $(q_1, \pi, q_2) \in \delta$ which rearrange the contents of registers in $[n_r+1, n]$ according to $\pi \in$ Mix: if the automaton is in state q_1 and the current register assignment is ρ, after the transition the automaton will move to q_2 and the new register assignment will be $\rho \circ \overline{\pi}$, where $\overline{\pi}(i) = i$ ($1 \leq i \leq n_r$) and $\overline{\pi}(i) = \pi(i)$ (otherwise). We write $L(\mathcal{A})$ for the set of words from $(\mathcal{L} \times \mathbb{T} \times$ Sto$)^*$ that are accepted by $\mathcal{A}$ by final state.

Given $\Gamma = \{x_1 : \theta_1, \cdots, x_m : \theta_m\}$ and θ, let us write $P^1_{\Gamma \vdash \theta}$ for the set of plays of length 1 over $[\![\Gamma \vdash \theta]\!]$. Recall that each of them will have the form ι^{Σ_0}, where $\iota \in I_\Gamma$, i.e. $\iota = (l_1, \cdots, l_m)$ with $l_i \in \mathcal{L}_{\theta_i}$. Moreover, the names in ι^{Σ_0} coincide with those of $\mathrm{dom}(\Sigma_0) = \nu(\Sigma_0)$. We order them by use of register assignments and set:
$$I^+_{\Gamma \vdash \theta} = \{(\iota^{\Sigma_0}, \rho_0) \mid \iota^{\Sigma_0} \in P^1_{\Gamma \vdash \theta}, \nu(\rho_0) = \nu(\Sigma_0), \exists k. \rho_0([1, k]) = \nu(\iota)\}$$
For brevity, we shall write each element $(\iota^{\Sigma_0}, \rho_0) \in I^+_{\Gamma \vdash \theta}$ as $\iota^{\Sigma_0}_{\rho_0}$.

Lemma 15. *Let $\Gamma \vdash C : \theta$ be a GRef☺-term in canonical form. For each $j = \iota^{\Sigma_0}_{\rho_0} \in I^+_{\Gamma \vdash \theta}$, there exists a deterministic $(|\nu(\iota)|, m_j)$-automaton $\mathcal{A}_j$ with initial register assignment ρ_0 such that $\bigcup_{w \in L(\mathcal{A}_j)} \mathrm{ext}(\iota^{\Sigma_0} w) = \mathrm{comp}([\![\Gamma \vdash C : \theta]\!]) \cap P^{\iota^{\Sigma_0}}_{\Gamma \vdash \theta}$, where $P^{\iota^{\Sigma_0}}_{\Gamma \vdash \theta}$ is the set of plays over $[\![\Gamma \vdash \theta]\!]$ that start from ι^{Σ_0}.*

Proof. We build upon the techniques developed in [10] insofar as the non-nominal part of the constructions and pointers are concerned. The essence of the nominal approach is revealed in the construction of the automaton, say $\mathcal{A}'$, corresponding to let $x^{\mathsf{ref}^2} \gamma = \mathrm{ref}(y^{\mathsf{ref}} \gamma)$ in C. This is done inductively, starting from the automaton for C, call it $\mathcal{A}$, in which x appears in the initial assignment with name a. Passing to $\mathcal{A}'$ then amounts to omitting a from all transitions in $\mathcal{A}$ (along with parts of the store that can only be reached through a) as long as a has not been played in a P-transition (scenarios in which this happens in an O-transition are discarded). At that point, we convert the transition into one which creates a fresh name and then proceed as $\mathcal{A}$. The construction is complicated by the fact that, while a is being omitted from the store, we need to keep track of its value inside the state and the value may be a chain of other hidden names.

As in [10], the hardest construction is that of the automaton for let $y = z(\lambda x.C)$ in C'. Because of the structure of the arenas involved, the automaton needs to be designed so that it can alternate between plays in C and C'. More specifically, from designated states in C, we need to allow for jumps to C' and vice versa. In [10], such jumps involve the stack so that the well-bracketing condition is preserved: each call is matched to the appropriate return.[3] In our setting the jumps involve the stack also in a more crucial way: when jumping to C', the automaton stores its register assignment to the stack so that, once control returns to C, the state can be recovered and computation can resume from where it had been interrupted. Such interleaving of computations from C and C' requires frequent rearrangements of registers to make sure the respective register assignments of C and C' are simulated in the single register assignment of the automaton we construct. For this, we follow a two-step construction which involves first introducing a notion of automaton operating on two distinct register assignments that do not interfere with each other, which we next reduce to an (n_r, n)-automaton.   □

Lemma 16. *Let* $\Gamma \vdash C_1, C_2 : \theta$ *be* GRef☺*-terms in canonical form. For each* $j = \iota_{\rho_0}^{\Sigma_0} \in I_{\Gamma\vdash\theta}^+$*, there exists a deterministic* $(|\nu(\iota)|, n_j)$*-automaton* $\mathcal{B}_j$ *with initial register assignment* ρ_0 *such that* $L(\mathcal{B}_j) = \emptyset$ *iff* $\mathsf{comp}(\llbracket \Gamma \vdash C_1 : \theta \rrbracket) \cap P_{\Gamma\vdash\theta}^{\iota\Sigma_0} \subseteq \llbracket \Gamma \vdash C_2 : \theta \rrbracket$.

Proof. Let $\mathcal{A}_j^1$, $\mathcal{A}_j^2$ be the automata obtained from the previous Lemma for C_1, C_2 respectively. Because our automata use visibly pushdown stacks and rely on the same partitioning of tags, we can synchronize them using a single stack and check whether any complete play from $P_{\Gamma\vdash\theta}^{\iota\Sigma_0}$ represented by $\mathcal{A}_j^1$ is also represented by $\mathcal{A}_j^2$. Note that this is not a direct inclusion check, because the automata represent plays via extensions and the representations are not uniquely determined. Consequently, $\mathcal{B}_j$ synchronizes $\mathcal{A}_j^1, \mathcal{A}_j^2$ and also explores each possible way of relating names that are present in stores. Once a clash is detected (e.g. different store values, global freshness vs existing name), $\mathcal{B}_j$ continues to simulate $\mathcal{A}_j^1$ only to see whether the offending scenario extends to a complete play of C_1. If so, it enters an accepting state.   □

Note that, although $I_{\Gamma\vdash\theta}^+$ is an infinite set, there exists a finite subset $J \subseteq I_{\Gamma\vdash\theta}^+$ such that $\{\mathcal{A}_j\}_{j\in J}$ already captures $\mathsf{comp}(\llbracket \Gamma \vdash C : \theta \rrbracket)$, because up to name-permutation there are only finitely many initial moves. Consequently, we only need finitely many of them to check whether $\Gamma \vdash C_1 \sqsubseteq C_2$. By Lemma 16, to achieve this we need to be able to decide the emptiness problem for (n_r, n)-automata. To show that it is indeed decidable we consider register pushdown automata [6] over infinite alphabets. They are similar to (n_r, n)-automata in that they are equipped with registers and a stack. The only significant differences are that they process single names in a computational step and do not have the ability to generate globally fresh names. The first difficulty is easily overcome by decomposing transitions of our automata into a bounded number of steps (the existence of the bound follows from the fact that symbolic stores in our transition function are bounded). To deal with freshness we need a separate argument.

---

[3] Jumps to C' are made by P-calls of specific type in C, and returns by corresponding O-returns.

Lemma 17. *The emptiness problem for register pushdown automata extended with fresh-symbol generation is decidable.*

Proof. Register pushdown automata have the ability to generate (locally fresh) symbols not present in the current registers. However, using them directly as a substitute for global freshness is out of question, because it would lead to spurious accepting runs due to the presence of the stack. To narrow the gap, we observe that if the automata could generate locally fresh symbols that are *in addition* not present on the stack then, for the purposes of emptiness testing, this stronger generative power, which we refer to as *quasi freshness*, could stand in for global freshness. It turns out that a run involving quasi fresh symbols can be simulated through local freshness. Whenever a quasi fresh name is generated, we will generate a locally fresh name annotated with a tag indicating its supposed quasi-freshness. The tags will accompany such names as they are being pushed from registers and popped back into them. Whenever we find that a tagged name in a register coincides with a name on top of the stack, we will interrupt the computation as this will indicate that the supposedly quasi-fresh symbol is not quasi fresh. If no violations are detected through tags, we can show that annotated locally fresh names in an accepting run may well be replaced by quasi-fresh ones.   □

Finally, combining Proposition 6 with Lemmata 15-17 we obtain:

Theorem 18. *Program approximation (and thus program equivalence) is decidable for* GRef☺*-terms.*

References

1. Abramsky, S., Ghica, D., Murawski, A.S., Ong, C.-H.L., Stark, I.: Nominal games and full abstraction for the nu-calculus. In: LICS, pp. 150–159 (2004)
2. Abramsky, S., McCusker, G.: Call-by-Value Games. In: Nielsen, M. (ed.) CSL 1997. LNCS, vol. 1414, pp. 1–17. Springer, Heidelberg (1998)
3. Ahmed, A., Dreyer, D., Rossberg, A.: State-dependent representation independence. In: POPL, pp. 340–353 (2009)
4. Alur, R., Madhusudan, P.: Visibly pushdown languages. In: STOC, pp. 202–211 (2004)
5. Benton, N., Leperchey, B.: Relational Reasoning in a Nominal Semantics for Storage. In: Urzyczyn, P. (ed.) TLCA 2005. LNCS, vol. 3461, pp. 86–101. Springer, Heidelberg (2005)
6. Cheng, E.Y.C., Kaminski, M.: Context-free languages over infinite alphabets. Acta Inf. 35(3), 245–267 (1998)
7. Dreyer, D., Neis, G., Birkedal, L.: The impact of higher-order state and control effects on local relational reasoning. In: ICFP, pp. 143–156 (2010)
8. Gabbay, M.J., Pitts, A.M.: A new approach to abstract syntax with variable binding. Formal Asp. Comput. 13, 341–363 (2002)
9. Honda, K., Yoshida, N.: Game-theoretic analysis of call-by-value computation. TCS 221(1-2), 393–456 (1999)
10. Hopkins, D., Murawski, A.S., Ong, C.-H.L.: A Fragment of ML Decidable by Visibly Pushdown Automata. In: Aceto, L., Henzinger, M., Sgall, J. (eds.) ICALP 2011, Part II. LNCS, vol. 6756, pp. 149–161. Springer, Heidelberg (2011)
11. Jeffrey, A., Rathke, J.: Towards a theory of bisimulation for local names. In: LICS (1999)
12. Kozen, D.: Automata and Computability. Springer (1997)

13. Koutavas, V., Wand, M.: Small bisimulations for reasoning about higher-order imperative programs. In: POPL, pp. 141–152 (2006)
14. Laird, J.: A game semantics of names and pointers. APAL 151, 151–169 (2008)
15. Murawski, A.S.: Functions with local state: regularity and undecidability. TCS 338 (2005)
16. Murawski, A.S., Tzevelekos, N.: Full Abstraction for Reduced ML. In: de Alfaro, L. (ed.) FOSSACS 2009. LNCS, vol. 5504, pp. 32–47. Springer, Heidelberg (2009)
17. Murawski, A.S., Tzevelekos, N.: Algorithmic Nominal Game Semantics. In: Barthe, G. (ed.) ESOP 2011. LNCS, vol. 6602, pp. 419–438. Springer, Heidelberg (2011)
18. Murawski, A.S., Tzevelekos, N.: Game semantics for good general references. In: LICS (2011)
19. Pitts, A.M., Stark, I.D.B.: Operational reasoning for functions with local state. In: Gordon, Pitts (eds.) Higher-Order Operational Techniques in Semantics, pp. 227–273. CUP (1998)
20. Reddy, U.S., Yang, H.: Correctness of data representations involving heap data structures. Sci. Comput. Program. 50(1-3), 129–160 (2004)
21. Sangiorgi, D., Kobayashi, N., Sumii, E.: Environmental bisimulations for higher-order languages. ACM Trans. Program. Lang. Syst. 33(1), 5:1–5:69 (2011)
22. Tzevelekos, N.: Full abstraction for nominal general references. LMCS 5(3:8) (2009)

Two-Level Game Semantics, Intersection Types, and Recursion Schemes[*]

C.-H. Luke Ong[1] and Takeshi Tsukada[2,3]

[1] Department of Computer Science, University of Oxford
[2] Graduate School of Information Science, Tohoku University
[3] JSPS Research Fellow

Abstract. We introduce a new cartesian closed category of *two-level* arenas and innocent strategies to model intersection types that are refinements of simple types. Intuitively a property (respectively computation) on the upper level refines that on the lower level. We prove *Subject Expansion*—any lower-level computation is closely and canonically tracked by the upper-level computation that lies over it—which is a measure of the robustness of the two-level semantics. The game semantics of the type system is *fully complete*: every winning strategy is the denotation of some derivation. To demonstrate the relevance of the game model, we use it to construct new semantic proofs of non-trivial algorithmic results in higher-order model checking.

1 Introduction

The recent development of *higher-order model checking*—the model checking of trees generated by higher-order recursion schemes (HORS) against (alternating parity) tree automata—has benefitted much from ideas and methods in semantics. Ong's proof [1] of the decidability of the monadic second-order (MSO) theories of trees generated by HORS was based on game semantics [2]. Using HORS as an intermediate model of higher-order computation, Kobayashi [3] showed that safety properties of functional programs can be verified by reduction to the model checking of HORS against trivial automata (i.e. Büchi tree automata with a trivial acceptance condition). His model checking algorithm is based on an intersection-type-theoretic characterisation of the trivial automata acceptance problem of trees generated by HORS.[1] This type-theoretic approach was subsequently refined and extended to characterise alternating parity tree automata [5], thus yielding a new proof of Ong's MSO decidability result. (Several other proofs [6,7] of the result have since been published.)

This paper was motivated by a desire to understand the connexions between the game-semantic proof [1] and the type-based proof [3,5] of the MSO decidability result. As a first step in clarifying their relationship, we construct a *two-level game semantics* to model intersection types that are refinements of simple types. Given a set Q of

---

[*] A full version with proofs is available at http://www.cs.ox.ac.uk/people/luke.ong/personal/publications/icalp12.pdf.

[1] Independently, Salvati [4] has proposed essentially the same intersection type system for the simply-typed λ-calculus without recursion from a different perspective.

© Springer-Verlag Berlin Heidelberg 2012

colours (modelling the states of an automaton), we introduce a cartesian closed category whose objects are triples (A, U, K) called *two-level arenas*, where A is a Q-*coloured arena* (modelling intersection types), K is a standard arena (modelling simple types), and U is a colour-forgetting function from A-moves to K-moves which preserves the justification relation. A map of the category from (A, U, K) to (A', U', K') is a pair of innocent and colour-reflecting strategies, $\sigma : A \longrightarrow A'$ and $\bar\sigma : K \longrightarrow K'$, such that the induced colour-forgetting function maps plays of σ to plays of $\bar\sigma$. This captures the intuition that the upper-level computation represented by σ refines (or is more constrained than) the lower-level computation represented by $\bar\sigma$, a semantic framework reminiscent of two-level denotational semantics in abstract interpretation as studied by Nielson [8]. Given triples $\mathcal{A}_1 = (A_1, U_1, K)$ and $\mathcal{A}_2 = (A_2, U_2, K)$ that have the same base arena K, their *intersection* $\mathcal{A}_1 \wedge \mathcal{A}_2$ is $(A_1 \times A_2, [U_1, U_2], K)$. Building on the two-level game semantics, we make the following contributions.

(i) How good is the two-level game semantics? Our answer is *Subject Expansion* (Theorem 3), which says intuitively that any computation (reduction) on the lower level can be closely and canonically tracked by the higher-level computation that lies over it. Subject Expansion clarifies the relationship between the two levels; we think it is an important measure of the robustness (and, as we shall see, the reason for the usefulness) of the game semantics.

(ii) We put the two-level game model to use by modelling Kobayashi's intersection type system [3]. Derivations of intersection-type judgements, which we represent by the terms of a new proof calculus, are interpreted by *winning strategies* i.e. compact and total (in addition to innocent and colour-reflecting). We prove that the interpretation is *fully complete* (Theorem 5): every winning strategy is the denotation of some derivation.

(iii) Finally, to demonstrate the usefulness and relevance of the two-level game semantics, we apply it to construct new semantic proofs of three non-trivial *algorithmic* results in higher-order model checking: (a) characterisation of trivial automata acceptance (existence of an accepting run-tree) by a notion of typability [3], (b) minimality of the type environment induced by traversal tree [1], and (c) completeness of GTRecS, a game-semantics based practical algorithm for model checking HORS against trivial automata [9].

Outline. In Section 2, the idea of two-level structure is explained informally. We present coloured arenas, two-level arenas, innocent strategies and related game-semantic notions in Section 3, culminating in the Subject Expansion Theorem. In Section 4 we construct a fully complete two-level game model of Kobayashi's intersection type system. Finally, Section 5 applies the game model to reason about algorithmic problems in higher-order model checking.

2 Two Structures of Intersection Type System

This section presents the intuitions behind the two levels. We explain that two different structures are naturally extracted from a derivation in an intersection type system. Here we use term representation for explanation. Two-level game semantics will be developed in the following sections based on this idea.

$$\dfrac{\dfrac{g : \tau}{g : p_1 \wedge p_3 \to q_1} \quad \dfrac{\dfrac{x : \sigma \quad x : \sigma}{x : p_1 \quad x : p_3}}{x : p_1 \wedge p_3}}{g\, x : q_1} \qquad \dfrac{\dfrac{g : \tau}{g : p_2 \to q_2} \quad \dfrac{x : \sigma}{x : p_2}}{g\, x : q_2}$$

$$g\, x : q_1 \wedge q_2$$

Fig. 1. A type derivation of the intersection type system. Here type environment $\Gamma = \{g : ((p_1 \wedge p_3) \to q_1) \wedge (p_2 \to q_2),\ x : p_1 \wedge p_2 \wedge p_3\}$ is omitted.

$$\dfrac{\dfrac{g : \tau'}{\mathsf{p}_1(g) : p_1 \times p_3 \to q_1} \quad \dfrac{\dfrac{x : \sigma' \quad x : \sigma}{\mathsf{p}_1(x) : p_1 \quad \mathsf{p}_3(x) : p_3}}{\langle \mathsf{p}_1(x), \mathsf{p}_2(x) \rangle : p_1 \times p_3}}{\mathsf{p}_1(g)\ \langle \mathsf{p}_1(x), \mathsf{p}_2(x) \rangle : q_1} \qquad \dfrac{\dfrac{g : \tau}{\mathsf{p}_2(g) : p_2 \to q_2} \quad \dfrac{x : \sigma}{\mathsf{p}_2(x) : p_2}}{\mathsf{p}_2(g)\ \mathsf{p}_2(x) : q_2}$$

$$\langle \mathsf{p}_1(g)\ \langle \mathsf{p}_1(x), \mathsf{p}_2(x) \rangle,\ \mathsf{p}_2(g)\ \mathsf{p}_2(x) \rangle : q_1 \times q_2$$

Fig. 2. A type derivation of the product type system, which corresponds to Fig. 1. Here $\Gamma' = \{g : ((p_1 \times p_3) \to q_1) \times (p_2 \to q_2),\ x : p_1 \times p_2 \times p_3\}$ is omitted.

The intersection type constructor $\wedge$ of an intersection type system is characterised by the following typing rules.[2]

$$\dfrac{\Gamma \vdash t : \tau_1 \quad \Gamma \vdash t : \tau_2}{\Gamma \vdash t : \tau_1 \wedge \tau_2} \qquad \dfrac{\Gamma \vdash t : \tau_1 \wedge \tau_2}{\Gamma \vdash t : \tau_1} \qquad \dfrac{\Gamma \vdash t : \tau_1 \wedge \tau_2}{\Gamma \vdash t : \tau_2}$$

At first glance, they resemble the rules for products. Let $\langle t_1, t_2 \rangle$ be a pair of t_1 and t_2 and p_i be the projection to the ith element (for $i \in \{1, 2\}$).

$$\dfrac{\Gamma \vdash t_1 : \tau_1 \quad \Gamma \vdash t_2 : \tau_2}{\Gamma \vdash \langle t_1, t_2 \rangle : \tau_1 \times \tau_2} \qquad \dfrac{\Gamma \vdash t : \tau_1 \times \tau_2}{\Gamma \vdash \mathsf{p}_1(t) : \tau_1} \qquad \dfrac{\Gamma \vdash t : \tau_1 \times \tau_2}{\Gamma \vdash \mathsf{p}_2(t) : \tau_2}$$

When we ignore terms and replace $\times$ by $\wedge$, the rules in the two groups coincide. In fact, they are so similar that a derivation of the intersection type system can be transformed to a derivation of the product type system by replacing $\wedge$ by $\times$ and adjusting terms to the rules for product. See Figures 1 and 2 for example. This is the first structure behind an intersection-type derivation, which we call the *upper-level structure*.

However the upper-level structure alone does not capture all features of the intersection type system: specifically some derivations of the product type system have no corresponding derivation in the intersection type system. For example, while the type judgement $x : p_1, y : p_2 \vdash \langle x, y \rangle : p_1 \times p_2$ is derivable, no term inhabits the judgement $x : p_1, y : p_2 \vdash ? : p_1 \wedge p_2$.

Terms in the rules explain this gap. We call them *lower-level structures*. To construct a term of type $\tau_1 \times \tau_2$, it suffices to find *any* two terms t_1 of type τ_1 and t_2 of type τ_2. However to construct a term of type $\tau_1 \wedge \tau_2$, we need to find a term t that has both type τ_1 and type τ_2. Thus a product type derivation has a corresponding intersection

---

[2] In the type system in Section 4, these rules are no longer to be independent rules, but a similar argument stands.

type derivation only if for all pairs $\langle t_1, t_2 \rangle$ appearing at the derivation, the respective structures of t_1 and t_2 are "coherent".

For example, let us examine the derivation in Figure 2, which contains two pair constructors. One appears at $\langle \mathsf{p}_1(x), \mathsf{p}_3(x) \rangle : p_1 \times p_3$. Here the left argument $\mathsf{p}_1(x) : p_1$ and the right argument $\mathsf{p}_3(x) : p_3$ are "coherent" in the sense that they are the same except for details such as types and indexes of projections. In other words, by forgetting such details, $\mathsf{p}_1(x) : p_1$ and $\mathsf{p}_3(x) : p_3$ become the same term x. The other pair appears at the root and the "forgetful" map maps both the left and right arguments to $g\ x$.

This interpretation decomposes an intersection type derivation into three components: a derivation in the simple type system with product (the upper-level structure), a term (the lower-level structure) and a "forgetful" map from the upper-level structure to the lower-level structure. Since recursion schemes are simply typed, we can assume a term to also be simply typed for our purpose. Hence the resulting two-level structure consists of two derivations in the simple type system with a map on nodes from one to the other.

3   Two-Level Game Semantics

For sets A and B, we write $A + B$ for disjoint union and $A \times B$ for cartesian product. We first introduce some basic notions in game semantics [2].

We introduce Q-coloured arenas and innocent strategies, which are models of the simply-typed λ-calculus with multiple base types ranging over Q.

Definition 1 (Coloured Arena). For a set Q of symbols, a *Q-coloured arena A* is a quadruple $(M_A, \vdash_A, \lambda_A, c_A)$, where (i) M_A is a set of *moves*, (ii) $\vdash_A \subseteq M_A + (M_A \times M_A)$ is a *justification relation*, (iii) $\lambda_A : M_A \to \{P, O\}$, and (iv) $c_A : M_A \to Q$ is a *colouring*. We write $\vdash_A m$ for $m \in (\vdash_A)$ and $m \vdash_A m'$ for $(m, m') \in (\vdash_A)$. The justification relation must satisfy the conditions:

- For each $m \in M_A$, either $\vdash_A m$ or $m' \vdash_A m$ for a unique $m' \in M_A$.
- If $\vdash_A m$, then $\lambda_A(m) = O$. If $m \vdash_A m'$, then $\lambda_A(m) \neq \lambda(m')$.

Fer a Q-coloured arena A, the set $Init_A \subseteq M_A$ of *initial moves* of A is $\{m \in M_A \mid \vdash_A m\}$. A move $m \in M_A$ is called an *O-move* if $\lambda_A(m) = O$ and a *P-move* if $\lambda_A(m) = P$.

A *justified sequence* of a Q-coloured arena A is a sequence of moves such that each element except the first is equipped with a *justification pointer* to some previous move. A *play* of an arena A is a justified sequence s that satisfies: (i) Well-openness, (ii) Alternation, (iii) Justification, (iv) Visibility. A *P-strategy* (or a *strategy*) σ of an arena A is a prefix-closed subset of plays of A that satisfies Determinacy, Contingent Completeness, and

Colour Reflection. Only the opponent can change the colour i.e. for every O-move m and P-move m', if $s \cdot m \cdot m' \in \sigma$, then $c(m) = c(m')$.

A strategy σ is *innocent* just if for every pair of plays $s \cdot m, s' \cdot m' \in \sigma$ ending with P-moves m and m', $\ulcorner s \urcorner = \ulcorner s' \urcorner$ implies $\ulcorner s \cdot m \urcorner = \ulcorner s' \cdot m' \urcorner$, writing $\ulcorner s \urcorner$ to mean the *P-view* of s. Further σ is *winning* just if the following hold:

Compact. The domain $\mathrm{dom}(f_\sigma)$ of the view function of σ is a finite set.

Total. If $s \cdot m \in \sigma$ for an O-move m, then $s \cdot m \cdot m' \in \sigma$ for some P-move m'.

Given Q-coloured arenas A and B, the *product* $A \times B$ and *function space arena* $A \Rightarrow B$ are standard. We define a category whose objects are Q-coloured arenas; maps from A to B are innocent strategies of the arena $A \Rightarrow B$. The category is cartesian closed, and is thus a model of the simply-typed lambda calculus with (indexed) products.

Theorem 1. *For every set Q, the category of Q-coloured arenas and innocent strategies is cartesian closed with the product $A \times B$ and function space $A \Rightarrow B$.*

Definition 2 (Two-Level Arenas). An *two-level arena* based on Q is a triple $\mathcal{A} = (A, U, K)$, where A is a Q-colored arena, K is a $\{o\}$-colored arena (i.e. an ordinary arena, which we call the *base arena* of $\mathcal{A}$) and U is a map from M_A to M_K that satisfies: (i) $\lambda_A(m) = \lambda_K(U(m))$ (ii) If $m \vdash_A m'$ then $U(m) \vdash_K U(m')$; and if $\vdash_A m$ then $\vdash_K U(m)$.

For a justified sequence $s = m_1 \cdot m_2 \cdots m_k$, we write $U(s)$ to mean the justified sequence $U(m_1) \cdot U(m_2) \cdots U(m_k)$ whose justification pointers are induced by those of s. Let $\mathcal{A} = (A, U, K)$ be a two-level arena. It is easy to see that if s is a play of A, then $U(s)$ is a play of K.

For a strategy σ of A, $U(\sigma) := \{U(s) \mid s \in \sigma\}$ is a set of plays of K, which is not necessarily a strategy, since $U(s)$ may not satisfy determinacy. (Recall that some upper-level structure has no corresponding lower-level structure.)

Definition 3 (Strategies). A *strategy* of a two-level arena (A, U, K) is a pair $(\sigma, \bar{\sigma})$ of strategies of A and K respectively such that $U(\sigma) \subseteq \bar{\sigma}$; it is *innocent* just if σ and $\bar{\sigma}$ are innocent as strategies of A and K respectively.

Let $\mathcal{A}_i = (A_i, U_i, K_i)$ where $i = 1, 2$ be two-level arenas. We define product, function space and intersection constructions as follows.

Product. $\mathcal{A}_1 \times \mathcal{A}_2 := (A_1 \times A_2, U, K_1 \times K_2)$, where $U : (M_{A_1} + M_{A_2}) \to (M_{K_1} + M_{K_2})$ is defined as $U_1 + U_2$.

Function Space. $\mathcal{A}_1 \Rightarrow \mathcal{A}_2 := (A_1 \Rightarrow A_2, U, K_1 \Rightarrow K_2)$, where $U : ((M_{A_1} \times Init_{A_2}) + M_{A_2}) \to ((M_{K_1} \times Init_{K_2}) + M_{K_2})$ is defined as $U_1 \times U_2 + U_2$.

Intersection. *Provided* $K_1 = K_2 = K$, define $\mathcal{A}_1 \wedge \mathcal{A}_2 := (A_1 \times A_2, U, K)$, where $U : (M_{A_1} + M_{A_2}) \to M_K$ is defined as $[U_1, U_2]$.

We can now define a category whose objects are two-level arenas, and maps $\mathcal{A}_1 \to \mathcal{A}_2$ are innocent strategies of $\mathcal{A}_1 \Rightarrow \mathcal{A}_2$. The composite of $(\sigma_1, \bar{\sigma}_1) : \mathcal{A}_1 \Rightarrow \mathcal{A}_2$ and $(\sigma_2, \bar{\sigma}_2) : \mathcal{A}_2 \Rightarrow \mathcal{A}_3$ is defined as $(\sigma_1; \sigma_2, \bar{\sigma}_1; \bar{\sigma}_2) : \mathcal{A}_1 \Rightarrow \mathcal{A}_3$. Let $\top$ be the terminal object in the category of Q-coloured arenas.

Theorem 2. *(i) Let $\mathcal{A}_i = (A_i, U_i, K)$. Then $\mathcal{A}_1 \wedge \mathcal{A}_2$ is the pullback of*

$$\mathcal{A}_1 \xrightarrow{(!_{A_1}, \mathrm{id}_K)} (\top, \emptyset, K) \xleftarrow{(!_{A_2}, \mathrm{id}_K)} \mathcal{A}_2.$$ *(ii) The category of two-level arenas and innocent strategies is cartesian closed.*

Finally we introduce *Subject Expansion* which is a property that relates the two levels of game semantics. The name originates from a characteristic property of (intersection) type systems, which states that for any terms t and t', type environment Γ and type τ, if $t \longrightarrow t'$ and $\Gamma \vdash t' : \tau$, then $\Gamma \vdash t : \tau$. Subject expansion plays a central rôle in completeness of intersection type systems.

In two-level game semantics, it seems best to formulate Subject Expansion as a kind of factorisation theorem, stated as follows.

Theorem 3 (Subject Expansion). *Let $\mathcal{A}_i = (A_i, U_i, K_i)$ be a two-level arena for $i = 1, 2$ and K be a base arena. If*

$$\mathcal{A}_1 \xrightarrow{\;(\sigma,\bar{\sigma})\;} \mathcal{A}_2 \qquad and \qquad K_1 \xrightarrow[\;\bar{\sigma}_1 \searrow \; \underset{K}{\circlearrowleft} \; \nearrow \bar{\sigma}_2\;]{\;\bar{\sigma}\;} K_2$$

(a map of two-level arenas)   (maps of base arenas)

then there are a two-level arena $\mathcal{A}$ whose base arena is K and strategies $\sigma_1 : A_1 \to A$ and $\sigma_2 : A \to A_2$ such that

$$\mathcal{A}_1 \xrightarrow[\;(\sigma_1,\bar{\sigma}_1) \searrow \; \underset{\mathcal{A}}{\circlearrowleft} \; \nearrow (\sigma_2,\bar{\sigma}_2)\;]{\;(\sigma,\bar{\sigma})\;} \mathcal{A}_2$$

Moreover, there is a canonical triple $(\sigma_1, \mathcal{A}, \sigma_2)$: for every triple $(\sigma'_1, \mathcal{A}', \sigma'_2)$ that satisfies the requirement, there exists a mapping φ from moves of $\mathcal{A}$ to moves of $\mathcal{A}'$ such that $[\mathrm{id}_{\mathcal{A}_1}, \varphi](\sigma_1) \subseteq \sigma'_1$ and $[\varphi, \mathrm{id}_{\mathcal{A}_2}](\sigma_2) \subseteq \sigma'_2$.

As an application, we shall use Subject Expansion to prove the completeness of the intersection type system for recursion scheme model checking (Theorem 6).

4   Interpretation of Intersection Types

In this section, we interpret Kobayashi's intersection type system [3] in the two-level game model, and show that the interpretation is *fully complete* i.e. every winning strategy is the denotation of some derivation.

We consider the standard Church-style simply-typed lambda calculus. However, to avoid confusion with intersection types, we henceforth refer to simple types as *kinds*, defined by $\kappa ::= o \mid \kappa_1 \to \kappa_2$. Let Δ be a *kind environment* i.e. a set of variable-kind bindings, $x : \kappa$. We write $\Delta \vdash t :: \kappa$ to mean t has kind κ under the environment Δ. Fix a set Q of symbols, ranged over by q. *Intersection pre-types* are defined by $\tau, \sigma ::= q \mid \tau \to \sigma \mid \bigwedge_{i \in I} \tau_i$. The *well-kindedness relation* $\tau :: \kappa$ is defined by the following rules.

$$\frac{}{q :: o} \qquad \frac{\tau_i :: \kappa \quad (\text{for all } i \in I) \qquad \sigma :: \kappa'}{(\bigwedge_{i \in I} \tau_i) \to \sigma :: \kappa \to \kappa'}$$

An *intersection type* is an intersection pre-type τ such that $\tau :: \kappa$ for some κ.

An *(intersection) type environment* Γ is a set of variable-type bindings, $x : \bigwedge_{i \in I} \tau_i$. We write $\Gamma :: \Delta$ just if $x : \bigwedge_{i \in I} \tau_i \in \Gamma$ implies that for some κ, $x : \kappa \in \Delta$ and $\tau_i :: \kappa$ for all $i \in I$. *Valid typing sequents* are defined by induction over the following rules.

$$\frac{}{\Gamma, x : \bigwedge_{i \in I} \tau_i \vdash x : \tau_i} \qquad \frac{\Gamma, x : \bigwedge_{i \in I} \tau_i \vdash t : \sigma \qquad \tau_i :: \kappa \quad (\text{for all } i \in I)}{\Gamma \vdash \lambda x^\kappa.t : (\bigwedge_{i \in I} \tau_i) \to \sigma}$$

$$\frac{\Gamma \vdash t_1 : (\bigwedge_{i \in I} \tau_i) \to \sigma \qquad \Gamma \vdash t_2 : \tau_i \quad \text{(for all } i \in I)}{\Gamma \vdash t_1\, t_2 : \sigma}$$

Lemma 1. *If* $\Delta \vdash t :: \kappa$ *and* $\Gamma :: \Delta$ *and* $\Gamma \vdash t : \tau$, *then* $\tau :: \kappa$.

For notational convenience, we use a Church-style simply-kinded lambda calculus with (indexed) product as a term representation of derivations. The raw terms are defined as follows.

$$M ::= \mathsf{p}_i(x) \mid \lambda x^{\bigwedge_{i \in I} \tau_i}.M \mid M_1\, M_2 \mid \prod_{i \in I} M_i$$

where I is a finite indexing set. We omit I and simply write $\lambda x^{\bigwedge_i \tau_i}$ and so on if I is clear from the context or unimportant. We say a term M is *well-formed* just if for every application subterm $M_1\, M_2$ of M, M_2 has the form $\prod_{i \in I} N_i$. We consider only well-formed terms. By abuse of notation, we write $\top$ for $\prod \emptyset$.

We give a type system for terms of the calculus, which ressemble the intersection type system, but is syntax directed, i.e., a term completely determines the structure of a derivation.

$$\frac{}{\Gamma, x : \bigwedge_{i \in I} \tau_i \Vdash \mathsf{p}_i(x) : \tau_i} \qquad \frac{\Gamma \Vdash M_i : \tau_i \qquad \tau_i :: \kappa \qquad \text{(for all } i)}{\Gamma \Vdash \prod_i M_i : \bigwedge_i \tau_i}$$

$$\frac{\Gamma \Vdash M_1 : (\bigwedge_i \tau_i) \to \sigma \qquad \Gamma \Vdash M_2 : \bigwedge_i \tau_i}{\Gamma \Vdash M_1\, M_2 : \sigma} \qquad \frac{\Gamma, x : \bigwedge_{i \in I} \tau \Vdash M : \sigma}{\Gamma \Vdash \lambda x^{\bigwedge_{i \in I} \tau_i}.M : (\bigwedge_{i \in I} \tau_i) \to \sigma}$$

We call a term-in-context $\Gamma \Vdash M : \tau$ a *proof term*. Observe that a proof term is essentially a typed lambda term with (indexed) product. Here an intersection type $\tau_1 \wedge \cdots \wedge \tau_n$ is interpreted as a product type $\tau_1 \times \cdots \times \tau_n$ and a proof term $M_1 \sqcap \cdots \sqcap M_n$ is a tuple $\langle M_1, \ldots, M_n \rangle$. Then all variables are bound to tuples and a proof term $\mathsf{p}_i(x)$ is a projection into the ith element.

Unfortunately, not all the proof terms correspond to a derivation of the intersection type system. For example, $\lambda f^{(q_1 \wedge q_2) \to p}.\lambda x^{q_1}.\lambda y^{q_2}.f(\mathsf{p}(x) \sqcap \mathsf{p}(y))$ is a proof term of the type $((q_1 \wedge q_2) \to p) \to q_1 \to q_2 \to p$, but there is no inhabitant of that type. In the intersection type system, $t : \tau \wedge \sigma$ only if $t : \tau$ and $t : \sigma$ for the same term t, but the proof term $\mathsf{p}(x) \sqcap \mathsf{p}(y)$ violates the requirement.

We introduce a judgement $M :: t$ that means the structure of M coincides with the structure of t. By definition, $\top :: t$ for every term t.

$$\mathsf{p}_i(x) :: x \qquad\qquad \lambda x^{\bigwedge_i \tau_i}.M :: \lambda x^\kappa.t := M :: t \wedge \forall i.\tau_i :: \kappa$$
$$\prod_i M_i :: t := \forall i.M_i :: t \qquad M_1\, M_2 :: t_1\, t_2 := M_1 :: t_1 \wedge M_2 :: t_2$$

We write $[\Gamma :: \Delta] \vdash [M :: t] : [\tau :: \kappa]$ just if $\Gamma :: \Delta$, $M :: t$, $\tau :: \kappa$, $\Delta \vdash t :: \kappa$ and $\Gamma \Vdash M : \tau$. Let t be a term such that $\Delta \vdash t :: \kappa$. It is easy to see that there is a one-to-one correspondence between a derivation of $\Gamma \vdash t : \tau$ and a proof term M such that $[\Gamma :: \Delta] \vdash [M :: t] : [\tau :: \kappa]$.

Example 1. Let $Q = \{q_1, q_2\}$ and take $\theta \to (q_1 \wedge q_2) \to q_1 :: (o \to o) \to o \to o$ where $\theta = (q_1 \to q_1) \wedge (q_2 \to q_1) \wedge (q_1 \wedge q_2 \to q_1)$ and terminal $f : q_1 \to q_2$. Set $M := \lambda x^\theta\, y^{q_1 \wedge q_2}.\mathsf{p}_2(x)(f^{q_1 \to q_2}(\mathsf{p}_1(x)(\mathsf{p}_3(x)(\mathsf{p}_1(y) \sqcap \mathsf{p}_2(y)))))$. Then we have $M :: \lambda xy.x(f(x(x\, y))$.

A two-level arena represents a proof of well-kindedness, $\tau :: \kappa$. The interpretation is straightforward since we have arena constructors $\Rightarrow$ and $\wedge$:

$$[\![q :: o]\!] := ([\![q]\!], U, [\![o]\!]) \qquad [\![(\bigwedge_{i \in I} \tau_i) \rightarrow \sigma :: \kappa \rightarrow \kappa']\!] := (\bigwedge_{i \in I}[\![\tau_i :: \kappa]\!]) \Rightarrow [\![\sigma :: \kappa']\!]$$

where $[\![q]\!]$ is a Q-coloured arena with a single move of the colour q, $[\![o]\!]$ is a $\{o\}$-coloured arena with a single move, and U maps the unique move of $[\![q]\!]$ to the unique move of $[\![o]\!]$. Let $\Gamma = x_1 : \bigwedge_{i \in I_1} \tau_i^1, \ldots, x_n : \bigwedge_{i \in I_n} \tau_i^n$ be a type environment with $\Gamma :: \Delta$ where $\Delta = x_1 : \kappa_1, \ldots, x_n : \kappa_n$. Then $[\![\Gamma :: \Delta]\!] := \prod_{j \leq n}(\bigwedge_{i \in I_j}[\![\tau_i^j :: \kappa_i]\!])$.

A proof $[\Gamma :: \Delta] \vdash [M :: t] : [\tau :: \kappa]$, which is equivalent to a derivation of $\Gamma \vdash t : \tau$, is interpreted as a strategy of the two-level arena $[\![\Gamma :: \Delta]\!] \Rightarrow [\![\tau :: \kappa]\!]$, defined by the following rules (for simplicity, we write $[\![M :: t]\!]$ instead of $[\![[\Gamma :: \Delta] \vdash [M :: t] : [\tau :: \kappa]]\!]$):

$$[\![\mathsf{p}_i(x) :: x]\!] := \pi_x; \mathsf{p}_i \qquad\qquad [\![M_1 \, M_2 :: t_1 \, t_2]\!] := \langle [\![M_1 :: t_1]\!], [\![M_2 :: t_2]\!] \rangle; \mathbf{eval}$$
$$[\![\textstyle\prod_i M_i :: t]\!] := \textstyle\prod_i [\![M_i :: t]\!] \qquad\qquad [\![\lambda x.M :: \lambda x.t]\!] := \Lambda([\![M :: t]\!])$$

where π_x is the projection $[\![(\Gamma, x : \bigwedge_i \tau_i) :: (\Delta, x : \kappa)]\!] \longrightarrow [\![\bigwedge_i \tau_i :: \kappa]\!]$ and for strategies $\sigma_i : [\![\Gamma :: \Delta]\!] \longrightarrow [\![\tau_i :: \kappa]\!]$ indexed by i, the strategy $\prod_i \sigma_i : [\![\Gamma :: \Delta]\!] \longrightarrow \bigwedge_i [\![\tau_i :: \kappa]\!]$ is the canonical map of the pullback.

Lemma 2 (Componentwise Interpretation). *Let* $[\Gamma :: \Delta] \vdash [M :: t] : [\tau :: \kappa]$ *be a derivation. Then* $[\![M :: t]\!] = ([\![M]\!], [\![t]\!])$.

Theorem 4 (Adequacy). *Let* $[\Gamma :: \Delta] \vdash [M_1 :: t_1] : [\tau :: \kappa]$ *and* $[\Gamma :: \Delta] \vdash [M_2 :: t_2] : [\tau :: \kappa]$ *be two proofs such that* $[\![M_1 :: t_1]\!] =_{\beta\eta} [\![M_2 :: t_2]\!]$. *Then* $[\![M_1 :: t_1]\!] = [\![M_2 :: t_2]\!]$.

Theorem 5 (Definability). *Let* $(\sigma, \bar{\sigma}) : [\![\Gamma :: \Delta]\!] \rightarrow [\![\tau :: \kappa]\!]$ *be a winning strategy. There is a derivation* $[\Gamma :: \Delta] \vdash [M :: t] : [\tau :: \kappa]$ *such that* $(\sigma, \bar{\sigma}) = [\![M :: t]\!]$.

Proof. (Sketch) By the standard argument of definability [2], we have a proof term M and a simply-typed lambda term t such that $[\![M]\!] = \sigma : [\![\Gamma]\!] \longrightarrow [\![\tau]\!]$ and $[\![t]\!] = \bar{\sigma} : [\![\Delta]\!] \longrightarrow [\![\kappa]\!]$, where $[\![\cdot]\!]$ is the standard interpretation of typed lambda terms (here intersection $\wedge$ in Γ and τ is interpreted as a product). If $[\Gamma :: \Delta] \vdash [M :: t] : [\tau :: \kappa]$ is a valid derivation, by Lemma 2, we have $[\![M :: t]\!] = (\sigma, \bar{\sigma})$ as required. Thus it suffices to show that $M :: t$, which can be shown by an easy induction. $\qquad\square$

We can use Church-style type-annotated terms in β-normal η-long form, called *canonical terms*, to represent winning strategies, which are terms-in-context of the form: $\Gamma \Vdash \mathsf{p}_i(x) \, M_1 \cdots M_n : q$ where $\Gamma = \cdots, x : \bigwedge_i \alpha_i, \cdots$ and $\alpha_i = \tau_1 \rightarrow \cdots \rightarrow \tau_n \rightarrow q$, and for each $k \in \{1, \ldots, n\}$,

$$M_k = \prod_{j \in J_k} \lambda y_{kj1}^{\tau_{kj1}} \ldots y_{kjr}^{\tau_{kjr}}.N_{kj} : \bigwedge_{j \in J_k} \beta_{kj} = \tau_k$$

such that for each $j \in J_k$, $\beta_{kj} = \tau_{kj1} \rightarrow \cdots \rightarrow \tau_{kjr} \rightarrow q_{kj}$ with $r = r_{kj}$ and $\Gamma, y_{kj1} : \tau_{kj1}, \cdots, y_{kjr} : \tau_{kjr} \Vdash N_{kj} : q_{kj}$ is a canonical term. (We assume that canonical terms are proof terms that represent derivations.)

By definition, canonical terms are not λ-abstractions. We call terms-in-context such as M_k above canonical terms in (partially) *curried form*; they have the shape $\Gamma \Vdash \lambda \bar{x}.M : \tau_1 \to \cdots \to \tau_n \to q$. Note that in case $n = 0$, the curried form retains an outermost "dummy lambda" $\Gamma \Vdash \lambda.M : q$. With this syntactic convention, we obtain a tight correspondence between syntax and semantics.

Lemma 3. *Let $\tau :: \kappa$. There is a one-to-one correspondence between winning strategies over the two-level arena $[\![\tau :: \kappa]\!]$ and canonical terms in curried form of the shape $\emptyset \Vdash M : \tau$ (with η-long β-normal simply-typed term t such that $M :: t$).*

A strategy $(\sigma, \bar{\sigma})$ of $\mathcal{A} = (A, U, K)$ is *P-full* (respectively *O-full*) just if every P-move (respectively O-move) of A occurs in σ. Suppose $(\sigma, \bar{\sigma})$ is a winning strategy of $[\![\tau :: \kappa]\!]$. Then: (i) If $(\sigma, \bar{\sigma})$ is P-full, then it is also O-full. (ii) There is a subtype $\tau' :: \kappa$ of τ such that $(\sigma, \bar{\sigma})$ is winning and P-full over $[\![\tau' :: \kappa]\!]$.

A derivation $[\Gamma :: \Delta] \vdash [M :: t] : [\tau :: \kappa]$ is *relevant* just if for each abstraction subterm $\lambda x^{\bigwedge_{i \in I} \tau_i}.M'$ of M and $i \in I$, M' has a free occurrence of $\mathsf{p}_i(x)$.

Lemma 4. $[\Gamma :: \Delta] \vdash [M :: t] : [\tau :: \kappa]$ *is relevant iff $[\![M :: t]\!]$ is P-full.*

5   Applications to HORS Model-Checking

Fix a ranked alphabet Σ and a HORS $G = \langle \Sigma, \mathcal{N}, S, \mathcal{R} \rangle$ we first give the game semantics $[\![G]\!]$ of G (see [1] for a definition of HORS). Let $\mathcal{N} = \{ F_1 : \kappa_1, \ldots, F_n : \kappa_n \}$ with $F_1 = S$ (start symbol), and $\Sigma = \{ a_1 : r_1, \ldots, a_m : r_m \}$ where each $r_i = ar(a_i)$, the arity of a_i. Writing $[\![\Sigma]\!] := \prod_{i=1}^m [\![o^{r_i} \to o]\!]$ and $[\![\mathcal{N}]\!] := \prod_{i=1}^n [\![\kappa_i]\!]$, the *game semantics* of G, $[\![G]\!] : [\![\Sigma]\!] \longrightarrow [\![o]\!]$, is the composite

$$[\![\Sigma]\!] \xrightarrow{\Lambda(\mathbf{g})} ([\![\mathcal{N}]\!] \Rightarrow [\![\mathcal{N}]\!]) \xrightarrow{Y} [\![\mathcal{N}]\!] \xrightarrow{\{S::o\}} [\![o]\!]$$

in the cartesian closed category of o-coloured arenas and innocent strategies, where $\mathbf{g} = \langle g_1, \ldots, g_n \rangle : [\![\Sigma]\!] \times [\![\mathcal{N}]\!] \longrightarrow [\![\mathcal{N}]\!]$ with $g_i = [\![\Sigma \cup \mathcal{N} \vdash \mathcal{R}(F_i) :: \kappa_i]\!]$ and $\Lambda(\text{-})$ is currying; Y is the standard fixpoint strategy (see [2, §7.2]); and $\{ S :: o \} = \pi_1 : [\![\mathcal{N}]\!] \longrightarrow [\![o]\!]$ is the projection map.

Remark 1. Since the set of P-views of $[\![G]\!]$ coincide with the branch language[3] of the *value tree* of G (i.e. the Σ-labelled tree generated by G; see [1]) and an innocent strategy is determined by its P-views, we identify the map $[\![G]\!]$ with the value tree of G.

Now fix a trivial automaton $\mathcal{B} = \langle Q, \Sigma, q_I, \delta \rangle$. We extend the game-semantic account to express the run tree of $\mathcal{B}$ over the value tree $[\![G]\!]$ in the category of Q-based two-level arenas and innocent strategies. First set

$$[\![\delta :: \Sigma]\!] := \prod_{a \in \Sigma} \bigwedge_{(q, a, \bar{q}) \in \delta} [\![q_1 \to \cdots \to q_{ar(a)} \to q :: o^{ar(a)} \to o]\!] = ([\![\delta]\!], U, [\![\Sigma]\!])$$

---

[3] Let m be the maximum arity of the symbols in Σ, and write $[m] = \{ 1, \cdots, m \}$. The *branch language* of $t : dom(t) \longrightarrow \Sigma$ consists of (i) $(f_1, d_1)(f_2, d_2) \cdots$ if there exists $d_1 d_2 \cdots \in [m]^\omega$ s.t. $t(d_1 \cdots d_i) = f_{i+1}$ for every $i \in \omega$; and (ii) $(f_1, d_1) \cdots (f_n, d_n) f_{n+1}$ if there exists $d_1 \cdots d_n \in [m]^*$ s.t. $t(d_1 \cdots d_i) = f_{i+1}$ for $0 \leq i \leq n$, and the arity of f_{n+1} is 0.

where $[\![\delta]\!]$ is the Q-coloured arena $\prod_{a\in\Sigma}\prod_{(q,a,\bar{q})\in\delta}[\![q_1 \to \cdots \to q_{ar(a)} \to q]\!]$ and $\bar{q} = q_1, q_2, \ldots, q_{ar(a)}$.

A *run tree* of $\mathcal{B}$ over $[\![G]\!]$ is just an innocent strategy $(\rho, [\![G]\!])$ of the arena $[\![\delta :: \Sigma]\!] \Rightarrow [\![q_I :: o]\!] = ([\![\delta]\!] \Rightarrow [\![q_I]\!], V, [\![\Sigma]\!] \Rightarrow [\![o]\!])$. Every P-view $\bar{p} \in [\![G]\!]$ has a unique "colouring" i.e. a P-view $p \in \rho$ such that $V(p) = \bar{p}$. This associates a colour (state) with each node of the value tree, which corresponds to a run tree in the concrete presentation.

Characterisation by Complete Type Environment. Using G and $\mathcal{B}$ as before, Kobayashi [3] showed that $[\![G]\!]$ is accepted by $\mathcal{B}$ if, and only if, there is a *complete type environment* Γ, meaning that (i) $S : q_I \in \Gamma$, (ii) $\Gamma \vdash \mathcal{R}(F) : \theta$ for each $F : \theta \in \Gamma$. As a first application of two-level arena games, we give a semantic counterpart of the characterisation. Let $\Gamma = \{ F_1 : \bigwedge_{j\in I_1} \tau_{1j} :: \kappa_1, \ldots, F_n : \bigwedge_{j\in I_n} \tau_{nj} :: \kappa_n \}$ be a type environment of G. Set $[\![\Gamma :: \mathcal{N}]\!] := \prod_{i=1}^{n} \bigwedge_{j\in I_i} [\![\tau_{ij} :: \kappa_i]\!] = ([\![\Gamma]\!], U_1, [\![\mathcal{N}]\!])$ where $[\![\Gamma]\!] := \prod_{i=1}^{n} \prod_{j\in I_i} [\![\tau_{ij}]\!]$.

Theorem 6. *Using Σ, G and $\mathcal{B}$ as before, $[\![G]\!]$ is accepted by $\mathcal{B}$ if, and only if, there exists Γ such that (i) $S : q_I \in \Gamma$, and (ii) there exists a strategy σ (say) of the Q-coloured arena $[\![\delta]\!] \times [\![\Gamma]\!] \Rightarrow [\![\Gamma]\!]$ such that $(\sigma, \mathbf{g})$ defines a winning strategy of the two-level arena $([\![\delta :: \Sigma]\!] \times [\![\Gamma :: \mathcal{N}]\!]) \Rightarrow [\![\Gamma :: \mathcal{N}]\!]$. (Thanks to Theorem 5, (ii) is equivalent to: $\Gamma \vdash \mathcal{R}(F) : \theta$ for each $F : \theta$ in Γ; hence Γ is complete.)*

Proof. (Sketch) We use Subject Expansion (Theorem 3) to prove the left-to-right direction. To prove the right-to-left, consider the composite

$$[\![\delta :: \Sigma]\!] \xrightarrow{\Lambda(\sigma, \mathbf{g})} ([\![\Gamma :: \mathcal{N}]\!] \Rightarrow [\![\Gamma :: \mathcal{N}]\!]) \xrightarrow{(Y,Y)} [\![\Gamma :: \mathcal{N}]\!] \xrightarrow{(\{S:q_I\}, \{S::o\})} [\![q_I :: o]\!]. \quad \square$$

Minimality of Traversals-induced Typing Using the same notation as before, interaction sequences from $\mathbf{Int}(\Lambda(\mathbf{g}), \mathbf{fix}) \subseteq \mathrm{Int}([\![\Sigma]\!], [\![\mathcal{N}]\!] \Rightarrow [\![\mathcal{N}]\!], [\![o]\!])$ form a tree, which is (in essence) the *traversal tree* in the sense of Ong [1].

Prime types, which are intersection types of the form $\theta = \bigwedge_{i\in I_1} \theta_{1i} \to \cdots \to \bigwedge_{i\in I_n} \theta_{ni} \to q$, are equivalent to *variable profiles* (or simply *profiles*) [1]. Precisely θ corresponds to profile $\widehat{\theta} := (\{ \widehat{\theta_{1i}} \mid i \in I_1 \}, \cdots, \{ \widehat{\theta_{ni}} \mid i \in I_n \}, q)$. We write profiles of ground kind as q, rather than (q). Henceforth, we shall use prime types and profiles interchangeably.

Tsukada and Kobayashi [10] introduced (a kind-indexed family of) binary relations $\leq_\kappa$ between profiles of kind κ, and between sets of profiles of kind κ, by induction over the following rules.

(i) If for all $\theta \in A$ there exists $\theta' \in A'$ such that $\theta \leq_\kappa \theta'$ then $A \leq_\kappa A'$.
(ii) If $A_i \leq_{\kappa_i} A'_i$ for each i then $(A_1, \ldots, A_n; q) \leq_{\kappa_1 \to \cdots \to \kappa_n \to o} (A'_1, \ldots, A'_n, q)$.

A *profile annotation* (or simply *annotation*) of the traversal tree $\mathbf{Int}(\Lambda(\mathbf{g}), \mathbf{fix})$ is a map of the nodes (which are move-occurrences of $M_{[\![\Sigma]\!]} + M_{[\![\mathcal{N}]\!]\Rightarrow[\![\mathcal{N}]\!]} + M_{[\![o]\!]}$) of the tree to profiles. We say that an annotation of the traversal tree is *consistent* just if whenever a move m, of kind $\kappa_1 \to \cdots \to \kappa_n \to o$ and simulates q, is annotated with profile $(A_1, \cdots, A_n, q')$, then (i) $q' = q$, (ii) for each i, A_i is a set of profiles of kind κ_i, (iii) if m' is annotated with θ and i-points to m, then $\theta \in A_i$. Now consider *annotated moves*, which are moves paired with their annotations, written (m, θ). We say that a profile annotation is *innocent* just if whenever $u_1 \cdot (m_1, \theta_1)$ and $u_2 \cdot (m_2, \theta_2)$ are even-length paths in the annotated traversal tree such that $\ulcorner u_1 \urcorner = \ulcorner u_2 \urcorner$, then $m_1 = m_2$ and $\theta_1 = \theta_2$.

Every consistent (and innocent) annotation α of an (accepting) traversal tree gives rise to a typing environment, written Γ_α, which is the set of bindings $F_i : \theta$ where $i \in \{1, \ldots, n\}$ and θ is the profile that annotates an occurrence of an initial move of $[\![\kappa_i]\!]$. Note that Γ_α is finite because there are only finitely many types of a given kind. We define a relation between annotations: $\alpha_1 \leq \alpha_2$ just if for each occurrence m of a move of kind κ in the traversal tree, $\alpha_1(m) \leq_\kappa \alpha_2(m)$.

Theorem 7. *(i) Let α be a consistent and innocent annotation of a traversal tree. Then Γ_α is a complete type environment.*

(ii) There is $\leq$-minimal consistent and innocent annotation, written $\alpha_{\min}$. Then $\Gamma_{\alpha_{\min}} \leq \Gamma_\alpha$ meaning that for all $F : \theta \in \Gamma_{\alpha_{\min}}$ there exists $F : \theta' \in \Gamma_\alpha$ such that $\theta \leq \theta'$.

(iii) Every complete type environment Γ determines a consistent and innocent annotation α_Γ of the traversal tree.

Game-Semantic Proof of Completeness of GTRecS. GTRecS [9] is a higher-order model checker proposed by Kobayashi. Although GTRecS is inspired by game-semantics, the formal development of the algorithm is purely type-theoretical and no concrete relationship to game semantics is known. Here we give a game-semantic proof of completeness of GTRecS based on two-level arena games.

The novelty of GTRecS lies in a function on type bindings, named **Expand**. For a set Γ of nonterminal-type bindings, **Expand**(Γ) is defined as

$$\Gamma \cup \bigcup \{ \Gamma' \cup \{F_i : \tau'\} \mid \Gamma \preceq_P \Gamma' \wedge \Gamma' \vdash \mathcal{R}(F_i) : \tau' \wedge \Gamma \preceq_O \{F_i : \tau'\} \},$$

where $\Gamma' \vdash \mathcal{R}(F_i) : \tau'$ is relevant. Here for types τ_1 and τ_2, $\tau_1 \preceq_P \tau_2$ if the arena $[\![\tau_2]\!]$ is obtained by adding only proponent moves to $[\![\tau_1]\!]$. For example, $(\bigwedge \emptyset) \to q \preceq_P ((\bigwedge \emptyset) \to q') \to q$ but $(\bigwedge \emptyset) \to q \npreceq_P (q'' \to q') \to q$, since q' is at the proponent position and q'' at the opponent position. $\Gamma \preceq_P \Gamma'$ is defined as $\forall F : \tau' \in \Gamma'. \exists F : \tau \in \Gamma. \tau \preceq_P \tau'$. Similarly, $\tau \preceq_O \tau'$ and $\Gamma \preceq_O \Gamma'$ are defined.

Our goal is to analyse **Expand** game theoretically. The result is Lemma 5, which states that **Expand** overapproximates one step interaction of two strategies, σ and **fix**. Completeness of GTRecS is a corollary of Lemma 5.

Assume that G is typable and fix a type environment $\Gamma = \{F_i : \bigwedge_j \tau_{i,j} \mid F_i \in \mathcal{N}\}$ such that $\vdash G : \Gamma$. Let $\sigma : [\![\delta]\!] \longrightarrow (\Gamma^1 \Rightarrow \Gamma^2)$ (here we use superscripts to distinguish occurrences of Γ) be the winning strategy that is induced from the derivation of $\vdash G : \Gamma$. The strategy **fix** $: ([\![\Gamma^1]\!] \Rightarrow [\![\Gamma^2]\!]) \longrightarrow [\![q_I]\!]$ is defined as the composite of $([\![\Gamma]\!]^1 \Rightarrow [\![\Gamma]\!]^2) \xrightarrow{Y} [\![\Gamma]\!] \xrightarrow{\{S:q_I\}} [\![q_I]\!]$. For $n \in \{1, 2, \ldots\}$, the *nth approximation of* **fix** is defined by $\lfloor \textbf{fix} \rfloor_n = \{s \in \textbf{fix} \mid |s| \leq 2n + 1\}$. Thus $\lfloor \textbf{fix} \rfloor_n$ is a strategy that behaves like **fix** until the nth interaction, but stops after that.

Let $n \in \{0, 1, 2, \ldots\}$. The n-th approximation of **fix** induces approximation of arenas and type environments. An arena $\lfloor \Gamma^1 \Rightarrow \Gamma^2 \rfloor_n$ is defined as the restriction of

$\Gamma^1 \Rightarrow \Gamma^2$ that consists of only moves appearing at $\mathbf{Int}(\sigma, \lfloor \mathbf{fix} \rfloor_n)$. The arena $\lfloor \Gamma^1 \Rightarrow \Gamma^2 \rfloor_n$ is decomposed as $\prod_{i,j} (\lfloor \Gamma^1 \rfloor_{n,i,j} \Rightarrow \lfloor \tau_{i,j} \rfloor_n)$. Let $\lfloor \Gamma^1 \rfloor_n$ be the union of variable-type bindings corresponding to $\bigcup_{i,j} \lfloor \Gamma^1 \rfloor_{n,i,j}$ and $\lfloor \Gamma^2 \rfloor_n$ be the set of type bindings $\{F_i : \lfloor \tau_{i,j} \rfloor_n\}_{i,j}$.

Lemma 5. $\lfloor \Gamma^1 \rfloor_n \cup \lfloor \Gamma^2 \rfloor_n \subseteq \mathbf{Expand}^n(\{S : q_0\})$.

Conclusions and Further Directions. Two-level arena games are an accurate model of intersection types. Thanks to Subject Expansion, they are a useful semantic framework for reasoning about higher-order model checking.

For future work, we aim to (i) consider properties that are closed under disjunction and quantifications, and (ii) study a call-by-value version of intersection games. In orthogonal directions, it would be interesting to (iii) construct an intersection game model for untyped recursion schemes [10], and (iv) build a CCC of intersection games parameterised by an alternating parity tree automaton, thus extending our semantic framework to mu-calculus properties.

Acknowledgement. This work is partially supported by Kakenhi 22 · 3842 and EPSRC EP/F036361/1. We thank Naoki Kobayashi for encouraging us to think about game-semantic proofs and for insightful discussions.

References

1. Ong, C.H.L.: On model-checking trees generated by higher-order recursion schemes. In: LICS, pp. 81–90. IEEE Computer Society (2006)
2. Hyland, J.M.E., Ong, C.H.L.: On full abstraction for PCF: I, II, and III. Inf. Comput. 163(2), 285–408 (2000)
3. Kobayashi, N.: Types and higher-order recursion schemes for verification of higher-order programs. In: Shao, Z., Pierce, B.C. (eds.) POPL, pp. 416–428. ACM (2009)
4. Salvati, S.: On the membership problem for non-linear abstract categorial grammars. Journal of Logic, Language and Information 19(2), 163–183 (2010)
5. Kobayashi, N., Ong, C.H.L.: A type system equivalent to the modal mu-calculus model checking of higher-order recursion schemes. In: LICS, pp. 179–188. IEEE Computer Society (2009)
6. Hague, M., Murawski, A.S., Ong, C.H.L., Serre, O.: Collapsible pushdown automata and recursion schemes. In: LICS, pp. 452–461 (2008)
7. Salvati, S., Walukiewicz, I.: Krivine Machines and Higher-Order Schemes. In: Aceto, L., Henzinger, M., Sgall, J. (eds.) ICALP 2011, Part II. LNCS, vol. 6756, pp. 162–173. Springer, Heidelberg (2011)
8. Nielson, F.: Two-level semantics and abstract interpretation. Theor. Comput. Sci. 69(2), 117–242 (1989)
9. Kobayashi, N.: A Practical Linear Time Algorithm for Trivial Automata Model Checking of Higher-Order Recursion Schemes. In: Hofmann, M. (ed.) FOSSACS 2011. LNCS, vol. 6604, pp. 260–274. Springer, Heidelberg (2011)
10. Tsukada, T., Kobayashi, N.: Untyped Recursion Schemes and Infinite Intersection Types. In: Ong, L. (ed.) FOSSACS 2010. LNCS, vol. 6014, pp. 343–357. Springer, Heidelberg (2010)

An Automata-Theoretic Model
of Idealized Algol
(Extended Abstract)

Uday S. Reddy[1] and Brian P. Dunphy[2]

[1] University of Birmingham
u.s.reddy@birmingham.ac.uk
[2] University of Illinois at Urbana-Champaign

Abstract. In this paper, we present a new model of class-based Algol-like programming languages inspired by automata-theoretic concepts. The model may be seen as a variant of the "object-based" model previously proposed by Reddy, where objects are described by their observable behaviour in terms of events. At the same time, it also reflects the intuitions behind state-based models studied by Reynolds, Oles, Tennent and O'Hearn where the effect of commands is described by state transformations. The idea is to view stores as automata, capturing not only their states but also the allowed state transformations. In this fashion, we are able to combine both the state-based and event-based views of objects. We illustrate the efficacy of the model by proving several test equivalences and discuss its connections to the previous models.[1]

1 Introduction

Imperative programming languages provide *information hiding* via local variables accessible only in their declaring scope. This is exploited in object-oriented programming in a fundamental way. The use of such information hiding in everyday programming can be said to have revolutionized the practice of software development.

Meyer and Sieber [10] pointed out that the traditional semantic models for imperative programs do not capture such information hiding. Rapid progress was made in the 1990's to address the problem. O'Hearn and Tennent [15] proposed a model using relational parametricity to capture the independence of data representations. Reddy [19] proposed an alternative event-based model which hides data representations entirely, and this was adapted to full Idealized Algol in [12]. Both the models have been proved fully abstract for second-order types of Idealized Algol (though this does not cover "passive" or "read-only" types such as expressions) [12,13]. Abramsky, McCusker and Honda [2,1] refined the event-based model using games semantics and proved it fully abstract for full higher-order types.

---

[1] A full version of this paper, along with appendices, is available electronically at http://www.cs.bham.ac.uk/~udr/papers/icalp2012.

A. Czumaj et al. (Eds.): ICALP 2012, Part II, LNCS 7392, pp. 337–350, 2012.
© Springer-Verlag Berlin Heidelberg 2012

Despite all this progress, the practical application of these models for program reasoning had stalled. As we shall see, "second-order functions" in Idealized Algol only correspond to basic functions (almost zero-order functions) in the object-oriented setting. The event-based model is a bit removed from the normal practice in program reasoning, while the applicability of the parametricity model for genuine higher-order functions has not been investigated. In fact, Pitts and Stark [18] showed an "awkward example" in a bare bones ML-like language (discussed in Sec. 2), which could not be proved using their formulation of the parametricity technique.

The present work began in the late 90's with the motivation of bridging the gap between state-based parametricity models and the event-based models, because they clearly had complementary strengths. These investigations led to an automata theory-inspired framework where both states and events play a role [20]. However, it was noticed that the basic ingredients of the model were already present in the early work of Reynolds [23]. The subsequent work focused on formalizing the category-theoretic foundations of the framework, documented in [6], but the applications of the framework remained unexplored.

The interest in the approach has been renewed with two parallel developments in recent work. Amal Ahmed, Derek Dreyer and colleagues [5] began to investigate reasoning principles for higher-order ML-like languages where similar ideas have reappeared. In the application of Separation Logic to concurrency, a technique called "deny-guarantee reasoning" has been developed [4] where, again, a combination of states and events is employed. With this paper, we hope to provide a denotational semantic foundation for such techniques and stimulate further work in this area.

We see denotational semantic models as giving an abstract characterisation of what kind of computational entities are expressible in a programming language. Once an accurate semantic model is developed, it can be used for proving observational equivalences, constructing reasoning principles or programming logics as well as for supporting informal reasoning routinely carried out by programmers.

2   Motivation

In this section, we informally motivate the ideas behind the new semantic model.

The two existing classes of semantic models for imperative programs are state-based parametricity models [13,15] and event/game-based models [1,2,19]. Both of them were first presented for Algol-like languages, and later adapted to object-oriented programs [21]. The new model here may be seen as a refinement of the state-based model using ideas from the event-based model. Like the state-based model, it will be extensional (no intermediate computation steps appear in denotations) but it borrows ideas from the event-based model to capture irreversible state change.

In the state-based model, an object is described as a state machine with a state set Q, an initial state $q_0 \in Q$, and the effect of the methods on the object state. Abstractly, such a structure may be thought of as belonging to the type $\exists_Q Q \times F(Q)$, where $F(Q)$ is a type denoting the method signature of the object. The existential type [11] is the type of "abstract types," i.e., structures with hidden representations.

For example, two forms of counter objects with methods for reading and incrementing can be semantically described by:

$$M = \langle Q = Int, 0, \{val = \lambda n.\, n,\ inc = \lambda n.\, n + 1\}\rangle$$
$$M' = \langle Q' = Int, 0, \{val = \lambda n.\, (-n),\ inc = \lambda n.\, n - 1\}\rangle$$

Here, val is given by a function of type $Q \to Int$ and the effect of inc is given by a function of type $Q \to Q$. (We are ignoring the issues of divergence and recursion.) The information hiding properties of the programming language would allow the objects to be used solely by their methods without direct access to the internal state. This gives rise to a notion of "behavioral equivalence" for objects, which occurs when they differ only in the hidden internal state but have the same observable behavior. The behavioral equivalence of the two implementations of counters can be established by exhibiting a *simulation relation* between the state sets:

$$n\ [R]\ n' \iff n \geq 0 \wedge n' = -n \tag{1}$$

and showing that all the operations "preserve" the simulation relation. Once this is done, we regard M and M' as "similar" (denoted $M \sim M'$). *Behavioural equivalence* is the reflexive-transitive closure $\sim^*$ of similarity. So, we can argue that two objects are behaviorally equivalent not only when they are similar, but also when there is a series of similarity proofs $M = M_1 \sim M_2 \sim \cdots \sim M_n = M'$.

State-based models of this kind struggle to capture the *irreversibility* of state change. The action of incrementing the counter overwrites the old state of the counter and it is not possible to go back to that state. If we pass a counter object to a procedure, we can be sure that, after the procedure returns, the value in the counter could be no less that what it was before the call. The state-based models do not have such a direction of time and often contain a variety of "snap-back" operators in their mathematical domains which violate the direction of time [13,15,12,19]. (The event-based and games models, in contrast, do capture the direction of time, which is crucial for their full abstraction properties.) Offsetting this technical deficiency, the state-based models have the advantage of being highly intuitive and familiar from traditional reasoning principles of programs.

In this paper, we define a new model that combines the advantages of the state-based and event-based models. For this purpose, we turn to algebraic automata theory [7]. We use an abstract form of semiautomata called *transformation monoids* in place of simple state sets to describe the internal structure of objects. A model of objects can now be given in *four* parts: a state set Q, a monoid of state transformations $T \subseteq [Q \to Q]$, an initial state $q_0 \in Q$, and the effect of the methods on the object state using the state transformations in T. Abstractly, such a structure may be regarded as belonging to a type such as $\exists_{(Q,T)} Q \times F(Q,T)$. The essential difference from the state-based model is the addition of the T component in modeling the internal store, which represents the *allowed* transformations of the store regarded as a state machine. If the full monoid of transformations $T(Q) = [Q \to Q]$ is used as the T component and simulation relations do the same, the model reduces to the state-based model.

For example, the automata-theoretic representation of the two counter objects is:

$$N = \langle Q = Int, \ T = Int^+, \ 0, \ \{val = \lambda n. \, n, \ inc = \lambda n. \, n + 1\} \rangle$$
$$N' = \langle Q' = Int, \ T' = Int^-, \ 0, \ \{val = \lambda n. -n, \ inc = \lambda n. \, n - 1\} \rangle$$

Here, $Int^+ = \{\lambda n. \, n + k \mid k \geq 0\}$ and $Int^- = \{\lambda n. \, n - k \mid k \geq 0\}$ are the set of allowed transformations. Note that they are monoids. The type of val is $Q \to Int$ as before, but the type of inc is T. Any state change operations in methods are interpreted in T, so that they are restricted to the *allowed* transformations of the state machine.

Proving the equivalence of the two state machines requires us to exhibit *two relations*: a relation R_Q between the state sets and a relation R_T between the state transformations:

$$\begin{aligned} n \begin{bmatrix} R_Q \end{bmatrix} n' &\iff n \geq 0 \wedge n' = -n \\ a \begin{bmatrix} R_T \end{bmatrix} a' &\iff \forall n, n'. \, a(n) \geq n \wedge a'(n') - n' = -(a(n) - n) \end{aligned} \qquad (2)$$

The two relations have to satisfy some coherence conditions, which are detailed in Sec. 4. Using these relations, it is easy to prove, for instance, that a procedure that takes a counter as an argument can only increase the value of the counter (as visible from the outside). The transformation components in the state machines provide a direction of time, which is absent in the purely state-based model.

While simulation relations are useful for proving the equivalence of two implementations of classes, they form an instance of a general theory of relational parametricity which works for relations of arbitrary arity [8]. The case of "unary relations" gives us a new notion of *invariants*. Our theory therefore posits that invariants of classes again come in two parts: one on state sets and one on state transformations. The invariants for counter objects represented by N are:

$$P_Q(n) \iff n \geq 0 \qquad P_T(a) \iff \forall n. \, a(n) \geq n \qquad (3)$$

State invariants are well-known from traditional reasoning methods, while the invariant properties of transformations might be called "action invariants" and are relatively unknown. (See, however, the "history invariants" of [9].)

The recent work on reasoning about state has focused on higher-order procedures and higher-order state, in particular the work of Ahmed, Dreyer and colleagues [5]. This work has brought home the fact that the traditional theory of Algol-like languages fails to be abstract for higher-order procedures. [2] We illustrate the problem with an example from Pitts and Stark [18], which

---

[2] O'Hearn and Reynolds [13] proved that their state-based model was fully abstract for second-order Algol types. Translated to object-oriented languages, this amounts to saying that they can prove the equivalence of classes whose methods take at best value-typed, i.e., state-independent, arguments. If the methods take higher-type arguments, e.g., other procedures, then the full abstraction results do not apply.

was termed an "awkward example" in their paper. Consider the following class, written in the IA+ language [21]:

$$RC = \textbf{class} : \textbf{comm} \to \textbf{comm}$$
$$\textbf{local } \text{Var[int] } x; \ \textbf{init } x := 0;$$
$$\textbf{meth } \{m = \lambda c.\ x := 1; c;\ test(x = 1)\}$$
$$\textbf{where } test(b) \triangleq \textbf{if } b \textbf{ then skip else diverge}$$

This class provides a single method of type **comm** $\to$ **comm**. (Recall that Idealized Algol is a call-by-name language.) The problem is to argue that the method always terminates. Intuitively, one might expect that this should always be the case because the local variable x is only available inside the class. So, calling c cannot change x and the test should succeed. However, the reasoning is unsound. Consider the following usage of class RC:

$$\textbf{new } RC\ p.\ (p.m\ (p.m\ \textbf{skip}))$$

When the outer call to $p.m$ is executed, it sets x to 1 and calls its argument $c \equiv p.m$ **skip**. Since the argument in turn calls $p.m$, it has the effect of setting x to 1. So, the argument that c does not have "access" to x is not sound. The phenomenon exhibited in this example is termed a "reentrant callback" [3] and it is a common technique used in constructing object-oriented programs.

A more sophisticated argument for the termination of RC's method notes that the *only change* that a call to c can make to x is setting it to 1. However, as noted by Dreyer et al. [5], this cannot be proved by exhibiting invariants on states. The state-invariant for the class only states that x can be 0 or 1. All the previous state-based denotational models [13,15] fail for this example.

In our framework, we start by defining a two-part invariant for the class:

$$P_Q(x) \iff x = 0 \lor x = 1 \qquad P_T(a) \iff a = (\lambda n.\ n) \lor a = (\lambda n.\ 1) \qquad (4)$$

To maintain P_T as an "invariant", the method m must restrict its actions to those satisfying P_T, while *assuming* that the argument c does so as well. So, by assumption, the call to c will either leave x unchanged $(\lambda n.\ n)$ or set it to 1 $(\lambda n.\ 1)$. In either case, the value of x at the end of c will be 1. So, the method always terminates. Other examples discussed by Dreyer et al. [5] can be verified similarly, as long as they fit within our framework — with only ground-typed state and no control effects. We show some of the details in Sec. 5.

3   The Semantic Framework

The programming language we use in this paper is the language IA+ described in [21], which represents Idealized Algol [23] extended with classes.

Recall that Idealized Algol is a call-by-name simply typed lambda calculus (with full higher-order procedures including the potential for aliasing and interference), with base types supporting imperative programming. These base types include **val**[δ] for data values of type δ, **exp**[δ] for (state-reading) expressions yielding data values of type δ, and **comm** for (state-transforming) commands. Here, δ ranges over "data types" such as **int** and **bool**.

To support classes, we use a type constructor **cls** so that **cls** θ is the type of classes whose method suite is of type θ. So, θ is the "interface type" of the class. The language comes with a family of predefined classes Var[δ] for assignable variables of type δ, whose interface type is a record type of the form:

$$\text{Var}[\delta] : \textbf{cls} \,\{\text{get} : \textbf{exp}[\delta], \text{put} : \textbf{val}[\delta] \to \textbf{comm}\}$$

In essence, a variable is treated as an object with a "get" method that reads the state of the variable and "put" method that changes the state to a given value. User-defined classes are available using terms of the form

$$\textbf{class} : \theta$$
$$\textbf{local}\,C\,x;\ \textbf{init}\,A;\ \textbf{meth}\,M$$

where C is another class, x is a locally bound identifier for the "instance variable," A is a command for initializing the instance variable, and M is a term of type θ serving as the suite of "methods" for the objects of this class. For simplicity of exposition, we only consider "constant classes" in the main body of the paper, which are defined by *closed terms* of type **cls** θ.

New instances of classes are created in commands using terms of the form **new** C o. B, whose effect is to create an instance of class C, bind it to o and execute a command B. So, thinking of the binding o. B as a function, **new** is effectively a constant of type: **cls** $\theta \to (\theta \to \textbf{comm}) \to \textbf{comm}$.

Possible World Semantics

Semantics of Algol-like languages is normally given using a category-theoretic possible world semantics, where the "worlds" represent types of stores [23] (also called "store shapes" [25].) A simple form of worlds can be just sets of locations, but it is the point of this paper to propose a more abstract treatment of stores.

We use letters $W, X, Y, Z, \ldots$ for worlds. A morphism $f : W \to X$ represents the idea that X is a larger world than W or, to put another way, a W-typed store can be extracted from an X-typed store via the morphism f. (Note the reversal of direction in the second statement.) To capture relational parametricity and data abstraction, we also assume relations between worlds, denoted as $R : W \leftrightarrow W'$. Formally, these pieces of data should form a *parametricity graph of categories* [6], denoted **W**.

We leave this structure unspecified for the time being, except to note that it should be able to specify for each world W, a set of states $\mathcal{Q}_W$ and a set of state transformations $\mathcal{T}_W$, the latter of which is a submonoid of the set of partial functions $\mathcal{Q}_W \rightharpoonup \mathcal{Q}_W$.

Each programming language type θ is interpreted as a *parametricity graph-functor* (PG-functor) of the form $[\![\theta]\!] : \mathbf{W} \to \mathbf{DCPO}$ where **DCPO** is the parametricity graph of directed-complete partial orders with continuous functions as morphisms and directed-complete relations as edges. (These functors should satisfy the "Oles condition" that they should factor through the embedding **CPO**$_\perp$ $\hookrightarrow$ **DCPO**. This obtains a cartesian closed category of functors [12].) We use readable notation for these functors $[\![\textbf{comm}]\!] = \textsc{Comm}$, $[\![\textbf{exp}[\delta]]\!] = \textsc{Exp}_\delta$, etc.

The functors needed for interpreting IA+ are shown in a schematic form below:

$$\textsc{Comm}(X) = \mathcal{T}_X$$
$$\textsc{Exp}_\delta(X) = [\mathcal{Q}_X \rightharpoonup [\![\delta]\!]]$$
$$\textsc{Var}_\delta(X) = \textsc{Exp}_\delta(X) \times [[\![\delta]\!] \to \textsc{Comm}(X)]$$
$$(F \times G)(X) = F(X) \times G(X) \tag{5}$$
$$(F \Rightarrow G)(X) = \forall_{h:Z \leftarrow X} [F(Z) \to G(Z)]$$
$$(\textsc{cls}\, F)(X) = \exists_Z (\mathcal{Q}_Z)_\perp \times F(Z)$$

Note that commands for a store type X are interpreted as state transformations for X, expressions for store type X are interpreted as partial functions from states of X to values, and variables are interpreted as pairs of get and put operations. The $\forall$ and $\exists$ operators in the last two lines are formally the parametric limit and parametric colimit constructions [6], but they can also be understood intuitively as the types of polymorphic functions and abstract types respectively. The meaning of a procedure of type $F \Rightarrow G$ at store type X allows for it to be called in a larger store Z (which might be obtained by allocating new local variables) and maps arguments of type F in the larger store Z to results of type G in the store Z. The parametric limit interpretation of $\forall$ ensure that this will be done *uniformly* in the store type Z, depending only on the fact that it is a larger store than X. The interpretation of a class at a store type X ignores X (because we are considering only *constant classes*), specifies a new store Z for the representation of the objects of this class, and provides an initial state and an interpretation of the methods on store Z. The intuitions behind this interpretation are essentially standard [15,13,21].

The focus of this paper is on defining a suitable category (or, rather, a parametricity graph) **W** for modelling stores. We do this using automata-theoretic ideas in the following sections.

4   Stores as Automata

In algebraic automata theory [7], we find three related notions. A *semiautomaton* is a triple (Q, Σ, α) where Q is a set (of "states"), Σ is a set (of "events"), $\alpha : \Sigma \to [Q \rightharpoonup Q]$ provides an interpretation of the events as state transformations. A more general notion is a *monoid action* (Q, M, α) where the set Σ of events is generalized to a monoid M and α is given to be compatible with the monoid structure. A special case of a monoid action is a *transformation monoid* (Q, T) where T is a submonoid of the monoid of state transformations $[Q \rightharpoonup Q]$ and, hence, α is implicit. In this work, we choose transformation monoids as the basis for our modelling of stores and leave its generalization to monoid actions to future work.

The monoid structure of T is obtained by the sequential composition of transformations (written as $a \cdot b$) and the unit transformation (written as "null"). The monoid structure supports sequential composition of commands. To support divergence and iteration, we require that T be a monoid in $\mathbf{CPO}_\perp$ (the category of pointed cpo's and strict continuous functions). We refer to such a monoid as a *complete ordered monoid*. In addition, commands in imperative languages also have the ability to read the current state and tailor their actions accordingly. Consider if-then-else, for example. Reynolds [23] addressed this issue in his early work, and postulated an operation called the "diagonal." Consequently, we name the resulting structures after Reynolds.

Definition 1. *A* Reynolds transformation monoid *is a triple* $X = (\mathcal{Q}_X, \mathcal{T}_X, read_X)$ *where* $\mathcal{Q}_X$ *is a set,* $\mathcal{T}_X$ *is a submonoid of the monoid of transformations* $[\mathcal{Q}_X \rightharpoonup \mathcal{Q}_X]$ *and the two components are jointly closed under the operation* $read_X : (\mathcal{Q}_X \to \mathcal{T}_X) \to \mathcal{T}_X$ *given by:* $read_X(p) = \lambda x. \, p(x)(x)$.

The $read_X$ operation maps a state-dependent transformation p into a normal transformation, which first reads the current state (x), uses it to satisfy the state-dependence of p, and finally executes the resulting action. Given a transformation monoid, it is always possible to close it under the read operation by adding enough elements to $\mathcal{T}_X$. We call it the "read-closure" of the transformation monoid. A full transformation monoid $(Q, T(Q))$ is always a Reynolds transformation monoid.

A *relation of Reynolds transformation monoids* $R : X \leftrightarrow X'$ is a pair $R = (R_Q, R_T)$ where $R_Q : \mathcal{Q}_X \leftrightarrow \mathcal{Q}_{X'}$ is a normal set-theoretic relation and $R_T : \mathcal{T}_X \leftrightarrow \mathcal{T}_{X'}$ is a complete ordered monoid relation (relation compatible with the units, multiplication, least elements and sup's of directed sets) such that:

$$R_T \subseteq [R_Q \rightharpoonup R_Q]$$
$$read_X \, [[R_Q \to R_T] \to R_T] \, read_{X'}$$

These conditions ensure that parametric transformations will include all the normal operations of imperative programming: sequential composition, if-then-else, assignments, and iteration.

A *morphism of Reynolds transformation monoids* $f : W \to X$ is a pair $f = (\phi_f, \tau_f)$ where $\phi_f : \mathcal{Q}_X \to \mathcal{Q}_W$ is a function and $\tau_f : \mathcal{T}_W \to \mathcal{T}_X$ is a complete ordered monoid morphism such that the pair $(\langle \phi_f \rangle^\smile, \langle \tau_f \rangle)$ is a relation of Reynolds transformation monoids. The notation $\langle - \rangle$ stands for the graph of a function treated as a relation and $(-)^\smile$ stands for the converse of a relation.

Note that ϕ_f and τ_f go in opposite directions. Computationally, the intuition is that, when X is a larger store than W, it *extends* and possibly *constrains* the states of the current store. So, it is possible to recover the state information at the level of the current store W via the function ϕ_f. On the other hand, the actions possible in the current store should continue to be possible in the larger store, which is modelled by the function τ_f. In the terminology of algebraic automata theory, X "covers" W.

A *relation-preservation square* of Reynolds transformation monoids f $[R \to S]$ f' exists iff ϕ_f $[S_Q \to R_Q]$ $\phi_{f'}$ and τ_f $[R_T \to S_T]$ $\tau_{f'}$.

This data constitutes a cpo-enriched reflexive graph **RTM**. (See [6,15] for the background on reflexive graphs.)

Lemma 1. **RTM** *is a cpo-enriched parametricity graph, i.e., it is relational, fibred and satisfies the identity condition.*

The move from state-based models to automata-based models has already paid a rich dividend, because the reflexive graphs of worlds in [13,15] are not parametricity graphs. A parametricity graph has a *subsumption* map whereby each morphism $f : X \to Y$ is "subsumed" by an edge $\langle f \rangle : X \leftrightarrow Y$. This is given by $\langle f \rangle = [f, \mathrm{id}_Y] I_Y$. In the case of **RTM**, this gives $\langle f \rangle = (\langle \phi_f \rangle^{\smile}, \langle \tau_f \rangle)$. The subsumption maps lead to a strong theory of relational parametricity which includes naturality as a *special case*. These properties are not available in the models of [13,14], where naturality is an independent condition from parametricity, indicating that their theory parametricity is not strong enough.

Examples of morphisms. The *expansion* of a full transformation monoid $(Q, T(Q))$ with additional state components represented by a set Z, and leading to a larger world $(Q \times Z, T(Q \times Z))$, is represented by a morphism $\times Z :$ $(Q, T(Q)) \to (Q \times Z, T(Q \times Z))$. The ϕ component of the morphism is a projection and the τ component expands actions $a \in T(Q)$ to $T(Q \times Z)$ by leaving the Z components unchanged. A *state change restriction* morphism for a Reynolds transformation monoid $(\mathcal{Q}_X, \mathcal{T}_X)$ restricts the state transformations to a submonoid $T' \subseteq \mathcal{T}_X$. The morphism $f : (\mathcal{Q}_X, T') \to (\mathcal{Q}_X, \mathcal{T}_X)$ has the identity for the ϕ component and an injection for the τ component. A *passivity restriction* morphism is an extreme case of state change restriction morphism that prohibits all state changes: $p_X : (\mathcal{Q}_X, \mathbf{0}_X) \to (\mathcal{Q}_X, \mathcal{T}_X)$ where $\mathbf{0}_X$ is the complete ordered monoid containing the unit transformation null_X and all its approximations.

The parametricity graph **RTM** has a symmetric monoidal structure. For any Reynolds transformation monoids $X = (\mathcal{Q}_X, \mathcal{T}_X)$ and $Y = (\mathcal{Q}_Y, \mathcal{T}_Y)$ representing two separate stores of locations (along with allowed transformations), thee is another one $X \star Y$ that corresponds to their combined store. The unit for $\star$ is $\mathbf{I} = (\mathbf{1}, \mathbf{0_1})$, which represents the empty store. This is in fact the initial object in the parametricity graph **RTM**.

5   Semantics

As noted in Section 3, the semantics of the programming language is given in the functor category $\mathbf{W} \to \mathbf{DCPO}$ (restricted to the functors satisfying the "Oles condition") where $\mathbf{W}$ is a parametricity graph of worlds. We now choose $\mathbf{W} = \mathbf{RTM}$, the parametricity graph of Reynolds transformation monoids.

The interpretation of types is as in (5). The type expressions $F(X)$ shown there have an associated relation action $F(R)$ and an action on morphisms $F(f)$ so as to form PG-functors. The details are shown in the full paper.

Using these functor actions $[\![\theta]\!](f : X \to Y)$, we can upgrade any value d of type θ at world X to a future world Y. When the morphism $f : X \to Y$ is clear from the context, we often use the short-hand notation $d{\uparrow}_X^Y \triangleq [\![\theta]\!](f)(d)$ to denote such upgrading.

The meaning of a term M with typing: $x_1 : \theta_1, \ldots, x_n : \theta_n \vdash M : \theta$ is a parametric transformation of type $[\![M]\!] : (\prod_{x_i} [\![\theta_i]\!]) \to [\![\theta]\!]$. This means that, for each world (Reynolds transformation monoid) X, $[\![M]\!]_X$ is a continuous function of type $(\prod_{x_i} [\![\theta_i]\!](X)) \to [\![\theta]\!](X)$ such that all relations are preserved, i.e., for any relation $R : X \leftrightarrow X'$, the meanings $[\![M]\!]_X$ and $[\![M]\!]_{X'}$ are related by $(\prod_{x_i} [\![\theta_i]\!](R)) \to [\![\theta]\!](R)$. To the extent that IA+ is a simply typed lambda calculus, this is standard [6,15]. The details are shown in the full paper.

Lemma 2. *All the combinators of Idealized Algol are parametric transformations.*

Theorem 1. *The meaning of every IA+ term $x_1 : \theta_1, \ldots, x_n : \theta_n \vdash M : \theta$ is a parametric transformation of type $(\prod_{x_i} [\![\theta_i]\!]) \to [\![\theta]\!]$.*

Example Equivalences

We note at the outset that the example equivalences that can be validated in previous state-based models such as those of [10,15] continue to hold in our model. We focus on the new equivalences validated here.

In the following discussion, we use the information notation $a{\uparrow}_Z^Y$ for the "upgrading" of a value $a \in F(Z)$ to $F(Y)$, using a morphism $h : Z \to Y$. The morphism h is left implicit in the notation, but it will be clear from the context.

Example 1 (Pitts and Stark). Consider the following classes:

$C_1 = $ **class** : **comm** $\to$ **comm**
    local Var[int] x;
    init $x := 0$;
    meth $\lambda c. x := 1; c; test(x = 1)$

$C_2 = $ **class** : **comm** $\to$ **comm**
    local Var[int] x;
    init $x := 0$;
    meth $\lambda c. c$

where $test(b) = $ **if** b **then skip else diverge**.

The meanings of the classes should be semantic values of type: $\exists_Z ((\mathcal{Q}_Z)_\bot \times \forall_{g:Y \leftarrow Z} \text{COMM}(Y) \to \text{COMM}(Y))$. Equality at this type is behavioral equivalence of structures, as mentioned in Sec. 2. The semantic definition provides two structures M_1 and M_2 as the meanings of the two classes. We then construct a behavioural equivalence proof of the form $M_1 \sim M_1' \sim M_2' \sim M_2$, where M_1' and M_2' are pared down versions of M_1 and M_2 that record only the transformations used by the objects.

We first note how such paring down can be done. The initial structure M_1 obtained for class C_1 is as follows:

$$\mathcal{Q}_Z = \mathit{Int} \qquad \mathcal{T}_Z = T(\mathit{Int})$$
$$\mathit{init}_1 = 0 \qquad \mathit{meth}_1 = \Lambda g : Y \leftarrow Z. \lambda c : \text{COMM}(Y).$$
$$\text{put}(1){\uparrow}_Z^Y \cdot c \cdot \text{check}(1){\uparrow}_Z^Y$$

where $\text{put}(k) = \lambda n.\, k$ and $\text{check}(k) = \text{read } \lambda n.\, \text{if } n = k \text{ then null else } \bot$.
We can notice that the object only uses states in the subset $\{\, n \mid n \geq 0\,\}$ and transformations in the subset $\{\text{put}(1)\}$. The minimal complete ordered monoid containing the latter is $\{\bot, \text{null}_{Z_2}, \text{put}(1)\}$. So we can define the pared-down structure for C_1 as follows:

$$\mathcal{Q}_{Z_1} = \mathit{Int} \qquad \mathcal{T}_{Z_1} = \text{read-closure of } \{\bot, \text{null}_{Z_1}, \text{put}(1)\}$$
$$\mathit{init}_1 = 0 \qquad \mathit{meth}_1 = \varLambda g\colon Y \leftarrow Z_1.\, \lambda c\colon \text{COMM}(Y).$$
$$\text{put}(1){\uparrow}^{Y}_{Z_1} \cdot c \cdot \text{check}(1){\uparrow}^{Y}_{Z_1}$$

It is easy to see that the two structures are similar using a simulation relation P of the form:

$$P_Q = \{\, (n,n) \mid n \geq 0\,\} \qquad P_T = \{(\bot, \bot),\, (\text{null}_Z, \text{null}_{Z_1}),\, (\text{put}(1), \text{put}(1))\}$$

For the class C_2, we have a similar pared-down structure M_2':

$$\mathcal{Q}_{Z_2} = \{0\} \qquad \mathcal{T}_{Z_2} = \text{read-closure of } \{\bot, \text{null}_{Z_2}\}$$
$$\mathit{init}_2 = 0 \qquad \mathit{meth}_2 = \varLambda g\colon Y \leftarrow Z_2.\, \lambda c\colon \text{COMM}(Y).\, c$$

To demonstrate that the two classes are equal, we exhibit a relation $S\colon Z_1 \leftrightarrow Z_2$ given by:

$$S_Q = \{\, (n,0) \mid n \geq 0\,\} \qquad S_T = \{(\bot, \bot),\, (\text{null}_{Z_1}, \text{null}_{Z_2}),\, (\text{put}(1), \text{null}_{Z_2})\}$$

The preservation properties to be verified are:

$$\mathit{init}_1 \left[(S_Q)_\bot\right] \mathit{init}_2 \qquad \mathit{meth}_1 \left[(\text{COMM} \Rightarrow \text{COMM})(S)\right] \mathit{meth}_2$$

Note that $(\text{COMM} \Rightarrow \text{COMM})S = \forall_{R \leftarrow S} \left[\text{COMM}(R) \rightarrow \text{COMM}(R)\right] = \forall_{R \leftarrow S} \left[R_T \rightarrow R_T\right]$. So, the relationship to be proved between the two method suites is:

$$\forall g_1\colon Z_1 \rightarrow Y_1.\, \forall g_2\colon Z_2 \rightarrow Y_2.\, g_1 \left[S \rightarrow R\right] g_2 \Longrightarrow$$
$$\forall c_1, c_2.\, c_1 \left[R_T\right] c_2 \Longrightarrow \mathit{meth}_1[g_1](c_1) \left[R_T\right] \mathit{meth}_2[g_2](c_2)$$

To show the conclusion, we need $\text{put}(1){\uparrow}^{Y_1}_{Z_1} \left[R_T\right] \text{null}{\uparrow}^{Y_2}_{Z_2}$, $c_1 \left[R_T\right] c_2$, and $\text{check}(1){\uparrow}^{Y_1}_{Z_1} \left[R_T\right] \text{null}{\uparrow}^{Y_2}_{Z_2}$. The first condition follows from $\text{put}(1) \left[S_T\right] \text{null}_{Z_2}$, and the second by assumption. For the third condition, we need to argue that the state in Z_1 read by $\text{check}(1)$ is 1. This follows from the fact that the projection of c_1 to Z_1 could only be null or $\text{put}(1)$. Hence, we have the required property. ∎

Example 2 (Thamsborg, Ahmed, Dreyer and Rossberg). Consider the following classes:

$C_1 = \textbf{class} : \textbf{comm} \rightarrow \textbf{comm}$
    local Var[int] x;
    init $x := 0$;
    meth $\lambda c.\, x := 0;\, c;\, x := 1;\, c;\, \mathit{test}(x = 1)$

$C_2 = \textbf{class} : \textbf{comm} \rightarrow \textbf{comm}$
    local Var[int] x;
    init $x := 0$;
    meth $\lambda c.\, c;\, c$

where $test(b) = $ **if** b **then skip else diverge**.

This is similar to the previous example, except that we have two calls to c in the method of C_1, interspersed by *different* assignments to x. We construct a 3-step similarity proof $M_1 \sim M_1' \sim M_2' \sim M_2$ as earlier. The differences from the previous example in the structures M_1' and M_2' are as follows:

$$\mathcal{T}_{Z_1} = \text{read-closure of } \{\overline{\bot}, \text{null}_{Z_1}, \text{put}(0), \text{put}(1)\}$$
$$meth_1 = \Lambda g\colon Y \leftarrow Z_1.\ \lambda c\colon \text{COMM}(Y).$$
$$\text{put}(0){\uparrow}_{Z_1}^{Y} \cdot c \cdot \text{put}(1){\uparrow}_{Z_1}^{Y} \cdot c \cdot \text{check}(1){\uparrow}_{Z_1}^{Y}$$
$$\mathcal{T}_{Z_2} = \text{read-closure of } \{\overline{\bot}, \text{null}_{Z_2}\}$$
$$meth_2 = \Lambda g\colon Y \leftarrow Z_2.\ \lambda c\colon \text{COMM}(Y).\ c \cdot c$$

Note that we have put(0) among the transitions in $\mathcal{T}_{Z_1}$, unlike in the previous example. However, we can use exactly the *same* relation $S : Z_1 \leftrightarrow Z_2$ for showing the equivalence of M_1' and M_2':

$$S_Q = \{\,(n, 0) \mid n \geq 0\,\} \qquad S_T = \{(\overline{\bot}, \overline{\bot}),\ (\text{null}_{Z_1}, \text{null}_{Z_2}),\ (\text{put}(1), \text{null}_{Z_2})\}$$

The simulation property $meth_1 \left[(\text{COMM} \Rightarrow \text{COMM})(S) \right] meth_2$ involves the condition:

$$\forall g_1\colon Z_1 \to Y_1.\ \forall g_2\colon Z_2 \to Y_2.\ g_1 \left[S \to R \right] g_2 \Longrightarrow$$
$$\forall c_1, c_2.\ c_1 \left[R_T \right] c_2 \Longrightarrow meth_1[g_1](c_1) \left[R_T \right] meth_2[g_2](c_2)$$

Assuming $c_1 \left[R_T \right] c_2$, we first argue that $meth_1[g_1](c_1)$ and $meth_2[g_2](c_2)$ are related by $R_Q \rightharpoonup R_Q$. Starting from related initial states n and 0, the first action in $meth_1$ is put(0), which changes the local state to 0. Calling c_1 has the effect of either null_{Z_1} or put(1) on x. So, x is either 0 or 1, both of which are related to 0 by R_Q. The next action put(1) overrides the previous effect and changes the local state to 1. The second call to c_1 again has the effect of either null_{Z_1} or put(1), with the result that the local state continues to be 1 and, so, check(1) succeeds. Thus, the overall effect of $meth_1$ is to set the local state to 1, i.e., a put(1) action, and two calls to c_1 for the effects on the non-local state. This is related to $c_2 \cdot c_2$ in $meth_2$ by the R_T relation. ∎

Dreyer et al. [5] characterize actions such as put(0) in $meth_1$ as "private transitions" because their effect is not visible at the end of method calls. Their proof method involves formalizing such private transitions and distinguishing them from "public transitions." Note that no special treatment is needed in our semantics to capture the idea of private transitions. To draw a rough correspondence, we might regard the transitions included in the $\mathcal{T}_Z$ components as the "private transitions" and those related by S_T relations as the "public transitions." But these two can vary in each step of a behavioral equivalence proof. So, we do not see a fixed notion of private versus public transitions in our setting.

6   Conclusion

We have outlined a new denotational semantic model for class-based Algol-like languages, which combines the advantages of the existing models. Similar to the state-based models, it is able to represent the effect of operations as state transformations. At the same time, it also represents stores as rudimentary form of objects, whose state changes are treated from the outside in a modular fashion. Further, this modeling allows one to prove observational equivalences of programs that were not possible in the previous models. This work complements that of Ahmed, Dreyer and colleagues [5] who use an *operational approach* to develop similar reasoning principles.

In principle, this work could have been done any time after 1983, because Reynolds used a similar framework for his semantics in [23] and formulated relational parametricity in [24]. We can only speculate why it wasn't done. The alternative model invented by Oles [17] was considered equivalent to the Reynolds's model and it appeared to be simpler as well as more general. However, sharp differences between the two models become visible as soon as relational parametricity is considered. This fact was perhaps not appreciated in the intervening years.

In terms of further work to be carried out, we have not addressed the issues of dynamic storage (pointers) but we expect that the prior work in parametricity semantics [22] will be applicable. We have not considered higher-order store, i.e., storing procedures in variables. This problem is known to be hard in the framework of functor category models and it may take some time to get resolved. More exciting work awaits to be done in applying these ideas to study program reasoning, including Specification Logic, Separation Logic, Rely-guarantee and Deny-guarantee reasoning techniques [4].

References

1. Abramsky, S., Honda, K., McCusker, G.: A fully abstract game semantics for general references. In: LICS 1998, pp. 334–344 (1998)
2. Abramsky, S., McCusker, G.: Linearity, sharing and state. In: Algol-like Languages [16], ch. 20
3. Barnett, M., Naumann, D.A.: Friends Need a Bit More: Maintaining Invariants Over Shared State. In: Kozen, D. (ed.) MPC 2004. LNCS, vol. 3125, pp. 54–84. Springer, Heidelberg (2004)
4. Dodds, M., Feng, X., Parkinson, M., Vafeiadis, V.: Deny-Guarantee Reasoning. In: Castagna, G. (ed.) ESOP 2009. LNCS, vol. 5502, pp. 363–377. Springer, Heidelberg (2009)
5. Dreyer, D., Neis, G., Birkedal, L.: The impact of higher-order state and control effects on local relational reasoning. In: ICFP (2010)
6. Dunphy, B.P., Reddy, U.S.: Parametric limits. In: LICS, pp. 242–253. IEEE (July 2004)
7. Eilenberg, S.: Automata, Languages, and Machines, vol. A and B. Academic Press (1974)
8. Fiore, M.P., Jung, A., Moggi, E., O'Hearn, P.W., Riecke, J., Rosolini, G., Stark, I.: Domains and denotational semantics: History, accomplishments and open problems. EATCS 59, 227–256 (1996)
9. Rustan, K., Leino, M., Schulte, W.: Using History Invariants to Verify Observers. In: De Nicola, R. (ed.) ESOP 2007. LNCS, vol. 4421, pp. 80–94. Springer, Heidelberg (2007)

10. Meyer, A.R., Sieber, K.: Towards fully abstract semantics for local variables. In: POPL, pp. 191–203. ACM (1988); Reprinted as Chapter 7 of [16]

11. Mitchell, J.C., Plotkin, G.D.: Abstract types have existential types. TOPLAS 10(3), 470–502 (1988)

12. O'Hearn, P.W., Reddy, U.S.: Objects, interference and the Yoneda embedding. Theoretical Computer Science 228(1), 211–252 (1999)

13. O'Hearn, P.W., Reynolds, J.C.: From Algol to polymorphic linear lambda-calculus. JACM 47(1), 167–223 (2000)

14. O'Hearn, P.W., Tennent, R.D.: Semantical analysis of specification logic, Part 2. Inf. Comput. 107(1), 25–57 (1993); Reprinted as Chapter 14 of [16]

15. O'Hearn, P.W., Tennent, R.D.: Parametricity and local variables. JACM 42(3), 658–709 (1995); Reprinted as Chapter 16 of [16]

16. O'Hearn, P.W., Tennent, R.D.: Algol-like Languages, vol. 2. Birkhäuser, Boston (1997)

17. Oles, F.J.: Type algebras, functor categories and block structure. In: Nivat, M., Reynolds, J.C. (eds.) Algebraic Methods in Semantics, pp. 543–573. Cambridge Univ. Press (1985)

18. Pitts, A.M., Stark, I.D.B.: Operational reasoning for functions with local state. In: Gordon, A.M., Pitts, A.M. (eds.) Higher Order Operational Techniques in Semantics, pp. 227–274. Cambridge Univ. Press, Cambridge (1998)

19. Reddy, U.S.: Global state considered unnecessary: An introduction to object-based semantics. J. Lisp and Symbolic Computation 9, 7–76 (1996); Reprinted as Chapter 19 of [16]

20. Reddy, U.S.: Parametricity and naturality in the semantics of Algol-like languages. Electronic manuscript, University of Birmingham (December 1998), http://www.cs.bham.ac.uk/~udr/

21. Reddy, U.S.: Objects and classes in Algol-like languages. Inf. Comput. 172, 63–97 (2002)

22. Reddy, U.S., Yang, H.: Correctness of data representations involving heap data structures. Sci. of Comput. Prog. 50(1-3), 129–160 (2004)

23. Reynolds, J.C.: The essence of Algol. In: de Bakker, J.W., van Vliet, J.C. (eds.) Algorithmic Languages, pp. 345–372. North-Holland (1981); Reprinted as Chapter 3 of [16]

24. Reynolds, J.C.: Types, abstraction and parametric polymorphism. In: Mason, R.E.A. (ed.) Information Processing 1983, pp. 513–523. North-Holland, Amsterdam (1983)

25. Tennent, R.D.: Denotational semantics. In: Abramsky, S., Gabbay, D.M., Maibaum, T.S.E. (eds.) Handbook of Logic in Computer Science, vol. 3, pp. 169–322. Oxford University Press (1994)

Towards a Unified Theory
of Operational and Axiomatic Semantics[*]

Grigore Roşu[1,2] and Andrei Ştefănescu[1]

[1] University of Illinois at Urbana-Champaign, USA
{grosu,stefane1}@illinois.edu
[2] Alexandru Ioan Cuza University, Iaşi, Romania

Abstract. This paper presents a nine-rule *language-independent* proof system that takes an operational semantics as axioms and derives program reachability properties, including ones corresponding to Hoare triples. This eliminates the need for language-specific Hoare-style proof rules to verify programs, and, implicitly, the tedious step of proving such proof rules sound for each language separately. The key proof rule is *Circularity*, which is coinductive in nature and allows for reasoning about constructs with repetitive behaviors (e.g., loops). The generic proof system is shown sound and has been implemented in the MatchC verifier.

1 Introduction

An operational semantics defines a formal executable model of a language typically in terms of a transition relation $cfg \Rightarrow cfg'$ between program configurations, and can serve as a formal basis for language understanding, design, and implementation. On the other hand, an axiomatic semantics defines a proof system typically in terms of Hoare triples $\{\psi\}$ code $\{\psi'\}$, and can serve as a basis for program reasoning and verification. Operational semantics are well-understood and comparatively easier to define than axiomatic semantics for complex languages. More importantly, operational semantics are typically executable, and thus testable. For example, we can test them by executing the same test suites that compiler testers use, as has been done with the operational semantics of C [6]. Thus, we can build confidence in and eventually trust them.

The state-of-the-art in mechanical program verification (see, e.g., [1,9,13,15,19,27]) is to describe the trusted operational semantics in a powerful logical framework or language, say $\mathcal{L}$, and then to use the capabilities of $\mathcal{L}$ (e.g., induction) to verify programs. To avoid proving low-level and program-specific lemmas, Hoare-style proof rules (or consequences of them such as weakest-precondition or strongest-postcondition procedures) are typically also formalized and proved sound in $\mathcal{L}$ w.r.t. the given operational semantics. Despite impressive mechanical theorem proving advances in recent years, language designers still perceive the operational and axiomatic semantics as two distinct endeavors, and proving their formal relationship as a burden. With few notable exceptions, real languages are rarely given both semantics. Consequently, many program verifiers end up building upon possibly unsound axiomatic semantics.

---

[*] Full version of this paper, with proofs, available at http://hdl.handle.net/2142/30827.

A. Czumaj et al. (Eds.): ICALP 2012, Part II, LNCS 7392, pp. 351–363, 2012.
© Springer-Verlag Berlin Heidelberg 2012

The above lead naturally to the idea of a *unified theory* of programming, in the sense of [12], where various semantic approaches coexists with systematic relationships between them. The disadvantage of the approach in [12] is that one still needs two or more semantics of the same language. Another type of a unified theory could be one where we need only *one* semantics of the language, the theory providing the necessary machinery to achieve the same benefits as in each individual semantics, at the same cost. In the context of operational and axiomatic semantics, such a theory would have the following properties: (1) it is as executable, testable, and simple as operational semantics, so it can be used to define sound-by-construction models of programming languages; and (2) it is as good for program reasoning and verification as axiomatic semantics, so no other semantics for verification purposes—and, implicitly, no tedious soundness proofs—are needed. Such a unified theory could be, for example, a language-independent Hoare-logic-like framework taking the operational semantics rules as axioms. To understand why this is not easy, consider the Hoare logic rule for while in a C-like language:

$$\frac{\mathcal{H} \vdash \{\varphi \wedge e \neq 0\}\, s\, \{\varphi\}}{\mathcal{H} \vdash \{\varphi\}\, \mathtt{while}(e)\, s\, \{\varphi \wedge e = 0\}}$$

This proof rule is far from being language-independent. It heavily relies on the C-like semantics of this particular while construct ($e = 0$ means e is false, $e \neq 0$ means e is true). If by mistake we replace $e \neq 0$ with $e = 1$, then we get a wrong Hoare logic. This problem is amplified by its lack of executability/testability, which is why each Hoare logic needs to be proved sound w.r.t. a trusted semantics for each language separately.

We present the first steps towards such a unified theory of operational and axiomatic semantics. Our result is a sound and language-independent proof system for *matching logic reachability rules* $\varphi \Rightarrow \varphi'$ between *patterns*. A pattern φ specifies a set of program configurations that *match* it. A rule $\varphi \Rightarrow \varphi'$ specifies reachability: configurations matching φ eventually transit to ones matching φ'. Patterns were used in [21] to define language-specific axiomatic semantics (e.g., the while proof rule was similar to the one above). Our new approach is much closer to a unified theory. Although we support a limited number of operational semantics styles (including the popular reduction semantics with evaluation contexts [7]), our new proof system is language-independent.

Matching logic reachability rules generalize the basic elements of both operational and axiomatic semantics. Transitions between configurations, upon which operational semantics build, are instances of one-step reachability rules. Also, Hoare triples $\{\psi\}$ code $\{\psi'\}$ can be viewed as reachability rules between a pattern holding code with constraints ψ and a pattern holding the empty code and constraints ψ'. The proof system that we propose in this paper takes an operational semantics given as a set of reachability rules, say $\mathcal{A}$ (axioms), and derives other reachability rules. The key proof rule of our system is *Circularity*, which has a coinductive nature:

$$\frac{\mathcal{A} \vdash \varphi \Rightarrow^+ \varphi'' \qquad \mathcal{A} \cup \{\varphi \Rightarrow \varphi'\} \vdash \varphi'' \Rightarrow \varphi'}{\mathcal{A} \vdash \varphi \Rightarrow \varphi'}$$

It deductively and language-independently captures the various circular behaviors that appear in languages, due to loops, recursion, etc. Circularity adds new reachability rules to $\mathcal{A}$ during the proof derivation process, which can be used in their own proof! The

---

IMP language syntax

$PVar ::=$ program variables

$Exp ::= PVar \mid Int \mid Exp$ op Exp

$Stmt ::=$ skip $\mid PVar := Exp \mid Stmt; Stmt$
  $\mid$ if(Exp) $Stmt$ else $Stmt$
  $\mid$ while(Exp) $Stmt$

IMP evaluation contexts syntax

$Context ::= \blacksquare$
  $\mid \langle Context, State \rangle$
  $\mid Context$ op $Exp \mid Int$ op $Context$
  $\mid PVar := Context \mid Context; Stmt$
  $\mid$ if($Context$) $Stmt$ else $Stmt$

IMP operational semantics

lookup $\langle C, \sigma \rangle[x] \Rightarrow \langle C, \sigma \rangle[\sigma(x)]$

op i_1 op $i_2 \Rightarrow i_1$ op_{Int} i_2

asgn $\langle C, \sigma \rangle[x := i] \Rightarrow \langle C, \sigma[x \leftarrow i] \rangle[\text{skip}]$

seq skip; $s_2 \Rightarrow s_2$

cond$_1$ if(i) s_1 else $s_2 \Rightarrow s_1$ if $i \neq 0$

cond$_2$ if(0) s_1 else $s_2 \Rightarrow s_2$

while while(e) $s \Rightarrow$
  if(e) s; while(e) s else skip

Fig. 1. IMP language syntax and operational semantics based on evaluation contexts

correctness of this proof circularity is given by the fact that progress is required to be made (indicated by $\Rightarrow^+$ in $\mathcal{A} \vdash \varphi \Rightarrow^+ \varphi''$) before a circular reasoning step is allowed.

Sections 2 and 3 recall operational semantics and matching logic patterns. Sections 4 and 5 contain our novel theoretical notions and contribution, the sound proof system for matching logic reachability. Section 6 discusses MatchC, an automated program verifier based on our proof system. Section 7 discusses related and future work, and concludes.

2 Operational Semantics, Reduction Rules, and Transition Systems

Here we recall basic notions of operational semantics, reduction rules, and transition systems, and introduce our notation and terminology for these. We do so by means of a simple imperative language, IMP. Fig. 1 shows its syntax and an operational semantics based on evaluation contexts. IMP has only integer expressions. When used as conditions of if and while, zero means false and any non-zero integer means true (like in C). Expressions are formed with integer constants, program variables, and conventional arithmetic constructs. For simplicity, we only assume a generic binary operation, op. IMP statements are the variable assignment, if, while and sequential composition.

Various operational semantics styles define programming languages (or calculi, or systems, etc.) as sets of rewrite or reduction rules of the form "$l \Rightarrow r$ if b", where l and r are program configurations with variables constrained by the boolean condition b. One of the most popular such operational approaches is reduction semantics with evaluation contexts [7], with rules "$C[t] \Rightarrow C'[t']$ if b", where C is the evaluation context which reduces to C' (typically $C = C'$), t is the redex which reduces to t', and b is a side condition. Another approach is the chemical abstract machine [4], where l is a chemical solution that reacts into r under condition b. The $\mathbb{K}$ framework [23] is another, based on plain (no evaluation contexts) rewrite rules. Several large languages have been given such semantics, including C [6] (whose definition has about 1200 such rules).

Here we chose to define IMP using reduction semantics with evaluation contexts. Note, however, that our subsequent results work with any of the aforementioned operational approaches. The program configurations of IMP are pairs $\langle$code, $\sigma \rangle$, where code is a program fragment and σ is a state term mapping program variables into integers. As usual, we assume appropriate definitions for the integer and map domains available, together with associated operations like arithmetic operations (i_1 op_{Int} i_2, etc.) on the integers and lookup ($\sigma(x)$) or update ($\sigma[x \leftarrow i]$) on the maps.

The IMP definition in Fig. 1 consists of seven reduction rule schemas between program configurations, which make use of first-order variables: σ is a variable of sort *State*; x is a variable of sort *PVar*; i, i_1, i_2 are variables of sort *Int*; e is a variable of sort *Exp*; s, s_1, s_2 are variables of sort *Stmt*. A rule mentions a context (containing a code context and a state) and a redex which together form a configuration, and reduces the said configuration by rewriting the redex and possibly the context. As notational shortcut, a context is not mentioned if not used; e.g., the rule **op** stands in fact for the rule $\langle C, \ \sigma \rangle[i_1 \, \mathsf{op} \, i_2] \ \Rightarrow \ \langle C, \ \sigma \rangle[i_1 \, op_{Int} \, i_2]$. The code context meta-variable C allows one to instantiate a schema into reduction rules, one for each valid redex of each code fragment. For example, with C set to $x := \blacksquare \, \mathsf{op} \, y$, the **op** rule schema becomes the rule $\langle x := (i_1 \, \mathsf{op} \, i_2) \, \mathsf{op} \, y, \ \sigma \rangle \Rightarrow \langle x := (i_1 \, op_{Int} \, i_2) \, \mathsf{op} \, y, \ \sigma \rangle$.

We can therefore regard the operational semantics of IMP above as a set of reduction rules of the form "$l \Rightarrow r$ if b", where l and r are configurations with variables constrained by the boolean condition b. The subsequent results work with such sets of reduction rules and are agnostic to the particular underlying operational semantics style.

Let S (for "semantics") be a set of reduction rules like above, and let Σ be the underlying signature; also, let *Cfg* be a distinguished sort of Σ (for "configurations"). S yields a transition system on any Σ-algebra/model $\mathcal{T}$, no matter whether $\mathcal{T}$ is a term model or not. Let us fix an arbitrary model $\mathcal{T}$, which we may call a *configuration model*; as usual, $\mathcal{T}_{Cfg}$ denotes the elements of $\mathcal{T}$ of sort *Cfg*, which we call *configurations*:

Definition 1. *S induces a **transition system** $(\mathcal{T}, \Rightarrow_S^{\mathcal{T}})$ as follows: $\gamma \Rightarrow_S^{\mathcal{T}} \gamma'$ for some $\gamma, \gamma' \in \mathcal{T}_{Cfg}$ iff there is some rule "$l \Rightarrow r$ if b" in S and some $\rho : Var \rightarrow \mathcal{T}$ such that $\rho(l) = \gamma$, $\rho(r) = \gamma'$ and $\rho(b)$ holds (Var is the set of variables appearing in rules in S and we used the same ρ for its homomorphic extension to terms l, r and predicates b).*

$(\mathcal{T}, \Rightarrow_S^{\mathcal{T}})$ is a conventional transition system, i.e., a set together with a binary relation on it (in fact, $\Rightarrow_S^{\mathcal{T}} \subseteq \mathcal{T}_{Cfg} \times \mathcal{T}_{Cfg}$), and captures precisely how the language defined by S operates. We use it in Section 5 to define and prove the soundness of our proof system.

3 Matching Logic Patterns

Here we recall the notion of a matching logic pattern from [21]. Note that this section is the only overlap between [21] and this paper. The objective there was to use patterns as a state specification logic to give language-specific axiomatic (non-operational) semantics. The approach and proof system in this paper (Sections 4 and 5) are quite different: they are language-independent and use the operational semantics rules as axioms.

We assume the reader is familiar with basic concepts of algebraic specification and first-order logic. Given an algebraic signature Σ, we let $\mathcal{T}_\Sigma$ denote the initial Σ-algebra of ground terms (i.e., terms without variables) and let $\mathcal{T}_\Sigma(X)$ denote the free Σ-algebra of terms with variables in X. $\mathcal{T}_{\Sigma,s}(X)$ denotes the set of Σ-terms of sort s. These notions extend to algebraic specifications. Many mathematical and computing structures can be defined as initial Σ-algebras: boolean algebras, natural/integer/rational numbers, monoids, groups, rings, lists, sets, bags (or multisets), mappings, trees, queues, stacks, etc. CASL [17] and Maude [5] use first-order and algebraic specifications as underlying semantic infrastructure; we refer the reader to [5, 17] for examples. Here we only need maps, to represent program states. We use the notation $Map_{PVar, Int}$ for the sort corresponding to maps taking program variables into integers. We use an infix "$\mapsto$" for map entries and (an associative and commutative) comma "," to separate them.

Matching logic is parametric in configurations, or more precisely in a configuration model. Next, we discuss the configuration signature of IMP, noting that different languages or calculi have different signatures. The same machinery works for all.

Fig. 2 shows the configuration syntax of IMP. The sort *Syntax* is a generic sort for "code". Thus, terms of sort *Syntax* correspond to program fragments. States are terms of sort *State*, mapping program variables to integers. A program configuration is a term $\langle code, \sigma \rangle$ of sort *Cfg*, with code a term of sort *Syntax* and σ of sort *State*.

$PVar$ ::= IMP identifiers
Int ::= integer numbers
$Syntax$ ::= IMP syntax
$State$ ::= $Map_{PVar, Int}$
Cfg ::= $\langle Syntax, State \rangle$

Fig. 2. IMP configurations

Let Σ be the algebraic signature associated to some desired configuration syntax. Then a Σ-algebra gives a configuration *model*, namely a universe of concrete configurations. From here on we assume that Σ is a fixed signature and $\mathcal{T}$ a fixed configuration model. Note that $\mathcal{T}$ can be quite large (including models of integers, maps, etc.). We assume that Σ has a distinguished sort *Cfg* and *Var* is a sort-wise infinite set of variables.

Definition 2. *Matching logic extends the syntax of first order logic with equality (abbreviated FOL) by adding Σ-terms with variables, called **basic patterns**, as formulae:*

$$\varphi ::= ... \text{ conventional FOL syntax } \mid \quad \mathcal{T}_{\Sigma,Cfg}(Var)$$

*Matching logic formulae are also called **patterns**.*

Let $\psi, \psi_1, \psi', ...$ range over conventional FOL formulae (without patterns), $\pi, \pi_1, \pi', ...$ over basic patterns, and $\varphi, \varphi_1, \varphi', ...$ over any patterns. Matching logic satisfaction is *(pattern) matching* in $\mathcal{T}$. The satisfaction of the FOL constructs is standard. We extend FOL's valuations to include a $\mathcal{T}$ configuration, to be used for matching basic patterns:

Definition 3. *We define the relation $(\gamma, \rho) \models \varphi$ over configurations $\gamma \in \mathcal{T}_{Cfg}$, valuations $\rho : Var \to \mathcal{T}$ and patterns φ as follows (among the FOL constructs, we only show $\exists$):*

$(\gamma, \rho) \models \exists X \, \varphi$ iff $(\gamma, \rho') \models \varphi$ for some $\rho' : Var \to \mathcal{T}$ with $\rho'(y) = \rho(y)$ for all $y \in Var \backslash X$

$(\gamma, \rho) \models \pi$ iff $\gamma = \rho(\pi)$ where $\pi \in \mathcal{T}_{\Sigma,Cfg}(Var)$

We write $\models \varphi$ when $(\gamma, \rho) \models \varphi$ for all $\gamma \in \mathcal{T}_{Cfg}$ and all $\rho : Var \to \mathcal{T}$.

The pattern below matches the IMP configurations holding the code computing the sum of the natural numbers from 1 to n, and the state mapping the program variables s, n into integers s, n, such that n is positive. We use typewriter fonts for program variables and *italic* fonts for mathematical variables.

$$\exists s \, (\langle \text{ s:=0; while(n>0)(s:=s+n; n:=n-1)}, \ (\text{s} \mapsto s, \text{n} \mapsto n) \rangle \wedge n \geq_{Int} 0)$$

Note that s is existentially quantified, while n is not. That means that n can be further constrained if the pattern above is put in some larger context. Similarly, the pattern

$$\langle \text{skip}, (\text{s} \mapsto n *_{Int} (n +_{Int} 1)/_{Int} 2, \text{n} \mapsto 0) \rangle$$

is satisfied (with the same ρ, that is, the same n) by all the final configurations reachable (in IMP's transition system) from the configurations specified by the previous pattern.

Next, we show how matching logic formulae can be translated into FOL formulae, so that its satisfaction becomes FOL satisfaction in the model of configurations, $\mathcal{T}$.

Definition 4. *Let $\square$ be a fresh Cfg variable. For a pattern φ, let $\varphi^\square$ be the FOL formula replacing basic patterns $\pi \in \mathcal{T}_{\Sigma,Cfg}(Var)$ with equalities $\square = \pi$. If $\rho : Var \to \mathcal{T}$ and $\gamma \in \mathcal{T}_{Cfg}$ then let $\rho^\gamma : Var \cup \{\square\} \to \mathcal{T}$ be the mapping $\rho^\gamma(x) = \rho(x)$ for $x \in Var$ and $\rho^\gamma(\square) = \gamma$.*

With the notation in Definition 4, $(\gamma,\rho) \models \varphi$ iff $\rho^\gamma \models_{\mathrm{FOL}} \varphi^\square$, and $\models \varphi$ iff $\mathcal{T} \models_{\mathrm{FOL}} \varphi^\square$. Therefore, matching logic is a methodological fragment of the FOL theory of $\mathcal{T}$. Thus, we can actually use conventional theorem provers or proof assistants for pattern reasoning.

4 Matching Logic Reachability

Patterns were used in [21] to specify state properties in axiomatic semantics. Unfortunately, that approach shares a major disadvantage with other axiomatic approaches: the target language needs to be given a new semantics. Axiomatic semantics are less intuitive than operational semantics, are not easily executable, and thus are hard to test and need to be proved sound. What we want is *one* formal semantics of a language, which should be both executable and suitable for program verification. In this section we introduce the notion of *matching logic reachability*, and show that it captures operational semantics and can be used to specify reachability properties about programs. Assume an arbitrary but fixed configuration signature Σ and a model $\mathcal{T}$, like in Sections 2 and 3.

Definition 5. *A (matching logic) **reachability rule** is a pair $\varphi \Rightarrow \varphi'$, where φ, called the **left-hand side (LHS)**, and φ', called the **right-hand side (RHS)**, are matching logic patterns (which can have free variables). A (matching logic) **reachability system** is a set of reachability rules. A reachability system S induces a **transition system** $(\mathcal{T}, \Rightarrow_S^{\mathcal{T}})$ on the configuration model $\mathcal{T}$, as follows: $\gamma \Rightarrow_S^{\mathcal{T}} \gamma'$ for some $\gamma, \gamma' \in \mathcal{T}_{Cfg}$ iff there is some rule $\varphi \Rightarrow \varphi'$ in S and some $\rho : Var \to \mathcal{T}$ such that $(\gamma,\rho) \models \varphi$ and $(\gamma',\rho) \models \varphi'$. Configuration $\gamma \in \mathcal{T}_{Cfg}$ **terminates** in $(\mathcal{T}, \Rightarrow_S^{\mathcal{T}})$ iff there is no infinite $\Rightarrow_S^{\mathcal{T}}$-sequence starting with γ. A rule $\varphi \Rightarrow \varphi'$ is **well-defined** iff for any $\gamma \in \mathcal{T}_{Cfg}$ and $\rho : Var \to \mathcal{T}$ with $(\gamma,\rho) \models \varphi$, there is some $\gamma' \in \mathcal{T}_{Cfg}$ with $(\gamma',\rho) \models \varphi'$. The reachability system S is **well-defined** iff each rule is well-defined, and is **deterministic** iff $(\mathcal{T}, \Rightarrow_S^{\mathcal{T}})$ is deterministic.*

As mentioned in Section 2, various operational semantics styles define languages as sets of rules "$l \Rightarrow r$ if b", where l and r are Cfg terms with variables constrained by boolean condition b (a predicate over the variables in l and r). These reduction rules are just special matching logic reachability rules: "$l \Rightarrow r$ if b" can be viewed as the reachability rule $l \wedge b \Rightarrow r$, as both specify the same transitions $\gamma \Rightarrow_S^{\mathcal{T}} \gamma'$ between configurations $\gamma, \gamma' \in \mathcal{T}_{Cfg}$. This is because Definition 3 implies that $(\gamma,\rho) \models l \wedge b$ and $(\gamma',\rho) \models r$ (like in Definition 5) iff $\rho(l) = \gamma$, $\rho(r) = \gamma'$ and $\rho(b)$ holds (like in Definition 1). Note that well-definedness makes sense in general (since, e.g., φ' can be *false*), but that matching logic rules of the form $l \wedge b \Rightarrow r$ are always well-defined (pick γ' to be $\rho(r)$).

Hence, any operational semantics defined using reduction rules like above *is* a particular matching logic reachability system. If we relax the meaning of $\varphi \Rightarrow \varphi'$ from "one step" to "zero, one or more steps", then reachability rules can in fact express arbitrary reachability program properties. Consider for example the IMP code fragment "`s:=0; while(n>0)(s:=s+n; n:=n-1)`", say SUM. We can express its semantics as:

$$\langle \text{SUM}, (\text{s} \mapsto s, \text{n} \mapsto n) \rangle \wedge n \geq_{Int} 0 \Rightarrow \langle \text{skip}, (\text{s} \mapsto n *_{Int} (n +_{Int} 1)/_{Int}2, \text{n} \mapsto 0) \rangle$$

This says that any configuration γ holding SUM and some state binding program variables s and n to integers s and respectively positive n, eventually transits to a configuration γ' whose code is consumed, s is bound to the sum of numbers up to n and n is 0. Note that $s, n \in Var$ are free logical variables in this rule, so they are instantiated the same way in γ and γ', while s and n are program variables, that is, constants of sort *PVar*.

As shown in [25], to *any* IMP Hoare triple $\{\psi\}$ code $\{\psi'\}$ we can associate the following reachability rule: $\exists X_{\text{code}}(\langle \text{code}, \sigma_{X_{\text{code}}} \rangle \wedge \psi_X) \Rightarrow \exists X_{\text{code}}(\langle \text{skip}, \sigma_{X_{\text{code}}} \rangle \wedge \psi'_X)$, where X is a set containing a logical integer variable x for each variable x appearing in the Hoare triple, $X_{\text{code}} \subseteq X$ is the subset corresponding to program variables appearing in code, $\sigma_{X_{\text{code}}}$ binds each program variable x to its logical variable x, and ψ_X, ψ'_X are the formulae obtained from ψ, ψ' by replacing each variable x with its corresponding logical variable x and each arithmetic operation op with its corresponding domain operation op_{Int}. As an example, consider the Hoare triple specifying the semantics of SUM above, namely $\{\text{n} = \text{oldn} \wedge \text{n} \geq 0\}$ SUM $\{\text{s} = \text{oldn*(oldn+1)/2} \wedge \text{n} = 0\}$. The `oldn` variable is needed to remember the initial value of n. Hoare logic makes no theoretical distinction between program and logical variables, nor between program expression constructs and logical expression constructs. The corresponding matching logic reachability rule is

$$\exists s, n \, (\langle \text{SUM}, (\text{s} \mapsto s, \text{n} \mapsto n) \rangle \wedge n = oldn \wedge n \geq_{Int} 0)$$
$$\Rightarrow \exists s, n \, (\langle \text{skip}, (\text{s} \mapsto s, \text{n} \mapsto n) \rangle \wedge s = oldn *_{Int} (oldn +_{Int} 1)/_{Int}2 \wedge n = 0)$$

While this reachability rule mechanically derived from the Hoare triple is more involved than the one we originally proposed, it is not hard to see that they specify the same pairs of configurations γ, γ'. The proof system in Section 5 allows one to formally show them equivalent. Therefore, in the case of IMP, we can use reachability rules as an alternative to Hoare triples for specifying program properties. Note that reachability rules are strictly more expressive than Hoare triples, as they allow any code in their RHS patterns, not only `skip`. Replacing Hoare logic reasoning by matching logic reachability reasoning using translations like above is discouraged, as one would be required to still provide a Hoare logic for the target language. A strong point of our approach is that one does *not* have to go through this tedious step. We implemented the proof system in Section 5 and verified many programs (see Section 6) with it, using only the operational semantic rules, without having to prove any Hoare logic proof rules as lemmas. The reason we show this translation here is only to argue that the matching logic reachability rules are expressive.

Next, we define the semantic validity of matching logic reachability rules. Recall the semantic validity (i.e., partial correctness) of Hoare triples: $\models \{\psi\}$ code $\{\psi'\}$ iff for all states s, if $s \models \psi$ and code executed in state s terminates in state s', then $s' \models \psi'$. This elegant definition has the luxury of relying on another (typically operational or denotational) semantics of the language, which provides the notions of "execution", "termination" and "state". Here all these happen in the same semantics, given as a reachability system. The closest matching logic element to a "state" is a ground configuration in $\mathcal{T}_{Cfg}$, which also includes the code. Since the transition system $(\mathcal{T}, \Rightarrow_S^{\mathcal{T}})$ is meant to give all the operational behaviors of the defined language, we introduce the following

Definition 6. *Let S be a reachability system and $\varphi \Rightarrow \varphi'$ a reachability rule. We define $S \models \varphi \Rightarrow \varphi'$ iff for all $\gamma \in \mathcal{T}_{Cfg}$ and $\rho : Var \to \mathcal{T}$ such that γ terminates in $(\mathcal{T}, \Rightarrow_S^{\mathcal{T}})$ and $(\gamma, \rho) \models \varphi$, there exists some $\gamma' \in \mathcal{T}_{Cfg}$ such that $\gamma \Rightarrow_S^{\star\mathcal{T}} \gamma'$ and $(\gamma', \rho) \models \varphi'$.*

Intuitively, $S \models \varphi \Rightarrow \varphi'$ specifies reachability, in the sense that "any terminating ground configuration that matches φ transits, on some possible execution path, to a configuration that matches φ'". If S is deterministic (Definition 5), then "some path" becomes equivalent to "all paths", and thus $\varphi \Rightarrow \varphi'$ captures partial correctness.

Note that, if S is well-defined, then $S \models \varphi \Rightarrow \varphi'$ holds for all $\varphi \Rightarrow \varphi' \in S$ (well-defined axioms are semantically valid). Also, as already mentioned, φ' does not need to have an empty (skip) code cell. If it does, then so does γ' in the definition above, and, in the case of IMP, γ' is unique and thus we recover the Hoare validity as a special case.

Taking S to be the operational semantics of IMP in Section 2, $S \models \varphi \Rightarrow \varphi'$ can be proved for the two example reachability rules $\varphi \Rightarrow \varphi'$ for SUM in this section. Unfortunately, such proofs are tedious, involving low-level details about the IMP transition system and induction. What we want is an abstract proof system for deriving matching logic reachability rules, which does not refer to the low-level transition system.

5   Sound Proof System

Fig. 3 shows our nine-rule language-independent proof system for deriving matching logic reachability rules. We start with a set of reachability rules representing an operational semantics of the target language, and then either "execute" programs or derive program properties in a generic manner, without relying on the specifics of the target language except for using its operational semantics rules as axioms. In particular, no auxiliary lemmas corresponding to Hoare logic rules are proved or needed. Initially, $\mathcal{A}$ contains the operational semantics of the target language. The first group of rules (Reflexivity, Axiom, Substitution, Transitivity) has an operational nature and derives concrete and (linear) symbolic executions; any executable semantic framework has similar rules (see, e.g., rewriting logic [16]). The second group of rules (Case Analysis, Logic Framing, Consequence and Abstraction) has a more deductive nature and is inspired from the subset of language-independent rules of Hoare logic [11]. The first group combined with Case Analysis and Logic Framing enables the derivation of all symbolic execution paths. The Circularity proof rule is new and captures the various circular behaviors that appear in languages, due to loops, recursion, jumps, etc.

Let $\mathcal{A} \vdash \varphi \Rightarrow^+ \varphi'$ be the derivation relation obtained by dropping the Reflexivity rule from the proof system in Fig. 3. $\mathcal{A} \vdash \varphi \Rightarrow^+ \varphi'$ says that a configuration satisfying φ can transit in one or more operational semantics steps to one satisfying φ'. The Circularity rule in Fig. 3 says that we can derive the sequent $\mathcal{A} \vdash \varphi \Rightarrow \varphi'$ whenever we can derive

Rules of operational nature	Rules of deductive nature

Reflexivity :

$$\frac{\cdot}{\mathcal{A} \vdash \varphi \Rightarrow \varphi}$$

Axiom :

$$\frac{\varphi \Rightarrow \varphi' \in \mathcal{A}}{\mathcal{A} \vdash \varphi \Rightarrow \varphi'}$$

Substitution :

$$\frac{\mathcal{A} \vdash \varphi \Rightarrow \varphi' \quad \theta : Var \to \mathcal{T}_{\Sigma}(Var)}{\mathcal{A} \vdash \theta(\varphi) \Rightarrow \theta(\varphi')}$$

Transitivity :

$$\frac{\mathcal{A} \vdash \varphi_1 \Rightarrow \varphi_2 \quad \mathcal{A} \vdash \varphi_2 \Rightarrow \varphi_3}{\mathcal{A} \vdash \varphi_1 \Rightarrow \varphi_3}$$

Case Analysis :

$$\frac{\mathcal{A} \vdash \varphi_1 \Rightarrow \varphi \quad \mathcal{A} \vdash \varphi_2 \Rightarrow \varphi}{\mathcal{A} \vdash \varphi_1 \vee \varphi_2 \Rightarrow \varphi}$$

Logic Framing :

$$\frac{\mathcal{A} \vdash \varphi \Rightarrow \varphi' \quad \psi \text{ is a (patternless) FOL formula}}{\mathcal{A} \vdash \varphi \wedge \psi \Rightarrow \varphi' \wedge \psi}$$

Consequence :

$$\frac{\models \varphi_1 \to \varphi_1' \quad \mathcal{A} \vdash \varphi_1' \Rightarrow \varphi_2' \quad \models \varphi_2' \to \varphi_2}{\mathcal{A} \vdash \varphi_1 \Rightarrow \varphi_2}$$

Abstraction :

$$\frac{\mathcal{A} \vdash \varphi \Rightarrow \varphi' \quad X \cap FreeVars(\varphi') = \emptyset}{\mathcal{A} \vdash \exists X \varphi \Rightarrow \varphi'}$$

Rule for circular behavior

Circularity :
$$\frac{\mathcal{A} \vdash \varphi \Rightarrow^+ \varphi'' \quad \mathcal{A} \cup \{\varphi \Rightarrow \varphi'\} \vdash \varphi'' \Rightarrow \varphi'}{\mathcal{A} \vdash \varphi \Rightarrow \varphi'}$$

Fig. 3. Matching logic reachability proof system

the rule $\varphi \Rightarrow \varphi'$ by starting with one or more steps in $\mathcal{A}$ and continuing with steps which can involve both rules from $\mathcal{A}$ and the rule to be proved itself, $\varphi \Rightarrow \varphi'$. The first step can for example be a loop unrolling step in the case of loops, or a function invocation step in the case of recursive functions, etc. The use of the claimed properties in their own proofs in Circularity is reminiscent of *circular coinduction* [22]. Like in circular coinduction, where the claimed properties can only be used in some special contexts, Circularity also disallows their unrestricted use: it only allows them to be guarded by a trusted, operational step. It would actually be unsound to drop the operational-step-guard requirement: for example, if $\mathcal{A}$ contained $\varphi_1 \Rightarrow \varphi_2$ then $\varphi_2 \Rightarrow \varphi_1$ could be "proved" in a two-step transitivity, using itself, the rule in $\mathcal{A}$ and then itself again.

Theorem 1. *Soundness (see [26] for the proof) Let S be a well-defined matching logic reachability system (typically corresponding to an operational semantics), and let $S \vdash \varphi \Rightarrow \varphi'$ be a sequent derived with the proof system in Fig. 3. Then $S \models \varphi \Rightarrow \varphi'$.*

Hence, proof derivations are sound w.r.t. the (transition system generated by the) operational semantics. The well-definedness requirement is acceptable (operational semantics satisfy it) and needed (otherwise not even the axioms satisfy $S \models \varphi \Rightarrow \varphi'$).

We next illustrate our proof system by means of examples. The proof below may seem low level when compared to the similar proof done using Hoare logic. However, note that it is quite mechanical, the user only having to provide the invariant (φ_{inv}). The rest is automatic and consists of applying the operational reduction rules whenever they match, except for the circularities which are given priority; when the redex is an if, a Case Analysis is applied. Our current MatchC implementation can prove it automatically, as well as much more complex programs (see Section 6). Although the paper Hoare logic proofs for simple languages like IMP may look more compact, note that in general they make assumptions which need to be addressed in implementations, such as that expressions do not have side effects, or that substitution is available and atomic, etc.

Consider the SUM code (Section 4) "s:=0; while(n>0)(s:=s+n; n:=n-1)", and the property given as a matching logic reachability rule, say $\mu^1_{SUM} \equiv (\varphi_{LHS} \Rightarrow \varphi_{RHS})$:

$$\langle SUM, (s \mapsto s, n \mapsto n) \rangle \wedge n \geq_{Int} 0 \Rightarrow \langle skip, (s \mapsto n *_{Int} (n +_{Int} 1)/_{Int}2, n \mapsto 0) \rangle$$

Let us formally derive this property using the proof system in Fig. 3. Let S be the operational semantics of IMP in Fig. 1 and let φ_{inv} be the pattern

$$\langle LOOP, (s \mapsto (n -_{Int} n') *_{Int} (n +_{Int} n' +_{Int} 1)/_{Int}2, n \mapsto n') \rangle \wedge n' \geq_{Int} 0$$

where LOOP is "while (n>0) (s := s+n; n := n-1)". We derive $S \vdash \mu^1_{SUM}$ by Transitivity with $\mu_1 \equiv (\varphi_{LHS} \Rightarrow \exists n' \varphi_{inv})$ and $\mu_2 \equiv (\exists n' \varphi_{inv} \Rightarrow \varphi_{RHS})$. By Axiom **asgn** (Fig. 1, within the SUM context) followed by Substitution with $\theta(\sigma) = (s \mapsto s, n \mapsto n)$, $\theta(x) = s$ and $\theta(i) = 0$ followed by Logic Framing with $n \geq_{Int} 0$, we derive $\varphi_{LHS} \Rightarrow \langle skip; LOOP, (s \mapsto 0, n \mapsto n) \rangle \wedge n \geq_{Int} 0$. This "operational" sequence of Axiom, Substitution and Logic Framing is quite common; we abbreviate it ASLF. Further, by ASLF with **seq** and Transitivity, we derive $\varphi_{LHS} \Rightarrow \langle LOOP, (s \mapsto s, n \mapsto n) \rangle \wedge n \geq_{Int} 0$. $S \vdash \mu_1$ now follows by Consequence. We derive $S \vdash \mu_2$ by Circularity with $S \vdash \exists n' \varphi_{inv} \Rightarrow^+ \varphi_{if}$ and $S \cup \{\mu_2\} \vdash \varphi_{if} \Rightarrow \varphi_{RHS}$, where φ_{if} is the formula obtained from φ_{inv} replacing its code with "if (n>0) (s := s+n; n := n-1; LOOP) else skip". ASLF (**while**) followed by Abstraction derive $S \vdash \exists n' \varphi_{inv} \Rightarrow^+ \varphi_{if}$. For the other, we use Case Analysis with $\varphi_{if} \wedge n' \leq_{Int} 0$ and $\varphi_{if} \wedge n' >_{Int} 0$. ASLF (**lookup$_n$**, **op$_>$**, **cond$_2$**) together with some Transitivity and Consequence steps derive $S \cup \{\mu_2\} \vdash \varphi_{if} \wedge n' \leq_{Int} 0 \Rightarrow \varphi_{RHS}$ (μ_2 is not needed in this derivation). Similarly, ASLF (**lookup$_n$**, **op$_>$**, **cond$_1$**, **lookup$_n$**, **lookup$_s$**, **op$_+$**, **asgn**, **seq**, **lookup$_n$**, **op$_-$**, **asgn**, **seq**, and μ_2) together with Transitivity and Consequence steps derive $S \cup \{\mu_2\} \vdash \varphi_{if} \wedge n' >_{Int} 0 \Rightarrow \varphi_{RHS}$. This time μ_2 is needed and it is interesting to note how. After applying all the steps above and the LOOP fragment of code is reached again, the pattern characterizing the configuration is

$$\langle LOOP, (s \mapsto (n -_{Int} n') *_{Int} (n +_{Int} n' +_{Int} 1)/_{Int}2 +_{Int} n', n \mapsto n' -_{Int} 1) \rangle \wedge n' >_{Int} 0$$

The circularity μ_2 can now be applied, via Consequence and Transitivity, because this formula implies $\exists n' \varphi_{inv}$ (indeed, pick the existentially quantified n' to be $n' -_{Int} 1$).

We can similarly derive the other reachability rule for SUM in Section 4, say μ^2_{SUM}. Instead, let us prove the stronger result that the two matching logic reachability rules are equivalent, that is, that $\mu^1_{SUM} \vdash \mu^2_{SUM}$ and $\mu^2_{SUM} \vdash \mu^1_{SUM}$. Using conventional FOL reasoning and the Consequence proof rule, one can show μ^2_{SUM} equivalent to the reachability rule
$$\exists s \left(\langle SUM, (s \mapsto s, n \mapsto oldn) \rangle \wedge oldn \geq_{Int} 0 \right) \Rightarrow \langle skip, (s \mapsto oldn *_{Int} (oldn +_{Int} 1)/_{Int}2, n \mapsto 0) \rangle$$
The equivalence to μ^1_{SUM} now follows by applying the Substitution rule with $oldn \mapsto n$ and Consequence and, respectively, Substitution with $n \mapsto oldn$ and Abstraction.

In the sister paper [25] (dedicated to completeness), we show that any property derived using IMP's Hoare logic proof system can also be derived using our proof system in Fig. 5, of course modulo the representation of Hoare triples as matching logic reachability rules described in Section 4. For example, a Hoare logic proof step for while is translated in our proof system into an Axiom step (with **while** in Fig. 1), a Case Analysis (for the resulting if statement), a Circularity (as part of the positive case, when the while statement is reached again), an Abstraction (to add existential quantifiers for the logical variables added as part of the translation), and a few Transitivity steps.

6   Implementation in MatchC

A concern to a verification framework based on operational semantics is that it may not be practical, due to the amount of required user involvement or to the amount of low-level details that need to be provided. To test the practicality of matching logic reachability, we picked a fragment of C and implemented a proof-of-concept program verifier for it based on matching logic, named MatchC. Our C fragment includes functions, structures, pointers and I/O primitives. MatchC uses reachability rules for program specifications and its implementation is directly based on the proof system in Fig. 3.

MatchC has verified various programs manipulating lists and trees, performing arithmetic and I/O operations, and implementing sorting algorithms, binary search trees, AVL trees, and the Schorr-Waite graph marking algorithm. Users only provide the program specifications (reachability rules) as annotations, in addition to the unavoidable formalizations of the used mathematical domains. The rest is automatic. For example, it takes MatchC less than 2 seconds to verify Schorr-Waite for full correctness. The Matching Logic web page, http://fsl.cs.uiuc.edu/ml, contains an online interface to run MatchC, where users can try more than 50 existing examples (or type their own).

Let S be the reachability system giving the language semantics, and let C be the set of reachability rules corresponding to user-provided specifications (properties that one wants to verify). MatchC derives the rules in C using the proof system in Fig. 3. It begins by applying Circularity for each rule in C and reduces the task to deriving sequents of the form $S \cup C \vdash \varphi \Rightarrow \varphi'$. To prove them, it symbolically executes φ with steps from $S \cup C$ searching for a formula that implies φ'. An SMT solver (Z3 [18]) is invoked to solve the side conditions of the rules. Whenever the reduction semantics rule for a conditional statement cannot apply because its condition is symbolic, a Case Analysis is applied and formula split into a disjunction. Rules in C are given priority; thus, if each loop and function is given a specification then MatchC will always terminate (Z3 cutoff is 5s).

A previous version of MatchC, based on the proof system in [21], was discussed in [24]. The new implementation based on the proof system in Fig. 3 will be presented in detail elsewhere. We here only mean to highlight the practical feasibility of our approach.

7   Conclusion, Additional Related Work, and Future Work

To our knowledge, the proof system in Fig. 3 is the first of its kind. We can now define only *one* semantics of the target language, which is operational and thus well-understood and comparatively easier to define than an axiomatic semantics. Moreover, it is testable using existing rewrite engines or functional languages incorporating pattern matching (e.g., Haskell). For example, we can test it by executing program benchmarks that compiler testers use, like in [6]. Then, we take this semantics and use it *as is* for program verification. We not only skip completely the tedious step of having to prove the relationship between operational and an axiomatic semantics of the same language, but we can also change the language at will (or fix semantic bugs), without having to worry about doing that in two different places and maintaining the soundness proofs.

The idea of regarding a program as a specification transformer to analyze programs in a forwards-style goes back to Floyd in 1967 [8]. However, unlike ours, Floyd's rules are language-specific, not executable, and introduce quantifiers. Dynamic logic [2, 10] extends FOL with modal operators to embed program fragments within program specifications. Like in matching logic, in dynamic logic programs and specifications also coexist in the same logic. However, unlike in matching logic, one still needs to define an alternative (dynamic logic) language semantics, with language-specific proof rules.

Leroy and Grall [14] use the coinductive interpretation of a standard big-step semantics as a semantic foundation both for terminating and for non-terminating evaluation. Our approach is different in that: (1) our proof system can reason about reachability between arbitrary formulae, rather than just the evaluation of programs to values, and (2) although Circularity has a coinductive flavor, we take the inductive interpretation of our proof system and obtain soundness by appropriate guarding of the circular rules.

Symbolic execution is a popular technique for program analysis and verification, for example automatic testing and model-checking in Java Path Finder (JPF) [20], or strongest postcondition computation in separation logic [3]. The first six proof rules in Fig. 3 can be used to symbolically execute a pattern formula φ. Different techniques are used to make symbolic execution efficient and scalable for different programming languages, like path merging in JPF. Our approach is orthogonal, in that our proof system can be used to justify those techniques based on the language operational semantics.

We believe our proof system can be extended to work with SOS-style conditional reduction rules. Concurrency and non-determinism were purposely left out; these are major topics which deserve full attention. General relative completeness and total correctness also need to be addressed. Like other formal semantics, matching logic can also be embedded into higher-level formalisms and theorem provers, so that proofs of relationships to other semantics can be mechanized, and even programs verified. Ultimately, we would like to have a generic verifier taking an operational semantics as input, together with an extension allowing users to provide pattern annotations, and to yield an automated program verifier based on the proof system in Fig. 3.

Acknowledgements. We thank Santiago Escobar, Klaus Havelund, Jose Meseguer, Peter Olveczky, Madhusudan Parthasarathy, Vlad Rusu, the members of the $\mathbb{K}$ team, and the anonymous reviewers for their valuable comments on drafts of this paper. The work in this paper was supported in part by NSA contract H98230-10-C-0294, by NSF grant CCF-0916893 and by (Romanian) SMIS-CSNR 602-12516 contract no. 161/15.06.2010.

References

1. Appel, A.W.: Verified Software Toolchain. In: Barthe, G. (ed.) ESOP 2011. LNCS, vol. 6602, pp. 1–17. Springer, Heidelberg (2011)
2. Beckert, B., Hähnle, R., Schmitt, P.H. (eds.): Verification of Object-Oriented Software. LNCS (LNAI), vol. 4334. Springer, Heidelberg (2007)
3. Berdine, J., Calcagno, C., O'Hearn, P.W.: Symbolic Execution with Separation Logic. In: Yi, K. (ed.) APLAS 2005. LNCS, vol. 3780, pp. 52–68. Springer, Heidelberg (2005)
4. Berry, G., Boudol, G.: The chemical abstract machine. Th. Comp. Sci. 96(1), 217–248 (1992)
5. Clavel, M., Durán, F., Eker, S., Lincoln, P., Martí-Oliet, N., Meseguer, J., Talcott, C.: All About Maude. LNCS, vol. 4350. Springer, Heidelberg (2007)
6. Ellison, C., Rosu, G.: An executable formal semantics of C with applications. In: POPL, pp. 533–544. ACM (2012)

7. Felleisen, M., Findler, R.B., Flatt, M.: Semantics Engineering with PLT Redex. MIT (2009)
8. Floyd, R.W.: Assigning meaning to programs. In: Symposia in Applied Mathematics, vol. 19, pp. 19–32. AMS (1967)
9. George, C., Haxthausen, A.E., Hughes, S., Milne, R., Prehn, S., Pedersen, J.S.: The RAISE Development Method. BCS Practitioner Series. Prentice Hall (1995)
10. Harel, D., Kozen, D., Tiuryn, J.: Dynamic logic. In: Handbook of Philosophical Logic, pp. 497–604 (1984)
11. Hoare, C.A.R.: An axiomatic basis for computer programming. Commun. ACM 12(10), 576–580 (1969)
12. Hoare, C.A.R., Jifeng, H.: Unifying Theories of Programming. Prentice Hall (1998)
13. Jacobs, B.: Weakest pre-condition reasoning for java programs with JML annotations. J. Log. Algebr. Program. 58(1-2), 61–88 (2004)
14. Leroy, X., Grall, H.: Coinductive big-step operational semantics. Inf. Comput. 207(2), 284–304 (2009)
15. Liu, H., Moore, J.S.: Java Program Verification via a JVM Deep Embedding in ACL2. In: Slind, K., Bunker, A., Gopalakrishnan, G.C. (eds.) TPHOLs 2004. LNCS, vol. 3223, pp. 184–200. Springer, Heidelberg (2004)
16. Meseguer, J.: Conditioned rewriting logic as a united model of concurrency. Theor. Comput. Sci. 96(1), 73–155 (1992)
17. Mosses, P.D. (ed.): CASL Reference Manual. LNCS, vol. 2960. Springer, Heidelberg (2004)
18. de Moura, L., Bjørner, N.: Z3: An Efficient SMT Solver. In: Ramakrishnan, C.R., Rehof, J. (eds.) TACAS 2008. LNCS, vol. 4963, pp. 337–340. Springer, Heidelberg (2008)
19. Nipkow, T.: Winskel is (almost) right: Towards a mechanized semantics textbook. Formal Aspects of Computing 10, 171–186 (1998)
20. Pasareanu, C.S., Mehlitz, P.C., Bushnell, D.H., Gundy-Burlet, K., Lowry, M.R., Person, S., Pape, M.: Combining unit-level symbolic execution and system-level concrete execution for testing NASA software. In: ISSTA, pp. 15–26. ACM (2008)
21. Roşu, G., Ellison, C., Schulte, W.: Matching Logic: An Alternative to Hoare/Floyd Logic. In: Johnson, M., Pavlovic, D. (eds.) AMAST 2010. LNCS, vol. 6486, pp. 142–162. Springer, Heidelberg (2011)
22. Roşu, G., Lucanu, D.: Circular Coinduction: A Proof Theoretical Foundation. In: Kurz, A., Lenisa, M., Tarlecki, A. (eds.) CALCO 2009. LNCS, vol. 5728, pp. 127–144. Springer, Heidelberg (2009)
23. Rosu, G., Serbanuta, T.F.: An overview of the K semantic framework. J. Log. Algebr. Program. 79(6), 397–434 (2010)
24. Rosu, G., Stefanescu, A.: Matching logic: A new program verification approach (NIER track). In: ICSE, pp. 868–871. ACM (2011)
25. Rosu, G., Stefanescu, A.: From Hoare logic to matching logic reachability. In: FM (to appear, 2012)
26. Rosu, G., Stefanescu, A.: Towards a unified theory of operational and axiomatic semantics. Tech. Rep., Univ. of Illinois (May 2012), http://hdl.handle.net/2142/30827
27. Sasse, R., Meseguer, J.: Java+ITP: A verification tool based on Hoare logic and algebraic semantics. Electr. Notes Theor. Comput. Sci. 176(4), 29–46 (2007)

Loader and Urzyczyn Are Logically Related

Sylvain Salvati[1], Giulio Manzonetto[2], Mai Gehrke[3], and Henk Barendregt[4]

[1] INRIA Bordeaux — Sud-Ouest, Talence, France
sylvain.salvati@labri.fr
[2] LIPN, CNRS UMR 7030, Université Paris-Nord, Villetaneuse, France
giulio.manzonetto@lipn.univ-paris13.fr
[3] LIAFA, CNRS, Université Paris-Diderot, Paris, France
mai.gehrke@liafa.jussieu.fr
[4] Radboud University, Intelligent Systems, Nijmegen, The Netherlands
henk@cs.ru.nl

Abstract. In simply typed λ-calculus with one ground type the following theorem due to Loader holds. (i) Given the full model $\mathcal{F}$ over a finite set, the question whether some element $f \in \mathcal{F}$ is λ-definable is undecidable. In the λ-calculus with intersection types based on countably many atoms, the following is proved by Urzyczyn. (ii) It is undecidable whether a type is inhabited.

Both statements are major results presented in [3]. We show that (i) and (ii) follow from each other in a natural way, by interpreting intersection types as continuous functions logically related to elements of $\mathcal{F}$. From this, and a result by Joly on λ-definability, we get that Urzyczyn's theorem already holds for intersection types with at most two atoms.

Keywords: λ-calculus, λ-definability, inhabitation, undecidability.

Introduction

Consider the simply typed λ-calculus on simple types $\mathbb{T}^0$ with one ground type 0. Recall that a hereditarily finite full model of simply typed λ-calculus is a collection of sets $\mathcal{F} = (\mathcal{F}_A)_{A \in \mathbb{T}^0}$ such that $\mathcal{F}_0 \neq \emptyset$ is finite and $\mathcal{F}_{A \to B} = \mathcal{F}_B^{\mathcal{F}_A}$ (i.e. the set of functions from $\mathcal{F}_A$ to $\mathcal{F}_B$) for all simple types A, B. An element $f \in \mathcal{F}_A$ is λ-*definable* whenever, for some closed λ-term M having type A, we have $[\![M]\!] = f$, where $[\![M]\!]$ denotes the interpretation of M in $\mathcal{F}$. The following question, raised by Plotkin in [7], is known as the Definability Problem:

> DP: "Given an element f of any hereditarily finite full model,
> is f λ-definable?"

A natural restriction considered in the literature [5,6] is the following:

> DP$_n$: "Given an element f of $\mathcal{F}_n$, is f λ-definable?"

where $\mathcal{F}_n$ (for $n \geq 1$) denotes the unique (up to isomorphism) full model whose ground set $\mathcal{F}_0$ has n elements. Statman's conjecture stating that DP is decidable

A. Czumaj et al. (Eds.): ICALP 2012, Part II, LNCS 7392, pp. 364–376, 2012.
© Springer-Verlag Berlin Heidelberg 2012

[9] was refuted by Loader [6], who proved in 1993 (but published in 2001) that DP_n is undecidable for every $n > 6$. Such a result was then strengthened by Joly, who showed in [5] that DP_n is undecidable for all $n > 1$.

Theorem 1. *1.* **(Loader)** *The Definability Problem is undecidable.*
2. **(Loader/Joly)** DP_n *is undecidable for every* $n > 6$ *(resp.* $n > 1$*).*

Consider now the λ-calculus endowed with the intersection type system CDV (Coppo-Dezani-Venneri [4]) based on a countable set $\mathbb{A}$ of atomic types. Recall that an intersection type σ is *inhabited* if $\vdash_\wedge M : \sigma$ for some closed λ-term M. The Inhabitation Problem for this type theory is formulated as follows:

IHP: "Given an intersection type σ, is σ inhabited?"

We will also be interested in the following restriction of IHP:

IHP_n: "Given an intersection type σ with at most n atoms, is σ inhabited?"

In 1999, Urzyczyn [10] proved that IHP is undecidable for suitable intersection types, called "game types" in [3, §17E], and thus for the whole CDV. His idea was to prove that solving IHP for a game type σ is equivalent to winning a suitable "tree game" G. An arbitrary number of atoms may be needed since, in the Turing-reduction, the actual amount of atoms in σ is determined by G.

Theorem 2 (Urzyczyn).

1. The Inhabitation Problem is undecidable.
2. The Inhabitation Problem for game types is undecidable.

The undecidability of DP and that of IHP are major results presented thoroughly in [3, §4A] and [3, §17E]. In the proof these problems are reduced to well-known undecidable problems (and eventually to the Halting problem). However, the instruments used to achieve these results are very different — the proof by Loader proceeds by reducing DP to the two-letter word rewriting problem, while the proof by Urzyczyn reduces IHP to the emptiness problem for queue automata (through a series of reductions). The fact that these proofs are different is not surprising since the two problems, at first sight, really *look* unrelated.

Our main contribution is to show that ·DP and IHP are actually Turing-equivalent, by providing a perhaps unexpected link between the two problems. The key ideas behind our constructions are the following. Every intersection $\alpha_1 \wedge \cdots \wedge \alpha_k$ of atoms can be viewed as a set $\{\alpha_1, \ldots, \alpha_k\}$, and every arrow type $\sigma \to \tau$ as a (continuous) step function. Moreover, Urzyczyn's game types follow the structure of simple types. Combining these ingredients we build a continuous model $\mathcal{S} = (\mathcal{S}_A)_{A \in \mathbb{T}^0}$ over a finite set of atomic types, which constitutes a "bridge" between intersection type systems and full models of simply typed λ-calculus. Then, exploiting very natural semantic logical relations, we can study the continuous model, cross the bridge and infer properties of the full model. Our constructions allow us to obtain the following Turing-reductions (recall that if the problem P_1 is undecidable and $P_1 \leq_T P_2$, then also P_2 is undecidable):

$$\Lambda: \qquad M,N,P ::= x \mid MN \mid \lambda x.M, \text{ where } x \in \text{Var}$$
$$\mathbb{T}^0: \qquad A,B,C ::= 0 \mid A \to B$$
$$\mathbb{T}^{\mathbb{A}}_{\wedge}: \qquad \gamma,\rho,\sigma,\tau ::= \alpha \mid \sigma \to \tau \mid \sigma \wedge \tau, \text{ where } \alpha \in \mathbb{A}$$

(a) Sets Λ of λ-terms, $\mathbb{T}^0$ of simple types, $\mathbb{T}^{\mathbb{A}}_{\wedge}$ of intersection types over $\mathbb{A}$.

$$\sigma \leq \sigma \;(\text{refl}) \qquad\qquad \sigma \wedge \tau \leq \sigma \;(\text{incl}_L) \qquad\qquad \sigma \wedge \tau \leq \tau \;(\text{incl}_R)$$

$$(\sigma \to \tau) \wedge (\sigma \to \tau') \leq \sigma \to (\tau \wedge \tau') \quad (\to_\wedge)$$

$$\frac{\sigma \leq \gamma \quad \gamma \leq \tau}{\sigma \leq \tau} \;(\text{trans}) \qquad \frac{\sigma \leq \tau \quad \sigma \leq \tau'}{\sigma \leq \tau \wedge \tau'} \;(\text{glb}) \qquad \frac{\sigma' \leq \sigma \quad \tau \leq \tau'}{\sigma \to \tau \leq \sigma' \to \tau'} \;(\to)$$

(b) Rules defining the subtyping relation $\leq$ on intersection types $\mathbb{T}^{\mathbb{A}}_{\wedge}$.

$$\frac{}{x_1 : \sigma_1, \ldots, x_n : \sigma_n \vdash_\wedge x_i : \sigma_i} \;(\text{ax}) \qquad \frac{\Gamma \vdash_\wedge M : \tau \to \sigma \quad \Gamma \vdash_\wedge N : \tau}{\Gamma \vdash_\wedge MN : \sigma} \;(\to_E)$$

$$\frac{\Gamma, x : \sigma \vdash_\wedge M : \tau}{\Gamma \vdash_\wedge \lambda x.M : \sigma \to \tau} \;(\to_I) \quad \frac{\Gamma \vdash_\wedge M : \sigma \quad \Gamma \vdash_\wedge M : \tau}{\Gamma \vdash_\wedge M : \sigma \wedge \tau} \;(\wedge_I) \quad \frac{\Gamma \vdash_\wedge M : \sigma \quad \sigma \leq \tau}{\Gamma \vdash_\wedge M : \tau} \;(\leq)$$

(c) Rules defining the intersection type system CDV.

Fig. 1. Definition of terms, types, subtyping and derivation rules for CDV. The rules for simply typed λ-calculus are obtained from those in (c) leaving out $(\wedge_I)$ and $(\leq)$.

(i) Inhabitation Problem for game types $\leq_T$ Definability Problem,
(ii) Definability Problem $\leq_T$ Inhabitation Problem (cf. [8]),
(iii) DP$_n$ $\leq_T$ IHP$_n$ (cf. [8]).

Therefore, by (i) and (ii) we get that the undecidability of DP and IHP follows from each other. Moreover, by (iii) and Theorem 1(2) we conclude that IHP$_n$ is undecidable whenever $n > 1$, which is a new result refining Urzyczyn's one.

1   Preliminaries: Some Syntax, Some Semantics

To make this article more self-contained, this section summarizes some definitions and results that we will use later in the paper. Given a set X, we write $\mathcal{P}(X)$ for the set of all subsets of X, and $Y \subseteq_f X$ if Y is a finite subset of X.

1.1   Typed Lambda Calculi

We take untyped λ-calculus for granted together with the notions of closed λ-term, α-conversion, (β-)normal form and strong normalization. We denote by Var the set of variables and by Λ the set of λ-terms. Hereafter, we consider λ-terms up to α-conversion and we adopt Barendregt's variable convention.

We mainly focus on two particular typed λ-calculi (see [3] for more details).

The simply typed λ-calculus *à la Curry* over a single atomic type 0. The set $\mathbb{T}^0$ of *simple types* $A, B, C, \ldots$ is defined in Figure 1(a). *Simple contexts* Δ are partial functions from Var to $\mathbb{T}^0$; we write $\Delta = x_1 : A_1, \ldots, x_n : A_n$ for the function of domain $\{x_1, \ldots, x_n\}$ such that $\Delta(x_i) = A_i$ for i in $[1; n]$. We write $\Delta \vdash M : A$ if M *has type* A *in* Δ, and we say that such an M is *simply typable*.

The intersection type system CDV over an infinite set $\mathbb{A}$ of atomic types. This system was first introduced by Coppo, Dezani and Venneri [4] to characterize strongly normalizable λ-terms. The set $\mathbb{T}^{\mathbb{A}}_{\wedge}$ of *intersection types* is given in Figure 1(a) and it is partially ordered by the subtyping relation $\leq$ defined in Figure 1(b). We denote by $\simeq$ the equivalence generated by $\leq$. As usual, we may write $\bigwedge_{i=1}^{n} \sigma_i \to \tau_i$ for $(\sigma_1 \to \tau_1) \wedge \cdots \wedge (\sigma_n \to \tau_n)$.

Contexts $\Gamma = x_1 : \tau_1, \ldots, x_n : \tau_n$ are handled as in the simply typed case. We write $\Gamma \vdash_{\wedge} M : \sigma$ if the judgment can be proved using the rules of Figure 1(c).

As a matter of notation, given two sets Y, Z of intersection types, we let $Y^{\wedge} = \{\sigma_1 \wedge \cdots \wedge \sigma_n \mid \sigma_i \in Y$ for $i \in [1; n]\}$ and $Y \to Z = \{\tau \to \sigma \mid \tau \in Y, \sigma \in Z\}$.

We now present some well known properties of CDV. For their proofs, we refer to [4], [3, Thm. 14.1.7] and [3, Thm. 14.1.9] respectively.

Theorem 3. *A λ-term M is typable in CDV iff M is strongly normalizable.*

Theorem 4 (β-soundness). *For all $k \geq 1$, if $\bigwedge_{i=1}^{k} \sigma_i \to \rho_i \leq \gamma_1 \to \gamma_2$ then there is a non-empty subset $K \subseteq [1; k]$ such that $\gamma_1 \leq \bigwedge_{i \in K} \sigma_i$ and $\bigwedge_{i \in K} \rho_i \leq \gamma_2$.*

Theorem 5 (Inversion Lemma). *The following properties hold:*

1. *$\Gamma \vdash_{\wedge} x : \sigma$ iff $\Gamma(x) \leq \sigma$,*
2. *$\Gamma \vdash_{\wedge} MN : \sigma$ iff there is ρ such that $\Gamma \vdash_{\wedge} M : \rho \to \sigma$ and $\Gamma \vdash_{\wedge} N : \rho$,*
3. *$\Gamma \vdash_{\wedge} \lambda x.M : \sigma$ iff there is $n \geq 1$ such that $\sigma = \bigwedge_{i=1}^{n} \sigma_i \to \sigma'_i$ for some σ_i, σ'_i,*
4. *$\Gamma \vdash_{\wedge} \lambda x.M : \sigma \to \tau$ iff $\Gamma, x : \sigma \vdash_{\wedge} M : \tau$.*

1.2 Type Structures Modelling the Simply Typed Lambda Calculus

A *typed applicative structure* $\mathcal{M}$ is a pair $((\mathcal{M}_A)_{A \in \mathbb{T}^0}, \bullet)$ where each $\mathcal{M}_A$ is a structure whose carrier is non-empty, and $\bullet$ is a function that associates to every $d \in \mathcal{M}_{A \to B}$ and every $e \in \mathcal{M}_A$ an element $d \bullet e$ in $\mathcal{M}_B$. From now on, we shall write $d \in \mathcal{M}$ to denote $d \in \mathcal{M}_A$ for some A. We say that $\mathcal{M}$ is: *hereditarily finite* if every $\mathcal{M}_A$ has a finite carrier; *extensional* whenever, for all $A, B \in \mathbb{T}^0$ and $d, d' \in \mathcal{M}_{A \to B}$, we have that $d \bullet e = d' \bullet e$ for every $e \in \mathcal{M}_A$ entails $d = d'$.

A *valuation in $\mathcal{M}$* is any map $\nu_{\mathcal{M}}$ from Var to elements of $\mathcal{M}$. A valuation $\nu_{\mathcal{M}}$ *agrees* with a simple context Δ when $\Delta(x) = A$ implies $\nu_{\mathcal{M}}(x) \in \mathcal{M}_A$. Given a valuation $\nu_{\mathcal{M}}$ and an element $d \in \mathcal{M}$, we write $\nu_{\mathcal{M}}[x := d]$ for the valuation $\nu'_{\mathcal{M}}$ that coincides with $\nu_{\mathcal{M}}$, except for x, where $\nu'_{\mathcal{M}}$ takes the value d. When there is no danger of confusion we may omit the subscript $\mathcal{M}$ and write ν.

A *valuation model $\mathcal{M}$* is an extensional typed applicative structure such that the clauses below define a total interpretation function $[\![\cdot]\!]^{\mathcal{M}}_{(\cdot)}$ which maps derivations $\Delta \vdash M : A$ and valuations ν agreeing with Δ to elements of $\mathcal{M}_A$:

- $[\![\Delta \vdash x : A]\!]_{\nu}^{\mathcal{M}} = \nu(x)$,
- $[\![\Delta \vdash NP : A]\!]_{\nu}^{\mathcal{M}} = [\![\Delta \vdash N : B \to A]\!]_{\nu}^{\mathcal{M}} \bullet [\![\Delta \vdash P : B]\!]_{\nu}^{\mathcal{M}}$,
- $[\![\Delta \vdash \lambda x.N : A \to B]\!]_{\nu}^{\mathcal{M}} \bullet d = [\![\Delta, x : A \vdash N : B]\!]_{\nu[x:=d]}^{\mathcal{M}}$ for every $d \in \mathcal{M}_A$.

When the derivation (resp. the model) is clear from the context we may simply write $[\![M]\!]_{\nu}^{\mathcal{M}}$ (resp. $[\![M]\!]_{\nu}$). For M closed, we simplify the notation further and write $[\![M]\!]$ since its interpretation is independent from the valuation.

The full model over a set $X \neq \emptyset$, denoted by $\mathrm{Full}(X)$, is the valuation model $((\mathcal{F}_A)_{A \in \mathbb{T}^0}, \bullet)$ where $\bullet$ is functional application, $\mathcal{F}_0 = X$ and $\mathcal{F}_{A \to B} = \mathcal{F}_B^{\mathcal{F}_A}$.

The continuous model over a cpo $(D, \leq)$, written $\mathrm{Cont}(D, \leq)$, is the valuation model $((\mathcal{D}_A, \sqsubseteq_A)_{A \in \mathbb{T}^0}, \bullet)$ such that $\bullet$ is functional application and:

- $\mathcal{D}_0 = D$ and $f \sqsubseteq_0 g$ iff $f \leq g$,
- $\mathcal{D}_{A \to B} = [\mathcal{D}_A \to \mathcal{D}_B]$ consisting of the monotone functions from $\mathcal{D}_A$ to $\mathcal{D}_B$ with the pointwise partially ordering $\sqsubseteq_{A \to B}$.

We will systematically omit the subscript A in $\sqsubseteq_A$ when clear from the context.

Note that both $\mathrm{Full}(X)$ and $\mathrm{Cont}(D, \leq)$ are extensional. Moreover, whenever X (resp. D) is finite $\mathrm{Full}(X)$ (resp. $\mathrm{Cont}(D, \leq)$) is hereditarily finite.

Logical relations have been extensively used in the study of semantic properties of λ-calculus (see [2, §4.5] for a survey). As we will see in Sections 4 and 5 they constitute a powerful tool for relating different valuation models.

Definition 1. *Given two valuation models $\mathcal{M}, \mathcal{N}$, a logical relation $\mathscr{R}$ between $\mathcal{M}$ and $\mathcal{N}$ is a family $\{\mathscr{R}_A\}_{A \in \mathbb{T}^0}$ of binary relations $\mathscr{R}_A \subseteq \mathcal{M}_A \times \mathcal{N}_A$ such that for all $A, B \in \mathbb{T}^0$, $f \in \mathcal{M}_{A \to B}$ and $g \in \mathcal{N}_{A \to B}$ we have:*

$$f \ \mathscr{R}_{A \to B} \ g \ \text{iff} \ \forall h \in \mathcal{M}_A, h' \in \mathcal{N}_A[h \ \mathscr{R}_A \ h' \Rightarrow f(h) \ \mathscr{R}_B \ g(h')].$$

Given $f \in \mathcal{M}_A$ we define $\mathscr{R}_A(f) = \{g \in \mathcal{N}_A \mid f \ \mathscr{R}_A \ g\}$ and, for $Y \subseteq \mathcal{M}_A$, $\mathscr{R}_A(Y) = \bigcup_{f \in Y} \mathscr{R}_A(f)$. Similarly, for $g \in \mathcal{N}_A$ and $Z \subseteq \mathcal{N}_A$ we have $\mathscr{R}_A^-(g) = \{f \in \mathcal{N}_A \mid f \ \mathscr{R}_A \ g\}$ and $\mathscr{R}_A^-(Z) = \bigcup_{g \in Z} \mathscr{R}_A^-(g)$.

It is well known that a logical relation $\mathscr{R}$ is univocally determined by the value of $\mathscr{R}_0$, and that the fundamental lemma of logical relations holds [2, §4.5].

Lemma 1 (Fundamental Lemma). *Let $\mathscr{R}$ be a logical relation between $\mathcal{M}$ and $\mathcal{N}$ then, for all closed M having simple type A, we have $[\![M]\!]^{\mathcal{M}} \ \mathscr{R}_A \ [\![M]\!]^{\mathcal{N}}$.*

2 Uniform Intersection Types and CDV$^\omega$

A useful approach to prove that a general decision problem is undecidable, is to identify a "sufficiently difficult" fragment of the problem. For instance, Urzyczyn in [10] shows the undecidability of inhabitation for a proper subset $\mathcal{G}$ of intersection types called *game types* in [3, §17E]. Formally, $\mathcal{G} = \mathbb{A} \cup \mathcal{B} \cup \mathcal{C}$ where:

$$\mathcal{A} = \mathbb{A}^\wedge, \mathcal{B} = (\mathcal{A} \to \mathcal{A})^\wedge, \mathcal{C} = (\mathcal{D} \to \mathcal{A})^\wedge \text{ for } \mathcal{D} = \{\sigma \wedge \tau \mid \sigma, \tau \in (\mathcal{B} \to \mathcal{A})\}.$$

(Recall that the notations $Y^\wedge$ and $Y \to Z$ were introduced in Subsection 1.1.) In our case we focus on intersection types that are *uniform* with simple types, in the sense that such intersection types follow the structure of the simple types.

Let us fix an arbitrary set $X \subseteq \mathbb{A}$. We write $\mathbb{T}_\wedge^X$ for the set of intersection types based on X.

Definition 2. *The set $\Xi_X(A)$ of intersection types uniform with $A \in \mathbb{T}^0$ is defined by induction on A as follows:*

$$\Xi_X(0) = X^\wedge, \qquad\qquad \Xi_X(B \to C) = (\Xi_X(B) \to \Xi_X(C))^\wedge.$$

When there is little danger of confusion, we simply write $\Xi(A)$ for $\Xi_X(A)$.

It turns out that game types are all uniform: $\mathcal{A} \subseteq \Xi_\mathbb{A}(0)$, $\mathcal{B} \subseteq \Xi_\mathbb{A}(0 \to 0)$ and $\mathcal{D} \subseteq \Xi_\mathbb{A}((0 \to 0) \to 0)$ thus $\mathcal{C} \subseteq \Xi_\mathbb{A}(((0 \to 0) \to 0) \to 0)$. Therefore the inhabitation problem for uniform intersection types over $\mathbb{A}$ is undecidable too.

Theorem 6 (Urzyczyn revisited). *The problem of deciding whether a type $\sigma \in \bigcup_{A \in \mathbb{T}^0, X \subseteq_f \mathbb{A}} \Xi_X(A)$ is inhabited in CDV is undecidable.*

For technical reasons, that will be clarified in the next section, we need to introduce the system CDV^ω over $\mathbb{A} \cup \{\omega\}$, a variation of CDV where intersection types are extended by adding a distinguished element ω at ground level.

In this framework, the set $\Xi_{X \cup \{\omega\}}(A)$ of *intersection types with ω uniform with A* will be denoted by $\Xi_X^\omega(A)$, or just $\Xi^\omega(A)$ when X is clear. We write ω_A for the type in $\Xi^\omega(A)$ defined by $\omega_0 = \omega$ and $\omega_{B \to C} = \omega_B \to \omega_C$.

The system CDV^ω over $\mathbb{T}_\wedge^{\mathbb{A} \cup \{\omega\}}$, whose judgments are denoted by $\Gamma \vdash_\wedge^\omega M : \sigma$, is generated by adding the following rule to the definition of $\leq$ in Figure 1(b):

$$\frac{\sigma \in \Xi_\mathbb{A}^\omega(A)}{\sigma \leq \omega_A} \ (\leq_A)$$

Therefore CDV^ω is different from the usual intersection type systems with ω. By construction, for every $A \in \mathbb{T}^0$, the type ω_A is a maximal element of $\Xi^\omega(A)$. Using [3, Thm. 14A.7], we easily get that the Inversion Lemma (Theorem 5) still works for CDV^ω, while the β-soundness holds in the following restricted form.

Recall that $\simeq$ stands for the equivalence generated by $\leq$.

Theorem 7 (β-soundness for CDV^ω). *Let $k \geq 1$. Suppose $\gamma_1 \to \gamma_2 \not\simeq \omega_A$ for all $A \in \mathbb{T}^0$ and $\bigwedge_{i=1}^k \sigma_i \to \rho_i \leq \gamma_1 \to \gamma_2$, then there is a non-empty subset $K \subseteq [1; k]$ such that $\gamma_1 \leq \bigwedge_{i \in K} \sigma_i$ and $\bigwedge_{i \in K} \rho_i \leq \gamma_2$.*

We now provide some useful properties of uniform intersection types.

Lemma 2. *Let $\sigma \in \Xi^\omega(A)$ and $\tau \in \Xi^\omega(A')$. Then we have that $\sigma \leq \tau$ entails $A = A'$.*

To distinguish arbitrary contexts from contexts containing uniform intersection types (with or without ω) we introduce some terminology.

We say that a context Γ is a *Ξ-context* (resp. *Ξ^ω-context*) if it ranges over uniform intersection types (resp. with ω). A Ξ-context (resp. Ξ^ω-context) $\Gamma = x_1 : \sigma_1, \ldots, x_n : \sigma_n$ is *uniform with* $\Delta = x_1 : A_1, \ldots, x_n : A_n$ if every σ_i belongs to $\Xi(A_i)$ (resp. to $\Xi^\omega(A_i)$).

Lemma 3. *Let $\rho \in \mathbb{T}_{\wedge}^{A\cup\{\omega\}}$, $\tau \in \Xi^{\omega}(B)$ and Γ be a Ξ^{ω}-context. Then we have that $\Gamma, x : \tau \vdash_{\wedge}^{\omega} xN_1 \cdots N_k : \rho$ iff there are $A, A_1, \ldots, A_k \in \mathbb{T}^0$ and $\sigma \in \Xi^{\omega}(A)$ and $\tau_i \in \Xi^{\omega}(A_i)$ for i in $[1;k]$ such that $B = A_1 \to \cdots \to A_k \to A$ and:*

1. *$\sigma \leq \rho$,*
2. *$\Gamma, x : \tau \vdash_{\wedge}^{\omega} xN_1 \cdots N_k : \sigma$,*
3. *$\tau \leq \tau_1 \to \cdots \to \tau_k \to \sigma$,*
4. *$\Gamma, x : \tau \vdash_{\wedge}^{\omega} N_i : \tau_i$ for all i in $[1;k]$.*

Furthermore, if Γ is a Ξ-context, $\rho \in \mathbb{T}_{\wedge}^{A}$ and $\tau \in \Xi(B)$, then σ and the τ_i for i in $[1;k]$ may also be chosen as uniform intersection types without ω (while the type judgments $\vdash_{\wedge}^{\omega}$ still need to be in CDV$^{\omega}$).

Theorem 8 (Uniform Inversion Lemma for CDV$^{\omega}$). *Let $\sigma \in \Xi^{\omega}(A)$ and Γ be a Ξ^{ω}-context. Then we have that (where we suppose that each term in a type judgment is in normal form):*

1. *$\Gamma \vdash_{\wedge}^{\omega} x : \sigma$ iff $\Gamma(x) \leq \sigma$,*
2. *$\Gamma \vdash_{\wedge}^{\omega} MN : \sigma$ iff there exist $B \in \mathbb{T}^0$ and $\tau \in \Xi^{\omega}(B)$ such that $\Gamma \vdash_{\wedge}^{\omega} M : \tau \to \sigma$ and $\Gamma \vdash_{\wedge}^{\omega} N : \tau$,*
3. *$\Gamma \vdash_{\wedge}^{\omega} \lambda x.N : \sigma$ iff $A = B \to C$ and there are $\tau_i \in \Xi^{\omega}(B), \tau_i' \in \Xi^{\omega}(C)$ such that $\sigma = \bigwedge_{i=1}^{n} \tau_i \to \tau_i'$ and $\Gamma, x : \tau_i \vdash_{\wedge}^{\omega} N : \tau_i'$ for all i in $[1;n]$.*

Corollary 1. *For M a normal λ-term, $\sigma \in \Xi^{\omega}(A)$ and Γ a Ξ^{ω}-context uniform with Δ, we have that $\Gamma \vdash_{\wedge}^{\omega} M : \sigma$ entails $\Delta \vdash M : A$.*

Proof. A simple consequence of the Uniform Inversion Lemma (with Lemma 2 when M is a variable). □

The corollary above does not generalize to arbitrary λ-terms as the following example illustrates. Let $M = \lambda zy.y$ and $N = \lambda x.xx$, then we have that $\vdash_{\wedge}^{\omega} MN : \alpha \to \alpha \in \Xi^{\omega}(0 \to 0)$ since $\vdash_{\wedge}^{\omega} N : \gamma$ and $\vdash_{\wedge}^{\omega} M : \gamma \to \alpha \to \alpha$ where $\gamma = (\beta \wedge (\beta \to \beta)) \to \beta$. However N is not simply typable, hence neither is MN. Note that, while we consider only uniform intersection types, we do not restrict the intersection type systems so that the type γ still may be used in a deduction.

CDV and CDV$^{\omega}$ are equivalent on normal forms in the following sense.

Lemma 4. *For every normal $M \in \Lambda$, for every Ξ-context $\Gamma = x_1 : \tau_1, \ldots, x_n : \tau_n$ uniform with $\Delta = x_1 : A_1, \ldots, x_n : A_n$, and for every $\sigma \in \Xi(A)$ we have:*

$$\Gamma \vdash_{\wedge} M : \sigma \iff \Gamma \vdash_{\wedge}^{\omega} M : \sigma.$$

Proof. ($\Rightarrow$) Trivial, as CDV is a subsystem of CDV$^{\omega}$.

($\Leftarrow$) We proceed by induction on the structure of M. The cases where M is a variable or a λ-abstraction can be treated thanks to Theorem 5 for CDV$^{\omega}$ and the induction hypothesis. Concerning the case where $M = x_i N_1 \cdots N_k$, from the ω-free version of Lemma 3, we have that $A_i = B_1 \to \cdots \to B_k \to A$, there exist $\tau_1, \ldots, \tau_k$ respectively in $\Xi(B_1), \ldots, \Xi(B_k)$ such that $\tau_i \leq \tau_1 \to \cdots \to \tau_k \to \sigma$ and $\Gamma \vdash_{\wedge}^{\omega} N_i : \tau_i$ for each i in $[1;k]$. Therefore, by the induction hypothesis, we have that for every i in $[1;k]$, $\Gamma \vdash_{\wedge} N_i : \tau_i$ which entails that $\Gamma \vdash_{\wedge} M : \sigma$. □

3   The Continuous Model over $\mathcal{P}(X)$

Hereafter we consider fixed an arbitrary set $X \subseteq_f \mathbb{A}$. We are going to represent uniform intersection types based on $X \cup \{\omega\}$, as elements of the continuous model $\mathcal{S}$ over $\mathcal{P}(X)$, ordered by set-theoretical inclusion.

Let $\mathcal{S} = \{(\mathcal{S}_A, \sqsubseteq_A)\}_{A \in \mathbb{T}^0} = \text{Cont}(\mathcal{P}(X), \subseteq)$. Each $\mathcal{S}_A$ is a finite join-semilattice and thus a complete lattice. We denote the join by $\sqcup$ and the bottom by $\perp_A$.

Given $f \in \mathcal{S}_A, g \in \mathcal{S}_B$ we write $f \mapsto g$ for the corresponding *step function*:

$$(f \mapsto g)(h) = \begin{cases} g & \text{if } f \sqsubseteq_A h, \\ \perp_B & \text{otherwise.} \end{cases}$$

For all A we define a function $\iota_A : \Xi^\omega(A) \to \mathcal{S}_A$ by induction on A as follows.

Definition 3. *For $\alpha \in X$ and $\sigma, \tau \in \Xi^\omega(0)$ we let $\iota_0(\alpha) = \{\alpha\}$, $\iota_0(\omega) = \perp_0 = \emptyset$, $\iota_0(\sigma \wedge \tau) = \iota_0(\sigma) \sqcup \iota_0(\tau)$. For $\sigma, \tau \in \Xi^\omega(A \to B)$ we define:*

$$\iota_{A \to B}(\sigma \to \tau) = \iota_A(\sigma) \mapsto \iota_B(\tau), \qquad \iota_{A \to B}(\sigma \wedge \tau) = \iota_{A \to B}(\sigma) \sqcup \iota_{A \to B}(\tau).$$

Remark 1. Given $\sigma \in \Xi^\omega(A)$, we have that $\sigma \simeq \omega_A$ entails $\iota_A(\sigma) = \perp_A$.

Thanks to the presence of the maximal element ω_A, the correspondence between $\Xi^\omega(A)$ and $\mathcal{S}_A$ is actually very faithful (in the sense of Corollary 2).

Lemma 5. *Let $h = \bigsqcup_{i=1}^n f_i \mapsto g_i$, then for every f we have:*

(i) $h(f) = \bigsqcup_{i \in K} g_i$ where $K = \{i \in [1; n] \mid f_i \sqsubseteq f\}$.
(ii) $h \sqsubseteq f$ iff $g_i \sqsubseteq f(f_i)$ for all $1 \le i \le n$.

Lemma 6. *Step functions are generators:* $\forall f \in \mathcal{S}_{A \to B}$, $f = \bigsqcup_{g \in \mathcal{S}_A} g \mapsto f(g)$.

Proof. Let $h = \bigsqcup_{g \in \mathcal{S}_A} g \mapsto f(g)$. We need to prove that, for every $g \in \mathcal{S}_A$, $f(g) = h(g)$. From Lemma 5(i), we have that $h(g) = \bigsqcup_{g' \sqsubseteq g} f(g')$. Since f is monotone, we have that for every $g' \sqsubseteq g$, $f(g') \sqsubseteq f(g)$ and therefore $\bigsqcup_{g' \sqsubseteq g} f(g') \sqsubseteq f(g)$. Since obviously $f(g) \sqsubseteq \bigsqcup_{g' \sqsubseteq g} f(g')$, we obtain $f(g) = \bigsqcup_{g' \sqsubseteq g} f(g') = h(g)$. $\square$

Lemma 7. *For all $A \in \mathbb{T}^0$, $\sigma, \tau \in \Xi^\omega(A)$ we have $\sigma \le \tau$ iff $\iota_A(\tau) \sqsubseteq \iota_A(\sigma)$.*

Proof. We proceed by induction on A. In case $A = 0$, the equivalence is clear since $\mathcal{P}(X)$ is the free $\sqcup$-semilattice with bottom over X and $\Xi^\omega(0)/\simeq$ is the free $\wedge$-semilattice with top over X.

In case $A = B \to C$, we have two subcases. Case 1, $\tau \simeq \omega_D$ for some $D \in \mathbb{T}^0$. Then by Lemma 2 we get $D = A$, by Remark 1 we get $\iota_A(\tau) = \perp_A$ and the equivalence follows since both $\sigma \le \tau$ and $\iota_A(\tau) \sqsubseteq \iota_A(\sigma)$ hold. Case 2, $\sigma = \bigwedge_{i=1}^n \sigma_i \to \sigma_i'$, $\tau = \bigwedge_{j=1}^m \tau_j \to \tau_j'$ and $\tau \not\simeq \omega_D$ for any $D \in \mathbb{T}^0$. By Remark 1 we can assume, without loss of generality, that for every j in $[1; m]$ we have $\tau_j \to \tau_j' \not\simeq \omega_D$ for all $D \in \mathbb{T}^0$. (Indeed for those k such that $\tau_k \to \tau_k' \simeq \omega_D$ one reasons as in Case 1.) We now prove the equivalence for this case.

($\Rightarrow$) If $\sigma \leq \tau$, then by β-soundness, for every j in $[1; m]$, there is K_j included in $[1; n]$ such that $\tau_j \leq \bigwedge_{i \in K_j} \sigma_i$ and $\bigwedge_{i \in K_j} \sigma'_i \leq \tau'_j$. By the induction hypothesis:

$$(1) \quad \bigsqcup_{i \in K_j} \iota_B(\sigma_i) \sqsubseteq \iota_B(\tau_j) \qquad (2) \quad \iota_C(\tau'_j) \sqsubseteq \bigsqcup_{i \in K_j} \iota_C(\sigma'_i)$$

We now prove that, for every $f \in \mathcal{S}_B$, $\iota_A(\tau)(f) \sqsubseteq \iota_A(\sigma)(f)$. From Lemma 5(i), we get $\iota_A(\tau)(f) = \bigsqcup_{j \in J} \iota_C(\tau'_j)$ where $J = \{j \in [1; m] \mid \iota_B(\tau_j) \sqsubseteq f\}$. By definition of J, we have that $\bigsqcup_{j \in J} \iota_B(\tau_j) \sqsubseteq f$ so, by (1), we obtain $\bigsqcup_{j \in J, i \in K_j} \iota_B(\sigma_i) \sqsubseteq f$. Therefore by Lemma 5(i), we get $\bigsqcup_{j \in J, i \in K_j} \iota_C(\sigma'_i) \sqsubseteq \iota_A(\sigma)(f)$ and, using (2), we obtain $\iota_A(\tau)(f) \sqsubseteq \iota_A(\sigma)(f)$. As a conclusion we have $\iota_A(\tau) \sqsubseteq \iota_A(\sigma)$.

($\Leftarrow$) If $\iota_A(\tau) \sqsubseteq \iota_A(\sigma)$, then we have in particular $\iota_A(\tau)(\iota_B(\tau_j)) \sqsubseteq \iota_A(\sigma)(\iota_B(\tau_j))$ for each $j \in [1, m]$. From Lemma 5(i), we have that $\iota_A(\tau)(\iota_B(\tau_j)) = \bigsqcup_{i \in I_j} \iota_C(\tau'_i)$ where $I_j = \{i \in [1; m] \mid \tau_i \leq \tau_j\}$. Since $\tau_j \leq \tau_j$ we must have $j \in I_j$ and therefore, we obtain $\iota_C(\tau'_j) \sqsubseteq \iota_A(\tau)(\iota_B(\tau_j))$. So, again by Lemma 5(i), we have that $\iota_A(\sigma)(\iota_B(\tau_j)) = \bigsqcup_{k \in K_j} \iota_C(\sigma'_k)$ where $K_j = \{k \in [1; n] \mid \tau_j \leq \sigma_k\}$. Thus we get $\iota_C(\tau'_j) \sqsubseteq \bigsqcup_{k \in K_j} \iota_C(\sigma'_k)$ and hence, by the induction hypothesis, $\bigwedge_{k \in K_j} \sigma'_k \leq \tau'_j$. Now, by definition of K_j, we also have $\tau_j \leq \bigwedge_{k \in K_j} \sigma_k$. As we can find such a K_j for every j in $[1; m]$, we can finally conclude that $\sigma \leq \tau$. $\qquad \square$

Corollary 2. *The map ι_A is an order-reversing bijection on $\Xi^\omega(A)/\simeq$.*

Proof. If $\tau \leq \sigma$ and $\sigma \leq \tau$, then Lemma 7 implies that $\iota_A(\tau) = \iota_A(\sigma)$. From this it ensues that ι_A is an order-reversing injection. To prove that it is actually a bijection, we need to show that ι_A is surjective. We proceed by induction on A. Clearly when $A = 0$, ι_A is surjective. If $A = B \to C$ then we get from the induction hypothesis that ι_B and ι_C are bijections between $\Xi^\omega(B)/\simeq$ and $\mathcal{S}_B$, and between $\Xi^\omega(C)/\simeq$ and $\mathcal{S}_C$, respectively. Now, given f in $\mathcal{S}_A$, we define $\tau_f \in \Xi^\omega(A)$ to be $\bigwedge_{g \in \mathcal{S}_B} \iota_B^{-1}(g) \to \iota_C^{-1}(f(g))$. But, $\iota_{A \to B}(\tau_f) = \bigsqcup_{g \in \mathcal{S}_B} g \mapsto f(g)$ which is equal to f by Lemma 6. $\qquad \square$

The above results are related to Stone duality for intersection types (cf. [1]).

Proposition 1. *Let M be a normal term such that $x_1 : A_1, \ldots, x_n : A_n \vdash M : A$. Then for all $\tau_i \in \Xi^\omega(A_i)$, $\sigma \in \Xi^\omega(A)$ the following two sentences are equivalent:*

1. $x_1 : \tau_1, \ldots, x_n : \tau_n \vdash_\wedge^\omega M : \sigma$,
2. $\iota_A(\sigma) \sqsubseteq [\![M]\!]_\nu^{\mathcal{S}}$, *for all valuations ν such that $\nu(x_i) = \iota_{A_i}(\tau_i)$.*

Proof. Let $\Delta = x_1 : A_1, \ldots, x_n : A_n$ and $\Gamma = x_1 : \tau_1, \ldots, x_n : \tau_n$.
$(1 \Rightarrow 2)$ We proceed by structural induction on M.

- In case $M = x_i$, then $\tau_i \leq \sigma$ and, by Lemma 7, $\iota_{A_i}(\sigma) \sqsubseteq \iota_{A_i}(\tau_i) = [\![x_i]\!]_\nu$.
- In case $M = NP$, then, from Theorem 8(2), there are $B \in \mathbb{T}^0$ and $\tau \in \Xi^\omega(B)$ such that $\Gamma \vdash_\wedge^\omega N : \tau \to \sigma$ and $\Gamma \vdash_\wedge^\omega P : \tau$. By induction $\iota_{B \to A}(\tau \to \sigma) \sqsubseteq [\![N]\!]_\nu$ and $\iota_B(\tau) \sqsubseteq [\![P]\!]_\nu$, thus, $\iota_A(\sigma) = \iota_{B \to A}(\tau \to \sigma)(\iota_B(\tau)) \sqsubseteq [\![N]\!]_\nu(\iota_B(\tau))$ and, by monotonicity, $[\![N]\!]_\nu(\iota_B(\tau)) \sqsubseteq [\![N]\!]_\nu([\![P]\!]_\nu) = [\![NP]\!]_\nu$. From this we finally get $\iota_A(\sigma) \sqsubseteq [\![NP]\!]_\nu$.

– In case $M = \lambda x.N$, then by Theorem 8(3) we have that $A = B \to C$ and, for all $j \in [1;n]$, there are $\sigma_j \in \Xi^\omega(B)$, $\sigma'_j \in \Xi^\omega(C)$ such that $\sigma = \bigwedge_{j=1}^n \sigma_j \to \sigma'_j$ and $\Gamma, x : \sigma_j \vdash^\omega_\wedge N : \sigma'_j$. Thus, by induction hypothesis, we get $\iota_C(\sigma'_j) \sqsubseteq [\![N]\!]_{\nu[x:=\iota_B(\sigma_j)]}$. From Lemma 5(ii) it ensues that $\iota_A(\sigma) \sqsubseteq [\![M]\!]_\nu$.

$(2 \Rightarrow 1)$ It suffices to establish by induction that $[\![M]\!]_\nu = \iota_A(\sigma)$, for all ν such that $\nu(x_i) = \iota_{A_i}(\tau_i)$, entails $\Gamma \vdash^\omega_\wedge M : \sigma$. Indeed, if τ is such that $\iota_A(\tau) \sqsubseteq [\![M]\!]_\nu$ then by Lemma 7 and $\sigma \le \tau$ we obtain, using the subsumption rule, that $\Gamma \vdash^\omega_\wedge M : \tau$.

– If $M = x_i$, then $[\![x_i]\!]_\nu = \iota_{A_i}(\tau_i) = \iota_A(\sigma)$ and $\sigma \simeq \tau_i$. Thus $\Gamma \vdash^\omega_\wedge x_i : \sigma$.
– If $M = NP$, then there is B such that $\Delta \vdash N : B \to A$ and $\Delta \vdash P : B$. By Corollary 2, there are $\tau \in \Xi^\omega(B \to A), \rho \in \Xi^\omega(B)$ such that $[\![N]\!]_\nu = \iota_{B \to A}(\tau)$ and $[\![P]\!]_\nu = \iota_B(\rho)$. The induction hypothesis implies that $\Gamma \vdash^\omega_\wedge N : \tau$ and $\Gamma \vdash^\omega_\wedge P : \rho$ are derivable. By hypothesis we know that $[\![M]\!]_\nu = \iota_A(\sigma)$. From Lemma 5(ii), since $\iota_A(\sigma) = [\![M]\!]_\nu = [\![N]\!]_\nu([\![P]\!]_\nu) = \iota_{B \to A}(\tau)(\iota_B(\rho))$, we have $\iota_B(\rho) \mapsto \iota_A(\sigma) \sqsubseteq \iota_{B \to A}(\tau)$ and thus, by Lemma 7, $\tau \le \rho \to \sigma$. Hence $\Gamma \vdash^\omega_\wedge N : \rho \to \sigma$ is derivable, which implies that $\Gamma \vdash^\omega_\wedge M : \sigma$ is derivable.
– If $M = \lambda x.N$, then $A = B \to C$. By Corollary 2 we can choose, for every $g \in S_B$, $\sigma_g \in \Xi^\omega(B)$ such that $\iota_B(\sigma_g) = g$ and $\tau_g \in \Xi^\omega(C)$ such that $\iota_C(\tau_g) = [\![N]\!]_{\nu[x:=g]} = [\![M]\!]_\nu(g)$. By the induction hypothesis, for every $g \in S_B$, we have $\Gamma, x : \sigma_g \vdash^\omega_\wedge N : \tau_g$. Therefore, $\Gamma \vdash^\omega_\wedge M : \sigma_g \to \tau_g$ and $\Gamma \vdash^\omega_\wedge M : \bigwedge_{g \in S_B} \sigma_g \to \tau_g$. By definition $\iota_A(\bigwedge_{g \in S_B} \sigma_g \to \tau_g) = \bigsqcup_{g \in S_b} \iota_B(\sigma_g) \mapsto \iota_C(\tau_g) = \bigsqcup_{g \in S_b} g \mapsto [\![M]\!]_\nu(g)$ which is equal, by Lemma 6, to $[\![M]\!]_\nu$.   □

4   Inhabitation Reduces to Definability

We now prove that the undecidability of the Definability Problem follows from the undecidability of the inhabitation problem (for game types) in CDV. A preliminary version of this result was announced in the invited paper [8].

The proof we present here is obtained by linking via a suitable logical relation $\mathscr{I}$ the continuous model S built in the previous section and $\mathcal{F} = \{\mathcal{F}_A\}_{A \in \mathbb{T}^0} = \mathrm{Full}(\mathcal{P}(X))$, where $X \subseteq_f A$. Let $\mathscr{I}$ be the logical relation between S and $\mathcal{F}$ generated by taking the identity at ground level (indeed $S_0 = \mathcal{F}_0 = \mathcal{P}(X)$).

Lemma 8. $\mathscr{I}$ *is a logical retract, i.e. at every level $A \in \mathbb{T}^0$ we have $\forall f_1, f_2 \in S_A$, $\mathscr{I}_A(f_1) \cap \mathscr{I}_A(f_2) \ne \emptyset$ iff $f_1 = f_2$. Equivalently, both next statements hold:*

(i) for all $f \in S_A$ there is $g \in F_A$ such that $f \mathscr{I}_A g$,
(ii) for all $f, f' \in S_A, g \in F_A$ if $f \mathscr{I}_A g$ and $f' \mathscr{I}_A g$ then $f = f'$.

Proof. We prove the main statement by induction on A, then both items follow. The base case $A = 0$ is trivial, so we consider the case $A = B \to C$.
$(\Rightarrow)$ By definition of $\mathscr{I}_A(f_1), \mathscr{I}_A(f_2)$ we have:

$$\mathscr{I}_A(f_1) \cap \mathscr{I}_A(f_2) = \{h \mid \forall g \in S_B, \forall k \in \mathscr{I}_B(g), h(k) \in \mathscr{I}_C(f_1(g)) \cap \mathscr{I}_C(f_2(g))\}.$$

Now, $\mathscr{I}_A(f_1) \cap \mathscr{I}_A(f_2) \ne \emptyset$ entails $\mathscr{I}_C(f_1(g)) \cap \mathscr{I}_C(f_2(g)) \ne \emptyset$ for all $g \in S_B$. By induction, this holds when $f_1(g) = f_2(g)$ for all $g \in S_B$, i.e. when $f_1 = f_2$.

($\Leftarrow$) If $f_1 = f_2$ then $\mathscr{I}_A(f_1) = \{h \mid \forall g \in \mathcal{S}_B, \forall k \in \mathscr{I}_B(g), h(k) \in \mathscr{I}_C(f_1(g))\}$. To prove $\mathscr{I}_A(f_1) \neq \emptyset$, we build a relation $h \subseteq \mathcal{F}_B \times \mathcal{F}_C$ that is actually functional and belongs to it. Fix any $d \in \mathcal{F}_C$ and, for every $g \in \mathcal{S}_B$, an element $r_g \in \mathscr{I}_C(f_1(g))$ which exists by induction hypothesis. Define h as the smallest relation such that $(k, r_g) \in h$ if $k \in \mathscr{I}_B(g)$, and $(k, d) \in h$ if $k \notin \bigcup_{g \in \mathcal{S}_B} \mathscr{I}_B(g)$. As, by induction hypothesis, $\mathscr{I}_B(g_1)$ and $\mathscr{I}_B(g_2)$ are disjoint for all $g_1 \neq g_2$ then h is functional. By construction, $h \in \mathscr{I}_C(f_1(g))$. $\qquad\square$

As a consequence we get, for every subset $S \subseteq \mathcal{S}_A$, that $\mathscr{I}_A^-(\mathscr{I}_A(S)) = S$. Given $f \in \mathcal{S}_A$ we write $f\uparrow$ for its upward closure in $\mathcal{S}_A$: $\{f' \in \mathcal{S}_A \mid f \sqsubseteq f'\}$.

Proposition 2. *Let $\sigma \in \Xi(A)$. For every normal λ-term M having type A we have $\vdash_\wedge M : \sigma$ iff $[\![M]\!]^{\mathcal{F}} \in \mathscr{I}_A(\iota_A(\sigma)\uparrow)$.*

Proof. We have the following computable chain of equivalences:

$$\vdash_\wedge M : \sigma \iff \vdash_\wedge^\omega M : \sigma, \qquad \text{by Lemma 4,}$$
$$\iff [\![M]\!]^{\mathcal{S}} \in \iota_A(\sigma)\uparrow, \qquad \text{by Proposition 1,}$$
$$\iff [\![M]\!]^{\mathcal{F}} \in \mathscr{I}_A(\iota_A(\sigma)\uparrow), \qquad \text{by Lemma 1 plus Lemma 8.} \qquad\square$$

Theorem 9. *The undecidability of the Definability Problem follows by a reduction from the one of the Inhabitation Problem for game types, Theorem 2(2).*

Proof. Suppose by contradiction that DP is decidable. We want to decide whether $\sigma \in \bigcup_{A \in \mathbb{T}^0, X \subseteq_f \mathbb{A}} \Xi_X(A)$ is inhabited in CDV. By Theorem 3 and Corollary 1 we can focus on normal simply typed λ-terms. Now we can take the set Y of all atoms in σ, compute the simple type A such that $\sigma \in \Xi_Y(A)$, and effectively construct the finite set $\mathscr{I}_A(\iota_A(\sigma)\uparrow) \subseteq \text{Full}(Y)$. If DP is decidable, then we can also decide with finitely many tests whether there is a λ-definable $f \in \mathscr{I}_A(\iota_A(\sigma)\uparrow)$. By Proposition 2 such an f exists if and only if σ is inhabited. This yields a reduction of IHP for game types (hence for uniform types, Theorem 6) to DP. $\qquad\square$

5   Definability Reduces to Inhabitation

In this section we prove the converse of Theorem 9, namely that the undecidability of inhabitation follows directly from the undecidability of λ-definability in the full model $\mathcal{F} = \text{Full}(X)$ over a fixed set $X \subseteq_f \mathbb{A}$. The main idea is a simple embedding of the elements of $\mathcal{F}$ into the uniform intersection types.

Also in this proof the continuous model $\mathcal{S} = \text{Cont}(\mathcal{P}(X), \subseteq)$ will play a key role. (Remark that the ground set of $\mathcal{S}$ is still $\mathcal{P}(X)$, while $\mathcal{F}$ is now over X.) We start by defining an injection $\varphi_A : \mathcal{F}_A \to \mathcal{S}_A$ by induction on A:

- if $A = 0$, then $\varphi_A(f) = \{f\}$,
- if $A = B \to C$, then $\varphi_A(f) = \bigsqcup_{g \in \mathcal{F}_B} \varphi_B(g) \mapsto \varphi_C(f(g))$.

Now, given f in $\mathcal{F}_A$ we define an intersection type ξ_f in $\Xi(A)$ as follows:

- if $A = 0$, then $\xi_f = f$,
- if $A = B \to C$, then $\xi_f = \bigwedge_{g \in \mathcal{F}_B} \xi_g \to \xi_{f(g)}$.

Lemma 9. *For every f in $\mathcal{F}_A$, we have $\varphi_A(f) = \iota_A(\xi_f)$.*

We consider the logical relation $\mathcal{J}$ between the full model $\mathcal{F}$ and the continuous model $\mathcal{S}$ generated by $\mathcal{J}_0 = \{(f, F) \mid f \in F \subseteq \mathcal{F}_0\}$.

Lemma 10. *For every $f \in \mathcal{F}_A$ and $g \in \mathcal{S}_A$ we have $f \ \mathcal{J}_A \ g$ iff $\varphi_A(f) \sqsubseteq g$.*

Proof. By induction on A, the case $A = 0$ being obvious. Let $A = B \to C$.

($\Rightarrow$) Suppose $f \ \mathcal{J}_A \ g$. We want to prove that $\varphi_A(f) \sqsubseteq g$. That is, for all $h \in \mathcal{S}_B$, we have $\varphi_A(f)(h) \sqsubseteq g(h)$. Let $h \in \mathcal{S}_B$, then by definition of φ_A, we have $\varphi_A(f)(h) = \bigsqcup\{\varphi_C(f(k)) \mid \varphi_B(k) \sqsubseteq h, k \in \mathcal{F}_B\}$. But $\varphi_B(k) \sqsubseteq h$ implies $k \ \mathcal{J}_B \ h$ by induction hypothesis, which implies that $f(k) \ \mathcal{J}_C \ g(h)$ since $f \ \mathcal{J}_A \ g$. Now using the induction hypothesis for C, we get $\varphi_C(f(k)) \sqsubseteq g(h)$. That is, $\varphi_A(f)(h)$ is a supremum of things all of which are below $g(h)$, thus $\varphi_A(f)(h) \sqsubseteq g(h)$.

($\Leftarrow$) Suppose $\varphi_A(f) \sqsubseteq g$. Let $h \in \mathcal{F}_B$ and $h' \in \mathcal{S}_B$ with $h \ \mathcal{J}_B \ h'$, that is, by the induction hypothesis, with $\varphi_B(h) \sqsubseteq h'$. We want to show that $f(h) \ \mathcal{J}_C \ g(h')$ or, equivalently, again by the induction hypothesis, that $\varphi_C(f(h)) \sqsubseteq g(h')$. Now, by definition, $\varphi_A(f)(h') = \bigsqcup\{\varphi_C(f(k)) \mid \varphi_B(k) \sqsubseteq h', k \in \mathcal{F}_B\}$, and by assumption $h \in \mathcal{F}_B$ and $\varphi_B(h) \sqsubseteq h'$, so $\varphi_C(f(h)) \sqsubseteq \varphi_A(f)(h')$. On the other hand, $\varphi_A(f) \sqsubseteq g$ as functions on $\mathcal{S}_A$ and $h' \in \mathcal{S}_B$, so $\varphi_A(f)(h') \sqsubseteq g(h')$. By transitivity of the order we obtain $\varphi_C(f(h)) \sqsubseteq g(h')$ as required. $\square$

Proposition 3. *Given f in $\mathcal{F}_A$, we have $[\![M]\!]^{\mathcal{F}} = f$ iff $\vdash_\wedge M : \xi_f$.*

Proof. We have the following computable chain of equivalences:

$$
\begin{aligned}
[\![M]\!]^{\mathcal{F}} = f &\iff f \ \mathcal{J}_A \ [\![M]\!]^{\mathcal{S}}, && \text{by Lemma 1,} \\
&\iff \varphi(f) \sqsubseteq [\![M]\!]^{\mathcal{S}}, && \text{by Lemma 10,} \\
&\iff \iota_A(\xi_f) \sqsubseteq [\![M]\!]^{\mathcal{S}}, && \text{by Lemma 9,} \\
&\iff \vdash_\wedge M : \xi_f, && \text{by Proposition 1.} \qquad\square
\end{aligned}
$$

Therefore f is definable iff ξ_f is inhabited. This yields a reduction of the Definability Problem (resp. DP_n) to the Inhabitation Problem (resp. IHP_n).

Theorem 10. *1. The undecidability of IHP_n for all $n > 1$ follows by a reduction from the undecidability of DP_n for all $n > 1$, Theorem 1(2).*
2. The undecidability of the Inhabitation Problem follows by a reduction from the undecidability of the Definability Problem, Theorem 1(1).

Acknowledgements. We are grateful to Antonio Bucciarelli for interesting discussions. This work is partly supported by NWO Project 612.000.936 CALMOC, ANR 2010 BLAN 0202 01 FREC and ANR 2010 BLAN 0202 02 FREC.

References

1. Abramsky, S.: Domain theory in logical form. In: Symposium on Logic and Computer Science (LICS 1987), pp. 47–53. IEEE Computer Science Press (1987)

2. Amadio, R., Curien, P.-L.: Domains and lambda-calculi. Cambridge Tracts in Theoretical Computer Science, vol. (46). Cambridge University Press (1998)
3. Barendregt, H.P., Dekkers, W., Statman, R.: Lambda calculus with types (to appear), Draft available at http://www.cs.ru.nl/~henk/book.pdf
4. Coppo, M., Dezani-Ciancaglini, M., Venneri, B.: Functional characters of solvable terms. Mathematical Logic Quarterly 27(2-6), 45–58 (1981)
5. Joly, T.: Encoding of the Halting Problem into the Monster Type & Applications. In: Hofmann, M.O. (ed.) TLCA 2003. LNCS, vol. 2701, pp. 153–166. Springer, Heidelberg (2003)
6. Loader, R.: The undecidability of lambda definability. In: Logic, Meaning and Computation: Essays in Memory of Alonzo Church, pp. 331–342 (2001)
7. Plotkin, G.: Lambda definability and logical relations. Memorandum SAI-RM-4, School of Artificial Intelligence. University of Edinburgh (1973)
8. Salvati, S.: Recognizability in the Simply Typed Lambda-Calculus. In: Ono, H., Kanazawa, M., de Queiroz, R. (eds.) WoLLIC 2009. LNCS, vol. 5514, pp. 48–60. Springer, Heidelberg (2009)
9. Statman, R.: Completeness, invariance and λ-definability. The Journal of Symbolic Logic 47(1), 17–26 (1982)
10. Urzyczyn, P.: The emptiness problem for intersection types. The Journal of Symbolic Logic 64(3), 1195–1215 (1999)

Languages of Profinite Words and the Limitedness Problem

Szymon Toruńczyk

University of Warsaw

Abstract. We present a new framework for the limitedness problem. The key novelty is a description using profinite words, which unifies and simplifies the previous approaches, allowing a seamless extension of the theory of regular languages. We also define a logic over profinite words, called MSO+inf and show that the satisfiability problem of MSO+$\mathbb{B}$ reduces to the satisfiability problem of our logic.

1 Introduction

This paper is an attempt to establish a natural framework for problems related to the limitedness problem. A notable example of such a problem is the decidability of the logic MSO+$\mathbb{B}$.

Fig. 1. A distance automaton over the input alphabet $\{a, b\}$

The *limitedness problem* was introduced by Hashiguchi [7] on his way to solving the famous star height problem. In its basic form, it concerns *distance automata*, i.e. nondeterministic automata, whose transitions are additionally labeled by nonnegative, integer weights, such as the one depicted in Figure 1. A distance automaton is *limited* if there exists a bound n such that every accepted word has some accepting run whose sum of weights is bounded by n. Thus the *limitedness problem* is a decision problem which asks whether a given automaton is limited. The automaton in the example is not limited: the words $a, a^2, a^3, \ldots$ require accepting runs of ever larger weights.

The *logic MSO+$\mathbb{B}$* was introduced by Bojańczyk in his dissertation (see also [2]) in relation with a problem concerning modal μ-calculus. It is an extension of the usual MSO logic – over infinite trees or words – by the quantifier $\mathbb{B}$, defined so that the formula $\mathbb{B}X.\varphi(X)$ holds if and only if all the sets of positions X satisfying the formula φ in the given model have a commonly bounded size. A typical language of infinite words defined in this logic is:

$$L_B = \{a^{n_1} b a^{n_2} b \ldots : \quad \text{the sequence } n_1, n_2, \ldots \text{ is bounded}\}.$$

Note that this language is not ω-regular, as its complement does not contain any ultimately periodic word. As a far-reaching project (see [3] for a survey),

A. Czumaj et al. (Eds.): ICALP 2012, Part II, LNCS 7392, pp. 377–389, 2012.
© Springer-Verlag Berlin Heidelberg 2012

Bojańczyk posed the question of decidability of satisfiability of the logic MSO+$\mathbb{B}$ over infinite trees. Still, it is not even known to be decidable over infinite words.

A syntactic fragment of the logic MSO+$\mathbb{B}$ was proved decidable in [4]. The key tool used in that paper is a model of automata called ωB-automata, recognizing ωB-regular languages such as L_B. Later, Colcombet discovered that limitedness of distance automata can be easily decided using their results concerning ωB-automata. The link with the limitedness problem has been exploited in [5], where Colcombet defined B-automata and developed his theory of regular cost functions and stabilization semigroups. B-automata directly generalize distance automata, by allowing more than one counter which, moreover, can be reset.

Our contribution is a topological framework which builds upon the theory of Colcombet. Our extension is mainly not by means of new results, but by means of a new language which unifies and simplifies significantly many notions and proofs from the literature.

As a starting point, we see that B-automata naturally define languages of *profinite* words. The set of profinite words has a rich algebraic and topological structure, which we find very useful in the context of limitedness. For example, consider the distance automaton from Fig. 1. There is a *profinite* word, denoted a^ω (not to be confused with the infinite word) witnessing that the automaton is *not* limited – this word is defined as the limit of the sequence of finite words $a, a^{2!}, a^{3!}, \ldots$ We say that this profinite word does *not* belong to the language of this automaton; the language of this automaton consists of profinite words which only have finitely many a's, such as b or $b^\omega a$.

We call the class of languages of profinite words defined by B-automata *B-regular languages*. Using our framework, we reformulate and reprove the result of Colcombet [5,6], by characterizing B-regular languages in terms of logic, regular expressions and semigroups. These characterizations generalize the main results of the papers [10,12,8,1,4]. The description in terms of semigroups immediately implies decidability of the limitedness problem for B-automata, which, in our framework is simply the question of language universality. In particular, together with Kirsten's elegant reduction of the star height problem to the limitedness problem, our framework provides yet another proof of decidability of the star height problem. The result also implies decidability of a more general problem – limitedness of Boolean combinations of B-automata.

On top of these characterizations, we provide a novel finite-index characterization of B-regular languages, à la the Myhill-Nerode theorem. Altogether, these characterizations demonstrate that the class of B-regular languages is robust, and that our framework is suitable.

Lastly, we show that our framework is suited for dealing with the satisfiability problem for MSO+$\mathbb{B}$ over infinite words – we prove that this problem can be reduced to the satisfiability problem of a new logic MSO+inf over profinite words, which we introduce here. In fact, our reduction is very general, and works for very many logics. The proof extends Büchi's ideas, and consists of two key ingredients: convergent Ramsey factorizations of infinite words, and a model of deterministic automata over infinite words with a profinite acceptance condition.

Related work. Several proofs of decidability of the limitedness problem exist [7,10,12,8,1,5]. Our proof builds on ideas from all of these papers, and simplifies them significantly. Hashigushi's #-expressions acquire a new, concrete meaning in our framework, as simply defining profinite words. We extend Leung's insight of considering the compact topological semigroup of all matrices over the tropical semiring, to considering the profinite semigroup. Also, Leung introduced finite versions of his topological semigroups, which are predecessors of stabilization semigroups of Colcombet. The factorization forests of Simon play a key role in the main technical part of our proof. The proof of Kirsten applies to a model very similar to B-automata, but with a hierarchical constraint on the counter operations. Kirsten generalized Leung's proof, providing further instances of stabilization semigroups; however, the topological insights of Leung disappeared, as he no longer considered compact topological semigroups.

Colcombet used ideas from [4] and of Kirsten in [5], where he developed his theory of *regular cost functions* – equivalence classes of number-valued functions defined by B-automata. Regular cost functions also have equivalent descriptions in terms of regular expressions, logic and semigroups. The crucial discovery of [5] is the abstract notion of *stabilization semigroups* and their tight relationship with B-automata. Our topological insights shed new light on this relationship.

On a general level, and also on the level of proof structure, our approach resembles, and is inspired by the approach of Colcombet. Most notions considered in this paper have analogues in the work of Colcombet. We outline the key differences. As we deal with languages which are subsets of a topological semigroup, many classical notions naturally lift to our setting – such as recognizable subsets, Myhill-Nerode equivalence, homomorphisms. In Colcombet's framework, cost functions are not sets, and have no apparent algebraic nor topological structure. Because of this, the natural notions mentioned above do not exist, or have non-obvious definitions – an example is the notion of *compatible mapping* [5], which corresponds to our ∞-*homomorphism*. There is no notion of a Boolean combination of cost functions. As a result, cost functions are not suited for the study of the full logic MSO+$\mathbb{B}$. On a technical level, the proofs in [5,6] deal with the relative notions of "big" vs. "small" values, and this relativity needs to be carefully controlled. In our more abstract setting, we deal with the absolute notions of infinite vs. finite, and computations involve usual set-theoretic equalities. This vastly simplifies the proofs.

2   Preliminaries

In this section, we recall the definitions of B- and S-automata and of profinite words. Let us fix a finite alphabet A; finite words are assumed to be elements of A^+. In the examples, we will more concretely assume the alphabet $A = \{a, b\}$. By $\mathbb{N}$ we denote $\{0, 1, 2, \ldots\}$, and by $\overline{\mathbb{N}}$ we denote $\mathbb{N} \cup \{\omega\}$. We treat $\overline{\mathbb{N}}$ as a compact metric space, in which $d(m, n) = |2^{-m} - 2^{-n}|$, where is defined $2^{-\omega}$ as 0.

B-automata and S-automata (implicit in [4], defined in [5]) are nondeterministic automata over finite words, equipped with a finite number of counters. There are two counter operations available for each counter: *inc* increases the current value of the counter by 1 and *reset* sets the value to 0. A transition of a B- or S-automaton may trigger any sequence of operations on its counters. If the operation *reset* is performed in a run ρ on a counter which currently stores a value n, then we say that n *is a reset value* in the considered run ρ. The two models – B- and S-automata – differ in the semantics of the functions they define.

First, consider a B-automaton $\mathcal{A}$. Since $\mathcal{A}$ is nondeterministic, there might be many runs over a single word. For a particular run ρ, we define the *value* of ρ as its maximal reset value. Next, the *valuation* $f_{\mathcal{A}}(w)$ of an input word w under the automaton $\mathcal{A}$ is the minimum of the values of all accepting runs ρ over w:

$$f_{\mathcal{A}}(w) = \min_{\rho} \max\{n : \text{ in the run } \rho, \text{ the value } n \text{ is a reset value}\}.$$

Note that min ranges only over the accepting runs ρ of $\mathcal{A}$. We assume $\max(\emptyset) = 0$ and $\min(\emptyset) = \omega$, so if $\mathcal{A}$ has no accepting run over w, then $f_{\mathcal{A}}(w) = \omega$.

If $\mathcal{A}$ is an S-automaton, the definition of a valuation $f_{\mathcal{A}}(w)$ of an input word w is completely dual – simply swap min with max in the formula above.

Example 1 (The running example). Let $\mathcal{A}$ be the B-automaton with one counter which is depicted in the left-hand side of the figure below.

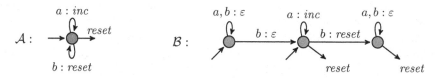

We declare that the automaton resets its counter after reading the entire word – this extra feature can be easily eliminated using nondeterminism. Then,

$$f_{\mathcal{A}}(w) = \max\{n_1, n_2, \ldots, n_k\} \qquad \text{for} \qquad w = a^{n_1} b a^{n_2} \ldots b a^{n_k}.$$

Now consider the S-automaton $\mathcal{B}$ depicted in the right-hand side of the figure. It has one counter, which is also assumed to be reset at the end of the run. The reader can check that each accepting run of $\mathcal{B}$ over an input word w counts the length of some block of a's in w, and that $f_{\mathcal{B}}(w)$ is the length of the largest such block. Therefore, $f_{\mathcal{B}}$ and $f_{\mathcal{A}}$ are precisely the same function from A^+ to $\overline{\mathbb{N}}$.

Example 2. Let $\mathcal{A}$ be a finite nondeterministic automaton. If we view $\mathcal{A}$ as a B-automaton with no counters, the induced function assigns 0 to any word accepted by $\mathcal{A}$ and ω to any rejected word. Dually, if we treat $\mathcal{A}$ as an S-automaton, the induced function assigns ω to any accepted word, and 0 to any rejected word.

A B- or S-automaton is said to be *limited* if the function $f_{\mathcal{A}}$ has finite range (which may nevertheless contain the value ω). The *limitedness problem* for B- or S-automata is then to decide whether a given B- or S-automaton is limited. The automata in the example are not limited, since $f_{\mathcal{A}}(a^n) = n$ for any $n \in \mathbb{N}$.

Profinite words should be thought of as limits of sequences of finite words, with respect to *all* regular languages. A formal definition follows (see e.g. [11] for more details). We say that an infinite sequence $w_1, w_2, \ldots \in A^+$ of finite words *ultimately* belongs to a language $L \subseteq A^+$ if almost all the words $w_1, w_2, \ldots$ belong to L. A sequence of words is *convergent* if for any regular language L, the sequence ultimately belongs to L or ultimately belongs to the complement of L. Every constant sequence is convergent. The sequence $a, a^{2!}, a^{3!}, \ldots$ is also convergent – by a pumping argument for regular languages – but the sequence $a, a^2, a^3, \ldots$ is not convergent, since the regular language $(aa)^+$ only contains every other of its elements. Two convergent sequences are *equivalent* if they belong ultimately to precisely the same regular languages. In other words, interleaving one sequence with the other yields a convergent sequence. An equivalence class of convergent sequences is a *profinite word*. We denote profinite words by $x, y, \ldots$, and the set of all profinite words by $\widehat{A^+}$. Note that the set of finite words A^+ naturally embeds into the set of profinite words $\widehat{A^+}$, via constant convergent sequences.

The set of profinite words forms a semigroup: if $w_1, w_2, \ldots$ and $v_1, v_2, \ldots$ are two convergent sequences, then the sequence $w_1 v_1, w_2 v_2, \ldots$ is also convergent. The *ω-power* of a convergent sequence $w_1, w_2, w_3, \ldots$ is the sequence $w_1^{1}, w_2^{2!}, w_3^{3!}, \ldots$, which also turns out to be convergent. Therefore, the *ω-power* of a profinite word x is well-defined, and denoted x^ω.

The set of profinite words carries a compact metric: the distance between two profinite words x, y is $\frac{1}{n}$, where n is the smallest size – measured as size of the minimal automaton – of a regular language L such that x ultimately belongs to L and y does not. This metric is compatible with the notion of convergence defined above. In particular, the set A^+ of finite words is dense in the set of profinite words, $\widehat{A^+}$. Multiplication and the ω-power are continuous mappings over $\widehat{A^+}$. The closure $\overline{L}$ in $\widehat{A^+}$ of any regular language $L \subseteq A^+$ turns out to be both closed and open, i.e. *clopen* in $\widehat{A^+}$. Conversely, any clopen subset of $\widehat{A^+}$ is of the form $\overline{L}$ for some regular language L, so clopen sets correspond precisely to regular languages. Any open set in $\widehat{A^+}$ is a (possibly infinite) union of clopen sets.

3 Languages of Profinite Words

In this section we discuss several ways of describing subsets – or *languages* – of profinite words: via automata, regular expressions and logic.

B- and S-regular languages. The essential idea underlying our theory is to consider B- and S-automata as processing not only finite words, but also profinite words. Let $\mathcal{A}$ be a B- or S-automaton. The following fact relies on the simple observation that for each $n \in \mathbb{N}$, the language $\{w \in A^+ : f_\mathcal{A}(w) < n\}$ is regular.

Fact 1. *Let $w_1, w_2, \ldots$ be a convergent sequence of finite words. Then, the sequence $f_\mathcal{A}(w_1), f_\mathcal{A}(w_2), \ldots$ is convergent in $\overline{\mathbb{N}} = \mathbb{N} \cup \{\omega\}$.*

Therefore, it makes sense to define, for any $x \in \widehat{A^+}$,

$$\widehat{f_\mathcal{A}}(x) \overset{def}{=} \lim_{n \to \infty} f_\mathcal{A}(w_n),$$

where $w_1, w_2, \ldots$ is any sequence of finite words which converges to x.

It is straightforward to show that $\widehat{f_A}$ is a well-defined continuous function from $\widehat{A^+}$ to $\overline{\mathbb{N}}$. By density of A^+ in $\widehat{A^+}$, the continuous extension of f_A to $\widehat{A^+}$ is unique, so we will further identify f_A with $\widehat{f_A}$.

Similarly to the idea underlying cost functions [5], we do not care about the exact values of the function f_A (this would quickly lead to undecidability, as demonstrated by Krob [9]). What we care about is over which sequences of words, f_A grows indefinitely. By continuity of f_A and compactness of $\widehat{A^+}$, this is encoded in the set

$$\{x \in \widehat{A^+} : f_A(x) = \omega\}.$$

This is a closed set as the inverse image of a point under a continuous mapping.

This motivates the following definitions. For an S-automaton $\mathcal{A}$, we define the set $L(\mathcal{A})$ consisting of all profinite words x such that $f_A(x) = \omega$. For a B-automaton $\mathcal{A}$, we define $L(\mathcal{A})$ dually, as the language of all profinite words x such that $f_A(x) < \omega$. In either case, we call $L(\mathcal{A})$ the *language recognized by* $\mathcal{A}$. The reason why the definitions differ is that S-automata try to maximize, while B-automata try to minimize the value of a run. We call a language $L \subseteq \widehat{A^+}$ *B-regular* (respectively, *S-regular*), if it is recognized by a B-automaton (respectively, S-automaton). Note that S-regular languages are closed, and B-regular languages are open subsets of $\widehat{A^+}$. In particular, a language is both B- and S-regular if and only if it is clopen.

Example 3. Let $\mathcal{A}$ be the B-automaton from Example 1, computing the largest block of a's. Then $L(\mathcal{A})$ is the language of all profinite words for which every block of a's has uniformly bounded length:

$$L(\mathcal{A}) = \{x \in \widehat{A^+} : f_A(x) < \omega\} = \bigcup_{n \in \mathbb{N}} \{x \in \widehat{A^+} : x \text{ has no infix } a^n\}.$$

It is not difficult to show (using compactness and continuity of multiplication) that a profinite word has arbitrarily long blocks of a's if and only if it contains a^ω as an infix. (We say that u is an *infix* of v if $v = v_1 \cdot u \cdot v_2$ for some, potentially empty, profinite words v_1, v_2.) Therefore, if $\mathcal{B}$ is the S-automaton from Example 1 (recall that $f_A = f_B$), we deduce that

$$L(\mathcal{B}) = \{x \in \widehat{A^+} : f_B(x) = \omega\} = \widehat{A^+} - L(\mathcal{A}) = \{x_1 \cdot a^\omega \cdot x_2 : x_1, x_2 \in \widehat{A^+}\}.$$

Limitedness. Assume that we want to test for limitedness of a B-automaton $\mathcal{A}$. It is easy to reduce the general case to the case when the underlying finite automaton accepts all finite words (consider the disjoint union of $\mathcal{A}$ with the finite automaton accepting precisely the words rejected by $\mathcal{A}$). Then, limitedness is simply the universality problem, as an easy compactness argument shows:

Fact 2. *A B-automaton $\mathcal{A}$ which accepts all finite words is limited iff* $L(\mathcal{A}) = \widehat{A^+}$.

Closure properties. As usual with nondeterministic automata, both classes – of B- and S-regular languages – are closed under language projection, and also under union and intersection. They are not, however, closed under complements:

the complement of the B-regular language $L(\mathcal{A})$ from the previous example is not B-regular, since it is not an open set. However, this complement is an S-regular language, as it is equal to $L(\mathcal{B})$. More generally, we will see the deep result that complements of B-regular languages are S-regular, and vice versa.

The Logic MSO+inf. We introduce the logic MSO+inf over profinite words. First, we define its base fragment, the logic MSO. A formula of this logic describes a set of profinite words. Usually, one sees finite words as models whose elements are positions of the word, and so a formula of MSO speaks about sets of positions of the word. However, in profinite words, "positions" are not well-defined. To define the logic MSO over profinite words, we view the constructs of MSO as operations on languages of profinite words. We describe how to interpret the second-order existential quantifier $\exists$; for the other constructs, the idea is even simpler. We view the quantifier $\exists$ as language projection. What language do we project? A formula $\varphi(X)$ beneath a quantifier $\exists$ defines a language L_φ over the extended alphabet $A \times \{0, 1\}$. For example, $\varphi(X) = a(X) \wedge \text{singleton}(X)$ defines the language L_φ of those profinite words over $A \times \{0, 1\}$, which contain precisely one symbol $(a, 1)$ and no other symbols with a 1 on the second coordinate. We define the language of the formula $\exists X.\varphi(X)$ as the projection of the language L_φ, forgetting about the second coordinate. Therefore, $\exists X.a(X) \wedge \text{singleton}(X)$ describes the set of profinite words with at least one letter a.

With similar ideas, it is easy to interpret all the usual constructs of MSO as language operations: the Boolean connectives $\wedge, \vee, \neg$, the binary predicates $<, \in$ and the unary predicates $a(X)$, per each letter $a \in A$. This way, we define the semantic of the MSO logic over profinite words. This logic describes precisely the class of clopen sets. To go beyond that, we add a predicate $\inf(X)$ which holds in a profinite word over $A \times \{0, 1\}$ if it has infinitely many 1's on the second coordinate. This is a closed, but not open property of profinite words over the alphabet $A \times \{0, 1\}$, so it is not definable in MSO. We denote the logic MSO extended by the quantifier inf by MSO+inf and distinguish the syntactic fragment MSO+inf$^+$ (resp., MSO+inf$^-$) where the predicate inf appears only under an even (resp. odd) number of negations.

Example 4. Consider the S-regular language $L(\mathcal{B})$ from Example 3: "there is an infinite block of a's". It can be described by the following formula of MSO+inf$^+$:

$$\exists X. \inf(X) \ \wedge \ \forall x, y, z. (x \in X \ \wedge \ z \in X \ \wedge \ (x < y < z) \implies (y \in X \wedge a(y))).$$

This example can be easily extended, yielding the following.

Proposition 3. *B-regular languages are definable in* MSO+inf$^-$ *and S-regular languages are definable in* MSO+inf$^+$. *The translations are effective.*

B- and S-regular Expressions. For describing concisely languages of profinite words we consider the usual regular expressions, extended by two new operations: *finite* iteration, denoted $L^{<\infty}$, and *infinite* iteration, denoted L^∞, apart from the usual, *unrestricted* iteration L^*. Formally, we define *profinite sequences* of profinite words as profinite words over the alphabet A extended by an additional *separator* symbol †. A profinite word $x \in \widehat{A^+}$ is an *element* of a profinite sequence $\hat{x}$

if $\dagger x \dagger$ is an infix of $\dagger \hat{x} \dagger$. The *concatenation* of $\hat{x}$ is obtained by removing the symbols $\dagger$. We define L^{∞} (resp. $L^{<\infty}$ and L^{*}) as the set of concatenations of profinite sequences containing infinitely (resp. finitely, arbitrarily) many separators, and whose elements belong to L. *B-regular expressions* can only use the exponents $<\infty$ and $*$; *S-regular expressions* can only use the exponents ∞ and $*$.

Example 5. The B-regular expression $(a^{<\infty} b)^{*} a^{<\infty}$ describes precisely the language accepted by the B-automaton $\mathcal{A}$ from Example 3 – "every block of a's has a finite length". The S-regular expression $(a + b)^{*} a^{\infty} (a + b)^{*}$ describes precisely the complement of $L(\mathcal{A})$, i.e. the language accepted by the S-automaton $\mathcal{B}$.

Mimicking the standard translation from regular expressions to automata we get:

Proposition 4. *A language defined by a B-/S-regular expression is B-/S-regular.*

Relationship with Theory of Colcombet. We recall the *domination relation* [5]. For $f, g \colon A^{+} \to \overline{\mathbb{N}}$ we write $f \preccurlyeq g$ if for every set $K \subseteq A^{+}$, $f(K)$ is bounded whenever $g(K)$ is bounded. A *cost function* is an equivalence class of $\preccurlyeq$. A *regular* cost function is any equivalence class containing some function $f_{\mathcal{A}}$ induced by a B-automaton $\mathcal{A}$. A function $f \colon A^{+} \to \overline{\mathbb{N}}$ defines a closed language L_{f} of profinite words:

$$L_{f} \stackrel{def}{=} \{ \lim_{n \to \infty} w_{n} : \quad w_{1}, w_{2}, \ldots \in A^{+} \text{ convergent, with } f(w_{n}) \text{ unbounded} \}.$$

Clearly, if $f \preccurlyeq g$ then $L_{f} \subseteq L_{g}$. The mapping $f \mapsto L_{f}$ induces a bijective correspondence Φ between certain cost functions – equivalence classes of uniformly continuous functions – and *closed* languages of profinite words. This correspondence relates the semantics of many objects considered in this paper and in [5,6]. For example, a B-automaton $\mathcal{A}$ defines both a B-regular language $L_{\mathcal{A}}$ and a regular cost function $f_{\mathcal{A}}$. It is easy to see that $f_{\mathcal{A}}$ and $\widehat{A^{+}} - L_{\mathcal{A}}$ are linked via Φ. Similarly, *cost MSO* corresponds via Φ to MSO+inf^{-}. However, not every language definable in the full logic MSO+inf corresponds to some cost function.

4 Recognizable Languages

In this section, we present the key technical tool – recognition by homomorphisms. First we discover the appropriate algebraic structure of the recognizers.

Syntactic Congruence. Just as multiplication is intimately related with regular languages, multiplication together with the ω-power over $\widehat{A^{+}}$ turn out to be of central importance for B- and S-regular languages. For notational reasons, we view $(\widehat{A^{+}}, \cdot, \omega)$ as an algebra over the signature $\langle \cdot, \# \rangle$, where the ω-power of $\widehat{A^{+}}$ plays the role of the operation $\#$ of the signature. Let $L \subseteq \widehat{A^{+}}$. Its $\langle \cdot, \# \rangle$-*syntactic congruence* $\simeq_{L}$ is the coarsest equivalence relation over $\widehat{A^{+}}$ which preserves multiplication, the ω-power, and membership in L.

Example 6. Let $L = (a^{<\infty} b)^* a^{<\infty}$ be the language of the B-automaton which computes the maximal length of a block of a's. It is easy to see that the equivalence classes of $\simeq_L$ (and hence also of $\simeq_K$, for $K = \widehat{A^+} - L$) are:

$$a^{<\infty}, \qquad (a^{<\infty} b)^+ a^{<\infty}, \qquad (a+b)^* a^{\infty} (a+b)^*.$$

Stabilization Semigroups. We consider languages $L \subseteq \widehat{A^+}$ whose $\langle \cdot , \# \rangle$-syntactic congruence has a finite index. Such a set yields a finite $\langle \cdot , \# \rangle$-*syntactic algebra*, i.e. the quotient $S_L = \widehat{A^+}/\simeq_L$. Since $\simeq_L$ is a congruence, the syntactic algebra is equipped with two operations – the usual multiplication, and *stabilization*, denoted $\#$, which stems from the ω-power in the profinite semigroup. The syntactic algebra also naturally inherits the quotient topology from $\widehat{A^+}$, which is usually non-Hausdorff, i.e. there might be singleton sets which are not closed. However, if L is a closed or open language, then the quotient topology is T_0, i.e. if $s \in \overline{\{t\}}$ and $t \in \overline{\{s\}}$ for $s, t \in S_L$, then $s = t$. Multiplication and stabilization in S_L are continuous with respect to the topology, and also satisfy several properties which are easily derived from the properties of multiplication and the ω-power over $\widehat{A^+}$. Namely, for $s, t \in S$ and idempotent $e \in S$,

$$\begin{aligned} s \cdot (t \cdot s)^{\#} &= (s \cdot t)^{\#} \cdot s & s^{\#} \cdot s^{\#} &= s^{\#} \\ (s^{\#})^{\#} &= s^{\#} & e \cdot e^{\#} &= e^{\#} \\ (s^n)^{\#} &= s^{\#} \quad \text{for } n = 1, 2, 3 \ldots & e^{\#} &\in \overline{\{e\}}. \end{aligned}$$

A *stabilization semigroup* is a T_0 topological space S equipped with two continuous operations $\cdot$ and $\#$ satisfying the above axioms, apart from associativity of $\cdot$. The above notion of stabilization semigroup corresponds precisely – via a simple correspondence – to the original notion defined by Colcombet [5].

Example 7. Let S_L denote the quotient set induced by the language L from Example 6. As noted there, S_L consists of three equivalence classes, which we denote by $[a]$, $[b]$ and $[a^{\omega}]$, respectively. Multiplication, stabilization and topology over S_L flow from the properties of the three equivalence classes: multiplication is commutative and each element is idempotent, $[a^{\omega}]$ is the zero element and $[a]$ is the neutral element; stabilization maps $[a]$ to $[a^{\omega}]$ and s to s otherwise; $[a^{\omega}]$ is contained in the closure of $[a]$ and in the closure of $[b]$.

Recognizability. We consider an analogue of the notion of recognizability by semigroups in the classical theory. Recall that a subset $L \subseteq \widehat{A^+}$ is *recognizable* if there is a mapping $\alpha \colon A \to S$ to a finite discrete semigroup such that for the induced homomorphism $\hat{\alpha} \colon \widehat{A^+} \to S$ we have $L = \hat{\alpha}^{-1}(F)$ for some $F \subseteq S$.

Instead of semigroups, we deal with stabilization semigroups. A *homomorphism* $\hat{\alpha}$ from $\widehat{A^+}$ to a stabilization semigroup S is required to preserve multiplication and map the ω-power in $\widehat{A^+}$ to stabilization in S. We use a notion of invariance of $\hat{\alpha}$ under *infinite substitutions*, which intuitively means that if a profinite word x is factorized into a profinite sequence of factors, and each factor x_i is replaced by some other factor y_i with $\hat{\alpha}(x_i) = \hat{\alpha}(y_i)$, then, for the resulting concatenation y of the factors y_i, $\hat{\alpha}(x) = \hat{\alpha}(y)$. We say that such a homomorphism $\hat{\alpha} \colon \widehat{A^+} \to S$ is an ∞-*homomorphism*. The following result plays a pivotal role in the theory, and its proof is difficult comparing to the classical case. It can be deduced – although not immediately – from a theorem of Colcombet.

Theorem 5 (cf. Theorem 2 in [5]). *Let* $\alpha \colon A \to S$ *be any mapping from a finite alphabet* A *to a finite stabilization semigroup* S. *Then there exists a unique* ∞-*homomorphism* $\hat{\alpha} \colon \widehat{A^+} \to S$ *extending* α. *The mapping* $\hat{\alpha}$ *is continuous. Its image is the subset of* S *generated from* $\alpha(A)$ *by the operations* $\langle \cdot , \# \rangle$.

Note that the extension $\hat{\alpha}$ is not necessarily the *unique* continuous homomorphic extension of α. We call $\hat{\alpha}$ the ∞-homomorphism *induced* by α. We say that a language $L \subseteq \widehat{A^+}$ is *recognized* by $\hat{\alpha} \colon \widehat{A^+} \to S$ if $L = \hat{\alpha}^{-1}(F)$ for some $F \subseteq S$; if additionally F is closed (resp. open) in S, we say that L is $\downarrow$-*recognizable* (resp. $\uparrow$-*recognizable*). Note that a recognizable set is described in a finite manner by $\alpha \colon A \to S$ and $F \subseteq S$. It is crucial that the image of $\hat{\alpha}$ can be computed from α.

Example 8. Let S be the stabilization semigroup $\widehat{A^+}/\simeq_L$ from the previous example, whose elements are $[a], [b], [a^\omega]$. Let $\alpha \colon A \to S$ map a to $[a]$ and b to $[b]$. We will check that the quotient mapping $\alpha_L \colon \widehat{A^+} \to S$ is the ∞-homomorphism induced by α. We argue that α_L is invariant under infinite substitutions. Consider a profinite word x, and choose some factorization of x. Replace each factor by some other factor, with the same image under α_L. Schematically:

$$
\begin{array}{ccccccc}
x = & aaa & aaba & aaa & \cdots & ab^\omega a & baaab \\
& \downarrow & \downarrow & \downarrow & \cdots & \downarrow & \downarrow \\
y = & aaaaa & (ab)^\omega & aaaaaaa & \cdots & aaaaabaaaa & aaaaabaaa
\end{array}
$$

Intuitively, it is clear that if the original word x contains no infinite block of a's, then no such block can appear in the resulting word y either. Hence, $\alpha_L(y) = \alpha_L(x)$.

The proof of Theorem 5 extends the idea of Simon's factorization trees to profinite words and stabilization semigroups, which we shortly describe. Start with any profinite word x. We want to determine the *type* of x, i.e. $\hat{\alpha}(x)$. If x is a single letter a, then its type is $\alpha(a)$. If not, we try to factorize x into a profinite sequence of factors, for which the type can be determined. We use three rules:

- If $x = x_1 \cdot x_2$, and $\hat{\alpha}(x_1) = s_1$, $\hat{\alpha}(x_2) = s_2$, then $\hat{\alpha}(x) = s_1 \cdot s_2$,
- If x factorizes into finitely many factors, each of idempotent type e, then $\hat{\alpha}(x) = e$,
- If x factorizes into infinitely many factors, each of idempotent type e, then $\hat{\alpha}(x) = e^\#$.

We prove by induction on $|S|$ that in a finite number of steps, depending only on $|S|$, using the above three rules, any profinite word x can be iteratively split into single letters. Moreover, we prove that the resulting type does not depend on the chosen "factorization tree". The proof of existence of factorization trees is similar to the proof of Simon's theorem, and proceeds by induction on the size of S. The proof of uniqueness requires the use of the axioms of stabilization semigroups. It is similar to a proof of an analogous statement in [6]. An important difference is that there only finite words have factorization trees, and their output is unique only in an asymptotic way.

The standard Cartesian-product construction yields several closure properties for recognizable languages. For closure under projection, we use two enhanced variants of the powerset construction, similar to constructions from [6].

Proposition 6. *Recognizable languages are closed under Boolean combinations. $\downarrow$-recognizable (resp. $\uparrow$-recognizable) languages are closed under unions and intersections. Complements of $\downarrow$-recognizable languages are $\uparrow$-recognizable and vice versa. $\downarrow$-recognizable and $\uparrow$-recognizable languages are closed under projections.*

Applying inductively the above to formulas of MSO+inf$^-$ or MSO+inf$^+$ we get:

Corollary 1. *Languages definable in MSO+inf$^-$ are $\downarrow$-recognizable, and languages definable in MSO+inf$^+$ are $\uparrow$-recognizable. The translations are effective.*

5   The Main Results

The main theorem collects the notions and results listed above, proving the equivalence of several characterizations. Except for the last item – the finite-index characterization – it can be derived relatively easily from [5,6].

Theorem 7. *Let $L \subseteq \widehat{A^+}$ and $K = \widehat{A^+} - L$ be its complement. The following conditions 1-9 are equivalent:*

1. *L is defined by a B-regular expression,* 5. *K is defined by an S-regular expression,*
2. *$L = L(\mathcal{A})$ for some B-automaton $\mathcal{A}$,* 6. *$K = L(\mathcal{B})$ for some S-automaton $\mathcal{B}$,*
3. *L is definable in MSO+inf$^-$,* 7. *K is definable in MSO+inf$^+$,*
4. *L is $\uparrow$-recognizable,* 8. *K is $\downarrow$-recognizable,*
   9. *The $\langle \cdot, \# \rangle$-syntactic congruence of K has finite index and $K = \overline{K \cap A^{\langle \cdot, \omega \rangle}}$.*

Above, $A^{\langle \cdot, \omega \rangle}$ denots the set of ω-terms – profinite words which can be generated from A by applying multiplication and the ω-power. They are analogues of ultimately periodic words in the theory of ω-regular languages. It follows that a B- or S-regular language is determined by its ω-terms, similarly as an ω-regular language is determined by its ultimately periodic words.

By the last part of Theorem 5, the image of an ∞-homomorphism to a finite stabilization semigroup can be computed by a fixed point calculation. Hence, emptiness of recognizable languages is decidable. This proves the following.

Theorem 8. *Emptiness of Boolean combinations of B-regular languages is decidable. In particular, the limitedness problem is decidable for B-automata.*

The above result extends the decidability results of Hashiguchi and Kirsten. As emptiness of Boolean combinations reduces to inclusion testing, it is equivalent to the main result of [5] – that the domination relation is decidable for B-automata.

6   From Infinite Words to Profinite Words

We describe a connection between ω-words (i.e. mappings from $\mathbb{N}$ to A) and profinite words. Recall that any ω-regular language can be presented as a finite union of languages of the form $U \cdot V^\omega$, where $U, V \subseteq A^+$ are regular languages of finite words. We generalize this observation, and provide a meta-reduction between the satisfiability problems for logics over ω-words to corresponding logics over profinite words. The proof resembles Büchi's original proof of decidability of MSO. Instead of the usual Ramsey lemma, we use the following observation.

> For any ω-word $w \in A^\omega$ there is a factorization $w = u_0 \cdot u_1 \cdot u_2 \cdots$ such that the sequence $u_0, u_1, u_2, \ldots \in A^+$ is convergent to some $u_\infty \in \widehat{A^+}$.

Its proof is an easy, repeated application of the usual Ramsey lemma.

Let $V \subseteq \widehat{A^+}$ be a language of profinite words, and $\varepsilon > 0$ a real number. Consider the following language of infinite words $V_\varepsilon^\omega \subseteq A^\omega$:

$$V_\varepsilon^\omega \overset{def}{=} \{v_1 \cdot v_2 \cdot v_3 \cdots : \forall_n v_n \in A^+, \exists v_\infty \in V^* : \lim_{n \to \infty} v_n = v_\infty, \forall_n d(v_n, v_\infty) < \varepsilon\}.$$

For a regular language $U \subseteq A^+$ of finite words, we say that the expression $U \cdot V^\omega$ is *well-formed* if the language $U \cdot V_\varepsilon^\omega$ does not depend on the choice of $\varepsilon < 1/n$, where n is the size of the minimal automaton recognizing U. In this case, we define the language $U \cdot V^\omega$ as $U \cdot V_\varepsilon^\omega$. For example, the expression $(a+b)^* \cdot (a^{<\infty}b)^\omega$ is well-formed and describes the language L_B from the introduction. For a class $\mathcal{L}$ of languages of profinite words, we define a class of languages of infinite words:

$$\omega\mathcal{L} \overset{def}{=} \left\{ \begin{array}{l} \textit{finite unions of languages defined by well-formed} \\ \textit{expressions } U \cdot V^\omega \textit{ with } U \subseteq A^+ \textit{ regular and } V \in \mathcal{L} \end{array} \right\}$$

In the following theorem, by REGULAR, B-REGULAR, S-REGULAR, MSO+inf we denote the corresponding classes of languages of profinite words. To each we apply the map $\mathcal{L} \mapsto \omega\mathcal{L}$ described above, yielding classes of languages of infinite words. The proof is very general, and also applies to other classes of languages.

Theorem 9. *Every ω-regular language is in ωREGULAR. Every ωB-regular language is in ωB-REGULAR. Every ωS-regular language is in ωS-REGULAR. Every MSO+$\mathbb{B}$ definable language is in ωMSO+inf. The translations are effective.*

The reduction described above allows to transfer results from profinite words to ω-words. For instance, the main results of [4], concerning ωB- and ωS-regular languages, follow from the results in our paper. More importantly, we get:

Corollary 2. *The satisfiability problem for the logic MSO+$\mathbb{B}$ over ω-words reduces to the satisfiability problem for the logic MSO+inf over profinite words.*

We mention that by refining our Theorem 9, Skrzypczak [13] proved that a language of infinite words which is both ωB-regular and ωS-regular must in fact be ω-regular – reflecting the immediate, analogous fact for profinite words.

Conclusion.   We   presented   a   new   proof   and   framework for the limitedness problem. We rise the question of decidability of the logic MSO+inf over profinite words.

Acknowledgements. This paper is based on my PhD thesis, supervised by Mikołaj Bojańczyk. I am very grateful to Thomas Colcombet for many useful comments.

Due to space limitations, most details are deferred to the appendix [14].

References

1. Abdulla, P.A., Krcal, P., Yi, W.: R-Automata. In: van Breugel, F., Chechik, M. (eds.) CONCUR 2008. LNCS, vol. 5201, pp. 67–81. Springer, Heidelberg (2008)
2. Bojańczyk, M.: A Bounding Quantifier. In: Marcinkowski, J., Tarlecki, A. (eds.) CSL 2004. LNCS, vol. 3210, pp. 41–55. Springer, Heidelberg (2004)
3. Bojańczyk, M.: Beyond ω-regular languages. In: STACS, pp. 11–16 (2010)
4. Bojańczyk, M., Colcombet, T.: Bounds in ω-regularity. In: Logic in Computer Science, pp. 285–296 (2006)
5. Colcombet, T.: The Theory of Stabilisation Monoids and Regular Cost Functions. In: Albers, S., Marchetti-Spaccamela, A., Matias, Y., Nikoletseas, S., Thomas, W. (eds.) ICALP 2009, Part II. LNCS, vol. 5556, pp. 139–150. Springer, Heidelberg (2009)
6. Colcombet, T.: Regular cost functions, Part I: Logic and algebra over words (2011) (submitted)
7. Hashiguchi, K.: Limitedness theorem on finite automata with distance functions. Journal of Computer and System Sciences 24, 233–244 (1982)
8. Kirsten, D.: Distance desert automata and the star height problem. Theoretical Informatics and Applications 39(3), 455–511 (2005)
9. Krob, D.: The Equality Problem for Rational Series with Multiplicities in the Tropical Semiring is Undecidable. In: Kuich, W. (ed.) ICALP 1992. LNCS, vol. 623, pp. 101–112. Springer, Heidelberg (1992)
10. Leung, H.: On the topological structure of a finitely generated semigroup of matrices. Semigroup Forum 37, 273–278 (1988)
11. Pin, J.-É.: Profinite methods in automata theory. In: Albers, S., Marion, J.-Y. (eds.) STACS. LIPIcs, vol. 3, pp. 31–50. Schloss Dagstuhl - Leibniz-Zentrum fuer Informatik, Germany (2009)
12. Simon, I.: On semigroups of matrices over the tropical semiring. ITA 28(3-4), 277–294 (1994)
13. Skrzypczak, M.: Separation property for ωB- and ωS-regular languages (submitted, 2012)
14. Toruńczyk, S.: Languages of profinite words and the limitedness problem (2012), http://www.mimuw.edu.pl/~szymtor/papers/BS-icalp.pdf

The Complexity
of Mean-Payoff Automaton Expression*

Yaron Velner

The Blavatnik School of Computer Science, Tel Aviv University, Israel

Abstract. Quantitative languages are extension of Boolean languages that assign to each word a real number. With quantitative languages, systems and specifications can be formalized more accurately. For example, a system may use a varying amount of some resource (e.g., memory consumption, or power consumption) depending on its behavior, and a specification may assign a maximal amount of available resource to each behavior, or fix the long-run average available use of the resource.

Mean-payoff automata are finite automata with numerical weights on transitions that assign to each infinite path the long-run average of the transition weights. Mean-payoff automata forms a class of quantitative languages that is not robust, since it is not closed under the basic algebraic operations: min, max, sum and numerical complement. The class of *mean-payoff automaton expressions*, recently introduced by Chatterjee et al., is currently the only known class of quantitative languages that is robust, expressive and decidable. This class is defined as the closure of mean-payoff automata under the basic algebraic operations. In this work, we prove that all the classical decision problems for mean-payoff expressions are PSPACE-complete. Our proof improves the previously known 4EXPTIME upper bound. In addition, our proof is significantly simpler, and fully accessible to the automata-theoretic community.

1 Introduction

In algorithmic verification of reactive systems, the system is modeled as a finite-state transition system, and requirements are captured as languages of infinite words over system observations [15,17]. The classical verification framework only captures *qualitative* aspects of system behavior, and in order to describe *quantitative* aspects, for example, consumption of resources such as CPU and energy, the framework of *quantitative languages* was proposed [6].

Quantitative languages are a natural generalization of Boolean languages that assign to every word a real number instead of a Boolean value. With such languages, quantitative specifications can be formalized. In this model, an implementation L_A satisfies (or refines) a specification L_B if $L_A(w) \leq L_B(w)$ for all words w.

This notion of refinement is a *quantitative generalization of language inclusion*, and it can be used to check for example if for each behavior, the long-run

---

* Fuller version with detailed proofs is available at [18].

A. Czumaj et al. (Eds.): ICALP 2012, Part II, LNCS 7392, pp. 390–402, 2012.
© Springer-Verlag Berlin Heidelberg 2012

average response time of the system lies below the specified average response requirement. The other classical decision problems such as emptiness, universality, and language equivalence have also a natural quantitative extension. For example, the *quantitative emptiness problem* asks, given a quantitative language L and a rational threshold ν, whether there exists some word w such that $L(w) \geq \nu$, and the *quantitative universality problem* asks whether $L(w) \geq \nu$ for all words w. We also consider the notion of *distance* between two quantitative languages L_A and L_B, defined as $\sup_{w \in \Sigma^\omega} |L_A(w) - L_B(w)|$.

The model of mean-payoff automaton is a popular approach to express quantitative properties; in this model, a *payoff* (or a weight) is associated with each transition of the automaton, the mean-payoff of a finite run is simply the average of the payoff of the transitions in the run, and the mean-payoff of an infinite run is the limit, as the length of the run tends to infinity.

In this work, we study the computational complexity of the classical decision problems for the class of quantitative languages that are defined by *mean-payoff expression*. An expression is either a deterministic[1] mean-payoff automaton, or it is the max, min or sum of two mean-payoff expressions. This class, introduced in [4], is robust as it is closed under the max, min, sum and the numerical complement operators [4].

The decidability of the classical decision problems, as well as the computability of the distance problem, was first established in [4]; in this paper we describe alternative proofs for these results. Our proofs offer the following advantages: First, the proofs yield PSPACE complexity upper bounds, which match corresponding PSPACE lower bounds; in comparison to 4EXPTIME upper bounds achieved in [4]. Second, our proofs reside only in the frameworks of graph theory and basic linear-programing, which are common practices among the automata-theoretic community, whereas a substantial part of the proofs in [4] resides in the framework of computational geometry.

Our proofs are based on a reduction from the emptiness problem to the feasibility problem for a set of linear inequalities; for this purpose, inspired by the proofs in [19], we establish a connection between the emptiness problem and the problem of finding a multi-set of cycles, with certain properties, in a directed graph. The reduction also reveals how to compute the maximum value of an expression, and therefore the decidability of all mentioned problems is followed almost immediately.

This paper is organized as follows: In the next section we formally define the class of mean-payoff expressions; in Section 3 we describe a PSPACE algorithm that computes the maximum value of an expression that does not contain the max operator; in Section 4 we show PSPACE algorithm for all the classical problems and prove their corresponding PSPACE lower bounds. Due to lack of space, in some cases the proofs are omitted, and in other cases only sketches of the proofs are presented. The full proofs are available at [18].

---

[1] We note that the restriction to deterministic automata is inherent; for nondeterministic automata all the decision problems are undecidable [4].

2   Mean-Payoff Automaton Expression

In this section we present the definitions of mean-payoff expressions from [4].

Quantitative Languages. A *quantitative language* L over a finite alphabet Σ is a function $L : \Sigma^\omega \to \mathbb{R}$. Given two quantitative languages L_1 and L_2 over Σ, we denote by $\max(L_1, L_2)$ (resp., $\min(L_1, L_2)$, $\text{sum}(L_1, L_2)$ and $-L_1$) the quantitative language that assigns $\max(L_1(w), L_2(w))$ (resp., $\min(L_1(w), L_2(w))$, $L_1(w) + L_2(w)$, and $-L_1(w)$) to each word $w \in \Sigma^\omega$. The quantitative language $-L$ is called the *complement* of L.

Cut-point Languages. Let L be a quantitative language over Σ. Given a threshold $\nu \in \mathbb{R}$, the *cut-point* language defined by (L, ν) is the language $L^{\geq \nu} = \{w \in \Sigma^\omega | L(w) \geq \nu\}$.

Weighted Automata. A *(deterministic) weighted automaton* is a tuple $A = \langle Q, q_I, \Sigma, \delta, wt \rangle$, where (i) Q is a finite set of states, $q_I \in Q$ is the initial state, and Σ is a finite alphabet; (ii) $\delta \subseteq Q \times \Sigma \times Q$ is a set of transitions such that for every $q \in Q$ and $\sigma \in \Sigma$ the size of the set $\{q' \in Q | (q, \sigma, q') \in \delta\}$ is exactly 1; and (iii) $wt : \delta \to \mathbb{Q}$ is a *weight function*, where $\mathbb{Q}$ is the set of rationals.

The Product of Weighted Automata. The *product* of the weighted automata $A_1, \ldots, A_n$ such that $A_i = \langle Q_i, q_I^i, \Sigma, \delta_i, wt_i \rangle$ is the multidimensional weighted automaton $\mathcal{A} = A_1 \times \cdots \times A_n = \langle Q_1 \times \cdots \times Q_n, (q_I^1, \ldots, q_I^n), \Sigma, \delta, wt \rangle$ such that $t = ((q_1, \ldots, q_n), \sigma, (q_1', \ldots, q_n')) \in \delta$ if $t_i = (q_i, \sigma, q_i') \in \delta_i$ for all $i \in \{1, \ldots, n\}$, and $wt(t) = (wt_1(t_1), \ldots, wt_n(t_n)) \in \mathbb{Q}^n$. We denote by $\mathcal{A}_i$ the projection of the automaton $\mathcal{A}$ to dimension i.

Words and Runs. A *word* $w \in \Sigma^\omega$ is an infinite sequence of letters from Σ. A *run* of a weighted automaton A over an infinite word $w = \sigma_1 \sigma_2 \ldots$ is the (unique) infinite sequence $r = q_0 \sigma_1 q_1 \sigma_2 \ldots$ of states and letters such that $q_0 = q_I$, and $(q_i, \sigma_{i+1}, q_{i+1}) \in \delta$ for all $i \geq 0$. We denote by $wt(w) = wt(r) = v_0 v_1 \ldots$ the sequence of weights that occur in r where $v_i = wt(q_i, \sigma_{i+1}, q_{i+1})$ for all $i \geq 0$.

Quantitative Language of Mean-Payoff Automata. The *mean-payoff value* (or limit average) of a sequence $\bar{v} = v_0 v_1 \ldots$ of real numbers is either $LimInfAvg(\bar{v}) = \liminf_{n \to \infty} \frac{1}{n} \cdot \sum_{i=0}^{n-1} v_i$; or $LimSupAvg(\bar{v}) = \limsup_{n \to \infty} \frac{1}{n} \cdot \sum_{i=0}^{n-1} v_i$. The quantitative language $\underline{A}$ of a weighted automaton A is defined by $\underline{A}(w) = LimInfAvg(wt(w))$; analogously the quantitative language $\overline{A}$ is defined by $\overline{A}(w) = LimSupAvg(wt(w))$. In the sequel we also refer to the quantitative language $\underline{A}$ as the *LimInfAvg automaton* A, and analogously the *LimSupAvg automaton* A is the quantitative language $\overline{A}$.

Mean-Payoff Automaton Expressions. A *mean-payoff automaton expression* E is obtained by the following grammar rule:

$$E ::= \underline{A} | \overline{A} | \max(E, E) | \min(E, E) | \text{sum}(E, E)$$

where A is a *deterministic* (one-dimensional) weighted automaton. The quantitative language L_E of a mean-payoff automaton expression E is $L_E = \underline{A}$ (resp., $L_E = \overline{A}$) if $E = \underline{A}$ (resp., if $E = \overline{A}$), and $L_E = \text{op}(L_{E_1}, L_{E_2})$ if $E = \text{op}(E_1, E_2)$

for op $\in$ {max, min, sum}. We shall, by convenient abuse of notation, interchangeably use E to denote both the expression and the quantitative language of the expression (that is, E will also denote L_E). An expression E is called an *atomic expression* if $E = \underline{A}$ or $E = \overline{A}$, where A is a weighted automaton.

It was established in [4] (and it follows almost immediately by the construction of the class) that the class of mean-payoff automaton expressions is closed under max, min, sum and numerical complement.

Decision Problems and Distance. We consider the following classical decision problems for a quantitative language defined by a mean-payoff expression. Given a quantitative language L and a threshold $\nu \in \mathbb{Q}$, the *quantitative emptiness* problem asks whether there exists a word $w \in \Sigma^\omega$ such that $L(w) \geq \nu$, and the *quantitative universality problem* asks whether $L(w) \geq \nu$ for all words $w \in \Sigma^\omega$.

Given two quantitative languages L_1 and L_2, the *quantitative language-inclusion problem* asks whether $L_1(w) \leq L_2(w)$ for all words $w \in \Sigma^\omega$, and the *quantitative language-equivalence problem* asks whether $L_1(w) = L_2(w)$ for all words $w \in \Sigma^\omega$. Finally, the *distance* between L_1 and L_2 is $D_{\sup}(L_1, L_2) = \sup_{w \in \Sigma^\omega} |L_1(w) - L_2(w)|$; and the corresponding computation problem is to compute the value of the distance.

Maximum Value of Expression. Given an expression E, its *supremum value* is the real number $\sup_{w \in \Sigma^\omega} E(w)$. While it is obvious that such supremum exists, it was proved in [4] that a maximum value also exists (that is, there exists $w' \in \Sigma^\omega$ s.t $E(w') = \sup_{w \in \Sigma^\omega} E(w)$). Hence the *maximum value* of the expression E is $\sup_{w \in \Sigma^\omega} E(w)$ or equivalently $\max_{w \in \Sigma^\omega} E(w)$.

Encoding of Expressions and Numbers. An expression E is encoded by the tuple $(\langle E \rangle, \langle A_1 \rangle, \ldots, \langle A_k \rangle)$, where $\langle E \rangle$ is the expression string and $A_1, \ldots, A_k$ are the weighted automata that occur in the expression, w.l.o.g we assume that each automaton occur only once. A rational number is encoded as a pair of integers, where every integer is encoded in binary.

3   PSPACE Algorithm for Computing the Maximum Value of max-free Expressions

In this section we consider only max-free expressions, which are expressions that contain only the min and sum operators. We will present a PSPACE algorithm that computes the maximum value of such expressions; computing the maximum value amounts to computing the maximum threshold for which the expression is nonempty; for this purpose we present four intermediate problems (and solutions), each problem is presented in a corresponding subsection below. The first problem asks whether an intersection of cut-point languages of *LimSupAvg* automata is empty; the second problem asks the same question for *LimInfAvg* automata; the third problem asks if an arbitrary intersection of cut-point languages of *LimSupAvg* and *LimInfAvg* automata is empty; and the last problem asks whether a max-free expression is empty. We will first present a naive solution for these problems; the solution basically lists all the simple cycles in the product automaton of the automata that occur in the expression; it then constructs linear constraints, with coefficients that depend on the weight vectors of the simple cycles, which their feasibility corresponds to the non-emptiness of the expression.

In the fifth subsection we will analyze the solution for the max-free emptiness problem; we will show a PSPACE algorithm that solves the problem; and we will bound the number of bits that are needed to encode the maximum threshold for which the expression is nonempty (recall that such maximal threshold is the maximum value of the expression); this will yield a PSPACE algorithm for computing the maximum value of a max-free expression.

In subsections 3.1-3.4 we shall assume w.l.o.g that the product automaton of all the automata that occur in the expression is a strongly connected graph; this can be done since in these subsections we do not refer to the complexity of the presented procedures.

3.1   The Emptiness Problem for Intersection of *LimSupAvg* Automata

In this subsection we consider the problem where k weighted automata $A_1, \ldots, A_k$ and a rational threshold vector $\bar{r} = (r_1, \ldots, r_k)$ are given, and we need to decide whether there exists an infinite word $w \in \Sigma^\omega$ such that $\overline{A_i}(w) \geq r_i$ for all $i \in \{1, \ldots, k\}$; equivalently, whether the intersection $\bigcap_{i=1}^k \overline{A_i}^{\geq r_i}$ is nonempty.

Informally, we prove that there is such w iff for every $i \in \{1, \ldots, k\}$, there is a word w_i such that $\overline{A_i}(w_i) \geq r_i$.

Formally, let $\mathcal{A} = A_1 \times \cdots \times A_k$ be the product automaton of the automata $A_1, \ldots, A_k$. Recall that an infinite word corresponds to an infinite path in $\mathcal{A}$, and that w.l.o.g we assume that the graph of $\mathcal{A}$ is strongly connected. Let $C_1, C_2, \ldots, C_n$ be the simple cycles that occur in $\mathcal{A}$. The next lemma claims that it is enough to find one cycle with average weight r_i for every dimension i.

Lemma 1. *There exists an infinite path π in $\mathcal{A}$ such that $\overline{A_i}(\pi) \geq r_i$, for all $i \in \{1, \ldots, k\}$, iff for every $i \in \{1, \ldots, k\}$ there exists a simple cycle C_i in $\mathcal{A}$, with average weight at least r_i*

Proof. The direction from left to right is easy: Since for every $i \in \{1, \ldots, k\}$ there exists a path π_i (namely π) such that $\overline{A_i}(\pi_i) \geq r_i$ it follows that there exists a simple cycle in $\mathcal{A}$ with average at least r_i in dimension i. (This fact is well-known for one-dimensional weighted automata, e.g., see [20], and hence it is true for the projection of $\mathcal{A}$ to the i-th dimension.)

For the converse direction, we assume that for every $i \in \{1, \ldots, k\}$ there exists a simple cycle C_i in $\mathcal{A}$ with average weight at least r_i. Informally, we form the path π by following the edges of the cycle C_i until the average weight in dimension i is sufficiently close to r_i, and then we do likewise for dimension $1 + (i \pmod k)$, and so on. (Full proof is given in [18].)   □

Lemma 1 shows that the emptiness problem for intersection of *LimSupAvg* automata can be naively solved by an exponential time algorithm that constructs the product automaton and checks if the desired cycles exist.

3.2   The Emptiness Problem for Intersection of *LimInfAvg* Automata

In this subsection we consider the problem where k weighted automata $A_1, \ldots, A_k$ and a rational threshold vector $\bar{r} = (r_1, \ldots, r_k)$ are given, and we need to decide whether there exists an infinite word $w \in \Sigma^\omega$ such that $A_i(w) \geq r_i$ for all $i \in \{1, \ldots, k\}$; or equivalently, whether the intersection $\bigcap_{i=1}^k A_i^{\geq r_i}$ is nonempty.

For *LimInfAvg* automata, the componentwise technique we presented in the previous subsection will not work; to solve the emptiness problem for the intersection of such automata we need the notion of $\bar{r}$ multi-cycles.

$\bar{r}$ **multi-cycles.** Let G be a directed graph equipped with a multidimensional weight function $wt : E \to \mathbb{Q}^k$, and let $\bar{r}$ be a vector of rationals. A *multi-cycle* is a multi-set of simple cycles; the length of a multi-cycle $\mathcal{C} = \{C_1, \ldots, C_n\}$, denoted by $|\mathcal{C}|$, is $\sum_{i=1}^n |C_i|$. A multi-cycle $\mathcal{C} = \{C_1, \ldots, C_n\}$ is said to be an $\bar{r}$ *multi-cycle* if $\frac{1}{|\mathcal{C}|} \sum_{j=1}^n wt(C_j) \geq \bar{r}$, that is, if the average weight of the multi-cycle, in every dimension i, is at least r_i.

In the sequel we will establish a connection between the problem of finding an $\bar{r}$ multi-cycle and the emptiness problem for intersection of *LimInfAvg* automata.

A polynomial time algorithm that decides if an $\bar{r}$ multi-cycle exists is known [12]; in this work however, it is sufficient to present the naive way for finding such multi-cycles; for this purpose we construct the following set of linear constraints: Let $\mathbb{C}$ denote the set of all simple cycles in $\mathcal{A}$; for every $c \in \mathbb{C}$ we define a variable X_c; we define the $\bar{r}$ *multi-cycle constraints* to be:

$$\sum_{c \in \mathbb{C}} X_c wt(c) \geq \bar{r} \; ; \; \sum_{c \in \mathbb{C}} |c| X_c = 1 \; ; \text{ and for every } c \in \mathbb{C}: X_c \geq 0$$

In the next lemma we establish the connection between the feasibility of the $\bar{r}$ multi-cycle constraints and the existence of an $\bar{r}$ multi-cycle.

Lemma 2. *The automaton $\mathcal{A}$ has an $\bar{r}$ multi-cycle iff the corresponding $\bar{r}$ multi-cycle constraints are feasible.*

Proof. The direction from left to right is immediate, indeed if we define X_c as the number of occurrences of cycle c in the witness $\bar{r}$ multi-cycle divided by the length of that multi-cycle, then we get a solution for the set of constraints.

In order to prove the converse direction, it is enough to notice that if the constraints are feasible then they have a rational solution. Let $\overline{X}$ be such rational solution, and let N be the least common multiple of all the denominators of the elements of $\overline{X}$; by definition, the multi-set that contains NX_c copies of the cycle c is an $\bar{r}$ multi-cycle.   □

In the following lemma we establish a connection between the problem of finding an $\bar{r}$ multi-cycle and the emptiness problem for intersection of *LimInfAvg* automata.

Lemma 3. *There exists an infinite path π in $\mathcal{A}$ such that $\underline{A}_i(\pi) \geq r_i$, for all $i \in \{1, \ldots, k\}$, iff the graph of $\mathcal{A}$ contains an $\bar{r} = (r_1, \ldots, r_k)$ multi-cycle.*

Proof. To prove the direction from right to left, we show, in the following lemma, that if an $\bar{r}$ multi-cycle does not exist, then for every infinite path there is a dimension i for which $\underline{A}_i(\pi) < r_i$.

Lemma 4. *Let $G = (V, E)$ be a directed graph equipped with a weight function $wt : E \to \mathbb{Q}^k$, and let $\bar{r} \in \mathbb{Q}^k$ be a threshold vector. If G does not have an $\bar{r}$ multi-cycle, then there exist constants $\epsilon_G > 0$ and $m_G \in \mathbb{N}$ such that for every finite path π there is a dimension i for which $wt_i(\pi) \leq m_G + (r_i - \epsilon_G)|\pi|$.*

Lemma 4 implies that if an $\bar{r}$ multi-cycle does not exists, then for every infinite path π there exist a dimension i and an infinite sequence of indices $j_1 < j_2 < j_3 \ldots$ such that the average weight of the prefix of π, of length j_m, is at most $r_i - \frac{\epsilon_G}{2}$, for all $m \in \mathbb{N}$. Hence by definition $LimInfAvg_i(\pi) < r_i$.

In order to prove the converse direction, let us assume that G has an $\bar{r}$ multi-cycle $\mathcal{C} = C_1, \ldots, C_n$, such that the cycle C_i occurs m_i times in $\mathcal{C}$. We obtain the witness path π in the following way (we demonstrate the claim for $n = 2$): let π_{12} be a path from C_1 to C_2 and π_{21} be a path from C_2 to C_1 (recall that the graph is strongly connected), we define

$$\pi = C_1^{m_1} \pi_{12} C_2^{m_2} \pi_{21} (C_1^{m_1})^2 \pi_{12} (C_2^{m_2})^2 \pi_{21} \ldots (C_1^{m_1})^\ell \pi_{12} (C_2^{m_2})^\ell \pi_{21} \ldots$$

Informally, the long-run average weight of the path π is determined only by the cycles $C_1^{m_1}$ and $C_2^{m_2}$, since the effect of the paths π_{12} and π_{21} on the average weight of a prefix of π becomes negligible as the length of the prefix tends to infinity. Thus $\underline{A}_i(\pi) \geq r_i$ for every dimension i, which concludes the proof of Lemma 3 □

Lemma 3 and Lemma 2 immediately give us the following naive algorithm for the emptiness problem for intersection of $LimInfAvg$ automata: First, construct the product automaton; second, list all the simple cycles in the product automaton; third, construct the $\bar{r}$ multi-cycle constraints and check for their feasibility.

When the automata and the threshold vector are clear from the context, we shall refer to the $\bar{r}$ multi-cycle constraints, which are constructed from the intersection of the given $LimInfAvg$ automata and the threshold $\bar{r}$, as the *lim-inf constraints*.

3.3 The Emptiness Problem for Intersection of *LimSupAvg* and *LimInfAvg* Automata

In this subsection we consider the problem where $2k$ weighted automata $A_1, \ldots, A_k, B_1, \ldots, B_k$ and two k-dimensional rational threshold vectors $\overline{r^a}$ and $\overline{r^b}$ are given, and we need to decide whether there exists an infinite word $w \in \Sigma^\omega$ such that $\underline{A}_i(w) \geq r_i^a$ and $\overline{B}_i(w) \geq r_i^b$ for all $i \in \{1, \ldots, k\}$; or equivalently, whether the intersection $(\bigcap_{i=1}^k \underline{A}_i^{\geq r_i^a}) \cap (\bigcap_{i=1}^k \overline{B}_i^{\geq r_i^b})$ is nonempty.

Our solution will be a result of the following two lemmata. The first lemma claims that there is a word that satisfies all the conditions iff there are words $w_1, \ldots, w_k$ such that w_j satisfies all the lim-inf conditions and the lim-sup condition for the automaton B_j.

Lemma 5. *There exists an infinite word w for which $\underline{A_i}(w) \geq r_i^a$ and $\overline{B_i}(w) \geq r_i^b$ for all $i \in \{1, \ldots, k\}$ iff there exist k infinite words $w_1, w_2, \ldots w_k$ such that for every $j \in \{1, \ldots, k\}$:*

$$\underline{A_i}(w_j) \geq r_i^a \text{ for all } i \in \{1, \ldots, k\} \text{ ; and } \overline{B_j}(w_j) \geq r_j^b$$

The second lemma shows that the emptiness problem for an intersection of lim-inf automata and one lim-sup automaton can be reduced to the emptiness problem for an intersection of lim-inf automata.

Lemma 6. *The intersection $\overline{B_1}^{\geq r_1^b} \cap (\bigcap_{i=1}^k \underline{A_i}^{\geq r_i^a})$ is nonempty iff the intersection $\underline{B_1}^{\geq r_1^b} \cap (\bigcap_{i=1}^k \underline{A_i}^{\geq r_i^a})$ is nonempty.*

Due to Lemma 5 and 6 we can solve the emptiness problem for intersection of lim-inf and lim-sup automata in the following way: First, we construct the product automata $\mathcal{A}^i = A_1 \times \cdots \times A_k \times B_i$ for all $i \in \{1, \ldots, k\}$ and list all the simple cycles that occur in it; second, we construct the threshold vector $\overline{r^i} = (r_1^a, \ldots, r_k^a, r_i^b)$ and check if the graph of $\mathcal{A}^i$ has an $\overline{r^i}$ multi-cycle, for all $i \in \{1, \ldots, k\}$. Due to Lemma 5 and 6 the intersection is nonempty iff every $\mathcal{A}^i$ has an $\overline{r^i}$ multi-cycle, that is, if the $\overline{r^i}$ multi-cycle constraints are feasible.

Recall that the existence of an $\overline{r^i}$ multi-cycle in the graph of $\mathcal{A}^i$ is equivalent to the feasibility of the corresponding lim-inf constraints for $\mathcal{A}^i$ and $\overline{r^i}$; in the sequel, we will refer to the set of constraints $\bigcup_{i=1}^k \{$lim-inf constraints for $\mathcal{A}^i$ and $\overline{r^i}\}$ as the min-*only constraints*. (As we use them to decide the emptiness of expressions that contain only the min operator.)

In this subsection we proved that the emptiness of the intersection of *LimInfAvg* and *LimSupAvg* automata is equivalent to the feasibility of the corresponding min-only constraints.

3.4   The Emptiness Problem for max-free Expressions

In this subsection we solve the emptiness problem for max-free expressions. The solution we present is a reduction to the emptiness problem for an intersection of lim-inf and lim-sup automata with a threshold vector that satisfies certain linear constraints; the reduction yields a naive double-exponential complexity upper-bound for the problem, which we will improve in the succeeding subsection.

The reduction is based on the next simple observation.

Observation 1. *The expression $E = E_1 + E_2$ is nonempty with respect to the rational threshold ν iff there exist two thresholds $\nu_1, \nu_2 \in \mathbb{R}$ such that (i) The intersection of the cut-point languages $E_1^{\geq \nu_1}$ and $E_2^{\geq \nu_2}$ is nonempty; and (ii) $\nu_1 + \nu_2 \geq \nu$.*

If $E = E_1 + E_2$ and E_1 and E_2 are min-only expressions then we decide the emptiness of E in the following way: we combine the min-only constraints for the expressions E_1 and E_2 with respect to arbitrary thresholds r_1 and r_2 (that is, r_1 and r_2 are variables in the constraints), note that these are still linear constraints; we then check the feasibility of the constraints subject to $r_1 + r_2 \geq \nu$. (Note that as all the constraints are linear, this can be done by linear programming.)

The next lemma shows that in the general case, the emptiness problem for an arbitrary max-free expression and a threshold ν can be reduced, in polynomial time, to the emptiness problem for an intersection of lim-inf and lim-sup automata with respect to threshold vectors r^a and r^b subject to certain linear constraints on r^a and r^b.

Lemma 7. *Let E be a max-free expression with atomic expressions $e_1, \ldots, e_k$, and let ν be a rational threshold, then there exist a $2k \times 2k$ matrix M_E and a $2k$-dimensional vector $\overline{b_\nu}$, with rational coefficients, and computable in polynomial time (from E and ν) such that:*

> *The expression E is nonempty (with respect to ν) iff there exists a $2k$-dimensional vector of reals $\overline{r}$ such that the intersection $\bigcap_{i=1}^{k} e_i^{\geq r_i}$ is nonempty **and** $M_E \times \overline{r} \geq \overline{b_\nu}$.*

Instead of formally proving the correctness of Lemma 7, we provide a generic example that illustrates the construction of the matrix M_E and the vector $\overline{b_\nu}$.

Example 1. *Let $E = \min(\underline{A_1}, (\underline{A_2} + \overline{A_3})) + \min(\overline{A_4}, \underline{A_5})$. Then for every $\nu \in \mathbb{R}$, each the following condition is equivalent to $E^{\geq \nu} \neq \emptyset$.*

- $\exists r_6, r_7 \in \mathbb{R}$ *such that* $L^{\geq r_6}_{\min(\underline{A_1}, (\underline{A_2} + \overline{A_3}))} \cap L^{\geq r_7}_{\min(\overline{A_4}, \underline{A_5})} \neq \emptyset$ *and* $r_6 + r_7 \geq \nu$.
- $\exists r_1, r_4, r_5, r_6, r_7, r_8$ *such that* $\underline{A_1}^{\geq r_1} \cap L^{\geq r_8}_{\underline{A_2} + \overline{A_3}} \cap \overline{A_4}^{\geq r_4} \cap \underline{A_5}^{\geq r_5} \neq \emptyset$ *and* $r_1 \geq r_6, r_8 \geq r_6, r_4 \geq r_7, r_5 \geq r_7$ *and* $r_6 + r_7 \geq \nu$.
- $\exists r_1, r_2, r_3, r_4, r_5, r_6, r_7, r_8$ *such that* $\underline{A_1}^{\geq r_1} \cap \underline{A_2}^{\geq r_2} \cap \overline{A_3}^{\geq r_3} \cap \overline{A_4}^{\geq r_4} \cap \underline{A_5}^{\geq r_5} \neq \emptyset$ *and* $r_1 \geq r_6, r_2 + r_3 \geq r_8, r_8 \geq r_6, r_4 \geq r_7, r_5 \geq r_7$ *and* $r_6 + r_7 \geq \nu$.

The reader should note that we associate every variable r_i either with a sub-expression or with an atomic expression; as we assume that each atomic expression occurs only once, the number of variables is at most $2k$.

Hence we can solve the emptiness problem for a rational threshold ν and a max-free expression E, which contains the atomic expressions $e_1, \ldots, e_k$ in the following way: First, we construct the matrix M_E and the vector $\overline{b_\nu}$; second, we construct the min-only constraints for the intersection $\bigcap_{i=1}^{k} e_i^{\geq r_i}$ and check for their feasibility subject to the constraints $M_E \times r \geq \overline{b_\nu}$.

In the sequel we will refer to the min-only constraints along with the $M_{E,\nu} \times \overline{r} \geq \overline{b_\nu}$ constraints as the max-*free constraints*; we will show that even though the size of the constraints is double-exponential, there is a PSPACE algorithm that decides their feasibility (when the input is E and ν).

3.5   PSPACE Algorithm for the Emptiness Problem of max-free Expressions

In this subsection we will present a PSPACE algorithm that for given max-free expression E and rational threshold ν, decides the feasibility of the max-free constraints; as shown in subsection 3.4, such algorithm also solves the emptiness problem for max-free expressions. Informally, we will show that if the max-free constraints are feasible then they have a *short* solution, and that a short solution can be verified by a polynomial-space machine; hence the problem is in NPSPACE, and due to Savitch Theorem, also in PSPACE.

The next lemma describes key properties of max-free constraints, which we will use to obtain the PSPACE algorithm.

Lemma 8. *For every* max-*free expression E:*

1. *For every threshold ν, the* max-*free constraints have at most $O(k^2)$ constraints, where k is the number of automata that occur in E, that are not of the form of $x \geq 0$, where x is a variable.*
2. *There exists a bound t, polynomial in the size of the expression, such that for every threshold ν, the* max-*free constraints are feasible iff there is a solution that assigns a nonzero value to at most t variables.*
3. *There exists a bound t, polynomial in the size of the expression, such that the maximum threshold $\nu \in \mathbb{R}$, for which the* max-*free constraints are feasible, is a rational and can be encoded by at most t bits. (In particular such maximum ν exists.)*

Recall that a rational solution for the max-free constraints corresponds to vectors of thresholds and a set of multi-cycles, each multi-cycle with an average weight that matches its corresponding threshold vector; by Lemma 8(2) the number of different simple cycles that occur in the witness multi-cycles set is at most t. We also observe that if a multi-set of cycles (that are not necessarily simple) with average weight vector $\overline{\nu}$ exists, then a $\overline{\nu}$ multi-cycle (of simple cycles) also exists, since we can decompose every non-simple cycle to a set of simple cycles; thus, a $\overline{\nu}$ multi-cycle exists iff there exists a multi-set of *short* cycles, where the length of each cycle in the multi-set is at most the number of vertices in the graph (note that in particular, every simple cycle is short).

Hence, we can decide the feasibility of the max-free constraints in the following way: First, we guess t weight vectors of t short cycles that occur in the same strongly connected component (SCC) of the product automaton of all the automata that occur in the expression; second, we construct the $O(k^2)$ constraints of the max-free constraints and assign zero values to all the variables of the non-chosen cycles; third, we check the feasibility of the formed $O(k^2)$ constraints, where each constraint has at most $t + 1$ variables.

Note that we can easily perform the last two steps in polynomial time (as the values of the weight vector of every short cycle can be encoded by polynomial number of bits); hence, to prove the existence of a PSPACE algorithm, it is enough to show how to encode (and verify by a polynomial-space machine) t average weight vectors of t short cycles that belong to one SCC of the product automaton. Informally, the encoding scheme is based on the facts that every vertex in the product automaton is a k-tuple of states, and that a path is a sequence of alphabet symbols; the verification is done by simulating the k automata in parallel, and since the size of the witness string should be at most exponential, we can do it with a polynomial-size tape. (More details are given in [18].)

To conclude, we proved that there is a PSPACE algorithm that decides the feasibly of the max-free constraints, and therefore the next lemma follows.

Lemma 9. *The emptiness problem for* max-free *expressions is in PSPACE.*

Lemma 9 along with Lemma 8(3) imply a PSPACE algorithm that computes the maximum value of a max-free expression; the next lemma formally states this claim.

Lemma 10. *(i) The maximum value of a* max-free *expression is a rational value that can be encoded by polynomial number of bits (in particular, every expression has a maximum value); and (ii) The maximum value of a* max-free *expression is PSPACE computable.*

4   The Complexity of Mean-Payoff Expression Problems

In this section we will prove PSPACE membership, and PSPACE hardness, for the classical mean-payoff expression problems; the key step in the proof of the PSPACE membership is the next theorem, which extends Lemma 10 to arbitrary expressions (as opposed to only max-free expressions).

Theorem 1. *(i) The maximum value of an expression is a rational value that can be encoded by polynomial number of bits (in particular, every expression has a maximum value); and (ii) The maximum value of an expression is PSPACE computable.*

Proof (of Theorem 1). Informally, we prove that if the number of max operators in the expression E is $m > 0$, then we can construct in linear time two expressions E_1 and E_2, each with at most $m - 1$ max operators and of size at most $|E|$, such that $E = \max(E_1, E_2)$; hence, in order to compute the maximum value of E, we recursively compute the maximum values of E_1 and E_2, and return the maximum of the two values; note that if the expression is max-free (that is, if $m = 0$), then thanks to Lemma 10, the maximum value is PSPACE computable and can be encoded by polynomial number of bits. (Formal proof is given in [18].)   □

The PSPACE membership of the classical problems follows almost trivially from Theorem 1; for example, given two expressions E_1 and E_2, we can determine whether for all $w \in \Sigma^\omega$, $E_1(w) \geq E_2(w)$, by checking if the maximum value of the expression $F = (E_2 - E_1)$ is non-positive, and the distance between E_1 and E_2 is the maximum value of the expression $F = \max(E_1 - E_2, E_2 - E_1)$.

The PSPACE lower bounds for the decision problems are obtained by reductions from the emptiness problem for intersection of regular languages (see proofs in [18]), which is PSPACE-hard [14].

Thus, we get the main result of this paper:

Theorem 2. *For the class of mean-payoff automaton expressions, the quantitative emptiness, universality, language inclusion, and equivalence problems are PSPACE-complete, and the distance is PSPACE computable.*

5   Conclusion and Future Work

We proved tight complexity bounds for all classical decision problems for mean-payoff expressions and for the distance computation problem. Future work is to investigate the decidability of games with mean-payoff expression winning condition.

Acknowledgements. The author would like to thank Prof. Alexander Rabinovich for his helpful comments. This research was partially supported by the Israeli Centers of Research Excellence (I-CORE) program, (Center No. 4/11).

References

1. Alur, R., Degorre, A., Maler, O., Weiss, G.: On Omega-Languages Defined by Mean-Payoff Conditions. In: de Alfaro, L. (ed.) FOSSACS 2009. LNCS, vol. 5504, pp. 333–347. Springer, Heidelberg (2009)
2. Bojanczyk, M.: Beyond omega-regular languages. In: STACS, pp. 11–16 (2010)
3. Boker, U., Chatterjee, K., Henzinger, T.A., Kupferman, O.: Temporal specifications with accumulative values. In: LICS, pp. 43–52 (2011)
4. Chatterjee, K., Doyen, L., Edelsbrunner, H., Henzinger, T.A., Rannou, P.: Mean-Payoff Automaton Expressions. In: Gastin, P., Laroussinie, F. (eds.) CONCUR 2010. LNCS, vol. 6269, pp. 269–283. Springer, Heidelberg (2010)
5. Chatterjee, K., Doyen, L., Henzinger, T.A.: Expressiveness and closure properties for quantitative languages. Logical Methods in Computer Science (2010)
6. Chatterjee, K., Doyen, L., Henzinger, T.A.: Quantitative languages. ACM Trans. Comput. Log. 11(4) (2010)
7. Chatterjee, K., Ghosal, A., Henzinger, T.A., Iercan, D., Kirsch, C.M., Pinello, C., Sangiovanni-Vincentelli, A.: Logical reliability of interacting real-time tasks. In: DATE 2008, pp. 909–914. ACM (2008)
8. Droste, M., Gastin, P.: Weighted automata and weighted logics. Theor. Comput. Sci., 69–86 (2007)
9. Droste, M., Kuich, W., Vogler, H.: Handbook of Weighted Automata. Springer Publishing Company, Incorporated (2009)
10. Droste, M., Kuske, D.: Skew and Infinitary Formal Power Series. In: Baeten, J.C.M., Lenstra, J.K., Parrow, J., Woeginger, G.J. (eds.) ICALP 2003. LNCS, vol. 2719, pp. 426–438. Springer, Heidelberg (2003)
11. Droste, M., Rahonis, G.: Weighted Automata and Weighted Logics on Infinite Words. In: Ibarra, O.H., Dang, Z. (eds.) DLT 2006. LNCS, vol. 4036, pp. 49–58. Springer, Heidelberg (2006)

12. Kosaraju, S., Sullivan, G.: Detecting cycles in dynamic graphs in polynomial time. In: STOC, pp. 398–406. ACM (1988)
13. Kupferman, O., Lustig, Y.: Lattice Automata. In: Cook, B., Podelski, A. (eds.) VMCAI 2007. LNCS, vol. 4349, pp. 199–213. Springer, Heidelberg (2007)
14. Lange, K.J., Rossmanith, P.: The Emptiness Problem for Intersections of Regular Languages. In: Havel, I.M., Koubek, V. (eds.) MFCS 1992. LNCS, vol. 629, pp. 346–354. Springer, Heidelberg (1992)
15. Manna, Z., Pnueli, A.: The temporal logic of reactive and concurrent systems. Springer-Verlag New York, Inc. (1992)
16. Murty, K.G.: Linear Programming. Wiley, New York (1983)
17. Pnueli, A.: The temporal logic of programs. In: FOCS, pp. 46–57 (1977)
18. Velner, Y.: The complexity of mean-payoff automaton expression. CoRR, abs/1106.3054 (2012)
19. Velner, Y., Rabinovich, A.: Church Synthesis Problem for Noisy Input. In: Hofmann, M. (ed.) FOSSACS 2011. LNCS, vol. 6604, pp. 275–289. Springer, Heidelberg (2011)
20. Zwick, U., Paterson, M.: The complexity of mean payoff games on graphs. Theor. Comput. Sci. 158(1&2), 343–359 (1996)

On the Locality of Some NP-Complete Problems[*]

Leonid Barenboim[**]

Department of Computer Science, Ben-Gurion University of the Negev,
POB 653, Beer-Sheva 84105, Israel
leonidba@cs.bgu.ac.il

Abstract. We consider the distributed message-passing $\mathcal{LOCAL}$ model. In this model a communication network is represented by a graph where vertices host processors, and communication is performed over the edges. Computation proceeds in synchronous rounds. The running time of an algorithm is the number of rounds from the beginning until all vertices terminate. Local computation is free. An algorithm is called *local* if it terminates within a constant number of rounds. The question of what problems can be computed locally was raised by Naor and Stockmeyer [16] in their seminal paper in STOC'93. Since then the quest for problems with local algorithms, and for problems that cannot be computed locally, has become a central research direction in the field of distributed algorithms [9,11,13,17].

We devise the first local algorithm for an *NP-complete* problem. Specifically, our randomized algorithm computes, with high probability, an $O(n^{1/2+\epsilon} \cdot \chi)$-coloring within $O(1)$ rounds, where $\epsilon > 0$ is an arbitrarily small constant, and χ is the chromatic number of the input graph. (This problem was shown to be NP-complete in [21].) On our way to this result we devise a constant-time algorithm for computing $(O(1), O(n^{1/2+\epsilon}))$-network-decompositions. Network-decompositions were introduced by Awerbuch et al. [1], and are very useful for solving various distributed problems. The best previously-known algorithm for network-decomposition has a polylogarithmic running time (but is applicable for a wider range of parameters) [15]. We also devise a $\Delta^{1+\epsilon}$-coloring algorithm for graphs with sufficiently large maximum degree Δ that runs within $O(1)$ rounds. It improves the best previously-known result for this family of graphs, which is $O(\log^* n)$ [19].

1 Introduction

1.1 The Model

We consider the distributed message-passing model. This model, widely known as the $\mathcal{LOCAL}$ model, was formalized by Linial in a seminal paper in FOCS'87 [14]. In this model a communication network is represented by an n-vertex graph $G = (V, E)$ of maximum degree $\Delta = \Delta(G)$. The vertices of the graph host processors,

---

[*] A full version of this paper with full proofs is available online [2].

[**] The author is supported by the Adams Fellowship Program of the Israel Academy of Sciences and Humanities.

and communication is performed over the edges. Each vertex has a distinct identity number (henceforth, ID) of size $O(\log n)$ bits. The model is synchronous, meaning that computation proceeds in discrete rounds. In each round vertices are allowed to perform unbounded local computation, and send messages to their neighbors that arrive before the beginning of the next round. The input for a distributed algorithm is the underlying network. However, initially each vertex knows only the number of vertices n, and the IDs of its neighbors. Within r rounds, a vertex can learn the topology of its r-hop-neighborhood. For a given problem on graphs, a vertex has to compute only its part in the output. For example, for vertex coloring problems, each vertex has to compute only its color. However, the union of outputs of all vertices must constitute a correct solution. The running time of a distributed algorithm is the number of rounds from the beginning until the last vertex terminates. Local computation is free, and is not taken into account. This is motivated by the study of the ability of each vertex to arrive to a solution based on coordination only with close vertices.

1.2   Problems and Results

A legal vertex coloring is an assignment of colors to vertices, such that each pair of neighbors are assigned distinct colors. Vertex coloring problems are among the most fundamental and extensively studied problems in the field of distributed algorithms. Many variations have been studied. The most common variation is the $(\Delta + 1)$-coloring problem. The goal of this problem is computing a legal vertex coloring using at most $\Delta + 1$ colors. This problem has a very simple greedy solution in the sequential setting. Specifically, each vertex performs a color selection based on its 1-hop neighborhood. (The sequential time of the algorithm is linear.) However, in the distributed setting it becomes much more complicated. It is impossible to compute a solution based on an $O(1)$-hop-neighborhood [14]. The best currently-known deterministic distributed algorithms require $O(\Delta + \log^* n)$ [4,9], and $2^{O(\sqrt{\log n})}$ [18] rounds. The best currently-known randomized algorithm requires $O(\sqrt{\log n} + \log \Delta)$ rounds [19]. Moreover, Linial [14] proved that any distributed algorithm requires $\Omega(\log^* n)$ rounds for computing $(\Delta+1)$-, and even Δ^2-, coloring. (On the other hand, $O(\Delta^2)$-coloring is currently known to have a distributed algorithm that requires $O(\log^* n)$ rounds [14].) On special graph families it is often possible to employ fewer than $\Delta + 1$ colors. However, such algorithms provably cannot terminate within a small number of rounds. In particular, for graphs with arboricity[1] a, coloring a graph with $O(a)$ colors can be performed in $O(a^\epsilon \cdot \log n)$ rounds [4], for an arbitrarily small constant $\epsilon > 0$. However, any algorithm for this task requires $\Omega(\log n / \log a)$ rounds [3].

In the current paper we focus on problems which are hard in the sequential setting, as opposed to the problems mentioned above. Nevertheless, we show that NP-Complete vertex-coloring problems can be solved within $O(1)$ rounds in the distributed $\mathcal{LOCAL}$ setting. We devise a randomized algorithm that computes an $O(n^{1/2+\epsilon}\chi)$-coloring, within a constant number of rounds, with high probability.

---

[1] The *arboricity* is the minimum number of forests that cover the graph edges.

(χ is the chromatic number of the input graph, and $\epsilon > 0$ is an arbitrarily small constant.) Computing $O(n^{1-\epsilon}\chi)$-coloring (and, in particular, $O(n^{1/2+\epsilon}\chi)$-coloring) is known to be NP-complete [21]. To the best of our knowledge, prior to our work NP-complete problems could be solved only within $O(Diam(G))$ rounds. (In the $\mathcal{LOCAL}$ model every computable problem can be solved within $O(Diam(G))$ rounds, since all vertices can learn the topology of the entire input graph.)

The question of what problems can be solved within a constant number of rounds is one of the most fundamental questions in the field of distributed algorithms. It was raised around twenty years ago in the seminal paper of Naor and Stockmeyer, titled "What can be computed locally?" [16]. In this paper, an algorithm that requires $O(1)$ rounds is called *a local algorithm*. Despite a very intensive research in this direction that was conducted in the last twenty years, few problems with local algorithms on general graphs are known. (On the other hand, on constant-diameter graphs, any problem can be solved locally. Therefore, in the current setting, this question is meaningful only with respect to families of graphs with superconstant diameter.) Specifically, there are known local algorithms for computing weak-colorings [16], Δ-forests-decomposition [17], edge-defective-colorings [9], and dominating-set approximation [12,10,13]. (All these problems have simple sequential solutions.) On the other hand, many problems provably cannot be computed locally. In particular, minimum vertex cover, minimum dominating set, maximum independent set, maximum matching, maximal independent set, and maximal matching require $\Omega(\sqrt{\log n})$ rounds [11]. (The first three problems are NP-complete. The last three problems have polynomial sequential solutions. In particular, the last two problems have very simple greedy sequential algorithms.) Also, it is known that Δ^k-coloring, for any constant k requires $\Omega(\log^* n)$ rounds. Thus, discovering non-trivial (and, especially, hard) problems that can be computed within $O(1)$ rounds is of significant interest.

In our $O(n^{1/2+\epsilon}\chi)$-coloring algorithm the vertices perform NP-complete local computations in each round. Therefore, this algorithm has mainly theoretical interest. (However, we stress that unless P=NP, it is impossible to solve NP-complete problems by using polynomial local computation per round.) On the other hand, in addition to $O(n^{1/2+\epsilon}\chi)$-coloring, we devise local algorithms for several problems, in which vertices perform polynomial local computations. Therefore, these algorithms may be useful in practice. Specifically, we devise local algorithms for $(O(1), O(n^{1/2+\epsilon}))$-network-decomposition, and for $\Delta^{1+\epsilon}$-coloring graphs with large degree. The best previously known algorithm for network-decomposition requires $O(\log^2 n)$ rounds, but it employs different parameters for the decomposition [15]. The best previously known $\Delta^{1+\epsilon}$-coloring algorithm requires $O(\log^* n)$ time [19]. We elaborate on these problems in Section 1.4.

1.3 The Difficulty of Solving NP-Complete Problems Locally

Discovering NP-complete problems that can be solved locally in the distributed setting is interesting for the following reasons. Although the sequential setting and the distributed $\mathcal{LOCAL}$ setting are considerably different each from another, it is plausible that many NP-complete problems are also difficult to compute in the distributed setting. Despite that each vertex has an unbounded local computational power, the vertices have limited knowledge about the input graph. On the other hand, NP-complete problems usually define global constraints, which makes the computation difficult in the occasion of partial input knowledge. Consider, for example, the maximum clique problem. Consider a graph G in which two cliques K_4 and K_3 are connected by a path of length $\Theta(n)$, and, therefore, are at distance $\Theta(n)$ each from another. All vertices of K_4 must decide that they belong to the maximum clique. However, if the vertices of K_4 aware only of their $o(n)$-hop-neighborhood, then they cannot distinguish G from another graph G' that connects K_4 and K_5. Therefore, they cannot always arrive to a correct solution if the number of rounds is $o(n)$. Similar phenomenon occurs in additional problems.

Another example of a difficulty that NP-complete problems arise in the distributed setting can be found in the area of local decision and verification. (See, e.g., [8]). In this area the vertices are required to verify locally the correctness of the output, and at least one vertex needs to react in case of an incorrect output. While some simple sequential problems, such as maximal independent set and maximal matching are locally verifiable, NP-complete problems are more difficult for distributed verification because of their global constraints.

1.4   Our Techniques

Our main technical contribution is devising *network-decomposition* algorithms that require $O(1)$ rounds. Roughly speaking, a network decomposition is a partition of the vertices into clusters of bounded diameter, such that the supergraph formed by contracting clusters into single vertices has bounded chromatic number. (See Section 2 for a formal definition.) Network-decompositions are among the most useful structures in the field of distributed graph algorithms. Once an appropriate network decomposition is computed, it becomes possible to solve efficiently a variety of problems. These problems include vertex colorings, edge colorings, maximal independent set, maximal matching, and additional problems. The best currently-known deterministic $(\Delta + 1)$-coloring algorithms and maximal independent set algorithms employ network-decompositions [18]. Since the best currently-known network-decomposition algorithms require quite a large number of rounds (superlogarithmic for deterministic algorithms, and polylogarithmic for randomized ones) the quest for efficient network-decomposition algorithms is of great interest.

We devise a novel partitioning technique that allows computing network-decomposition with cluster diameter $O(1)$ and supergraph chromatic number $O(n^{1/2+\epsilon})$. Using a randomized algorithm we partition the vertex set of the input graph into subsets. Each subset has its own helpful properties that allow computing the network-decomposition efficiently. Specifically, one of the subsets contains a small dominating set, with high probability. We show that small dominating sets are very useful for computing network-decompositions. Another subset in the partition has bounded maximum degree, with high probability. This is very useful as well, since we can compute a $\Delta^{1+\epsilon}$-coloring in constant number of rounds on such graphs. Such a coloring is, in particular, a network-decomposition. Once we compute network-decompositions of the subsets, we merge the results to achieve a unified network decomposition of the input graph.

1.5 Related Work

Cole and Vishkin [6] and Goldberg and Plotkin [7] devised deterministic 3-coloring algorithms for paths, cycles and trees that require $O(\log^* n)$ rounds. Awerbuch, Goldberg, Luby, and Plotkin [1] devised a deterministic network-decomposition algorithm that requires $2^{O(\sqrt{\log n \log \log n})}$ rounds. It was later improved by Panconesi and Srinivasan [18], who achieved running time of $2^{O(\sqrt{\log n})}$ rounds. Schneider and Wattenhofer [20] devised a randomized coloring algorithm that produces, for a wide range of graphs, a $(1 - 1/O(\chi))\Delta$-coloring within $O(\log \chi + \log^* n)$ time.

To the best of our knowledge, the hardest problem that could be solved locally prior to our work is computing a constant approximation of minimum dominating sets (MDS) on planar graphs [13]. Although computing MDS on planar graphs is NP-complete, the constant approximation for this problem presented in [13] can be computed in polynomial time in the sequential setting.

2 Preliminaries

Unless the base value is specified, all logarithms in this paper are to base 2.

The graph $G' = (V', E')$ is a *subgraph* of $G = (V, E)$, denoted $G' \subseteq G$, if $V' \subseteq V$ and $E' \subseteq E$. For a subset $V' \subseteq V$, the graph $G(V')$ denotes the subgraph of G induced by V'. The *degree* of a vertex v in a graph $G = (V, E)$, denoted $deg_G(v)$, is the number of edges incident to v. The *distance* between a pair of vertices $u, v \in V$, denoted $dist_G(u, v)$, is the length of the shortest path between u and v in G. A vertex u such that $(u, v) \in E$ is called a *neighbor* of v in G. The *neighborhood* of v in G, denoted $\Gamma_G(v)$, is the set of neighbors of v in G. If the graph G can be understood from context, then we omit the underscript G. The *r-hop-neighborhood* of v in G is $\Gamma_G^r(v) = \{u \mid dist_G(u, v) \le r\}$. If the graph G can be understood from context, we use the shortcut $\Gamma_r(v)$ for $\Gamma_G^r(v)$. The maximum degree of a vertex in G, denoted $\Delta(G)$, is defined by $\Delta(G) = \max_{v \in V} deg(v)$. The *diameter* of G is the maximum distance between a pair of vertices in G. A *dominating set* $U \subseteq V$ satisfies that for each $v \in V$, either $v \in U$, or there is a neighbor of v in U. The *chromatic number* $\chi(G)$ of a graph G is the minimum number of colors that can be used in a legal coloring of the vertices of G. The *Minimum-Coloring problem* is the problem of computing a legal coloring of the vertices of G using $\chi(G)$ colors. For a graph $G = (V, E)$, a function $f : V \to \mathbb{N}$ is called a *label assignment*. For a graph G with a label assignment f, a connected component of vertices with the same label forms a *cluster*. More formally, a *cluster* is a connected component $U \subseteq V$, such that for each $u, v \in U$, it holds that $f(u) = f(v)$, and for each $u \in U, v \in \Gamma(u) \setminus U$, it holds that $f(u) \neq f(v)$. A (d, c)-*network-decomposition* of a graph G is an assignment of labels from the

set $\{1, 2, ..., c\}$ to vertices of V, such that each cluster has diameter at most d.
An algorithm succeeds *with high probability* if it succeeds with probability
$1 - 1/n^k$, for an arbitrarily large constant $k \geq 1$.

3   Approximating Minimum-Coloring Using Network Decompositions

In this section we show how to approximate Minimum-Coloring on a graph with
a given (d, c)-network-decomposition. First, we provide a high-level description
of the algorithm. Suppose that we are given a graph G, and a label assignment
$f : V \rightarrow \{1, 2, ..., c\}$, such that each cluster has diameter at most d. We c-
approximate Minimum-Coloring in the following way. First, for each cluster $U \subseteq$
V, we compute a Minimum-Coloring $\varphi_U : U \rightarrow \{1, 2, ..., \chi(G(U))\}$. Next, we
compute a new color $\varphi(v)$ for each $v \in V$. Let W be the cluster of v. (Notice
that by definition, each vertex belongs to exactly one cluster.) We set $\varphi(v) =$
$\varphi_W(v) \cdot c + f(v) - 1$. Intuitively, the color $\varphi(w)$ can be seen as the ordered pair
$\langle \varphi_W(v), f(v) \rangle$. The coloring φ is returned by the algorithm. In the sequel we
show that φ is a c-approximation of Minimum-Coloring of G.

Next, we provide a detailed description of a distributed algorithm that employs the high-level idea described above. The algorithm is called *Procedure
Approximate*. Similarly to all distributed algorithms that we will describe, it defines the behavior of each vertex $v \in V$ in each round. Procedure Approximate
accepts as input the label $f(v)$ of v, and the number of labels c. In the first
stage of the procedure, v collects the entire topology of the cluster W that v
belongs to. It is widely known (see, e.g., [1,18]) that collecting the topology of
an r-hop-neighborhood of a vertex v can be performed in r rounds. Therefore,
each vertex can collect the topology of its $(d + 1)$-neighborhood $\Gamma_{d+1}(v)$ within
$(d + 1)$-rounds. Since the diameter of the cluster W of v is at most d, it holds
that $W \subseteq \Gamma_{d+1}(v)$. Hence v learns the topology of W within $(d + 1)$-rounds.

In the second stage, Procedure Approximate computes a Minimum-Coloring
of the cluster W of v. To this end, it employs a deterministic algorithm that
performs exhaustive search locally. Specifically, for $i = 1, 2, ...$, the algorithm
goes over all possible (either legal or illegal) colorings of $G(W)$ with i colors. For
each coloring it checks whether it is legal or not, and terminates in the first time
a legal coloring is found. Observe that this technique guarantees that all vertices
that belong to the same cluster W compute the same coloring. Indeed, all vertices
$w \in W$ have learnt the entire topology of W, and perform an exhaustive search
on $G(W)$ locally. Since all the vertices perform exactly the same deterministic
algorithm that runs on the input $G(W)$, the output is identical for all vertices in
W. Denote by φ_W the coloring returned by the exhaustive search. Each vertex v
sets $\varphi(v) = \varphi_W(v) \cdot c + f(v) - 1$, and terminates. This completes the description
of Procedure Approximate. Next we analyze its correctness and running time.

Lemma 3.1. *Procedure Approximate invoked on a graph G with a (d, c)-network-
decomposition requires $(d + 1)$-rounds.*

Proof. The only stage of Procedure Approximate that is not performed locally is the stage that collects the information of a cluster of diameter d. This requires $(d+1)$-rounds. □

Lemma 3.2. *Procedure Approximate invoked on a graph G with a (d,c)-network-decomposition computes a c-approximate Minimum-Coloring of G.*

Proof. (Sketch) First we need to prove that Procedure Approximate computes a legal coloring. This proof can be found in the full version of this paper [2].

Next, we prove that the computed coloring is a c-approximate Minimum-Coloring of G. Observe that for any $W \subseteq V$ it holds that $\chi(G(W)) \leq \chi(G)$. Indeed, a legal coloring of G using $\chi(G)$ colors restricted to W is, in particular, a legal coloring of $G(W)$. Therefore, $G(W)$ can be legally colored with at most $\chi(G)$ colors. Consequently, for each cluster $W \subseteq V$, and each vertex $v \in W$, it holds that $1 \leq \varphi_W(v) \leq \chi(G(W)) \leq \chi(G)$. Therefore, $\varphi(v) = \varphi_W(v) \cdot c + f(v) - 1 \leq \chi(G) \cdot c + c - 1$. On the other hand, since $\varphi_W(v) \geq 1$ and $f(v) \geq 1$, it holds that $\varphi(v) \geq c$. Therefore, φ employs at most $\chi(G) \cdot c + c - 1 - c + 1 = \chi(G) \cdot c$ colors. We remark that all vertices $v \in V$ should subtract $c - 1$ from $\varphi(v)$ to achieve a color in the range $\{1, 2, ..., \chi(G) \cdot c\}$. □

Linial and Saks [15], devised a randomized algorithm for computing $(O(\log n), O(\log n))$-network-decomposition within $O(\log^2 n)$ rounds. If one is willing to spend that much time, then Lemmas 3.1 - 3.2 in conjunction with the algorithm of Linial and Saks allow computing an $O(\log n \cdot \chi(G))$-coloring of an input graph G within $O(\log^2 n)$ rounds. In the sequel, we devise coloring algorithms that employ more colors, but require a constant number of rounds.

4   Computing Network Decompositions

4.1   Partitioning Procedure

In this section we devise an algorithm for computing $(O(1), O(n^{1/2+\epsilon}))$-network-decompositions, for an arbitrarily small constant $\epsilon > 0$. Using this algorithm in conjunction with Lemmas 3.1 - 3.2 we obtain an $O(n^{1/2+\epsilon})$-approximation algorithm for Minimum-Coloring of G. This algorithm terminates within $O(1)$ rounds. The algorithm for computing $(O(1), O(n^{1/2+\epsilon}))$-network-decompositions is called *Procedure Decompose*. The main idea of the algorithm is partitioning the vertex set V into two subsets A and B that satisfy certain helpful properties. Specifically, the induced subgraph $G(A)$ contains a dominating set D of A, such that the size of D is sufficiently small. The set D consists of $O(n^{1/2})$ vertices. The set B, on the other hand, satisfies a different property. Specifically, the maximum degree of $G(B)$ is bounded by $O(n^{1/2} \log n)$. In the sequel we show how to compute network-decompositions of $G(A)$ and $G(B)$, and how to combine them to achieve the desired network-decomposition of G. In this section we devise an algorithm for computing A and B within $O(1)$ rounds.

The algorithm for computing a partition of V into two subsets A and B is called *Procedure Partition*. Procedure Partition is a randomized algorithm that works in the following way. Each vertex $v \in V$ holds a local Boolean variable v_m. We say that a vertex marks itself if it sets $v_m = true$. The vertex v is unmarked if and only if $v_m = false$. Initially, all vertices are unmarked. The steps that each vertex $v \in V$ performs are described below.

Procedure Partition
1. v marks itself with probability $1/n^{1/2}$, independently of other vertices.
2. **if** v is marked, **then** it sends a 'marked' message to all its neighbors.
3. **if** v is marked or v has a marked neighbor, **then** v joins the set A.
   else v joins the set B.

Step 1 and 3 of Procedure Partition are performed locally, and step 2 requires one communication round. Therefore, the procedure requires $O(1)$ rounds. Next, we prove that Procedure Partition partitions V into the subsets A and B that satisfy the properties mentioned above.

Lemma 4.1. *The set A contains a dominating set D of A, with size $|D| = O(n^{1/2})$, with high probability.*

Proof. The set A contains all the vertices that are marked during the execution of the algorithm, and all their neighbors. Denote by D the set of vertices that are marked. The set D is a dominating set of A. We show that with high probability, $|D| = O(n^{1/2})$. Let X_v denote the random indicator variable, such that $X_v = 1$ if v marks iself, and $X_v = 0$ otherwise. Let $X = \sum_{v \in V} X_v$ be the sum of n indicator variables. Let $\gamma > 0$ be an arbitrarily small constant. The expected number of marked vertices is $\mathbb{E}(X) = n \cdot 1/n^{1/2} = n^{1/2}$. Hence, by the Chernoff bound for upper tails, it holds that $Pr[X > (1 + \gamma)\mathbb{E}(X)] \leq \left(\frac{e^\gamma}{(1+\gamma)^{1+\gamma}} \right)^{\mathbb{E}(X)}$. Set $\gamma = 1$. It holds that $Pr[X > 2\mathbb{E}(X)] \leq (e/4)^{n^{1/2}} \leq 1/n^k$, for an arbitrarily large constant k, and sufficiently large n.   □

Lemma 4.2. *The subgraph $G(B)$ induced by B has maximum degree $O(n^{1/2} \log n)$, with high probability. (See the full version of this paper [2] for the proof.)*

4.2   Network-Decompositions in Graphs with Bounded Degree

In this section we device an algorithm that allows computing an $(O(1), n^{1/2+\epsilon})$-network-decomposition of B. (We postpone the description of the algorithm for A to Section 4.3.) The algorithm we devise to be used for B is quite general. It computes a legal coloring of the underlying graph, rather than a network-decomposition. However, a legal coloring using ℓ colors is, in particular, an $(O(1), \ell)$-network-decomposition. Moreover, our algorithm colors any graph with maximum degree $\Delta \geq n^\mu$, using $\Delta^{1+\epsilon}$ colors, for arbitrarily small constants $\epsilon, \mu > 0$. For graphs with maximum degree smaller than n^μ, our algorithm produces an $O(n^{\mu+\epsilon \cdot \mu})$-coloring. Observe that applying this algorithm on $G(B)$ results in an $O((n^{1/2} \log n)^{1+\epsilon})$-coloring of $G(B)$, which is an $O(n^{1/2+\epsilon})$-coloring.

The algorithm is called *Procedure Color*. It accepts as input a parameter Δ which is an upper bound of the maximum degree of the underlying graph, such that $\Delta \geq n^\mu$. The procedure performs a constant number of rounds. Each round consists of the following steps.

Procedure Color
1. v draws uniformly at random a color q_v from the range $\{1, 2, ..., \lceil \Delta^{1+\epsilon} \rceil\}$, and sends q_v to all neighbors.
2. **if** q_v is different from the colors of all neighbors of v (including those that have already terminated)
   then v sets q_v as its final color, informs its neighbors, and terminates.
3. **else** v discards the color q_v.

Observe that once the procedure terminates in all vertices, each vertex v holds a color q_v that is different from the colors of all its neighbors. To prove this, let i denote the round in which a vertex v has terminated. All neighbors u of v that have terminated before v, have selected a final color q_u before round i. Thus, these colors do not change in round i and afterwards. Therefore, $q_v \neq q_u$ for each neighbor u of v that has terminated before v. (Otherwise, v would not terminate in round i.) By the same argument, we conclude that for each neighbor u that terminates after v, it holds that $q_u \neq q_v$. It is left to show that for all neighbors w of v that terminates in round i, it holds that $q_v \neq q_w$. Assume for contradiction that the colors that v and w select in round i are identical. Then nor v nor w terminate in round i. This is a contradiction. Therefore, if all vertices terminate, the produced coloring is legal.

Next, we analyze the performance of Procedure Color in case that it is executed for a single round. (Not necessarily the first one.)

Lemma 4.3. *Suppose that Procedure Color is executed in round i, for $i \geq 1$, by a vertex v. The probability that v does not terminate in round i is at most $1/\Delta^\epsilon$.*

Proof. Assume without loss of generality that in round i the vertex v selects a color after all its neighbors do so. The number of different colors selected by all neighbors of v is at most Δ. The vertex v selects a color from the range $\{1, 2, ..., \lceil \Delta^{1+\epsilon} \rceil\}$. Therefore, the probability that it selects a color that is identical to a color of a neighbor is at most $\Delta/\Delta^{1+\epsilon} = 1/\Delta^\epsilon$. $\qquad\square$

Next, we analyze the probability that a vertex does not terminate within i rounds, for an integer constant $i > 0$. The probability that a vertex does not terminate in the first round is at most $1/\Delta^\epsilon$, by Lemma 4.3. The probability that a vertex does not terminates in round i, conditioned on that it does not terminate within rounds $1, 2, ..., i-1$, is at most $1/\Delta^\epsilon$ as well. Therefore, the probability that a vertex does not terminate within i rounds is $\hat{\rho} = (1/\Delta^\epsilon)^i$. For an arbitrarily large constant k, set $i = \lceil k/(\mu \cdot \epsilon) \rceil$. It holds that $\hat{\rho} = (1/\Delta^\epsilon)^i \leq (1/n^{\mu \cdot \epsilon})^i \leq 1/n^k$. By the union bound, the probability that there exists a vertex that does not terminate is at most $n \cdot 1/n^k = 1/n^{k-1}$. Therefore, all vertices terminate with high probability. We summarize this in the following theorem.

Theorem 4.4. *After a constant number of rounds, Procedure Color computes a legal $\Delta^{1+\epsilon}$-coloring of an input graph of maximum degree at most $\Delta \geq n^{\mu}$, for arbitrarily small constants $\epsilon, \mu > 0$, with high probability.*

Corollary 4.5. *For a graph with maximum degree $O(n^{1/2} \log n)$, an $(O(1), n^{1/2+\epsilon})$-network-decomposition can be computed in $O(1)$ rounds, with high probability.*

4.3 Network-Decompositions in Graphs with Small Dominating Set

In this section we devise an algorithm for computing $(O(1), n^{1/2+\epsilon})$-network-decomposition for graph that contain a dominating set of size $O(n^{1/2})$. This algorithm can be used for A. The algorithm is called *Procedure Dominate*. It accepts as input a dominating set D of the underlying graph $G' = (V', E')$, such that D contains $O(n^{1/2})$ vertices. Our ultimate goal would be assigning unique labels from the range $\{1, 2, ..., O(n^{1/2})\}$, to the vertices of D. If we would be able to do so, then each vertex in $V' \setminus D$ could select a label of (an arbitrary) neighbor that belongs to D. Consequently, the diameter of each cluster would be at most 2. Indeed, all vertices with the same label in $V' \setminus D$ are connected to a common vertex in D, and all vertices in D have distinct labels.

It is currently unknown whether it is possible to compute labels for vertices in D as described above, within a constant number of rounds. We address a problem with somewhat weaker requirements. This is, however, sufficient for computing the desired network-decomposition. Specifically, we require that the labels assigned to vertices of D are taken from the range $\{1, 2, ..., O(n^{1/2+\epsilon})\}$. Also, we do not require that all the labels are unique. Instead, we require that each vertex $v \in D$ selects a label that is distinct from the labels of vertices in $\Gamma_{G'}^3(v) \cap D$. In particular, such a labeling constitutes a distance-3 coloring of D. This way, once vertices from D select appropriate labels, and vertices from $V' \setminus D$ select a label of an arbitrary neighbor from D, we obtain an $(O(1), n^{1/2+\epsilon})$-network-decomposition. We will prove this claim shortly, but first we describe the algorithm in a more detail. In each iteration, each vertex $v \in D$ performs the following steps. These steps are performed for k iterations, where $k > 0$ is an integer constant to be determined later. The vertices of $V' \setminus D$ do not perform any steps in these k iterations.

Procedure Dominate
(Performed by vertices $v \in D$)
1. v draws uniformly at random a label l_v from the set $\{1, 2, ..., \lfloor n^{1/2+\epsilon} \rfloor\}$.
2. v collects the topology of $\Gamma_{G'}^3(v)$, including labels.
3. **if** l_v is distinct from all labels in $\Gamma_{G'}^3(v) \cap D$
   then v sets l_v as its final label, informs its neighbors, and terminates.
4. **else** v discards the label l_v.

Observe that each iteration of Procedure Dominate requires four rounds. Three rounds are required for collecting the topology of $\Gamma_{G'}^3(v)$, and one round is required for informing the neighbors about a final selection of l_v. We will prove that after k iterations, for a sufficiently large constant k, all vertices $v \in D$ terminate, with high probability. In iteration $k + 1$, all vertices $u \in V' \setminus D$ select a final label l_u, such that $l_u = l_w$, for an arbitrary $w \in \Gamma_{G'}(v) \cap D$. This completes the description of Procedure Dominate. We analyze its correctness and running time below.

Lemma 4.6. *After a constant number of iterations of executing Procedure Dominate, all vertices $v \in D$ terminate, with high probability.*

Proof. Observe that for each $v \in D$, the number of vertices in $\Gamma_{G'}^3(v) \cap D$ is $O(n^{1/2})$, since $|D| = O(n^{1/2})$. Therefore, the probability that a vertex does not terminate after a single iteration is $O(n^{1/2})/n^{1/2+\epsilon} = 1/\Omega(n^\epsilon)$. The probability that a vertex does not terminate after k iterations is $1/\Omega(n^{\epsilon \cdot k})$. For an arbitrarily large constant k', there exist a sufficiently large constant k, such that $1/\Omega(n^{\epsilon \cdot k}) < 1/n^{k'}$. Therefore, a vertex terminates after a constant number of iterations, with high probability. By the union bound, all vertices in D terminate within k' iterations, with probability at least $1 - 1/n^{k'-1}$. $\qquad\square$

Lemma 4.7. *Procedure Dominate computes an $(O(1), O(n^{1/2+\epsilon}))$-network- decomposition, with high probability. (See full version of this paper [2] for proof.)*

Using Lemma 4.7 we can compute an $(O(1), O(n^{1/2+\epsilon}))$-network-decomposition of $G(A)$. Using Corollary 4.5 we can compute an $(O(1), O(n^{1/2+\epsilon}))$-network-decomposition of $G(B)$. It is left to show how to combine these two network-decompositions to obtain a unified network-decomposition of G. To this end, each vertex $u \in A$, with a label l_u, computes a new label $l'_u = l_u \cdot 2$. Each vertex $v \in B$, with a label l_v, computes a new label $l'_v = l_v \cdot 2 + 1$. Consequently, for each $v \in A, u \in B$, it holds that $l'_v \neq l'_u$. Consider a cluster in G with respect to the new labeling. All the vertices in the cluster have the same new label. Hence all of them have the same old label, as well. Therefore, either all of them belong to A, or all of them belong to B. Therefore, the diameter of the cluster is $O(1)$. Since the number of labels in A and in B is $O(n^{1/2+\epsilon})$ the total number of new labels is $O(n^{1/2+\epsilon})$ as well.

Finally, observe that all the procedures can be combined to produce an $O(n^{1/2+\epsilon} \cdot \chi(G))$-coloring of G from scratch. To this end, the vertices first compute the value $t = \lfloor k \cdot n^{1/2} \cdot \log n \rfloor$, where k is the constant hidden in the O-notation in Lemma 4.2. (Recall that all vertices know n.) Then, they invoke Procedure Partition. Consequently, each vertex knows whether it belongs to A or to B. Moreover, the vertices in A know whether they belong to the dominating set D or not. (The vertices that belong to D are marked.) Next, the vertices of A execute Procedure Dominate. The vertices of B execute Procedure Color with the value t as input. Consequently, the desired network-decompositions of $G(A)$ and $G(B)$ are computed. Then they are combined into a unified $(O(1), O(n^{1/2+\epsilon})$-decomposition of G. Next, an $O(n^{1/2+\epsilon} \cdot \chi(G))$-coloring is computed using Procedure Approximate in $O(1)$ rounds. (See Lemmas 3.1 -3.2.) The input for Procedure Approximate, $t' = O(n^{1/2+\epsilon})$, can be computed locally by each vertex. We summarize this discussion in the following theorem.

Theorem 4.8. *For any graph G with n vertices, and an arbitrarily small positive constant $\epsilon > 0$, with high probability, we can compute within $O(1)$ rounds:*
(1) *An $(O(1), O(n^{1/2+\epsilon}))$-network-decomposition of G.*
(2) *An $O(n^{1/2+\epsilon} \cdot \chi(G))$-coloring of G.*

Acknowledgements. The author is grateful to Michael Elkin for fruitful discussions and very helpful remarks.

References

1. Awerbuch, B., Goldberg, A.V., Luby, M., Plotkin, S.: Network decomposition and locality in distributed computation. In: Proc. 30th IEEE Symp. on Foundations of Computer Science, pp. 364–369 (October 1989)
2. Barenboim, L.: On the locality of some NP-complete problems (2012), http://arXiv.org/abs/1204.6675
3. Barenboim, L., Elkin, M.: Sublogarithmic distributed MIS algorithm for sparse graphs using Nash-Williams decomposition. In: Proc. 27th ACM Symp. on Principles of Distributed Computing, pp. 25–34 (2008)
4. Barenboim, L., Elkin, M.: Distributed $(\Delta + 1)$-coloring in linear (in Δ) time. In: Proc. 41st ACM Symp. on Theory of Computing, pp. 111–120 (2009)
5. Barenboim, L., Elkin, M.: Deterministic distributed vertex coloring in polylogarithmic time. In: Proc. 29th ACM Symp. on Principles of Distributed Computing, pp. 410–419 (2010)
6. Cole, R., Vishkin, U.: Deterministic coin tossing with applications to optimal parallel list ranking. Information and Control 70(1), 32–53 (1986)
7. Goldberg, A., Plotkin, S.: Efficient parallel algorithms for $(\Delta + 1)$-coloring and maximal independent set problem. In: Proc. 19th ACM Symp. on Theory of Computing, pp. 315–324 (1987)
8. Fraigniaud, P., Korman, A., Peleg, D.: Local Distributed Decision. In: Proc. 52nd IEEE Symp. on Foundations of Computer Science, pp. 708–717 (2011)
9. Kuhn, F.: Weak graph colorings: distributed algorithms and applications. In: Proc. 21st ACM Symp. on Parallel Algorithms and Architectures, pp. 138–144 (2009)
10. Kuhn, F., Moscibroda, T., Wattenhofer, R.: The price of being near-sighted. In: Proc. 17th ACM-SIAM Symp. on Discrete Algorithms, pp. 980–989 (2006)
11. Kuhn, F., Moscibroda, T., Wattenhofer, R.: Local Computation: Lower and Upper Bounds (2010), http://arXiv.org/abs/1011.5470
12. Kuhn, F., Wattenhofer, R.: Constant-time distributed dominating set approximation. In: 22nd ACM Symp. Principles of Distributed Computing, pp. 25–32 (2003)
13. Lenzen, C., Oswald, Y., Wattenhofer, R.: What Can Be Approximated Locally? Case Study: Dominating Sets in Planar Graphs. In: Proc. 20th ACM Symp. on Parallelism in Algorithms and Architectures, pp. 46–54 (2008); See also TIK report number 331, ETH Zurich (2010)
14. Linial, N.: Distributive Graph Algorithms-Global Solutions from Local Data. In: Proc. 28th IEEE Symp. on Foundations of Computer Science, pp. 331–335 (1987)
15. Linial, N., Saks, M.: Low diameter graph decompositions. Combinatorica 13(4), 441–454 (1993)
16. Naor, M., Stockmeyer, L.: What can be computed locally? In: Proc. 25th ACM Symp. on Theory of Computing, pp. 184–193 (1993)

17. Panconesi, A., Rizzi, R.: Some simple distributed algorithms for sparse networks. Distributed Computing 14(2), 97–100 (2001)
18. Panconesi, A., Srinivasan, A.: On the complexity of distributed network decomposition. Journal of Algorithms 20(2), 581–592 (1995)
19. Schneider, J., Wattenhofer, R.: A New Technique For Distributed Symmetry Breaking. In: 29th ACM Symp. Principles of Distributed Computing, pp. 257–266 (2010)
20. Schneider, J., Wattenhofer, R.: Distributed Coloring Depending on the Chromatic Number or the Neighborhood Growth. In: Kosowski, A., Yamashita, M. (eds.) SIROCCO 2011. LNCS, vol. 6796, pp. 246–257. Springer, Heidelberg (2011)
21. Zuckerman, D.: Linear Degree Extractors and the Inapproximability of Max Clique and Chromatic Number. Theory of Computing 3(1), 103–128 (2007)

Growing Half-Balls: Minimizing Storage and Communication Costs in CDNs*

Reuven Bar-Yehuda[1], Erez Kantor[1,**], Shay Kutten[1,***], and Dror Rawitz[2]

[1] Technion, Haifa 32000, Israel
reuven@cs.technion.ac.il, erez.kantor@gmail.com, kutten@ie.technion.ac.il
[2] Tel-Aviv University, Tel-Aviv 69978, Israel
rawitz@eng.tau.ac.il

Abstract. The Dynamic Content Distribution problem addresses the trade-off between storage and delivery costs in modern virtual Content Delivery Networks (CDNs). That is, a video file can be stored in multiple places so that the request of each user is served from a location that is near by to the user. This minimizes the delivery costs, but is associated with a storage cost. This problem is NP-hard even in grid networks. In this paper, we present a constant factor approximation algorithm for grid networks. We also present an $O(\log \delta)$-competitive algorithm, where δ is the normalized diameter of the network, for general networks with general metrics. We show a matching lower bound by using a reduction from online undirected STEINER TREE. Our algorithms use a rather intuitive approach that has an elegant representation in geometric terms.

1   Introduction

This paper reports research that is been performed as a part of Net-HD [1], an Israeli consortium of Industry and academia, designing a Content Delivery Network (CDN) over the Internet, for delivering video on demand. The idea is to save bandwidth, by storing movies' copies in various nodes in the network. The project is in a rather advanced stage, and a prototype has already been demonstrated. Some of the results reported here were used in its design.

In Net-HD, each CDN is supposed to be controlled by an Internet Service Provider. Thus, the network structure is known to the CDN operator. When a request for a movie arrive at some node, v, the CDN selects may select *any* other node u (as in a P2P network) which currently has a copy of that movie, and instructs u send a copy to v. Serving from a near by node saves the cost of delivering from a remote server. On the other hand, this incurs a storage cost at u; hence the system may save by deleting copies sometimes. On the other hand, when the demand grows in some localities, the system may decide to add copies. The problem is to optimize the trade-off between the above two costs for each sequence of requests.

---

* Supported in part by the Net-HD MAGNET Consortium.

** Supported by Eshkol fellowship, the Ministry of Science and Technology, Israel.

*** Supported in part by the ISF and by the Technion Gordon Center.

A. Czumaj et al. (Eds.): ICALP 2012, Part II, LNCS 7392, pp. 416–427, 2012.
© Springer-Verlag Berlin Heidelberg 2012

The decisions (in the case of Net-HD, but also in many other cases) are made by a central location that knows the topology of the network as explained above. It also knows the locations of the copies. This is feasible since the control messages ordering u to deliver a copy to v consumes negligible bandwidth, compared to the amount consumed by the copy of the movie.

CDNs, especially for video, have been growing rapidly, becoming a major application of the Internet. In May 2011, Youtube passed the barrier of three billion requests per day, three times as much as it had a year earlier. Netflix, that rents movies over the internet, is claimed to account for 20% of peek U.S. bandwidth use [4]. Some other big players (or emerging players) in renting movies online are Amazon, Hulu, Apple, Blockbuster, even Facebook (with Warner Bros [15]). Video traffic, not including file sharing P2P traffic (such as Bit Torrent, eMule, or even Rapidshare), was estimated to be 25% of the Internet traffic in October 2010 [2]. The sum of all forms of video was predicted to exceed 91% of the global consumer traffic by 2014 [3]. Reducing video traffic, even over the commercial (and legal) CDNs would make a very significant contribution to providing services over the Internet. (In fact, it has been claimed that some Internet Service Providers suffer from the video load so much that they limit the amount of bandwidth they give to YouTube (perform throttling [14]); this is in spite of the fact that YouTube (Google) transports most of this traffic over its own network, and passes it to ISPs relatively close to the users).

We strive to reduce the traffic, but also to save on the cost of storage space. Indeed, storage is rather cheap, so normally, there is no need to evict a copy of one movie from a node when bringing a copy of another movie. However, when the ISP (or the caching provider) sees a fast growth in the demand at some node, it installs additional storage, incurring costs. Alternatively, storage can be rented from various cloud computing providers and from other CDNs even today. Additional storage renting by location will probably arise when demand grows, similar to the case currently with fiber and links. Another kind of cost materializes in the form of slower servers (since an overloaded server must use a slower secondary storage, or, alternatively, may need to postpone some space consuming tasks). Finally, we note that algorithms that minimize storage cost may help to approximate the more traditional caching, where a charge for space is not assumed, but rigid space bounds are. (e.g., paging algorithms).

Our Results. We present a constant factor approximation algorithm for grid networks. We comment that it is rather easy to show that this problem is NP-hard in general graphs (using a reduction from the STEINER TREE problem [13]) and even on grids (using [12]). (However, as shown in [9], it is not possible to translate upper bounds to online undirected spanning tree algorithms to solve the problem addressed here). We present an $O(\log \delta)$-competitive algorithm, where δ is the normalized diameter of the network, for general networks with general metrics. In the full version, we show a matching lower bound using a reduction from the Steiner tree problem. Our algorithms use a rather intuitive approach ("growing half balls") that has an elegant representation in geometric terms.

Related Work. In [18,16], Papadimitriou et al. addressed a special case of this problem, were the servers (who can store copies but cannot issue requests) are arranged in a line network. All the requests there are issued only by *users*, who are connected to one of the endpoints of the line. Moreover, it is assumed there that every user does issue exactly a single request. (The results for the algorithms in the current paper hold also when storage and requests are allowed anywhere, and when each node may issue any number of requests). They present an optimal solution for that problem, as well as a ratio 2 approximation for a tree topology of the servers. (Again, only users issued requests, each user issued exactly one request, and users were attached only to the leaves). In [17], they show that allowing the storage and delivery costs to change in time (which we do not allow) makes the problem NP-Hard even if the tree is a star, and that the time invariant case is NP-Hard on general networks. They also state that the above approximation method would work also for the time varying costs, for the above network (of servers tree with single request users at the leaves).

They also connected the problems to the problem of Steiner Trees in semi directed graphs "space-time" graphs; those are the Cartesian products of the network graphs with the set of requests' times; the directed edges are along the time axis. This paper (as well as [9] mentioned below) use these "space-time" graphs. They gave a (greedy) 2-approximation on trees for the problem where the costs did not change with time. For the more general problem, with costs that vary with time, they gave a polynomial algorithm for the special case of a line network plus a star at one of its ends, where all requests must be at the nodes of the star.

A related problem of dynamic servers with costs associated both with delivery and with storage (or "rental") was studied in [9]. There, the main results are for the case that the deletion of a server is associated with a cost. Hence, it is hard to compare the competitiveness results given there and given here. (Their constraint on deletions may force the adversary off line algorithm to work harder, thus allowing for a better competitive factor). Indeed, we present a (rather straightforward) lower bound of $\Omega(\log \delta)$ for the competitive ratio of our problem on a general graph, while their lower bound is $\Omega(\frac{\log \log \delta}{\log \log \log \delta})$. They also show that it is not obvious how to modify any existing Steiner tree algorithm (or, even closer, a Rectilinear Steiner Arborescence algorithm (RSA)) to solve the problem studied here. Specifically, they show that there exists instances where the cost of an RSA solution is higher in the order of magnitude than the cost of a solution to the no-deletion cost problem; the converse is also true.

Our problem, but on a line network, is also closely related to the symmetric version of the Rectilinear Steiner Arborescence (StRSA) problem, for which a PTAS is known to exist [20]. It may or may not be possible to extend this to the grid networks we approximated. However, the problem solved in [20] was for the continuous case, while our motivation dictates addressing the discrete case.

A static problem with a trade-off between bandwidth and storage on tree networks was introduced in [19]. A generalization and a polynomial algorithm to find an optimal solution appeared in [10] (still for the static case).

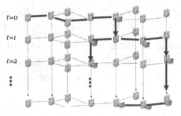

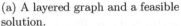

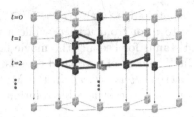

(a) A layered graph and a feasible solution.

(b) An h-ball with radius $\rho = 2$.

Fig. 1. An example of a layered graph, a feasible solution and an h-ball. Monitors represent movie requests.

Online migration and replication were tasks suggested in [8], but without dealing with the storage cost. A known online problem that resembles our problem somewhat is that of file allocation, see [7,6]. (In databases and systems contexts, see, e.g. [11] for a survey of early work). Like that problem, the one studied here can be viewed as a combination of the two tasks suggested by [8]. The similarity is in the possibility to replicate a file to a near by location to save in the cost of reading it. However, the pressure to reduce the number of replicas is different, yielding different algorithms. There, when writing a file, one needs to write *all* the replicas. Hence, the algorithms there are motivated to reduce the number of replicas that are "far away" from each other. On the other hand, our algorithms are more likely to eliminate replicas *especially* when they are *not* "far away" (since then bringing a copy later, when needed, will be cheap).

2   Preliminaries

Problem Definition. Given a graph $G = (V, E)$, construct another graph $\mathcal{G} = (\mathcal{V}, \mathcal{E})$, intuitively, by "layering" multiple copies of G, one per time unit. Connect each node in each copy to the same node in the next copy. (See Fig. 1a.) More formally, the node set $\mathcal{V}$ contains a *node replica* (sometimes called just a *replica*) $v(t)$ of every $v \in V$, for every time step $t \in \mathbb{N}$. Denote $V(t) = \{v(t) : v \in V\}$ and $E(t) = \{(v(t), u(t)) : (u, v) \in E\}$, for every $t \in \mathbb{N}$. It follows that $\mathcal{V} = \bigcup_t V(t)$. On the other hand, the set of edges contains not only $\mathcal{H} \triangleq \bigcup_t E(t)$, but also a set of directed edges $\mathcal{A} \triangleq \{(v(t), v(t+1)) : v \in V, t \in \mathbb{N}\}$. Pictorially, we term the edges in $\mathcal{H}$ (intuitively, edges of a copy of G) *horizontal edges*, and directed edges in $\mathcal{A}$ (connecting different copies of G) *vertical edges*. (For easy distinction, term the latter *arcs*). Following Fig. 1a, we say that $V(t)$ (and the nodes in $V(t)$) are *higher* than $V(t+1)$ (and its nodes).

In the MOVIE ALLOCATION problem we are given a graph $G = (V, E)$, an *origin* node $v_0 \in V$ (who is said to hold a movie copy initially), and a set of *requests* $\mathcal{R} \subseteq \mathcal{V}$. We assume that the storage cost is normalized to 1 and we are given a cost function $c : E \to \mathbb{R}_+$. A feasible solution is a subset of edges $\mathcal{F} \subseteq \mathcal{E}$ such that for every $v(t) \in \mathcal{R}$, there exists a path in $\mathcal{F}$ from $v_0(0)$ to $v(t)$.

A horizontal edge $(v(t), u(t)) \in \mathcal{F} \cap \mathcal{H}$ stands for sending a copy of the movie from v to u, or from u to v, while a vertical (directed) edge $(v(t), v(t+1)) \in \mathcal{F} \cap \mathcal{A}$ stands for keeping the movie in v's cache at time step t. See example in Fig. 1a. The cost of a vertical edge is 1, while the cost of an A horizontal edge $(v(t), u(t)) \in \mathcal{H}$ is $c(v, u)$. In the MOVIE ALLOCATION problem our goal is to find a minimum cost feasible solution.

Online Model. In the online version of MOVIE ALLOCATION we need to make two types of decisions at each time t, before seeing future requests: (1) At which nodes to cache a copy of the movie for future use (that is, add vertical edges); (2) From which current (time t) cache to serve the requests (by adding horizontal edges) that arrive at time t. We assume that the online algorithm may replicate the movie for efficient delivery, but at least one copy of the movie must remain in the network at all times. Alternatively, the system (but not the algorithm) can have the option to delete the movie altogether—this decision is then made known to the algorithm. We argue that this is a natural assumption. Normally, content is not destroyed, but is kept somewhere at least for archival purposes, or because somebody will eventually request. (The heavy tail nature of requests for Internet content, means that it is rather typical that "not so popular" content is sometimes requested nevertheless). If this content resides somewhere in the world and can be brought into the system for some payment, then our algorithms can be adapted to handle this case nevertheless.

Anyhow, without this assumption, no competitive algorithm exists. It is easy to see that an online algorithm cannot eliminate the last copy in any case, so it spends unboundedly more than an offline algorithm who may delete it. A similar assumption was made in the case of the file allocation problem [7,6].

Definitions and Notation. Denote $n = |V|$. Let $\mathcal{R} = \{v_1(t_1), \ldots, v_N(t_N)\}$, where $t_j \leq t_{j+1}$, for every j. Namely, we assume that the requests are sorted in a non-decreasing order of time step. Define $\mathcal{R}_t = \mathcal{R} \cap V(t)$. When G is the line network, we assume that $V = \{0, \ldots, n-1\}$, where 0 is the origin. We write for two requests, $i_j(t_j), i_k(t_k)$ that $i_j(t_j) \prec i_k(t_k)$ if either $t_j < t_k$ or $t_j = t_k$ and $i_j < i_k$. This total order implies that if $\mathcal{R} \neq \emptyset$ then it has a minimal request.

Given two replicas $v(t)$ and $u(s)$, let $\mathcal{P}[v(t), u(s)]$ be the set of edges that comprises some arbitrary shortest path from $v(t)$ to $u(s)$, and let $d(v(t), u(s)) \triangleq \sum_{e \in \mathcal{P}[v(t), u(s)]} c(e)$. This path is unique in some important cases, e.g. if $u = v$, or if G is a line and $s = t$.

In our analysis, OPT is the set of edges in some optimal solution whose cost is $c(\text{OPT})$. In the unit costs case, where $c(e) = 1$, for every e, the cost of the optimal solution is denoted by $|\text{OPT}|$.

Half Balls. Our algorithms and their analysis are based on the following notion. Intuitively, a *half-ball*, or an *h-ball*, of *radius* $\rho \in \mathbb{N}$ centered at a node $v(t)$ contains every node from which there exists a path of length ρ to $v(t)$. See Fig. 1b. We term them *half-balls* to emphasize that there is no path from a replica $u(t+1)$ to a replica $v(t)$. Recall that we assume that the cost $c(e)$ of an

---

Algorithm 1. Store $(v_j(t_j))$

1: Let u be the closest node with a copy of the movie at time t_j, and let $\rho_j = d(u, v_j)/4$.
2: Satisfy request $v_j(t_j)$ via $u(t_j)$.
3: Hold movie until $t_j + 2\rho_j$.   ▷ v_j keeps its copy
4: Hold movie as long as this is the only copy.

---

edge $e \in E$ can be of any (positive) value. For convenience of description, we define h-balls for the unit cost case. This eases the definition since in this case each edge is either completely inside the h-ball, or completely outside of it.

Definition 1 (H-ball). *Given a node $v(t) \in \mathcal{V}$ and a radius $\rho \in \mathbb{N}$, the h-ball of radius ρ centered at $v(t)$ is defined as $\mathcal{V}[v(t), \rho] = \{u(s) : d(u(s), v(t)) \leq \rho\}$. Also, $\mathcal{E}[v(t), \rho]$ denotes the set of edges of the h-ball. Such is every edge $(v_1(t_1), v_2(t_2)) \in \mathcal{E}$ if both its endpoints $v_1(t_1), v_2(t_2) \in \mathcal{V}[v(t), \rho]$.*

A simple property of the h-balls implies a bound on the optimum:

Lemma 1. *Let $\mathcal{R} = \{v_1(t_1), \ldots, v_k(t_N)\}$ be a set of requests, and let $\rho_1, \ldots, \rho_N$ be a series of radii, such that $\rho_j \leq d(v_j(t_j), v_0(0))$, for every j. If every edge from $\mathcal{E}$ is contained in at most α h-balls from $\mathcal{E}(v_1(t_1), \rho_1), \ldots, \mathcal{E}(v_k(t_k), \rho_k)$, then $|\text{OPT}| \geq \frac{1}{\alpha}\sum_j \rho_j$.*

Proof. Let $\mathcal{F}$ be a optimal solution. $\mathcal{F}$ must contain a path from $v_0(0)$ to $v_j(t_j)$, for every j, hence $|\mathcal{F} \cap \mathcal{E}(v_j(t_j), \rho_j)| \geq \rho_j$. Since each edge is contained in at most α h-balls, we have $|\mathcal{F}| \geq \frac{1}{\alpha}\sum_j |\mathcal{F} \cap \mathcal{E}(v_j(t_j), \rho_j)| \geq \frac{1}{\alpha}\sum_j \rho_j$.   □

A similar result can be shown for general costs. Observe that in this case $\mathcal{E}(v_1(t_j), \rho_j)$ may contain edge parts.

3   Online Algorithm for General Networks

Let us now present an $O(\log \delta)$-competitive online algorithm for general networks, and analyze it using h-balls. For ease of exposition we analyze the algorithm for unit costs. A matching lower bound is given in the full version.

Algorithm **Store** is rather simple: upon the arrival of a request at a node v at time t, the request is satisfied by obtaining a movie from the closest node u that has a copy of the movie. The movie is saved in v's cache for a period of time that is proportional to the distance between u and and v. If several requests arrive at the same time, we process them one by one in an arbitrary order.

To upper bound the competitive ratio, consider the h-balls $\mathcal{V}[v_1(t_1), \rho_1], \ldots, \mathcal{V}[v_n(t_n), \rho_n]$ induced by Algorithm **Store**. We bound the number of h-balls that contain the same edge, and then use Lemma 1.

Lemma 2. *If $\mathcal{V}[v_j(t_j), \rho_j] \cap \mathcal{V}[v_k(t_k), \rho_k] \neq \emptyset$, then either $\rho_k \geq 2\rho_j$ or $\rho_k \leq \frac{1}{2}\rho_j$.*

Proof. W.l.o.g., assume that $t_j \le t_k$. If $t_k > t_j + 2\rho_j$, then $\mathcal{V}[v_j(t_j), \rho_j] \cap \mathcal{V}[v_k(t_k), \rho_k] \neq \emptyset$ only if $\rho_k \ge 2\rho_j$. If $t_k \le t_j + 2\rho_j$ and $d(v_j, v_k) \le 2\rho_j$, we have that $\rho_k \le d(v_j, v_k)/4 \le \frac{1}{2}\rho_j$, since the request at $v_k(t_k)$ can be satisfied by receiving a copy from v_j who has a copy of the movie at time t_k. It remains to consider the case where $t_k \le t_j + 2\rho_j$ and $d(v_j, v_k) > 2\rho_j$. If $\mathcal{V}[v_j(t_j), \rho_j]$ and $\mathcal{V}[v_k(t_k), \rho_k]$ intersect, they must do so at time t_j. It follows that there is a node v such that $d(v_j, v) \le \rho_j$ and $d(v_k, v) \le \rho_k$, hence, $d(v_j, v_k) \le \rho_j + \rho_k$. On the other hand, $4\rho_k \le d(v_j, v_k)$, since v_j has a copy of the movie at time t_k. This leads to a contradiction, since $\rho_j + \rho_k \ge \frac{2d(v_j, v_k)}{2} > \frac{2\rho_j + 4\rho_k}{2} > \rho_j + \rho_k$. Hence, $\mathcal{V}[v_j(t_j), \rho_j]$ and $\mathcal{V}[v_k(t_k), \rho_k]$ do not intersect in this case.   □

Theorem 1. *Algorithm* **Store** *is an $O(\log \delta)$-competitive online algorithm for* MOVIE ALLOCATION.

Proof. Break the execution of the algorithm into phases. Two consecutive phases are separated by an interval in which some node v_ℓ holds the movie after $t_\ell + 2\rho_\ell$ (Line 4 is invoked). Let t'_ℓ denote the time at which the movie was removed from v_ℓ's cache. Clearly, **Store** is optimal within time interval $[t_\ell, t'_\ell]$, since every solution must hold at least one copy of the movie during $[t_\ell, t'_\ell]$.

Within the phases, **Store** pays at most $6\rho_j$ for servicing the jth request and for storing the movie at v_j. On the other hand, Lemma 2 implies that each edge in $\mathcal{E}$ is contained in at most $\log \frac{\max_i \rho_i}{\min_i \rho_i}$ h-balls from $\mathcal{E}(v_1(t_1), \rho_1), \ldots, \mathcal{E}(v_N(t_N), \rho_N)$. Thus, by Lemma 1 we know that $\sum_j \rho_j \le \log \frac{\max_i \rho_i}{\min_i \rho_i} \cdot \text{OPT} \le \log \delta \cdot \text{OPT}$. Hence, $|\textbf{Store}(\mathcal{R})| \le \sum_\ell (t'_\ell - t_\ell) + \sum_j 6\rho_j = O(\log \delta) \cdot \text{OPT}$.   □

4   Approximation Algorithm for the Grid Network

In this section, we present an $O(1)$-approximation algorithm for MOVIE ALLOCATION in the grid network. We describe the algorithm for the special case of unit costs. The description of the general case is omitted for lack of space.

Let us now expand the method of h-balls, to obtain a constant factor approximation also in grid networks. When using the ℓ_∞ norm (and the d_∞ distance), an h-ball centered at $v(t)$ with radius ρ is a half-cube or a *box*. See Fig. 2a. We use $\mathcal{V}_\infty[v(t), \rho]$ and $\mathcal{E}_\infty[v(t), \rho]$ to denote nodes and edges of a box of radius ρ centered at $v(t)$. If $d(v'(t'), v(t)) = \rho$, we say that $v'(t')$ is on the *border* of the h-ball. An edge $(v_1(t_1), v_2(t_2)) \in \mathcal{E}_\infty[v(t), \rho]$ is said to be on the border of the h-ball if both $v_1(t_1)$ and $v_2(t_2)$ are on the border of the h-ball. $\mathcal{E}_\infty(v(t), \rho)$ denotes the set of edges in $\mathcal{E}[v(t), \rho]$ that are not border edges. Note that Lemma 1 still holds for this kind of h-balls, since $\mathcal{E}[v(t), \rho] \subseteq \mathcal{E}_\infty[v(t), \rho]$.

Given a set of boxes $\mathcal{V}_\infty[v_1(t_1), \rho_1], \ldots, \mathcal{V}_\infty[v_{j-1}(t_{j-1}), \rho_{j-1}]$ in $\mathcal{G}$, we *grow* a box centered at $v_j(t_j)$, where $t_j \ge \max_{k=1}^{j-1} t_k$, by increasing its radius ρ_j until it collides with another box. We consider two possible collision types depending on the intersection.

Border Collision: There exists a collision node on the border of $\mathcal{V}_\infty[v_k(t_k), \rho_k]$.
There exists a node $u \in V$ such that $d_\infty(v_j, u) = \rho_j$ and $d_\infty(v_k, u) = \rho_k$.

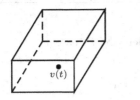
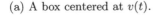

(a) A box centered at $v(t)$.

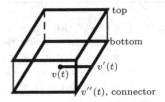

(b) A skeleton of a box centered at $v(t)$.

Fig. 2. Example of a box and its skeleton. The box is given on the left, and the box with its skeleton is given on the right. The thick line represent the skeleton.

Non-border Collision: The top of $V_\infty[v_j(t_j), \rho_j]$ is strictly contained in the bottom of $V_\infty[v_k(t_k), \rho_k]$.

For every $u \in V$ such that $d_\infty(v_j, u) \leq \rho_j$ we have that $d_\infty(v_k, u) < \rho_k$.

Algorithm **Skeleton** constructs a solution by assembling parts from each box. Note that the process of growing boxes induces a tree structure on the requests, namely, a request j is said to be the child of request k, if ρ_j was determined by a collision with the box of request k. (If there is more than one option, pick one arbitrarily.) The following definition is useful in describing parts of a box that are used for connecting the box to its parent box and children boxes. Specifically, the *skeleton* of a box $V_\infty(v(t), \rho)$ is defined as follows. (See example in Fig. 2b).

$$S[v(t), \rho] = P[v(t), v'(t)] \ \cup$$

bottom:
$$\{(u(t), u'(t)) : d_\infty(v, u) = d_\infty(v, u') = \rho\} \ \cup \tag{1}$$

connector:
$$P[v''(t), v''(t - \rho)] \ \cup \tag{2}$$

top:
$$\{(w(t - \rho), w'(t - \rho)) : d_\infty(v, w) = d_\infty(v, w') = \rho\} \tag{3}$$

where v' is some node such that $d(v, v') = \rho$, and v'' is some node such that $d_\infty(v, v'') = \rho$. Observe that the only vertical part of the skeleton is part (2). We refer to (1) and (3) as the bottom $\underline{S}[v(t), \rho]$ and the top $\overline{S}[v(t), \rho]$ of the skeleton, respectively. (Note that the top [respectively, the bottom] of the skeleton is a subset of the top [resp., the bottom] of the box. Node v'' is called the *connector* of the skeleton, since it is used to connect the bottom and the top of the skeleton. Node v' is called the *border gateway*, since it is used to connect the center of the box (node $v(t)$) to the border of the bottom of the skeleton.

Algorithm **Skeleton** is the first phase of the algorithm. It grows a box for each request and then adds the skeleton of the box to an edge multiset $\mathcal{F}$. We say a "multiset" to stress that all skeleton edges of a request are put in $\mathcal{F}$; we do not take advantage of cases where one of these edges already belongs to $\mathcal{F}$ (e.g. when they belong to the skeleton of a higher box). **Skeleton** copes only with border collisions. Later, we show how to transform $\mathcal{F}$ into a solution that copes with non-border collisions as well.

Algorithm 2. Skeleton $(\mathcal{G}, v_0, \mathcal{R})$

1: $\mathcal{F} \leftarrow \emptyset$
2: **for** $j = 1 \rightarrow N$ **do**
3:     Let ρ_j be the collision radius of $v_j(t_j)$, and let k be j's parent
4:     **if** $\exists u$ such that $d_\infty(u, v_j) = \rho_j$ and $d_\infty(u, v_k) = \rho_k$ **then** let $v_j'' = u$
          else let v_j'' be some node such that $d_\infty(v_j'', v_j) = \rho_j$
5:     $\mathcal{F} \leftarrow \mathcal{F} \cup \mathcal{S}[v_j(t_j), \rho_j]$ using v_j'' as the connector
6: **end for**
7: **return** $\mathcal{F}$

Each request adds $O(1)$ simple straight paths to the solution $\mathcal{F}$. Hence, it is easy to show that the running time of the algorithm is polynomial.

The set of edges returned by Algorithm **Skeleton** may be an infeasible solution. However, it is feasible if no non-border collision occurs during execution. This is proved in the following lemma. The lemma after that, analyzes the competitiveness (in that case).

Lemma 3. *Let $v_j(t_j)$ be the jth request, and let k be j's parent. If ρ_j was determined by a border collision, then there is a path from $v_k(t_k)$ to $v_j(t_j)$ in $\mathcal{F}$.*

Proof. First, observe that $d_\infty(v_j'', v_k) = \rho_k$ if a border collision occurs. Hence, $v_j(t_j)$ can be reached from $v_k(t_k)$ using the following path. First, go from $v_k(t_k)$ to $v_k'(t_k)$ (i.e. go on the skeleton of the higher box to the border of its bottom). Then, go from $v_k'(t_k)$ to $v_j''(t_k)$ (proceed on the bottom part of the higher skeleton to the connector node). From $v_j''(t_k)$, go to $v_j''(t_j)$ (stay on the connector node until the later time of the bottom of the lower box). Proceed (on the bottom part of the skeleton $\mathcal{S}[v_j(t_j), \rho_j)]$) to $v_j'(t_j)$ (the border gateway of the bottom of the lower box). Finally, go from $v_j'(t_j)$ to $v_j(t_j)$. $\qquad \square$

Lemma 4. $|\mathcal{F}| \leq 18 \cdot \text{OPT}$.

Proof. $|\mathcal{S}[v_j(t_j), \rho]| \leq 18\rho_j$, since parts (1) and (3) of $v_j(t_j)$'s skeleton contain up to $8\rho_j$ edges each, while the rest contain a total of $2\rho_j$ edges. The lemma follows from Lemma 1. $\qquad \square$

Coping with Non-border Collisions. Let us now show how to modify the above multiset $\mathcal{F}$, such that $\mathcal{F}$ contains a path from every request to its parent, even in the case of non-border collisions. The modification consists of two actions: (1) replacing certain edges of $\mathcal{F}$ by earlier (higher) copies of these edges, and (2) adding $O(|\text{OPT}|)$ new edges.

We describe the action of edge replacement using the notion of *edge ascension*. An edge $(v(t), u(t)) \in \mathcal{H} \cap \mathcal{F}$ is said to be ascended if it is replaced by the edge $(v(t'), u(t'))$, where $t' \leq t$ is the maximal time such that $(v(t'), u(t'))$ is located on the bottom of some box $\mathcal{V}_\infty[v_j(t_j), \rho_j]$. Pictorially, the edge is elevated until it reaches the bottom of some box. It may be that an edge is already located on the bottom of some box, in this case, $t' = t$. If no such time t' exists, then we say that the ascension failed and the edge is deleted.

Consider the jth request $v_j(t_j)$. Since $\mathcal{F}$ is a multiset, every edge in the top of $\overline{S}[v_j(t_j), \rho_j]$ has a copy (in $\mathcal{F}$) that was not used to connect a request to its parent (in Lemma 3). We try to ascend every edge of $\overline{S}[v_j(t_j), \rho_j]$, for every j, resulting in a revised (but not yet feasible) solution $\mathcal{F}'$.

Observation 1. $|\mathcal{F}'| \leq |\mathcal{F}|$.

Observation 2. Let $v_j(t_j)$ be a child request of $v_k(t_k)$, whose collision radius was determined by a non-border collision. Then, $\overline{S}[v_j(t_j), \rho_j] \subseteq \mathcal{F}'$.

Proof. Since we are dealing with a non-border collision, the edges in $\overline{S}[v_j(t_j), \rho_j]$ are located on the bottom of $\mathcal{V}_\infty[v_k(t_k), \rho_k]$ and therefore are not moved during the ascension process. $\qquad\square$

Consider a request $v_k(t_k)$ and let $u_1(s_1), \ldots, u_q(s_q)$ be those among its children, whose collision radii were determined by non-border collisions. We describe an edge set $\mathcal{F}_k$ such that $\mathcal{F}' \cup \mathcal{F}_k$ contains a path from $v_k(t_k)$ to $u_j(s_j)$ for every j. Let G_k be the subgraph of G that is induced by node set $V_k = \{v : d_\infty(v_k, v) \leq \rho_k\}$. We define a weight function on $E_k = E \cap (V_k \times V_k)$: $w(x, y) = 0$, if $(x(t_k), y(t_k)) \in \mathcal{F}'$, and $w(x, y) = 1$, otherwise. We now use Arora's PTAS for the STEINER TREE problem [5] to compute F_k, which is a $(1 + \varepsilon)$-approximate Steiner tree that connects $u_1'', \ldots, u_q''$ to v_k in G_k, where u_j'' is the connector of the skeleton of $u_j(s_j)$. Translating this set of edges of G to a set of edges in $\mathcal{G}$, we define $\mathcal{F}_k = \{(x(t_k), y(t_k)) : (x, y) \in F_k\}$.
    A subset $\mathcal{T} \subseteq \mathcal{E} \setminus \mathcal{F}'$ is called a *completion* of $\mathcal{F}'$ if $\mathcal{F}' \cup \mathcal{T}$ is a feasible solution.

Lemma 5. $\bigcup_k \mathcal{F}_k$ is a completion of $\mathcal{F}'$.

Proof. First, border collision children are connected to their parents by $\mathcal{F}'$. Let $u_j(s_j)$ be a non-border collision child of $v_k(t_k)$. $\mathcal{F}_k$ contains a path from $v_k(t_k)$ to $u_j''(t_k)$ and $\mathcal{F}'$ contains a path from $u_j''(t_k)$ to $u_j(s_j)$ that is located on the skeleton of $u_j(s_j)$. The lemma follows since the edges in $\overline{S}[u_j(s_j), \sigma_j]$ are not ascended (Observation 2). $\qquad\square$

It remains to bound the completion's. For that, we show that any completion $\mathcal{T}$ of $\mathcal{F}'$ can be transformed into some $\mathcal{T}'$ that consists of Steiner trees, one in each $E_k(t_k)$, for every k, without increasing its cost. (Unfortunately, it was too difficult to show a similar claim for $\mathcal{F}$, which is the reason we defined $\mathcal{F}'$).

Lemma 6. $|\bigcup_k \mathcal{F}_k| \leq (1 + \varepsilon)\mathrm{OPT}$.

Proof. We show that there exists an optimal completion of $\mathcal{F}'$, denoted by $\mathcal{T}'$, that consists of Steiner trees, one in $E_k(t_k)$, for every k. Since any feasible solution is also a completion of $\mathcal{F}'$ it follows that $|\bigcup_k \mathcal{F}_k| = \sum_k |F_k| \leq \sum_k (1 + \varepsilon)|T_k| \leq (1 + \varepsilon)|\mathcal{T}'| \leq (1 + \varepsilon)\mathrm{OPT}$, where T_k is an optimal Steiner tree in G_k.
    It remains to show that there exists an optimal completion of $\mathcal{F}'$ that consists of Steiner trees, one in $E_k(t_k)$, for every k. Let $\mathcal{T}$ be an optimal completion of $\mathcal{F}'$. We modify $\mathcal{T}$ and obtain another optimal completion $\mathcal{T}'$ of $\mathcal{F}'$ that consists of Steiner trees in G_k, for all k. We perform the following operations on $\mathcal{T}$:

1. We perform edge ascension on edges in $(\mathcal{T} \cap \mathcal{H}) \setminus \bigcup_j \mathcal{E}_\infty[v_j(t_j), \rho_j]$, namely on all horizontal edges in $\mathcal{T}$ that are not located within some box.

2. For every connected component $\mathcal{C}$ of $\mathcal{T} \cap \mathcal{E}_\infty[v_j(t_j), \rho_j]$, for some j, we do the following. Let $x(\tau)$ be the root of $\mathcal{C}$. If $\tau = t_j - \rho_j$ and $\mathcal{C}$ contains a node at t_j, then add a shortest path from $v_j(t_j)$ to $x(t_j)$.

3. We perform $edge\ descension$ on edges in $(\mathcal{T} \cap \mathcal{H}) \cap \bigcup_j \mathcal{E}_\infty[v_j(t_j), \rho_j]$, namely we move all horizontal edges in $\mathcal{T}$ that are located within some box $\mathcal{E}_\infty[v_j(t_j), \rho_j]$ to time t_j.

4. We remove the edges in $(\mathcal{T} \cap \mathcal{A}) \cap \bigcup_j \mathcal{E}_\infty[v_j(t_j), \rho_j]$.

The resulting set of edges is denoted by $\mathcal{T}'$.

Observe that in Operations 1 and 3, we move edges, and therefore we do not increase the number of edges. Operation 2 adds at most ρ_j edges for every component $\mathcal{C}$ that satisfies the conditions, while Operation 4 deletes at least ρ_j edges for every such component. It follows that $|\mathcal{T}'| \le |\mathcal{T}|$.

It remains to show that $\mathcal{T}'$ is a feasible completion of $\mathcal{F}'$. Consider a request $v_k(t_k)$ and let $u_1(s_1), \ldots, u_q(s_q)$ be those among its children, whose collision radii $\sigma_1, \ldots, \sigma_q$ were determined by non-border collisions. We distinguish between two types of children of $v_k(t_k)$ according to the last node $x_j(\tau_j)$ on the path from $v_0(0)$ to a child $u_j(s_j)$ in $\mathcal{F}' \cup \mathcal{T}$ that satisfies one of the following two conditions:

1. $d_\infty(x_j, v_k) = \rho_k$ and $\tau_j \in [t_k - \rho_k, s_j]$.
2. $d_\infty(x_j, v_k) < \rho_k$ and $\tau_j = t_k - \rho_k$.

Consider a Type 1 child $u_j(s_j)$. Due to Operations 1 and 3, the path from $x_j(\tau_j)$ to $u_j(s_j)$ is moved to time t_k. However, some edges may be missing due to the following reasons:

1. An edge ascension may be blocked by a box that is located beneath the box of $v_k(t_k)$.
2. An edge may be contained in a box located beneath the box of $v_k(t_k)$, and in this case it descends.

However, the missing parts in the path from $x_j(\tau_j)$ to $u_j(s_j)$ at time t_k are replaced by the ascended skeleton edges that are contained in $\mathcal{F}'$. Since $x_j(t_k)$ is located on the bottom part of the skeleton of $v_k(t_k)$, $\mathcal{T}'$ contains a path that connects $v_k(t_k)$ to the top part of the skeleton of $u_j(s_j)$, and therefore to $u_j(s_j)$.

A Type 2 child is handled similarly with one difference. Node $x_j(t_k)$ is not located on the bottom part of the skeleton of $v_k(t_k)$. However, it is connected to $v_k(t_k)$ due to Operation 2. The lemma follows since $\mathcal{T}'$ induces a Steiner tree in G_k, for all k. □

It is not hard to show that the running time of **Skeleton**, as well as of the computation of the completion, is polynomial. Hence, Lemmas 4 and 6 lead us to the main result of this section:

Theorem 2. *There exists a polynomial-time $(19 + \varepsilon)$-approximation algorithm for* MOVIE ALLOCATION, *for every $\varepsilon > 0$.*

References

1. Net-HD Consortium, http://www.nethd.org.il
2. Cisco visual networking index: Usage study. Cisco Visual Networking Index (October 25, 2010), http://www.cisco.com/en/US/solutions/collateral/ns341/ns525/ns537/ns705/Cisco_VNI_Usage_WP.html
3. Hyperconnectivity and the approaching zettabyte era. Cisco Visual Networking Index (June 2, 2010), http://www.cisco.com/en/US/solutions/~collateral/ns341/ns525/ns537/ns705/~ns827/VNI_Hyperconnectivity_WP.html
4. Abell, J.C.: Netflix instant account for 20 percent of peek U.S. bandwidth use. Wire Magazine (October 21, 2010)
5. Arora, S.: Polynomial time approximation schemes for euclidean traveling salesman and other geometric problems. Journal of the ACM 45(5), 753–782 (1998)
6. Awerbuch, B., Bartal, Y., Fiat, A.: Competitive distributed file allocation. In: 25th ACM STOC, pp. 164–173 (1993)
7. Bartal, Y., Fiat, A., Rabani, Y.: Competitive algorithms for distributed data management. In: 24th ACM STOC, pp. 39–50 (1992)
8. Black, D., Sleator, D.: Competitive algorithms for replication and migration problems. Tech. Rep. CMU-CS-89-201, CMU (1989)
9. Charlikar, M., Halperin, D., Motwani, R.: The dynamic servers problem. In: 9th Annual Symposium on Discrete Algorithms, pp. 410–419 (1998)
10. Cidon, I., Kutten, S., Soffer, R.: Optimal allocation of electronic content. Computer Networks 40(2), 205–218 (2002)
11. Dowdy, L.W., Foster, D.V.: Comparative models of the file assignment problem. ACM Comp. Surveys 14(2), 287–313 (1982)
12. Garey, M.R., Graham, R.L., Johnson, D.S.: The complexity of computing steiner minimal trees. SIAM J. on Applied Math. 32(4), 835–859 (1977)
13. Hwang, F.K., Richards, D.S.: Steiner tree problems. Networks 22(1), 55–89 (1992)
14. Jones, D.: Proof that telstra is throtteling youtube bandwidth in australia, http://www.eevblog.com, http://www.youtube.com/watch?v=9iDRynyBl0c
15. Kopytoff, V.G.: Shifting online, Netflix faces new competition. New York Times (Business Day, Technology) (September 26, 2010)
16. Papadimitriou, C., Ramanathan, S., Rangan, P.: Information caching for delivery of personalized video programs for home entertainment channels. In: IEEE International Conf. on Multimedia Computing and Systems, pp. 214–223 (1994)
17. Papadimitriou, C., Ramanathan, S., Rangan, P.: Optimal Information Delivery. In: Staples, J., Katoh, N., Eades, P., Moffat, A. (eds.) ISAAC 1995. LNCS, vol. 1004, pp. 181–187. Springer, Heidelberg (1995)
18. Papadimitriou, C., Ramanathan, S., Rangan, P., Sampathkumar, S.: Multimedia information caching for personalized video-on demand. Computer Communications 18(3), 204–216 (1995)
19. Schaffa, F., Nussbaumer, J.P.: On bandwidth and storage tradeoffs in multimedia distribution networks. In: IEEE INFOCOM, pp. 1020–1026 (1995)
20. Xiuzhen Cheng, B.D., Lu, B.: Polynomial time approximation scheme for symmetric rectilinear steiner arborescence problem. J. Global Optim. 21, 385–396 (2001)

Super-Fast Distributed Algorithms for Metric Facility Location

Andrew Berns, James Hegeman, and Sriram V. Pemmaraju*

Department of Computer Science
The University of Iowa
Iowa City, Iowa 52242-1419, USA
{andrew-berns,james-hegeman,sriram-pemmaraju}@uiowa.edu

Abstract. This paper presents a distributed $O(1)$-approximation algorithm in the $\mathcal{CONGEST}$ model for the metric facility location problem on a size-n clique network that has an expected running time of $O(\log \log n \cdot \log^* n)$ rounds. Though metric facility location has been considered by a number of researchers in low-diameter settings, this is the first sub-logarithmic-round algorithm for the problem that yields an $O(1)$-approximation in the setting of non-uniform facility opening costs. Since the facility location problem is specified by $\Omega(n^2)$ bits of information, any fast solution in the $\mathcal{CONGEST}$ model must be truly distributed. Our paper makes three main technical contributions. First, we show a new lower bound for metric facility location. Next, we demonstrate a reduction of the distributed metric facility location problem to the problem of computing an $O(1)$-ruling set of an appropriate spanning subgraph. Finally, we present a sub-logarithmic-round (in expectation) algorithm for computing a 2-ruling set in a spanning subgraph of a clique. Our algorithm accomplishes this by using a combination of randomized and deterministic sparsification.

1 Introduction

This paper explores the design of "super-fast" distributed algorithms in settings in which bandwidth constraints impose severe restrictions on the volume of information that can quickly reach an individual node. As a starting point for our exploration, we consider networks of diameter one (i.e., cliques) so as to focus on bandwidth constraints only and avoid penalties imposed by network distance between nodes. We assume the standard $\mathcal{CONGEST}$ model [19], which is a synchronous message-passing model in which each node in a size-n network can send a message of size $O(\log n)$ along each incident communication link in each round. By "super-fast" algorithms we mean algorithms whose running time is strictly sub-logarithmic, in any sense - deterministic, in expectation, or with high probability (w.h.p.). Several researchers have previously considered the design of such "super-fast" algorithms; see [10,12,18] for good examples of relevant concepts. The working hypothesis is that in low-diameter (communication) settings, where congestion, rather than network distance, is the main bottleneck,

---
* This work is supported in part by National Science Foundation grant CCF 0915543

A. Czumaj et al. (Eds.): ICALP 2012, Part II, LNCS 7392, pp. 428–439, 2012.
© Springer-Verlag Berlin Heidelberg 2012

we should be able to design algorithms that are much faster than algorithms for "local" problems in high-diameter settings.

The focus of this paper is the *distributed facility location* problem, which has been considered by several researchers [6,14,16,17] in low-diameter settings. We first describe the sequential version of the problem. The input to the facility location problem consists of a set of *facilities* $\mathcal{F} = \{x_1, x_2, \ldots, x_m\}$, a set of *clients* $\mathcal{C} = \{y_1, y_2, \ldots, y_n\}$, an *opening cost* f_i associated with each facility x_i, and a *connection cost* $D(x_i, y_j)$ between each facility x_i and client y_j. The goal is to find a subset $F \subseteq \mathcal{F}$ of facilities to *open* so as to minimize the facility opening cost plus connection costs, i.e.,

$$FacLoc(F) := \sum_{x_i \in F} f_i + \sum_{y_j \in \mathcal{C}} D(F, y_j)$$

where $D(F, y_j) := \min_{x_i \in F} D(x_i, y_j)$. Facility location is an old and well-studied problem in operations research [1,3,4,8,21] that arises in contexts such as locating hospitals in a city or locating distribution centers in a region.

The *metric facility location* problem is an important special case of facility location in which the connection costs satisfy the following "triangle inequality:" for any $x_i, x_{i'} \in \mathcal{F}$ and $y_j, y_{j'} \in \mathcal{C}$, $D(x_i, y_j) + D(y_j, x_{i'}) + D(x_{i'}, y_{j'}) \geq D(x_i, y_{j'})$. The facility location problem, even in its metric version, is NP-complete and finding approximation algorithms for the problem has been a fertile area of research. A series of constant-factor approximation algorithms have been proposed for the metric facility location problem, with a steady improvement in the constant specifying the approximation factor. See [11] for a recent 1.488-approximation algorithm. This result is near-optimal because it is known [7] that the metric facility location problem has no polynomial-time algorithm yielding an approximation guarantee better than 1.463 unless $NP \subseteq DTIME(n^{O(\log \log n)})$. For non-metric facility location, a simple greedy algorithm yields an $O(\log n)$-approximation, and this is also optimal (to within a constant factor) because it is easy to show that the problem is at least as hard as set cover.

More recently, the facility location problem has also been used as an abstraction for the problem of locating resources in wireless networks [5,15]. Motivated by this application, several researchers have considered the facility location problem in a distributed setting. In [14,16,17], the underlying communication network is a complete bipartite graph with $\mathcal{F}$ and $\mathcal{C}$ forming the bipartition. At the beginning of the algorithm, each node, whether it is a facility or a client, has knowledge of the connection costs between itself and all nodes in the other part. In addition, the facilities know their opening costs. In [6], the underlying communication network is a clique. Each node in the clique may choose to open as a facility, and each node that does not open will connect to an open facility. Note that all of the aforementioned work assumes the $\mathcal{CONGEST}$ model of distributed computation. The facility location problem considered in [15] assumes that the underlying communication network is a *unit disk graph* (UDG), and also considers the $\mathcal{LOCAL}$ model. While such a network can be of high diameter relative to the number of nodes in the network, this paper [15] reduces the UDG facility location problem to a low-diameter facility location-type problem and uses this in the eventual solution.

None of the prior papers, however, achieve near-optimal approximation (i.e., constant-factor in the case of metric facility location and $O(\log n)$ in the case of non-metric facility location) in *sub-logarithmic* rounds. While [6] does present a *constant-round*, constant-factor approximation to metric facility location on a clique, it is only for the special case of *uniform* metric facility location, i.e., when all facility opening costs are identical. The question that drives this paper, then, is the following: Can we achieve a distributed constant-factor approximation algorithm for the metric facility location problem in the clique setting in strictly sub-logarithmic time? One can ask similar questions in the bipartite setting and for non-metric facility location as well, but as a first step we focus on the metric version of the facility location problem on a clique.

Distributed facility location is challenging even in low-diameter settings because the input consists of $\Theta(m \cdot n + m)$ pieces of information, distributed across the network, which cannot quickly be delivered to a single node (or even a small number of nodes) due to the bandwidth constraints of the $\mathcal{CONGEST}$ model. Therefore, any fast distributed algorithm for the problem must be truly distributed and needs to take advantage of the available bandwidth and of structural properties of approximate solutions. Also worth noting is that even though tight lower bounds on the running times of distributed approximation algorithms have been established [9], none of these bounds extend to low-diameter settings.

2   Results

Main result. The main result of this paper is an $O(1)$-approximation algorithm for metric facility location on a clique which has an expected running time of $O(\log \log n \cdot \log^* n)$ rounds. If the metric satisfies additional properties (e.g., it has constant doubling dimension), then we obtain an $O(\log^* n)$-round $O(1)$-approximation to the problem. Our results are achieved via a combination of techniques that include (i) a new lower bound for the optimal cost of metric facility location and (ii) a sparsification technique combining randomization with a deterministic subroutine that repeatedly leverages the available bandwidth to process sparse subgraphs.

2.1   Overview of Technical Contributions

We start by describing the distributed facility location problem on a clique, as in [6]. Let (P, D) be a discrete metric space with point set $P = \{p_1, p_2, \ldots, p_n\}$. Let f_i be the opening cost of p_i. Now view the metric space (P, D) as a completely connected size-n network $C = (P, E)$ with each point p_i represented by a node, which we will also call p_i. Each node p_i knows f_i and the connection costs (distances) $D(p_i, p_j)$ for all $p_j \in P$. The problem is to design a distributed algorithm that runs on C in the $\mathcal{CONGEST}$ model and produces a subset $X \subseteq P$ such that each node $p_i \in X$ opens and provides services as a facility and each node $p_i \notin X$ connects to the nearest open node. The goal is to guarantee that $FacLoc(X) \leq \alpha \cdot OPT$, where OPT is the cost of an optimal solution to the given instance of facility location and α is some constant. We call this the CLIQUEFACLOC problem. Of course, we also want our algorithm to be "super-fast" (in some sense), i.e., terminate in $o(\log n)$ rounds.

Our paper makes three main technical contributions.

- **A new lower bound for metric facility location.** For $p \in P$, let $B(p, r)$ denote the set of points $q \in P$ satisfying $D(p, q) \leq r$. For each p_i, let r_i be the non-negative real number satisfying

$$\sum_{q \in B(p_i, r_i)} (r_i - D(p_i, q)) = f_i$$

As observed by Mettu and Plaxton [13], r_i exists and is unique. Bădoiu et al. proved in [2] that $\sum_{i=1}^{n} r_i$ is a constant-factor approximation for OPT in the case of *uniform* facility opening costs. This fact plays a critical role in the design of the constant-round, constant-factor approximation algorithm of Gehweiler et al. [6] for the special case of CLIQUEFACLOC in which all facility opening costs are identical. However, the sum $\sum_{i=1}^{n} r_i$ can be arbitrarily large compared to OPT when the f_i's are allowed to vary. (Consider an example consisting of only two nodes, one of whose opening costs is large in comparison to the other and to the distance between them.) Therefore, we apply the idempotent transformation

$$r_i \rightarrow \bar{r}_i = \min_{1 \leq j \leq n} \{D(p_i, p_j) + r_j\},$$

and use $\bar{r}_i$ instead of r_i to derive a lower bound. Note that for any i, $\bar{r}_i \leq r_i$. We show later that $\sum_{i=1}^{n} \bar{r}_i$ bounds the optimal cost OPT from below (to within a constant factor) in the general case of non-uniform facility opening costs.

- **Reduction to an $O(1)$-ruling set.** Our next contribution is an $O(1)$-round reduction of the distributed facility location problem on a clique to the problem of computing an $O(1)$-ruling set of a specific spanning subgraph. Let $C' = (P, E')$ be a spanning subgraph of C. A subset $Y \subseteq P$ is said to be *independent* if no two nodes in Y are neighbors in C'. An independent set Y is a *maximal independent set* (MIS) if no superset $Y' \supset Y$ is independent in C'. An independent set Y is β-ruling if every node in P is at most β hops along edges in C' from some node in Y. Clearly, an MIS is a 1-ruling set. We describe an algorithm that approximates distributed facility location on a clique by first computing a spanning collection of subgraphs $C_1, C_2, C_3, \ldots$ in $O(1)$ rounds. Then we show that a solution to the facility location problem (i.e., a set of nodes to open) can be obtained by computing a β-ruling set for each of the subgraphs C_j, $j \geq 1$, and combining the solutions in a certain way. We show that combining the β-ruling sets can also be done in $O(1)$ rounds. The parameter β affects the approximation factor of the computed solution and enforcing $\beta = O(1)$ ensures that the solution to facility location is an $O(1)$-approximation.

– **An $O(1)$-ruling set via a combination of randomized and deterministic sparsification.** We present an expected-$O(\log\log n \cdot \log^* n)$-round algorithm for computing a 2-ruling set of a given spanning subgraph C' of a clique C. We start by describing a deterministic "subroutine" that takes a subset $Z \subseteq P$ as input and computes an MIS of $C'[Z]$ (i.e., the subgraph of C' induced by Z) in c rounds if $C'[Z]$ has at most $c \cdot n$ edges. This is achieved via a simple load-balancing scheme that communicates the entire subgraph $C'[Z]$ to all nodes in c rounds. We then show how to use randomization repeatedly to peel off subgraphs with linearly many edges for processing by the aforementioned subroutine. In this manner, the entire graph C' can be processed using a number of calls, to the deterministic subroutine, which is $O(\log\log n \cdot \log^* n)$ in expectation.

3   Reduction to the $O(1)$-Ruling Set Problem

3.1   A New Lower Bound for Non-uniform Metric Facility Location

In this subsection we show that $\sum_{i=1}^{n} \bar{r}_i$ is a constant-factor lower bound to the optimal cost OPT. To facilitate this, we recall a definition from Mettu and Plaxton [13]. The $charge(\cdot, \cdot)$ of a node p_i with respect to a collection of (open) facilities X (also known as a *configuration*) is defined by

$$charge(p_i, X) = D(p_i, X) + \sum_{p_j \in X} \max\{0, r_j - D(p_j, p_i)\}$$

where $D(p_i, X) = \min_{p_j \in X} D(p_i, p_j)$. Mettu and Plaxton showed that the cost of a configuration X, $FacLoc(X)$, is precisely equal to the sum of the charges with respect to X, i.e. $\sum_{i=1}^{n} charge(p_i, X)$ [13].

The Mettu-Plaxton configuration F_{MP} is derived from the Mettu-Plaxton algorithm [13]. This algorithm, referred to as the MP-algorithm, is a sequential, greedy algorithm for facility location on a clique in which facilities open (sequentially) precisely when there is no already-open facility within distance $2 \cdot r_i$ [13]. Our algorithm borrows from the core ideas of the MP-algorithm.

The F_{MP} configuration was shown to have a cost at most three times OPT [13]. So for any configuration X,

$$FacLoc(X) \geq \frac{1}{3}FacLoc(F_{MP}) = \frac{1}{3}\sum_{i=1}^{n} charge(p_i, F_{MP})$$

We now present the following lemma, which relates $FacLoc(X)$ (for any X) to $\sum_{i=1}^{n} \bar{r}_i$.

Lemma 1. $FacLoc(X) \geq (\sum_{i=1}^{n} \bar{r}_i)/6$ *for any configuration X.*

3.2   Algorithm

We present our facility location algorithm in Algorithm 1. Our distributed algorithm consists of three stages. We use the notations $G[H]$ and $E[G]$ to refer to the subgraph induced by H and the edge set of G, respectively.

Algorithm 1. FACILITYLOCATION

Input: A discrete metric space of nodes (P, D), with opening costs;
a sparsity parameter s
Assumption: Each node knows its own opening cost and the distances from itself to other nodes
Output: A subset of nodes (a *configuration*) to be declared open

1.   Each node p_i computes and broadcasts its value r_i; $r_0 := \min_i r_i$.
2.   Each node computes a partition of the network into classes H_k with
     $$3^k \cdot r_0 \leq r_j < 3^{k+1} \cdot r_0 \text{ for } p_j \in H_k.$$
3.   Each node determines its neighbors within its own class;
     for $p_i, p_j \in H_k$, $(p_i, p_j) \in E[G[H_k]]$ if and only if $D(p_i, p_j) \leq r_i + r_j$.
4.   All nodes now use procedure RULINGSET($\bigcup_k G[H_k]$, s) to determine, for
     each k, a sparse set $T_k \subseteq H_k$.
5.   Each node broadcasts its membership status with respect to the sparse set
     T_k of its own class.
6.   A node $p_i \in H_k$ declares itself to be open if:
     (i) p_i is a member of the sparse set $T_k \subseteq H_k$, and
     (ii) There is no node p_j belonging to a class $H_{k'}$, with $k' < k$,
         such that $D(p_i, p_j) \leq 2r_i$.
7.   Each node broadcasts its status (open or not), and nodes connect.

Stage 1 (Steps 1-2). Each node knows its own opening cost and the distance to other nodes, so node p_i computes r_i and broadcasts that value to all others. Once this is complete, each node knows all of the r_i values. Next, every node computes a partition of the network into groups whose r_i values vary by at most a factor of 3 (Step 2). Specifically, let $r_0 := \min_{1 \leq j \leq n} \{r_j\}$, and define the class H_k to be the set of nodes p_i such that $3^k \cdot r_0 \leq r_i < 3^{k+1} \cdot r_0$. Every node computes the class into which each node in the network, including itself, falls.

Stage 2 (Steps 3-5). We now focus our attention on class H_k. Suppose $p_i, p_j \in H_k$. We define p_i and p_j to be *adjacent* in class H_k if $D(p_i, p_j) \leq r_i + r_j$. Each node in H_k can determine its neighbors in H_k. We refer to the graph on nodes in H_k induced by this adjacency condition as $G[H_k]$. Next, consider the graph (on all n nodes) $\bigcup_k G[H_k]$ which is the union of all induced graphs $G[H_k]$. A sparse set of nodes in $\bigcup G[H_k]$ determines, for each k, a sparse set $T_k \subseteq H_k$. Therefore, apply procedure RULINGSET() to $\bigcup_k G[H_k]$. We describe a super-fast-expected-time implementation of RULINGSET() in Section 4. After a sparse set has been constructed for each class H_k, each node broadcasts its membership status.

Stage 3 (Steps 6-7). Finally, a node p_i in class H_k opens if (i) $p_i \in T_k$, and (ii) there is no node $p_j \in B(p_i, 2r_i)$ of a class $H_{k'}$ with $k' < k$. Open facilities declare themselves via broadcast, and every node connects to the nearest open facility.

3.3   Analysis

We show that our algorithm produces an $O(s)$-approximation to CLIQUEFA-CLOC in $O(\mathcal{T}(n, s))$ communication rounds, where $\mathcal{T}(n, s)$ is the running time (in rounds) of procedure RULINGSET().

Communication rounds. Stage 1 requires exactly one round of communication, to broadcast r_i values. Stage 2 requires $O(\mathcal{T}(n, s))$ rounds to compute the s-sparse subsets $\{T_k\}_k$, and an additional round to broadcast membership status. Stage 3 requires one round, in order to inform others of a nodes decision to open or not. Thus, the running time of our algorithm in communication rounds is $O(\mathcal{T}(n, s))$.

Cost approximation. Let F be the set of nodes opened by our algorithm. We analyze $FacLoc(F)$ by bounding $charge(p_i, F)$ for each p_i. Recall that $FacLoc(F) = \sum_{i=1}^n charge(p_i, F)$. Since $charge(p_i, F)$ is the sum of two terms, $D(p_i, F)$ and $\sum_{p_j \in F} \max\{0, r_j - D(p_j, p_i)\}$, we bound each separately by a $O(s)$-multiple of $\bar{r}_i$.

Algorithm 2. RULINGSET

Input: An undirected graph $G = (V, E)$; a sparsity parameter s
Assumptions: Each node has knowledge of its neighbors in G; each node can send a message to any other node (not just along edges of G)
Output: An independent s-ruling set $T \subseteq G$

The sparse subset $T_k \subseteq H_k$ has the property that for any node $p_i \in H_k$, $D(p_i, T_k) \le 2 \cdot 3^{k+1} r_0 \cdot s$, where s is the sparsity parameter passed to procedure RULINGSET(). Also, for no two members of T_k is the distance between them less than $2 \cdot 3^k r_0$. Note that here we are using distances from the metric D of (P, D).

Now, in our cost analysis, we consider a node $p_i \in H_k$. To bound $D(p_i, F)$, observe that either $p_i \in T_k$, or else there exists a node $p_j \in T_k$ such that $D(p_i, p_j) \le 2 \cdot 3^{k+1} r_0 \cdot s \le 6r_i \cdot s$. Also, if a node $p_j \in T_k$ does not open, then there exists another node $p_{j'}$ in a class $H_{k'}$, with $k' < k$, such that $D(p_j, p_{j'}) \le 2r_j$.

We are now ready to bound the components of $charge(p_i, F)$.

Lemma 2. *For all i, $D(p_i, F) \le (81s + 81) \cdot \bar{r}_i$.*

Lemma 3. *For all i, $\sum_{p_j \in F} \max\{0, r_j - D(p_j, p_i)\} \le 9\bar{r}_i$.*

Combining the two previous lemmas gives

$$FacLoc(F) = \sum_{i=1}^{n} charge(p_i, F) = \sum_{i=1}^{n} \left[D(p_i, F) + \sum_{p_j \in F} \max\{0, r_j - D(p_j, p_i)\} \right]$$

$$\leq \sum_{i=1}^{n} [(81s + 81) \cdot \bar{r}_i + 9\bar{r}_i] \leq (81s + 90) \cdot \sum_{i=1}^{n} \bar{r}_i$$

Theorem 1. *Algorithm 1 (*FACILITYLOCATION*) computes an $O(s)$-factor approximation to* CLIQUEFACLOC *in $O(\mathcal{T}(n, s))$ rounds, where $\mathcal{T}(n, s)$ is the running time of the* RULINGSET*() procedure called with argument s.*

4   Computing a 2-Ruling Set

The facility location algorithm in Section 3 depends on being able to efficiently compute an independent β-ruling set, for small β, of an arbitrary spanning subgraph C' of a clique C. This section describes how to compute an (independent) 2-ruling set of C' in a number of rounds which is $O(\log \log n \cdot \log^* n)$ in expectation.

4.1   A Useful Subroutine

We first present a deterministic subroutine for efficiently computing a maximal independent set of a sparse, induced subgraph of C'. For a subset $M \subseteq P$, we use $C'[M]$ to denote the subgraph of C' induced by M and $E[M]$ and $e[M]$ to denote the set and number (respectively) of edges in $C'[M]$. The subroutine we present below computes an MIS of $C'[M]$ in $e[M]/n$ rounds. Later, we use this repeatedly in situations where $e[M] = O(n)$.

We assume that nodes in P have unique identifiers and can therefore be totally ordered according to these. Let $\rho_i \in \{0, 1, \ldots, n - 1\}$ denote the rank of node p_i in this ordering. Imagine (temporarily) that edges are oriented from lower-rank nodes to higher-rank nodes and let $\mathcal{E}(p_i)$ denote the set of outgoing edges incident on p_i. Let d_i denote $|\mathcal{E}(p_i)|$, the outdegree of p_i, and let $D_i = \sum_{j:\rho_j < \rho_i} d_j$ denote the outdegree sum of lower-ranked nodes.

The subroutine shares the entire topology of $C'[M]$ with all nodes in the network. To do this efficiently, we map each edge $e \in E[M]$ to a node in P. Information about e will be sent to the node to which e is mapped. Each node will then broadcast information about all edges that have been mapped to it. See Algorithm 3.

Theorem 2. *Algorithm 3 computes an MIS L of $C'[M]$ in $\frac{e[M]}{n} + O(1)$ rounds.*

4.2   Algorithm

We are now ready to present an algorithm for computing a 2-ruling set of C' which is "super-fast" in expectation. We show that this algorithm has an expected running time of $O(\log \log n \cdot \log^* n)$ rounds. The algorithm proceeds in

---

Algorithm 3. Deterministic MIS for Sparse Graphs

Input: A subset of nodes $M \subseteq P$
Output: An MIS L of $C'[M]$
Algorithm executed by node p_i

1. Broadcast ID.
2. Calculate and broadcast d_i.
3. Assign a distinct label $\ell(e)$ from $\{D_i, D_i + 1, \ldots, D_i + d_i - 1\}$ to each incident outgoing edge e.
4. Send each outgoing edge e to a node p_j with rank $\rho_j = (\ell(e) \mod n)$.
5. Broadcast all edges received in previous step, one per round.
6. Compute $C'[M]$ from received edges and use a deterministic algorithm to locally compute MIS L.

Stages. In Stage i, $i = 1, 2, \ldots$ we process nodes whose degrees (in graph C') lie in the range $[n^{1/2^i}, n^{1/2^{i-1}})$. At the end of Stage i, every node has degree less than $n^{1/2^i}$; thus the algorithm consists of $O(\log \log n)$ Stages.

Each Stage consists of *Phases*. Consider the Stage in which we process the set $S(d)$ of nodes whose degrees are in the range $[d, d^2)$. In each Phase of this Stage, $|S(d)|$ decreases. To understand the rate at which this occurs, consider the function $t(k)$ defined recursively by $t(0) = 1$, $t(k+1) = e^{\sqrt{t(k)}}$, for all $k \geq 0$. This is a rapidly-growing function that reaches n in $O(\log^* n)$ steps. At the beginning of Phase k, $|S(d)| \leq n/t(k)$, and as Phase k proceeds, $S(d)$ shrinks. Phase k ends when $|S(d)| \leq \frac{n}{t(k+1)}$ (loop starting in Line 6). Because of the rate at which $t(k)$ grows, each Stage consists of $O(\log^* n)$ Phases.

Each Phase consists of *Iterations*. Consider the Stage in which we process nodes with degrees in $[d, d^2)$, and then consider an Iteration in Phase k of this Stage. In this Iteration, nodes in $S(d)$ join a set M independently with probability $q = \sqrt{t(k)}/d$ (Line 8). Nodes not in $S(d)$ join M with probability $1/\sqrt{d}$ (Line 9). The probability q is set such that the expected number of edges in $C'[M]$ is bounded above by $2n$. Once the set M is picked, we use Algorithm 3 to process $C'[M]$ in $O(1)$ rounds, and then we delete M and its neighborhood $N(M)$ (Lines 10-12). This ends an Iteration. In expectation, only a constant number of Iterations are needed to complete a Phase. Because the size of $S(d)$ decreases during a Phase, we can raise q (Line 15) while still ensuring that the expected number of edges in $C'[M]$ is $\leq 2n$. Within a Stage, q is increased until it reaches $1/\sqrt{d}$. As well, by the time q reaches $1/\sqrt{d}$, $S(d)$ will have diminished such that $|E[C'[S(d)]]| = O(n)$. The $S(d)$ remnant can then be processed in $O(1)$ time (Lines 18-20) to finish the Stage. See Algorithm 4.

4.3   Analysis

Lemma 4. *Algorithm 4 computes a 2-ruling set of C'.*

Algorithm 4. Super-Fast 2-Ruling Set

Input: A spanning subgraph C' of the clique C
Output: A 2-ruling set $T \subseteq P$ of C'

1. $i = 1; d := \sqrt{n}$ (equal to $n^{1/2^i}$); $T := \emptyset$
2. **while** $d > 10$ **do**
        Start of Stage i:
3. $k := 0; q := \frac{1}{d}$ (equal to $\sqrt{t(k)}/d$);
4. $S(d) := \{p \in P : deg_{C'}(p) \geq d\}$; $lastPhase := false$;
5.   **while** ($true$) **do**
            Start of Phase k:
6.   **while** ($|S(d)| > \frac{n}{t(k+1)}$ **and** $\neg lastPhase$)
                    or ($|S(d)| > \frac{n}{e^{\sqrt{d}}}$ **and** $lastPhase$) **do**
                Start of Iteration
7. $M := \emptyset$
8.   Add each $p \in S(d)$ to M with probability q.
9.   Add each $p \in P \setminus S(d)$ to M with probability $\frac{1}{\sqrt{d}}$.
10.   Compute an MIS L on M using Algorithm 3.
11. $T := T \cup L$
12.   Remove $(M \cup N(M))$ from C'.
13. $S(d) := \{p \in P : deg_{C'}(p) \geq d\}$
                End of Iteration
14.   **if** $lastPhase$ **then break**;
15. $q := \frac{\sqrt{t(k+1)}}{d}$; $k := k + 1$;
16.   **if** $q > \frac{1}{\sqrt{d}}$ **then**
17. $q := \frac{1}{\sqrt{d}}$; $lastPhase := true$;
            End of Phase
18. $M := S(d)$; Compute an MIS L on M using Algorithm 3.
19. $T := T \cup L$
20.   Remove $(M \cup N(M))$ from C'.
21. $d := n^{1/2^{i+1}}$; $i := i + 1$
        End of Stage
22. $M := C'$; Compute an MIS L on M using Algorithm 3.
23. $T := T \cup L$
24. Output T.

---

438   A. Berns, J. Hegeman, and S.V. Pemmaraju

Lemma 5. *For any $d \geq 0$, the smallest k for which $t(k) \geq d$ is $O(\log^* d)$.*

Lemma 6. *Consider Phase k in Stage i. Let $d = n^{1/2^i}$. Then the maximum degree (in C') of a node during Phase k is less than d^2. Furthermore, $|S(d)| \leq n/t(k)$.*

Lemma 7. *In any Iteration, the expected number of edges in the subgraph (of C') induced by M is bounded above by $2n$.*

Lemma 8. *Fix a Stage i, and suppose that $t(k) \leq d$. Then the expected number of Iterations (Lines 7-13) required in Phase k before $|S(d)| \leq \frac{n}{t(k+1)}$ is $O(1)$.*

Lemma 9. *Fix a Stage i, and suppose that $t(k) \leq d < t(k+1)$. Then the expected number of Iterations required in Phase $k+1$ before $|S(d)| \leq \frac{n}{e^{\sqrt{d}}}$ is $O(1)$, and at the end of Phase $k+1$, the number of edges in $C'[S(d)]$ is $\tilde{O}(n)$.*

Theorem 3. *Algorithm 4 computes a 2-ruling set of C' and has an expected running time of $O(\log \log n \cdot \log^* n)$ communication rounds.*

5   Conclusions

Using Algorithm 4 as a specific instance of the procedure RULINGSET() for $s = 2$ and combining Theorems 1 and 3 leads us to Theorem 4. We also note that under special circumstances an $O(1)$-ruling set of a spanning subgraph of a clique can be computed even more quickly. For example, if the subgraph of C induced by the nodes in class H_k is growth-bounded for each k, then we can use the Schneider-Wattenhofer [20] result to compute an MIS for $G[H_k]$ in $O(\log^* n)$ rounds (in the $\mathcal{CONGEST}$ model). It is easy to see that if the metric space (P, D) is Euclidean with constant dimension or even has constant doubling dimension, H_k would be growth-bounded for each k.

Theorem 4. *There exists an algorithm that solves the CLIQUEFACLOC problem with an expected running time of $O(\log \log n \cdot \log^* n)$ communication rounds.*

Theorem 5. *The CLIQUEFACLOC problem can be solved in $O(\log^* n)$ rounds on a metric space of constant doubling dimension.*

References

1. Balinski, M.: On finding integer solutions to linear programs. In: Proceedings of IBM Scientific Computing Symposium on Combinatorial Problems, pp. 225–248 (1966)
2. Bădoiu, M., Czumaj, A., Indyk, P., Sohler, C.: Facility Location in Sublinear Time. In: Caires, L., Italiano, G.F., Monteiro, L., Palamidessi, C., Yung, M. (eds.) ICALP 2005. LNCS, vol. 3580, pp. 866–877. Springer, Heidelberg (2005)
3. Cornuejols, G., Nemhouser, G., Wolsey, L.: Discrete Location Theory. Wiley (1990)

4. Eede, M.V., Hansen, P., Kaufman, L.: A plant and warehouse location problem. Operational Research Quarterly 28(3), 547–554 (1977)
5. Frank, C.: Facility Location. In: Wagner, D., Wattenhofer, R. (eds.) Algorithms for Sensor and Ad Hoc Networks. LNCS, vol. 4621, pp. 131–159. Springer, Heidelberg (2007)
6. Gehweiler, J., Lammersen, C., Sohler, C.: A distributed O(1)-approximation algorithm for the uniform facility location problem. In: Proceedings of the Eighteenth Annual ACM Symposium on Parallelism in Algorithms and Architectures, SPAA 2006, pp. 237–243. ACM Press, New York (2006)
7. Guha, S., Khuller, S.: Greedy strikes back: Improved facility location algorithms. In: Proceedings of the Ninth Annual ACM-SIAM Symposium on Discrete Algorithms, pp. 649–657. Society for Industrial and Applied Mathematics (1998)
8. Hamburger, M.J., Kuehn, A.A.: A heuristic program for locating warehouses. Management Science 9(4), 643–666 (1963)
9. Kuhn, F., Moscibroda, T., Wattenhofer, R.: Local Computation: Lower and Upper Bounds. CoRR abs/1011.5470 (2010)
10. Lenzen, C., Wattenhofer, R.: Brief announcement: Exponential speed-up of local algorithms using non-local communication. In: Proceeding of the 29th ACM SIGACT-SIGOPS Symposium on Principles of Distributed Computing, pp. 295–296. ACM (2010)
11. Li, S.: A 1.488 Approximation Algorithm for the Uncapacitated Facility Location Problem. In: Aceto, L., Henzinger, M., Sgall, J. (eds.) ICALP 2011, Part II. LNCS, vol. 6756, pp. 77–88. Springer, Heidelberg (2011)
12. Lotker, Z., Patt-Shamir, B., Pavlov, E., Peleg, D.: Minimum-weight spanning tree construction in O(log log n) communication rounds. SIAM J. Comput. 35(1), 120–131 (2005)
13. Mettu, R.R., Plaxton, C.G.: The online median problem. SIAM J. Comput. 32(3), 816–832 (2003)
14. Moscibroda, T., Wattenhofer, R.: Facility location: distributed approximation. In: Proceedings of the Twenty-Fourth Annual ACM Symposium on Principles of Distributed Computing, pp. 108–117. ACM Press, New York (2005)
15. Pandit, S., Pemmaraju, S.V.: Finding Facilities Fast. In: Garg, V., Wattenhofer, R., Kothapalli, K. (eds.) ICDCN 2009. LNCS, vol. 5408, pp. 11–24. Springer, Heidelberg (2009)
16. Pandit, S., Pemmaraju, S.V.: Return of the primal-dual: distributed metric facility location. In: Proceedings of the 28th ACM Symposium on Principles of Distributed Computing, PODC 2009, pp. 180–189. ACM Press, New York (2009)
17. Pandit, S., Pemmaraju, S.V.: Rapid randomized pruning for fast greedy distributed algorithms. In: Proceedings of the 29th ACM SIGACT-SIGOPS Symposium on Principles of Distributed Computing, pp. 325–334. ACM (2010)
18. Patt-Shamir, B., Teplitsky, M.: The round complexity of distributed sorting: extended abstract. In: PODC, pp. 249–256. ACM Press (2011)
19. Peleg, D.: Distributed computing: a locality-sensitive approach, vol. 5. Society for Industrial and Applied Mathematics (2000)
20. Schneider, J., Wattenhofer, R.: A log-star distributed maximal independent set algorithm for growth-bounded graphs. In: Proceedings of the Twenty-Seventh ACM Symposium on Principles of Distributed Computing, pp. 35–44. ACM (2008)
21. Stollsteimer, J.F.: A working model for plant numbers and locations. Journal of Farm Economics 45(3), 631–645 (1963)

Preventing Unraveling in Social Networks: The Anchored k-Core Problem

Kshipra Bhawalkar[1], Jon Kleinberg[2], Kevin Lewi[1], Tim Roughgarden[1], and Aneesh Sharma[3]

[1] Stanford University, Stanford, CA, USA
[2] Cornell University, Ithaca, NY, USA
[3] Twitter, Inc.

Abstract. We consider a model of user engagement in social networks, where each player incurs a cost to remain engaged but derives a benefit proportional to the number of engaged neighbors. The natural equilibrium of this model corresponds to the k-core of the social network — the maximal induced subgraph with minimum degree at least k.

We study the problem of "anchoring" a small number of vertices to maximize the size of the corresponding anchored k-core — the maximal induced subgraph in which every non-anchored vertex has degree at least k. This problem corresponds to preventing "unraveling" — a cascade of iterated withdrawals. We provide polynomial-time algorithms for general graphs with $k = 2$, and for bounded-treewidth graphs with arbitrary k. We prove strong inapproximability results for general graphs and $k \geq 3$.

1 Introduction

A defining property of social networks — where nodes represent individuals, and edges represent friendships — is that the behavior of an individual is influenced by that of his or her friends. In particular, they often exhibit positive "network effects", where the utility of an individual is increasing in the number of friends that behave in a certain way. For example, empirical work has determined that individuals are more likely to contribute useful content to a social network if their friends do [4]. Increasingly, empirical studies suggest that the influence of interactions in social network extends to behavior outside of these networks, as well [9]. An obvious question, studied from a system-building perspective in [10], is how to design or modify social networks to maximize the participation and engagement of its users.

For concreteness, consider scenarios where each individual of a social network has two strategies, to "engage" or to "drop out". Being engaged could mean contributing to a public good (like network content), signing up for a new social network feature, adopting one technology instead of another, and so on. We assume that a player is more likely to be engaged if many friends are. For this Introduction, we focus on our most basic model. We first describe our model via

A. Czumaj et al. (Eds.): ICALP 2012, Part II, LNCS 7392, pp. 440–451, 2012.
© Springer-Verlag Berlin Heidelberg 2012

a process of cascading withdrawals, and then formulate it using a simultaneous-move game.

Assume that all individuals are initially engaged, and for a parameter k, a node remains engaged if and only if at least k friends are engaged. For example, engagement could represent active participation in the social network, which is worthwhile to an individual if and only if at least k friends are also actively participating. Or, dropping out could represent the abandonment of an incumbent product in favor of a newly arrived competitor; when the number of one's friends using the old product falls below k, one switches to the new product.

In this basic model, it is clear that all individuals with less than k friends will drop out. These initial withdrawals can be contagious, spreading to individuals with many more than k friends. See Figure 1 for an example of this phenomenon. In general, when such iterated withdrawals die out, the remaining engaged individuals correspond to a well-known concept in graph theory — the k-core of the original social network, which by definition is the (unique) maximal induced subgraph with minimum degree at least k. Alternatively, the k-core is the (unique) result of iteratively deleting nodes that have degree less than k, in any order.

Schelling [17, P.214] describes this type of "unraveling" in typically picturesque language, by contrasting the cycle with the line (with $k = 2$). He imagines people sitting with reading lamps, each of whom can get additional partial illumination from the lamps of their neighbor(s):

> In some cases the arrangement matters. If everybody needs 100 watts to read by and a neighbor's bulb is equivalent to half one's own, and everybody has a 60-watt bulb, everybody can read as long as he and both his neighbors have their lights on. Arranged in a circle, everybody will keep his light on if everybody else does (and nobody will if his neighbors do not); arranged in a line, the people at the ends cannot read anyway and the whole thing unravels.

A Game-Theoretic Formulation. The k-core can be seen as the maximal equilibrium in a natural game-theoretic model; it has been studied previously in this guise in the social sciences literature [5,6,16]. Concretely, imagine that each node in a social network G is considering whether to remain engaged in a social activity. We suppose that each node v in G incurs an (integer) cost of $k > 0$ for the effort it must spend to remain engaged. Node v also obtains a benefit of 1 from each neighbor w who is engaged; this reflects the idea that the benefit from participation in the activity comes from interaction with neighbors in the social network.

If each node makes its decision simultaneously, we can model the situation as a simultaneous-move game in which the nodes are the players, and v's possible strategies are to remain engaged or to drop out. For a choice of strategies σ by each player, let S_σ be the set of players who choose to remain engaged. The payoff of v is 0 if it drops out, and otherwise it is v's degree in the induced subgraph $G[S_\sigma]$ minus k. Note that we can talk about sets of engaged nodes and strategy profiles interchangeably.

There is a natural structure to the set of pure Nash equilibria in this game: σ is an equilibrium if and only if $G[S_\sigma]$ has minimum degree k (so that no engaged player wants to drop out), and no node in $V - S_\sigma$ has k or more neighbors in S_σ (so that no player who has dropped out wants to remain engaged). There will generally be multiple equilibria — for example, if G has minimum degree at least k, then $S_\sigma = \emptyset$ and $S_\sigma = V$ define two of possibly many equilibria. There is always a unique maximal equilibrium σ^*, in the sense that S_{σ^*} contains S_σ for all other equilibria σ. This maximal equilibrium is easily seen to correspond to the k-core of G — the unique maximal set S^* of minimum internal degree at least k in G.

Chwe [5,6] and Saaskilahti [16] argue that it is reasonable to assume that this maximal equilibrium will be selected in an actual play of the game, since it optimizes the welfare of all the players simultaneously (as well as the provider of the service, whose goal is to attract a large audience). That is, all incentives are aligned to coordinate on this equilibrium.

The Anchored k-Core Problem. The unraveling described in Schelling's line example is often undesirable, and could represent the end of a social network, a product, or a public good. When and how can such unraveling be prevented? For instance, in Schelling's example, the solution is clear: giving the two readers at the ends an extra lamp yields persistent illumination for all.

We formalize this problem as the *anchored k-core* problem. In the most basic version of the problem, the input is an undirected graph and two parameters $k, b \in \{1, 2, \ldots, n\}$, where b denotes a budget. Solutions correspond to subsets of at most b vertices, which are said to be *anchored*. Anchored vertices remain engaged no matter what their friends do — for example, due to external incentives like rewards for participation, or rebates for using a product. The anchored k-core, corresponding to the final subgraph of engaged individuals, is computed like the k-core, except that anchored vertices are never deleted. That is, unanchored vertices with degree less than k are deleted iteratively, in any order. In Schelling's line example, anchoring the two endpoints causes the anchored 2-core to be equal to the entire network. Another example, with $b = 2$ and $k = 3$, is displayed in Figure 2.

(a) The original graph G (b) The 3-core of G

Fig. 1. k-core for $k = 3$ on an example graph G

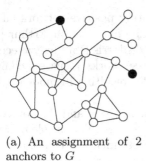

(a) An assignment of 2 anchors to G

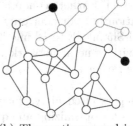

(b) The vertices saved in an anchored 3-core (with budget 2) of G

Fig. 2. Anchored 3-core with budget 2 on G

Summarizing, we have seen that cascades of withdrawals can cause an unraveling of engagement, but that such cascades can sometimes be prevented by anchoring (i.e., rendering non-strategic) a small number of individuals. The goal of this paper is to study systematically the optimization problem of anchoring a given number of individuals to maximize the amount of engagement in a social network — to maximize the size of the resulting anchored k-core. Solving this problem identifies the individuals whose participation is most crucial to the overall health of a community.

Our Results. We first study general graphs, where we identify a "phase transition" in the computational complexity of the anchored k-core problem with respect to the parameter k. (The problem is interesting for all $k \geq 2$.) First, we prove that the anchored 2-core problem is solvable in polynomial (even near-linear) time. Second, we prove that the anchored k-core problem admits no non-trivial polynomial-time approximation algorithm for every $k \geq 3$. Precisely, we prove that it is NP-hard to distinguish between instances in which $\Omega(n)$ vertices are in the optimal anchored k-core, and those in which the optimal anchored k-core has size only $O(b)$. This inapproximability result holds even for a natural resource augmentation version of the problem. We also prove, for every $k \geq 3$, that the problem is $W[2]$-hard with respect to the budget parameter b.

Our negative results motivate studying the anchored k-core problem in restricted classes of graphs, and here we provide positive results. For arbitrary k, we show that the anchored k-core problem can be solved exactly in polynomial time in graphs with bounded treewidth. Our polynomial-time algorithm extends to many natural variations of the problem: for directed graphs, for non-uniform anchoring costs, for vertex-specific values of k, and others.

Further Related Work. Our mathematical model of user engagement in social networks appears to be new, although it is related to a number of previous works in the social sciences literature. Saaskilahti's model [16] is the closest to ours — the payoff structure in [16] includes ours as a special case, though only a few special network topologies are considered (the complete graph, the cycle, and the star). Earlier economic models that capture positive network effects of participation but consider only the complete graph are given in Arthur [1] and Katz and Shapiro [11]. Blume [2], Ellison [8], and Morris [14] analyze economic models with general network topolgies, but these works focus on competing behaviors rather than on positive network effects, resulting in models different from ours.

The papers cited above focus on equilibrium analysis and do not consider algorithms for optimizing an equilibrium, as we do here. The problem of identifying influential nodes of a social network in order to incite cascades, introduced by Kempe et al. [12] and studied further in [13,15], shares some of the spirit of the optimization problem studied in the present work.

2   General Graphs

In this section, we investigate the anchored k-core problem on general graphs. We will see that the problem can be solved exactly for $k = 2$ but becomes intractable for $k \geq 3$. Also, we'll show that one can't (under suitable complexity assumptions) substantially improve upon the brute-force algorithm that tries all $\binom{n}{b}$ subsets of placements for anchors unless FPT = W[2].

We first make the following observation. Let G be a graph for which we would like to compute the anchored k-core with budget b. We construct a new graph from G, which we will call RemoveCore(G), in the following manner: Compute the set of vertices of the k-core, C_k, and remove all edges between pairs of vertices $u, v \in C_k$. We also imagine an anchor placed at each vertex $v^* \in C_k$ (without actually subtracting from the budget).

Proposition 1. *An assignment of anchors has size z in RemoveCore(G) if and only if it has size z in G.*

Intuitively, each $v^* \in C_k$ would remain in the graph without the assistance of any anchors. So, we can think of these vertices as already being anchored, and the structure within C_k no longer affects the anchored k-core of G. Thus, it is enough to devise a solution for the graph RemoveCore(G) in order to obtain a solution for the anchored k-core problem on G.

2.1   Anchored 2-Core

The RemoveCore procedure greatly simplifies the structure of the input graph G for the case of $k = 2$.

Proposition 2. *For every input graph G, with $k = 2$, RemoveCore(G) is a forest, where each tree in the forest contains at most one member of the k-core.*

For each tree in the graph, we will call a tree *rooted* if there exists a member of the k-core in the tree, and *non-rooted* otherwise. Let $\mathcal{R}$ and $\mathcal{S}$ be the set of rooted and non-rooted trees, respectively.

An Efficient Optimal Algorithm. We now show how to solve the anchored 2-core problem for any budget b. We describe this algorithm intuitively and present it more explicitly as Algorithm 2.1 below. First, find two vertices $v_1, v_2 \in \mathcal{R}$ such that placing an anchor at v_1 maximizes the number of vertices saved across all placements of a single anchor in $\mathcal{R}$, and v_2 does the same assuming that v_1 has already been placed. Next, find $v_3, v_4 \in \mathcal{S}$ such that placing anchors at v_3 and v_4 simultaneously maximizes the number of vertices saved across all placements of two anchors in $\mathcal{S}$. Let $c_\mathcal{R}(v_1)$, $c_\mathcal{R}(v_2)$, and $c_\mathcal{S}(v_3, v_4)$ denote the number of vertices saved by placing v_1, v_2, and v_3 and v_4 (together), respectively. If $c_\mathcal{R}(v_1) + c_\mathcal{R}(v_2) > c_\mathcal{S}(v_3, v_4)$, or $b = 1$, then place an anchor at v_1 and decrease b by 1. Otherwise, place anchors at v_3 and v_4 and decrease b by 2. After the anchor placement, re-run RemoveCore on the graph (now with the saved vertices as part of the k-core) and repeat the process of determining $\{v_1, v_2, v_3, v_4\}$ until $b = 0$.

Algorithm 2.1. An efficient, exact algorithm for anchored 2-core

---
$G \leftarrow$ RemoveCore(G) $//$ G is now a forest
$S \leftarrow \emptyset$
while $b > 0$ **do**
    Partition the trees of G into sets $\mathcal{R}$ and $\mathcal{S}$
    $v_1 \leftarrow$ a vertex furthest from root of trees in $\mathcal{R}$
    $v_2 \leftarrow$ a vertex to be furthest from root of trees in $\mathcal{R}$ after v_1 is anchored
    $(v_3, v_4) \leftarrow$ a pair of vertices on the endpoints of a longest path across trees in $\mathcal{S}$
    if $c_\mathcal{R}(v_1) + c_\mathcal{R}(v_2) > c_\mathcal{S}(v_3, v_4)$ or $b = 1$ **then**
        $S \leftarrow S \cup \{v_1\}$, $b \leftarrow b - 1$ $//$ Place an anchor on v_1
    else
        $S \leftarrow S \cup \{v_3, v_4\}$, $b \leftarrow b - 2$ $//$ Place an anchor on v_3 and v_4
        $G \leftarrow$ RemoveCore(G) $//$ G modified due to anchoring vertices

---

Theorem 1. *Algorithm 2.1 yields an anchored 2-core of maximum size.*

A Faster Implementation. The above algorithm runs in time $O(n^2)$, since both the RemoveCore step and finding the maximum path across all trees takes time $O(n)$, and this must be repeated for each anchor placed. However, there is an implementation of the algorithm that runs in time $O(m + n \log n)$ through the use of priority queues, and is detailed in the full version.

Corollary 1. *There is an $O(m + n \log n)$ time exact algorithm for the anchored 2-core problem.*

2.2 Inapproximability of $k \geq 3$

The natural next step is to determine the complexity of the anchored k-core problem for $k \geq 3$. Note that every solution to the anchored k-core problem has objective function value in the range $[b, n]$. In this section we show that for $k \geq 3$, it is NP-hard to approximate the optimal anchored k-core within a factor of $O(n^{1-\epsilon})$.

A Preliminary Construction. Our reduction is from the Set Cover problem. Fix an instance I of set cover with n sets $S_1, \ldots, S_n$ and m elements $\{e_1, \ldots, e_m\} = \bigcup_{i=1}^{n} S_i$. We first give the construction only for instances of set cover such that for all i, $|S_i| \leq k - 1$. Then, we show how to lift this restriction while still obtaining the same results.

We now define a corresponding anchored k-core instance. Let H be an arbitrarily large graph where every vertex has degree k except for a single vertex with degree $k-1$ — call this vertex $t(H)$. Now, consider the graph consisting of a set of nodes $\{v_1, \ldots, v_n\}$ associated with sets $S_1, \ldots, S_n$ and a set $B = \{H_1, \ldots, H_m\}$ consisting of m disjoint copies H_j of H, where each copy of H is associated with an element e_j. There is an edge between v_i and $t(H_j)$ if and only if $e_j \in S_i$.

Lemma 1. *For instances of set cover with maximum set size at most $k - 1$, there is a set cover of size z if and only if there exists an assignment in the corresponding anchored k-core instance using only z anchors such that all vertices in B are saved.*

Proof. Notice that the H_j's are designed such that if there exists some i such that v_i is adjacent to $t(H_j)$, then all vertices in H_j will be saved. Thus, if there is a set cover $\mathcal{C}$ of size z, then one can place the z anchors at v_i for all i such that $S_i \in \mathcal{C}$ and hence save all vertices in B. For the converse, we see that it is enough to restrict attention to assignments with anchors placed on v_i's. Since we are assuming that $|S_i| < k$ for all sets, each v_i will not be saved unless anchored. Thus, we must anchor some vertex adjacent to $t(H_j)$ for each copy H_j, which corresponds precisely to a set cover of size z.

Now, define $\mathsf{tree}(d, y)$ to be a perfect d-ary tree (each node has exactly d children) with exactly y leaves, if $y \geq d$, and a single root node with y leaves when $y < d$. To lift the restriction on the maximum set size, we replace each instance of v_i with $\mathsf{tree}(k - 1, |S_i|)$, and if $y_1, \ldots, y_{|S_i|}$ are the leaves, then for each $e_\ell \in S_i$, we contract the pairs of vertices (y_ℓ, e_ℓ).

Lemma 2. *For every instance of Set Cover, there is a set cover of size z if and only if there exists an assignment for the corresponding anchored k-core instance using only z anchors such that all vertices in B are saved.*

The Reduction from Set Cover. At this point, we have already shown that obtaining an optimal solution for the anchored k-core problem for $k \geq 3$ is NP-hard. Now, there exists a way to arrange the edges between each copy of H such that if there exists some vertex of B which is not saved, then *the majority* of the vertices will not be saved, either. Since this construction is somewhat complicated, we defer the details to the full version. From here on, we refer to the full construction as the graph $G(c, I)$, where I is an instance of Set Cover and c is an arbitrarily large constant.

Recall the decision problem for (unweighted) Set Cover: Given a collection of sets $\mathcal{C}$ which contain elements from a universe of size m, does there exist a set cover of size at most ℓ? We are able to show that $G(c, I)$ has the following two properties:

1. If I is a yes-instance, then there exists an assignment of ℓ anchors such that at least $km^{c+1}\ell$ vertices are saved.
2. If I is a no-instance, then no assignment of ℓ anchors can save more than $km\ell$ anchors.

Since c can be arbitrarily large, we can ensure that $km^{c+1}\ell = \Omega(n)$, where n is the number of vertices in the anchored k-core instance. We can therefore conclude with the following theorem and corollary.

Theorem 2. *It is NP-hard to distinguish between instances of anchored k-core where the optimal solution has value $\Omega(n)$ versus when the optimal solution has value $O(b)$.*

Corollary 2. *It is NP-hard to approximate the anchored k-core problem on general graphs within an $O(n^{1-\epsilon})$ factor.*

Resource Augmentation Extensions. Suppose we are interested in comparing the performance of an algorithm given a budget of $O(b \cdot \alpha)$ anchors against the optimal assignment given only b anchors, for some $\alpha > 1$. In the full version, we augment the original reduction above to yield the following result.

Corollary 3. *For $\alpha = o(\log n)$, unless $P = NP$, there does not exist a polynomial time algorithm which, given $O(b \cdot \alpha)$ anchors, finds a solution within an $O(n^{1-\epsilon})$ multiplicative factor of the optimal solution with b anchors.*

W[2]-Hardness with Respect to Budget. We can also show that the anchored k-core problem is not in FPT with respect to the budget parameter b. We establish this result via a reduction from the Dominating Set problem, and the proof can be found in the full version. Recall that the decision version of Dominating Set is as follows: given a budget ℓ, determine if there a subset S of vertices such that $|S| \leq \ell$ and each vertex in $V \setminus S$ is adjacent to a vertex in S. As shown in [7], this problem is W[2]-hard. In the full version, we establish that if the anchored k-core problem can be solved in time $O(f(b) \cdot \text{poly}(n))$ for any $k \geq 3$, then Dominating Set is in FPT with respect to ℓ, and hence FPT = W[2].

Theorem 3. *For every $k \geq 3$, the anchored k-core problem is W[2]-hard with respect to the parameter b.*

3   Graphs with Bounded Treewidth

Although we see that the anchored k-core problem is hopelessly inapproximable on general graphs, we next give polynomial time exact algorithms for graphs with bounded treewidth. The treewidth of a graph is defined as the minimum width over all tree decompositions of the graph, where the width of a tree decomposition is one more than the size of the largest node in the tree decomposition (see [3] for a tutorial and survey on treewidth). In this section, we present an algorithm that runs in time $O(f(k,w)\cdot b^2)\cdot\mathsf{poly}(n)$, where $f(k,w) = (3(k+1)^2)^w$, using $w-1$ as the graph's treewidth. To distinguish the vertices of a tree decomposition from the vertices of the original graph, we will call the elements of a tree decomposition nodes, and the elements of the original graph will remain as vertices. We will use the concept of nice tree decompositions for graphs, defined in [3] — the idea that a tree decomposition can be converted into another tree decomposition (a "nice" one) of the same treewidth and $O(n)$ nodes, but with the special property that each node comes in one of four types:

- Leaf Node: Only one vertex is associated with this node
- Introduce Node: The node has a single child, and if X is the set of vertices associated with this node and Y is the child, then $X = Y \cup \{v\}$ for some v.
- Forget Node: The node has a single child, and if X is the set of vertices associated with this node and Y is the child, then $X = Y \setminus \{v\}$ for some v.
- Join Node: The node has two children, and if X is the set of vertices associated with this node, Y and Z are its children, then $X = Y = Z$.

We show how to solve a generalization of the anchored k-core problem. Let $\mathsf{threshold}(v)$ represent the *threshold* of v, which is the minimum number of neighbors that v requires in order to remain in the k-core (assuming v is not anchored). Traditionally, in k-core, we use $\mathsf{threshold}(v) = k$ for all $v \in V(G)$. Since we are now considering a situation where the threshold function varies across vertices, we will instead use k to denote the maximum threshold across all vertices.

For a fixed assignment of anchors, a vertex is either *anchored*, *not saved*, or *indirectly saved* (not anchored, but saved). We show that the categorization of vertices into these three types is enough to capture the complexity of the problem on graphs with bounded treewidth. Define a *fixture* f of a tree decomposition T to be an assignment of these three types to a subset of $G[r(T)]$, the vertices of G that are associated with the root node of T. We say that an assignment A of anchors to vertices *satisfies* a fixture f if under the assignment A, the type of each vertex designated by the fixture agrees with the type induced by the assignment A. Define a threshold alteration m (which we will simply call an *alteration*) to be a setting of the thresholds of some subset $S \subseteq V(G)$ so that for each $v \in S$, m reduces the threshold of v by some integer in the interval $[0, \mathsf{threshold}(v)]$. We use the notation $m(T)$ to denote the tree obtained by lowering the thresholds of all vertices as prescribed by m.

3.1   The Algorithm

We first define a subroutine: $\mathsf{Solve}(T)$ for a tree decomposition T, also outlined in Algorithm 3.1. The output of $\mathsf{Solve}(T)$ will be, for all fixtures f and all alterations m (within $r(T)$), and $\hat{b} \in [1, b]$, the table $\mathsf{Solutions}(m(T), f, \hat{b})$. Each entry of this table of solutions, $\mathsf{Solutions}_T[f][m][b]$, will describe an optimal assignment of b anchors which satisfies the fixture f on the graph $G[T]$, the vertices of G that are associated with the nodes of T, under alteration m. Note that if no such

Algorithm 3.1. Solve(T): The main subroutine used in the exact algorithm for graphs with bounded treewidth.

(Solutions$_{T_1}$, Solutions$_{T_2}$) ← (Solve(T_1), Solve(T_2))
for all fixtures f, alterations m, and budgets b **do**
  if $r(T)$ is a leaf node **then**
    Solutions$_T[f][m][b]$ ← the result dicated by the fixture f
  if $r(T)$ is a forget node **then**
    S ← {fixtures f' of $T_1 : \forall v \in G[r(T)], f(v) = f'(v)$}
    Solutions$_T[f][m][b]$ ← max$_{f' \in S}$ Solutions$_{T_1}[f'][m][b]$
  if $r(T)$ is an introduce node **then**
    Set f' to be such that $\forall v \in G[rT_1)], f'(v) = f(v)$
    Let v be the sole element of $G[r(T)] \setminus G[r(T_1)]$
    Set m' so that $\forall u \in N(v), m'(u) = m(u) - 1$ and $\forall u \notin N(v), m'(u) = m(u)$
    if $f(v)$ is anchored **then**
      Solutions$_T[f][m][b]$ ← Solutions$_{T_1}[f'][m'][b-1]$
    if $f(v)$ is not saved **then**
      Solutions$_T[f][m][b]$ ← Solutions$_{T_1}[f'][m][b]$
    if $f(v)$ is indirectly saved **then**
      Solutions$_T[f][m][b]$ ← Solutions$_{T_1}[f'][m'][b]$
  if $r(T)$ is a join node **then**
    $\hat{b}$ ← $b - |\{v : f(v) \text{ is anchored}\}|$
    t ← max$_{i \in [0,\hat{b}], \hat{m} \in [0,k]^w}$ (Solutions$_{T_1}[f][\hat{m}][i]$ + Solutions$_{T_2}[f][m - \hat{m}][\hat{b} - i]$)
    Solutions$_T[f][m][b]$ ← t
return Solutions$_T$

---

assignment that satisfies the stated restrictions can exist, then the output of the entry is ⊥. (This could occur if, for example, the fixture f requires 3 nodes to be anchored yet $b < 3$.)

Thus, Solve(T) will output at most $(3(k + 1))^w \cdot b$ solutions. The 3^w term is due to the fact that there are 3^w possible fixtures, and the $(k + 1)^w$ term comes from the fact that each vertex has threshold at most k, and so there are at most $k + 1$ choices for lowering the thresholds of each vertex in the root node (within the range $[0, k]$). We now show that given the outputs of Solve(T_i) for each child subtree T_i, we can compute Solve(T). This will be done through a case analysis on the node type of the root node, denoted by $r(T)$.

If $r(T)$ is a leaf node, then Solutions($m(T), f, b$) is trivial to determine for all alterations m, fixtures f, and budgets b, as there is only one vertex associated with $r(T)$ and there are no other nodes in T (and so, an anchor is placed on this vertex if it is anchored under f and $b \geq 1$). Otherwise, let T_1 and T_2 denote the child subtrees of $r(T)$. If $r(T)$ is a forget node, then let v be the vertex that is associated with the child node but not in $G[r(T)]$. To find Solutions($m(T), f, b$) for all f, m, and b, we simply compute the maximum solution over possible choices of the fixture type of v and the threshold alteration induced by the partial fixture of the other vertices in $G[r(T)]$.

If $r(T)$ is an introduce node, then let v be the vertex in $G[r(T)]$ that is not associated with the child node. For fixtures f such that v is anchored, we simply subtract one from the budget for T_1 and retrieve the optimal solution there under the induced partial fixture. This is obtained from the output of Solutions$(m'(T), f, b)$, where $m'(u) = m(u) - 1$ if $u \in N(v)$ and $m'(u) = m(u)$ otherwise. If v is indirectly saved, we do not change the budget but obtain the optimal solution under the induced partial fixture for Solutions$(m'(T), f, b)$. If v is not saved, we use the optimal solution on T_1 under the induced fixture f. Finally, if $r(T)$ is a join node, then let $i \in [1, b]$. Also, for a fixture f, let S be the set of all vertices in the root node that are indirectly saved. Note that $|S| \leq w$. We iterate over all of the at most $(k+1)^w$ possibilities of dividing up the thresholds for each $v \in R$ between T_1 and T_2. We then iterate over all pairs of solutions OPT$(m(T_1), f, i)$ and OPT$(m(T_2), f, b - i)$ and take the maximum over such i.

This approach is repeated in a bottom-up manner until we have covered the entire tree. Finally, we simply take the output of Solve(T) (the original tree decomposition T of the entire graph) and find the optimal solution corresponding to the tree on b anchors by taking the maximum value over all fixtures. The formal proof of correctness is deferred to the full version.

Generalizations. This dynamic programming approach allows for several generalizations to the anchored k-core problem on graphs with bounded treewidth, all of which can be solved exactly and run in time $O(f(k, w) \cdot \text{poly}(n, b))$. For example, one can assign weights to vertices. Also, as we have already seen, the vertex thresholds can be non-uniform. Furthermore, the edges of the graph can be directed, and each arc $a = (u, v)$ can have a weight $w(a)$ such that ensuring that u is in the graph contributes a value of $w(a)$ to the threshold of v.

4   Concluding Remarks

There remain several attractive open problems related to the anchored k-core problem, especially on restricted classes of graphs. Is there a PTAS for planar graphs? What can be said about the problem on random graphs? Can the running time of our polynomial-time algorithm for bounded treewidth graphs be improved? Is there a linear-time algorithm for the anchored 2-core problem in general graphs?

References

1. Brian Arthur, W.: Competing technologies, increasing returns, and lock-in by historical events. The Economic Journal 99(394), 116–131 (1989)
2. Blume, L.: The statistical mechanics of strategic interaction. Games and Economic Behavior 5, 387–424 (1993)
3. Bodlaender, H.L., Koster, A.M.C.A.: Combinatorial optimization on graphs of bounded treewidth. Comput. J., 255–269 (2008)
4. Burke, M., Marlow, C., Lento, T.: Feed me: motivating newcomer contribution in social network sites. In: CHI (2009)

5. Chwe, M.S.-Y.: Structure and strategy in collective action. American Journal of Sociology 105(1), 128–156 (1999)
6. Chwe, M.S.-Y.: Communication and coordination in social networks. Review of Economic Studies 67, 1–16 (2000)
7. Downey, R.G., Fellows, M.R., McCartin, C., Rosamond, F.: Parameterized approximation of dominating set problems. Information Processing Letters 109, 68–70 (2008)
8. Ellison, G.: Learning, local interaction, and coordination. Econometrica 61, 1047–1071 (1993)
9. Ellison, N.B., Steinfield, C., Lampe, C.: The Benefits of Facebook Friends: Social Capital and College Students' Use of Online Social Network Sites. Journal of Computer-Mediated Communication (2007)
10. Farzan, R., Dabbish, L.A., Kraut, R.E., Postmes, T.: Increasing commitment to online communities by designing for social presence. In: CSCW (2011)
11. Katz, M.L., Shapiro, C.: Network externalities, competition, and compatibility. American Economic Review 75(3), 424–440 (1985)
12. Kempe, D., Kleinberg, J., Tardos, É.: Maximizing the spread of influence through a social network. In: KDD, pp. 137–146 (2003)
13. Kempe, D., Kleinberg, J., Tardos, É.: Influential Nodes in a Diffusion Model for Social Networks. In: Caires, L., Italiano, G.F., Monteiro, L., Palamidessi, C., Yung, M. (eds.) ICALP 2005. LNCS, vol. 3580, pp. 1127–1138. Springer, Heidelberg (2005)
14. Morris, S.: Contagion. Review of Economic Studies 67, 57–78 (2000)
15. Mossel, E., Roch, S.: On the submodularity of influence in social networks. In: STOC, pp. 128–134 (2007)
16. Saaskilahti, P.: Monopoly pricing of social goods. Technical report, University Library of Munich, Germany (2007)
17. Schelling, T.C.: Micromotives and Macrobehavior. Norton (1978)

Edge Fault Tolerance on Sparse Networks[*]

Nishanth Chandran[1,**], Juan Garay[2], and Rafail Ostrovsky[3,***]

[1] Microsoft Research, Redmond
nish@microsoft.com
[2] AT&T Labs – Research
garay@research.att.com
[3] Departments of Computer Science and Mathematics, UCLA
rafail@cs.ucla.edu

Abstract. Byzantine agreement, which requires n processors (nodes) in a completely connected network to agree on a value dependent on their initial values and despite the arbitrary, possible malicious behavior of some of them, is perhaps the most popular paradigm in fault-tolerant distributed systems. However, partially connected networks are far more realistic than fully connected networks, which led Dwork, Peleg, Pippenger and Upfal [STOC'86] to formulate the notion of *almost-everywhere (a.e.) agreement* which shares the same aim with the original problem, except that now not all pairs of nodes are connected by reliable and authenticated channels. In such a setting, agreement amongst all correct nodes cannot be guaranteed due to possible poor connectivity with other correct nodes, and some of them must be given up. The number of such nodes is a function of the underlying communication graph and the adversarial set of nodes.

In this work we introduce the notion of *almost-everywhere agreement with edge corruptions* which is exactly the same problem as described above, except that we additionally allow the adversary to completely control some of the communication channels between two correct nodes— i.e., to "corrupt" edges in the network. While it is easy to see that an a.e. agreement protocol for the original node-corruption model is also an a.e. agreement protocol tolerating edge corruptions (albeit for a reduced fraction of edge corruptions with respect to the bound for node corruptions), no polynomial-time protocol is known in the case where a constant fraction of the edges can be corrupted and the degree of the network is sub-linear.

---

[*] A full version of this paper, entitled "Almost-Everywhere Secure Computation with Edge Corruptions," is available at http://eprint.iacr.org/2012/221.

[**] Part of this work was done at UCLA.

[***] Supported in part by NSF grants 0830803, 09165174, 1065276, 1118126 and 1136174, US-Israel BSF grant 2008411, OKAWA Foundation Research Award, IBM Faculty Research Award, Xerox Faculty Research Award, B. John Garrick Foundation Award, Teradata Research Award, and Lockheed-Martin Corporation Research Award. This material is based upon work supported by the Defense Advanced Research Projects Agency through the U.S. Office of Naval Research under Contract N00014-11-1-0392. The views expressed are those of the author and do not reflect the official policy or position of the Department of Defense or the U.S. Government.

A. Czumaj et al. (Eds.): ICALP 2012, Part II, LNCS 7392, pp. 452–463, 2012.
© Springer-Verlag Berlin Heidelberg 2012

We make progress on this front, by constructing graphs of degree $O(n^\epsilon)$ (for arbitrary constant $0 < \epsilon < 1$) on which we can run a.e. agreement protocols tolerating a constant fraction of adversarial edges. The number of given-up nodes in our construction is μn (for some constant $0 < \mu < 1$ that depends on the fraction of corrupted edges), which is asymptotically optimal. We remark that allowing an adversary to corrupt edges in the network can be seen as taking a step closer towards guaranteeing a.e. agreement amongst honest nodes even on adversarially chosen communication networks, as opposed to earlier results where the communication graph is specially constructed.

In addition, building upon the work of Garay and Ostrovsky [Eurocrypt'08], we obtain a protocol for *a.e. secure computation* tolerating edge corruptions on the above graphs.

Keywords: Fault tolerance, almost-everywhere agreement, bounded-degree network, secure multiparty computation.

1   Introduction

Byzantine agreement [12,10] is perhaps the most popular paradigm in fault-tolerant distributed systems. It requires n parties (processors) to agree upon a common value that is dependent on their inputs, even when some of them may behave arbitrarily. More specifically, n parties $P_1, \cdots, P_n$, each holding input v_i, must run a protocol such that at the end of the protocol, all "honest" (i.e., not misbehaving) parties output the same value and if $v_i = v$ for all the honest parties, then the honest parties output v. Traditionally, protocols for Byzantine agreement assume that any two of the n parties share a reliable and authenticated channel which they use for communication. However, for protocols executed over large networks such as the Internet, in which nodes are typically connected by a communication graph of small degree, this assumption is unreasonable.

In light of this, the seminal work of Dwork, Peleg, Pippenger, and Upfal [5] considered the problem of reaching agreement over networks that are *not* fully connected, and even of low degree, where every party shares a reliable and authenticated channel only with a few of the other $n - 1$ parties. More formally, in the Dwork *et al.* formulation, the n parties (or nodes) are connected by a communication network $\mathcal{G}$. Nodes that are connected by an edge in $\mathcal{G}$ share a reliable and authentic channel, but other nodes must communicate via paths in the graphs that may not be available to them (due to the adversarial "corruption" of some of the nodes). Naturally, in such a setting, one may not be able to guarantee agreement amongst all honest parties; for example, one cannot hope to be able to communicate at all with an honest party whose neighbors are all adversarial. Given this fact—and ubiquitously—Dwork *et al.* termed the new problem *almost-everywhere (a.e.) agreement*, wherein the number of such abandoned nodes (which henceforth will be called "doomed") introduces another parameter of interest, in addition to the degree of the communication graph (which we wish to minimize), and the number of adversarial nodes that can be tolerated (which we wish to maximize) in reaching agreement.

Indeed, in [5], Dwork *et al.* provide a.e. agreement protocols for various classes of low-degree graphs and bounds on the number of adversarial nodes as well as doomed nodes. For example, they construct a graph of constant degree and show an agreement protocol on this graph tolerating a $\frac{\alpha}{\log n}$ fraction of corrupted nodes (for constant $0 < \alpha < 1$), guaranteeing agreement amongst $(1 - \alpha - \mu)n$ of the honest nodes (for constant $0 < \mu < 1$). In another construction, they give a graph of degree $\mathcal{O}(n^\epsilon)$ (for constant $0 < \epsilon < 1$) and show an agreement protocol on this graph tolerating a constant α ($0 < \alpha < 1$) fraction of corrupted nodes, and again guaranteeing agreement amongst $(1 - \alpha - \mu)n$ nodes. In a subsequent and remarkable result, Upfal [13] constructed a constant-degree graph and showed the existence of an a.e. agreement protocol on this graph tolerating a constant fraction of corrupted nodes, while giving up a constant fraction of the honest nodes. Unfortunately, the protocol of [13] runs in exponential time (in n). More recently, Chandran, Garay, and Ostrovsky [3] constructed a graph of degree $\mathcal{O}(\log^k n)$ (for constant $k > 1$) and show an agreement protocol on this graph tolerating a constant fraction of corrupted nodes, while giving up only $\mathcal{O}(\frac{n}{\log n})$ honest nodes.

Edge corruptions. All existing work on a.e. agreement considers the case where only nodes may be corrupted and misbehave, while communication on all edges is assumed to be perfectly reliable and authentic. In many settings, however, such a guarantee might again be unrealistic and a bit optimistic. Think for example, of the communication subsystem of a networked computer (e.g., network interface controller card) being infected by malicious software designed to disrupt or alter operation. This would affect the communication between honest parties. Or worse, of a scenario where the secret keys shared by two parties in the system is compromised, yet the parties themselves are honest.

In this work, we address this situation and endow the adversary with additional powers which allow him, in addition to corrupting nodes, to corrupt some of the *edges* in the network—i.e., we consider *a.e. agreement with edge corruptions.* When he does (corrupt an edge), he is able to completely control the communication channel between the two honest nodes, from simply preventing them to communicate, to injecting arbitrary messages that the receiving end will accept as valid. As in the node-only corruption case, in this case also some of the honest nodes in the network must be abandoned. As further motivation towards considering edge corruptions, we remark that allowing an adversary to corrupt edges in the network moves us a step closer towards guaranteeing a.e. agreement on adversarially chosen communication networks. In an ideal scenario, we would like to construct a.e. computation protocols on arbitrary adversarially chosen communication networks. Unfortunately, this is impossible in general[1].

---

[1] The adversary could simply design networks where several nodes have extremely poor connectivity and hence corrupting a few edges could create several disconnected components in the network of small size.

Table 1. *Agreement against edge corruptions from agreement against node corruptions. (Expl., Det., and Rand. denote Explicit, Deterministic and Randomized, resp.)*

Reference	Graph degree	Frac. of corrupt edges	Graph/Protocol	Running time
[5]	$\mathcal{O}(n^\epsilon)$	$\frac{\alpha}{n^\epsilon}$; adaptive	Expl./Det.	Polynomial
[5]	$\mathcal{O}(1)$	$\frac{\alpha}{\log n}$; adaptive	Expl./Det.	Polynomial
[13]	$\mathcal{O}(1)$	α; adaptive	Expl./Det.	Exponential
[9]	$\mathcal{O}(\log^k n)$	$\frac{\alpha}{\log^k n}$; static	Expl./Rand.	Polynomial
[3]	$\mathcal{O}(\log^k n)$	$\frac{\alpha}{\log^k n}$; adaptive	Expl./Det.	Polynomial
[This work]	$\mathcal{O}(n^\epsilon)$	α; adaptive	Rand./Det.	Polynomial

However, we can take a step in this direction by allowing the adversary to corrupt edges in the network that we design, thereby "modifying" the network.

Observe that an a.e. agreement protocol for node corruptions can be readily transformed into an a.e. agreement protocol for edge corruptions, albeit for a reduced fraction of edge corruptions. More specifically, let d be the maximum degree of any node in a graph $\mathcal{G}$ on n nodes that admits an a.e. agreement protocol Π amongst $p < n$ nodes, in the presence of x corrupt nodes. Then, it is easy to see that $\mathcal{G}$ admits an a.e. agreement protocol Π' amongst p nodes in the presence of x corrupt edges[2]. However, this means that the graph will only admit an agreement protocol for an $\frac{x}{nd}$ fraction of corrupted edges, as opposed to an $\frac{x}{n}$ fraction of corrupted nodes in the former case. Therefore, the result that we get for the case of edge corruptions using this method is asymptotically weaker than in the case of node corruptions (except when d is a constant). Unfortunately, by applying this method, none of the existing protocols for a.e. agreement against node corruptions give us an a.e. agreement protocol tolerating a constant fraction of edge corruptions. This is depicted in Table 1, where we outline the results obtained via this approach using the results on a.e. agreement for node corruptions from the works of [5,13,9,3], and compare them with the results we obtain in this work. In all the results listed in the table, $0 < \alpha < 1$ is a constant, $0 < \epsilon < 1$ can be any arbitrary constant, and $k > 1$ is a constant.

Note that all the previous results (except for the result obtained as a corollary to [13], in which the protocol's running time is exponential) cannot handle the case where we have a *constant* fraction of corrupted edges. In this work, we are precisely interested in this case. Specifically, we construct the first a.e. agreement protocol on graphs with sub-linear degree that can tolerate a constant fraction of edge corruptions. We remark here that while the above graph constructions are deterministic, we construct our graph (upon which we obtain an a.e. agreement protocol tolerating constant fraction of corrupted edges) probabilistically, and our result holds with high probability. However, a graph satisfying the conditions required for our a.e. agreement protocol to be successful can be sampled with probability $1 - \mathsf{neg}(n)$, where $\mathsf{neg}(n)$ denotes a function that is negligible in n, and furthermore, one can also efficiently check if the graph thus sampled satisfies the necessary conditions for our protocol.

---

[2] To simulate an adversary corrupting edge (u, v), simply corrupt either node u or v.

Our results and techniques. In this work, we show that for every constant ϵ, $0 < \epsilon < 1$, there exists a graph, call it $\mathcal{G}_{\text{main}}$, on n nodes, with maximum degree $d_m = \mathcal{O}(n^\epsilon)$, and such that it admits an a.e. agreement protocol that guarantees agreement amongst a $\gamma_m n$ fraction of honest nodes (for some constant $0 < \gamma_m < 1$), even in the presence of an α_m fraction of corrupted edges (i.e., at most $\frac{\alpha_m n d_m}{2}$ corrupted edges), for some constant $0 < \alpha_m < 1$. Our protocol works against an adversary that is *adaptive* (i.e., the adversary can decide which edges to corrupt on the fly during the protocol after observing messages of honest parties) and *rushing* (i.e., in every round, the adversary can decide on its messages after seeing the messages from the honest parties). We now outline the high-level ideas behind our construction:

1. The first step in our construction is to build a graph with higher degree, $\mathcal{O}(\sqrt{n} \log n)$, on which we can have an a.e. agreement protocol tolerating a constant fraction of corrupted edges.

    – To do this, we first observe a property of a graph that is sufficient for such a construction (besides, obviously, every node having degree $\mathcal{O}(\sqrt{n} \log n)$), namely, that any two nodes in the graph have $\mathcal{O}(\log^2 n)$ number of paths of length 2 between them.

    – Second, we observe that the Erdős-Renyi random graph $\mathcal{G}(n, \frac{\log n}{\sqrt{n}})$ satisfies the above two properties with high probability. That is, graph $\mathcal{G}$ on n nodes satisfying the above two properties can be easily sampled by putting an edge (u, v) in $\mathcal{G}$, independently, with probability $p = \frac{\log n}{\sqrt{n}}$.

    – Once we have a graph satisfying these properties, the construction is fairly straightforward: to obtain reliable communication between any two nodes, say, u and v, u simply sends the message to all nodes in the network via all the paths of length 2, and all the nodes then send the message to v, again via all their paths of length 2. One can then show that if v takes a simple majority of the received values, then a constant fraction of the nodes can communicate reliably even in the presence of a constant fraction of corrupted edges. (As shown in [5], reliable pairwise communication is sufficient to obtain an a.e. agreement protocol amongst those nodes.)

2. Next, we show how to construct a graph, $\mathcal{G}'$, recursively from $\mathcal{G} \leftarrow G(n, \frac{\log n}{\sqrt{n}})$ above such that the new graph is of size n^2 and its degree at most twice that of $\mathcal{G}$, and yet we can have an agreement protocol on $\mathcal{G}'$ tolerating a constant fraction of corrupted edges.

    – We construct $\mathcal{G}'$ by taking n "copies" of $\mathcal{G}$ to form n "clouds," and then connecting the clouds using another copy of $\mathcal{G}$. We connect two clouds by connecting the ith node in one with the ith node in the other.

    – Now our hope is to be able to simulate the communication between two nodes u and v in the following way: u will send the message to all nodes in its cloud (call this cloud C_u). Cloud C_u will then send the message to cloud C_v (the cloud which v is a part of). Finally, v will somehow receive the message from cloud C_v.

- The problem with this approach is that we need to have a protocol that will allow two clouds to communicate reliably. But clouds themselves are comprised of nodes, some of which might be corrupted or doomed; hence, the transmission from cloud C_u to cloud C_v might end up being unreliable. To get over this problem, we make use of a specific type of agreement protocol known as *differential agreement* [6], which, informally and whenever possible, allows parties to agree on the majority value of the honest parties' inputs. Careful application of this protocol allows us to perform a type of "error-correction" of the message when it is being transferred from one cloud to another.

- Combining the above techniques leads us to our main result, an a.e. agreement protocol on graphs of degree $\mathcal{O}(n^\epsilon)$ (for all constants $0 < \epsilon < 1$), tolerating a constant fraction of corrupted edges, while giving up μn honest nodes (for a constant $0 < \mu < 1$).

In [8], Garay and Ostrovsky considered the problem of (unconditional, or information-theoretic) secure multi-party computation (MPC) [1,4] in the context of partially connected networks with adversarial nodes. By applying our new a.e. agreement protocol to the construction in [8], we obtain an a.e. MPC protocol tolerating both node and edge corruptions for graphs of degree $\mathcal{O}(n^\epsilon)$ and same parameters as above.

Due to lack of space, a more detailed overview of related work, as well as supplementary material and proofs, are presented in the full version [2].

2 Model, Definitions and Building Blocks

Let $\mathcal{G} = (\mathcal{V}, \mathcal{E})$ denote a graph with n nodes (i.e., $|\mathcal{V}| = n$). The nodes of the graph represent the processors (parties, "players") participating in the protocol, while the edges represent the communication links connecting them. We assume a synchronous network and that the protocol communication is divided into rounds. In every round, all parties can send a message on all of their communication links (i.e., on all edges incident on the node representing the party); these messages are delivered before the next round.

An adversary $\mathcal{A}$ can "corrupt" a set of nodes (as in taking over them and completely control their behavior), $\mathcal{T}_{\text{nodes}} \subset \mathcal{V}$, as well as a set of edges, $\mathcal{T}_{\text{edges}} \subset \mathcal{E}$, in the network such that $|\mathcal{T}_{\text{nodes}}| \le t_n$ and $|\mathcal{T}_{\text{edges}}| \le t_e$. $\mathcal{A}$ has unbounded computational power and can corrupt both nodes and edges *adaptively* (that is, the adversary can decide which nodes and edges to corrupt on the fly during the course of the protocol, after observing the messages from honest parties). Furthermore, $\mathcal{A}$ is *rushing*, meaning that it can decide the messages to be sent by adversarial parties (or on adversarial edges) in a particular round after observing the messages sent by honest parties in the same round.

Almost-everywhere agreement. The problem of *almost-everywhere agreement* ("a.e. agreement" for short) was introduced by Dwork, Peleg, Pippenger and Upfal [5] in the (traditional) context of node corruptions. A.e. agreement "gives up" some of the non-faulty nodes in the network from reaching agreement, which is unavoidable due to their poor connectivity with other non-faulty nodes. We refer to the given-up nodes as *doomed* nodes; the honest nodes for which we guarantee agreement are referred to as *privileged* nodes. Let the set of doomed nodes be denoted by $\mathcal{X}$ and the set of privileged nodes by $\mathcal{P}$; note that the sets $\mathcal{P}$ and $\mathcal{X}$ are a function of the set of corrupted nodes ($\mathcal{T}_{\text{nodes}}$) and the underlying graph. Let $|\mathcal{X}| = x$ and $|\mathcal{P}| = p$. Clearly, we have $p + x + t = n$. We present the formal definition of a.e. agreement, and state the relevant results from Dwork *et al.* [5], in the full version.

Differential agreement. We now present a tool that will be used in our recursive construction in Section 3.2. Fitzi and Garay [6] introduced the problem of δ-*differential agreement* (also, "consensus") developing on the so-called "strong consensus" problem [11], in which every party begins with an input v from a domain $\mathcal{D}$.[3] We describe the problem below and state the results from [6]. In the standard Byzantine agreement problem [12,10], n parties attempt to reach agreement on some value v (for simplicity, we assume $v \in \{0, 1\}$). Let c_v denote the number of honest parties whose initial value is v, and δ be a non-negative integer. δ-*differential agreement* is defined as follows:

Definition 1. *A protocol for parties* $\{P_1, P_2, \cdots, P_n\}$, *each holding initial value* v_i, *is a* δ-differential agreement *protocol if the following conditions hold for any adversary* $\mathcal{A}$ *that corrupts a set* $\mathcal{T}_{\text{nodes}}$ *of parties with* $|\mathcal{T}_{\text{nodes}}| \leq t_n$:

- AGREEMENT: *All honest parties output the same value.*
- δ-DIFFERENTIAL VALIDITY: *If the honest parties output* v, *then* $c_v + \delta \geq c_{\bar{v}}$.

Theorem 1. [6] *In a synchronous, fully connected network,* δ-*differential agreement is impossible if* $n \leq 3t_n$ *or* $\delta < t_n$. *On the other hand, there exists an efficient (i.e., polynomial-time) protocol that achieves* t_n-*differential agreement for* $n > 3t_n$ *in* $t_n + 1$ *rounds.*

We will use $\text{DA}(n, t_n, \delta)$ to denote a δ-differential agreement protocol for a fully connected network tolerating up to t_n faulty processors.

The edge corruption model. In this work we additionally allow the adversary to corrupt edges on the network graph—the set $\mathcal{T}_{\text{edges}} \subset \mathcal{E}$, $|\mathcal{T}_{\text{edges}}| \leq t_e$. We will bound this quantity, as well as the total number of nodes that the adversary can corrupt, and attempt to construct a network graph $\mathcal{G}$ of small (sublinear) degree on which a significant number of honest nodes can still reach agreement. We now give some definitions and make some remarks about a.e. agreement and a.e. secure computation for this setting.

We first observe that since we are working with (asymptotically) regular graphs, obtaining an a.e. (agreement, MPC) protocol in the presence of a constant fraction of corrupted edges will also imply a protocol in the presence of a

---

[3] In contrast to standard Byzantine agreement, the validity condition in the strong consensus problem states that the output value v must have been the input of some honest party P_i (which is implicit in the case of binary Byzantine agreement).

constant fraction of corrupted edges and a constant fraction of corrupted nodes nodes, as every corrupted node can be "simulated" by corrupting all the edges incident on this node. Thus, we will henceforth consider only adversarial edges and assume that all the nodes are honest.

As in the case of a.e. agreement on sparse networks in the presence of adversarial nodes, a.e. agreement in the presence of adversarial edges also "gives up" certain honest nodes in the network, which, as argued before, is unavoidable due to their poor connectivity with other honest nodes. Let the set of such doomed nodes be denoted by $\mathcal{X}$ and the set of privileged nodes by $\mathcal{P}$. Note that the sets $\mathcal{P}$ and $\mathcal{X}$ are a function of both the set of corrupted edges ($\mathcal{T}_{edges}$) and the underlying graph. Let $|\mathcal{X}| = x$ and $|\mathcal{P}| = p$; we let the fraction of corrupt edges be α_e. The definition of a.e. agreement with corrupted edges, in particular, now readily follows in the same manner.

Next, we remark that the problem of a.e. agreement for edge corruptions also reduces to that of constructing a reliable message transmission protocol between any two nodes $u, v \in \mathcal{P}$, in particular those which are not directly connected by an edge in $\mathcal{E}$. Furthermore, Garay and Ostrovsky showed that, given such a channel between two nodes u and $v \in \mathcal{P}$, plus some additional paths, most of which (i.e., all but one) might be corrupted, it is possible to construct a (unidirectional) *secure* (i.e., private and reliable) channel between them. The construction is via a protocol known as *secure message transmission by public discussion* (SMT-PD) [8,7]. In turn, from the protocol for a secure channel, an a.e. MPC protocol amongst the nodes in $\mathcal{P}$, satisfying the same notion of security as in [8], readily follows We refer the reader to the full version for details on the definitions of these primitives in the presence of edge corruptions and a.e. MPC (for node-corruptions).

Finally, we remark that one can define the notion of *a.e. differential agreement* (for edge corruptions) in the same manner as a.e. agreement by replacing the set of honest parties with the set of privileged parties in Definition 1 (i.e., by treating doomed parties also as adversarial). Furthermore, note that one can also obtain an a.e. differential agreement protocol (for edge corruptions) from the construction of a reliable message transmission protocol between any two nodes $u, v \in \mathcal{P}$: simply, execute a standard differential agreement protocol and replace every communication between nodes with an execution of the message transmission protocol. We will use $\mathsf{AE\text{-}DA}(n, t_n, \delta)$ to denote an a.e. δ-differential agreement protocol for a partially connected network where the number of privileged parties is $n - t_n$.

3   A.E. Agreement on Low-Degree Networks

In this section we construct a graph in which the maximum degree of any node is low, and yet, there exists a set of nodes (of size a constant times the total number of nodes), such that all nodes in this set can reach agreement even when a constant fraction of the edges in the graph are corrupted. First, our goal will be to construct a graph $\mathcal{G} = (\mathcal{V}, \mathcal{E})$ on n nodes with maximum degree d, and a protocol for reliable message transmission scheme, $\mathsf{TS}_{u,v}^{\mathcal{G}}(m)$, with the following properties. Let the set of edges that are corrupted by an adversary be denoted by $\mathcal{T}_{\text{edges}} \subset \mathcal{E}, |\mathcal{T}_{\text{edges}}| \leq \alpha nd$. We shall show that there exists a set of nodes $\mathcal{P} \subseteq \mathcal{V}$, such that $|\mathcal{P}| \geq \gamma n$, and any two nodes $u, v \in \mathcal{P}$ can communicate using $\text{TS}^{\mathcal{G}}_{u,v}(m)$. This will then be sufficient to obtain a protocol for a.e. agreement (as in the work of Dwork $et\ al.$ [5]), as well as a.e. secure computation (using the techniques from Garay and Ostrovsky [8]). Our graph will have maximum degree $\mathcal{O}(n^\epsilon)$, for arbitrary constants $0 < \epsilon < 1$, such that $|\mathcal{P}| \geq \gamma n$, for constant $0 < \gamma < 1$.

We begin this section by constructing such a message transmission scheme on a graph of larger degree, $\mathcal{O}(\sqrt{n}\log n)$, and then show how to use that construction to obtain a scheme on a graph of maximum degree $\mathcal{O}(n^\epsilon)$.

3.1   A.e. Agreement on $\mathcal{O}(\sqrt{n}\log n)$-Degree Graphs

We now show how to construct a graph of maximum degree $\mathcal{O}(\sqrt{n}\log n)$, and then present a protocol for a remote message transmission protocol between any two nodes $u, v \in \mathcal{P}$, tolerating a constant fraction of corrupted edges. For simplicity, we will assume that all messages in our protocols are binary. We remark that this restriction can be easily removed.

Let $\mathcal{G} = (\mathcal{V}, \mathcal{E})$ denote a graph on n nodes, d_v the degree of vertex $v \in \mathcal{V}$, and $\mathsf{Paths}_2(u, v)$ the set of all paths between any two vertices $u, v \in \mathcal{V}$ of length exactly 2. Let $\mathcal{G}$ satisfy the following two properties:

1. $\frac{\sqrt{n}\log n}{2} \leq d_v \leq 2\sqrt{n}\log n$ for all $v \in \mathcal{V}$; and

2. $|\mathsf{Paths}_2(u, v)| \geq \frac{\log^2 n}{2}$ for all $u, v \in \mathcal{V}$.

We will construct our transmission scheme on any graph $\mathcal{G}$ satisfying the above properties. We first observe that such a graph is easy to construct probabilistically. Consider the Erdős-Renyi random graph $\mathcal{G}(n, p)$, with $p = \frac{\log n}{\sqrt{n}}$; that is, construct the graph $\mathcal{G}$ such that there is an edge between every pair of nodes u and v, independently with probability $p = \frac{\log n}{\sqrt{n}}$ (for simplicity, we allow self-edges). Then, except with negligible (in n) probability, $\mathcal{G}(n, p)$ satisfies the conditions that we require of graph $\mathcal{G}$. (For completeness, we provide the proof of this in the full version.) We sometimes denote this process by $\mathcal{G} \leftarrow G(n, p)$. We now present two lemmas for graph $\mathcal{G}$ satisfying the two properties above.

Lemma 1. *In graph* $\mathcal{G}$, *no edge participates in more than* $4\sqrt{n}\log n$ *paths of length exactly 2* ($\mathsf{Paths}_2(u, v)$) *between any two vertices* $u, v \in \mathcal{V}$.

Let $0 < \alpha_e, \alpha_n < 1$ be constants denoting the fraction of corrupt edges and corrupt nodes in the graph, respectively. Note that if we are able to design a protocol that can tolerate $\alpha_e\sqrt{n}(n-1)\log n + 2\alpha_n\sqrt{n}(n-1)\log n$ edge corruptions, then we will automatically get a protocol that can tolerate an α_e fraction of corrupt edges and an α_n fraction of corrupt nodes. Hence, let $\alpha = \alpha_e + 2\alpha_n$; we will construct a protocol that can tolerate an α fraction of corrupt edges (and no corrupt nodes). The next lemma bounds the number of nodes in $\mathcal{G}$ with poor connectivity.

Lemma 2. *Let $\mathcal{Y}_u$ denote the set of nodes v such that the fraction of paths in* $\mathsf{Paths}_2(u, v)$ *with no corrupt edges is* $\leq \frac{1}{2}$. *We say that a node* $u \in V$ *is doomed if* $|\mathcal{Y}_u| \geq \frac{n}{4}$. *Then, in graph* $\mathcal{G}$, *at most* $64\alpha n$ *nodes are doomed.*

The set of privileged nodes $\mathcal{P}$ in $\mathcal{G}$ will simply be the nodes that are not doomed. By Lemma 2 above, we have that $|\mathcal{P}| \geq (1 - 64\alpha)n = \gamma n$ (for some constant $0 < \gamma < 1$). We now present the construction of a message transmission protocol between any two nodes $u, v \in \mathcal{P}$:

$\mathsf{TS}^{\mathcal{G}}_{u,v}(m)$

1.   For every node $w \in V$, u sends m over all paths in $\mathsf{Paths}_2(u, w)$.

2.   Every node $w \in V$, upon receiving m over the different paths, takes the majority of the values received, and sends this value to v over all paths in $\mathsf{Paths}_2(w, v)$.

3.   For every w, v takes the majority value of all messages received over $\mathsf{Paths}_2(w, v)$ as the message received from w. Then, v takes the majority (over all w) of the received values as the value sent by u.

We now show that if nodes u and v are not doomed, then the protocol described above is a reliable message transmission protocol.

Lemma 3. *Let* $u, v \in \mathcal{P}$ *(i.e., any two nodes in* $\mathcal{G}$ *that are not doomed), Then, after an execution of* $\mathsf{TS}^{\mathcal{G}}_{u,v}(m)$, *$v$ outputs m with probability 1.*

3.2   A.e. Agreement on $\mathcal{O}(n^\epsilon)$-Degree Graphs

In this section we present our main technical result: we show how to recursively increase the number of nodes in graph $\mathcal{G}$ from the previous section, while not increasing its degree (asymptotically), and show how to implement a reliable message transmission on such graphs (this will in turn lead to a.e. agreement protocols on such graphs with the same parameters). We will do this in two steps. Let $\gamma = (1 - 64\alpha)$. We will first show the following:

Lemma 4. *Let* $\mathcal{G}$ *be a graph on n nodes with maximum degree d. Furthermore, let* $\mathcal{G}$ *be such that it admits a reliable message transmission protocol,* $\mathsf{TS}^{\mathcal{G}}_{u,v}(\cdot)$, *between any two nodes u and v from a set of size at least γn nodes even in the presence of αnd corrupt edges. Then, there exists a graph $\mathcal{G}'$ on n^2 nodes of maximum degree $2d$, such that $\mathcal{G}'$ admits a reliable message transmission protocol between any two nodes u and v from a set of size at least $\gamma^2 n^2$ nodes even in the presence of $\alpha^2 n^2 d$ corrupt edges.*

In the full version [2], we show how to apply the $\mathcal{G}'$ construction from $\mathcal{G}$ recursively to obtain the desired result on graphs of degree $\mathcal{O}(n^\epsilon)$.

Construction of $\mathcal{G}'$. We construct $\mathcal{G}'$ as follows. Take n copies of graph $\mathcal{G}$; we will call each copy a *cloud*, and denote them $C_1, \cdots, C_n$. Connect the n clouds using another copy of graph $\mathcal{G}$. We do this by connecting the ith node in cloud C_j to the ith node in cloud C_k by an edge, whenever there is an edge between j and k in $\mathcal{G}$. We will call such a collection of edges between clouds C_j and C_k as a *cloud-edge*. Note that the maximum degree of any node in $\mathcal{G}'$ is $2d$.

We now describe how a node u in cloud C_j will communicate with a node v in cloud C_k—call this protocol $\mathsf{TS}^{\mathcal{G}'}_{u,v}(m)$. To do this, we will first describe how two clouds that share a cloud-edge will communicate. Let every node $i \in C_j$ hold a value m_i as input (note that every node need not hold the same value m_i) and assume cloud C_j wishes to communicate with cloud C_k. We describe a protocol such that, assuming a large-enough fraction of nodes in C_j hold the same input value, say m, then at the end of this protocol's execution a large-enough fraction of nodes in cloud C_k will output m. We call this protocol $\mathsf{CloudTransmit}_{C_j,C_k}(m_i)$. Let δ be such that $64\alpha n < \delta < (\gamma - 130\alpha)n$.

$\underline{\mathsf{CloudTransmit}_{C_j,C_k}(m_i)}$

1. For every node $1 \leq i \leq n$, the ith node in C_j sends m to the ith node in C_k through the edge connecting these two nodes.

2. The nodes in C_k execute a.e. differential agreement protocol $\mathsf{AE\text{-}DA}(n, 64\alpha n , \delta)$ using the value they received from their counterpart node in C_j as input. (Recall that the existence of protocol $\mathsf{TS}^{\mathcal{G}}_{u,v}(m)$ between privileged nodes in $\mathcal{G}$ guarantees that one can construct an a.e. differential agreement protocol; see the full version for more details on constructing an a.e. agreement protocol on $\mathcal{G}$.)

3. Each node takes the output of protocol $\mathsf{AE\text{-}DA}(n, 64\alpha n, \delta)$ as its output in this protocol.

We are now ready to describe $\mathsf{TS}^{\mathcal{G}'}_{u,v}(m)$:

$\underline{\mathsf{TS}^{\mathcal{G}'}_{u,v}(m)}$

1. u sends m to i for all nodes i in cloud C_j using $\mathsf{TS}^{\mathcal{G}}_{u,i}(m)$ from Section 3.1. The ith node in C_j receives message m_i.

2. Clouds C_j and C_k now execute protocol $\mathsf{TS}^{\mathcal{G}}_{C_j,C_k}(m_i)$ over the graph $\mathcal{G}$ connecting the n clouds[4]. Whenever cloud C_w is supposed to send a message to C_z according to the protocol, they use protocol $\mathsf{CloudTransmit}_{C_w,C_z}(\cdot)$ over the cloud-edge connecting C_w and C_z.

3. Node $v \in C_k$ takes its output in the protocol $\mathsf{TS}^{\mathcal{G}}_{C_j,C_k}(m_i)$ as the value sent by u.

We prove the correctness of the transmission scheme above through a series of lemmas. At a high level, our proof goes as follows. We will call a cloud C_j as good if it does not have too many corrupt edges within it (that is, corrupt edges of the form (u,v) with both u and v in C_j); otherwise we will call the cloud, bad. We first show that an adversary cannot create too many bad clouds. Next, we define what it means for a cloud-edge between two clouds C_j and C_k to be good (informally, the cloud-edge is good if both C_j and C_k are good clouds and there are sufficient number of edges connecting privileged nodes in C_j and C_k). We then show that the adversary cannot create too many bad cloud-edges.

---

[4] We again use m_i as the input argument, since the input values to nodes in C_j might be different.

Next, we show that two good clouds can communicate reliably across a good cloud-edge. Finally, we show that there exists a large set of clouds such that any two privileged nodes in any two clouds from this set, can communicate reliably. From this, the proof of Lemma 4 readily follows. We refer to the full version [2] of the paper for the complete proof of correctness of the transmission scheme.

We now arrive at our main result by applying the construction of $\mathcal{G}'$ from $\mathcal{G}$ recursively, a constant number of times, beginning with graph $\mathcal{G} \leftarrow G(n, \sqrt{n} \log n)$. That is, we obtain a transmission scheme on $\mathcal{O}(n^\epsilon)$ degree graphs and then show obtain our a.e. agreement protocol as well as the a.e. MPC protocol on the same graph. That is, we show:

Theorem 2. *For all sufficiently large n and all constant $0 < \epsilon < 1$, there exists a graph $\mathcal{G}_{main} = (\mathcal{V}, \mathcal{E})$ with maximum degree $\mathcal{O}(n^\epsilon)$, and a set of nodes $\mathcal{P} \subseteq \mathcal{V}$, with $|\mathcal{P}| \geq \mu n$ (for constant $0 < \mu < 1$) such that the nodes in $\mathcal{P}$ can execute an a.e. agreement protocol and a secure multi-party computation protocol (satisfying the security definition of [8]), even in the presence of of an α fraction of edge corruptions in $\mathcal{G}_{main}$ (for some constant $0 < \alpha < 1$).*

References

1. Ben-Or, M., Goldwasser, S., Wigderson, A.: Completeness theorems for non-cryptographic fault-tolerant distributed computation. In: STOC 1988 (1988)
2. Chandran, N., Garay, J., Ostrovsky, R.: Almost-everywhere secure computation with edge corruptions. Cryptology ePrint Archive, Report 2012/221, http://eprint.iacr.org/
3. Chandran, N., Garay, J., Ostrovsky, R.: Improved Fault Tolerance and Secure Computation on Sparse Networks. In: Abramsky, S., Gavoille, C., Kirchner, C., Meyer auf der Heide, F., Spirakis, P.G. (eds.) ICALP 2010, Part II. LNCS, vol. 6199, pp. 249–260. Springer, Heidelberg (2010)
4. Chaum, D., Crépeau, C., Damgård, I.: Multiparty unconditionally secure protocols (abstract). In: STOC 1988 (1988)
5. Dwork, C., Peleg, D., Pippenger, N., Upfal, E.: Fault tolerance in networks of bounded degree (preliminary version). In: STOC 1986 (1986)
6. Fitzi, M., Garay, J.: Efficient player-optimal protocols for strong and differential consensus. In: PODC 2003 (2003)
7. Garay, J., Givens, C., Ostrovsky, R.: Secure Message Transmission by Public Discussion: A Brief Survey. In: Chee, Y.M., Guo, Z., Ling, S., Shao, F., Tang, Y., Wang, H., Xing, C. (eds.) IWCC 2011. LNCS, vol. 6639, pp. 126–141. Springer, Heidelberg (2011)
8. Garay, J.A., Ostrovsky, R.: Almost-Everywhere Secure Computation. In: Smart, N.P. (ed.) EUROCRYPT 2008. LNCS, vol. 4965, pp. 307–323. Springer, Heidelberg (2008)
9. King, V., Saia, J., Sanwalani, V., Vee, E.: Towards secure and scalable computation in peer-to-peer networks. In: FOCS 2006 (2006)
10. Lamport, L., Shostak, R., Pease, M.: The Byzantine generals problem. ACM Transactions on Programming Languages and Systems 4(3) (1982)
11. Neiger, G.: Distributed consensus revisited. Information Processing Letters 49(4), 195–201 (1994)
12. Pease, M., Shostak, R., Lamport, L.: Reaching agreement in the presence of faults. Journal of the ACM (1980)
13. Upfal, E.: Tolerating linear number of faults in networks of bounded degree. In: PODC 1992 (1992)

Incentive Ratios of Fisher Markets

Ning Chen[1], Xiaotie Deng[2], Hongyang Zhang[3], and Jie Zhang[4]

[1] Division of Mathematical Sciences, Nanyang Technological University, Singapore
ningc@ntu.edu.sg
[2] Department of Computer Science, University of Liverpool, UK
xiaotie@liv.ac.uk
[3] Department of Computer Science, Shanghai Jiao Tong University, China
hongyang90@gmail.com
[4] Department of Computer Science, Aarhus University, Denmark
csjiezhang@gmail.com

Abstract. In a Fisher market, a market maker sells m items to n potential buyers. The buyers submit their utility functions and money endowments to the market maker, who, upon receiving submitted information, derives market equilibrium prices and allocations of its items. While agents may benefit by misreporting their private information, we show that the percentage of improvement by a unilateral strategic play, called incentive ratio, is rather limited—it is less than 2 for linear markets and at most $e^{1/e} \approx 1.445$ for Cobb-Douglas markets. We further prove that both ratios are tight.

1 Introduction

The Internet and world wide web have created a possibility for buyers and sellers to meet at a marketplace where pricing and allocations can be determined more efficiently and effectively than ever before. Market equilibrium, which ensures optimum fairness and efficiency, has become a paradigm for practical applications. It is well known that a market equilibrium always exists given mild assumptions on the utility functions of participating individuals [3].

However, there has been a major criticism on the market equilibrium in that it has not taken strategic behaviors of buyers and sellers into consideration: In a Fisher market, a market equilibrium price vector and associated allocations, computed in terms of utility functions and money endowments of participating individuals, may change even if one participant has a change in its utility function or endowment. Hence, an individual may misreport its private information to divert to a favorable outcome.

This phenomenon was first formally described by Adsul et al. [1] for linear and Chen et al. [5] for Leontief utility functions. Existence of such manipulations may impede potential uses of market equilibrium as a solution mechanism. To overcome such limitations, we explore the effect of a participant's incentive on the market equilibrium mechanism. We adopt the notion of *incentive ratio* [5] as the factor of the largest utility gains that a participant may achieve by behaving

A. Czumaj et al. (Eds.): ICALP 2012, Part II, LNCS 7392, pp. 464–475, 2012.
© Springer-Verlag Berlin Heidelberg 2012

strategically in the full information setting (the formal definition is referred to Section 2). The ratio characterizes the extent to which utilities can be increased by manipulations of individuals. Similar ideas have been applied in auctions under the concept of approximate strategic-proofness such as in [19,14].

While the big space of manipulations suggest that one may substantially increase his utility by behaving strategically, surprisingly, it was shown in [5] that for any Leontief utility market, the incentive ratio is upper bounded by 2. In this paper, we further study the incentive ratios of two other important functions: linear and Cobb-Douglas utilities [2]. For both utility models, manipulations do help to improve one's obtained utility. The following example shows such a case in a Cobb-Douglas market (a similar example for linear markets can be found in [1]).

Example 1. In a Cobb-Douglas market, there are two items with unit supply each and two buyers with endowments $e_1 = \frac{1}{2}, e_2 = \frac{1}{2}$ and utility functions $u_1(x, y) = x^{\frac{1}{4}} y^{\frac{3}{4}}, u_2(x, y) = x^{\frac{3}{4}} y^{\frac{1}{4}}$, respectively. When both buyers bid their utility functions and endowments truthfully, the equilibrium price is $\mathbf{p} = (\frac{1}{2}, \frac{1}{2})$, and the equilibrium allocations are $(\frac{1}{4}, \frac{3}{4})$ and $(\frac{3}{4}, \frac{1}{4})$; their utilities are $u_1 = u_2 = (\frac{1}{4})^{\frac{1}{4}} (\frac{3}{4})^{\frac{3}{4}}$. If buyer 1 strategically reports $u_1'(x, y) = x^{\frac{1}{2}} y^{\frac{1}{2}}$, then the equilibrium price is $\mathbf{p}' = (\frac{5}{8}, \frac{3}{8})$, and the equilibrium allocations are $(\frac{2}{5}, \frac{2}{3})$ and $(\frac{3}{5}, \frac{1}{3})$; their utilities are $u_1' = (\frac{2}{5})^{\frac{1}{4}} (\frac{2}{3})^{\frac{3}{4}}$ and $u_2' = (\frac{3}{5})^{\frac{3}{4}} (\frac{1}{3})^{\frac{1}{4}}$. Hence, $u_1' > u_1$ and the first buyer gets a strictly larger utility.

Our main results are the following, which bound the incentive ratios of linear and Cobb-Douglas markets.

Theorem. For any linear utility market, the incentive ratio is less than 2; and for any Cobb-Douglas utility market, the incentive ratio is at most $e^{1/e} \approx 1.445$. Both ratios are tight.

Our results give a further evidence for the solution concept of market equilibrium to be used in practical applications—while one may improve his utility by (complicated) manipulations, the increment is reasonably bounded by a small constant. Therefore, in a marketplace especially with incomplete information and a large number of participants, identifying a manipulation strategy is rather difficult and may be worthless. This echoes the results of, e.g., [18,13], saying that, in certain marketplaces, the fraction of participants with incentives to misreport their bids approaches zero as the market becomes large.

Our proof for the incentive ratio of linear markets is built on a reduction from fractional equilibrium allocations to an instance with integral equilibrium allocations, preserving the incentive ratio. We then, using the seminal Karush, Kuhn, Tucker (KKT) condition, show that if any participant is able to improve his utility by a factor of at least 2, everyone else can simultaneously obtain a utility increment. This, at a high level view, contradicts the market equilibrium condition of the original setting.

For Cobb-Douglas utility markets, our proof lies on a different approach by revealing interconnections of the incentive ratios of markets with different sizes.

In particular, we prove that the incentive ratio is independent of the number of buyers, by showing a reduction from any n-buyer market to a 2-buyer market, and vice versa. This result implies that the size of a market is not a factor to affect the largest possible utility gain by manipulations. Given this property, we restrict on a market with 2 buyers to bound the incentive ratio.

1.1 Related Work

Eisenberg and Gale [11] introduced a convex program to capture market equilibria of Fisher markets with linear utilities. Their convex program can be solved in polynomial time using the ellipsoid algorithm [12] and interior point algorithm [21]. Devanur et al. [9] gave the first combinatorial polynomial time algorithm for computing a Fisher market equilibrium with linear utility functions. The first strongly polynomial time algorithm for this problem was recently given by Orlin [17]. For Cobb-Douglas markets, Eaves [10] gave the necessary and sufficient conditions for existence of a market equilibrium, and gave an algorithm to compute a market equilibrium in polynomial time. Other computational studies on different market equilibrium models and utilities can be found in, e.g., [7,6,4,12,8] and the references within.

The concept of incentive ratio, which quantifies the benefit of unilateral strategic plays from a single participant, is in spirit similar to approximate truthfulness [19,14]. In the study of incentive ratio, the focus is on classic market designs with a stable outcome. It makes no attempt to consider the mechanism itself, but rather, focuses on individual's strategic plays and measures his benefit due to the incentive incompatibility of the mechanism.

Organization. In Section 2, we define the market equilibrium mechanism model and the notion of incentive ratio. In Section 3 and 4, we consider linear and Cobb-Douglas utility markets and derive matching incentive ratios, respectively. We conclude our work in Section 5.

2 Preliminary

In a Fisher market M, there are a set of n buyers and a set of m divisible items of unit quantity each for sale. We denote by $[n] = \{1, 2, \cdots, n\}$ and $[m] = \{1, 2, \cdots, m\}$ the set of buyers and items, respectively. Each buyer i has an initial cash endowment $e_i > 0$, which is normalized to be $\sum_{i=1}^{n} e_i = 1$, and has a utility function $u_i(x_i)$, where $x_i = (x_{i1}, \ldots, x_{im}) \in [0, 1]^m$ is an allocation vector denoting the amount that i receives from each item j.

An outcome of the market is a tuple $(\mathbf{p}, \mathbf{x})$, where $\mathbf{p} = (p_1, \ldots, p_m)$ is a price vector of all items and $\mathbf{x} = (x_1, x_2, \ldots, x_n)$ is an allocation vector. An outcome is called a *market equilibrium* if the following conditions hold: (i) All items are sold out, i.e., $\sum_{i=1}^{n} x_{ij} = 1$ for $j \in [m]$, and (ii) each buyer gets an allocation that maximizes its utility under the constraint $\sum_{j \in [m]} x_{ij} p_j \leq e_i$ for the given price vector. Such an equilibrium solution exists under a mild condition [3] on the utility functions.

One extensively studied class of utility functions is that of Constant Elasticity of Substitution (CES) functions [20]: For each i, its utility function $u_i(x_i) = \left(\sum_{j=1}^{m} \alpha_{ij} x_{ij}^\rho\right)^{\frac{1}{\rho}}$, where $-\infty < \rho < 1$ and $\rho \neq 0$, and $\alpha_i = (\alpha_{i1}, \ldots, \alpha_{im}) \geq 0$ is a given vector associated with buyer i. The CES utility functions allow us to model a wide range of realistic preferences of buyers, and have been shown to derive, in the limit, a number of special classes. In this paper, we will consider linear and Cobb-Douglas utility functions, which are derived when $\rho \to 1$ and $\rho \to 0$, respectively.

2.1 Incentive Ratio

Notice that a market equilibrium output crucially depends on the utility functions and endowments that buyers hold. This implies, in particular, that if a buyer manipulates his function or endowment, the outcome would be changed, and the buyer may possibly obtain a larger utility. This phenomenon has been observed for, e.g., linear [1] and Leontief functions [5]. One natural question is that how much such benefits can be obtained from manipulations with respect to the given market equilibrium rule (i.e., mechanism).

To this end, Chen et al. [5] defined the notion of incentive ratio to characterize such utility improvements by manipulations. Formally, in a given market M, for each buyer $i \in [n]$, let $u_i(\cdot)$ be his true private utility function and U_i be the space of utility functions that i can feasibly report; note that $u_i \in U_i$. Define $U = U_1 \times U_2 \times \cdots \times U_n$ and $U_{-i} = U_1 \times \cdots \times U_{i-1} \times U_{i+1} \times \cdots \times U_n$. Another private information that every buyer holds is his endowment e_i. For a given input, a vector of utility functions $(u_1, \ldots, u_n) \in U$ and a vector of endowments $(e_1, \ldots, e_n)$, we denote by $x_i(u_1, \ldots, u_n; e_1, \ldots, e_n)$ the equilibrium allocation of buyer i. In the market equilibrium mechanism, a buyer i can report any utility function $u_i' \in U_i$ and endowment $e_i' \in \mathbb{R}^+$. The *incentive ratio* of buyer i in the market M is defined to be

$$\zeta_i^M = \max_{u_{-i} \in U_{-i}, e_{-i} \in \mathbb{R}^{+n-1}} \frac{\max_{u_i' \in U_i, e_i' \in \mathbb{R}^+} u_i(x_i(u_i', u_{-i}; e_i', e_{-i}))}{u_i(x_i(u_i, u_{-i}; e_i, e_{-i}))}.$$

In the above definition, the numerator is the largest possible utility of buyer i when he unilaterally changes his bid.[1],[2] The incentive ratio of the market with respect to a given space of utility functions U is defined as $\zeta^M = \max_{i \in [n]} \zeta_i^M$. Incentive ratio quantifies the benefit of strategic behaviors of each individual buyer.

---

[1] Note that for some utility functions, an equilibrium allocation may not be unique. This may lead to different true utilities for a given manipulation bid. Our definition of incentive ratio is the strongest in the sense that it bounds the largest possible utility in all possible equilibrium allocations, which include, of course, the best possible allocation.

[2] Practically, a buyer can bid any endowment e_i'. However, reporting a larger budget results in a deficit in a resulting equilibrium, and thus, a negative utility. We therefore assume without loss of generality that $e_i' \leq e_i$ for all buyers.

For the considered linear and Cobb-Douglas (or any other CES) utilities, the true utility functions are characterized by the parameters $(\alpha_1, \alpha_2, \ldots, \alpha_n)$, where each $\alpha_i = (\alpha_{i1}, \ldots, \alpha_{im})$. The definition of incentive ratio can be simplified as follows

$$\zeta^{CES} = \max_{i \in [n]} \max_{\alpha_{-i}, e_{-i}} \frac{\max_{\alpha'_i, e'_i} u_i(x_i(\alpha'_i, \alpha_{-i}; e'_i, e_{-i}))}{u_i(x_i(\alpha_i, \alpha_{-i}; e_i, e_{-i}))}.$$

3   Linear Utility Functions

In this section, we will consider incentive ratio for linear utility functions, i.e., $u_i(x_i) = \sum_{j \in [m]} \alpha_{ij} x_{ij}$. Our main result in this section is the following.

Theorem 1. *The incentive ratio of linear markets is*

$$\zeta^{linear} < 2.$$

Consider a given linear market M and an arbitrary input scenario (α, e) where every buyer i bids utility vector $\alpha_i = (\alpha_{ij})_{j \in [m]}$ and endowment e_i. Let (p, x) be a market equilibrium of the instance (α, e). Let $r_i = \frac{u_i(x_i)}{e_i}$ be the bang-per-buck of buyer i, where $x_i = (x_{ij})_{j \in [m]}$ is the allocation of buyer i in the equilibrium. The Karush, Kuhn, Tucker (KKT) condition [16] implies that for any item j, if $x_{ij} > 0$ then $\frac{\alpha_{ij}}{p_j} = r_i$, and if $x_{ij} = 0$ then $\frac{\alpha_{ij}}{p_j} \leq r_i$.

Consider any fixed buyer, say i^*, and a scenario when all other buyers keep their bids and i^* unilaterally changes his bid to $\alpha'_{i^*} = (\alpha'_{i^*j})_{j \in [m]}$ and e'_{i^*}. Denote the resulting instance by $(\alpha', e') = (\alpha'_{i^*}, \alpha_{-i^*}; e'_{i^*}, e_{-i^*})$ and its equilibrium by (p', x'). For each buyer $i \in [n]$, define

$$c_i = \frac{u_i(x'_i)}{u_i(x_i)} = \frac{\sum_{j \in [m]} \alpha_{ij} x'_{ij}}{\sum_{j \in [m]} \alpha_{ij} x_{ij}}$$

to be the factor of utility changes of the buyer. Note that c_{i^*} gives the factor of how much more utility that i^* can get by manipulation. For the new setting (α', e'), the utility of buyer i is changed by a factor of c_i; thus, his bang-per-buck is changed by a factor of c_i as well, i.e., becomes $r_i c_i$.

Lemma 1. $c_{i^*} < 2.$

Note that our discussions do not rely on any specific initial instance (α, e) and manipulation (α', e'). Thus, the above lemma immediately implies that the incentive ratio of buyer i^* is less than 2, i.e., $\zeta_{i^*}^M < 2$. The same argument holds for all other buyers. Therefore, $\zeta^{linear} < 2$ and Theorem 1 follows. In the remaining of this section, we will prove this lemma.

In our proof, we assume that all input utility coefficients and endowments are rational numbers; thus, the computed equilibrium is composed of rational numbers as well. To simplify the proof, we first reduce an equilibrium with fractional allocations to an instance with integral equilibrium allocations, preserving the factor c_i of utility gains.

Proposition 1. *In the given market M, there exist another two linear market instances, where one is derived from the other by one strategic play of a buyer, such that they admit $\{0,1\}$-integral equilibrium allocations and c_i remains unchanged for all buyers.*

By the above claim, in the following we assume without loss of generality that the two equilibrium allocations $\mathbf{x}$ and $\mathbf{x}'$ are $\{0,1\}$-integral. That is, for any $i \in [n]$ and $j \in [m]$, $x_{ij}, x'_{ij} \in \{0,1\}$. Let $S_i = \{j \in [m] \mid x_{ij} = 1\}$ and $S'_i = \{j \in [m] \mid x'_{ij} = 1\}$ be the sets of items allocated to buyer i in the two allocations, respectively.

Proposition 2. *For any buyer i, $\sum_{j \in S'_i} p_j \geq c_i e_i$.*

Proof. Since the allocations x_i and x'_i are integral, we have

$$c_i = \frac{u_i(x'_i)}{u_i(x_i)} = \frac{\sum_{j \in S'_i} \alpha_{ij}}{\sum_{j \in S_i} \alpha_{ij}}$$

For $j \in S_i$, we have $\frac{\alpha_{ij}}{p_j} = r_i$. Thus, $\sum_{j \in S_i} \alpha_{ij} = \sum_{j \in S_i} r_i p_j = r_i e_i$. For $j \in S'_i$, we have $\frac{\alpha_{ij}}{p_j} \leq r_i$. Thus, $\sum_{j \in S'_i} \alpha_{ij} \leq r_i \cdot \sum_{j \in S'_i} p_j$. Therefore,

$$\sum_{j \in S'_i} p_j \geq \frac{1}{r_i} \sum_{j \in S'_i} \alpha_{ij} = \frac{c_i}{r_i} \sum_{j \in S_i} \alpha_{ij} = \frac{c_i}{r_i} r_i e_i = c_i e_i.$$

The claim follows. $\qquad\square$

Proposition 3. *Consider any buyer $i \neq i^*$ and any item $j \in [m]$. If $j \in S_i$, then $c_i \geq \frac{p_j}{p'_j}$; if $j \in S'_i$, then $\frac{p_j}{p'_j} \geq c_i$.*

Proof. Note that the bids of buyer i in the two scenarios (α, e) and (α', e') are the same. If $j \in S_i$, we have $\frac{\alpha_{ij}}{p_j} = r_i$ and $\frac{\alpha_{ij}}{p'_j} \leq r_i c_i$; therefore, $c_i \geq \frac{p_j}{p'_j}$. If $j \in S'_i$, we have $\frac{\alpha_{ij}}{p'_j} = r_i c_i$ and $\frac{\alpha_{ij}}{p_j} \leq r_i$; thus, $\frac{p_j}{p'_j} \geq c_i$. $\qquad\square$

Finally, we are ready to prove Lemma 1.

Proof (of Lemma 1). Assume to the contrary that $c_{i^*} \geq 2$. By Proposition 2, $\sum_{j \in S'_{i^*}} p_j \geq c_{i^*} e_{i^*}$. Since $\sum_{j \in S_{i^*} \cap S'_{i^*}} p_j \leq \sum_{j \in S_{i^*}} p_j = e_{i^*}$, we have

$$\sum_{j \in S'_{i^*} \setminus S_{i^*}} p_j \geq (c_{i^*} - 1) \cdot e_{i^*}$$

Further, we have $\sum_{j \in S'_{i^*} \setminus S_{i^*}} p'_j \leq \sum_{j \in S'_{i^*}} p'_j = e'_{i^*}$. Hence,

$$\frac{\sum_{j \in S'_{i^*} \setminus S_{i^*}} p_j}{\sum_{j \in S'_{i^*} \setminus S_{i^*}} p'_j} \geq \frac{(c_{i^*} - 1) \cdot e_{i^*}}{e'_{i^*}} \triangleq \Delta$$

This implies there exists $j \in S'_{i^*} \setminus S_{i^*}$ such that $\frac{p_j}{p'_j} \geq \Delta$. Let R denote the set of all such items in $S'_{i^*} \setminus S_{i^*}$. From the above discussion, we have $R \neq \emptyset$.

Consider the following iterative procedure:

1. Initialize: Let $A = \{i^*\}$ and $T = \{i \mid S_i \cap R \neq \emptyset\}$ (note that $T \neq \emptyset$).
2. Do the following until $T = \emptyset$:

   - Pick an arbitrary $i \in T$.
   - Let $T \leftarrow T \setminus \{i\}$ and $A \leftarrow A \cup \{i\}$.
   - Let $T \leftarrow T \cup \{k \notin A \cup T \mid S_k \cap (S_i' \setminus S_i) \neq \emptyset\}$.

Intuitively, in each iteration, we find all buyers that win some items from the set $S_i' \setminus S_i$ in the equilibrium allocation $\mathbf{x}$. Our main observation is that, for all buyers ever added to T in the procedure, their utility gains by manipulations are at least Δ. We prove this by induction on the iterations.

In the initialization step, this fact is true for all buyers in T: For any $i \in T$, by the definition of T and Proposition 3, there exists an item $j \in S_i \cap R$ such that $c_i \geq \frac{p_j}{p_j'}$. By the definition of R, $\frac{p_j}{p_j'} \geq \Delta$. Therefore, we have $c_i \geq \Delta$. Next, we consider the induction step. For any buyer k added to T during the procedure, since $S_k \cap (S_i' \setminus S_i) \neq \emptyset$, let j be an item in $S_k \cap (S_i' \setminus S_i)$. By Proposition 3, we have $c_k \geq \frac{p_j}{p_j'} \geq c_i$. Since i used to be in T, by induction hypothesis, $c_i \geq \Delta$; hence, $c_k \geq \Delta$. Therefore, for any buyer k ever added to T in the process, $c_k \geq \Delta$.

Note that the iterative procedure must terminate as every buyer can be added into T at most once. We consider the subset A at the end of the procedure, which includes all buyers ever added into T. Note that since the initial $T \neq \emptyset$, $A \setminus \{i^*\} \neq \emptyset$. By the rule of updating T, we know that all items in R and S_i' (for all $i \in A \setminus \{i^*\}$) are bought by buyers in the set A in the equilibrium allocation $\mathbf{x}$. That is,

$$\left(\bigcup_{i \in A \setminus \{i^*\}} S_i' \right) \cup R \cup (S_{i^*}' \cap S_{i^*}) \subseteq \bigcup_{i \in A} S_i.$$

Further, by the definition of R, we have $R \subseteq S_{i^*}'$, and thus,

$$\sum_{j \in (S_{i^*}' \setminus S_{i^*}) \setminus R} p_j < \Delta \cdot \sum_{j \in (S_{i^*}' \setminus S_{i^*}) \setminus R} p_j' \leq \Delta \cdot e_{i^*}' = (c_{i^*} - 1) \cdot e_{i^*}$$

Therefore,

$$\sum_{j \in \bigcup_{i \in A} S_i'} p_j = \sum_{j \in (S_{i^*}' \setminus S_{i^*}) \setminus R} p_j + \sum_{j \in \left(\bigcup_{i \in A \setminus \{i^*\}} S_i' \right) \cup R \cup (S_{i^*}' \cap S_{i^*})} p_j$$

$$< (c_{i^*} - 1) \cdot e_{i^*} + \sum_{j \in \left(\bigcup_{i \in A} S_i \right)} p_j$$

$$\leq (c_{i^*} - 1) \cdot e_{i^*} + \sum_{i \in A} e_i \tag{1}$$

On the other hand, we have

$$
\sum_{j \in \bigcup_{i \in A} S_i'} p_j = \sum_{i \in A} \sum_{j \in S_i'} p_j
$$

$$
\geq \sum_{i \in A} c_i e_i \qquad \text{(Proposition 2)}
$$

$$
\geq c_{i^*} e_{i^*} + \sum_{i \in A \setminus \{i^*\}} \Delta \cdot e_i \qquad (\forall\, i \in A \setminus \{i^*\},\, c_i \geq \Delta)
$$

$$
= c_{i^*} e_{i^*} + \sum_{i \in A \setminus \{i^*\}} \frac{(c_{i^*} - 1) \cdot e_{i^*}}{e_{i^*}'} \cdot e_i
$$

$$
\geq c_{i^*} e_{i^*} + \sum_{i \in A \setminus \{i^*\}} (c_{i^*} - 1) \cdot e_i \qquad (e_{i^*} \geq e_{i^*}')
$$

This contradicts formula (1), as $c_{i^*} \geq 2$ and $A \setminus \{i^*\}$ is nonempty. $\square$

The proved ratio $\zeta^{\text{linear}} < 2$ is tight, as the following example shows.

Example 2. There are three items and two buyers with utilities and endowments:
$u_1 = \left(\frac{1+\varepsilon}{2-2\varepsilon^2-\varepsilon^3}, \frac{1-\varepsilon-2\varepsilon^2-\varepsilon^3}{2-2\varepsilon^2-\varepsilon^3}, 0 \right)$, $u_2 = \left(\varepsilon^2, \varepsilon, 1 - \varepsilon - \varepsilon^2 \right)$, $e_1 = \varepsilon + \varepsilon^2$, and $e_2 = 1 - \varepsilon - \varepsilon^2$. When both buyers bid truthfully, the equilibrium price is $\mathbf{p} = (\varepsilon + \varepsilon^2, \frac{\varepsilon(1-\varepsilon-\varepsilon^2)}{1-\varepsilon^2}, \frac{(1-\varepsilon-\varepsilon^2)^2}{1-\varepsilon^2})$, and equilibrium allocations are $x_1 = (1, 0, 0)$ and $x_2 = (0, 1, 1)$; the utility of the first buyer is $u_1 = \frac{1+\varepsilon}{2-2\varepsilon^2-\varepsilon^3}$. When the first buyer bids $u_1' = u_2$, the equilibrium price becomes $\mathbf{p}' = \left(\varepsilon^2, \varepsilon, 1 - \varepsilon - \varepsilon^2 \right)$, and the best equilibrium allocations are $x_1' = (1, 1, 0)$ and $x_2' = (0, 0, 1)$; the utility of the first buyer becomes $u_1' = 1$. Thus, the utility gain is $\frac{u_1'}{u_1} = \frac{2-2\varepsilon^2-\varepsilon^3}{1+\varepsilon}$, which approaches to 2 when ε is arbitrarily small.

4   Cobb-Douglas Utility Functions

In a Cobb-Douglas market, buyers' utility functions are of the form $u_i(x_i) = \prod_{j \in [m]} x_{ij}^{\alpha_{ij}}$, where $\sum_{j=1}^{m} \alpha_{ij} = 1$, for all $i \in [n]$. To guarantee the existence of a market equilibrium, we assume that each item is desired by at least one buyer, i.e., $\alpha_{ij} > 0$ for some i. This, together with the fact that each buyer desires at least one item (followed by the fact that $\alpha_{ij} > 0$ for some j), implies that a market equilibrium always exists [15].

For a given Cobb-Douglas market with reported utilities $(\alpha_{ij})_{j \in [m]}$ and endowment e_i from each buyer i, market equilibrium prices and allocations are unique and can be computed by the following equations [10].

$$
p_j = \sum_{i=1}^{n} e_i \alpha_{ij} \tag{2}
$$

$$
x_{ij} = \frac{e_i \alpha_{ij}}{\sum_{i=1}^{n} e_i \alpha_{ij}} \tag{3}
$$

Based on these characterizations, in the following we will analyze the incentive ratio of Cobb-Douglas markets.

4.1   Manipulation on Endowments

Note that the private information of a buyer is composed of two parts: money endowment and utility function. In this section, we show that a buyer will never misreport his endowment.

Lemma 2. *In any Cobb-Douglas market, bidding endowments truthfully is a dominant strategy for all buyers.*

By the above result, in the following discussions, we assume that all buyers report their endowments truthfully, and will only consider their strategic behaviors on utility functions.

4.2   Reductions on Market Sizes

We first show that the incentive ratio of Cobb-Douglas markets is independent of the number of buyers. Let

$$\zeta(n) = \max\left\{\zeta^{M_n} \mid M_n \text{ is a Cobb-Douglas market with } n \text{ buyers}\right\}$$

be the largest incentive ratio of all markets with n buyers. Note that $\zeta(1) = 1$.

Theorem 2. *For Cobb-Douglas markets, incentive ratio is independent of the number of buyers, i.e., $\zeta(n) = \zeta(n')$ for any $n > n' \geq 2$.*

The claim follows from the following two lemmas.

Lemma 3. *For any n-buyer market M_n, there is a 2-buyer market M_2 such that $\zeta^{M_2} \geq \zeta^{M_n}$. This implies that $\zeta(2) \geq \zeta(n)$.*

Proof. Consider a market M_n with n buyers; assume without loss of generality that the first buyer defines the maximal incentive ratio, i.e., $\zeta^{M_n} = \zeta_1^{M_n}$. Given M_n, we will construct a market M_2 with two buyers as follows.

- Input of M_n: n buyers $[n] = \{1,\ldots,n\}$, each with an endowment e_i and a utility function $u_i = \prod_{j=1}^{m} x_{ij}^{\alpha_{ij}}$.
- Construction of M_2: 2 buyers $1^*, 2^*$.
  - For 1^*, endowment $e_{1^*} = e_1$, utility function $u_{1^*} = \prod_{j=1}^{m} x_{1^*j}^{\alpha_{1^*j}}$, where $\alpha_{1^*j} = \alpha_{1j}$.
  - For 2^*, endowment $e_{2^*} = 1 - e_1$, utility function $u_{2^*} = \prod_{j=1}^{m} x_{2^*j}^{\alpha_{2^*j}}$, where $\alpha_{2^*j} = \sum_{i=2}^{n} \frac{e_i}{1-e_1}\alpha_{ij}$.

For the constructed M_2, we can easily verify it is a well defined Cobb-Douglas market, i.e., $e_{1^*} + e_{2^*} = 1$, $\sum_j \alpha_{1^*j} = 1$ and $\sum_j \alpha_{2^*j} = \sum_j \sum_{i=2}^{n} \frac{e_i}{1-e_1}\alpha_{ij} = 1$. The above reduction is based on unifying buyers $2,\ldots,n$ in M_n into one buyer 2^* in M_2. We will prove that the incentive ratio of buyer 1^* in M_2 is the same as buyer 1 in M_n, i.e., $\zeta_{1^*}^{M_2} = \zeta_1^{M_n}$. This immediately implies that $\zeta^{M_2} \geq \zeta_{1^*}^{M_2} = \zeta_1^{M_n} = \zeta^{M_n}$, and thus, $\zeta(2) \geq \zeta(n)$.

Let $\mathbf{p} = (p_j)_{j \in [m]}$ and $\mathbf{p}^* = (p_j^*)_{j \in [m]}$ be the equilibrium prices of markets M_n and M_2 respectively. Further, let $x_1 = (x_{1j})_{j \in [m]}$ and $x_{1^*}^* = (x_{1^*j}^*)_{j \in [m]}$ denote the equilibrium allocations of buyer 1 in M_n and buyer 1^* in M_2, respectively. Since for any $j \in [m]$,

$$p_j = \sum_{i=1}^n e_i \alpha_{ij} = e_{1^*} \alpha_{1^*j} + (1 - e_1) \sum_{i=2}^n \frac{e_i}{1 - e_1} \alpha_{ij} = e_{1^*} \alpha_{1^*j} + e_{2^*} \alpha_{2^*j} = p_j^*,$$

we obtain

$$x_{1j} = \frac{e_1 \alpha_{1j}}{\sum_{i=1}^n e_i \alpha_{ij}} = \frac{e_{1^*} \alpha_{1^*j}}{e_{1^*} \alpha_{1^*j} + e_{2^*} \alpha_{2^*j}} = x_{1^*j}^*.$$

Hence,

$$u_1(\mathbf{x}_1) = \prod_{j=1}^m x_{1j}^{\alpha_{1j}} = \prod_{j=1}^m (x_{1^*j}^*)^{\alpha_{1^*j}} = u_{1^*}(\mathbf{x}_{1^*}^*)$$

Denote by $\alpha_1' = (\alpha_{1j}')_{j \in [m]}$ the best response of buyer 1 in M_n. By the same argument as above, buyer 1^* can get the same utility in M_2 as buyer 1 in M_n by reporting $\alpha_1' = (\alpha_{1j}')_{j \in [m]}$. That is, $u_1' = u_{1^*}'$. This implies that $\zeta_{1^*}^{M_2} = \zeta_1^{M_n}$ and completes the proof of the claim. $\qquad\square$

We can have a similar reduction from any 2-buyer market to an n-buyer market.

Lemma 4. *For any 2-buyer market M_2, there is an n-buyer market M_n such that $\zeta^{M_n} \geq \zeta^{M_2}$. This implies $\zeta(n) \geq \zeta(2)$.*

4.3 Incentive Ratio

Theorem 3. *The incentive ratio of Cobb-Douglas markets is*

$$\zeta^{\text{Cobb-Douglas}} \leq e^{1/e} \approx 1.445.$$

Proof. According to Theorem 2, it suffices to consider the case with 2 buyers. We consider two scenarios: For the fixed bid vector $(\alpha_{2j})_{j \in [m]}$ of buyer 2, buyer 1 bids $(\alpha_{1j})_{j \in [m]}$ and $(\alpha_{1j}')_{j \in [m]}$, respectively, with resulting equilibrium allocations $\mathbf{x}_1 = (x_{1j})_{j \in [m]}$ and $\mathbf{x}_1' = (x_{1j}')_{j \in [m]}$. Then,

$$\zeta = \frac{u_1(\mathbf{x}_1')}{u_1(\mathbf{x}_1)} = \frac{\prod_{j=1}^m x_{1j}'^{\alpha_{1j}}}{\prod_{j=1}^m x_{1j}^{\alpha_{1j}}} = \frac{\prod_{j=1}^m \left(\frac{e_1 \alpha_{1j}'}{e_1 \alpha_{1j}' + e_2 \alpha_{2j}}\right)^{\alpha_{1j}}}{\prod_{j=1}^m \left(\frac{e_1 \alpha_{1j}}{e_1 \alpha_{1j} + e_2 \alpha_{2j}}\right)^{\alpha_{1j}}}$$

$$= \prod_{j=1}^m \left(\frac{\alpha_{1j}'(e_1 \alpha_{1j} + e_2 \alpha_{2j})}{\alpha_{1j}(e_1 \alpha_{1j}' + e_2 \alpha_{2j})}\right)^{\alpha_{1j}} = \prod_{j=1}^m \left(\frac{\alpha_{1j}' \alpha_{1j} + \frac{e_2}{e_1} \alpha_{2j} \alpha_{1j}'}{\alpha_{1j}' \alpha_{1j} + \frac{e_2}{e_1} \alpha_{2j} \alpha_{1j}}\right)^{\alpha_{1j}} \triangleq \prod_{j=1}^m R_j$$

where R_j, $j = 1, \ldots, m$, is the j-th term of the above formula.

For each item j, it is easy to see that $R_j > 1$ if and only if $\alpha'_{1j} > \alpha_{1j}$. Let $S = \{j \mid R_j > 1\}$, and $r_j = R_j^{1/\alpha_{1j}}$ for $j \in S$ (note that r_j is well defined for those items in S as $\alpha_{1j} > 0$). Therefore, $r_j > 1$ if and only if $R_j > 1$. Further, when $r_j > 1$, one can see that $\alpha'_{1j} \geq r_j \alpha_{1j}$. This implies that $\sum_{j \in S} r_j \alpha_{1j} \leq \sum_{j \in S} \alpha'_{1j} \leq 1$. Hence,

$$\zeta \leq \prod_{j \in S} r_j^{\alpha_{1j}} \leq \prod_{j \in S} e^{\frac{r_j \alpha_{1j}}{e}} \leq e^{\frac{\sum_{j \in S} r_j \alpha_{1j}}{e}} \leq e^{1/e}.$$

Note that the above second inequality follows from the fact that for any $x, y \geq 0$, $x^y \leq e^{xy/e}$, which can be verified easily. Therefore, the theorem follows. $\qquad \Box$

The upper bound established in the above claim is tight, which can be seen from the following example.

Example 3. There are 2 items and 2 buyers, with endowments e_1 and e_2, respectively. Let $n = \frac{e_2}{e_1}$, $\epsilon_1 = \frac{1}{n}$, and $\epsilon_2 = \frac{1}{n^3}$. The utilities vectors of the two buyers are defined as $u_1 = (\frac{1}{e}, 1 - \frac{1}{e})$ and $u_2 = (1 - \epsilon_2, \epsilon_2)$. When buyer 1 bids $u'_1 = (1 - \epsilon_1, \epsilon_1)$ instead of u_1, the incentive ratio is given by the following formula:

$$\left(\frac{\frac{1 - \frac{1}{n}}{e} + n(1 - \frac{1}{n^3})(1 - \frac{1}{n})}{\frac{1 - \frac{1}{n}}{e} + n(1 - \frac{1}{n^3})\frac{1}{e}} \right)^{\frac{1}{e}} \cdot \left(\frac{\frac{1}{n}(1 - \frac{1}{e}) + n \cdot \frac{1}{n^3} \cdot \frac{1}{n}}{\frac{1}{n}(1 - \frac{1}{e}) + n \cdot \frac{1}{n^3} \cdot (1 - \frac{1}{e})} \right)^{1 - \frac{1}{e}}$$

When n goes to infinity, the limit of the left factor is $e^{1/e}$, and the limit of the right factor is 1. Therefore, the ratio approaches $e^{1/e}$.

5   Conclusions

We study two important class of utility functions in Fisher markets: linear and Cobb-Douglas utilities, and show that their incentive ratios are bounded by 2 and $e^{1/e}$, respectively. It is interesting to explore the incentive ratios of other CES functions. In particular, is the incentive ratio bounded by a constant for any CES function? Another interesting direction is to characterize Nash equilibria in the market mechanism. In the full version paper, we give a sufficient and necessary condition for truthful bidding being a Nash equilibrium in Cobb-Douglas markets. It is intriguing to have such characterizations for other CES functions.

Acknowledgements. The last author acknowledges support from the Danish National Research Foundation and the National Science Foundation of China (grant 61061130540) for the Sino-Danish Center for the Theory of Interactive Computation, and from the Center for research in Foundations of Electronic Markets (CFEM), supported by the Danish Council for Strategic Research.

References

1. Adsul, B., Babu, C.S., Garg, J., Mehta, R., Sohoni, M.: Nash Equilibria in Fisher Market. In: Kontogiannis, S., Koutsoupias, E., Spirakis, P.G. (eds.) SAGT 2010. LNCS, vol. 6386, pp. 30–41. Springer, Heidelberg (2010)
2. Arrow, K., Chenery, H., Minhas, B., Solow, R.: Capital-Labor Substitution and Economic Efficiency. The Review of Economics and Statistics 43(3), 225–250 (1961)
3. Arrow, K., Debreu, G.: Existence of an Equilibrium for a Competitive Economy. Econometrica 22, 265–290 (1954)
4. Chen, N., Deng, X., Sun, X., Yao, A.C.-C.: Fisher Equilibrium Price with a Class of Concave Utility Functions. In: Albers, S., Radzik, T. (eds.) ESA 2004. LNCS, vol. 3221, pp. 169–179. Springer, Heidelberg (2004)
5. Chen, N., Deng, X., Zhang, J.: How Profitable Are Strategic Behaviors in a Market? In: Demetrescu, C., Halldórsson, M.M. (eds.) ESA 2011. LNCS, vol. 6942, pp. 106–118. Springer, Heidelberg (2011)
6. Codenotti, B., Varadarajan, K.R.: Efficient Computation of Equilibrium Prices for Markets with Leontief Utilities. In: Díaz, J., Karhumäki, J., Lepistö, A., Sannella, D. (eds.) ICALP 2004. LNCS, vol. 3142, pp. 371–382. Springer, Heidelberg (2004)
7. Deng, X., Papadimitriou, C., Safra, S.: On the Complexity of Equilibria. In: STOC 2002, pp. 67–71 (2002)
8. Devanur, N., Kannan, R.: Market Equilibria in Polynomial Time for Fixed Number of Goods or Agents. In: FOCS 2008, pp. 45–53 (2008)
9. Devanur, N., Papadimitriou, C., Saberi, A., Vazirani, V.: Market Equilibrium via a Primal-Dual Algorithm for a Convex Program. JACM 55(5) (2008)
10. Eaves, B.C.: Finite Solution of Pure Trade Markets with Cobb-Douglas Utilities. Mathematical Programming Study 23, 226–239 (1985)
11. Eisenberg, E., Gale, D.: Consensus of Subjective Probabilities: The Pari-Mutuel Method. Annals of Mathematical Statistics 30, 165–168 (1959)
12. Jain, K.: A polynomial Time Algorithm for Computing an Arrow-Debreu Market Equilibrium for Linear Utilities. SIAM Journal on Computing 37(1), 303–318 (2007)
13. Kojima, F., Pathak, P.: Incentives and Stability in Large Two-Sided Matching Markets. American Economic Review 99(3), 608–627 (2009)
14. Kothari, A., Parkes, D., Suri, S.: Approximately-Strategyproof and Tractable Multiunit Auctions. Decision Support Systems 39(1), 105–121 (2005)
15. Maxfield, R.: General Equilibrium and the Theory of Directed Graphs. Journal of Mathematical Economics 27(1), 23–51 (1997)
16. Nisan, N., Roughgarden, T., Tardos, E., Vazirani, V.: Algorithmic Game Theory. Cambridge University Press (2007)
17. Orlin, J.: Improved Algorithms for Computing Fisher's Market Clearing Prices. In: STOC 2010, pp. 291–300 (2010)
18. Roberts, D., Postlewaite, A.: The Incentives for Price-Taking Behavior in Large Exchange Economies. Econometrica 44, 113–127 (1976)
19. Schummer, J.: Almost Dominant Strategy Implementation, MEDS Department, Northwestern University, Discussion Papers 1278 (1999)
20. Solov, R.: A Contribution to the Theory of Economic Growth. Quarterly Journal of Economics 70, 65–94 (1956)
21. Ye, Y.: A Path to the Arrow-Debreu Competitive Market Equilibrium. Mathematical Programming 111(1-2), 315–348 (2008)

Computational Complexity of Traffic Hijacking under BGP and S-BGP*

Marco Chiesa[1], Giuseppe Di Battista[1], Thomas Erlebach[2], and Maurizio Patrignani[1]

[1] Dept. of Computer Science and Automation, Roma Tre University
[2] Dept. of Computer Science, University of Leicester

Abstract. Harmful Internet hijacking incidents put in evidence how fragile the Border Gateway Protocol (BGP) is, which is used to exchange routing information between Autonomous Systems (ASes). As proved by recent research contributions, even S-BGP, the secure variant of BGP that is being deployed, is not fully able to blunt traffic attraction attacks. Given a traffic flow between two ASes, we study how difficult it is for a malicious AS to devise a strategy for hijacking or intercepting that flow. We show that this problem marks a sharp difference between BGP and S-BGP. Namely, while it is solvable, under reasonable assumptions, in polynomial time for the type of attacks that are usually performed in BGP, it is NP-hard for S-BGP. Our study has several by-products. E.g., we solve a problem left open in the literature, stating when performing a hijacking in S-BGP is equivalent to performing an interception.

1 Introduction and Overview

On 24th Feb. 2008, Pakistan Telecom started an unauthorized announcement of prefix 208.65.153.0/24 [13]. This announcement was propagated to the rest of the Internet, which resulted in the *hijacking* of YouTube traffic on a global scale. Incidents like this put in evidence how fragile is the *Border Gateway Protocol (BGP)* [10], which is used to exchange routing information between Internet Service Providers (ISPs). Indeed, performing a hijacking attack is a relatively simple task. It suffices to issue a BGP announcement of a victim prefix from a border router of a malicious (or unaware) *Autonomous System (AS)*. Part of the traffic addressed to the prefix will be routed towards the malicious AS rather than to the intended destination. A mischievous variation of the hijacking is the *interception* when, after passing through the malicious AS, the traffic is forwarded to the correct destination. This allows the rogue AS to eavesdrop or even modify the transit packets.

In order to cope with this security vulnerability, a variant of BGP, called S-BGP [8], has been proposed, that requires a PKI infrastructure both to validate the correctness of the AS that originates a prefix and to allow an AS to sign its announcements to other ASes. In this setting an AS cannot forge announcements that do not derive from announcements received from its neighbors. However,

---

* An extended version of the paper is available on the arXiv Web site.

A. Czumaj et al. (Eds.): ICALP 2012, Part II, LNCS 7392, pp. 476–487, 2012.
© Springer-Verlag Berlin Heidelberg 2012

Table 1. Complexity of finding a HIJACK strategy in different settings

	AS-paths of any length	Bounded AS-path length	Bounded AS-path length and AS degree
Origin-spoofing	NP-hard (Thm. 1)	P(Thm. 2)	P
S-BGP	NP-hard	NP-hard (Thm. 3)	P(Thm. 4)

[4] contains surprising results: (i) simple hijacking strategies are tremendously effective and (ii) finding a strategy that maximizes the amount of traffic that is hijacked is NP-hard both for BGP and for S-BGP.

In this paper we tackle the hijacking and interception problems from a new perspective. Namely, given a traffic flow between two ASes, how difficult is it for a malicious AS to devise a strategy for hijacking or intercepting at least that specific flow? We show that this problem marks a sharp difference between BGP and S-BGP. Namely, while it is polynomial time solvable, under reasonable assumptions, for typical BGP attacks, it is NP-hard for S-BGP. This gives new complexity related evidence of the effectiveness of the adoption of S-BGP. Also, we solve an open problem [4], showing when every hijack in S-BGP results in an interception. Tab. 1 summarizes our results. Rows correspond to different settings for a malicious AS m. The origin-spoofing setting (Sect. 2) corresponds to a scenario where m issues BGP announcements pretending to be the owner of a prefix. Its degree of freedom is to choose a subset of its neighbors for such a bogus announcement. This is the most common type of hijacking attack to BGP [1]. In S-BGP (Sect. 3) m must enforce the constraints imposed by S-BGP, which does not allow to pretend to be the owner of a prefix that is assigned to another AS. Columns of Tab. 1 correspond to different assumptions about the Internet. In the first column we assume that the longest *valley-free* path (i.e. a path enforcing certain customer-provider constraints) in the Internet can be of arbitrary length. This column has a theoretical interest since the length of the longest path (and hence valley-free path) observed in the Internet remained constant even though the Internet has been growing in terms of active AS numbers during the last 15 years [7]. Moreover, in today's Internet about 95% of the ASes is reached in 3 AS hops [7]. Hence, the second column corresponds to a quite realistic Internet, where the AS-path length is bounded by a constant. In the third column we assume that the number of neighbors of m is bounded by a constant. This is typical in the periphery of the Internet. A "P" means that a Polynomial-time algorithm exists. Since moving from left to right the setting is more constrained, we prove only the rightmost NP-hardness results, since they imply the NP-hardness results to their left. Analogously, we prove only the leftmost "P" results.

1.1 A Model for BGP Routing

As in previous work on interdomain hijacking [4], we model the Internet as a graph $G = (V, E)$. A vertex in V is an *Autonomous System (AS)*. Edges in E are *peerings* (i.e., connections) between ASes. A vertex owns one or more *prefixes*, i.e., sets of contiguous IP numbers. The routes used to reach prefixes are spread and selected via BGP. Since each prefix is handled independently by BGP, we focus on a single prefix π, owned by a destination vertex d.

BGP allows each AS to autonomously specify which paths are forbidden (*import policy*), how to choose the best path among those available to reach a destination (*selection policy*), and a subset of neighbors to whom the best path should be announced (*export policy*). BGP works as follows. Vertex d initializes the routing process by sending *announcements* to (a subset of) its neighbors. Such announcements contain π and the *path* of G that should be traversed by the traffic to reach d. In the announcements sent from d such a path contains just d. We say that a path $P = (v_n \ \ldots \ v_0)$ is *available* at vertex v if v_n announces P to v. Each vertex checks among its available paths that are not filtered by the import policy, which is the best one according to its selection policy, and then it announces that path to a set of its neighbors in accordance with the export policy. Further, BGP has a loop detection mechanism, i.e., each vertex v ignores a route if v is already contained in the route itself.

Policies are typically specified according to two types of relationships [6]. In a *customer-provider* relationship, an AS that wants to access the Internet pays an AS which sells this service. In a *peer-peer* relationship two ASes exchange traffic without any money transfer between them. Such commercial relationships between ASes are represented by orienting a subset of the edges of E. Namely, edge $(u, v) \in E$ is directed from u to v if u is a customer of v, while it is undirected if u and v are peers. A path is *valley-free* if provider-customer and peer-peer edges are only followed by provider-customer edges.

The Gao-Rexford [3] Export-all (*GR-EA*) conditions are commonly assumed to hold in this setting [4]. **GR1**: G has no directed cycles that would correspond to unclear customer-provider roles. **GR2**: Each vertex $v \in V$ sends an announcement containing a path P to a neighbor n only if path $(n \ v)P$ is valley-free. Otherwise, some AS would provide transit to either its peers or its providers without revenues. **GR3**: A vertex prefers paths through customers over those provided by peers and paths through peers over those provided by providers. **Shortest Paths**: Among paths received from neighbors of the same class (customers, peers, and provider), a vertex chooses the shortest ones. **Tie Break**: If there are multiple such paths, a vertex chooses according to some tie break rule. As in [4], we assume that the one whose next hop has lowest AS number is chosen. Also, as in [2], to tie break equal class and equal length simple paths $P_1^u = (u \ v)P_1^v$ and $P_2^u = (u \ v)P_2^v$ at the same vertex u from the same neighbor v, if v prefers P_1^v over P_2^v, then u prefers P_1^u over P_2^u. This choice is called *policy consistent* in [2] and it can be proven that it has the nice property of making the entire set of policies considered in this paper policy consistent. **NE policy**: a vertex always exports a path except when GR2 forbids it to do so.

Since we assume that the GR-EA conditions are satisfied, then a (partially directed) graph is sufficient to fully specify the policies of the ASes. Hence, in the following a *BGP instance* is just a graph.

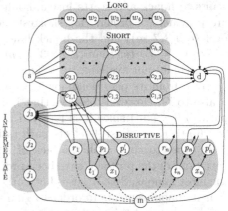

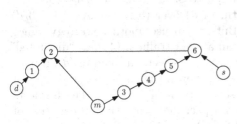

Fig. 1. A network for Examples 1 and 2

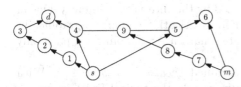

Fig. 2. A network for Example 3

Fig. 3. Reduction of a constrained 3-SAT problem to the HIJACK problem when m has S-BGP cheating capabilities

1.2 Understanding Hacking Strategies

We consider the following problem. A BGP instance with three specific vertices, d, s, and m are given, where such vertices are: the AS originating a prefix π, a source of traffic for π, and an attacker, respectively. All vertices, but m, behave correctly, i.e., according to the BGP protocol and GR-EA conditions. Vertex m is interested in two types of attacks: *hijacking* and *interception*. In the hijacking attack m's goal is to attract to itself at least the traffic from s to d. In the interception attack m's goal is to be traversed by at least the traffic from s to d.

In Fig. 1 $(2,6)$ is peer-to-peer and the other edges are customer-provider. Prefix π is owned and announced by d. According to BGP, the traffic from s to d follows $(s\ 6\ 2\ 1\ d)$. In fact, 2 selects $(1\ d)$. Vertex 6 receives a unique announcement from d (it cannot receive an announcement with $(5\ 4\ 3\ m\ 2\ 1\ d)$ since it is not valley-free). By cheating, (**Example 1**) m can deviate the traffic from s to d attracting traffic from s. In fact, if m pretends to be the owner of π and announces it to 2, then 2 prefers, for shortest-path, $(2\ m)$ over $(2\ 1\ d)$. Hence, the traffic from s to d is received by m following $(s\ 6\ 2\ m)$. A hijack!

Observe that m could be smarter (**Example 2**). Violating GR2, it can announce $(2\ 1\ d)$ to 3. Since each of 3, 4 and 5 prefers paths announced by customers (GR3), the propagation of this path is guaranteed. Therefore, 6 has two available paths, namely, $(2\ 1\ d)$ and $(5\ 4\ 3\ m\ 2\ 1\ d)$. The second one is preferred because 5 is a customer of 6, while 2 is a peer of 6. Hence, the traffic from s to d is received by m following path $(s\ 6\ 5\ 4\ 3\ m)$. Since after passing through m the traffic reaches d following $(m\ 2\ 1\ d)$ this is an interception.

Fig. 2 allows to show a negative example (**Example 3**). According to BGP, the traffic from s to d follows $(s\ 4\ d)$. In fact, s receives only paths $(4\ d)$ and $(1\ 2\ 3\ d)$, both from a provider, and prefers the shortest one. Suppose that m

wants to hijack and starts just announcing π to 6. Since all the neighbors of s are providers, s prefers, for shortest path, (4 d) over (5 6 m) (over (1 2 3 d) over (4 9 8 7 m)) and the hijack fails. But m can use another strategy. Since (s 5 6 m) is shorter than (s 1 2 3 d), m can attract traffic if (4 d) is "disrupted" and becomes not available at s. This happens if 4 selects, instead of (d), a path received from its peer neighbor 9 (m may announce that it is the originator of π also to 7). However, observe that if 4 selects path (4 9 8 7 m) then 5 selects path (5 9 8 7 m) since it is received from a peer and stops the propagation of (s 5 6 m). Hence, s still selects path (s 1 2 3 d) and the hijack fails.

In order to cope with the lack of any security mechanism in BGP, several variations of the protocol have been proposed by the Internet community. One of the most famous, S-BGP, uses both origin authentication and cryptographically-signed announcements in order to guarantee that an AS announces a path only if it has received this path in the past.

The attacker m has more or less constrained *cheating capabilities*. 1. With the *origin-spoofing* cheating capabilities m can do the typical BGP announcement manipulation. I.e., m can pretend to be the origin of prefix π owned by d, announcing this to a subset of its neighbors. 2. With the *S-BGP* cheating capabilities m must comply with the S-BGP constraints. I.e.: (a) m cannot pretend to be the origin of prefix π; and (b) m can announce a path (m u)P only if u announced P to m in the past. However, m can still announce paths that are not the best to reach d and can decide to announce different paths to different neighbors. In Example 2, m has S-BGP cheating capabilities.

In this paper we study the computational complexity of the HIJACK and of the INTERCEPTION problems. The HIJACK problem is formally defined as follows. **Instance:** A BGP instance G, a source vertex s, a destination vertex d, a manipulator vertex m, and a cheating capability for m. **Question:** Does there exist a set of announcements that m can simultaneously send to its neighbors, according to its cheating capability, that produces a stable state for G where the traffic from s to d goes to m? The INTERCEPTION problem is defined in the same way but changing "the traffic from s to d goes to m" to "the traffic from s to d passes through m before reaching d".

BGP policies can be so complex that there exist configurations that do not allow to reach any stable routing state (see, e.g., [5]). A routing state is *stable* if there exists a time t such that after t no AS changes its selected path. If the GR-EA conditions are satisfied [3], then a BGP network always converges to a stable state. However, there is a subtle issue to consider in attacks. As we have seen in the examples, m can deliberately ignore the GR-EA conditions. Anyway, the following lemma makes it possible, in our setting, to study the HIJACK and the INTERCEPTION problem ignoring stability related issues.

Lemma 1. *Let G be a BGP instance and suppose that at a certain time a manipulator m starts announcing steadily any set of arbitrary paths to its neighbors. Routing in G converges to a stable state.*

The existence of a stable state (pure Nash equilibrium) in a game where one player can deviate from a standard behavior has been proved, in a different setting in [2]. Such a result and Lemma 1 are somehow complementary since the export policies they consider are more general than Export-All, while the convergence to the stable state is not guaranteed (even if such a stable state is always reachable from any initial state).

2   Checking If an Origin-Spoofing BGP Attack Exists

In this section, we show that, in general, it is hard to find an attack strategy if m has an origin-spoofing cheating capability (Theorem 1), while the problem turns to be easier in a realistic setting (Theorem 2).

We introduce some notation. A *ranking function* determines the level of preference of paths available at vertex v. If P_1, P_2 are available at v and P_1 is *preferred* over P_2 we write $P_1 <^v_\lambda P_2$. The *concatenation* of two nonempty paths $P = (v_k \ v_{k-1} \ \dots \ v_i)$, $k \geq i$, and $Q = (v_i \ v_{i-1} \ \dots \ v_0)$, $i \geq 0$, denoted as PQ, is the path $(v_k \ v_{k-1} \ \dots \ v_{i+1} \ v_i \ v_{i-1} \ \dots \ v_0)$. Also, let P be a valley-free path from vertex v. We say that P *is of class* 3, 2, or 1 if its first edge connects v with a customer, a peer, or a provider of v, respectively. We also define a function f^v for each vertex v, that maps each path from v to the integer of its class. Given two paths P and P' available at v if $f^v(P) > f^v(P')$ we say that the class of P *is better than* the class of P'. In stable routing state S, a path $P = (v_1 \ \dots \ v_n)$ is *disrupted at vertex* v_i by a path P' if there exists a vertex v_i of P such that v_i selects path P'. Also, if P' is preferred over $(v_i \ \dots \ v_n)$ because of the GR3 condition, we say that path P *is disrupted by a path of a better class*. Otherwise, if P' is preferred over $(v_i \ \dots \ v_n)$ because of the shortest-paths criterion, we say that P is disrupted by a path of the *same class*.

A hijacking can be obviously found in exponential time by a simple brute force approach which simulates every possible attack strategy and verifies its effectiveness. The following result in the case the Internet graph has no bound constraints may be somehow expected.

Theorem 1. *If the manipulator has origin-spoofing cheating capabilities, then problem* HIJACK *is NP-hard.*

Surprisingly, in a more realistic scenario, where the length of valley-free paths is bounded by a constant k, we have that in the origin-spoofing setting an attack strategy can be found in polynomial time $(n^{O(k)}$, where n is the number of vertices of G). Let N be the set of neighbors of m. Indeed, the difficulty of the HIJACK problem in the origin-spoofing setting depends on the fact that m has to decide to which of the vertices in N it announces the attacked prefix π, which leads to an exponential number of possibilities. However, when the longest valley-free path in the graph is bounded by a constant k, it is possible to design a polynomial-time algorithm based on the following intuition, that will be formalized below. Suppose m is announcing π to a subset $A \subseteq N$ of its neighbors and path $p = (z \ \dots \ n \ m)$ is available at an arbitrary vertex z of the graph. Let n_1, n_2 be two vertices of $N \setminus A$. If p is disrupted (is not disrupted) by better class both when π is announced either to n_1 or to n_2, then p is disrupted (is not

Algorithm 1. Algorithm for the HIJACK problem where m has origin-spoofing capabilities and the longest valley-free path in the graph is bounded.

1: **Input:** instance of HIJACK problem, m has origin-spoofing cheating capabilities;
2: **Output:** an attack pattern if the attack exists, fail otherwise;
3: let P_{sm} be the set of all valley-free paths from s to m;
4: **for all** p_{sm} in P_{sm} **do**
5:     let w be the vertex of p_{sm} adjacent to m; let A be a set of vertices and initialize A to $\{w\}$; let N be the set of the neighbors of m;
6:     **for all** n in $N \setminus \{w\}$ **do**
7:         **if** there is no path p through (m, n) to a vertex x of p_{sm} such that $f^x(p) > f^x(p_{xm})$, where p_{xm} is the subpath of p_{sm} from x to m **then**
8:             insert n into A
9:     **if** the attack succeeds when m announces π only to the vertices in A **then**
10:            **return** A
11: **return** fail

---

disrupted) by better class when π is announced to both n_1 and n_2. This implies that once m has a candidate path p^* for attracting traffic from s, it can check independently to which of its neighbors it can announce π without disrupting p^* by better class, which guarantees that a path from m to z longer than p cannot be selected at z.

In order to prove Theorem 2, we introduce the following lemmata that relate attacks to the structure of the Internet.

Lemma 2. *Consider a valley-free path $p = (v_n \; ... \; v_1)$ and consider an attack of m such that v_1 announces a path p_{v_1} to v_2 to reach prefix π and p is possibly disrupted only by same class. Vertex v_n selects a path $p_n \leq_\lambda^{v_n} pp_{v_1}$.*

Lemma 3. *Consider a successful attack for m and let p_{sm} be the path selected at s. Let p_{sd} be a valley-free path from s to d such that it does not traverse m and such that $p_{sd} <_\lambda^s p_{sm}$. Path p_{sd} is disrupted by a path of better class.*

Lemma 4. *Let $p = (v_n \; ... \; v_1)$ be a valley-free path. Consider an attack where v_1 announces a path p_1 to v_2. Vertex v_n selects a path of class at least $f^{v_n}(p)$.*

Theorem 2. *If the manipulator has origin-spoofing cheating capabilities and the length of the longest valley-free path is bounded by a constant, then problem HIJACK is in P.*

Proof. We tackle the problem with Alg. 1. First, observe that line 9 tests if a certain set of announcements causes a successful attack and, in that case, it returns the corresponding set of neighbors to whom m announces prefix π. Hence, if Alg. 1 returns without failure it is trivial to see that it found a successful attack. Suppose now that there exists a successful attack a^* from m that is not found by Alg. 1. Let p_{sm}^* be the path selected by s in attack a^*. Let A^* be the set of neighbors of m that receives prefix π from m in the successful attack.

Consider the iteration of the Alg. 1 where path p_{sm}^* is analyzed in the outer loop. At the end of the iteration Alg. 1 constructs a set A of neighbors of m. Let a be an attack from m where m announces π only to the vertices in A.

First, we prove that $A^* \subseteq A$. Suppose by contradiction that there exists a vertex $n \in A^*$ that is not contained in A. It implies that there exists a valley-free path p through (m, n) to a vertex x of p^*_{sm} such that $f^x(p) > f^x(p_{xm})$, where p_{xm} is the subpath of p^*_{sm} from x to m. Since m announces π to n, by Lemma 4, we have that x selects a path p' of class at least $f^x(p)$, that is a contradiction since p^*_{sm} would be disrupted by better class. Hence, $A^* \subseteq A$.

Now, we prove that attack a is a successful attack for m. Consider a valley-free path p_{sd} from s to d that does not traverse m and is preferred over p^*_{sm}. By Lemma 3 it is disrupted by better class in attack a^*. By Lemma 4, since $A^* \subseteq A$, we have that also in a path p_{sd} is disrupted by better class. Let x be the vertex adjacent to s in p_{sd}. Observe that, vertex s cannot have an available path $(s\ x)p$ to d such that $(s\ x)p <^s_\lambda p^*_{sm}$, because $(s\ x)p$ must be disrupted by better class.

Moreover, consider path p^*_{sm}. Since in a^* path p^*_{sm} is not disrupted by better class by a path to d, by Lemma 4, there does not exist a path p'_{xd} from a vertex x of p^*_{sm} to d of class higher than p_{xm}, where p_{xm} is the subpath of p^*_{sm} from x to m. Hence, path p^*_{sm} cannot be disrupted by better class by a path to d. Also, observe that for each $n \in A$ there is no path p through (m, n) to a vertex x of p^*_{sm} such that $f^x(p) > f^x(p_{xm})$, where p_{xm} is the subpath of p_{sm} from x to m. Hence, p^*_{sm} can be disrupted only by same class. By Lemma 2, we have that s selects a path p such that $p \leq^s_\lambda p^*_{sm}$. Since path p cannot be a path to d, attack a is successful. This is a contradiction since we assumed that Alg. 1 failed.

Finally, since the length of the valley-free paths is bounded, the iterations of the algorithm where paths in P_{sm} are considered require a number of steps that is polynomial in the number of vertices of the graph. Also, the disruption checks can be performed in polynomial time by using the algorithm in [11].   □

3   S-BGP Gives Hackers Hard Times

We open this section by strengthening the role of S-BGP as a security protocol. Indeed, S-BGP adds more complexity to the problem of finding an attack strategy (Theorem 3). After that we also provide an answer to a conjecture posed in [4] about hijacking and interception attacks in S-BGP when a single path is announced by the manipulator. In this case, we prove that every successful hijacking attack is also an interception attack (Theorem 5).

Theorem 3. *If the manipulator has S-BGP cheating capabilities and the length of the longest valley-free path is bounded by a constant, then problem* HIJACK *is NP-hard.*

Proof. We reduce from a version of 3-SAT where each variable appears at most three times and each positive literal at most once [9]. Let F be a logical formula in conjunctive normal form with variables $X_1 \ldots X_n$ and clauses $C_1 \ldots C_h$. We build a BGP instance G (see Fig. 3) consisting of 4 structures: INTERMEDIATE, SHORT, LONG, and DISRUPTIVE. The LONG structure is a directed path of length 6 with edges (s, w_1), (w_1, w_2), $\ldots$, (w_4, w_5), and (w_5, d). The INTERMEDIATE structure consists of a valley-free path joining m and s. It has length 4 and it is composed by a directed path $(s\ j_3\ j_2\ j_1)$, and a directed edge (m, j_1). The SHORT structure has h directed paths from s to d. Each path has length at most 4 and has edges $(s, c_{i,1}), (c_{i,1}, c_{i,2}), \ldots, (c_{i,v(C_i)}, d)$ $(1 \leq i \leq h)$, where $v(C_i)$ is the size of C_i. The DISRUPTIVE structure contains, for each variable X_i vertices, r_i, t_i, x_i, p_i and p_i'. Vertices, r_i, t_i, and x_i, are reached via long directed paths from m and are connected by (t_i, p_i), (x_i, p_i), (x_i, p_i'), (r_i, j_3), (p_i, j_3), and (p_i, d). Finally, suppose X_i occurs in clause C_j with a literal in position l. If the literal is negative the undirected edge $(p_i, c_{j,l})$ is added, otherwise, edges $(p_i, c_{j,l})$, $(r_i, c_{j,l})$, $(c_{j,l}, j_3)$, and undirected edge $(p_i', c_{j,l})$ are added. An edge connects m to d. Vertices s, d, and m have source, destination, and manipulator roles, respectively.

Intuitively, the proof works as follows. The paths that allow traffic to go from s to m are only those passing through the DISRUPTIVE structure and the one in the INTERMEDIATE structure. Also, the path through the INTERMEDIATE structure is shorter than the one through the LONG structure, which is shorter than those through the DISRUPTIVE structure. If m does not behave maliciously, s receives only paths traversing the SHORT structure and the LONG structure. In this case s selects one of the paths in the SHORT structure according to its tie break policy. If m wants to attract traffic from s, then: (i) path $(j_3 \; j_2 \; j_1 \; m \; d)$ must be available at s and (ii) all paths contained in the SHORT structure must be disrupted by a path announced by m. If (i) does not hold, then s selects the path contained in either the LONG structure or the SHORT structure. If (ii) does not hold, then s selects a path contained in the SHORT structure.

Our construction is such that the 3-SAT formula is true iff m can attract the traffic from s to d. To understand the relationship with the 3-SAT problem, consider the behavior of m with respect to variable X_1 (see Fig. 3) that appears with a positive literal in the first position of clause C_1, a negative literal in the first position of C_2 and a negative literal in the second position of C_h.

First, we explore the possible actions that m can perform in order to disrupt paths in the SHORT structure. Since m has S-BGP cheating capabilities, m is constrained to propagate only the announcements it receives. If m does not behave maliciously, m receives path (d) from d and paths P_{r_1}, P_{t_1}, and P_{x_1} from r_1, t_1, and x_1, respectively. These paths have the following properties: P_{r_1} contains vertex $c_{1,1}$ that is contained in the path of the SHORT structure that corresponds to clause C_1; paths P_{t_1} and P_{x_1} both contain vertex p_1 and do not contain vertex $c_{1,1}$ since p_1 prefers $(p_1 \; d)$ over $(p_1 \; c_{1,1} \; c_{1,2} \; c_{1,3} \; d)$.

Now, we analyze what actions are not useful for m to perform an attack. If m issues any announcement towards t_1 or r_1 the path traversing the INTERMEDIATE structure is disrupted by better class. Also, if m sends a path P_{r_1}, P_{t_1}, or P_{x_1} towards r_j, t_j, or x_j, with $j = 2, \ldots, n$, the path traversing the INTERMEDIATE structure is disrupted by better class. Also, if m sends $(m \; d)$ to x_1, then the path traversing the INTERMEDIATE structure is disrupted from $c_{1,1}$ by better class. If m sends P_{x_1} to x_1, then it is discarded by x_1 because of loop detection. In each of these cases m cannot disrupt any path traversing the SHORT structure without disrupting the path traversing the INTERMEDIATE structure. Hence, m can disrupt path in the SHORT structure without disrupting the path traversing the INTERMEDIATE structure only announcing P_{r_1} and P_{t_1} from m towards x_1.

If path P_{t_1} is announced to x_1, then p_1 discards that announcement because of loop detection and path $(s\ c_{1,1}\ c_{1,2}\ c_{1,3}\ d)$ is disrupted from p'_1 by better class. Also, the path through the INTERMEDIATE structure remains available because the announcement through p'_1 cannot reach j_3 from $c_{1,1}$, otherwise valley-freeness would be violated. Hence, announcing path P_{t_1}, corresponds to assigning true value to variable X_1, since the only path in the SHORT structure that is disrupted is the one that corresponds to the clause that contains the positive literal of X_1.

If path P_{r_1} is announced to x_1, then $c_{1,1}$ discards that announcement because of loop detection and both paths $(s\ c_{2,1}\ c_{2,2}\ c_{2,3}\ d)$ and $(s\ c_{h,1}\ c_{h,2}\ c_{h,3}\ d)$ are disrupted by better class from p_1. Also, the path through the INTERMEDIATE structure remains available because the announcement through p_1 cannot reach j_3 from $c_{2,1}$ or $c_{h,2}$, otherwise valley-freeness would be violated. Hence, announcing path P_{r_1}, corresponds to assigning false value to variable X_1, since the only paths in the SHORT structure that are disrupted are the ones that correspond to the clauses that contain a negative literal of X_1.

Hence, announcing path P_{t_1} (P_{r_1}) from m to x_1 corresponds to assigning the true (false) value to variable X_1. As a consequence, m can disrupt every path in the SHORT structure without disrupting the path in the INTERMEDIATE structure iff formula F is satisfiable.   □

Theorem 4. *If the manipulator has S-BGP cheating capabilities and its degree is bounded by a constant, then problem* HIJACK *is in P.*

To study the relationship between hijacking and interception we introduce the following technical lemma.

Lemma 5. *Let G be a BGP instance, let m be a vertex with S-BGP cheating capabilities, and let $d \neq m$ be any vertex of G. Let N' be the set of the neighbors to whom m is announcing a path. Consider a vertex v admitting a class c valley-free path p to d such that either $p = (v\ \dots\ n\ m\ \dots\ d)$ to d, where $n \in N'$, or p does not contain m. Vertex v has an available path of class c or better to d, irrespective of the paths propagated by m to N'.*

Theorem 5. *Let N' be the set of the neighbors to whom a manipulator m with S-BGP cheating capabilities is announcing a path. If m announces the same path to all the vertices in N', then every successful hijacking attack is also a successful interception attack. If m announces different paths to different vertices, then the hijacking may not be an interception.*

Proof. We prove the following more technical statement that implies the first part of the theorem. Let G be a BGP instance, let m be a vertex with S-BGP cheating capabilities. Let p be a path available at m in the stable state S reached when m behaves correctly. Suppose that m starts announcing p to any subset of its neighbors. Let S' be the corresponding routing state. Path p remains available at vertex m in S'. The truth of the statement implies that m can forward the traffic to d by exploiting p.

Suppose for a contradiction that path p is disrupted in S' when m propagates it to a subset of its neighbors. Let x be the first vertex of p that prefers a different

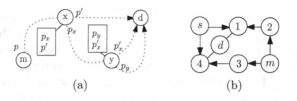

Fig. 4. Proof of Theorem 5. (a) The order of paths into the boxes represents the preference of the vertices. (b) The BGP instance used in the example.

path p_x (p is disrupted by p_x) in S' and let p' be the subpath of p from vertex d to x (see Fig. 4a). Observe that p is not a subpath of p_x as x cannot select a path that passes through itself. Since p_x is not available at x in S, let y be the vertex in p_x closest to d that selects a path p_y that is preferred over p'_x in S, where p'_x is the subpath of p_x from y to d.

We have two cases: either $f^x(p_x) > f^x(p')$ or $f^x(p_x) = f^x(p')$ (i.e., p_x is preferred to p' by better or by same class).

Suppose that $f^x(p_x) > f^x(p')$. By Lemma 5, since there exists a valley-free path p_x from x to d that does not traverse m, then x has an available path of class at least $f^x(p_x)$. Hence, x cannot select path p' in S, a contradiction.

Suppose that $f^x(p_x) = f^x(p')$. Two cases are possible: either p_y contains x or not. In the first case either $f^y(p_y) > f^y(p'_x)$ or $f^y(p_y) = f^y(p'_x)$. If $f^y(p_y) > f^y(p'_x)$, then we have that $f^y(p_y) \leq f^x(p') = f^x(p_x) \leq f^y(p'_x)$, a contradiction. If $f^y(p_y) = f^y(p'_x)$, we have that $|p'_x| < |p_x| \leq |p'| < |p_y|$. A contradiction since a longer path is preferred.

The second case ($f^x(p_x) = f^x(p')$ and p_y does not contain x) is more complex. We have that $|p'| \geq |p_x|$. Also, by Lemma 5, since p_y and p'_x do not pass through m, then y has an available path of class at least $\max\{f^y(p_y), f^y(p'_x)\}$. As y alternatively chooses p_y and p'_x we have that $f^y(p_y) = f^y(p'_x)$, which implies that $|p'_x| \geq |p_y|$. Denote by p_{xy} the subpath $(v_m \ldots v_0)$ of p_x, where $v_0 = y$ and $v_m = x$. Consider routing in state S. Two cases are possible: either $p_{xy}p_y$ is available at x or not. In the first case, since $|p'| \geq |p_x| = |p_{xy}p'_x| \geq |p_{xy}p_y|$, we have a contradiction because p' would not be selected in S. In the second case, we will prove that for each vertex $v_h \neq x$ in p_{xy} we have that $|p_h| \leq |(v_h \ldots v_0)p_y|$, where p_h is the path selected by v_h in S. This implies that $|(v_m v_{m-1})p_{m-1}| \leq |p_{xy}p_y| \leq |p_x| \leq |p'|$ and this leads to a contradiction. In fact, if $|(v_m v_{m-1})p_{m-1}| < |p'|$, then we have a contradiction because p' would not be selected in S. Otherwise, if $|(v_m v_{m-1})p_{m-1}| = |p'|$, we have that $|p_x| = |p'|$. Then, x prefers p_x over p' because of tie break. We have a contradiction since also $(v_m v_{m-1})p_{m-1}$ is preferred over p' because of tie break in S.

Finally, we prove that for each vertex $v_h \neq x$ in p_{xy} we have that $|p_h| \leq |(v_h \ldots v_0)p_y|$. This trivially holds for $v_0 = y$. We prove that if it holds for v_i then it also holds for v_{i+1}. If v_{i+1} selects $(v_{i+1} v_i)p_i$, then the property holds. Otherwise, $(v_{i+1} v_i)p_i$ is disrupted either by better class or by same class by a path p_{i+1}. In the first case, by Lemma 5, a path of a class better than $(v_{i+1} \ldots v_0)p'_x$ is available at v_{i+1} and so v_{i+1} cannot select $(v_{i+1} \ldots v_0)p'_x$ in S', a contradiction.

In the second case, we have that $|p_{i+1}| \leq |(v_{i+1}\ v_i)p_i| \leq |(v_{i+1}\ \ldots\ v_0)p_y|$. The second inequality comes from the induction hypothesis.

This concludes the first part of the proof. For proving the second part we show an example where m announces different paths to different neighbors and the resulting hijacking is not an interception. Consider the BGP instance in Fig. 4b. In order to hijack traffic from s, vertices 1 and 4 must be hijacked. Hence, m must announce $(m\ 3\ 4\ d)$ to 2 and $(m\ 2\ 1\ d)$ to 3. However, since $(3\ 4\ d)$ and $(2\ 1\ d)$ are no longer available at m the interception fails.   □

4   Conclusions and Open Problems

Given a communication flow between two ASes we studied how difficult it is for a malicious AS m to devise a strategy for hijacking or intercepting that flow. This problem marks a sharp difference between BGP and S-BGP. Namely, while in a realistic scenario the problem is computationally tractable for typical BGP attacks it is NP-hard for S-BGP. This gives new evidence of the effectiveness of the adoption of S-BGP. It is easy to see that all the NP-hardness results that we obtained for the hijacking problem easily extend to the interception problem. Further, we solved a problem left open in [4], showing when performing a hijacking in S-BGP is equivalent to performing an interception.

Several problems remain open: 1. We focused on a unique m. How difficult is it to find a strategy involving several malicious ASes [4]? 2. In [12] it has been proposed to disregard the AS-paths length in the BGP decision process. How difficult is it to find an attack strategy in this different model?

References

1. IP hijacking (2012), http://en.wikipedia.org/wiki/IP_hijacking
2. Engelberg, R., Schapira, M.: Weakly-Acyclic (Internet) Routing Games. In: Persiano, G. (ed.) SAGT 2011. LNCS, vol. 6982, pp. 290–301. Springer, Heidelberg (2011)
3. Gao, L., Rexford, J.: Stable internet routing without global coordination. In: Proc. SIGMETRICS (2000)
4. Goldberg, S., Schapira, M., Hummon, P., Rexford, J.: How secure are secure interdomain routing protocols? In: Proc. SIGCOMM (2010)
5. Griffin, T., Shepherd, F.B., Wilfong, G.: The stable paths problem and interdomain routing. IEEE/ACM Trans. on Networking 10(2), 232–243 (2002)
6. Huston, G.: Interconnection, peering, and settlements. In: Proc. INET (1999)
7. Huston, G.: AS6447 BGP routing table analysis report (2012), http://bgp.potaroo.net/as6447/
8. Kent, S., Lynn, C., Seo, K.: Secure border gateway protocol (S-BGP) (2000)
9. Papadimitriou, C.M.: Computational complexity (1994)
10. Rekhter, Y., Li, T., Hares, S.: A Border Gateway Protocol 4 (BGP-4). RFC 4271
11. Sami, R., Schapira, M., Zohar, A.: Searching for stability in interdomain routing. In: Proc. INFOCOM (2009)
12. Schapira, M., Zhu, Y., Rexford, J.: Putting BGP on the right path: a case for next-hop routing. In: HotNets (2010)
13. Underwood, T.: Pakistan hijacks YouTube (2008), http://www.renesys.com/blog/2008/02/pakistan_hijacks_youtube_1.shtml

Efficiency-Revenue Trade-Offs in Auctions

Ilias Diakonikolas[1,*], Christos Papadimitriou[1,**], George Pierrakos[1,**],
and Yaron Singer[2,***]

[1] UC Berkeley, EECS
{ilias,christos,georgios}@cs.berkeley.edu
[2] Google, Inc.
yaron@cs.berkeley.edu

Abstract. When agents with independent priors bid for a single item, Myerson's optimal auction maximizes expected revenue, whereas Vickrey's second-price auction optimizes social welfare. We address the natural question of *trade-offs* between the two criteria, that is, auctions that optimize, say, revenue under the constraint that the welfare is above a given level. If one allows for randomized mechanisms, it is easy to see that there are polynomial-time mechanisms that achieve any point in the trade-off (the *Pareto curve*) between revenue and welfare. We investigate whether one can achieve the same guarantees using *deterministic* mechanisms. We provide a negative answer to this question by showing that this is a (weakly) NP-hard problem. On the positive side, we provide polynomial-time deterministic mechanisms that approximate with arbitrary precision any point of the trade-off between these two fundamental objectives for the case of two bidders, even when the valuations are correlated arbitrarily. The major problem left open by our work is whether there is such an algorithm for three or more bidders with independent valuation distributions.

1 Introduction

Two are the fundamental results in the theory of auctions. First, Vickrey observed that there is a simple way to run an auction so that social welfare (efficiency) is maximized: The second-price (Vickrey) auction is optimally efficient, independently of how bidder valuations are distributed. However, the whole point of the Vickrey auction is to deliberately sacrifice auctioneer revenue in order to achieve efficiency. If auctioneer revenue is to be maximized, Myerson showed in 1980 that, when the bidders' valuations are distributed independently, a straightforward auction (essentially, a clever reduction to Vickrey's auction via an ingenious transformation of valuations) achieves this.

These two criteria, social welfare and revenue, are arguably of singular and paramount importance. It is therefore a pity that they seem to be at loggerheads: It is not hard to establish that optimizing any one of these two criteria can be very suboptimal with respect to the other. In other words, there is a substantial *trade-off* between these two

---

* Supported by a Simons Postdoctoral Fellowship.
** Research supported by NSF grant CC-0964033 and by a Google University Research Award.
*** Research supported by a Microsoft Graduate Fellowship and a Facebook Graduate Fellowship.

A. Czumaj et al. (Eds.): ICALP 2012, Part II, LNCS 7392, pp. 488–499, 2012.
© Springer-Verlag Berlin Heidelberg 2012

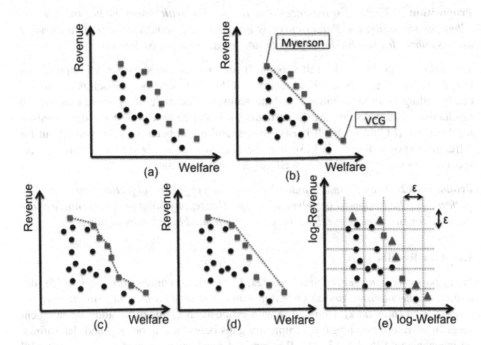

Fig. 1. The Pareto points of the bi-criterion auction problem are shown as squares (a); the Pareto points may be far off the line connecting the two extremes (b), and may be non-convex (c). The Pareto points of randomized auctions comprise the upper boundary of the convex closure of the Pareto points (d). Even though the Pareto set may be exponential in size, for any $\epsilon > 0$, there is always a polynomially small set of ϵ-Pareto points, the triangular points in (e), that is, points that are not dominated by other solutions by more than ϵ in any dimension. We study the problem of computing such a set in polynomial time.

important and natural objectives. *What are the various intermediate (Pareto) points of this trade-off? And can each such point be computed — or all such points summarized somehow — in polynomial time?* This is the fundamental problem that we consider in this paper. See Figure 1 (a) for a graphical illustration.

The problem of exploring the revenue/welfare trade-off in auctions turns out to be a rather sophisticated problem, defying several naive approaches. One common-sense approach is to simply randomize between the optima of the two extremes, Vickrey's and Myerson's auctions. This can produce very poor results, since it only explores the straight line joining the two extreme points, which can be very far from the true trade-off (Figure 1 (b)). A second common-sense approach is the so-called *slope search*: To explore the trade-off space, just optimize the objective "revenue + λ· welfare" for various values of $\lambda > 0$. By modifying Myerson's auction this objective can indeed be optimized efficiently, as it was pointed out seven years ago by Likhodedov and Sandholm [19]. The problem is that the trade-off curve may not be convex (Figure 1 (c)), and hence the algorithm of [19] can miss vast areas of trade-offs:

Proposition 1. *There exist instances with two bidders with monotone hazard rate distributions for which the Pareto curve is not convex; in contrast the Pareto curve is always convex for one bidder with a monotone hazard rate distribution.*

The proof is deferred to the full version. It follows that the slope search approach of [19] is incorrect. However, the correctness of the slope search approach is restored if one is willing to settle for randomized mechanisms: The trade-off space of randomized mechanisms is always convex (in particular, it is the convex hull of the deterministic mechanisms, (Figure 1 (d)). It is easy to see (and it had been actually worked out for different purposes already in [23]) that the optimum randomized mechanism with respect to the metric "revenue + λ· welfare" is easy to calculate.

Proposition 2. *The optimum randomized mechanism for the objective "revenue + λ· welfare" can be computed in polynomial time. Hence, any point of the revenue/welfare trade-off for randomized mechanisms can be computed in polynomial time.*

1.1 Our Results

In this paper we consider the problem of exploring the revenue/welfare trade-off for deterministic mechanisms, and show that it is an intractable problem in general, even for two bidders (Theorem 2). Comparing with Proposition 2, this result adds to the recent surge in literature pointing out complexity gaps between randomized and deterministic mechanisms [26, 12, 13, 11]. Randomized mechanisms are of course a powerful and useful analytical concept, but it is deterministic mechanisms and auctions that we are chiefly interested in. Hence such complexity gaps are meaningful and onerous. We also show that there are instances for which the set of Pareto optimal mechanisms has exponential size.

On the positive side, we show that the problem can be solved for two bidders, even for correlated valuations (Theorem 4). By "solved" we mean that any trade-off point can be approximated with arbitrarily high precision in polynomial time in both the input and the precision — that is to say, by an FPTAS. It also means (by results in [28]) that an approximate summary of the trade-off (the ϵ-Pareto curve), of polynomial size (Figure 1(e)), can be computed in polynomial time. The derivation of the two-bidders algorithm (see Section 4.1) is quite involved. We first find a pseudo-polynomial dynamic programming algorithm for the problem of finding a mechanism with welfare (resp. revenue) *exactly* a given number. This algorithm is very different from the one in [26] for optimal auctions in the two bidder case, but it exploits the same feature of the problem, namely its planar nature. We then recall Theorem 4 of [28] (Section 2) which establishes a connection between such pseudo-polynomial algorithms for the exact problems and FPTAS for the trade-off problem. However, the present problem violates several key assumptions of that theorem, and a custom reduction to the exact problem is needed.

Unfortunately for three or more bidders the above approach no longer works; this is not surprising since, as it was recently shown in [26], just maximizing revenue is an APX-hard problem in the correlated case. The main problem left open in this work is whether there is an FPTAS for three or more bidders with *independent* valuation distributions.

We also look at another interesting case of the n-bidder problem, in which the valuation distributions have support two. This case is of some methodological interest because, in general, n-dimensional problems of this sort in mechanism design have not been characterized computationally, because of the difficulty related to the exponential size of the solution sought; binary-valued bidders have served as a first step towards the understanding of auction problems in the past, for example in the study of optimal *multi-object* auctions [3]. We show that the trade-off problem is in PSPACE and (weakly) NP-hard (Theorem 5).

1.2 Related Work

Although [19] appears to be the only previous paper explicitly treating optimal auction design as a multi-objective optimization problem, there has been substantial work in studying the relation of the two objectives. The most prominent paper in the area is that of Bulow and Klemperer [4] who show that the revenue benefits of adding one extra bidder and running the efficiency-maximizing auction surpasses those of running the revenue-maximizing auction. In [2] the authors show that for valuations drawn independently from the same monotone hazard rate distribution, an analogous theorem holds for efficiency: by adding $\Theta(\log n)$ extra bidders and running Myerson's auction, one gets at least the efficiency of Vickrey's auction. This paper also shows that for these distributions both the welfare and the revenue ratios between Vickrey and Myerson's auctions are bounded by $1/e$: in our terms this implies that the extreme points of the Pareto curve lie within a constant factor of each other and so constant factor approximations are trivial; we note that no such constant ratios are known for more general distributions (not even for the case of regular distributions), assuming of course that the ratio between all bidders' maximum and minimum valuation is arbitrary. This kind of revenue and welfare ratios are also studied in [29] for keyword auctions (multi-item auctions), and in [24] for single-item english auctions and valuations drawn from a distribution with bounded support. In [1] the authors present some tight bounds for the efficiency loss of revenue-optimal mechanisms, which depend on the number of bidders and the size of the support. Finally, and very recently, [7] gives simple auctions (in particular, second-price auctions with appropriately chosen reserve prices) that simultaneously guarantee a 20% fraction of both the optimal revenue and the optimal social welfare, when bidders' valuations are drawn independently from (possibly different) regular distributions: in multiobjective optimization parlance, their auctions belong to the *knee* of the Pareto curve. In this work (Section 4) we provide an algorithm for approximating *any* point of the Pareto curve within arbitrary precision, albeit sacrificing the simplicity of the auction format.

2 Preliminaries

2.1 Bayesian Mechanism Design

We are interested in auctioning a single, indivisible item to n bidders. We assume every bidder i has a private valuation v_i for the item and that her valuation is drawn from some discrete probability distribution over support of size h_i with probability density function $f_i(\cdot)$. We use v_i^k and f_i^k, $k = 1, \ldots, h_i$, to denote the k-th smallest element in the support of bidder i and its probability mass respectively.

Formally an auction consists of an allocation rule $x_i(v_1, \ldots, v_n)$, the probability of bidder i getting allocated the item, and a payment rule $p_i(v_1, \ldots, v_n)$ which is the price paid by bidder i. In this paper we focus our attention on deterministic mechanisms so that $x_i(\cdot) \in \{0, 1\}$. We demand from our auctions to satisfy the two standard constraints of ex-post incentive compatibility (IC) and individual rationality (IR); it is well known [25] that any such auction has the following special form: if we fix the valuation of all bidders except for bidder i, then there is a threshold value $t_i(v_{-i})$, such that bidder i only gets the item for values $v_i \geq t_i(v_i)$ and pays $t_i(v_{-i})$. In particular one can show that, for the discrete setting and for the objectives of welfare and revenue we are interested in, we can wlog assume that the threshold values t_i of any Pareto optimal auction will always be on the support of bidder i.

Relying on the above characterization, we will describe our mechanisms using the concept of an *allocation matrix* A: a $h_1 \times \ldots \times h_n$ matrix where entry $(i_1, \ldots, i_n)$ corresponds to the tuple $(v_1^{i_1}, \ldots, v_n^{i_n})$ of bidder's valuations. Each entry takes values from $\{0, 1, \ldots, n\}$ indicating which bidder gets allocated the item for the given tuple of valuations, with 0 indicating that the auctioneer keeps the item. In order for an allocation matrix to correspond to a valid (ex-post IC and IR) auction a necessary and sufficient condition is the following *monotonicity constraint*: if $A[i_1, \ldots, i_j, \ldots, i_n] = j$ then $A[i_1, \ldots, k, \ldots, i_n] = j$ for all $k \geq i_j$. Notice that the payment of the bidder who gets allocated the item can be determined as the least value in his support for which he still gets the item, keeping the values of the other bidders fixed; moreover, when there is only a constant number of bidders, the allocation matrix provides a polynomial representation of an auction.

2.2 Multi-objective Optimization

Trade-offs are present everywhere in life and science — in fact, one can argue that optimization theory studies the very special and degenerate case in which we happen to be interested in only one objective. There is a long research tradition of *multi-objective* or *multi-criterion optimization*, developing methodologies for computing the trade-off points (called the *Pareto set*) of optimization problems with many objectives, see for example [18, 14, 20]. However, there is a computational awkwardness about this problem: Even for simple cases, such as bicriterion shortest paths, the Pareto set (the set of all undominated feasible solutions) can be exponential, and thus it can never be polynomially computed. In 2000, Papadimitriou and Yannakakis [28] identified a sense in which this is a meaningful problem: They showed that there is *always* a set of solutions of polynomial size that are *approximately* undominated, within arbitrary precision; a multi-objective problem is considered tractable if such a set can be computed in polynomial time. Since then, much progress has been made in the algorithmic theory of multi-objective optimization [30, 10, 9, 17, 6, 5, 8], and much methodology has been developed, some of which has been applied to mechanism design before [16]. In this paper we use this methodology for studying Bayesian auctions under the two criteria of expected revenue and social welfare.

The BI-CRITERION AUCTION **problem.** We want to design deterministic auctions that perform favorably with respect to (expected) social welfare, defined as SW $=$ $\mathbb{E}[\sum_i x_i v_i]$ and (expected) revenue, defined as Rev $= \mathbb{E}[\sum_i p_i]$. Based on the aforementioned characterization with allocation matrices, we can view an auction as a feasible solution to a combinatorial problem. An instance specifies the number n of bidders and for each bidder its distribution on valuations. The size of the instance is the number of bits needed to represent these distributions. We map solutions (mechanisms) to points (x, y) in the plane, where we use the x-axis for the welfare and the y-axis for the revenue. The objective space is the set of such points.

Let $p, q \in \mathbb{R}^2_+$. We say that p dominates q if $p \geq q$ (coordinate-wise). We say that p ϵ-covers q ($\epsilon \geq 0$) if $p \geq q/(1+\epsilon)$. Let $A \subseteq \mathbb{R}^2_+$. The Pareto set of A, denoted by $P(A)$, is the subset of undominated points in A (i.e. $p \in P(A)$ iff $p \in A$ and no other point in A dominates p). We say that $P(A)$ is *convex* if it contains no points that are dominated by convex combinations of other points. Given a set $A \subseteq \mathbb{R}^2_+$ and $\epsilon > 0$, an ϵ-*Pareto set* of A, denoted by $P_\epsilon(A)$, is a subset of points in A that ϵ-cover all vectors in A. Given two mechanisms M, M' we define domination between them according to the 2-vectors of their objective values. This naturally defines the Pareto set and approximate Pareto sets for our auction setting.

As shown in [28], for every instance and $\epsilon > 0$, there exists an ϵ-Pareto set of polynomial size. The issue is one of efficient computability. There is a simple necessary and sufficient condition, which relates the efficient computability of an ϵ-Pareto set to the following *GAP Problem*: given an instance I, a (positive rational) 2-vector $b = (W_0, R_0)$, and a rational $\delta > 0$, either return a mechanism M whose 2-vector dominates b, i.e. $\text{SW}(M) \geq W_0$ and $\text{Rev}(M) \geq R_0$, or report that there does *not* exist any mechanism that is better than b by at least a $(1 + \delta)$ factor in both coordinates, i.e. such that $\text{SW}(M) \geq (1 + \delta) \cdot W_0$ and $\text{Rev}(M) \geq (1 + \delta) \cdot R_0$. There is an FPTAS for constructing an ϵ-Pareto set iff there is an FPTAS for the GAP Problem [28].

Remark 1. Even though our exposition focuses on discrete distributions, our results easily extend to continuous distributions as well. As in [26], given a sufficiently smooth continuous density (say Lipschitz-continuous), whose support lies in a finite interval $[\underline{v}, \overline{v}]$,[1] we can appropriately discretize (while preserving the optimal values within $O(\epsilon)$) and run our algorithms on the discrete approximations.

From Exact to Bi-criterion. We will make essential use of a result from [28] reducing the multi-objective version of a linear optimization problem A to its exact version: Let A be a discrete linear optimization problem whose objective function(s) have *non-negative* coefficients. The *exact version* of a A is the following problem: Given an instance x of A, and a positive rational C, is there a feasible solution with objective function value *exactly* C? For such problems, a pseudo-polynomial algorithm for the exact version of implies an FPTAS for the multi-objective version:

Theorem 1 ([28]). *Let A be a* linear *multi-objective problem whose objective functions have* non-negative *coefficients: If there exists a pseudo-polynomial algorithm for the exact version of A, then there exists an FPTAS for constructing an approximate Pareto curve for A.*

---

[1] This is the standard approach in economics, see for example [22].

To obtain our main algorithmic result (Theorem 4), we design a pseudo-polynomial algorithm for the exact version of the BI-CRITERION AUCTION problem and apply Theorem 1 to deduce the existence of an FPTAS. However, it is not obvious why BI-CRITERION AUCTION satisfies the condition of the theorem, since in the standard representation of the problem as a linear problem, the objective functions typically have negative coefficients. We show however (Lemma 2) that there exists an alternate representation with monotonic linear functions.

3   The Complexity of Pareto Optimal Auctions

Our main result in this section is that – in contrast with randomized auctions – designing deterministic Pareto optimal auctions under welfare and revenue objectives is an intractable problem; in particular, we show that, even for 2 bidders [2] whose distributions are independent and regular, the problem of maximizing one criterion subject to a lower bound on the other is (weakly) NP-hard.

Theorem 2. *For two bidders with independent regular distributions, it is NP-hard to decide whether there exists an auction with welfare at least W and revenue at least R.*

Proof (Sketch). Due to space constraints, in this version of the paper we only provide the reduction for the exact problem for the welfare objective; quite simple and intuitive, it also captures the main idea in the (significantly more elaborate) proof for the bi-criterion problem, which can be found in the full version.

The reduction is from the Partition problem: we are given a set $B = \{b_1, \ldots, b_k\}$ of k positive integers, and we wish to determine whether it is possible to partition B into two subsets with equal sum. We assume that $b_i \geq b_{i+1}$ for all i. Consider the rescaled values $b_i' := b_i / (10k \cdot T)$, where $T = \sum_{i=1}^{k} b_i$, and the set $B' = \{b_1', \ldots, b_k'\}$. It is clear that there exists a partition of B iff there exists a partition of B'.

We construct an instance of the auction problem with two bidders whose independent valuations v_r (row bidder) and v_c (column bidder) are uniformly distributed over supports of size k. (To avoid unnecessary clutter in the expressions, we assume w.l.o.g – by linearity – that the "probability mass" of all elements in the support is equal to 1, as opposed to $1/k$.) The valuation distribution for the row bidder is supported on the set $\{1, 2, \ldots, k\}$, while the column bidder's valuation comes from the set $\{1 + b_1', 2 + b_2', \ldots, k + b_k'\}$. Since $b_i' \geq b_{i+1}'$ and $\sum_{i=1}^{k} b_i' = 1/(10k)$, it is straightforward to verify that both distributions are indeed regular (the proof is deferred to the full version).

The main idea of the proof is this: appropriately *isolate* a subset of 2^k feasible mechanisms whose welfare values encode the sum of values $\sum_{i \in S} b_i'$ for all possible subsets $S \subseteq [k]$. The existence of a mechanism with a specified welfare value would then reveal the existence of a partition. Formally, we prove that there exists a Partition of B' iff there exists a feasible mechanism M^* with (expected) welfare

$$\text{SW}(M^*) = (2/3) \cdot (k-1)k(k+1) + (1/2) \cdot k(k+1) + \sum_{i=2}^{k} (i-1)b_i' + 1/(20k) \quad (1)$$

---

[2] Note that for a single bidder, one can enumerate all feasible mechanisms in linear time.

Consider the allocation matrix of a feasible mechanism. Recall that a mechanism is feasible iff its allocation matrix satisfies the monotonicity constraint. The main claim is that *all mechanisms that could potentially satisfy (1) must allocate the item to the highest bidder, except potentially for the outcomes* $(v_r = i, v_c = i + b'_i)$ *(i.e. the ones corresponding to entries on the secondary diagonal of the matrix) when the item can be allocated to either bidder.* Denote by $\mathcal{R}$ the aforementioned subclass of mechanisms. The above claim follows from the next lemma, which shows that mechanisms in $\mathcal{R}$ maximize welfare (see full version for the proof):

Lemma 1. *We have* $\max_{M \notin \mathcal{R}} \mathrm{SW}(M) < \min_{M \in \mathcal{R}} \mathrm{SW}(M) < \mathrm{SW}(M^*)$.

To complete the proof, observe that all 2^k mechanisms in $\mathcal{R}$ satisfy monotonicity, hence are feasible. Also note that there is a natural bijection between subsets $S \subseteq [k]$ and these mechanisms: we include i in S iff on input $(v_r = i, v_c = i + b'_i)$ the item is allocated to the column bidder. Denote by $M(S)$ the mechanism in $\mathcal{R}$ corresponding to subset S under this mapping; we will compute the welfare of $M(S)$. Note that the contribution of each entry of the allocation matrix (input) to the welfare equals the valuation of the bidder who gets the item for that input. By the definition of $\mathcal{R}$, for the entries above the secondary diagonal, the row bidder gets the item (since her valuation is strictly larger than that of the column bidder – this is evident since $\max_i b'_i < 1/(10k)$). Therefore, the contribution of these entries to the welfare equals $\sum_{i=2}^{k} i(i-1) = (1/3)(k-1)k(k+1)$. Similarly, for the entries below the diagonal, the column bidder gets the item and their contribution to the welfare is $\sum_{i=2}^{k}(i+b'_i)(i-1) = (1/3)(k-1)k(k+1) + \sum_{i=2}^{k}(i-1)b'_i$. Finally, for the diagonal entries, if $S \subseteq [k]$ is the subset of indices for which the column bidder gets the item, the welfare contribution is $\sum_{i \in S}(i+b'_i) + \sum_{i \in [k]\setminus S} i = k(k+1)/2 + \sum_{i \in S} b'_i$. Hence, we have:

$$\mathrm{SW}(M(S)) = (2/3) \cdot (k-1)k(k+1) + (1/2) \cdot k(k+1) + \sum_{i=2}^{k}(i-1)b'_i + \sum_{i \in S} b'_i \quad (2)$$

Recalling that $\sum_{i=1}^{k} b'_i = 1/(10k)$, (1) and (2) imply that there exists a partition of B' iff there exists a feasible mechanism satisfying (1). This completes the proof sketch. (See the full version of the paper for the much more elaborate proof of the general case.) □

We can also prove that the size of the Pareto curve can be exponentially large (in other words, the problem of computing the entire curve is exponential even if $P = NP$). The construction is given in the full version.

Theorem 3. *There exists a family of two-bidder instances for which the size of the Pareto curve for* BI-CRITERION AUCTION *grows exponentially.*

4 An FPTAS for 2 Bidders

In this section we give our main algorithmic result:

Theorem 4. *For two bidders, there is an FPTAS to approximate the Pareto curve of the* BI-CRITERION AUCTION *problem, even for arbitrarily correlated distributions.*

In the proof, we design a pseudo-polynomial algorithm for the exact version of the problem (for both the welfare and revenue objectives) and then appeal to Theorem 1. There is a difficulty, however, in showing that the problem satisfies the assumptions of Theorem 1, because in the most natural linear representation of the problem, the coefficients for revenue, coinciding with the virtual valuations, may be negative, thus violating the hypothesis of Theorem 1.

We use the following alternate representation: Instead of considering the contribution of each entry (bid tuple) of the allocation matrix separately, we consider the revenue and welfare resulting from all the *single-bidder mechanisms* (pricings) obtained by fixing the valuation of the other bidder.

Definition 1. *Let $r_1^{i_1,i_2}$ and $w_1^{i_1,i_2}$ be the (contribution to the) revenue and welfare from bidder 1 of the pricing which offers bidder 1 a price of $v_1^{i_1}$ when bidder 2's value is $v_2^{i_2}$:*
$$r_1^{i_1,i_2} = \sum_{j \geq i_1} v_1^{i_1} \cdot f(j, i_2) \text{ and } w_1^{i_1,i_2} = \sum_{j \geq i_1} v_1^{j} \cdot f(j, i_2), \text{ where } f(\cdot, \cdot) \text{ is the}$$
joint (possibly non-product) valuation distribution. (The quantities $r_2^{i_1,i_2}$ and $w_2^{i_1,i_2}$ are defined analogously.)

Lemma 2. *The BI-CRITERION AUCTION problem can be expressed in a way that satisfies the conditions of Theorem 1.*

Proof. We consider variables x_{ij}, y_{ij}, $i \in [h_1]$, $j \in [h_2]$. The x_{ij}'s are defined as follows: $x_{ij} = 1$ iff $A[i, j] = 1$ and $A[i', j] \neq 1$ for all $i' < i$. I.e. $x_{ij} = 1$ iff the (i, j)-th entry of A is allocated to bidder 1 and, for this fixed value of j, i is the smallest index for which bidder 1 gets allocated; symmetrically, $y_{ij} = 1$ iff $A[i, j] = 2$ and $A[i, j'] \neq 2$ for all $j' < j$. It is easy to see that the feasibility constraints are linear in these variables. We can also express the objectives as linear functions with non-negative coefficients as follows:
$$\text{Rev}(x, y) = \sum_{i=1}^{h_1} \sum_{j=1}^{h_2} x_{ij} r_1^{i,j} + \sum_{i=1}^{h_1} \sum_{j=1}^{h_2} y_{ij} r_2^{i,j}$$
$$\text{SW}(x, y) = \sum_{i=1}^{h_1} \sum_{j=1}^{h_2} x_{ij} w_1^{i,j} + \sum_{i=1}^{h_1} \sum_{j=1}^{h_2} y_{ij} w_2^{i,j} \qquad \square$$

4.1 An Algorithm for the Exact Version of BI-CRITERION AUCTION

The main idea behind our algorithm, inspired by the characterization of Lemma 2, is to consider the contribution from each bidder (fixing the value of the other) independently, by going over all (linearly many) single-bidder mechanisms for both bidders. The challenging part is to combine the individual single-bidder mechanisms into a single two-bidder mechanism and to this end we employ dynamic programming:

Assume that both bidders have valuations of support size h; the subproblems we consider in our dynamic program correspond to settings where we condition that the valuation of each bidder is drawn from an upwards closed subset of his original support. Formally, let $M[i, j, W]$ be True iff there exists an auction that uses the valuations $(v_1^i, \ldots, v_1^h)$ and $(v_2^j, \ldots, v_2^h)$ and has welfare exactly W. In what follows $N_{i,j}$ is the normalization factor for valuations (jointly) drawn from $(v_1^i, \ldots, v_1^h)$ and $(v_2^j, \ldots, v_2^h)$, namely $N_{i,j} = \sum_{k \geq i, l \geq j} f(v_1^k, v_2^l)$.

Lemma 3. *We can update the quantity $M[i, j, W]$ as follows:*

$$M[i, j, W] = \bigvee_{k \geq j} M\left[i + 1, j, \left(W \cdot N_{i,j} - w_2^{i,k}\right) \cdot N_{i+1,j}^{-1}\right]$$

$$\vee \bigvee_{k \geq i} M\left[i, j + 1, \left(W \cdot N_{i,j} - w_1^{k,j}\right) \cdot N_{i,j+1}^{-1}\right]$$

$$\vee \bigvee_{\substack{k > i \\ l > j}} M\left[i + 1, j + 1, \left(W \cdot N_{i,j} - w_1^{k,j} - w_2^{i,l}\right) \cdot N_{i+1,j+1}^{-1}\right]$$

Proof. Let $A[i \ldots h, j \ldots h]$ be the allocation matrix of the auction that results from the above update rule, fixing i and j. We start by noting that any allocation matrix A can have one of the following four forms:

F1: There exist i' and j' such that $A[i, j'] = 1$ and $A[i', j] = 2$.
F2: There exists i' such that $A[i', j] = 2$ but there is no j' such that $A[i, j'] = 1$.
F3: There exists j' such that $A[i, j'] = 1$ but there is no i' such that $A[i', j] = 2$.
F4: There exist no i' and j' such that $A[i, j'] = 1$ or $A[i', j] = 2$.

Because of monotonicity it follows immediately that no allocation matrix of form F1 can be valid, and the other three forms correspond to the three terms of the recurrence; finally note that for any such form, say F2, the first term of the update rule for $M[i, j, W]$ runs over all possible pricings for bidder 1 (keeping the value of bidder 2 at v_2^j) and checks whether they induce the required welfare. □

We omit the straightforwards details of how the above recurrence can be efficiently implemented as a pseudo-polynomial dynamic programming algorithm. The algorithm for deciding whether there exists an auction with revenue exactly R is identical to the above by simply replacing R (the revenue target value) for W and $r_j^{i_1, i_2}$ for $w_j^{i_1, i_2}$.

5   The Case of n Bidders

When the number n of bidders is part of the input, the allocation matrix is no longer a polynomially succinct representation of a mechanism. In fact, it is by no means clear whether BI-CRITERION AUCTION is even in NP in this case: we next show that for the case of n binary bidders, the problem is NP-hard and in $PSPACE$:

Theorem 5. *For n binary-valued bidders* BI-CRITERION AUCTION *is (weakly) NP-hard and in PSPACE.*

Proof (Sketch). For simplicity, we prove both results for the exact version of the problem for welfare; the bi-objective case follows by a straightforward but tedious generalization.

   The NP-hardness reduction is from Partition. Let $B = \{b_1, \ldots, b_k\}$ be a set of positive rationals; we can assume by rescaling that $\sum_{i=1}^k b_i = 1/100$. We construct an instance of the auction problem as follows: there are k bidders, with uniform distributions (again we will assume unit masses for simplicity) over the following supports

$\{l_i, h_i\}, i = 1 \ldots n$, where $l_i < h_i$. We set $l_i = b_i$ and demand that $\{h_i\}_{i=1,\ldots,n}$ forms a super-increasing sequence (i.e. $h_{i+1} > \sum_{j=1}^{i} h_j$), with $h_1 > \max_i b_i$. The claim is that there exists a partition of B iff there exists an auction with welfare equal to $\sum_{i=1}^{k} h_i + (1/2) \sum_{i=1}^{k} b_i$. To see this notice that – since the sequence $\{h_i\}_{i=1\ldots n}$ is super-increasing – any mechanism with the above welfare value must must allocate to bidder i for *exactly* one valuation tuple (v_i, v_{-i}) where $v_i = h_i$; the corresponding contribution to the welfare from this case is h_i. Monotonicity then implies that this auction can allocate to bidder i for *at most* one valuation tuple (v_i, v_{-i}) where $v_i = l_i$; the corresponding contribution to the welfare from this case is b_i. We therefore get a bijection between subsets of B and mechanisms, by including an element b_i in the set S iff bidder i gets allocated the item for some valuation tuple (v_i, v_{-i}) where $v_i = l_i$, and the claim follows.

For the PSPACE upper bound, we start by noting that the problem of computing an auction with welfare (or revenue) *exactly* W, can be formulated as the problem of computing a matching of weight exactly W in a particular type of bipartite graphs (first pointed out in [12], see also the full version of the paper) with a number of nodes that is exponential in the number of bidders. The EXACT MATCHING problem is known to be solvable in RNC [21]; since our input provides an exponentially succinct representation of the constructed graph, we are interested in the so-called *succinct version* of the problem [15, 27]. By standard techniques, the succinct version of EXACT MATCHING in our setting is solvable in PSPACE, and the theorem follows.   □

We conjecture the above upper bound to be tight (i.e. the problem is actually PSPACE-complete) even for n bidders with arbitrary supports.

6   Open Questions

Is there is an FPTAS for 3 bidders? We conjecture that there is, and in fact for any constant number of bidders. Of course, the approach of our FPTAS for 2 bidders cannot be generalized, since it works for the correlated case, which is APX-complete for 3 or more bidders. We have derived two different dynamic programming-based PTAS's for the uncorrelated problem, but so far, despite a hopeful outlook, we have failed to generalize them to 3 bidders. Finally, we conjecture that for n bidders the problem is significantly harder, namely PSPACE-complete and inapproximable.

On a different note, it would be interesting to see if we can get better approximations for some special types of distributions; we give one such type of result in the full version of the paper. Are there improved approximation guarantees for more general kinds of distributions and n bidders?

References

[1] Abhishek, V., Hajek, B.E.: Efficiency loss in revenue optimal auctions. In: CDC (2010)
[2] Aggarwal, G., Goel, G., Mehta, A.: Efficiency of (revenue-)optimal mechanisms. In: EC (2009)
[3] Armstrong, M.: Optimal multi-object auctions. Review of Economic Studies 67(3), 455–481

[4] Bulow, J., Klemperer, P.: Auctions versus negotiations. American Economic Review 86(1), 180–194 (1996)

[5] Chekuri, C., Vondrak, J., Zenklusen, R.: Multi-budgeted matchings and matroid intersection via dependent rounding. In: SODA (2011)

[6] Daskalakis, C., Diakonikolas, I., Yannakakis, M.: How good is the chord algorithm? In: SODA (2010)

[7] Daskalakis, C., Pierrakos, G.: Simple, Optimal and Efficient Auctions. In: Chen, N., Elkind, E., Koutsoupias, E. (eds.) WINE 2011. LNCS, vol. 7090, pp. 109–121. Springer, Heidelberg (2011)

[8] Diakonikolas, I.: Approximation of Multiobjective Optimization Problems. PhD thesis, Columbia University (2010)

[9] Diakonikolas, I., Yannakakis, M.: Succinct Approximate Convex Pareto Curves. In: SODA (2008)

[10] Diakonikolas, I., Yannakakis, M.: Small approximate pareto sets for biobjective shortest paths and other problems. SIAM J. Comput. 39, 1340–1371 (2009)

[11] Dobzinski, S.: An impossibility result for truthful combinatorial auctions with submodular valuations. In: STOC (2011)

[12] Dobzinski, S., Fu, H., Kleinberg, R.D.: Optimal auctions with correlated bidders are easy. In: STOC (2011)

[13] Dughmi, S., Roughgarden, T., Yan, Q.: From convex optimization to randomized mechanisms: toward optimal combinatorial auctions. In: STOC (2011)

[14] Ehrgott, M.: Multicriteria optimization. Springer (2005)

[15] Galperin, H., Wigderson, A.: Succinct representations of graphs. Information and Control 56(3), 183–198 (1983)

[16] Grandoni, F., Krysta, P., Leonardi, S., Ventre, C.: Utilitarian mechanism design for multiobjective optimization. In: SODA (2010)

[17] Grandoni, F., Ravi, R., Singh, M.: Iterative Rounding for Multi-Objective Optimization Problems. In: Fiat, A., Sanders, P. (eds.) ESA 2009. LNCS, vol. 5757, pp. 95–106. Springer, Heidelberg (2009)

[18] Climacao, E.J.: Multicriteria Analysis. Springer (1997)

[19] Likhodedov, A., Sandholm, T.: Mechanism for optimally trading off revenue and efficiency in multi-unit auctions. In: EC (2003)

[20] Miettinen, K.M.: Nonlinear Multiobjective Optimization. Kluwer (1999)

[21] Mulmuley, K., Vazirani, U.V., Vazirani, V.V.: Matching is as easy as matrix inversion. In: STOC (1987)

[22] Myerson, R.B.: Optimal auction design. Mathematics of Operations Research 6, 58–73

[23] Myerson, R.B., Satterthwaite, M.A.: Efficient mechanisms for bilateral trading. Journal of Economic Theory 29(2), 265–281 (1983)

[24] Neeman, Z.: The effectiveness of english auctions. Games and Economic Behavior 43(2), 214–238 (2003)

[25] Nisan, N., Roughgarden, T., Tardos, É., Vazirani, V.V.: Algorithmic Game Theory (2007)

[26] Papadimitriou, C.H., Pierrakos, G.: On optimal single-item auctions. In: STOC (2011)

[27] Papadimitriou, C.H., Yannakakis, M.: A note on succinct representations of graphs. Information and Control 71(3), 181–185 (1986)

[28] Papadimitriou, C.H., Yannakakis, M.: On the approximability of trade-offs and optimal access of web sources. In: FOCS (2000)

[29] Roughgarden, T., Sundararajan, M.: Is efficiency expensive? In: 3rd Workshop on Sponsored Search (2007)

[30] Vassilvitskii, S., Yannakakis, M.: Efficiently computing succinct trade-off curves. Theoretical Computer Science 348, 334–356 (2005)

Deterministic Network Exploration by Anonymous Silent Agents with Local Traffic Reports

Yoann Dieudonné[1,*] and Andrzej Pelc[2,**]

[1] MIS, Université de Picardie Jules Verne, France
[2] Département d'informatique, Université du Québec en Outaouais,
Gatineau, Québec, Canada

Abstract. A team consisting of an unknown number of mobile agents, starting from different nodes of an unknown network, possibly at different times, have to explore the network: every node must be visited by at least one agent and all agents must eventually stop. Agents are anonymous (identical), execute the same deterministic algorithm and move in synchronous rounds along links of the network. They are silent: they cannot send any messages to other agents or mark visited nodes in any way. In the absence of any additional information, exploration with termination of an arbitrary network in this weak model is impossible. Our aim is to solve the exploration problem giving to agents very restricted *local traffic reports*. Specifically, an agent that is at a node v in a given round, is provided with three bits of information, answering the following questions: Am I alone at v? Did any agent enter v in this round? Did any agent exit v in this round? We show that this small information permits to solve the exploration problem in arbitrary networks. More precisely, we give a deterministic terminating exploration algorithm working in arbitrary networks for all initial configurations that are not *perfectly symmetric*, i.e., in which there are agents with different *views* of the network. The algorithm works in time polynomial in the (unknown) size of the network. A deterministic terminating exploration algorithm working for all initial configurations in arbitrary networks does not exist.

Keywords: exploration, deterministic algorithm, anonymous mobile agent, network, graph.

1 Introduction

The Background. A team consisting of an unknown number of mobile agents, starting from different nodes of an unknown network, possibly at different times, have to explore the network: every node must be visited by at least one agent

* A part of this research was done during this author's stay at the Research Chair in Distributed Computing of the Université du Québec en Outaouais as a postdoctoral fellow.
** Supported in part by NSERC discovery grant and by the Research Chair in Distributed Computing of the Université du Québec en Outaouais.

A. Czumaj et al. (Eds.): ICALP 2012, Part II, LNCS 7392, pp. 500–512, 2012.
© Springer-Verlag Berlin Heidelberg 2012

and all agents must eventually stop. The latter requirement is called *termination*. The exploration task is fundamental for many network problems. For example, software agents may need to collect data located at nodes of the network, or they may have to visit all nodes in order to check if they are functional.

The Model and the Problem. The network is modeled as an undirected connected graph, referred to hereafter as a graph. We seek exploration algorithms that do not rely on the knowledge of node labels, and can work in anonymous graphs as well (cf. [4,18]). The importance of designing such algorithms is motivated by the fact that, even when nodes are equipped with distinct labels, agents may be unable to perceive them because of limited sensory capabilities, or nodes may refuse to reveal their labels, e.g., due to security or privacy reasons. On the other hand, we assume that edges incident to a node v have distinct labels in $\{0, \ldots, d-1\}$, where d is the degree of v. Thus every undirected edge $\{u,v\}$ has two labels, which are called its *port numbers* at u and at v. Port numbering is *local*, i.e., there is no relation between port numbers at u and at v. Note that in the absence of port numbers, edges incident to a node would be indistinguishable for agents and thus exploration would be often impossible, as the adversary could prevent any agent from taking some edge incident to the current node and thus block access to a part of the graph.

Agents start from different nodes of the graph and traverse its edges in synchronous rounds. In each round every agent is at some node. When an agent is at a node v in a given round $t-1$, in round t it may either remain at v, or it may be at an adjacent node w. In the latter case we say that in round t the agent exited (left) node v and entered node w. When an agent enters a node, it learns its degree and the port of entry.

Agents are anonymous (identical) and they execute the same deterministic algorithm. They are silent: they cannot send messages to other agents, communicate with agents currently in the same node or mark visited nodes or traversed edges in any way. They cannot observe other nodes or agents (even those that are simultaneously at the same node). Agents that cross each other on an edge, traversing it simultaneously in different directions, do not notice this fact. In the beginning, the adversary places agents in some nodes of the graph, each agent in a different node. Such a placement is called an *initial configuration*. The adversary wakes up some of the agents at possibly different times. A dormant agent, not woken up by the adversary, is woken up by the first agent that visits its starting node, if such an agent exists. Every agent starts executing the algorithm in the round of its wake-up. Agents do not know the topology of the graph, they do not know any upper bound on its size and they do not know the size of the team. We assume that the memory of the agents is unlimited: from the computational point of view they are modeled as Turing machines.

In the absence of any additional information, deterministic terminating exploration of an arbitrary network in this very weak model is impossible. Indeed, since agents are totally oblivious of each other (even when they meet, they do not notice this event), they act as if they were alone in the network and hence exploration with stop is impossible even in the class of oriented rings (i.e., rings in which ports 0,1 at all nodes are situated clockwise). If the adversary wakes up all agents simultaneously, each agent would perform the same sequence of clockwise/counterclockwise moves in all rings and every agent would have to stop after time t since wake-up. For a team whose most distant agents are at distance d, at most $d + 2t + 1$ nodes could be visited, and hence exploration would be incorrect for sufficiently large rings.

Since some additional information is needed to solve the exploration problem, our aim is to solve it giving to agents as little information as possible. The information each agent gets, called *local traffic report*, has two features: it is very small in size and has local character both regarding space and time, i.e., it concerns only the node in which the agent is currently located and only the current round. Specifically, an agent that is at a node v in a given round, is provided with three bits of information, answering the following questions: Am I alone at v in this round? Did any agent enter v in this round? Did any agent exit v in this round? (Notice that if the agent entered node v in round t, the second answer is always "yes", but it is informative if the agent did not move in round t.) On the basis of these three answers, together with the degree of the current node, the last port of entry and all information recorded in the agent's memory to date, the agent either decides to remain in the current node or to leave it by a given port. Notice that the agent does not know how many agents are together with it in a given round, or how many agents left or entered and by which ports. So even the local information available to the agent is very modest: it is restricted to the above three bits provided by the node in which the agent is currently located. Notice also that this information is not personalized: each agent residing at the same node in a given round receives the same three bits (which can be simply posted at each node in any round in which there is some agent at this node). Nevertheless we show that this restricted information permits to solve the exploration problem in arbitrary networks. The time of an exploration algorithm is the number of rounds between the wake-up of the first agent and the last round in which some agent moves.

Our Results. We give a deterministic terminating exploration algorithm working in arbitrary graphs, for all initial configurations that are not *perfectly symmetric*, i.e., in which there are agents with different *views* [1] of the graph. The algorithm works in time polynomial in the (unknown) size of the graph. Removing the restriction to non perfectly symmetric configurations is impossible: a (terminating) deterministic exploration algorithm working for all initial configurations in arbitrary graphs does not exist.

Due to space constraints, many lemmas and proofs are omitted and will appear in the full version of this paper.

Related Work. Algorithms for graph exploration by mobile agents (often called robots) have been intensely studied in recent literature. A lot of research is concerned with the case of a single robot exploring a labeled graph. In [1,3,4,8,13] the robot explores strongly connected directed graphs and it can move only in

---

[1] See Section 2 for a precise definition of the view.

the direction from head to tail of an edge, not vice-versa. In particular, [8] investigates the minimum time of exploration of directed graphs, and [1,13] give improved algorithms for this problem in terms of the deficiency of the graph (i.e., the minimum number of edges to be added to make the graph Eulerian). Many papers, e.g., [9,11,12,16] study the scenario where the explored graph is undirected and the robot can traverse edges in both directions. In [9] the authors investigate the problem of how the availability of a map influences the efficiency of exploration. In some papers, additional restrictions on the moves of the robot are imposed. It is assumed that the robot has either a restricted tank [2], forcing it to periodically return to the base for refueling, or that it is tethered, i.e., attached to the base by a rope or cable of restricted length [12].

Exploration of anonymous graphs presents different difficulties. In this case it is impossible to explore arbitrary graphs by a single robot, if no marking of nodes is allowed. Hence the scenario adopted in [3,4] allows the use of *pebbles* which the robot can drop on nodes to recognize already visited ones, and then remove them and drop in other places. The authors concentrate attention on the minimum number of pebbles allowing efficient exploration and mapping of arbitrary directed n-node graphs.

In all the above papers, except [4] which deals with randomized algorithms, exploration is performed by a single robot. Deterministic exploration by many robots usually assumed that moves of the robots are centrally coordinated. In [15], approximation algorithms are given for the collective exploration problem in arbitrary graphs. On the other hand, in [14] the authors study the problem of distributed collective exploration of trees of unknown topology. However, the robots performing exploration start in the same node and can directly communicate with each other. Exploration of arbitrary anonymous graphs by robots communicating through *whiteboards* has been studied in [7].

Some of the techniques from this paper were previously used in [10], where the aim was to gather anonymous agents in one node of the graph.

2   Preliminaries

Throughout the paper, the number of nodes of a graph is called its size. In this section we recall four procedures known from the literature, that will be used as building blocks in our algorithm. The aim of the first two procedures is graph exploration, i.e., visiting all nodes of the graph (cf., e.g., [5,17]). The first of these procedures assumes an upper bound n on the size of the graph and the second one makes no assumptions on the size but it is performed by an agent using a fixed token placed at the starting node of the agent. In our applications the roles of the token and of the exploring agent will be played by agents. The first procedure works in time polynomial in the known upper bound n on the size of the graph, and the second in time polynomial in the size of the graph. Moreover, at the end of the second procedure the agent is with the token and has a complete map of the graph with all port numbers marked. We call the first procedure $EXPLO(n)$ and the second procedure EST, for *exploration with a stationary token.* We denote by $T(EXPLO(n))$ (respectively $T(EST(n))$) the maximum time of execution of the procedure $EXPLO(n)$ (respectively procedure EST) in a graph of size at most n.

Before describing the third procedure we define the following notion from [19]. Let G be a graph and v a node of G, of degree k. The *view* from v is the infinite rooted tree $\mathcal{V}(v)$ with labeled ports, defined recursively as follows. $\mathcal{V}(v)$ has the root x_0 corresponding to v. For every node v_i, $i = 1, \ldots, k$, adjacent to v, there is a neighbor x_i in $\mathcal{V}(v)$ such that the port number at v corresponding to edge $\{v, v_i\}$ is the same as the port number at x_0 corresponding to edge $\{x_0, x_i\}$, and the port number at v_i corresponding to edge $\{v, v_i\}$ is the same as the port number at x_i corresponding to edge $\{x_0, x_i\}$. Node x_i, for $i = 1, \ldots, k$, is now the root of the view from v_i.

The third procedure, described in [6], permits a single anonymous agent starting at node v to find a positive integer $S(v)$, called the *signature* of the agent, such that $\mathcal{V}(v) = \mathcal{V}(w)$ if and only if $S(v) = S(w)$. This procedure, called $SIGN(n)$ works for any graph of known upper bound n on its size, and its running time is polynomial in n. After the completion of $SIGN(n)$ the agent is back at its starting node. We denote by $T(SIGN(n))$ the maximum time of execution of the procedure $SIGN(n)$ in a graph of size at most n.

Finally, the fourth procedure is for gathering two agents in a graph of unknown size. It is due to Ta-Shma and Zwick [18] and relies on the fact that agents have distinct labels. (Using it as a building block in our scenario of anonymous agents is one of the difficulties that we need to overcome.) Each agent knows its own label (which is a parameter of the algorithm) but not the label of the other agent. We will call this procedure $TZ(\ell)$, where ℓ is the label of the executing agent. In [18], the authors give a polynomial P in two variables, increasing in each of the variables, such that, if there are agents with distinct labels ℓ_1 and ℓ_2 operating in a graph of size n, appearing at their starting positions in possibly different times, then they will meet after at most $P(n, |\ell|)$ rounds since the appearance of the later agent, where ℓ is the smaller label. Also, if an agent with label ℓ_i performs $TZ(\ell_i)$ for $P(n, |\ell_i|)$ rounds and the other agent is inert during this time, the meeting is guaranteed.

An initial configuration is called *perfectly symmetric* if the views from all initial positions of agents are identical. For example, any initial configuration in an oriented ring (i.e., a ring in which ports 0,1 at all nodes are situated clockwise) is perfectly symmetric. The following result justifies that, in order to guarantee exploration by a team of anonymous agents in arbitrary graphs, we restrict attention to initial configurations that are not perfectly symmetric.

Proposition 1. *There is no deterministic terminating algorithm that accomplishes exploration for all perfectly symmetric configurations in arbitrary graphs.*

Of course, it is possible to explore a graph starting from any *particular* (even perfectly symmetric) configuration. Such a dedicated algorithm would know the size n of the graph and hence procedure $EXPLO(n)$ could be used by each agent separately. This is not our goal: we want to give a *universal* exploration algorithm that works for arbitrary graphs, and hence, in view of Proposition 1, we restrict attention to initial configurations that are not perfectly symmetric.

We mentioned in the introduction that in the absence of any additional information, such as, e.g., our local traffic reports, exploration for all graphs is impossible. However, the simple example given there concerns oriented rings, where all configurations are perfectly symmetric. We now know that even with local traffic reports, exploring all graphs starting from all possible configurations (including perfectly symmetric) is impossible. So a question could be raised of whether in the weaker model without additional information it is possible to perform exploration starting from all non-perfectly symmetric configurations, as we will do with local traffic reports. The answer to this more refined question is also negative. Let M be the modification of our model which results from removing local traffic reports.

Proposition 2. *There is no deterministic terminating algorithm accomplishing exploration for all non-perfectly symmetric configurations in the model M.*

3   An Auxiliary Algorithm: Exploration with Tokens

In this section we describe an auxiliary algorithm which is a generalization of the procedure EST from [5]. While EST is an exploration of an arbitrary unknown graph by a single agent with a stationary token placed at its initial position, we will show an algorithm to explore an arbitrary unknown graph by an arbitrary team of anonymous identical agents with identical stationary tokens each placed at the initial position of one agent. Note that such an algorithm is not an immediate consequence of exploration by one agent with one stationary token. Indeed, since all tokens are identical, an agent cannot be sure if a token it currently visits is its own token, or a token of another agent. Thus, if each agent applies exploration with one token and treats every visited token as its own, it may fail to visit all nodes of the graph.

We first briefly describe the procedure EST based on [5] that will be subsequently adapted to our needs. The agent constructs a BFS tree rooted at its starting node r marked by the stationary token. In this tree it marks port numbers at all nodes. A path in a BFS tree is the sequence of ports from an ancestor to a descendant. The length $|\alpha|$ of a path α is the corresponding number of edges. During the BFS traversal, some nodes are added to the BFS tree. In the beginning the root is added. Whenever a node w is added to the BFS tree, all its neighbors are visited by the agent. For each neighbor v of w, the agent verifies if v is equal to some node u previously added to the tree. To do this, the agent travels from v using the reversal $\bar{q}$ of the path q from r to u. If at the end of this backtrack it meets the token, then $v = u$. In this case v is not added to the tree as a neighbor of w and is called w-*rejected*. If not, then $v \neq u$. This verification is done for all nodes that are already in the tree. If v is different from all these nodes, then it is added to the tree.

Now consider the scenario where, instead of a single agent with a token, there is an arbitrary team of anonymous identical agents with identical stationary tokens, each placed at the initial position of one agent. The adversary places each agent with its token in some round, possibly different for different agents. Each agent with its token is placed at a different node. Each agent is woken up at the time of its placement by the adversary. Upon wake-up each agent applies the above described procedure EST, with the following additional provision: neighbors $v_1, \ldots, v_i$ of a node w that has been added to the BFS tree constructed by the agent, are visited in increasing order of ports at w corresponding to the edges $\{w, v_j\}$. Call this procedure EST^*. The following proposition implies that the team of agents performing the above procedure explores the entire graph.

Proposition 3. *Let v be any node of a graph G. Consider any placement of all agents (with their tokens) in G, located at nodes $w_1, \ldots, w_k$, performing procedure EST^* in G. Let T be the set of all paths from nodes w_i to v (coded as sequences of ports). Let $Min(T)$ be the set of paths in T of the smallest length. The node v is visited by the agent whose initial position corresponds to the lexicographically smallest path in $Min(T)$.*

4   The Main Algorithm

In this section we present our main result, which is an exploration algorithm executed by an unknown team of identical anonymous agents, working in an arbitrary unknown graph, for any initial configuration which is not perfectly symmetric. Unlike in the auxiliary algorithm presented in the previous section, our model does not allow using real tokens or marking nodes in any way. In fact, some agents will play the role of tokens and others the role of explorers. The algorithm works in time polynomial in the (unknown) size of the graph.

Idea of the Algorithm
At a high level the idea of the algorithm is the following. Upon wake-up, agents start walking in the graph with the aim of meeting other agents. One of the goals of a meeting is to assign to one of the meeting agents the role of a token and to the others the role of explorers. This is done in order to simulate our auxiliary algorithm and thus complete exploration.

There are two difficulties in implementing this idea. The first is to guarantee that each agent meets another agent. In this part we will use the assumption that the initial configuration is not perfectly symmetric. The second difficulty is much more serious. Since agents do not have direct means of communication and they become aware of each other only through the obtained local traffic reports, already breaking symmetry between agents that have met, in order to assign them the roles of token and explorer, is a significant challenge that has to be overcome by carefully orchestrated moves of the agents, that will be reflected in the reports. Afterwards, an agent that plays the role of a token, must inform other agents of its presence. Again, this can be achieved only by specific moves of such an agent, reflected in the reports. (Otherwise, an explorer might be unable to differentiate meeting a token from meeting another explorer.) Thus even agents playing the role of a token cannot be completely inert: they will move from one node, called their *home* to an adjacent node, called their *office*. This, however, creates a new difficulty. When an explorer enters the home of an agent playing the role of a token, the latter may be at its office. Hence we need to organize the moves of agents in such a way that the home of each "token" will be recognized by explorers and it will in fact be this home (rather than the agent whose home it is) that will play the role of a "genuine" token from the simulated auxiliary algorithm.

We now give a detailed description of the algorithm.

Algorithm Team-Exploration

For clarity of the description we introduce five states in which an agent can be (depending on the content of its memory). These states are cruiser, token, explorer, fighter and ghost. For simplicity we will use the expressions, e.g., "agent A is a cruiser", instead of "agent A is in state cruiser" and "agent A becomes an explorer in round t" instead of "agent A transits to state explorer in round t". The latter expression means that the agent was not in state explorer in round $t-1$ but it is in this state in round t.

The algorithm consists in prescribing, for an agent A in each of those states, under which conditions the agent in a given state leaves the current node and by which port, as well as in describing transitions between states.

We will use boolean variables $alone(v,t)$, $exit(v,t)$ and $enter(v,t)$ whose values are defined as follows.

– $alone(v,t)$ is true if and only if an agent, located at node v in round t, is alone in this round,
– $exit(v,t)$ is true if and only if some agent left node v in round t,
– $enter(v,t)$ is true if and only if some agent entered node v in round t.

In order to get the value of $alone(v,t)$, $exit(v,t)$, $enter(v,t)$, an agent must be at node v in round t. A typical piece of information triggering the move of an agent in round t is of the form $enter(v,t-2) \wedge \neg alone(v,t-1) \wedge \neg enter(v,t-1)$. In order to get this information, an agent must be at node v both in round $t-2$ and in round $t-1$. In this case we will say that the agent *learns* the formula $enter(v,t-2) \wedge \neg alone(v,t-1) \wedge \neg enter(v,t-1)$. (Round numbers are not global and are for description only.)

In the beginning of the algorithm each agent is a cruiser and remains in this state at least until wake-up.

State cruiser. [Intuition. The goal of a cruiser is to meet another agent, either another cruiser or a token, in order to become either a token or an explorer in the first case (passing through the state fighter) or to become an explorer in the second case).]

When agent A is woken up then, as long as at the current node v in the current round t the report is $alone(v,t)$, agent A acts as follows. It works in phases numbered $1, 2, \ldots$. In phase n the agent supposes that the graph has size n. Agent A performs $EXPLO(n)$ in order to visit all nodes and wake up all still dormant agents, if the assumption was correct. Then agent A performs the procedure $SIGN(n)$ finding the current signature of its initial position v, called the *label*

ℓ_n of agent A. Let L_n be the maximum possible label of an agent in phase n (Note that L_n is polynomial in n.) Then agent A performs $TZ(\ell_n)$ for Δ_n rounds, where $\Delta_n = T(EXPLO(n)) + T(SIGN(n)) + 2P(n, L_n) + \Sigma_{i=1}^{n-1}Q_i$, for $n \geq 2$, and $\Delta_1 = 0$. In this formula, Q_i defined as $T(EXPLO(i)) + T(SIGN(i)) + \Delta(i)$ is an upper bound on the duration of phase i. If $alone(v, t)$ is the report given to agent A in all rounds of phase n, agent A starts phase $n + 1$. As soon as the report is $\neg alone(v, t)$ in some round t of phase k, agent A interrupts this phase and executes the following protocol in every round $t' \geq t$ while it stays at v, until the protocol specifies that it leaves the node, or that it changes its state.

---

if agent A learns $enter(v, t' - 2) \wedge \neg enter(v, t' - 1) \wedge \neg alone(v, t' - 1) \wedge exit(v, t')$
then in round $t' + 1$ agent A leaves node v by port 0 and becomes an **explorer**
else

    if agent A learns $alone(v, t')$ **then** in round $t' + 1$ agent A remains a **cruiser** and leaves node v continuing the phase k of the above procedure where it was interrupted in round t
    else

        if agent A learns $enter(v, t' - 3) \wedge \neg enter(v, t' - 2) \wedge \neg alone(v, t' - 2) \wedge$
        $\neg exit(v, t' - 1) \wedge \neg enter(v, t' - 1) \wedge \neg enter(v, t') \wedge \neg alone(v, t')$
        then in round $t' + 1$ agent A remains at node v and becomes a **fighter**

State fighter. [Intuition. **Cruisers** become **fighters** at a node in order to break symmetry between them and assign to one of them state **token** and to all others at the same node state **explorer**.]

When agent A becomes a **fighter** in round t, it has not moved in round t and is at some node v. There are two possible cases. Either v is the initial position of A and A has never yet moved from v, or A entered v by some port. In the second case, let d be the degree of node v, let s be the last round in which agent A entered node v before becoming a **fighter**, let e be the port number by which agent A entered node v in round s, and let i be the number of rounds since round s in which agent A learned that some agent entered node v.

Agent A executes the following protocol in every round $t' \geq t$ while it stays at v, until the protocol specifies that it leaves the node, or changes its state.

---

if agent A learns $exit(v, t')$
then in round $t' + 1$ agent A leaves node v by port 0 and becomes an **explorer**
else

    if agent A learns $enter(v, t')$
    then in round $t' + 1$ agent A remains at node v and becomes a **cruiser**
    else

        if ((v is the initial position of A, and A has never yet moved from v)
        or $t' = t + i \cdot d + e$)
        then in round $t' + 1$ agent A leaves v by port 0 and becomes a **token**

The role of the above protocol is to prevent two **fighters** from leaving node v simultaneously. We will later prove this property and use it to show that two **tokens** cannot have the same home.

State explorer. [Intuition. **Explorers** simulate procedure EST^* treating homes of **tokens** as nodes containing tokens in the simulated procedure. Recognizing homes of **tokens** is a significant challenge that is overcome by using carefully orchestrated moves of agents reflected in the report given to **explorers**.]

When agent A becomes an **explorer** in round t, leaving node w in this round, it starts simulating procedure EST^* from node w as its starting position. This procedure tells the agent by which port it must leave any current node. However, to carry out the simulation, agent A must decide at each node, whether in the simulated procedure EST^* the current node contains a token or not. Call the current node, in which A is located, *red* in the first case and *white* in the second case. We will later show that a node is red (i.e., in the simulated procedure EST^* it contains a token) if and only if it is the home of an agent in state **token** in the simulation.

As long as the report given to agent A is $alone(v, t)$, the agent considers the current node v as white and continues the simulation leaving the current node by the port prescribed by procedure EST^*. As soon as the report is $\neg alone(v, t)$ in some round t of the simulation of EST^*, agent A interrupts this simulation and executes the following protocol in every round $t' \geq t$ while it stays at v, until the protocol specifies that it leaves the node.

---

if agent A learns $[enter(v, t'-2) \wedge \neg enter(v, t'-1) \wedge \neg alone(v, t'-1)] \vee alone(v, t')$
then
    if the simulation of EST^* is finished
    then in round $t'+1$ agent A leaves node v by port 0 and becomes a **ghost**
    else in round $t'+1$ agent A remains an **explorer** and leaves node v
    continuing the simulation of EST^* where it was interrupted in round t

---

When A starts the simulation of EST^* becoming an **explorer** in some round, it decides that the node that it leaves in this round is red. In order to continue the simulation of EST^* when the above protocol indicates it, agent A must decide if the node which it is about to leave in round $t'+1$ is red or white. This decision is made in round t' using the following rule:
if agent A learns $enter(v, t'-2) \wedge \neg enter(v, t'-1) \wedge \neg alone(v, t'-1) \wedge exit(v, t')$
then A decides in round t' that v is red
else A decides in round t' that v is white

State token. [Intuition. The role of a **token** is to help other agents to identify its home. This is crucial in order to carry out a correct simulation of EST^* by **explorers**. In order to do it, a **token** moves between its home and its office in a carefully prescribed manner.]

When agent A becomes a **token** in round s, leaving node w by port 0 in this round, node w is called the *home* of **token** A and node w' which A enters in round s is called the *office* of **token** A. When agent A is at its office w' in round t', then it leaves it and enters its home w in round $t'+1$. When agent A is at its home w in some round t, then it executes the following protocol in every round $t' \geq t$ while it stays at w, until the protocol specifies that it leaves this node.

> **if** agent A learns $enter(w, t' - 1) \wedge \neg alone(w, t') \wedge \neg enter(w, t')$
> **then** agent A leaves w by port 0 and enters its office w' in round $t' + 1$.

State ghost. [Intuition. An agent transits to state **ghost** from state **explorer** after finishing the simulation of EST^*. The goal of a **ghost** is to find a node which is not a home of a **token** and stop at this node. Stopping at a home of a **token** would entice this **token** to move perpetually, thus potentially precluding termination.]

When agent A becomes a **ghost**, it works in phases numbered $1, 2, \ldots$. In phase n the agent performs procedure $EXPLO(n)$. At each step of each phase agent A stops at the current node v in round t in which it enters this node and it executes the following protocol in every round $t' \geq t$ while it stays at v, until the protocol specifies that it leaves this node.

> **if** agent A learns $enter(v, t' - 2) \wedge \neg enter(v, t' - 1) \wedge \neg alone(v, t' - 1) \wedge exit(v, t')$
> **then** in round $t' + 1$ agent A leaves node v continuing the procedure $EXPLO(n)$
> where it was interrupted in round t
> **if** agent A learns $\neg enter(v, t' - 3) \wedge \neg enter(v, t' - 2) \wedge \neg enter(v, t' - 1) \wedge exit(v, t')$
> **then** in round $t' + 1$ agent A leaves node v by port 0 and becomes an **explorer**

5   Analysis of the Algorithm

In this section we show that Algorithm Team-Exploration terminates in polynomial time and that it performs correct exploration, provided that the initial configuration is not perfectly symmetric.

Main Theorem
Consider an arbitrary graph G and an initial configuration of agents in G that is not perfectly symmetric. Algorithm Team-Exploration terminates in time polynomial in the size of the graph, starting from this configuration. At the termination of the algorithm, each node of the graph is visited by some agent.

Plan of the Proof
In order to prove that Algorithm Team-Exploration is correct, we show that it is a faithful simulation of the procedure EST^* from the auxiliary scenario. In this simulation homes of **tokens** play the role of "genuine" tokens, and **explorers** play the role of agents executing the simulated procedure EST^*. By Proposition 3 we know that in the auxiliary scenario in which each agent is equipped with a "genuine" token, each node is visited by some agent at the time when all agents terminate the execution of procedure EST^*. Hence, if the simulation is faithful, all nodes will be visited upon completion of Algorithm Team-Exploration as well. To prove this faithfulness, we need to guarantee the following facts:

1. at least one agent becomes a **token**, similarly as at least one "genuine" token appears in EST^*.

2. **explorers** appear together with **tokens**, similarly as agents appear together with "genuine" tokens in EST^*.
3. **explorers** correctly recognize homes of **tokens**, similarly as agents recognize "genuine" tokens in EST^*.
4. an **explorer** always terminates the simulation of EST^*.

Guaranteeing the above facts is the main difficulty of the proof of correctness.

The proof that Algorithm Team-Exploration terminates in polynomial time is split into two parts. First we show that the number of rounds in which some agent moves is polynomial in the size of the graph. This implies termination but is not enough to estimate time, as time intervals between two such rounds could potentially be large. Hence in the second part we prove that the largest time interval between two rounds when some agent moves is also polynomial in the size of the graph.

References

1. Albers, S., Henzinger, M.R.: Exploring unknown environments. SIAM J. Comput. 29, 1164–1188 (2000)
2. Awerbuch, B., Betke, M., Rivest, R., Singh, M.: Piecemeal graph learning by a mobile robot. In: Proc. 8th Conf. on Comput. Learning Theory, pp. 321–328 (1995)
3. Bender, M.A., Fernandez, A., Ron, D., Sahai, A., Vadhan, S.: The power of a pebble: exploring and mapping directed graphs. In: Proc. STOC 1998, pp. 269–278 (1998)
4. Bender, M.A., Slonim, D.: The power of team exploration: Two robots can learn unlabeled directed graphs. In: Proc. FOCS 1994, pp. 75–85 (1994)
5. Chalopin, J., Das, S., Kosowski, A.: Constructing a Map of an Anonymous Graph: Applications of Universal Sequences. In: Lu, C., Masuzawa, T., Mosbah, M. (eds.) OPODIS 2010. LNCS, vol. 6490, pp. 119–134. Springer, Heidelberg (2010)
6. Czyzowicz, J., Kosowski, A., Pelc, A.: How to meet when you forget: Log-space rendezvous in arbitrary graphs. In: Proc. PODC 2010, pp. 450–459 (2010)
7. Das, S., Flocchini, P., Kutten, S., Nayak, A., Santoro, N.: Map construction of unknown graphs by multiple agents. Theoretical Computer Science 385, 34–48 (2007)
8. Deng, X., Papadimitriou, C.H.: Exploring an unknown graph. Journal of Graph Theory 32, 265–297 (1999)
9. Dessmark, A., Pelc, A.: Optimal graph exploration without good maps. Theoretical Computer Science 326, 343–362 (2004)
10. Dieudonné, Y., Pelc, A.: Deterministic gathering of anonymous agents in arbitrary networks (2011), http://arxiv.org/abs/1111.0321
11. Diks, K., Fraigniaud, P., Kranakis, E., Pelc, A.: Tree exploration with little memory. Journal of Algorithms 51, 38–63 (2004)
12. Duncan, C.A., Kobourov, S.G., Kumar, V.S.A.: Optimal constrained graph exploration. In: Proc. 12th Ann. ACM-SIAM Symp. on Discrete Algorithms (SODA 2001), pp. 807–814 (2001)
13. Fleischer, R., Trippen, G.: Exploring an Unknown Graph Efficiently. In: Brodal, G.S., Leonardi, S. (eds.) ESA 2005. LNCS, vol. 3669, pp. 11–22. Springer, Heidelberg (2005)

14. Fraigniaud, P., Gasieniec, L., Kowalski, D., Pelc, A.: Collective tree exploration. Networks 48, 166–177 (2006)
15. Frederickson, G.N., Hecht, M.S., Kim, C.E.: Approximation algorithms for some routing problems. SIAM J. Comput. 7, 178–193 (1978)
16. Gasieniec, L., Pelc, A., Radzik, T., Zhang, X.: Tree exploration with logarithmic memory. In: Proc. SODA 2007, pp. 585–594 (2007)
17. Reingold, O.: Undirected connectivity in log-space. Journal of the ACM 55 (2008)
18. Ta-Shma, A., Zwick, U.: Deterministic rendezvous, treasure hunts and strongly universal exploration sequences. In: Proc. SODA 2007, pp. 599–608 (2007)
19. Yamashita, M., Kameda, T.: Computing on Anonymous Networks: Part I-Characterizing the Solvable Cases. IEEE Trans. Parallel Distrib. Syst. 7, 69–89 (1996)

A QPTAS for ε-Envy-Free Profit-Maximizing Pricing on Line Graphs

Khaled Elbassioni

Max-Planck-Institut für Informatik, Saarbrücken, Germany
elbassio@mpi-inf.mpg.de

Abstract. We consider the problem of pricing edges of a line graph so as to maximize the profit made from selling intervals to single-minded customers. An instance is given by a set E of n edges with a limited supply for each edge, and a set of m clients, where each client j specifies one interval of E she is interested in and a budget B_j which is the maximum price she is willing to pay for that interval. An envy-free pricing is one in which every customer is allocated (possibly empty) interval maximizing her utility. Recently, Grandoni and Rothvoss (SODA 2011) gave a polynomial-time approximation scheme (PTAS) for the unlimited supply case with running time $(nm)^{O((\frac{1}{\varepsilon})^{\frac{1}{\varepsilon}})}$. By utilizing the known hierarchical decomposition of doubling metrics, we give a PTAS with running time $(nm)^{O(\frac{1}{\varepsilon^2})}$. We then consider the limited supply case, and the notion of ε-envy-free pricing in which a customer gets an allocation maximizing her utility within an additive error of ε. For this case we develop an approximation scheme with running time $(nm)^{O(\frac{\log^4 \max_e H_e}{\varepsilon^3})}$, where $H_e = \frac{B_{\max}(e)}{B_{\min}(e)}$ is the maximum ratio of the budgets of any two customers demanding edge e. This yields a PTAS in the uniform budget case, and a quasi-PTAS for the general case.

1 Introduction

Consider the situation when there is a set of n items, available in limited or unlimited supply, and a set of m customers interested in buying subsets of these items. We restrict our attention to the so called *single-minded (SM)* customers in which each client is only interested in a single subset of the items, and declares her maximum budget for purchasing that subset. From the seller's point of view, a question that naturally arises is: How to assign prices to the items so as to maximize the revenue made from selling these items to customers that can afford them? Clearly, with limited supply, customers compete for the available items, and a pricing scheme will be acceptable by the customers if any customer who can afford her preferred subset is guaranteed to get it.

Guruswami et al. [GHK+05] initiated the study of the computational complexity of such *envy-free* pricing schemes in the general combinatorial auction setting. Envy-freeness is a natural fairness condition that corresponds to a stable situation in which each customer is maximally happy with her allocation under the announced prices, and has no reason to seek a different allocation. Recently, there has been a growing interest in studying the approximability of such pricing problems (see e.g. [AH06, BB06, BK07, CS08, DFHS06, GvLSU06, GR11, GVR06, GHK+05], motivated by their practical

A. Czumaj et al. (Eds.): ICALP 2012, Part II, LNCS 7392, pp. 513–524, 2012.
© Springer-Verlag Berlin Heidelberg 2012

applications and, also by their connections to other areas such as profit-maximizing truthful mechanisms (see e.g. [AH06, GHK+06, GHK+05]).

Recognizing the generality of the problem, Guruswami et al. [GHK+05] proceeded to study various more-tractable special cases. In particular, they identified the case when the items are edges on a path and the customer-sets correspond to subpaths (or intervals) on this path, as one of the simplest versions in the *single-minded* setting, which they called the *highway problem*. This version can be motivated from a scheduling perspective, and was shown to be strongly NP-hard in [ERRS09] even in the unlimited supply case. A number of results for special cases were given in [BB06, BK06, GHK+05] and a quasi-polynomial-time approximation scheme (QPTAS) in the *non-envy-free* setting was given in [ESZ07]. Very recently, Grandoni and Rothvoss settled the complexity of the problem in the *unlimited* supply case by giving a PTAS [GR11]. For the *envy-free limited supply* case, the best known result is due to Cheung and Swamy [CS08], who used LP-based techniques to get an $O(\log c_{\max})$-approximation in polynomial time, where $c_{\max}$ is the maximum item supply.

Our contribution. We use the connection to hierarchical decompositions of doubling metrics [Tal04] to obtain an improved PTAS for the unlimited supply version of the highway problem. Somewhat surprisingly, a (seemingly trivial) one dimensional-version of such decomposition turns out to be sufficient to give a PTAS with running time $\left(\frac{nm}{\varepsilon}\right)^{O\left(\frac{1}{\varepsilon^2}\right)}$, improving on the PTAS in [GR11], whose running time $(nm)^{O\left(\left(\frac{1}{\varepsilon}\right)^{\frac{1}{\varepsilon}}\right)}$ is doubly exponential in $\frac{1}{\varepsilon}$. Our method also extends to other problems treated in [GR11], such as the maximum feasible subsystem problem with interval matrices. Our second result, which can be thought of as the main contribution of this paper, is to give a QPTAS for obtaining a pricing satisfying an ε-approximate version of the envy-free pricing notion. Specifically, we call a pricing to be ε-*envy-free* if a winner (that is, a customer who gets allocated a non-empty subset) can afford her allocated subset, while a loser cannot afford a $(1 + \varepsilon)$-factor of the price of her preferred subset. Extending our first PTAS, we develop an approximation scheme for this version with running time $\left(\frac{nm}{\varepsilon}\right)^{O\left(\frac{\log^4 \max_e H_e}{\varepsilon^3}\right)}$, where $H_e := \frac{B_{\max}(e)}{B_{\min}(e)}$ is the maximum ratio of the budgets of any two customers demanding edge e. This yields a PTAS in the uniform budget case (when all budgets are the same, and in fact, in much more general settings), and a QPTAS for the general case, and shows that the problem is unlikely to be *APX*-hard; in contrast, one should note that, even in the *uniform* budget, *unlimited* supply case, the general SM-pricing problem (where bundles are not necessarily intervals) is hard to approximate within a factor of $O(\log^\epsilon m)$ and $O(n^\epsilon)$, for some $\epsilon > 0$, under some plausible complexity assumptions [DFHS06, Bri08].

Overview of the technique. The general idea is to use the hierarchical decomposition of Talwar [Tal04], applied to a line semi-metric (which has doubling dimension 1) defined by the optimal pricing p^* on the set of vertices of the given line. At the highest level, this decomposition divides the set of points on the line into a constant number of clusters. Then each cluster is divided further into lower-level clusters and so on. The diameters of the clusters shrink geometrically (see Figure 1).

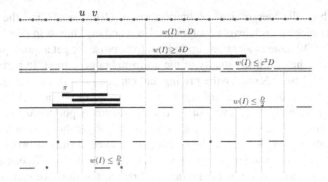

Fig. 1. Hierarchical decomposition. Thick black intervals correspond to customer path declarations; Thin black intervals are the intervals defined by the hierarchical decomposition; red intervals define the refinement of the interval with weight D in the refined decomposition. π is the capacity profile consumed by the intervals crossing edge $[u, v]$. We use $\delta = \varepsilon$.

This decomposition guarantees that, for any $\varepsilon > 0$, we can recover a fraction of at least $1 - \varepsilon$ of the optimal profit, from customers whose intervals are "large" in price w.r.t. the diameter of the cluster of the decomposition where they belong. Here, an interval I is considered large if the price $w(I) = p^*(I)$ is at least εD, where D is the diameter (or price) of the cluster to which I belongs (e.g. the thick black interval in Figure 1). It follows that we need only to estimate the prices in each cluster within a factor of ε^2 to ensure that we make only an error of at most $(1 + \varepsilon)$ relative to the price $p^*(I)$ for any large interval I. This leads us to the notion of *refined hierarchical decomposition*, where each cluster of diameter D is further divided into more intervals each with price at most $\varepsilon^2 D$ (e.g. the red intervals in Figure 1). Having shown such structural result on near-optimal solutions, we can build a dynamic programming (DP) table that guesses all such decompositions at each level. This allows us to handle the unlimited supply case. We remark that this approach resembles in essence the one in [GR11]; however, our decomposition is more efficient since it decomposes each interval into $O(1/\varepsilon^2)$ intervals at each level, in contrast to $O((1/\varepsilon)^{1/\varepsilon})$ intervals at each level in [GR11].

To extend this basic approach to the *capacitated* variant, we need to use one more idea from the QPTAS for the unsplittable flow problem developed in [BCES06]. Namely, we have to restrict the capacity requirement at each level of the decomposition to a profile admitting a *constant-size* description (e.g. the dashed profile π in Figure 1).

There is a number of technical hurdles that we have to overcome in applying these techniques:

– At a given level of the decomposition, there is a number of "forced winners" who have to be included in the solution because they can afford the price of their intervals (otherwise we violate the envy-freeness condition). Unlike the unforced winners, the capacity requirement of the forced winners might not be restricted to a profile with a constant-size description, and thus, if we were to guess such requirements, we would need an exponential number of guesses. We resolve this issue by noting that these forced winners, at each level, can be represented, more or less, as a function of a pricing profile which has a constant-size description.

- The depth of the hierarchical decomposition is $\Omega(\log(nm))$ and the number of possible choices at each level is polynomial in n and m. Thus as in [GR11], we have to build the DP table in a bottom-up fashion (otherwise we get a quasi-poly running time even in the unlimited supply or the uniform budget cases). In particular, each DP entry will be indexed by the pricing and capacity profiles at the current level, and in addition, by the *residual capacity* needed by all winners at higher levels. Since the total number of levels in the decomposition is polylogarithmic, the total description size of this residual capacity is also polylogarithmic, which may lead to a quasi-polynomial blow-up in the running time, even in the uniform budget case. We overcome this problem by showing that, due to the envy-freeness condition and the geometric nature of the hierarchical decomposition, we can describe this residual capacity in terms of the profiles at *only a constant number* of levels.
- As mentioned above, the near-optimal pricing is obtained by considering only large intervals. However, we have to guarantee that small intervals also satisfy the envy-freeness condition. We resolve this issue by first showing that we may assume it is only encountered by customers whose budgets are sufficiently small. Then a small increase in the prices at the end will scare away such customers, without increasing by much the price of the large intervals.

Some related work. The highway problem and its generalization to trees, the *tollbooth problem*, were introduced in [GHK+05] as special cases of the more general single-minded pricing problem, where intervals are replaced by arbitrary sets of items. For the unlimited supply version of the general SM-pricing problem, there exist $O(\log m + \log n)$-approximation algorithms [BK06, GHK+05], and almost matching inapproximability results of $O(\log^{\epsilon} m)$ and $O(n^{\epsilon})$, for some $\epsilon > 0$ [DFHS06, Bri08]. The unlimited-supply highway problem was shown to be strongly NP-hard in [ERRS09], and a PTAS was given in [GR11]; for the tollbooth problem, the currently best approximation factor is $O(\log n/\log\log n)$ [GS10] and the problem is known to be APX-hard [GHK+05]. For the limited supply setting, a pseudo-polynomial algorithm was given in [GHK+05] for the envy-free version of tollbooth problem on rooted trees, when all paths start from the root. A QPTAS for the non-envy free version of the highway problem is given in [ESZ07].

Probably the main difficulty with dealing with such pricing problems in general is the lack of good upper bounds on the optimal solution. The only obvious upper bound is the *maximum social welfare*, which is the maximum profit that can be obtained assuming each winner pays her full budget. The currently best result for the envy-free pricing version of both the tollbooth and highway problems uses this upper bound [CS08] and achieves an approximation factor of $\log c_{\max}$, where $c_{\max}$ is the maximum supply of an item. On the other hard, this gap of $\log c_{\max}$ is unavoidable using this upper bound, since there is a simple example achieving this gap. In that sense, our result shows that one can get substantial improvements in the approximation ratio, if the envy freeness condition is replaced by ε-envy-freeness.

2   The Highway Problem

Let $V = \{0, 1, \ldots, n\}$ and $E = \{e_1, \ldots, e_n\}$, with $e_i = \{i - 1, i\}$, for $i = 1, \ldots, n$. In the *highway problem*, denoted henceforth by HP, we are given a set of m

(*single-minded*) customers: customer j is interested in buying an interval (or a path) I_j on V, and declares a *budget* $B(I_j) \in \mathbb{R}_+$, which is the maximum amount of money she is welling to pay if she gets that interval. Our task is to assign a price $p(e) \in \mathbb{R}_+$ for each edge $e \in E$, and to find a set of winners $\mathcal{W} \subseteq [m]$ (those that get allocated non-empty intervals), so as to maximize $p(\mathcal{W}) := \sum_{j \in \mathcal{W}} p(I_j)$ subject to the capacity constraints: $|\{j \in \mathcal{W} : e \subseteq I_j\}| \leq c(e)$, for all $e \in E$, and the *envy-freeness* condition

$$p(I_j) \leq B(I_j) \qquad \text{for all } j \in \mathcal{W} \tag{1}$$
$$p(I_j) \geq B(I_j) \qquad \text{for all } j \notin \mathcal{W}. \tag{2}$$

Given $\varepsilon > 0$, we say that the pricing $p : E \to \mathbb{R}_+$ is ε-*envy-free* if it satisfies (1) and

$$p(I_j) \geq (1 - \varepsilon)B(I_j) \qquad \text{for all } j \notin \mathcal{W}. \tag{3}$$

A 0-envy-free (resp., 1-envy-free) pricing will be simply called an envy-free (resp., non-envy-free) pricing.

3   Hierarchical Decomposition for Line Semi-metrics

3.1   Basic Definitions

A semi-metric that can be embedded into the line is called a *line semi-metric*. Formally, there is a line graph $G = (V, E)$ with non-negative weights $w : E \to \mathbb{R}_+$, and an ordering on V, say $V = \{u_0, u_1, \ldots, u_n\}$, such that the distance between u_i and u_j, $d(u_i, u_j) = w([u_i, u_j]) := \sum_{k=i+1}^{j} w(e_k)$, for all i, j, where[1] $e_k = [u_{k-1}, u_k] \in E$. Henceforth we will denote such a line semi-metric by (V, w), where we assume w.l.o.g. that $V = \{0, 1, \ldots, n\}$. An interval $I \subseteq V$ is path in G, and we denote the set of all such intervals by $\mathbb{I}(V) \subseteq 2^V$. For an interval $I \in \mathbb{I}(V)$, we denote respectively by $\mathrm{lt}(I)$ and $\mathrm{rt}(I)$ the left and right-end points of I, that is $I = [\mathrm{lt}(I), \mathrm{rt}(I)]$. For a collection of intervals $\mathcal{J} \subseteq \mathbb{I}(V)$ and an edge $e \in E$, we denote by $\mathcal{J}(e)$ the subset of intervals in $\mathcal{J}$ containing e, and write $w(\mathcal{J}) := \sum_{J \in \mathcal{J}} w(J)$. We use $w_{\min} = \min_{i: w[i,i+1] > 0} \{w[i, i+1]\}$, $w_{\max} = w([0, n])$, and $\mathrm{Ar}_w = \frac{w_{\max}}{w_{\min}}$ for the *aspect ratio* of the metric.

3.2   Basic Hierarchical Decomposition for Line Semi-metrics

The decomposition presented in this section is based on the (trivial) fact that a line semi-metric has doubling dimension 1.

Let (V, w) be a line semi-metric. Two intervals $I = [i, j]$ and $I' = [k, l]$ are said to be *edge-disjoint* if $|I \cap I'| \leq 1$, and *consecutive* if $k = j + 1$. For a consecutive set of intervals $\mathcal{I} = \{[u_i, v_i] : i = 1 \ldots, k\}$, we denote by $\mathcal{E}(\mathcal{I}) \subseteq E$ the set of edges in between, that is, $\mathcal{E}(\mathcal{I}) = \{[v_i, u_{i+1}] : i = 1, \ldots, k - 1\}$. As usual, we denote by $w(I) = d(i, j)$ the length (or weight) of interval $I = [i, j]$.

---

[1] For convenience, we will sometimes treat edges also as intervals, and allow an interval to be a single point.

We consider distances of geometrically decreasing scales. Recall the relevant distances are between $w_{\min}$ and $w_{\max}$. We consider powers of 2, and have $L = L_w :=$ $\lceil \log_2 \mathsf{Ar}_w \rceil$ distance scales. We let $D_L := w_{\max}$, and $D_{i-1} := \frac{D_i}{2}$, for $1 \leq i \leq L$ (so that $D_0 \leq w_{\min}$). For $I \in \mathbb{I}(V)$, define $i(I) = i_w(I)$ to be the largest integer $i \in \{0, 1, \ldots, L-1\}$ such that $D_i < w(I)$, if $w(I) > w_{\min}$, and set $i(I) = 0$, otherwise. Let $\tau(I) = \tau_w(I) := D_{i(I)}$; thus $\tau(I) < w(I) \leq 2\tau(I)$, unless $w(I) \leq w_{\min}$.

For a line semi-metric, the hierarchical decomposition given in [Tal04] for doubling metrics specializes to the following.

Definition 1 ($((\ell, l)$-bounded hierarchical decomposition for line semi-metrics). *Given a line semi-metric (V, w) and integers ℓ and l, an (ℓ, l)-bounded hierarchical decomposition $\mathbf{T} = (\mathcal{N}, \mathbf{r}, \mathcal{C})$ is defined by a rooted tree $\mathbf{T}$ with root $\mathbf{r}$ and set of nodes $\mathcal{N}$; each node $\mathbf{v} \in \mathcal{N}$ is associated with an interval $I(\mathbf{v}) \in \mathbb{I}(V)$, such that*

(HD1) the depth of $\mathbf{T}$ is at most l;

(HD2) for any non-leaf node $\mathbf{v} \in \mathcal{N}$, the set $\mathcal{C}(\mathbf{v})$ of children of $\mathbf{v}$ has size between 2 and ℓ;

(HD3) $I(\mathbf{r}) = [0, n]$ and $w(I(\mathbf{v})) = 0$ if and only if $\mathbf{v} \in \mathcal{N}$ is a leaf;

(HD4) for each $\mathbf{v} \in \mathcal{N}$, the set $\{I(\mathbf{u}) : \mathbf{u} \in \mathcal{C}(\mathbf{v})\}$ consists of disjoint consecutive intervals that cover the (vertices of the) whole interval $I(\mathbf{v})$. Moreover, for all $\mathbf{u} \in \mathcal{C}(\mathbf{v})$, $w(I(\mathbf{u})) < \tau(I(\mathbf{v}))$.

Given such a hierarchical decomposition $\mathbf{T} = (\mathcal{N}, \mathbf{r}, \mathcal{C})$, we denote by $\mathbf{T}(\mathbf{v})$ the subtree of $\mathbf{T}$ rooted at $\mathbf{v}$, and let $\mathcal{E}(\mathbf{v}) := \mathcal{E}(\{I(\mathbf{u}) : \mathbf{u} \in \mathcal{C}(\mathbf{v})\})$. Two intervals I and J are said to be crossing, denoted by $I \perp J$, if $I \cap J \neq \emptyset$ and neither of them is contained in the other. We will say that an interval $I \in \mathbb{I}(V)$ *is split by the decomposition* at node $\mathbf{v} \in \mathcal{N}$ if $I \subseteq I(\mathbf{v})$ but $I \perp I(\mathbf{u})$ for some $\mathbf{u} \in \mathcal{C}(\mathbf{v})$. Correspondingly, we say that I is split at level i of the decomposition if I is split at node $\mathbf{v}$ where $i(I(\mathbf{v})) = i$. Note that an interval can be split by $\mathbf{T}$ at no more than one node. Let us write $\mathcal{J}(\mathbf{v})$ for the subset of intervals $J \subseteq I(\mathbf{v})$ that are split at node $\mathbf{v}$, and $\mathcal{J}[\mathbf{T}(\mathbf{v})]$ for the union $\bigcup_{\mathbf{u}} \mathcal{J}(\mathbf{u})$ over all nodes $\mathbf{u}$ of the subtree $\mathbf{T}(\mathbf{v})$. When considering the highway problem, we may represent a set of intervals $\mathcal{J}$ by the set of indices $\mathcal{W} := \{j \in [m] : I_j \in \mathcal{J}\}$ of their corresponding customers, and write $\mathcal{W}(\mathbf{v})$, $\mathcal{W}[\mathbf{T}(\mathbf{v})]$, for $\mathcal{J}(\mathbf{v})$, $\mathcal{J}[\mathbf{T}(\mathbf{v})]$, etc.

Proposition 1. *For any line semi-metric (V, w), there exits a randomized $(8, L_w)$-bounded hierarchical decomposition $\mathbf{T}$, such that*

(HD5) for any $J \in \mathbb{I}(V)$ and any $i \in \{0, 1, \ldots, L-1\}$, $\Pr[J$ is split at level $i] \leq \frac{16w(J)}{D_i}$.

3.3   Refined Hierarchical Decomposition for Line Semi-metrics

Given sets of edges $F, F' \subseteq E$, and sets of pairwise edge-disjoint intervals $\mathcal{H}, \mathcal{H}' \subseteq \mathbb{I}(V)$, we will say that $(F', \mathcal{H}')$ is a *refinement* of $(F, \mathcal{H})$ if: $F \subseteq F'$, and every interval $I \in \mathcal{H}$ is the (edge-disjoint) union of intervals and edges from $F' \cup \mathcal{H}'$. For our purposes, we will need to work with a more refined hierarchical decomposition, defined as follows.

Definition 2 $((\ell, l, \eta)$-refined hierarchical decomposition for line semi-metrics).
Given a line semi-metric (V, w), integers ℓ and l, and $\eta \in (0, 1)$, an (ℓ, l, η)-refined hierarchical decomposition $\mathbf{T} = (\mathcal{N}, \mathbf{r}, \mathcal{C})$ is an (ℓ, l)-bounded hierarchical decomposition $\mathbf{T}$, such that for each non-leaf node $\mathbf{v} \in \mathcal{N}$, the interval $I(\mathbf{v})$ is partitioned into the edge-disjoint union of a set of intervals $\mathcal{H}(\mathbf{v})$ and a set of edges $F(\mathbf{v})$, with the following further properties:

(HD6) $|\mathcal{H}(\mathbf{v})| \leq \frac{2}{\eta}$ and $|F(\mathbf{v})| \leq \frac{1}{\eta}$;

(HD7) $w(I) \leq \eta \cdot w(I(\mathbf{v}))$ for all $I \in \mathcal{H}(\mathbf{v})$ and $w(e) > \eta \cdot w(I(\mathbf{v}))$ for all $e \in F(\mathbf{v})$;

(HD8) $(\bigcup_{\mathbf{u} \in \mathcal{C}(\mathbf{v})} F(\mathbf{u}) \cup \mathcal{E}(\{I(\mathbf{u}) : \mathbf{u} \in \mathcal{C}(\mathbf{v})\}), \bigcup_{\mathbf{u} \in \mathcal{C}(\mathbf{v})} \mathcal{H}(\mathbf{u}))$ is a refinement of $(F(\mathbf{v}), \mathcal{H}(\mathbf{v}))$.

Refined hierarchical decompositions are a compact way for approximating the distance in the underlying metric.

Definition 3 (Approximate distance w.r.t. a refined hierarchical decomposition).
Given an (ℓ, l, η)-refined hierarchical decomposition $\mathbf{T}$ for a line semi-metric (V, w), the approximate distance $\mathrm{val}_w([u, v])$ between points $u, v \in V$ is defined as follows. If $J = [u, v]$ is not split by $\mathbf{T}$, then $\mathrm{val}_w(J) = w(J) = 0$; otherwise, let $\mathbf{v}$ be the node of $\mathbf{T}$ at which J is split, then $\mathrm{val}_w(J) := \sum_{I \in F(\mathbf{v}) \cup \mathcal{H}(\mathbf{v}),\ I \subseteq J} w(I)$.

Definition 4 (δ-aligned interval). *Let $\mathbf{T}$ be a hierarchical decomposition for a line semi-metric (V, w). For $\delta > 0$, an interval $J \in \mathbb{I}(V)$ is said to be δ-aligned w.r.t. $\mathbf{T}$ if*

$$J \text{ is split by } \mathbf{T} \text{ at node } \mathbf{v} \Rightarrow w(J) \geq \delta \cdot w(I(\mathbf{v})).$$

For a set $\mathcal{J}$ of intervals, we denote by $\mathcal{J}[\mathbf{T}(\mathbf{v}), \delta]$ the subset of intervals in $\mathcal{J}[\mathbf{T}(\mathbf{v})]$ that are δ-aligned w.r.t. the subtree of $\mathbf{T}$ rooted at node $\mathbf{v}$, and will write $\mathcal{J}[\mathbf{v}, \delta]$ for the subset of intervals in $\mathcal{J}(\mathbf{v})$ that are δ-aligned w.r.t. $\mathbf{T}$.

Proposition 2. *Let $\mathbf{T}$ be an (ℓ, l, η)-refined hierarchical decomposition for a line semi-metric (V, w). If $J \in \mathbb{I}(V)$ is split at node $\mathbf{v}$ in $\mathbf{T}$ then $\mathrm{val}_w(J) \leq w(J) \leq \mathrm{val}_w(J) + 2\eta \cdot w(I(\mathbf{v}))$. If J is δ-aligned w.r.t. $\mathbf{T}$, then $(1 - \frac{2\eta}{\delta})w(J) \leq \mathrm{val}_w(J) \leq w(J)$.*

The key fact on which the whole approach is based is the following.

Lemma 1. *Let $w : E \to \mathbb{R}_+$ be a non-negative weight function on E. Then for any $\delta, \eta > 0$, there exists an $(8, L_w, \eta)$-refined hierarchical decomposition $\mathbf{T}$ of (V, w) such that, for any set of intervals $\mathcal{J} \subseteq \mathbb{I}(V)$, $w(\mathcal{J}[\mathbf{T}, \delta]) \geq (1 - 64\delta)w(\mathcal{J})$.*

4 A Dynamic Program for ε-Envy Free Pricing

In the following we fix $\varepsilon \in (0, 1)$. For a subset $\mathcal{M} \subseteq [m]$ and a pricing $p : E \to \mathbb{R}_+$, we denote by $p(\mathcal{M})$ the sum $\sum_{j \in \mathcal{M}} p(I_j)$. Given an ε-envy-free pricing p, the set of winners, under p, is obtained by solving the following packing integer program: $\mathrm{OPT}_{p,\varepsilon} := \max\{\sum_{j \in [m]} p(I_j)x_j : \sum_{j\, :\, e \in I_j} x_j \leq c(e), \forall e \in E, x_j = 1\ \forall j \text{ s.t. } p(I_j) < (1 - \varepsilon)B(I_j), x_j = 0\ \forall j \text{ s.t. } p(I_j) > B(I_j), x_j \in \{0, 1\}\ \forall j \in [m]\}$. Note that this IP can be solved as an LP since the constraint matrix is totally unimodular. By $\mathcal{W}_{p,\varepsilon} := \{j : x_j^* = 1\}$, we denote the set of winners in an optimal solution x^* of this IP, that is, $\mathrm{OPT}_{p,\varepsilon} = p(\mathcal{W}_{p,\varepsilon})$. For simplicity, we will write $\mathrm{OPT} := \mathrm{OPT}_{p^*,0}$, where p^* is an optimal envy-free pricing.

4.1 Restricting the Search Space for Pricing

Let ε be a given constant. We also let $\delta > 0$ and $\eta \in (0,1)$ be constants that will be fixed later. For $e \in E$, define $B_{\max}(e) := \max_{j \in [m]:\ e \in I_j} B(I_j)$ and $B_{\min} :=$ $\min_{j \in [m],\ e \in I_j} B(I_j)$, $H_e := \frac{B_{\max}(e)}{B_{\min}(e)}$, and $H := \max_e H_e$.

Throughout the paper, we will use the following notation: $P := \frac{mn^2(1+\varepsilon)}{\varepsilon^2}$ and $P' := \lceil \log_2 (nP) \rceil$. The following proposition is obtained by standard techniques.

Proposition 3. *We may assume w.l.o.g. that the given instance of* HP *satisfies the following conditions.* $B_{\min} := 1$ *and* $H \leq B_{\max} \leq \frac{mn^2(1+\varepsilon)}{\varepsilon^2}$. *There exists an ε-envy-free pricing* $\tilde{p} : E \mapsto \mathbb{R}_+$ *for which* $\tilde{p}(e) \in \{0,1,\ldots,P\}$, *for every* $e \in E$, *and* $\mathrm{OPT}_{\tilde{p},\varepsilon} \geq (1 - 2\varepsilon)\mathrm{OPT}$.

Applying Lemma 1 to the ε-envy-free pricing $\tilde{p}$ given by Proposition 3, we obtain the following structural result.

Corollary 1 (Structure of near-optimal solution in the unlimited supply case). *There exists a $(8, P', \eta)$-refined hierarchical decomposition* $\mathbf{T}$ *of* $(V, \tilde{p})$ *such that* $\tilde{p}(\mathcal{W}_{\tilde{p},\varepsilon}[\mathbf{T}, \delta]) \geq (1 - 64\delta - 2\varepsilon)\mathrm{OPT}$. *For any j such that I_j is δ-aligned w.r.t.* $\mathbf{T}$, *it holds that* $(1 - \frac{2\eta}{\delta})\tilde{p}(I_j) \leq \mathrm{val}_{\tilde{p}}(I_j) \leq \tilde{p}(I_j)$.

Refined hierarchical decompositions give rise to restricted pricing profiles, defined at each node of the decomposition. We define this formally as follows.

Definition 5 (ε-pricing profile). *Given a line graph $G = (V, E)$, an ε-pricing profile* $\mathcal{R} = (I, F, \mathcal{H}, \mathcal{E}, p)$ *is given by (where η is a function of ε)*

- *an interval $I \in \mathbb{I}(V)$, sets of edges $F, \mathcal{E} \subseteq I$, a set of pairwise edge-disjoint intervals $\mathcal{H} \subseteq \mathbb{I}(V)$, s.t. $I = \cup_{J \in \mathcal{H}} J \cup F$, $|F| \leq \frac{1}{\eta}$, $|\mathcal{E}| \leq \ell := 8$ and $|\mathcal{H}| \leq \frac{2}{\eta}$;*
- *a pricing function $p : \mathbb{I}(V) \to \mathbb{R}_+$, partially-defined on $\{I\} \cup \mathcal{H} \cup \{\{e\} : e \in F \cup \mathcal{E}\}$, such that: $p(J) \leq \eta \cdot p(I)$ for all $J \in \mathcal{H}$ and $p(e) > \eta \cdot p(I)$ for all $e \in F$, $p(I) = \sum_{J \in F \cup \mathcal{H}} p(J)$, and $p(J) \in \{0,1,2,\ldots,\lceil nP \rceil\}$ for $J \in \{I\} \cup \mathcal{H} \cup F$.*

Given a refined hierarchical decomposition $\mathbf{T}$ of $(V, \tilde{p})$, we can associate with each node $\mathbf{v}$ of $\mathbf{T}$, an ε-pricing profile $\mathcal{R} = (I, F, \mathcal{H}, \mathcal{E}, p)$, where $I = I(\mathbf{v})$, $\mathcal{H} = \mathcal{H}(\mathbf{v})$, $F = F(\mathbf{v})$, $\mathcal{E} = \mathcal{E}(\mathbf{v})$ and p is the restriction of $\tilde{p}$ on $\{I\} \cup \mathcal{H} \cup \{\{e\} : e \in F \cup \mathcal{E}\}$.

An ε-pricing profile $\mathcal{R}$ is said to be consistent with a number of ε-pricing profiles $\mathcal{R}_1, \ldots, \mathcal{R}_k$, if there is a refined hierarchical decomposition $\mathbf{T}$ such that $\mathcal{R}$ is the pricing profile corresponding to a node $\mathbf{v}$ in $\mathbf{T}$, while $\mathcal{R}_1, \ldots, \mathcal{R}_k$ are the pricing profiles corresponding to the children of $\mathbf{v}$.

We omit the proof of the following theorem from this extended abstract.

Theorem 1. *There is a PTAS, for* HP *with unlimited supply, that runs in time* $\left(\frac{nm}{\varepsilon}\right)^{O(\frac{1}{\varepsilon^2})}$.

4.2 Restricting the Search Space for the Consumed Capacities

We recall the following definition from [BCES06].

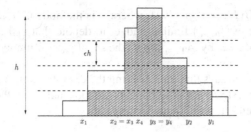

Fig. 2. A profile and its ε-restriction

Definition 6. (ε-*capacity profile*) *Let* $e = [i - 1, i]$ *be an edge of* E, $h, 1/\epsilon \in \mathbb{Z}_+$, *with* $h \leq c(e)$, *and set* $d = \max\{\epsilon h, 1\}$. *Let* $x = (x_1, \ldots, x_{h/d})$ *and* $y = (y_1, \ldots, y_{h/d})$ *be two vectors of points in* V, *such that* $x_1 \leq x_2 \leq \cdots \leq x_{h/d} \leq i - 1 < i \leq y_{h/d} \leq \cdots \leq y_2 \leq y_1$. *Then the vector* $(\ell_1, \ldots, \ell_n)$, *where*

$$\ell_i = \begin{cases} 0, & \text{for } i \leq x_1 \text{ and } i > y_1 \\ j \cdot d, & \text{for } x_j < i \leq x_{j+1} \text{ and } y_{j+1} < i \leq y_j \\ h, & \text{for } x_{h/d} < i \leq y_{h/d}, \end{cases}$$

is said to be an ε-(*restricted*) *capacity profile* with peak e and *height* h.

For an interval I we denote by $c \mid_I := (c(e) : e \subseteq I)$ the restriction of the capacity vector on I. For a collection of intervals $\mathcal{J} \subseteq \mathbb{I}(V)$ and any edge $e \in E$, define a profile $\mathrm{prof}(\mathcal{J}(e)) \in \mathbb{Z}_+^E$ by: $\mathrm{prof}(\mathcal{J}(e))_{e'} = |\mathcal{J}(e) \cap \mathcal{J}(e')|$, for $e' \in E$. Lemma 3.1 in [BCES06] states that any such profile can be sufficiently accurately approximated by an ε-capacity profile (see Figure 2). Applying this to the set of intervals split at a given node of the decomposition, we obtain the following result.

Corollary 2. *Let* $p : E \to \mathbb{Z}_+$ *be a pricing of* E *and* **T** *be a hierarchical decomposition for* (V, p). *Consider a node* **v** *of* **T**. *Write* $\mathcal{E}(\mathbf{v}) := \{e_1, \ldots, e_k\}$ *(numbered, say, from left to right), and let* $\mathcal{J}$ *be a given subset of* δ-*aligned intervals w.r.t.* **T**, *such that* J *is split at* **v** *for all* $J \in \mathcal{J}$. *For,* $i = 1, \ldots, k$, *write* $\mathcal{J}_i := \mathcal{J}(e_i) \setminus \bigcup_{j=1}^{i-1} \mathcal{J}(e_j)$ *and* $h_i := |\mathcal{J}_i|$. *Then there exist* εδ-*capacity profiles* $\pi_1, \ldots, \pi_k$ *with peaks* $e_1, \ldots, e_k$ *and heights* $h_1, \ldots, h_k$, *respectively, and subsets* $\mathcal{J}_i' \subseteq \mathcal{J}_i$, *such that, for* $i = 1, \ldots, k$,
  (i) $\mathrm{prof}(\mathcal{J}_i') \leq \pi_i \leq \mathrm{prof}(\mathcal{J}_i)$,
  (ii) $p(\mathcal{J}_i') \geq (1 - 4\varepsilon)p(\mathcal{J}_i)$, *and hence* $p(\bigcup_{i=1}^k \mathcal{J}_i') \geq (1 - 4\varepsilon)p(\mathcal{J})$.

Remark 1. There is a polynomial-time procedure, $\mathrm{PACK}(\mathcal{J}, \mathcal{R}, (\pi_1, \ldots, \pi_k))$, that finds for any given subset of intervals $\mathcal{J}$, ε-pricing profile $\mathcal{R} = (I, F, \mathcal{H}, \mathcal{E}, p)$, and ε-capacity profiles $\pi_1, \ldots, \pi_k$, a maximum packing $\mathcal{J}_i' \subseteq \mathcal{J}_i$ (as defined in Corollary 2) into π_i, for $i = 1, \ldots, k$, with maximum value w.r.t. $\mathcal{R}$.

Let ε be the desired accuracy. In the following, we fix $\varepsilon_0 := \frac{\varepsilon}{8(\log_2 \lceil \varepsilon H \rceil + 1)}$, $\delta := \varepsilon_0$, $\varepsilon'' := \varepsilon_0^2$, $\eta := \varepsilon_0^3$, $\varepsilon' := \frac{2\eta}{\delta}$ and $\varepsilon_a := \varepsilon_0^3$. Consider an ε'-envy-free pricing $\tilde{p} : E \to \mathbb{R}_+$, and the corresponding refined hierarchical decomposition **T** given by corollary 1. Given $\varepsilon_b, \varepsilon_c \geq 0$, we say that a customer j has an $[\varepsilon_b, \varepsilon_c]$-limited budget (at **v**) w.r.t.

$\mathbf{T}$, if $\varepsilon_b \tilde{p}(I(\mathbf{v})) \le B(I_j) \le \varepsilon_c \tilde{p}(I(\mathbf{v}))$, where $\mathbf{v}$ is the node in $\mathbf{T}$ at which I_j is split. (Similarly, we will use $[\varepsilon_b, \varepsilon_c)$-limited budget to denote that $\varepsilon_b \tilde{p}(I(\mathbf{v})) \le B(I_j) < \varepsilon_c \tilde{p}(I(\mathbf{v}))$, etc.) We denote by $\mathcal{A}_{\mathbf{T},[\varepsilon_b,\varepsilon_c]}(\mathbf{v})$ the set of $[\varepsilon_b, \varepsilon_c]$-limited budget customers at node $\mathbf{v}$ w.r.t. $\mathbf{T}$.

The following lemma says that, in satisfying the envy-freeness condition, we do not have to worry about customers whose budgets are too small w.r.t. the price determined by the location of their intervals in the decomposition.

Lemma 2. *Let $\varepsilon, \varepsilon'', \eta, \delta > 0$ be given constants and $W \subseteq [m]$ be a set of customers satisfying the capacity constraints. Let $\mathbf{T}$ be an (ℓ, l, η)-refined hierarchical decomposition of (V, p), where p partialy satisfies the ε-envy-free pricing conditions (1) and (3) w.r.t. W: $p(I_j) \le B(I_j)$ for all $j \in W$ and $p(I_j) \ge (1 - \varepsilon)B(I_j)$ for all $j \in [m] \setminus (W \cup W')$, where W' is the set of $(0, \varepsilon'']$-limited budget customers in $[m]$. Then there is an $\varepsilon + \varepsilon'''$-envy-free pricing p', such that $\mathrm{OPT}_{p',\varepsilon+\varepsilon'''} \ge (1 - \varepsilon''')p(W[\mathbf{T}, \delta])$, where $\varepsilon''' = (\ell - 1)(2\log_2 \frac{\varepsilon'' H}{\delta} + 1)[\varepsilon'' + \frac{2\varepsilon''}{\delta}]$.*

For an ε-pricing profile $\mathcal{R} = (I, F, \mathcal{H}, \mathcal{E}, p)$, we let $\pi_p(\mathcal{R}) := \mathrm{prof}(\mathcal{M})$ be the capacity consumed by the set of customers $\mathcal{M} := \mathcal{M}(\mathcal{R}) = \mathcal{M}_1(\mathcal{R}) \cup \mathcal{M}_2(\mathcal{R}) \cup \mathcal{M}_3(\mathcal{R}) \subseteq \mathcal{M}_0(\mathcal{R})$, where $\mathcal{M}_0(\mathcal{R}) := \{j : I_j \subseteq I, I_j \supseteq e \text{ for some } e \in \mathcal{E}\}$, and

$$\mathcal{M}_1(\mathcal{R}) := \{j \in \mathcal{M}_0(\mathcal{R}) : \delta p(I) \le \mathrm{val}_p(I_j) < (1 - \varepsilon')^2 B(I_j)\}$$
$$\mathcal{M}_2(\mathcal{R}) := \{j \in \mathcal{M}_0(\mathcal{R}) : \mathrm{val}_p(I_j) + 2\eta \cdot p(I) < (1 - \varepsilon')B(I_j) \text{ and } \mathrm{val}_p(I_j) < \delta p(I)\}$$
$$\mathcal{M}_3(\mathcal{R}) := \{j \in \mathcal{M}_0(\mathcal{R}) : \mathrm{val}_p(I_j) \le \varepsilon''(1 - \frac{4\eta}{\varepsilon''})p(I) \text{ and } B(I_j) \ge \varepsilon'' p(I)\}.$$

The set $\mathcal{M}(\mathcal{R})$ approximates the set of forced winners (whose budgets are above the price) at a given node $\mathbf{v}$ of $\mathbf{T}$, where $\mathcal{R}$ approximates the pricing profile. For a vector $\pi := (\pi_1, \ldots, \pi_k)$ of $\varepsilon\delta$-capacity profiles, let $\mathcal{M}(\mathcal{R}, \pi) := \mathrm{PACK}(\mathcal{M}_4(\mathcal{R}), \mathcal{R}, \pi))$ be as defined in Remark 1, where

$$\mathcal{M}_4(\mathcal{R}) = \{j \in \mathcal{M}_0(\mathcal{R}) : B(I_j) \ge \mathrm{val}_p(I_j) \ge (1 - \varepsilon')^2 B(I_j) \text{ and } \mathrm{val}_p(I_j) \ge \delta p(I)\}.$$

Thus, $\mathcal{M}(\mathcal{R}, \pi)$ represents the set of non-forced winners at $\mathbf{v}$.

Lemma 3 (Structure of near-optimal solution in the limited supply case). *There exists a capacitated $(8, P', \eta)$-refined hierarchical decomposition $\mathbf{T}$ of $(V, \tilde{p})$, and a subset of winners $W := W' \cup W''$, $W' \cap W'' = \emptyset$, such that*

(C1) $\mathrm{prof}(W) \le c$ and $\tilde{p}(W[\mathbf{T}, \delta]) \ge (1 - \frac{64\delta}{1-\varepsilon'} - \frac{4\eta}{\delta} - 4\varepsilon)\mathrm{OPT}$;

(C2) for each node $\mathbf{v} \in \mathcal{N}(\mathbf{T})$: there exists an ε-pricing profile $\mathcal{R}(\mathbf{v}) := (I = I(\mathbf{v}), F = F(\mathbf{v}), \mathcal{H} = \mathcal{H}(\mathbf{v}), \mathcal{E} = \mathcal{E}(\mathbf{v}), p)$ consistent with $\tilde{p}$, and a vector of $k \le \ell - 1$ $\varepsilon\delta$-capacity profiles $\pi(\mathbf{v}) := (\pi_1, \ldots, \pi_k)$, with peaks at the edges in $\mathcal{E}(\mathbf{v})$, such that (i) $W'(\mathbf{v}) := \mathcal{M}(\mathcal{R}(\mathbf{v}))$; and (ii) $W''(\mathbf{v}) := \mathcal{M}(\mathcal{R}(\mathbf{v}), \pi(\mathbf{v}))$;

(C3) for each node $\mathbf{v} \in \mathcal{N}(\mathbf{T})$, there exist ancestors $\mathbf{v}^L, \mathbf{v}^R, \mathbf{u}_1, \ldots, \mathbf{u}_d$ of $\mathbf{v}$, where $0 \le d \le 4\log_2 \frac{H}{(1-\varepsilon')\varepsilon_a} + 8$ and possibly $\mathbf{v}^L = \mathbf{v}^R = \mathbf{v}$, s.t. the following holds: (i) $j \notin W$ for all j such that $(I_j \not\subseteq I(\mathbf{v}^L)$ and $I_j \supseteq [\mathrm{lt}(I(\mathbf{v})), \mathrm{lt}(I(\mathbf{v})) + 1])$ or $(I_j \not\subseteq I(\mathbf{v}^R)$ and $I_j \supseteq [\mathrm{rt}(I(\mathbf{v})) - 1, \mathrm{rt}(I(\mathbf{v}))])$; and (ii) let $\mathcal{J}_1(\mathbf{v})$ be the set of winners $j \in W(\mathbf{v}^L) \cup W(\mathbf{v}^R) \cup \bigcup_{i=1}^{d} W(\mathbf{u}_i)$ s..t. I_j overlaps

with $I(\mathbf{v})$ *(that is,* $|I_j \cap I(\mathbf{v})| \geq 2$*), and let* $\mathcal{J}_2(\mathbf{v})$ *be the set of customers* j *that are not split at any of the nodes* $\mathbf{v}^L, \mathbf{v}^R, \mathbf{u}_1, \ldots, \mathbf{u}_d$*, s.t.* $(I_j \subseteq I(\mathbf{v}^L)$ *and* $I_j \supseteq [\mathrm{lt}(I(\mathbf{v})), \mathrm{lt}(I(\mathbf{v})) + 1])$ *or* $(I_j \subseteq I(\mathbf{v}^R)$ *and* $I_j \supseteq [\mathrm{rt}(I(\mathbf{v})) - 1, \mathrm{rt}(I(\mathbf{v}))])$*, then* $\bigcup_{\mathbf{u} \ ancestor \ of \ \mathbf{v}} \mathcal{W}(\mathbf{u}) \subseteq \mathcal{J}_1(\mathbf{v}) \cup \mathcal{J}_2(\mathbf{v})$ *and* $\mathrm{prof}(\mathcal{J}_1(\mathbf{v}) \cup \mathcal{J}_2(\mathbf{v}) \cup \mathcal{W}[\mathbf{T}(\mathbf{v})]) \mid_{I(\mathbf{v})} \leq c \mid_{I(\mathbf{v})}$.

4.3 Dynamic Program for the Limited Supply Case

Motivated by Lemma 3, we introduce the following definition.

Definition 7 (ε-**profile vector**). *Given* $\varepsilon > 0$*, an* ε-*profile vector is a tuple* $\lambda = (\bar{d}, \{\mathcal{R}(r)\}_{r \in [\bar{d}]}, \{\pi(r)\}_{r \in [\bar{d}]})$*, where* $\bar{d} \leq 4 \log_2 \frac{H}{(1-\varepsilon')\varepsilon_a} + 11$ *is a positive integer,* $\mathcal{R}(r) = (I^r, F^r, \mathcal{H}^r, \mathcal{E}^r, p^r)$ *is an* ε-*pricing profile, and* $\pi(r) = (\pi_1(r), \ldots, \pi_{k_r}(r))$ *is a vector of* $k_r \leq \ell - 1$ $\varepsilon\delta$-*capacity profiles with peaks in* $\mathcal{E}^r$.

In the above definition, $\bar{d}-1$ corresponds to the number of ancestors $\mathbf{v}^L, \mathbf{v}^R, \mathbf{u}_1, \ldots, \mathbf{u}_d$ of a given node $\mathbf{v}$ in $\mathbf{T}$, as specified in condition (C3) of Lemma 3; $\mathcal{R}(r)$ and $\pi(r)$, for $r \in \{2, \ldots, \bar{d}\}$, represent respectively the pricing and capacity profile vectors at these ancestors, with $h = 2, 3$ representing the profiles at $\mathbf{v}^L, \mathbf{v}^R$, respectively; and $\mathcal{R}(1)$ and $\pi(1)$ correspond to the profiles at $\mathbf{v}$ itself. With this interpretation in mind, we define the *residual capacity requirement* of λ as $c_\lambda := \sum_{r=2}^{\bar{d}}(\sum_{j=1}^{k_r} \pi_j(r) + \pi_p(\mathcal{R}(r))) + \mathrm{prof}(\mathcal{J}_2(\mathbf{v}))$, where $\mathcal{J}_2(\mathbf{v})$ is the set of customers defined in (C3) (which are completely determined by $\mathcal{R}(r)$ and $\pi(r)$, for $h = 2, \ldots, \bar{d}$).

We build a DP to search for a pricing $\tilde{p}$ and the corresponding decomposition and set of winners given by Lemma 3. First we guess (which means iterate over all possible values) $1 \leq L := L_{\tilde{p}} \leq P'$. As before, we let $D_L := 2^L$, and $D_{l-1} := \frac{D_l}{2}$, for $1 \leq l \leq L$, and $D_{-1} := 0$. Given an integer $-1 \leq l \leq L$ and an ε-pricing profile $\mathcal{R} = (I, F, \mathcal{H}, \mathcal{E}, p)$, we say that $\mathcal{R}$ is *l-valid* if $p(I) \leq D_l$; we say that λ *l-valid* if $\mathcal{R}(1)$ is *l-valid* and $(c_\lambda + \mathrm{prof}(\mathcal{M}(\mathcal{R}(1)) \cup \mathcal{M}(\mathcal{R}(1), \pi(1)))) \mid_{I(\mathbf{v})} \leq c \mid_{I(\mathbf{v})}$.

For an *l-valid* ε-profile vector $\lambda = (\bar{d}, \{\mathcal{R}(r)\}_{r \in [\bar{d}]}, \{\pi(r)\}_{r \in [\bar{d}]})$, define $\mathrm{DP}_l[\lambda] := \max_{\mathcal{J} \subseteq \{j: I_j \subseteq I\}} \mathrm{val}_{p'}(\mathcal{J})$ to be the maximum (approximate) profit obtainable, with an ε'-envy pricing $p' : I \rightarrow \mathbb{R}_+$, consistent with $\mathcal{R}(1) = (I, F, \mathcal{H}, \mathcal{E}, p)$, from the instance with intervals contained completely inside I, when

(i) $\mathcal{R}(1)$ is an (*l-valid*) ε-pricing profile associated with some node $\mathbf{v}$ in $\mathbf{T}$, and $\pi(1) = \pi(\mathbf{v})$, as defined by (C2) ;
(ii) the restriction of the capacity profile on $I(\mathbf{v})$, required by all winners at ancestors of $\mathbf{v}$, is bounded from above by $c_\lambda \mid_I$: $\mathrm{prof}(\bigcup_{\mathbf{u} \ ancestor \ of \ \mathbf{v}} \mathcal{W}(\mathbf{u})) \mid_{I(\mathbf{v})} \leq c_\lambda \mid_{I(\mathbf{v})}$.

Let us further denote by $\mathcal{P}_l[\lambda]$ the set of all $k \leq 8$ tuples of ε-profile vectors, consistent with λ. Then we have the following recurrence, which can be used in a bottom-up fashion to compute $\mathrm{DP}_L[\cdot]$: $\mathrm{DP}_l[\lambda] := \max_{(\lambda_1, \ldots, \lambda_k) \in \mathcal{P}_l[\lambda]} \sum_{i=1}^{k} \mathrm{DP}_{l-1}[\lambda_i] + \mathrm{val}_p(\mathrm{WS}^\lambda)$, where $\mathrm{WS}^\lambda := \mathcal{M}(\mathcal{R}(1)) \cup \mathcal{M}(\mathcal{R}(1), \pi(1))$ is the set of winners computed w.r.t. the profile $\lambda = (\bar{d}, \{\mathcal{R}(r)\}_{r \in [\bar{d}]}, \{\pi(r)\}_{r \in [\bar{d}]})$.

Finally, the maximum of $\mathrm{DP}_L[\lambda]$, over all ε-profile vectors λ with $\bar{d} = 1$, and the corresponding pricing scaled by $(1 - 2\varepsilon_0^2)$ can be found by following the optimal choices in the table. The final pricing p' is obtained by modifying the pricing $\tilde{p}$, according to Lemma 2.

Theorem 2. *There is a QPTAS, for* HP *with limited supply, that runs in time* $\left(\frac{nm}{\varepsilon}\right)^{O\left(\frac{\log^4 H}{\varepsilon^3}\right)}$.

References

[AH06] Aggarwal, G., Hartline, J.D.: Knapsack auctions. In: SODA, pp. 1083–1092 (2006)

[BB06] Balcan, M.F., Blum, A.: Approximation algorithms and online mechanisms for item pricing. In: EC, New York, NY, USA, pp. 29–35 (2006)

[BCES06] Bansal, N., Chakrabarti, A., Epstein, A., Schieber, B.: A quasi-PTAS for unsplittable flow on line graphs. In: STOC, pp. 721–729 (2006)

[BK06] Briest, P., Krysta, P.: Single-minded unlimited supply pricing on sparse instances. In: SODA, pp. 1093–1102 (2006)

[BK07] Briest, P., Krysta, P.: Buying cheap is expensive: Hardness of non-parametric multi-product pricing. In: SODA, pp. 716–725 (2007)

[Bri08] Briest, P.: Uniform Budgets and the Envy-Free Pricing Problem. In: Aceto, L., Damgård, I., Goldberg, L.A., Halldórsson, M.M., Ingólfsdóttir, A., Walukiewicz, I. (eds.) ICALP 2008, Part I. LNCS, vol. 5125, pp. 808–819. Springer, Heidelberg (2008)

[CS08] Cheung, M., Swamy, C.: Approximation algorithms for single-minded envy-free profit-maximization problems with limited supply. In: FOCS, pp. 35–44 (2008)

[DFHS06] Demaine, E.D., Feige, U., Hajiaghayi, M.T., Salavatipour, M.R.: Combination can be hard: approximability of the unique coverage problem. In: SODA, pp. 162–171 (2006)

[ERRS09] Elbassioni, K., Raman, R., Ray, S., Sitters, R.: On Profit-Maximizing Pricing for the Highway and Tollbooth Problems. In: Mavronicolas, M., Papadopoulou, V.G. (eds.) SAGT 2009. LNCS, vol. 5814, pp. 275–286. Springer, Heidelberg (2009)

[ESZ07] Elbassioni, K., Sitters, R., Zhang, Y.: A Quasi-PTAS for Profit-Maximizing Pricing on Line Graphs. In: Arge, L., Hoffmann, M., Welzl, E. (eds.) ESA 2007. LNCS, vol. 4698, pp. 451–462. Springer, Heidelberg (2007)

[GHK+05] Guruswami, V., Hartline, J.D., Karlin, A.R., Kempe, D., Kenyon, C., McSherry, F.: On profit-maximizing envy-free pricing. In: SODA, pp. 1164–1173 (2005)

[GHK+06] Goldberg, A.V., Hartline, J.D., Karlin, A.R., Saks, M., Wright, A.: Competitive auctions. Games and Economic Behavior 55(2), 242–269 (2006)

[GR11] Grandoni, F., Rothvoss, T.: Pricing on paths: A ptas for the highway problem. In: SODA, pp. 675–684 (2011)

[GS10] Gamzu, I., Segev, D.: A Sublogarithmic Approximation for Highway and Tollbooth Pricing. In: Abramsky, S., Gavoille, C., Kirchner, C., Meyer auf der Heide, F., Spirakis, P.G. (eds.) ICALP 2010, Part I. LNCS, vol. 6198, pp. 582–593. Springer, Heidelberg (2010)

[GvLSU06] Grigoriev, A., van Loon, J., Sitters, R., Uetz, M.: How to Sell a Graph: Guidelines for Graph Retailers. In: Fomin, F.V. (ed.) WG 2006. LNCS, vol. 4271, pp. 125–136. Springer, Heidelberg (2006)

[GVR06] Glynn, P.W., Van Roy, B., Rusmevichientong, P.: A nonparametric approach to multi-product pricing. Operations Research 54(1), 82–98 (2006)

[Tal04] Talwar, K.: Bypassing the embedding: algorithms for low dimensional metrics. In: STOC, pp. 281–290 (2004)

Minimizing Rosenthal Potential in Multicast Games*

Fedor V. Fomin[1], Petr Golovach[2], Jesper Nederlof[3], and Michał Pilipczuk[1]

[1] Department of Informatics, University of Bergen, Bergen, Norway
{fomin,michal.pilipczuk}@ii.uib.no
[2] School of Engineering and Computing Science, Durham University, Durham, UK
petr.golovach@durham.ac.uk
[3] Utrecht University, Utrecht, the Netherlands
j.nederlof@uu.nl

Abstract. A multicast game is a network design game modelling how selfish non-cooperative agents build and maintain one-to-many network communication. There is a special source node and a collection of agents located at corresponding terminals. Each agent is interested in selecting a route from the special source to its terminal minimizing the cost. The mutual influence of the agents is determined by a cost sharing mechanism, which evenly splits the cost of an edge among all the agents using it for routing. The existence of a Nash equilibrium for the game was previously established by the means of Rosenthal potential. Anshelevich et al. [FOCS 2004, SICOMP 2008] introduced a measure of quality of the best Nash equilibrium, the price of stability, as the ratio of its cost to the optimum network cost. While Rosenthal potential is a reasonable measure of the quality of Nash equilibra, finding a Nash equilibrium minimizing this potential is NP-hard.

In this paper we provide several algorithmic and complexity results on finding a Nash equilibrium minimizing the value of Rosenthal potential. Let n be the number of agents and G be the communication network. We show that

- For a given strategy profile s and integer $k \geq 1$, there is a local search algorithm which in time $n^{O(k)} \cdot |G|^{O(1)}$ finds a better strategy profile, if there is any, in a k-exchange neighbourhood of s. In other words, the algorithm decides if Rosenthal potential can be decreased by changing strategies of at most k agents;
- The running time of our local search algorithm is essentially tight: unless $FPT = W[1]$, for any function $f(k)$, searching of the k-neighbourhood cannot be done in time $f(k) \cdot |G|^{O(1)}$.

The key ingredient of our algorithmic result is a subroutine that finds an equilibrium with minimum potential in $3^n \cdot |G|^{O(1)}$ time. In other words, finding an equilibrium with minimum potential is fixed-parameter tractable when parameterized by the number of agents.

* This work is supported by EPSRC (EP/G043434/1), Royal Society (JP100692), European Research Council (ERC) grant "Rigorous Theory of Preprocessing", reference 267959, and Nederlandse Organisatie voor Wetenschappelijk Onderzoek (NWO), project: 'Space and Time Efficient Structural Improvements of Dynamic Programming Algorithms'.

A. Czumaj et al. (Eds.): ICALP 2012, Part II, LNCS 7392, pp. 525–536, 2012.
© Springer-Verlag Berlin Heidelberg 2012

1   Introduction

Modern networks are often designed and used by non-cooperative individuals with diverse objectives. A considerable part of Algorithmic Game Theory focuses on optimization in such networks with selfish users [2,6,9,13,14,16,21,22].

In this paper we study the conceptually simple but mathematically rich cost-sharing model introduced by Anshelevich et al. [3,4], see also [15, Chapter 12]. In a variant of the cost-sharing game, which was called by Chekuri et al. the *multicast game* [5], the network is represented by a weighted directed graph with a distinguished source node r, and a collection of n agents located at corresponding terminals. Each agent is trying to select a cheapest route from r to its terminal. The mutual influence of the players is determined by a cost sharing mechanism identifying how the cost of each edge in the network is shared among the agents using this edge. When h agents use an edge e of cost c_e, each of them has to pay c_e/h. This is a very natural cost sharing formula which is also the outcome of the Shapley value.

The multicast game belongs to the widely studied class of congestion games. This class of games was defined by Rosenthal [20], who also proved that every congestion game has a Nash equilibrium. Rosenthal showed that for every congestion game it is possible to define a potential function which decreases if a player improves its selfish cost. Best-response dynamics in these games always lead to a set of paths that forms a Nash equilibrium. Furthermore, every local minimum of Rosenthal potential corresponds to a Nash equilibrium and vice versa. However, while we know that the multicast game always has a Nash equilibrium, the number of iterations in best-response dynamics achieving an equilibrium can be exponential (see [3, Theorem 5.1]), and it is an important open question if any Nash equilibrium can be found in polynomial time. The next step in the study of Rosenthal potential was done by Anshelevich et al. [3], who showed that Rosenthal potential can be used not only for proving the existence of a Nash equilibrium but also to estimate the quality of equilibrium. Anshelevich et al. defined the price of stability, as the ratio of the best Nash equilibrium cost and the optimum network cost, the *social optimum*. In particular, the cost of a Nash equilibrium minimizing Rosenthal potential is within $\log n$-factor of the social optimum, and thus the global minimum of the potential brings to a cheap equilibrium. The computational complexity of finding a Nash equilibrium achieving the bound of $\log n$ relative to the social optimum is still open, while computing the minimum of the Rosenthal potential is NP-hard [3,5].

Our results. In this paper we analyze the following local search problem. Given a strategy profile s, we are interested if a profile with a smaller value of Rosenthal potential can be found in a k-exchange neighbourhood of s, which is the set of all profiles that can be obtained from s by changing strategies of at most k players. Our motivation to study this problem is two-fold.

- If we succeed in finding some Nash equilibrium, say by implementing best-response dynamics, which is still far from the social optimum, it is

an important question if the already found equilibrium can be used to find a
better one efficiently. Local search heuristic in this case is a natural approach.
- Since the number of iterations in best-response dynamics scenario can be
  exponential (see [3, Theorem 5.1]), it can be useful to combine the best-
  response dynamics with a heuristic that at some moments tries to make
  "larger jumps", i.e., instead of decreasing Rosenthal potential by changing
  strategy of one player, to decrease the potential by changing in one step
  strategies of several players.

Let us remark that the number of paths, and thus strategies, every player can
select from, is exponential, so the size of the search space also can be exponential. Since the size of k-exchange neighbourhood is exponential, it is not clear
a priori, if searching of a smaller value of Rosenthal potential in a k-exchange
neighbourhood of a given strategy profile can be done in polynomial time. We
show that for a fixed k, the local search can be performed in polynomial time.
The running time of our algorithm is $n^{O(k)} \cdot |G|^{O(1)}$, where n is the total number
of players[1]. As a subroutine, our algorithm uses a fixed-parameter algorithm
computing the minimum of Rosenthal potential in time $3^n \cdot |G|^{O(1)}$. We find this
auxiliary algorithm to be interesting in its own. It is known that for a number
of local search algorithms, exploration of the k-exchange neighbourhood can be
done by fixed-parameter tractable (in k) algorithms [10,17,23]. We show that,
unfortunately, this is not the case for the local search of Rosenthal potential
minimum. We use tools from Parameterized Complexity, to show that the running time of our local search algorithm is essentially tight: unless $FPT = W[1]$,
searching of the k-neighbourhood cannot be done in time $f(k) \cdot |G|^{O(1)}$ for any
function $f(k)$.

2 Preliminaries

Multicast game and Rosenthal potential. A network is modeled by a directed
$G = (V, E)$ graph. There is a special *root* or *source* node $r \in V$. There are n
multicast users, *players*, and each player has a specified *terminal* node t_i (several
players can have the same terminals). A strategy s^i for player i is a path P_i from
r to t_i in G. We denote by Π the set of players and by S^i the finite set of
strategies of player i, which is the set of all paths from r to t_i. The joint strategy
space $S = S^1 \times S^2 \times \cdots \times S^n$ is the Cartesian product of all the possible strategy
profiles. At any given moment, a strategy profile (or a configuration) of the game
$s \in S$ is the vector of all the strategies of the players, $s = (s^1, \ldots, s^n)$. Notice
that for a given strategy profile s, several players can use paths that go through
the same edge. For each edge $e \in E$ and a positive integer h, we have a cost
$c_e(h) \in \mathbb{R}$ of the edge e for each player who uses a path containing e, provided
that exactly h players share e. With each player i, we associate the cost function
c^i mapping a strategy profile $s \in S$ to real numbers, i.e., $c^i : S \to \mathbb{R}$. For a

---

[1] The number of arithmetic operations used by our algorithms does not depend on the
size of the input weights, i.e. the claimed running times are in the unit-cost model.

strategy profile $s \in S$, let $n_e(s)$ be the number of players using the edge e in s. Then the cost the i-th player has to pay is

$$c^i(s) = \sum_{e \in E(P_i)} c_e(n_e(s)),$$

and the total cost of s is

$$c(s) = \sum_{i=1}^{n} c^i(s).$$

The *potential* of a strategy profile $s \in S$, or equivalently, the set of paths $(P_1, \ldots, P_n)$, is

$$\Phi(s) = \sum_{e \in \cup_{i=1}^{n} E(P_i)} \sum_{h=1}^{n_e(s)} c_e(h). \tag{1}$$

In this paper, we are especially interested in the case where the cost of every edge is split evenly between the players sharing it, i.e, the payment of player i for edge e is $c_e(h) = \frac{c_e}{h}$ for $c_e \in \mathbb{R}$. Respectively, *Rosenthal potential* of a strategy profile $s \in S$ is

$$\Phi(s) = \sum_{e \in \cup_{i=1}^{n} E(P_i)} c_e \cdot \mathcal{H}(n_e(s)),$$

where $\mathcal{H}(h) = 1 + 1/2 + 1/3 + \cdots + 1/h$ is the h-th Harmonic number.

For a strategy profile $s \in S$ and $i \in \{1, 2, \ldots, n\}$, we denote by s^{-i} the strategy profile of the players $j \neq i$, i.e. $s^{-i} = (s^1, \ldots, s^{i-1}, s^{i+1}, \ldots s^n)$. We use $(s^{-i}, \bar{s}^i)$ to denote the strategy profile identical to s, except that the ith player uses strategy $\bar{s}^i$ instead of s^i. Similarly, for a subset of players Π_0, we define $s^{-\Pi_0}$, the profile of players $j \notin \Pi_0$. For $\sigma \in \times_{i \in \Pi_0} S^i$, we denote by $(s^{-\Pi_0}, \sigma)$ the strategy profile obtained from s by changing the strategies of players in Π_0 to σ.

A strategy profile $s \in S$ is a *Nash equilibrium* if no player $i \in \Pi$ can benefit from unilaterally deviating from his action to another action, i.e.,

$$\forall i \in \Pi \text{ and } \forall \bar{s}^i \in S^i, \ c^i(s^{-i}, \bar{s}^i) \geq c^i(s).$$

The crucial property of Rosenthal potential Φ is that each step performed by a player improving his payoff also decreases Φ. Consequently, if Φ admits a minimal value in strategy profile, this strategy profile is a Nash equilibrium.

Parameterized complexity. We briefly review the relevant concepts of parameterized complexity theory that we employ. For deeper background on the subject see the books by Downey and Fellows [7], Flum and Grohe [12], and Niedermeier [19].

In the classical framework of P *vs* NP, there is only one measurement (the overall input size) that frames the distinction between efficient and inefficient algorithms, and between tractable and intractable problems. Parameterized complexity is essentially a two-dimensional sequel, where in addition to the overall input size n, a secondary measurement k (the *parameter*) is introduced, with the aim of capturing the contributions to problem complexity due to such things as typical input structure, sizes of solutions, goodness of approximation, etc. Here, the parameter is deployed as a measurement of the amount of current solution modification allowed in a local search step. The parameter can also represent an aggregrate of such bounds.

The central concept in parameterized complexity theory is the concept of *fixed-parameter tractability* (FPT), that is solvability of the parameterized problem in time $f(k) \cdot n^{O(1)}$. The importance is that such a running time isolates all the exponential costs to a function of only the parameter.

The main hierarchy of parameterized complexity classes is

$$FPT \subseteq W[1] \subseteq W[2] \subseteq \cdots \subseteq W[P] \subseteq XP.$$

The formal definition of classes $W[t]$ is technical, and, in fact, irrelevant to the scope of this paper. For our purposes it suffices to say that a problem is in a class if it is FPT-reducible to a complete problem in this class. Given two parameterized problems Π and Π', an *FPT reduction* from Π to Π' maps an instance (I, k) of Π to an instance (I', k') of Π' such that

(1) $k' = h(k)$ for some computable function h,
(2) (I, k) is a YES-instance of Π if and only if (I', k') is a YES-instance of Π', and
(3) the mapping can be computed in FPT time.

Hundreds of natural problems are known to be complete for the aforementioned classes, and $W[1]$ is considered the parameterized analog of NP, because the k-STEP HALTING PROBLEM for nondeterministic Turing machines of unlimited nondeterminism (trivially solvable by brute force in time $O(n^k)$) is complete for $W[1]$. Thus, the statement $FPT \neq W[1]$ serves as a plausible complexity assumption for proving intractability results in parameterized complexity. INDEPENDENT SET, parameterized by solution size, is a more combinatorial example of a problem complete for $W[1]$. We refer the interested reader to the books by Downey and Fellows [7] or Flum and Grohe [12] for a more detailed introduction to the hierarchy of parameterized problems.

Local Search. Local search algorithms are among the most common heuristics used to solve computationally hard optimization problems. The common method of local search algorithms is to move from solution to solution by applying local changes. Books [1,18] provide a nice introduction to the wide area of local search. Best-response dynamics in congestion games corresponds to local search in 1-exchange neighbourhood minimizing Rosenthal potential Φ; improving moves for players decrease the value of the potential function. For strategy profiles $s_1, s_2 \in S$, we define the Hamming distance $D(s_1, s_2) = |s_1 \triangle s_2|$ between s_1 and s_2, that is the number of players implementing different strategies in s_1 and s_2. We study the following parameterized version of the local search problem for multicast.

We define *arena* as a directed graph G with root vertex r, a multiset of target vertices $t_1, \ldots, t_\ell$ and for every edge e of the graph a cost function $c_e : \mathbb{Z}^+ \to \mathbb{R}^+ \cup \{0\}$ such that $c_e(h) \geq c(h+1)$ for $h \geq 1$.

---

p-LOCAL SEARCH ON POTENTIAL Φ **Parameter: k**
Input: An arena consisting of graph G, vertices $r, (t_1, \ldots, t_\ell)$ and cost functions c_e, a strategy profile s, and an integer $k \geq 0$
Problem: Decide whether there is a strategy profile s' such that $\Phi(s') < \Phi(s)$ and $D(s, s') \leq k$, where Φ is as defined in (1).

---

3   Minimizing Rosenthal Potential

The aim of this section is to prove the following theorem.

Theorem 1. *The p-LOCAL SEARCH ON POTENTIAL Φ problem is solvable in time*

$$\binom{|\Pi|}{k} \cdot 3^k \cdot |G|^{O(1)}.$$

Let us remark that in particular, if Φ is Rosenthal's potential, and hence the cost functions are of the special type $c_e(h) = \frac{c_e}{h}$, the p-LOCAL SEARCH ON POTENTIAL Φ problems can be solved within the running time of Theorem 1.

We need some additional terminology. Let G be a directed graph. We say that a subdigraph T of G is an *out-tree* if T is a directed tree with only one vertex r of in-degree zero (called the *root*). The vertices of T of out-degree zero are called *leaves*. We also say that a strategy profile s^* is *optimal* if it gives the minimum value of the potential, i.e., for any other strategy profile s, $\Phi(s) \geq \Phi(s^*)$. Let $s = (P_1, \ldots, P_{|\Pi|})$ be a strategy profile and $C \geq 1$ be an integer. We say that s *uses C arcs* if the union T of the paths P_i consists of C arcs.

If edge-sharing is profitable, then we can make the following observation about the structure of optimal strategies.

Lemma 1. *Let C be an integer such that there is a strategy profile using at most C arcs. Let $s = (P_1, \ldots, P_{|\Pi|})$ be a strategy profile using at most C arcs such that*

(i) Among all profiles using at most C arcs, s is optimal. In other words, for any profile s' using at most C arcs, we have $\Phi(s') \geq \Phi(s)$.
(ii) Subject to (i), S uses the minimum number of arcs.

Then the union T of the paths P_i, $i \in \{1, \ldots, |\Pi|\}$, is an out-tree rooted in r.

Proof. Targeting towards a contradiction, let us assume that $T = \cup_{i=1}^{|\Pi|} P_i$ is not an out-tree. Then there are paths $P_i, P_j, i, j \in \{1, \ldots, |\Pi|\}$, that have a common vertex $v \neq r$ such that the (r, v)-subpaths P_i^v and P_j^v of P_i and P_j respectively enter v by different arcs.

We show first that

$$\sum_{e \in E(P_i^v)} c_e(n_e(s)) > \sum_{e \in E(P_j^v)} c_e(n_e(s)). \tag{2}$$

cannot occur. Assume that (2) holds. We claim that then the i-th player can improve his strategy and, consequently, Φ can be decreased, which will contradict the optimality of s. Denote by P the (r, t_i)-walk obtained from P_i by replacing path P_i^v by P_j^v. Notice that P is not necessarily a path. Let P' be a (r, t_i)-path in P and let us construct the new strategy profile $s' = (s^{-i}, P')$. This profile uses at most C arcs. By non-negativity of $c_e(h)$, the new cost for the i-th player is equal to

$$\sum_{e \in E(P')} c_e(n_e(s')) = \sum_{e \in E(P') \cap E(P_i)} c_e(n_e(s)) + \sum_{e \in E(P') \setminus E(P_i)} c_e(n_e(s) + 1) \le$$

$$\le \sum_{e \in E(P) \cap E(P_i)} c_e(n_e(s)) + \sum_{e \in E(P) \setminus E(P_i)} c_e(n_e(s) + 1).$$

Since for each $e \in E$ and $h \ge 1$, we have $c_e(h) \ge c_e(h+1)$,

$$\sum_{e \in E(P) \setminus E(P_i)} c_e(n_e(s) + 1) \le \sum_{e \in E(P) \setminus E(P_i)} c_e(n_e(s)).$$

Therefore,

$$\sum_{e \in E(P')} c_e(n_e(s')) \le \sum_{e \in E(P)} c_e(n_e(s)).$$

By (2), we have

$$\sum_{e \in E(P)} c_e(n_e(s)) < \sum_{e \in E(P_i)} c_e(n_e(s)),$$

and the claim that player i can improve follows.

Hence,

$$\sum_{e \in E(P_i^v)} c_e(n_e(s)) \le \sum_{e \in E(P_j^v)} c_e(n_e(s)).$$

By the same arguments as above, we can replace P_j by a (r, t_j)-path P in the walk obtained from P_j by the replacement of P_j^v by P_i^v without increasing Φ. Moreover, we can repeat this operation for each path P_h, $h \ne i$, that enters v by an arc that is different from the arc in P_i. The number of arcs used by the paths in the modified strategy profiles is at most the number of arcs used in s, and thus is at most C. It remains to observe that we obtain a strategy profile where v has in-degree one in the union of paths. But it contradicts the choice of s, since we obtain a strategy profile that uses less arcs. Hence, T is an out-tree rooted in r. □

We use Lemma 1 to find an optimal strategy profile using the approach proposed by Dreyfus and Wagner [8] for the STEINER TREE problem.

Theorem 2. *Given an arena as input, the minimum value of a potential Φ can be found in time $3^{|\Pi|} \cdot |G|^{O(1)}$. The algorithm can also construct the corresponding optimal strategy profile s^* within the same time complexity.*

Proof. We give a dynamic programming algorithm. For simplicity, we only describe how to find the minimum of Φ, but it is straightforward to modify the algorithm to obtain the corresponding strategy profile.

Let $T = \{t_1, \ldots, t_{|\Pi|}\}$ be the multiset of terminals. We construct partial solutions for subsets $X \subseteq T$. Also, while at the end we are interested in the answer for the source r, our partial solutions are constructed for all vertices of G. For a vertex $u \in V(G)$ and a multiset $X \subseteq T$, let Γ_u^X denote the version of the game, in which only players associated with X build paths from u to their respective terminals. Therefore, we are interested in the game Γ_r^T. For a non-negative integer m, we define $\Psi(u, X, m)$ as the minimum value of the potential $\Phi(s)$ in the game Γ_u^X, taken over all strategy profiles s such that the union of paths in s contains at most m arcs (we say that s *uses* arc e if it is contained in some path from s). We assume that $\Psi(u, X, m) = +\infty$ if there are no feasible strategy profiles. Notice that by Lemma 1, the number of arcs used in an optimal strategy in the original problem is at most $|V(G)| - 1$. Hence, our aim is to compute $\Psi(r, T, |V(G)| - 1)$.

Clearly, $\Psi(u, \emptyset, m) = 0$ for all $u \in V$ and $m \geq 0$. For non-empty X and $m = 0$, $\Psi(u, X, m) = 0$ if all terminals in X are equal to u, and $\Psi(u, X, m) = +\infty$ otherwise. We need the following claim.

Claim. For $X \neq \emptyset$ and $m \geq 1$, $\Psi(u, X, m)$ satisfies the following equation:

$$\Psi(u, X, m) = \min\{\Psi(u, X, m-1),$$

$$\Psi(u, X \setminus Y, m_1) + \Psi(v, Y, m_2) + \sum_{h=1}^{|Y|} c_{(u,v)}(h)\}, \tag{3}$$

where the minimum is taken over all arcs $(u, v) \in E(G)$, $\emptyset \neq Y \subseteq X$, and $m_1, m_2 \geq 0$ such that $m_1 + m_2 = m - 1$; it is assumed that $\Psi(u, X, m) = \Psi(u, X, m-1)$ if the out-degree of u is zero.

Proof. Let $\psi = \min\{\Psi(u, X, m-1), \Psi(u, X \setminus Y, m_1) + \Psi(v, Y, m_2) + \sum_{h=1}^{|Y|} c_{(u,v)}(h)\}$. We prove that $\Psi(u, X, m) = \psi$ by first showing that $\Psi(u, X, m) \geq \psi$, and then that $\Psi(u, X, m) \leq \psi$. Without loss of generality assume that $X = \{t_1, \ldots, t_\ell\} \subseteq T$, where $\ell = |X|$.

If $\Psi(u, X, m) = +\infty$, then $\Psi(u, X, m) \geq \psi$. Suppose that $\Psi(u, X, m) \neq +\infty$ and consider a strategy $s^* = (P_1, \ldots, P_\ell)$ in the game Γ_u^X which is optimal among those using at most m arcs and, subject to this condition, the number of used arcs is minimum; in particular, s^* has potential $\Psi(u, X, m)$. By Lemma 1, $H = \cup_{i=1}^{\ell} P_i$ is an out-tree rooted in u. If $|E(H)| < m$, then $\Psi(u, X, m) =$

$\Psi(u, X, m - 1) \geq \psi$. Assume that $|E(H)| = m$. As $m \geq 1$, vertex u has an out-neighbor v in H. Denote by H_1 and H_2 the components of $H - (u, v)$, where H_1 is an out-tree rooted in u and H_2 is an out-tree rooted in v. Let $Y \subseteq X$ be the multiset of terminals in H_2 and let $m_1 = |E(H_1)|$, $m_2 = |E(H_2)|$. Notice that exactly $|Y|$ players are using the arc (u, v) in s^* and Y is nonempty. Then $\Psi(u, X, m) \geq \Psi(u, X \setminus Y, m_1) + \Psi(v, Y, m_2) + \sum_{h=1}^{|Y|} c_{(u,v)}(h) \geq \psi$.

Now we prove that $\Psi(u, X, m) \leq \psi$. If $\psi = \Psi(u, X, m - 1)$ then the claim is trivial, so let v, Y, m_1 and m_2 be such that $\psi = \Psi(u, X \setminus Y, m_1) + \Psi(v, Y, m_2) + \sum_{h=1}^{|Y|} c_{(u,v)}(h)$. Assume without loss of generality that $Y = \{t_1, \ldots, t_{\ell'}\}$ for some $\ell' \leq \ell$. If $\Psi(u, X \setminus Y, m_1) = +\infty$ or $\Psi(v, Y, m_2) = +\infty$, then the inequality is trivial. Suppose that $\Psi(u, X \setminus Y, m_1) \neq +\infty$ and $\Psi(v, Y, m_2) \neq +\infty$. Consider a strategy s_1^* in the game $\Gamma_u^{X \setminus Y}$ that is optimal among those using at most m_1 arcs, and a strategy s_2^* in the game Γ_v^Y that is optimal among those using at most m_2 arcs. Of course, the potential of s_1^* is equal to $\Psi(u, X \setminus Y, m_1)$, while the potential of s_2^* is equal to $\Psi(u, Y, m_2)$. We construct the strategy profile s in the game Γ_u^X as follows. For each terminal $t_j \in X \setminus Y$, the players use the (u, t_j)-path from s_1^*. For any $t_j \in Y$, the players use the (v, t_j)-path from s_2^* after accessing v from u via the arc (u, v), unless u already lies on this (v, t_j)-path, in which case they simply use the corresponding subpath of the (v, t_j)-path. Note that s uses at most $m_1 + m_2 + 1 = m$ arcs. Because for every $e \in E(G)$ and every $h \geq 1$, we have that $c_e(h) \geq 0$, and $c_e(h) \geq c_e(h + 1)$, we infer that $\Phi(s) \leq \psi$, as possible overlapping of arcs used in s_1^*, s_2^* and the arc (u, v) can only decrease the potential of s. Since $\Psi(u, X, m) \leq \Phi(s)$, this implies that $\Psi(u, X, m) \leq \psi$. $\square$

In order to finish the proof of Theorem 2, we need to observe that using the recurrence (3) one can compute the value $\Psi(r, T, |V(G)| - 1)$ in time $3^{|\Pi|} \cdot |G|^{O(1)}$. $\square$

We use Theorem 2 to construct algorithm for p-LOCAL SEARCH ON POTENTIAL Φ and to conclude with the proof of Theorem 1.

Proof (of Theorem 1). Consider an instance of p-LOCAL SEARCH ON POTENTIAL Φ. Let $T = \{t_1, \ldots, t_{|\Pi|}\}$ be the multiset of terminals and let s be a strategy profile. Recall that p-LOCAL SEARCH ON POTENTIAL Φ asks whether at most k players can change their strategies in such a way that the potential decreases. Observe that we can assume that *exactly* k players are going to change their strategies because some of these players can choose their old strategies. There are $\binom{|\Pi|}{k}$ possibilities to choose a set of k players $\Pi_0 \subseteq \Pi$. We consider all possible choices and for each set Π_0, we check whether the players from this set can apply some strategy to decrease Φ.

Denote by $X \subseteq T$ the multiset of terminals of the players from Π_0, and let $s' = s^{-\Pi_0}$. We compute the potential $\Phi(s')$ for this strategy profile. Now we redefine the cost of edges as follows: for each $e \in E(G)$ and $h \geq 1$, $c_e'(h) = c_e(n_e(s') + h)$. Clearly, $c_e'(h) \geq 0$ and $c_e'(h) \geq c_e'(h + 1)$. Let Φ' be the potential for these edge costs. We find the minimum value of $\Phi'(s^*)$ for the set of players Π_0 and the corresponding terminals X. It remains to observe that $\Phi(s') + \Phi'(s^*) = \min\{\Phi(s'') \mid s'' = (s^{-\Pi_0}, \sigma), \sigma \in \prod_{i \in \Pi_0} S^i\}$. By Theorem 2, we can find $\Phi'(s^*)$ in time $3^k \cdot |G|^{O(1)}$ and the claim follows. $\square$

4 Intractability of Local Search for Rosenthal Potential

This section is devoted to the proof of the following theorem. Due to space restriction, the most technical part of the proof, i.e., the proof of the completeness of the reduction, will appear in the full version of the paper.

Theorem 3. p-LOCAL SEARCH ON POTENTIAL Φ, where Φ is Rosenthal potential for multicasting game, is $W[1]$-hard.

Proof. (Construction and Soundness) We provide an FPT reduction from the MULTICOLOURED CLIQUE problem, which is known to be $W[1]$-hard [11].

MULTICOLOURED CLIQUE   **Parameter:** k
Input: An undirected graph H with vertices partitioned into k sets $V_1, V_2, \ldots, V_k$, such that H does not contain edges connecting vertices from the same part V_i.
Problem: Is there a clique C in G of size k?

Observe that by the assumption on the structure of H, the clique C has to contain exactly one vertex from every part V_i.

We take an instance (H, k) of MULTICOLOURED CLIQUE and construct an instance $(G, s, k(k-1))$ of p-LOCAL SEARCH ON POTENTIAL Φ. First, we provide the construction of the new instance; then, we discuss its soundness and completeness. During the reduction we assume k to be large enough; for constant k we solve the instance (H, k) in polynomial time by a brute-force search and output a trivial YES or NO instance of p-LOCAL SEARCH ON POTENTIAL Φ.

Construction. First create the root vertex r. For every $u \in V_i$, we create k vertices: $\bar{u}$ and $u_1, \ldots, u_{i-1}, u_{i+1}, \ldots, u_k$. Denote $F_u = \{u_1, \ldots, u_{i-1}, u_{i+1}, \ldots, u_k\}$. We connect them in the following manner: we construct one arc $(r, \bar{u})$ with cost $R = k^2$, and for all $j \in \{1, 2, \ldots, i-1, i+1, \ldots, k\}$ we construct arc $(\bar{u}, u_j)$ with cost 0. With every vertex u_j for all $u \in V(H)$ we associate a player that builds a path from r to u_j. In the initial strategy profile s, each of $(k-1)|V(H)|$ players builds a path that leads to his vertex via the corresponding vertex $\bar{u}$. Observe that the potential of this strategy profile is equal to $|V(H)| \cdot R \cdot \mathcal{H}(k-1)$.

We now construct the part of the graph that is responsible for the choice of the clique. We create a *pseudo-root* r' and an arc (r, r') with cost

$$W = \frac{1}{\mathcal{H}(k(k-1))} \left(k \cdot R \cdot \mathcal{H}(k-1) - \frac{3}{2}\binom{k}{2} - \varepsilon \right),$$

where $\varepsilon = \frac{k-1}{k^5}$. Note that $W \geq 1$ for sufficiently large k. For every edge $uv \in E(H)$, where $u \in V_i$ and $v \in V_j$, $i \neq j$, we create a vertex x_{uv}, arc (r', x_{uv}) of cost 1, and arcs (x_{uv}, u_j), (x_{uv}, v_i) of cost 0. This concludes the construction.

Before we proceed with the formal proof of the theorem, let us give some intuition behind the construction. From a clique C in H, we can derive a common strategy of $k(k-1)$ players assigned to vertices from $\bigcup_{u \in V(C)} F_u$, who can agree

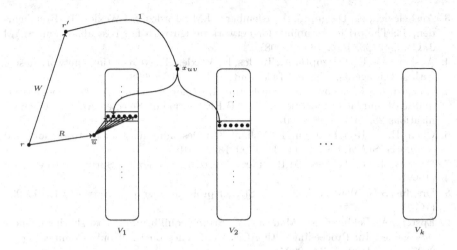

Fig. 1. Graph G

to jointly rebuild their paths via the pseudo-root r'. The "cost of entrance" for remodelling the strategy in this manner is paying for the expensive arc (r, r'); however, this can amortised by sharing cheap arcs (r', x_{uv}) for $uv \in E(C)$. The costs have been chosen so that only the maximum possibility of sharing, which corresponds to a clique in H, can yield a decrease of the potential.

Soundness. Assume that C is a clique in H with k vertices. Let us remind, that in the initial strategy profile s each player is using the corresponding arc $(r, \overline{u})$ for his path. We want to show that H contains a clique of size k. We construct the new strategy profile s' by changing strategies of $k(k-1)$ players as follows. For every $uv \in E(C)$, where $u \in V_i$ and $v \in V_j$, $i \neq j$, the players associated with vertices u_j and v_i reroute their paths so that in s' they lead via r' and x_{uv} to respective targets. In comparison to the profile s, the new profile s':

- has congestion withdrawn from arcs $(r, \overline{u})$ for $u \in V(C)$—this decreases the potential by $k \cdot R \cdot \mathcal{H}(k-1)$;
- has congestion introduced to arcs (r, r') and (r', x_{uv}) for $uv \in E(C)$—this increases the potential by $W \cdot \mathcal{H}(k(k-1)) + \frac{3}{2}\binom{k}{2}$.

Therefore, $\Phi(s') = \Phi(s) - k \cdot R \cdot \mathcal{H}(k-1) + W \cdot \mathcal{H}(k(k-1)) + \frac{3}{2}\binom{k}{2} = \Phi(s) - \varepsilon < \Phi(s)$.

The proof of the completeness of the reduction is deferred to the full version of the paper due to space constraints. $\square$

References

1. Aarts, E.H.L., Lenstra, J.K.: Local Search in Combinatorial Optimization. Princeton University Press (1997)
2. Albers, S.: On the value of coordination in network design. SIAM J. Comput. 38, 2273–2302 (2009)

3. Anshelevich, E., Dasgupta, A., Kleinberg, J.M., Tardos, É., Wexler, T., Roughgarden, T.: The price of stability for network design with fair cost allocation. SIAM J. Comput. 38, 1602–1623 (2008)
4. Anshelevich, E., Dasgupta, A., Tardos, É., Wexler, T.: Near-optimal network design with selfish agents. Theory of Computing 4, 77–109 (2008)
5. Chekuri, C., Chuzhoy, J., Lewin-Eytan, L., Naor, J., Orda, A.: Non-cooperative multicast and facility location games. IEEE Journal on Selected Areas in Communications 25, 1193–1206 (2007)
6. Chen, H.-L., Roughgarden, T., Valiant, G.: Designing network protocols for good equilibria. SIAM J. Comput. 39, 1799–1832 (2010)
7. Downey, R.G., Fellows, M.R.: Parameterized complexity. Springer, New York (1999)
8. Dreyfus, S.E., Wagner, R.A.: The Steiner problem in graphs. Networks 1, 195–207 (1972)
9. Epstein, A., Feldman, M., Mansour, Y.: Strong equilibrium in cost sharing connection games. In: Proceedings 8th ACM Conference on Electronic Commerce (EC 2007), pp. 84–92. ACM (2007)
10. Fellows, M., Fomin, F.V., Lokshtanov, D., Rosamond, F., Saurabh, S., Villanger, Y.: Local search: Is brute-force avoidable? In: Proceedings of the 21st International Joint Conference on Artificial Intelligence (IJCAI 2009), pp. 486–491. AAAI (2009)
11. Fellows, M.R., Hermelin, D., Rosamond, F.A., Vialette, S.: On the parameterized complexity of multiple-interval graph problems. Theor. Comput. Sci. 410, 53–61 (2009)
12. Flum, J., Grohe, M.: Parameterized Complexity Theory. Springer, Berlin (2006)
13. Gupta, A., Kumar, A., Pál, M., Roughgarden, T.: Approximation via cost sharing: Simpler and better approximation algorithms for network design. J. ACM 54, 11 (2007)
14. Gupta, A., Srinivasan, A., Tardos, É.: Cost-sharing mechanisms for network design. Algorithmica 50, 98–119 (2008)
15. Kleinberg, J., Tardos, E.: Algorithm design. Addison-Wesley, Boston (2005)
16. Koutsoupias, E., Papadimitriou, C.H.: Worst-case equilibria. Computer Science Review 3, 65–69 (2009)
17. Marx, D.: Local search. Parameterized Complexity Newsletter 3, 7–8 (2008)
18. Michiels, W., Aarts, E.H.L., Korst, J.: Theoretical Aspects of Local Search. Springer (2007)
19. Niedermeier, R.: Invitation to fixed-parameter algorithms. Oxford University Press (2006)
20. Rosenthal, R.W.: A class of games possessing pure-strategy Nash equilibria. Internat. J. Game Theory 2, 65–67 (1973)
21. Roughgarden, T., Sundararajan, M.: Quantifying inefficiency in cost-sharing mechanisms. J. ACM 56 (2009)
22. Roughgarden, T., Tardos, É.: How bad is selfish routing? J. ACM 49, 236–259 (2002)
23. Szeider, S.: The parameterized complexity of k-flip local search for sat and max sat. Discrete Optimization 8, 139–145 (2011)

Multiparty Proximity Testing with Dishonest Majority from Equality Testing*

Ran Gelles, Rafail Ostrovsky, and Kina Winoto

University of California, Los Angeles
{gelles,rafail}@cs.ucla.edu, kwinoto@ucla.edu

Abstract. Motivated by the recent widespread emergence of location-based services (LBS) over mobile devices, we explore efficient protocols for *proximity-testing*. Such protocols allow a group of friends to discover if they are all close to each other in some physical location, without revealing their individual locations to each other. We focus on hand-held devices and aim at protocols with very small communication complexity and a small constant number of rounds.

The proximity-testing problem can be reduced to the *private equality testing* (PET) problem, in which parties find out whether or not they hold the same input (drawn from a low-entropy distribution) without revealing any other information about their inputs to each other. While previous works analyze the 2-party PET special case (and its LBS application), in this work we consider highly-efficient schemes for the *multiparty* case with no honest majority. We provide schemes for both a direct-communication setting and a setting with a honest-but-curious mediating server that does not learn the users' inputs. Our most efficient scheme takes 2 rounds, where in each round each user sends only a couple of ElGamal ciphertexts.

1 Introduction

The ubiquity of mobile devices has led to the rise of *Location-Based Services* (LBS), services that depend on one's current location [1]. An interesting location-based service is *proximity testing*, in which users determine if they are in proximity to one another. This has plenty of possible applications. For instance, assume you and your friends visit a mall, identified by a GROUPON®-like company who sends you a group-coupon for, say, one of the mall's restaurants. While several products and applications based on proximity have recently appeared (such as magnetU®, etc.), a great deal of concern arises from privacy issues [17,13,29,14,26]. A recent work by Narayanan, Thiagarajan, Lakhani, Hamburg, and Boneh [20] provides a private scheme for proximity testing by reducing it to private equality testing (PET). The idea is to quantize one's approximate locations (say, GPS position) into centers of overlapping hexagonal cells, and then run several PET instances using the quantized locations (see discussion in [20]).

---

* Full version of the paper is available on-line [9].

A. Czumaj et al. (Eds.): ICALP 2012, Part II, LNCS 7392, pp. 537–548, 2012.
© Springer-Verlag Berlin Heidelberg 2012

In an equality testing scheme each of the users A and B holds a private input out of a set $\mathcal{D}$ of possible inputs (X_A and X_B respectively), and the aim is to jointly perform a private computation of the predicate $X_A \stackrel{?}{=} X_B$, without revealing the private inputs to each other. The PET task is a special (and more simple) case of private set intersection (which is, in turn, a special case of secure multiparty computation (MPC)), in which each user holds a set of private inputs, and the output is the intersection of all input sets (in some variants, the protocol only outputs the size of the intersection). A PET scheme can always be realized using a private set-intersection scheme if each user's set includes only his private input X_i. Since we focus on the special case of PET rather than the general set-intersection, we are able to achieve secure protocols with improved efficiency, specifically, less communication rounds.

Our Contributions. In this work we provide efficient, constant round protocols for n-party PET, secure against a coalition of up to $n - 1$ malicious adversaries.

We begin by extending the 2-party schemes of Narayanan et al. [20] and Lipmaa [18] to an arbitrary number of users, obtaining a 2-round n-party PET scheme that uses a trusted server, as well as a 2-round n-party PET scheme that does not rely on a mediating server.

We also consider security against *arbitrary* computationally-bounded adversaries. While it is relatively simple to design a scheme resistant to semi-honest adversaries, this is not the case with malicious adversaries. In addition to privacy issues, malicious adversaries sometimes jeopardize the soundness of the protocol. We show schemes that are secure against a malicious adversary (with abort). To the best of our knowledge, the schemes we present here are the first n-party PETs proven secure against malicious adversaries with no honest majority (excluding general-purpose MPC schemes).

Our Techniques. Our first n-party PET scheme (based on [20]) assumes an honest-but-curious server, with the additional requirement that the server does not learn the users' inputs. Each party sends its input to the server, masked with a random one-time pad (OTP), known to all users except the server. Using the linearity of OTP, and using randomness shared with some of the users, the server is able to compute a message which is fixed if all the parties share the same private input, and uniformly distributed otherwise.

Our second scheme (based on ideas from [18,8,20]) does not require a trusted server. The users replace the role of the server by "summing" the inputs themselves. In order to keep the inputs private yet still be able to manipulate them, we replace the OTP with homomorphic encryptions. We also add a layer of secret sharing [2] to prevent intermediate results from leaking information.

In order to obtain privacy against a malicious-adversary, it is common to use zero-knowledge proof-of-knowledge (ZKPoK) in which a prover convinces a verifier that he knows some secret, without leaking any information about the secret. When considering the multiparty case, the parties either conduct a 2-party ZKPoK sequentially, so the number of rounds becomes linear in the number of parties, or they perform multiple instances of a 2-party ZKPoK in parallel, which requires proving its security for parallel composition. To this end we use a variant which allows a single prover to perform a ZKPoK scheme with n verifiers, namely, a 1-Prover n-Verifier ZKPoK scheme. On top of being zero-knowledge, the prover must convince an honest verifier that he knows some secret, even if the rest of the verifiers collude with the prover to fool the honest verifier. This task can be performed in a constant number of rounds; in addition, it also reduces the communication—the prover broadcasts a single message to all the verifiers rather than engaging in n instances of the same protocol. We show how to construct a 1-Prover n-Verifier ZKPoK from a 2-party ZKPoK (a Σ-protocol) and a trapdoor commitment scheme. For concreteness, our PET scheme uses a 1-Prover n-Verifier ZKPoK for discrete log based on a Σ-Protocol by Schnorr [25] and a trapdoor-commitment based on Pedersen's commitment [24].

Our alternative PET scheme utilizes a different underlying primitive, namely, a password-authenticated key-exchange (PAKE) [15]. Specifically, we show how to realize a 2-round PET secure against malicious adversaries, using oracle access to an underlying PAKE subroutine. The PAKE is used to establish an encryption key, separately between each pair of users. If each party's private input is the same, they will end up sharing the same key (otherwise their keys will mismatch). The parties then use the keys to send random shares to each other, such that the sum of the shares is, say, 0. All these shares are then sent to Alice, who adds them up—the sum will be 0 if every pair of users used matching keys (otherwise, the sum will be random).

Related Work. 2-party PET was inspired by Yao's *socialist millionaire*'s problem [28], and examined for "real-life" problems by Fagin, Naor and Winkler [7]. The work of Boudot, Schoenmakers and Traoré [4] gives a rigorous security proof for this problem in the *random oracle model*. The work of Naor and Pinkas [19] draws a connection between PET and Oblivious Transfer (OT). This connection is extended by Lipmaa [18], providing private OT protocols based on homomorphic encryption, and showing how to realize an efficient (2-party) PET scheme using the same methods. Lipmaa also provides a security proof for PET in the semi-honest model. Tonicelli, Machado David and de Morais Alves [27] present an efficient UC-secure scheme for a 2-party PET.

Damgård, Fitzi, Kiltz, Nielsen and Toft [6] present an unconditionally secure scheme (in the secret sharing model) for multiparty PET with $O(1)$ rounds (this translates to 7 rounds in the plain model), however their result requires an honest majority. Nishide and Ohta [21] extend this result and achieve a 5-round scheme (in the plain model), again, assuming honest majority.

2 Preliminaries, Model and Privacy Definition

Let us begin by describing several notations and cryptographic primitives we use throughout. We let κ be the security parameter, and assume that any function implicitly depends on this parameter (especially when we write *neg* to describe a negligible function in κ, i.e., $neg < 1/poly(\kappa)$ for large enough κ). We say that two ensembles of distributions $\{X_\kappa\}_{\kappa \in \mathbb{N}}$, $\{Y_\kappa\}_{\kappa \in \mathbb{N}}$ are *computationally indistinguishable* and write $\{X_\kappa\} \overset{c}{\equiv} \{Y_\kappa\}$ if for any probabilistic polynomial-time (PPT) algorithm C, for large enough κ, $|\Pr[C(1^\kappa, X_\kappa) = 1] - \Pr[C(1^\kappa, Y_\kappa) = 1]| < neg$.

Semantically Secure Public-Key Encryption. An encryption scheme (GEN, ENC, DEC) is called *semantically secure* if for any PPT adversary Adv, and any two messages m_0, m_1 of equal length chosen by the adversary, and pk generated by $GEN(1^\kappa)$ it holds that $\Pr[\mathsf{Adv}(1^\kappa, pk, ENC_{pk}(m_b)) = b] < \frac{1}{2} + neg$, over the coin tosses of the algorithms and the choice of $b \in \{0, 1\}$. See [10] for a formal definition, as well as the analog definition for semantically secure *private-key* encryptions systems.

Homomorphic Encryption. Informally, we say that an encryption $ENC_k(m, r)$ that takes message m, randomness r and key k is *additively homomorphic* if $ENC_k(m_1, r_1) \cdot ENC_k(m_2, r_2) = ENC_k(m_1 + m_2, r_1 \circ r_2)$ for some deterministic binary operation $\circ$. See e.g. [22] for a formal definition. We focus on an homomorphic variant of the ElGamal scheme: let G be a cyclic group of order q in which the Decisional Diffie-Hellman (DDH) problem is difficult [3], and $g \in G$ a generator. A random number $x \in \mathbb{Z}_q^*$ is chosen as private key, and the public key is $(g, h) = (g, g^x)$. To encrypt a message m with the above public key, one randomly picks $r \in \mathbb{Z}_q$ and computes $ENC_{(g,h)}(m, r) = (g^r, h^r \cdot h^m)$. The decryption of a ciphertext $c = (a, b)$ is given by $DEC_x(c) = b/a^x = h^m$. It is easy to verify that this variant is additively homomorphic. Note that the decryption outputs h^m rather than m, however, for a low-entropy $\mathcal{D}$, m can be found by searching the entire domain.

Another homomorphic encryption system is the *Paillier Encryption system* [23], whose security relies on the decisional composite residuosity assumption. Although our schemes are specified using the ElGamal system, they can be implemented using variants of the Paillier encryption, or using any other semantically secure additive homomorphic encryption.

Private Equality Testing. We now define the ideal functionality of an asymmetric Private Equality Testing (PET) scheme. An n-party asymmetric-PET scheme is a distributed protocol between a specific party A (Alice) and another $n - 1$ parties, denoted $B_1, B_2, \ldots, B_{n-1}$. Each one of the users holds a private input $X_A, X_{B_1}, X_{B_2}, \ldots, X_{B_{n-1}}$, respectively. For ease of notation, we denote the private inputs as $\bar{X} = (X_A, X_1, X_2, \ldots, X_{n-1})$. At the end of the computation Alice obtains the value

$$\mathcal{F}_{ideal}(X_A, X_1, \ldots, X_{n-1}) = \begin{cases} 1 & \text{if } X_A = X_1 = \cdots = X_{n-1} \\ 0 & \text{otherwise} \end{cases}$$

and the rest of the parties obtain $\perp$. That is, the protocol is asymmetric: Alice learns whether or not all private inputs are equal, while the other parties learn no information at all. Henceforth, we refer to asymmetric-PET simply as PET.

We say that some realization π of a PET scheme is *complete* if it answers correctly for all positive instances, $\Pr[\pi(\bar{X}) = \mathcal{F}_{ideal}(\bar{X}) = 1] = 1$. Similarly, we say that a scheme is ε-*sound* if, for negative instances, the protocol answers affirmatively with probability at most ε, over the randomness of the parties, $\max_{\bar{X}} \Pr[\pi(\bar{X}) = 1 \wedge \mathcal{F}_{ideal}(\bar{X}) = 0] \leq \varepsilon$.

Adversaries. In this work we consider both semi-honest (also known as honest-but-curious) and malicious adversaries, using standard notions (for instance, see [10]). While in the *semi-honest adversary model* parties are assumed to follow the protocol, in the *malicious adversary model* the corrupted parties might deviate from the protocol as desired, as well as intercept and block messages from other users (excluding broadcast messages). The corruption model is static.

The trusted server is always honest (even in the malicious-adversary model) and never colludes with other parties. However, differently from the standard notion of trusted server we assume that the server is curious (passive adversary) and require that it cannot learn the honest users' inputs.

Privacy. The privacy model we consider here is a stand alone security against a coalition of up to $n - 1$ adversaries (this is also known as *secure function evaluation*). Informally, we would like the private inputs $\bar{X}$ to remain hidden, regardless of the adversary's actions. For instance, even if $n - 1$ parties collude together, they should have no advantage in finding the input of the nth player, other than just guessing it. We distinguish two types of attacks: *on-line* and *off-line* [12]. In an *on-line* attack, the adversary actively participates in a PET scheme, while in an off-line attack, the adversary tries to extract data from a PET instance's transcript, possibly by checking for inconsistencies between a given PET transcript and each one of the values in the (low-entropy) dictionary $\mathcal{D}$. Throughout this paper we assume $|\mathcal{D}|$ is polynomial in the security parameter.

To formalize the notion of privacy, we use the standard notion of *simulatability* in the *real/ideal paradigm*. Namely, we require that for any PPT adversary Adv running π with any set of honest parties, there exists a PPT simulator Sim such that for any private inputs, Sim's output and Adv's view are computationally indistinguishable, yet Sim only has access to the ideal functionality $\mathcal{F}_{ideal}$,[1] (rewindable) black-box access to Adv, and the trusted setup parameters of the protocol.

Assume $P_1, \ldots, P_m$ are the honest parties $m \leq n$, where each P_i has a unique role $\text{role}_i \in \{A, B_1, \ldots, B_n\}$ and a private input X_{P_i}. Denote by $View_{\bar{X}}(\text{Adv} \Leftrightarrow \pi)$ the view of the adversary running π with $P_1, \ldots, P_m$, with inputs $\bar{X}$. The view contains all the messages and randomness of the corrupt parties. Denote by $Output_{\bar{X}}(\text{Sim}^{\text{Adv}} \Leftrightarrow \mathcal{F}_{ideal})$ the output of a simulator which has rewindable black-box access to Adv, running $\mathcal{F}_{ideal}$ with the honest parties, with inputs $\bar{X}$. Note that both the adversary and the simulator are given the inputs of the corrupt parties $X_{P_{m+1}}, \ldots, X_{P_n}$, but are oblivious to the inputs of the honest parties $X_{P_1}, \ldots, X_{P_m}$.

---

[1] Note that due to the asymmetric nature of the protocol, Sim has access to $\mathcal{F}_{ideal}$ only when it simulates A.

Definition 1. *We say that an Equality Testing protocol π is **private** if for any set of $m \leq n$ honest parties $P_1, \ldots, P_m$ acting as $role_1, \ldots, role_m$, with unique $role_i \in \{A, B_1, \ldots, B_{n-1}\}$, and for any PPT Adv, there exists a PPT simulator Sim such that for any set of inputs $\bar{X}$*

$$View_{\bar{X}}(\text{Adv} \Leftrightarrow \pi) \overset{c}{\equiv} Output_{\bar{X}}(\text{Sim}^{\text{Adv}} \Leftrightarrow \mathcal{F}_{ideal}) \qquad \text{if } A \text{ is corrupt, and}$$

$$View_{\bar{X}}(\text{Adv} \Leftrightarrow \pi) \overset{c}{\equiv} Output_{\bar{X}}(\text{Sim}^{\text{Adv}}) \qquad\qquad \text{otherwise.}$$

Recall that all of our definitions use an implicit security parameter κ. Specifically, the above distributions are in fact the ensembles $\{View_{\bar{X}}(\text{Adv} \Leftrightarrow \pi)\}_\kappa$ etc., and all the PPT machines run in polynomial time in κ.

In order to achieve a private PET, some assumptions are necessary. The following claim suggests that a minimal requirement for the honest parties is to resist impersonation attacks.

Claim. Assuming messages can be blocked, a private PET scheme allows each user to identify the sender of a received message.

To see this, assume a private PET with $m \geq 2$ honest parties in which an honest party, (wlog) P_1, cannot identify the origin of a received message. Let Adv run the PET scheme with P_1, acting as the other $n-1$ users (blocking the messages of the other honest users). Since P_1 cannot validate the origin of his received messages, he proceeds using Adv's messages as the legitimate messages. Then, Adv is able to guess X_{P_1} via a trivial guessing attack: he picks a value $x \in \mathcal{D}$ and runs the PET scheme, setting all the inputs to x. If the PET scheme returns 1, Adv outputs x; otherwise Adv outputs a uniform guess from $\mathcal{D} - \{x\}$. This attack succeeds with probability at least $2/|\mathcal{D}|$ (assuming a uniform distribution on $\mathcal{D}$), regardless of the inputs of the other honest parties, which contradicts the privacy of the scheme.

3 n-Party PET with an Honest-But-Curious Server

We now provide a private equality testing protocol for an arbitrary number of users, assuming an honest-but-curious server. The main idea is that user A performs a generalization of the 2-party scheme [20] with each one of the users (B_1 to B_{n-1}), however, replies are sent to the server S, which incorporates them all into a single reply that is sent to A. Informally, if all private inputs are the same, then their sum is exactly n times the value of a single input. The server can add up the values, deduct n times Alice's value, and multiply the result with a random number. That way, the result is 0 if all users sent the same value, and random otherwise. One can see that this scheme cannot have perfect soundness, since it might be that the inputs sum up to the expected value, without being equal.

We assume a secure channel between each party and the trusted server. The scheme is given in Protocol 1.

Theorem 1. *Protocol 1 is complete, $1/p$-sound and private against a semi honest adversary and a semi honest server.*

Protocol 1. A private n-party PET with a trusted server S

(trusted) setup : Let p be a prime number, known to all parties (p depends on the security parameter). All the values below are over $\mathbb{Z}_p$.

Let k_A be a key known to all the users, excluding the server. In addition, for every $i \in \{1, \ldots, n-1\}$, A and B_i share $k_{A,i}$. Each B_i shares randomness $r_{S,i}$ with the server. Each user has a secure private channel with the server.

1. Alice sends $m_A = X_A + k_A$ to the server.
2. Each B_i sends $m_i = r_{S,i}(X_i + k_A) + k_{A,i}$ to the server.
3. The server computes $\mathsf{res} = \sum_i (r_{S,i} m_A - m_i)$, and sends res to party A.
4. Alice's output is the boolean predicate $\mathsf{res} \overset{?}{=} \sum_i -k_{A,i}$.

Theorem 2. *Protocol 1 is complete, $(\varepsilon + neg)$-sound and private against a malicious adversary and a semi-honest server, where ε is the probability of guessing X_A.*

Due to space limitation we defer the proofs to the full version of this paper [9].

4 Homomorphic PET Schemes

We now show multiparty PET schemes that do not use a trusted server. Generalizing 2-party PET scheme into the multiparty case is not straightforward: if Alice performs a 2-party PET separately with each one of the B_is, she can learn which of the B_is are located next to her, and which are not, thus this trivial extension is not private.

In order to achieve privacy in the multiparty case, each user[2] i performs a secret sharing of a random value s_i, which is used to mask the 2-party PET result between the user and Alice. The users sum all the shares they have received from all the other users and obtain a secret sharing of $\sum_i s_i$. Then, the users send all the shares to Alice, who can reconstruct $\sum_i s_i$. Since the value of any individual s_i remains secret, Alice has no use of the separate 2-party PET results, and she must "sum" them up in order to deduct the sum of the secret shares. This prevents Alice from learning any information about the users separately, and guarantees privacy.

We assume secure channels between any two users, and let G be a cyclic group with prime order q, in which DDH is hard. We let g be a fixed generator of G. The scheme is described in Protocol 2. Note that the protocol requires only 2 rounds of communication, where a single round means simultaneous mutual communication between all the parties.

Theorem 3. *Protocol 2 is private, complete and $O(1/q)$-sound against a semi-honest adversary.*

See detailed proof in the full version of this paper [9].

---

[2] We slightly abuse the notations here, so that $i, j \in \{A, B_1, \ldots, B_{n-1}\}$. We further abuse the notations and refer to a general user (except Alice) as B_i.

Protocol 2. A private n-party homomorphic PET against semi-honest coalitions

1. Each party $i = A, B_1, \ldots, B_{n-1}$, randomly picks $s_i \in \mathbb{Z}_q$ and computes an (n,n)-secret-sharing of s_i, obtaining $s_{iA}, s_{iB_1}, \ldots, s_{iB_{n-1}}$ such that $\sum_j s_{ij} = s_i$. The share s_{ij} is sent to user j over a secure channel.
2. Alice randomly picks (a private key) $x \in \mathbb{Z}_q^*$, and publishes her public key $(g, h) = (g, g^x)$. Alice encodes her input as h^{X_A} and broadcasts (or sends to each party over the secure channel) an ElGamal encryption of her encoded input, that is, she sends $(g^{r_A}, h^{X_A + r_A})$ with a random $r_A \in \mathbb{Z}_q$.
3. Each party $i = B_1, \ldots, B_{n-1}$ receives the message (a_A, b_A) and performs the following. The user randomly picks two numbers r_i, t_i and computes $a_i = a_A^{r_i} \cdot g^{t_i}$ and $b_i = b_A^{r_i} \cdot h^{t_i - r_i X_i + s_i}$.
4. Each user computes the sum of all the shares he has (including his own, s_{ii}), $\hat{s}_i = \sum_j s_{ji}$, and sends Alice the message $m_i = (a_i, b_i, \hat{s}_i)$ over the secure channel.
5. Alice receives a message $(u_1, u_2, u_3)_i$ from each of the other parties. She decrypts each message using her private key so that for every user i she gets $\mathsf{res}_i = (u_2)_i / (u_1)_i^x = h^{r_i(X_A - X_i) + s_i}$. Alice also computes $\sigma = \sum_i (u_3)_i + \hat{s}_A$.
6. Alice's output is the boolean predicate $h^\sigma \overset{?}{=} h^{s_A} \prod_i \mathsf{res}_i$.

4.1   Private Homomorphic PET against Malicious Adversaries

The above scheme is not private when the adversary is malicious, mainly due to the following two issues. First, a malicious A can send different messages to different parties (e.g., different values of X_A with different B_is); yet, this can easily be resolved if A uses a broadcast channel. The other issue is that a malicious A might ignore the input X_A, and the simulator might not be able to know the value X_A actually used. Moreover, instead of choosing a secret key x and publishing a public-key pair (g, g^x), the adversary might publish and use a public key for which he does not know the secret key, so there is no hope for the simulator to learn X_A from A's encrypted message (Step 2).

A standard technique to overcome this issue is to use a zero-knowledge proof of knowledge (ZKPoK) scheme, in which party A proves she knows the secret key x corresponding to the public key (g, g^x), yet without revealing any information about x to the other parties. This also allows the simulator to extract the secret x, which in turn lets it extract the value X_A from A's message.

Party A is required to convince all the B_is that she knows the secret x, which requires performing the ZKPoK scheme $n - 1$ times, separately with each B_i. If this is done sequentially, the round complexity increases from constant to linear in n. On the other hand, if $n - 1$ instances are to be performed in parallel, the ZKPoK must be secure under parallel composition. While the round complexity remains the same for such composition, the communication complexity multiplies by $n - 1$. Although there is no hope to obtain sub-linear communication for parallel composition, the fact that A plays the same role in all the instances allows us to "compose" the separate instances so that all the B_is collaborate to act as a "single" verifier. This way, A performs a single ZKPoK instance over a broadcast channel, and we save a factor of $n - 1$ in A's outgoing communication. We call such a composed scheme a 1-Prover n-Verifier ZKPoK.

Definition 2 (1PnV-ZKPoK protocol[3]). *Let R be some binary relation and let $\kappa()$ be a function from bitstrings to the interval $[0..1]$. An interactive protocol between a single prover $P(x, w)$ and n verifiers $V_1(x), \ldots, V_n(x)$ is called a 1-Prover n-Verifier Zero-Knowledge Proof of Knowledge (1PnV-ZKPoK) protocol for a relation R with knowledge error κ, if for any $(x, w) \in R$ the following holds:*

1. *(Completeness) The honest Prover on input $(x, w) \in R$ interacting with honest verifiers, causes all the verifiers to accept with probability 1.*
2. *(Zero-knowledge) For any PPT verifiers $V_1', \ldots, V_n'$ there exists a PPT simulator that has rewindable black-box access to $V_1', \ldots, V_n'$ and produces a transcript T that is indistinguishable from a transcript of $V_1', \ldots, V_n'$ with the real prover P.*
3. *(Extended extractability) There exists a PPT extractor Ext such that the following holds. For any prover P' and any verifiers $V_1', \ldots, V_{n-1}'$ (that might collude with P'), let $\epsilon(x)$ be the probability (the honest) V_n accepts x. There exists a constant c such that whenever $\epsilon(x) > \kappa(x)$, Ext outputs a correct w in expected time at most $\frac{|x|^c}{\epsilon(x) - \kappa(x)}$.*

Note that while the zero-knowledge property is standard, when considering the proof-of-knowledge (extractability) property, the prover is allowed to collude with several of the verifiers, in order to fool an honest verifier. In this case Ext is given black-box access to the corrupt parties, possibly, including other verifiers.

In the full version [9] we show how to construct a 1PnV-ZKPoK from a Σ-Protocol and a (perfectly-hiding) trapdoor commitment. Specifically, we show:

Lemma 1. *There exists a 5-move 1PnV-ZKPoK scheme for the discrete logarithm relation with negligible knowledge error.*

We obtain a PET protocol private against malicious parties by combining Protocol 2 with a 1PnV-ZKPoK for discrete log. The complete scheme takes 6 rounds and is given in Protocol 3 (see a more detailed description in [9]).

Protocol 3. n-party PET with Homomorphic Encryption and ZKPoK

---

1. Alice randomly picks (a private key) $x \in \mathbb{Z}_q^*$, and publishes her public key $(g, h) = (g, g^x)$.
2. Party A performs a 1P(n-1)V-ZKPoK scheme for discrete log with all of the B_is where each of the B_is acts as a verifier. For any party, if the ZKPoK scheme fails then the party broadcasts an abort message and aborts.
3. All parties continue with Protocol 2 with (g, g^x) being the public key.

Theorem 4. *Protocol 3 is private against any coalition of malicious adversaries.*

Due to space limitation we defer the proof to the full version of our paper.

---

[3] We consider only PPT parties, thus this is, in fact, a ZK *argument*.

5 Realizing PET via PAKE

In this section we construct a PET scheme based on a password-authenticated key-exchange (PAKE) primitive (see for instance [11,15,5] and references therein). PAKE is performed between two users, A and B, where each user $i \in \{A, B\}$ possesses a password pw_i drawn from a low-entropy dictionary $\mathcal{D}$. The distributed computation $\mathsf{PAKE}(pw_A, pw_B)$ outputs a key k_i for each user i, such that k_i is uniformly distributed and $k_A = k_B$ if $pw_A = pw_B$.

We show that an n-party PET scheme can be realized using a chosen-plaintext unforgeable semantically-secure encryption scheme along with a secure 2-party PAKE, using one's private input as one's password. Our construction is defined assuming a secure two-party PAKE as a sub-protocol. Specifically, Let $2\mathsf{PAKE}^{\mathcal{D},N} : (pw_1, pw_2) \to (k_1, k_2)$ be the ideal functionality that accepts two passwords and outputs two numbers $k_1, k_2 \in Z_N$ such that k_1, k_2 are uniformly chosen and $k_1 = k_2$ if $pw_1 = pw_2$ and $pw_1, pw_2 \in \mathcal{D}$.

The main idea is to use $2\mathsf{PAKE}^{\mathcal{D},N}$ to form secure channels between each pair of users, using their private input X_i as the $2\mathsf{PAKE}^{\mathcal{D},N}$ password. Unless the users share the same input, they would use different keys to encrypt and decrypt messages, and "randomize" the transmitted messages. The encryption scheme is *chosen-plaintext unforgeable* (see [16]) which guarantees that for any x the adversary cannot forge a valid encryption of x with non-negligible probability.

The scheme works as follows. Each user performs an (n, n)-secret sharing of the value 0 and sends the shares to the rest of the $n - 1$ users, encrypted with a chosen-plaintext unforgeable semantically-secure encryption, using the PAKE generated key. Next, each user adds all the shares he has received, so the users still jointly hold a secret-sharing of 0. Finally, all the parties send Alice their accumulated shares. The scheme is described in Protocol 4.

Protocol 4. A realization of an n-party PET via $2\mathsf{PAKE}^{\mathcal{D},N}$

(trusted) setup : Assume $\mathbb{Z}_N$ is some fixed finite field, where N depends on κ. Assume a symmetric-key semantically secure $1/N$-chosen-plaintext unforgeable encryption scheme (GEN, ENC, DEC), and oracle access to $2\mathsf{PAKE}^{\mathcal{D},N}$.

1. Each user $i \in (A, B_1, B_2, \ldots, B_{n-1})$ runs $2\mathsf{PAKE}^{\mathcal{D},N}$ with any other user $j \neq i$, using his private input X_i as his PAKE password. Let $2\mathsf{PAKE}^{\mathcal{D},N}(X_i, X_j) = (sk_{ij}, sk_{ji})$.
2. Each user $i \in (A, B_1, B_2, \ldots, B_{n-1})$, randomly picks n values $s_{ij} \in \mathbb{Z}_N$ such that $\sum_j s_{ij} = 0$ and sends $ENC_{sk_{ij}}(s_{ij})$ to user $j \in (A, B_1, B_2, \ldots, B_{n-1})$.
3. User i, upon receiving the message m_j from user j, computes $\tilde{s}_{ji} = DEC_{sk_{ij}}(m_j)$. Next, each user $i \neq A$ sends the value $ENC_{sk_{iA}}(s_{ii} + \sum_{j \neq i} \tilde{s}_{ji})$ to Alice.
4. Alice decrypts the message from user i using the key sk_{Ai}, to obtain $\tilde{\tilde{s}}_i$. Alice computes $\mathsf{res} = s_{AA} + \sum_{i \neq A}(\tilde{\tilde{s}}_i + \tilde{s}_{iA})$, the sum of all the shares she has received (including her own).
5. Alice's output is the value of the boolean predicate $\mathsf{res} \overset{?}{=} 0$.

In the case that all users hold the same password (and thus use matching encryption keys), Alice reconstructs the value 0 and learns that all the parties have the same private input. Otherwise, shares sent over channels with mis-matching keys are decrypted to a random value. This prevents Alice from reconstructing 0, and she only learns that at least one party has a different private input.

Theorem 5. *Protocol 4 is $(\varepsilon + neg)$-sound and private in the malicious adversary model, where ε is the probability of guessing the password of party A.*

We defer the proof to the full version of this paper [9].

Acknowledgments. We thank Alan Roytman, Sanjam Garg and Akshay Wadia for useful discussions, and thank Serge Fehr for pointing out trapdoor commitment based ZKPoK schemes to us. We also thank the anonymous reviewers for comments and suggestions.

RG is supported in part by DARPA and NSF grants 0830803, 09165174, 1065276, 1118126 and 1136174. RO is supported in part by NSF grants 0830803, 09165174, 1065276, 1118126 and 1136174, US-Israel BSF grant 2008411, OKAWA Foundation Research Award, IBM Faculty Research Award, Xerox Faculty Research Award, B. John Garrick Foundation Award, Teradata Research Award, and Lockheed-Martin Corporation Research Award. This material is based upon work supported by the Defense Advanced Research Projects Agency through the U.S. Office of Naval Research under Contract N00014-11-1-0392. The views expressed are those of the author and do not reflect the official policy or position of the Department of Defense or the U.S. Government.

References

1. Axel, K.: Location-Based Services: Fundamentals and Operation. John Wiley & Sons, Hoboken (2005)
2. Ben-Or, M., Goldwasser, S., Wigderson, A.: Completeness theorems for non-cryptographic fault-tolerant distributed computation. In: STOC 1988, pp. 1–10 (1988)
3. Boneh, D.: The Decision Diffie-Hellman Problem. In: Buhler, J.P. (ed.) ANTS 1998. LNCS, vol. 1423, pp. 48–63. Springer, Heidelberg (1998)
4. Boudot, F., Schoenmakers, B., Traoré, J.: A fair and efficient solution to the socialist millionaires' problem. Discrete Applied Mathematics 111(1-2), 23–36 (2001)
5. Canetti, R., Halevi, S., Katz, J., Lindell, Y., MacKenzie, P.: Universally Composable Password-Based Key Exchange. In: Cramer, R. (ed.) EUROCRYPT 2005. LNCS, vol. 3494, pp. 404–421. Springer, Heidelberg (2005)
6. Damgård, I., Fitzi, M., Kiltz, E., Nielsen, J.B., Toft, T.: Unconditionally Secure Constant-Rounds Multi-party Computation for Equality, Comparison, Bits and Exponentiation. In: Halevi, S., Rabin, T. (eds.) TCC 2006. LNCS, vol. 3876, pp. 285–304. Springer, Heidelberg (2006)
7. Fagin, R., Naor, M., Winkler, P.: Comparing information without leaking it. Commun. ACM 39, 77–85 (1996)
8. Freedman, M.J., Nissim, K., Pinkas, B.: Efficient Private Matching and Set Intersection. In: Cachin, C., Camenisch, J.L. (eds.) EUROCRYPT 2004. LNCS, vol. 3027, pp. 1–19. Springer, Heidelberg (2004)
9. Gelles, R., Ostrovsky, R., Winoto, K.: Multiparty Proximity Testing with Dishonest Majority from Equality Testing. In: Czumaj, A., et al. (eds.) ICALP 2012, Part II. LNCS, vol. 7392, pp. 537–548. Springer, Heidelberg (2012)

10. Goldreich, O.: Foundations of cryptography. Basic applications, vol. II. Cambridge University Press, New York (2004)
11. Goldreich, O., Lindell, Y.: Session-key generation using human passwords only. Journal of Cryptology 19, 241–340 (2006)
12. Gong, L., Lomas, M., Needham, R., Saltzer, J.: Protecting poorly chosen secrets from guessing attacks. IEEE Journal on Selected Areas in Communications 11(5), 648–656 (1993)
13. Gruteser, M., Grunwald, D.: Anonymous usage of location-based services through spatial and temporal cloaking. In: MobiSys 2003, pp. 31–42 (2003)
14. Kalnis, P., Ghinita, G., Mouratidis, K., Papadias, D.: Preventing location-based identity inference in anonymous spatial queries. IEEE Transactions on Knowledge and Data Engineering 19(12), 1719–1733 (2007)
15. Katz, J., Ostrovsky, R., Yung, M.: Efficient Password-Authenticated Key Exchange Using Human-Memorable Passwords. In: Pfitzmann, B. (ed.) EUROCRYPT 2001. LNCS, vol. 2045, pp. 475–494. Springer, Heidelberg (2001)
16. Katz, J., Yung, M.: Unforgeable Encryption and Chosen Ciphertext Secure Modes of Operation. In: Schneier, B. (ed.) FSE 2000. LNCS, vol. 1978, pp. 284–299. Springer, Heidelberg (2001)
17. Li, K.A., Sohn, T.Y., Huang, S., Griswold, W.G.: Peopletones: a system for the detection and notification of buddy proximity on mobile phones. In: MobiSys 2008, pp. 160–173 (2008)
18. Lipmaa, H.: Verifiable Homomorphic Oblivious Transfer and Private Equality Test. In: Laih, C.-S. (ed.) ASIACRYPT 2003. LNCS, vol. 2894, pp. 416–433. Springer, Heidelberg (2003)
19. Naor, M., Pinkas, B.: Oblivious transfer and polynomial evaluation. In: STOC 1999, pp. 245–254 (1999)
20. Narayanan, A., Thiagarajan, N., Lakhani, M., Hamburg, M., Boneh, D.: Location Privacy via Private Proximity Testing. In: NDSS 2011 (2011)
21. Nishide, T., Ohta, K.: Multiparty Computation for Interval, Equality, and Comparison Without Bit-Decomposition Protocol. In: Okamoto, T., Wang, X. (eds.) PKC 2007. LNCS, vol. 4450, pp. 343–360. Springer, Heidelberg (2007)
22. Ostrovsky, R., Skeith III, W.E.: A Survey of Single-Database Private Information Retrieval: Techniques and Applications. In: Okamoto, T., Wang, X. (eds.) PKC 2007. LNCS, vol. 4450, pp. 393–411. Springer, Heidelberg (2007)
23. Paillier, P.: Public-Key Cryptosystems Based on Composite Degree Residuosity Classes. In: Stern, J. (ed.) EUROCRYPT 1999. LNCS, vol. 1592, pp. 223–238. Springer, Heidelberg (1999)
24. Pedersen, T.P.: Non-interactive and Information-Theoretic Secure Verifiable Secret Sharing. In: Feigenbaum, J. (ed.) CRYPTO 1991. LNCS, vol. 576, pp. 129–140. Springer, Heidelberg (1992)
25. Schnorr, C.P.: Efficient signature generation by smart cards. Journal of Cryptology 4, 161–174 (1991)
26. Šikšnys, L., Thomsen, J.R., Šaltenis, S., Yiu, M.L., Andersen, O.: A Location Privacy Aware Friend Locator. In: Mamoulis, N., Seidl, T., Pedersen, T.B., Torp, K., Assent, I. (eds.) SSTD 2009. LNCS, vol. 5644, pp. 405–410. Springer, Heidelberg (2009)
27. Tonicelli, R., David, B.M., de Morais Alves, V.: Universally Composable Private Proximity Testing. In: Boyen, X., Chen, X. (eds.) ProvSec 2011. LNCS, vol. 6980, pp. 222–239. Springer, Heidelberg (2011)
28. Yao, A.: Protocols for secure computations. In: SFCS 1982, pp. 160–164 (1982)
29. Zhong, G., Goldberg, I., Hengartner, U.: Louis, Lester and Pierre: Three Protocols for Location Privacy. In: Borisov, N., Golle, P. (eds.) PET 2007. LNCS, vol. 4776, pp. 62–76. Springer, Heidelberg (2007)

Anonymous Card Shuffling and Its Applications to Parallel Mixnets

Michael T. Goodrich[1] and Michael Mitzenmacher[2]

[1] Dept. of Computer Science, University of California, Irvine
goodrich@ics.uci.edu
[2] Dept. of Computer Science, Harvard University
michaelm@eecs.harvard.edu

Abstract. We study the question of how to shuffle n cards when faced with an opponent who knows the initial position of all the cards *and* can track every card when permuted, *except* when one takes $K < n$ cards at a time and shuffles them in a private buffer "behind your back," which we call *buffer shuffling*. The problem arises naturally in the context of parallel mixnet servers as well as other security applications. Our analysis is based on related analyses of load-balancing processes. We include extensions to variations that involve corrupted servers and adversarially injected messages, which correspond to an opponent who can peek at some shuffles in the buffer and who can mark some number of the cards. In addition, our analysis makes novel use of a sum-of-squares metric for anonymity, which leads to improved performance bounds for parallel mixnets and can also be used to bound well-known existing anonymity measures.

1 Introduction

Suppose an honest player, Alice, is playing cards with a card shark, Bob, who has a photographic memory and perfect vision. Not trusting Bob to shuffle, Alice insists on shuffling the deck for each hand they play. Unfortunately, Bob will only agree to this condition if he gets to scan through the deck of n cards before she shuffles, so that he sees each card and its position in the deck, and if he also gets to watch her shuffle. It isn't hard to realize that, even though several well-known card shuffling algorithms, like random riffle shuffling [1], top-to-random shuffling [4], and Fisher-Yates shuffling [10], are great at placing cards in random order, they are terrible at obscuring that order from someone like Bob who has memorized the initial ordering of the cards and is watching Alice's every move. Thus, these algorithms on their own are of little use to Alice. What she needs is a way to shuffle that can place cards in random order in a way that hides that order from Bob. We refer to this as the *anonymous shuffling* problem. Our goal in this paper is to show that, as long as Alice has a private buffer where she can shuffle a subset of the cards, she can solve the anonymous shuffling problem.

Our main motivation for studying the anonymous shuffling problem in this paper comes from the problem of designing efficient parallel mixnets. A *parallel mix network* (or *mixnet*) is a distributed mechanism for connecting a set of n inputs with a set of n outputs in a way that hides the way the inputs and outputs are connected. This connection hiding is achieved by routing the n inputs as messages through a set of M *mix*

A. Czumaj et al. (Eds.): ICALP 2012, Part II, LNCS 7392, pp. 549–560, 2012.
© Springer-Verlag Berlin Heidelberg 2012

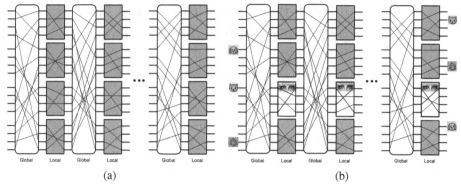

Fig. 1. (a) A parallel mixnet with $n = 16$ inputs and $M = 4$ mix servers. Shaded boxes illustrate mix servers, whose internal permutations are hidden from the adversary. The adversary is allowed to see the global permutation performed in each round. (b) A corrupted parallel mixnet, where $s = 3$ servers are not colluding with the adversary, who has injected $f = 3$ fake messages into the network.

servers in a series of synchronized rounds. In each round, the n inputs are randomly assigned to servers so that each server is assigned $K = n/M$ messages. Then, each server randomly permutes the messages it receives and performs an encryption operation so that it is computationally infeasible for an eavesdropper watching the inputs and outputs of any (honest) server to determine which inputs are matched to the outputs. The mixnet repeats this process for a specific number of rounds. The goal of the adversary in this scenario is to determine (that is, link) one or more of the input messages with their corresponding outputs, while the mixnet shuffles so as to reduce the linkability between inputs and outputs to an acceptably small level. (See Figure 1a.)

Each of the servers is assumed to run the mixnet protocol correctly, which is enforced using cryptographic primitives and public sources of randomness (e.g., see [2,6,9,12,13]). In some cases, we also allow for a *corrupted* parallel mixnet, where some number $s \geq 1$ of the servers behave properly, but the remaining $M - s$ servers collude with the adversary so as to reveal how they are internally permuting the messages they receive. In addition, the adversary may also be allowed to inject some number, $f < n$, of fake messages that are marked in a way that allows the adversary to determine their placement at any point in the process, including in the final output ordering. (See Figure 1b.) In this paper, we are interested in studying a class of algorithms for anonymous shuffling, to show how the analysis of these algorithms can lead to improved protocols for uncorrupted and corrupted parallel mixnets.

1.1   Previous Related Work

Chaum [3] introduced the concept of mix networks for achieving anonymity in messaging, and this work has led to a host of other papers on the topic (e.g., see [12,13]).

Golle and Juels [6] study the parallel mixing problem, where mix servers process messages synchronously in parallel rounds, and discuss the cryptographic primitives sufficient to support parallel mixnet functionality. Their scheme has a total mixing time of $2n(M - s + 1)/M$ and a number of parallel mixing rounds that is $2(M - s + 1)$, assuming that M^2 divides n. It achieves a degree of anonymity "close" to $n - f$, using a specialized anonymity measure, Anon_t, that they define (which we discuss in more detail in Section 2). Note that if s is, say, $M/2$, then their protocol requires as many rounds as the number of servers, which diminishes the potential benefits of a *parallel* mixnet. In particular, their approach uses sequences of round-robin permutations (cyclic shifts) rather than the standard parallel mixing protocol described above and illustrated in Figure 1. Even then, Borisov [2] shows that their scheme can leak linkages between inputs and outputs (as can the standard parallel mixing protocol if the number of rounds is too small) if the vast majority of inputs are fake messages introduced by the adversary. Thus, it is reasonable to place realistic limits on how large f can be, such as $f \leq n/2$, and require that the number of parallel mixing rounds is high enough to guarantee a high degree of anonymity. Klonowski and Kutylowski [9] also study the anonymity of parallel mixnets, characterizing it in terms of variation distance (which we discuss in the next section) for honest servers and for the case of a single corrupted server. They do not consider an adversary who can inject fake messages, however, and they only treat the case when M^2 is much less than n.

Goodrich, Mitzenmacher, Ohrimenko, and Tamassia [7] study a simple variant of the anonymous shuffling problem with no corrupted servers or fake cards, addressing a problem similar to parallel mixing in the context of oblivious storage. They show that when the number of cards per server each round is $K = n^{1/c}$, then $c + 1$ rounds are sufficient to hide any specific initial card, so that the adversary can guess its location with probability only $1/n + o(1/n)$. The current work provides a much more general and detailed result, using much more robust techniques.

Our techniques are based on work in dynamic load balancing by Ghosh and Muthukrishnan [5]. In their setting, tasks are balanced in a dynamic network by repeatedly choosing random matchings and balancing tasks across each edge. Here, we extend this work by choosing random subcollections of K cards and balancing weights, corresponding to probabilites of a specific card being one of those K, among the K cards via the shuffling.

1.2 Our Results

We study the problem of analyzing parallel mixnets in terms of a buffer-based solution to the anonymous shuffling problem, assuming, as with other works on parallel mixnets [2,6,7,9], that cryptographic primitives exist to enforce re-encryption for each mix server, along with public sources of randomness and permutation verification so that servers must correctly follow the mixing protocol even if corrupted.

In the *buffer shuffling* algorithm [7,9], Alice repeatedly performs a series of shuffling rounds, as in the parallel mixnet paradigm. That is, each round begins with Alice performing a random shuffle that places the cards in random order (albeit in a way that the adversary, Bob, can see). Then she splits the ordered cards into M piles, with each pile getting $K = n/M$ cards. Finally, she randomly shuffles each pile, using a private buffer that Bob cannot see into. Once she has completed her private shuffles, she stacks up her piles, which become the working deck for the next round. She repeats these rounds until she is satisfied that the deck is sufficiently shuffled for the adversary. Note that during her shuffling, Bob can see cards go in and out of her buffer, but he cannot normally see

Table 1. Summary of our results. We compare with the solutions of Golle and Juels [6], Klonowski and Kutylowski [9], and Goodrich *et al.* [7]. The server restriction column refers to the parameter c in the inequality $K \geq n^{1/c}$. The bounds of "medium" and "high" for corruption tolerance are made more precise in the statement of the theorems.

Solution	Corruption Tolerance	Server Restriction	Allows for Corrupt Servers	Allows for Fake Messages	Rounds
GJ [6]	high	$c = 2$ only	$\checkmark$	$\checkmark$	$O(M)$
KK [9]	only one	$c = 2$ only	$\checkmark$	–	$O(\log n)$
GMOT [7]	none	const. c	–	–	$O(1)$
Our Theorem 2	medium	const. c	$\checkmark$	–	$O(1)$
Our Theorem 3	high	const. c	$\checkmark$	$\checkmark$	$O(\log n)$

cards while they are in the buffer. As we describe in more detail shortly, Alice's goal is to prevent Bob from being able to track a card; that is, Bob should only be able to guess the location of a card with probability $1/n + o(1/n)$, where generally we take the $o(1/n)$ term to be $O(1/n^b)$ for some $b \geq 1$.

To characterize the power of the adversary in the parallel mixnet framework, we consider buffer shuffling in a context where, for $M - s$ specific uses of Alice's buffer within each round, Bob is allowed to see how the cards are shuffled inside it. Likewise, we assume he is allowed to mark $f < n$ of the cards in a way that lets him determine their position in the deck at any time. We provide a novel analysis of this framework, and show how this analysis can be used to design improved methods for designing parallel mixnets. For instance, we show that buffer shuffling achieves our goal with $O(1)$ rounds even if the number of servers is relatively large and that buffer shuffling can be performed in $O(\log n)$ rounds even for high degrees of compromise. We summarize our results and how they compare with the previous related work in Table 1.

2   Anonymity Measures

We can model the anonymous shuffling problem in terms of probability distributions. Without loss of generality, we can assume that the initial ordering of cards is $[n] = (1, 2, \ldots, n)$. After Alice performs t rounds of shuffling, let $w_i(t, c)$ denote the probability from the point of view of the adversary that the card in position i at time t is the card numbered c, and let $W(i, t)$ denote the distribution defined by these probabilities (we may drop the i and t if they are clear from the context). The ideal is for this probability to be $1/n$ for all i and c, which corresponds to the uniform distribution, U.

A natural way to measure anonymity is to use a distance metric to determine how close the distribution W is to U, for any particular card i or in terms of a maximum taken over all the cards. The goal is for this metric to converge to 0 quickly as a function of t.

Maximum difference. The *maximum-difference* metric, which is also known as the L_∞ metric, specialized to measure the distance between W and U, is

$$\alpha(t) = \max_{i,c} |w_i(t,c) - 1/n|.$$

As mentioned above, the goal is to minimize $\alpha(t)$, getting close to 0 as quickly as possible.

Note that, in the case of buffer shuffling, the formula for $\alpha(t)$ can be simplified. In particular, since Alice starts each round with a random permutation, $w_i(t,c) = w_i(t,1)$. Thus, in our case, we can drop the c and focus on $w_i(t)$, the probability that the i-th card is 1. In this case, we can simplify the definition as

$$\alpha(t) = \max_i |w_i(t) - 1/n|.$$

The Anon *measure of Golle and Juels.* In the context of parallel mixing, Golle and Juels [6] define a measure for anonymity, which, using the above notation, would be defined as follows:

$$\mathsf{Anon}_t = \min_i \left(\max_c w_i(t,c) \right)^{-1},$$

which they try to maximize. Note that $\max_c w_i(t,c) \geq 1/n$ for all i, so, to be consistent with the goals for other anonymity measures, which are all based on minimizations, we can use the following Anon_t' definition for an anonymity measure equivalent to that of Golle and Juels:

$$\mathsf{Anon}_t' = \max_i \left(\max_c w_i(t,c) \right) = \max_{i,c} w_i(t,c) = (\mathsf{Anon}_t)^{-1}.$$

The Anon_t' measure is not an actual distance metric, with respect to U, however, since its smallest value is $1/n$, not 0. In addition, it is biased towards the knowledge gained by the adversary for positive identifications and can downplay knowledge gained by ruling out possibilities. To see this, note that, if we let W^+ denote all the $w_i(t,c)$'s that are at least $1/n$ and W^- denote all the $w_i(t,c)$ values less than $1/n$, then

$$\alpha(t) = \max\{ \max_{w_i(t,c) \in W^+} \{w_i(t,c) - 1/n\}, \max_{w_i(t,c) \in W^-} \{1/n - w_i(t,c)\}\}$$

$$= \max\{\mathsf{Anon}_t' - 1/n, \max_{w_i(t,c) \in W^-} \{1/n - w_i(t,c)\}\}.$$

Therefore, we prefer to use anonymity measures that are based on metrics and are unbiased measures of the distance from W to the uniform distribution, U.

Variation Distance. Li *et al.* [11] introduce a notion of anonymity called *threshold closeness* or *t-closeness*. For categorical data, as in card shuffling and mixnets, this metric amounts to the *variation distance* between the W-distribution defined by Alice's shuffling method and the (desired) uniform distribution, U, where each card occurs with probability $1/n$ (see also [9]). In particular, this metric would be defined as follows for buffer shuffling:

$$\beta(t) = \frac{1}{2} \sum_{i=1}^{n} |w_i(t) - 1/n|,$$

which is the same as half the L_1 distance between the W-distribution and the uniform distribution, U. As with other distance metrics, the goal is to minimize $\beta(t)$.

The sum-of-squares metric. For this paper, we have chosen to focus on a metric for anonymity that is derived from a simple measure that is well-known for its sensitivity to outliers (which are undesirable in the context of anonymity). In this, the *sum-of-squares* metric, we take the sum of the squared differences between the given distribution and our desired ideal. In the context of buffer shuffling, this would be defined as follows:

$$\Phi(t) = \sum_{i=1}^{n}(w_i(t) - 1/n)^2,$$

which can be further simplified as follows:

$$\Phi(t) = \sum_{i=1}^{n}(w_i^2(t) - 2w_i(t)/n + 1/n^2)$$

$$= \left(\sum_{i=1}^{n} w_i^2(t)\right) - 1/n.$$

This amounts to the square of the L_2-distance between the W-distribution and the uniform distribution, U. The goal is to minimize $\Phi(t)$.

Relationships between anonymity measures. Another benefit of the $\Phi(t)$ metric is that it can be used to bound other metrics and measures for anonymity, by well-known relationships among the L_p norms. For instance, we can derive upper bounds for other metrics (which we leave as exercises for the interested reader), such as

$$\alpha(t) \leq \Phi(t)^{1/2} \quad \text{and} \quad \beta(t) \leq (n\Phi(t))^{1/2}/2.$$

In addition, even though Anon_t' is not a metric, we can derive the following bound for it, since $\text{Anon}_t' \leq \alpha(t) + 1/n$:

$$\text{Anon}_t' \leq \Phi(t)^{1/2} + 1/n.$$

So, for the remainder of this paper, we focus primarily on the $\Phi(t)$ metric.

3 Algorithms and Analysis

Our parallel mixing algorithm repeats the following steps:

1. Shuffle the cards, placing them according to a uniform permutation.
2. Under this ordering, divide the cards up into consecutive groups of $K = n/M$ cards.[1]
3. For each group of K cards, shuffle their cards randomly, hidden from the adversary.

---

[1] As we also study, we could alternatively assign each card uniformly at random to one of the $M = n/K$ piles, with each group getting $K = n/M$ cards in expectation.

We refer to each repetition of the above steps as a *round*. In the parallel mixnet setting, each group of K cards would be shuffled at a different server.

Let $w_i(t)$ be the probability that the ith card after t rounds is the first card from time 0 from the point of view of the adversary. (We drop the dependence on t where the meaning is clear.) Initially, $w_1 = 1$, and $w_2 \ldots w_n$ are all 0. Motivated by [5], let $\Phi(t)$ be a potential function $\Phi(t) = (\sum w_i(t)^2) - \frac{1}{n}$, based on the sum-of-squares metric, and let $\Delta\Phi(t) = \Phi(t) - \Phi(t+1)$. (Again, we drop the explicit dependence on t where suitable.)

Our first goal is to prove the following theorem.

Theorem 1. *A non-corrupted parallel mixnet, designed as described above, has* $\mathbf{E}[\Phi(t)] \leq K^{-t}$. *In particular, such a mixnet, with* $K \geq n^{1/c}$, *can mix messages in* $t = bc$ *rounds so that the expected sum-of-squares error,* $\mathbf{E}[\Phi(t)]$, *between card-assignment probabilities and the uniform distribution is at most* $1/n^b$, *for any fixed* $b \geq 1$.

Before proving this theorem, we note some implications. From Theorem 1 and Markov's inequality, using $t = 2bc$ rounds, we can bound the probability that $\Phi(t) > 1/n^b$ to be at most $1/n^b$, for any fixed $b \geq 1$. So, taking $b = 2$ implies $\alpha_t \leq 1/n$ with probability $1 - 1/n^2$, taking $b = 3$ implies $\beta_t \leq 1/n$ with probability $1 - 1/n^3$, and taking $b = 2$ implies $\text{Anon}'_t < (n-1)^{-1}$, with probability $1 - 1/n^2$, which achieves the anonymity goal of Golle and Juels [6] (who only treat the case $c = 2$). Therefore, a constant number of rounds suffices for anonymously shuffling the inputs in a parallel mixnet, provided servers can internally mix $K \geq n^{1/c}$ items, for some constant $c \geq 1$.

We now move to the proof. Let $\Delta\Phi^*$ represent how the potential changes when a group of K cards is shuffled during a round. For clarity, we examine the cases of $K = 2$ and 3 before the general case.

- For 2 cards with incoming weights w_i and w_j (outgoing weights are the average):

$$\Delta\Phi^* = w_i^2 + w_j^2 - 2((w_i + w_j)/2)^2$$
$$= (w_i - w_j)^2/2.$$

- For 3 cards with incoming weights w_i, w_j, and w_k:

$$\Delta\Phi^* = w_i^2 + w_j^2 + w_k^2 - 3((w_i + w_j + w_k)/3)^2$$
$$= (w_i - w_j)^2/3 + (w_j - w_k)^2/3 + (w_k - w_i)^2/3.$$

- For K cards with weights $w_{i1}, w_{i2}, \ldots, w_{iK}$:

$$\Delta\Phi^* = \sum_{k=1}^{K} w_{ik}^2 - K\left(\frac{\sum_{k=1}^{K} w_{ik}}{K}\right)^2$$
$$= \frac{1}{K} \sum_{1 \leq j < k \leq K}^{K} (w_{ij} - w_{ik})^2.$$

We now proceed to bound $\mathbf{E}[\Phi(t)]$ by making use of $\Delta\Phi$.

$$
\mathbf{E}[\Delta\Phi] = \frac{1}{K} \sum_{1 \le i < j \le n} \Pr\left((i,j) \text{ are in the same set of } K \text{ cards}\right) (w_i - w_j)^2
$$
$$
= \frac{K-1}{K(n-1)} \sum_{i<j} (w_i - w_j)^2
$$
$$
= \frac{K-1}{2K(n-1)} \sum_{1 \le i,j \le n} (w_i - w_j)^2.
$$

Also,

$$
\mathbf{E}[\Delta\Phi/\Phi] = \frac{K-1}{2K(n-1)} \frac{\sum_{i,j} \left((w_i - 1/n) - (w_j - 1/n)\right)^2}{\sum_k (w_k - 1/n)^2}.
$$

Let $x_i = w_i - 1/n$ to get

$$
\mathbf{E}[\Delta\Phi/\Phi] = \frac{K-1}{2K(n-1)} \frac{\sum_{i,j} (x_i - x_j)^2}{\sum_k x_k^2}.
$$

Interestingly, when $K = n$, we should have mixing in one step, so in this case $\mathbf{E}[\Delta\Phi/\Phi]$ should be 1. Notice if that is the case, then perhaps surprisingly the above expression is independent of the actual x_i values, and then we have immediately:

$$
\mathbf{E}[\Delta\Phi/\Phi] = \frac{n(K-1)}{K(n-1)}.
$$

We can in fact confirm this easily. Since $\sum_k x_k = 0$, we have

$$
\sum_{i,j}(x_i - x_j)^2 = \sum_{i,j}(x_i - x_j)^2 + 2\left(\sum_k x_k\right)^2 = 2n \sum_k x_k^2,
$$

and cancellation gives the desired result.

This analysis also gives us fast convergence to the uniform distribution in the general case. Let $\gamma = \frac{n(K-1)}{K(n-1)}$, and note $\gamma \le 1$. In particular,

$$
1 - \gamma = \frac{n-K}{K(n-1)} < \frac{1}{K}.
$$

Also note $\Phi(0) < 1$. So we have

$$
\mathbf{E}[\Phi(t+1)] = (1-\gamma)\mathbf{E}[\Phi(t)],
$$

and a simple induction yields

$$
\mathbf{E}[\Phi(t)] = (1-\gamma)^t \Phi(0) \le K^{-t}.
$$

The rest of the theorem follows easily.

3.1 Extensions to Mixnets with Corrupted Servers

In the case of there being corrupted servers, Bob will know the permutation for the cards assigned to each such server. In terms of the analysis, we can treat the permutation for each corrupted server as the identity operation, since Bob can simply undo that permutation. Let us suppose, then, that there are $M = n/K$ servers, so that each obtains K cards in each round, and that $1 \leq s \leq n/K$ servers are uncorrupted. Following our previous analysis, we find

$$
\begin{aligned}
\mathbf{E}[\Delta\Phi] &= \frac{1}{K} \sum_{1 \leq i < j \leq n} \Pr\left((i,j) \text{ are in the same uncorrupted server}\right) (w_i - w_j)^2 \\
&= \frac{(K-1)}{K(n-1)} \frac{s}{n/K} \sum_{i<j} (w_i - w_j)^2 \\
&= \frac{s(K-1)}{2n(n-1)} \sum_{1 \leq i,j \leq n} (w_i - w_j)^2.
\end{aligned}
$$

Again, based on our previous analysis, we have

$$
\mathbf{E}[\Delta\Phi/\Phi] = \frac{s(K-1)}{n-1}.
$$

Now let $\gamma' = \frac{s(K-1)}{(n-1)}$; if, for example, $s = \epsilon \frac{n-1}{K-1}$ then $1 - \gamma' = 1 - \epsilon$. In that case,

$$
\mathbf{E}[\Phi(t)] = (1 - \epsilon)^t.
$$

Theorem 2. *A corrupted parallel mixnet, designed as described above, with $s \geq \epsilon(n-1)/(K-1)$ non-corrupted servers, for $\epsilon \geq 1/2$, can mix messages in $t = b\log n$ rounds so that the expected sum-of-squares error, $\mathbf{E}[\Phi(t)]$, between card-assignment probabilities and the uniform distribution is at most $1/n^b$, for any fixed $b \geq 1$. Likewise, if there are at most $\frac{n^{-1/c}(n-1)}{K-1} - \frac{n-K}{K(K-1)}$ corrupted servers, with $K \geq n^{1/c}$ for some constant $c \geq 1$, then in $t = bc$ rounds it is also the case that $\mathbf{E}[\Phi(t)]$ is at most $1/n^b$, for any fixed $b \geq 1$.*

Thus, by Markov's inequality, using $t = 2b\log n$ or $t = 2bc$ rounds, depending on the number of uncorrupted servers, s, we can bound the probability that $\Phi(t) > 1/n^b$ to itself be at most $1/n^b$, for any fixed $b \geq 1$.

As an instructive specific example, suppose $K = M = \sqrt{n}$, and there are a constant z servers that are corrupted. Then

$$
1 - \epsilon = 1 - (\sqrt{n} - z)\frac{K-1}{n-1} = 1 - \frac{\sqrt{n} - z}{\sqrt{n}+1} = \frac{z+1}{\sqrt{n}+1}.
$$

Hence, in this specific case,

$$
\mathbf{E}[\Phi(t)] = \left(\frac{z+1}{\sqrt{n}+1}\right)^t < \frac{(z+1)^t}{n^{t/2}},
$$

and for any constant b after $4b$ rounds we have that $\Phi(t) \leq n^{-b}$ with probability $P(n^{-b})$.

As our expressions become less clean in our remaining settings, we state a general theorem which can be applied to these settings in a straightforward way:

Theorem 3. *Given a parallel mixnet with corrupted servers or adversarially generated inputs, let $\gamma = \mathbf{E}[\Delta\Phi/\Phi]$ in that setting. Then in $t = b\log_{1/(1-\gamma)} n$ rounds the expected sum-of-squares error, $\mathbf{E}[\Phi(t)]$, between card-assignment probabilities and the uniform distribution is at most $1/n^b$, for any fixed $b \geq 1$. In particular, if $\gamma \geq 1/2$, at most $b\log n$ rounds are required; if $\gamma \geq 1 - n^{-1/c}$, at most bc rounds are required.*

3.2 Extensions to Mixnets with Corrupted Inputs

For the case of corrupted inputs, Bob will be able to track those cards throughout the shuffle process. In the shuffling setting, we can think of some number of the cards as being marked—no matter what we do, Bob knows the locations of those cards. In terms of the analysis, we can treat this in the following way: when we have a group of K cards, it is as though we are shuffling only $K' \leq K$ cards, where K' is the number of unmarked cards in the collection of K cards. Let us suppose that $f \leq n - 2$ cards are marked. Note that we may think of w_i as being 0 for any cards in a marked position; alternatively, without loss of generality, let us calculate at each step as though w_i is non-zero only for $i = 1$ to $n - f$. (Think of w_i as being the appropriate value for the ith unmarked card.) Note that, for consistency, we must have

$$\Phi(t) = \left(\sum_{i=1}^{n-f} w_i(t)^2\right) - \frac{1}{n-f}.$$

Following our previous analysis, we find

$$\mathbf{E}[\Delta\Phi] = \sum_{K'=2}^{K} \frac{1}{K'} \sum_{1 \leq i < j \leq n-f} \Pr\left(\begin{matrix}(i,j) \text{ are in the same set of } K' \text{ out} \\ \text{of } K \text{ unmarked cards}\end{matrix}\right)(w_i - w_j)^2$$

$$= \sum_{K'=2}^{K} \frac{\binom{n-f}{K'}\binom{f}{K-K'}}{\binom{n}{K}} \frac{K'-1}{K'(n-1)}\sum_{i<j}(w_i - w_j)^2$$

$$= \sum_{K'=2}^{K} \frac{\binom{K}{K'}\binom{n-K}{n-f-K'}}{\binom{n}{f}} \frac{K'-1}{K'(n-1)}\sum_{i<j}(w_i - w_j)^2$$

$$= \frac{1}{2(n-1)\binom{n}{f}}\left(\sum_{K'=2}^{K}\binom{K}{K'}\binom{n-K}{n-f-K'}\frac{K'-1}{K'}\right)\sum_{1 \leq i,j \leq n-f}(w_i - w_j)^2.$$

We can then compute $\mathbf{E}[\Delta\Phi/\Phi]$ as

$$\frac{1}{2(n-1)\binom{n}{f}}\left(\sum_{K'=2}^{K}\binom{K}{K'}\binom{n-K}{n-f-K'}\frac{K'-1}{K'}\right)\frac{\sum_{1 \leq i,j \leq n-f}(x_i - x_j)^2}{\sum_{1 \leq k \leq n-f}x_k^2}.$$

Following the same computations as previously, we have

$$\mathbf{E}[\Delta\Phi/\Phi] = \frac{n-f}{(n-1)\binom{n}{f}} \left(\sum_{K'=2}^{K} \binom{K}{K'} \binom{n-K}{n-f-K'} \frac{K'-1}{K'} \right).$$

Note the $n - f$ term in the numerator in place of an n.

Now let ν equal the right hand side above; then we have

$$\mathbf{E}[\Phi(t)] = (1 - \nu)^t.$$

In particular, it is clear that $\nu \leq \frac{n(K-1)}{K(n-1)}$, so the convergence of $\mathbf{E}[\Phi(t)]$ to 0 happens more slowly than in the case without corrupted inputs, as expected. Nevertheless, we can still derive a theorem analogous to Theorem 2 using Theorem 3 and the above characterization of $\mathbf{E}[\Phi(t)]$. We omit a full restatement for space reasons.

3.3 Extensions to Mixnets with Corrupted Servers and Inputs

One nice aspect of our analysis is that combinations of corrupted servers and inputs are entirely straightforward. In this setting, we have

$$\mathbf{E}[\Delta\Phi] = \sum_{K'=2}^{K} \frac{1}{K'} \sum_{1 \leq i < j \leq n-f} \Pr \left(\begin{array}{l} (i,j) \text{ are in the same set of } K' \text{ out} \\ \text{of } K \text{ unmarked cards at an uncor-} \\ \text{rupted server} \end{array} \right) (w_i - w_j)^2$$

$$= \sum_{K'=2}^{K} \frac{s}{n/K} \frac{\binom{n-f}{K'}\binom{f}{K-K'}}{\binom{n}{K}} \frac{K'-1}{K'(n-1)} \sum_{i<j} (w_i - w_j)^2$$

$$= \frac{sK}{n(n-1)} \sum_{K'=2}^{K} \frac{\binom{K}{K'}\binom{n-K}{n-f-K'}}{\binom{n}{f}} \frac{K'-1}{K'} \sum_{i<j} (w_i - w_j)^2$$

$$= \frac{sK}{2n(n-1)\binom{n}{f}} \left(\sum_{K'=2}^{K} \binom{K}{K'} \binom{n-K}{n-f-K'} \frac{K'-1}{K'} \right) \sum_{1 \leq i,j \leq n-f} (w_i - w_j)^2.$$

Hence

$$\mathbf{E}[\Delta\Phi/\Phi] = \frac{sK(n-f)}{n(n-1)\binom{n}{f}} \left(\sum_{K'=2}^{K} \binom{K}{K'} \binom{n-K}{n-f-K'} \frac{K'-1}{K'} \right).$$

Given this bound, we can then derive a theorem analogous to Theorem 2 for the case when mix servers can be corrupted and the adversary can inject fake messages using Theorem 3. We omit a full restatement for space reasons.

In a more complete version of this paper [8], we extend this analysis further to show how to derive bounds for the case when the assignment of messages to servers is done uniformly at random rather than in way that assigns exactly $K = n/M$ messages per server.

4   Conclusion and Open Problems

In this paper, we have provided a comprehensive analysis of buffer shuffling and shown that this leads to improved algorithms for achieving anonymity and unlinkability in parallel mixnets. An interesting direction for future research could be to extend this analysis to other topologies, including hypercubes and expander graphs.

Acknowledgments. This research was supported in part by the National Science Foundation under grants 0713046, 0721491, 0724806, 0847968, 0915922, 0953071, 0964473, 1011840, and 1012060, and by the ONR under grant N00014-08-1-1015. We would like to thank Justin Thaler for several helpful comments regarding this paper.

References

1. Aldous, D., Diaconis, P.: Shuffling cards and stopping times. Amer. Math. Monthly 93(5), 333–348 (1986)
2. Borisov, N.: An Analysis of Parallel Mixing with Attacker-Controlled Inputs. In: Danezis, G., Martin, D. (eds.) PET 2005. LNCS, vol. 3856, pp. 12–25. Springer, Heidelberg (2006)
3. Chaum, D.L.: Untraceable electronic mail, return addresses, and digital pseudonyms. Commun. ACM 24(2), 84–90 (1981)
4. Diaconis, P., Fill, J.A., Pitman, J.: Analysis of top to random shuffles. Combinatorics, Probability and Computing 1(02), 135–155 (1992)
5. Ghosh, B., Muthukrishnan, S.: Dynamic load balancing in parallel and distributed networks by random matchings. In: 6th ACM Symp. Par. Alg. & Arch. (SPAA), pp. 226–235 (1994)
6. Golle, P., Juels, A.: Parallel mixing. In: 11th ACM Conf. on Comp. and Comm. Security (CCS), pp. 220–226 (2004)
7. Goodrich, M.T., Mitzenmacher, M., Ohrimenko, O., Tamassia, R.: Practical oblivious storage. In: 2nd ACM Conf. on Data and App. Sec. & Priv. (CODASPY), pp. 1–10 (2011)
8. Goodrich, M., Mitzenmacher, M.: Anonymous card shuffling and its applications to parallel mixnets. Arxiv preprint (2012)
9. Klonowski, M., Kutyłowski, M.: Provable Anonymity for Networks of Mixes. In: Barni, M., Herrera-Joancomartí, J., Katzenbeisser, S., Pérez-González, F. (eds.) IH 2005. LNCS, vol. 3727, pp. 26–38. Springer, Heidelberg (2005)
10. Knuth, D.E.: Art of Computer Programming, 3rd edn. Seminumerical Algorithms, vol. 2. Addison-Wesley (1997)
11. Li, N., Li, T., Venkatasubramanian, S.: t-closeness: Privacy beyond k-anonymity and l-diversity. In: 23rd IEEE Int. Conf. on Data Engineering (ICDE), pp. 106–115 (2007)
12. Ren, J., Wu, J.: Survey on anonymous communications in computer networks. Computer Communications 33(4), 420–431 (2010)
13. Sampigethaya, K., Poovendran, R.: A survey on mix networks and their secure applications. Proc. of the IEEE 94(12), 2142–2181 (2006)

Byzantine Agreement with a Rational Adversary

Adam Groce, Jonathan Katz*, Aishwarya Thiruvengadam, and Vassilis Zikas**

Department of Computer Science, University of Maryland
{agroce,jkatz,aish,vzikas}@cs.umd.edu

Abstract. Traditionally, cryptographers assume a "worst-case" adversary who can act arbitrarily. More recently, they have begun to consider *rational* adversaries who can be expected to act in a utility-maximizing way. Here we apply this model for the first time to the problem of *Byzantine agreement* (BA) and the closely related problem of *broadcast*, for natural classes of utilities. Surprisingly, we show that many known results (e.g., equivalence of these problems, or the impossibility of tolerating $t \geq n/2$ corruptions) do not hold in the rational model. We study the feasibility of information-theoretic (both perfect and statistical) BA assuming complete or partial knowledge of the adversary's preferences. We show that perfectly secure BA is possible for $t < n$ corruptions given complete knowledge of the adversary's preferences, and characterize when statistical security is possible with only partial knowledge. Our protocols have the added advantage of being more efficient than BA protocols secure in the traditional adversarial model.

1   Introduction

The problem of *Byzantine agreement* (BA) was introduced by Lamport, Shostak, and Pease [15] as an abstraction of their earlier work on distributed computation among fallible processors [18]. The problem comes in two flavurs, called *consensus* and *broadcast*. In consensus, we have n players each having an input and it is required that they all agree on an output y (consistency), where if all correct players have the same input x then $y = x$ (correctness). In broadcast, only one player (the *sender*) has input, and the requirements are that all players should agree on an output y (consistency), such that if the sender correctly follows the protocol then y is equal to its input (correctness).

In the original work of Lamport et al. [15] BA was motivated by the so-called *Byzantine generals problem*: The generals of the Byzantine army, along with their troops, have encircled an enemy city. Each general is far away from the rest and messengers are used for communication. The generals must agree upon a common plan (to attack or to retreat), though one or more of the generals may be traitors who will attempt to foil the plan. The good generals do not know who the traitors are. If the good generals agree upon the plan unanimously, the

---
* Supported in part by NSF grants #5-23126 and #5-24541.
** Supported in part by a fellowship from the Swiss National Science Foundation (Project No. PBEZP2-134445).

A. Czumaj et al. (Eds.): ICALP 2012, Part II, LNCS 7392, pp. 561–572, 2012.
© Springer-Verlag Berlin Heidelberg 2012

plan will succeed. The traitors, however, may choose to coordinate in a manner that would mislead the good generals into disagreement.

In the formal definition of the problem, the destructive behavior of the traitors is modeled by assuming that they are corrupted by a (central) adversary who coordinates them and tries to break the security. Breaking the security of BA corresponds to violating some of the aforementioned properties, i.e, correctness or consistency (or both). But imagine that the generals had some additional preferences. Say, for example, the traitors wanted to cause inconsistency among the honest generals but, if they can't do that, they would at least prefer the honest generals retreat rather than attack. Or maybe they just want as few generals to attack as possible. In short, it is realistic to assume that the traitors are not acting arbitrarily, but instead have a clear set of preferences, and prefer some outcomes over others.

In this paper, we model the above by assuming *rational* adversaries. A rational adversary prefers a particular outcome of the protocol and may deviate in an attempt to achieve its preference. This is different from the traditional malicious adversary setting, wherein the only goal of the adversary is to break the security of a protocol. We investigate feasibility of rational Byzantine agreement (RBA) under various assumptions regarding the adversary's preferences. Interestingly, in addition to providing a conceptually simple way of capturing realistic situations like the one described above, the model yields significant differences with respect the traditional (non-rational) BA setting. In particular, many properties — even some well-established impossibility results — that are taken for granted in the traditional model are no longer true in the rational setting.

1.1   Our Results

We present, for the first time, a definition of Byzantine agreement taking into account rational behavior on the part of the adversary. In our work, we adopt a somewhat different approach than that taken in some other work blending game theory and cryptography (see below): rather than treating *all* players as rational, we assume that some players are honest and will follow the protocol without question, while other players (those controlled by the adversary) are rational and will attempt to alter the outcome so as to increase their utility.

We study rational broadcast and Byzantine agreement for a natural class of adversarial utility functions defined by the adversary's preferences over the possible outcomes: agreement on 0, agreement on 1, and disagreement. Interestingly, many of the statements that are considered self-evident in the BA literature break down in the rational setting. Examples include the impossibility of consensus for $t \geq n/2$, the usefulness of setups for statistical (and computational) security, as well as the reduction of consensus to broadcast for $t < n/2$. We also study of feasibility of RBA for all possible orderings on the adversary's preferences in the following two cases: (1) the utility function of the adversary is known, and (2) only the adversary's preference between agreement and disagreement is known (but among the possible outcomes for agreement, it is not known which one is more preferred).

1.2 Related Work

Byzantine agreement and broadcast have been studied extensively, and we limit ourselves to a discussion of the main results. Early work showed that (without any setup), Byzantine agreement is possible if and only if the number of corrupted parties t is strictly less than $1/3$ of the total number of parties n. The situation changes when a trusted setup allowing digital signatures (e.g., a public-key infrastructure (PKI)) is assumed. Such a setup does not make a difference for perfect security as there is always some (possibly negligible) probability that the adversary guesses the secret keys of the honest parties and breaks the security of the BA protocol. However, as shown in [18], computationally secure broadcast can be achieved for arbitrary $t < n$ if it is assumed that honest players can sign the messages they send. The same bound was shown to be achievable for the case of statistical security, assuming a setup for information-theoretically secure (pseudo-)signatures [4,19]. It follows from the definition (for a traditional, worst-case adversary) that consensus is impossible in any setting when $t \geq n/2$.

There has recently been a significant amount of interest in bridging cryptographic and game-theoretic models and definitions; see, e.g., [12,13,10,1,17,3,11]. We refer to [14] for a (now slightly outdated) survey. Most closely related to our own work, the BAR model [2] was developed to capture Byzantine, altruistic, and rational behavior in a distributed system; protocols that tolerate a combination of Byzantine and rational nodes were proposed for reliable broadcast [6], state machine replication, and gossip-based multicast [16]. In contrast to these works, our model considers some nodes to be rational and the rest to be honest. In work done concurrently with our own, Bei et al. [5] have shown that rational consensus is impossible in the presence of colluding rational agents and crash failures. Their model assumes that each player is individually rational, and wants to strategically manipulate the protocol according to his own set of preferences. In addition, some of these agents could have crash failures. This is incomparable to our model where we assume that some players honestly follow the protocol while the rest are under the control of a centralized, rational adversary.

2 Byzantine Agreement

We briefly review the traditional definitions of broadcast and consensus. We let $P_1, \ldots, P_n$ denote the parties running the protocol, and let t be (an upper bound on) the number of deviating parties. We let v_i and w_i denote the input and output, respectively, of P_i. We assume the standard network model, where all pairs of players have (authenticated) point-to-point channels. We assume synchronous communication and allow a computationally unbounded adversary. We refer to a (static) adversary who corrupts up to t parties as a t-adversary.

Definition 1 (Consensus). *Each player P_i initially has input v_i. A protocol is a* perfectly secure consensus protocol *if it satisfies the following properties:*

1. *(Consistency): All honest players output the same value w.*
2. *(Correctness): If all honest players begin with the same input value, i.e.*
   $v_i = v$ *for all i, then* $w = v$.

Definition 2 (Broadcast). *We refer to player* P_1 *as the* sender, *who is transmitting his input* v_1 *to the remaining* $n - 1$ *receivers. A protocol is a perfectly secure* broadcast *protocol if it satisfies the following properties:*

1. *(Consistency): All honest players output the same value w.*
2. *(Correctness): If the sender* P_1 *is honest then* $w = v_1$.

The definitions of BA for statistical and computational security are obtained by requiring that the corresponding properties are satisfied except with negligible probability in the presence of an unbounded and a computationally bounded adversary, respectively.

Detectable Broadcast. A useful primitive in our constructions is *detectable broadcast*, which was defined by Fitzi et al. in [7] as a relaxation of the definition of broadcast. Informally, a detectable broadcast guarantees that at the protocol termination, either a successful realization of broadcast has been achieved or all honest parties have agreed that the protocol has been aborted.

The formal definition for detectable broadcast [7] is as follows. For simplicity, we describe the definition for the case where the input is a bit.

Definition 3 (Detectable broadcast). *A protocol for detectable broadcast must satisfy the following properties:*

1. *(Correctness): All honest players either abort or accept and output 0 or 1. If any honest player aborts, so does every honest player. If no honest players abort, then the output satisfies the security conditions of broadcast (according to Definition 2).*
2. *(Completeness): If all players are honest, all players accept (and therefore achieve broadcast without error).*
3. *(Fairness): If any honest player aborts then the adversary receives no information about the sender's bit.*

A protocol for detectable broadcast was presented in [8] that satisfies the above definition except with some negligible error probability in the presence of an unbounded adversary (i.e., with statistical security) corrupting arbitrary many parties $(t < n)$. The protocol for detectable broadcast given by [8] requires $t + 5$ rounds and $O(n^8 (\log n + k)^3)$ total bits of communication, where k is a security parameter and $t < n$.

3   Rational Byzantine Agreement

We next define our model of a rational adversary and within it the definitions of rational BA. A *rational adversary* is characterized by some utility function which describes his preference over possible outcomes of the protocol execution. In the following we describe generic definitions of security in the presence of such an adversary; subsequently, we specify a natural class of utilities for an adversary attacking a BA protocol. Towards the end of the section, we also study the relation between the traditional and the rational definition of BA.

The Adversary's Utility. In any analysis of security against rational adversaries, one needs to define the adversaries' behavior. The first step to doing this is to define their utility, which provides a method for deciding which outcomes an adversary (or any other rational player) prefers to which others. We present a definition of utility that we believe is natural, reasonable, and can be worked with easily. In particular, we consider real utility, i.e., the utility is described by real numbers associated with particular outcomes. For simplicity, we limit ourselves to protocols that are attempting to broadcast or agree on a single bit. The adversary's utility is defined on the following events: (1) All honest players output (agree on) 1, (2) all honest players output (agree on) 0, and (3) honest players have disagreeing output. In particular we define the utility function of the adversary as follows: For values $u_0, u_1, u_2 \in \mathbb{R}$:

$$U[\text{agreement on } 0] := u_0, \ U[\text{agreement on } 1] := u_1, \text{ and } U[\text{disagreement}] := u_2$$

For simplicity we assume that the values u_0, u_1, and u_2 are distinct, but all our proofs go through even if some of them are equal. We assume that rational players will choose from the strategies available to them the one that results in the most preferred outcome. However, since strategies and the protocol can be randomized, a particular set of strategies will imply not a particular outcome but a particular distribution over outcomes. The utility of a distribution is then the expected value of the utility of an outcome drawn from that distribution.

Definition 4 (Utility). *An* expectation utility *is a utility that conforms to the following condition. Using D_z to represent the probability distribution where outcome $z \in Z$ occurs with probability 1, we require that $U(D) = E[U(D_z)|z \leftarrow D]$.*

The above utility function corresponds, of course, to a substantial simplification of possible outcomes. For example, some sorts of disagreement could be preferred over any unanimous output while other types are disliked. Nevertheless, these outcomes capture a meaningful portion of potential outcome variation. In order to maximize the strength of our results, we assume that the adversary knows the inputs of the honest players (which are disclosed very early in most protocols anyway) and can therefore choose its strategy to maximize utility for that particular input set.

Definition. We assume that all corrupted players are colluding. Equivalently, there is a single adversary that directs the actions of up to t (non-adaptively) corrupted players. The other players are honest, meaning that rather than acting according to their selfish interests they simply run the protocol as specified. This means that the "game" we are considering actually only has one player. We are essentially considering what is a Nash equilibrium strategy for the adversary. However, the Nash equilibrium of a one-player game is simply a straightforward utility-optimization, so we leave out the complexities of Nash equilibria in our definition. When we refer to a "strategy" we mean simply a function that takes as input the view of the adversary so far and outputs its next message/action.

Definition 5 (Perfect security). *A protocol for broadcast or consensus is perfectly secure against rational adversaries controlling t players with utility U if for every t-adversary there is a strategy S such that for any choice of input for honest players*

1. *(S is tolerable): S induces a distribution of final outputs D in which no security condition is violated with nonzero probability, and*
2. *(S is Nash): For any strategy $S' \neq S$ with induced output distribution D': $U(D) \geq U(D')$.*

In addition to this standard notion, we will be considering a definition following from statistical equilibria. Here we introduce a security parameter k. The strategy sees the security parameter at the beginning of the game and can alter its behavior based on that parameter. We require not that the security-respecting strategy be perfectly optimal but that it is within a negligible distance to optimal. This means that the incentive to deviate could be made arbitrarily small, and would get extremely small very quickly as the security parameter is raised.

Definition 6 (Statistical security). *A protocol for broadcast or consensus is statistically secure against rational adversaries controlling t players with utility U if for every t-adversary there is a strategy S such that for any choice of input for honest players S induces a distribution of final outputs D_k when the security parameter is k and the following properties hold:*

1. *(S is tolerable): no security condition is violated with nonzero probability in D_k for any k, and*
2. *(S is statistical Nash): for any strategy $S' \neq S$ with induced output distributions D'_k there is a negligible function $negl(\cdot)$ such that $U(D_k) + negl(k) > U(D'_k)$.*

Remark (Statistical tolerability and honestly perfect protocols). The above definition requires that the strategy S is *perfectly* tolerable. One could weaken this definition to require *statistical tolerability*, i.e., require that the tolerability property is satisfied except with some negligible probability. We argue that this does not make a difference for any protocol which, assuming no party is corrupted, satisfies the properties of BA with perfect security (we refer to such protocols as *honestly perfect*). Indeed, for an honestly perfect protocol there exists a strategy S^H, namely the strategy corresponding to honestly executing the protocol, which is perfectly tolerable. Let D_k^H denote the distribution induced by S_k^H and D_k denote the distribution induced by the optimal strategy S from the above definition where we require that S is only statistically tolerable. The statistical tolerability of S implies that $U(D_k) = U(D_k^H) \pm negl(k)$. This, combined with the fact that S is statistical Nash, implies that $U(D_k^H) + negl(k) \geq U(D_k')$ for all D_k'. Hence, D_k^H is statistically Nash and perfectly tolerable.

We note that a *computational* security definition could also be considered. Such a definition is equivalent to the statistical case except that the strategy function is required to be computable in polynomial time (in k). We do not consider computational security in this work. However, all our statements about feasibility with statistical security hold also for computational security.

Relation to the Traditional Definition. In the following we show that rational BA reduces to traditional BA. The proof is based on the observation that if a protocol is secure according to the traditional definition of BA, then in RBA every adversarial strategy is Nash.

Theorem 1 (BA implies RBA). *If protocol Π perfectly securely realizes traditional consensus (resp., broadcast) in the presence of a (non-rational) t-adversary, then Π is perfectly secure for consensus (resp., broadcast) against rational t-adversaries with utility U. The statement holds also for statistical security assuming the protocol π is honestly perfect.*[1]

4   Rational Byzantine Agreement: Basic Results

In this section, we motivate the study of feasibility of rational BA by demonstrating that some of the results that are taken for granted in the traditional BA literature become invalid in the rational setting.

The Traditional Impossibility of Consensus Fails. It is well-known that when $t \geq n/2$, there exists no consensus protocol which tolerates a t-adversary, even when the parties have access to a broadcast channel. The idea of the proof is the following: Consider the setting where the first $n/2$ of the parties have input 0, and the remaining have input 1. Assume the following adversarial scenarios: (A) the adversary corrupts the first $n/2$ or (B) the adversary corrupts the last $n/2$ parties; in both scenarios the adversary has the corrupted parties execute their correct protocol. In Scenario A, the honest parties should all output 1, whereas in Scenario B they should output 0. Consider now a third scenario (Scenario C) where the adversary does not corrupt any party. Because this Scenario is indistinguishable from Scenario B, the first half of parties should output 0; however, because Scenario C is indistinguishable from Scenario A, the second half of parties should output 1, which leads to contradiction.

We show that in the rational setting this impossibility does not, in general, hold: Consider a rational adversary with utility $u_2 > u_1 > u_0$. Then, as the following lemma suggests, assuming a (traditional) broadcast channel, there exists a consensus protocol tolerating arbitrary many parties, i.e., $t < n$, even with perfect security. The protocol, denoted as Π' works as follows:

---

[1] Recall that a BA protocol is honestly perfect if it satisfies the perfect security definition in the absence of an adversary.

Protocol $\Pi'(v_1, \ldots, v_n)$
1. Every party P_i broadcasts his input v_i.
2. If all parties broadcast the same value then output it, otherwise output 0.

The idea of the proof is that the adversary will never try to introduce an inconsistency, as if he does so he will be punished with his worst preferred outcome (i.e., 0).

Lemma 1. *The protocol Π' described above is (perfectly) secure for consensus against rational t-adversaries with $t < n$ and utilities satisfying $u_2 > u_1 > u_0$.*

The Traditional Reduction of Consensus to Broadcast Fails. Traditional consensus and broadcast are known to be equivalent assuming $t < n/2$ parties are corrupted. The idea is the following: assuming consensus, broadcast can be achieved by the sender sending his input to every party and then invoking consensus on the received values. Similarly, assuming broadcast (and $t < n/2$) consensus can be achieved by having every party broadcast his input and taking the majority of the broadcasted values he receives to be his output.

Surprisingly, the above straightforward reduction of consensus to broadcast does not transfer through to the rational setting. Informally, the reason for the failure of the reduction is the inherent incomposability issue that appears in most rational security models. In particular, for the case of the above reduction, it is possible that when attacking a consensus protocol that uses broadcast protocols as subroutines, that an adversary can achieve a *desired* outcome in the consensus protocol by violating the security of the broadcast subroutines in ways that would seem, on their own, *undesirable*.

Due to space limitations, we refer the reader to the full version of this work for a concrete description of our counterexample and the corresponding analysis.

Luckily, the reduction in the other direction is still successful. The proof can be found in the full version of this paper.

Theorem 2. *Assume that a consensus protocol exists that is secure against rational adversaries with a particular utility. A protocol can be constructed for broadcast that is secure against rational adversaries with the same utility.*

Equivalence of Statistical and Perfect Security (Setup-Independent) Perhaps one of the most unexpected differences between traditional and rational BA is the fact that in the rational setting (with real utilities), a setup does not offer anything with respect to feasibility in the information theoretic setting, as perfect security is possible for $t < n$. This is in contrast to the traditional BA where a setup is known to bring the exact bound for statistical complexity from $t < n/3$ (for both consensus and broadcast) down to $t < n/2$ for consensus and $t < n$ for broadcast. The following theorem states that the two levels of information theoretic security, i.e., perfect and statistical security, are equivalent in the rational setting. The idea for reducing perfect to statistical security is the following: because the values u_0, u_1, and u_2 are real numbers, in any statistical protocol one can fix the security parameter to be large enough, so that the adversary does not any more have an incentive to cheat, which will lead to a perfectly secure protocol. A proof can be found in the full version of this work.

Theorem 3. *There exists a statistically secure protocol for rational consensus (resp., broadcast) tolerating some adversary $\mathcal{A}$ if and only if there exists a perfectly secure protocol for rational consensus (resp., broadcast) tolerating $\mathcal{A}$.*

5   Feasibility Assuming Complete Knowledge

In this section, we give a complete characterization of feasibility of RBA for information-theoretic security. Note that as implied by Theorem 3, the bound for statistical and perfect security is the same. In fact, this bound is $t < n$ independent of the adversary's preference. This is stated in the following theorem. Due to space limitations, we only describe the idea of the proof and sketch the main argument for some of the cases, and refer to the full version of this work for a complete handling of all the cases.

Theorem 4. *There exists a protocol for perfectly secure Byzantine agreement tolerating a rational t-adversary, where the utilities $u_0, u_1, u_2 \in \mathbb{R}$ are known, tolerating arbitrarily many corruptions, i.e., $t < n$. The statement holds both for broadcast and consensus.*

Proof (sketch). The general idea is, as in the proof of Lemma 1, to force the adversary play the strategy which guarantees the security of the protocol by having the protocol punish him in case he does not. Note that, by Theorem 2, it suffices to describe consensus protocols; furthermore, by Theorem 3, it suffices to achieve statistical security. The proof considers two cases: (1) the adversary's most favorable choice is *not* disagreement, and (2) the adversary's most favorable choice is disagreement. In the first case, the following consensus protocol works:

1. Every player P_i sends his input v_i to every player P_j.
2. For every P_j: if the same value was received from every P_i then output it, otherwise output the bit b' which the adversary prefers least (i.e., $u_{b'} < \min\{u_2, u_{1-b'}\}$).

Intuitively, the above protocol is secure, as when there is pre-agreement among the honest parties, then the adversary has no incentive to destroy it, and when there is disagreement they will output b'.

The somewhat more involved setting occurs when the top choice of the adversary is to force disagreement. We consider the following consensus protocol:

1. Each P_i uses detectable broadcast [7] to broadcast his input v_i.
2. If any abort occurs or there is disagreement among the broadcasted value, then output the adversary's least preferred bit b' (i.e., $u_{b'} < \min\{u_2, u_{1-b'}\}$). Otherwise output the value broadcasted by all parties.

The fact that the above protocol is (statistically) secure is argued as follows: Consistency follows trivially from the consistency of detectable broadcast; furthermore, because the adversary has no incentive to break the detectable broadcast protocol, and, by the security of detectable broadcast, when it does not abort it satisfies the correctness property, the following adversarial strategy is a Nash equilibrium: allow all honest senders to broadcast their input and have the corrupted senders broadcast $1 - b'$.

6   Feasibility with Partial Knowledge

So far we have been considering only cases where the protocol is designed with full knowledge of the adversary's preferences, but it is also possible to consider cases where the adversary's preferences are not fully known. The goal is to guarantee security against any adversary that has preferences consistent with the limited information that we have. If no information about the adversary's preferences is known, this reduces to the traditional setting of a malicious adversary. If some information exists, however, the situation can be more interesting.

In the full-information settings we have been considering up to this point, statistical and perfect security are provably equivalent (Theorem 3). This result does not hold in the case of partial information. In fact, we can give protocols and impossibility results that prove that no such equivalence holds when setup is allowed. Similarly, we give results that show that when no setup is allowed, consensus and broadcast are not equivalent. In order to show these results, we consider the situation where it is known whether or not the adversary wishes to create disagreement between the parties, but it is not known what the adversary's preferences are among different potential agreeing outputs.

Disagreement is the Adversary's Most Favorable Option. If we consider the setting where disagreement is known to be the adversary's most preferred outcome, all the impossibility proofs from the traditional world apply. We can therefore deduce that the bounds for both broadcast and consensus are the same as in the traditional setting. We state this formally below and refer the reader to the full version of this work for the proof. (This is a tight bound, since it matches the possibility result from the malicious setting, which of course also applies in any rational setting.)

Theorem 5. *Assuming $n \geq 3$, there does not exist a perfectly or statistically secure rational consensus protocol that tolerates any t-adversary with $t \geq n/3$ and disagreement as the most-preferred outcome.*

Disagreement is the Adversary's Least Favorable Option. Finally, we consider the case where the adversary wants to avoid disagreement, but has unknown preferences on the other outcomes. This case is interesting because it provides an instance where what is possible is provably different for broadcast than it is for consensus, in contrast to the traditional setting of a malicious adversary (with perfect security). The proof of the following theorem can be found in the full version.

Theorem 6. *Assuming $n \geq 3$, there exists a perfectly secure rational consensus protocol tolerating any t-adversary with disagreement as the least-preferred outcome if and only if $t < n/2$. The statement holds also for statistical security.*

We complete our characterization by looking at feasibility of broadcast for the case where disagreement is the adversary's least-preferred outcome. As shown in the following theorem, in that case broadcast can be achieved, by the trivial multi-send protocol, tolerating an arbitrary number of corruptions. Combined with Theorem 6, this proves that in this setting perfectly secure broadcast is easier to achieve than consensus.

Theorem 7. *There exists a perfectly secure rational broadcast protocol tolerating any t-adversary for $t < n$ with disagreement as the least-preferred outcome. The statement holds for statistical security as well.*

Proof. We use the following perfectly secure protocol: The sender sends his input to every party who outputs the value received from the sender. If $t = n - 1$ the security conditions are trivially satisfied. For the case where there are at least two honest players, we consider two cases. In the first case, the sender is honest. As a result, all honest players are sent the correct output and no error is made. In the second case, the sender is not honest. In this case, the adversary would not have the sender send disagreeing messages to honest parties, since disagreement is the least-preferred outcome. However, because the sender is dishonest, any agreeing output from the honest parties is consistent with the security conditions, so no security violation can occur. Statistical security follows immediately, since it is a weaker definition.

Acknowledgments. We thank Dov Gordon for collaboration during the early stages of this work [9].

References

1. Abraham, I., Dolev, D., Gonen, R., Halpern, J.: Distributed computing meets game theory: robust mechanisms for rational secret sharing and multiparty computation. In: PODC 2006, pp. 53–62. ACM Press (2006)
2. Aiyer, A.S., Alvisi, L., Clement, A., Dahlin, M., Martin, J.-P., Porth, C.: BAR fault tolerance for cooperative services. In: SOSP 2005, pp. 45–58. ACM (2005)
3. Asharov, G., Canetti, R., Hazay, C.: Towards a Game Theoretic View of Secure Computation. In: Paterson, K.G. (ed.) EUROCRYPT 2011. LNCS, vol. 6632, pp. 426–445. Springer, Heidelberg (2011)
4. Baum-Waidner, B., Pfitzmann, B., Waidner, M.: Unconditional Byzantine Agreement With Good Majority. In: Jantzen, M., Choffrut, C. (eds.) STACS 1991. LNCS, vol. 480, pp. 285–295. Springer, Heidelberg (1991)
5. Bei, X., Chen, W., Zhang, J.: Distributed consensus resilient to both crash failures and strategic manipulations, arXiv 1203.4324 (2012)
6. Clement, A., Li, H.C., Napper, J., Martin, J.-P., Alvisi, L., Dahlin, M.: BAR primer. In: DSN 2008, pp. 287–296. IEEE Computer Society (2008)

7. Fitzi, M., Gisin, N., Maurer, U., von Rotz, O.: Unconditional Byzantine Agreement and Multi-party Computation Secure against Dishonest Minorities from Scratch. In: Knudsen, L.R. (ed.) EUROCRYPT 2002. LNCS, vol. 2332, pp. 482–501. Springer, Heidelberg (2002)

8. Fitzi, M., Gottesman, D., Hirt, M., Holenstein, T., Smith, A.: Detectable Byzantine agreement secure against faulty majorities. In: PODC 2002, pp. 118–126. ACM Press (2002)

9. Gordon, S.D., Katz, J.: Byzantine agreement with a rational adversary. Rump session presentation, Crypto 2006 (2006)

10. Gordon, S.D., Katz, J.: Rational Secret Sharing, Revisited. In: De Prisco, R., Yung, M. (eds.) SCN 2006. LNCS, vol. 4116, pp. 229–241. Springer, Heidelberg (2006)

11. Groce, A., Katz, J.: Fair Computation with Rational Players. In: Pointcheval, D., Johansson, T. (eds.) EUROCRYPT 2012. LNCS, vol. 7237, pp. 81–98. Springer, Heidelberg (2012)

12. Halpern, J., Teague, V.: Rational secret sharing and multiparty computation: Extended abstract. In: STOC 2004, pp. 623–632. ACM Press (2004)

13. Izmalkov, S., Micali, S., Lepinski, M.: Rational secure computation and ideal mechanism design. In: FOCS 2005, pp. 585–595. IEEE Computer Society Press (2005)

14. Katz, J.: Bridging Game Theory and Cryptography: Recent Results and Future Directions. In: Canetti, R. (ed.) TCC 2008. LNCS, vol. 4948, pp. 251–272. Springer, Heidelberg (2008)

15. Lamport, L., Shostak, R.E., Pease, M.C.: The Byzantine generals problem. ACM Trans. Programming Language Systems 4(3), 382–401 (1982)

16. Li, H.C., Clement, A., Wong, E.L., Napper, J., Roy, I., Alvisi, L., Dahlin, M.: Bar gossip. In: OSDI 2006, pp. 191–204. USENIX Association (2006)

17. Ong, S.J., Parkes, D.C., Rosen, A., Vadhan, S.: Fairness with an Honest Minority and a Rational Majority. In: Reingold, O. (ed.) TCC 2009. LNCS, vol. 5444, pp. 36–53. Springer, Heidelberg (2009)

18. Pease, M., Shostak, R.E., Lamport, L.: Reaching agreement in the presence of faults. Journal of the ACM 27(2), 228–234 (1980)

19. Pfitzmann, B., Waidner, M.: Unconditional Byzantine Agreement for any Number of Faulty Processors. In: Finkel, A., Jantzen, M. (eds.) STACS 1992. LNCS, vol. 577, pp. 339–350. Springer, Heidelberg (1992)

Random Hyperbolic Graphs:
Degree Sequence and Clustering*
(Extended Abstract)

Luca Gugelmann[1], Konstantinos Panagiotou[2], and Ueli Peter[1]

[1] Institute of Theoretical Computer Science, ETH Zurich, 8092 Zurich, Switzerland
[2] Department of Mathematics, University of Munich, 80333 Munich, Germany

Abstract. Recently, Papadopoulos, Krioukov, Boguñá and Vahdat [IN-FOCOM'10] introduced a random geometric graph model that is based on hyperbolic geometry. The authors argued empirically and by some preliminary mathematical analysis that the resulting graphs have many of the desired properties for models of large real-world graphs, such as high clustering and heavy tailed degree distributions. By computing explicitly a maximum likelihood fit of the Internet graph, they demonstrated impressively that this model is adequate for reproducing the structure of such with high accuracy.

In this work we initiate the rigorous study of random hyperbolic graphs. We compute exact asymptotic expressions for the expected number of vertices of degree k for all k up to the maximum degree and provide small probabilities for large deviations. We also prove a constant lower bound for the clustering coefficient. In particular, our findings confirm rigorously that the degree sequence follows a power-law distribution with controllable exponent and that the clustering is nonvanishing.

1 Introduction

Modeling the topology of large networks is a fundamental problem that has attracted considerable attention in the last decades. Networks provide an abstract way of describing relationships and interactions between elements of complex and heterogeneous systems. Examples include technological networks like the World Wide Web or the Internet, biological networks like the human brain, and social networks which describe various kinds of interactions between individuals.

An accurate mathematical model can have enormous impact on several research areas. From the viewpoint of computer science, an obvious benefit is that it could enable us to design more efficient algorithms that exploit the underlying structures. Moreover, the process of modeling may suggest and reveal novel types of qualitative network features, which become patterns to look for in datasets. Finally, an appropriate model allows us to generate artificial instances, which

---

* See http://arxiv.org/abs/1205.1470 for the full version.

A. Czumaj et al. (Eds.): ICALP 2012, Part II, LNCS 7392, pp. 573–585, 2012.
© Springer-Verlag Berlin Heidelberg 2012

resemble realistic instances to a high degree, for simulation purposes. Unfortunately, from today's point of view, a significant proportion of the current literature is devoted only to experimental studies of properties of real-world networks, and there has been only little rigorous mathematical work.

A reasonable model for real-world networks must be able, when setting the parameters appropriately, to replicate the salient features of the real-world graphs under consideration. Moreover, the model should be mathematically tractable and simple enough to be of use in large scale simulations. There are plenty of models that satisfy the first criterion, but are hard to analyze from a mathematical viewpoint. On the other hand, there exists a plethora of analytically tractable models, which unfortunately do not yet replicate satisfactory enough the properties that are observed in large networks.

In this work we initiate the rigorous study of a class of models for large networks, the so-called *random hyperbolic graphs*. Such graphs were shown empirically to have startling similarities with several real-world networks, and in particular with the Internet graph. Before we describe the model and our results, let us proceed with considering some properties of large networks in more detail.

Properties of large networks. Since the 60's, the study of networks of various kinds has grown into a significant research area. One of the initiators in this field, the sociologist Stanley Milgram, investigated the network that is obtained from the relationships among people [15,20]. In his work he discovered what is nowadays known as the *small-world phenomenon*, which postulated that the distance between two random people is on average between five and six.

Another property that is found in many networks addresses the degree distribution. In a celebrated paper, Faloutsos et. al. [11] observed that the Internet exhibits a so-called scale-free nature: the degree sequence follows approximately a *power-law* distribution, which means that the number of vertices of degree k is proportional to some inverse power of k, for all sufficiently large k. This sets such a network dramatically apart from e.g. a typical Erdős-Rényi random graph and stirred significant interest in exploring the causes of this phenomenon. From today's viewpoint, it is well-known that many graphs have a *heavy-tailed* degree distribution, which may be close to a power-law or a log-normal or a combination of these distributions (see [16] and references therein).

A third distinctive feature of large real-world graphs is the appearance *clustering* [17,19,21]. The network average of the probability that two neighbors of a random vertex are also directly connected is called the clustering coefficient. Measured clustering coefficients for social networks are typically tens of percent, and similar values have been measured for many other networks as well, including technological and biological ones.

Models of large networks. Perhaps the first step towards a random graph model for real-world networks was made by Watts and Strogatz [21] in 1998, who addressed the small-world phenomenon and clustering and gave reasons for its emergence. However, the degree distribution of the generated graphs follows a Poisson distribution, and thus is not heavy tailed. Barabási and Albert proposed

[3] that the cause for power-law degree distributions is preferential attachment: the networks evolve continuously by the addition of new vertices, and each new vertex chooses its neighbors with a probability that is proportional to their current degree. This model was shown by Bollobás et al. [4] to produce power-law degree distributions, but on the other hand it generates graphs that typically have a vanishing clustering coefficient [5]. Nevertheless, the Barabási-Albert model was the beginning of a vast series of proposed models that suggested mechanisms according to which a network can evolve (see e.g. [1,6,7,8,9,13,14] for a non-exhaustive but representative list). Unfortunately, all this processes fail in producing a power-law degree distribution and a nonvanishing clustering coefficient simultaneously.

Hyperbolic random graphs. Recently, a very promising model was introduced by Papadopoulos, Krioukov, Boguñá and Vahdat [18]. The authors demonstrated impressively that complex scale-free network topologies with high clustering coefficients emerge naturally from hyperbolic metric spaces. Their model, which we will denote by *random hyperbolic graph*, consists in its simplest variant of the uniform distribution of n vertices within a disk of radius $R = R(n)$ in the hyperbolic plane, where two vertices are connected if their hyperbolic distance is at most R. The authors show via simulations and some preliminary theoretical analysis that the generated graphs exhibit a power-law degree distribution, whose exponent can be tweaked via model parameters. Further, the authors indicate that with a slightly more complex model they can also control the clustering of the generated graphs to bring it in line with real-world networks.

Our contribution. Regarding the experiments just described, it seems at least fair to say that random hyperbolic graphs provide an attractive model that has a high potential for describing the characteristics of many real-world networks. Moreover, a simple formulation and a strong affinity to random geometric graphs indicate that this model might be mathematically tractable. In this work we show that this is indeed the case and initiate thereby the rigorous study of hyperbolic random graphs. First, we prove a constant lower bound on the clustering coefficient of hyperbolic random graphs which confirms the claimed high clustering. We then show that the expected degree distribution indeed follows a power-law *across all scales*, i.e., even up to the *maximum degree*. Note that in the seminal papers [18] and [12] the degree distribution was also considered, however only for constant degrees and without any error guarantees. In addition, we prove small probabilities for large deviations, i.e. we show that sampling from this distribution returns with high probability a graph with the desired properties, which is crucial for validating experimental results. We also compute tight bounds for the average and maximum degree that hold with high probability.

There are many models for which either a power-law degree sequence [1,4,7] or a large clustering coefficient [21] has been proven. But this is the first model that provably satisfies both properties. Note also that while there are some models (see for example [4] and [7]) for which a power-law degree distribution up to polynomially large degrees can be showed, to the best of our knowledge this is the first rigorous proof that the degree distribution of a random graph model is scale-free *up to the maximum degree*. Further, our results reveal some fundamental combinatorial properties of the model, thus setting the groundwork for further theoretical investigations. We strongly believe that these facts together with the nice combinatorial structure of the model make it attractive for the theoretical computer science and random graph community.

2   Model and Results

Let us begin this section with a few facts about the geometry of hyperbolic planes. We will restrict ourselves to the most basic notions, and refer the reader to e.g. [2] and many references therein for an extensive introduction.

First of all, there are many equivalent representations of the hyperbolic plane, each one highlighting different aspects of the underlying geometry. We will consider here the so-called *native* representation, which was described by Papadopoulos et. al. in [18], as it is most convenient for defining the model of random hyperbolic graphs.

One basic feature of the hyperbolic plane is that it is isotropic, meaning that the geometry is the same regardless of direction. In other words, we can distinguish an arbitrary point, which we call the *center* or the *origin*. In the native representation of the hyperbolic plane we will use polar coordinates (r, θ) to specify the position of any vertex v, where the radial coordinate r equals the hyperbolic distance of v from the origin. Given this notation, the distance d of two vertices with coordinates (r, θ) and (r', θ') can be computed by solving the equation

$$\cosh(d) = \cosh(r)\cosh(r') - \sinh(r)\sinh(r')\cos(\theta - \theta'), \qquad (1)$$

where $\cosh(x) = (e^x + e^{-x})/2$ and $\sinh(x) = (e^x - e^{-x})/2$. For our purposes we will denote from now on by $d(r, r', \theta - \theta')$ the solution of (1) for d.

The crucial difference between the Euclidean and the hyperbolic plane is that the latter contains in a well-defined sense more "space". More specifically, a circle with radius r has in the Euclidean plane a length of $2\pi r$, while its length in the hyperbolic plane is $2\pi \sinh(r) = \Omega(e^r)$. In other words, a circle in the hyperbolic plane has a length that is exponential in its radius as opposed to linear.

Based on the above facts, the authors of [18] defined a model of random geometric graphs that in its simplest version consists of the uniform distribution of n points into a hyperbolic disk of radius $R = R(n)$ around the origin. Two points in this disk are connected by an edge only if they are at hyperbolic distance at most R from each other, as defined in (1). More precisely, note that the total area of a circle of radius r equals $2\pi(\cosh(r)-1)$. To choose the n points uniformly at random in the hyperbolic disk of radius R it suffices to choose for each a polar coordinate (r, θ) such that θ is chosen uniformly at random in the interval, say, $(-\pi, \pi]$ and its radial coordinate r is drawn according to the distribution with density function $\sinh(r)/(\cosh(R) - 1)$, where $0 \le r \le R$. To add flexibility to the model, the authors of [18] use a slightly different density function for the radial coordinate: $\alpha \sinh(\alpha r)/(\cosh(\alpha R) - 1)$, where $\alpha > 1/2$. For $\alpha < 1$ this favors points closer to the center, while for $\alpha > 1$ points with radius closer to R are favored. For $\alpha = 1$ this corresponds to the uniform distribution.

Let us now proceed to a formal definition of the model. With all the above notation at hand, the random hyperbolic graph $G_{\alpha,C}(n)$ with n vertices and parameters α and C is defined as follows.

Definition 1 (Random Hyperbolic Graph $G_{\alpha,C}(n)$). *Let $\alpha > 1/2$, $C \in \mathbb{R}$, $n \in \mathbb{N}$, and set $R = 2\log n + C$. The random hyperbolic graph $G_{\alpha,C}(n)$ has the following properties.*

- *The vertex set V of $G_{\alpha,C}(n)$ is $V = \{1, \ldots, n\}$.*
- *Every $v \in V$ is equipped with random polar coordinates (r_v, θ_v), where $r_v \in [0, R]$ has density $p(r) := \alpha \frac{\sinh(\alpha r)}{\cosh(\alpha R) - 1}$ and θ_v is drawn uniformly from $[-\pi, \pi]$.*
- *The edge set of $G_{\alpha,C}(n)$ is given by $\{\{u, v\} \subset \binom{V}{2} : d(r_u, r_v, \theta_u - \theta_v) \leq R\}$.*

The restrictions in the model parameters, especially the condition $\alpha > 1/2$ and the definition of R will become clear in the sequel. Informally speaking, the choice of R guarantees that the resulting graph has a bounded average degree (depending on α and C only). If $\alpha \leq 1/2$, then the degree sequence is so heavy tailed that this is impossible.

Let us mention at this point that in [18] an even more general model was also proposed. There, each pair of vertices is connected with a probability that may depend on the hyperbolic distance of those vertices. In particular, this probability is large if the vertices have distance $\leq R$, and becomes quickly smaller when the distance is larger than R. We will not treat this model here.

Let us next describe the results that we show for $G_{\alpha,C}(n)$. First of all, we study the clustering of hyperbolic random graphs. The local clustering coefficient of a vertex v is defined by

$$\bar{c}_v = \begin{cases} 0 & \text{if } \deg(v) < 2 \\ \frac{|\{\{u_1, u_2\} \in E \mid u_1, u_2 \in \Gamma(v)\}|}{\binom{\deg(v)}{2}} & \text{else} \end{cases}, \tag{2}$$

where $\Gamma(v) := \{u \mid \{u, v\} \in E\}$. The global clustering coefficient of a graph $G = (V, E)$ is the average over all local clustering coefficients

$$\bar{c}(G) = \frac{1}{n} \sum_{v \in V} \bar{c}_v. \tag{3}$$

Our first theorem gives a constant lower bound on the global clustering coefficient which holds with high probability.

Theorem 2. *Let $\alpha > 1/2$, $C \in \mathbb{R}$ and $\bar{c} = \bar{c}(G_{\alpha,C}(n))$. Then $\mathbb{E}[\bar{c}] = \Theta(1)$ and with high probability $\bar{c} = (1 + o(1))\mathbb{E}[\bar{c}]$.*

Then we study the degree sequence, and provide sharp bounds for the number of vertices of degree k.

Theorem 3. *Let $\alpha > 1/2$ and $C \in \mathbb{R}$. There is a $\delta > 0$ such that with high probability, for all $0 \leq k \leq n^{\delta'}$, where $\delta' < \delta$, the fraction of vertices of degree exactly k in $G_{\alpha,C}(n)$ is*

$$(1 + o(1)) \frac{2\alpha e^{-\alpha C}}{k!} \left(\frac{2\alpha}{\pi(\alpha - 1/2)}\right)^{2\alpha} \left(\Gamma(k - 2\alpha) - \int_0^\xi t^{k-2\alpha-1} e^{-t} dt\right), \quad (4)$$

where $\Gamma(x) = \int_0^\infty t^{x-1} e^{-t} dt$ denotes the Gamma function and $\xi = \frac{2\alpha}{\pi(\alpha-1/2)} e^{-C/2}$. If $n^\delta \leq k \leq \frac{n^{1/2\alpha}}{\log n}$ then with high probability the fraction of vertices of degree at least k in $G_{\alpha,C}(n)$ is

$$(1 + o(1)) \left(\frac{2\alpha}{\pi(\alpha - 1/2)}\right)^{2\alpha} e^{-\alpha C} k^{-2\alpha}. \quad (5)$$

This result demonstrates that the degree sequence of $G_{\alpha,C}(n)$ is a power-law with exponent $2\alpha + 1 > 2$. To see this, note that for sufficiently large k we have that $\Gamma(k - 2\alpha)/k! = \Theta(k^{-2\alpha-1})$, i.e., (4) and (5) imply that the number of vertices of degree k in $G_{\alpha,C}(n)$ is $(1 + o(1))c_{\alpha,C} k^{-2\alpha-1} n$, for some $c_{\alpha,C} > 0$.

Our next result gives bounds for the average degree of $G_{\alpha,C}(n)$.

Theorem 4. *Let $\alpha > 1/2$ and $C \in \mathbb{R}$. Then the average degree of $G_{\alpha,C}(n)$ is $(1 + o(1)) \frac{2\alpha^2 e^{-C/2}}{\pi(\alpha-1/2)^2}$.*

Note that Theorem 3 and Theorem 4 confirm the results in [18].

Finally, we give sharp bounds for the maximum degree in $G_{\alpha,C}(n)$.

Theorem 5. *Let $\alpha > 1/2$ and $C \in \mathbb{R}$. Then the maximum vertex degree of $G_{\alpha,C}(n)$ is with high probability $n^{1/(2\alpha)+o(1)}$.*

Due to space constraints in this extended abstract we only show the proof for Theorem 2. The proofs of theorems 4 and 5 are straightforward applications of Lemma 7 in the next section. The proof for Theorem 3 is more involved and requires considering vertices with low and high degree separately. All missing proofs can be found in the full version of this paper.

3   Properties of the Model

Recall that according to Definition 1 the mass of a point $p = (r, \theta)$ is $f(r) = \frac{\alpha \sinh(\alpha r)}{2\pi(\cosh(\alpha R)-1)}$, and does only depend on the radial coordinate of p. Accordingly, we define the probability measure $\mu(S)$ of a point set S as

$$\mu(S) = \int_S f(y) dy. \quad (6)$$

A vertex located at (θ, r) is connected to all vertices with coordinates (θ', r') such that $d(r, r', \theta - \theta') \leq R$. The *ball* of radius x around (r, θ) is defined by

$$B_{r,\theta}(x) = \{(r', \theta') \mid d(r, r', \theta - \theta') \leq x\}. \quad (7)$$

Since in the definition of our model two points are connected if and only if they are at distance at most R, we will typically consider the intersection $B_{r_1,\theta_1}(R) \cap B_{0,0}(R)$ which corresponds to the point set in which all vertices are connected to a fixed vertex at (r_1, θ_1). By (6) we can determine the probability measure of such a set by integrating $f(y)$ over all points in the set. In our specific case we achieve this by integrating first over all $y \in [0, R]$ and then over all θ such that $d(r_1, y, \theta_1 - \theta) \leq R$. As $f(y)$ does not depend on θ we are only interested in the range of θ for which this inequality is satisfied. One extremal of $(\theta_1 - \theta)$ for which it is satisfied is clearly

$$\theta_{r_1}(y) = \arg \max_{0 \leq \phi \leq \pi} \{d(r_1, y, \phi) \leq R\} = \arccos\left(\frac{\cosh(r_1)\cosh(y) - \cosh(R)}{\sinh(r_1)\sinh(y)}\right). \tag{8}$$

Because of symmetry of the cosine the other extremal is $-\theta_{r_1}(y)$ and we therefore have to integrate from $-\theta_{r_1}(y)$ to $\theta_{r_1}(y)$ as all those angles θ satisfy $d(r_1, y, \theta) \leq R$. Therefore we arrive at the following expression:

$$\mu(B_{r,\theta}(R) \cap B_{0,0}(R)) = \int_0^R \int_{-\theta_r(y)}^{\theta_r(y)} f(y)\,d\theta\,dy = 2\int_0^R \int_0^{\theta_r(y)} f(y)\,d\theta\,dy. \tag{9}$$

Note that $\mu(B_{r,\theta}(x) \cap B_{0,0}(R))$ does not depend on θ and therefore we shorten it to $\mu(B_r(x) \cap B_0(R))$. Before we commence with the more technical part of this section, we quickly refresh the following basic estimates of $\cosh(x)$ and $\sinh(x)$. For all $x \geq 0$

$$\frac{e^x}{2} \leq \cosh(x) \leq e^x \quad \text{and} \quad \frac{e^x}{3} \overset{(x \geq 1/2 \ln 3)}{\leq} \sinh(x) \leq \frac{e^x}{2}. \tag{10}$$

We first state a technical lemma that gives almost tight bounds on $\theta_r(y)$.

Lemma 6. Let $0 \leq r \leq R$ and $y \geq R - r$. Then

$$\theta_r(y) = 2e^{\frac{R-r-y}{2}}\left(1 + \Theta\left(e^{R-r-y}\right)\right).$$

Our second lemma gives precise estimates for the measures of several useful combinations of balls. It is an important ingredient of proofs of our theorems. Note that the claimed formulas look a bit overloaded on the first sight; however, the derived bounds make it applicable for different purposes as we demonstrate in the later proofs.

Lemma 7. For any $0 \leq r \leq R$ and any $0 \leq x \leq R$ we have

$$\mu(B_0(x)) = e^{-\alpha(R-x)}(1 + o(1)) \tag{11}$$

$$\mu(B_r(R) \cap B_0(R)) = \frac{2\alpha e^{-r/2}}{\pi(\alpha - 1/2)}\left(1 \pm O\left(e^{-(\alpha-1/2)r} + e^{-r}\right)\right). \tag{12}$$

Further, for $x \leq R - r$

$$\mu((B_r(R) \cap B_0(R)) \setminus B_0(x)) = \frac{2\alpha e^{-r/2}}{\pi(\alpha - 1/2)}\left(1 \pm O\left(e^{-(\alpha-1/2)r} + e^{-r}\right)\right), \tag{13}$$

while for $x \geq R - r$ it holds that

$$\mu\big((B_r(R) \cap B_0(R)) \setminus B_0(x)\big) =$$
$$= \frac{2\alpha e^{-r/2}}{\pi(\alpha - 1/2)} \left(1 - \left(1 + \frac{\alpha - 1/2}{\alpha + 1/2} e^{-2\alpha x}\right) e^{-(\alpha - 1/2)(R - x)}\right).$$
$$\cdot \left(1 \pm O\big(e^{-r} + e^{-r-(R-x)(\alpha - 3/2)}\big)\right). \quad (14)$$

Before we continue, let us give an intuitive description of the statement. Let us in particular consider (12), as the subsequent equations are refinements of it. Equation (12) states that the mass of the intersection of $B_r(R)$ and $B_0(R)$ is, up to constants and error terms, equal to $e^{-r/2}$. Recall that in $G_{\alpha,C}(n)$ every point in $B_r(R) \cap B_0(R)$ is connected to the point p with radial coordinate r and $\theta = 0$. Thus, the degree of p is a binomial distribution with parameters n and $e^{-r/2}$. In particular, if r is small, then the expected degree of p is large, and on the other hand, if $r = 2 \log n \approx R$, then the expected degree of p is $O(1)$. In other words, the closer a vertex is located to the border of the disc, the smaller its degree will be, and (12) allows us to quantify precisely the dependence.

The next lemma confirms the somewhat intuitive fact that a ball around a point has a higher measure the closer the point is located to the center of the disk.

Lemma 8. *For all $0 \leq r_0 \leq R$, all $r_0 \leq r \leq R$ and all $0 \leq x \leq R$*

$$\mu(B_0(R) \cap B_{r_0}(R)) \geq \mu(B_0(R) \cap B_r(R))$$

and

$$\mu(B_0(R) \cap B_{r_0}(R) \setminus B_0(x)) \geq \mu(B_0(R) \cap B_r(R) \setminus B_0(x)).$$

4   Proofs of the Main Results

Before we give the proofs for our theorems, let us briefly describe a technique that we use to show concentration for the clustering coefficient and the degree sequence. We will apply an Azuma-Hoeffding-type large deviation inequality (see Lemma 9 below) to show that the sum of the local clustering coefficients $X := \sum_{v \in V} c_v$ and the number of vertices of degree k are concentrated around its expectation.

In a typical setting, such concentration inequalities require some kind of *Lipschitz condition* that is satisfied by the function under consideration. In our specific setting, the functions are X and the number of vertices of degree k, and it is required to provide a bound for the maximum effect that any vertex has. However, the only a priori bound that can be guaranteed is that for example the number of vertices of degree k can change by at most n, as a vertex may connect or not connect to any other vertex. To make the situation worse, this bound is even tight, since a vertex can be placed at the center of disc, i.e., if it has radial coordinate equal to 0.

We will overcome this obstacle as follows. Instead of counting the total number of vertices of degree k and the sum of all the local clustering coefficients, we will consider only vertices that lie far away from the center of the disc, i.e., which have radial coordinate larger than βR, for some appropriate $\beta > 0$. Moreover, we will consider only vertices such that all their neighbors have a large radial coordinate as well. This restriction will allow us to bound the maximum effect on the target function, as with high probability all these vertices do not have too large degree.

More formally, we proceed as follows. We partition the vertex set of $G_{\alpha,C}(n)$ into two sets. The *inner set* $I = I(\beta)$ contains all vertices of radius at most βR while the *outer set* $O = O(\beta)$ contains all vertices of radius larger than βR.

We will use the following large deviation inequality. Let f be a function on the random variables $X_1, \ldots, X_n$ that take values in some set A_i. We say that f is Lipschitz with coefficients $c_1 \ldots c_n$ and bad event $\mathcal{B}$ if for all $x, y \in A$

$$\left| \mathbb{E}[f | X_1, \ldots X_{i-1}, X_i = x, \overline{\mathcal{B}}] - \mathbb{E}[f | X_1, \ldots X_{i-1}, X_i = y, \overline{\mathcal{B}}] \right| \leq c_i.$$

(We denote by $\overline{\mathcal{B}}$ the complement of $\mathcal{B}$.) Then the following estimates are true.

Theorem 9 (Theorem 7.1 in [10]). *Let f be a function of n independent random variables $X_1, \ldots, X_n$, each X_i taking values in a set A_i, such that $\mathbb{E}[f]$ is bounded. Assume that $m \leq f(X_1, \ldots, X_n) \leq M$. Let $\mathcal{B}$ any event, and let c_i be the Lipschitz coefficients of f. Then*

$$\Pr[\, |f - \mathbb{E}[f]\,| \, > t + (M - m)\Pr[\mathcal{B}]] \leq 2e^{-2t^2/\sum_i c_i^2} + 2\Pr[\mathcal{B}].$$

In our setting, the random variables of interest are usually functions of $X_1, \ldots, X_n$, where X_i denotes the coordinates of the ith vertex. For the clustering coefficient and the degree sequence a coordinate change can not have a large effect on the random variable as long as the degree of the corresponding vertex is small. The following lemma states that all vertices in $O(\beta)$ have not too high degrees.

Lemma 10. *Let $\alpha > 1/2$ and $0 < \beta < 1$. There is a constant $c > 0$ such that the following is true. Let*

$$\mathcal{B} := \left\{ \text{there is a vertex in } O(\beta) = B_0(R) \setminus B_0(\beta R) \text{ with degree at least } cn^{1-\beta} \right\}.$$

Then $\Pr[\mathcal{B}] = e^{-\Omega(n^{1-\beta})}.$

4.1 The Clustering Coefficient

Recall the definition of the local and global clustering coefficient in (2) and (3). We will need the following technical statement, which gives an estimate for the measure of the intersection of the balls around two coordinates (r_1, θ_1) and (r_2, θ_2) if their angle difference $\theta := |\theta_1 - \theta_2|$ is very small.

Lemma 11. *Let $\beta > 1/2$, $\beta R \leq x \leq R$, $r_1 \geq r_2 \geq x$ and $0 \leq \theta \leq e^{-r_2/2} - e^{-r_1/2}$. Then*

$$\mu\left(B_0(R) \cap B_{r_1,0}(R) \cap B_{r_2,\theta}(R) \setminus B_0(x)\right) = \mu\left(B_0(R) \cap B_{r_1}(R) \setminus B_0(x)\right). \quad (15)$$

Proof. It follows from similar observations like the ones that lead to (9) that

$$\mu\left(B_0(R) \cap B_{r_1,0}(R) \cap B_{r_2,\theta}(R) \setminus B_0(x)\right) = \int_x^R \int_{\max\{-\theta_{r_1}(y),\theta-\theta_{r_2}(y)\}}^{\min\{\theta_{r_1}(y),\theta+\theta_{r_2}(y)\}} f(y)d\phi dy.$$

For $\theta \leq e^{-r_2/2} - e^{-r_1/2}$ and $r_1 \geq r_2$, using Lemma 6, it can be verified that

$$\max\{-\theta_{r_1}(y), \theta - \theta_{r_2}(y)\} = -\theta_{r_1}(y) \text{ and } \min\{\theta_{r_1}(y), \theta + \theta_{r_2}(y)\} = \theta_{r_1}(y). \quad \square$$

Proof (Theorem 2). Let $\beta := 2/3$ and for a graph $G = (V, E)$ let

$$X := \sum_{\substack{v \in V \\ \deg(v) \geq 2}} \frac{|\{\{u_1, u_2\} \in E \mid u_1, u_2 \in \Gamma(v)\}|}{\binom{\deg(v)}{2}}$$

and

$$Y := \sum_{\substack{v \in O(\beta) \\ \deg(v) \geq 2}} \frac{|\{\{u_1, u_2\} \in E \mid u_1, u_2 \in \Gamma(v) \cap O(\beta)\}|}{\binom{\deg(v)}{2}}.$$

Clearly, $\bar{c} = \frac{X}{n}$ and $X \geq Y$. It therefore suffices to derive a constant lower bound on $\mathbb{E}[Y]$ and to show that Y is concentrated around its expectation. Let $\mathbb{E}[Y_r \mid \mathcal{E}]$ be the expected value of $\frac{|\{\{u_1,u_2\} \in E \mid u_1,u_2 \in \Gamma(v) \cap O(\beta)\}|}{\binom{\deg(v)}{2}}$ for a vertex v with radius r conditioned on the event $\mathcal{E}$ that the vertex has degree at least 2. We observe that $\mathbb{E}[Y_r \mid \mathcal{E}]$ is exactly the probability that two randomly chosen neighbors u_1 and u_2 of a vertex v at radius r are connected. In order to derive this probability for a fixed vertex v at radius r, let us suppose that u_1 is at coordinate $(y, \phi) \in B_0(R) \cap B_r(R) \setminus B_0(\beta R)$. Note that by (8) these coordinates satisfy $\beta R \leq y \leq R$ and $-\theta_r(y) \leq \phi \leq \theta_r(y)$. Moreover, the probability for the event that u_1 is at (y, ϕ) is given by

$$\frac{f(y)}{\mu(B_0(R) \cap B_r(R))}.$$

The vertex u_2 is connected to u_1 in such a way that $\{u_1, u_2\}$ contributes to Y_r only if u_2 lies in the intersection of the balls $B_0(R) \cap B_r(R) \setminus B_0(\beta R)$ and $B_0(R) \cap B_y(R) \setminus B_0(\beta R)$. Therefore, the contribution of u_2 to $\mathbb{E}[Y_r \mid \mathcal{E}]$, given the coordinates of u_2, is

$$\frac{\mu(B_{r,0}(R) \cap B_{y,\phi}(R) \cap B_0(R) \setminus B_0(\beta R))}{\mu(B_0(R)) \cap B_r(R))}.$$

Note that we choose u_2 uniformly from all neighbors of v which makes it possible that $u_1 = u_2$. However, since the degree of v is at least 2 this event happens with probability at most $1/2$. Putting all the above facts together implies that

$$\mathbb{E}[Y_r \mid \mathcal{E}] \geq \frac{1}{2} \int_{\beta R}^{R} \int_{-\theta_r(y)}^{\theta_r(y)} \frac{f(y)\mu(B_{r,0}(R) \cap B_{y,\phi}(R) \cap B_0(R) \setminus B_0(\beta R))}{(\mu(B_0(R) \cap B_r(R)))^2} d\phi dy.$$

Since the term in the integral above does not depend on the angle, we can replace the integral from $-\theta_r(y)$ to $\theta_r(y)$ by twice the integral from 0 to $\theta_r(y)$. Further, we derive a lower bound on that term by integrating the radius only from r to R. Observe also that for $r \leq y \leq R$ the upper boundary of the angle, $\theta_r(y) \overset{(\text{Lem. }6)}{=} (1 + o(1))2e^{\frac{R-r-y}{2}}$, is at least $\xi := e^{-\frac{r}{2}} - e^{-\frac{y}{2}}$ and that therefore the term above is at least

$$\frac{\int_r^R \int_0^\xi f(y) \cdot \mu(B_{r,0}(R) \cap B_{y,\phi}(R) \cap B_0(R) \setminus B_0(\beta R)) d\phi dy}{(\mu(B_0(R) \cap B_r(R) \setminus B_0(\beta R)))^2}.$$

Applying Lemma 11, this integral simplifies to

$$\frac{1}{(\mu(B_0(R) \cap B_r(R) \setminus B_0(\beta R)))^2} \int_r^R \int_0^\xi f(y) \cdot \mu(B_0(R) \cap B_y(R) \setminus B_0(\beta R)) d\phi dy$$

$$\overset{(\text{Lem. }7), (10)}{\geq} \frac{\alpha - 1/2}{12} e^{r - \alpha R} \int_r^R (e^{-r/2} - e^{-y/2}) e^{y(\alpha - 1/2)} d\phi dy$$

$$= \frac{1}{12} \left(e^{r/2 - \alpha R} \left[e^{y(\alpha - 1/2)} \right]_r^R - \frac{\alpha - 1/2}{(\alpha - 1)} e^{r - \alpha R} \left[e^{y(\alpha - 1)} \right]_r^R \right)$$

$$\geq \frac{1}{24} \left(e^{-(R-r)/2} + \frac{1}{2(\alpha - 1)} e^{-\alpha(R-r)} - \frac{\alpha - 1/2}{\alpha - 1} e^{-(R-r)} \right).$$

Let v_R be a vertex at radius R. It follows from Lemma 8 that the degree distribution of every vertex in the graph dominates the degree distribution of v_R. Therefore, for any $v \in V$

$$\Pr[\mathcal{E}] = \Pr[\deg(v) \geq 2] \geq \Pr[\deg(v_R) \geq 2] \geq \Pr[\deg(v_R) = 2]$$

$$= \binom{n-1}{2} (\mu(B_0(R) \cap B_R(R)))^2 (1 - \mu(B_0(R) \cap B_R(R)))^{n-3}$$

$$\geq n^2 e^{-R} \frac{\alpha^2}{\pi^2(\alpha - 1/2)^2} e^{-\frac{2\alpha e^{-R/2}}{\pi(\alpha - 1/2)} n} = e^{-C} \frac{\alpha^2}{\pi^2(\alpha - 1/2)^2} e^{-\frac{2\alpha e^{-C/2}}{\pi(\alpha - 1/2)}}.$$

By integrating $\mathbb{E}[Y_r \mid \mathcal{E}]$ over all $\beta R \leq r \leq R$ and multiplying with $\Pr[\mathcal{E}]$, $f(y)$ and n we get the expected value of Y

$$\mathbb{E}[Y] = n \int_{\beta R}^{R} f(y) \cdot \Pr[\mathcal{E}] \cdot \mathbb{E}[\overline{c_r} \mid \mathcal{E}] dr \geq \frac{n \cdot e^{-C} \alpha^2 e^{-\frac{2\alpha e^{-C/2}}{\pi(\alpha - 1/2)}}}{600\pi^3(\alpha - 1/2)(\alpha + 1)(\alpha + 1/2)}.$$

$$\tag{16}$$

Thus, $\mathbb{E}[Y] = \Theta(n)$.

Set $f := Y, t := n^{6/7}$ and 'bad' event $\mathcal{B}$ and c as stated in Lemma 10. Note that each coordinate change can influence f by at most $c_i := cn^{1-\beta} + 1$ as long as $\bar{\mathcal{B}}$ holds. It therefore follows from Theorem 9 and Lemma 10 that $\Pr\left[X \leq \mathbb{E}[Y] - n^{6/7} - Pr[\mathcal{B}]\right] = o(1)$ and therefore that the clustering coefficient is with high probability at least $\mathbb{E}[Y]/n = \Theta(1)$. □

References

1. Aiello, W., Chung, F., Lu, L.: Random evolution in massive graphs. In: Proceedings of the 42nd IEEE Symposium on Foundations of Computer Science, pp. 510–519. IEEE (2001)
2. Anderson, J.W.: Hyperbolic geometry, 2nd edn. Springer (2005)
3. Barabási, A.L., Albert, R.: Emergence of Scaling in Random Networks. Science 286(5439), 509–512 (1999)
4. Bollobás, B., Riordan, O., Spencer, J., Tusnády, G.: The degree sequence of a scale-free random graph process. Random Structures and Algorithms 18(3), 279–290 (2001)
5. Bollobás, B., Riordan, O.: Mathematical results on scale-free random graphs. In: Handbook of Graphs and Networks, pp. 1–34 (2002)
6. Borgs, C., Chayes, J., Daskalakis, C., Roch, S.: First to market is not everything: an analysis of preferential attachment with fitness. In: Proceedings of the 39th ACM Symposium on Theory of Computing, pp. 135–144. ACM (2007)
7. Buckley, P., Osthus, D.: Popularity based random graph models leading to a scale-free degree sequence. Discrete Mathematics 282(1-3), 53–68 (2004)
8. Chierichetti, F., Kumar, R., Lattanzi, S., Panconesi, A., Raghavan, P.: Models for the compressible web. In: Proceedings of the 50th IEEE Symposium on Foundations of Computer Science, pp. 331–340. IEEE (2009)
9. Cooper, C., Frieze, A.: A general model of web graphs. Random Structures & Algorithms 22(3), 311–335 (2003)
10. Dubhashi, D., Panconesi, A., Press, C.U.: Concentration of measure for the analysis of randomized algorithms. Cambridge University Press (2009)
11. Faloutsos, M., Faloutsos, P., Faloutsos, C.: On power-law relationships of the internet topology. ACM SIGCOMM Computer Communication Review 29, 251–262 (1999)
12. Krioukov, D., Papadopoulos, F., Kitsak, M., Vahdat, A., Boguñá, M.: Hyperbolic Geometry of Complex Networks. Physical Review E 82(3) (2010)
13. Lattanzi, S., Sivakumar, D.: Affiliation networks. In: Proceedings of the 41st ACM Symposium on Theory of Computing, pp. 427–434. ACM (2009)
14. Leskovec, J., Kleinberg, J., Faloutsos, C.: Graphs over time: densification laws, shrinking diameters and possible explanations. In: Proceedings of the 11th ACM SIGKDD International Conference on Knowledge Discovery in Data Mining, pp. 177–187. ACM (2005)
15. Milgram, S.: The small world problem. Psychology Today 2(1), 60–67 (1967)
16. Mitzenmacher, M.: A brief history of generative models for power law and lognormal distributions. Internet Mathematics 1(2), 226–251 (2004)
17. Newman, M., Park, J.: Why social networks are different from other types of networks. Physical Review E 68(3), 036122 (2003)

18. Papadopoulos, F., Krioukov, D., Boguñá, M., Vahdat, A.: Greedy forwarding in dynamic scale-free networks embedded in hyperbolic metric spaces. In: Proceedings of the 29th IEEE International Conference on Computer Communications, INFOCOM 2010, pp. 2973–2981 (2010)
19. Serrano, M., Boguñá, M.: Clustering in complex networks. i. general formalism. Physical Review E 74(5), 056114 (2006)
20. Travers, J., Milgram, S.: An experimental study of the small world problem. Sociometry, 425–443 (1969)
21. Watts, D.J., Strogatz, S.H.: Collective dynamics of small-world networks. Nature 393(6684), 440–442 (1998)

Topology-Aware VM Migration in Bandwidth Oversubscribed Datacenter Networks

Navendu Jain[1], Ishai Menache[1], Joseph (Seffi) Naor[2,*],
and F. Bruce Shepherd[3,*]

[1] Microsoft Research, Redmond, WA
[2] Technion, Haifa, Israel
[3] McGill University

Abstract. Virtualization can deliver significant benefits for cloud computing by enabling VM migration to improve utilization, balance load and alleviate hotspots. While several *mechanisms* exist to migrate VMs, few efforts have focused on optimizing migration *policies* in a multi-rooted tree datacenter network. The general problem has multiple facets, two of which map to generalizations of well-studied problems: (1) Migration of VMs in a bandwidth-oversubscribed tree network generalizes the maximum multicommodity flow problem in a tree, and (2) Migrations must meet load constraints at the servers, mapping to variants of the matching problem – generalized assignment and demand matching. While these problems have been individually studied, a new fundamental challenge is to *simultaneously* handle the packing constraints of server load and tree edge capacities. We give approximation algorithms for several versions of this problem, where the objective is to alleviate a maximal number of hot servers. In the full version of this work [5], we empirically demonstrate the effectiveness of these algorithms through large scale simulations on real data.

1 Introduction

Virtual machine (VM) technology has emerged as a key building block for cloud computing. The idea is to provide a layer of abstraction over resources of a physical server and multiplex them among its hosted VMs. Virtualization provides several benefits such as performance isolation, security, ease-of-management, and flexibility of running applications in a user-customized environment. A typical datacenter comprises tens of thousands of servers hosting VMs organized in racks, clusters or containers e.g., 40-80 servers per rack. These racks are interconnected in a network organized as a spanning tree topology with a high bandwidth oversubscription [4]. As a result, the cost to move data between servers is lowest within the same rack, relatively higher within neighboring racks, and significantly higher between far away ones [4].

---

* Work done in part while visiting Microsoft Research. Work supported in part by the Technion-Microsoft Electronic Commerce Research Center, by ISF grant 954/11 and by NSERC Discovery and Accelerator Grants.

A. Czumaj et al. (Eds.): ICALP 2012, Part II, LNCS 7392, pp. 586–597, 2012.
© Springer-Verlag Berlin Heidelberg 2012

In a cloud computing setup, the VM load may significantly fluctuate due to time-of-day effects, flash crowds, incremental application growth, and varying resource demand of co-located VMs [11]. This risks the creation of hotspots that can degrade the quality of service (QoS) of hosted applications, e.g., long response delays or low throughput. Therefore, to mitigate hotspots at runtime, cloud platforms provide live migration which transparently moves an entire VM (with its memory/disk state, processor registers, OS, and applications) from an overloaded server to an underloaded one with near-zero downtime, an important feature when live services are being hosted. This VM migration framework represents both a new opportunity and challenge to enable agile and dynamic resource management in data centers [2, 6, 10–13].

While several *mechanisms* exist for live VM migration (e.g., VMware VMotion, Windows Hyper-V, and Xen XenMotion), there remains a need for optimized, computationally-efficient migration policies. In particular, two key questions need to be answered in any chosen policy:

Q1. Which VMs to migrate from overloaded servers? First, we need to identify which VMs to move from a hotspot so as to reduce server load below a specified threshold. There are various strategies, e.g., selecting VMs till the load falls below the threshold, either in descending order of load and size, or at random. While the former may minimize the number of migrated VMs, the latter may move relatively more VMs while avoiding high migration cost scenarios.

Q2. Which servers to migrate the VMs to? Second, we need to select target servers so as to optimize the placement of selected VMs. In particular, the reconfiguration cost (e.g., bandwidth and latency) to migrate a VM from a hotspot to a target server depends on the network topology between servers. Specifically, in a bandwidth oversubscribed datacenter network, data movement to far nodes risks a long reconfiguration time window, typically proportional to both the network distance and the migrated data volume. Further, VM migrations may interfere with foreground network traffic, risking performance degradation of running applications.

Unfortunately, prior efforts have given little attention to address these challenges. Many cloud systems perform initial provisioning of VMs, but the user needs to detect the hotspot and re-provision VMs to a (likely) different server. Note that determining a new mapping of VMs over physical servers is NP-hard[1]. Several greedy heuristics have been proposed such as first-fit placement of overloaded VMs, applying a sequence of move and swap operations [11], and hottest-to-coldest in which the largest load VM is moved to the coldest server [13]. However, these techniques do not consider the network topology connecting the servers (thereby risking high migration costs in Q2). Others have advocated using stable matching techniques [12] applied to a system of cloud providers and consumers, each aiming to maximize their own benefit. However, the authors

---

[1] In fact, with general loads, even a single edge network captures the NP-hard knapsack problem, and even unit load versions on trees capture hard instances of edge-disjoint paths [10].

assume that VMs are already assigned for migration (thereby skipping Q1) and ignores edge capacity constraints or migration costs in the network connecting the servers. Thus, there is a clear need to develop *automated techniques* to optimize VM migration costs in bandwidth oversubscribed datacenter networks.

The Constrained Migration Model. Servers in a data center are interconnected in an undirected spanning tree network topology with servers as leaf nodes and switches and routers as internal nodes [4]. Each edge in the tree has a capacity measured as bits per time unit.

VMs or jobs (we use these terms interchangeably) are allocated to physical servers with each server typically hosting multiple VMs. Each VM is characterized by three parameters: (i) *transfer size* (typically 1-30 GB); we assume here that these are uniform (e.g., pre-compiled VM images and bounds on allocated RAM) and hence are all normalized to size 1, (ii) *computational load* (e.g., in CPU units); the server load is defined as the sum aggregate of the loads of the VMs hosted on it, and (iii) *value of migration* to prioritize migration of mission-critical VMs. Note that transfer size, load, and value are *independent* parameters.

The set of servers is logically partitioned into *hot* and *cold* servers. For simplicity, we refer to each cold server as a single core having *free* capacity. Exceeding this capacity risks performance degradation of its VMs and hosted applications therein. Each hot server has an *excess* load, quantifying the load reduction needed to meet application QoS.

Our core problem is the *constrained migration problem* (CoMP), formally defined in Section 2. Here we wish to compute a maximal set of hot servers that can be relieved by migrating a subset of their hosted VMs. In our context, in addition to load constraints imposed by server CPU capacity, migration patterns must also obey edge capacities inherent to the topology's bandwidth constraints.

Handling load and size constraints is a nontrivial task. To understand this challenge, we put CoMP in a wider context. First, consider a simplified version in which all server load constraints are ignored and assume that each hot server has already determined a *single* VM which it needs to migrate to alleviate overload. This becomes a maximum disjoint paths problem in a tree, generalizing the maximum matching problem, and is APX-hard even when the edges capacities belong to the set $\{1, 2\}$ [3]. Another special case of CoMP is when all routing constraints are ignored (i.e., tree capacities are set to ∞); i.e., we only have server load constraints. Again, even under the restriction that each server has selected a single job to migrate, this problem instance reduces to a generalized assignment (and maximum demand matching) problem [8, 9]. While these variants have been individually studied, we make initial headway on the new fundamental challenge to *simultaneously* handle packing constraints of server load and tree edge capacities. A longer version of this paper [5] elaborates further on the algorithmic context and its related work.

Algorithmic Contributions. We now give an overview of our results on CoMP. While we are unable to give theoretical bounds for CoMP in its full generality, we obtain approximation algorithms in the following three cases.

1) *Single hot server* (Section 3): We compute a set of VMs at a given hot server for migration to cold servers. Our algorithm either determines that no such set exists (so relieving the hotspot server is not possible) or computes a set of hosted VMs to migrate, that may incur a small additive violation in the load capacities at some of the destination cold servers.

2) *Multiple hot servers* – directed tree approach (Section 4): Our results for the single hot server generalize very nicely to the version of the problem on a directed tree in which the goal is to relieve a maximum number of hot servers. In particular, we provide a bi-criteria guarantee for this case. While the approximate solution for this case does not directly solve CoMP we use it as a building block for our system, called WAVE, briefly described below. A detailed description of WAVE is available in the accompanying technical report [5].

3) *Maximum throughput* – undirected tree approach (Section 5): We consider a maximum throughput relaxation of the problem, in which we aim to maximize the number of VMs that are migrated from hot servers. This version can closely approximate CoMP in scenarios where the set of jobs that are allowed to migrate in each hot server is very small. We extend the integer decomposition approach used in [1] to achieve an 8-approximation for this problem.

Techniques. The CoMP optimization problem is a *packing integer program*. Our solution for the multiple hot server problem is based on a two-phase algorithm. In Phase 1, we first solve a (standard) LP relaxation which fractionally routes the VMs from the hot servers to the cold servers. The key question is how to round this fractional solution without incurring too much loss. Here we devise a new approach - in Phase 2 we reduce ("round" in some sense) this fractional solution to a second LP which is well-structured. In particular, it is defined by a system of totally unimodular constraints, and hence its basic solutions are guaranteed to be integral [7]. Further, we show that its solutions simultaneously relieve the hot servers and satisfy tree edge capacities, at the expense of exceeding the load at each cold server by only a small additive constant.

WAVE - System Implementation and Evaluation. Based on our directed tree algorithms, we design and evaluate a system called WAVE (Workload Aware VM placEment) for mitigation of hotspots in data centers. WAVE uses a heuristic which iteratively examines hot servers on a rack by rack basis. Specifically, the heuristic invokes our directed tree approximation algorithm by migrating jobs away from hot servers in a single rack (see Figure 1). We iteratively process each rack separately and update the (residual) edge and load capacities when we finish its migrations after each rack iteration.

To evaluate WAVE, we conduct a detailed simulation study based on modeling data center workloads. Our results show that WAVE can efficiently and quickly alleviate single server hotspots and more complex multi-server hotspots via short migration paths, while still scaling to large data centers. The reader is referred to [5] for complete details.

We believe that the problems addressed in this paper open up a new set of challenging theoretical algorithmic questions. These are discussed in [5].

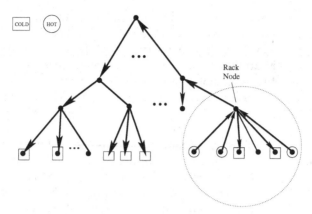

Fig. 1. Tree network topology. Edge orientation is away from hot servers in the specified rack.

2 The Constrained Migration Problem (CoMP)

We now define the optimization problem of VM migration in a bandwidth over-subscribed data center network. The servers are organized in a tree structure denoted by $T = (V, \mathcal{E})$ where V denotes the nodes and edges (or links) are denoted by $e \in \mathcal{E}$. We focus on the case where links are undirected (bidirectional), and state cases where we refer to the directed tree version. In datacenter operations, there may be a temporal budget during which migrations must be completed. To account for this, each link has a capacity denoted by c_e measured as bits per time unit. For practical interest, we logically allocate network capacity for two parts: (i) foreground application traffic and (ii) background VM migrations; we use c_e to denote the latter.

The set of servers is denoted by $\mathcal{K} \subseteq V$. This set is logically partitioned into *hot* and *cold* servers: $\mathcal{K} = \mathcal{K}_{hot} \cup \mathcal{K}_{cold}$. For simplicity, each cold server k denotes a single core having *free* capacity denoted by L_k. Exceeding this capacity risks performance degradation of its VMs and hosted applications therein; a multiplicative constant can be used to model multi-core systems. Multiple resources such as network, disk and memory can be handled similarly but we presently focus on the single resource case. Each hot server h has an *excess* demand denoted by L_h. This quantifies the reduction in load required to meet application QoS.

The set of VMs or jobs are (possibly pre-selected as *subject-to-migration* (STM)) numbered $j = 1, \ldots, N$. Each such job j is characterized by the following parameters: (1) Transfer size s_j, which is normalized to 1 for all VMs. (2) Computational load ℓ_j (e.g., in CPU units) for each job j. (3) The current (hot) server $sv(j)$ hosting VM j, and a subset $\mathrm{Dest}(j)$, called the *destination set*, of possible (cold) destinations to which j is allowed to migrate. (4) A value (or cost) of migration v_j.

Feasible Solutions. A feasible solution to CoMP specifies a collection $\mathcal{J}$ of jobs, and for each such job there is a target server $k(j) \in \mathcal{K}_{cold}$ for its migration. Obviously j is a job located on some hot server $sv(j)$ and $k(j) \in \mathrm{Dest}(j)$. For general networks, we would also specify a migration path for such a pair $(j, k(j))$, but this is uniquely determined in a tree topology. Obviously, the total migration of jobs from $\mathcal{J}$ should not exceed the capacity of any edge in T. In addition, if S_h is the set of jobs located at some hot server h, then $\sum(\ell_j : j \in \mathcal{J} \cap S_h) \geq L_h$. Similarly, for each cold server k, $\sum(\ell_j : j \in \mathcal{J} \cap S_k) \leq L_k$.

Objective Functions. The most natural objective function is simply maximizing the number of hot servers to decongest. We call this the *all-or-nothing decongestion version*. Also of interest is the *partial decongestion model*, where the objective is to migrate a maximum weight/number of migrating jobs; we call this the *maximum throughput version* (i.e., one achieves some benefit by partially decongesting servers).

Notation. In the sequel we can always scale link capacities and CPU units so that $s_{min}, \ell_{min} = 1$. We now clarify some notation related to a given undirected graph $H = (V, E)$ with node set V, and edge set E. For any $S \subseteq V(H)$ we denote by $\delta_H(S)$ the set of edges with exactly one endpoint in S. Sometimes we work with a tree T whose node set is also V. For an edge $e \in E(T)$, deleting it from T gives an obvious partition of V into two sets V_1 and V_2. The *fundamental cut* (in H induced by e) consists of $\delta_H(V_1)$. When there is no confusion, we use the notation Fund(e) to denote the edges in this cut.

3   Relieving a Single Hot Spot (Single Source CoMP)

In this section we consider the single hot server CoMP, i.e., computing a set of jobs at a single hot server h which can be migrated to cold servers, thus relieving the hot server. We describe an approximation algorithm which finds such a set of jobs if one exists, yet might incur a small additive violation in some destination cold server load capacities. Since we have a single hot server, the maximum throughput and all-or-nothing objective functions are equivalent here.

Our starting point is an LP formulation of CoMP. The optimum of the linear program is a *fractional* solution having the property that its value *opt* is an upper bound on the total load of jobs that can be feasibly migrated from h. We then show how to "round" this LP solution to migrate some of these jobs *integrally*. Our rounding process may incur a violation of the total load constraints at some destination cold servers by at most an additive term of ℓ_{max} ($\ell_{max} = \max_i \ell_i$).

Our overall method proceeds as follows. We first solve an LP relaxation. If it does not succeed in reducing the overload of hot server h, then we quit. Otherwise, we use the LP solution to produce a second LP with a nice structure, namely total unimodularity. This implies that all basic solutions for the second LP are integral. We can further prove that the feasible solutions for the second LP still relieve the server h, satisfy tree capacities, and exceed the load at each cold server by at most an additive term of ℓ_{max}.

3.1   Converting a Fractional Migration from a Single Hot Server

We think of our tree T as rooted at node h, i.e., edges are directed "away" from h. This is similar to the orientations in Figure 1, except that here we direct away from only one hot server (as opposed to multiple servers within a rack). We can assume all other leaves of T correspond to cold servers (or else delete them).

We first solve the following natural LP relaxation of CoMP. The maximization objective function guarantees that if there is a feasible solution which decongests h, then the LP optimal value will be at least L_h. That is, it ensures that we fractionally remove enough jobs (at least L_h worth), if that is possible given the constraints (we note that we modify this objective function in our empirical evaluation to incorporate penalties for long migration paths.)

Variables:

- C is the set of cold servers; J is the set of jobs on the hot server.
- $\text{Dest}(j)$ is the set of possible cold servers, for each $j \in J$.
- $x(jk)$ indicates the fractional amount of migration
  of job $j \in J$ to server $k \in C$.
- $z_j = \sum_{k \in \text{Dest}(j)} x(jk)$ indicates the total fractional migration
  (in $[0, 1]$) of job $j \in J$.

The LP objective: $OPT_{LP1} = \max \sum_{j \in J} \ell_j z_j$
Migration Constraints: for each job j: $\sum_{k \in \text{Dest}(j)} x(jk) \leq 1$
Flow constraints: for each edge $e \in T$: $\sum_{jk \in \text{Fund}(e)} x(jk) \leq c_e$
Load Constraints: for each $k \in C$: $\sum_{j:k \in \text{Dest}(j)} x(jk)\ell_j \leq L_k$
Non-Negativity: $x(jk), z_j \geq 0$

Thus, the LP generates a *fractional migration* vector (x^*, z^*) which aims to decongest the hot server. The components of the solution are: (1) $x^*(jk)$ representing what fraction of job j is migrated to server k (the flow from j to k), and (2) $z_j^* = \sum_k x^*(jk) \leq 1$ representing the total amount of job j on migration. If the solution satisfies $\sum_j \ell_j z_j^* \geq L = L_h$, then the hot server is relieved. We assume x^*, z^* are obtained by an LP solver.

3.2 Multiflows on a Directed Tree

We next recast the fractional migration problem as a *directed multiflow* problem on a tree. This yields another LP - the *Phase 2 LP* - which we show has integral optimal solutions (every basic solution is integral). We create the Phase 2 LP from a new directed tree T^* with extra leaves. For each job j, we create a new *job node* j and add a new leaf edge (j, h) from j to the server node h. These edges have capacity 1. We denote by V_J the set of new job nodes in this construction.

We also add new leaves at each cold server. To introduce these, it is convenient to define a *job-edge graph* $H = (V_J \cup \mathcal{K}_{cold}, E_{job})$. For each cold server $k \in \text{Dest}(j)$, we add a *job* edge (j, k) if job j is partially migrated to k in the fractional solution x^*. In this case, if $f = (j, k)$ is such an edge, then we also use ℓ_f to denote the load ℓ_j of j. Let E_{job} be the resulting set of job edges. These yield the bipartite demand graph.

Note that a feasible (integral) migration of h's jobs corresponds to choosing $M \subseteq E_{job}$ such that: (i) at most one job edge is chosen incident to each $j \in V_J$

(i.e., we do not try to migrate a job twice); (ii) for each edge $e \in T$, the number of job edges "crossing" e (i.e., its fundamental cut $Fund(e)$ in the job-edge graph H) is at most c_e (in other words, the total flow of jobs through e is at most c_e), and (iii) for each $k \in \mathcal{K}_{cold}$, $\sum_{f \in M \cap \delta_H(k)} \ell_f \leq L_k$ (i.e., the total load of jobs migrated to k is at most L_k).

The first two constraints are modeled purely as routing constraints within the tree, i.e., if we choose job edges which have a feasible routing in T^*, then (i) and (ii) hold. The last constraint is different from the first two, since routing a fraction x_{jk}^* of job j to server k induces a load of $\ell_j x_{jk}^*$, and not just x_{jk} which is the induced flow on tree edges (since all sizes are one). This constraint introduces a challenge due to different units for size and load. To address this challenge, we approximately model constraint (iii) as a flow constraint, allowing some additive server overload. To do this, we enlarge T^* with some cold server leaves.

For each cold server k, define its "fractional degree", $f(k)$, to be the total flow (not load) of jobs being migrated to k. We next create new leaf edges at k: $(k,1), (k,2), \ldots, (k, \lceil f(k) \rceil)$, each with capacity one. We call these *bucket leaves* at k. We now redirect the job edges of E_{job} terminating at k to bucket leaf nodes as follows. First, let $f_1, f_2, \ldots, f_p$ be the job edges currently terminating at k, where $f_i = (j_i, k)$. Without loss of generality, assume $\ell_1 \geq \ell_2 \geq \ldots \geq \ell_p$, and consider the fractional amounts $x^*(j_i k)$ that the LP routed from job j_i to server k. We greedily group the f_i's into $\lceil f(k) \rceil$ buckets as follows. Let s be the smallest value such that $\sum_{i=1}^{s} x^*(j_i k) \geq 1$. Then, we remove $f_1, f_2, \ldots, f_s$ from E_{job}, and add instead edges from each j_i to bucket leaf node 1. If the latter sum is strictly larger than 1, then we make two copies of f_s, and the second copy is redirected to leaf node 2. We then proceed to make our buckets $B_1, B_2, \ldots, B_{\lceil f(k) \rceil}$ of job edges in this obvious inductive fashion. Note that the total fractional weight of job edges into each k-leaf node can be viewed as exactly 1, except for the last bin whose total weight is $f(k) - \lfloor f(k) \rfloor$. Figure 2 gives a pictorial example of this operation.

Note that by construction of the multiflow problem, the fractional solution (x^*, z^*) immediately yields a feasible flow in T^*, fractionally migrating the same amount for each job.

Lemma 1. *There is a multiflow in the capacitated tree T^* which routes $x^*(jk)$ for each job j and server k.*

The Phase 2 LP has the following useful property.

Lemma 2. *Any integral solution to the multiflow problem on the expanded directed tree T^* corresponds to a migration which satisfies the above (i), (ii) and*

$$(iii') \quad \sum_{f \in M \cap \delta_H(k)} \ell_f \leq L_k + \ell_{\max}.$$

3.3 Total Unimodularity (TUM) of the Routing Constraints in T^*

Certain multiflow problems on directed trees have a special structure. Namely, let A be the $\{0,1\}$ matrix whose rows are indexed by directed edges of T^*, and

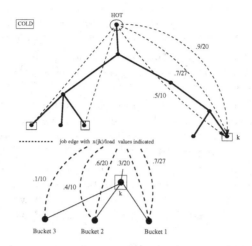

Fig. 2. Construction of buckets at cold server

whose columns are indexed by directed paths associated with our job edges E_{job} (where the paths are extended from job nodes to bucket leaves). For a job edge $f = (j, k_r)$ (where k_r denotes some leaf bucket node of cold server k), we put a 1 in row e and column f precisely if e is a directed arc on the path in T^* from j to k_r. It turns out (see [7]) that the resulting matrix A is a network matrix, and hence is *totally unimodular*, i.e., the determinant of every square submatrix is in $\{-1, 0, 1\}$. It follows (cf. [7]) that if $\max w^T y : Ay \leq b, 0 \leq y \leq 1$ has a fractional solution, for some integral capacities $b : E(T^*) \to \mathbf{Z}_+$, then it has an integral basic optimal solution. Since our original solution (x^*, z^*) induces a feasible fractional solution for the multiflow problem on T^*, by taking $w = l$, we must have an integral solution whose objective value is at least $\sum_j \ell_j z_j^*$. That is, we have an almost-feasible (up to (iii') in Lemma 1) integral migration that relieves the hot server. We now combine the pieces to obtain the following.

Theorem 1. *There is a polynomial time algorithm for single-source CoMP in (directed or undirected) trees with the following guarantee. If there is a feasible solution which decongests the single server h, then it finds a decongesting set of jobs which is feasible with respect to tree edge capacities, and violates the load at any cold server by at most an additive amount ℓ_{max}.*

4   All-or-Nothing with Multiple Servers in Directed Trees

The algorithm for resolving a single hot spot works partly due to the fact that the all-or-nothing objective function is equivalent to the maximum through-put objective function. Namely, by maximizing the number of migrating jobs we can determine whether we succeed in hitting the threshold L_h of server h. In this section we consider the all-or-nothing objective function with multiple hot servers. Hence the complicating knapsack constraints re-enter the picture.

We focus on the case where the tree topology is directed, which is also the version used within the system WAVE. We first need to introduce a new multiserver LP which models the relief of a multiple number of hot servers. In particular, its value *opt* is an upper bound on the number of hot servers we can relieve.

The Multiserver LP. We introduce an LP relaxation for multiple hot server migration. It has variables $x(jk)$ and z_j as before, and we also incorporate a variable $m_h \in [0,1]$ for each hot server $h \in \mathcal{K}_{hot}$. Variable m_h measures the (fractional) extent to which we reduce the overload of server h. This is modeled by including the constraints $\sum_{j \in Loc(h)} z_j \ell_j \geq m_h L_h$, and $0 \leq m_h \leq 1$. (Here $Loc(h)$ is the set of jobs on server h available for migration). We then solve this expanded LP with a new objective of maximizing $\sum_{h \in \mathcal{K}_{hot}} m_h$. Note that if *opt* is the optimal value, then this value is an upper bound on the total number of hot servers we could relieve in one round.

Ideally, we would convert a solution to the LP into a valid integral solution which relieves a constant factor $\Omega(opt)$ of servers, with some minimal violation of load constraints at cold servers. Since we consider migration paths in a directed tree, we still inherit a total unimodular structure but there are new difficulties. First, the objective function now uses variables m_h, and these no longer correspond to variables associated with the TUM matrix columns (i.e., the z_j or $x(jk)$ variables). More troubling (theoretically) are difficulties arising from the *all-or-nothing* objective function, i.e., that either a hot server is completely relieved or it is not. The multiserver LP may return solutions in which a large number of m_h's have a very small value, whereas we need to find an all-or-nothing subset where the m_h's are all equal to 1. Currently, our techniques essentially only apply to fractional solutions where we have $\Omega(opt)$ variables $m_h = 1$ (or close to 1). For further details on the theoretical bounds we can obtain through our two-phase LP approach, the reader is referred to [5].

An Iterative Online Heuristic. We employ the algorithmic approach described above within an iterative heuristic for mitigating hotspots across the whole datacenter. This is the basis for our system WAVE, which is reported in [5]. We proceed by addressing hotspot overloads in racks, one by one. Observe that migrations from a rack are *consistent*, in the sense that the resulting orientation on each tree edge is in the same direction. To see this, note that each server which is a child of some rack node is either hot or cold (or neither). Hence, direction of migration along such a leaf edge is determined (upwards from a hot server, or downwards to a cold server); see Figure 1. Moreover, any migrations beyond this rack are obviously all oriented away from the rack node. Thus the migration problem for each rack, has the necessary structure to apply the two-phase LP approach. For each rack, the second optimization yields feasible integral migrations, which are added to a batch scheduling of jobs to be migrated. We then update the tree-capacities and server CPU load capacities used by this migration, and iteratively select a new rack to mitigate its hot servers. We note that the LP approach allows one to penalize migration paths which are longer. Incorporating such penalty in the objective function leads to significant performance enhancements (more hot servers get migrated and shorter migration paths).

5   Maximum throughput and b-Matched Multiflows

Given the complications inherent to the multiserver LP, we tackle the maximum throughput objective as a (sometimes suitable) alternative. We first note that in the *directed* tree setting, the techniques from the single source algorithm apply directly in the absence of the multiserver LP complications. This is because our technique for binning jobs at the cold servers is oblivious to which hot server the job edge was migrating from. Hence, analogous to Theorem 1, we can migrate LP-OPT-worth of jobs without violating tree capacities, and violating cold server capacities by at most ℓ_{max}.

These techniques do not apply in the undirected setting; we address this now. We formulate maximum throughput CoMP in a slightly more general setting, to emphasize its connection to maximum multiflows in trees (MEDP). In MEDP, we have an edge-capacitated undirected tree $T = (V, E)$, $c : e \to \mathbf{Z}_+$, together with a collection of point-to-point demands $f = uv$ (each demand may also have a profit p_f). The *maximum multiflow problem* (MEDP) asks for a maximum weight (profit) subset of demands that can be simultaneously routed in T without violating edge capacities. A 2-approximation is known for the unweighted version of this problem [3] and 4-approximation for the weighted case [1].

We consider an extension of the above problem where each demand also comes with a load ℓ_f. In addition, each $v \in T$ also comes with a *capacity* (possibly infinite) $b(v)$. A *b-matched multiflow* is a subset of demands F that can be routed in T while satisfying its capacity constraints, and such that for each v: $\sum(\ell_f : f \in \delta(v) \cap F) \le b(v)$. The problem of finding a maximum b-matched multiflow obviously generalizes MEDP on trees. The reason that it slightly generalizes CoMP on trees is that we no longer have a partition of nodes into hot (supply) and cold (capacitated) servers.

We first establish an 8-approximation for maximum b-matched multiflow, with some additive error in the resource constraints $b(v)$. In fact, we show something stronger, a decomposition result using a technique from [1]. In that work, the authors call a set J of demand edges k-*routable* if when routing all these demands in T, the total flow through any edge is at most kc_e. They prove that any k-routable set J can be partitioned ("colored") into $4k$ sets, each of which is routable (this is the key step to obtaining a polytime 4-approximation for weighted MEDP in trees). With additional work, one can employ their result to obtain a similar decomposition for b-matched multiflows. Our results are the following (further details and proofs appear in [5]).

Theorem 2. *There is an 8-approximation for maximum weighted b-matched mutliflows (and hence maximum throughput CoMP) in undirected trees, if we allow an additive violation of ℓ_{max} at the knapsack constraints for nodes.*

One can also avoid the additive violation of ℓ_{max} with further degradation to the approximation ratio.

Theorem 3. *There is a polynomial time $O(1)$-approximation for maximum weighted b-matched multiflow problem (and hence maximum throughput CoMP) in undirected trees.*

Finally, in the more general *unsplittable flow* setting, each demand f imposes a flow of d_f when it is routed through T's edge capacities. This maximum unsplittable flow problem also has an $O(1)$-approximation on trees [1], assuming the no-bottleneck assumption: $d_{max} \leq c_{min}$. At this point, we have been unable to push these techniques, however, to give constant approximation results in the b-matched unsplittable flow setting.

References

1. Chekuri, C., Mydlarz, M., Shepherd, F.: Multicommodity demand flow in a tree and packing integer programs. ACM Transactions on Algorithms (TALG) 3(3), 27–es (2007)
2. Clark, C., Fraser, K., Hand, S., Hansen, J.G., Jul, E., Limpach, C., Pratt, I., Warfield, A.: Live migration of virtual machines. In: NSDI (2005)
3. Garg, N., Vazirani, V., Yannakakis, M.: Primal-dual approximation algorithms for integral flow and multicut in trees. Algorithmica 18(1), 3–20 (1997)
4. Greenberg, A., Hamilton, J., Jain, N., Kandula, S., Kim, C., Lahiri, P., Maltz, D., Patel, P., Sengupta, S.: VL2: a scalable and flexible data center network. In: SIGCOMM (2009)
5. Jain, N., Menache, I., Naor, S., Shepherd, F.: Topology-aware VM migration in bandwidth oversubscribed datacenter networks. Technical report (May 2012), http://research.microsoft.com/apps/pubs/?id=162997
6. Lagar-Cavilla, H.A., Whitney, J., Scannell, A., Patchin, P., Rumble, S.M., de Lara, E., Brudno, M., Satyanarayanan, M.: Snowflock: Rapid virtual machine cloning for cloud computing. In: Eurosys (2009)
7. Schrijver, A.: Theory of linear and integer programming. John Wiley & Sons (1998)
8. Shepherd, F., Vetta, A.: The demand-matching problem. Mathematics of Operations Research 32(3), 563 (2007)
9. Shmoys, D., Tardos, É.: An approximation algorithm for the generalized assignment problem. Mathematical Programming 62(1), 461–474 (1993)
10. Sundararaj, A., Sanghi, M., Lange, J., Dinda, P.: An optimization problem in adaptive virtual environments. ACM SIGMETRICS Performance Evaluation Review 33(2), 6–8 (2005)
11. Wood, T., Shenoy, P.J., Venkataramani, A., Yousif, M.S.: Black-box and gray-box strategies for virtual machine migration. In: NSDI (2007)
12. Xu, H., Li, B.: Egalitarian stable matching for VM migration in cloud computing. In: INFOCOM Workshops, pp. 631–636 (2011)
13. Xu, Y., Sekiya, Y.: Virtual machine migration strategy in federated cloud. In: Internet Conference (2010)

Counting Arbitrary Subgraphs in Data Streams[*]

Daniel M. Kane[1], Kurt Mehlhorn[2], Thomas Sauerwald[2], and He Sun[2,3]

[1] Department of Mathematics, Stanford University, USA
[2] Max Planck Institute for Informatics, Germany
[3] Institute for Modern Mathematics and Physics, Fudan University, China

Abstract. We study the subgraph counting problem in data streams. We provide the first non-trivial estimator for approximately counting the number of occurrences of an *arbitrary* subgraph H of constant size in a (large) graph G. Our estimator works in the turnstile model, i.e., can handle both edge-insertions and edge-deletions, and is applicable in a distributed setting. Prior to this work, only for a few non-regular graphs estimators were known in case of edge-insertions, leaving the problem of counting general subgraphs in the turnstile model wide open. We further demonstrate the applicability of our estimator by analyzing its concentration for several graphs H and the case where G is a power law graph.

Keywords: data streams, subgraph counting, network analysis.

1 Introduction

Counting (small) subgraphs in massive graphs is one of the fundamental tasks in algorithm design and has various applications, including analyzing the connectivity of networks, uncovering the structural information of large graphs, and indexing graph databases. The current best known algorithm for the simplest non-trivial version of the problem, counting the number of triangles, is based on matrix multiplication, and is infeasible for massive graphs. To overcome this, we consider the problem in the data streaming setting, where the edges come sequentially and the algorithm is required to approximate the number of subgraphs without storing the whole graph.

Formally in this problem, we are given a set of items $s_1, s_2, \ldots$ in a data stream. These items arrive sequentially and represent edges of an underlying graph $G = (V, E)$. Two standard models [14] in this context are the *Cash Register Model* and the *Turnstile Model*. In the cash register model, each item s_i represents one edge and these arrived items form a graph G with edge set $E := \bigcup \{s_i\}$, where $E = \emptyset$ initially. The turnstile model generalizes the cash register model and is applicable to dynamic situations. Specifically, each item s_i in the turnstile model is of the form (e_i, sign_i), where e_i is an edge of G and $\text{sign}_i \in \{+, -\}$ indicates that e_i is inserted to or deleted from G. That is, after reading the ith item, $E \leftarrow E \cup \{e_i\}$ if $\text{sign}_i = +$, and $E \leftarrow E \setminus \{e_i\}$ otherwise.

---

[*] This material is based upon work supported by the National Science Foundation under Award No. 1103688.

A. Czumaj et al. (Eds.): ICALP 2012, Part II, LNCS 7392, pp. 598–609, 2012.
© Springer-Verlag Berlin Heidelberg 2012

In a more general distributed setting, there are k distributed sites, each receiving a stream S_i of elements over time, and every S_i is processed by a local host. When the number of subgraphs is asked for, these k hosts cooperate to give an approximation for the underlying graph formed by $\bigcup_{i=1}^{k} S_i$.

Our Results & Techniques. We present the first sketch for counting *arbitrary* subgraphs of constant size in data streams. While most of the previous algorithms are based on sampling techniques and cannot be extended to count subgraphs with complex structures, our algorithm can approximately count arbitrary (possibly directed) subgraphs. Moreover, our algorithm runs in the turnstile model and is applicable in the distributed setting.

More formally, for any fixed subgraph H of constant size, we present an algorithm that $(1 \pm \varepsilon)$-approximates the number of occurrences of H in G, denoted by $\#H$. That is, for any constant $0 < \varepsilon < 1$, with probability at least $2/3$ the output Z of our algorithm satisfies $Z \in [(1 - \varepsilon) \cdot \#H, (1 + \varepsilon) \cdot \#H]$. For several families of graphs G and H, our algorithm achieves a $(1 \pm \varepsilon)$-approximation for the number of subgraphs H in G within sublinear space. Our result generalizes previous work which can only count cycles in the turnstile model [10, 11], and answers the 11th open problem in the 2006 IITK Workshop on Algorithms for Data Steams [12].

We further consider counting stars in power law graphs, which include many practical networks. We show that $O\left(\frac{1}{\varepsilon^2} \cdot \log n\right)$ bits suffice to get a $(1 \pm \varepsilon)$-approximation for counting stars S_k, while the exact counting needs $n \cdot \log n$ bits of space. Our main results are summarized in Table 1.

Our sketch relies on a novel approach of designing random vectors that are based on different combinations of complex numbers. By using different roots of unity and random mappings from vertices in G to complex numbers, we obtain an unbiased estimator for $\#H$. This partially answers Problem 4 of the survey by Muthukrishnan [14], which asks for suitable applications of complex-valued hash functions in data streaming algorithms. Apart from counting subgraphs in streams, we believe that our new approach will have more applications.

Discussion. To demonstrate that for a large family of graphs G our algorithm achieves a $(1 \pm \varepsilon)$-approximation within sublinear space, we consider Erdős-Rényi random graphs $G = G(n, p)$, where each edge is placed independently with a fixed probability $p \geq (1 + \varepsilon) \cdot \ln(n)/n$. Random graphs are of interest for the performance of our algorithm, as the independent appearance of the edges in $G = G(n, p)$ reduces the number of particular patterns. In other words, if our algorithm has low space complexity for counting a subgraph H in $G(n, p)$, then the space complexity is even lower for counting a more frequently occurring subgraph in a real-world graph G which has the same density as $G(n, p)$.

Regarding the space complexity of our algorithm on random graphs, assume for instance that the subgraph H is a P_3 or S_3 (i.e., a path or a star with three edges). The expected number of occurrences of such a graph is of order $n^4 p^3 \gg 1$. It can be shown by standard techniques (cf. [1, Section 4.4]) that the number of occurrences is also of this order with probability $1 - o(1)$ as $n \to \infty$. Assuming

Table 1. Space requirement for $(1 \pm \varepsilon)$-approximately counting an undirected and connected graph H with $k = O(1)$ edges. Here δ and Δ denote the minimum and maximum degree, respectively. Space complexity is measured in terms of bits.

Conditions	Space Complexity	Reference
any graph G any graph H	$O\!\left(\frac{1}{\varepsilon^2} \cdot \frac{m^k \cdot \Delta(G)^k}{(\#H)^2} \cdot \log n\right)$	Theorem 7
any graph G H with $\delta(H) \geqslant 2$	$O\!\left(\frac{1}{\varepsilon^2} \cdot \frac{m^k}{(\#H)^2} \cdot \log n\right)$	Theorem 7
any graph G stars S_k	$O\left(\frac{n^{1-1/(2k)}}{\varepsilon^2} \cdot \left(\frac{n^{3/2-1/(2k)} \cdot \Delta(G)^{2k}}{(\#S_k)^2} + 1\right) \cdot \log n\right)$	Theorem 8
Power law graph G stars S_k	$O\left(\frac{1}{\varepsilon^2} \cdot \log n\right)$	Theorem 9

that this event occurs, Theorem 7 along with the facts that $m = \Theta(n^2 p)$ and $\Delta(G) = \Theta(np)$ implies a $(1 \pm \varepsilon)$-approximation algorithm for P_3 (or S_3) with space complexity $O(\frac{1}{\varepsilon^2} \cdot n \cdot \log n)$. For stars S_k with any constant k, the result from Theorem 8 yields a $(1 \pm \varepsilon)$-approximation algorithm in space $O\left(\frac{1}{\varepsilon^2} \cdot \sqrt{n} \cdot \log n\right)$. Finally, for any cycle with $k = O(1)$ edges, Theorem 7 gives an algorithm with space complexity $O(\frac{1}{\varepsilon^2} \cdot p^{-k} \cdot \log n)$, which is sublinear for sufficiently large values of p, e.g., $p = \omega(n^{-1/k})$.

Related Work. Bar-Yossef, Kumar and Sivakumar were the first to study the subgraph counting problem in data streams and presented an algorithm for counting triangles [3]. After that, the problem of counting triangles in data streams was studied extensively [4, 6, 10, 16]. The problem of counting other subgraphs was also addressed in the literature. Buriol et al. [7] considered the problem of estimating clustering indexes in data streams. Bordino et al. [5] extended the technique of counting triangles [6] to all subgraphs on three and four vertices. Manjunath et al. [11] presented an algorithm for counting cycles of constant size in data streams. Among these results, only two algorithms [10, 11] work in the turnstile model and these only hold for cycles.

Apart from designing algorithms in the streaming model, the subgraph counting problem has been studied extensively. Alon et al. [2] presented an algorithm for counting given-length cycles. Gonen et al. [9] showed how to count stars and other small subgraphs in sublinear time. In particular, several small subgraphs in a network, named network motifs, have been identified as the simple building blocks of complex biological networks and the distribution of their occurrences could reveal answers to many important biological questions [13, 17].

Notation. Let $G = (V, E)$ be an undirected graph without self-loops and multiple edges. The set of vertices and edges are represented by $V[G]$ and $E[G]$, respectively. We will assume that $V[G] = \{1, \ldots, n\}$ and n is known in advance.

For any vertex $u \in V[G]$, the degree of u is denoted by $\deg(u)$. The maximum and minimum degree of G are denoted by $\Delta(G)$ and $\delta(G)$, respectively.

Given two directed graphs H_1 and H_2, we say that H_1 is *homomorphic* to H_2 if there is a mapping $\varphi : V[H_1] \to V[H_2]$ such that $(u, v) \in E[H_1]$ implies $(\varphi(u), \varphi(v)) \in E[H_2]$. Graphs H_1 and H_2 are said to be *isomorphic* if there is a bijection $\varphi : V[H_1] \to V[H_2]$ such that $(u, v) \in E[H_1]$ iff $(\varphi(u), \varphi(v)) \in E[H_2]$. Let $\text{auto}(H)$ be the number of automorphisms of graph H.

For any graph H, we call a subgraph H_1 of G that is not necessarily induced an *occurrence* of H, if H_1 is isomorphic to H. Let $\#(H, G)$ be the number of occurrences of H in G. When reference to G is clear, we may also write $\#H$. A kth root of unity is any number of the form $e^{2\pi i \cdot j / k}$, where $0 \leqslant j < k$. For $p, q \in \mathbb{N}$ define $\left(e^{2\pi i \cdot j / k} \right)^{p/q}$ as $e^{2\pi i \cdot (jp)/(kq)}$.

2   An Unbiased Estimator for Counting Subgraphs

We present a framework for counting general subgraphs. Suppose that H is a fixed graph with t vertices and k edges, and we want to count the number of occurrences of H in G. For the notation, we denote vertices of H by a, b and c, and vertices of G by u, v and w, respectively. Let the degree of vertex a in H be $\deg_H(a)$. We equip the edges of H with an arbitrary orientation, as this is necessary for the further analysis. Therefore, each edge in H together with its orientation can be expressed as $\overrightarrow{ab}$ for some $a, b \in V[H]$. For simplicity and with slight abuse of notation we will use H to denote such an oriented graph.

At a high level, our estimator maintains k complex-valued variables $Z_{\overrightarrow{ab}}(G)$, where $\overrightarrow{ab} \in E[H]$, and these variables are set to be zero initially. For every arriving edge $\{u, v\} \in E[G]$ we update each $Z_{\overrightarrow{ab}}(G)$ according to

$$Z_{\overrightarrow{ab}}(G) \leftarrow Z_{\overrightarrow{ab}}(G) + \mathcal{M}_{\overrightarrow{ab}}(u, v) + \mathcal{M}_{\overrightarrow{ab}}(v, u) \ ,$$

where $\mathcal{M}_{\overrightarrow{ab}} : V[G] \times V[G] \to \mathbb{C}$ is defined with respect to edge $\overrightarrow{ab} \in E[H]$ and can be computed in constant time. Hence

$$Z_{\overrightarrow{ab}}(G) = \sum_{\{u,v\} \in E[G]} \mathcal{M}_{\overrightarrow{ab}}(u, v) + \mathcal{M}_{\overrightarrow{ab}}(v, u) \ .$$

Intuitively $\mathcal{M}_{\overrightarrow{ab}}(u, v)$ gives $\{u, v\}$ the orientation $\overrightarrow{uv}$ and maps $\overrightarrow{uv}$ to $\overrightarrow{ab}$, and $\mathcal{M}_{\overrightarrow{ab}}(u, v) + \mathcal{M}_{\overrightarrow{ab}}(v, u)$ is used to express two different orientations of edge $\{u, v\}$. For every query for $\#(H, G)$, the estimator simply outputs the real part of $\alpha \cdot \prod_{\overrightarrow{ab} \in E[H]} Z_{\overrightarrow{ab}}(G)$, where $\alpha \in \mathbb{R}^+$ is a scaling factor. For any k edges $(u_1, v_1), \ldots, (u_k, v_k)$ in G and k edges $\overrightarrow{a_1 b_1}, \ldots, \overrightarrow{a_k b_k}$ in H, we want $\alpha \cdot \prod_{i=1}^{k} \mathcal{M}_{\overrightarrow{a_i b_i}}(u_i, v_i)$ to be one if these edges $(u_1, v_1), \ldots, (u_k, v_k)$ form an occurrence of H, and zero otherwise.

More formally, each $\mathcal{M}_{\overrightarrow{ab}}(u, v)$ is defined according to the degree of vertices a, b in graph H and consists of the product of three types of random variables $Q, X_c(w)$ and $Y(w)$, where $c \in V[H]$ and $w \in V[G]$:

- Variable Q is a random τth root of unity, where $\tau := 2^t - 1$.
- For vertex $c \in V[H]$ and $w \in V[G]$, function $X_c(w)$ is a random $\deg_H(c)$th root of unity, and for each vertex $c \in V[H]$, $X_c : V[G] \to \mathbb{C}$ is chosen independently and uniformly at random from a family of $4k$-wise independent hash functions. Variables Q and $X_c(\cdot)$ for $c \in V[H]$ are chosen independently.
- For every $w \in V[G]$, $Y(w)$ is a random element from $S := \{1, 2, 4, 8, \ldots, 2^{t-1}\}$ as part of a $4k$-wise independent hash function. Variables $X_c(\cdot)$ for $c \in V[H]$, $Y(\cdot)$ and Q are chosen independently.

Given the notations above, we define each function $\mathcal{M}_{\overrightarrow{ab}}$ as

$$\mathcal{M}_{\overrightarrow{ab}}(u, v) := X_a(u) \, X_b(v) \, Q^{\frac{Y(u)}{\deg_H(a)}} \, Q^{\frac{Y(v)}{\deg_H(b)}} \ .$$

Estimator 1 gives the formal description of the update and query procedures.

Estimator 1. Counting $\#(H, G)$

<u>Step 1 (Update):</u> When an edge $e = \{u, v\} \in E[G]$ arrives, update each $Z_{\overrightarrow{ab}}$ w.r.t.

$$Z_{\overrightarrow{ab}}(G) \leftarrow Z_{\overrightarrow{ab}}(G) + \mathcal{M}_{\overrightarrow{ab}}(u, v) + \mathcal{M}_{\overrightarrow{ab}}(v, u). \tag{1}$$

<u>Step 2 (Query):</u> When $\#(H, G)$ is required, output the real part of

$$\frac{t^t}{t! \cdot \text{auto}(H)} \cdot Z_H(G) \ , \tag{2}$$

where Z_H is defined by

$$Z_H(G) := \prod_{\overrightarrow{ab} \in E[H]} Z_{\overrightarrow{ab}}(G) \ . \tag{3}$$

Estimator 1 is applicable in a quite general setting: First, the estimator runs in the turnstile model. For simplicity the update procedure above is only described for the edge-insertion case. For every item of the stream that represents an edge-deletion, we replace "+" by "−" in (1). Second, our estimator also works in the distributed setting, where every local host maintains variables $Z_{\overrightarrow{ab}}$ for $\overrightarrow{ab} \in E[H]$, and does the update for every arriving item in the local stream. When the output is required, these variables located at different hosts are summed up and we return the estimated value according to (3). Third, the estimator above can be revised easily to count the number of directed subgraphs in a directed graphs. Since in this case we need to change the constant of (2) accordingly, in the rest of our paper we only focus on the case of counting undirected graphs.

3 Analysis of the Estimator

Let us first explain the intuition behind our estimator. By definition we have

$$Z_H(G) = \prod_{\overrightarrow{ab} \in E[H]} Z_{\overrightarrow{ab}}(G) = \prod_{\overrightarrow{ab} \in E[H]} \sum_{\{u,v\} \in E[G]} \left(\mathcal{M}_{\overrightarrow{ab}}(u, v) + \mathcal{M}_{\overrightarrow{ab}}(v, u) \right) \ .$$

Since H has k edges, $Z_H(G)$ is a product of k terms and each term is a sum over all edges of G each with two possible orientations. Hence, in the expansion of $Z_H(G)$ any k-tuple $(e_1, \ldots, e_k) \in E^k[G]$ contributes 2^k different terms to $Z_H(G)$ and each term corresponds to a certain orientation of $(e_1, \ldots, e_k)$. Let $\overrightarrow{T} = (\overrightarrow{e_1}, \ldots, \overrightarrow{e_k})$ be an arbitrary orientation of $(e_1, \ldots, e_k)$ and $G_{\overrightarrow{T}}$ be the directed graph induced from $\overrightarrow{T}$.

At a high level, we use three types of variables to test if $G_{\overrightarrow{T}}$ is isomorphic to H. These variables play different roles, as described below. (i) For $c \in V[H]$ and $w \in V[G]$, we have $\mathbb{E}[X_c^i(w)] \neq 0$ $(1 \leqslant i \leqslant \deg_H(c))$ iff $i = \deg_H(c)$. Random variables $X_c(w)$ guarantee that $G_{\overrightarrow{T}}$ contributes to $\mathbb{E}[Z_H(G)]$ only if $G_{\overrightarrow{T}}$ is homomorphic to H. (ii) Through function $Y : V[G] \to S$ every vertex $u \in V_{\overrightarrow{T}}$ maps to one element $Y(u)$ in S randomly. If $|V_{\overrightarrow{T}}| = |S| = t$, then with constant probability, vertices in $V_{\overrightarrow{T}}$ map to different t numbers in S. Otherwise, $|V_{\overrightarrow{T}}| < t$ and vertices in $V_{\overrightarrow{T}}$ cannot map to different t elements. Since Q is a random τth root of unity, $\mathbb{E}[Q^i] \neq 0$ $(1 \leqslant i \leqslant \tau)$ iff $i = \tau$, where $\tau = \sum_{\ell \in S} \ell$. The combination of Q and Y guarantees that $G_{\overrightarrow{T}}$ contributes to $\mathbb{E}[Z_H(G)]$ only if graph H and $G_{\overrightarrow{T}}$ have the same number of vertices. Combining (i) and (ii), only subgraphs isomorphic to H contribute to $\mathbb{E}[Z_H(G)]$.

Lemma 1 ([8]). *For any $c \in V[H]$ let X_c be a randomly chosen $\deg_H(c)$th root of unity. Then for any $1 \leqslant i \leqslant \deg_H(c)$, it holds that*

$$\mathbb{E}[X_c^i] = \begin{cases} 1, & i = \deg_H(c) \ , \\ 0, & 1 \leqslant i < \deg_H(c) \ . \end{cases}$$

In particular, $\mathbb{E}[X_c] = 1$ if $\deg_H(c) = 1$.

Lemma 2. *Let R be a primitive τth root of unity and $k \in \mathbb{N}$. Then*

$$\sum_{\ell=0}^{\tau-1} (R^k)^\ell = \begin{cases} \tau, & \tau \mid k \ , \\ 0, & \tau \nmid k \ . \end{cases}$$

Lemma 3. *Let $x_i \in \mathbb{Z}_{\geqslant 0}$ and $\sum_{i=0}^{t-1} x_i = t$. Then $2^t - 1 \mid \sum_{i=0}^{t-1} 2^i \cdot x_i$ if and only if $x_0 = \cdots = x_{t-1} = 1$.*

Based on the three lemmas above, we prove that $Z_H(G)$ is an unbiased estimator for $\#(H, G)$.

Theorem 4. *Let H be a graph with t vertices and k edges. Assume that variables $X_c(w), Y(w)$ for $c \in V[H], w \in V[G]$ and Q are as defined above. Then*

$$\mathbb{E}[Z_H(G)] = \frac{t! \cdot \mathrm{auto}(H)}{t^t} \cdot \#(H, G) \ .$$

Proof. Let $(e_1, \ldots, e_k) \in E^k(G)$ and $\overrightarrow{T} = (\overrightarrow{e_1}, \ldots, \overrightarrow{e_k})$ be an arbitrary orientation of $(e_1, \ldots, e_k)$, where $\overrightarrow{e_i} = \overrightarrow{u_i v_i}$. Consider the expansion of $Z_H(G)$ below:

$$Z_H(G) = \prod_{\overrightarrow{ab} \in E[H]} Z_{\overrightarrow{ab}}(G) = \prod_{\overrightarrow{ab} \in E[H]} \sum_{\{u,v\} \in E[G]} \left(M_{\overrightarrow{ab}}(u, v) + M_{\overrightarrow{ab}}(v, u) \right) \ .$$

The term corresponding to $(\vec{e_1}, \ldots, \vec{e_k})$ in the expansion of $Z_H(G)$ is

$$\prod_{i=1}^{k} \mathcal{M}_{\overrightarrow{a_i b_i}}(u_i, v_i) = \prod_{i=1}^{k} X_{a_i}(u_i)\, X_{b_i}(v_i)\, Q^{\frac{Y(u_i)}{\deg_H(a_i)}}\, Q^{\frac{Y(v_i)}{\deg_H(b_i)}}, \tag{4}$$

where $\overrightarrow{a_i b_i}$ is the ith edge of H (where we assume any order) and $\overrightarrow{u_i v_i}$ is the ith edge in $\vec{T}$. We show that the expectation of (4) is non-zero if and only if the graph induced by $\vec{T}$ is an occurrence of H in G. Moreover, if the expectation of (4) is non-zero, then its value is a constant.

For any vertex w of G and any vertex c of H, let

$$\theta_{\vec{T}}(c, w) := \big|\{i : (u_i = w \text{ and } a_i = c) \text{ or } (v_i = w \text{ and } b_i = c)\}\big|$$

be the number of edges in $\vec{T}$ with head (or tail) w mapping to the edges in H with head (or tail) c. Since every vertex c of H is incident to $\deg_H(c)$ edges, for any $c \in V[H]$ it holds that $\sum_{w \in V_{\vec{T}}} \theta_{\vec{T}}(c, w) = \deg_H(c)$. By the definition of $\theta_{\vec{T}}$, we can rewrite (4) as

$$\left(\prod_{c \in V[H]} \prod_{w \in V_{\vec{T}}} X_c^{\theta_{\vec{T}}(c,w)}(w) \right) \cdot \left(\prod_{c \in V[H]} \prod_{w \in V_{\vec{T}}} Q^{\frac{\theta_{\vec{T}}(c,w)Y(w)}{\deg_H(c)}} \right).$$

Therefore $Z_H(G)$ is equal to

$$\sum_{\substack{e_1,\ldots,e_k \\ e_i \in E[G]}} \sum_{\vec{T}=(\vec{e_1},\ldots,\vec{e_k})} \left(\prod_{c \in V[H]} \prod_{w \in V_{\vec{T}}} X_c^{\theta_{\vec{T}}(c,w)}(w) \right) \cdot \left(\prod_{c \in V[H]} \prod_{w \in V_{\vec{T}}} Q^{\frac{\theta_{\vec{T}}(c,w)Y(w)}{\deg_H(c)}} \right),$$

where the first summation is over all k-tuples of edges in $E[G]$ and the second summation is over all their possible orientations. By linearity of expectations of these random variables and the assumption that $X_c(\cdot)$ for $c \in V[H]$, $Y(\cdot)$, and Q have sufficient independence, we have

$$\mathbb{E}[Z_H(G)]$$

$$= \sum_{\substack{e_1,\ldots,e_k \\ e_i \in E[G]}} \sum_{\vec{T}=(\vec{e_1},\ldots,\vec{e_k})} \left(\prod_{c \in V[H]} \mathbb{E}\left[\prod_{w \in V_{\vec{T}}} X_c^{\theta_{\vec{T}}(c,w)}(w) \right] \right) \cdot \mathbb{E}\left[\prod_{\substack{c \in V[H] \\ w \in V_{\vec{T}}}} Q^{\frac{\theta_{\vec{T}}(c,w)Y(w)}{\deg_H(c)}} \right].$$

Let

$$\alpha_{\vec{T}} := \underbrace{\left(\prod_{c \in V[H]} \mathbb{E}\left[\prod_{w \in V_{\vec{T}}} X_c^{\theta_{\vec{T}}(c,w)}(w) \right] \right)}_{A} \cdot \underbrace{\mathbb{E}\left[\prod_{c \in V[H]} \prod_{w \in V_{\vec{T}}} Q^{\frac{\theta_{\vec{T}}(c,w)Y(w)}{\deg_H(c)}} \right]}_{B}.$$

We will next show that $\alpha_{\overrightarrow{T}}$ is either zero or a nonzero constant independent of $\overrightarrow{T}$. The latter is the case if and only if G_T, the undirected graph induced from the edge set $\overrightarrow{T}$, is an occurrence of H in G.

We consider the product A at first. Assume that $A \neq 0$. Using the same technique as [11], we construct a homomorphism from H to $G_{\overrightarrow{T}}$. Remember that: (i) For any $c \in V[H]$ and $w \in V_{\overrightarrow{T}}$, we have $\theta_{\overrightarrow{T}}(c, w) \leqslant \deg_H(c)$, and (ii) $\mathbb{E}\left[X_c^i(w)\right] \neq 0$ iff $i \in \{0, \deg_H(c)\}$. Therefore for any fixed $\overrightarrow{T}$ and $c \in V[H]$, it holds that $\mathbb{E}\left[\prod_{w \in V_{\overrightarrow{T}}} X_c^{\theta_{\overrightarrow{T}}(c,w)}(w)\right] \neq 0$ iff $\theta_{\overrightarrow{T}}(c, w) \in \{0, \deg_H(c)\}$ for all w. Now assume that $\mathbb{E}\left[\prod_{w \in V_{\overrightarrow{T}}} X_c^{\theta_{\overrightarrow{T}}(c,w)}(w)\right] \neq 0$ for every $c \in V[H]$. Then $\theta_{\overrightarrow{T}}(c, w) \in \{0, \deg_H(c)\}$ for all $c \in V[H]$ and $w \in V[G]$. Since $\sum_w \theta_{\overrightarrow{T}}(c, w) = \deg_H(c)$ for any $c \in V[H]$, there is a unique vertex $w \in V_{\overrightarrow{T}}$ such that $\theta_{\overrightarrow{T}}(c, w) = \deg_H(c)$. Define $\varphi : V[H] \to V_{\overrightarrow{T}}$ as $\varphi(c) = w$ for the vertex w satisfying $\theta_{\overrightarrow{T}}(c, w) = \deg_H(c)$. Then φ is a homomorphism, i.e. $(a, b) \in E[H]$ implies $(\varphi(a), \varphi(b)) \in E[G_{\overrightarrow{T}}]$. Hence $A \neq 0$ implies H is homomorphic to $G_{\overrightarrow{T}}$, and

$$\prod_{c \in V[H]} \mathbb{E}\left[\prod_{w \in V_{\overrightarrow{T}}} X_c^{\theta_{\overrightarrow{T}}(c,w)}(w)\right] = \prod_{c \in V[H]} \mathbb{E}\left[X_c^{\deg_H(c)}(\varphi(c))\right] = 1 . \tag{5}$$

Second we consider the product B. Our task is to show that, under the condition $A \neq 0$, $G_{\overrightarrow{T}}$ is an occurrence of H if and only if $B \neq 0$. Observe that

$$\mathbb{E}\left[\prod_{c \in V[H]} \prod_{w \in V_{\overrightarrow{T}}} Q^{\frac{\theta_{\overrightarrow{T}}(c,w)Y(w)}{\deg_H(c)}}\right] = \mathbb{E}\left[Q^{\sum_{c \in V[H]} \sum_{w \in V_{\overrightarrow{T}}} \frac{\theta_{\overrightarrow{T}}(c,w)Y(w)}{\deg_H(c)}}\right] .$$

Case 1: Assume that $G_{\overrightarrow{T}}$ is an occurrence of H in G. Then $|V_{\overrightarrow{T}}| = |V[H]|$ and the function φ constructed above is a bijection, which implies that

$$\sum_{c \in V[H]} \sum_{w \in V_{\overrightarrow{T}}} \frac{\theta_{\overrightarrow{T}}(c,w)Y(w)}{\deg_H(c)} = \sum_{c \in V[H]} Y(\varphi(c)) = \sum_{w \in V_{\overrightarrow{T}}} Y(w) .$$

Without loss of generality, let $V_{\overrightarrow{T}} = \{w_1, \ldots, w_t\}$. By considering all possible choices for $Y(w_1), \ldots, Y(w_t)$, denoted by $y(w_1), \ldots, y(w_t) \in S$, and independence between Q and $Y(w)$, where $w \in V[G]$, we have

$$B = \sum_{j=0}^{\tau-1} \sum_{y(w_1),\ldots,y(w_t) \in S} \frac{1}{\tau} \left(\prod_{i=1}^{t} \mathbf{Pr}\left[Y(w_i) = y(w_i)\right]\right) \cdot \exp\left(\frac{2\pi i j}{\tau} \sum_{\ell=1}^{t} y(w_\ell)\right)$$

$$= \sum_{j=0}^{\tau-1} \sum_{\substack{y(w_1),\ldots,y(w_t) \in S \\ \vartheta := y(w_1)+\cdots+y(w_t), \tau | \vartheta}} \frac{1}{\tau} \left(\frac{1}{t}\right)^t \exp\left(\frac{2\pi i}{\tau} \cdot \vartheta \cdot j\right) +$$

$$\sum_{j=0}^{\tau-1} \sum_{\substack{y(w_1),\ldots,y(w_t) \in S \\ \vartheta := y(w_1)+\cdots+y(w_t), \tau \nmid \vartheta}} \frac{1}{\tau} \left(\frac{1}{t}\right)^t \exp\left(\frac{2\pi i}{\tau} \cdot \vartheta \cdot j\right) .$$

Applying Lemma 2 with $R = \exp\left(\frac{2\pi i}{\tau}\right)$, the second summation is zero. Hence by Lemma 3 we have

$$B = \sum_{\substack{y(w_1),\ldots,y(w_t)\in S \\ \tau | y(w_1)+\cdots+y(w_t)}} \left(\frac{1}{t}\right)^t = \sum_{\substack{y(w_1),\ldots,y(w_t)\in S \\ y(w_1)+\cdots+y(w_t)=\tau}} \left(\frac{1}{t}\right)^t = \left(\frac{1}{t}\right)^t \cdot t! = \frac{t!}{t^t} .$$

Case 2: Assume that $G_{\vec{\mathcal{T}}}$ is not an occurrence of H in G and let $V_{\vec{\mathcal{T}}} = \{w_1,\ldots,w_{t'}\}$, where $t' < t$. Then there is a vertex $w \in V_{\vec{\mathcal{T}}}$ and different $b,c \in V[H]$, such that $\varphi(b) = \varphi(c) = w$. As before we have

$$\sum_{c\in V[H]} \sum_{w\in V_{\vec{\mathcal{T}}}} \frac{\theta_{\vec{\mathcal{T}}}(c,w)Y(w)}{\deg_H(c)} = \sum_{c\in V[H]} Y(\varphi(c)) .$$

By Lemma 3, $\tau \nmid \sum_{c\in V[H]} Y(\varphi(c))$. Hence

$$B = \sum_{j=0}^{\tau-1} \sum_{\substack{y(w_1),\ldots,y(w_{t'})\in S \\ \vartheta := \sum_{c\in V[H]} y(\varphi(c))}} \frac{1}{\tau} \left(\frac{1}{t}\right)^{t'} \exp\left(\frac{2\pi i}{\tau} \cdot \vartheta \cdot j\right) = 0 ,$$

where the last equality follows from Lemma 2 with $R = \exp\left(\frac{2\pi i}{\tau}\right)$.

Let $\mathbf{1}_{G_{\vec{\mathcal{T}}} \equiv H}$ be the indicator variable that is one if $G_{\vec{\mathcal{T}}}$ and H are isomorphic and zero otherwise. By the definition of graph automorphism and (5),

$$\mathbb{E}[Z_H(G)] = \sum_{\substack{e_1,\ldots,e_k \\ e_i \in E[G]}} \sum_{\vec{\mathcal{T}}=(\vec{e_1},\ldots,\vec{e_k})} \frac{t!}{t^t} \cdot \left(\mathbf{1}_{G_{\vec{\mathcal{T}}} \equiv H}\right) = \frac{t! \cdot \mathrm{auto}(H)}{t^t} \cdot \#(H,G) . \qquad \square$$

We can use a similar technique to analyze the variance of $Z_H(G)$ and apply Chebyshev's inequality on complex-valued random variables to upper bound the number of trials required for a $(1 \pm \varepsilon)$-approximation. Since $Z_H(G)$ is complex-valued, we need to upper bound $Z_H(G) \cdot \overline{Z_H(G)}$, which relies on the number of subgraphs of $2k$ edges in G with certain properties.

Lemma 5. *Let G be a graph with m edges and H be any graph with k edges (possibly with multiple edges), where k is a constant. The following statements hold: (i) If $\delta(H) \geqslant 2$, then $\#(H,G) = O\left(m^{k/2}\right)$; (ii) If every connected component of H contains at least two edges, then $\#(H,G) = O\left(m^{k/2} \cdot (\Delta(G))^{k/2}\right)$.*

Lemma 6. *Let G be any graph with m edges, H be any graph with k edges for a constant k. Random variables $X_c(w)$ $(c \in V[H], w \in V[G])$ and Q are defined as above. Then the following statements hold:*

1. *If $\delta(H) \geqslant 2$, then $\mathbb{E}\left[Z_H(G) \cdot \overline{Z_H(G)}\right] = O\left(m^k\right)$.*
2. *Let H be a connected graph with $k \geqslant 2$ edges and $\mathcal{H}$ be the set of all subgraphs H' in G with the following properties: (i) H' has $2k$ edges, and (ii) every connected component of H' contains at least two edges. Then $\mathbb{E}\left[Z_H(G) \cdot \overline{Z_H(G)}\right] = O\left(|\mathcal{H}|\right)$.*

By using Chebyshev's inequality, we can get a $(1 \pm \varepsilon)$-approximation by running independent copies of our estimator in parallel and returning the average of the output of these copies. This leads to our main result for counting the number of occurrences of H.

Theorem 7. *Let G be any graph with m edges and H be any graph with $k = O(1)$ edges. For any constant $0 < \varepsilon < 1$, there is an algorithm to $(1 \pm \varepsilon)$-approximate $\#(H, G)$ using (i) $O\left(\frac{1}{\varepsilon^2} \cdot \frac{m^k}{(\#H)^2} \cdot \log n\right)$ bits if $\delta(H) \geqslant 2$, or (ii) using $O\left(\frac{1}{\varepsilon^2} \cdot \frac{m^k \cdot (\Delta(G))^k}{(\#H)^2} \cdot \log n\right)$ bits for any H.*

Discussion. Statement (i) of Theorem 7 extends the main result of [11, Theorem 1] which requires H to be a cycle. Note that a naïve sampling-based approach would choose a random k-tuple of edges and require $m^k/(\#H)$ space. Theorem 7 improves upon this approach, in particular if the graph G is sparse and the number of occurrences of H is a growing function in n.

4   Extensions

We have developed a general framework for counting arbitrary subgraphs of constant size. For several typical applications we can further improve the space complexity by grouping the sketches or using certain properties of the underlying graph G. For the ease of the discussion we only focus on counting stars.

Grouping Sketches. The space complexity in Theorem 7 relies on the number of edges that the sketch reads. To reduce the variance, a natural way is to use multiple copies of the sketches, and every sketch is only responsible for the updates of the edges from a certain subgraph.

To formulate this intuition, we partition $V = \{1, \ldots, n\}$ into $g := n^{1-1/(2k)}$ subsets $\mathcal{V}_1, \ldots, \mathcal{V}_g$, and $\mathcal{V}_i := \{j : (i-1) \cdot n^{1/(2k)} + 1 \leqslant j \leqslant i \cdot n^{1/(2k)}\}$. Without loss of generality we assume that $n^{1/(2k)} \in \mathbb{N}$. Associated with every $\mathcal{V}_i$, we maintain a sketch $\mathcal{C}_i$, whose description is shown in Estimator 2. For every arriving edge $e = \{u, v\}$ in the stream, we update sketch $\mathcal{C}_i$ if $u \in \mathcal{V}_i$ or $v \in \mathcal{V}_i$. Since (i) the central vertex of every occurrence of S_k is in exactly one subset $\mathcal{V}_i$, and (ii) every edge adjacent to one vertex in $\mathcal{V}_i$ is taken into account by sketch $\mathcal{C}_i$, every occurrence of S_k in G is only counted by one sketch $\mathcal{C}_i$.

Estimator 2. Counting $\#(S_k, G|_{\mathcal{V}_i})$, update procedure

---

Step 1 (Update): When an edge $e = \{u, v\} \in E[G]$ arrives, update each variable $Z_{\overrightarrow{ab}}$:
(a) If $u \in \mathcal{V}_i$ and $v \in \mathcal{V}_i$, then

$$Z_{\overrightarrow{ab}}(G) \leftarrow Z_{\overrightarrow{ab}}(G) + \mathcal{M}_{\overrightarrow{ab}}(u, v) + \mathcal{M}_{\overrightarrow{ab}}(v, u).$$

(b) If $u \in \mathcal{V}_i$ and $v \in \partial \mathcal{V}_i$, then $Z_{\overrightarrow{ab}}(G) \leftarrow Z_{\overrightarrow{ab}}(G) + \mathcal{M}_{\overrightarrow{ab}}(u, v)$.
(c) If $u \in \partial \mathcal{V}_i$ and $v \in \mathcal{V}_i$, then $Z_{\overrightarrow{ab}}(G) \leftarrow Z_{\overrightarrow{ab}}(G) + \mathcal{M}_{\overrightarrow{ab}}(v, u)$.

More formally, let $\widetilde{\#}(S_k, G|_{\mathcal{V}_i})$ be the number of S_k whose central vertex is in $\mathcal{V}_i$. It holds that $\#(S_k, G) = \sum_{i=1}^{g} \widetilde{\#}(S_k, G|_{\mathcal{V}_i})$. This indicates that if every $\mathcal{C}_i$ is unbiased for $\widetilde{\#}(S_k, G|_{\mathcal{V}_i})$, then we can use the sum of returned values from different $\mathcal{C}_i$'s to approximate $\#(S_k, G)$.

Theorem 8. *Let G be a graph with n vertices. For any constants $0 < \varepsilon < 1$ and k, there is an algorithm to $(1 \pm \varepsilon)$-approximate $\#(S_k, G)$ with space complexity*

$$O\left(\frac{n^{1-1/(2k)}}{\varepsilon^2} \cdot \left(\frac{n^{3/2-1/(2k)} \cdot \Delta(G)^{2k}}{(\#S_k)^2} + 1\right) \cdot \log n\right).$$

Let us consider graphs G with $\Delta(G)/\delta(G) = o(n^{1/(4k)})$ and $\delta(G) \geqslant k$. Since $\#(S_k, G) = \Omega\left(n \cdot \delta(G)^k\right)$, Theorem 8 implies that $o\left(\frac{1}{\varepsilon^2} \cdot n \cdot \log n\right)$ bits suffice to give a $(1 \pm \varepsilon)$-approximation.

Counting on Power Law Graphs. Besides organizing the sketches into groups, the space complexity can be also reduced by using the structural information of the underlying graph G. One important property shared by many biological, social or technological networks is the so-called *Power Law* degree distribution, i.e., the number of vertices with degree d, denoted by $f(d) := |\{v \in V : \deg(v) = d\}|$, satisfies $f(d) \sim d^{-\beta}$, where $\beta > 0$ is the power law exponent. For many networks, experimental studies indicate that β is between 2 and 3, see [15].

Formally, we use the following model based on the cumulative degree distribution. For given constants $\sigma \geqslant 1$ and $d_{\min} \in \mathbb{N}$, we say that G has an approximate power law degree distribution with exponent $\beta \in (2,3)$, if $\sum_{d=k}^{n-1} f(d) \in \left[\lfloor \sigma^{-1} \cdot n \cdot k^{-\beta+1} \rfloor, \sigma \cdot n \cdot k^{-\beta+1}\right]$ for any $k \geqslant d_{\min}$. Our result on counting stars on power law graphs is as follows.

Theorem 9. *Assume that G has an approximate power law degree distribution with exponent $\beta \in (2,3)$. Then, for any two constants $0 < \varepsilon < 1$ and k, we can $(1 \pm \varepsilon)$-approximate $\#(S_k, G)$ using $O\left(\frac{1}{\varepsilon^2} \cdot \log n\right)$ bits.*

References

[1] Alon, N., Spencer, J.: The Probabilistic Method, 3rd edn. Wiley-Interscience Series in Discrete Mathematics and Optimization. John Wiley & Sons (2008)

[2] Alon, N., Yuster, R., Zwick, U.: Finding and counting given length cycles. Algorithmica 17(3), 209–223 (1997)

[3] Bar-Yossef, Z., Kumar, R., Sivakumar, D.: Reductions in streaming algorithms, with an application to counting triangles in graphs. In: Proc. 13th Symp. on Discrete Algorithms (SODA), pp. 623–632 (2002)

[4] Becchetti, L., Boldi, P., Castillo, C., Gionis, A.: Efficient semi-streaming algorithms for local triangle counting in massive graphs. In: Proc. 14th Intl. Conf. Knowledge Discovery and Data Mining (KDD), pp. 16–24 (2008)

[5] Bordino, I., Donato, D., Gionis, A., Leonardi, S.: Mining large networks with subgraph counting. In: Proc. 8th Intl. Conf. on Data Mining (ICDM), pp. 737–742 (2008)

[6] Buriol, L.S., Frahling, G., Leonardi, S., Marchetti-Spaccamela, A., Sohler, C.: Counting triangles in data streams. In: Proc. 25th Symp. Principles of Database Systems (PODS), pp. 253–262 (2006)

[7] Buriol, L.S., Frahling, G., Leonardi, S., Sohler, C.: Estimating Clustering Indexes in Data Streams. In: Arge, L., Hoffmann, M., Welzl, E. (eds.) ESA 2007. LNCS, vol. 4698, pp. 618–632. Springer, Heidelberg (2007)

[8] Ganguly, S.: Estimating Frequency Moments of Data Streams Using Random Linear Combinations. In: Jansen, K., Khanna, S., Rolim, J.D.P., Ron, D. (eds.) RANDOM 2004 and APPROX 2004. LNCS, vol. 3122, pp. 369–380. Springer, Heidelberg (2004)

[9] Gonen, M., Ron, D., Shavitt, Y.: Counting stars and other small subgraphs in sublinear-time. SIAM J. Disc. Math. 25(3), 1365–1411 (2011)

[10] Jowhari, H., Ghodsi, M.: New Streaming Algorithms for Counting Triangles in Graphs. In: Wang, L. (ed.) COCOON 2005. LNCS, vol. 3595, pp. 710–716. Springer, Heidelberg (2005)

[11] Manjunath, M., Mehlhorn, K., Panagiotou, K., Sun, H.: Approximate Counting of Cycles in Streams. In: Demetrescu, C., Halldórsson, M.M. (eds.) ESA 2011. LNCS, vol. 6942, pp. 677–688. Springer, Heidelberg (2011)

[12] McGregor, A.: Open Problems in Data Streams and Related Topics, IITK Workshop on Algorithms For Data Sreams (2006), http://www.cse.iitk.ac.in/users/sganguly/data-stream-probs.pdf

[13] Milo, R., Shen-Orr, S., Itzkovitz, S., Kashtan, N., Chklovskii, D., Alon, U.: Network motifs: Simple building blocks of complex networks. Science 298(5594), 824–827 (2002)

[14] Muthukrishnan, S.: Data Streams: Algorithms and Applications. Foundations and Trends in Theoretical Computer Science 1(2) (2005)

[15] Newman, M.E.J.: The structure and function of complex networks. SIAM Review 45, 167–256 (2003)

[16] Pagh, R., Tsourakakis, C.E.: Colorful triangle counting and a mapreduce implementation. Inf. Process. Lett. 112(7), 277–281 (2012)

[17] Wong, E., Baur, B., Quader, S., Huang, C.: Biological network motif detection: principles and practice. Briefings in Bioinformatics, 1–14 (June 2011)

k-Chordal Graphs: From Cops and Robber to Compact Routing via Treewidth[*]

Adrian Kosowski[1], Bi Li[2,3], Nicolas Nisse[2], and Karol Suchan[4,5]

[1] CEPAGE, INRIA, LaBRI, Talence, France
[2] MASCOTTE, INRIA, I3S(CNRS/UNS), Sophia Antipolis, France
[3] CAS & AAMS, Beijing, China
[4] FIC, Universidad Adolfo Ibáñez, Santiago, Chile
[5] WMS, AGH - University of Science and Technology, Krakow, Poland

Abstract. *Cops and robber games* concern a team of cops that must capture a robber moving in a graph. We consider the class of k-chordal graphs, i.e., graphs with no induced cycle of length greater than k, $k \geq 3$. We prove that $k - 1$ cops are always sufficient to capture a robber in k-chordal graphs. This leads us to our main result, a new structural decomposition for a graph class including k-chordal graphs.

We present a quadratic algorithm that, given a graph G and $k \geq 3$, either returns an induced cycle larger than k in G, or computes a *tree-decomposition* of G, each *bag* of which contains a dominating path with at most $k-1$ vertices. This allows us to prove that any k-chordal graph with maximum degree Δ has treewidth at most $(k-1)(\Delta-1)+2$, improving the $O(\Delta(\Delta-1)^{k-3})$ bound of Bodlaender and Thilikos (1997). Moreover, any graph admitting such a tree-decomposition has small hyperbolicity.

As an application, for any n-node graph admitting such a tree-decomposition, we propose a *compact routing scheme* using routing tables, addresses and headers of size $O(\log n)$ bits and achieving an additive stretch of $O(k \log \Delta)$. As far as we know, this is the first routing scheme with $O(k \log \Delta + \log n)$-routing tables and small additive stretch for k-chordal graphs.

Keywords: Treewidth, chordality, compact routing, cops and robber games.

1 Introduction

Because of the huge size of real-world networks, an important current research effort concerns exploiting their structural properties for algorithmic purposes. Indeed, in large-scale networks, even algorithms with polynomial-time in the size of the instance may become unpractical. So, it is important to design algorithms depending only quadratically or linearly on the size of the network when its topology is expected to satisfy some properties. Among these properties, the *chordality* of a graph is the length of its longest induced (i.e., chordless) cycle. The

---

[*] Partially supported by programs Fondap and Basal-CMM, Anillo ACT88 and Fondecyt 11090390 (K.S.), FP7 STREP EULER (N.N.).

A. Czumaj et al. (Eds.): ICALP 2012, Part II, LNCS 7392, pp. 610–622, 2012.
© Springer-Verlag Berlin Heidelberg 2012

hyperbolicity of a graph reflects how the metric (distances) of the graph is close to the metric of a tree. A graph has hyperbolicity $\leq \delta$ if, for any $u, v, w \in V(G)$ and for any shortest paths P_{uv}, P_{vw}, P_{uw} between these three vertices, any vertex in P_{uv} is at distance at most δ from $P_{vw} \cup P_{uw}$ [Gro87]. Intuitively, in a graph with small hyperbolicity, any two shortest paths between the same pair of vertices are close to each other. Several recent works take advantage of such structural properties of large-scale networks for algorithm design (e.g., routing [KPBV09]). Indeed, Internet-type networks have a so-called high clustering coefficient (see e.g. [WS98]), leading to the existence of very few long chordless cycles, whereas their low (logarithmic) diameter implies a small hyperbolicity [dMSV11].

Another way to study tree-likeness of graphs is by *tree-decompositions*. Introduced by Robertson and Seymour, such decompositions play an important role in design of efficient algorithms. Roughly speaking, a tree-decomposition maps each vertex of a graph to a subtree of the *decomposition tree* in a way that the subtrees assigned to adjacent vertices intersect [Bod98]. The nodes of the decomposition tree are called *bags*, and the size of a bag is the number of vertices assigned to it (assigned subtrees intersect the bag). The *width* of a tree-decomposition is the maximum size over its bags, and the *treewidth* of a graph is the smallest width over its tree-decompositions. By using dynamic programming based on a tree-decomposition, many NP-hard problems have been shown to be linear time solvable for graph with bounded treewidth [CM93]. In particular, there are linear-time algorithms to compute an optimal tree-decomposition of a graph with bounded treewidth [BK96]. However, from the practical point of view, this approach has several drawbacks. First, all above-mentioned algorithms are linear in the size of the graph but (at least) exponential in the treewidth. Moreover, due to the high clustering coefficient of large-scale networks, their treewidth is expected to be large [dMSV11]. Hence, to face these problems, it is important to focus on the structure of the bags of the tree-decomposition, instead of trying to minimize their size. For instance, several works study the diameter of the bags [DG07]. In this work, we consider tree-decompositions in which each bag admits a particular small dominating set. Such decompositions turn out to be applicable to a large family of graphs (including k-chordal graphs).

1.1 Our Results

Our results on tree decomposition are inspired by a study of the so called *cops and robber games*. The aim of such a game is to capture a robber moving in a graph, using as few cops as possible. This problem has been intensively studied in the literature, allowing for a better understanding of the structure of graphs [BN11].

Outline of the paper. We start by presenting our results for the cops and robber problem in Section 2. Next, using these results, in Section 3 we provide a new type of efficiently computable tree-decomposition which we call *good tree decomposition*. Our tree decomposition turns out to be applicable to many real-world graph classes (including k-chordal graphs), and has several algorithmic applications. Finally, we focus on the applications of this decomposition to the *compact routing problem*, a research area in which tree decompositions have already proved useful [Dou05]. The objective of compact routing is to provide a scheme for finding a path from a sender node to a known destination, taking routing decisions for the packet at every step using only very limited information stored at each node. In Section 4, we show how to use our tree decomposition to minimize the additive stretch of the routing scheme (i.e., the difference between the length of a route computed by the scheme and that of a shortest path connecting the same pair of nodes) in graphs admitting with k-good tree-decomposition for any given integer $k \geq 3$ (including k-chordal graphs), assuming logarithmic size of packet headers and routing tables stored at each node.

The necessary terminology concerning cops and robber games, tree decompositions, and compact routing, is introduced in the corresponding sections.

Main contributions. Our main contribution is the design of a $O(m^2)$ algorithm that, given a m-edge graph G and an integer $k \geq 3$, either returns an induced cycle of length at least $k + 1$ in G or computes a tree-decomposition of G with each bag having a dominating path of order $\leq k - 1$. That is, each bag of our tree-decomposition contains a chordless path with at most $k - 1$ vertices, such that any vertex in the bag is either in the path or adjacent to some vertex of the path. If G, with maximum degree Δ, admits such a decomposition, then G has treewidth at most $(k - 1)(\Delta - 1) + 2$, tree-length at most k and hyperbolicity at most $\lfloor \frac{3}{2}k \rfloor$. In particular, this shows that the treewidth of any k-chordal graph is upper-bounded by $O(k \cdot \Delta)$, improving the exponential bound of [BT97]. The proposed algorithm is mainly derived from our proof of the fact that $k - 1$ cops are sufficient to capture a robber in k-chordal graphs.

Our tree-decomposition may be used efficiently for solving problems using dynamic programming in graphs of small chordality and small maximum degree. In this paper, we focus on different application. We present a compact routing scheme that uses our tree-decomposition and that achieves an additive stretch $\leq 2k(\lceil \log \Delta \rceil + \frac{5}{2}) - 5$ with routing tables, addresses and message headers of $O(\max\{k \cdot \log \Delta, \log n\})$ bits. An earlier approach of Dourisboure achieved stretch $k + 1$, but with routing tables of size $O(\log^2 n)$.

1.2   Related Work

Chordality and hyperbolicity. Chordality and hyperbolicity are both parameters measuring the tree-likeness of a graph. Some papers consider relations between them [WZ11]. In particular, the hyperbolicity of a k-chordal graph is at most k, but the difference may be arbitrary large (take a $3 \times n$-grid). The seminal definition of Gromov hyperbolicity is the following. A graph G is d-hyperbolic provided that for any vertices $x, y, u, v \in V(G)$, the two larger of the three sums $d(u, v) + d(x, y), d(u, x) + d(v, y)$ and $d(u, y) + d(v, x)$ differ by at most $2d$ [Gro87]. This definition is equivalent to the one we use in this paper, using so called *thin triangles*, up to a constant ratio. No algorithm better than the $O(n^4)$-brute force algorithm (testing all 4-tuples in G) is known to compute Gromov hyperbolicity of n-node graphs. The problem of computing the chordality of a graph G is NP-complete since it may be related to computing a longest cycle in the graph obtained from G after subdividing all edges once. Finding the longest induced path is $W[2]$-complete [CF07] and the problem is Fixed Parameter Tractable in planar graphs [KK09].

Treewidth. It is NP-complete to decide whether the treewidth of a graph G is at most k [ACP87]. The treewidth problem is polynomially solvable in chordal graphs, cographs, circular arc graphs, chordal bipartite graphs, etc. [Bod98]. Bodlaender and Thilikos proved that the treewidth of a k-chordal graph with maximum degree Δ is at most $\Delta(\Delta-1)^{k-3}$ which implies that treewidth is polynomially computable in the class of graphs with chordality and maximum degree bounded by constants [BT97]. They also proved that the treewidth problem is NP-complete for graphs with small maximum degree [BT97].

Compact routing. In [AGM+08], a universal name-independent routing scheme with stretch linear in k and $n^{1/k}polylog(n)$ space is provided. There are weighted trees for which every name-independent routing scheme with space less than $n^{1/k}$ requires stretch at least $2k+1$ and average stretch at least $k/4$ [AGD06]. Subsequently, the interest of the scientific community was turned toward specific properties of graphs. Several routing schemes have been proposed for particular graph classes: e.g., trees [FG01], bounded doubling dimension [AGGM06], excluding a fixed graph as a minor [AG06], etc. The best compact routing scheme in k-chordal graphs (independent from the maximum degree) is due to Dourisboure and achieves a stretch of $k+1$ using routing tables of size $O(\log^2 n)$ bits [Dou05]. A routing scheme achieving stretch $k-1$ with a distributed algorithm for computing routing tables of size $O(\Delta \log n)$ bits has been proposed in [NSR12].

2   A Detour through Cops and Robber Games

Let us first fix some notations. G denotes a simple connected undirected graph with vertex set V and edge set E. $n = |V|$ is the *order* of G and $m = |E|$ is the *size* of G. $V(H)$ and $E(H)$ denotes the vertex and edge set of H, respectively. The set of vertices adjacent to $v \in V$ is denoted $N(v)$ and called *open neighborhood* of v. $N[v] = N(v) \cup \{v\}$ is the *closed neighborhood* of v. We extend this notation to write $N[U] = \cup\{N[u] \mid u \in U\}$ and $N(U) = N[U] \setminus U$. $d_G(v) = |N_G(v)|$ is the *degree* of v, Δ denotes the maximum degree among the vertices of G. The graph obtained from G by *removing an edge* $\{x,y\}$ is denoted $G \setminus \{x,y\}$; the result of *removing a vertex* v and all adjacent edges is denoted $G \setminus \{v\}$. Like above, we extend this to denote removing sets of vertices or edges. For $U \subset V$, $G[U]$ is the subgraph of G *induced* by U. It can be obtained as the result of removing from G the vertices in $V \setminus U$, denoted by $G \setminus (V \setminus U)$. Given two paths $P = (p_1, \ldots, p_k)$ and $Q = (q_1, \ldots, q_r)$, we denote by (P, Q) the path present in the graph induced by $V(P) \cup V(Q)$ - to make descriptions more concise, we omit the detail of reversing P or Q if necessary.

Let us formally define a cops' strategy to capture a robber. Given a graph G, a player starts by placing $k \geq 1$ cops on some vertices in V, then a visible robber is placed on one vertex v in V. Alternately, the robber can move to a vertex x in $N(v)$, and the cop-player may move each cop along an edge. The robber is captured if, at some step, a cop occupies the same vertex. The *cop-number* of a graph G, denoted by $cn(G)$, is the fewest number of cops required to capture a robber in G (see the recent book [BN11]). Bounds on the cop-number have been provided for various graph classes and provided many nice structural results [BN11]. We consider the class of k-chordal graphs.

Theorem 1. *Let $k \geq 3$. For any k-chordal connected graph G, $cn(G) \leq k - 1$, and there exists a strategy where all $k - 1$ cops always occupy a chordless path.*

Proof. Let $v \in V$ be any vertex and place all cops at it. Then, the robber chooses a vertex. Now, at some step, assume that the cops are occupying $\{v_1, \cdots, v_i\}$ which induce a chordless path, $i \leq k - 1$, and it is the turn of the cops (initially $i = 1$). Let $N = \cup_{j \leq i} N[v_j]$, if the robber occupies a vertex in N, it is captured during the next move. Else, let $R \neq \emptyset$ be the connected component of $G \setminus N$ occupied by the robber. Finally, let S be the set of vertices in N that have some neighbor in R. Clearly, while R is not empty, then so does S.

Now, there are two cases to be considered. If $N(v_1) \cap S \subseteq \cup_{1 < j \leq i} N[v_j]$. This case may happen only if $i > 1$. Then, "remove" the cop(s) occupying v_1. That is, the cops occupying v_1 go to v_2. Symmetrically, if $N(v_i) \cap S \subseteq \cup_{1 \leq j < i} N[v_j]$, then the cops occupying v_i go to v_{i-1}. Then, the cops occupy a shorter chordless path while the robber is still restricted to R.

Hence, there is $u \in (N(v_1) \cap S) \setminus (\cup_{1 < j \leq i} N[v_j])$ and $v \in (N(v_i) \cap S) \setminus (\cup_{1 \leq j < i} N[v_j])$. First, we show that this case may happen only if $i < k - 1$. Indeed, otherwise, let P be a shortest path between such u and v with all internal vertices in R (possibly, P is reduced to an edge). Such a path exists by definition of S. Then, $(v_1, \cdots, v_i, v, P, u)$ is a chordless cycle of length at least $i + 2$. Since G is k-chordal, this implies that $i + 2 \leq k$. Then, one cop goes to $v = v_{i+1}$ while all the vertices in $\{v_1, \cdots, v_i\}$ remain occupied. Since $v \in S$, it has some neighbor in R, and then, the robber is restricted to occupy R' the connected component of $G \setminus (N \cup N[v])$ which is strictly contained in R.

Therefore, proceeding as described above strictly reduces the area of the robber (i.e., R) after $< k$ steps and then the robber is eventually captured.   □

Note that previous Theorem somehow extends the model in [CN05] where the authors consider the game when two cops always remaining at distance at most 2 from each other must capture a robber. It is possible to improve the previous result in case $k = 4$. Indeed, for any 4-chordal connected graph G, then $cn(G) \leq 2$. Due to lack of space, the proof is omitted and can be found in [KLNS12].

Theorem 1 relies on chordless paths P in G such that $N[P]$ is a separator of G, i.e., there exist vertices a and b of G such that all paths between a and b intersect $N[P]$. In next section, we show how to adapt this to compute particular tree-decompositions.

3   Structured Tree-Decomposition

In this section, we present our main contribution, that is, an algorithm that, given a n-node graph G and an integer $k \geq 3$, either returns an induced cycle of length at least $k+1$ in G or computes a tree-decomposition of G with interesting structural properties. First, we need some definitions.

A *tree-decomposition* of a graph $G = (V, E)$ is a pair $(\{X_i | i \in I\}, T = (I, M))$, where T is a tree and $\{X_i | i \in I\}$ is a family of subsets, called bags, of vertices of G such that (1) $V = \cup_{i \in I} X_i$; (2) $\forall \{uv\} \in E$ there is $i \in I$ such that $u, v \in X_i$; and (3) $\forall v \in V$, $\{i \in I | v \in X_i\}$ induces a (connected) subtree of T. The *width* of a tree-decomposition is the size of its largest bag minus 1 and its ℓ-*width* is the maximum diameter of the subgraphs induced by the bags. The *treewidth* denoted by $tw(G)$, resp., *tree-length* denoted by $tl(G)$, of a graph G is the minimum width, resp., ℓ-width, over all possible tree-decompositions of G [DG07].

Let $k \geq 2$. A k-caterpillar is a graph that has a dominating set, called *backbone*, which induces a chordless path of order at most $k - 1$. That is, any vertex of a k-caterpillar either belongs to the backbone or is adjacent to a vertex of the backbone. A tree-decomposition is said to be k-*good* if each of its bags induces a k-caterpillar.

Theorem 2. *There is a $O(m^2)$-algorithm that takes a graph G of size m and an integer $k \geq 3$ as inputs and: either returns an induced cycle of length at least $k + 1$; or returns a k-good tree-decomposition of G;*

Proof. The proof is by induction on $|V(G)| = n$. We prove that either we find an induced cycle larger than k, or for any chordless path $P = (v_1, \ldots, v_i)$ with $i \leq k - 1$, there is a k-good tree-decomposition for G with one bag containing $N_G[P]$. Obviously, it is true if $|V(G)| = 1$. Now we assume that it is true for any graph G with n' nodes, $1 \leq n' < n$, and we show it is true for n-node graphs.

Let G be a connected n-node graph, $n > 1$. Let $P = (v_1, \ldots, v_i)$ be any chordless path with $i \leq k - 1$ and let $N = N_G[P]$, $N_j = N_G[v_j]$ for $j = 1, \ldots, i$ and $G' = G \setminus N$. There are three cases to be considered:

Case 1. $G' = \emptyset$. In this case, we have $G = N$. The desired tree-decomposition consists of one node, corresponding to the bag N.

Case 2. G' is disconnected. Let $C_1, \ldots, C_r, r \geq 2$, be the connected components of G' For any $j \leq r$, let G_j be the graph induced by $C_j \cup N$. Note that any induced cycle in G_j, $j \leq r$, is an induced cycle in G. By the induction hypothesis, either there is an induced cycle $\mathcal{C}$ larger than k in G_j, then $\mathcal{C}$ is also an induced cycle larger than k in G, or our algorithm computes a k-good tree-decomposition TD_j of G_j with one bag X_j containing N. To obtain the k-good tree-decomposition of G, we combine the TD_j's, $j \leq r$, by adding a bag $X = N$ adjacent to all the bags X_j for $j = 1, \ldots, r$. It is easy to see that this tree-decomposition satisfies our requirements.

Case 3. G' is connected. We consider the order of the path $P = (v_1, \ldots, v_i)$. In the following proof, first we prove that if the order of path P, $i = k - 1$, then we can find either an induced cycle larger than k or the required tree-decomposition for G. Subsequently, we prove it is also true for path with length $i < k - 1$ by reversed induction on i. More precisely, if $i < k-1$, either we find directly the desired cycle or tree-decomposition, or we show that there exists a vertex v_{i+1} such that $V(P) \cup \{v_{i+1}\}$ induces a chordless path P' of order $i + 1$. By reverse induction on i we can find either an induced cycle larger than k or a k-good tree-decomposition of G with one bag containing $N_G[V(P')] \supseteq N_G[V(P)]$.

(a) If $i = k - 1$, then we consider the following two cases.

 - Assume first that there is $u \in N_G(V(P)) \cup \{v_1, v_i\}$ (in particular, $u \notin V(P) \setminus \{v_1, v_i\}$) such that $N_G(u) \subseteq N_G[V(P) \setminus \{u\}]$. Let $\tilde{G} = G \setminus \{u\}$. Then $\tilde{G}$ is a graph with $n' = n - 1$ vertices. By the induction hypothesis on $n' < n$, the algorithm either finds an induced cycle larger than k in $\tilde{G}$, then it is also the one in G; Otherwise our algorithm computes a k-good tree-decomposition $\widetilde{TD}$ of $\tilde{G}$ with one bag $\tilde{X}$ containing $N_{\tilde{G}}[V(P) \setminus \{u\}]$. To obtain the required tree-decomposition of G, we just add vertex u into the bag $\tilde{X}$. The tree-decomposition is still k-good.

 - Otherwise, there exist two distinct vertices $v_0 \in N_G(v_1)$ and $v_{i+1} \in N_G(v_i)$ and there are vertices $u_1, u_2 \in V(G')$ (possibly $u_1 = u_2$) such that $\{v_0, u_1\} \in E(G)$ and $\{v_{i+1}, u_2\} \in E(G)$. If $\{v_0, v_{i+1}\} \in E(G)$, (P, v_0, v_{i+1}) is an induced cycle with $k + 1$ vertices. Otherwise, let Q be a shortest path between u_1 and u_2 in G' (Q exists since G' is connected). So $(P, v_{i+1}, u_2, Q, u_1, v_0)$ is an induced cycle with at least $k + 1$ vertices in G.

(b) If $i < k - 1$, we proceed by reverse induction on i. Namely, assume that, for any chordless path Q with $i + 1$ vertices, our algorithm either finds an induced cycle larger than k in G or computes a k-good tree-decomposition of G with one bag containing $N[Q]$. Note that the initialization of the induction holds for $i = k - 1$ as described in case (a). We show it still holds for a chordless path with i vertices. We consider the following two cases.

 - Either there is $u \in N_G(V(P)) \cup \{v_1, v_i\}$ (in particular, $u \notin V(P) \setminus \{v_1, v_i\}$) such that $N_G(u) \subseteq N_G[V(P) \setminus \{u\}]$. That is, we are in the same case as the first item of (a). We proceed as above and the result holds by induction on n.

 - Or there is $w \in N_G(v_1) \cup N_G(v_i) \setminus V(P)$ such that (P, w) is chordless (i.e., w is a neighbor of v_1 or v_i but not both). Therefore, we apply the induction hypothesis (on i) on $P' = (P, w)$. By the assumption on i, either our algorithm returns an induced cycle larger than k or it computes a k-good tree-decomposition of G with one bag containing $N_G[V(P')] \supseteq N_G[V(P)]$.

Let us analyze the algorithm and its complexity. Let G be a m-edge n-node graph with maximum degree Δ. The algorithm proceeds in steps of $O(m)$ time, each time considering one vertex. We prove that at each step (but the first one), at least one edge will be *considered* and that all edges are considered at most once. This implies a time-complexity of $O(m^2)$ for the algorithm. Due to lack of space, the proof of the time-complexity is omitted and can be found in [KLNS12]. □

Due to lack of space, the proofs of the following consequences of Theorem 2 are omitted and can be found in [KLNS12].

Theorem 3. *Let G be a graph that admits a k-good tree-decomposition. Then $tw(G) \leq (k-1)(\Delta - 1) + 2$ where Δ is its maximum degree, $tl(G) \leq k$, and hyperbolicity at most $\lfloor \frac{3}{2}k \rfloor$.*

Corollary 1. *Any k-chordal graph G with maximum degree Δ has treewidth at most $(k-1)(\Delta - 1) + 2$, tree-length at most k and hyperbolicity at most $\lfloor \frac{3}{2}k \rfloor$.*

Corollary 2. *There is an algorithm that, given a m-edge graph G and $k \geq 3$, states that either G has chordality at least $k+1$ or G has hyperbolicity at most $\lfloor \frac{3}{2}k \rfloor$, in time $O(m^2)$.*

4   Application of k-Good Tree-Decompositions for Routing

In this section, we propose a compact routing scheme for any n-node graph G that admits a k-good tree-decomposition (including k-chordal graphs). here here

4.1   Model and Performance of the Routing Scheme

We propose a *labelled* routing scheme which means that we are allowed to give one identifier, $name(v)$, of $O(\log n)$ bits to any vertex v of G. Moreover, following [FG01], we consider the *designer-port* model, which allows us to choose the permutation of ports (assign a label of $\log d_v$ bits to any edge incident to v in $V(G)$). Finally, to any node $v \in V(G)$, we assign a routing table, denoted by $Table(v)$, where local information of $O(k \cdot \log \Delta + \log n)$ bits is stored. Any message has a *header* that contains the address $name(t)$ of the destination t, three modifiable integers $pos \in \{-1, 0, \cdots, k-1\}$, $cnt_d, cnt'_d \in \{-1, 0, \cdots, \Delta + 1\}$, one bit $start$ and some memory, called $path$, of size $O(k \cdot \log \Delta)$ bits. $start$ and $path$ change only once.

Following our routing scheme, a node v that receives a message uses its header, $name(v)$, $Table(v)$ and the port-numbers of the edges incident to v to compute its new header and to choose the edge $e = \{v, u\}$ over which it relays the message. Then, the node u knows that the message arrived from v. The length of the path followed by a message from a source $s \in V(G)$ to a destination $t \in V(G)$, using the routing scheme, is denoted by $|P(s,t)|$, and the *stretch* of the scheme is $\max_{s,t \in V(G)} |P(s,t)| - d(s,t)$ where $d(s,t)$ is the distance between s and t in G.

To design our routing scheme, we combine the compact routing scheme in trees of [FG01] together with the k-good tree-decomposition. Roughly, the scheme consists in following the paths in a BFS-tree F of G, using the scheme in [FG01], and uses one bag of the tree-decomposition as a short-cut between two branches of F. Intuitively, if the source s and the destination t are "far apart", then there is a bag X of the tree-decomposition that separates s and t in G. The message follows the path in F to the root of F until it reaches X, then an exhaustive search is done in X until the message finds an ancestor y of t, and finally it follows the path from y to t in F using the scheme of [FG01]. The remaining part of this Section is devoted to the proof of the next Theorem that summarizes the performances of our routing scheme.

Theorem 4. *For any n-node m-edge graph G with maximum degree Δ and with a k-good tree-decomposition, there is a labelled routing scheme $\mathcal{R}$ with the following properties. $\mathcal{R}$ uses addresses of size $O(\log n)$ bits, port-numbers of size $O(\log \Delta)$ bits and routing tables of size $O(k \cdot \log \Delta + \log n)$ bits. The routing tables, addresses and port-numbers can be computed in time $O(m^2)$. Except the address of the destination (not modifiable), the header of a message contains $O(k \cdot \log \Delta)$ modifiable bits. The header and next hop is computed in time $O(1)$ at each step of the routing. Finally, the additive stretch is $\leq 2k(\lceil \log \Delta \rceil + \frac{5}{2}) - 5$.*

4.2   Data Structures

Routing in Trees [FG01]. Since, we use the shortest path routing scheme proposed in [FG01] for trees, we start by recalling some of the data structures they use. Let F be a tree rooted in $r \in V(F)$. For any $v \in V(F)$, let F_v be the subtree of F rooted in v and let $w_F(v) = |V(F_v)|$ be the *weight* of v. Consider a Depth-First-Search (DFS) traversal of F, starting from r, and guided by the weight of the vertices, i.e., at each vertex, the DFS visits first the largest subtree, then the second largest subtree, and so on. For any $v \in V(F)$, let $Id_F(v) \in \{1, \cdots, n\}$ be the preordering rank of v in the DFS. It is important to note that, for any $u, v \in V(F)$, $v \in V(F_u)$ if and only if $Id_F(u) \leq Id_F(v) \leq Id_F(u) + w_F(u) - 1$.

For any $v \in V(F)$ and any e incident to v, the edge e receives a *port-number* $p_F(e, v)$ at v as follows. $p_F(e, v) = 0$ if $v \neq r$ and e leads to the parent of v in F, i.e., e is the first edge on the path from v to r. Otherwise, let $(u_1, \cdots, u_d)$ be the children of v ($d = d_F v$ if $v = r$ and $d = d_F v - 1$ otherwise) ordered by their weight, i.e., such that $w_F(u_1) \geq \cdots \geq w_F(u_d)$. Then, let $p_F(\{u_i, v\}, v) = i$, for any $i \leq d$. Finally, each vertex $v \in V(F)$ is assigned a routing table $RT_F(v)$ and an address $\ell_F(v)$ of size $O(\log n)$ bits allowing a shortest path routing in trees (see details in [FG01]).

Our Data Structures. Let G be a graph with the k-good tree-decomposition $(T = (I, M), \{X_i | i \in I\})$. Let $r \in V(G)$. Let F be a Breadth-First-Search(BFS) tree of G rooted at r. Let T be rooted in $b \in I$ such that $r \in X_b$.

We use (some of) the data structures of [FG01] for both trees F and T. More precisely, for any $v \in V(G)$, let $Id_F(v), w_F(v), \ell_F(v)$ and $RT_F(v)$ be defined as above for the BFS-tree F. Moreover, we add $d_F(v)$ to store the degree of v in the tree F . $p_{e,v} = p_F(e, v)$ for edges that belong to F are defined as above, the ports $> d_F(v)$ will be assigned to edges that do not belong to F. With $d_F(v)$ at hand, the ports that correspond to edges in F can be easily distinguished from ports assigned to edges in $\overline{F}$.

For any $i \in I$, let $Id_T(i)$ and $w_T(i)$ be defined as above for the tree T. For any $v \in V(G)$, let $B_v \in I$ be the bag of T containing v (i.e., $v \in X_{B_v}$) that is closest to the root b of T. To simplify the notations, we set $Id_T(v) = Id_T(B_v)$ and $w_T(v) = w_T(B_v)$. These structures will be used to decide "where" we are in the tree-decomposition when the message reaches $v \in V(G)$.

Let $\overline{F} = G \setminus E(F)$. For any $v \in V(G)$, let $(u_1, \cdots, u_d) = N_{\overline{F}}(v)$ be the neighborhood of v in $\overline{F}$ ordered such that $Id_F(u_1) < \cdots < Id_F(u_d)$. We assign $p_{e_i,v} = d_F(v) + i$, where $e_i = \{v, u_i\}$, for each u_i in this order. This ordering will allow to decide whether one of the vertices in $N_{\overline{F}}(v)$ is an ancestor of a given node t in time $O(\log \Delta)$ by binary search.

Finally, for any $i \in I$, let $P_i = (v_1, \cdots, v_\ell)$ be the backbone of B_i with $\ell \leq k-1$ (recall we consider a k-good tree decomposition). Let $(e_1, \cdots, e_{\ell-1})$ be the set of edges of P_i in order. We set $Backbone_i = (p_{e_1,v_1}, p_{e_1,v_2}, p_{e_2,v_2}, \cdots, p_{e_{\ell-1},v_\ell})$. For any $v \in V(G)$ such that $Id_T(v) = i \in I$, if $v = v_j \in P_i$, then $back(v) = (\emptyset, j)$ and if $v \notin P_i$, let $back(v) = (p_{e,v}, j)$ where $e = \{v, v_j\}$ and v_j $(j \leq \ell)$ is the neighbor of v in P_i with j minimum. This information will be used to cross a bag (using its backbone) of the tree-decomposition.

Now, for every $v \in V(G)$, we define the address $name(v) = \langle \ell_F(v), Id_T(v) \rangle$. Note that, in particular, $\ell_F(v)$ contains $Id_F(v)$. We also define the routing table of v as $Table(v) = \langle RT_F(v), w_T(v), Backbone(v), back(v) \rangle$.

Next table summarizes all these data structures.

	notation	description
$name(v)$	$\ell_F(v)$	the address of v in tree F [FG01]
	$Id_T(v)$	the identifier of the highest bag B_v containing v in T
	$RT_F(v)$	the routing table used of v for routing in F [FG01]
	$d_F(v)$	the degree of v in F
$Table(v)$	$w_T(v)$	the weight of the subtree of T rooted in B_v
	$Backbone(v)$	information to navigate in the backbone of B_v
	$back(v)$	information to reach the backbone of B_v from v

Clearly, $name(v)$ has size $O(\log n)$ bits and $Table(v)$ has size $O(k \cdot \log \Delta + \log n)$ bits. Moreover, any edge e incident to v receives a port-number $p_{e,v}$ of size $O(\log \Delta)$ bits.

4.3 Routing Algorithm in k-Good Tree-Decomposable Graphs

Let us consider a message that must be sent to some destination $t \in V(G)$. Initially, the header of the message contains $name(t)$, the three counters pos, cnt_d, $cnt'_d = -1$, the bit $start = 0$ and the memory $path = \emptyset$. Let $v \in V(G)$ be the current node where the message stands. First, using $Id_F(t)$ in $name(t)$, $Id_F(v)$ in $name(v)$ and $w_F(v)$ in $RT_F(v) \in Table(v)$, it is possible to decide in constant time if v is an ancestor of t in F. Similarly, using $Id_T(t)$ in $name(t)$, $Id_T(v)$ in $name(v)$ and $w_T(v)$ in $Table(v)$, it is possible to decide if the highest bag B_v containing v is an ancestor of B_t in T. There are several cases to be considered.

- If v is an ancestor of t in F, then using the protocol of [FG01] the message is passed to the child w of v that is an ancestor of t in F towards t. Recursively, the message arrives at t following a shortest path in G, since F is a BFS-tree.

- Else, if $path = \emptyset$, then
  - if neither B_v is an ancestor of B_t in T nor $B_t = B_v$, then the message follows the edge leading to the parent of v in F, i.e., the edge with port-number $p_{e,v} = 0$. Note that the message will eventually reach w that either is an ancestor of t in F or B_w is an ancestor of B_t in T, since the message follows a shortest path to the root r of F and B_r is the ancestor of any bag in T.
  - Else, an ancestor of t belongs to B_v since either $B_v = B_t$, or B_v is an ancestor of B_t. Therefore, since T is a tree-decomposition, B_v has to contain a vertex on the shortest path from t to r in F. Now the goal of it is to explore the bag B_v using its backbone $P = (v_1, \cdots, v_\ell)$ $(\ell < k)$, until the message finds an ancestor of t in F.
    The aim of this case is to put the message on the backbone, and then explore the backbone using $Backbone(v)$ copied in $path$ in the header of the message. Using $back(v) = (p, j) \in Table(v)$, pos is set to j. If $p = \emptyset$ then the message already is on the backbone. Otherwise, the message is sent over the port p. The idea is to explore the neighborhoods of vertices on the backbone, starting from v_1. Note that at in what follows $path \neq \emptyset$ and $pos \neq -1$.
- Else, if $start = 0$, then the message is at $v = v_j \in P$ and pos indicates the value of j. Moreover, in the field $path$ of the header, there are the port-numbers allowing to follow P. If $pos > 1$ then $pos = j - 1$ is set and the message follows the corresponding port-number to reach v_{j-1}. Otherwise, $start$ is set to 1, $cnt_d = d_F(v_1)$ and $cnt'_d = d_G(v_1) + 1$.
- Else, if $start = 1$, then the exploration of a bag containing an ancestor of t has begun. The key point is that any ancestor w of t in F is such that $Id_F(w) \leq Id_F(t) \leq Id_F(w) + w_F(w) - 1$. Using this property, for each vertex v_j of the backbone $P = (v_1, \cdots, v_\ell)$, the message visits v_j and its parent in F, and explores $N_{\overline{F}}(v_j)$ by binary search. Notice that the other neighbors of v_j are its descendants in F, so if t has an ancestor among them, then v_j also is an ancestor of t.
  - If $cnt_d = cnt'_d - 1$, the neighborhood of the current node $v = v_j$, where $j = pos$, has already been explored and no ancestor of t has been found. In that case, using $path$, the message goes to v_{j+1} the next vertex in the backbone. $pos = j + 1$ is set.
  - Otherwise, let $pn = \lfloor \frac{cnt'_d + cnt_d}{2} \rfloor$. The message takes port-number pn from v towards vertex w. If w is an ancestor of t, we go to the first case of the algorithm. Otherwise, the message goes back to $v = v_j$. This is possible since the node knows the port over which the message arrives. Moreover, if $Id_F(t) > Id_F(w) + w_F(w) - 1$, then cnt_d is set to pn and cnt'_d is set to pn otherwise.

The fact that the message eventually reaches its destination follows from the above description. Moreover, the computation of the next hop and the modification of the header clearly takes time $O(1)$. Due to lack of space, the proof of the stretch, described in next lemma, is omitted and can be found in [KLNS12].

Lemma 1. *Our routing scheme has stretch $\leq 2k(\lceil \log \Delta \rceil + \frac{5}{2}) - 5$.*

5   Conclusion and Further Work

It would be interesting to reduce the $O(k \cdot \log \Delta)$ stretch due to the dichotomic search phase of our routing scheme. Another interesting topic concerns the computation of tree-decompositions not trying to minimize the size of the bag but imposing some specific algorithmically useful structure.

References

[ACP87]   Arnborg, S., Corneil, D.G., Proskurowski, A.: Complexity of finding embeddings in a k-tree. SIAM J. Alg. Discrete Methods 8, 277–284 (1987)

[AG06]   Abraham, I., Gavoille, C.: Object location using path separators. In: PODC, pp. 188–197. ACM (2006)

[AGD06]   Abraham, I., Gavoille, C., Malkhi, D.: On space-stretch trade-offs: Lower bounds. In: SPAA, pp. 217–224 (2006)

[AGGM06] Abraham, I., Gavoille, C., Goldberg, A.V., Malkhi, D.: Routing in networks with low doubling dimension. In: ICDCS, p. 75 (2006)

[AGM+08] Abraham, I., Gavoille, C., Malkhi, D., Nisan, N., Thorup, M.: Compact name-independent routing with minimum stretch. ACM T. Alg. 4(3) (2008)

[BK96]   Bodlaender, H.L., Kloks, T.: Efficient and constructive algorithms for the pathwidth and treewidth of graphs. J. Algorithms 21(2), 358–402 (1996)

[BN11]   Bonato, A., Nowakovski, R.: The game of Cops and Robber on Graphs. American Math. Soc. (2011)

[Bod98]   Bodlaender, H.L.: A partial k-arboretum of graphs with bounded treewidth. Theor. Comput. Sci. 209(1-2), 1–45 (1998)

[BT97]   Bodlaender, H.L., Thilikos, D.M.: Treewidth for graphs with small chordality. Disc. Ap. Maths 79(1-3), 45–61 (1997)

[CF07]   Chen, Y., Flum, J.: On parameterized path and chordless path problems. In: CCC, pp. 250–263 (2007)

[CM93]   Courcelle, B., Mosbah, M.: Monadic second-order evaluations on tree-decomposable graphs. TCS 109, 49–82 (1993)

[CN05]   Clarke, N.E., Nowakowski, R.J.: Tandem-win graphs. Discrete Mathematics 299(1-3), 56–64 (2005)

[DG07]   Dourisboure, Y., Gavoille, C.: Tree-decompositions with bags of small diameter. Discrete Mathematics 307(16), 2008–2029 (2007)

[dMSV11] de Montgolfier, F., Soto, M., Viennot, L.: Treewidth and hyperbolicity of the internet. In: NCA, pp. 25–32. IEEE Comp. Soc. (2011)

[Dou05]   Dourisboure, Y.: Compact routing schemes for generalised chordal graphs. J. of Graph Alg. and App. 9(2), 277–297 (2005)

[FG01]   Fraigniaud, P., Gavoille, C.: Routing in Trees. In: Yu, Y., Spirakis, P.G., van Leeuwen, J. (eds.) ICALP 2001. LNCS, vol. 2076, pp. 757–772. Springer, Heidelberg (2001)

[Gro87]   Gromov, M.: Hyperbolic groups. Essays in Group Theory 8, 75–263 (1987)

[KK09]   Kobayashi, Y., Kawarabayashi, K.: Algorithms for finding an induced cycle in planar graphs and bounded genus graphs. In: 20th Annual ACM-SIAM Symp. on Discrete Alg. (SODA), pp. 1146–1155. SIAM (2009)

[KLNS12] Kosowski, A., Li, B., Nisse, N., Suchan, K.: k-chordal graphs: from cops and robber to compact routing via treewidth, Report, INRIA-RR7888 (2012), http://www-sop.inria.fr/members/Bi.Li/RR-7888.pdf

[KPBV09] Krioukov, D.V., Papadopoulos, F., Boguñá, M., Vahdat, A.: Greedy forwarding in scale-free networks embedded in hyperbolic metric spaces. SIGMETRICS Performance Evaluation Review 37(2), 15–17 (2009)

[NSR12] Nisse, N., Suchan, K., Rapaport, I.: Distributed computing of efficient routing schemes in generalized chordal graphs. TCS (to appear, 2012)

[WS98] Watts, D.J., Strogatz, S.: Collective dynamics of 'small-world' networks. Nature 393(6684), 440–442 (1998)

[WZ11] Wu, Y., Zhang, C.: Hyperbolicity and chordality of a graph. Electr. J. Comb. 18(1) (2011)

Contention Issues in Congestion Games*

Elias Koutsoupias and Katia Papakonstantinopoulou

University of Athens, Athens, Greece
{elias,katia}@di.uoa.gr

Abstract. We study time-dependent strategies for playing congestion games. The players can time their participation in the game with the hope that fewer players will compete for the same resources. We study two models: the boat model, in which the latency of a player is influenced only by the players that start at the same time, and the conveyor belt model in which the latency of a player is affected by the players that share the system, even if they started earlier or later; unlike standard congestion games, in these games the order of the edges in the paths affect the latency of the players. We characterize the symmetric Nash equilibria of the games with affine latencies of networks of parallel links in the boat model and we bound their price of anarchy and stability. For the conveyor belt model, we characterize the symmetric Nash equilibria of two players on parallel links. We also show that the games of the boat model are themselves congestion games. The same is true for the games of two players for the conveyor belt model; however, for this model the games of three or more players are not in general congestion games and may not have pure equilibria.

Keywords: Algorithmic game theory, price of anarchy, congestion games, contention.

1 Introduction

In the last dozen years, the concepts of the price of anarchy (PoA) and stability (PoS) have been successfully applied to many classes of games, most notably to congestion games and its relatives [17,24,21]. In congestion games, the players compete for a set of resources, such as facilities or links; the cost of each player depends on the number of players using the same resources; the assumption is that each resource can be shared among the players, but with a cost. Another interesting class of games are the contention games [12] in which the players again compete for resources, but the resources cannot be shared. If more than one players attempt to share a resource at the same time, the resource becomes unavailable and the players have to try again later. There are however interesting games that lie between the two extreme cases of the congestion and contention games. For example, the game that users play for dealing with congestion on a network seems to lie in between—the TCP congestion control policy is a strategy

---

* Partially supported by the ESF-NSRF research program Thales (AGT).

A. Czumaj et al. (Eds.): ICALP 2012, Part II, LNCS 7392, pp. 623–635, 2012.
© Springer-Verlag Berlin Heidelberg 2012

of this game. Timing is part of the strategy of the players (as in contention games) and the latency of a path depends on how many players use its edges (as in congestion games).

In this work, we attempt to abstract away the essential features of these games, to model them, and to study their properties, their Nash equilibria, and their price of anarchy and stability. The games that we consider are essentially congestion games with the addition of time dimension. The difference with congestion games is that players now don't simply select which path to use, but they also decide *when* to initiate the transmission.

Consider a link or facility e of a congestion game with latency function ℓ_e. In the congestion game the latency that a player experiences on the link is $\ell_e(k)$, where k is the number of players that use the link. In our model however, in which the players can also decide when to start, the latency needs to be redefined. We define and study two latency models for the links:

The boat model: in which only the group of players that start together affect the latency of the group: imagine that one boat departs from the source of the link at every time step; all players that decide to start at time t enter the boat which takes them to their destination; the speed of the boat depends only on the number of players in the boat and it is independent of the players on the other boats.

The conveyor belt model: in which the latency of a player depends on the number of other players using the link at the same time regardless if they started earlier or later. Specifically, the link is like a conveyor belt from the source to the destination; the speed of the belt at every time depends on the number of people on it. An interesting variant of this model is when the player is affected only by the players that have been already in the link but not by the players that follow; we don't study this model in this work.

Notice that in the boat model, the order in which the players finish a link may differ from the order in which they start. This, for example, can happen when a player starts later but with a smaller group of people. This cannot happen in the conveyor belt model.

In this work, we consider

- non-adaptive strategies, in which the players decide on their strategy in advance. Their pure strategies consist of a path and a starting time.
- symmetric strategies.

Intuitively, in the boat model, the aim of the players is to select a path with small latency and to avoid other players that start at the same time. In the conveyor belt model the aim is similar but the players try to avoid other players that start *near* the same time.

Related work. Contention resolution in communication networks is a problem that has attracted the interest of diverse communities of Computer Science. Its significance comes from the fact that contention is inherent in many critical network applications. One of them is the design of multiple access protocols for communication networks, such as Slotted Aloha: According to it, a source transmits a packet through the network, as soon as this packet is available. If a collision takes place, that is, another source attempted to transmit simultaneously, the source waits for some random number of time slots and attempts to retransmit at the beginning of the next slot. The increase of users of the network incurs a large number of collisions and subsequently poor utilization of the system's resources.

During the last four décades many more refined multiple access protocols have been proposed to increase the efficiency of Aloha, the vast majority of which assume that the agents follow the protocol, even if they might prefer not doing so. Recently, slotted Aloha has been studied from a game-theoretic point of view, trying to capture the selfish nature of its users. Part of this work has been done by Altman et al. [2,3]. The authors model slotted Aloha as a game among the transmitters who aim at transmitting their stochastic flow, using the retransmission probability that maximizes their throughput [2] or minimizes their delay [3]. They show that the system possesses symmetric equilibria and that its throughput deteriorates with larger number of players or arrival rate of new packets. Things get better considering a cost for each transmission though. Another slotted Aloha game is studied by MacKenzie and Wicker [18]. Here the agents aim at minimizing the time spent for unsuccessful transmissions before each successful one, while each transmission incurs some cost to the player. Their game possesses a symmetric equilibrium, and some of its instantiations possess equilibria that achieve the maximum possible throughput of Aloha.

Much of the prior game-theoretic work considers transmission protocols that always transmit with the same fixed probability. In [12] and [11] the authors consider more complex protocols (multi-round games), where a player's transmission probability is allowed to be an arbitrary function of his play history and the sequence of feedback he has received, and propose asymptotically optimal protocols. In [12], the authors propose a protocol which is a Nash equilibrium and has constant price of stability, i.e., all agents will successfully transmit within time proportional to their number. This protocol assumes that the cost of any single transmission is zero. In [11] the case of non-zero transmission cost is addressed, and a protocol is proposed where after each time slot, the number of attempted transmissions is returned as feedback to the users.

There is a lot of work on game theoretic issues of packet switching. For example, [15] considers the game in which users select their transmission rate, [1] considers TCP-like games in which the strategies of the players are the parameters of the AIMD (additive increase / multiplicative decrease) algorithm, and [13] considers game-theoretic issues of congestion control. All these works are concerned with the steady or long term version of the problems and they don't consider time-dependent strategies in the spirit of this work.

Routing in networks by selfish agents is another area that has been extensively studied based on the notion of the price of anarchy (PoA) [17] and the price of stability (PoS) [5]. The PoA and the PoS compare the social cost of the worst-case and best-case equilibrium to the social optimum. Selfish routing is naturally modeled as a congestion game. The class of congestion or potential games [22,20] consists of the games where the cost of each player depends on the resources he uses and the number of players using each resource. The effect of selfishness in infinite congestion games was first studied in [24] and of finite congestion games in [8,6].

The above results concern classical networks or static flows on networks. Perhaps the closest in spirit to our work are the recent attempts to study game-theoretic issues of dynamic flows, or more precisely, of flows over time. In [16], the authors consider selfish selection of routing paths when users have to wait in a FIFO queue before using every edge of their paths; the waiting time is not part of their strategy, but depends on the traffic in front of them. The same model is assumed by [19] who considers the Braess' paradox for flows over time. More results appeared in [7] which gives an efficiently computable Stackelberg strategy for which the competitive equilibrium is not much worse than the optimal, for two natural measures of optimality: total delay of the players and time taken to route a fixed amount of flow to the sink. In a slightly different model, [4] considers game-theoretic issues of discrete-time models in which the latency of each edge depends on its history. All these papers consider non-atomic congestion games. In a different direction which involves atomic games, [14] considers temporal congestion games that are based on coordination mechanisms [10] and congestion games with time-dependent costs.

All these models share with this work the interest in game-theoretic issues of timing in routing, but they differ in an essential ingredient: in our games, timing is the most important part of the players strategy, while in the previous work, time delays exist because of the interaction of the players; in particular, *in all these models the strategy of the players is to select only a path*, while in our games the strategy is essentially the timing. We view our model as a step towards understanding games related to TCP congestion control; this does not seem to be in the research agenda of game-theoretic issues of flows over time.

Short description of results. We first study structural properties of the boat and conveyor belt games. In the next section, we characterize the symmetric Nash equilibria and the optimal symmetric solution of the boat model game for parallel links of affine latency functions and any number of players. From these we get that the price of anarchy and stability is very low $3\sqrt{2}/4 \approx 1.06$. We also study the class of conveyor belt games. These are more complicated games and here we consider only two players and arbitrary latency functions (for two players the class of affine and the class of arbitrary latency functions are identical). The price of anarchy and stability is (for large latencies) again approximately $3\sqrt{2}/4 \approx 1.06$. This is the price of anarchy we computed for the boat model, but the relation is not as straightforward as it may appear: in the boat model we take the limit as the number of players tends to infinity, while in the conveyor model, we take the limit as the latencies tend to infinity.

To our knowledge, these games differ significantly from the classes of congestion games that have been studied before. Also, the techniques developed for bounding the PoA and the PoS for congestion games do not seem to be applicable in our setting. In particular, the smoothness analysis arguments [8,23,9] do not seem to apply because we consider symmetric equilibria. In fact, the focus and difficulty of our analysis is to characterize the Nash equilibria and not to bound the PoA (or PoS).

The decision to study only symmetric strategies is based on the assumption that these games are played by many players with no coordination among them. We consider this work as a step towards the study of real-life situations such as the TCP congestion control mechanism in which the players are essentially indistinguishable and therefore symmetric.

In all the games that we study, there exists a unique symmetric equilibrium. For this type of equilibria, the definition of the price of anarchy is uncomplicated: We simply take the ratio of the cost of one player over the cost of one player *of the symmetric optimal solution*. Since there is a unique Nash equilibrium, the price of stability is equal to the price of anarchy.

Due to the space limitations, some proofs are omitted but can be found in the full version of the paper. Moreover, the structural properties of the two models are presented in short here, and are described in detail in the full version of this work.

Formally, the games that we study here are the following: Let G be a *network congestion game* with n players and latency functions $\ell_e(k)$ on its link e. We define two new games based on G, the boat model game and the conveyor belt game. The pure strategies of both new games of every player consist of one strategy (path) of the original game and one non-negative time step $t \in \mathbb{Z}_0^+$. Their difference lies in the cost of the pure strategies.

In the boat model, the cost of a player is simply $t + \sum_{e \in P} \ell_e(n_t(e))$, where $n_t(e)$ denotes the set of players that also start at time t and use edge e. In the conveyor belt model the cost is more complicated. It depends on the notion of work: in a time interval $[t, t + \Delta t]$ in which player i uses link e, it completes work $\Delta t/\ell_e(k)$, where k is the number of players using the same link during this time interval. A player finishes a link when it completes total work of 1 for this link; the player then moves to the next link of its path.

The following theorem describes the nature of the time-dependent games.

Theorem 1. *All boat games are congestion games. In contrast, only the 2-player conveyor belt games are congestion games, and for 3 or more players there are games that have no pure equilibria. Furthermore, the order of using the facilities in conveyor belt games is important: a reordering of the edges of a path can result in a different game.*

2   Nash Equilibria of the Boat Model

In this section, we first consider symmetric Nash equilibria of n players for the boat model of parallel links. We also compute the optimal non-selfish solution and estimate the PoA.

Nash equilibria computation. A pure strategy for a player is to select a link e and a time t. A mixed strategy is given by probabilities $p_{e,t}$ with $\sum_{e,t} p_{e,t} = 1$: the player uses link e at time step t with probability $p_{e,t}$. A set of probabilities $p_{e,t}$ is a Nash equilibrium when a player has no incentive to change it to some other values q. To find the Nash equilibria, we first estimate the latency $d_{e,t}$ when the player selects pure strategy (e,t):

$$d_{e,t} = t + \sum_{k=0}^{n-1} \binom{n-1}{k} p_{e,t}^k (1 - p_{e,t})^{n-1-k} \ell_e(k+1). \tag{1}$$

Let $d = \min_{e,t} d_{e,t}$ denote the minimum value. Then the probabilities define a symmetric mixed Nash equilibrium if and only if $p_{e,t} > 0$ implies $d = d_{e,t}$.

To find the Nash equilibria, the first crucial step is to show that the probabilities in every link must be non-increasing in t. This is shown by the following lemma which holds for arbitrary latency functions, not only for affine ones:

Lemma 1. *If for every edge e the latencies $\ell_e(k)$ are non-decreasing in k, then every symmetric Nash equilibrium is a non-increasing sequence of probabilities:* $p_{e,t} \geq p_{e,t+1}.$

(The proof is omitted and can be found in the full version of the paper.)

We define the support of the Nash equilibrium to be the set of strategies that have minimum latency: $S_e = \{t : d_{e,t} = d\}$. Alternatively, we could have defined the support to be the set of strategies with non-zero probability at the Nash equilibrium; the two notions are similar but not identical in some cases. Notice the convention $d_{e,h_e+1} > d = d_{e,h_e}$, in the definition of the support. The last lemma shows that the support S_e of every link e is of the from $\{0, \ldots, h_e\}$ for some integer h_e.

We now focus on affine latency functions, $\ell_e(k) = a_e k + b_e$, for which the cost $d_{e,t}$ in (1) takes a simple closed form:

$$d_{e,t} = t + a_e + b_e + (n-1) a_e p_{e,t}, \tag{2}$$

which shows that the probabilities of the Nash equilibria are of the form:

$$p_{e,t} = \begin{cases} \frac{d - a_e - b_e - t}{(n-1)a_e} & \text{for } t \leq h_e \\ 0 & \text{otherwise} \end{cases} \tag{3}$$

Observe that at every Nash equilibrium $p_{e,t}$, the non-zero probabilities decrease linearly with t. These probabilities are determined by the cost d of each player and the integers h_e (one for each link). In fact, the parameters h_e are very tightly related with the cost d of each player:

Theorem 2. *There is a unique symmetric Nash equilibrium with support $S_e = \{t : 0 \leq t \leq h_e = \lfloor d - a_e - b_e \rfloor\}$, where d is the expected cost of every player; its probabilities are given by*

$$p_{e,t} = \begin{cases} \frac{d - a_e - b_e - t}{(n-1)a_e} & \text{for } t \leq d - a_e - b_e \\ 0 & \text{otherwise} \end{cases}$$

The expected cost $L_{NE} = d$ of every player is the unique solution of the equation

$$\sum_e \frac{(\lfloor d - a_e - b_e \rfloor + 1)\,(2(d - a_e - b_e) - \lfloor d - a_e - b_e \rfloor)}{2(n-1)a_e} = 1. \qquad (4)$$

Its value is approximately

$$d \approx \frac{\sum_e \frac{a_e + b_e}{2(n-1)a_e} + \sqrt{\left(\sum_e \frac{a_e + b_e}{2(n-1)a_e}\right)^2 + \left(\sum_e \frac{1}{2(n-1)a_e}\right)\left(1 - \sum_e \frac{(a_e + b_e)^2}{2(n-1)a_e}\right)}}{\sum_e \frac{1}{2(n-1)a_e}}, \qquad (5)$$

and as n tends to infinity this tends to $\sqrt{\frac{2n}{\sum_e a_e^{-1}}}$.

(The proof is omitted and can be found in the full version of the paper.)

The optimal setting. Let us now consider the optimal symmetric protocol. With similar reasoning, the expected latency of a player is

$$L_{OPT} = \sum_e \sum_{t=0}^{\infty} p_{e,t}\left(t + \sum_{k=0}^{n-1}\binom{n-1}{k}p_{e,t}^k (1 - p_{e,t})^{n-1-k}\,\ell_e(k+1)\right) = \sum_e \sum_{t=0}^{\infty} p_{e,t}\,d_{e,t}$$

We seek the probabilities $p_{e,t}$ with $\sum_{e,t} p_{e,t} = 1$ which minimize the above expression. We again focus on affine latencies. With $\ell_e(k) = ak + b$, the above expression has the following compact form $L_{OPT} = \sum_e \sum_{t=0}^{\infty} p_{e,t}\,(t + a_e + b_e + (n-1)\,a_e\,p_{e,t})$. We minimize this subject to $\sum_{e,t} p_{e,t} = 1$. Using a Lagrange multiplier and taking derivatives, we get that the minimum occurs when the probabilities have the form $p_{e,t} = (\lambda - a_e - b_e - t)/(2(n-1)a_e)$, for some constant λ, and $p_{e,t} = 0$ when $\lambda - a_e - b_e - t \leq 0$. This means that they decrease linearly with t until $c_e = \lambda - a_e - b_e$, when they become 0 and they remain 0 from that point on. Thus, the form of the optimal probabilities resembles the form of the Nash equilibrium probabilities; the only difference is that the optimal probabilities drop slower to 0 (the factors are $2(n-1)a_e$ and $(n-1)a_e$ respectively). A similar bicriteria relation between the Nash equilibria and the optimal solution has been observed in simple congestion games before [24]. Taking into account the constant term also we get,

Lemma 2. *The set of probabilities of the optimal solution for latencies $\ell_e(k) = a_e k + b_e$ is a Nash equilibrium for latencies $\ell_e(k) = 2a_e k + (b_e - a_e)$.*

Therefore the probabilities of the optimal solution are:

$$p_{e,t} = \begin{cases} \frac{\lambda - a_e - b_e - t}{2(n-1)a_e} & t \leq h_e^* \\ 0 & \text{otherwise} \end{cases} \qquad (6)$$

where $h_e^* = \lfloor \lambda - a_e - b_e \rfloor$, and the value of λ is the unique solution of the equation $\sum_{e,t} p_{e,t} = 1$. Thus, λ is determined by an equation similar to (4) (they essentially differ only in the denominator):

$$\sum_e \frac{(\lfloor \lambda - a_e - b_e \rfloor + 1)(2(\lambda - a_e - b_e) - \lfloor \lambda - a_e - b_e \rfloor)}{4(n-1)a_e} = 1. \qquad (7)$$

From the probabilities we can compute L_{OPT}. Observe that the optimal case differs from the Nash equilibrium case of the previous subsection in that the parameters λ and L_{OPT} are distinct (while in the Nash equilibrium case they are identical—equal to d).

As in the case of the Nash equilibrium, it is useful to define $\eta_e^* = \lambda - a_e - b_e$. We can then compute the optimal latency: $L_{OPT} =$

$$\sum_e \sum_{t=0}^{\infty} p_{e,t}\, (t + a_e + b_e + (n-1)\, a_e\, p_{e,t}) = \sum_e \frac{(h_e^* + 1)\,(6\eta_e^*(\eta_e^* + 2a_e) - h_e^*(2h_e^* + 6a + 1))}{24(n-1)a_e}.$$

To get an approximate estimate as n tends to infinity, we observe that λ is approximately given by $\sum_e \frac{\lambda^2}{4(n-1)a_e} \approx 1$ which implies $\lambda \approx 2\sqrt{\frac{n}{\sum_e a_e^{-1}}}$. From this, we can find an approximate value for L_{OPT}:

$$L_{OPT} \approx \frac{\eta_e^{*3}}{6(n-1)a_e} \approx \sum_e \frac{\lambda^3}{6(n-1)a_e} = \frac{4}{3}\sqrt{\frac{n}{\sum_e a_e^{-1}}}$$

The price of anarchy. Comparing the value of L_{OPT} to the cost d of the Nash equilibrium, we see that the PoA and the PoS of the boat model on parallel links with affine latency functions tends to $\frac{3\sqrt{2}}{4} \approx 1.06$, as the number of players n tends to infinity (while the parameters of the network remain fixed).

Theorem 3. *For every fixed set of parallel links with positive a_e and b_e, the PoA (and PoS) tends to $3\sqrt{2}/4 \approx 1.06$, as the number of players n tends to infinity.*

However, for fixed number of players and because of the integrality of h and h^*, the situation is more complicated. Figure 1 shows the PoA for typical values of a_e and n, for one link. The situation is captured by the following theorem:

Theorem 4. *For one link and fixed number of players n, the PoA is maximized when $a_e = 1/(n-1)$ and $b_e = 0$. For these values the NE is pure $(p_{e,0} = 1)$, but the optimal symmetric solution is given by the probabilities $p_{e,0} = 3/4$ and $p_{e,1} = 1/4$. For these strategies we get $L_{NE} = d = n/(n-1)$, $L_{OPT} = (7n+1)/(8(n-1))$, and $PoA = 8n/(7n+1)$.*

To compare the cost L_{NE} and L_{OPT} we first investigate the solutions of the equations (4) and (7) as functions of a_e; since we care about the worst-case PoA, we can safely assume that $b_e = 0$ because $b_e \geq 0$ is added to both the numerator and the denominator of the PoA.

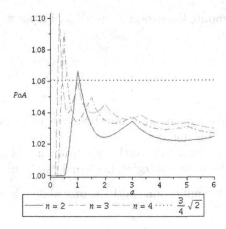

Fig. 1. PoA of the single-link boat games

For every nonnegative integer k, let us define $A_k = \frac{k(k+1)}{2(n-1)}$ which are the values of a_e where the value $d - a_e$ becomes integral (equal to k). The following lemma gives the solution of (4) for the intervals $[A_k, A_{k+1})$ where the integral part of $d - a_e$ is constant. It also extends it to the optimal cost.

Lemma 3. For $a_e \in [A_k, A_{k+1})$, $L_{NE} = \frac{n+k}{k+1} a_e + \frac{k}{2}$.
For $a_e \in [A_k/2, A_{k+1}/2)$, $L_{OPT} = \frac{n+k}{k+1} a_e + \frac{k}{2} - \frac{k(k+1)(k+2)}{48(n-1)a_e}$.

Proof. We first show that for $a_e \in [A_k, A_{k+1})$ the value of d given by equation (4) satisfies $\lfloor d - a_a \rfloor = k$. It suffices to show that the value $x = d - a_e - k$ is in $[0, 1)$.

3   Nash Equilibria of the Conveyor Belt Model

We now turn our attention to the conveyor belt model, which is more complicated than the boat model. In the conveyor belt model each link is like a conveyor belt whose speed depends on the number of players on it. *We only consider the case of 2 players in this section.* The cost $c_e(t, t')$ of a player for pure strategies (e, t) when the other player starts using link e at time step t' is computed using
$$f_i = t_i + \ell_e(1) + \max\left(0, (\ell_e(2) - \ell_e(1))\left(1 - \frac{|t_2 - t_1|}{\ell_e(1)}\right)\right)$$ where t_i, f_i are the start and finish times of player i respectively (see the full version for details).

To simplify the discussion, we assume that $\ell_e(1)$ is an integer; this does not seem to really change the nature of equilibria, except perhaps when $\ell_e(1) < 1$ which does not seem a very interesting case.

Nash equilibria computation. Consider a symmetric Nash equilibrium with probabilities $p_{e,t}$, the same for every player. It is a Nash equilibrium when a player has no incentive to change his probabilities to different values. To find the Nash equilibria, we first compute the expected cost $d_{e,t}$ of a player when he plays pure strategy (e, t):

$$d_{e,t} = \sum_{t'} c_e(t, t') = t + \ell_e(1) + (\ell_e(2) - \ell_e(1)) \sum_{r=-\ell_e(1)}^{\ell_e(1)} \left(1 - \frac{|r|}{\ell_e(1)}\right) p_{e,t+r} \quad (8)$$

The probabilities define a symmetric mixed Nash equilibrium when probability $p_{e,t} > 0$ implies $d_{e,t} = d = \min_{e,t} d_{e,t}$.

We are interested in symmetric Nash equilibria, that is equilibria that occur when all players use the same strategies. Let's first establish a very intuitive fact:

Claim. If at the Nash equilibrium, positive probability is allocated to edge e, then $p_{e,0} > 0$.

(*The proof is omitted and can be found in the full version of the paper.*)

The next lemma shows that the support $S_e = \{t : d_{e,t} = d\}$ of every mixed Nash equilibrium is of the form $\{0, \ldots, \hat{h}_e\}$ for some $\hat{h}_e$.

Lemma 4. *If for some t there exists $s \geq t$ with $p_{e,t} > 0$, then t is in the support S_e, i.e. $d_{e,t} = d$.*

(*The proof is omitted and can be found in the full version of the paper.*)

The previous lemma establishes that the support S_e starts at 0 and is contiguous. With this, we can now determine the exact structure of Nash equilibria.

Theorem 5. *The Nash equilibria of the conveyor belt game of two players in parallel links have probabilities*

$$p_{e,t} = \begin{cases} \frac{d - \ell_e(1) - t}{\ell_e(2) - \ell_e(1)} & t \leq d - \ell_e(1) \text{ and } \frac{t}{\ell_e(1)} \in \mathbb{Z}^+ \\ 0 & otherwise \end{cases} \quad (9)$$

where d is the expected cost of each player and it is the unique solution of the equation

$$\sum_e \frac{(\lfloor \eta_e \rfloor + 1)(2\eta_e - \lfloor \eta_e \rfloor)}{2^{\frac{\ell_e(2) - \ell_e(1)}{\ell_e(1)}}} = 1, \quad (10)$$

where $\eta_e = d/\ell_e(1) - 1$.

Proof. Consider some $0 < t < \hat{h}_e$. Then from the definition of $d_{e,t}$ we can compute $d_{e,t+1} - 2d_{e,t} + d_{e,t-1} = \frac{\ell_e(2) - \ell_e(1)}{\ell_e(1)} (p_{e,t-\ell_e(1)} - 2p_{e,t} + p_{e,t+\ell_e(1)})$. Since for $t \in \{1, \ldots, \hat{h}_e - 1\}$, all $t - 1$, t and $t + 1$ are in the support S_e, we have that $d_{e,t-1} = d_{e,t} = d_{e,t+1}$. In turn, this gives that the right-hand side is 0 and we get that $p_{e,t+\ell_e(1)} - p_{e,t} = p_{e,t} - p_{e,t-\ell_e(1)}$; this shows that if we consider times that differ by $\ell_e(1)$, the probabilities drop linearly and more specifically that for integers k, x: $p_{e,k\ell_e(1)+x} - p_{e,x} = k(p_{e,x+\ell_e(1)} - p_{e,x})$.

This linearity allows us to conclude that $p_{e,t} = 0$ for every t which is not a multiple of $\ell_e(1)$. To see this consider some $x \in \{1, \ldots, \ell_e(1) - 1\}$ and the sequence $p_{e,x-\ell_e(1)}, p_{e,x}, p_{e,x+\ell_e(1)}, \ldots, p_{e,x+k\ell_e(1)}$. This sequence is linear and starts with a 0 (since $x - \ell_e(1) < 0$) and ends again in 0 (if we take k such that $\hat{h}_e < x + k\ell_e(1) \le \hat{h}_e + \ell_e(1)$).

The above reasoning does not apply to the value $x = 0$, because $p_{e,t+\ell_e(1)} - p_{e,t} = p_{e,t} - p_{e,t-\ell_e(1)}$ only for $t \in \{1, \ldots, \hat{h}_e - 1\}$. To summarize, the NE with support $\{0, \ldots, \hat{h}_e\}$ have non-zero probabilities only on the multiples of $\ell_e(1)$. This means that either the players start together, or they do not overlap, which is *exactly the property of the boat model*. It follows, that for one link, the Nash equilibrium is identical to the Nash equilibrium of the boat game with time step expanded to $\ell_e(1)$. For more than one link, the time steps in each link are different, because $\ell_e(1)$ are different. Nevertheless the analysis of the boat model carries over to the conveyor belt model.

The proof now is essentially the same with the boat model, but with the extra restriction that the time steps are not the same in all links. Since the probabilities are non-zero only at integral multiples of $\ell_e(1)$, the latency becomes $d_{e,t} = t + \ell_e(1) + (\ell_e(2) - \ell_e(1))p_{e,t}$ when t is an integral multiple of $\ell_e(1)$. It follows that the probabilities are as in (9). The cost d is determined by the equation $\sum_{e,t} p_{e,t} = 1$. Using the expressions for the probabilities, this equation is equivalent to (10). This is identical to the equation for d for the boat model and the argument about the uniqueness of the solution carries over.

The Optimal setting. Let's now consider the optimal symmetric protocol. With reasoning similar to that in the boat model and omitting the details (which can be found in the full version of the paper), we get that the minimum occurs when

$$\lambda = t + \ell_e(1) + 2(\ell_e(2) - \ell_e(1)) \sum_{r=-\ell_e(1)}^{\ell_e(1)} \left(1 - \tfrac{|r|}{\ell_e(1)}\right) p_{e,t+r}, \qquad (11)$$

for some λ. The factor 2 in the last term comes from the convolution in the L_{OPT} expression. We notice again the bicriteria property.

Lemma 5. *The probabilities of the optimal solution for two players in the conveyor belt model of parallel links with latencies $\ell_e(k)$ is a Nash equilibrium for latencies $\ell'_e(k) = 2\ell_e(k) - \ell_e(1)$.*

Proof. By comparing equations (8) and (11) that determine the Nash equilibria and the optimal solution, we see that the latencies must satisfy $\ell'_e(1) = \ell_e(1)$ and $\ell'_e(2) - \ell'_e(1) = 2(\ell_e(2) - \ell_e(1))$, which can be expressed as in the lemma.

Since the conveyor belt Nash equilibrium and optimal solution are very similar to the ones of the boat model, the analysis of the price of anarchy is similar, and their expressions can be approximated well as the latencies $\ell_e(k)$ tend to infinity. For one link, the cost d of the Nash equilibrium is approximately $\sqrt{2\ell_e(1)(\ell_e(2) - \ell_e(1))}$ while the optimal cost is $\frac{4}{3}\sqrt{2\ell_e(1)(\ell_e(2) - \ell_e(1))}$, which shows that the price of anarchy tends to $3\sqrt{24} \approx 1.06$, again. Since this is not sufficiently different than the boat model, we omit the details.

References

1. Akella, A., Seshan, S., Karp, R., Shenker, S., Papadimitriou, C.: Selfish behavior and stability of the internet: a game-theoretic analysis of TCP. In: Proceedings of the 2002 SIGCOMM Conference, pp. 117–130. ACM (2002)
2. Altman, E., El Azouzi, R., Jiménez, T.: Slotted aloha as a game with partial information. Comput. Netw. 45, 701–713 (2004)
3. Altman, E., Barman, D., El Azouzi, R., Jiménez, T.: A game theoretic approach for delay minimization in slotted ALOHA. In: IEEE International Conference on Communications (2004)
4. Anshelevich, E., Ukkusuri, S.: Equilibria in Dynamic Selfish Routing. In: Mavronicolas, M., Papadopoulou, V.G. (eds.) SAGT 2009. LNCS, vol. 5814, pp. 171–182. Springer, Heidelberg (2009)
5. Anshelevich, E., Dasgupta, A., Kleinberg, J.M., Tardos, É., Wexler, T., Roughgarden, T.: The price of stability for network design with fair cost allocation. In: Proceedings of FOCS 2004, pp. 295–304 (2004)
6. Awerbuch, B., Azar, Y., Epstein, A.: Large the price of routing unsplittable flow. In: Proceedings of STOC 2005, pp. 57–66 (2005)
7. Bhaskar, U., Fleischer, L., Anshelevich, E.: A stackelberg strategy for routing flow over time. In: Proceedings of SODA 2011, pp. 192–201. SIAM (2011)
8. Christodoulou, G., Koutsoupias, E.: The price of anarchy of finite congestion games. In: Proceedings of STOC 2005, pp. 67–73. ACM (2005)
9. Christodoulou, G., Koutsoupias, E., Spirakis, P.G.: On the Performance of Approximate Equilibria in Congestion Games. In: Fiat, A., Sanders, P. (eds.) ESA 2009. LNCS, vol. 5757, pp. 251–262. Springer, Heidelberg (2009)
10. Christodoulou, G., Koutsoupias, E., Nanavati, A.: Coordination mechanisms. Theor. Comput. Sci. 410(36), 3327–3336 (2009)
11. Christodoulou, G., Ligett, K., Pyrga, E.: Contention Resolution under Selfishness. In: Abramsky, S., Gavoille, C., Kirchner, C., Meyer auf der Heide, F., Spirakis, P.G. (eds.) ICALP 2010, Part II. LNCS, vol. 6199, pp. 430–441. Springer, Heidelberg (2010)
12. Fiat, A., Mansour, Y., Nadav, U.: Efficient contention resolution protocols for selfish agents. In: Proceedings of SODA 2007, pp. 179–188 (2007)
13. Garg, R., Kamra, A., Khurana, V.: A game-theoretic approach towards congestion control in communication networks. ACM SIGCOMM Computer Communication Review 32(3), 47–61 (2002)
14. Hoefer, M., Mirrokni, V.S., Röglin, H., Teng, S.-H.: Competitive Routing over Time. In: Leonardi, S. (ed.) WINE 2009. LNCS, vol. 5929, pp. 18–29. Springer, Heidelberg (2009)
15. Kesselman, A., Leonardi, S., Bonifaci, V.: Game-Theoretic Analysis of Internet Switching with Selfish Users. In: Deng, X., Ye, Y. (eds.) WINE 2005. LNCS, vol. 3828, pp. 236–245. Springer, Heidelberg (2005)
16. Koch, R., Skutella, M.: Nash Equilibria and the Price of Anarchy for Flows over Time. In: Mavronicolas, M., Papadopoulou, V.G. (eds.) SAGT 2009. LNCS, vol. 5814, pp. 323–334. Springer, Heidelberg (2009)
17. Koutsoupias, E., Papadimitriou, C.: Worst-Case Equilibria. In: Meinel, C., Tison, S. (eds.) STACS 1999. LNCS, vol. 1563, pp. 404–413. Springer, Heidelberg (1999)
18. MacKenzie, A.B., Wicker, S.B.: Stability of multipacket slotted aloha with selfish users and perfect information. In: Proceedings of IEEE INFOCOM, pp. 1583–1590 (2003)

19. Macko, M., Larson, K., Steskal, Ľ.: Braess's Paradox for Flows over Time. In: Kontogiannis, S., Koutsoupias, E., Spirakis, P.G. (eds.) SAGT 2010. LNCS, vol. 6386, pp. 262–275. Springer, Heidelberg (2010)
20. Monderer, D., Shapley, L.S.: Potential games. Games and Economic Behavior 14, 124–143 (1996)
21. Nisan, N., Roughgarden, T., Tardos, É., Vazirani, V.V.: Algorithmic game theory. Cambridge Univ. Pr. (2007)
22. Rosenthal, R.W.: A class of games possessing pure-strategy Nash equilibria. International Journal of Game Theory 2, 65–67 (1973)
23. Roughgarden, T.: Intrinsic robustness of the price of anarchy. In: Proceedings of STOC 2009, pp. 513–522. ACM, New York (2009)
24. Roughgarden, T., Tardos, É.: How bad is selfish routing? J. ACM 49(2), 236–259 (2002)

Online Mechanism Design
(Randomized Rounding on the Fly)[*]

Piotr Krysta[1] and Berthold Vöcking[2]

[1] Department of Computer Science, University of Liverpool
p.krysta@liverpool.ac.uk
[2] Department of Computer Science, RWTH Aachen University
voecking@cs.rwth-aachen.de

Abstract. We study incentive compatible mechanisms for combinatorial auctions (CAs) in an online model with sequentially arriving bidders, where the arrivals' order is either random or adversarial. The bidders' valuations are given by demand oracles. Previously known online mechanisms for CAs assume that each item is available at a certain multiplicity $b > 1$. Typically, one assumes $b = \Omega(\log m)$, where m is the number of different items.

We present the first online mechanisms guaranteeing competitiveness for any multiplicity $b \geq 1$. We introduce an online variant of oblivious randomized rounding enabling us to prove competitive ratios that are close to or even beat the best known offline approximation factors for various CAs settings. Our mechanisms are universally truthful, and they significantly improve on the previously known mechanisms.

1 Introduction

Efficient allocation of resources to different entities is a fundamental problem in computer science and economics. Incentive compatible, or truthful, mechanisms address this problem by aggregating the preferences of different participants in such a way that it is best for each of them to truthfully report its true valuations for different bundles of resources. In many practical situations decisions about the allocation of resources must be made online without or with limited knowledge about the future. Examples are the provision of resources like CPU time or bandwidth to a set of tasks that changes over time, or the on-demand allocation of radio spectrum to a fluctuating set of mobile users.

We study truthful mechanisms for combinatorial auctions (CAs) in an online model with sequentially arriving bidders. There are m indivisible items, and each bidder comes with a valuation function specifying the values of different bundles (subsets) for her. An online mechanism serves the bidders one by one in their arrival order. We consider two kinds of arrivals: bidders might appear in an order specified by a random permutation or, alternatively, by an adversary. The former kind of arrivals is used in the classical secretary setup [13,16,3], the latter is used in the standard competitive analysis of online algorithms [7]. For each bidder, the mechanism has to make a decision about the bidder's allocation and payment without knowing the valuations of bidders arriving later.

---

[*] This work has been supported by EPSRC grants EP/F069502/1 and EP/G069239/1.

A. Czumaj et al. (Eds.): ICALP 2012, Part II, LNCS 7392, pp. 636–647, 2012.
© Springer-Verlag Berlin Heidelberg 2012

On the first view, the online requirement seems to be an additional complication added to mechanism design. Obviously, VCG based approaches cannot be used in the online setting as they require to compute an optimal offline solution. However, considering bidders one after the other yields an alternative way for achieving truthfulness by setting a price for each bidder and letting the bidder pick her utility maximizing allocation. Bartal et al. [5] use this approach to devise approximation algorithms for CAs with items of multiplicity b, i.e., each item can be sold up to b times. They present a polynomial-time truthful mechanism with an approximation factor of $O(b \cdot m^{1/(b-2)})$, for $b \geq 3$, with respect to the social welfare. Although their algorithm is not fully online since it uses an offline estimate of the maximum bid, this work exemplifies that serving bidders one after the other is a promising approach for designing truthful mechanisms.

The argument that online algorithms are a promising way to achieve truthfulness is also supported by Awerbuch et al. [1]. They show that every online algorithm satisfying certain natural assumptions can be turned into a truthful online mechanism. They present a revenue extraction method with guaranteed expected revenue of the derived online mechanism: Given a c-competitive online algorithm and assuming that the ratio between the largest and the smallest bid is bounded by a parameter $\mu > 1$, they obtain a truthful online mechanism with competitive factor $O(c + \log \mu)$ with respect to the revenue. The problem however is that previous work does not provide online algorithms for CAs with reasonable competitive ratios, unless one assumes a relatively high multiplicity b. In particular, all examples of variants of CAs in [1] assume $b = \Omega(\log m)$.

Essentially all previous approaches to online resource allocation, e.g., [2,9,8], are based on the following idea: The online algorithm sets prices for the items and sells the items to the bidders at these prices. Initially, all items have the same price. The prices for different items increase over time at different "speed" depending on the "popularity" of the corresponding items. Whenever an item is sold, its price is increased in a multiplicative way. Thus, the algorithm learns the "right" item's price by selling copies of the item. That is why a relatively high multiplicity leads to a good competitive factor.

We give the first online mechanisms with good competitive ratios for various kinds of CAs and small multiplicities. In particular, our analysis covers the case $b = 1$. We combine the online allocation of items using increasing prices with the concept of oblivious randomized rounding [15]. The main idea is that we relax the constraints by assuming that each item is available at a certain "virtual multiplicity". This enables the online algorithm to learn prices by assigning virtual copies to the bidders. Randomized rounding is used to decide which of the virtual copies does actually correspond to the real item. The randomized rounding procedure is applied online, i.e., immediately after the virtual copy is assigned to the bidder, and it is oblivious, i.e., the rounding probability does not depend on the valuations. The latter property is crucial for truthfulness.

1.1 Formal Description of the Problem

Combinatorial auctions (CAs). In a combinatorial auction (CA), a set U of m items (goods) shall be allocated to n bidders. Each $e \in U$ has multiplicity $b \geq 1$, i.e., it can be assigned to at most b bidders. Bidders receive subsets (bundles) of U. Each bidder i has a valuation function $v_i : 2^U \to \mathbb{R}_{\geq 0}$ satisfying standard assumptions: $v_i(\cdot)$ is non-decreasing (*free disposal*) and $v_i(\emptyset) = 0$ (*normalization*). The set of feasible allocations is $A = \left\{ S = (S_1, \ldots, S_n) \in (2^U)^n \mid \forall e \in U : \sum_{i=1}^n |S_i \cap \{e\}| \leq b \right\}$.

The valuation of bidder i for allocation S is denoted by $v_i(S)$. Let us explicitly point out that v_i depends only on S_i and not on $S_{-i} = (S_1, \ldots, S_{i-1}, S_{i+1}, \ldots, S_n)$, that is, $v_i(S) = v_i(S_i)$ (*no externalities*). For more details please refer to [6].

Besides general valuation functions, we consider so-called XOS functions, introduced in [19]. A valuation function $v : 2^U \to \mathbb{R}_{\geq 0}$ is XOS if there exists additive (i.e., linear) valuation functions $w_j : 2^U \to \mathbb{R}_{\geq 0}$, $j = 1, 2, 3, \ldots$, such that for every $S \subseteq U$ we have $v(S) = \max_j \{w_j(S)\}$. It is known that v is XOS if and only if it is *fractionally subadditive*, i.e., for every *fractional cover* defined by a collection of sets $T_k \subseteq S$ and numbers $0 \leq \alpha_k \leq 1$ satisfying $\sum_{k : e \in T_k} \alpha_k \geq 1$ (for every $e \in S$), it holds $v(S) \leq \sum_k \alpha_k v(T_k)$, see [15].

Valuation functions are assumed to be given by a black box (oracle) that can be queried by the mechanism. The black box answers *demand queries*: Given a price vector $(p_e)_{e \in U}$, what is the utility maximizing subset S_i and its valuation $v_i(S_i)$? The utility for S_i under price vector $p = (p_e)_{e \in U}$ is $u_i(p, S_i) = v_i(S_i) - \sum_{e \in S_i} p_e$. Our algorithms use restricted demand oracles $D_i(U', p)$, for $U' \subseteq U$, returning the utility maximizing subset $S_i' \subseteq U'$. A restricted demand oracle can be simulated by an unrestricted demand oracle by setting prices of items $U \setminus U'$ to a sufficiently large number.

A deterministic mechanism for CAs is a pair (f, q) with $f : V^n \to A$ being a social choice function and $q = (q_1, \ldots, q_n)$, $q_i : V^n \to \mathbb{R}$ being a payment scheme, where V is the set of all possible valuation functions. The social choice function determines the allocation and the payment scheme specifies the payments requested from the bidders. *Incentive compatibility.* A mechanism is *(deterministically) truthful*, or *incentive compatible*, if it is a dominant strategy for each bidder to report his true valuation, i.e., for all $i \in [n]$, all v_i, v_i' and all declarations v_{-i} of the other bidders except bidder i ($v_{-i} = (v_1, \ldots, v_{i-1}, v_{i+1}, \ldots, v_n)$), i's utility when bidding v_i is not smaller than i's utility when bidding v_i', i.e., $v_i(f(v_i, v_{-i})) - q_i(v_i, v_{-i}) \geq v_i(f(v_i', v_{-i})) - q_i(v_i', v_{-i})$. A randomized mechanism is a probability distribution over deterministic mechanisms. It is *universally truthful* if each of these deterministic mechanisms is truthful. A weaker notion is *truthfulness in expectation* where a bidder's expected utility is maximized when revealing his true valuations. Our mechanisms are universally truthful.

We achieve truthfulness by considering bidders one by one in a given order and offering items at fixed prices to each bidder. In particular, if $U_i \subseteq U$ is the set of items that the mechanism decides to offer to bidder i, then the mechanism defines prices $p = (p_e)_{e \in U}$ and computes S_i by calling the demand oracle $D_i(U_i, p)$. Prices offered to bidder i do not depend on the i's valuation function v_i. This approach yields truthfulness by the *direct characterization* of truthful mechanisms (see, e.g., [5], or [20]).

Online models. We assume that bidders arrive one by one, let us say in the order $1, 2, \ldots, n$. The mechanism needs to compute the allocation S_i and the payment q_i for bidder i without knowing the valuations of the bidders $i + 1, \ldots, n$. We distinguish two models of arrivals: In the random arrival model (secretary model) bidders arrive in an order specified by a random permutation. In the adversarial arrival model, bidders arrive in arbitrary (adversarial) order. The objective is to find an allocation $S \in A$ maximizing the *social welfare* $v(S) = \sum_{i=1}^n v_i(S_i)$.

The *competitive ratio* R of a possibly randomized online algorithm is defined as follows. Let Σ be the set of all arrival sequences of n bidders with valuations for m items. For $\sigma \in \Sigma$, let $S(\sigma)$ be the allocation computed by the algorithm and $opt(\sigma)$ an optimal allocation for σ. Then $R = \sup_{\sigma \in \Sigma} v(opt(\sigma))/\mathbf{E}\left[v(S(\sigma))\right]$. In the adversarial model, R cannot be bounded if the valuations are unbounded. Therefore, we assume that the maximum valuation of every bidder is lower bounded by 1 and upper bounded by a parameter $\mu \geq 1$. In the random arrival model we assume unbounded valuations.

1.2 Our Contribution

We present the first online mechanisms guaranteeing competitiveness for any multiplicity $b \geq 1$. Our results are based on a novel algorithmic approach combining the online allocation of items with the concept of oblivious randomized rounding: For each bidder, the online mechanism makes a single call to a demand oracle and assigns the bundle selected by the oracle with a probability independent of bidders' valuations. This way, we obtain universally truthful mechanisms that achieve competitive ratios being close to or even beating the best known approximation factors for the corresponding offline setting. We first present our results for the secretary model (randomly ordered arrivals):

- At first, we consider CAs with general valuations. For any multiplicity $b \geq 1$, we obtain a competitive ratio of $O(m^{1/(b+1)} \log(bm))$. To our knowledge, this is the first result for online optimization with multiplicity $b = o(\log m)$.
- At second, we consider CAs with general valuations for bundles of size at most d. For any multiplicity $b \geq 1$, we obtain a competitive ratio $O(d^{1/b} \log(bm))$. Previously, a competitive factor $O(d^2)$ was known only for multiplicity $b = 1$ [17].
- At third, we consider CAs with XOS valuations with $b = 1$, i.e., only one copy per item is available. We achieve a competitive factor of $O(\log m)$. To our knowledge, this is the first result about online optimization for submodular or XOS valuations.

Our online algorithms in the secretary model can be used for offline optimization as well. They correspond to polynomial-time universally truthful offline mechanisms calling each bidder's demand oracle only once. The strength of our approach becomes visible especially in the case of submodular and XOS valuations where we can improve on the best previously known approximation ratio which was $O(\log m \log \log m)$ [10] with a universally truthful mechanism; this mechanism due to Dobzinski [10] works for more general subadditive valuations, but it is the best known (offline) universally truthful mechanism for XOS bidders. Previously there was known an $O(\log m/\log \log m)$-approximate mechanism for more general subadditive bidders [11] but it is only truthful in expectation and is presented within a polynomial communication model. Another $O(\log m/\log \log m)$-approximate truthful in expectation mechanism for subadditive bidders was given by Feige [14][1]. Dobzinski et al. [12] designed a universally truthful $O(\log^2 m)$-approximate mechanism for XOS bidders.

---

[1] In fact this mechanism uses an even weaker notion than truthfulness in expectation, as truthfulness maximizes the expected utility only approximately.

For general valuations, we obtain a universally truthful mechanism with an approximation ratio deviating only by the factor $O(\log m)$ from the complexity theoretic lower bound $\Omega(m^{1/(b+1)})$ ([6]) (in terms of m), and from the best known (non-truthful) approximation upper bound $O(d^{1/b})$ [4] (in terms of d). In fact, the framework of Lavi and Swamy [18] yields mechanisms with approximation factors that match this lower bound. However, these mechanisms are only truthful in expectation. Only, for the case $b = 1$, a universally truthful $O(\sqrt{m})$-approximation mechanism is known [12,10].

An appealing feature of our mechanisms, compared to other randomized mechanisms for CAs, e.g., [10,12,18], is their simplicity in both the algorithmic approach and interaction with the bidders, e.g., they call each bidder's oracle only once. Our mechanisms use the random order of bidders' arrivals only for the purpose of extracting an estimate of the bids' range. If the arrival order is adversarial then one needs to assume there is an a priori defined range of the bids as otherwise one cannot obtain a competitive online algorithm. Thus, we assume that the ratio of the largest to the smallest maximum valuation over all bidders is bounded by a parameter $\mu \geq 1$, cf. Section 1.1.

Assuming that bidders arrive in adversarial order, we obtain similar competitive factors as for bidders arriving in random order: The competitive ratios listed above hold in the adversarial arrival model when replacing the $\log bm$ term in each of the bounds by the term $\log \mu bm$. Let us finally point out that the framework in [1] can be applied to our algorithms such that these competitive ratios hold not only with respect to social welfare but also with respect to revenue.

We would also like to mention that due to the generality of the CA model we study, our results also apply to mechanism design for network routing problems, where each bidder wants to connect two or more of her network sites by purchasing collections of edges/links in a network; for a detailed description of this setting see, e.g., [18].

2   The Overselling MPU Algorithm

We will describe and analyze here a multiplicative price-update algorithm similar to these used in many places in the literature, see, e.g., [1,5]. Assume for simplicity that the supplies of all goods are the same, that is, $b_e = b$ for each good $e \in U$, for some integer b; our approach can be generalized to allowing for arbitrary goods' supplies b_e.

The difference in the algorithm here, as compared to previous work, is that we will allow the algorithm to oversell the items and thus violate the items' supply b by a factor of $O(\log bm)$, which enables us to prove that the algorithm provides a constant approximation to the optimum with respect to the original supplies. Our final mechanism will then correct this possible infeasibility in a randomized fashion. The analysis of the final mechanism will be based in parts on lemmas that we prove in this section.

The algorithm is deterministic. It considers bidders in any given (adversarial) order. We assume that the algorithm is given two parameters $\mu \geq 1$ and $L \geq 0$, such that L is a lower bound on the maximum valuation among all bidders and bundles such that there is at most one bidder whose valuation exceeds μL. Observe that $L \leq v(opt)$.

The algorithm allocates a set S_i to each bidder i. It proceeds as follows. It starts with some relatively small, uniform price $p_0 = \frac{L}{4bm}$ of each item. When considering bidder i, the algorithm uses the demand oracle $D_i(U_i, p)$ to pick a set S_i, i.e., it chooses a utility maximizing subset with respect to the current price vector p over a subset $U_i \subseteq U$.

Algorithm 1. Overselling MPU algorithm

---

1 For each good $e \in U$ do $p_e^1 := p_0$.
2 For each bidder $i = 1, 2, \ldots, n$ do
3   Set $S_i := D_i(U_i, p^i)$, for a suitable $U_i \subseteq U$.
4   Update for each good $e \in S_i$: $p_e^{i+1} := p_e^i \cdot 2^{1/b}$.

Then the prices of the elements in the set S_i are increased by a factor $r = 2^{1/b}$ and the next bidder is considered. For a detailed description see Algorithm 1.

Bidder i's payment for bundle S_i is $p^i = \sum_{e \in S_i} p_e^i$. Let ℓ_e^i be the number of copies of item e allocated to all bidders preceding bidder i and $\ell_e^* = \ell_e^{n+1}$ denote the total allocation of item e to all bidders. Let, moreover, $p_e^* = p_0 \cdot r^{\ell_e^*}$ be item e's price at the end of the algorithm. Observe that the call to the demand oracle in line 3 of Algorithm 1 does not specify how to choose the sets U_i, for $1 \leq i \leq n$. We say that the demand oracle of a bidder is *unrestricted* if $U_i = U$, and it is *restricted* if $U_i \subsetneq U$. Indeed, the bound on the approximation ratio that we will give in this section holds only for unrestricted demand oracles. However, two of the lemmas that we prove in this section hold for (possibly) restricted demand oracles, i.e., for an arbitrarily chosen collection $U_1, \ldots, U_n$. These lemmas will be re-used in subsequent sections in which we present randomized algorithms that ensure feasibility by using restricted demand oracles in combination with oblivious randomized rounding.

Lemma 1 below holds for (possibly) restricted demand oracles and gives an upper bound on the factor by which the algorithm exceeds the number of available copies.

Lemma 1. *For any choice of $U_1, \ldots, U_n \subseteq U$, allocation S maps at most sb copies of each item to the bidders, where $s = \log(4\mu bm) + \frac{2}{b}$, where the log's base is 2.*

Proof. Consider any item $e \in U$. Suppose, after some step, $\lceil sb - 2 \rceil \geq b\log(4\mu bm)$ copies of item e have been assigned to bidders. Then the price of e is larger than $p_0 \cdot 2^{\log(4\mu bm)} \geq \mu L$. After this step, the algorithm might give further copies of e only to bidders whose maximum valuation exceeds μL. By the definition of μ, there is at most one such bidder that receives at most one copy of b. Hence, at most $\lceil sb - 1 \rceil \leq sb$ copies of e are assigned, which proves the lemma.   □

Next we prove two lemmas that give lower bounds on the social welfare $v(S) = \sum_{i=1}^n v_i(S_i)$ of the sets $S_1, \ldots, S_n$ picked by Algorithm 1. The first lower bound holds for (possibly) restricted demand oracles, the second lower bound holds only when using unrestricted demand oracles. Let $v(opt)$ denote the optimal social welfare with respect to the original supplies b, and recall that p_e^* denotes the final price of item $e \in U$.

Lemma 2. *For any choice of $U_1, \ldots, U_n \subseteq U$, $v(S) \geq b \sum_{e \in U} p_e^* - bmp_0$.*

Proof. Let $r = 2^{1/b}$. As bidders are individually rational, $v_i(S_i) \geq \sum_{e \in S_i} p_e^i$. Hence,

$$v(S) \geq \sum_{i=1}^n \sum_{e \in S_i} p_e^i = \sum_{i=1}^n \sum_{e \in S_i} p_0 r^{\ell_e^i} = p_0 \sum_{e \in U} \sum_{k=0}^{\ell_e^* - 1} r^k = p_0 \sum_{e \in U} \frac{r^{\ell_e^*} - 1}{r - 1} .$$

Now applying $p_e^* = p_0 r^{\ell_e^*}$ and $1/(r-1) = 1/(2^{1/b}-1) \geq b$ gives the lemma. $\quad\square$

Lemma 3. *For $U_1 = \cdots = U_n = U$, $v(S) \geq v(opt) - b \sum_{e \in U} p_e^*$.*

Proof. Consider any feasible allocation $T = (T_1, \ldots, T_n)$ assigning at most b of each item. As the algorithm uses an unrestricted demand oracle, we have $v_i(S_i) - \sum_{e \in S_i} p_e^i \geq v_i(T_i) - \sum_{e \in T_i} p_e^i$ and, hence,

$$v_i(S_i) \geq v_i(T_i) - \sum_{e \in T_i} p_e^i. \tag{1}$$

As $p_e^* \geq p_e^i$, for every i and e, this implies $v_i(S_i) \geq v_i(T_i) - \sum_{e \in T_i} p_e^*$. Summing over all bidders gives $v(S) = \sum_{i=1}^n v_i(S_i) \geq \sum_{i=1}^n v_i(T_i) - \sum_{i=1}^n \sum_{e \in T_i} p_e^* \geq v(T) - b \sum_{e \in U} p_e^*$, where the latter equation employs that T is feasible so that each item is given to at most b sets. Taking for T_i to be the bundle assigned to bidder i in an optimal solution *opt* and summing this inequality for all bidders implies the lemma. $\quad\square$

Combining the above lemmas yields the following result for the algorithm.

Theorem 1. *Algorithm 1 with $p_0 = \frac{L}{4bm}$ produces an infeasible allocation S in which up to $b \log(4\mu bm) + 2$ copies of each item are assigned to the bidders. If $U_1 = \cdots = U_n = U$, then $v(S) \geq \frac{1}{4} v(opt)$.*

Proof. The bound on the number of copies follows from Lemma 1. Lemma 2 gives $b \sum_{e \in U} p_e^* \leq v(S) + bmp_0$. Substituting this upper bound on the sum of the final prices into the equation in Lemma 3 gives $v(S) \geq v(opt) - v(S) - bmp_0$, which yields $2v(S) \geq v(opt) - bmp_0$. Now because $v(opt) \geq L$, we have $p_0 = \frac{L}{4bm} \leq \frac{v(opt)}{4bm}$. This gives $bmp_0 \leq \frac{1}{4} v(opt)$ and, hence, $v(S) \geq \frac{1}{4} v(opt)$. $\quad\square$

Let us stress that the solution produced by Algorithm 1 is infeasible. In the following section, we will show how to parametrize the demand oracles, that is, how to choose the subsets $U_1, \ldots, U_n$ such that a feasible solution is obtained.

3   The MPU Algorithm with Oblivious Randomized Rounding

We present here a variant of the MPU algorithm not exceeding the number of available copies. The idea is to provisionally assign bundles S_i of virtual copies to the bidders following the multiplicative price-update policy of Algorithm 1. The number of virtual copies exceeds the available number of real copies by a factor $O(\log(\mu bm))$, which corresponds to the upper bound of the overselling factor of Algorithm 1 shown in Lemma 1.

Algorithm 2 uses a randomized rounding procedure to decide which of the provisionally assigned bundles of virtual copies are actually converted into final bundles R_i of real copies. More precisely, for each bidder i, the algorithm sets $R_i = S_i$ with probability q and $R_i = \emptyset$ with probability $1 - q$, where $0 \leq q \leq 1$ is a parameter. For different settings of combinatorial auctions we will specify different values of q.

The algorithm outputs the allocation R. The payment for R_i is $p^i = \sum_{e \in R_i} p_e^i$. Observe that the definition of U_i in line 3 of the algorithm ensures that, for every item

---

Algorithm 2. MPU algorithm with oblivious randomized rounding

---

1 For each good $e \in U$ do $p_e^1 := p_0, b_e^1 := b$.
2 For each bidder $i = 1, 2, \ldots, n$ do
3   Set $S_i := D_i(U_i, p^i)$, for $U_i = \{e \in U \mid b_e^i > 0\}$.
4   Update for each good $e \in S_i$: $p_e^{i+1} := p_e^i \cdot 2^{1/b}$.
5   With probability q set $R_i := S_i$ else $R_i := \emptyset$.
6   Update for each good $e \in R_i$: $b_e^{i+1} := b_e^i - 1$.

---

e, the number of copies of e in $R_1 \cup \ldots \cup R_n$ does not exceed the number b of available copies. Hence, the final allocation R is feasible. Furthermore, if the probability q in step 5 is set to 0, then the provisional allocation S computed by Algorithm 2 is identical to the allocation S computed by Algorithm 1 with unrestricted demand oracles. Obviously, however, one needs to set $q > 0$ as the final allocation R is empty, otherwise.

Now, large parts of the analysis of Algorithm 1 hold for arbitrary chosen sets $U_1, \ldots, U_n \subseteq U$. If we set $q > 0$ then the properties of S from Lemmas 1 and 2 still hold, only Lemma 3 uses unrestricted demand oracles called with $U_1 = \cdots = U_n = U$ and, hence, cannot directly be applied to Algorithm 2 if $q > 0$. In particular, (1) in the proof of this lemma fails when using restricted demand oracles as done by Algorithm 2.

Lemma 4 shows that Algorithm 2 with $0 < q \leq 1$ yields an $O(\frac{1}{q})$-approximation of the optimal social welfare, given a relaxed stochastic version of (1). Observe that prices p_e^i are random variables that depend on the random coin flips for bidders $1, \ldots, i-1$.

Lemma 4. *Suppose the probability $q > 0$ in Algorithm 2 is chosen sufficiently small such that, for any $1 \leq i \leq n$ and any bundle $T \subseteq U$,*

$$\mathbf{E}\left[v_i(T \cap U_i)\right] \geq \frac{1}{2} v_i(T) \ . \tag{2}$$

Then $\mathbf{E}\left[v(S)\right] \geq \frac{1}{8}v(opt)$ and $\mathbf{E}\left[v(R)\right] \geq \frac{q}{8}v(opt)$.

Proof. Fix any bidder i. Consider any feasible allocation $T_1, \ldots, T_n$. The set S_i is chosen by utility maximizing demand oracle $D_i(U_i, p^i)$, thus $v_i(S_i) \geq v_i(T_i \cap U_i) - \sum_{e \in T_i \cap U_i} p_e^i$, for any outcome of the algorithm's random coin flips; which implies $\mathbf{E}\left[v_i(S_i)\right] \geq \mathbf{E}\left[v_i(T_i \cap U_i)\right] - \sum_{e \in T_i \cap U_i} \mathbf{E}\left[p_e^i\right]$. By the assumption in the lemma, this gives $\mathbf{E}\left[v_i(S_i)\right] \geq \frac{1}{2}v_i(T_i) - \sum_{e \in T_i} \mathbf{E}\left[p_e^i\right]$.

As before, let p_e^* denote final price of item e. As $p_e^* \geq p_e^i$, we obtain $2 \cdot \mathbf{E}\left[v_i(S_i)\right] \geq v_i(T_i) - 2 \cdot \sum_{e \in T_i} \mathbf{E}\left[p_e^*\right]$. Summing over all bidders gives thus

$$2\,\mathbf{E}\left[v(S)\right] = 2\sum_{i=1}^{n} \mathbf{E}\left[v_i(S_i)\right] \geq \sum_{i=1}^{n} v_i(T_i) - 2\sum_{i=1}^{n}\sum_{e \in T_i} \mathbf{E}\left[p_e^*\right] \ .$$

Let us now assume that $T_1, \ldots, T_n$ is the optimal allocation. This way, we obtain

$$2\,\mathbf{E}\left[v(S)\right] \geq v(opt) - 2\sum_{i=1}^{n}\sum_{e \in T_i} \mathbf{E}\left[p_e^*\right] \geq v(opt) - 2\,b\sum_{e \in U} \mathbf{E}\left[p_e^*\right],$$

where the latter equation uses that each item is given to at most b sets.

Now we proceed analogously to the proof of Theorem 1. Lemma 2 shows that $b \sum_{e \in U} p_e^* \leq v(S) + bmp_0$, for any outcome of the random coin flips of the algorithm. Hence, $b \sum_{e \in U} \mathbf{E}[p_e^*] \leq \mathbf{E}[v(S)] + bmp_0$. Substituting this upper bound into the equation above and using the fact that $p_0 = \frac{L}{4bm} \leq \frac{v(opt)}{4bm}$ gives

$$2 \cdot \mathbf{E}[v(S)] \geq v(opt) - 2 \cdot \mathbf{E}[v(S)] - 2bmp_0 \geq v(opt) - 2\mathbf{E}[v(S)] - \tfrac{1}{2}v(opt).$$

Hence $4 \cdot \mathbf{E}[v(S)] \geq \tfrac{1}{2}v(opt)$, so $\mathbf{E}[v(S)] \geq \tfrac{1}{8}v(opt)$.

Finally, we observe that the random variable $v_i(S_i)$ depends on random coin flips made by the algorithm when considering bidders $1, \ldots, i-1$. The probability whether bidder i receives the provisional bundle S_i or the empty set is independent of the random flips made for the preceding bidders. Thus, $\mathbf{E}[v(R_i)] = q\mathbf{E}[v(S_i)]$ as $\mathbf{Pr}[R_i = S_i] = q$. By linearity of expectation $\mathbf{E}[v(R)] = q\mathbf{E}[v(S)]$, so $\mathbf{E}[v(R)] \geq \tfrac{q}{8}v(opt)$.   □

Lemma 4 provides a recipe for devising online algorithm: Depending on the auction setting, choose $q > 0$ in such a way that (2) can be established. The larger the value of q, the better the competitive ratio. We apply this recipe to three different CAs settings.

3.1   Analysis for CAs with Bundles of Bounded Cardinality

We study here the approximation factor achieved by Algorithm 2 for CAs with general valuations. Lemma 5 below proves that Assumption (2) in Lemma 4 holds for $q^{-1} = O(d^{1/b} \log m)$. The probability in Lemma 5 is with respect to the random coin flips for bidders $1, \ldots, i-1$ which influence price vector p^i, set U_i and, hence, the choice of S_i.

Lemma 5. *Let us consider general CAs with bundles of maximum cardinality $d \geq 1$ and multiplicity $b \geq 1$, and let $q^{-1} = 2ed^{1/b}(\log(4\mu bm) + \tfrac{2}{b})$. For any $1 \leq i \leq n$ and any bundle $T \subseteq U$ of at most d items, $\mathbf{E}[v_i(T \cap U_i)] \geq \tfrac{1}{2}v_i(T)$.*

Proof. Consider bidder i. By Lemma 1, item $e \in U$ is contained in at most $\ell = b \cdot \log(4\mu bm) + 2$ of the provisional bundles $S_1, \ldots, S_{i-1}$. Each of these ℓ bundles is turned into a final bundle with probability $q = b/(2ed^{1/b}\ell)$. Now consider $e \in T$. Observe, $e \in U_i$ if e was sold $\leq b-1$ times, that is, at most $b-1$ of its provisional bundles became final. The probability that $e \notin U_i$ is thus

$$\binom{\ell}{b} \cdot q^b \leq \left(\frac{e\ell}{b}\right)^b \cdot \left(\frac{b}{2ed^{1/b}\ell}\right)^b = \frac{1}{2d} \ .$$

Because $|T| \leq d$, by the union bound we have $\mathbf{Pr}[T \cap U_i = T] \geq \tfrac{1}{2}$. As a consequence, $\mathbf{E}[v_i(T \cap U_i)] \geq \tfrac{1}{2}v_i(T)$.   □

Combining Lemma 4 and 5 yields

Theorem 2. *Algorithm 2 with q as defined above is $O(d^{1/b} \log(\mu bm))$-competitive for general CAs with bundles of maximum cardinality $d \geq 1$ and multiplicity $b \geq 1$.*   □

3.2 Algorithm for General CAs with Multiplicity

For CAs with general valuations and bundles of unbounded cardinality, we introduce a mechanism that flips a fair coin to choose one out of two algorithms. If the coin shows head, then Algorithm 2 is executed with parameter q set analogously to the case of bundles of cardinality at most $d = \lfloor m^{\frac{b}{b+1}} \rfloor$, that is, with $q^{-1} = 2ed^{1/b}(\log(4\mu bm)+\frac{2}{b})$. If the coin shows tail, then the mechanism only considers bundles of full cardinality m. This setting corresponds to an auction where each bidder wants to buy only a single super item (corresponding to the full bundle U) which is available in b copies. The b copies of the super item are sold by calling the algorithm of Section 3.1 with $m = 1$ (single item) and $d = 1$ (bundles of cardinality 1).

Theorem 3. *The algorithm as defined above is an $O(m^{1/(b+1)} \log(\mu bm))$-competitive algorithm for general CAs with multiplicity $b \geq 1$.*

Due to space limitations, we only sketch the analysis of this theorem. Let opt_S (opt_L) denote an offline optimum of the original CA problem assuming that only sets of size at most (at least) d can be allocated to the bidders. We show, if $v(opt_S) \geq v(opt_L)$, then the head algorithm is $O(m^{1/(b+1)} \log(\mu bm))$-competitive with respect to $v(opt_S)$. Otherwise, the tail algorithm is $O(m^{1/(b+1)} \log(\mu b))$-competitive with respect to $v(opt_L)$. This implies the theorem as $v(opt) \leq v(opt_L) + v(opt_S)$.

Suppose $v(opt_S) \geq v(opt_L)$ and consider the head algorithm. Analogously to the proof of Lemma 4, one can show that $\mathbf{E}\left[v(S)\right] \geq \frac{1}{4}v(opt_S) - \frac{1}{8}L$ with L being a lower bound on the maximum valuations among all bidders. By our assumptions $L \leq v(opt_L) \leq v(opt_S)$. Hence, $\mathbf{E}\left[v(S)\right] \geq \frac{1}{8}v(opt_S)$, which gives $\mathbf{E}\left[v(R)\right] \geq \frac{q}{8}v(opt_S)$. That is, the head algorithm achieves a competitive ratio of $\frac{8}{q} = O(d^{1/b} \log(\mu bm)) = O(m^{1/(b+1)} \log(\mu bm))$ with respect to opt_S.

Now suppose $v(opt_L) > v(opt_S)$ and consider the tail algorithm. This algorithm assigns only full bundles. The optimal allocation for full bundles, called opt_F gives the full bundle to the b bidders with highest valuations. The number of bundles allocated by opt_L is at most $\frac{bm}{d}$ as the total number of copies is bm and bundles have size at least d. Comparing the numbers of allocated bundles implies $opt_F \geq \frac{d}{m} opt_L$. The tail algorithms sells b copies of the full bundle by calling the algorithm of Section 3.1 with $m = 1$ and $d = 1$. It is $O(\log(\mu b))$-competitive with respect to opt_F. Hence, the competitive ratio with respect to opt_L is $O(\frac{m}{d} \log(\mu b)) = O(m^{1/(b+1)} \log(\mu b))$.

3.3 Analysis for CAs with XOS Valuations

In this section, we consider the approximation factor that is achieved by Algorithm 2 for CAs with XOS valuations. W.l.o.g., we assume $b = 1$. The following lemma proves that Assumption (2) in Lemma 4 holds for $q^{-1} = O(\log m)$. The probability in the lemma is with respect to the random coin flips for the bidders $1, \dots, i-1$.

Lemma 6. *Let $q^{-1} = 2(\log(4\mu m) + 2)$. For any $1 \leq i \leq n$ and any bundle $T \subseteq U$, it holds $\mathbf{E}\left[v_i(T \cap U_i)\right] \geq \frac{1}{2}v_i(T)$.*

Proof. Consider bidder i. By Lemma 1, $e \in U$ is contained in at most $\ell = \log(4\mu m)+2$ of the provisional bundles $S_1, \dots, S_{i-1}$. Each of these ℓ bundles is turned into a final bundle with probability $q = 1/(2\ell)$. Thus, by the union bound, the probability that at least one of these ℓ bundles is turned into the final bundle is at most $\ell q = \frac{1}{2}$. Hence, the probability that none of the bundles containing e is turned into a final bundle is at least $\frac{1}{2}$. That is, $\mathbf{Pr}\,[e \in U_i] \geq \frac{1}{2}$, for every $e \in U$.

The set U_i and, hence, also the set $T \cap U_i$ depend on the random coin flips of the algorithm. For a subset $K \subseteq T$, let $\alpha(K)$ denote the probability that $T \cap U_i = K$. For $e \in T$, $\sum_{K \subseteq T, K \ni e} \alpha(K) \geq \mathbf{Pr}\,[\cup_{K \subseteq T, K \ni e}(T \cap U_i = K)] = \mathbf{Pr}\,[e \in U_i] \geq \frac{1}{2}$. Now define $\beta(K) = \min\{1, 2\alpha(K)\}$. Then $\sum_{K \subseteq T, K \ni e} \beta(K) \geq 1$. That is, the subsets $K \subseteq T$ and the numbers $0 \leq \beta(K) \leq 1$ are a fractional cover of T. We obtain

$$\mathbf{E}\,[v_i(T \cap U_i)] = \sum_{K \subseteq T} \alpha(K)v_i(K) \geq \frac{1}{2} \sum_{K \subseteq T} \beta(K)v_i(K) \geq \frac{1}{2} v_i(T) \ ,$$

where the last inequality uses that XOS valuations are fractionally subadditive.   □

Combining the lemma above with Lemma 4 yields

Theorem 4. *Algorithm 2 with q as defined above is $O(\log \mu m)$-competitive for CAs with XOS valuations.*   □

4   Truthful Mechanisms for Different Kind of Arrivals

In the previous sections, we assumed that the algorithm is given two parameters $\mu \geq 1$ and $L \geq 0$, such that L is a lower bound on the maximum valuation among all bidders and bundles such that there is at most one bidder whose valuation exceeds μL. We will now show how these parameters can be obtained in a truthful manner when bidders arrive in random and adversarial order, respectively.

Mechanism for random arrivals. We use the random order of arrivals only for one purpose: The first $n/2$ bidders correspond to a random sample of the bidders. These bidders are excluded from the allocation, that is, they do not get assigned a bundle but are used only for statistical purposes. Each of these bidders is asked for her maximum bid by querying her demand oracle for the utility maximizing bundle assuming prices are set to 0. Let L denote the maximum among these valuations.

With probability $\frac{1}{2}$, the second largest bidder, i.e., the one whose maximum valuation for a bundle is the second largest valuation among all bidders, belongs to the first half of arrivals. Thus, L and, hence, μL, for $\mu = 1$, is an upper bound on the valuations of all but the largest bidder. Excluding the second largest bidder from our consideration, the optimum social welfare among the other bidders is at least $v(opt)/2$. Thus, with probability at least $\frac{1}{2}$, the social welfare among the bidders arriving in the second half is at least $v(opt)/4$.

Summarizing, with probability at least $\frac{1}{4}$, the parameters L and $\mu = 1$ satisfy the assumptions in Section 2 and the second half of the arrivals carries a social welfare $\geq v(opt)/4$. Thus, applying Algorithm 2 to bidders in the second half, yields universally truthful mechanisms with competitive ratios as in Theorems 2, 3 and 4 with $\mu = 1$.

Mechanism for adversarial arrivals. For bidders arriving in adversarial order, we use that valuations have a bounded domain. By our assumptions, a bidder's maximum valuation over all bundles is assumed to be at least 1 and at most $\mu \geq 1$. This corresponds to the assumptions made in Section 2 about the parameter μ for the Algorithms 1 and 2. Thus, we obtain universally truthful mechanisms for the different settings of combinatorial auctions considered in the Theorems 2, 3 and 4.

Observe that truthfulness of the resulting mechanisms follows from the fact that they obey the direct characterization of truthfulness mentioned in Section 1.1.

References

1. Awerbuch, B., Azar, Y., Meyerson, A.: Reducing truth-telling online mechanisms to online optimization. In: STOC, pp. 503–510 (2003)
2. Awerbuch, B., Azar, Y., Plotkin, S.A.: Throughput-competitive on-line routing. In: FOCS, pp. 32–40 (1993)
3. Babaioff, M., Immorlica, N., Kleinberg, R.: Matroids, secretary problems, and online mechanisms. In: SODA, pp. 434–443 (2007)
4. Bansal, N., Korula, N., Nagarajan, V., Srinivasan, A.: On k-Column Sparse Packing Programs. In: Eisenbrand, F., Shepherd, F.B. (eds.) IPCO 2010. LNCS, vol. 6080, pp. 369–382. Springer, Heidelberg (2010)
5. Bartal, Y., Gonen, R., Nisan, N.: Incentive compatible multi unit combinatorial auctions. In: The Proc. of the 9th TARK, pp. 72–87 (2003)
6. Blumrosen, L., Nisan, N.: Combinatorial auctions. In: Nisan, N., Roughgarden, T., Tardos, E., Vazirani, V. (eds.) Algorithmic Game Theory (2007)
7. Borodin, A., El-Yaniv, R.: Online computation and competitive analysis. Cambridge University Press (1998)
8. Buchbinder, N., Naor, J.: Online Primal-Dual Algorithms for Covering and Packing Problems. In: Brodal, G.S., Leonardi, S. (eds.) ESA 2005. LNCS, vol. 3669, pp. 689–701. Springer, Heidelberg (2005)
9. Buchbinder, N., Naor, J.: Improved bounds for online routing and packing via a primal-dual approach. In: FOCS, pp. 293–304 (2006)
10. Dobzinski, S.: Two Randomized Mechanisms for Combinatorial Auctions. In: Charikar, M., Jansen, K., Reingold, O., Rolim, J.D.P. (eds.) RANDOM 2007 and APPROX 2007. LNCS, vol. 4627, pp. 89–103. Springer, Heidelberg (2007)
11. Dobzinski, S., Fu, H., Kleinberg, R.: Truthfulness via proxies. CoRR, abs/1011.3232 (2010)
12. Dobzinski, S., Nisan, N., Schapira, M.: Truthful randomized mechanisms for combinatorial auctions. In: STOC, pp. 644–652 (2006)
13. Dynkin, E.B.: The optimum choice of the instant for stopping a markov process. Sov. Math Dokl. 4 (1963)
14. Feige, U.: On maximizing welfare when utility functions are subadditive. In: STOC, pp. 41–50 (2006)
15. Feige, U.: On maximizing welfare when utility functions are subadditive. SIAM J. Comput. 39(1), 122–142 (2009)
16. Kleinberg, R.D.: A multiple-choice secretary algorithm with applications to online auctions. In: SODA, pp. 630–631 (2005)
17. Korula, N., Pál, M.: Algorithms for Secretary Problems on Graphs and Hypergraphs. In: Albers, S., Marchetti-Spaccamela, A., Matias, Y., Nikoletseas, S., Thomas, W. (eds.) ICALP 2009, Part II. LNCS, vol. 5556, pp. 508–520. Springer, Heidelberg (2009)
18. Lavi, R., Swamy, C.: Truthful and near-optimal mechanism design via linear programming. In: Proc. of FOCS, pp. 595–604 (2005)
19. Lehmann, B., Lehmann, D.J., Nisan, N.: Combinatorial auctions with decreasing marginal utilities. In: ACM Conference on Electronic Commerce, pp. 18–28 (2001)
20. Nisan, N.: Introduction to mechanism design (for computer scientists). In: Nisan, N., Roughgarden, T., Tardos, E., Vazirani, V. (eds.) Algorithmic Game Theory (2007)

Online Packing with Gradually Improving Capacity Estimations and Applications to Network Lifetime Maximization

Marcel Ochel, Klaus Radke, and Berthold Vöcking*

Department of Computer Science, RWTH Aachen University, Germany

Abstract. We introduce a general model for online packing problems with applications to lifetime optimization of wireless sensor networks. Classical approaches for lifetime maximization make the crucial assumption that battery capacities of sensor nodes are known a priori. For real-world batteries, however, the capacities are only vaguely known. To capture this aspect, we introduce an adversarial online model where estimates become more and more accurate over time, that is, when using the corresponding resources. Our model is based on general linear packing programs and we assume the remaining capacities to be always specified by lower and upper bounds that may deviate from each other by a fixed factor α.

We analyze the algorithmic consequences of our model and provide a general $\ln \alpha/\alpha$-competitive framework. Furthermore, we show a complementary upper bound of $O(1/\sqrt{\alpha})$.

1 Introduction

The devices participating in ad-hoc networks are typically assumed to have a limited energy supply, e. g. battery-driven nodes. By assigning specific transmission powers to the network nodes, different communication structures can be established. This flexibility allows to control the energy requirements for communication by adaptively choosing power assignments.

One of the most prominent problems in this context is network lifetime maximization [7,8,14–16]. For given communication requirements and radio characteristics a transmission scheme has to be found which maximizes the number of communication operations until the first network node runs out of energy. This can be seen as a packing of specific communication structures due to energy constraints [7,15].

To the best of our knowledge, all formulations of network lifetime assume perfect a priori knowledge about the status and future bahaviour of the energy supplies of each single network node. In real-world applications, however, this seems an unrealistic assumption. The measured residual energy after a (partial) execution of the transmission scheme may deviate from the initially predicted

---

* This work has been supported by DFG through UMIC Research Centre, RWTH Aachen University.

A. Czumaj et al. (Eds.): ICALP 2012, Part II, LNCS 7392, pp. 648–659, 2012.
© Springer-Verlag Berlin Heidelberg 2012

ideal behaviour. In addition to the non-perfect discharge characteristics of batteries (for a brief introduction see e.g. [18]), the energy budget can also be influenced by varying communication costs caused from temporal interference (e.g. by rain or transient noise). To exploit the updates on the residual energy levels during the execution of a transmission scheme, we introduce a new online model for general linear packing problems that can be used to increase the lifetime of wireless ad-hoc networks.

Abstracting from various technological details, we might assume that at any time an upper and a lower bound on the remaining capacity of every resource can be obtained and these bounds deviate at most by a factor of α, for a given parameter $\alpha \geq 1$. This way, while the guarantee on the relative error regarding the remaining capacity stays constant, the estimations on the true, initial capacity get more and more accurate with depletion.

We analyze the algorithmic consequences of our model and provide a general framework that exploits the gradually improving capacity estimates. Furthermore, we show a complementary upper bound on the best possible competitive ratio.

1.1 Model for the Competitive Analysis

The proposed online model for packing problems with gradually improving estimations on the packing constraints considers general linear packing programs of the form $\max p^\top x$ subject to $Ax \leq c$ and $x \geq 0$, with $p \in \mathbb{R}^n_{\geq 0}$, $A \in \mathbb{R}^{m \times n}_{\geq 0}$, and $c \in \mathbb{R}^m_{\geq 0}$. The coefficients in the vector c are called *(true) capacities*.

In our model, there are two players, an *online algorithm* and an *adversary*. The algorithm aims to find a feasible solution to the packing program maximizing the objective function $p^\top x$. It is equipped with the coefficients in A and p but only receives estimates on the capacities. The estimates are given in form of lower and upper bounds specified by the adversary.

In particular, there is a parameter $\alpha \geq 1$ that controls the quality of these estimates. The algorithm may always allocate a portion of the *remaining capacities*, i.e. of the capacities that have not been used previously. At the beginning of each allocation step the online algorithm receives an upper and a lower bound on the remaining capacities from the adversary. The only restriction is that the two bounds deviate at most by the factor α. Based on this imprecise information, the algorithm has to decide which amounts of the capacities shall be used in the current step.

We consider online algorithms that run in stages. Let N denote the number of stages. Initially, the online player starts with the solution $x^{(0)} = 0 \in \mathbb{R}^n_{\geq 0}$. It can increase the variables x from stage to stage. Each stage $s \in \{1, \dots, N\}$ proceeds as follows:

- The adversary reveals bound vectors $\ell^{(s)}, u^{(s)} \in \mathbb{R}^m_{\geq 0}$ on the remaining slack such that for each constraint $i \in \{1, \dots, m\}$ we have

$$\ell_i^{(s)} \leq c_i - (Ax^{(s-1)})_i \leq u_i^{(s)} \quad \text{and} \quad u_i^{(s)} \leq \alpha \cdot \ell_i^{(s)}.$$

– Based on p, A, $\ell^{(s)}$ and $u^{(s)}$ the online algorithm determines its additional allocations, i. e. it determines a $y^{(s)} \in \mathbb{R}^n_{\geq 0}$ and sets $x^{(s)} := x^{(s-1)} + y^{(s)}$.

We assume that the online algorithm has to guarantee that all solutions $x^{(1)}, \ldots, x^{(N)}$ are feasible. Note that we did not restrict the adversary in a way assuming that the most recently presented lower bound on the slack yields the best lower bound on the true capacity. Those presented in earlier stages might translate into better lower bounds on the true capacity.

To simplify the model, we assume w.l.o.g. that $u^{(s)} = \alpha \cdot \ell^{(s)}$ as by telling any better upper bound the adversary would provide more information than necessary. In other words, the adversary specifies only a lower bound vector satisfying $\ell^{(s)}_i \leq c_i - (Ax^{(s-1)})_i \leq \alpha\ell^{(s)}_i$. We will call such lower bounds α-sharp.

We evaluate online algorithms in terms of their *competitive ratio* [5], i. e. the worst-case ratio between the objective value obtained by the online algorithm and the best possible objective value under the true constraints.

1.2   New Results

We begin with analyzing a natural greedy approach in Section 2: In each stage the algorithm GREEDY computes the best allocation with respect to the currently known lower bounds on the capacities. We show that this algorithm, which at first might look promising, only achieves the trivial competitive ratio of $1/\alpha$.

In Section 3, we give a new algorithm which is called EQUITABLE as it uses only an equitable portion of the capacities in each stage. Our main result shows that its competitive ratio is at least $(1 - \epsilon) \cdot \frac{\ln(\alpha)}{\alpha - 1}$ on general packing problems if the algorithm runs for $N \geq \frac{\alpha - 1}{\alpha^\epsilon - 1}$ stages. This proves that already for a small number of stages, i. e. $N \geq \frac{\alpha^2}{\ln \alpha}$, the algorithm is $\frac{\ln(\alpha)}{\alpha}$-competitive. In the context of network lifetime problems this is a favorable property as the efficiency gain is not overshadowed by the communication overhead needed to implement a multistage approach. In addition, we give an upper bound for the competitive ratio of EQUITABLE showing that our analysis is tight.

The presented framework is applicable to all network lifetime problems that can be formulated in terms of linear packing programs. In fact, in every stage the known algorithms are used without any modification and the LP formulation is just needed for the analysis. For hard problem variants, where the underlying packing LPs are only solvable approximately, the competitiveness of our framework decreases by the approximation factor of the blackbox algorithm.

The specific algorithm called in the stages of our framework might return fractional solutions. We show that under reasonable conditions they can be converted to integral solutions via a randomized rounding technique similar to [20].

Finally, in Section 4, we analyze the limits of online algorithms using gradually improving capacity estimations. We prove an upper bound on the competitive ratio of any (randomized) online algorithm showing that there does not exist any online algorithm achieving a competitive ratio of $\frac{4}{\sqrt{\alpha}}$ or better. This upper bound is shown for an instance of the problem of packing arborescences (spanning trees) in edge-capacitated graphs as it appears in the context of the lifetime maximization problems discussed in Section 1.3.

1.3   Lifetime Maximization for Sensor Networks

A sensor network consists of a number of sensor nodes that monitor their environment and a base station where the data is periodically collected. Therefore the nodes have to communicate their data to the base station. As they are equipped with only a limited energy supply they do not transmit it directly but via convergecast operations, i. e. they send the data to a nearby node and the various transmissions form a spanning tree which is directed towards the base station. Note that the energy consumption of wireless transmissions grows superlinearly with the distance, i. e. quadratic or larger.

The network lifetime is the number of sensor polls until the first node runs out of energy. Just assuming that the remaining energy of each sensor node can always be estimated up to an uncertainty factor α, it can easily be seen that relying on a precomputed schedule can lead to a lifetime which is α times smaller than an optimal (ex-post) schedule. Since the problem is intrinsically online (once a communication has been executed the corresponding energy consumption can not be undone), we apply the model described in Section 1.1 and formulate the lifetime maximization problem in terms of online packing.

Online Maximum Lifetime Convergecast (OMLC): Given a sensor network we consider the corresponding weighted digraph $G = (V, E)$ where the vertex set V consists of the sensor nodes and one designated base node $b \in V$. The edges E and the weight function $w : E \to \mathbb{R}^+$ are defined according to the possible transmissions and their respective energy consumption. Due to interference and energy issues it is generally assumed that nodes can not transmit across the whole network but only within a specific radius. Hence, every node u is only adjacent to all those nodes $N(u)$ for which the necessary transmission power is below a threshold $t_{max}(u)$, i. e. $w((u,v)) \leq t_{max}(u), \forall v \in N(u)$.

The goal is to find a packing of convergecast trees $T \in \mathcal{T}_b$ rooted at b with maximum total multiplicity subject to the a priori unknown node capacities $c_{unknown} : V \to \mathbb{R}^+$. These capacities are revealed online by α-sharp lower bounds $\ell_u^{(s)}$ for every stage $s \in \mathbb{N}$.

$$\max \sum_{T \in \mathcal{T}_b} x_T$$

$$\text{s.t.} \quad \sum_{\substack{T \in \mathcal{T}_b : \\ (u,v) \in E(T)}} w((u,v)) \cdot x_T \leq c_{unknown}(u) \qquad \forall u \in V \setminus \{b\}$$

$$x_T \geq 0 \qquad \qquad \forall T \in \mathcal{T}_b$$

Note that we took the fractional variant of Maximum Lifetime Convergecast for our online formulation. Furthermore, the entries in each row do not exceed the corresponding transmission threshold $t_{max}(u)$. By scaling the constraints we can assume the entries to be within the interval $[0; 1]$ and since batteries are amply dimensioned for multiple transmissions, we can still assume the capacities to be large, which guarantees that always an integral $(1 - \epsilon)$-approximate solution can be derived from a fractional one by randomized rounding (see Lemma 3).

Although the above linear program has an exponential number of variables, it can be optimally solved in polynomial time by a flow-based reduction to the efficiently solvable [9] problem of packing arborescences in edge-capacitated graphs (similar to the one in [15] for single recipient maximum lifetime broadcast).

Note that the OMLC problem is only one simple example from the vast domain of network lifetime problems. The online algorithm EQUITABLE presented in this paper works for general packing problems and, hence, the results can be applied to different variants of lifetime maximization problems which might also incorporate a more data-centric view considering the amount of transmitted data in each gathering. Even though most lifetime maximization problems are hard to solve and admit only approximation algorithms [8, 15], those can be directly combined with the presented algorithmic framework (see Corollary 1).

1.4   Related Work

Optimization under uncertainty has been studied extensively in the computer science community. The predominent approach is stochastic programming [4, 10, 12, 19]. There, one assumes a stochastic model on the uncertain parameters and seeks solutions that optimize an objective function in expectation. The problems are often studied in the form of two-stage or multi-stage recourse models where corrective actions at higher prices are admitted as more and more of the random events occur. In our case though, we do not make any assumptions on the probability distribution of the outcomes and corrective actions are not allowed.

The general (offline) packing problem occurs in many applications and a lot of research has been done to obtain near optimal solutions in various cases [3, 11, 17]. Most of them are Lagrangian-relaxation algorithms and rely in spirit on the well known multiplicative weights update method.

Online packing has often been considered in the context of call control [1, 2] or online-flow [13]. In the form of abstract linear programming the online model of Buchbinder and Naor [6] is the most prominent one. They assume that the capacities of the constraints are known in advance but not the profit vector and the constraint matrix. In every round a new component of the solution vector is introduced along with the corresponding column of the constraint matrix, i. e. the packing constraints are gradually revealed. Furthermore, the online player can only increase a component of the solution vector in its corresponding round. This model is closely related to our work although it bears fundamental differences as can easily be seen by the bounds on the best possible algorithms.

2   Competitiveness of the GREEDY-Algorithm

The naive algorithm for solving a packing problem $\mathcal{P}$ of the form $\max p^\top x$ subject to $Ax \leq c$, $x \geq 0$ in our model works as follows. Take the initial lower bounds

$l^{(1)}$ on the slack vector and solve the (completely known) linear packing problem $\mathcal{P}(l^{(1)})$ defined as $\max p^\top x$ subject to $Ax \leq l^{(1)}$, $x \geq 0$. Since the optimal value of a packing LP is linear in the constraint capacities and since the components of $l^{(1)}$ are at most a factor of α smaller than the true capacities c, this algorithm is $1/\alpha$-competitive.

Of course it is better to iterate this behaviour. The GREEDY-algorithm uses the lower bounds on the new slacks again as the capacities for an offline problem, i.e. it solves the problem $\mathcal{P}(l^{(2)})$, and repeats this until the first constraint becomes tight. Nevertheless, the algorithm can be tricked into performing badly.

Lemma 1. *GREEDY is only* $O(\frac{1}{\alpha})$*-competitive.*

Proof. Consider the following linear program which we call (1,k)-PACK.

$$\begin{aligned} \max \quad & x_1 + x_2 \\ \text{s.t.} \quad & k \cdot x_1 + x_2 \leq c_1 \\ & x_1 + k \cdot x_2 \leq c_2 \\ & x_1, x_2 \geq 0 \end{aligned}$$

By adding the two constraints it is clear that the profit of any solution to this LP is at most $(c_1 + c_2)/(k+1)$. Let the initial lower bounds be $\ell^{(1)} = (1; 1)$. In the first stage GREEDY computes a solution with profit at most $2/(k+1)$ and blocks capacity equal to 1 on both constraints. At the beginning of the second stage the lower bound on the remaining slacks turns out to be $\ell^{(2)} = (0; \alpha - 1)$. Note that this also determines the original capacities to be $c = (1; \alpha)$. As the first constraint is already tight GREEDY is not able to perform any further improvements and stops with a profit of at most $2/(k+1)$. For $k = \alpha$ the optimal offline solution achieves a profit of exactly 1, which results in the claimed worst-case factor. $\square$

3 The EQUITABLE-Algorithm

In this section we give an algorithm which is $\frac{\ln(\alpha)}{\alpha}$-competitive for every packing problem defined in our online model. It avoids the drawbacks of GREEDY by restricting itself in each stage to a budget which ensures that in each forthcoming stage the solution can only improve.

Algorithm EQUITABLE
Input: Online packing LP $\mathcal{P}$ of the form $\max p^\top x$ subject to $Ax \leq c$, $x \geq 0$, with unknown c.

1. Fix the number N of stages to be large enough (e.g. $N := \sqrt{\alpha} + 1$).
2. Set $x^{(0)} = y^{(0)} := 0 \in \mathbb{R}^n_{\geq 0}$ and $\bar{\ell}^{(0)} = 0 \in \mathbb{R}^m_{\geq 0}$.
3. For $s := 1$ to N do
   (a) Probe the lower bounds $\ell^{(s)} = (\ell_1^{(s)}, \ldots, \ell_m^{(s)})$ on the remaining capacities.
   (b) For each constraint j set $\bar{\ell}_j^{(s)} := \max\{\ell_j^{(s)}, \bar{\ell}_j^{(s-1)} - (Ay^{(s-1)})_j\}$ to be the best known lower bound on the remaining slack.

(c) Define the current capacity budget for stage s to be

$$b^{(s)} := \frac{\bar{\ell}^{(s)}}{N - (s-1)}.$$

(d) Compute an optimal solution $y^{(s)}$ for the offline linear program $\max p^\top x$ subject to $Ax \leq b^{(s)}$, $x \geq 0$, and set $x^{(s)} := x^{(s-1)} + y^{(s)}$.

4. Output solution $x^{(N)}$ (if wanted use the remaining slack with GREEDY).

The definition of $b^{(s)}$ is deliberately chosen to guarantee monotonicity in the vector components as will be shown in the proof of Lemma 2. Furthermore, since the adversary might return new bounds on the remaining capacities that are considerably worse than what can be estimated from the last allocation we have to keep track of the currently best known lower bounds. This is done by the vector $\bar{\ell}$. Also notice that the algorithm needs relatively few stages to guarantee a decent competitive factor.

3.1   Competitiveness of EQUITABLE

When given a packing problem $\mathcal{P}$ of the form $\max p^\top x$ subject to $Ax \leq c$ and $x \geq 0$ we denote by $\mathcal{P}(c')$ the problem induced by substituting the capacities c by the vector c'. $\mathcal{P}^{\text{OPT}}$ shall denote the optimal objective value of $\mathcal{P}$.

By linearity of the objective function the output $x = x^{(N)}$ computed by the algorithm for $\mathcal{P}$ has value $\mathcal{P}^{\text{ALG}} := \sum_{s=1}^{N} \mathcal{P}^{\text{OPT}}(b^{(s)})$.

As the lower bounds provided in every stage may deviate from the true slacks by a factor of α, it might first seem impossible to obtain a better competitive ratio than GREEDY. However, by allowing EQUITABLE only to use a particular fraction of the remaining capacities, we can show that these budgets are monotonically increasing from stage to stage. Hence, the marginal profits obtained in every stage are also increasing. Moreover, this restrictive behavior guarantees that EQUITABLE, in contrast to GREEDY, cannot be tricked into getting stuck too early and enables the algorithm to somehow correct previous allocations when the estimates become better in later stages.

Lemma 2. *The marginal profit computed by EQUITABLE in stage $s+1$ can be lower bounded by*

$$\mathcal{P}^{OPT}(b^{(s+1)}) \geq \frac{\mathcal{P}^{OPT}}{s + \alpha \cdot (N - s)}.$$

Proof. First, let us observe that the vector $b^{(s)}$ monotonically increases in every component: By definition we know $\bar{\ell}^{(s+1)} \geq \bar{\ell}^{(s)} - Ay^{(s)}$ and since $Ay^{(s)} \leq b^{(s)}$, we get

$$b^{(s+1)} = \frac{\bar{\ell}^{(s+1)}}{N-s} \geq \frac{\bar{\ell}^{(s)} - b^{(s)}}{N-s} = b^{(s)}.$$

The key to comparing the overall optimal solution with the partial solutions achieved in every stage of the algorithm lies in the simple fact that $Ax^{(s)} + \alpha \cdot \ell^{(s+1)}$ provides an upper bound on c. Furthermore, $Ax^{(s)}$ does not exceed $\sum_{u=1}^{s} b^{(u)}$ which by monotonicity of the $b^{(u)}$ is smaller than $s \cdot b^{(s)}$.

Combining these facts we get

$$\mathcal{P}^{\mathrm{OPT}}(c) \le \mathcal{P}^{\mathrm{OPT}}(Ax^{(s)} + \alpha \cdot \ell^{(s+1)})$$
$$\le \mathcal{P}^{\mathrm{OPT}}(s \cdot b^{(s+1)} + \alpha \cdot \bar{\ell}^{(s+1)}) = \mathcal{P}^{\mathrm{OPT}}((s + \alpha \cdot (N - s)) \cdot b^{(s+1)}).$$

Because of the linearity of the packing problem this proves the lemma.   □

The bound of Lemma 2 significantly increases in later stages. Exploiting this, we get the following result.

Theorem 1. *The profit achieved by algorithm EQUITABLE with a total number of $N \ge \frac{\alpha - 1}{\alpha^\epsilon - 1}$ stages is at least*

$$\mathcal{P}^{\mathrm{ALG}} \ge (1 - \epsilon) \cdot \frac{\ln(\alpha)}{\alpha - 1} \cdot \mathcal{P}^{\mathrm{OPT}}.$$

Proof. For each stage we add the bounds given by Lemma 2,

$$\mathcal{P}^{\mathrm{ALG}} = \sum_{s=0}^{N-1} \mathcal{P}^{\mathrm{OPT}}(b^{(s+1)}) \ge \sum_{s=0}^{N-1} \frac{\mathcal{P}^{\mathrm{OPT}}}{s + \alpha \cdot (N - s)}.$$

We separate the first summand and interpret the rest as a Riemann sum,

$$\frac{\mathcal{P}^{\mathrm{ALG}}}{\mathcal{P}^{\mathrm{OPT}}} \ge \frac{1}{\alpha N} + \int_{x=0}^{N-1} \frac{1}{x + \alpha \cdot (N - x)} \, dx$$
$$= \frac{1}{\alpha N} + \frac{\ln(\alpha N) - \ln(N - 1 + \alpha)}{\alpha - 1} = \frac{\ln(\alpha) - \ln(1 + \frac{\alpha - 1}{N})}{\alpha - 1} + \frac{1}{\alpha N}.$$

Since for $N \ge \frac{\alpha - 1}{\alpha^\epsilon - 1}$ we have $\ln(1 + \frac{\alpha - 1}{N}) \le \epsilon \ln(\alpha)$, the claim follows.   □

Not surprisingly, with an increasing number of stages the performance guarantee of the algorithm improves and converges to $\frac{\ln(\alpha)}{\alpha - 1} \cdot \mathcal{P}^{\mathrm{OPT}}$. Theorem 1 shows that already for $N = \sqrt{\alpha} + 1$ the algorithm is $\frac{\ln(\alpha)}{2(\alpha - 1)}$-competitive. In order to get a competitive ratio strictly greater than $\frac{\ln(\alpha)}{\alpha}$ (i.e. $\epsilon \le \frac{1}{\alpha}$) it is sufficient to have a total number of $N \ge \frac{\alpha^2}{\ln(\alpha)}$ stages.

Note that the proof of Theorem 1 adapts to the situation where the online algorithm can only approximate an optimal solution of the packing problem $\mathcal{P}(b^{(s)})$.

Corollary 1. *If for $r \le 1$ in every stage s of the algorithm EQUITABLE, a feasible r-approximate solution for $\mathcal{P}(b^{(s)})$ is computed, the overall solution is $((1 - \epsilon) \cdot \frac{\ln(\alpha)}{\alpha - 1} \cdot r)$-competitive after a total number of $N \ge \frac{\alpha - 1}{\alpha^\epsilon - 1}$ many stages.*

This can also be useful when integral solutions are favored over fractional ones. If the capacities are large enough, randomized rounding can be used to get an $(1 - \delta)$-approximation to the optimal packing in every step without violating the capacity budgets.

Lemma 3 (proof omitted). *Let $0 < \delta \leq 1$ and assume $b_i^{(s)} \geq 6\log(em)/\delta^4$ for all $1 \leq i \leq m$ in the constraint-normalized packing LP. Then any optimal fractional solution x can be rounded to a feasible integral solution $\bar{x}$ with $p^\top \bar{x} \geq (1 - \delta)p^\top x$. The expected number of randomized rounding steps is at most 4.*

3.2   Upper Bound on the Competitiveness of EQUITABLE

We give an instance for the spanning tree packing problem where the competitiveness of the algorithm is $\Theta(\ln(\alpha)/\alpha)$ for an arbitrary number of stages, even when the remaining slack can be used optimally after finishing stage N.

For simplicity of notation we assume $\alpha \in \mathbb{N}$. Consider the graph $G = (V, E)$ defined by 2α nodes in the following way: two disjoint paths $P = (p_1, \ldots, p_\alpha)$ and $Q = (q_1, \ldots, q_\alpha)$, connected by additional edges induced by the path $R = (p_1, q_1, p_2, q_2, \ldots, p_\alpha, q_\alpha)$. The capacities of the edges $E_1 = E(R)$ belonging to R are c_1 while the capacities of the edges of $E_2 = E(P) \cup E(Q)$ are c_2.

Obviously, the value of any solution to instances of this form is bounded from above by the minimum of $\frac{1}{|V|-1}\sum_{e \in E} c_e$ and $\sum_{e \in E_1} c_e$. Indeed this bound is tight. If the capacities c_1 are not too small, i.e. while $c_1 \geq \frac{c_2}{2\alpha-1}$, then there is a fractional packing of spanning trees of total multiplicity $\frac{1}{|V|-1}\sum_{e \in E} c_e = c_1 + (1 - \frac{1}{2\alpha-1}) \cdot c_2$. This can be achieved by packing the $(2\alpha - 1)$ many spanning trees containing only a single edge of E_1 each with equally shared multiplicity together with the tree E_1 using the remaining capacity. For $c_1 < \frac{c_2}{2\alpha-1}$ this solution achieves a total multiplicity of $\sum_{e \in E_1} c_e$.

Thus, any optimal solution to instances of this form has to completely exhaust the capacities of edges belonging to E_1 in the latter case and even exhaust all edge capacities in the first case.

Now let $c_1 = 1$ and $c_2 = \alpha$ and define the corresponding lower bounds as functions in the constraint usage $\lambda_j = (Ax)_j$.

$$\ell_1(\lambda) = 1 - \lambda \quad \text{for } \lambda \leq 1 \qquad \qquad \ell_2(\lambda) = 1 - \frac{\lambda}{\alpha} \quad \text{for } \lambda \leq \alpha$$

These bounds are α-sharp and fulfill the premise of the following lemma.

Lemma 4 (proof omitted). *Consider the EQUITABLE algorithm with a total number of N stages. Let $q := \frac{\alpha-1}{\alpha} \cdot N$. When the lower bounds $\ell^{(1)}, \ldots, \ell^{(s)}$ are nonincreasing, then the total profit of the algorithm achieved up to any stage $s \leq \lceil q \rceil$ is bounded by $\mathcal{P}^{ALG \leq s} \leq (1 + \ln(\alpha)) \cdot \mathcal{P}^{OPT}(\ell^{(1)})$.*

So the achieved profit up to stage $\lceil q \rceil$ on this instance can be bounded by

$$\sum_{s=1}^{\lceil q \rceil} \mathcal{P}^{OPT}(b^{(s)}) \leq (1 + \ln(\alpha)) \cdot \frac{1}{|V|-1}\sum_{e \in E} \ell_e^{(1)} < 2 \cdot (1 + \ln(\alpha))$$

As the algorithm guarantees monotonicity of the budget vectors we know $b_1^{(s)} \geq b_1^{(1)} = 1/N$ for all s. Since in each stage the edges in E_1 are completely exhausted, the remaining capacity in those edges after $\lceil q \rceil$ many stages has dropped below $1/\alpha$, thus limiting the overall optimal solution achievable with the remaining capacities to at most $\sum_{e \in E_1} \frac{1}{\alpha} < 2$. Hence, $\mathcal{P}^{\text{ALG}} < 4 + 2 \ln(\alpha)$ while the optimal offline solution achieves a packing of multiplicity greater than α.

4   An $O(1/\sqrt{\alpha})$ Upper Bound on Packing Spanning Trees

4.1   A Deterministic Upper Bound

Definition 1 (Diamond Gadget $\hat{G}$). *Consider the complete bipartite graph* $\hat{G} = K_{2,n}$ *defined by* $V = A \cup B = \{u, w\} \cup \{v_1, \ldots, v_n\}$ *and* $E = A \times B$ *with* $n = \sqrt{\alpha}$. *For each node* $z \in B$ *the capacities of the incident edges* $\delta(z)$ *are defined by either setting edge* $\{u, z\}$ *to capacity* α *and edge* $\{w, z\}$ *to capacity* $\sqrt{\alpha}$ *or vice versa.*

The respective α-sharp lower bounds on the remaining capacities of an edge e given by the offline adversary will be according to ℓ' when e has capacity α, and $\bar{\ell}$ when e has capacity $\sqrt{\alpha}$, depending on the actual load λ_e on the edge.

$$\ell'(\lambda) = \begin{cases} \sqrt{\alpha} - \lambda & \text{for } 0 \leq \lambda < \sqrt{\alpha} - 1 \\ \alpha - \lambda & \text{for } \sqrt{\alpha} - 1 \leq \lambda \leq \alpha \end{cases} \quad \text{and} \quad \bar{\ell}(\lambda) = \sqrt{\alpha} - \lambda \quad \text{for } 0 \leq \lambda \leq \sqrt{\alpha}$$

Lemma 5. *For any diamond gadget there is an optimal (offline) algorithm achieving a spanning tree packing with total multiplicity α.*

Proof. Let $\bar{E}$ be the set of edges of capacity $\sqrt{\alpha}$ and $E' = E - \bar{E}$. For each of the $\sqrt{\alpha}$ many edges $\bar{e} \in \bar{E}$ just pack the spanning tree defined by $\{\bar{e}\} \cup E'$ with a multiplicity of $\sqrt{\alpha}$. □

Theorem 2. *For any deterministic online algorithm $\mathcal{A}$ a graph according to Definition 1 can be given, where $\mathcal{A}$ is not able to pack more than $3\sqrt{\alpha}$ many spanning trees.*

Proof. The lower bounds on all edges behave identical up to a load of $\sqrt{\alpha} - 1$. Hence, an online algorithm can not distinguish the different types of edges at the beginning. For every vertex $z \in B$ consider the one of its two incident edges where the load exceeds the threshold of $\sqrt{\alpha} - 1$ first. The offline adversary fixes the lower bound function for this edge $\bar{e}$ to $\ell_{\bar{e}} = \bar{\ell}$ and the function for the other edge $e' \in \delta(z) \setminus \{\bar{e}\}$ to $\ell_{e'} = \ell'$ thus revealing the capacities.

When lower bound functions are finally determined for all edges the additional amount of spanning trees that can be packed is at most $n = \sqrt{\alpha}$. This is easy to see, as n of the edges have a remaining capacity of less than $\bar{\ell}(\sqrt{\alpha} - 1) = 1$, however at least one of them is needed in any spanning tree.

How many spanning trees can be packed by an online algorithm before all of the lower bound functions are determined, then? For each set $\delta(z) = \{\bar{e}, e'\}$ the sum of the edge loads $\lambda_{\bar{e}} + \lambda_{e'}$ gets increased by at least one for every spanning tree in the packing solution. Thus, if the number of packed spanning trees exceeds $2\sqrt{\alpha}$, at least one of the edges of every set has blocked capactiy of more than $\sqrt{\alpha}$. This renders the solution infeasible if the last lower bound probing derived from blocked capacity not more than $\sqrt{\alpha} - 1$ for both edges of at least one of the sets $\delta(v)$. So the best possible overall online solution is bounded by $3\sqrt{\alpha}$.   □

Corollary 2. *No det. online algorithm is better than $O(1/\sqrt{\alpha})$-competitive.*

4.2   An Upper Bound on Randomized Algorithms

The proof of our randomized upper bound uses Yao's minimax principle. Towards this end, we construct a distribution of instances where an optimal spanning tree packing has multiplicity α but every deterministic online algorithm has an expected multiplicity of at most $4\sqrt{\alpha}$.

Definition 2 (Random graph $\hat{R}$). *The random graph $\hat{R}$ consists of M many diamond gadgets $\hat{G}_i = (\{u_i, w_i\} \cup B_i, E_i)$ according to Definition 1, connected by the path $P = (u_1, u_2, \ldots, u_M)$. The capacities of edges $e \in \bigcup_i E_i$ are defined by flipping an independent coin for each node $z \in \bigcup_i B_i$, to pick (uniformly at random) one of the two edges $\bar{e} \in E(z)$ and set its capacity to $\sqrt{\alpha}$. All other edges shall have capacity α.*

Note that each of the $2^{\sqrt{\alpha}}$ possible diamond gadgets is chosen with equal probability. By estimating the probability that any deterministic algorithm performs bad on at least one of the M many gadgets and thereby limiting the expected total multiplicity we get the following result.

Theorem 3 (proof omitted). *The spanning tree packing of every deterministic online algorithm has an expected multiplicity of at most $4\sqrt{\alpha}$ on the random graph $\hat{R}$ with $M \geq 2^{\sqrt{\alpha}-1}\ln(\alpha)$. So by Yao's principle the competitive ratio of any randomized online algorithm is at most $\frac{4}{\sqrt{\alpha}}$.*

5   Conclusion

We have introduced an adversarial model for analyzing online algorithms that have to deal with gradually sharpening constraints capturing that (battery) capacities in (network) optimization problems are typically not known a priori but estimates become more and more exact over time when using the corresponding resources. Our study shows that the natural iterative greedy approach fails to take advantage of the improving estimates, that is, it only achieves the trivial competitive ratio $1/\alpha$. Our main contribution, however, is a proof showing that a more conservative approach can exploit the gradually improving constraints: Algorithm EQUITABLE asymptotically approaches the competitive ratio $\ln \alpha/\alpha-1$ and, hence, improves significantly on the trivial performance guarantee. We complement this result by proving an upper bound on the best possible competitive ratio of order $O(1/\sqrt{\alpha})$. As open problems we ask for exact bounds on the competitive ratios for general packing problems and for interesting special cases like, e. g., packing spanning trees or packing paths (corresponding to max flow with gradually sharpening edge capacities).

References

1. Awerbuch, B., Azar, Y., Plotkin, S.A.: Throughput-competitive on-line routing. In: FOCS, pp. 32–40 (1993)
2. Azar, Y., Feige, U., Glasner, D.: A preemptive algorithm for maximizing disjoint paths on trees. Algorithmica 57(3), 517–537 (2010)
3. Bienstock, D., Iyengar, G.: Approximating fractional packings and coverings in o(1/epsilon) iterations. SIAM J. Comput. 35(4), 825–854 (2006)
4. Birge, J., Louveaux, F.: Introduction to stochastic programming. Springer (1997)
5. Borodin, A., El-Yaniv, R.: Online Computation and Competitive Analysis. Cambridge University Press (1998)
6. Buchbinder, N., Naor, J.: Online primal-dual algorithms for covering and packing. Math. Oper. Res. 34(2), 270–286 (2009)
7. Calinescu, G., Kapoor, S., Olshevsky, A., Zelikovsky, A.: Network Lifetime and Power Assignment in ad hoc Wireless Networks. In: Di Battista, G., Zwick, U. (eds.) ESA 2003. LNCS, vol. 2832, pp. 114–126. Springer, Heidelberg (2003)
8. Elkin, M., Lando, Y., Nutov, Z., Segal, M., Shpungin, H.: Novel algorithms for the network lifetime problem in wireless settings. Wireless Netw. 17, 397–410 (2011)
9. Gabow, H.N., Manu, K.S.: Packing algorithms for arborescences (and spanning trees) in capacitated graphs. Math. Program. 82, 83–109 (1998)
10. Garg, N., Gupta, A., Leonardi, S., Sankowski, P.: Stochastic analyses for online combinatorial optimization problems. In: SODA, pp. 942–951 (2008)
11. Garg, N., Könemann, J.: Faster and simpler algorithms for multicommodity flow and other fractional packing problems. In: FOCS, pp. 300–309 (1998)
12. Immorlica, N., Karger, D.R., Minkoff, M., Mirrokni, V.S.: On the costs and benefits of procrastination: approximation algorithms for stochastic combinatorial optimization problems. In: SODA, pp. 691–700 (2004)
13. Karp, R.M., Nierhoff, T., Tantau, T.: Optimal Flow Distribution Among Multiple Channels with Unknown Capacities. In: Goldreich, O., Rosenberg, A.L., Selman, A.L. (eds.) Theoretical Computer Science. LNCS, vol. 3895, pp. 111–128. Springer, Heidelberg (2006)
14. Michael, S.: Fast algorithm for multicast and data gathering in wireless networks. Information Processing Letters 107(1), 29–33 (2008)
15. Orda, A., Yassour, B.A.: Maximum-lifetime routing algorithms for networks with omnidirectional and directional antennas. In: MobiHoc, pp. 426–437 (2005)
16. Park, J., Sahni, S.: Maximum lifetime broadcasting in wireless networks. IEEE Transactions on Computers 54(9), 1081–1090 (2005)
17. Plotkin, S.A., Shmoys, D.B., Tardos, É.: Fast approximation algorithms for fractional packing and covering problems. In: FOCS, pp. 495–504 (1991)
18. Rao, R., Vrudhula, S., Rakhmatov, D.: Battery modeling for energy aware system design. Computer 36(12), 77–87 (2003)
19. Shmoys, D.B., Swamy, C.: Stochastic optimization is (almost) as easy as deterministic optimization. In: FOCS, pp. 228–237 (2004)
20. Srinivasan, A.: Approximation algorithms via randomized rounding: A survey. Series in Advanced Topics in Mathematics, pp. 9–71. Polish Scientific Publishers PWN (1999)

Distributed Algorithms for Network Diameter and Girth

David Peleg[1,*], Liam Roditty[2,**], and Elad Tal[2]

[1] Department of Computer Science and Applied Mathematics, The Weizmann
Institute of Science, Rehovot, Israel
david.peleg@weizmann.ac.il
[2] Department of Computer Science, Bar-Ilan University, Ramat-Gan, Israel
liamr@macs.biu.ac.il,elad.tal@gmail.com

Abstract. This paper considers the problem of computing the diameter
D and the girth g of an n-node network in the CONGEST distributed
model. In this model, in each synchronous round, each vertex can transmit a different short (say, $O(\log n)$ bits) message to each of its neighbors.
We present a distributed algorithm that computes the diameter of the
network in $O(n)$ rounds. We also present two distributed approximation
algorithms. The first computes a 2/3 multiplicative approximation of
the diameter in $O(D\sqrt{n}\log n)$ rounds. The second computes a $2 - 1/g$
multiplicative approximation of the girth in $O(D + \sqrt{gn}\log n)$ rounds.
Recently, Frischknecht, Holzer and Wattenhofer [11] considered these
problems in the CONGEST model but from the perspective of lower
bounds. They showed an $\tilde{\Omega}(n)$ rounds lower bound for exact diameter
computation. For diameter approximation, they showed a lower bound
of $\tilde{\Omega}(\sqrt{n})$ rounds for getting a multiplicative approximation of $2/3 + \varepsilon$.
Both lower bounds hold for networks with constant diameter. For girth
approximation, they showed a lower bound of $\tilde{\Omega}(\sqrt{n})$ rounds for getting a multiplicative approximation of $2 - \varepsilon$ on a network with constant
girth. Our exact algorithm for computing the diameter matches their
lower bound. Our diameter and girth approximation algorithms almost
match their lower bounds for constant diameter and for constant girth.

1 Introduction

Background and motivation. Network distance parameters play a key role in understanding the performance of distributed network algorithms. Distances affect
both the time and communication complexities of basic distributed tasks in networks, and various parameters quantifying the distance properties of the network
turn out to play a fundamental role in classifying distributed problems according
to their *locality* properties. (Intuitively, a problem is "local" if its solution does
not require communication between distant nodes.)

---

* Supported in part by grants from the Israel Science Foundation, the United-States
- Israel Binational Science Foundation and the Israel Ministry of Science.
** Work supported by the Israel Science Foundation (grant no. 822/10).

A. Czumaj et al. (Eds.): ICALP 2012, Part II, LNCS 7392, pp. 660–672, 2012.
© Springer-Verlag Berlin Heidelberg 2012

Perhaps the most prominent network parameter in that respect is the network *diameter*, hereafter denoted D, and the complexity of many distributed problems and algorithms is often expressed in terms of D, along with other parameters, such as the network size (number of nodes), hereafter denoted n. Among the problems whose time complexity is at least $\Omega(D)$ are calculating network-wide parameters such as the network size n or its diameter D, constructing various types of spanning trees, and more. At the same time, many of these problems can be solved in time $O(D)$, assuming a "convenient" distributed model (e.g., with synchronous communication and no failures). For instance, constructing a Breadth-First Search (BFS) tree from a given origin node can be achieved in time $O(D)$ by simple flooding, and once the tree is established, it can be used for finding n through a standard convergecast process (cf. [19]).

A natural and interesting question is whether the network diameter itself can be found in time $O(D)$. This question can be trivially answered in the affirmative assuming a synchronous communication model that allows arbitrarily large messages. In such a model (often known as the LOCAL model), it is possible to construct in parallel a BFS tree from every node of the network, calculate the depth of each of those trees, and then compute the maximum of those n values, all in time $O(D)$. A more realistic model, however, limits the size of a message allowed to be sent in a single communication round, in order to take into account congestion issues. In this model, often called the CONGEST model, the above algorithmic approach will clearly require time much higher than D. It is therefore interesting to ask whether the network diameter can still be computed in time $O(D)$ in the CONGEST model.

The first to raise this question were Frischknecht, Holzer and Wattenhofer [11], who answered it (perhaps surprisingly) in the negative, establishing a nontrivial lower bound of $\Omega(n)$ for the problem. That lower bound is proved by means of communication complexity techniques, and holds even for networks of constant diameter. Moreover, it is shown therein that even obtaining an approximation of the diameter by a factor of $2/3 + \epsilon$ may already require time that is sometimes higher than the network diameter, and specifically, at least $\Omega(\sqrt{n})$ time. (Note that approximating D by a factor of 2 is easily achievable in time $O(D)$, by computing a single BFS tree and returning its depth.)

Another distance parameter of considerable interest is the network *girth*, g, defined as the length of the shortest cycle in it. This natural graph parameter has many applications (e.g., in cycle packing [17] or computing a minimum cycle basis, which in itself has many uses [16]). Cycles are also important for classifying the locality of various distributed problems, as well as in the context of deadlock detection (cf. [6]). It is shown in [11] that even approximating the girth of a constant girth network by a factor of $2 - \epsilon$ in the CONGEST model requires $\Omega(\sqrt{n})$ time.

In this paper, we address the complementary question of establishing *upper* bounds on the complexity of computing or approximating the network diameter and girth in the CONGEST model, by developing fast distributed algorithms for those problems. In particular, we present a deterministic exact algorithm for computing the diameter, with time complexity of $O(n)$ rounds. This matches the lower bound of [11]. For diameter approximation, we present a randomized algorithm with time complexity $\tilde{O}(D\sqrt{n})$ rounds, that with high probability (i.e., prob. of $1 - 1/n^c$, for some constant c) has a multiplicative approximation factor of $2/3$. For girth approximation, we present a randomized algorithm with time complexity $\tilde{O}(D + \sqrt{ng})$ rounds that, w.h.p., yields a multiplicative approximation factor of $2 - 1/g$. The lower bounds of [11] are for constant diameter and girth, hence our approximation algorithms almost matches the lower bounds both for constant diameter and constant girth. At the end of both algorithms every vertex of the network has the same approximated value.

Related work. As mentioned earlier, it is straightforward to compute the network diameter by a distributed algorithm in $O(D)$ rounds with large messages. An efficient way to achieve that (with $O(n \log n)$ bit messages) is presented in [2].

The direct approach for computing the diameter is by first solving the *all-pairs shortest path (APSP)* problem. The APSP problem has been studied extensively, in both the distributed setting [3,5,12,15,23] and the sequential setting [7,10,25,29,24]. The algorithm of [15] is fast ($O(n)$ time) but involves using large messages. The algorithm of [3] requires time $O(n \log n)$ and uses short ($O(\log n)$ bits) messages, but it applies only to the special family of BHC graphs, whose topology is structured as a balanced hierarchy of clusters. Most other algorithms for APSP focus on optimizing the total message complexity, rather than the time complexity. The algorithm of [12] requires time $O(n^2)$.

In sequential settings, the exact diameter can be computed using any algorithm for computing the APSP. Similarly, an approximation of the diameter can be obtained by any algorithm that computes an approximation of the APSP (see for example [4,8,9]). Aingworth, Chekuri, Indyk and Motwani [1] presented an algorithm that approximates only the diameter in weighted directed graphs in $\tilde{O}(n^2 + m\sqrt{n})$ time. Their algorithm returns a path of length at least $\lfloor \frac{2}{3}D \rfloor$. It is possible to compute the exact diameter in time less than $O(mn)$ by using fast matrix multiplication [28,26].

The problem of computing the girth by a sequential algorithm has been studied since the 1970's. Itai and Rodeh [14] showed that the shortest cycle can be found in $O(nm)$ time by n applications of BFS, or in $O(n^\omega) \leq O(n^{2.3727})$ time using fast matrix multiplication [27].

Roditty and Tov [20] presented an $O(n^2 (\log n) \log nM)$ time algorithm with multiplicative approximation factor $4/3$ for undirected graphs with integral weights from the range $\{1, \ldots, M\}$. They improved a result of Lingas and Lundell [18] that obtained, in the same running time, a multiplicative approximation of 2.

Roditty and Vassilevska Williams [22] presented various approximation algorithms for the girth in unweighted undirected graphs with subquadratic running times. In particular, they presented an $\tilde{O}(n^{5/3})$ time algorithm with a multiplicative approximation factor of 2.

Roditty and Vassilevska Williams [21] showed that the girth problem in directed and undirected graphs with integral weights from the range $\{-M, \ldots, M\}$ can be solved in $\tilde{O}(Mn^\omega)$ time.

Recent developments. Independently, a similar algorithm to compute APSP and Diameter in time $O(n)$ appears at PODC 2012 [13].

In addition [13] presents an algorithm that computes a $1 + \varepsilon$ approximation of the diameter in $O(n/D + D)$ rounds. Combining their algorithm with ours results in an algorithm that computes $2/3$ approximation in $O(n^{3/4} + D)$ rounds.

For the girth [13] presents an algorithm that computes in $O(\min((n/g + D \cdot \log(D/g)), n))$ a $1 + \varepsilon$ approximation of the girth. Combining their algorithm with ours results in an algorithm that computes $2 - 1/g$ approximation of the girth in $O(n^{2/3} + D\log(D/g))$ rounds.

Model. We consider a synchronized network of processors represented by an undirected unweighted graph $G = (V, E)$, where vertices represent processors and edges represent bidirectional connections between processors. A processor is assumed to have unbounded computational power and a unique identifier from $\{1, \ldots, n\}$. Initially, every processor knows only its neighbors. In each round every vertex (processor) can send a message of B bits to each of its neighbors. Moreover, messages can be sent on both directions of an edge at the same round. This model is known as the CONGEST(B) model. When $B = \log n$ the model is denoted by CONGEST. In this paper we are interested in the number of rounds it takes a distributed algorithm that runs on the network processors and communicates using its links to complete a computational task such as computing the diameter of the network.

The distance between $u, v \in V$ is denoted by $d(u, v)$ and equals the length of the shortest path between u and v. The diameter of G is $D = \max_{u,v \in V} d(u, v)$. The girth of G, denoted g, is the length of the shortest cycle in G.

2   Exact All-Pairs of Shortest Paths and Diameter

In this section we present an exact distributed algorithm that computes for every vertex its distance to every other vertex in the graph. As a by-product, the algorithm also computes the network diameter. The algorithm completes the computation in $O(n)$ rounds.

The algorithm. The algorithm is composed of four stages.

Stage 1: Vertex v_0 initiates the process. It first builds a Breadth-First-Search (BFS) tree $BFS(v_0)$ for itself by standard flooding. It then uses this tree to synchronize the clocks of all vertices, so thereafter they all agree on time 0 and the current time. This is done by broadcasting a synchronization message "Time=t" over the tree, where v_0 initially sets $t = 0$, and each vertex getting the message increases t by 1 before forwarding it to its children in the tree.

Stage 2: The initiator v_0 runs a Depth-First-Search (DFS) process on the tree $BFS(v_0)$. In this process, each vertex v gets labelled by a number $\tau(v)$, which is the round in which the DFS visited v for the first time.

Stage 3: The initiator v_0 informs all vertices on a common round T_0 serving as the starting time for the combined process of building n BFS trees $BFS(v)$, one for each vertex $v \in V$. The individual starting times are staggered, and each vertex v starts building its $BFS(v)$ at time $T_0 + 2\tau(v)$. At the end of this stage, each $v \in V$ knows its depth in each $BFS(w)$, namely, $d(v, w)$, for every $w \in V$. As a result, each $v \in V$ knows the value

$$d(v) = \max_w \{ depth(v, BFS(w)) \} .$$

Stage 4: The vertices collectively calculate $D = \max_v \{d(v)\}$ by performing a standard convergecast process (cf. [19]) on $BFS(v_0)$.

Analysis. Let us first notice that the time labels assigned to the vertices satisfy the following.

Property 1. *For every $v, w \in V$, if $\tau(v) < \tau(w)$ then $d(v, w) \leq \tau(w) - \tau(v)$. Moreover, $\tau(v) < 2n$ for every $v \in V$.*

Before we analyze the number of rounds performed by the algorithm, we show that in stage 3, in which the n BFS processes are initiated on the network, no collisions occur, that is, given a vertex x, it cannot happen that two different BFS processes will have to send a message from x at the same round.

Lemma 1. *There are no collisions between messages of different BFS construction processes.*

Proof. Assume, for the sake of contradiction, that some collisions do occur, and the very first collision is between the BFS processes that construct the trees $BFS(v)$ and $BFS(w)$ of vertices v and w. Namely, there exists a vertex x such that at time t, the process constructing $BFS(v)$ needs to send a message M_v from x, and at the same time, the process constructing $BFS(w)$ needs to send a message M_w from x. As this is the very first collision to occur in stage 3, the message M_v must have traversed the way from v to x at full speed and thus reached x after exactly $d(v, x)$ time units. Similarly, M_w must have reached x after $d(w, x)$ time units. This implies that

$$T_0 + 2\tau(v) + d(v, x) = T_0 + 2\tau(w) + d(w, x) .$$

Assume, without loss of generality, that $\tau(v) < \tau(w)$. Then

$$d(v, x) - d(w, x) = 2\tau(w) - 2\tau(v) > 0 .$$

By the triangle inequality it follows that $d(v, x) \leq d(v, w) + d(w, x)$, or equivalently, that $d(v, x) - d(w, x) \leq d(v, w)$. Thus, we get that $0 < 2(\tau(w) - \tau(v)) \leq d(v, w)$, and hence $\tau(w) - \tau(v) < d(v, w)$. But since $\tau(v) < \tau(w)$, $d(v, w) \leq \tau(w) - \tau(v)$ by Property 1, leading to contradiction. $\square$

We are now ready to analyze the number of rounds performed by the algorithm. In stage 1, the root initiates the construction of a single BFS tree, which takes $O(D)$ rounds. In stage 2, the algorithm computes the time labels $\tau(v)$ via a DFS process, in $O(n)$ rounds. In stage 3, the algorithm initiates n BFS processes. Each of these processes takes $O(D)$ rounds. However, by Lemma 1 it follows that all BFS processes are independent (in the sense that the sets of edges on which they send messages at each round are disjoint), and therefore the messages of each process traverse the network without incurring any delays. Thus, the time it takes to complete this stage is $T_0 + \max_v\{\tau(v)\} + D$, which is $O(n)$. The last stage takes $O(D)$. We conclude that the running time is $O(n)$.

We get the following.

Theorem 2. *The algorithm computes in $O(n)$ rounds the diameter of the graph and the distances between any two vertices in the graph.* □

3   An Approximation Algorithm for the Diameter

In this section we show that the sequential algorithm of Aingworth, Chekuri, Indyk and Motwani [1], that computes in $\tilde{O}(m\sqrt{n} + n^2)$ time a 2/3 approximation of the diameter of directed graphs, can be implemented efficiently by a distributed algorithm in the CONGEST model. In particular, for any constant diameter our distributed algorithm obtains a 2/3 approximation of the diameter in $\tilde{O}(\sqrt{n})$ rounds.

The algorithm of Aingworth et al. [1] works as follows. For every vertex $u \in V$, let $N(u, s)$ denote the set of s vertices closest to u. (If two vertices have the same distance to u, then the one with the smallest id is considered to be closer.) First, the algorithm computes for every $u \in V$ a partial BFS tree $PB(u, s)$ that spans the vertices of $N(u, s)$. Let v_0 be the vertex whose partial BFS tree, $PB(v_0, s)$, is the deepest. For each of the s vertices $w \in PB(v_0, s)$, the algorithm constructs a full BFS tree $BFS(w)$. Next, the algorithm creates a new graph $\tilde{G}$ by connecting every vertex u to its s closest vertices, $N(u, s)$, with new edges. It then computes a dominating set M of size $\tilde{O}(n/s)$ on $\tilde{G}$ (one can verify that such a dominating set always exists in a graph of minimum degree s). For each vertex $w \in M$, the algorithm constructs a full BFS tree $BFS(w)$. Finally, the algorithm computes the depth of each of the BFS trees that were constructed, and returns their maximum as an estimate for the network diameter D. Aingworth et al. [1] showed that the maximum depth is at least $\lfloor \frac{2}{3}D \rfloor$.

The running time of the algorithm is easy to analyze. The partial BFS trees are computed in $O(ns^2)$ time. Then, s BFS trees are computed in $O(ms)$ time. A dominating set is computed in $O(n^2 + ns)$ time. Finally, $\tilde{O}(n/s)$ BFS trees are computed in $\tilde{O}(mn/s)$ time. Setting $s = \sqrt{n}$ we get $\tilde{O}(m\sqrt{n} + n^2)$ time.

We now describe how to implement this algorithm in the CONGEST model. We start with computing for every vertex u the set $N(u, s)$ consisting of its s closest vertices. (Recall that if two vertices have the same distance to u, then the one with the smaller id is considered to be closer.) The *depth* of the set $N(u, s)$ is the distance from u to the farthest vertex in $N(u, s)$.

The algorithm operates in *cycles*. The length of each cycle is s rounds. All vertices synchronize to start at the same time. At the ith cycle, every vertex v notifies its neighbors concerning vertices at distance $i - 1$ from it. The messages are sent according to the id of these vertices in ascending order. Each vertex transmits exactly s messages in total, throughout the entire process. During the ith cycle, every vertex gets messages from all its neighbors that can still transmit (i.e., that did not exhaust their quota of s messages), regarding the vertices that are at distance $i - 1$ from them. When the ith cycle ends, every vertex prepares its messages for the $(i + 1)$st cycle by examining the information from its neighbors and identifying all distance i vertices it has just learned of. The algorithm starts with every vertex sending a message to all its neighbors, saying that it is at distance 0 from itself. The algorithm ends when no vertex is transmitting.

We now prove the correctness of the algorithm.

Lemma 2. *Given s, the algorithm computes, for every $v \in V$, the set $N(v, s)$.*

Proof. Let $\Gamma(v, i)$ denote the set of vertices at distance i or less from v. Notice that for every vertex $v \in V$, and for every integer s, there exists some integer i_s such that $\Gamma(v, i_s - 1) \subset N(v, s) \subseteq \Gamma(v, i_s)$.

To prove the lemma, we prove by induction on i that for every vertex v, if $|\Gamma(v, i - 1)| < s$, then after the ith cycle ends, v knows all the vertices in $\Gamma(v, i) \cap N(v, s)$. This implies the lemma by considering the situation at vertex v after cycle i_s.

For the basis of the induction, we have to show that after the first cycle ends, every vertex v knows all its neighbors if its degree is at most s, and all the vertices of $N(v, s)$ otherwise. In fact, it is possible to prove a slightly stronger claim, namely, that after the first cycle ends, every vertex knows all its neighbors. This follows easily from the way that the algorithm starts: In the first cycle, every vertex sends a message to all its neighbors, telling them that it is at distance 0 from itself. Thus, every vertex gets messages from all its neighbors, and after the first cycle ends, it knows all of them.

For the induction hypothesis, assume that after the ith cycle ends, for every vertex v, if $|\Gamma(v, i - 1)| < s$, then v knows $\Gamma(v, i) \cap N(v, s)$. We prove that after the $(i + 1)$st cycle ends, for every vertex v, if $|\Gamma(v, i)| < s$, then v knows $\Gamma(v, i + 1) \cap N(v, s)$.

Consider a vertex v satisfying that $|\Gamma(v, i)| < s$. Assume, for the sake of contradiction, that some of the vertices of $\Gamma(v, i + 1) \cap N(v, s)$ are not reported to v during the $(i + 1)$st cycle, and let u be the smallest id vertex among them. Let w be the neighbor of v on the shortest path between v and u. Note that $u \in \Gamma(w, i) \cap N(w, s)$.

Also note that $\Gamma(w, i - 1) \subseteq \Gamma(v, i)$, and hence $|\Gamma(w, i - 1)| \leq |\Gamma(v, i)| < s$. It follows by the induction hypothesis that at the end of cycle i, w knows all the vertices of $\Gamma(w, i) \cap N(w, s)$, so in particular, it knows u. Moreover, as $|\Gamma(w, i - 1)| = r < s$, w must have reported all the vertices of $\Gamma(w, i - 1)$ to v by the end of cycle i.

On cycle $i+1$, w reports to v up to $s-r$ additional vertices from $\Gamma(w,i)$, yet it does not report u (which it already knows). This implies that w reports precisely $s-r$ vertices, and thus exhausts its quota, having reported to v about s vertices, all of whom are closer to v than u (or have equal distance and smaller id). But this means that u is not among the s vertices closest to v, i.e., $u \notin N(v,s)$, a contradiction.   □

We now analyze the running time of the algorithm. A vertex keeps on transmitting messages as long as it does not have all its s closest vertices. At the worst case scenario, its s closest vertices have depth D and it will stop transmitting only after D cycles end which equal to sD rounds. In additional $O(D)$ rounds, every vertex can convergecast the depth of its s closest vertices so we can pick the vertex with maximum depth. Let w be this vertex. We compute for each of the s closest vertices of w a complete BFS tree. The cost for this is $O(sD)$ rounds. To obtain a dominating set, we choose $\Theta(n/s \log n)$ vertices uniformly at random. With high probability, this set contains a vertex from the set of s closest vertices of each vertex. We compute a complete BFS tree for the sampled vertices in $\tilde{O}(Dn/s)$ rounds.

Thus, the total number of rounds is $\tilde{O}(Ds + nD/s)$ and if we set $s = \sqrt{n}$ we get $\tilde{O}(\sqrt{n}D)$ rounds.

Theorem 3. *There is a distributed algorithm that in $\tilde{O}(\sqrt{n}D)$ rounds computes, w.h.p, a $2/3$ approximation of the network diameter.*

4   Approximating the Girth

In this Section we present an algorithm that approximates the network girth. The estimation of our algorithm is at most $4\ell - 2$ when $g = 2\ell$, and at most $4\ell - 3$ when $g = 2\ell - 1$. Thus, our algorithm guarantees a $2 - 1/g$ multiplicative approximation. The time complexity of the algorithm is $\tilde{O}(D + \sqrt{n}g)$ rounds.

The algorithm operates in phases. Each phase is composed of $\Theta(s+(n/s) \log n)$ rounds. All vertices are synchronized to start at the same time. Before the first phase starts, a sampled vertex set S of size $\Theta(n/s \log n)$ is formed in a distributed manner. A vertex can transmit two types of messages, a *distance* message and a *cycle* message. Once a vertex receives a cycle message for the first time, it immediately transmits it and stops to transmit distance messages (if it gets two or more cycle messages at the same time, then it will transmit the one with the smallest cycle). At the ith phase, a vertex v transmits a distance message $\langle w, (i-1) \rangle$ to its neighbors if all the following conditions hold:

- v did not transmit a distance message in a previous phase $k < i$ of the form $\langle w, (k-1) \rangle$.
- v got in the previous phase, $i-1$, a message of the form $\langle w, (i-2) \rangle$.
- Either $w \in S$, or if $w \notin S$ then v has to transmit fewer than $\Theta(s+(n/s) \log n)$ distance messages in the ith phase, and in every phase $k < i$ it had to transmit fewer than $\Theta(s + n/s \log n)$ distance messages.

A cycle message has the form $\langle v, t \rangle$, which means that v is on a cycle of length at most t. At the beginning of the $(i+1)$st phase, a vertex v creates and transmits a cycle message and stops to transmit distance messages if it has not transmitted a cycle message in a previous phase, and one of the following conditions holds:

- v got in the ith phase a distance message of the form $\langle w, (i-1) \rangle$, and in the $(i-1)$th phase a distance message of the form $\langle w, (i-2) \rangle$, for some $w \in V$.
- v got in the ith phase two distance messages of the form $\langle w, (i-1) \rangle$, for some $w \in V$.

In the first case, v transmits a cycle message of the form $\langle v, 2i-1 \rangle$, and in the second case, v transmits a cycle message of the form $\langle v, 2i \rangle$. If both conditions are satisfied, then v transmits a cycle message of the form $\langle v, 2i-1 \rangle$. Notice that any vertex will create a cycle message at most once. Every vertex saves the cycle message that contains the smallest cycle that it has seen so far. When a vertex creates a new cycle message, it also saves it as the smallest cycle message it has seen so far. When a vertex receives a cycle message, it compares it to the saved cycle message, and if the new message contains a smaller cycle, then it saves the new message and immediately transmits it. The algorithm ends when no vertex is transmitting.

We now turn to discuss the correctness of the algorithm. The next Lemma is straightforward.

Lemma 3. *If v transmits a distance message $\langle w, (i-1) \rangle$ at the ith phase, then $d(w, v) = i - 1$.*

We now turn to show that if a vertex v creates a cycle message $\langle v, t \rangle$, then v is on a cycle of length at most t.

Lemma 4. *Let v be a vertex that at the beginning of the $(i+1)$st phase creates a cycle message of the form $\langle v, 2i \rangle$ (respectively, $\langle v, 2i-1 \rangle$). Then v is on a cycle of length at most $2i$ (resp., $2i-1$).*

Proof. A vertex v creates a cycle message at the beginning of the $(i+1)$st phase if it received two distance messages, $\langle w, t \rangle$ and $\langle w, t' \rangle$, where $t = i - 1$ and t' is either $i - 1$ or $i - 2$. In such a case, it must be that two different vertices u_1 and u_2 transmitted these distance messages, as each vertex notifies its neighbors exactly once on its distance from w. By Lemma 3 it follows that $d(u_1, w) = t$ and $d(u_2, w) = t'$. Thus, there are two paths between v and w, one that starts with the edge (v, u_1) and has length $t + 1$, and another that starts with the edge (v, u_2) and has length $t' + 1$. Let u be the first vertex (starting from v) that is common to these two paths. The portions of these paths between v and u form a simple cycle of length at most $t + t' + 2$, which is $2i$ in case $t' = i - 1$ and $2i - 1$ in case $t' = i - 2$. $\square$

We now show that at least one vertex of the graph creates a cycle message with a cycle length at most $2g - 2$ if g is even and $2g - 1$ if g is odd. We start with the case of even g.

Lemma 5. *Let $C = \{v_0, v_1, \ldots, v_{2\ell-1}\}$ be a shortest cycle in the graph. There exists a vertex that creates a cycle message of length at most $4\ell - 2$.*

Proof. Let P be the shortest path on the cycle C between v_0 and $v_{\ell-1}$ and let Q be the shortest path on the cycle C between v_0 and $v_{\ell+1}$. As C is a shortest cycle in the graph, P and Q are also shortest paths in the graph. If every vertex of P and Q transmitted the distance message containing its distance from v_0, then v_ℓ gets two distance messages $\langle v_0, \ell-1 \rangle$ at phase ℓ and creates a cycle message $\langle v_\ell, 2\ell \rangle$ at the beginning of phase $\ell+1$.

Assume that this is not the case. Thus, there is a vertex v_j, which w.l.o.g. is assumed to be on P, and thus $1 \le j \le \ell-1$, that gets a distance message of the form $\langle v_0, j-1 \rangle$ from v_{j-1} at phase j, and does not transmit a distance message $\langle v_0, j \rangle$ at phase $(j+1)$. As P is a shortest path in the graph, it cannot be that v_j already got a message of the form $\langle v_0, j'-1 \rangle$ for some $j' < j$. Also, it cannot be that v_j creates a cycle message at the beginning of phase $j+1$, as by Lemma 4, such a message implies that v_j is on a cycle of length at most $2j$, but $2j \le 2\ell-2$ and $g = 2\ell$.

Hence, v_j had to transmit more than $\Theta(s + (n/s)\log n)$ distance messages at some phase $2 \le k \le j+1$. W.h.p., one of these distance messages contains a vertex from S. Denote this vertex by w. The distance between v_j and w is $k-1$. All vertices always transmit distance messages that contain a sampled vertex, unless they already transmitting only cycle messages, thus, this message is transmitted by v_j. Let F be the shortest path on C between v_j and $v_{j+\ell-1}$, and let H be the shortest path on C between v_j and $v_{j+\ell+1}$.

We now distinguish between two cases. The first case is that the shortest path between w and every vertex of C goes through v_j. If at the $k+\ell-1$ phase $v_{\ell+j}$ receives two distance messages of the form $\langle w, \ell+k-2 \rangle$ from its neighbors $v_{j+\ell-1}$ and $v_{j+\ell+1}$, then at the beginning of the $(k+\ell)$th phase, $v_{j+\ell}$ creates a cycle message $\langle v_{\ell+j}, 2(\ell+k-1) \rangle$. As $k \le j+1 \le \ell$, we have $2(\ell+k-1) \le 4\ell-2$. If, however, at the $k+\ell-1$ phase $v_{\ell+j}$ does not receive two distance messages, then a vertex either on F or on H (or on both) started to transmit a cycle message at some phase between phases $k+1$ and $k+\ell-1$. Such a cycle has length at most $2(k+\ell-2) \le 4\ell-4$.

In the second case, there exists a vertex of C whose shortest path with w does not go through v_j. W.l.o.g., assume that this vertex is from $F \cup \{v_{\ell+j}\}$. Let v_r be the last vertex on F (starting from v_j) whose shortest path with w goes through v_j. Notice that $j \le r \le \ell+j-1$. If the distance between w and v_r is t, then the distance between w and v_{r+1} is either t or $t-1$. Now there are three possibilities. The first is that v_r gets two distance messages in phase t of the form $\langle w, t-1 \rangle$. It then creates a cycle message at the beginning of phase $t+1$ of the form $\langle v_r, 2t \rangle$. The second possibility is that v_r gets one distance message $\langle w, t-1 \rangle$ in phase t and one distance message $\langle w, t \rangle$ in phase $t+1$. It then creates a cycle message at the beginning of phase $t+2$ of the form $\langle v_r, 2t+1 \rangle$. The third possibility is that at least one of the distance messages does not reach v_r. Then it must be that v_r got a cycle message instead. Such a cycle message was created at phase $t+1$ or earlier, and the cycle is of length at most $2t$. As $t = k-1+r-j \le k-1+\ell+j-1-j \le \ell+k-2$ and $k \le \ell$, we get $2t+1 \le 4\ell-3$.

We conclude that a cycle message with a cycle of length at most $4\ell - 2$ is created by a vertex of the graph.   □

Next, we turn to the case of odd g.

Lemma 6. *Let* $C = \{v_0, v_2, \ldots, v_{2\ell-2}\}$ *be a shortest cycle in the graph. There exists a vertex in the graph that creates a cycle message of length at most* $4\ell - 3$.

The proof for odd g is very similar to the proof for even g, and is differed to the full version of this paper due to space limitations.

We now turn to analyze the time complexity of the algorithm.

Lemma 7. *After* $\tilde{O}(D + g(s + n/s))$ *rounds, every vertex of the graph has the desired approximation of the girth.*

Proof. By Lemmas 5 and 6, there exists a vertex v in the graph that creates a cycle message $\langle v, t \rangle$, where $t \leq 2g - 1$. Moreover, the way the algorithm is designed ensures that v will create this message after $O(g)$ phases from the beginning. Let v' be the vertex that creates the cycle message with the smallest cycle. Let this message be $\langle v', t' \rangle$. If $v' \neq v$, then $t' < t \leq 2g - 1$. By Lemma 4, there exists a cycle in the graph of length at most t', hence $t' \geq g$. This message will be created after $O(g)$ phases. As every vertex stops transmitting distance messages immediately after it receives a cycle message, and the smallest cycle message is always transmitted, this message will be distributed in $O(D)$ rounds.   □

A simple modification of the algorithm can improve the time complexity. Currently, every vertex $v \in V$ transmits a distance message $\langle v, 0 \rangle$ in the first phase. However, Lemmas 5 and 6 still guarantee that some vertex in the graph will create a cycle message with the desired approximation even if only a single vertex v of the cycle attaining the girth sends a distance message $\langle v, 0 \rangle$. As we do not really know which vertices occur on that cycle, we can create a sampled set R of $\tilde{O}(n/g)$ vertices, that w.h.p., will contain a vertex from every cycle of length g. Now, only vertices $v \in R$ send distance messages of the form $\langle v, 0 \rangle$. The sampled set $S \subset R$ is expected to be of size $\tilde{O}(n/(gs))$, and in every phase, only $\Theta(s + (n/(gs)) \log n)$ distance messages will be transmitted. The set S still ensures that, w.h.p., if a vertex v has more than $\Theta(s + (n/(gs)) \log n)$ distance messages to transmit in some phase, then it will have to transmit in this phase also a distance message that contains a vertex of S. If g is known, then within $\tilde{O}(D + g(s + n/(gs)))$ rounds, every vertex holds an approximation of the girth. Setting $s = \sqrt{n/g}$ results in $\tilde{O}(D + \sqrt{ng})$ rounds. As g is not known in advanced, an additional logarithmic factor must be added for performing a binary search.

The following Theorem stems from the above discussion.

Theorem 4. *There is a distributed algorithm that in* $\tilde{O}(D + \sqrt{ng})$ *rounds computes, w.h.p, a* $2 - 1/g$ *approximation of the girth.*

References

1. Aingworth, D., Chekuri, C., Indyk, P., Motwani, R.: Fast estimation of diameter and shortest paths (without matrix multiplication). SIAM J. Comput. 28(4), 1167–1181 (1999)
2. Almeida, P.S., Baquero, C., Cunha, A.: Fast distributed computation of distances in networks. Technical report (2011)
3. Antonio, J.K., Huang, G.M., Tsai, W.K.: A fast distributed shortest path algorithm for a class of hierarchically clustered data networks. IEEE Trans. Computers 41, 710–724 (1992)
4. Baswana, S., Kavitha, T.: Faster algorithms for approximate distance oracles and all-pairs small stretch paths. In: FOCS, pp. 591–602. IEEE Computer Society (2006)
5. Cicerone, S., D'Angelo, G., Di Stefano, G., Frigioni, D., Petricola, A.: Partially dynamic algorithms for distributed shortest paths and their experimental evaluation. J. Computers 2, 16–26 (2007)
6. Cidon, I., Jaffe, J.M., Sidi, M.: Local distributed deadlock detection by cycle detection and clustering. IEEE Trans. Software Eng. 13(1), 3–14 (1987)
7. Dijkstra, E.W.: A note on two problems in connection with graphs. Numer. Math., 269–271 (1959)
8. Dor, D., Halperin, S., Zwick, U.: All-pairs almost shortest paths. SIAM J. Comput. 29(5), 1740–1759 (2000)
9. Elkin, M.: Computing almost shortest paths. ACM Transactions on Algorithms 1(2), 283–323 (2005)
10. Floyd, R.W.: Algorithm 97: shortest path. Comm. ACM 5, 345 (1962)
11. Frischknecht, S., Holzer, S., Wattenhofer, R.: Networks cannot compute their diameter in sublinear time. In: Proc. 23rd ACM-SIAM Symp. on Discrete Algorithms, SODA (2012)
12. Haldar, S.: An 'all pairs shortest paths' distributed algorithm using $2n^2$ messages. J. Algorithms, 20–36 (1997)
13. Holzer, S., Wattenhofer, R.: Optimal distributed all pairs shortest paths and applications. In: Proc. 31st Annual ACM SIGACT-SIGOPS Symp. on Principles of Distributed Computing, PODC (2012)
14. Itai, A., Rodeh, M.: Finding a minimum circuit in a graph. SIAM J. Computing 7(4), 413–423 (1978)
15. Kanchi, S., Vineyard, D.: Time optimal distributed all pairs shortest path problem. Int. J. of Information Theories and Applications, 141–146 (2004)
16. Kavitha, T., Liebchen, C., Mehlhorn, K., Michail, D., Rizzi, R., Ueckerdt, T., Zweig, K.A.: Cycle bases in graphs characterization, algorithms, complexity, and applications. Computer Science Review 3(4), 199–243 (2009)
17. Krivelevich, M., Nutov, Z., Yuster, R.: Approximation algorithms for cycle packing problems. In: Proc. SODA, pp. 556–561 (2005)
18. Lingas, A., Lundell, E.-M.: Efficient approximation algorithms for shortest cycles in undirected graphs. Inf. Process. Lett. 109(10), 493–498 (2009)
19. Peleg, D.: Distributed Computing: A Locality-Sensitive Approach. SIAM (2000)
20. Roditty, L., Tov, R.: Approximating the girth. In: Proc. SODA, pp. 1446–1454 (2011)
21. Roditty, L., Vassilevska Williams, V.: Minimum weight cycles and triangles: Equivalences and algorithms. In: Proc. FOCS, pp. 180–189 (2011)

22. Roditty, L., Vassilevska Williams, V.: Subquadratic time approximation algorithms for the girth. In: SODA, pp. 833–845 (2012)
23. Segall, A.: Distributed network protocols. IEEE Trans. Inf. Th. IT-29, 23–35 (1983)
24. Seidel, R.: On the all-pairs-shortest-path problem in unweighted undirected graphs. JCSS 51, 400–403 (1995)
25. Warshall, S.: A theorem on boolean matrices. J. ACM 9(1), 11–12 (1962)
26. Vassilevska Williams, V.: Private communication
27. Vassilevska Williams, V.: Breaking the coppersmith-winograd barrier. In: STOC (2012)
28. Yuster, R.: Computing the diameter polynomially faster than apsp. CoRR, abs/1011.6181 (2010)
29. Zwick, U.: All pairs shortest paths using bridging sets and rectangular matrix multiplication. JACM 49(3), 289–317 (2002)

Author Index